Beginning ALGEBRA

Second Edition

Julie Miller
Daytona Beach Community College

Molly O'Neill
Daytona Beach Community College

Nancy Hyde
Formerly of Broward Community College

With Contributions
by **Mitchel Levy**

Boston Burr Ridge, IL Dubuque, IA New York San Francisco St. Louis
Bangkok Bogotá Caracas Kuala Lumpur Lisbon London Madrid Mexico City
Milan Montreal New Delhi Santiago Seoul Singapore Sydney Taipei Toronto

The McGraw·Hill Companies

BEGINNING ALGEBRA, SECOND EDITION

Published by McGraw-Hill, a business unit of The McGraw-Hill Companies, Inc., 1221 Avenue of the Americas, New York, NY 10020.

This book is printed on acid-free paper.

1 2 3 4 5 6 7 8 9 0 DOW/DOW 0 9 8 7 6

ISBN 978–0–07–302871–2
MHID 0–07–302871–1

ISBN 978-0-07-327083-8 (Annotated Instructor's Edition)
MHID 0-07-327083-0

Publisher: *Elizabeth J. Haefele*
Sponsoring Editor: *David Millage*
Director of Development: *David Dietz*
Developmental Editor: *Michelle Driscoll*
Marketing Manager: *Barbara Owca*
Senior Project Manager: *Vicki Krug*
Senior Production Supervisor: *Sherry L. Kane*
Lead Media Project Manager: *Stacy A. Patch*
Media Producer: *Amber M. Huebner*
Designer: *Laurie B. Janssen*
Cover Designer: *Asylum Studios*
Lead Photo Research Coordinator: *Carrie K. Burger*
Supplement Producer: *Melissa M. Leick*
Compositor: *Techbooks*
Typeface: *10/12 Times Ten Roman*
Printer: *R. R. Donnelley, Willard, OH*

Photo Credits: Page 3: © PhotoDisc/Getty R-F website; p. 30: © Corbis Website; p. 79: © Elena Rooraid/PhotoEdit; p. 87: © Vol. 44/Corbis CD; p. 99: © Corbis R-F Website; p. 162: © Vol. 44/Corbis CD; p. 171: © Judy Griesdieck/Corbis; p. 181: © Corbis R-F Website; p. 195: © Robert Brenner/PhotoEdit; p. 218: © David Young-Wolff/PhotoEdit; p. 237: © Tony Freeman/PhotoEdit; p. 251: © Vol. 107/Corbis CD; p. 283: © Susan Van Etten/Photo Edit; p. 340: © Jeff Greenberg/PhotoEdit; p. 365: © Vol. 132/Corbis; p. 367: © Linda Waymire; p. 402: © Vol. 26/Corbis; p. 442: © Corbis Website; p. 525: © Paul Morris/Corbis; p. 560: © EyeWire/Getty Images website; p. 589: Courtesy of NOAA; p. 602 left: © PhotoDisc Website; p. 602 right: © Vol. 59/Corbis; p. 635: © Dennis O'Clair Photography; p. 657: © Vol. 101/Corbis; p. 692: © PhotoDisc website; p. 710: © Masimo Listri/Corbis; p. 733: © Vol. 527/Corbis; p. 739: © Frank Whitney/Brand X; Pictures/PictureQuest.

Library of Congress Cataloging-in-Publication Data

Miller, Julie, 1962–
Beginning algebra / Julie Miller, Molly O'Neill, Nancy Hyde. — 2nd ed.
p. cm.
Includes indexes.
ISBN 978–0–07–302871–2 — ISBN 0–07–302871–1 (acid-free paper)
ISBN 978–0–07–327083–8 — ISBN 0–07–327083–0 (annotated instructor's ed. : acid-free paper)
1. Algebra—Textbooks. I. O'Neill, Molly, 1953– II. Hyde, Nancy. III. Title.

QA152.3.M55 2008
512.9—dc22 2006026558
CIP

www.mhhe.com

Contents

Dedication

To our students . . .

—Julie Miller —Molly O'Neill —Nancy Hyde

About the Authors

Julie Miller

Julie Miller has been on the faculty of the Mathematics Department at Daytona Beach Community College for 18 years, where she has taught developmental and upper-level courses. Prior to her work at DBCC, she worked as a software engineer for General Electric in the area of flight and radar simulation. Julie earned a bachelor of science in applied mathematics from Union College in Schenectady, New York, and a master of science in mathematics from the University of Florida. In addition to this textbook, she has authored several course supplements for college algebra, trigonometry, and precalculus, as well as several short works of fiction and nonfiction for young readers.

"My father is a medical researcher, and I got hooked on math and science when I was young and would visit his laboratory. I can remember using graph paper to plot data points for his experiments and doing simple calculations. He would then tell me what the peaks and features in the graph meant in the context of his experiment. I think that applications and hands-on experience made math come alive for me and I'd like to see math come alive for my students."

—Julie Miller

Molly O'Neill

Molly O'Neill is also from Daytona Beach Community College, where she has taught for 20 years in the Mathematics Department. She has taught a variety of courses from developmental mathematics to calculus. Before she came to Florida, Molly taught as an adjunct instructor at the University of Michigan–Dearborn, Eastern Michigan University, Wayne State University, and Oakland Community College. Molly earned a bachelor of science in mathematics and a master of arts and teaching from Western Michigan University in Kalamazoo, Michigan. Besides this textbook, she has authored several course supplements for college algebra, trigonometry, and precalculus and has reviewed texts for developmental mathematics.

"I differ from many of my colleagues in that math was not always easy for me. But in seventh grade I had a teacher who taught me that if I follow the rules of mathematics, even I could solve math problems. Once I understood this, I enjoyed math to the point of choosing it for my career. I now have the greatest job because I get to do math everyday and I have the opportunity to influence my students just as I was influenced. Authoring these texts has given me another avenue to reach even more students."

—Molly O'Neill

Nancy Hyde

Nancy Hyde served as a full-time faculty member of the Mathematics Department at Broward Community College for 24 years. During this time she taught the full spectrum of courses from developmental math through differential equations. She received a bachelor of science degree in math education from Florida State University and a master's degree in math education from Florida Atlantic University. She has conducted workshops and seminars for both students and teachers on the use of technology in the classroom. In addition to this textbook, she has authored a graphing calculator supplement for College Algebra.

"I grew up in Brevard County, Florida, with my father working at Cape Canaveral. I was always excited by mathematics and physics in relation to the space program. As I studied higher levels of mathematics I became more intrigued by its abstract nature and infinite possibilities. It is enjoyable and rewarding to convey this perspective to students while helping them to understand mathematics."

—Nancy Hyde

Mitchel Levy

Mitchel Levy of Broward Community College joined the team as the exercise consultant for the Miller/O'Neill/Hyde paperback series. Mitchel received his BA in mathematics in 1983 from the State University of New York at Albany and his MA in mathematical statistics from the University of Maryland, College Park in 1988. With over 17 years of teaching and extensive reviewing experience, Mitchel knows what makes exercise sets work for students. In 1987 he received the first annual "Excellence in Teaching" award for graduate teaching assistants at the University of Maryland. Mitchel was honored as the Broward Community College Professor of the year in 1994, and has co-coached the Broward math team to 3 state championships over 7 years.

"I love teaching all level of mathematics from Elementary Algebra through Calculus and Statistics."

—Mitchel Levy

Introducing . . .

The Miller/O'Neill/Hyde Series in Developmental Mathematics

Miller/O'Neill/Hyde casebound series

Beginning Algebra, 2/e

Intermediate Algebra, 2/e

Beginning and Intermediate Algebra, 2/e

Miller/O'Neill/Hyde worktext series

Basic College Mathematics

Introductory Algebra

Intermediate Algebra

Preface

From the Authors

First and foremost, we would like to thank the students and colleagues who have helped us prepare this text. The content and organization are based on a wealth of resources. Aside from an accumulation of our own notes and experiences as teachers, we recognize the influence of colleagues at Daytona Beach Community College as well as fellow presenters and attendees of national mathematics conferences and meetings. Perhaps our single greatest source of inspiration has been our students, who ask good, probing questions every day and challenge us to find new and better ways to convey mathematical concepts. We gratefully acknowledge the part that each has played in the writing of this book.

In designing the framework for this text, the time we have spent with our students has proved especially valuable. Over the years we have observed that students struggle consistently with certain topics. We have also come to know the influence of forces beyond the math, particularly motivational issues. An awareness of the various pitfalls has enabled us to tailor pedagogy and techniques that directly address students' needs and promote their success. These techniques and pedagogy are outlined here.

Active Classroom

First, we believe students retain more of what they learn when they are actively engaged in the classroom. Consequently, as we wrote each section of text, we also wrote accompanying worksheets called **Classroom Activities** to foster accountability and to encourage classroom participation. Classroom Activities resemble the examples that students encounter in the textbook. The activities can be assigned to individual students or to pairs or groups of students. Most of the activities have been tested in the classroom with our own students. In one class in particular, the introduction of Classroom Activities transformed a group of "clock watchers" into students who literally had to be ushered out of the classroom so that the next class could come in. The activities can be found in the *Instructor's Resource Manual*, which is available through MathZone.

Conceptual Support

While we believe students must practice basic skills to be successful in any mathematics class, we also believe concepts are important. To this end, we have included numerous writing questions and homework exercises that ask students to **"interpret the meaning in the context of the problem."** These questions make students stop and think, so they can process what they learn. In this way, students will learn underlying concepts. They will also form an understanding of what their answers mean in the contexts of the problems they solve.

Writing Style

Many students believe that reading a mathematics text is an exercise in futility. However, students who take the time to read the text and features may cast that notion aside. In particular, the **Tips** and **Avoiding Mistakes** boxes should prove especially enlightening. They offer the types of insights and hints that are usually only revealed during classroom lecture. On the whole, students should be very comfortable with the reading level, as the language and tone are consistent with those used daily within our own developmental mathematics classes.

Real-World Applications

Another critical component of the text is the inclusion of **contemporary real-world examples and applications.** We based examples and applications on information that students encounter daily when they turn on the news, read a magazine, or surf the Internet. We incorporated data for students to answer mathematical questions based on information in tables and graphs. When students encounter facts or information that is meaningful to them, they will relate better to the material and remember more of what they learn.

Study Skills

Many students in this course lack the basic study skills needed to be successful. Therefore, at the beginning of the homework exercises, we included a set of **Study Skills Exercises.** The exercises focus on one of nine areas: learning about the course, using the text, taking notes, completing homework assignments, test taking, time management, learning styles, preparing for a final exam, and defining **key terms.** Through completion of these exercises, students will be in a better position to pass the class and adopt techniques that will benefit them throughout their academic careers.

Language of Mathematics

Finally, for students to succeed in mathematics, they must be able to understand its language and notation. We place special emphasis on the skill of translating mathematical notation to English expressions and vice versa through **Translating Expressions Exercises.** These appear intermittently throughout the text. We also include key terms in the homework exercises and ask students to define these terms.

What Sets This Book Apart?

We believe that the thoughtfully designed pedagogy and contents of this textbook offer any willing student the opportunity to achieve success, opening the door to a wider world of possibilities.

While this textbook offers complete coverage of the beginning algebra curriculum, there are several concepts that receive special emphasis.

Problem Recognition

Problem recognition is an important theme carried throughout this edition and is integrated into the textbook in a number of different ways. First, we developed **Problem Recognition Exercises** that appear in selected chapters. The purpose of the Problem Recognition Exercises is to present a collection of problems that may look very similar to students, upon first glance, but are actually quite different in the manner of their individual solutions. By completing these exercises, students gain a greater awareness in problem recognition—that is, identifying a particular problem type upon inspection and applying the appropriate method to solve it.

We have carefully selected the content areas that seem to present the greatest challenge for students in terms of their ability to differentiate among various problem types. For these topics (listed below) we developed Problem Recognition Exercises.

Addition and Subtraction of Signed Numbers (page 88)
Properties of Exponents (page 404)
Operations on Polynomials (page 434)
Operations on Rational Expressions (page 552)
Comparing Rational Equations and Rational Expressions (page 571)
Operations on Radicals (page 650)

The problem recognition theme is also threaded throughout the book within the exercise sets. In particular, Section 6.6, "General Factoring Summary," is worth special mention. While not formally labeled "Problem Recognition," each factoring exercise requires students to label the *type* of factoring problem presented before attempting to factor the polynomial (see page 493, directions for problems 7–74).

The concept of problem recognition is also applied in every chapter, on a more micro level. We looked for opportunities within the section-ending Practice Exercises to include exercises that involve comparison between problem types. See, for example,

page 422, exercises 2–13
page 628, exercises 91 and 93 as well as 103 and 105
page 642, exercises 74–79 (contrast parts a and b)
page 649, exercises 51–58
page 649, exercises 59–62

We also included Mixed Exercises that appear in many of the Practice Exercise sets to give students opportunities to practice exercises that are not lumped together by problem type.

We firmly believe that students who effectively learn how to distinguish between various problems and the methods to solve them will be better prepared for Intermediate Algebra and courses beyond.

Design

While the content of a textbook is obviously critical, we believe that the design of the page is equally vital. We have often heard instructors describe the importance of "white space" in the design of developmental math textbooks and the need for simplicity, to prevent distractions. All of these comments have factored heavily in the page layout of this textbook. For example, we left ample space between exercises in the section-ending Practice Exercise sets to make it easier for students to read and complete homework assignments.

Similarly, we developed design treatments within sections to present the content, including examples, definitions, and summary boxes, in an organized and reader-friendly way. We also limited the number of colors and photos in use, in an effort to avoid distraction on the pages. We believe these considerations directly reflect the needs of our students and will enable them to navigate through the content successfully.

Chapter R

Chapter R is a reference chapter. We designed it to help students reacquaint themselves with the fundamentals of fractions, decimals, percents, and geometry. This chapter also addresses study skills and helpful hints to use the resources provided in the text and its supplements.

Factoring

Many years ago, we experimented in the classroom with our approach to factoring. We began factoring trinomials with the general case first, that is, with leading coefficient not equal to 1. This gave the students one rule for all cases, and it provided us with an extra class day for practice and group work. Most importantly, this approach forces students always to consider the leading coefficient, whether it be 1 or some other number. Thus, when students take the product of the inner terms and the product of the outer terms, the factors of the leading coefficient always come into play.

While we recommend presenting trinomials with leading coefficient other than 1 first, we want to afford flexibility to the instructors using this textbook. Therefore, we have structured our chapter on Factoring, Chapter 6, to offer instructors the option of covering trinomials with leading coefficient of 1 first (Section 6.2), or to cover trinomials with leading coefficient other than 1 first (Sections 6.3 and 6.4).

For those instructors who have never tried the method of presenting trinomials with leading coefficient other than 1 first, we suggest giving it a try. We have heard other instructors, who were at first resistant to this approach, remark how well it has worked for their students. In fact, one instructor told us, "I was skeptical, but I had the best exam results I have ever seen in factoring as a result of this approach! It has a big thumbs up from me and students alike."

Calculator Usage

The use of a scientific or a graphing calculator often inspires great debate among faculty who teach developmental mathematics. Our **Calculator Connections** boxes offer screen shots and some keystrokes to support applications where a calculator might enhance learning. Our approach is to use a calculator as a verification tool after analytical methods have been applied. The Calculator Connections boxes are self-contained units with accompanying exercises and can be employed or easily omitted at the recommendation of the instructor.

Calculator Exercises appear within the section-ending Practice Exercise sets where appropriate, but they are clearly noted as such and can be assigned or omitted at the instructor's discretion.

Listening to Students' and Instructors' Concerns

Our editorial staff has amassed the results of reviewer questionnaires, user diaries, focus groups, and symposia. We have consulted with a seven-member panel of beginning algebra instructors and their students on the development of this book. In addition, we have read hundreds of pages of reviews from instructors across the country. At McGraw-Hill symposia, faculty from across the United States gathered to discuss issues and trends in developmental mathematics. These efforts have involved hundreds of faculty and have explored issues such as content, readability, and even the aesthetics of page layout.

In our continuing efforts to improve our products, we invite you to contact us directly with your comments.

Julie Miller
millerj@dbcc.edu

Molly O'Neill
oneillm@dbcc.edu

Nancy Hyde
nhyde@montanasky.com

Acknowledgments and Reviewers

The development of this textbook would never have been possible without the creative ideas and constructive feedback offered by many reviewers. We are especially thankful to the following instructors for their valuable feedback and careful review of the manuscript.

Board of Advisors

Valarie Beaman-Hackle, *Macon State College*
Michael Everett, *Santa Ana College*
Teresa Hasenauer, *Indian River Community College*
Lori Holdren, *Manatee Community College–Bradenton*
Becky Hubiak, *Tidewater Community College*
Barbara Hughes, *San Jacinto College–Pasadena*
Mike Kirby, *Tidewater Community College*

Manuscript Reviewers

Cedric Atkins, *Charles Stewart Mott Community College*
David Bell, *Florida Community College*
Emilie Berglund, *Utah Valley State College*
Kirby Bunas, *Santa Rosa Junior College*
Annette Burden, *Youngstown State University*
Susan Caldiero, *Cosumnes River College*
Edie Carter, *Amarillo College*
Marcial Echenique, *Broward Community College*
Karen Estes, *St. Petersburg College*
Renu Gupta, *Louisiana State University–Alexandria*
Rodger Hergert, *Rock Valley College*
Linda Ho, *El Camino College*
Michelle Hollis, *Bowling Green Community College*
Laura Hoye, *Trident Technical College*
Randa Kress, *Idaho State University–Pocatello*
Donna Krichiver, *Johnson County Community College*
Karl Kruczek, *Northeastern State University*
Kathryn Lavelle, *Westchester Community College*
Jane Loftus, *Utah Valley State College*
Gary McCracken, *Shelton State Community College*
Karen Pagel, *Dona Ana Branch Community College*
Peter Remus, *Chicago State University*
Daniel Richbart, *Erie Community College–City Campus*

Special thanks go to Doris McClellan-Lewis for preparing the *Instructor's Solutions Manual* and the *Student's Solution Manual* and to Lauri Semarne for her work ensuring accuracy. Many, many thanks to Yolanda Davis, Andrea Hendricks, Patricia Jayne, and Cynthia Cruz for their work in the video series and to Kelly Jackson, an awesome teacher, for preparing the Instructor Notes. To Pat Steele, we appreciate your watchful eye over our manuscript.

Finally, we are forever grateful to the many people behind the scenes at McGraw-Hill, our publishing family. To Erin Brown, our lifeline on this project, without whom we'd be lost. To Liz Haefele for her passion for excellence and constant inspiration. To David Millage and Barb Owca for their countless hours of support and creative ideas promoting all of our efforts. To Jeff Huettman and Amber Huebner for the awesome technology so critical to student success, and finally to Vicki Krug for keeping the train on the track during production.

Most importantly, we give special thanks to all the students and instructors who use *Beginning Algebra* second edition in their classes.

Julie Miller Molly O'Neill Nancy Hyde

A COMMITMENT TO ACCURACY

You have a right to expect an accurate textbook, and McGraw-Hill invests considerable time and effort to make sure that we deliver one. Listed below are the many steps we take to make sure this happens.

OUR ACCURACY VERIFICATION PROCESS

First Round

Step 1: Numerous **college math instructors** review the manuscript and report on any errors that they may find, and the authors make these corrections in their final manuscript.

Second Round

Step 2: Once the manuscript has been typeset, the **authors** check their manuscript against the first page proofs to ensure that all illustrations, graphs, examples, exercises, solutions, and answers have been correctly laid out on the pages, and that all notation is correctly used.

Step 3: An outside, **professional mathematician** works through every example and exercise in the page proofs to verify the accuracy of the answers.

Step 4: A **proofreader** adds a triple layer of accuracy assurance in the first pages by hunting for errors, then a second, corrected round of page proofs is produced.

Third Round

Step 5: The **author team** reviews the second round of page proofs for two reasons: (1) to make certain that any previous corrections were properly made, and (2) to look for any errors they might have missed on the first round.

Step 6: A **second proofreader** is added to the project to examine the new round of page proofs to double check the author team's work and to lend a fresh, critical eye to the book before the third round of paging.

Fourth Round

Step 7: A **third proofreader** inspects the third round of page proofs to verify that all previous corrections have been properly made and that there are no new or remaining errors.

Step 8: Meanwhile, in partnership with **independent mathematicians,** the text accuracy is verified from a variety of fresh perspectives:

- The **test bank author** checks for consistency and accuracy as they prepare the computerized test item file.
- The **solutions manual author** works every single exercise and verifies their answers, reporting any errors to the publisher.
- A **consulting group of mathematicians,** who write material for the text's MathZone site, notifies the publisher of any errors they encounter in the page proofs.
- A video production company employing **expert math instructors** for the text's videos will alert the publisher of any errors they might find in the page proofs.

Final Round

Step 9: The **project manager,** who has overseen the book from the beginning, performs a **fourth proofread** of the textbook during the printing process, providing a final accuracy review.

⇒ What results is a mathematics textbook that is as accurate and error-free as is humanly possible, and our authors and publishing staff are confident that our many layers of quality assurance have produced textbooks that are the leaders of the industry for their integrity and correctness.

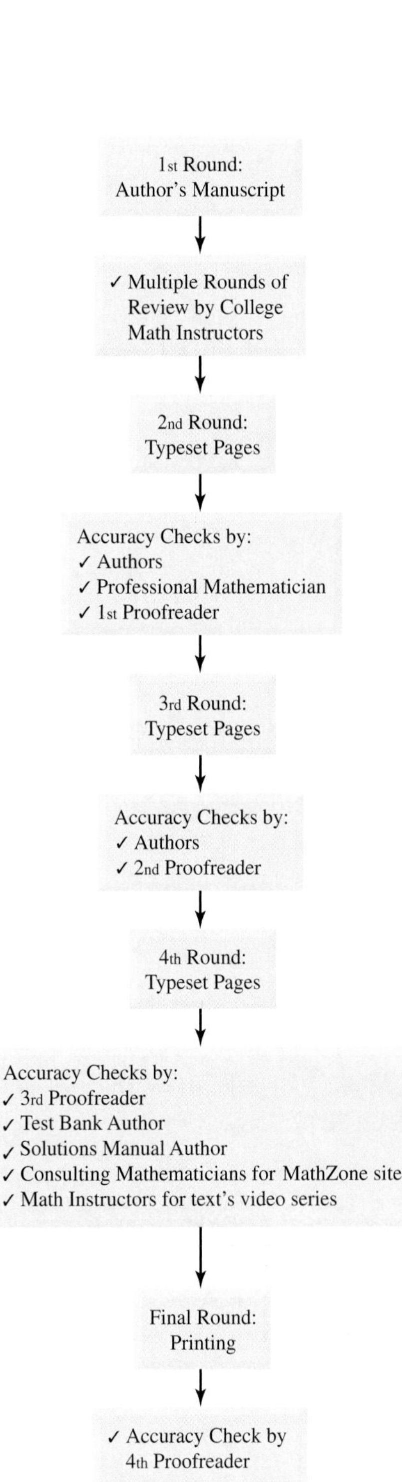

Guided Tour

Chapter Opener

Each chapter opens with a puzzle relating to the content of the chapter. Section titles are clearly listed for easy reference.

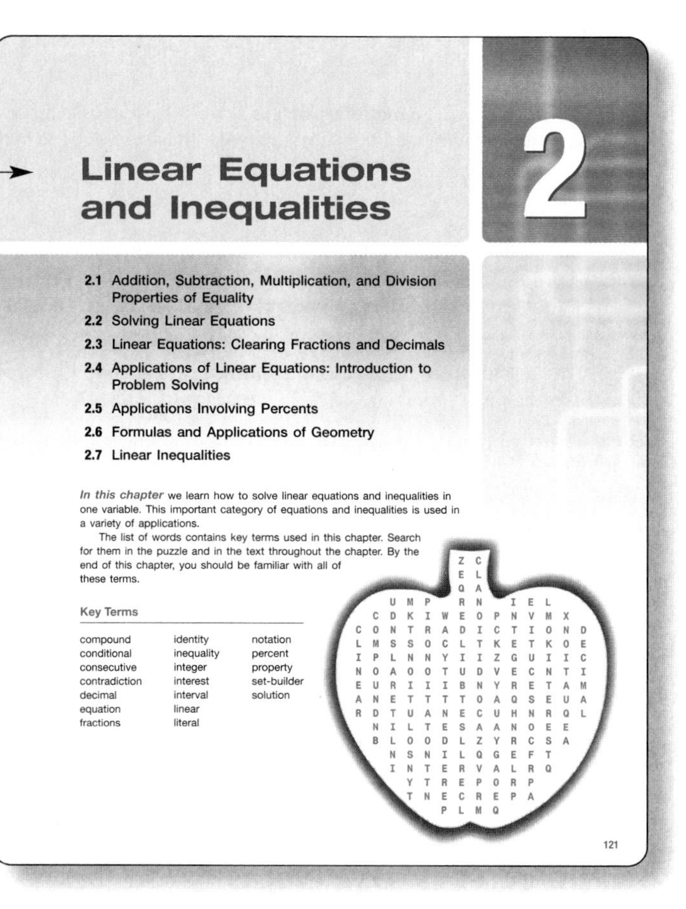

Linear Equations and Inequalities

2

2.1 Addition, Subtraction, Multiplication, and Division Properties of Equality
2.2 Solving Linear Equations
2.3 Linear Equations: Clearing Fractions and Decimals
2.4 Applications of Linear Equations: Introduction to Problem Solving
2.5 Applications Involving Percents
2.6 Formulas and Applications of Geometry
2.7 Linear Inequalities

In this chapter we learn how to solve linear equations and inequalities in one variable. This important category of equations and inequalities is used in a variety of applications.

The list of words contains key terms used in this chapter. Search for them in the puzzle and in the text throughout the chapter. By the end of this chapter, you should be familiar with all of these terms.

Key Terms

compound	identity	notation
conditional	inequality	percent
consecutive	integer	property
contradiction	interest	set-builder
decimal	interval	solution
equation	linear	
fractions	literal	

121

Concepts

A list of important learning concepts is provided at the beginning of each section. Each concept corresponds to a heading within the section and within the exercises, making it easy for students to locate topics as they study or as they work through homework exercises.

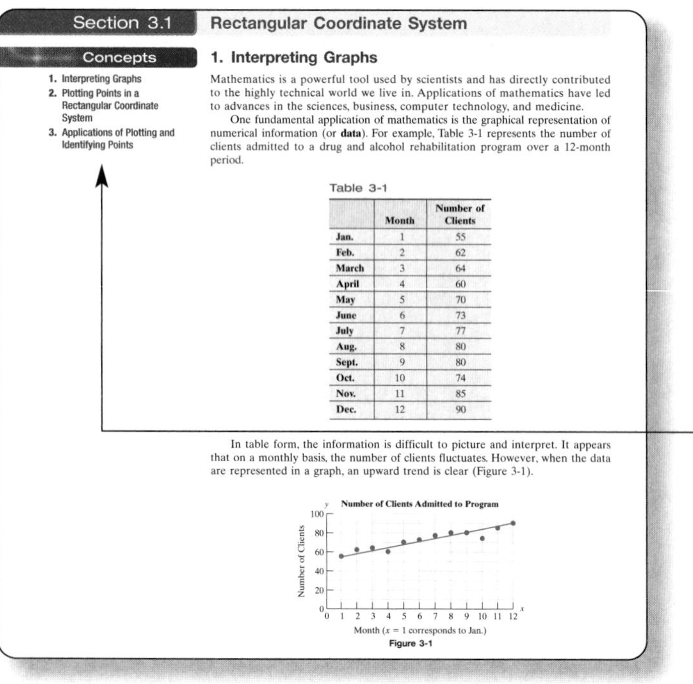

Section 3.1 Rectangular Coordinate System

Concepts

1. Interpreting Graphs
2. Plotting Points in a Rectangular Coordinate System
3. Applications of Plotting and Identifying Points

1. Interpreting Graphs

Mathematics is a powerful tool used by scientists and has directly contributed to the highly technical world we live in. Applications of mathematics have led to advances in the sciences, business, computer technology, and medicine.

One fundamental application of mathematics is the graphical representation of numerical information (or **data**). For example, Table 3-1 represents the number of clients admitted to a drug and alcohol rehabilitation program over a 12-month period.

Table 3-1

	Month	Number of Clients
Jan.	1	55
Feb.	2	62
March	3	64
April	4	60
May	5	70
June	6	73
July	7	77
Aug.	8	80
Sept.	9	80
Oct.	10	74
Nov.	11	85
Dec.	12	90

In table form, the information is difficult to picture and interpret. It appears that on a monthly basis, the number of clients fluctuates. However, when the data are represented in a graph, an upward trend is clear (Figure 3-1).

Figure 3-1

Worked Examples

Examples are set off in boxes and organized so that students can easily follow the solutions. Explanations appear beside each step and color-coding is used, where appropriate. For additional step-by-step instruction, students can run the "e-Professors" in MathZone. The e-Professors are based on worked examples from the text and use the solution methodologies presented in the text.

Example 5 Solving Linear Equations

Solve the equations:

a. $2 + 7x - 5 = 6(x + 3) + 2x$

b. $9 - (z - 3) + 4z = 4z - 5(z + 2) - 6$

Solution:

a. $2 + 7x - 5 = 6(x + 3) + 2x$

$-3 + 7x = 6x + 18 + 2x$ — **Step 1:** Add *like* terms on the left. Clear parentheses on the right.

$-3 + 7x = 8x + 18$ — Combine *like* terms.

$-3 + 7x - 7x = 8x - 7x + 18$ — **Step 2:** Subtract $7x$ from both sides.

$-3 = x + 18$ — Simplify.

$-3 - 18 = x + 18 - 18$ — **Step 3:** Subtract 18 from both sides.

$-21 = x$ — **Step 4:** Because the coefficient of the x term is already 1, there is no need to apply the multiplication or division property of equality.

$x = -21$

Step 5: The check is left to the reader.

b. $9 - (z - 3) + 4z = 4z - 5(z + 2) - 6$

$9 - z + 3 + 4z = 4z - 5z - 10 - 6$ — **Step 1:** Clear parentheses.

$12 + 3z = -z - 16$ — Combine *like* terms.

$12 + 3z + z = -z + z - 16$ — **Step 2:** Add z to both sides.

$12 + 4z = -16$

$12 - 12 + 4z = -16 - 12$ — **Step 3:** Subtract 12 from both sides.

$4z = -28$

$\frac{4z}{4} = \frac{-28}{4}$ — **Step 4:** Divide both sides by 4.

$z = -7$ — **Step 5:** The check is left for the reader.

Skill Practice Solve the equations.

6. $6y + 3 - y = 4(2y - 1)$ **7.** $3(3a - 4) - 4(5a + 2) = 7a - 2$

Skill Practice Exercises

Every worked example is followed by one or more Skill Practice exercises. These exercises offer students an immediate opportunity to work problems that mirror the examples. Students can then check their work by referring to the answers at the bottom of the page.

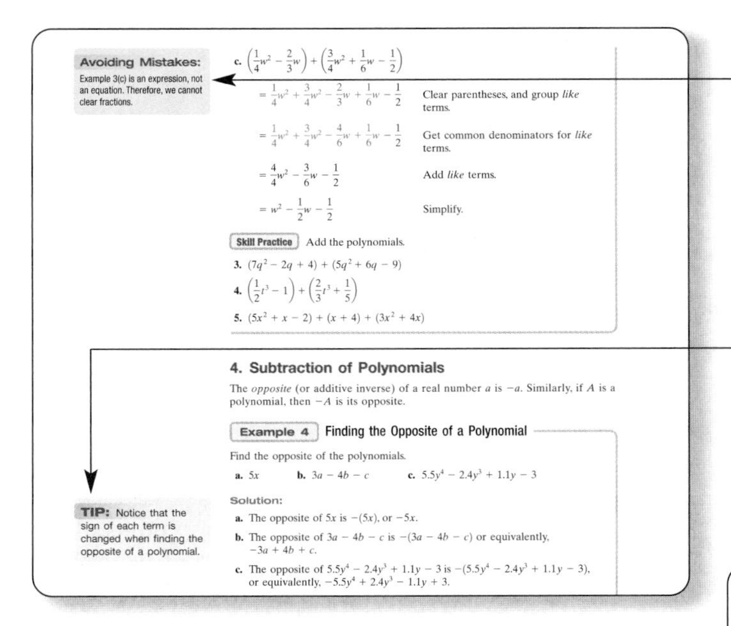

Avoiding Mistakes

Through notes labeled Avoiding Mistakes students are alerted to common errors and are shown methods to avoid them.

Tips

Tip boxes appear throughout the text and offer helpful hints and insight.

Problem Recognition Exercises

These exercises appear in selected chapters and cover topics that are inherently difficult for students. Problem recognition exercises give students the opportunity to practice the important, yet often overlooked, skill of distinguishing among different problem types.

Chapter 5 **Problem Recognition Exercises—Properties of Exponents**

Simplify completely. Assume that all variables represent nonzero real numbers.

1. t^3t^5 **2.** 2^32^5

3. $\frac{y^7}{y^2}$ **4.** $\frac{p^9}{p^3}$

5. $(r^2s^4)^2$ **6.** $(ab^3c^2)^3$

7. $\frac{w^4}{w^{-2}}$ **8.** $\frac{m^{-14}}{m^2}$

9. $\frac{y^{-7}x^4}{z^{-3}}$ **10.** $\frac{a^3b^{-6}}{c^{-8}}$

11. $(2.5 \times 10^{-3})(5.0 \times 10^5)$

12. $(3.1 \times 10^6)(4.0 \times 10^{-2})$

13. $\frac{4.8 \times 10^7}{6.0 \times 10^{-2}}$ **14.** $\frac{5.4 \times 10^{-2}}{9.0 \times 10^6}$

15. $\frac{1}{p^{-6}p^{-8}p^{-1}}$ **16.** p^6p^8p

21. $(2^5b^{-3})^{-3}$ **22.** $(3^{-2}y^3)^{-2}$

23. $\left(\frac{3x}{2y}\right)^{-4}$ **24.** $\left(\frac{6c}{5d^3}\right)^{-2}$

25. $(3ab^2)(a^2b)^3$ **26.** $(4x^2y^3)^3(xy^2)$

27. $\left(\frac{xy^2}{x^3y}\right)^4$ **28.** $\left(\frac{a^3b}{a^5b^3}\right)^5$

29. $\frac{(t^{-2})^3}{t^{-4}}$ **30.** $\frac{(p^3)^{-4}}{p^{-5}}$

31. $\left(\frac{2w^2x^3}{3y^0}\right)^3$ **32.** $\left(\frac{5a^0b^4}{4c^3}\right)^2$

33. $\frac{q^3r^{-2}}{s^{-1}t^5}$ **34.** $\frac{n^{-3}m^2}{p^{-3}q^{-1}}$

35. $\frac{(y^{-3})^2(y^5)}{(y^{-3})^{-4}}$ **36.** $\frac{(w^2)^{-4}(w^{-2})}{(w^5)^{-4}}$

37. $\left(\frac{-2a^2b^{-3}}{a^{-4}b^{-5}}\right)^{-3}$ **38.** $\left(\frac{-3x^{-4}y^3}{2x^5y^{-2}}\right)^{-2}$

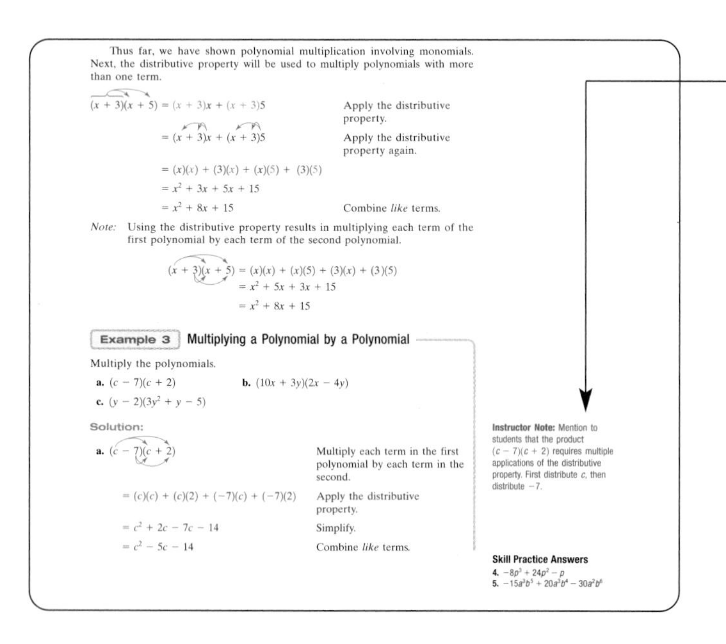

Thus far, we have shown polynomial multiplication involving monomials. Next, the distributive property will be used to multiply polynomials with more than one term.

$(x + 3)(x + 5) = (x + 3)x + (x + 3)5$ — Apply the distributive property.

$= (x + 3)x + (x + 3)5$ — Apply the distributive property again.

$= (x)(x) + (3)(x) + (x)(5) + (3)(5)$

$= x^2 + 3x + 5x + 15$

$= x^2 + 8x + 15$ — Combine *like* terms.

Note: Using the distributive property results in multiplying each term of the first polynomial by each term of the second polynomial.

$(x + 3)(x + 5) = (x)(x) + (x)(5) + (3)(x) + (3)(5)$
$= x^2 + 5x + 3x + 15$
$= x^2 + 8x + 15$

Example 3 Multiplying a Polynomial by a Polynomial

Multiply the polynomials.

a. $(c - 7)(c + 2)$ **b.** $(10x + 3y)(2x - 4y)$
c. $(y - 2)(3y^2 + y - 5)$

Solution:

a. $(c - 7)(c + 2)$ — Multiply each term in the first polynomial by each term in the second.

$= (c)(c) + (c)(2) + (-7)(c) + (-7)(2)$ — Apply the distributive property.

$= c^2 + 2c - 7c - 14$ — Simplify.

$= c^2 - 5c - 14$ — Combine *like* terms.

Instructor Note: Mention to students that the product $(c - 7)(c + 2)$ requires multiple applications of the distributive property. First distribute c, then distribute -7.

Skill Practice Answers
4. $-8p^3 + 24p^2 - p$
5. $-15a^3b^5 + 20a^3b^4 - 30a^2b^6$

Instructor Note (*AIE* only)

Instructor notes appear in the margins throughout each section of the *Annotated Instructor's Edition* (*AIE*). The notes may assist with lecture preparation in that they point out items that tend to confuse students, or lead students to err.

References to Classroom Activities (*AIE* only)

References are made to Classroom Activities at the beginning of each set of Practice Exercises in the *AIE*. The activities may be found in the *Instructor's Resource Manual*, which is available through MathZone and can be used during lecture or assigned for additional practice.

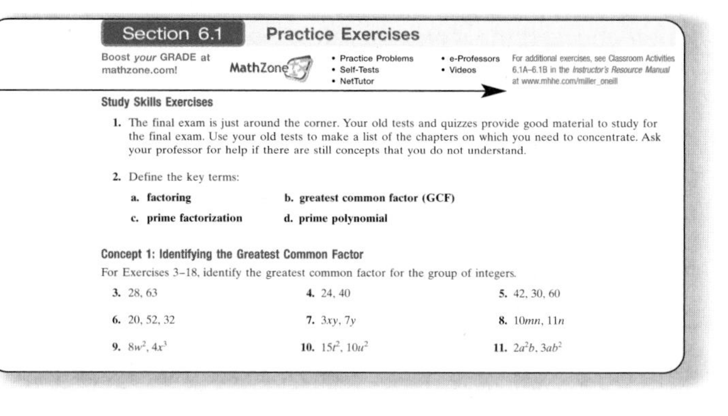

Section 6.1 Practice Exercises

Boost *your* GRADE at mathzone.com! MathZone
- Practice Problems
- Self-Tests
- NetTutor
- e-Professors
- Videos

For additional exercises, see Classroom Activities 6.1A–6.1B in the *Instructor's Resource Manual* at www.mhhe.com/miller_oneill

Study Skills Exercises

1. The final exam is just around the corner. Your old tests and quizzes provide good material to study for the final exam. Use your old tests to make a list of the chapters on which you need to concentrate. Ask your professor for help if there are still concepts that you do not understand.

2. Define the key terms:

a. factoring **b. greatest common factor (GCF)**
c. prime factorization **d. prime polynomial**

Concept 1: Identifying the Greatest Common Factor

For Exercises 3–18, identify the greatest common factor for the group of integers.

3. 28, 63 **4.** 24, 40 **5.** 42, 30, 60
6. 20, 52, 32 **7.** $3xy$, $7y$ **8.** $10mn$, $11n$
9. $8w^2$, $4x^3$ **10.** $15r^2$, $10u^2$ **11.** $2a^2b$, $3ab^2$

Section 6.2 Factoring Trinomials of the Form $x^2 + bx + c$ (Optional) 461

For Exercises 30–35, factor completely.

30. $-13x + x^2 - 30$ **31.** $12y - 160 + y^2$ **32.** $-18w + 65 + w^2$

33. $17t + t^2 + 72$ **34.** $22t + t^2 + 72$ **35.** $10q - 1200 + q^2$

36. Refer to page 457, write two important guidelines to follow when factoring trinomials.

For Exercises 37–48, factor completely. Be sure to factor out the GCF.

37. $3x^2 - 30x - 72$ **38.** $2z^2 + 4z - 198$ **39.** $8p^3 - 40p^2 + 32p$

40. $5w^4 - 35w^3 + 50w^2$ **41.** $y^4z^2 - 12y^3z^2 + 36y^2z^2$ **42.** $t^4u^2 + 6t^3u^2 + 9t^2u^2$

43. $-x^2 + 10x - 24$ **44.** $-y^2 - 12y - 35$ **45.** $-m^2 + m + 6$

46. $-n^2 + 5n + 6$ **47.** $-4 - 2c^2 - 6c$ **48.** $-40d - 30 - 10d^2$

Mixed Exercises

For Exercises 49–66, factor completely.

49. $x^3y^3 - 19x^2y^3 + 60xy^3$ **50.** $y^2z^5 + 17yz^5 + 60z^5$ **51.** $12p^2 - 96p + 84$

52. $5w^2 - 40w - 45$ **53.** $-2m^2 + 22m - 20$ **54.** $-3x^2 - 36x - 81$

55. $c^2 + 6cd + 5d^2$ **56.** $x^2 + 8xy + 12y^2$ **57.** $a^2 - 9ab + 14b^2$

58. $m^2 - 15mn + 44n^2$ **59.** $a^2 + 4a + 18$ **60.** $b^2 - 6a + 15$

61. $2q + q^2 - 63$ **62.** $-32 - 4t + t^2$ **63.** $x^2 + 20x + 100$

64. $z^2 - 24z + 144$ **65.** $t^2 + 18t - 40$ **66.** $d^2 + 2d - 99$

67. A student factored a trinomial as $(2x - 4)(x - 3)$. The instructor did not give full credit. Why?

68. A student factored a trinomial as $(y + 2)(5y - 15)$. The instructor did not give full credit. Why?

69. What polynomial factors as $(x - 4)(x + 13)$?

70. What polynomial factors as $(q - 7)(q + 10)$?

Expanding Your Skills

71. Find all integers, b, that make the trinomial $x^2 + bx + 6$ factorable.

72. Find all integers, b, that make the trinomial $x^2 + bx + 10$ factorable.

73. Find a value of c that makes the trinomial $x^2 + 6x + c$ factorable.

74. Find a value of c that makes the trinomial $x^2 + 8x + c$ factorable.

Practice Exercises

A variety of problem types appear in the section-ending Practice Exercises. Problem types are clearly labeled with either a heading or an icon for easy identification. References to MathZone are also found at the beginning of the Practice Exercises to remind students and instructors that additional help and practice problems are available. The core exercises for each section are organized by section concept. General references to examples are provided for blocks of core exercises. **Mixed Exercises** are also provided in some sections where no reference to concepts is offered.

Icon Key

The following key has been prepared for easy identification of "themed" exercises appearing within the Practice Exercises.

Student Edition

- Exercises Keyed to Video
- Calculator Exercises

AIE only

- Writing
- Translating Expressions
- Geometry

Study Skills Exercises appear at the beginning of selected exercise sets. They are designed to help students learn vocabulary and techniques to improve their study habits including exam preparation, note taking, and time management.

In the Practice Exercises, where appropriate, students are asked to define the **Key Terms** that are presented in the section. Assigning these exercises will help students to develop and expand their mathematical vocabularies.

Review Exercises also appear at the start of the Practice Exercises. The purpose of the Review Exercises is to help students retain their knowledge of concepts previously learned.

178 **Chapter 2** Linear Equations and Inequalities

Section 2.6 Practice Exercises

Boost *your* GRADE at mathzone.com! MathZone

- Practice Problems
- Self-Tests
- NetTutor
- e-Professors
- Videos

Study Skills Exercises

1. A good technique for studying for a test is to choose four problems from each section of the chapter and write each of them along with the directions on a 3×5 card. On the back, put the page number where you found that problem. Then shuffle the cards and test yourself on the procedure to solve each problem. If you find one that you do not know how to solve, look at the page number and do several of that type. Write four problems you would choose for this section.

2. Define the key term: **literal equation**

Review Exercises

For Exercises 3–7, solve the equation.

3. $3(2y + 3) - 4(-y + 1) = 7y - 10$

4. $-(3w + 4) + 5(w - 2) - 3(6w - 8) = 10$

5. $\frac{1}{2}(x - 3) + \frac{3}{4} = 3x - \frac{3}{4}$

6. $\frac{5}{6}x + \frac{1}{2} = \frac{1}{4}(x - 4)$

7. $0.5(y + 2) - 0.3 = 0.4y + 0.5$

Concept 1: Formulas

For Exercises 8–39, solve for the indicated variable.

8. $P = a + b + c$ for a **9.** $P = a + b + c$ for b **10.** $x = y - z$ for y

11. $c + d = e$ for d **12.** $p = 250 + q$ for q **13.** $y = 35 + x$ for x

14. $d = rt$ for t **15.** $d = rt$ for r **16.** $PV = nrt$ for t

17. $P_1V_1 = P_2V_2$ for V_1 **18.** $x - y = 5$ for x **19.** $x + y = -2$ for y

20. $3x + y = -19$ for y **21.** $x - 6y = -10$ for x **22.** $2x + 3y = 6$ for y

23. $5x + 2y = 10$ for y **24.** $-2x - y = 9$ for x **25.** $3x - y = -13$ for x

26. $4x - 3y = 12$ for y **27.** $6x - 3y = 4$ for y **28.** $ax + by = c$ for y

29. $ax + by = c$ for x **30.** $A = P(1 + rt)$ for t **31.** $P = 2(L + w)$ for L

32. $a = 2(b + c)$ for c **33.** $3(x + y) = z$ for x **34.** $Q = \frac{x + y}{2}$ for y

35. $Q = \frac{a - b}{2}$ for a **36.** $M = \frac{a}{S}$ for a **37.** $A = \frac{1}{3}(a + b + c)$ for c

38. $P = I^2R$ for R **39.** $F = \frac{GMm}{d^2}$ for m

Writing Exercises offer students an opportunity to conceptualize and communicate their understanding of algebra. These, along with the **Translating Expressions Exercises**, enable students to strengthen their command of mathematical language and notation and to improve their reading and writing skills.

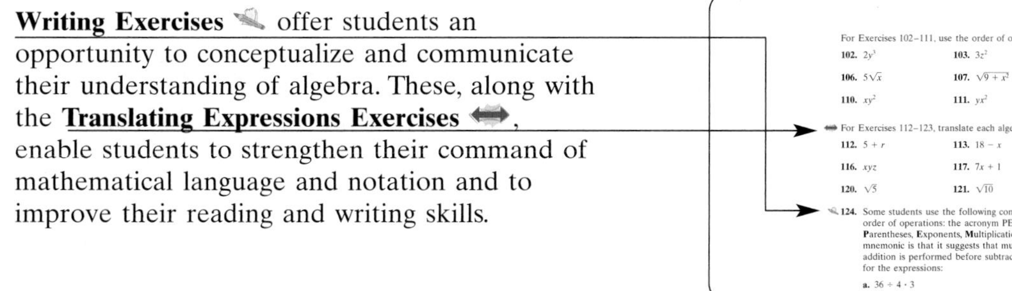

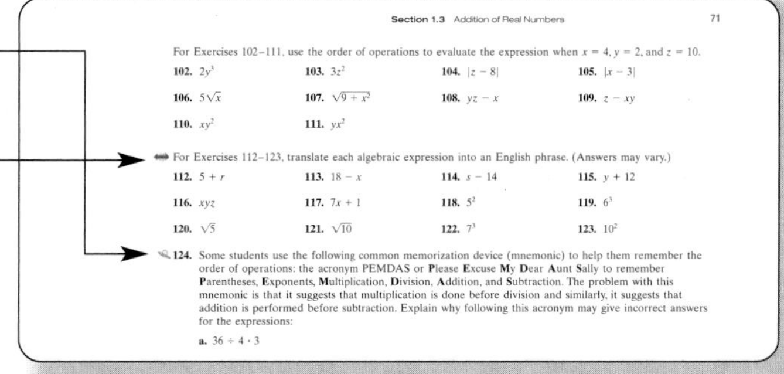

Section 1.3 Addition of Real Numbers 71

For Exercises 102–111, use the order of operations to evaluate the expression when $x = 4$, $y = 2$, and $z = 10$.

102. $2y^3$ **103.** $3z^2$ **104.** $|z - 8|$ **105.** $|x - 3|$

106. $5\sqrt{x}$ **107.** $\sqrt{9 + x^2}$ **108.** $yz - x$ **109.** $z - xy$

110. xy^2 **111.** yx^2

For Exercises 112–123, translate each algebraic expression into an English phrase. (Answers may vary.)

112. $5 + r$ **113.** $18 - x$ **114.** $s - 14$ **115.** $y + 12$

116. xyz **117.** $7x + 1$ **118.** 5^2 **119.** 6^3

120. $\sqrt{5}$ **121.** $\sqrt{10}$ **122.** 7^3 **123.** 10^2

124. Some students use the following common memorization device (mnemonic) to help them remember the order of operations: the acronym PEMDAS or **P**lease **E**xcuse **M**y **D**ear **A**unt **S**ally to remember **P**arentheses, **E**xponents, **M**ultiplication, **D**ivision, **A**ddition, and **S**ubtraction. The problem with this mnemonic is that it suggests that multiplication is done before division and similarly, it suggests that addition is performed before subtraction. Explain why following this acronym may give incorrect answers for the expressions:

a. $36 \div 4 \cdot 3$

Geometry Exercises appear throughout the Practice Exercises and encourage students to review and apply geometry concepts.

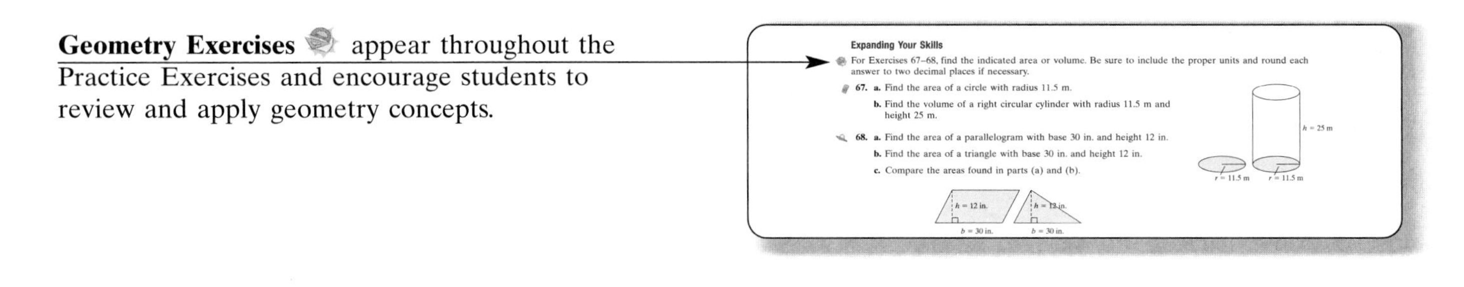
Expanding Your Skills

For Exercises 67–68, find the indicated area or volume. Be sure to include the proper units and round each answer to two decimal places if necessary.

67. a. Find the area of a circle with radius 11.5 m.

b. Find the volume of a right circular cylinder with radius 11.5 m and height 25 m.

68. a. Find the area of a parallelogram with base 30 in. and height 12 in.

b. Find the area of a triangle with base 30 in. and height 12 in.

c. Compare the areas found in parts (a) and (b).

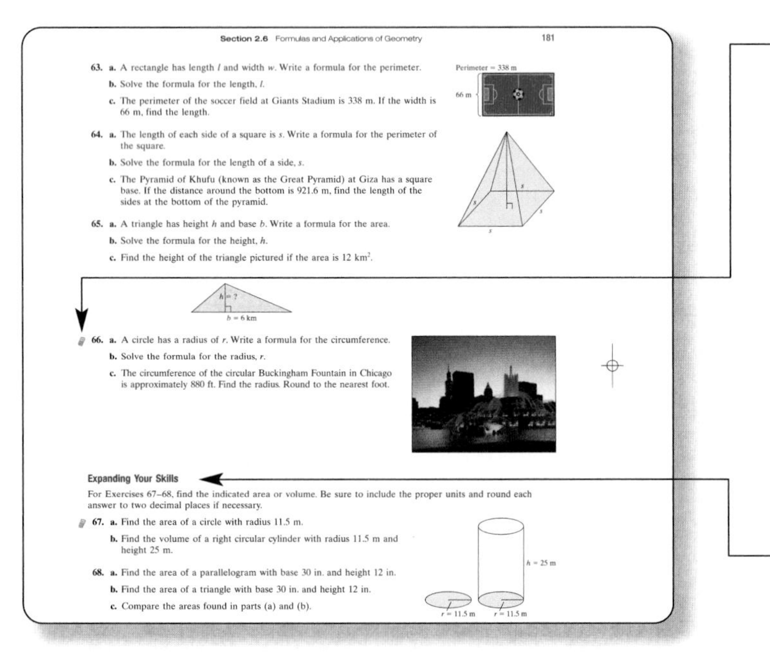
Section 2.6 Formulas and Applications of Geometry 181

63. a. A rectangle has length l and width w. Write a formula for the perimeter.

b. Solve the formula for the length, l.

c. The perimeter of the soccer field at Giants Stadium is 338 m. If the width is 66 m, find the length.

64. a. The length of each side of a square is s. Write a formula for the perimeter of the square.

b. Solve the formula for the length of a side, s.

c. The Pyramid of Khufu (known as the Great Pyramid) at Giza has a square base. If the distance around the bottom is 921.6 m, find the length of the sides at the bottom of the pyramid.

65. a. A triangle has height h and base b. Write a formula for the area.

b. Solve the formula for the height, h.

c. Find the height of the triangle pictured if the area is 12 km^2.

66. a. A circle has a radius of r. Write a formula for the circumference.

b. Solve the formula for the radius, r.

c. The circumference of the circular Buckingham Fountain in Chicago is approximately 880 ft. Find the radius. Round to the nearest foot.

Expanding Your Skills

For Exercises 67–68, find the indicated area or volume. Be sure to include the proper units and round each answer to two decimal places if necessary.

67. a. Find the area of a circle with radius 11.5 m.

b. Find the volume of a right circular cylinder with radius 11.5 m and height 25 m.

68. a. Find the area of a parallelogram with base 30 in. and height 12 in.

b. Find the area of a triangle with base 30 in. and height 12 in.

c. Compare the areas found in parts (a) and (b).

Calculator Exercises signify situations where a calculator would provide assistance for time-consuming calculations. These exercises were carefully designed to demonstrate the types of situations in which a calculator is a handy tool rather than a "crutch."

Exercises Keyed to Video are labeled with an icon to help students and instructors identify those exercises for which accompanying video instruction is available.

Applications based on real-world facts and figures motivate students and enable them to hone their problem-solving skills.

Expanding Your Skills exercises, found near the end of most Practice Exercises, challenge students' knowledge of the concepts presented.

Optional Calculator Connections are located at the ends of the sections and appear intermittently. They can be implemented at the instructor's discretion depending on the amount of emphasis placed on the calculator in the course. The Calculator Connections display keystrokes and provide an opportunity for students to apply the skill introduced.

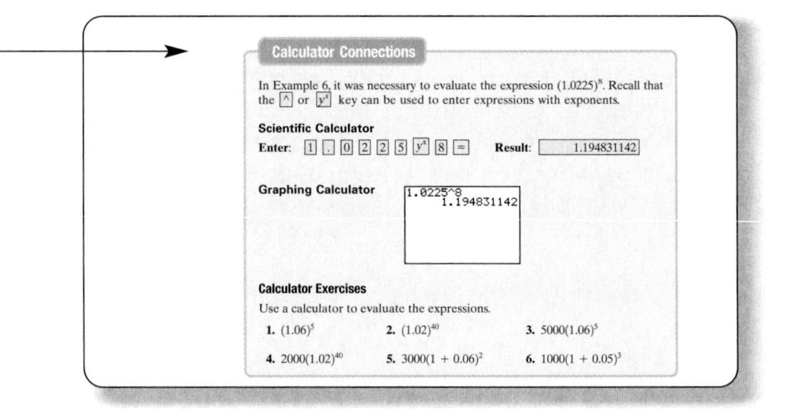
Calculator Connections

In Example 6, it was necessary to evaluate the expression $(1.0225)^8$. Recall that the [^] or [y^x] key can be used to enter expressions with exponents.

Scientific Calculator

Enter: [1] [.] [0] [2] [2] [5] [y^x] [8] [=] **Result:** 1.194831142

Graphing Calculator

1.0225^8
1.194831142

Calculator Exercises

Use a calculator to evaluate the expressions.

1. $(1.06)^5$ **2.** $(1.02)^{40}$ **3.** $5000(1.06)^5$

4. $2000(1.02)^{40}$ **5.** $3000(1 + 0.06)^2$ **6.** $1000(1 + 0.05)^3$

End-of-Chapter Summary and Exercises

The **Summary**, located at the end of each chapter, outlines key concepts for each section and illustrates those concepts with examples. Following the Summary is a set of **Review Exercises** that are organized by section. A **Chapter Test** appears after each set of Review Exercises. Chapters 2–9 also include **Cumulative Reviews** that follow the Chapter Tests. These end-of-chapter materials provide students with ample opportunity to prepare for quizzes or exams.

Chapter 6 SUMMARY

Section 6.1 Greatest Common Factor and Factoring by Grouping

Key Concepts

The **greatest common factor** (GCF) is the greatest factor common to all terms of a polynomial. To factor out the GCF from a polynomial, use the distributive property.

A four-term polynomial may be factorable by grouping.

Steps to Factoring by Grouping

1. Identify and factor out the GCF from all four terms.
2. Factor out the GCF from the first pair of terms. Factor out the GCF or its opposite from the second pair of terms.
3. If the two terms share a common binomial factor, factor out the binomial factor.

Examples

Example 1

$3x(a + b) - 5(a + b)$ Greatest common factor is $(a + b)$.

$= (a + b)(3x - 5)$

Example 2

$60xa - 30xb - 80ya + 40yb$

$= 10[6xa - 3xb - 8ya + 4yb]$ Factor out GCF.

$= 10[3x(2a - b) - 4y(2a - b)]$ Factor by grouping.

$= 10(2a - b)(3x - 4y)$

Chapter 6 Review Exercises

Section 6.1

For Exercises 1–4, identify the greatest common factor between each group of terms.

1. $15a^2b^4, 30a^3b, 9a^5b^3$ **2.** $3(x + 5), x(x + 5)$

3. $2c^3(3c - 5), 4c(3c - 5)$ **4.** $-2wyz, -4xyz$

For Exercises 5–10, factor out the greatest common factor.

5. $6x^2 + 2x^4 - 8x$ **6.** $11w^3y^3 - 44w^2y^5$

7. $-t^2 + 5t$ **8.** $-6u^2 - u$

9. $3b(b + 2) - 7(b + 2)$

10. $2(5x + 9) + 8x(5x + 9)$

26. When factoring a polynomial of the form $ax^2 - bx + c$, should the signs of the binomials be both positive, both negative, or different?

27. When factoring a polynomial of the form $ax^2 + bx + c$, should the signs of the binomials be both positive, both negative, or different?

28. When factoring a polynomial of the form $ax^2 + bx - c$, should the signs of the binomials be both positive, both negative, or different?

For Exercises 29–40, factor the trinomial using the trial-and-error method.

29. $2y^2 - 5y - 12$ **30.** $4w^2 - 5w - 6$

31. $10z^2 + 29z + 10$ **32.** $8z^2 + 6z - 9$

Chapter 6 Test

1. Factor out the GCF: $15x^4 - 3x + 6x^3$

2. Factor by grouping: $7a - 35 - a^2 + 5a$

3. Factor the trinomial: $6w^2 - 43w + 7$

4. Factor the difference of squares: $169 - p^2$

5. Factor the perfect square trinomial: $q^2 - 16q + 64$

6. Factor the sum of cubes: $8 + t^3$

For Exercises 7–25, factor completely.

7. $a^2 + 12a + 32$ **8.** $x^2 + x - 42$

9. $2y^2 - 17y + 8$ **10.** $6z^2 + 19z + 8$

11. $9t^2 - 100$ **12.** $v^2 - 81$

13. $3a^2 + 27ab + 54b^2$ **14.** $c^4 - 1$

15. $xy - 7x + 3y - 21$ **16.** $49 + p^2$

17. $-10u^2 + 30u - 20$ **18.** $12t^2 - 75$

19. $5y^2 - 50y + 125$ **20.** $21q^2 + 14q$

21. $2x^3 + x^2 - 8x - 4$ **22.** $y^3 - 125$

23. $m^2n^2 - 81$ **24.** $16a^2 - 64b^2$

25. $64x^3 - 27y^6$

For Exercises 26–30, solve the equation.

26. $(2x - 3)(x + 5) = 0$ **27.** $x^2 - 7x = 0$

28. $x^2 - 6x = 16$ **29.** $x(5x + 4) = 1$

30. $y^3 + 10y^2 - 9y - 90 = 0$

31. A tennis court has area of 312 yd^2. If the length is 2 yd more than twice the width, find the dimensions of the court.

32. The product of two consecutive odd integers is 35. Find the integers.

33. The height of a triangle is 5 in. less than the length of the base. The area is 42 in^2. Find the length of the base and the height of the triangle.

34. The hypotenuse of a right triangle is 2 ft less than three times the shorter leg. The longer leg is 3 ft less than three times the shorter leg. Find the length of the shorter leg.

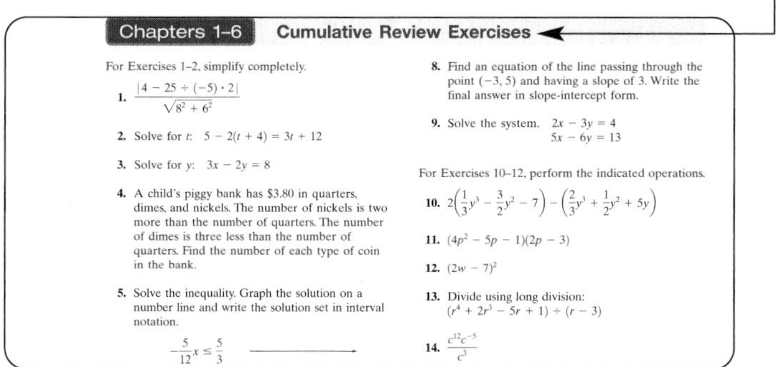

Chapters 1–6 Cumulative Review Exercises

For Exercises 1–2, simplify completely.

1. $\dfrac{|4 - 25 \div (-5) \cdot 2|}{\sqrt{8^2 + 6^2}}$

2. Solve for t: $5 - 2(t + 4) = 3t + 12$

3. Solve for y: $3x - 2y = 8$

4. A child's piggy bank has $3.80 in quarters, dimes, and nickels. The number of nickels is two more than the number of quarters. The number of dimes is three less than the number of quarters. Find the number of each type of coin in the bank.

5. Solve the inequality. Graph the solution on a number line and write the solution set in interval notation.

$-\dfrac{5}{12}x \le \dfrac{5}{3}$

8. Find an equation of the line passing through the point $(-3, 5)$ and having a slope of 3. Write the final answer in slope-intercept form.

9. Solve the system. $2x - 3y = 4$, $5x - 6y = 13$

For Exercises 10–12, perform the indicated operations.

10. $2\left(\frac{1}{3}y^3 - \frac{3}{2}y^2 - 7\right) - \left(\frac{2}{3}y^3 + \frac{1}{2}y^2 + 5y\right)$

11. $(4p^2 - 5p - 1)(2p - 3)$

12. $(2w - 7)^2$

13. Divide using long division: $(r^4 + 2r^3 - 5r + 1) \div (r - 3)$

14. $\dfrac{c^{12}c^{-5}}{c^3}$

Supplements

For the Instructor

Instructor's Resource Manual

The *Instructor's Resource Manual* (*IRM*), written by the authors, is a printable electronic supplement available through MathZone. The *IRM* includes discovery-based classroom activities, worksheets for drill and practice, materials for a student portfolio, and tips for implementing successful cooperative learning. Numerous classroom activities are available for each section of text and can be used as a complement to lecture or can be assigned for work outside of class. The activities are designed for group or individual work and take about 5–10 minutes each. With increasing demands on faculty schedules, these ready-made lessons offer a convenient means for both full-time and adjunct faculty to promote active learning in the classroom.

McGraw-Hill's **MathZone** is a complete online tutorial and course management system for mathematics and statistics, designed for greater ease of use than any other system available. Available with selected McGraw-Hill textbooks, the system enables instructors to **create and share courses and assignments** with colleagues and adjuncts with only a few clicks of the mouse. All assignments, questions, e-Professors, online tutoring, and video lectures are directly tied to **text-specific** materials.

MathZone courses are customized to your textbook, but you can edit questions and algorithms, import your own content, and **create** announcements and due dates for assignments.

MathZone has **automatic grading** and reporting of easy-to-assign, algorithmically generated homework, quizzing, and testing. All student activity within **MathZone** is automatically recorded and available to you through a **fully integrated grade book** that can be downloaded to Excel.

MathZone offers:

- **Practice exercises** based on the textbook and generated in an unlimited number for as much practice as needed to master any topic you study.
- **Videos** of classroom instructors giving lectures and showing you how to solve exercises from the textbook.
- **e-Professors** to take you through animated, step-by-step instructions (delivered via on-screen text and synchronized audio) for solving problems in the book, allowing you to digest each step at your own pace.
- **NetTutor,** which offers live, personalized tutoring via the Internet.

Instructor's Testing and Resource CD

This cross-platform CD-ROM provides a wealth of resources for the instructor. Among the supplements featured on the CD-ROM is a **computerized test bank** utilizing Brownstone Diploma® algorithm-based testing software to create customized exams quickly. This user-friendly program enables instructors to search for questions by topic, format, or difficulty level; to edit existing questions or to add new ones; and to scramble questions and answer keys for multiple versions of a single test. Hundreds of text-specific, open-ended, and multiple-choice questions are included in the question bank. Sample chapter tests are also provided.

ALEKS version 3.0 **ALEKS** (**A**ssessment and **LE**arning in **K**nowledge **S**paces) is an artificial-intelligence-based system for mathematics learning, available over the Web 24/7. Using unique adaptive questioning, ALEKS accurately assesses what topics each student knows and then determines exactly what each student is ready to learn next. ALEKS interacts with the students much as a skilled human tutor would, moving between explanation and practice as needed, correcting and analyzing errors, defining terms and changing topics on request, and helping them master the course content more quickly and easily. Moreover, the new ALEKS 3.0 now links to text-specific videos, multimedia tutorials, and textbook pages in PDF format. ALEKS also offers a robust classroom management system that enables instructors to monitor and direct student progress toward mastery of curricular goals. See www.highed.aleks.com.

New Connect2Developmental Mathematics Video Series!

Available on DVD and the MathZone website, these innovative videos bring essential Developmental Mathematics concepts to life! The videos take the concepts and place them in a real world setting so that students make the connection from what they learn in the classroom to real world experiences outside the classroom. Making use of 3D animations and lectures, Connect2Developmental Mathematics video series answers the age-old questions "Why is this important" and "When will I ever use it?" The videos cover topics from Arithmetic and Basic Mathematics through the Algebra sequence, mixing student-oriented themes and settings with basic theory.

Miller/O'Neill/Hyde Video Lectures on Digital Video Disk (DVD)

In the videos, qualified instructors work through selected problems from the textbook, following the solution methodology employed in the text. The video series is available on DVD or online as an assignable element of MathZone. The DVDs are closed-captioned for the hearing-impaired, are subtitled in Spanish, and meet the Americans with Disabilities Act Standards for Accessible Design. Instructors may use them as resources in a learning center, for online courses, and/or to provide extra help for students who require extra practice.

Annotated Instructor's Edition

In the *Annotated Instructor's Edition* (*AIE*), **answers to all exercises and tests appear adjacent to each exercise,** in a color used *only* for annotations. The *AIE* also contains **Instructor Notes** that appear in the margin. The notes may assist with lecture preparation. Also found in the *AIE* are icons within the Practice Exercises that serve to guide instructors in their preparation of homework assignments and lessons.

Instructor's Solutions Manual

The *Instructor's Solutions Manual* provides comprehensive, worked-out solutions to all exercises in the Chapter Openers; Practice Exercises; the Problem Recognition Exercises; the end-of-chapter Review Exercises; the Chapter Tests; and the Cumulative Review Exercises.

For the Student

www.mathzone.com

McGraw-Hill's MathZone is a powerful Web-based tutorial for homework, quizzing, testing, and multimedia instruction. Also available in CD-ROM format, MathZone offers:

Practice exercises based on the text and generated in an unlimited quantity for as much practice as needed to master any objective

Video clips of classroom instructors showing how to solve exercises from the text, step by step

e-Professor animations that take the student through step-by-step instructions, delivered on-screen and narrated by a teacher on audio, for solving exercises from the textbook; the user controls the pace of the explanations and can review as needed

NetTutor, which offers personalized instruction by live tutors familiar with the textbook's objectives and problem-solving methods.

Every assignment, exercise, video lecture, and e-Professor is derived from the textbook.

Student's Solutions Manual

The *Student's Solutions Manual* provides comprehensive, worked-out solutions to the odd-numbered exercises in the Practice Exercise sets; the Problem Recognition Exercises, the end-of-chapter Review Exercises, the Chapter Tests, and the Cumulative Review Exercises. Answers to the odd-and the even-numbered entries to the Chapter Opener puzzles are also provided.

New Connect2Developmental Mathematics Video Series!

Available on DVD and the MathZone website, these innovative videos bring essential Developmental Mathematics concepts to life! The videos take the concepts and place them in a real world setting so that students make the connection from what they learn in the classroom to real world experiences outside the classroom. Making use of 3D animations and lectures, Connect2Developmental Mathematics video series answers the age-old questions "Why is this important" and "When will I ever use it?" The videos cover topics from Arithmetic and Basic Mathematics through the Algebra sequence, mixing student-oriented themes and settings with basic theory.

Video Lectures on Digital Video Disk

The video series is based on exercises from the textbook. Each presenter works through selected problems, following the solution methodology employed in the text. The video series is available on DVD or online as part of MathZone. The DVDs are closed-captioned for the hearing impaired, are subtitled in Spanish, and meet the Americans with Disabilities Act Standards for Accessible Design.

NetTutor

Available through MathZone, NetTutor is a revolutionary system that enables students to interact with a live tutor over the Web. NetTutor's Web-based, graphical chat capabilities enable students and tutors to use mathematical notation and even to draw graphs as they work through a problem together. Students can also submit questions and receive answers, browse previously answered questions, and view previous sessions. Tutors are familiar with the textbook's objectives and problem-solving styles.

R

Reference

R.1 Study Tips

R.2 Fractions

R.3 Decimals and Percents

R.4 Introduction to Geometry

Chapter R is a reference chapter that provides a review of the basic operations on fractions, decimals, and percents. It also offers background on some of the important facts from geometry that will be used later in the text.

As you work through Section R.4, see if you recognize any of these important shapes and figures. Match each figure number on the left with a letter on the right. Then fill in the blanks below with the appropriate letter. The letters will form a word to complete the puzzle.

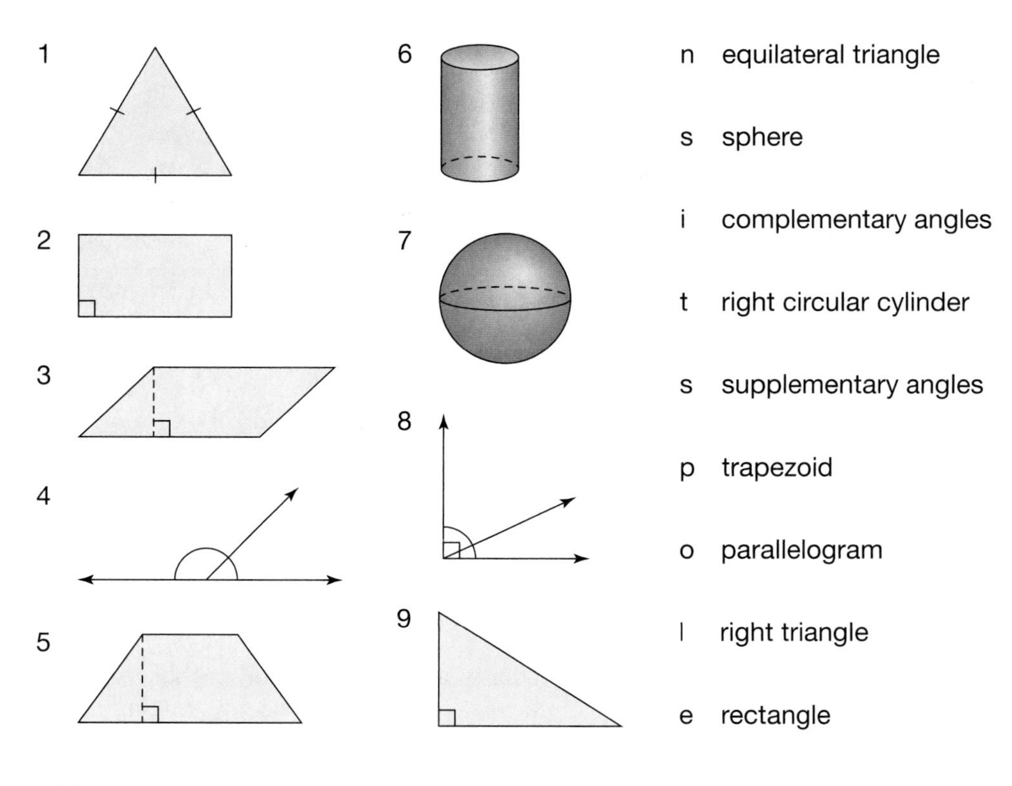

Without geometry, life would be __ __ __ __ __ __ __ __ __
5 3 8 1 6 9 2 7 4

Section R.1 Study Tips

Concepts

1. Before the Course
2. During the Course
3. Preparation for Exams
4. Where to Go for Help

In taking a course in algebra, you are making a commitment to yourself, your instructor, and your classmates. Following some or all of the study tips listed here can help you be successful in this endeavor. The features of this text that will assist you are printed in blue.

1. Before the Course

- Purchase the necessary materials for the course before the course begins or on the first day.
- Obtain a three-ring binder to keep and organize your notes, homework, tests, and any other materials acquired in the class. We call this type of notebook a portfolio.
- Arrange your schedule so that you have enough time to attend class and to do homework. A common rule of thumb is to set aside at least 2 hours for homework for every hour spent in class. That is, if you are taking a 4 credit-hour course, plan on at least 8 hours a week for homework. If you experience difficulty in mathematics, plan for more time. A 4 credit-hour course will then take *at least* 12 hours each week—about the same as a part-time job.
- Communicate with your employer and family members the importance of your success in this course so that they can support you.
- Be sure to find out the type of calculator (if any) that your instructor requires.

2. During the Course

- Read the section in the text *before* the lecture to familiarize yourself with the material and terminology.
- Attend every class, and be on time.
- Take notes in class. Write down all of the examples that the instructor presents. Read the notes after class, and add any comments to make your notes clearer to you. Use a tape recorder to record the lecture if the instructor permits the recording of lectures.
- Ask questions in class.
- Read the section in the text *after* the lecture, and pay special attention to the Tip boxes and Avoiding Mistakes boxes.
- Do homework every night. Even if your class does not meet every day, you should still do some work every night to keep the material fresh in your mind.
- Check your homework with the answers that are supplied in the back of this text. Correct the exercises that do not match, and circle or star those that you cannot correct yourself. This way you can easily find them and ask your instructor the next day.
- Write the definition and give an example of each Key Term found at the beginning of the Practice Exercises.
- The Problem Recognition Exercises provide additional practice distinguishing among a variety of problem types. Sometimes the most difficult part of learning mathematics is retaining all that you learn. These exercises are excellent tools for retention of material.
- Form a study group with fellow students in your class, and exchange phone numbers. You will be surprised by how much you can learn by talking about mathematics with other students.

- If you use a calculator in your class, read the Calculator Connections boxes to learn how and when to use your calculator.
- Ask your instructor where you might obtain extra help if necessary.

3. Preparation for Exams

- Look over your homework. Pay special attention to the exercises you have circled or starred to be sure that you have learned that concept.
- Read through the Summary at the end of the chapter. Be sure that you understand each concept and example. If not, go to the section in the text and reread that section.
- Give yourself enough time to take the Chapter Test uninterrupted. Then check the answers. For each problem you answered incorrectly, go to the Review Exercises and do all of the problems that are similar.
- To prepare for the final exam, complete the Cumulative Review Exercises at the end of each chapter, starting with Chapter 2. If you complete the cumulative reviews after finishing each chapter, then you will be preparing for the final exam throughout the course. The Cumulative Review Exercises are another excellent tool for helping you retain material.

4. Where to Go for Help

- At the first sign of trouble, see your instructor. Most instructors have specific office hours set aside to help students. Don't wait until after you have failed an exam to seek assistance.
- Get a tutor. Most colleges and universities have free tutoring available.
- When your instructor and tutor are unavailable, use the Student Solutions Manual for step-by-step solutions to the odd-numbered problems in the exercise sets.
- Work with another student from your class.
- Work on the computer. Many mathematics tutorial programs and websites are available on the Internet, including the website that accompanies this text: www.mhhe.com/miller_oneill

Section R.1 Practice Exercises

Boost *your* GRADE at mathzone.com!

- Practice Problems
- Self-Tests
- NetTutor
- e-Professors
- Videos

Concept 1: Before the Course

1. To be motivated to complete a course, it is helpful to have a very clear reason for taking the course. List your goals for taking this course.

2. Budgeting enough time to do homework and to study for a class is one of the most important steps to success in a class. Use the weekly calendar to help you plan your time for your studies this week. Also write other obligations such as the time required for your job, for your family, for sleeping, and for eating. Be realistic when estimating the time for each activity.

Time	Mon.	Tues.	Wed.	Thurs.	Fri.	Sat.	Sun.
7–8							
8–9							
9–10							
10–11							
11–12							
12–1							
1–2							
2–3							
3–4							
4–5							
5–6							
6–7							
7–8							
8–9							
9–10							

3. Taking 12 credit hours is the equivalent of a full-time job. Often students try to work too many hours while taking classes at school.

a. Write down how many hours you work per week and the number of credit hours you are taking this term.

number of hours worked per week ______________

number of credit hours this term ______________

b. The next table gives a recommended limit to the number of hours you should work for the number of credit hours you are taking at school. (Keep in mind that other responsibilities in your life such as your family might also make it necessary to limit your hours at work even more.) How do your numbers from part (a) compare to those in the table? Are you working too many hours?

Number of credit hours	Maximum number of hours of work per week
3	40
6	30
9	20
12	10
15	0

4. It is important to establish a place where you can study—someplace that has few distractions and is readily available. Answer the questions about the space that you have chosen.

a. Is there enough room to spread out books and paper to do homework?

b. Is this space available anytime?

c. Are there any distractions? Can you be interrupted?

d. Is the furniture appropriate for studying? That is, is there a comfortable chair and good lighting?

5. Organization is an important ingredient to success. A calendar or pocket planner is a valuable resource for keeping track of assignments and test dates. Write the date of the first test in this class.

Concept 2: During the Course

6. Taking notes can help in many ways. Good notes provide examples for reference as you do your homework. Also, taking notes keeps your mind on track during the lecture and helps make you an active listener. Here are some tips to help you take better notes in class.

a. In your next math class, take notes by drawing a vertical line about $\frac{3}{4}$ of the way across the paper as shown. On the left side, write down what your instructor puts on the board or overhead. On the right side, make your own comments about important words, procedures, or questions that you have.

b. Revisit your notes as soon as possible after class to fill in the missing parts that you recall from lecture but did not have time to write down.

c. Be sure that you label each page with the date, chapter, section, and topic. This will make it easier to study from when you study for a test.

7. Many instructors use a variety of styles to accommodate all types of learners. From the following list, try to identify the type of learner that best describes you.

Auditory Learner Do you learn best by listening to your instructor's lectures? Do you tape the lecture so that you can listen to it as many times as you need? Do you talk aloud when doing homework or study for a test?

Visual Learner Do you learn best by seeing problems worked out on the board? Do you understand better if there is a picture or illustration accompanying the problem? Do you take notes in class?

Tactile Learner Do you learn best with hands-on projects? Do you prefer having some sort of physical objects to manipulate? Do you prefer to move around the classroom in a lab situation?

Concept 3: Preparation for Exams

8. When taking a test, go through the test and work all the problems that you know first. Then go back and work on the problems that were more difficult. Give yourself a time limit for each problem (maybe 3 to 5 minutes the first time through the test). Circle the importance of each statement.

	not important	somewhat important	very important
a. Read through the entire test first.	1	2	3
b. If time allows, go back and check each problem.	1	2	3
c. Write out all steps instead of doing the work in your head.	1	2	3

9. One way to lessen test anxiety is to feel prepared for the exam. Check the ways that you think might be helpful in preparing for a test.

_____ Do the chapter test in the text.

_____ Get in a study group with your peers and go over the major topics.

_____ Write your own pretest.

_____ Use the online component that goes with your textbook.

_____ Write an outline of the major topics. Then find an example for each topic.

_____ Make flash cards with a definition, property, or rule on one side and an example on the other.

_____ Re-read your notes that you took in class.

_____ Write down the definitions of all key terms introduced in the chapter.

10. The following list gives symptoms of math anxiety.

- Experiencing loss of sleep and worrying about an upcoming exam.
- Mind becoming blank when answering a question or taking a test.
- Becoming nervous about asking questions in class.
- Experiencing anxiety that interferes with studying.
- Being afraid or embarrassed to let the instructor see your work.
- Becoming physically ill during a test.
- Having sweaty palms or shaking hands when asked a math question.

Have you experienced any of these symptoms? If so, how many and how often?

11. If you think that you have math anxiety, read the following list for some possible solutions. Check the activities that you can realistically try to help you overcome this problem.

______ Read a book on math anxiety.

______ Search the Web for help tips on handling math anxiety.

______ See a counselor to discuss your anxiety.

______ See your instructor to inform him or her about your situation.

______ Evaluate your time management to see if you are trying to do too much. Then adjust your schedule accordingly.

Concept 4: Where to Go for Help

12. Does your college offer free tutoring? If so, write down the room number and the hours of the tutoring center.

13. Does your instructor have office hours? If so, write down your professor's office number and office hours.

14. Is there a supplement to your text? If so, find out its price and where you can get it.

15. Find out how to access the online tutoring available with this text.

Section R.2 Fractions

Concepts

1. Basic Definitions
2. Prime Factorization
3. Simplifying Fractions to Lowest Terms
4. Multiplying Fractions
5. Dividing Fractions
6. Adding and Subtracting Fractions
7. Operations on Mixed Numbers

1. Basic Definitions

The study of algebra involves many of the operations and procedures used in arithmetic. Therefore, we begin this text by reviewing the basic operations of addition, subtraction, multiplication, and division on fractions and mixed numbers.

In day-to-day life, the numbers we use for counting are

the **natural numbers**: $1, 2, 3, 4, \ldots$ and

the **whole numbers**: $0, 1, 2, 3, \ldots$.

Whole numbers are used to count the number of whole units in a quantity. A fraction is used to express part of a whole unit. If a child gains $2\frac{1}{2}$ lb, the child has gained two whole pounds plus a portion of a pound. To express the additional half pound mathematically, we may use the fraction, $\frac{1}{2}$.

A Fraction and Its Parts

Fractions are numbers of the form $\frac{a}{b}$, where $\frac{a}{b} = a \div b$ and b does not equal zero.

In the fraction $\frac{a}{b}$, the **numerator** is a, and the **denominator** is b.

The denominator of a fraction indicates how many equal parts divide the whole. The numerator indicates how many parts are being represented. For instance, suppose Jack wants to plant carrots in $\frac{2}{5}$ of a rectangular garden. He can divide the garden into five equal parts and use two of the parts for carrots (Figure R-1).

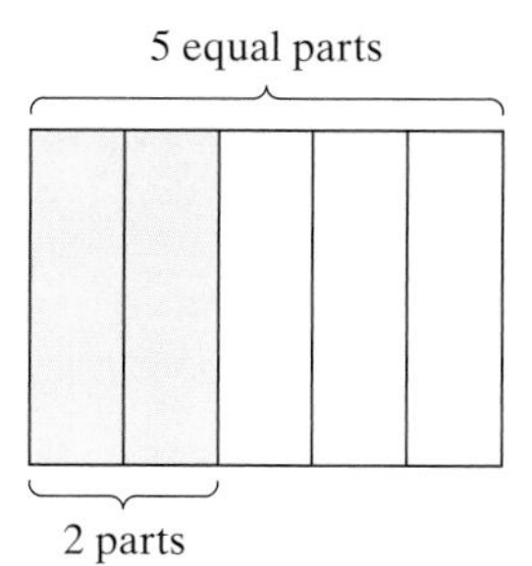

The shaded region represents $\frac{2}{5}$ of the garden.

Figure R-1

Definition of a Proper Fraction, an Improper Fraction, and a Mixed Number

1. If the numerator of a fraction is less than the denominator, the fraction is a **proper fraction**. A proper fraction represents a quantity that is less than a whole unit.
2. If the numerator of a fraction is greater than or equal to the denominator, then the fraction is an **improper fraction**. An improper fraction represents a quantity greater than or equal to a whole unit.
3. A **mixed number** is a whole number added to a proper fraction.

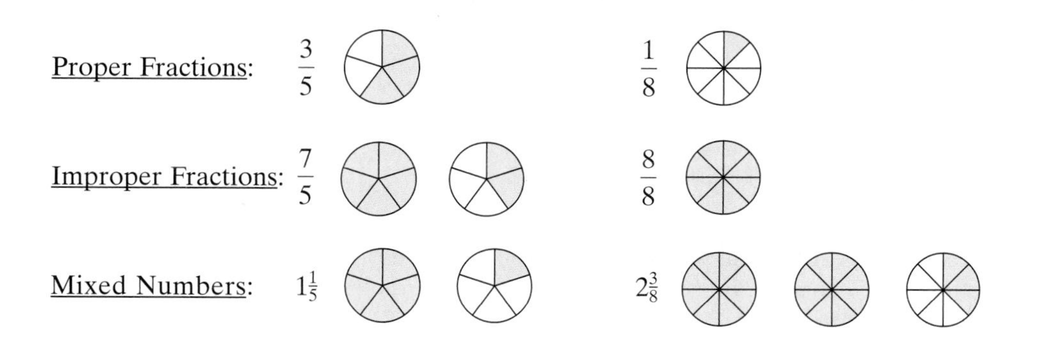

2. Prime Factorization

To perform operations on fractions it is important to understand the concept of a factor. For example, when the numbers 2 and 6 are multiplied, the result (called the **product**) is 12.

$$2 \times 6 = 12$$

factors product

The numbers 2 and 6 are said to be **factors** of 12. (In this context, we refer only to natural number factors.) The number 12 is said to be factored when it is written as the product of two or more natural numbers. For example, 12 can be factored in several ways:

$$12 = 1 \times 12 \qquad 12 = 2 \times 6 \qquad 12 = 3 \times 4 \qquad 12 = 2 \times 2 \times 3$$

A natural number greater than 1 that has only two factors, 1 and itself, is called a **prime number**. The first several prime numbers are 2, 3, 5, 7, 11, and 13. A natural number greater than 1 that is not prime is called a **composite number**. That is, a composite number has factors other than itself and 1. The first several composite numbers are 4, 6, 8, 9, 10, 12, 14, 15, and 16.

The number 1 is neither prime nor composite.

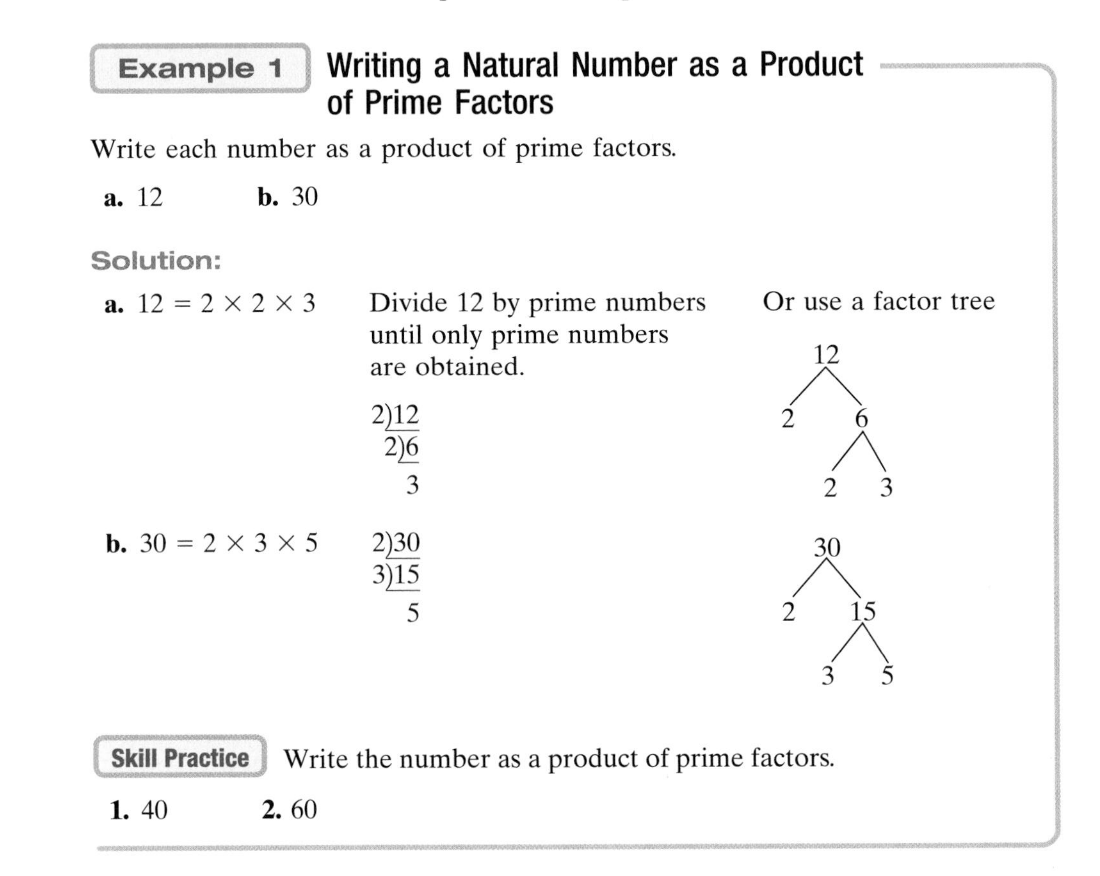

Example 1 **Writing a Natural Number as a Product of Prime Factors**

Write each number as a product of prime factors.

a. 12 **b.** 30

Solution:

a. $12 = 2 \times 2 \times 3$

Divide 12 by prime numbers until only prime numbers are obtained.

$2\overline{)12}$, $2\overline{)6}$, 3

Or use a factor tree

b. $30 = 2 \times 3 \times 5$

$2\overline{)30}$, $3\overline{)15}$, 5

Skill Practice Write the number as a product of prime factors.

1. 40 **2.** 60

3. Simplifying Fractions to Lowest Terms

The process of factoring numbers can be used to reduce or simplify fractions to lowest terms. A fractional portion of a whole can be represented by infinitely many fractions. For example, Figure R-2 shows that $\frac{1}{2}$ is equivalent to $\frac{2}{4}, \frac{3}{6}, \frac{4}{8}$, and so on.

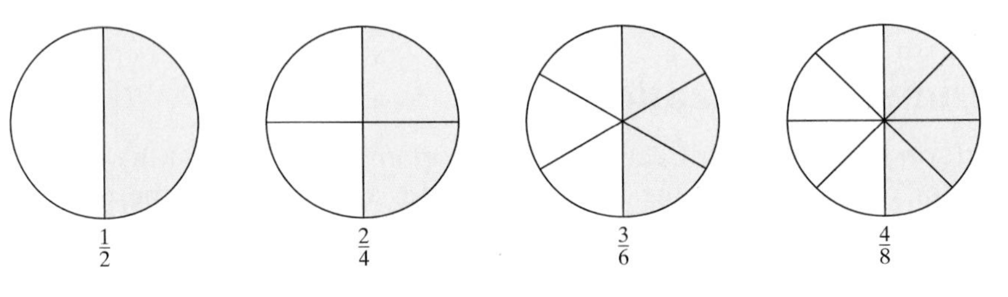

Figure R-2

Skill Practice Answers

1. $2 \times 2 \times 2 \times 5$
2. $2 \times 2 \times 3 \times 5$

The fraction $\frac{1}{2}$ is said to be in **lowest terms** because the numerator and denominator share no common factor other than 1.

To simplify a fraction to lowest terms, we use the following important principle.

The Fundamental Principle of Fractions

Consider the fraction $\dfrac{a}{b}$ and the nonzero number, c. Then,

$$\frac{a}{b} = \frac{a \div c}{b \div c}$$

The fundamental principle of fractions indicates that dividing both the numerator and denominator by the same nonzero number results in an equivalent fraction. For example, the numerator and denominator of the fraction $\frac{3}{6}$ both share a common factor of 3. To reduce $\frac{3}{6}$, we will divide both the numerator and denominator by the common factor, 3.

$$\frac{3}{6} = \frac{3 \div 3}{6 \div 3} = \frac{1}{2}$$

Before applying the fundamental principle of fractions, it is helpful to write the prime factorization of both the numerator and denominator. We can also use a slash, /, to denote division of common factors.

Example 2 Simplifying a Fraction to Lowest Terms

Simplify the fraction $\dfrac{30}{12}$ to lowest terms.

Solution:

From Example 1, we have $30 = 2 \times 3 \times 5$ and $12 = 2 \times 2 \times 3$. Hence,

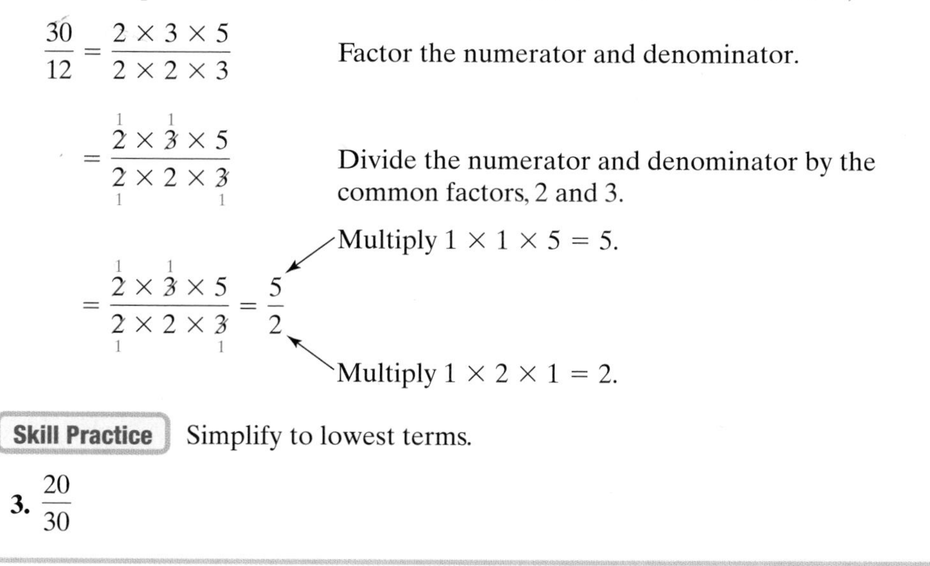

$$\frac{30}{12} = \frac{2 \times 3 \times 5}{2 \times 2 \times 3}$$ Factor the numerator and denominator.

$$= \frac{\overset{1}{\cancel{2}} \times \overset{1}{\cancel{3}} \times 5}{\underset{1}{\cancel{2}} \times 2 \times \underset{1}{\cancel{3}}}$$ Divide the numerator and denominator by the common factors, 2 and 3.

$$= \frac{\overset{1}{\cancel{2}} \times \overset{1}{\cancel{3}} \times 5}{\underset{1}{\cancel{2}} \times 2 \times \underset{1}{\cancel{3}}} = \frac{5}{2}$$

Multiply $1 \times 1 \times 5 = 5$.

Multiply $1 \times 2 \times 1 = 2$.

Skill Practice Simplify to lowest terms.

3. $\dfrac{20}{30}$

Skill Practice Answers

3. $\dfrac{2}{3}$

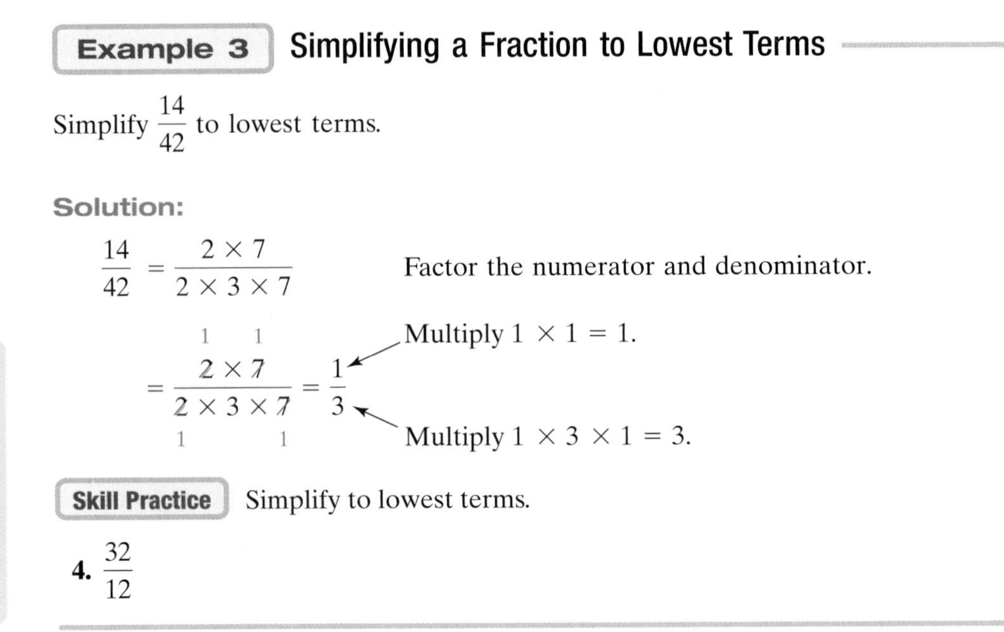

Example 3 Simplifying a Fraction to Lowest Terms

Simplify $\dfrac{14}{42}$ to lowest terms.

Solution:

$$\frac{14}{42} = \frac{2 \times 7}{2 \times 3 \times 7}$$ Factor the numerator and denominator.

$$= \frac{\overset{1}{\cancel{2}} \times \overset{1}{\cancel{7}}}{\underset{1}{\cancel{2}} \times 3 \times \underset{1}{\cancel{7}}} = \frac{1}{3}$$

Multiply $1 \times 1 = 1$.

Multiply $1 \times 3 \times 1 = 3$.

Skill Practice Simplify to lowest terms.

4. $\dfrac{32}{12}$

Avoiding Mistakes:

In Example 3, the common factors 2 and 7 in the numerator and denominator simplify to 1. It is important to remember to write the factor of 1 in the numerator. The simplified form of the fraction is $\frac{1}{3}$.

4. Multiplying Fractions

Multiplying Fractions

If b is not zero and d is not zero, then

$$\frac{a}{b} \times \frac{c}{d} = \frac{a \times c}{b \times d}$$

To multiply fractions, multiply the numerators and multiply the denominators.

Example 4 Multiplying Fractions

Multiply the fractions: $\dfrac{1}{4} \times \dfrac{1}{2}$

Solution:

$$\frac{1}{4} \times \frac{1}{2} = \frac{1 \times 1}{4 \times 2} = \frac{1}{8}$$ Multiply the numerators. Multiply the denominators.

Notice that the product $\frac{1}{4} \times \frac{1}{2}$ represents a quantity that is $\frac{1}{4}$ of $\frac{1}{2}$. Taking $\frac{1}{4}$ of a quantity is equivalent to dividing the quantity by 4. One-half of a pie divided into four pieces leaves pieces that each represent $\frac{1}{8}$ of the pie (Figure R-3).

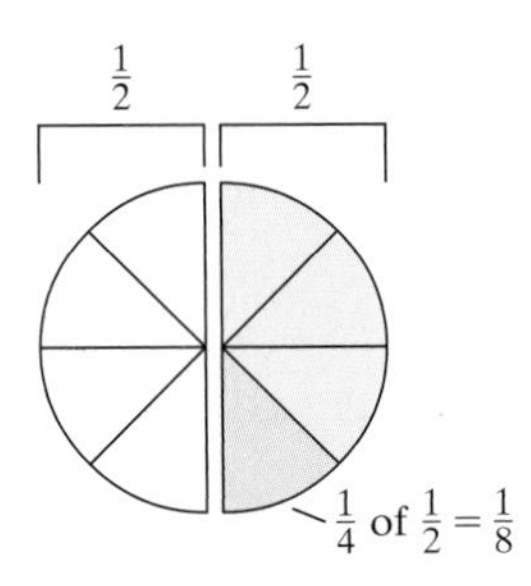

Figure R-3

Skill Practice Answers

4. $\dfrac{8}{3}$

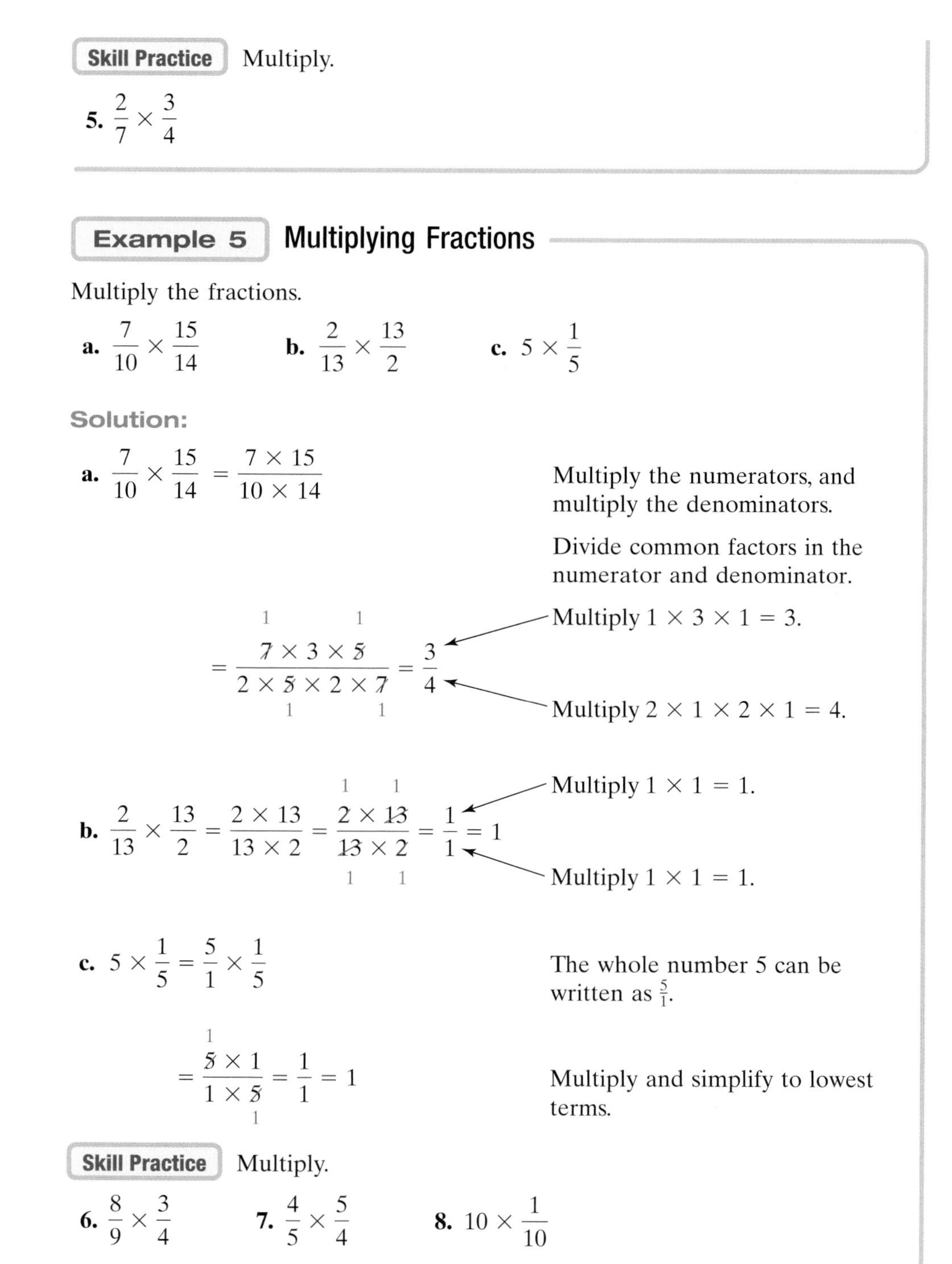

Skill Practice Multiply.

5. $\dfrac{2}{7} \times \dfrac{3}{4}$

Example 5 **Multiplying Fractions**

Multiply the fractions.

a. $\dfrac{7}{10} \times \dfrac{15}{14}$ **b.** $\dfrac{2}{13} \times \dfrac{13}{2}$ **c.** $5 \times \dfrac{1}{5}$

Solution:

a. $\dfrac{7}{10} \times \dfrac{15}{14} = \dfrac{7 \times 15}{10 \times 14}$ Multiply the numerators, and multiply the denominators.

Divide common factors in the numerator and denominator.

$= \dfrac{\overset{1}{\cancel{7}} \times 3 \times \overset{1}{\cancel{5}}}{2 \times \underset{1}{\cancel{5}} \times 2 \times \underset{1}{\cancel{7}}} = \dfrac{3}{4}$

Multiply $1 \times 3 \times 1 = 3$.

Multiply $2 \times 1 \times 2 \times 1 = 4$.

b. $\dfrac{2}{13} \times \dfrac{13}{2} = \dfrac{2 \times 13}{13 \times 2} = \dfrac{\overset{1}{\cancel{2}} \times \overset{1}{\cancel{13}}}{\underset{1}{\cancel{13}} \times \underset{1}{\cancel{2}}} = \dfrac{1}{1} = 1$

Multiply $1 \times 1 = 1$.

Multiply $1 \times 1 = 1$.

c. $5 \times \dfrac{1}{5} = \dfrac{5}{1} \times \dfrac{1}{5}$ The whole number 5 can be written as $\frac{5}{1}$.

$= \dfrac{\overset{1}{\cancel{5}} \times 1}{1 \times \underset{1}{\cancel{5}}} = \dfrac{1}{1} = 1$ Multiply and simplify to lowest terms.

Skill Practice Multiply.

6. $\dfrac{8}{9} \times \dfrac{3}{4}$ 7. $\dfrac{4}{5} \times \dfrac{5}{4}$ 8. $10 \times \dfrac{1}{10}$

5. Dividing Fractions

Before we divide fractions, we need to know how to find the reciprocal of a fraction. Notice from Example 5 that $\frac{2}{13} \times \frac{13}{2} = 1$ and $5 \times \frac{1}{5} = 1$. The numbers $\frac{2}{13}$ and $\frac{13}{2}$ are said to be reciprocals because their product is 1. Likewise the numbers 5 and $\frac{1}{5}$ are reciprocals.

Skill Practice Answers

5. $\dfrac{3}{14}$ 6. $\dfrac{2}{3}$ 7. 1 8. 1

The Reciprocal of a Number

Two nonzero numbers are **reciprocals** of each other if their product is 1. Therefore, the reciprocal of the fraction

$$\frac{a}{b} \text{ is } \frac{b}{a} \qquad \text{because} \qquad \frac{a}{b} \times \frac{b}{a} = 1$$

Number	Reciprocal	Product
$\frac{2}{13}$	$\frac{13}{2}$	$\frac{2}{13} \times \frac{13}{2} = 1$
$\frac{1}{8}$	$\frac{8}{1}$ (or equivalently 8)	$\frac{1}{8} \times 8 = 1$
$6\left(\text{or equivalently } \frac{6}{1}\right)$	$\frac{1}{6}$	$6 \times \frac{1}{6} = 1$

To understand the concept of dividing fractions, consider a pie that is half-eaten. Suppose the remaining half must be divided among three people, that is, $\frac{1}{2} \div 3$. However, dividing by 3 is equivalent to taking $\frac{1}{3}$ of the remaining $\frac{1}{2}$ of the pie (Figure R-4).

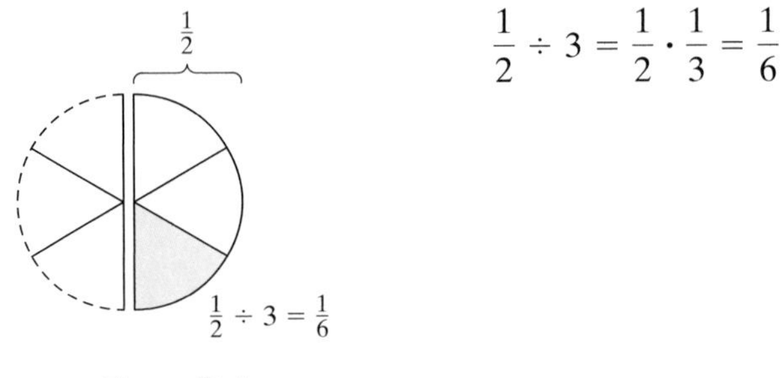

Figure R-4

This example illustrates that dividing two numbers is equivalent to multiplying the first number by the reciprocal of the second number.

Division of Fractions

Let a, b, c, and d be numbers such that b, c, and d are not zero. Then,

multiply

$$\frac{a}{b} \div \frac{c}{d} = \frac{a}{b} \times \frac{d}{c}$$

reciprocal

To divide fractions, multiply the first fraction by the reciprocal of the second fraction.

Example 6 Dividing Fractions

Divide the fractions.

a. $\dfrac{8}{5} \div \dfrac{3}{10}$ **b.** $\dfrac{12}{13} \div 6$

Solution:

a. $\dfrac{8}{5} \div \dfrac{3}{10} = \dfrac{8}{5} \times \dfrac{10}{3}$ Multiply by the reciprocal of $\frac{3}{10}$, which is $\frac{10}{3}$.

$= \dfrac{8 \times \overset{2}{\cancel{10}}}{\underset{1}{\cancel{5}} \times 3} = \dfrac{16}{3}$ Multiply and simplify to lowest terms.

b. $\dfrac{12}{13} \div 6 = \dfrac{12}{13} \div \dfrac{6}{1}$ Write the whole number 6 as $\frac{6}{1}$.

$= \dfrac{12}{13} \times \dfrac{1}{6}$ Multiply by the reciprocal of $\frac{6}{1}$, which is $\frac{1}{6}$.

$= \dfrac{\overset{2}{\cancel{12}} \times 1}{13 \times \underset{1}{\cancel{6}}} = \dfrac{2}{13}$ Multiply and simplify to lowest terms.

Skill Practice Divide.

9. $\dfrac{12}{25} \div \dfrac{8}{15}$ **10.** $\dfrac{9}{11} \div 3$

6. Adding and Subtracting Fractions

Adding and Subtracting Fractions

Two fractions can be added or subtracted if they have a common denominator. Let a, b, and c, be numbers such that b does not equal zero. Then,

$$\frac{a}{b} + \frac{c}{b} = \frac{a+c}{b} \quad \text{and} \quad \frac{a}{b} - \frac{c}{b} = \frac{a-c}{b}$$

To add or subtract fractions with the same denominator, add or subtract the numerators and write the result over the common denominator.

Example 7 Adding and Subtracting Fractions with the Same Denominator

Add or subtract as indicated.

a. $\dfrac{1}{12} + \dfrac{7}{12}$ **b.** $\dfrac{13}{5} - \dfrac{3}{5}$

Skill Practice Answers

9. $\dfrac{9}{10}$ **10.** $\dfrac{3}{11}$

TIP: The sum $\frac{1}{12} + \frac{7}{12}$ can be visualized as the sum of the pink and blue sections of the figure.

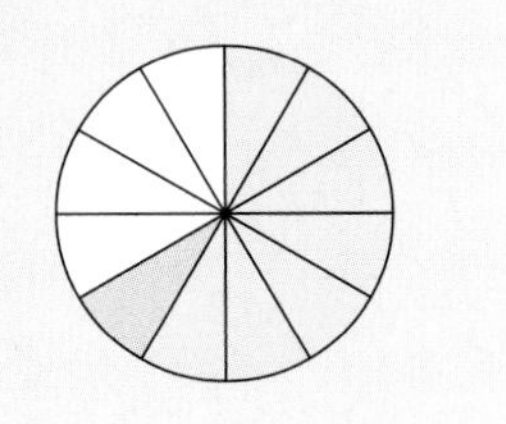

Solution:

a. $\dfrac{1}{12} + \dfrac{7}{12} = \dfrac{1 + 7}{12}$ Add the numerators.

$= \dfrac{8}{12}$

$= \dfrac{2}{3}$ Simplify to lowest terms.

b. $\dfrac{13}{5} - \dfrac{3}{5} = \dfrac{13 - 3}{5}$ Subtract the numerators.

$= \dfrac{10}{5}$ Simplify.

$= 2$ Simplify to lowest terms.

Skill Practice Add or subtract as indicated.

11. $\dfrac{2}{3} + \dfrac{5}{3}$ **12.** $\dfrac{5}{8} - \dfrac{1}{8}$

In Example 7, we added and subtracted fractions with the same denominators. To add or subtract fractions with different denominators, we must first become familiar with the idea of a least common multiple between two or more numbers. The **least common multiple (LCM)** of two numbers is the smallest whole number that is a multiple of each number. For example, the LCM of 6 and 9 is 18.

multiples of 6: 6, 12, 18, 24, 30, 36, . . .

multiples of 9: 9, 18, 27, 36, 45, 54, . . .

Listing the multiples of two or more given numbers can be a cumbersome way to find the LCM. Therefore, we offer the following method to find the LCM of two numbers.

Steps to Finding the LCM of Two Numbers

1. Write each number as a product of prime factors.
2. The LCM is the product of unique prime factors from *both* numbers. If a factor is repeated within the factorization of either number, use that factor the maximum number of times it appears in *either* factorization.

Example 8 Finding the LCM of Two Numbers

Find the LCM of 9 and 15.

Solution:

$9 = 3 \times 3$ and $15 = 3 \times 5$ Factor the numbers.

$\text{LCM} = 3 \times 3 \times 5 = 45$ The LCM is the product of the factors of 3 and 5, where 3 is repeated twice.

Skill Practice Find the LCM.

13. 10 and 25

Skill Practice Answers

11. $\dfrac{7}{3}$ **12.** $\dfrac{1}{2}$ **13.** 50

To add or subtract fractions with *different* denominators, we must first write each fraction as an equivalent fraction with a common denominator. A common denominator may be *any* common multiple of the denominators. However, we will use the least common denominator. The **least common denominator (LCD)** of two or more fractions is the LCM of the denominators of the fractions. The following steps outline the procedure to write a fraction as an equivalent fraction with a common denominator.

Writing Equivalent Fractions

To write a fraction as an equivalent fraction with a common denominator, multiply the numerator and denominator by the factors from the common denominator that are missing from the denominator of the original fraction.

Note: Multiplying the numerator and denominator by the *same* nonzero quantity will not change the value of the fraction.

Example 9 Writing Equivalent Fractions

a. Write the fractions $\frac{1}{9}$ and $\frac{1}{15}$ as equivalent fractions with the LCD as the denominator.

b. Subtract $\frac{1}{9} - \frac{1}{15}$.

Solution:

From Example 8, we know that the LCM for 9 and 15 is 45. Therefore, the LCD of $\frac{1}{9}$ and $\frac{1}{15}$ is 45.

a. $\frac{1}{9} = \frac{1 \times 5}{9 \times 5} = \frac{5}{45}$ Multiply numerator and denominator by 5. This creates a denominator of 45.

$\frac{1}{15} = \frac{1 \times 3}{15 \times 3} = \frac{3}{45}$ Multiply numerator and denominator by 3. This creates a denominator of 45.

b. $\frac{1}{9} - \frac{1}{15}$

$= \frac{5}{45} - \frac{3}{45}$ Write $\frac{1}{9}$ and $\frac{1}{15}$ as equivalent fractions with the same denominator.

$= \frac{2}{45}$ Subtract.

Skill Practice

14. Write the fractions $\frac{5}{8}$ and $\frac{5}{12}$ as equivalent fractions with the LCD as the denominator.

Example 10 Adding Fractions with Different Denominators

Suppose Nakeysha ate $\frac{1}{2}$ of an ice-cream pie, and her friend Carla ate $\frac{1}{3}$ of the pie. How much of the ice-cream pie was eaten?

Skill Practice Answers

14. $\frac{15}{24}, \frac{10}{24}$

Solution:

$\frac{1}{2} + \frac{1}{3}$ The LCD is $3 \times 2 = 6$.

$= \frac{1 \times 3}{2 \times 3} + \frac{1 \times 2}{3 \times 2}$ Multiply numerator and denominator by the missing factors.

$= \frac{3}{6} + \frac{2}{6}$

$= \frac{5}{6}$ Add the fractions.

Together, Nakeysha and Carla ate $\frac{5}{6}$ of the ice-cream pie.

By converting the fractions $\frac{1}{2}$ and $\frac{1}{3}$ to the same denominator, we are able to add *like*-size pieces of pie (Figure R-5).

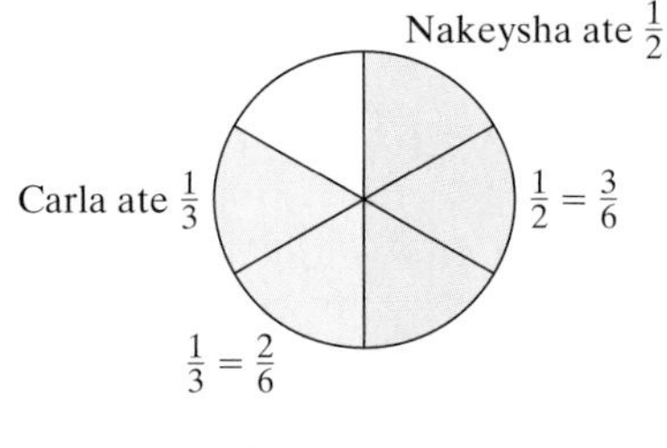

Figure R-5

Skill Practice

15. Adam painted $\frac{1}{4}$ of a fence and Megan painted $\frac{1}{6}$ of the fence. How much of the fence was painted?

Example 11 Adding and Subtracting Fractions

Simplify: $\frac{1}{2} + \frac{5}{8} - \frac{11}{16}$.

Solution:

$\frac{1}{2} + \frac{5}{8} - \frac{11}{16}$ Factor the denominators: $2 = 2$, $8 = 2 \times 2 \times 2$, and $16 = 2 \times 2 \times 2 \times 2$. The LCD is $2 \times 2 \times 2 \times 2 = 16$. Next write the fractions as equivalent fractions with the LCD.

$= \frac{1 \times 2 \times 2 \times 2}{2 \times 2 \times 2 \times 2} + \frac{5 \times 2}{8 \times 2} - \frac{11}{16}$ Multiply numerators and denominators by the factors missing from each denominator.

$= \frac{8}{16} + \frac{10}{16} - \frac{11}{16}$

$= \frac{8 + 10 - 11}{16}$ Add and subtract the numerators.

$= \frac{7}{16}$ Simplify.

Skill Practice Answers

15. Together, Adam and Megan painted $\frac{5}{12}$ of the fence.

Skill Practice Add.

16. $\frac{2}{3} + \frac{1}{2} + \frac{5}{6}$

7. Operations on Mixed Numbers

Recall that a mixed number is a whole number added to a fraction. The number $3\frac{1}{2}$ represents the sum of three wholes plus a half, that is, $3\frac{1}{2} = 3 + \frac{1}{2}$. For this reason, any mixed number can be converted to an improper fraction by using addition.

$$3\frac{1}{2} = 3 + \frac{1}{2} = \frac{6}{2} + \frac{1}{2} = \frac{7}{2}$$

TIP: A shortcut to writing a mixed number as an improper fraction is to multiply the whole number by the denominator of the fraction. Then add this value to the numerator of the fraction, and write the result over the denominator.

$3\frac{1}{2} \longrightarrow$ Multiply the whole number by the denominator: $3 \times 2 = 6$.
Add the numerator: $6 + 1 = 7$.
Write the result over the denominator: $\frac{7}{2}$.

To add, subtract, multiply, or divide mixed numbers, we will first write the mixed number as an improper fraction.

Example 12 **Operations on Mixed Numbers**

Perform the indicated operations.

a. $5\frac{1}{3} - 2\frac{1}{4}$ **b.** $7\frac{1}{2} \div 3$

Solution:

a. $5\frac{1}{3} - 2\frac{1}{4}$

$= \frac{16}{3} - \frac{9}{4}$ Write the mixed numbers as improper fractions.

$= \frac{16 \times 4}{3 \times 4} - \frac{9 \times 3}{4 \times 3}$ The LCD is 12. Multiply numerators and denominators by the missing factors from the denominators.

$= \frac{64}{12} - \frac{27}{12}$

$= \frac{37}{12}$ or $3\frac{1}{12}$ Subtract the fractions.

Skill Practice Answers

16. 2

TIP: An improper fraction can also be written as a mixed number. Both answers are acceptable. Note that

$$\frac{37}{12} = \frac{36}{12} + \frac{1}{12} = 3 + \frac{1}{12}, \text{ or } 3\frac{1}{12}$$

This can easily be found by dividing.

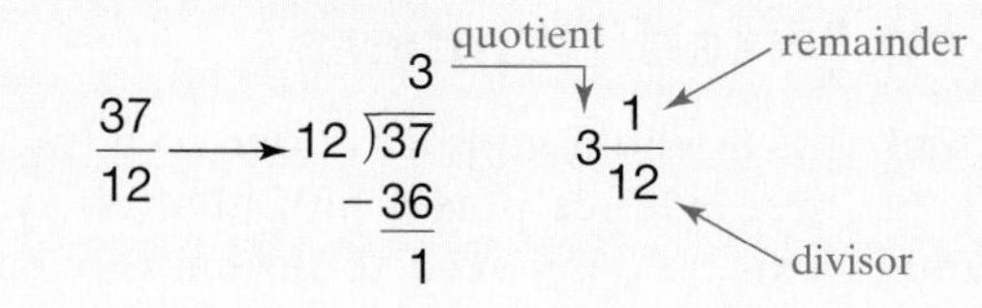

$$\frac{37}{12} \longrightarrow 12\overline{)37} \quad 3\frac{1}{12}$$
$$\begin{array}{r} 3 \\ 12\overline{)37} \\ -36 \\ \hline 1 \end{array}$$

Avoiding Mistakes:

Remember that when dividing (or multiplying) fractions, a common denominator is not necessary.

b. $7\frac{1}{2} \div 3$

$= \frac{15}{2} \div \frac{3}{1}$ Write the mixed number and whole number as fractions.

$= \frac{\overset{5}{\cancel{15}}}{2} \times \frac{1}{\underset{1}{\cancel{3}}}$ Multiply by the reciprocal of $\frac{3}{1}$, which is $\frac{1}{3}$.

$= \frac{5}{2}$ or $2\frac{1}{2}$ The answer may be written as an improper fraction or as a mixed number.

Skill Practice Perform the indicated operations.

17. $2\frac{3}{4} - 1\frac{1}{3}$ **18.** $3\frac{2}{3} \div 5\frac{5}{6}$

Skill Practice Answers

17. $\frac{17}{12}$ or $1\frac{5}{12}$ **18.** $\frac{22}{35}$

Section R.2 Practice Exercises

Boost *your* GRADE at mathzone.com!

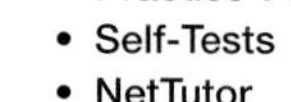

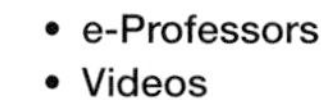

- Practice Problems
- Self-Tests
- NetTutor
- e-Professors
- Videos

Concept 1: Basic Definitions

For Exercises 1–8, identify the numerator and denominator of the fraction. Then determine if the fraction is a proper fraction or an improper fraction.

1. $\frac{7}{8}$ **2.** $\frac{2}{3}$ **3.** $\frac{9}{5}$ **4.** $\frac{5}{2}$

5. $\frac{6}{6}$ **6.** $\frac{4}{4}$ **7.** $\frac{12}{1}$ **8.** $\frac{5}{1}$

For Exercises 9–16, write a proper or improper fraction associated with the shaded region of each figure.

9. **10.**

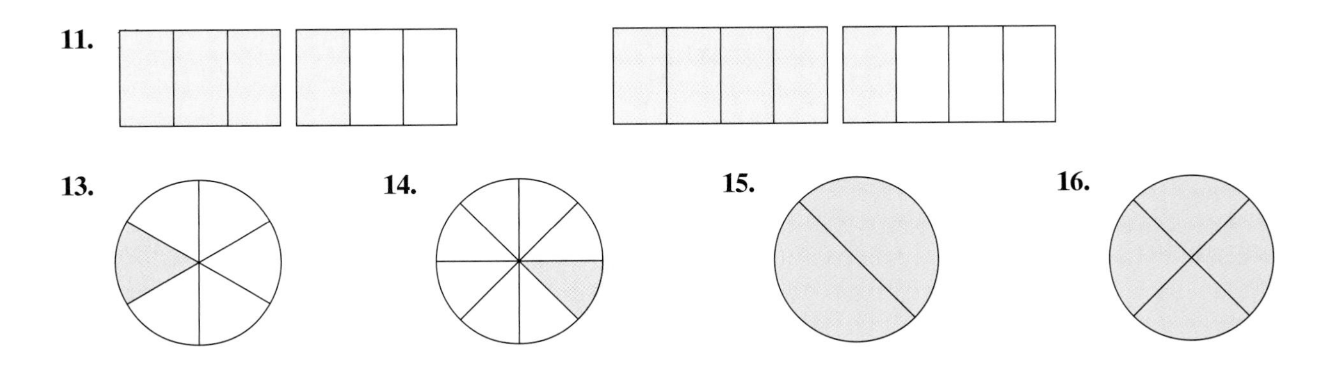

For Exercises 17–20, write both an improper fraction and a mixed number associated with the shaded region of each figure.

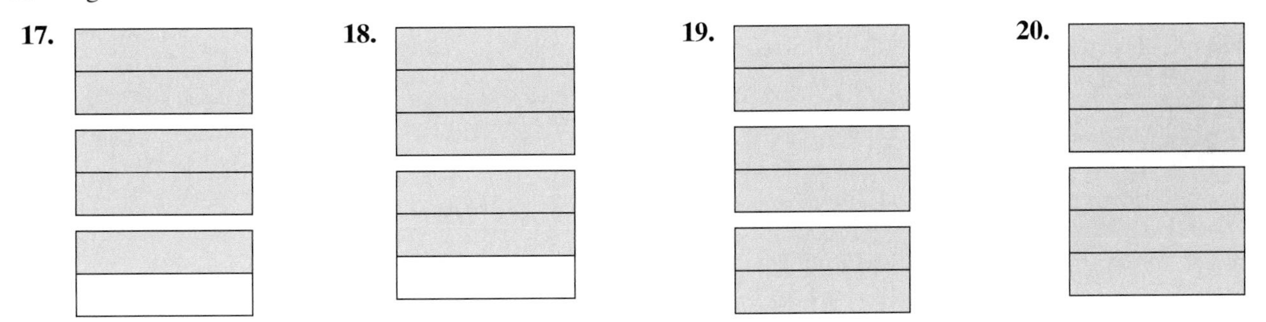

21. Explain the difference between the set of whole numbers and the set of natural numbers.

22. Explain the difference between a proper fraction and an improper fraction.

23. Write a fraction that simplifies to $\frac{1}{2}$. (Answers may vary.)

24. Write a fraction that simplifies to $\frac{1}{3}$. (Answers may vary.)

Concept 2: Prime Factorization

For Exercises 25–32, identify the number as either a prime number or a composite number.

25. 5 **26.** 9 **27.** 4 **28.** 2

29. 39 **30.** 23 **31.** 53 **32.** 51

For Exercises 33–40, write the number as a product of prime factors.

33. 36 **34.** 70 **35.** 42 **36.** 35

37. 110 **38.** 136 **39.** 135 **40.** 105

Concept 3: Simplifying Fractions to Lowest Terms

For Exercises 41–52, simplify each fraction to lowest terms.

41. $\frac{3}{15}$ **42.** $\frac{8}{12}$ **43.** $\frac{6}{16}$ **44.** $\frac{12}{20}$

45. $\frac{42}{48}$ **46.** $\frac{35}{80}$ **47.** $\frac{48}{64}$ **48.** $\frac{32}{48}$

49. $\frac{110}{176}$ **50.** $\frac{70}{120}$ **51.** $\frac{150}{200}$ **52.** $\frac{119}{210}$

Concepts 4–5: Multiplying and Dividing Fractions

For Exercises 53–54, determine if the statement is true or false. If it is false, rewrite as a true statement.

53. When multiplying or dividing fractions, it is necessary to have a common denominator.

54. When dividing two fractions, it is necessary to multiply the first fraction by the reciprocal of the second fraction.

For Exercises 55–66, multiply or divide as indicated.

55. $\frac{10}{13} \times \frac{26}{15}$ **56.** $\frac{15}{28} \times \frac{7}{9}$ **57.** $\frac{3}{7} \div \frac{9}{14}$ **58.** $\frac{7}{25} \div \frac{1}{5}$

59. $\frac{9}{10} \times 5$ **60.** $\frac{3}{7} \times 14$ **61.** $\frac{12}{5} \div 4$ **62.** $\frac{20}{6} \div 5$

63. $\frac{5}{2} \times \frac{10}{21} \times \frac{7}{5}$ **64.** $\frac{55}{9} \times \frac{18}{32} \times \frac{24}{11}$ **65.** $\frac{9}{100} \div \frac{13}{1000}$ **66.** $\frac{1000}{17} \div \frac{10}{3}$

67. Stephen's take-home pay is $1200 a month. If his rent is $\frac{1}{4}$ of his pay, how much is his rent?

68. Gus decides to save $\frac{1}{3}$ of his pay each month. If his monthly pay is $2112, how much will he save each month?

69. On a college basketball team, one-third of the team graduated with honors. If the team has 12 members, how many graduated with honors?

70. Shontell had only enough paper to print out $\frac{3}{5}$ of her book report before school. If the report is 10 pages long, how many pages did she print out?

71. Natalie has 4 yd of material with which she can make holiday aprons. If it takes $\frac{1}{2}$ yd of material per apron, how many aprons can she make?

72. There are 4 cups of oatmeal in a box. If each serving is $\frac{1}{3}$ of a cup, how many servings are contained in the box?

73. Gail buys 6 lb of mixed nuts to be divided into decorative jars that will each hold $\frac{3}{4}$ lb of nuts. How many jars will she be able to fill?

74. Troy has a $\frac{7}{8}$ in. nail that he must hammer into a board. Each strike of the hammer moves the nail $\frac{1}{16}$ in. into the board. How many strikes of the hammer must he make to drive the nail completely into the board?

Concept 6: Adding and Subtracting Fractions

For Exercises 75–78, add or subtract as indicated.

75. $\frac{5}{14} + \frac{1}{14}$ **76.** $\frac{9}{5} + \frac{1}{5}$ **77.** $\frac{17}{24} - \frac{5}{24}$ **78.** $\frac{11}{18} - \frac{5}{18}$

For Exercises 79–84, find the least common denominator for each pair of fractions.

79. $\frac{1}{6}, \frac{5}{24}$

80. $\frac{1}{12}, \frac{11}{30}$

81. $\frac{9}{20}, \frac{3}{8}$

82. $\frac{13}{24}, \frac{7}{40}$

83. $\frac{7}{10}, \frac{11}{45}$

84. $\frac{1}{20}, \frac{1}{30}$

For Exercises 85–100, add or subtract as indicated.

85. $\frac{1}{8} + \frac{3}{4}$

86. $\frac{3}{16} + \frac{1}{2}$

87. $\frac{3}{8} - \frac{3}{10}$

88. $\frac{12}{35} - \frac{1}{10}$

89. $\frac{7}{26} - \frac{2}{13}$

90. $\frac{11}{24} - \frac{5}{16}$

91. $\frac{7}{18} + \frac{5}{12}$

92. $\frac{3}{16} + \frac{9}{20}$

93. $\frac{3}{4} - \frac{1}{20}$

94. $\frac{1}{6} - \frac{1}{24}$

95. $\frac{5}{12} + \frac{5}{16}$

96. $\frac{3}{25} + \frac{8}{35}$

97. $\frac{1}{6} + \frac{3}{4} + \frac{5}{8}$

98. $\frac{1}{2} + \frac{2}{3} + \frac{5}{12}$

99. $\frac{4}{7} + \frac{1}{2} + \frac{3}{4}$

100. $\frac{9}{10} + \frac{4}{5} + \frac{3}{4}$

Concept 7: Operations on Mixed Numbers

For Exercises 101–123, perform the indicated operations.

101. $4\frac{3}{5} \div \frac{1}{10}$

102. $2\frac{4}{5} \div \frac{7}{11}$

103. $3\frac{1}{5} \times \frac{7}{8}$

104. $2\frac{1}{2} \times \frac{4}{5}$

105. $1\frac{2}{9} \div 6$

106. $2\frac{2}{5} \div \frac{2}{7}$

107. A board $26\frac{3}{8}$ in. long must be cut into three pieces of equal length. Find the length of each piece.

$26\frac{3}{8}$ in.

108. $2\frac{1}{8} + 1\frac{3}{8}$

109. $1\frac{3}{14} + 1\frac{1}{14}$

110. $1\frac{5}{6} - \frac{7}{8}$

111. $2\frac{1}{3} - \frac{5}{6}$

112. $1\frac{1}{6} + 3\frac{3}{4}$

113. $4\frac{1}{2} + 2\frac{2}{3}$

114. $1 - \frac{7}{8}$

115. $2 - \frac{3}{7}$

116. A futon, when set up as a sofa, measures $3\frac{5}{6}$ ft wide. When it is opened to be used as a bed, the width is increased by $1\frac{3}{4}$ ft. What is the total width of this bed?

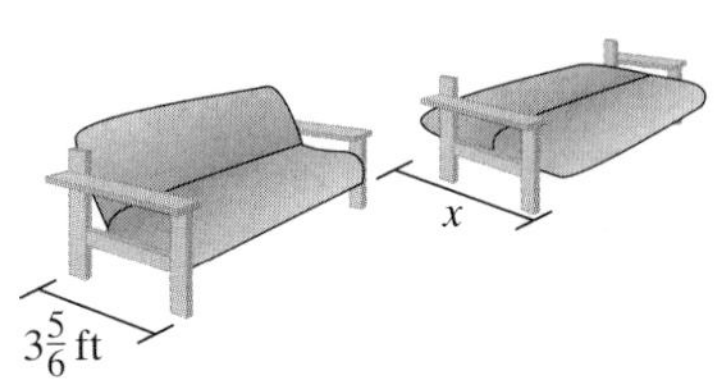

117. A plane trip from Orlando to Detroit takes $2\frac{3}{4}$ hr. If the plane traveled for $1\frac{1}{6}$ hr, how much time remains for the flight?

118. Antonio bought $\frac{1}{2}$ lb of turkey, $\frac{1}{3}$ lb of ham, and $\frac{3}{4}$ lb of roast beef. How much meat did Antonio buy?

119. José ordered two seafood platters for a party. One platter has $1\frac{1}{2}$ lb of shrimp, and the other has $\frac{3}{4}$ lb of shrimp. How many pounds does he have altogether?

120. Ayako took a trip to the store $5\frac{1}{2}$ miles away. If she rode the bus for $4\frac{5}{6}$ miles and walked the rest of the way, how far did she have to walk?

121. Average rainfall in Tampa, FL, for the month of November is $2\frac{3}{4}$ in. One stormy weekend $3\frac{1}{8}$ in. of rain fell. How many inches of rain over the monthly average is this?

122. Maria has 4 yd of material. If she sews a dress that requires $3\frac{1}{8}$ yd, how much material will she have left?

123. Pete started working out at the gym several months ago. His waist measured $38\frac{1}{2}$ in. when he began and is now $33\frac{3}{4}$ in. How many inches did he lose around his waist?

Section R.3 Decimals and Percents

Concepts

1. **Introduction to a Place Value System**
2. **Converting Fractions to Decimals**
3. **Converting Decimals to Fractions**
4. **Converting Percents to Decimals and Fractions**
5. **Converting Decimals and Fractions to Percents**
6. **Applications of Percents**

1. Introduction to a Place Value System

In a **place value** number system, each digit in a numeral has a particular value determined by its location in the numeral (Figure R-6).

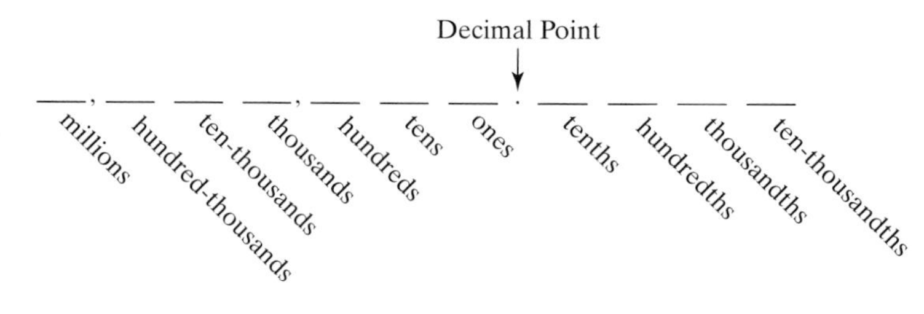

Figure R-6

For example, the number 197.215 represents

$$(1 \times 100) + (9 \times 10) + (7 \times 1) + \left(2 \times \frac{1}{10}\right) + \left(1 \times \frac{1}{100}\right) + \left(5 \times \frac{1}{1000}\right)$$

Each of the digits 1, 9, 7, 2, 1, and 5 is multiplied by 100, 10, 1, $\frac{1}{10}$, $\frac{1}{100}$, and $\frac{1}{1000}$, respectively, depending on its location in the numeral 197.215.

By obtaining a common denominator and adding fractions, we have

$$197.215 = 100 + 90 + 7 + \frac{200}{1000} + \frac{10}{1000} + \frac{5}{1000}$$

$$= 197 + \frac{215}{1000} \quad \text{or} \quad 197\frac{215}{1000}$$

Because 197.215 is equal to the mixed number $197\frac{215}{1000}$, we read 197.215 as one hundred ninety-seven *and* two hundred fifteen thousandths. The decimal point is read as the word *and*.

If there are no digits to the right of the decimal point, we usually omit the decimal point.

2. Converting Fractions to Decimals

In Section R.2, we learned that a fraction represents part of a whole unit. Likewise, the digits to the right of the decimal point represent a fraction of a whole unit. In this section, we will learn how to convert a fraction to a decimal number and vice versa.

Converting a Fraction to a Decimal

To convert a fraction to a decimal, divide the numerator of the fraction by the denominator of the fraction.

Example 1 Converting Fractions to Decimals

Convert each fraction to a decimal.

a. $\frac{7}{40}$ **b.** $\frac{2}{3}$ **c.** $\frac{13}{7}$

Solution:

a. $\frac{7}{40} = 0.175$

$$\begin{array}{r} 0.175 \\ 40\overline{)7.000} \\ \underline{40} \\ 300 \\ \underline{280} \\ 200 \\ \underline{200} \\ 0 \end{array}$$

The number 0.175 is said to be a **terminating decimal** because there are no nonzero digits to the right of the last digit, 5.

b. $\frac{2}{3} = 0.666\ldots$

$$\begin{array}{r} 0.666\ldots \\ 3\overline{)2.00000} \\ \underline{18} \\ 20 \\ \underline{18} \\ 20 \\ \underline{18} \\ 2\ldots \end{array}$$

The pattern 0.666 . . . continues indefinitely. Therefore, we say that this is a **repeating decimal.**

For a repeating decimal, a horizontal bar is often used to denote the repeating pattern after the decimal point.

Hence, $\frac{2}{3} = 0.\overline{6}$.

c. $\frac{13}{7} = 1.857142857142\ldots$

$$\begin{array}{r} 1.857142857142\ldots \\ 7\overline{)13.000000000000} \end{array}$$

The number 1.857142857142 . . . is also a repeating pattern, and in this case the pattern includes several digits. Therefore, we write $\frac{13}{7} = 1.\overline{857142}$.

Skill Practice Convert to a decimal.

1. $\frac{5}{8}$ **2.** $\frac{1}{6}$ **3.** $\frac{15}{11}$

Skill Practice Answers

1. 0.625 **2.** $0.1\overline{6}$ **3.** $1.\overline{36}$

3. Converting Decimals to Fractions

To convert a terminating decimal to a fraction, use the word name to write the number as a fraction or mixed number.

Example 2 **Converting Terminating Decimals to Fractions**

Convert each decimal to a fraction.

a. 0.0023 **b.** 50.06

Solution:

a. 0.0023 is read as twenty-three ten-thousandths. Thus,

$$0.0023 = \frac{23}{10{,}000}$$

b. 50.06 is read as fifty and six hundredths. Thus,

$$50.06 = 50\frac{6}{100}$$

$$= 50\frac{3}{50} \quad \text{Simplify the fraction to lowest terms.}$$

$$= \frac{2503}{50} \quad \text{Write the mixed number as a fraction.}$$

Skill Practice Convert to a fraction.

4. 0.107 **5.** 11.25

Repeating decimals also can be written as fractions. However, the procedure to convert a repeating decimal to a fraction requires some knowledge of algebra. Table R-1 shows some common repeating decimals and an equivalent fraction for each.

Table R-1

$0.\overline{1} = \frac{1}{9}$	$0.\overline{4} = \frac{4}{9}$	$0.\overline{7} = \frac{7}{9}$
$0.\overline{2} = \frac{2}{9}$	$0.\overline{5} = \frac{5}{9}$	$0.\overline{8} = \frac{8}{9}$
$0.\overline{3} = \frac{3}{9} = \frac{1}{3}$	$0.\overline{6} = \frac{6}{9} = \frac{2}{3}$	$0.\overline{9} = \frac{9}{9} = 1$

4. Converting Percents to Decimals and Fractions

The concept of percent (%) is widely used in a variety of mathematical applications. The word *percent* means "per 100." Therefore, we can write percents as fractions.

$6\% = \frac{6}{100}$ A sales tax of 6% means that 6 cents in tax is charged for every 100 cents spent.

$91\% = \frac{91}{100}$ The fact that 91% of the population is right-handed means that 91 people out of 100 are right-handed.

Skill Practice Answers

4. $\frac{107}{1000}$ **5.** $\frac{45}{4}$

The quantity $91\% = \frac{91}{100}$ can be written as $91 \times \frac{1}{100}$ or as 91×0.01.

Notice that the % symbol implies "division by 100" or, equivalently, "multiplication by $\frac{1}{100}$." Thus, we have the following rule to convert a percent to a fraction (or to a decimal).

Converting a Percent to a Decimal or Fraction

Replace the % symbol by $\div 100$ (or equivalently $\times \frac{1}{100}$ or $\times 0.01$).

Example 3 Converting Percents to Decimals

Convert the percents to decimals.

a. 78% **b.** 412% **c.** 0.045%

Solution:

a. $78\% = 78 \times 0.01 = 0.78$

b. $412\% = 412 \times 0.01 = 4.12$

c. $0.045\% = 0.045 \times 0.01 = 0.00045$

Skill Practice Convert the percent to a decimal.

6. 29% **7.** 3.5% **8.** 100%

TIP: Multiplying by 0.01 is equivalent to dividing by 100. This has the effect of moving the decimal point two places to the left.

Example 4 Converting Percents to Fractions

Convert the percents to fractions.

a. 52% **b.** $33\frac{1}{3}\%$ **c.** 6.5%

Solution:

a. $52\% = 52 \times \frac{1}{100}$ Replace the % symbol by $\frac{1}{100}$.

$= \frac{52}{100}$ Multiply.

$= \frac{13}{25}$ Simplify to lowest terms.

b. $33\frac{1}{3}\% = 33\frac{1}{3} \times \frac{1}{100}$ Replace the % symbol by $\frac{1}{100}$.

$= \frac{100}{3} \times \frac{1}{100}$ Write the mixed number as a fraction $33\frac{1}{3} = \frac{100}{3}$.

$= \frac{100}{300}$ Multiply the fractions.

$= \frac{1}{3}$ Simplify to lowest terms.

Skill Practice Answers
6. 0.29 **7.** 0.035 **8.** 1.00

c. $6.5\% = 6.5 \times \dfrac{1}{100}$ Replace the % symbol by $\frac{1}{100}$.

$= \dfrac{65}{10} \times \dfrac{1}{100}$ Write 6.5 as an improper fraction.

$= \dfrac{65}{1000}$ Multiply the fractions.

$= \dfrac{13}{200}$ Simplify to lowest terms.

Skill Practice Convert the percent to a fraction.

9. 30% **10.** $120\frac{1}{2}\%$ **11.** 25%

5. Converting Decimals and Fractions to Percents

To convert a percent to a decimal or fraction, we replace the % symbol by ÷ 100. To convert a decimal or percent to a fraction, we reverse this process.

Converting Fractions and Decimals to Percents

Multiply the fraction or decimal by 100%.

Example 5 Converting Decimals to Percents

Convert the decimals to percents.

a. 0.92 **b.** 10.80 **c.** 0.005

Solution:

a. $0.92 = 0.92 \times 100\% = 92\%$ Multiply by 100%.

b. $10.80 = 10.80 \times 100\% = 1080\%$ Multiply by 100%.

c. $0.005 = 0.005 \times 100\% = 0.5\%$ Multiply by 100%.

Skill Practice Convert the decimal to a percent.

12. 0.56 **13.** 4.36 **14.** 0.002

Example 6 Converting Fractions to Percents

Convert the fractions to percents.

a. $\dfrac{2}{5}$ **b.** $\dfrac{1}{16}$ **c.** $\dfrac{5}{3}$

Skill Practice Answers

9. $\dfrac{3}{10}$ **10.** $\dfrac{241}{200}$ **11.** $\dfrac{1}{4}$

12. 56% **13.** 436% **14.** 0.2%

Solution:

a. $\frac{2}{5} = \frac{2}{5} \times 100\%$ Multiply by 100%.

$= \frac{2}{5} \times \frac{100}{1}\%$ Write the whole number as a fraction.

$= \frac{2}{\cancel{5}_1} \times \frac{\cancel{100}^{20}}{1}\%$ Multiply fractions.

$= \frac{40}{1}\%$ or 40% Simplify.

b. $\frac{1}{16} = \frac{1}{16} \times 100\%$ Multiply by 100%.

$= \frac{1}{16} \times \frac{100}{1}\%$ Write the whole number as a fraction.

$= \frac{1}{\cancel{16}_4} \times \frac{\cancel{100}^{25}}{1}\%$ Multiply the fractions.

$= \frac{25}{4}\%$ Simplify.

$= \frac{25}{4}\%$ or 6.25% The value $\frac{25}{4}$ can be written in decimal form by dividing 25 by 4.

c. $\frac{5}{3} = \frac{5}{3} \times 100\%$ Multiply by 100%.

$= \frac{5}{3} \times \frac{100}{1}\%$ Write the whole number as a fraction.

$= \frac{500}{3}\%$ Multiply fractions.

$= \frac{500}{3}\%$ or $166.\overline{6}\%$ The value $\frac{500}{3}$ can be written in decimal form by dividing 500 by 3.

Skill Practice Convert the fraction to a percent.

15. $\frac{7}{8}$ **16.** $\frac{9}{5}$ **17.** $\frac{1}{9}$

6. Applications of Percents

Many applications involving percents involve finding a percent of some base number. For example, suppose a textbook is discounted 25%. If the book originally cost \$60, find the amount of the discount.

In this example, we must find 25% of \$60. In this context, the word *of* means multiply.

$$\begin{array}{ccccc} 25\% & \text{of} & \$60 & & \\ \downarrow & \downarrow & \downarrow & & \\ 0.25 & \times & 60 & = & 15 \end{array}$$

The amount of the discount is \$15.

Note that the *decimal form* of a percent is always used in calculations. Therefore, 25% was converted to 0.25 *before* multiplying by \$60.

Skill Practice Answers

15. 87.5% **16.** 180%
17. $11.\overline{1}\%$ or $11\frac{1}{9}\%$

Example 7 Applying Percentages

Shauna received a raise, so now her new salary is 105% of her old salary. Find Shauna's new salary if her old salary was $36,000 per year.

Solution:

The new salary is 105% of $36,000.

$1.05 \times 36{,}000 = 37{,}800$ The new salary is $37,800 per year.

Skill Practice

18. If 60% of the 35 students in an Algebra class are women, find the number of women.

Example 8 Applying Percentages

A couple leaves a 15% tip for a $32 dinner. Find the amount of the tip.

Solution:

The tip is 15% of $32.

$0.15 \times 32 = 4.80$ The tip is $4.80.

Skill Practice

19. The sales tax rate for one county is 6%. Find the amount of sales tax on a $52.00 fishing pole.

In some applications, it is necessary to convert a fractional part of a whole to a percent of the whole.

Example 9 Finding a Percentage

Union College in Schenectady, NY, accepts approximately 520 students each year from 3500 applicants. What percent does 520 represent? Round to the nearest tenth of a percent.

Solution:

$\frac{520}{3500} \approx 0.149$ Convert the fractional part of the total number of applicants to decimal form.

$= 0.149 \times 100\%$ Convert the decimal to a percent.

$= 14.9\%$ Simplify.

Approximately 14.9% of the applicants to Union College are accepted.

Skill Practice

20. Eduardo answered 66 questions correctly on a test with 75 questions. What percent of the questions does 66 represent?

Skill Practice Answers

18. 21 women **19.** $3.12
20. 88%

Calculator Connections

Calculators can display only a limited number of digits on the calculator screen. Therefore, repeating decimals and terminating decimals with a large number of digits will be truncated or rounded to fit the calculator display. For example, the fraction $\frac{2}{3} = 0.\overline{6}$ may be entered into the calculator as [2] [÷] [3]. The result may appear as 0.6666666667 or as 0.6666666666. The fraction $\frac{2}{11}$ equals the repeating decimal $0.\overline{18}$. However, the calculator converts $\frac{2}{11}$ to the terminating decimal 0.1818181818.

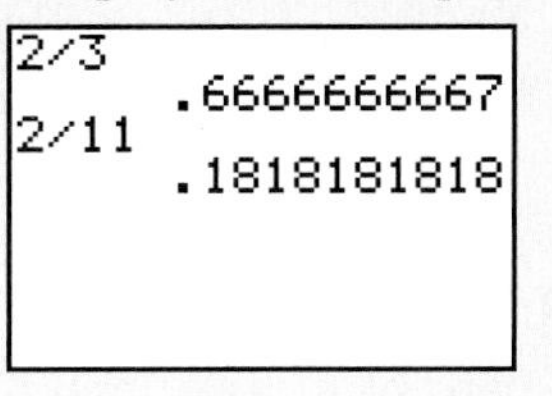

Calculator Exercises

Without using a calculator, find a repeating decimal to represent each of the following fractions. Then use a calculator to confirm your answer.

1. $\frac{4}{9}$

2. $\frac{7}{11}$

3. $\frac{3}{22}$

4. $\frac{5}{13}$

Section R.3 Practice Exercises

Boost *your* GRADE at mathzone.com!

- Practice Problems
- Self-Tests
- NetTutor
- e-Professors
- Videos

Concept 1: Introduction to a Place Value System

For Exercises 1–12, write the name of the place value for the underlined digit.

1. $4\underline{8}1.24$

2. $13\underline{4}5.42$

3. $2\underline{9}12.032$

4. $4\underline{2}08.03$

5. $2.\underline{3}81$

6. $8.\underline{2}49$

7. $21.41\underline{3}$

8. $82.7\underline{9}4$

9. $3\underline{2}.43$

10. $7\underline{8}.04$

11. $0.381\underline{9}2$

12. $0.897\underline{5}4$

13. The first 10 Roman numerals are: I, II, III, IV, V, VI, VII, VIII, IX, X. Is this numbering system a place value system? Explain your answer.

Concept 2: Converting Fractions to Decimals

For Exercises 14–21, convert each fraction to a terminating decimal or a repeating decimal.

14. $\frac{7}{10}$

15. $\frac{9}{10}$

16. $\frac{9}{25}$

17. $\frac{3}{25}$

18. $\frac{11}{9}$

19. $\frac{16}{9}$

20. $\frac{7}{33}$

21. $\frac{3}{11}$

Concept 3: Converting Decimals to Fractions

For Exercises 22–33, convert each decimal to a fraction or a mixed number.

22. 0.45 **23.** 0.65 **24.** 0.181 **25.** 0.273

26. 2.04 **27.** 6.02 **28.** 13.007 **29.** 12.003

30. $0.\overline{5}$ (*Hint:* Refer to Table R-1) **31.** $0.\overline{8}$ **32.** $1.\overline{1}$ **33.** $2.\overline{3}$

Concept 4: Converting Percents to Decimals and Fractions

For Exercises 34–43, convert each percent to a decimal and to a fraction.

34. The sale price is 30% off of the original price.

35. An HMO (health maintenance organization) pays 80% of all doctors' bills.

36. The building will be 75% complete by spring.

37. Chan plants roses in 25% of his garden.

38. The bank pays $3\frac{3}{4}\%$ interest on a checking account.

39. A credit union pays $4\frac{1}{2}\%$ interest on a savings account.

40. Kansas received 15.7% of its annual rainfall in 1 week.

41. Social Security withholds 5.8% of an employee's gross pay.

42. The world population in 2005 was 253% of the world population in 1950.

43. The cost of a home is 140% of its cost 10 years ago.

44. Explain how to convert a decimal to a percent.

45. Explain how to convert a percent to a decimal.

Concept 5: Converting Decimals and Fractions to Percents

For Exercises 46–57, convert the decimal to a percent.

46. 0.05 **47.** 0.06 **48.** 0.90 **49.** 0.70

50. 1.2 **51.** 4.8 **52.** 7.5 **53.** 9.3

54. 0.135 **55.** 0.536 **56.** 0.003 **57.** 0.002

For Exercises 58–69, convert the fraction to a percent.

58. $\frac{3}{50}$ **59.** $\frac{23}{50}$ **60.** $\frac{9}{2}$ **61.** $\frac{7}{4}$

62. $\frac{5}{8}$ **63.** $\frac{1}{8}$ **64.** $\frac{5}{16}$ **65.** $\frac{7}{16}$

66. $\frac{5}{6}$ **67.** $\frac{4}{15}$ **68.** $\frac{14}{15}$ **69.** $\frac{5}{18}$

Concept 6: Applications of Percents

70. A suit that costs \$140 is discounted by 30%. How much is the discount?

71. A community college has a 12% increase in enrollment over the previous year. If the enrollment last year was 10,800 students, how many students have enrolled this year?

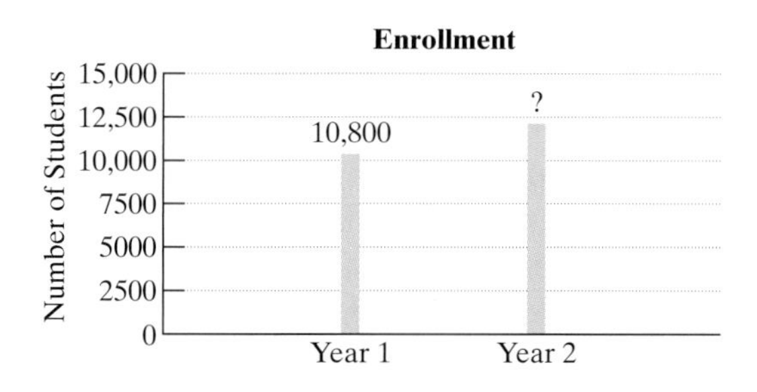

72. Louise completed 40% of her task that takes a total of 60 hours to finish. How many hours did she complete?

73. Tom's federal taxes amount to 27% of his income. If Tom earns \$12,500 per quarter, how much will he pay in taxes for that quarter?

74. A tip of \$7 is left on a meal that costs \$56. What percentage of the cost does the tip represent? Round to the nearest tenth of a percent.

75. José paid \$5.95 in sales tax on a textbook that costs \$85. Find the percent of the sales tax.

76. Sue saves \$37.50 each week out of her paycheck of \$625. What percent of her paycheck does her savings represent?

For Exercises 77–80, refer to the graph. The pie graph shows a family budget based on a net income of \$2400 per month.

77. Determine the amount spent on rent.

78. Determine the amount spent on car payments.

79. Determine the amount spent on utilities.

80. How much more money is spent than saved?

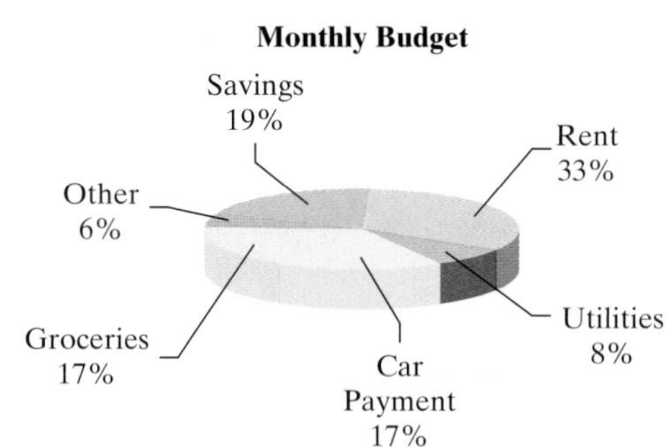

81. By the end of the year, Felipe will have 75% of his mortgage paid. If the mortgage was originally for \$90,000, how much will have been paid at the end of the year?

82. A certificate of deposit (CD) earns 8% interest in 1 year. If Mr. Patel has \$12,000 invested in the CD, how much interest will he receive at the end of the year?

83. On a state exit exam, a student can answer 30% of the problems incorrectly and still pass the test. If the exam has 40 problems, how many problems can a student answer incorrectly and still pass?

Section R.4 Introduction to Geometry

Concepts

1. Perimeter
2. Area
3. Volume
4. Angles
5. Triangles

1. Perimeter

In this section, we present several facts and formulas that may be used throughout the text in applications of geometry. One of the most important uses of geometry involves the measurement of objects of various shapes. We begin with an introduction to perimeter, area, and volume for several common shapes and objects.

Perimeter is defined as the distance around a figure. For example, if we were to put up a fence around a field, the perimeter would determine the amount of fencing. For a polygon (a closed figure constructed from line segments), the perimeter is the sum of the lengths of the sides. For a circle, the distance around the outside is called the **circumference**.

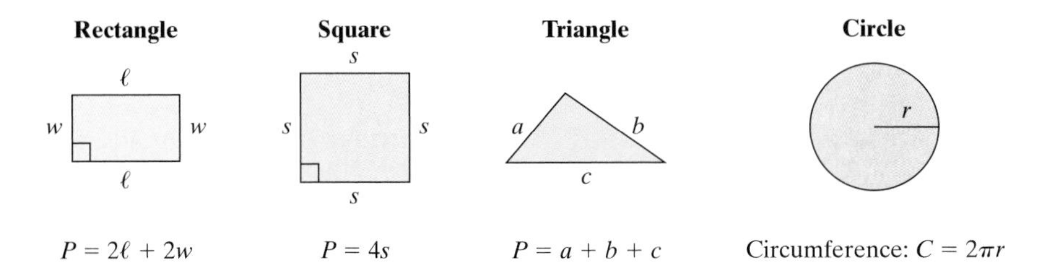

$d = 2r$

For a circle, r represents the length of a radius—the distance from the center to any point on the circle. The length of a diameter, d, of a circle is twice that of a radius. Thus, $d = 2r$. The number π is a constant equal to the ratio of the circumference of a circle divided by the length of a diameter, that is, $\pi = \frac{C}{d}$. The value of π is often approximated by 3.14 or $\frac{22}{7}$.

Example 1 Finding Perimeter and Circumference

Find the perimeter or circumference as indicated. Use 3.14 for π.

a. Perimeter of the polygon

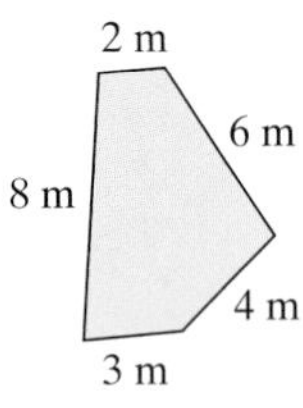

b. Perimeter of the rectangle

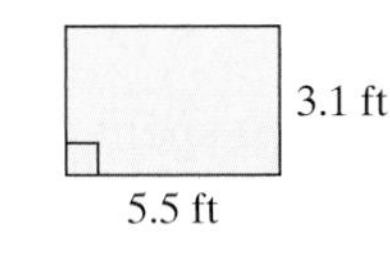

c. Circumference of the circle

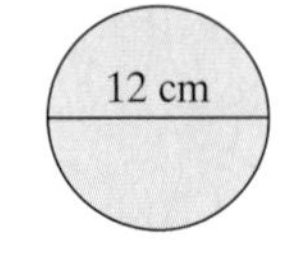

Solution:

a. $P = (8 \text{ m}) + (2 \text{ m}) + (6 \text{ m}) + (4 \text{ m}) + (3 \text{ m})$ Add the lengths of the sides.

$= 23 \text{ m}$ The perimeter is 23 m.

b. $P = 2\ell + 2w$

$= 2(5.5 \text{ ft}) + 2(3.1 \text{ ft})$ Substitute $\ell = 5.5$ ft and $w = 3.1$ ft.

$= 11 \text{ ft} + 6.2 \text{ ft}$

$= 17.2 \text{ ft}$ The perimeter is 17.2 ft.

c. $C = 2\pi r$

$= 2(3.14)(6 \text{ cm})$ Substitute 3.14 for π and $r = 6$ cm.

$= 6.28(6 \text{ cm})$

$= 37.68 \text{ cm}$ The circumference is 37.68 cm.

TIP: If a calculator is used to find the circumference of a circle, use the π key to get a more accurate answer than by using 3.14.

Skill Practice

1. Find the perimeter.

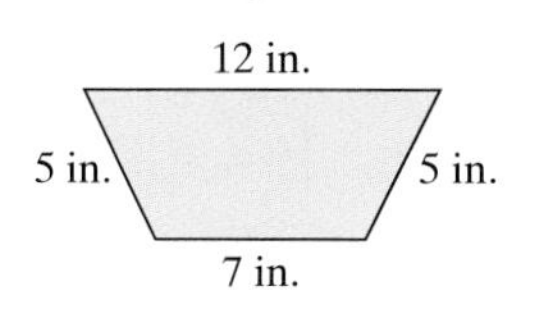

2. Find the circumference. Use 3.14 for π.

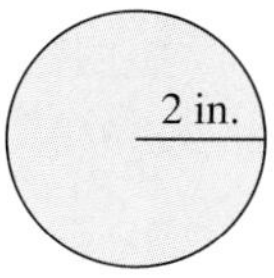

3. Find the perimeter of the square.

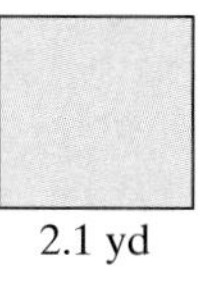

2. Area

The **area** of a geometric figure is the number of square units that can be enclosed within the figure. For example, the rectangle shown in Figure R-7 encloses 6 square inches (6 in.^{2}). In applications, we would find the area of a region if we were laying carpet or putting down sod for a lawn.

The formulas used to compute the area for several common geometric shapes are given here:

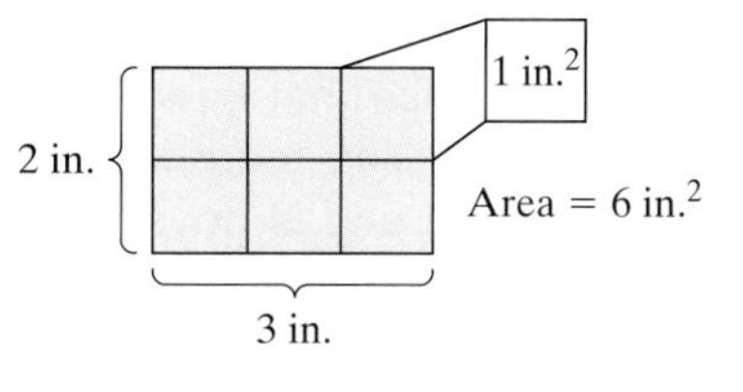

Figure R-7

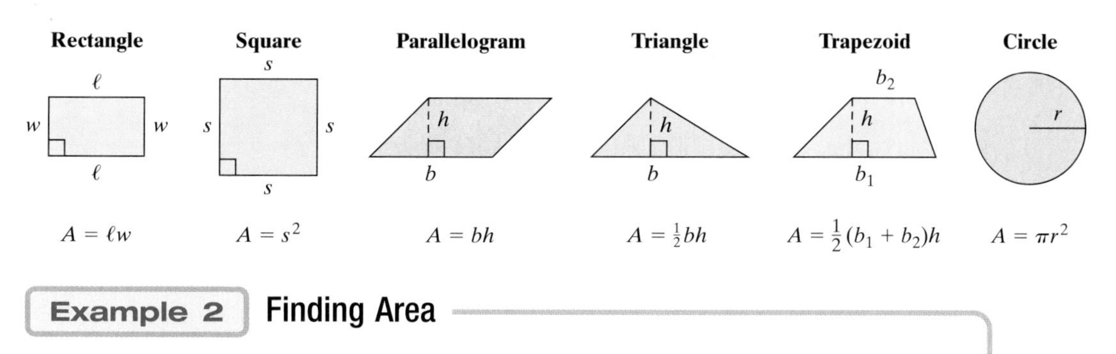

Example 2 Finding Area

Find the area enclosed by each figure.

a. **b.**

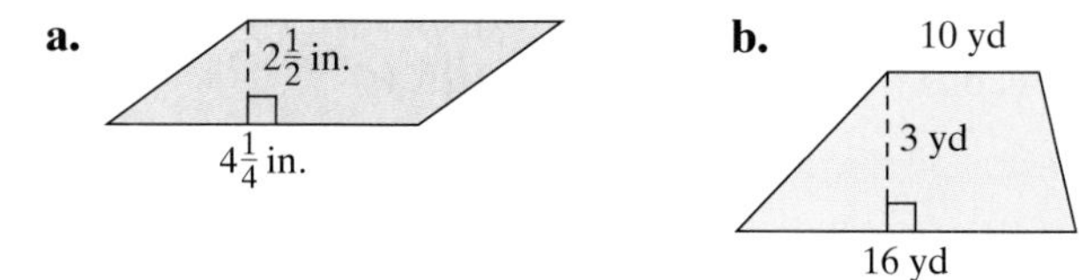

Skill Practice Answers

1. 29 in. **2.** 12.56 in.
3. 8.4 yd

> **TIP:** Notice that the units of area are given in square units such as square inches (in.^{2}), square feet (ft^2), square yards (yd^2), square centimeters (cm^2), and so on.

Solution:

a. $A = bh$ — The figure is a parallelogram.

$= (4\frac{1}{4}\text{ in.})(2\frac{1}{2}\text{ in.})$ — Substitute $b = 4\frac{1}{4}$ in. and $h = 2\frac{1}{2}$ in.

$= \left(\frac{17}{4}\text{ in.}\right)\left(\frac{5}{2}\text{ in.}\right)$

$= \frac{85}{8}\text{ in.}^2 \text{ or } 10\frac{5}{8}\text{ in.}^2$

b. $A = \frac{1}{2}(b_1 + b_2)h$ — The figure is a trapezoid.

$= \frac{1}{2}(16\text{ yd} + 10\text{ yd})(3\text{ yd})$ — Substitute $b_1 = 16$ yd, $b_2 = 10$ yd, and $h = 3$ yd.

$= \frac{1}{2}(26\text{ yd})(3\text{ yd})$

$= (13\text{ yd})(3\text{ yd})$

$= 39\text{ yd}^2$ — The area is 39 yd^2.

Skill Practice Find the area enclosed by the figure.

4. **5.**

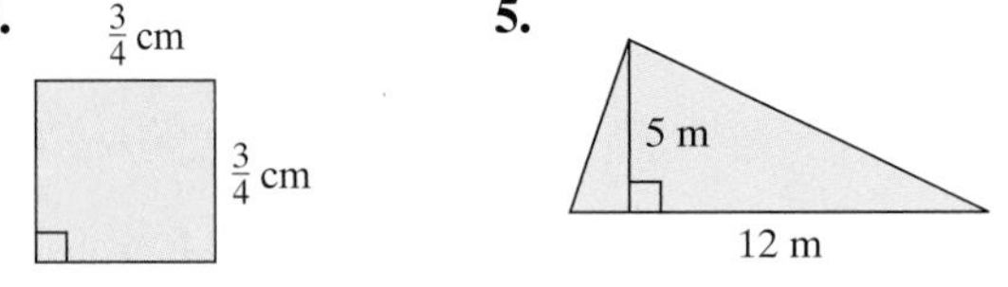

> **TIP:** Notice that several of the formulas presented thus far involve multiple operations. The order in which we perform the arithmetic is called the **order of operations** and is covered in detail in Section 1.2. We will follow these guidelines in the order given below:
>
> 1. Perform operations within parentheses first.
> 2. Evaluate expressions with exponents.
> 3. Perform multiplication or division in order from left to right.
> 4. Perform addition or subtraction in order from left to right.

Example 3 Finding Area of a Circle

Find the area of a circular fountain if the radius is 25 ft. Use 3.14 for π.

25 ft

Solution:

$A = \pi r^2$

$= (3.14)(25\text{ ft})^2$ — Substitute 3.14 for π and $r = 25$ ft.

$= (3.14)(625\text{ ft}^2)$ — Note: $(25\text{ ft})^2 = (25\text{ ft})(25\text{ ft}) = 625\text{ ft}^2$

$= 1962.5\text{ ft}^2$ — The area of the fountain is 1962.5 ft^2.

Skill Practice Answers

4. $\frac{9}{16}\text{ cm}^2$ **5.** 30 m^2

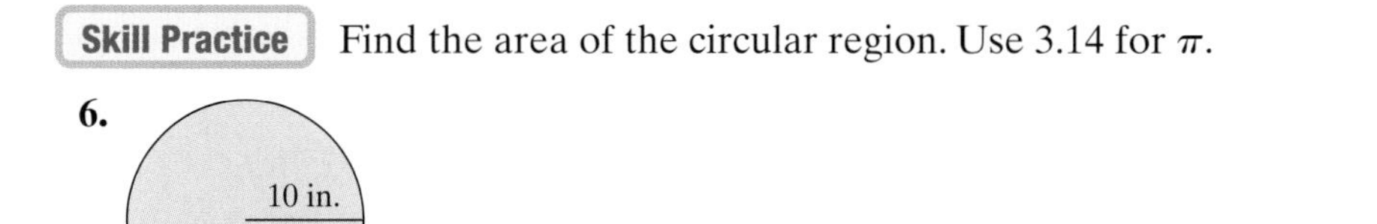

Skill Practice Find the area of the circular region. Use 3.14 for π.

6.

3. Volume

The **volume** of a solid is the number of cubic units that can be enclosed within a solid. The solid shown in Figure R-8 contains 18 cubic inches (18 in.^{3}). In applications, volume might refer to the amount of water in a swimming pool.

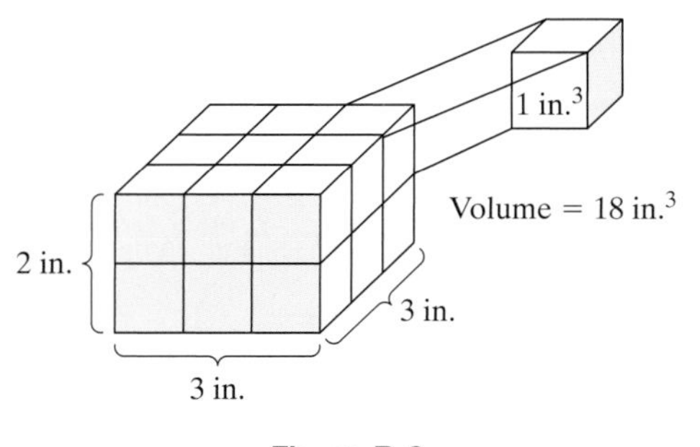

Figure R-8

The formulas used to compute the volume of several common solids are given here:

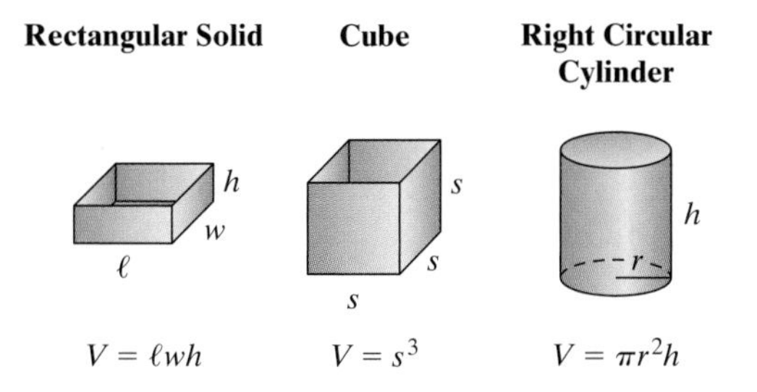

TIP: Notice that the volume formulas for the three figures just shown are given by the product of the area of the base and the height of the figure:

$V = \ell w h$	$V = s \cdot s \cdot s$	$V = \pi r^2 h$
Area of Rectangular Base	Area of Square Base	Area of Circular Base

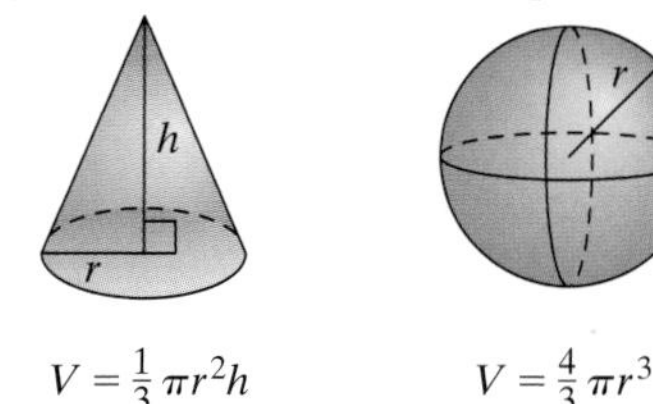

Skill Practice Answers

6. 314 in.^{2}

Example 4 Finding Volume

Find the volume of each object.

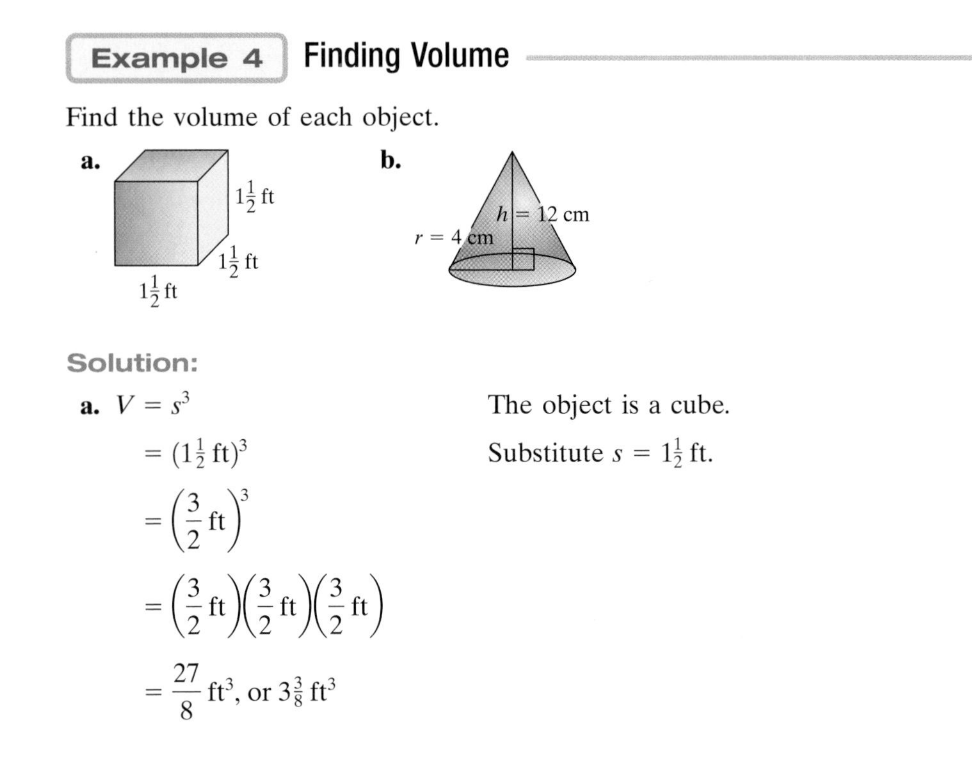

Solution:

a. $V = s^3$ — The object is a cube.

$= (1\frac{1}{2}\text{ ft})^3$ — Substitute $s = 1\frac{1}{2}$ ft.

$= \left(\frac{3}{2}\text{ ft}\right)^3$

$= \left(\frac{3}{2}\text{ ft}\right)\left(\frac{3}{2}\text{ ft}\right)\left(\frac{3}{2}\text{ ft}\right)$

$= \frac{27}{8}\text{ ft}^3$, or $3\frac{3}{8}\text{ ft}^3$

TIP: Notice that the units of volume are cubic units such as cubic inches (in.^3), cubic feet (ft^3), cubic yards (yd^3), cubic centimeters (cm^3), and so on.

b. $V = \frac{1}{3}\pi r^2 h$ — The object is a right circular cone.

$= \frac{1}{3}(3.14)(4\text{ cm})^2(12\text{ cm})$ — Substitute 3.14 for π, $r = 4$ cm, and $h = 12$ cm.

$= \frac{1}{3}(3.14)(16\text{ cm}^2)(12\text{ cm})$

$= 200.96\text{ cm}^3$

Skill Practice

7. Find the volume of the object.

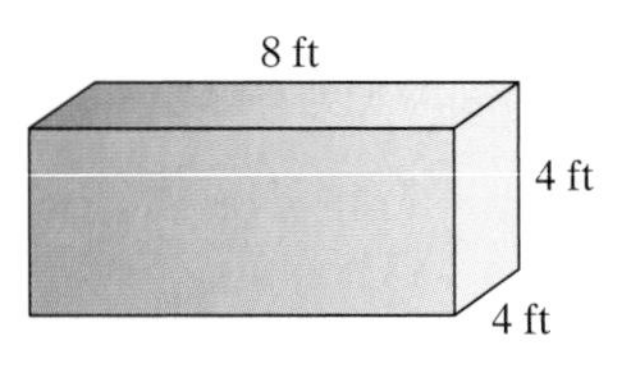

8. Find the volume of the object. Use 3.14 for π. Round to the nearest whole unit.

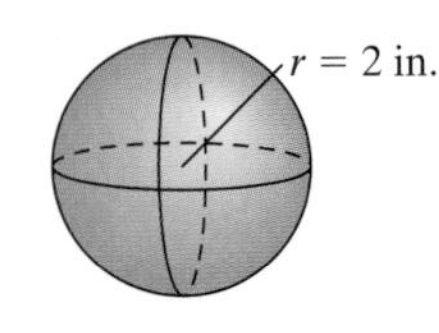

Skill Practice Answers

7. 128 ft^3 **8.** 33 in.^3

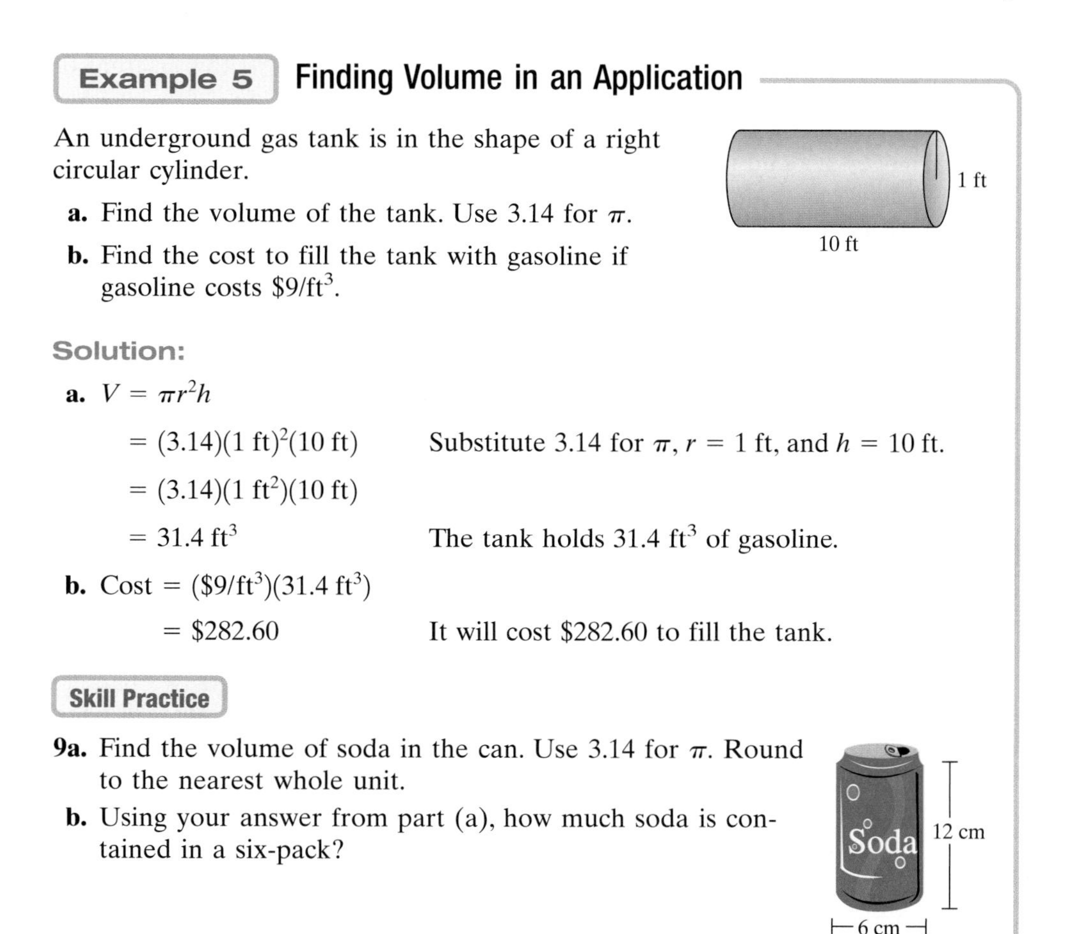

Example 5 **Finding Volume in an Application**

An underground gas tank is in the shape of a right circular cylinder.

a. Find the volume of the tank. Use 3.14 for π.

b. Find the cost to fill the tank with gasoline if gasoline costs $9/ft^3.

Solution:

a. $V = \pi r^2 h$

$= (3.14)(1 \text{ ft})^2(10 \text{ ft})$ Substitute 3.14 for π, $r = 1$ ft, and $h = 10$ ft.

$= (3.14)(1 \text{ ft}^2)(10 \text{ ft})$

$= 31.4 \text{ ft}^3$ The tank holds 31.4 ft^3 of gasoline.

b. $\text{Cost} = (\$9/\text{ft}^3)(31.4 \text{ ft}^3)$

$= \$282.60$ It will cost $282.60 to fill the tank.

Skill Practice

9a. Find the volume of soda in the can. Use 3.14 for π. Round to the nearest whole unit.

b. Using your answer from part (a), how much soda is contained in a six-pack?

4. Angles

Applications involving angles and their measure come up often in the study of algebra, trigonometry, calculus, and applied sciences. The most common unit used to measure an angle is the degree (°). Several angles and their corresponding degree measure are shown in Figure R-9.

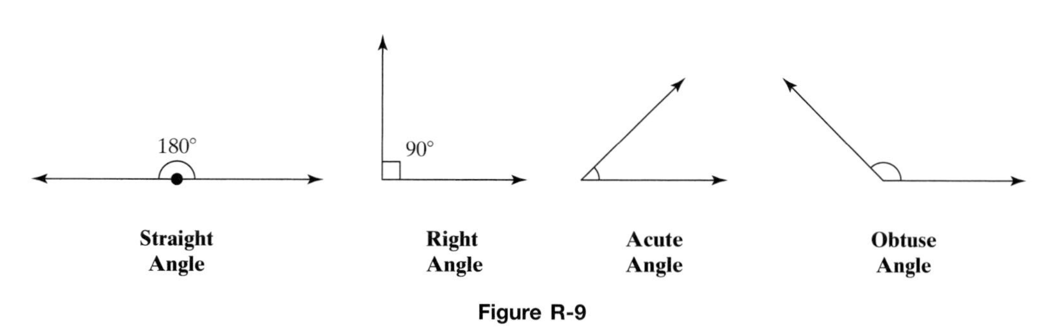

Figure R-9

- An angle that measures 90° is a **right angle** (right angles are often marked with a square or corner symbol, □).
- An angle that measures 180° is called a **straight angle**.
- An angle that measures between 0° and 90° is called an **acute angle**.
- An angle that measures between 90° and 180° is called an **obtuse angle**.
- Two angles with the same measure are **equal angles** (or **congruent angles**).

Skill Practice Answers

9a. 339 cm^3 **b.** 2034 cm^3

The measure of an angle will be denoted by the symbol m written in front of the angle. Therefore, the measure of $\angle A$ is denoted $m(\angle A)$.

- Two angles are said to be **complementary** if the sum of their measures is 90°.

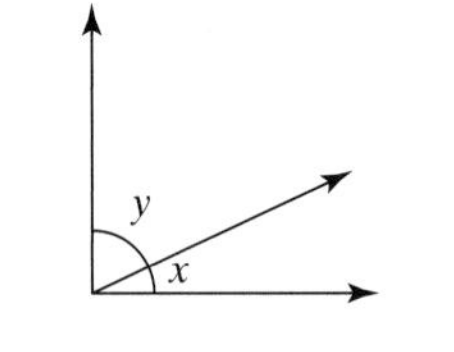

$m(\angle x) + m(\angle y) = 90°$

- Two angles are said to be **supplementary** if the sum of their measures is 180°.

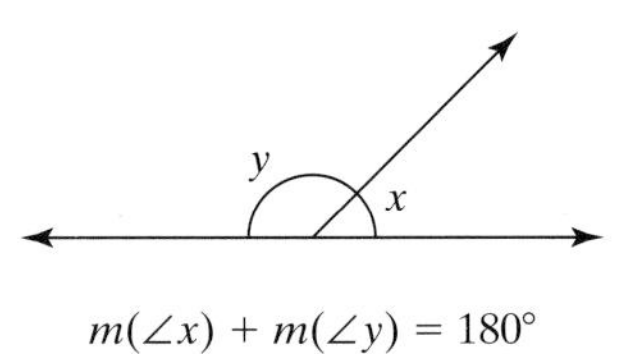

$m(\angle x) + m(\angle y) = 180°$

When two lines intersect, four angles are formed (Figure R-10). In Figure R-10, $\angle a$ and $\angle b$ are **vertical angles**. Another set of vertical angles is the pair $\angle c$ and $\angle d$. An important property of vertical angles is that the measures of two vertical angles are *equal*. In the figure, $m(\angle a) = m(\angle b)$ and $m(\angle c) = m(\angle d)$.

Parallel lines are lines that lie in the same plane and do not intersect. In Figure R-11, the lines L_1 and L_2 are parallel lines. If a line intersects two parallel lines, the line is called a **transversal**. In Figure R-11, the line m is a transversal and forms eight angles with the parallel lines L_1 and L_2.

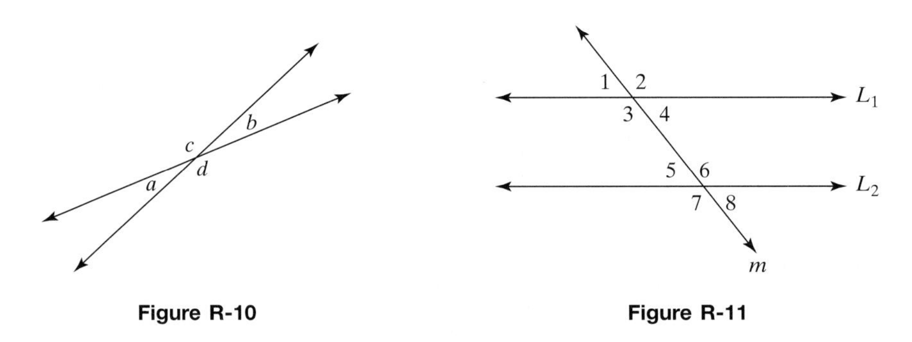

Figure R-10 **Figure R-11**

The angles 1–8 in Figure R-11 have special names and special properties.

Two angles that lie between the parallel lines (interior) and on alternate sides of the transversal are called **alternate interior angles**. Pairs of alternate interior angles are shown next.

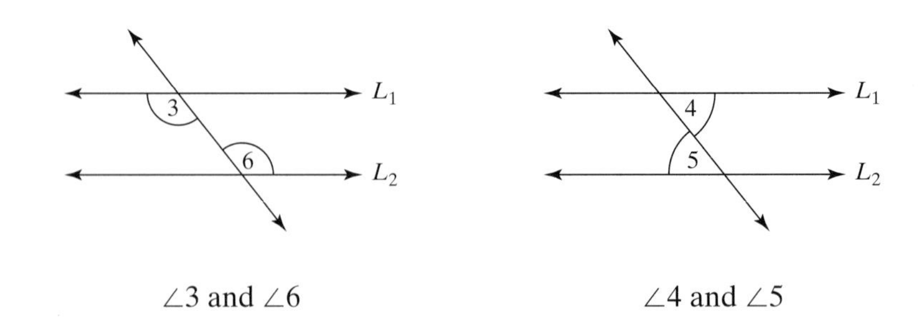

$\angle 3$ and $\angle 6$ $\angle 4$ and $\angle 5$

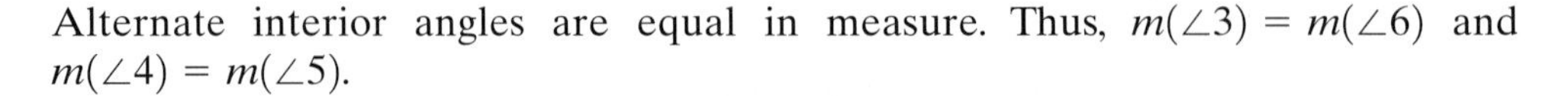

Alternate interior angles are equal in measure. Thus, $m(\angle 3) = m(\angle 6)$ and $m(\angle 4) = m(\angle 5)$.

Two angles that lie outside the parallel lines (exterior) and on alternate sides of the transversal are called **alternate exterior angles**. Pairs of alternate exterior angles are shown here.

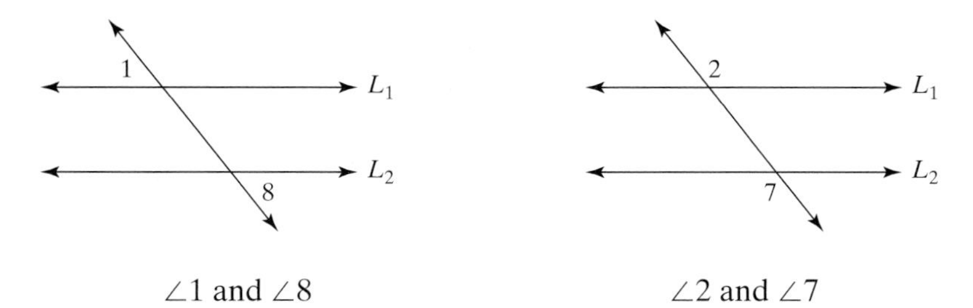

$\angle 1$ and $\angle 8$ $\qquad$ $\angle 2$ and $\angle 7$

Alternate exterior angles are equal in measure. Thus, $m(\angle 1) = m(\angle 8)$ and $m(\angle 2) = m(\angle 7)$.

Two angles that lie on the same side of the transversal such that one is exterior and one is interior are called **corresponding angles**. Corresponding angles cannot be adjacent to each other. Pairs of corresponding angles are shown here in color.

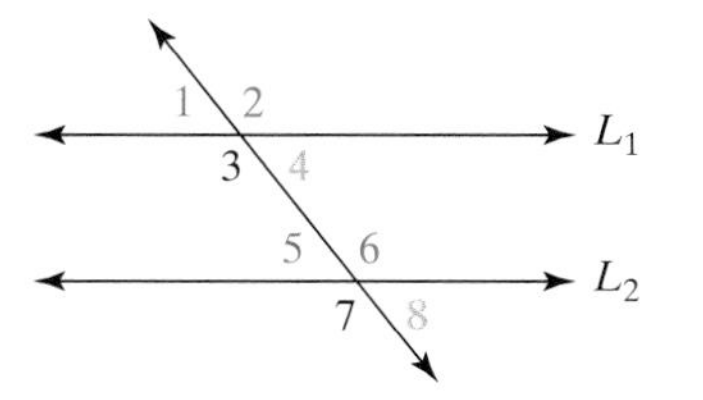

Corresponding angles are equal in measure. Thus, $m(\angle 1) = m(\angle 5)$, $m(\angle 2) = m(\angle 6)$, $m(\angle 3) = m(\angle 7)$, and $m(\angle 4) = m(\angle 8)$.

Example 6 Finding Unknown Angles in a Diagram

Find the measure of each angle and explain how the angle is related to the given angle of 70°.

a. $\angle a$ $\qquad$ **b.** $\angle b$

c. $\angle c$ $\qquad$ **d.** $\angle d$

Solution:

a. $m(\angle a) = 70°$ $\qquad$ $\angle a$ is a corresponding angle to the given angle of 70°.

b. $m(\angle b) = 70°$ $\qquad$ $\angle b$ and the given angle of 70° are alternate exterior angles.

c. $m(\angle c) = 70°$ $\qquad$ $\angle c$ and the given angle of 70° are vertical angles.

d. $m(\angle d) = 110°$ $\qquad$ $\angle d$ is the supplement of the given angle of 70°.

Skill Practice

10. Refer to the figure. Assume that lines L_1 and L_2 are parallel. Given that $m(\angle 3) = 23°$, find:

a. $m(\angle 2)$ $\qquad$ **b.** $m(\angle 4)$

c. $m(\angle 5)$ $\qquad$ **d.** $m(\angle 8)$

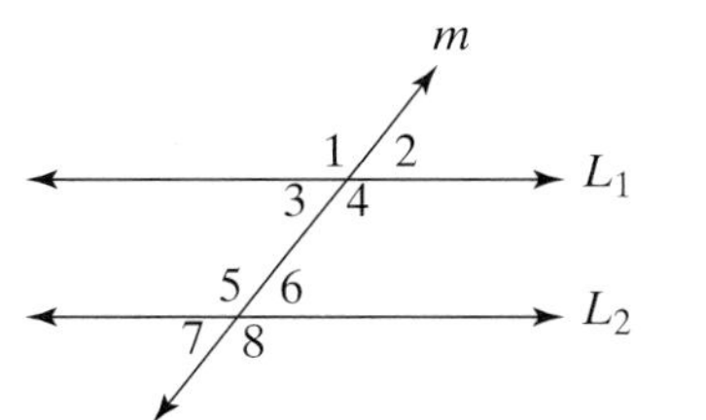

Skill Practice Answers

10a. 23° $\qquad$ **b.** 157°

c. 157° $\qquad$ **d.** 157°

5. Triangles

Triangles are categorized by the measures of the angles (Figure R-12) and by the number of equal sides or angles (Figure R-13).

- An **acute triangle** is a triangle in which all three angles are acute.
- A **right triangle** is a triangle in which one angle is a right angle.
- An **obtuse triangle** is a triangle in which one angle is obtuse.

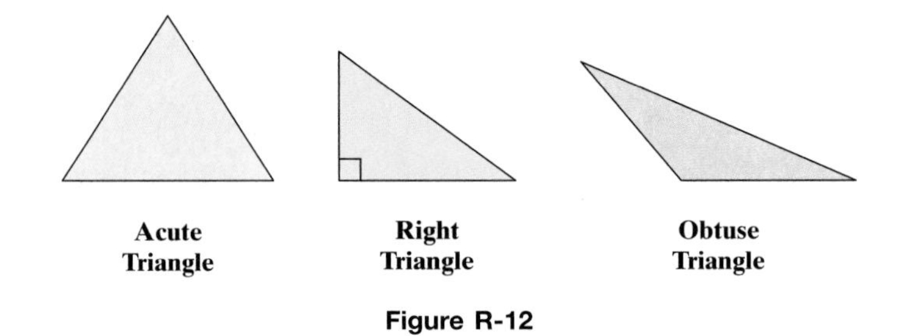

Figure R-12

- An **equilateral triangle** is a triangle in which all three sides (and all three angles) are equal in measure.
- An **isosceles triangle** is a triangle in which two sides are equal in measure (the angles opposite the equal sides are also equal in measure).
- A **scalene triangle** is a triangle in which no sides (or angles) are equal in measure.

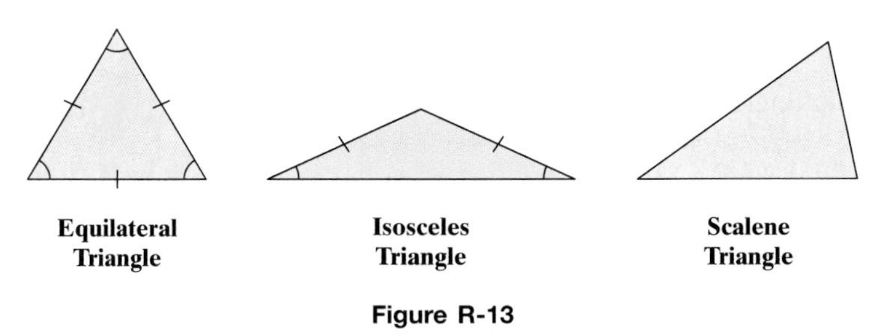

Figure R-13

The following important property is true for all triangles.

Sum of the Angles in a Triangle

The sum of the measures of the angles of a triangle is 180°.

Example 7 Finding Unknown Angles in a Diagram

Find the measure of each angle in the figure.

a. $\angle a$

b. $\angle b$

c. $\angle c$

d. $\angle d$

e. $\angle e$

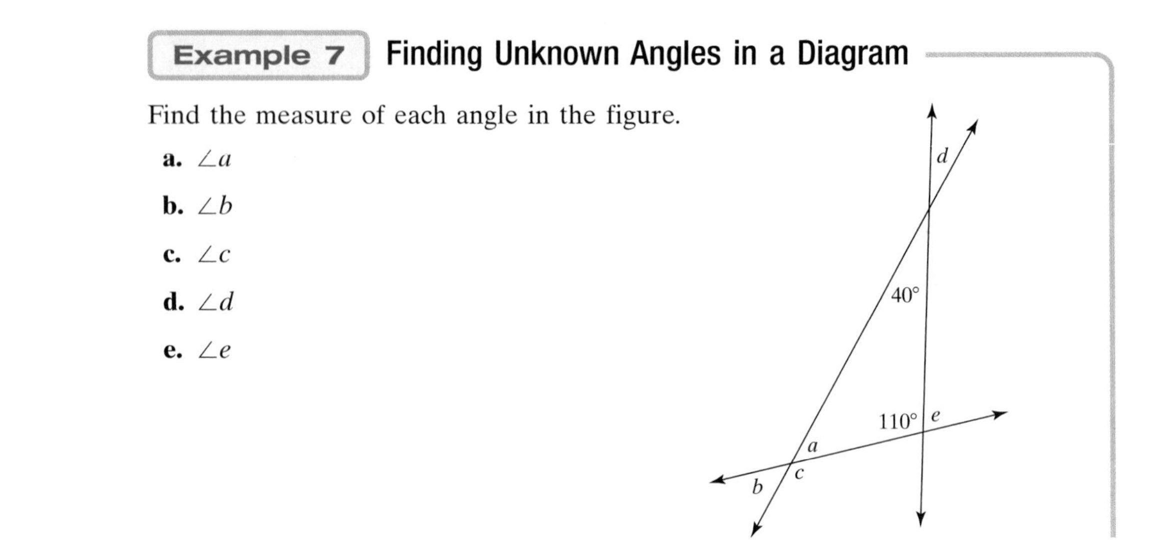

Solution:

a. $m(\angle a) = 30°$ — The sum of the angles in a triangle is 180°.

b. $m(\angle b) = 30°$ — $\angle a$ and $\angle b$ are vertical angles and are equal in measure.

c. $m(\angle c) = 150°$ — $\angle c$ and $\angle a$ are supplementary angles ($\angle c$ and $\angle b$ are also supplementary).

d. $m(\angle d) = 40°$ — $\angle d$ and the given angle of 40° are vertical angles.

e. $m(\angle e) = 70°$ — $\angle e$ and the given angle of 110° are supplementary angles.

Skill Practice

11. Refer to the figure. Find the measure of the indicated angle.

a. $\angle a$ **b.** $\angle b$ **c.** $\angle c$

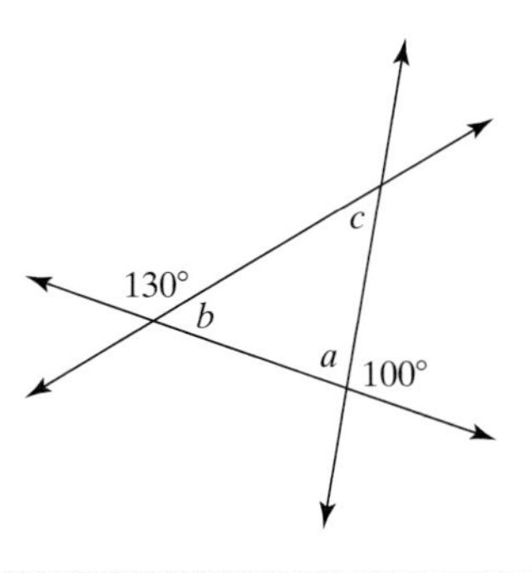

Skill Practice Answers

11a. 80° **b.** 50° **c.** 50°

Section R.4 Practice Exercises

Boost *your* GRADE at mathzone.com! MathZone

- Practice Problems
- Self-Tests
- NetTutor
- e-Professors
- Videos

Concept 1: Perimeter

1. Identify which of the following units could be measures of perimeter.

a. Square inches (in.^{2}) **b.** Meters (m) **c.** Cubic feet (ft^3)

d. Cubic meters (m^3) **e.** Miles (mi) **f.** Square centimeters (cm^2)

g. Square yards (yd^2) **h.** Cubic inches (in.^{3}) **i.** Kilometers (km)

For Exercises 2–5, find the perimeter of each figure.

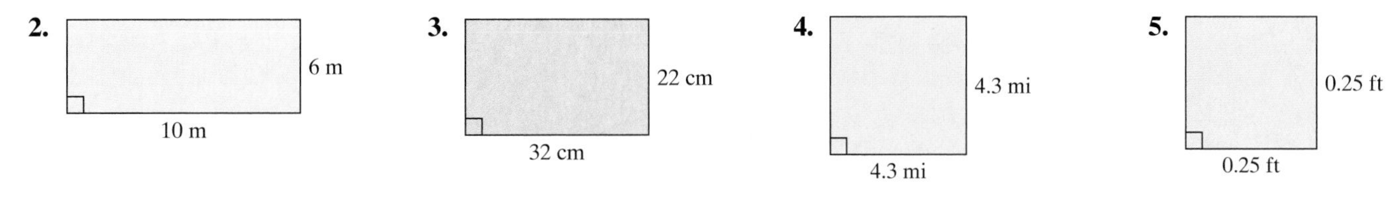

6. Identify which of the following units could be measures of circumference.

a. Square inches ($in.^2$) b. Meters (m) c. Cubic feet (ft^3)

d. Cubic meters (m^3) e. Miles (mi) f. Square centimeters (cm^2)

g. Square yards (yd^2) h. Cubic inches ($in.^3$) i. Kilometers (km)

For Exercises 7–10, find the perimeter or circumference. Use 3.14 for π.

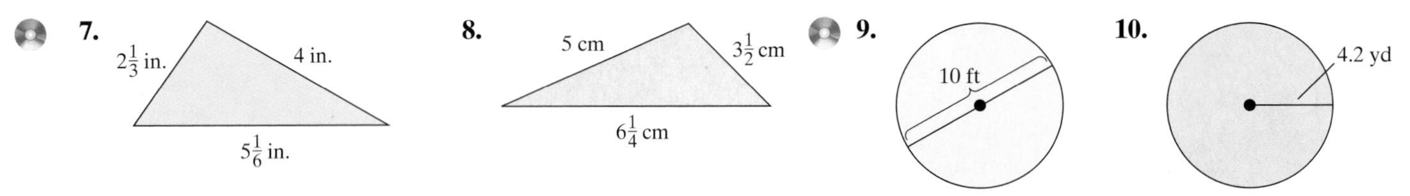

Concept 2: Area

11. Identify which of the following units could be measures of area.

a. Square inches ($in.^2$) b. Meters (m) c. Cubic feet (ft^3)

d. Cubic meters (m^3) e. Miles (mi) f. Square centimeters (cm^2)

g. Square yards (yd^2) h. Cubic inches ($in.^3$) i. Kilometers (km)

For Exercises 12–25, find the area. Use 3.14 for π.

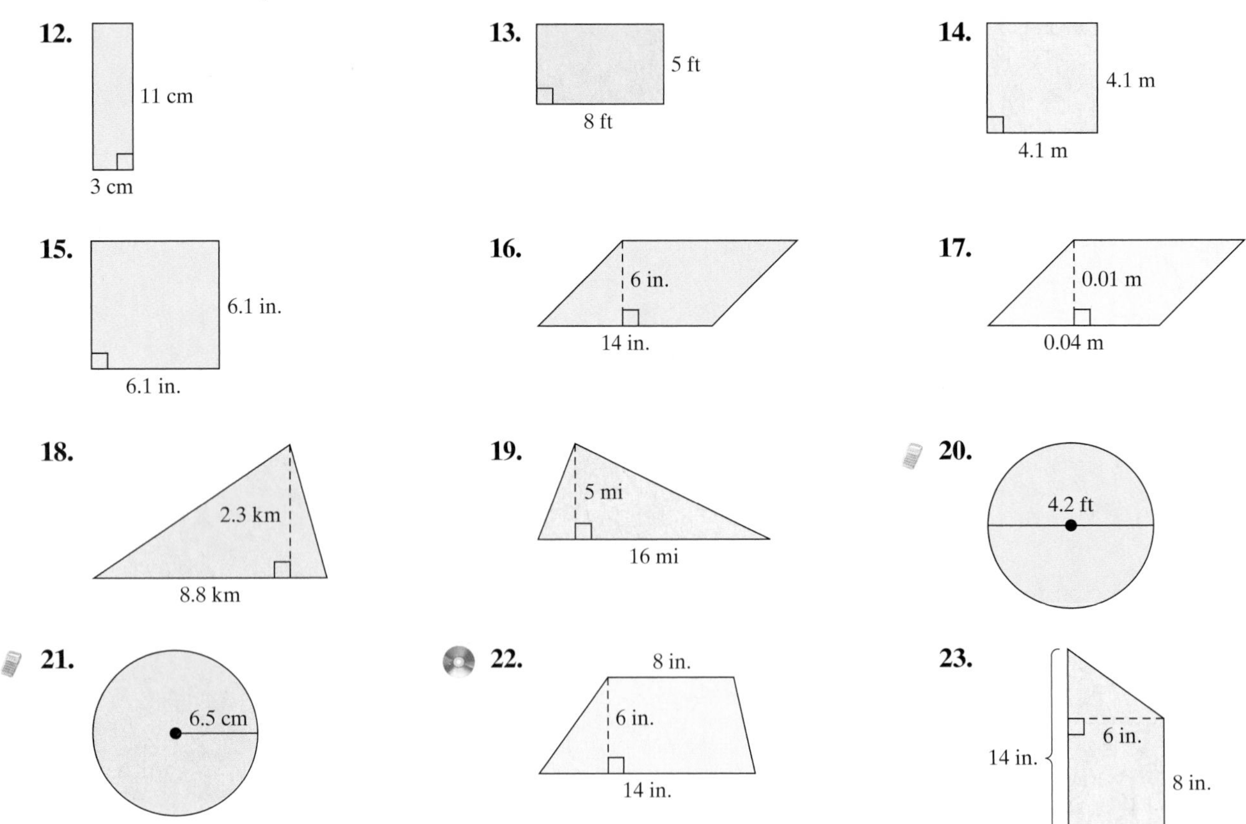

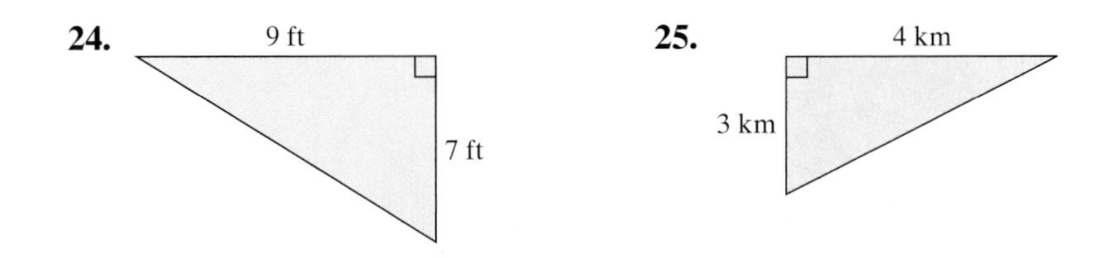

Concept 3: Volume

26. Identify which of the following units could be measures of volume.

a. Square inches (in.^2) **b.** Meters (m) **c.** Cubic feet (ft^3)

d. Cubic meters (m^3) **e.** Miles (mi) **f.** Square centimeters (cm^2)

g. Square yards (yd^2) **h.** Cubic inches (in.^3) **i.** Kilometers (km)

For Exercises 27–34, find the volume of each figure. Use 3.14 for π.

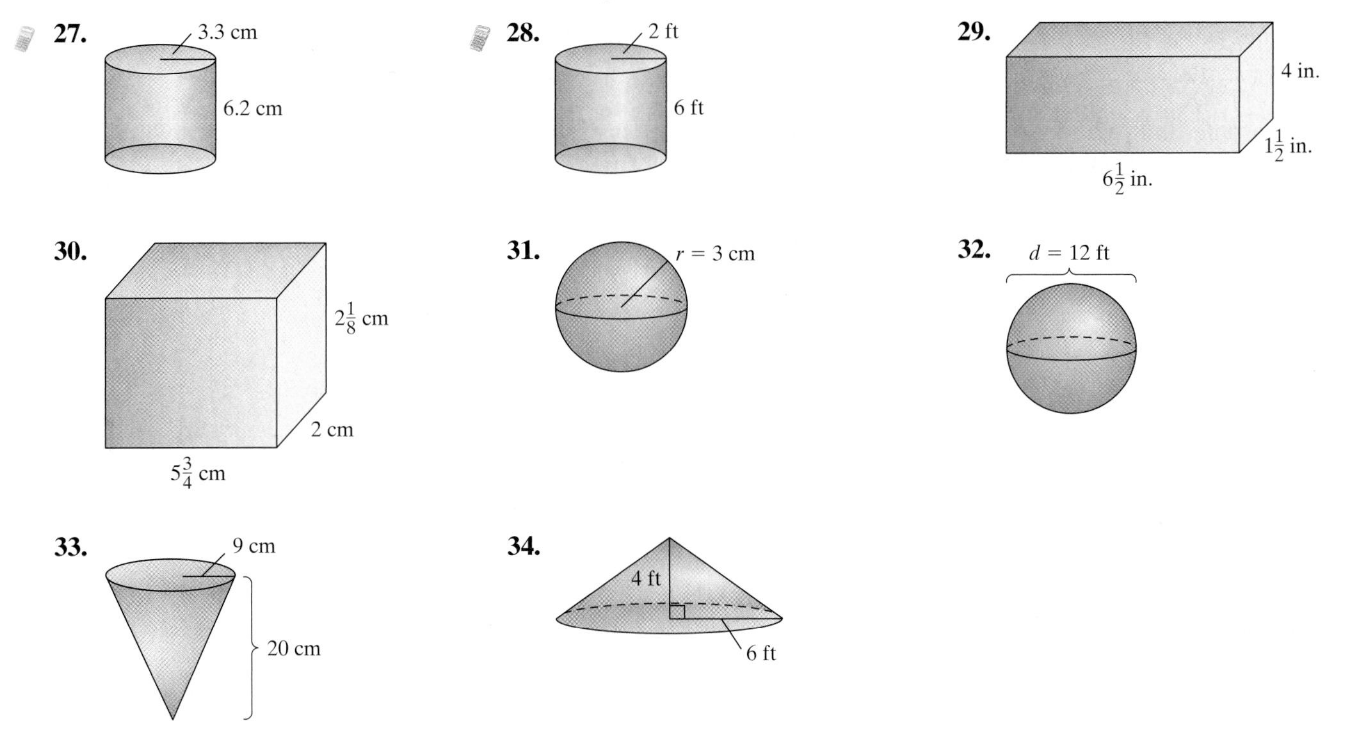

35. A florist sells balloons and needs to know how much helium to order. Each balloon is approximately spherical with a radius of 9 in. How much helium is needed to fill one balloon?

36. Find the volume of a spherical ball whose radius is 3 in. Use 3.14 for π.

37. Find the volume of a snow cone in the shape of a right circular cone whose radius is 3 cm and whose height is 12 cm. Use 3.14 for π.

38. A landscaping supply company has a pile of gravel in the shape of a right circular cone whose radius is 10 yd and whose height is 18 yd. Find the volume of the gravel. Use 3.14 for π.

39. Find the volume of a cube that is 3.2 ft on a side.

40. Find the volume of a cube that is 10.5 cm on a side.

Mixed Exercises: Perimeter, Area, and Volume

41. A wall measuring 20 ft by 8 ft can be painted for $50.

a. What is the price per square foot? Round to the nearest cent.

b. At this rate, how much would it cost to paint the remaining three walls that measure 20 ft by 8 ft, 16 ft by 8 ft, and 16 ft by 8 ft? Round to the nearest dollar.

42. Suppose it costs $320 to carpet a 16 ft by 12 ft room.

a. What is the price per square foot? Round to the nearest cent.

b. At this rate, how much would it cost to carpet a room that is 20 ft by 32 ft?

43. If you were to purchase fencing for a garden, would you measure the perimeter or area of the garden?

44. If you were to purchase sod (grass) for your front yard, would you measure the perimeter or area of the yard?

45. **a.** Find the area of a circular pizza that is 8 in. in diameter (the radius is 4 in.). Use 3.14 for π.

b. Find the area of a circular pizza that is 12 in. in diameter (the radius is 6 in.).

c. Assume that the 8-in. diameter and 12-in. diameter pizzas are both the same thickness. Which would provide more pizza, two 8-in. pizzas or one 12-in. pizza?

46. Find the area of a circular stained glass window that is 16 in. in diameter. Use 3.14 for π.

47. Find the volume of a soup can in the shape of a right circular cylinder if its radius is 3.2 cm and its height is 9 cm. Use 3.14 for π.

48. Find the volume of a coffee mug whose radius is 2.5 in. and whose height is 6 in. Use 3.14 for π.

Concept 4: Angles

For Exercises 49–56, answer True or False. If an answer is false, explain why.

49. The sum of the measures of two right angles equals the measure of a straight angle.

50. Two right angles are complementary.

51. Two right angles are supplementary.

52. Two acute angles cannot be supplementary.

53. Two obtuse angles cannot be supplementary.

54. An obtuse angle and an acute angle can be supplementary.

55. If a triangle is equilateral, then it is not scalene.

56. If a triangle is isosceles, then it is also scalene.

57. What angle is its own complement?

58. What angle is its own supplement?

59. If possible, find two acute angles that are supplementary.

60. If possible, find two acute angles that are complementary. Answers may vary.

61. If possible, find an obtuse angle and an acute angle that are supplementary. Answers may vary.

62. If possible, find two obtuse angles that are supplementary.

63. Refer to the figure.

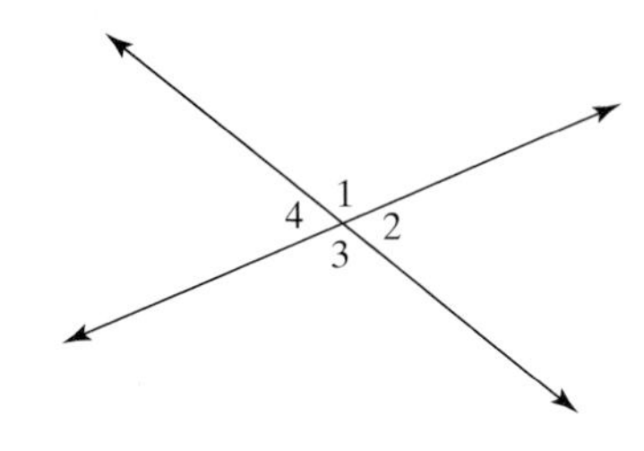

a. State all the pairs of vertical angles.

b. State all the pairs of supplementary angles.

c. If the measure of $\angle 4$ is 80°, find the measures of $\angle 1$, $\angle 2$, and $\angle 3$.

64. Refer to the figure.

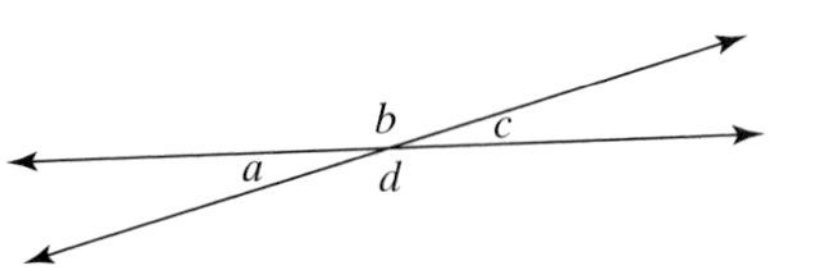

a. State all the pairs of vertical angles.

b. State all the pairs of supplementary angles.

c. If the measure of $\angle a$ is 25°, find the measures of $\angle b$, $\angle c$, and $\angle d$.

For Exercises 65–68, find the complement of each angle.

65. 33° **66.** 87° **67.** 12° **68.** 45°

For Exercises 69–72, find the supplement of each angle.

69. 33° **70.** 87° **71.** 122° **72.** 90°

For Exercises 73–80, refer to the figure. Assume that L_1 and L_2 are parallel lines cut by the transversal, n.

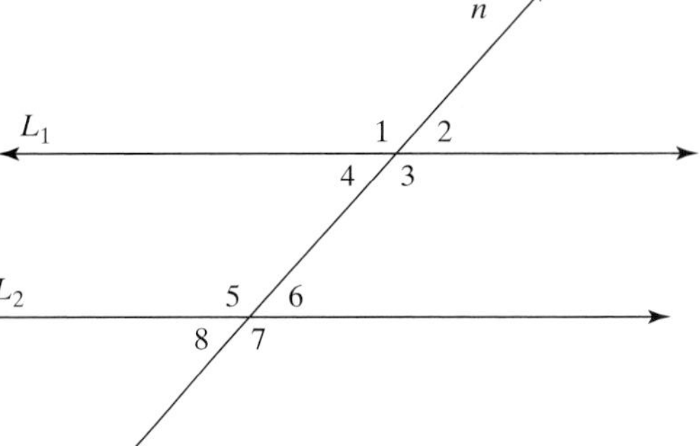

73. $m(\angle 5) = m(\angle$ ______) Reason: Vertical angles have equal measures.

74. $m(\angle 5) = m(\angle$ ______) Reason: Alternate interior angles have equal measures.

75. $m(\angle 5) = m(\angle$ ______) Reason: Corresponding angles have equal measures.

76. $m(\angle 7) = m(\angle$ ______) Reason: Corresponding angles have equal measures.

77. $m(\angle 7) = m(\angle$ ______) Reason: Alternate exterior angles have equal measures.

78. $m(\angle 7) = m(\angle$ ______) Reason: Vertical angles have equal measures.

79. $m(\angle 3) = m(\angle$ ______) Reason: Alternate interior angles have equal measures.

80. $m(\angle 3) = m(\angle$ ______) Reason: Vertical angles have equal measures.

81. Find the measures of angles a–g in the figure. Assume that L_1 and L_2 are parallel and that n is a transversal.

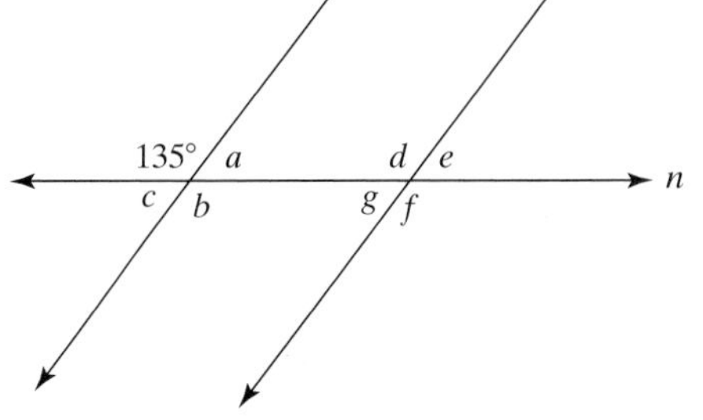

82. Find the measures of angles a–g in the figure. Assume that L_1 and L_2 are parallel and that n is a transversal.

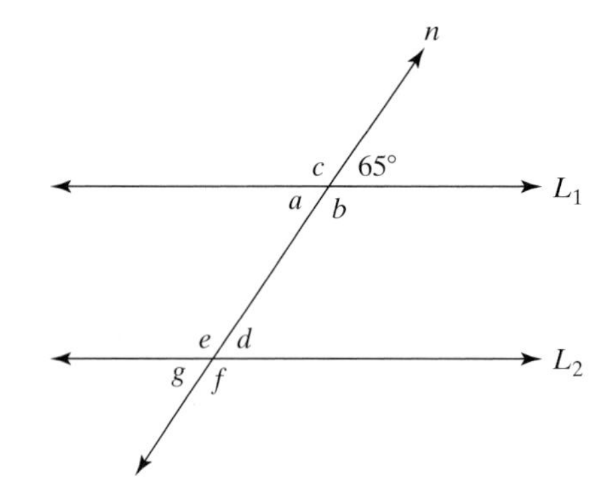

Concept 5: Triangles

For Exercises 83–86, identify the triangle as equilateral, isosceles, or scalene.

83. **84.** **85.** **86.**

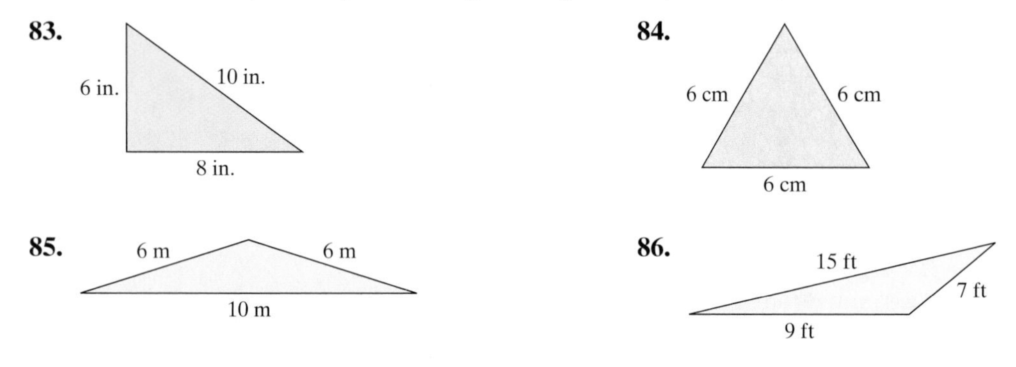

87. Can a triangle be both a right triangle and an obtuse triangle? Explain.

88. Can a triangle be both a right triangle and an isosceles triangle? Explain.

For Exercises 89–92, find the measures of the missing angles.

89. **90.** **91.** **92.**

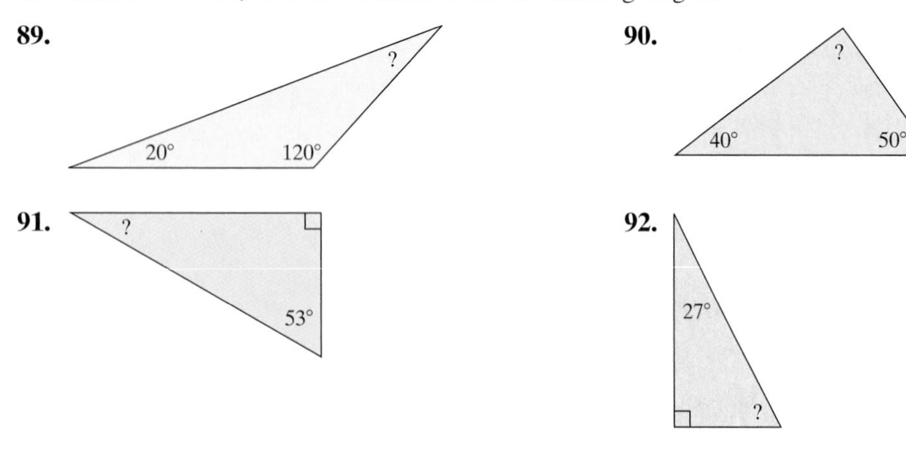

93. Refer to the figure. Find the measures of angles a–j.

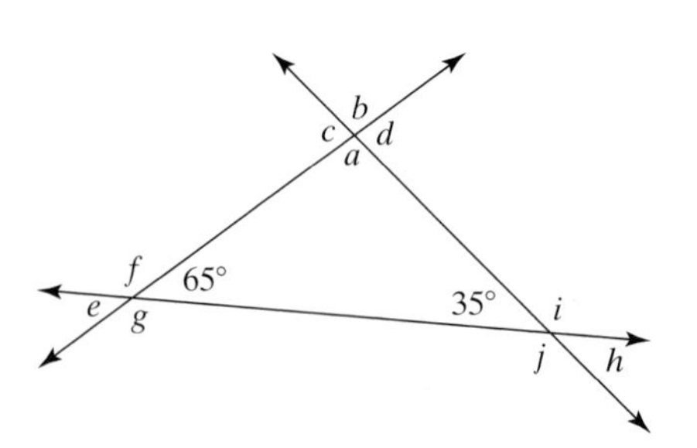

94. Refer to the figure. Find the measures of angles a–j.

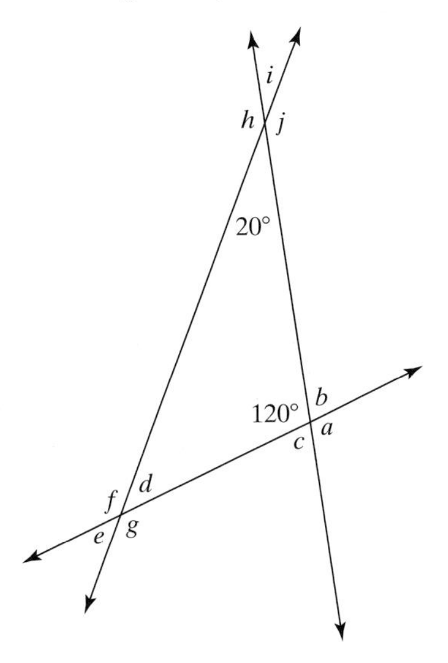

95. Refer to the figure. Find the measures of angles a–k. Assume that L_1 and L_2 are parallel.

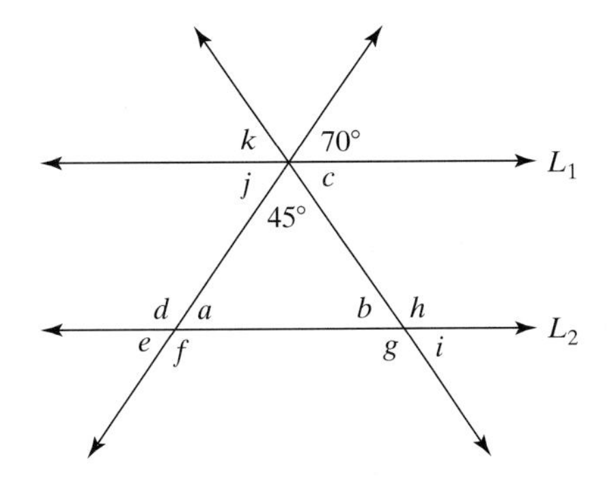

96. Refer to the figure. Find the measures of angles a–k. Assume that L_1 and L_2 are parallel.

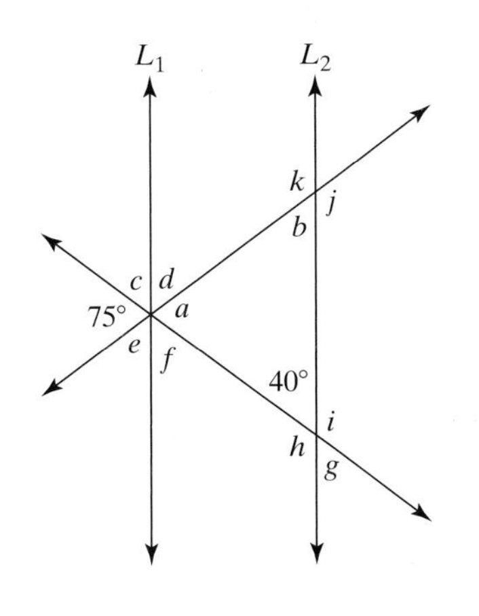

Expanding Your Skills

For Exercises 97–98, find the perimeter.

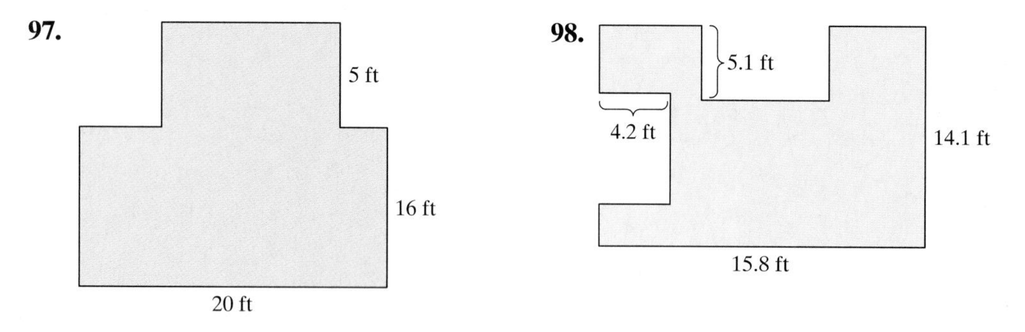

For Exercises 99–102, find the area of the shaded region. Use 3.14 for π.

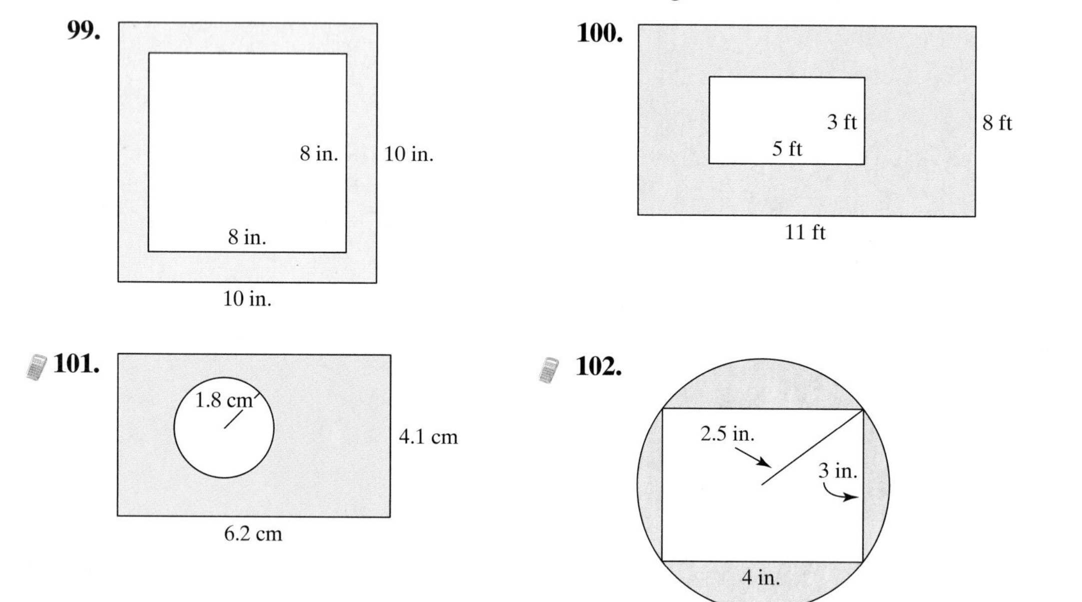

The Set of Real Numbers

1

In Chapter 1 we present operations on real numbers. To be successful in algebra, it is particularly important to understand the order of operations.

As you work through the chapter, pay attention to the order of operations. Then work through this puzzle. When you fill in the blanks, notice that negative signs may appear to the left of a horizontal number. They may also appear above a number written vertically.

Across

2. -12^2
5. $-3426 - 469 + 124 - 6421$
9. $-2.5 - 1.4 - 8.1 - 100$
11. $243 - |12 - 17| + |-600|$
12. $155 - 3(2 - 4^2)^2 - 4(1 - 3)$

Down

1. $(-12)^2$
3. $120 - \dfrac{\sqrt{25 - 9}}{-2}$
4. $600 \div [2^3 + 7 - (4 + 1)]$
6. $-\dfrac{4}{3} - \left(-\dfrac{11}{6}\right) + 15 - \dfrac{1}{2}$
7. $84(-12) - (-10{,}625)$
8. $\dfrac{120 - \sqrt{25 - 9}}{-2}$
10. $(-2.5)(-1.4)(-8.1)(-100)$

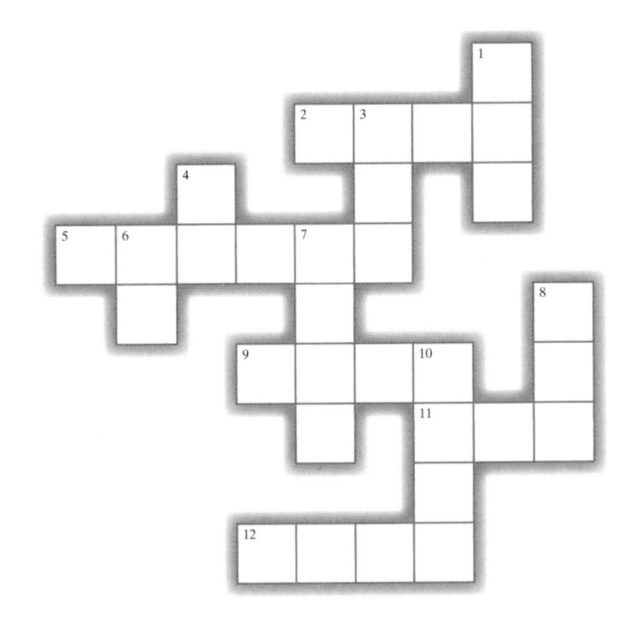

Section 1.1 Sets of Numbers and the Real Number Line

Concepts

1. Real Number Line
2. Plotting Points on the Number Line
3. Set of Real Numbers
4. Inequalities
5. Opposite of a Real Number
6. Absolute Value of a Real Number

1. Real Number Line

The numbers we work with on a day-to-day basis are all part of the set of **real numbers**. The real numbers encompass zero, all positive, and all negative numbers, including those represented by fractions and decimal numbers. The set of real numbers can be represented graphically on a horizontal number line with a point labeled as 0. Positive real numbers are graphed to the right of 0, and negative real numbers are graphed to the left. Zero is neither positive nor negative. Each point on the number line corresponds to exactly one real number. For this reason, this number line is called the **real number line** (Figure 1-1).

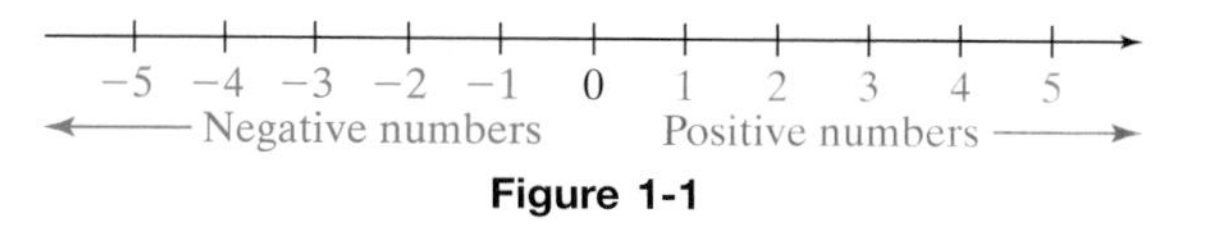

Figure 1-1

2. Plotting Points on the Number Line

Example 1 Plotting Points on the Real Number Line

Plot the points on the real number line that represent the following real numbers.

a. -3 **b.** $\frac{3}{2}$ **c.** -4.8 **d.** $\frac{16}{5}$

Solution:

a. Because -3 is negative, it lies three units to the left of zero.

b. The fraction $\frac{3}{2}$ can be expressed as the mixed number $1\frac{1}{2}$, which lies halfway between 1 and 2 on the number line.

c. The negative number -4.8 lies $\frac{8}{10}$ units to the left of -4 on the number line.

d. The fraction $\frac{16}{5}$ can be expressed as the mixed number $3\frac{1}{5}$, which lies $\frac{1}{5}$ unit to the right of 3 on the number line.

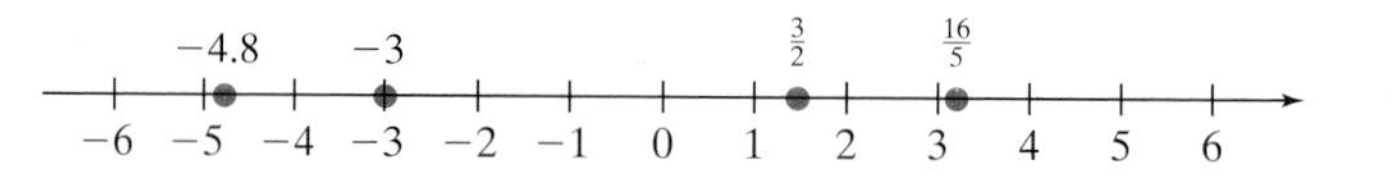

Skill Practice

1. Plot the numbers on a real number line.
$\{-1, -2.5, \frac{3}{4}, 4\}$

0

Skill Practice Answers

1.

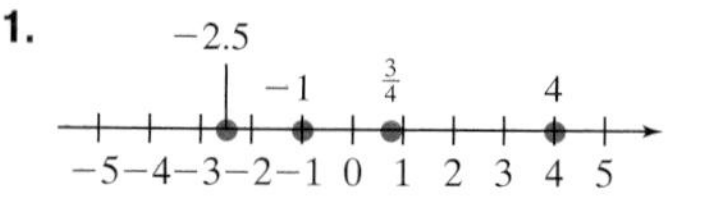

3. Set of Real Numbers

In mathematics, a well-defined collection of elements is called a set. "Well-defined" means the set is described in such a way that it is clear whether an element is in the set. The symbols { } are used to enclose the elements of the set. For example, the set {A, B, C, D, E} represents the set of the first five letters of the alphabet.

Several sets of numbers are used extensively in algebra and are subsets (or part) of the set of real numbers. These are:

The set of natural numbers
The set of whole numbers
The set of integers
The set of rational numbers
The set of irrational numbers

Definition of the Natural Numbers, Whole Numbers, and Integers

The set of **natural numbers** is $\{1, 2, 3, \ldots\}$
The set of **whole numbers** is $\{0, 1, 2, 3, \ldots\}$
The set of **integers** is $\{\ldots -3, -2, -1, 0, 1, 2, 3, \ldots\}$

Notice that the set of whole numbers includes the natural numbers. Therefore, every natural number is also a whole number. The set of integers includes the set of whole numbers. Therefore, every whole number is also an integer. The relationship among the elements of the natural numbers, whole numbers, and integers is shown in Figure 1-2.

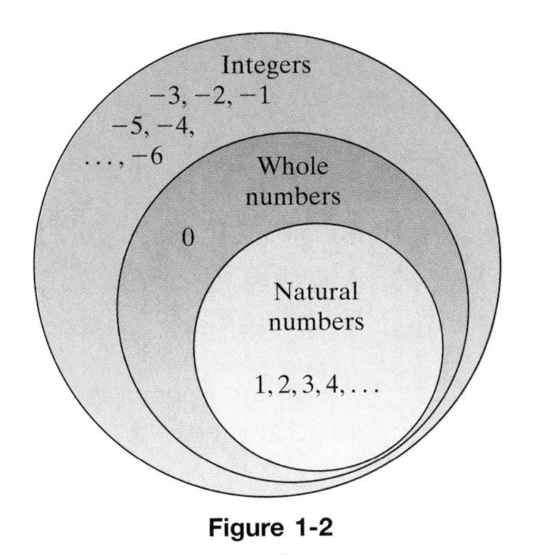

Figure 1-2

Fractions are also among the numbers we use frequently. A number that can be written as a fraction whose numerator is an integer and whose denominator is a nonzero integer is called a *rational number.*

Definition of the Rational Numbers

The set of **rational numbers** is the set of numbers that can be expressed in the form $\frac{p}{q}$, where both p and q are integers and q does not equal 0.

We also say that a rational number $\frac{p}{q}$ is a *ratio* of two integers, p and q, where q is not equal to zero.

Example 2 Identifying Rational Numbers

Show that the following numbers are rational numbers by finding an equivalent ratio of two integers.

a. $\dfrac{-2}{3}$ **b.** -12 **c.** 0.5 **d.** $0.\overline{6}$

Solution:

a. The fraction $\frac{-2}{3}$ is a rational number because it can be expressed as the ratio of -2 and 3.

b. The number -12 is a rational number because it can be expressed as the ratio of -12 and 1, that is, $-12 = \frac{-12}{1}$. In this example, we see that an integer is also a rational number.

c. The terminating decimal 0.5 is a rational number because it can be expressed as the ratio of 5 and 10. That is, $0.5 = \frac{5}{10}$. In this example we see that a terminating decimal is also a rational number.

d. The repeating decimal $0.\overline{6}$ is a rational number because it can be expressed as the ratio of 2 and 3. That is, $0.\overline{6} = \frac{2}{3}$. In this example we see that a repeating decimal is also a rational number.

Skill Practice Show that each number is rational by finding an equivalent ratio of two integers.

2. $\dfrac{3}{7}$ **3.** -5 **4.** 0.3 **5.** $0.\overline{3}$

TIP: Any rational number can be represented by a terminating decimal or by a repeating decimal.

Some real numbers, such as the number π, cannot be represented by the ratio of two integers. In decimal form, an irrational number is a nonterminating, nonrepeating decimal. The value of π, for example, can be approximated as $\pi \approx 3.14159265358979 32$. However, the decimal digits continue forever with no repeated pattern. Another example of an irrational number is $\sqrt{3}$ (read as "the positive square root of 3"). The expression $\sqrt{3}$ is a number that when multiplied by itself is 3. There is no rational number that satisfies this condition. Thus, $\sqrt{3}$ is an irrational number.

Definition of the Irrational Numbers

The set of **irrational numbers** is the set of real numbers that are not rational.

Note: An irrational number cannot be written as a terminating decimal or as a repeating decimal.

The set of real numbers consists of both the rational and the irrational numbers. The relationship among these important sets of numbers is illustrated in Figure 1-3:

Skill Practice Answers

2. ratio of 3 and 7
3. ratio of -5 and 1
4. ratio of 3 and 10
5. ratio of 1 and 3

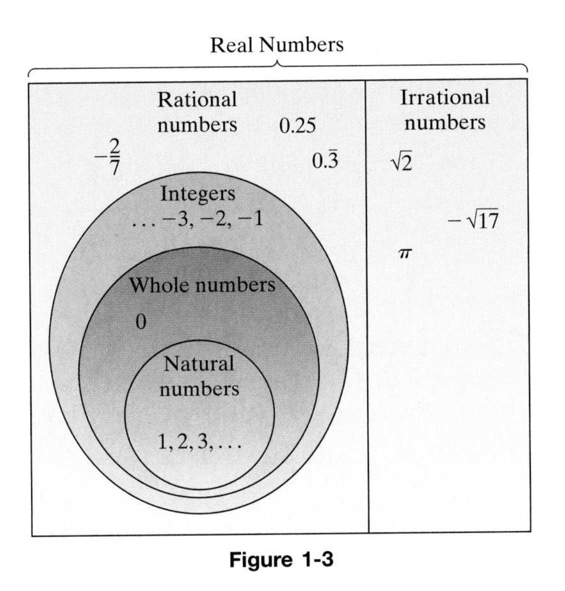

Figure 1-3

Example 3 Classifying Numbers by Set

Check the set(s) to which each number belongs. The numbers may belong to more than one set.

	Natural Numbers	Whole Numbers	Integers	Rational Numbers	Irrational Numbers	Real Numbers
5						
$\frac{-47}{3}$						
1.48						
$\sqrt{7}$						
0						

Solution:

	Natural Numbers	Whole Numbers	Integers	Rational Numbers	Irrational Numbers	Real Numbers
5	✔	✔	✔	✔ (ratio of 5 and 1)		✔
$\frac{-47}{3}$				✔ (ratio of −47 and 3)		✔
1.48				✔ (ratio of 148 and 100)		✔
$\sqrt{7}$					✔	✔
0		✔	✔	✔ (ratio of 0 and 1)		✔

Identify the sets to which each number belongs. Choose from: natural numbers, whole numbers, integers, rational numbers, irrational numbers, and real numbers.

6. -4 **7.** $0.\overline{7}$ **8.** $\sqrt{13}$ **9.** 12 **10.** 0

4. Inequalities

The relative size of two real numbers can be compared using the real number line. Suppose a and b represent two real numbers. We say that a is less than b, denoted $a < b$, if a lies to the left of b on the number line.

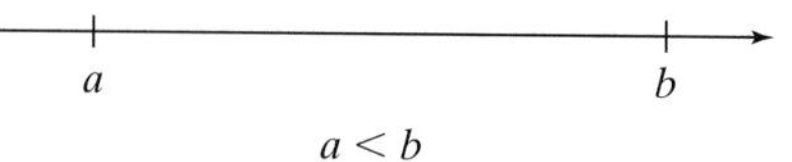

$a < b$

We say that a is greater than b, denoted $a > b$, if a lies to the right of b on the number line.

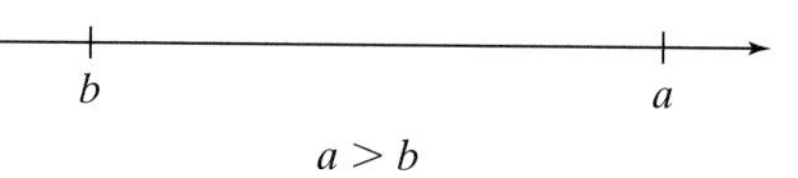

$a > b$

Table 1-1 summarizes the relational operators that compare two real numbers a and b.

Table 1-1

Mathematical Expression	Translation	Example
$a < b$	a is less than b.	$2 < 3$
$a > b$	a is greater than b.	$5 > 1$
$a \le b$	a is less than or equal to b.	$4 \le 4$
$a \ge b$	a is greater than or equal to b.	$10 \ge 9$
$a = b$	a is equal to b.	$6 = 6$
$a \ne b$	a is not equal to b.	$7 \ne 0$
$a \approx b$	a is approximately equal to b.	$2.3 \approx 2$

The symbols $<$, $>$, $\le$, $\ge$, and $\ne$ are called inequality signs, and the expressions $a < b$, $a > b$, $a \le b$, $a \ge b$, and $a \ne b$ are called **inequalities**.

Example 4 Ordering Real Numbers

The average temperatures (in degrees Celsius) for selected cities in the United States and Canada in January are shown in Table 1-2.

Table 1-2

City	Temp (°C)
Prince George, British Columbia	-12.1
Corpus Christi, Texas	13.4
Parkersburg, West Virginia	-0.9
San Jose, California	9.7
Juneau, Alaska	-5.7
New Bedford, Massachusetts	-0.2
Durham, North Carolina	4.2

Skill Practice Answers

6. integers, rational numbers, real numbers
7. rational numbers, real numbers
8. irrational numbers, real numbers
9. natural numbers, whole numbers, integers, rational numbers, real numbers
10. whole numbers, integers, rational numbers, real numbers

a. Plot a point on the real number line representing the temperature of each city.

b. Compare the temperatures between the following cities, and fill in the blank with the appropriate inequality sign: $<$ or $>$.

Solution:

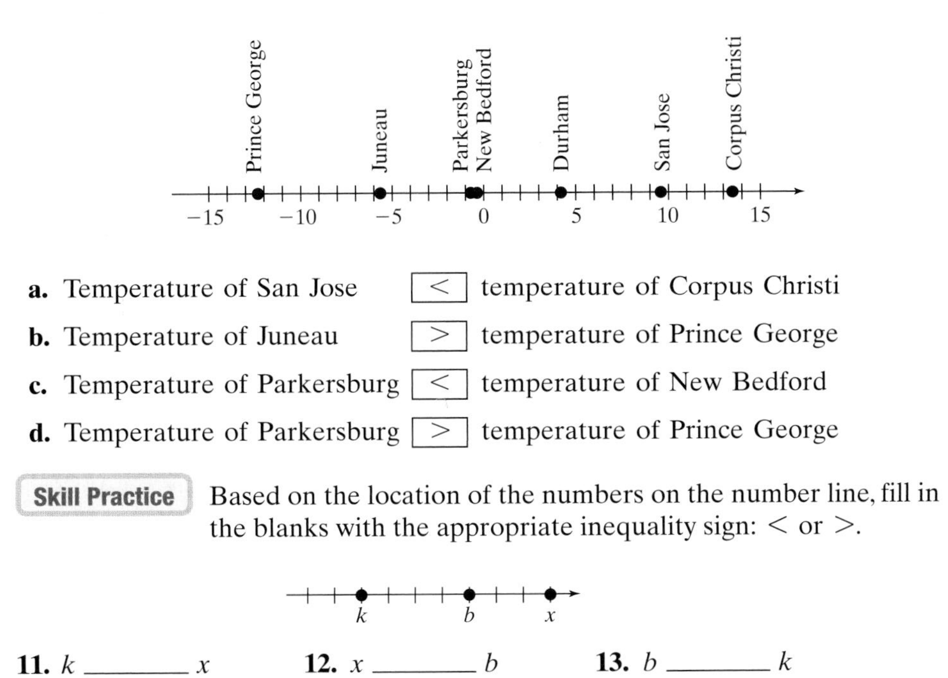

a. Temperature of San Jose $\boxed{<}$ temperature of Corpus Christi

b. Temperature of Juneau $\boxed{>}$ temperature of Prince George

c. Temperature of Parkersburg $\boxed{<}$ temperature of New Bedford

d. Temperature of Parkersburg $\boxed{>}$ temperature of Prince George

Skill Practice Based on the location of the numbers on the number line, fill in the blanks with the appropriate inequality sign: $<$ or $>$.

k b x

11. k ________ x **12.** x ________ b **13.** b ________ k

5. Opposite of a Real Number

To gain mastery of any algebraic skill, it is necessary to know the meaning of key definitions and key symbols. Two important definitions are the *opposite* of a real number and the *absolute value* of a real number.

Definition of the Opposite of a Real Number

Two numbers that are the same distance from 0 but on opposite sides of 0 on the number line are called **opposites** of each other. Symbolically, we denote the opposite of a real number a as $-a$.

Example 5 Finding the Opposite of a Real Number

a. Find the opposite of 5.

b. Find the opposite of $-\frac{4}{7}$.

c. Evaluate $-(0.46)$.

d. Evaluate $-(-\frac{11}{3})$.

Skill Practice Answers

11. $<$ **12.** $>$ **13.** $>$

Solution:

a. The opposite of 5 is -5.

b. The opposite of $-\frac{4}{7}$ is $\frac{4}{7}$.

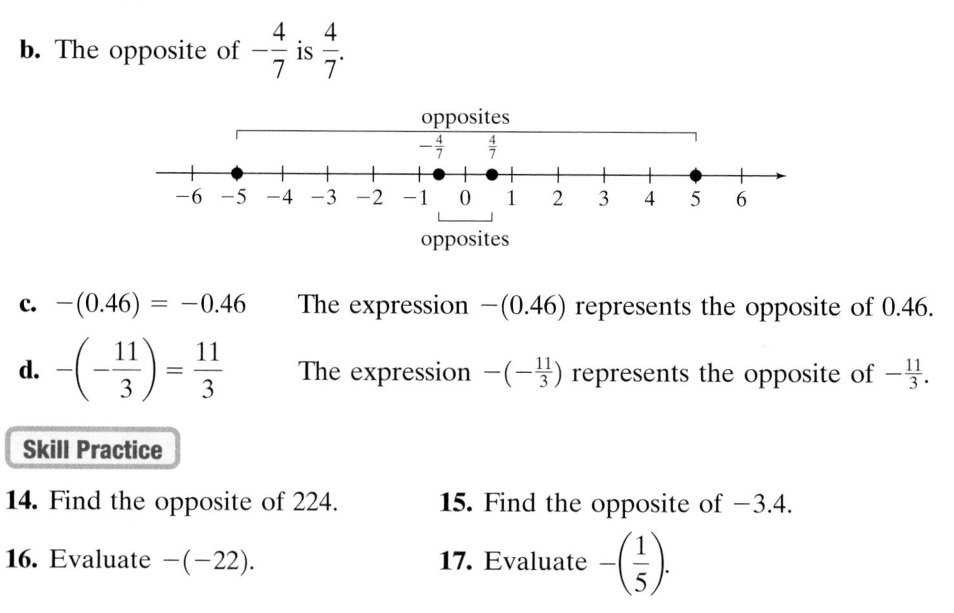

c. $-(0.46) = -0.46$ The expression $-(0.46)$ represents the opposite of 0.46.

d. $-\left(-\frac{11}{3}\right) = \frac{11}{3}$ The expression $-(-\frac{11}{3})$ represents the opposite of $-\frac{11}{3}$.

Skill Practice

14. Find the opposite of 224.

15. Find the opposite of -3.4.

16. Evaluate $-(-22)$.

17. Evaluate $-\left(\frac{1}{5}\right)$.

6. Absolute Value of a Real Number

The concept of absolute value will be used to define the addition of real numbers in Section 1.3.

Informal Definition of the Absolute Value of a Real Number

The **absolute value** of a real number a, denoted $|a|$, is the distance between a and 0 on the number line.

Note: The absolute value of any real number is nonnegative.

For example, $|3| = 3$ and $|-3| = 3$.

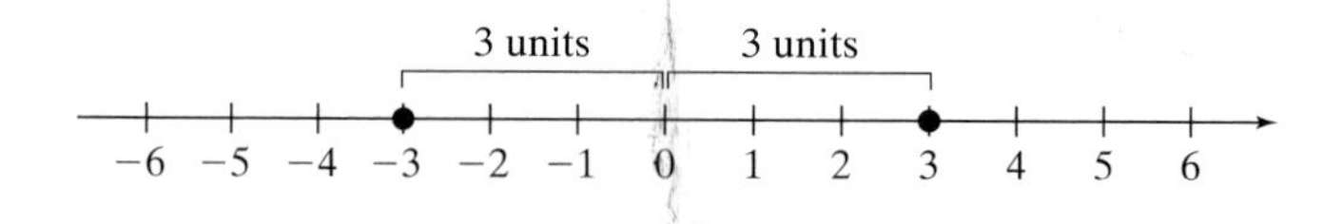

Example 6 Finding the Absolute Value of a Real Number

Evaluate the absolute value expressions.

a. $|-4|$ **b.** $|\frac{1}{2}|$ **c.** $|-6.2|$ **d.** $|0|$

Solution:

a. $|-4| = 4$ -4 is 4 units from 0 on the number line.

Skill Practice Answers

14. -224 **15.** 3.4

16. 22 **17.** $-\frac{1}{5}$

b. $\left|\frac{1}{2}\right| = \frac{1}{2}$ $\frac{1}{2}$ is $\frac{1}{2}$ unit from 0 on the number line.

c. $|-6.2| = 6.2$ -6.2 is 6.2 units from 0 on the number line.

d. $|0| = 0$ 0 is 0 units from 0 on the number line.

6.2 units $\frac{1}{2}$ unit

−7 −6 −5 −4 −3 −2 −1 0 1 2 3

4 units

Skill Practice Evaluate.

18. $|14|$ **19.** $|-99|$

20. $|0|$ **21.** $\left|-\frac{7}{8}\right|$

The absolute value of a number a is its distance from zero on the number line. The definition of $|a|$ may also be given symbolically depending on whether a is negative or nonnegative.

Definition of the Absolute Value of a Real Number

Let a be a real number. Then

1. If a is nonnegative (that is, $a \geq 0$), then $|a| = a$.
2. If a is negative (that is, $a < 0$), then $|a| = -a$.

This definition states that if a is a nonnegative number, then $|a|$ equals a itself. If a is a negative number, then $|a|$ equals the opposite of a. For example,

$|9| = 9$ Because 9 is positive, then $|9|$ equals the number 9 itself.

$|-7| = 7$ Because -7 is negative, then $|-7|$ equals the opposite of -7, which is 7.

Example 7 Comparing Absolute Value Expressions

Determine if the statements are true or false.

a. $|3| \leq 3$ **b.** $-|5| = |-5|$

Solution:

a. $|3| \leq 3$ True. The symbol $\leq$ means "less than *or* equal to." Since $|3|$ is equal to 3, then $|3| \leq 3$ is a true statement.

b. $-|5| = |-5|$ False. On the left-hand side, $-|5|$ is the opposite of $|5|$. Hence $-|5| = -5$. On the right-hand side, $|-5| = 5$. Therefore, the original statement simplifies to $-5 = 5$, which is false.

Skill Practice True or False.

22. $-4 < -4$ **23.** $|-17| = 17$

Skill Practice Answers

18. 14 **19.** 99 **20.** 0

21. $\frac{7}{8}$ **22.** False **23.** True

Calculator Connections

Scientific and graphing calculators approximate irrational numbers by using rational numbers in the form of terminating decimals. For example, consider approximating π and $\sqrt{3}$:

Scientific Calculator

Enter: [π] (or [2nd] [π]) **Result:** 3.141592654

Enter: [3] [$\sqrt{\ }$] **Result:** 1.732050808

Graphing Calculator

Enter: [2nd] [π] [ENTER]

Enter: [2nd] [$\sqrt{\ }$] [3] [ENTER]

```
π
          3.141592654
√(3)
          1.732050808
```

Note that when writing approximations, we use the symbol, $\approx$.

$$\pi \approx 3.141592654 \quad \text{and} \quad \sqrt{3} \approx 1.732050808$$

Section 1.1 Practice Exercises

Boost *your* GRADE at mathzone.com!

- Practice Problems
- Self-Tests
- NetTutor
- e-Professors
- Videos

Study Skills Exercises

1. In this text, we will provide skills for you to enhance your learning experience. In the first four chapters, each set of Practice Exercises will begin with an activity that focuses on one of the following areas: learning about your course, using your text, taking notes, doing homework, and taking an exam. In subsequent chapters, we will insert skills pertaining to the specific material in the chapter. In Chapter 6 we will give tips on studying for the final exam.

Each activity requires only a few minutes and will help you to pass this class and become a better math student. Many of these skills can be carried over to other disciplines and help you to become a model college student. To begin, fill in the following information:

a. Instructor's name

b. Days of the week that the class meets

c. The room number in which the class meets

d. Is there a lab requirement for this course?

If so, how often and what is the location of the lab?

e. Instructor's office number

f. Instructor's telephone number

g. Instructor's e-mail address

h. Instructor's office hours

2. Define the key terms:

a. real numbers **b. natural numbers** **c. whole numbers** **d. integers**

e. rational numbers **f. irrational numbers** **g. inequality** **h. opposite**

i. absolute value **j. real number line**

Concept 2: Plotting Points on the Number Line

3. Plot the numbers on a real number line: $\{1, -2, -\pi, 0, -\frac{5}{2}, 5.1\}$

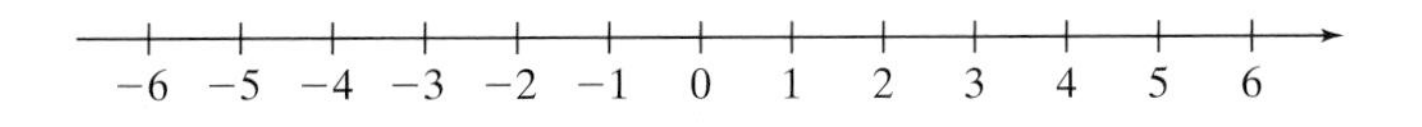

4. Plot the numbers on a real number line: $\{3, -4, \frac{1}{8}, -1.7, -\frac{4}{3}, 1.75\}$

−6 −5 −4 −3 −2 −1 0 1 2 3 4 5 6

Concept 3: Set of Real Numbers

For Exercises 5–20, describe each number as (a) a terminating decimal, (b) a repeating decimal, or (c) a nonterminating, nonrepeating decimal.

5. 0.29

6. 3.8

7. $\frac{1}{9}$

8. $\frac{1}{3}$

9. $\frac{1}{8}$

10. $\frac{1}{5}$

11. 2π

12. 3π

13. -0.125

14. -3.24

15. -3

16. -6

17. $0.\overline{2}$

18. $0.\overline{6}$

19. $\sqrt{6}$

20. $\sqrt{10}$

21. List all of the numbers from Exercises 5–20 that are rational numbers.

22. List all of the numbers from Exercises 5–20 that are irrational numbers.

23. List three numbers that are real numbers but not rational numbers.

24. List three numbers that are real numbers but not irrational numbers.

25. List three numbers that are integers but not natural numbers.

26. List three numbers that are integers but not whole numbers.

27. List three numbers that are rational numbers but not integers.

For Exercises 28–33, let $A = \{-\frac{3}{2}, \sqrt{11}, -4, 0.\overline{6}, 0, \sqrt{7}, 1\}$

28. Are all of the numbers in set A real numbers?

29. List all of the rational numbers in set A.

30. List all of the whole numbers in set A.

31. List all of the natural numbers in set A.

32. List all of the irrational numbers in set A.

33. List all of the integers in set A.

34. Plot the real numbers of set A on a number line. (*Hint:* $\sqrt{11} \approx 3.3$ and $\sqrt{7} \approx 2.6$)

−6 −5 −4 −3 −2 −1 0 1 2 3 4 5 6

Concept 4: Inequalities

35. The LPGA Samsung World Championship of women's golf scores for selected players are given in the table. Compare the scores and fill in the blanks with the appropriate inequality sign: $<$ or $>$.

LPGA Golfers	Final Score with Respect to Par
Annika Sorenstam	7
Laura Davies	-4
Lorie Kane	0
Cindy McCurdy	3
Se Ri Pak	-8

a. Kane's score ______ Pak's score.

b. Sorenstam's score ______ Davies' score.

c. Pak's score ______ McCurdy's score.

d. Kane's score ______ Davies' score.

36. The elevations of selected cities in the United States are shown in the figure. Compare the elevations and fill in the blanks with the appropriate inequality sign: $<$ or $>$. (A negative number indicates that the city is below sea level.)

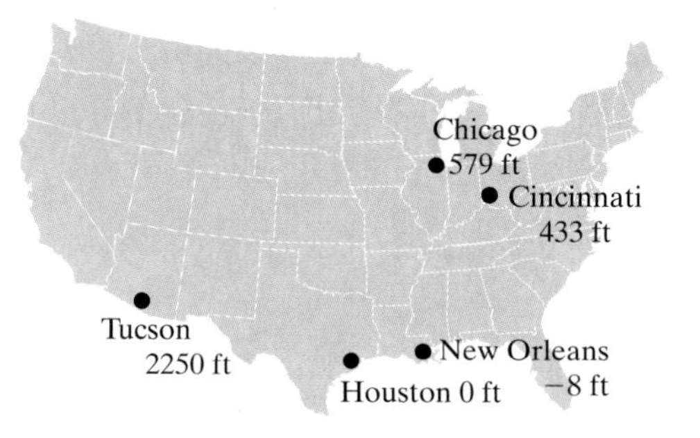

a. Elevation of Tucson _______ elevation of Cincinnati.

b. Elevation of New Orleans _______ elevation of Chicago.

c. Elevation of New Orleans _______ elevation of Houston.

d. Elevation of Chicago ______ elevation of Cincinnati.

Concept 5: Opposite of a Real Number

For Exercises 37–44, find the opposite of each number.

37. 18 **38.** 2 **39.** -6.1 **40.** -2.5

41. $-\frac{5}{8}$ **42.** $-\frac{1}{3}$ **43.** $\frac{7}{3}$ **44.** $\frac{1}{9}$

The opposite of a is denoted as $-a$. For Exercises 45–50, simplify.

45. $-(-3)$ **46.** $-(-5.1)$ **47.** $-\left(\frac{7}{3}\right)$ **48.** $-(-7)$

49. $-(-8)$ **50.** $-(36)$

Concept 6: Absolute Value of a Real Number

For Exercises 51–58, find the absolute value as indicated.

51. $|-2|$ **52.** $|-7|$ **53.** $|-1.5|$ **54.** $|-3.7|$

55. $-|-1.5|$ **56.** $-|-3.7|$ **57.** $\left|\frac{3}{2}\right|$ **58.** $\left|\frac{7}{4}\right|$

For Exercises 59–60, answer true or false. If a statement is false, explain why.

59. If n is positive, then $|n|$ is negative. **60.** If m is negative, then $|m|$ is negative.

For Exercises 61–84, determine if the statements are true or false. Use the real number line to justify the answer.

61. $5 > 2$ **62.** $8 < 10$ **63.** $6 < 6$ **64.** $19 > 19$

65. $-7 \geq -7$

66. $-1 \leq -1$

67. $\frac{3}{2} \leq \frac{1}{6}$

68. $-\frac{1}{4} \geq -\frac{7}{8}$

69. $-5 > -2$

70. $6 < -10$

71. $8 \neq 8$

72. $10 \neq 10$

73. $|-2| \geq |-1|$

74. $|3| \leq |-1|$

75. $\left|-\frac{1}{9}\right| = \left|\frac{1}{9}\right|$

76. $\left|-\frac{1}{3}\right| = \left|\frac{1}{3}\right|$

77. $|7| \neq |-7|$

78. $|-13| \neq |13|$

79. $-1 < |-1|$

80. $-6 < |-6|$

81. $|-8| \geq |8|$

82. $|-11| \geq |11|$

83. $|-2| \leq |2|$

84. $|-21| \leq |21|$

Expanding Your Skills

85. For what numbers, a, is $-a$ positive?

86. For what numbers, a, is $|a| = a$?

Order of Operations

Section 1.2

Concepts

1. Variables and Expressions
2. Evaluating Algebraic Expressions
3. Exponential Expressions
4. Square Roots
5. Order of Operations
6. Translations

1. Variables and Expressions

A **variable** is a symbol or letter, such as x, y, and z, used to represent an unknown number. **Constants** are values that do not vary such as the numbers 3, -1.5, $\frac{2}{7}$, and π. An algebraic **expression** is a collection of variables and constants under algebraic operations. For example, $\frac{3}{x}$, $y + 7$, and $t - 1.4$ are algebraic expressions.

The symbols used to show the four basic operations of addition, subtraction, multiplication, and division are summarized in Table 1-3.

Table 1-3

Operation	Symbols	Translation
Addition	$a + b$	**sum** of a and b a plus b b added to a b more than a a increased by b the total of a and b
Subtraction	$a - b$	**difference** of a and b a minus b b subtracted from a a decreased by b b less than a
Multiplication	$a \times b$, $a \cdot b$, $a(b)$, $(a)b$, $(a)(b)$, ab (*Note:* We rarely use the notation $a \times b$ because the symbol, $\times$, might be confused with the variable, x.)	**product** of a and b a times b a multiplied by b
Division	$a \div b$, $\frac{a}{b}$, a/b, $b\overline{)a}$	**quotient** of a and b a divided by b b divided into a ratio of a and b a over b a per b

2. Evaluating Algebraic Expressions

The value of an algebraic expression depends on the values of the variables within the expression.

Example 1 Evaluating an Algebraic Expression

Evaluate the algebraic expression when $p = 4$ and $q = \frac{3}{4}$.

a. $100 - p$ **b.** pq

Solution:

a. $100 - p$

$100 - (\quad)$	When substituting a number for a variable, use parentheses.
$= 100 - (4)$	Substitute $p = 4$ in the parentheses.
$= 96$	Subtract.

b. pq

$= (\quad)(\quad)$	When substituting a number for a variable, use parentheses.
$= (4)\left(\frac{3}{4}\right)$	Substitute $p = 4$ and $q = \frac{3}{4}$.
$= \frac{4}{1} \cdot \frac{3}{4}$	Write the whole number as a fraction.
$= \frac{12}{4}$	Multiply fractions.
$= 3$	Reduce to lowest terms.

Skill Practice Evaluate the algebraic expressions when $x = 5$ and $y = 2$.

1a. xy **b.** $20 - y$

3. Exponential Expressions

In algebra, repeated multiplication can be expressed using exponents. The expression $4 \cdot 4 \cdot 4$ can be written as

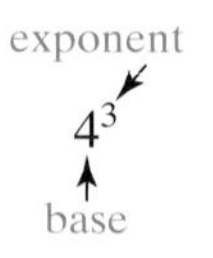

In the expression 4^3, 4 is the base, and 3 is the exponent, or power. The exponent indicates how many factors of the base to multiply.

Skill Practice Answers

1a. 10 **b.** 18

Definition of b^n

Let b represent any real number and n represent a positive integer. Then,

$$b^n = \underbrace{b \cdot b \cdot b \cdot b \ldots \cdot b}_{n \text{ factors of } b}$$

b^n is read as "b to the nth power."
b is called the **base,** and n is called the **exponent,** or **power.**
b^2 is read as "b squared," and b^3 is read as "b cubed."

The exponent, n, is a count of the number of times the base, b, is used as a factor.

Example 2 Evaluating Exponential Expressions

Translate the expression into words and then evaluate the expression.

a. 2^5 **b.** 5^2 **c.** $\left(\frac{3}{4}\right)^3$ **d.** 1^6

Solution:

a. The expression 2^5 is read as "two to the fifth power"
$2^5 = (2)(2)(2)(2)(2) = 32.$

b. The expression 5^2 is read as "five to the second power" or "five, squared"
$5^2 = (5)(5) = 25.$

c. The expression $(\frac{3}{4})^3$ is read as "three-fourths to the third power" or "three-fourths, cubed"

$$\left(\frac{3}{4}\right)^3 = \left(\frac{3}{4}\right)\left(\frac{3}{4}\right)\left(\frac{3}{4}\right) = \frac{27}{64}$$

d. The expression 1^6 is read as "one to the sixth power"
$1^6 = (1)(1)(1)(1)(1)(1) = 1.$

Skill Practice Evaluate.

2. 4^3 **3.** 2^4 **4.** $\left(\frac{2}{3}\right)^2$ **5.** $(1)^7$

4. Square Roots

The inverse operation to squaring a number is to find its **square roots**. For example, finding a square root of 9 is equivalent to asking "what number(s) when squared equals 9?" The symbol, $\sqrt{\ }$ (called a radical sign), is used to find the *principal* square root of a number. By definition, the principal square root of a number is nonnegative. Therefore, $\sqrt{9}$ is the nonnegative number that when squared equals 9. Hence $\sqrt{9} = 3$ because 3 is nonnegative and $(3)^2 = 9$. Several more examples follow:

$\sqrt{64} = 8$ Because $(8)^2 = 64$

$\sqrt{121} = 11$ Because $(11)^2 = 121$

$\sqrt{0} = 0$ Because $(0)^2 = 0$

Skill Practice Answers

2. 64 **3.** 16
4. $\frac{4}{9}$ **5.** 1

$\sqrt{\frac{1}{16}} = \frac{1}{4}$ Because $\frac{1}{4} \cdot \frac{1}{4} = \frac{1}{16}$

$\sqrt{\frac{4}{9}} = \frac{2}{3}$ Because $\frac{2}{3} \cdot \frac{2}{3} = \frac{4}{9}$

TIP: To simplify square roots, it is advisable to become familiar with the following squares and square roots.

$0^2 = 0 \rightarrow \sqrt{0} = 0$	$7^2 = 49 \rightarrow \sqrt{49} = 7$
$1^2 = 1 \rightarrow \sqrt{1} = 1$	$8^2 = 64 \rightarrow \sqrt{64} = 8$
$2^2 = 4 \rightarrow \sqrt{4} = 2$	$9^2 = 81 \rightarrow \sqrt{81} = 9$
$3^2 = 9 \rightarrow \sqrt{9} = 3$	$10^2 = 100 \rightarrow \sqrt{100} = 10$
$4^2 = 16 \rightarrow \sqrt{16} = 4$	$11^2 = 121 \rightarrow \sqrt{121} = 11$
$5^2 = 25 \rightarrow \sqrt{25} = 5$	$12^2 = 144 \rightarrow \sqrt{144} = 12$
$6^2 = 36 \rightarrow \sqrt{36} = 6$	$13^2 = 169 \rightarrow \sqrt{169} = 13$

5. Order of Operations

When algebraic expressions contain numerous operations, it is important to evaluate the operations in the proper order. Parentheses ( ), brackets [ ], and braces { } are used for grouping numbers and algebraic expressions. It is important to recognize that operations within parentheses and other grouping symbols must be done first. Other grouping symbols include absolute value bars, radical signs, and fraction bars.

Order of Operations

1. Simplify expressions within parentheses and other grouping symbols first. These include absolute value bars, fraction bars, and radicals. If imbedded parentheses are present, start with the innermost parentheses.
2. Evaluate expressions involving exponents and radicals.
3. Perform multiplication or division in the order that they occur from left to right.
4. Perform addition or subtraction in the order that they occur from left to right.

Example 3 Applying the Order of Operations

Simplify the expressions.

a. $17 - 3 \cdot 2 + 2^2$

b. $\frac{1}{2}\left(\frac{5}{6} - \frac{3}{4}\right)$

c. $25 - 12 \div 3 \cdot 4$

d. $6.2 - |-2.1| + \sqrt{15 - 6}$

e. $28 - 2[(6 - 3)^2 + 4]$

Solution:

a. $17 - 3 \cdot 2 + 2^2$

$= 17 - 3 \cdot 2 + 4$ Simplify exponents.

$= 17 - 6 + 4$ Multiply before adding or subtracting.

$= 11 + 4$ Add or subtract from left to right.

$= 15$

b. $\frac{1}{2}\left(\frac{5}{6} - \frac{3}{4}\right)$ Subtract fractions within the parentheses.

$= \frac{1}{2}\left(\frac{10}{12} - \frac{9}{12}\right)$ The least common denominator is 12.

$= \frac{1}{2}\left(\frac{1}{12}\right)$

$= \frac{1}{24}$ Multiply fractions.

c. $25 - 12 \div 3 \cdot 4$ Multiply or divide in order from left to right.

$= 25 - 4 \cdot 4$ Notice that the operation $12 \div 3$ is performed first (not $3 \cdot 4$).

$= 25 - 16$ Multiply $4 \cdot 4$ before subtracting.

$= 9$ Subtract.

d. $6.2 - |-2.1| + \sqrt{15 - 6}$

$= 6.2 - |-2.1| + \sqrt{9}$ Simplify within the square root.

$= 6.2 - (2.1) + 3$ Simplify the square root and absolute value.

$= 4.1 + 3$ Add or subtract from left to right.

$= 7.1$ Add.

e. $28 - 2[(6 - 3)^2 + 4]$

$= 28 - 2[(3)^2 + 4]$ Simplify within the inner parentheses first.

$= 28 - 2[(9) + 4]$ Simplify exponents.

$= 28 - 2[13]$ Add within the square brackets.

$= 28 - 26$ Multiply before subtracting.

$= 2$ Subtract.

Skill Practice Simplify the expressions.

6. $14 - 3 \cdot 2$

7. $\frac{13}{4} - \frac{1}{4}(10 - 2)$

8. $1 + 2 \cdot 3^2 \div 6$

9. $|-20| - (7 - 2)$

10. $60 - 5[(6 - 3) + 2^2]$

Skill Practice Answers

6. 8 **7.** $\frac{5}{4}$ **8.** 4

9. 15 **10.** 25

6. Translations

Example 4 Translating from English Form to Algebraic Form

Translate each English phrase to an algebraic expression.

a. The quotient of x and 5

b. The difference of p and the square root of q

c. Seven less than n

d. Seven less n

e. Eight more than the absolute value of w

Solution:

a. $\frac{x}{5}$ or $x \div 5$ — The quotient of x and 5

b. $p - \sqrt{q}$ — The difference of p and the square root of q

c. $n - 7$ — Seven less than n

d. $7 - n$ — Seven less n

e. $|w| + 8$ — Eight more than the absolute value of w

Avoiding Mistakes:

Recall that "a less than b" is translated as $b - a$. Therefore, the statement "seven less than n" must be translated as $n - 7$, not $7 - n$.

Skill Practice Translate each English phrase to an algebraic expression.

11. The product of 6 and y

12. The sum of b and the square root of c

13. Twelve less than x

14. Twelve less x

15. One more than two times x

Example 5 Translating from English Form to Algebraic Form

Translate each English phrase into an algebraic expression. Then evaluate the expression for $a = 6$, $b = 4$, and $c = 20$.

a. The product of a and the square root of b

b. Twice the sum of b and c

c. The difference of twice a and b

Solution:

a.

$a\sqrt{b}$	The product of a and the square root of b.
$= (\ \ )\sqrt{(\ \ )}$	Use parentheses to substitute a number for a variable.
$= (6)\sqrt{(4)}$	Substitute $a = 6$ and $b = 4$.
$= 6 \cdot 2$	Simplify the radical first.
$= 12$	Multiply.

Skill Practice Answers

11. $6y$ **12.** $b + \sqrt{c}$
13. $x - 12$ **14.** $12 - x$
15. $2x + 1$

b. $2(b + c)$	Twice the sum of b and c. To compute "twice the sum of b and c," it is necessary to take the sum first and then multiply by 2. To ensure the proper order, the sum of b and c must be enclosed in parentheses. The proper translation is $2(b + c)$.
$= 2((\quad) + (\quad))$	Use parentheses to substitute a number for a variable.
$= 2((4) + (20))$	Substitute $b = 4$ and $c = 20$.
$= 2(24)$	Simplify within the parentheses first.
$= 48$	Multiply.

c. $2a - b$	The difference of twice a and b.
$= 2(\quad) - (\quad)$	Use parentheses to substitute a number for a variable.
$= 2(6) - (4)$	Substitute $a = 6$ and $b = 4$.
$= 12 - 4$	Multiply first.
$= 8$	Subtract.

Skill Practice Translate each English phrase to an algebraic expression. Then evaluate the expression for $x = 3, y = 9, z = 10$.

16. The quotient of the square root of y and x

17. One-half the sum of x and y

18. The difference of z and twice x

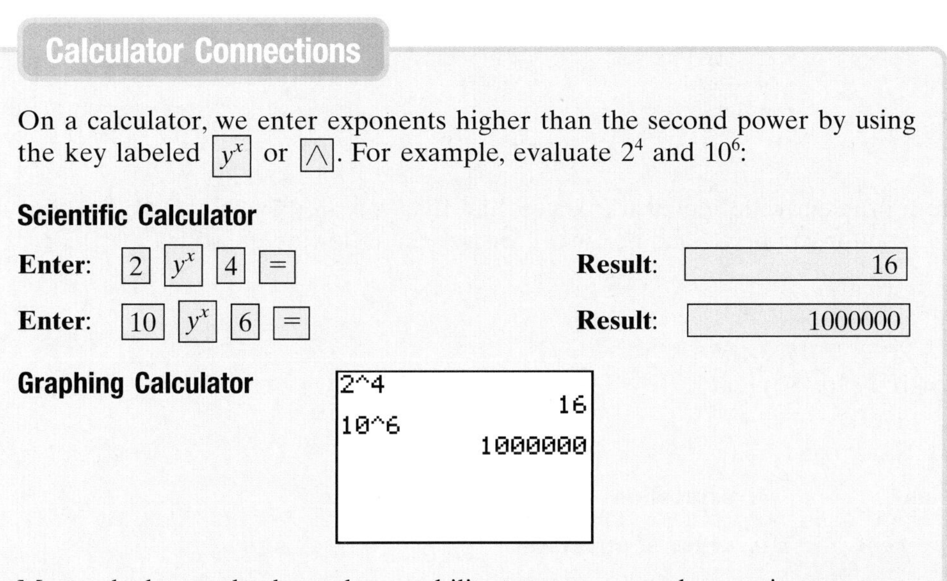

Calculator Connections

On a calculator, we enter exponents higher than the second power by using the key labeled [y^x] or [∧]. For example, evaluate 2^4 and 10^6:

Scientific Calculator

Enter: [2] [y^x] [4] [=] **Result:** 16

Enter: [10] [y^x] [6] [=] **Result:** 1000000

Graphing Calculator

Most calculators also have the capability to enter several operations at once. However, it is important to note that fraction bars and radicals require user-defined parentheses to ensure that the proper order of operations is followed. For example, evaluate the following expressions on a calculator:

a. $130 - 2(5 - 1)^3$ **b.** $\dfrac{18 - 2}{11 - 9}$ **c.** $\sqrt{25 - 9}$

Skill Practice Answers

16. $\dfrac{\sqrt{y}}{x}$; 1 **17.** $\dfrac{1}{2}(x + y)$; 6

18. $z - 2x$; 4

Scientific Calculator

Enter: [130] [−] [2] [×] [(] [5] [−] [1] [)] [y^x] [3] [=] Result: [2]

Enter: [(] [18] [−] [2] [)] [÷] [(] [11] [−] [9] [)] [=] Result: [8]

Enter: [(] [25] [−] [9] [)] [√] Result: [4]

Graphing Calculator

```
130-2*(5-1)^3
              2
(18-2)/(11-9)
              8
√(25-9)
              4
```

Calculator Exercises

Simplify the expressions without the use of a calculator. Then enter the expressions into the calculator to verify your answers.

1. $\dfrac{4+6}{8-3}$ **2.** $110 - 5(2+1) - 4$ **3.** $100 - 2(5-3)^3$

4. $3 + (4-1)^2$ **5.** $(12 - 6 + 1)^2$ **6.** $3 \cdot 8 - \sqrt{32 + 2^2}$

7. $\sqrt{18 - 2}$ **8.** $(4 \cdot 3 - 3 \cdot 3)^3$ **9.** $\dfrac{20 - 3^2}{26 - 2^2}$

Section 1.2 Practice Exercises

Boost *your* GRADE at mathzone.com!

MathZone

- Practice Problems
- Self-Tests
- NetTutor
- e-Professors
- Videos

Study Skills Exercises

1. Sometimes you may run into a problem with homework or you find that you are having trouble keeping up with the pace of the class. A tutor can be a good resource. Answer the following questions.

a. Does your college offer tutoring?

b. Is it free?

c. Where would you go to sign up for a tutor?

2. Define the key terms:

a. variable **b. constant** **c. expression** **d. base**

e. exponent **f. square root** **g. order of operations**

Review Exercises

3. Which of the following are rational numbers. $\left\{-4, 5.\overline{6}, \sqrt{29}, 0, \pi, 4.02, \dfrac{7}{9}\right\}$

4. Evaluate. $|-56|$

5. Evaluate. $|9.2|$

6. Find the opposite of 19.

7. Find the opposite of −34.2.

Concept 2: Evaluating Algebraic Expressions

For Exercises 8–19, evaluate the expressions for the given substitutions.

8. $y - 3$ when $y = 18$

9. $3q$ when $q = 5$

10. $\dfrac{15}{t}$ when $t = 5$

11. $8 + w$ when $w = 12$

12. $5 + 6d$ when $d = \dfrac{2}{3}$

13. $\dfrac{6}{5}h - 1$ when $h = 10$

14. $2(c + 1) - 5$ when $c = 4$

15. $4(4x - 1)$ when $x = \dfrac{3}{4}$

16. $p^2 + \dfrac{2}{9}$ when $p = \dfrac{2}{3}$

17. $z^3 - \dfrac{2}{27}$ when $z = \dfrac{2}{3}$

18. $5(x + 2.3)$ when $x = 1.1$

19. $3(2.1 - y)$ when $y = 0.5$

20. The area of a rectangle may be computed as $A = \ell w$, where ℓ is the length of the rectangle and w is the width. Find the area for the rectangle shown.

160 ft

360 ft

21. The perimeter of the rectangle from Exercise 20 may be computed as $P = 2\ell + 2w$. Find the perimeter.

22. The area of a trapezoid is given by $A = \frac{1}{2}(b_1 + b_2)h$, where b_1 and b_2 are the lengths of the two parallel sides and h is the height. Find the area of the trapezoid with dimensions shown in the figure.

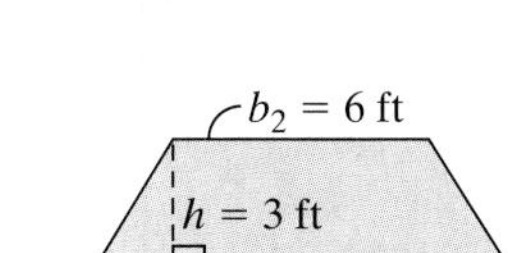

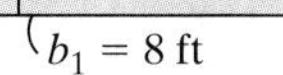

23. The volume of a rectangular solid is given by $V = \ell wh$, where ℓ is the length of the box, w is the width, and h is the height. Find the volume of the box shown in the figure.

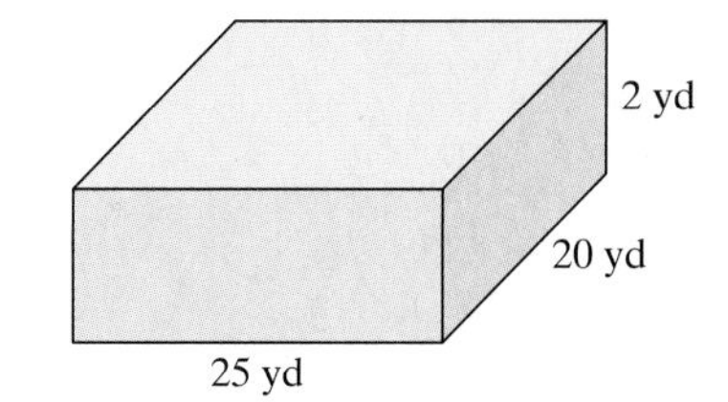

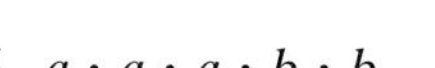

Concept 3: Exponential Expressions

For Exercises 24–31, write each of the products using exponents.

24. $\dfrac{1}{6} \cdot \dfrac{1}{6} \cdot \dfrac{1}{6} \cdot \dfrac{1}{6}$

25. $10 \cdot 10 \cdot 10 \cdot 10 \cdot 10 \cdot 10$

26. $a \cdot a \cdot a \cdot b \cdot b$

27. $7 \cdot x \cdot x \cdot y \cdot y$

28. $5c \cdot 5c \cdot 5c \cdot 5c \cdot 5c$

29. $3 \cdot w \cdot z \cdot z \cdot z \cdot z$

30. $8 \cdot y \cdot x \cdot x \cdot x \cdot x \cdot x$

31. $\dfrac{2}{3}t \cdot \dfrac{2}{3}t \cdot \dfrac{2}{3}t$

32. **a.** For the expression $5x^3$, what is the base for the exponent 3?

b. Does 5 have an exponent? If so, what is it?

33. **a.** For the expression $2y^4$, what is the base for the exponent 4?

b. Does 2 have an exponent? If so, what is it?

For Exercises 34–41, write each expression in expanded form using the definition of an exponent.

34. x^3

35. y^4

36. $(2b)^3$

37. $(8c)^2$

38. $10y^5$

39. x^2y^3

40. $2wz^2$

41. $3a^3b$

For Exercises 42–49, simplify the expressions.

42. 5^2

43. 4^3

44. $\left(\frac{1}{7}\right)^2$

45. $\left(\frac{1}{2}\right)^5$

46. $(0.25)^3$

47. $(0.8)^2$

48. 2^6

49. 13^2

Concept 4: Square Roots

For Exercises 50–61, simplify the square roots.

50. $\sqrt{81}$

51. $\sqrt{64}$

52. $\sqrt{4}$

53. $\sqrt{9}$

54. $\sqrt{100}$

55. $\sqrt{49}$

56. $\sqrt{16}$

57. $\sqrt{36}$

58. $\sqrt{\frac{1}{9}}$

59. $\sqrt{\frac{1}{64}}$

60. $\sqrt{\frac{25}{81}}$

61. $\sqrt{\frac{49}{100}}$

Concept 5: Order of Operations

For Exercises 62–89, use the order of operations to simplify the expressions.

62. $8 + 2 \cdot 6$

63. $7 + 3 \cdot 4$

64. $(8 + 2)6$

65. $(7 + 3)4$

66. $4 + 2 \div 2 \cdot 3 + 1$

67. $5 + 6 \cdot 2 \div 4 - 1$

68. $\frac{1}{4} \cdot \frac{2}{3} - \frac{1}{6}$

69. $\frac{3}{4} \cdot \frac{2}{3} + \frac{2}{3}$

70. $\frac{9}{8} - \frac{1}{3} \cdot \frac{3}{4}$

71. $\frac{11}{6} - \frac{3}{8} \cdot \frac{4}{3}$

72. $3[5 + 2(8 - 3)]$

73. $2[4 + 3(6 - 4)]$

74. $10 + |-6|$

75. $18 + |-3|$

76. $21 - |8 - 2|$

77. $12 - |6 - 1|$

78. $2^2 + \sqrt{9} \cdot 5$

79. $3^2 + \sqrt{16} \cdot 2$

80. $\sqrt{9 + 16} - 2$

81. $\sqrt{36 + 13} - 5$

82. $\frac{7 + 3(8 - 2)}{(7 + 3)(8 - 2)}$

83. $\frac{16 - 8 \div 4}{4 + 8 \div 4 - 2}$

84. $\frac{15 - 5(3 \cdot 2 - 4)}{10 - 2(4 \cdot 5 - 16)}$

85. $\frac{5(7 - 3) + 8(6 - 4)}{4[7 + 3(2 \cdot 9 - 8)]}$

86. $[4^2 \cdot (6 - 4) \div 8] + [7 \cdot (8 - 3)]$

87. $(18 \div \sqrt{4}) \cdot \{[(9^2 - 1) \div 2] - 15\}$

88. $48 - 13 \cdot 3 + [(50 - 7 \cdot 5) + 2]$

89. $80 \div 16 \cdot 2 + (6^2 - |-2|)$

Concept 6: Translations

For Exercises 90–101, translate each English phrase into an algebraic expression.

90. The product of 3 and x

91. The sum of b and 6

92. The quotient of x and 7

93. Four divided by k

94. The difference of 2 and a

95. Three subtracted from t

96. x more than twice y

97. Nine decreased by the product of 3 and p

98. Four times the sum of x and 12

99. Twice the difference of x and 3

100. Q less than 3

101. Fourteen less than t

For Exercises 102–111, use the order of operations to evaluate the expression when $x = 4$, $y = 2$, and $z = 10$.

102. $2y^3$ **103.** $3z^2$ **104.** $|z - 8|$ **105.** $|x - 3|$

106. $5\sqrt{x}$ **107.** $\sqrt{9 + x^2}$ **108.** $yz - x$ **109.** $z - xy$

110. xy^2 **111.** yx^2

For Exercises 112–123, translate each algebraic expression into an English phrase. (Answers may vary.)

112. $5 + r$ **113.** $18 - x$ **114.** $s - 14$ **115.** $y + 12$

116. xyz **117.** $7x + 1$ **118.** 5^2 **119.** 6^3

120. $\sqrt{5}$ **121.** $\sqrt{10}$ **122.** 7^3 **123.** 10^2

124. Some students use the following common memorization device (mnemonic) to help them remember the order of operations: the acronym PEMDAS or **P**lease **E**xcuse **M**y **D**ear **A**unt **S**ally to remember **P**arentheses, **E**xponents, **M**ultiplication, **D**ivision, **A**ddition, and **S**ubtraction. The problem with this mnemonic is that it suggests that multiplication is done before division and similarly, it suggests that addition is performed before subtraction. Explain why following this acronym may give incorrect answers for the expressions:

a. $36 \div 4 \cdot 3$

b. $36 - 4 + 3$

125. If you use the acronym **P**lease **E**xcuse **M**y **D**ear **A**unt **S**ally to remember the order of operations, what must you keep in mind about the last four operations?

126. Explain why the acronym **P**lease **E**xcuse **D**r. **M**ichael **S**mith's **A**unt could also be used as a memory device for the order of operations.

Expanding Your Skills

For Exercises 127–130, use the order of operations to simplify the expressions.

127. $\dfrac{\sqrt{\frac{1}{9}} + \frac{2}{3}}{\sqrt{\frac{4}{25}} + \frac{3}{5}}$ **128.** $\dfrac{5 - \sqrt{9}}{\sqrt{\frac{4}{9}} + \frac{1}{3}}$ **129.** $\dfrac{|-2|}{|-10| - |2|}$ **130.** $\dfrac{|-4|^2}{2^2 + \sqrt{144}}$

Addition of Real Numbers

Section 1.3

Concepts

1. **Addition of Real Numbers and the Number Line**
2. **Addition of Real Numbers**
3. **Translations**
4. **Applications Involving Addition of Real Numbers**

1. Addition of Real Numbers and the Number Line

Adding real numbers can be visualized on the number line. To add a positive number, move to the right on the number line. To add a negative number, move to the left on the number line. The following example may help to illustrate the process.

On a winter day in Detroit, suppose the temperature starts out at 5 degrees Fahrenheit (5°F) at noon, and then drops 12° two hours later when a cold front passes through. The resulting temperature can be represented by the expression

$5° + (-12°)$. On the number line, start at 5 and count 12 units to the left (Figure 1-4). The resulting temperature at 2:00 P.M. is $-7°F$.

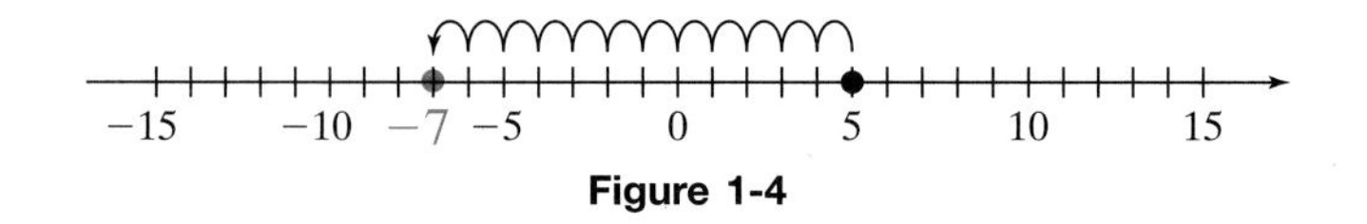

Figure 1-4

TIP: Note that we move to the left when we add a negative number, and we move to the right when we add a positive number.

Example 1 **Using the Number Line to Add Real Numbers**

Use the number line to add the numbers.

a. $-5 + 2$ **b.** $-1 + (-4)$ **c.** $4 + (-7)$

Solution:

a. $-5 + 2 = -3$

Start at -5, and count 2 units to the right.

b. $-1 + (-4) = -5$

Start at -1, and count 4 units to the left.

c. $4 + (-7) = -3$

Start at 4, and count 7 units to the left.

Skill Practice Use the number line to add the numbers.

1. $-2 + (-3)$ **2.** $5 + (-6)$ **3.** $-2 + 4$

2. Addition of Real Numbers

When adding large numbers or numbers that involve fractions or decimals, counting units on the number line can be cumbersome. Study the following example to determine a pattern for adding two numbers with the *same* sign.

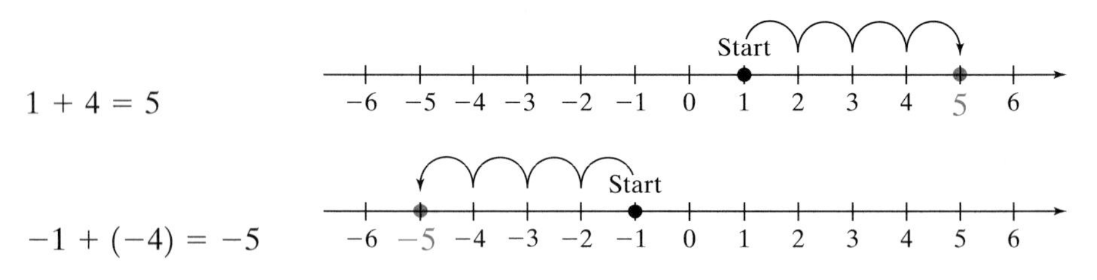

Adding Numbers with the *Same* Sign

To add two numbers with the *same* sign, add their absolute values and apply the common sign.

Skill Practice Answers

1. -5 **2.** -1 **3.** 2

Study the following example to determine a pattern for adding two numbers with *different* signs.

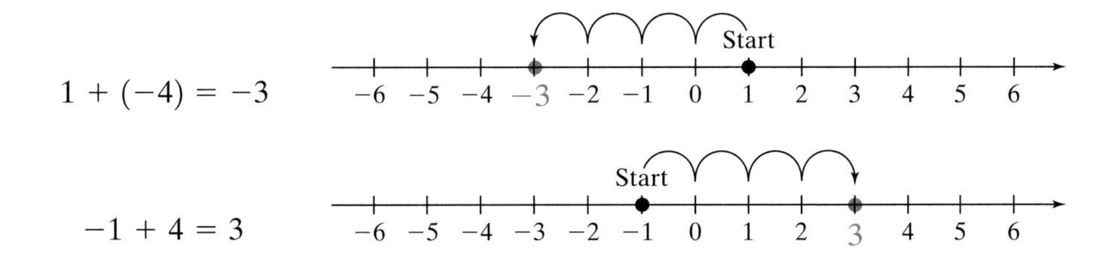

Adding Numbers with *Different* Signs

To add two numbers with *different* signs, subtract the smaller absolute value from the larger absolute value. Then apply the sign of the number having the larger absolute value.

Example 2 Adding Real Numbers with the Same Sign

Add.

a. $-12 + (-14)$ **b.** $7 + 63$ **c.** $-8.8 + (-3.7)$ **d.** $-\frac{4}{3} + \left(-\frac{6}{7}\right)$

Solution:

a. $-12 + (-14)$ — First find the absolute value of the addends. $|-12| = 12$ and $|-14| = 14$.

$= -(12 + 14)$ — Add their absolute values and apply the common sign (in this case, the common sign is negative).

common sign is negative

$= -26$ — The sum is -26.

b. $7 + 63$ — First find the absolute value of the addends. $|7| = 7$ and $|63| = 63$.

$= +(7 + 63)$ — Add their absolute values and apply the common sign (in this case, the common sign is positive).

common sign is positive

$= 70$ — The sum is 70.

c. $-8.8 + (-3.7)$ — First find the absolute value of the addends. $|-8.8| = 8.8$ and $|-3.7| = 3.7$.

$= -(8.8 + 3.7)$ — Add their absolute values and apply the common sign (in this case, the common sign is negative).

common sign is negative

$= -12.5$ — The sum is -12.5.

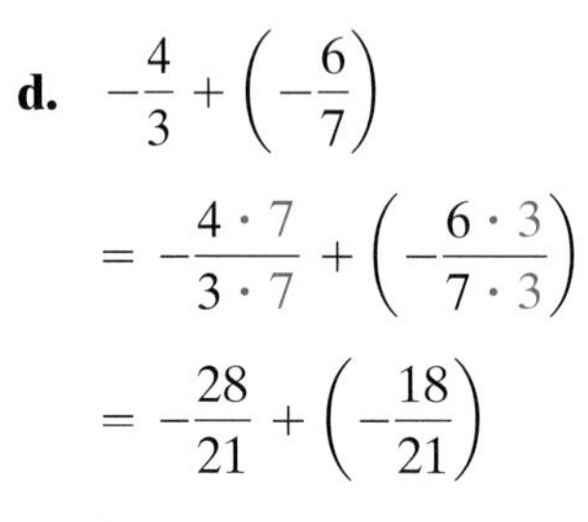

d. $-\frac{4}{3}+\left(-\frac{6}{7}\right)$ The least common denominator (LCD) is 21.

$= -\frac{4 \cdot 7}{3 \cdot 7}+\left(-\frac{6 \cdot 3}{7 \cdot 3}\right)$ Write each fraction with the LCD.

$= -\frac{28}{21}+\left(-\frac{18}{21}\right)$ Find the absolute value of the addends.

$\left|-\frac{28}{21}\right| = \frac{28}{21}$ and $\left|-\frac{18}{21}\right| = \frac{18}{21}$.

$= -\left(\frac{28}{21}+\frac{18}{21}\right)$ Add their absolute values and apply the common sign (in this case, the common sign is negative).

common sign is negative

$= -\frac{46}{21}$ The sum is $-\frac{46}{21}$.

Skill Practice Add the numbers.

4. $-5 + (-25)$ **5.** $7 + 12$

6. $-14.8 + (-9.7)$ **7.** $-\frac{1}{2}+\left(-\frac{5}{8}\right)$

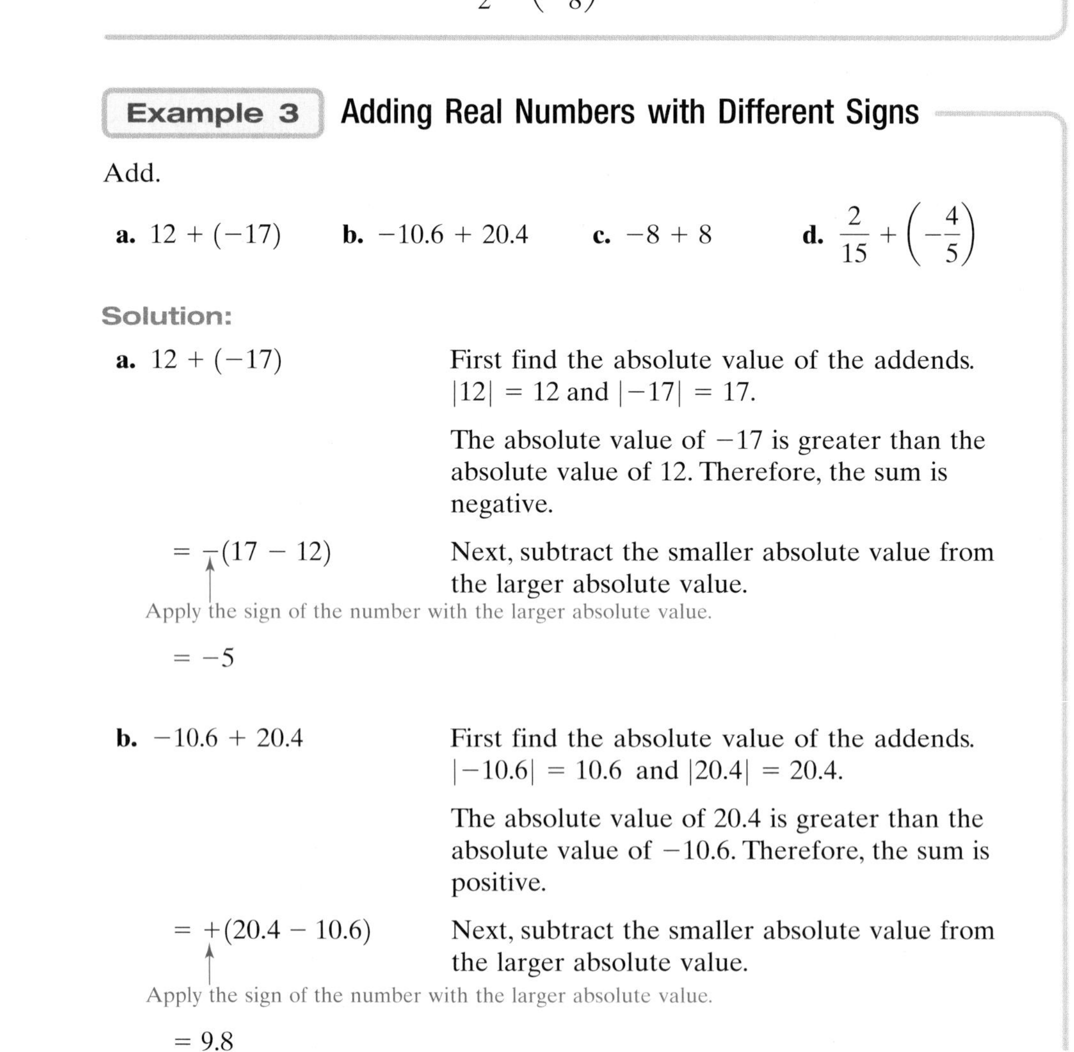

Example 3 Adding Real Numbers with Different Signs

Add.

a. $12 + (-17)$ **b.** $-10.6 + 20.4$ **c.** $-8 + 8$ **d.** $\frac{2}{15}+\left(-\frac{4}{5}\right)$

Solution:

a. $12 + (-17)$ First find the absolute value of the addends. $|12| = 12$ and $|-17| = 17$.

The absolute value of -17 is greater than the absolute value of 12. Therefore, the sum is negative.

$= -(17 - 12)$ Next, subtract the smaller absolute value from the larger absolute value.

Apply the sign of the number with the larger absolute value.

$= -5$

b. $-10.6 + 20.4$ First find the absolute value of the addends. $|-10.6| = 10.6$ and $|20.4| = 20.4$.

The absolute value of 20.4 is greater than the absolute value of -10.6. Therefore, the sum is positive.

$= +(20.4 - 10.6)$ Next, subtract the smaller absolute value from the larger absolute value.

Apply the sign of the number with the larger absolute value.

$= 9.8$

Skill Practice Answers

4. -30 **5.** 19

6. -24.5 **7.** $-\frac{9}{8}$

c. $-8 + 8$ — First find the absolute value of the addends. $|-8| = 8$ and $|8| = 8$.

$= (8 - 8)$ — The absolute values are equal. Therefore, their difference is 0. The number zero is neither positive nor negative.

$= 0$

d. $\frac{2}{15} + \left(-\frac{4}{5}\right)$ — The least common denominator is 15.

$= \frac{2}{15} + \left(-\frac{4 \cdot 3}{5 \cdot 3}\right)$ — Write each fraction with the LCD.

$= \frac{2}{15} + \left(-\frac{12}{15}\right)$ — Find the absolute value of the addends.

$$\left|\frac{2}{15}\right| = \frac{2}{15} \quad \text{and} \quad \left|-\frac{12}{15}\right| = \frac{12}{15}$$

The absolute value of $-\frac{12}{15}$ is greater than the absolute value of $\frac{2}{15}$. Therefore, the sum is negative.

$= -\left(\frac{12}{15} - \frac{2}{15}\right)$ — Next, subtract the smaller absolute value from the larger absolute value.

Apply the sign of the number with the larger absolute value.

$= -\frac{10}{15}$ — Subtract.

$= -\frac{2}{3}$ — Simplify by reducing to lowest terms. $-\frac{\overset{2}{\cancel{10}}}{\underset{3}{\cancel{15}}} = -\frac{2}{3}$

Skill Practice Add the numbers.

8. $-15 + 16$ **9.** $27.3 + (-18.1)$

10. $6.2 + (-6.2)$ **11.** $-\frac{9}{10} + \frac{2}{5}$

3. Translations

Example 4 **Translating Expressions Involving the Addition of Real Numbers**

Translate each English phrase into an algebraic expression. Then simplify the result.

a. The sum of -12, -8, 9, and -1

b. Negative three-tenths added to $-\frac{7}{8}$

c. The sum of -12 and its opposite

Solution:

a. $\underbrace{-12 + (-8)} + 9 + (-1)$ — The sum of -12, -8, 9, and -1.

$= \underbrace{-20 + 9} + (-1)$ — Add from left to right.

$= \underbrace{-11 + (-1)}$

$= -12$

Skill Practice Answers

8. 1 **9.** 9.2

10. 0 **11.** $-\frac{1}{2}$

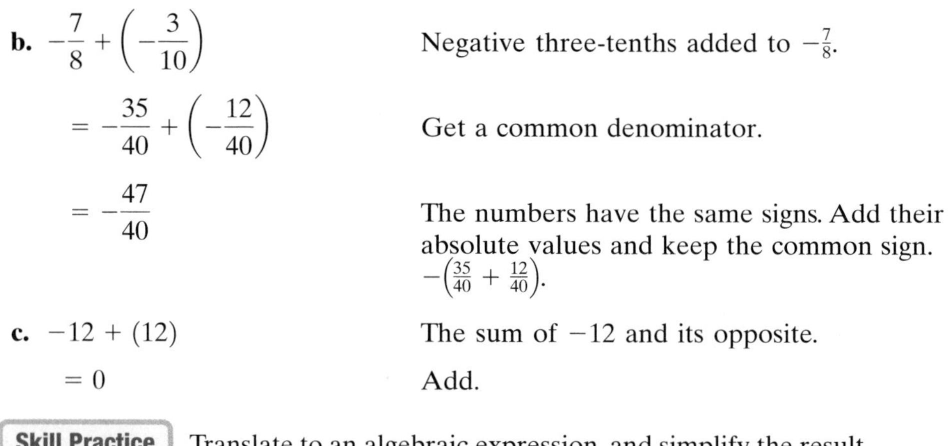

b. $-\frac{7}{8} + \left(-\frac{3}{10}\right)$ — Negative three-tenths added to $-\frac{7}{8}$.

$= -\frac{35}{40} + \left(-\frac{12}{40}\right)$ — Get a common denominator.

$= -\frac{47}{40}$ — The numbers have the same signs. Add their absolute values and keep the common sign. $-\left(\frac{35}{40} + \frac{12}{40}\right)$.

TIP: The sum of any number and its opposite is 0.

c. $-12 + (12)$ — The sum of -12 and its opposite.

$= 0$ — Add.

Skill Practice Translate to an algebraic expression, and simplify the result.

12. The sum of -10, 4, and -6

13. The sum of $-\frac{9}{4}$ and $\frac{11}{3}$

14. -60 added to its opposite

4. Applications Involving Addition of Real Numbers

Example 5 Adding Real Numbers in Applications

a. A running back on a football team gains 4 yards (yd). On the next play, the quarterback is sacked and loses 13 yd. Write a mathematical expression to describe this situation and then simplify the result.

b. A student has \$120 in her checking account. After depositing her paycheck of \$215, she writes a check for \$255 to cover her portion of the rent and another check for \$294 to cover her car payment. Write a mathematical expression to describe this situation and then simplify the result.

Solution:

a. $4 + (-13)$ — The loss of 13 yd can be interpreted as adding -13 yd.

$= -9$ — The football team has a net loss of 9 yd.

b. $\underbrace{120 + 215} + (-255) + (-294)$ — Writing a check is equivalent to adding a negative amount to the bank account.

$= \underbrace{335 + (-255)} + (-294)$ — Use the order of operations. Add from left to right.

$= 80 + (-294)$

$= -214$ — The student has overdrawn her account by \$214.

Skill Practice

15. A share of GE stock was priced at \$32.00 per share at the beginning of the month. After the first week, the price went up \$2.15 per share. At the end of the second week it went down \$3.28 per share. Write a mathematical expression to describe the price of the stock and find the price of the stock at the end of the 2-week period.

Skill Practice Answers

12. $-10 + 4 + (-6)$; -12

13. $-\frac{9}{4} + \frac{11}{3}$; $\frac{17}{12}$

14. $60 + (-60)$; 0

15. $32.00 + 2.15 + (-3.28)$; \$30.87 per share

Section 1.3 Practice Exercises

Boost *your* GRADE at mathzone.com! MathZone

- Practice Problems
- Self-Tests
- NetTutor
- e-Professors
- Videos

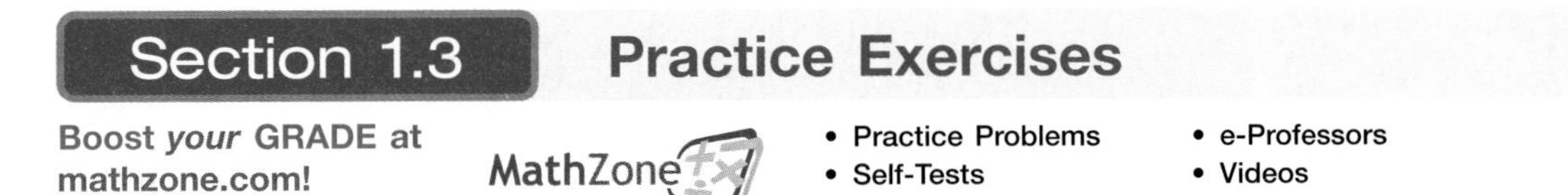

Study Skills Exercise

1. It is very important to attend class every day. Math is cumulative in nature, and you must master the material learned in the previous class to understand today's lesson. Because this is so important, many instructors tie attendance to the final grade. Write down the attendance policy for your class.

Review Exercises

Plot the points in set A on a number line. Then for Exercises 2–7 place the appropriate inequality ($<$, $>$) between the expressions.

$$A = \left\{-2, \frac{3}{4}, -\frac{5}{2}, 3, \frac{9}{2}, 1.6, 0\right\}$$

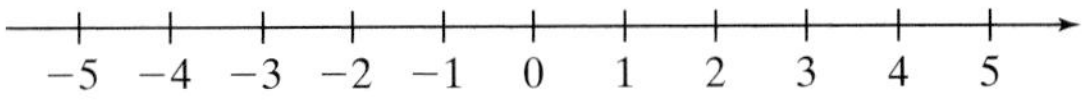

2. -2 ____ 0 **3.** $\frac{9}{2}$ ____ $\frac{3}{4}$ **4.** -2 ____ $-\frac{5}{2}$

5. 0 ____ $-\frac{5}{2}$ **6.** $\frac{3}{4}$ ____ 1.6 **7.** $\frac{3}{4}$ ____ $-\frac{5}{2}$

Concept 1: Addition of Real Numbers and the Number Line

For Exercises 8–15, add the numbers using the number line.

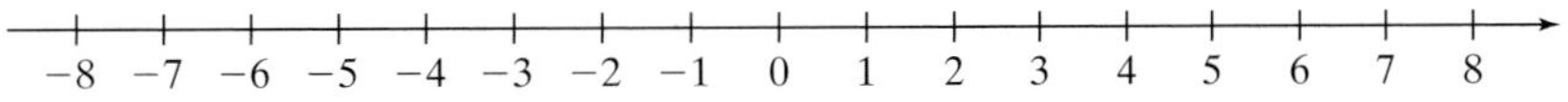

8. $-2 + (-4)$ **9.** $-3 + (-5)$ **10.** $-7 + 10$ **11.** $-2 + 9$

12. $6 + (-3)$ **13.** $8 + (-2)$ **14.** $2 + (-5)$ **15.** $7 + (-3)$

Concept 2: Addition of Real Numbers

For Exercises 16–43, add the integers.

16. $-19 + 2$ **17.** $-25 + 18$ **18.** $-4 + 11$ **19.** $-3 + 9$

20. $-16 + (-3)$ **21.** $-12 + (-23)$ **22.** $-2 + (-21)$ **23.** $-13 + (-1)$

24. $0 + (-5)$ **25.** $0 + (-4)$ **26.** $-3 + 0$ **27.** $-8 + 0$

28. $-16 + 16$ **29.** $11 + (-11)$ **30.** $41 + (-41)$ **31.** $-15 + 15$

32. $4 + (-9)$ **33.** $6 + (-9)$ **34.** $7 + (-2) + (-8)$ **35.** $2 + (-3) + (-6)$

36. $-17 + (-3) + 20$ **37.** $-9 + (-6) + 15$

38. $-3 + (-8) + (-12)$ **39.** $-8 + (-2) + (-13)$

40. $-42 + (-3) + 45 + (-6)$

41. $36 + (-3) + (-8) + (-25)$

42. $-5 + (-3) + (-7) + 4 + 8$

43. $-13 + (-1) + 5 + 2 + (-20)$

For Exercises 44–69, add the rational numbers.

44. $23.81 + (-2.51)$

45. $-9.23 + 10.53$

46. $-\frac{2}{7} + \frac{1}{14}$

47. $-\frac{1}{8} + \frac{5}{16}$

48. $\frac{2}{3} + \left(-\frac{5}{6}\right)$

49. $\frac{1}{2} + \left(-\frac{3}{4}\right)$

50. $-\frac{7}{8} + \left(-\frac{1}{16}\right)$

51. $-\frac{1}{9} + \left(-\frac{4}{3}\right)$

52. $-\frac{1}{4} + \frac{3}{10}$

53. $-\frac{7}{6} + \frac{7}{8}$

54. $-2.1 + \left(-\frac{3}{10}\right)$

55. $-8.3 + \left(-\frac{9}{10}\right)$

56. $\frac{3}{4} + (-0.5)$

57. $-\frac{3}{2} + 0.45$

58. $8.23 + (-8.23)$

59. $-7.5 + 7.5$

60. $-\frac{7}{8} + 0$

61. $0 + \left(-\frac{21}{22}\right)$

62. $-\frac{2}{3} + \left(-\frac{1}{9}\right) + 2$

63. $-\frac{1}{4} + \left(-\frac{3}{2}\right) + 2$

64. $-47.36 + 24.28$

65. $-0.015 + (0.0026)$

66. $516.816 + (-22.13)$

67. $87.02 + (-93.19)$

68. $-0.000617 + (-0.0015)$

69. $-5315.26 + (-314.89)$

70. State the rule for adding two numbers with different signs.

71. State the rule for adding two numbers with the same signs.

For Exercises 72–79, evaluate the expression for $x = -3$, $y = -2$, and $z = 16$.

72. $x + y + \sqrt{z}$

73. $2z + x + y$

74. $y + 3\sqrt{z}$

75. $-\sqrt{z} + y$

76. $|x| + |y|$

77. $z + x + |y|$

78. $-x + y$

79. $x + (-y) + z$

Concept 3: Translations

For Exercises 80–89, translate the English phrase into an algebraic expression. Then evaluate the expression.

80. The sum of -6 and -10

81. The sum of -3 and 5

82. Negative three increased by 8

83. Twenty-one increased by 4

84. Seventeen more than -21

85. Twenty-four more than -7

86. Three times the sum of -14 and 20

87. Two times the sum of -6 and 10

88. Five more than the sum of -7 and -2

89. Negative six more than the sum of 4 and -1

Concept 4: Applications Involving Addition of Real Numbers

90. The temperature in Minneapolis, Minnesota, began at $-5°$F (5° below zero) at 6:00 A.M. By noon the temperature had risen 13°, and by the end of the day, the temperature had dropped 11° from its noon time high. Write an expression using addition that describes the changes in temperature during the day. Then evaluate the expression to give the temperature at the end of the day.

91. The temperature in Toronto, Ontario, Canada, began at 4°F. A cold front went through at noon, and the temperature dropped 9°. By 4:00 P.M. the temperature had risen 2° from its noon time low. Write an expression using addition that describes the changes in temperature during the day. Then evaluate the expression to give the temperature at 4:00 P.M.

92. During a football game, the Nebraska Cornhuskers lost 2 yd, gained 6 yd, and then lost 5 yd. Write an expression using addition that describes the team's total loss or gain and evaluate the expression.

93. During a football game, the University of Oklahoma's team gained 3 yd, lost 5 yd, and then gained 14 yd. Write an expression using addition that describes the team's total loss or gain and evaluate the expression.

94. Yoshima has \$52.23 in her checking account. She writes a check for groceries for \$52.95.

a. Write an addition problem that expresses Yoshima's transaction.

b. Is Yoshima's account overdrawn?

95. Mohammad has \$40.02 in his checking account. He writes a check for a pair of shoes for \$40.96.

a. Write an addition problem that expresses Mohammad's transaction.

b. Is Mohammad's account overdrawn?

96. In the game show *Jeopardy*, a contestant responds to six questions with the following outcomes:

$$+\$100,\ +\$200,\ -\$500,\ +\$300,\ +\$100,\ -\$200$$

a. Write an expression using addition to describe the contestant's scoring activity.

b. Evaluate the expression from part (a) to determine the contestant's final outcome.

97. A company that has been in business for 5 years has the following profit and loss record.

Year	Profit/Loss (\$)
1	−50,000
2	−32,000
3	−5000
4	13,000
5	26,000

a. Write an expression using addition to describe the company's profit/loss activity.

b. Evaluate the expression from part (a) to determine the company's net profit or loss.

Section 1.4 Subtraction of Real Numbers

Concepts

1. Subtraction of Real Numbers
2. Translations
3. Applications Involving Subtraction
4. Applying the Order of Operations

1. Subtraction of Real Numbers

In Section 1.3, we learned the rules for adding real numbers. Subtraction of real numbers is defined in terms of the addition process. For example, consider the following subtraction problem and the corresponding addition problem:

$$6 - 4 = 2 \quad \Leftrightarrow \quad 6 + (-4) = 2$$

Start

−6 −5 −4 −3 −2 −1 0 1 2 3 4 5 6

In each case, we start at 6 on the number line and move to the left 4 units. That is, adding the opposite of 4 produces the same result as subtracting 4. This is true in general. To subtract two real numbers, add the opposite of the second number to the first number.

Subtraction of Real Numbers

If a and b are real numbers, then $a - b = a + (-b)$

$$10 - 4 = 10 + (-4) = 6$$
$$-10 - 4 = -10 + (-4) = -14$$
Subtracting 4 is the same as adding −4.

$$10 - (-4) = 10 + (4) = 14$$
$$-10 - (-4) = -10 + (4) = -6$$
Subtracting −4 is the same as adding 4.

Example 1 Subtracting Integers

Subtract the numbers.

a. $4 - (-9)$ **b.** $-6 - 9$ **c.** $-11 - (-5)$ **d.** $7 - 10$

Solution:

a. $4 - (-9)$

$= 4 + (9) = 13$

Change subtraction to addition. Take the opposite of −9.

b. $-6 - 9$

$= -6 + (-9) = -15$

Change subtraction to addition. Take the opposite of 9.

c. $-11 - (-5)$

$= -11 + (5) = -6$

Change subtraction to addition. Take the opposite of −5.

d. $7 - 10$

$= 7 + (-10) = -3$

Change subtraction to addition. Take the opposite of 10.

Skill Practice Subtract.

1. $1 - (-3)$ **2.** $-2 - 2$ **3.** $-6 - (-11)$ **4.** $8 - 15$

Skill Practice Answers

1. 4 **2.** −4
3. 5 **4.** −7

Example 2 **Subtracting Real Numbers**

a. $\frac{3}{20} - \left(-\frac{4}{15}\right)$ **b.** $-2.3 - 6.04$

Solution:

a. $\frac{3}{20} - \left(-\frac{4}{15}\right)$ The least common denominator is 60.

$\frac{9}{60} - \left(-\frac{16}{60}\right)$ Write equivalent fractions with the LCD.

$\frac{9}{60} + \left(\frac{16}{60}\right)$ Write in terms of addition.

$\frac{25}{60}$ Add.

$\frac{5}{12}$ Reduce to lowest terms.

b. $-2.3 - 6.04$

$-2.3 + (-6.04)$ Write in terms of addition.

-8.34 Add.

Skill Practice Subtract.

5. $-\frac{1}{6} - \frac{7}{12}$ **6.** $7.5 - (-1.5)$

2. Translations

Example 3 **Translating Expressions Involving Subtraction**

Write an algebraic expression for each English phrase and then simplify the result.

a. The difference of -7 and -5

b. 12.4 subtracted from -4.7

c. -24 decreased by the sum of -10 and 13

d. Seven-fourths less than one-third

Solution:

a. $-7 - (-5)$ The difference of -7 and -5

$= -7 + (5)$ Rewrite subtraction in terms of addition.

$= -2$ Simplify.

b. $-4.7 - 12.4$ 12.4 subtracted from -4.7

$= -4.7 + (-12.4)$ Rewrite subtraction in terms of addition.

$= -17.1$ Simplify.

TIP: Recall that "b subtracted from a" is translated as $a - b$. In Example 3(b), -4.7 is written first and then 12.4 is subtracted.

Skill Practice Answers

5. $-\frac{3}{4}$ **6.** 9

c. $-24 - (-10 + 13)$ — -24 decreased by the sum of -10 and 13

$= -24 - (3)$ — Simplify inside parentheses.

$= -24 + (-3)$ — Rewrite subtraction in terms of addition.

$= -27$ — Simplify.

TIP: Parentheses must be used around the sum of -10 and 13 so that -24 is decreased by the entire quantity $(-10 + 13)$.

d. $\frac{1}{3} - \frac{7}{4}$ — Seven-fourths less than one-third

$= \frac{1}{3} + \left(-\frac{7}{4}\right)$ — Rewrite subtraction in terms of addition.

$= \frac{4}{12} + \left(-\frac{21}{12}\right)$ — The common denominator is 12.

$= -\frac{17}{12}$ or $-1\frac{5}{12}$

Skill Practice Write an algebraic expression for each phrase and then simplify.

7. 8 less than -10
8. -72 subtracted from -82
9. The difference of 2.6 and -14.7
10. Two-fifths decreased by four-thirds

3. Applications Involving Subtraction

Example 4 Using Subtraction of Real Numbers in an Application

During one of his turns on *Jeopardy*, Harold selected the category "Show Tunes." He got the following results: the \$200 question correct; the \$400 question incorrect; the \$600 question correct; the \$800 question incorrect; and the \$1000 question correct. Write an expression that determines Harold's score. Then simplify the expression to find his total winnings for that category.

Solution:

$200 - 400 + 600 - 800 + 1000$

$= 200 + (-400) + 600 + (-800) + 1000$ — Rewrite subtraction in terms of addition.

$= -200 + 600 + (-800) + 1000$ — Add from left to right.

$= 400 + (-800) + 1000$

$= -400 + 1000$

$= 600$ — Harold won \$600 in that category.

Skill Practice Answers

7. $-10 - 8; -18$
8. $-82 - (-72); -10$
9. $2.6 - (-14.7); 17.3$
10. $\frac{2}{5} - \frac{4}{3}; -\frac{14}{15}$

Skill Practice

11. During Harold's first round on *Jeopardy*, he got the \$100, \$200, and \$400 questions correct but he got the \$300 and \$500 questions incorrect. Determine Harold's score for this round.

Example 5 Using Subtraction of Real Numbers in an Application

The highest recorded temperature in North America was 134°F, recorded on July 10, 1913, in Death Valley, California. The lowest temperature of −81°F was recorded on February 3, 1947, in Snag, Yukon, Canada.

Find the difference between the highest and lowest recorded temperatures in North America.

Solution:

$134 - (-81)$

$= 134 + (81)$ Rewrite subtraction in terms of addition.

$= 215$ Add.

The difference between the highest and lowest temperatures is 215°F.

Skill Practice

12. The record high temperature for the state of Montana occurred in 1937 and was 117°F. The record low occurred in 1954 and was −70°F. Find the difference between the highest and lowest temperatures.

4. Applying the Order of Operations

Example 6 Applying the Order of Operations

Simplify the expressions.

a. $-6 + \{10 - [7 - (-4)]\}$

b. $5 - \sqrt{35 - (-14)} - 2$

c. $\left(-\frac{5}{8} - \frac{2}{3}\right) - \left(\frac{1}{8} + 2\right)$

d. $-6 - |7 - 11| + (-3 + 7)^2$

Solution:

a. $-6 + \{10 - [7 - (-4)]\}$ Work inside the inner brackets first.

$= -6 + \{10 - [7 + (4)]\}$ Rewrite subtraction in terms of addition.

$= -6 + \{10 - (11)\}$ Simplify the expression inside braces.

$= -6 + \{10 + (-11)\}$ Rewrite subtraction in terms of addition.

$= -6 + (-1)$

$= -7$ Add.

Skill Practice Answers

11. −100, Harold lost \$100.
12. 187°F

b. $5 - \sqrt{35 - (-14)} - 2$ Work inside the radical first.

$= 5 - \sqrt{35 + (14)} - 2$ Rewrite subtraction in terms of addition.

$= 5 - \sqrt{49} - 2$

$= 5 - 7 - 2$ Simplify the radical.

$= 5 + (-7) + (-2)$ Rewrite subtraction in terms of addition.

$= -2 + (-2)$ Add from left to right.

$= -4$

c. $\left(-\frac{5}{8} - \frac{2}{3}\right) - \left(\frac{1}{8} + 2\right)$ Work inside the parentheses first.

$= \left[-\frac{5}{8} + \left(-\frac{2}{3}\right)\right] - \left(\frac{1}{8} + 2\right)$ Rewrite subtraction in terms of addition.

$= \left[-\frac{15}{24} + \left(-\frac{16}{24}\right)\right] - \left(\frac{1}{8} + \frac{16}{8}\right)$ Get a common denominator in each parentheses.

$= \left(-\frac{31}{24}\right) - \left(\frac{17}{8}\right)$ Add fractions in each parentheses.

$= \left(-\frac{31}{24}\right) + \left(-\frac{17}{8}\right)$ Rewrite subtraction in terms of addition.

$= -\frac{31}{24} + \left(-\frac{51}{24}\right)$ Get a common denominator.

$= -\frac{82}{24}$ Add.

$= -\frac{41}{12}$ Reduce to lowest terms.

d. $-6 - |7 - 11| + (-3 + 7)^2$ Simplify within absolute value bars and parentheses first.

$= -6 - |7 + (-11)| + (-3 + 7)^2$ Rewrite subtraction in terms of addition.

$= -6 - |-4| + (4)^2$

$= -6 - (4) + 16$ Simplify absolute value and exponent.

$= -6 + (-4) + 16$ Rewrite subtraction in terms of addition.

$= -10 + 16$ Add from left to right.

$= 6$

Skill Practice Simplify the expressions.

13. $\{-11 + [3 - (-4)]\} - 8$

14. $(12 - 5)^2 + \sqrt{4 - (-21)}$

15. $\left(-1 + \frac{1}{4}\right) - \left(\frac{3}{4} - \frac{1}{2}\right)$

16. $4 - 2\,|6 + (-8)| + (4)^2$

Skill Practice Answers

13. -12 **14.** 54
15. -1 **16.** 16

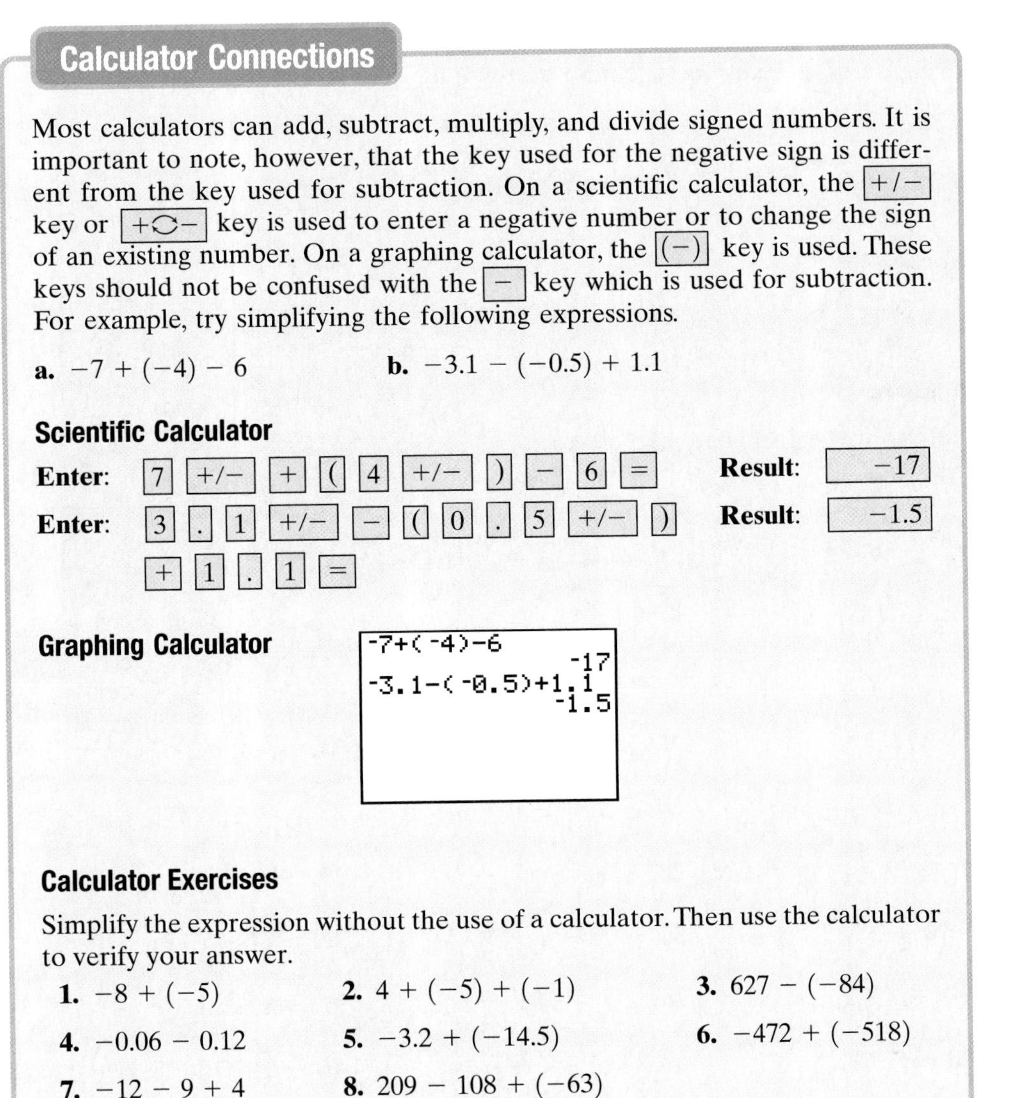

Calculator Connections

Most calculators can add, subtract, multiply, and divide signed numbers. It is important to note, however, that the key used for the negative sign is different from the key used for subtraction. On a scientific calculator, the [+/−] key or [+⟲−] key is used to enter a negative number or to change the sign of an existing number. On a graphing calculator, the [(−)] key is used. These keys should not be confused with the [−] key which is used for subtraction. For example, try simplifying the following expressions.

a. $-7 + (-4) - 6$ **b.** $-3.1 - (-0.5) + 1.1$

Scientific Calculator

Enter: [7] [+/−] [+] [(] [4] [+/−] [)] [−] [6] [=] **Result:** −17

Enter: [3] [.] [1] [+/−] [−] [(] [0] [.] [5] [+/−] [)] [+] [1] [.] [1] [=] **Result:** −1.5

Graphing Calculator

```
-7+(-4)-6
              -17
-3.1-(-0.5)+1.1
             -1.5
```

Calculator Exercises

Simplify the expression without the use of a calculator. Then use the calculator to verify your answer.

1. $-8 + (-5)$ **2.** $4 + (-5) + (-1)$ **3.** $627 - (-84)$

4. $-0.06 - 0.12$ **5.** $-3.2 + (-14.5)$ **6.** $-472 + (-518)$

7. $-12 - 9 + 4$ **8.** $209 - 108 + (-63)$

Section 1.4 Practice Exercises

Boost *your* GRADE at mathzone.com!

- Practice Problems
- Self-Tests
- NetTutor
- e-Professors
- Videos

Study Skills Exercise

1. Some instructors allow the use of calculators. Does your instructor allow the use of a calculator? If so, what kind?

Will you be allowed to use a calculator on tests or just for occasional calculator problems in the text?

Helpful Hint: If you are not permitted to use a calculator on tests, it is good to do your homework in the same way, without the calculator.

Review Exercises

For Exercises 2–5, translate each English phrase into an algebraic expression.

2. The square root of 6

3. The square of x

4. Negative seven increased by 10

5. Two more than $-b$

For Exercises 6–9, simplify the expression.

6. $4^2 - 6 \div 2$

7. $1 + 36 \div 9 \cdot 2$

8. $14 - |10 - 6|$

9. $|-12 + 7|$

Concept 1: Subtraction of Real Numbers

For Exercises 10–17, fill in the blank to make each statement correct.

10. $5 - 3 = 5 +$ ________

11. $8 - 7 = 8 +$ ________

12. $-2 - 12 = -2 +$ ________

13. $-4 - 9 = -4 +$ ________

14. $7 - (-4) = 7 +$ ________

15. $13 - (-4) = 13 +$ ________

16. $-9 - (-3) = -9 +$ ________

17. $-15 - (-10) = -15 +$ ________

For Exercises 18–41, subtract the integers.

18. $3 - 5$

19. $9 - 12$

20. $3 - (-5)$

21. $9 - (-12)$

22. $-3 - 5$

23. $-9 - 12$

24. $-3 - (-5)$

25. $-9 - (-5)$

26. $23 - 17$

27. $14 - 2$

28. $23 - (-17)$

29. $14 - (-2)$

30. $-23 - 17$

31. $-14 - 2$

32. $-23 - (-17)$

33. $-14 - (-2)$

34. $-6 - 14$

35. $-9 - 12$

36. $-7 - 17$

37. $-8 - 21$

38. $13 - (-12)$

39. $20 - (-5)$

40. $-14 - (-9)$

41. $-21 - (-17)$

For Exercises 42–63, subtract the real numbers.

42. $-\frac{6}{5} - \frac{3}{10}$

43. $-\frac{2}{9} - \frac{5}{3}$

44. $\frac{3}{8} - \left(-\frac{4}{3}\right)$

45. $\frac{7}{10} - \left(-\frac{5}{6}\right)$

46. $\frac{1}{2} - \frac{1}{10}$

47. $\frac{2}{7} - \frac{3}{14}$

48. $-\frac{11}{12} - \left(-\frac{1}{4}\right)$

49. $-\frac{7}{8} - \left(-\frac{1}{6}\right)$

50. $6.8 - (-2.4)$

51. $7.2 - (-1.9)$

52. $3.1 - 8.82$

53. $1.8 - 9.59$

54. $-4 - 3 - 2 - 1$

55. $-10 - 9 - 8 - 7$

56. $6 - 8 - 2 - 10$

57. $20 - 50 - 10 - 5$

58. $-36.75 - 14.25$

59. $-84.21 - 112.16$

60. $-112.846 + (-13.03) - 47.312$

61. $-96.473 + (-36.02) - 16.617$

62. $0.085 - (-3.14) + (0.018)$

63. $0.00061 - (-0.00057) + (0.0014)$

Concept 2: Translations

For Exercises 64–73, translate each English phrase into an algebraic expression. Then evaluate the expression.

64. Six minus -7

65. Eighteen minus -1

66. Eighteen subtracted from 3

67. Twenty-one subtracted from 8

68. The difference of -5 and -11

69. The difference of -2 and -18

70. Negative thirteen subtracted from -1

71. Negative thirty-one subtracted from -19

72. Twenty less than -32

73. Seven less than -3

Concept 3: Applications Involving Subtraction

74. On the game, *Jeopardy*, Jasper selected the category "The Last." He got the first four questions correct (worth \$200, \$400, \$600, and \$800) but then missed the last question (worth \$1000). Write an expression that determines Jasper's score. Then simplify the expression to find his total winnings for that category.

75. On Ethyl's turn in *Jeopardy*, she chose the category "Birds of a Feather." She already had \$1200 when she selected a Double Jeopardy question. She wagered \$500 but guessed incorrectly (therefore she lost \$500). On her next turn, she got the \$800 question correct. Write an expression that determines Ethyl's score. Then simplify the expression to find her total winnings.

76. In Ohio, the highest temperature ever recorded was 113°F and the lowest was -39°F. Find the difference between the highest and lowest temperatures. (*Source: Information Please Almanac*)

77. In Mississippi, the highest temperature ever recorded was 115°F and the lowest was -19°F. Find the difference between the highest and lowest temperatures. (*Source: Information Please Almanac*)

78. The highest mountain in the world is Mt. Everest, located in the Himalayas. Its height is 8848 meters (m) (29,028 ft). The lowest recorded depth in the ocean is located in the Marianas Trench in the Pacific Ocean. Its "height" relative to sea level is $-11{,}033$ m ($-36{,}198$ ft). Determine the difference in elevation, in meters, between the highest mountain in the world and the deepest ocean trench. (*Source: Information Please Almanac*)

79. The lowest point in North America is located in Death Valley, California, at an elevation of -282 ft (-86 m). The highest point in North America is Mt. McKinley, Alaska, at an elevation of 20,320 ft (6194 m). Find the difference in elevation, in feet, between the highest and lowest points in North America. (*Source: Information Please Almanac*)

Concept 4: Applying the Order of Operations

For Exercises 80–93, perform the indicated operations. Remember to perform addition or subtraction as they occur from left to right.

80. $6 + 8 - (-2) - 4 + 1$

81. $-3 - (-4) + 1 - 2 - 5$

82. $-1 - 7 + (-3) - 8 + 10$

83. $13 - 7 + 4 - 3 - (-1)$

84. $2 - (-8) + 7 + 3 - 15$

85. $8 - (-13) + 1 - 9$

86. $-6 + (-1) + (-8) + (-10)$

87. $-8 + (-3) + (-5) + (-2)$

88. $-6 - 1 - 8 - 10$

89. $-8 - 3 - 5 - 2$

90. $-\frac{13}{10} + \frac{8}{15} - \left(-\frac{2}{5}\right)$

91. $\frac{11}{14} - \left(-\frac{9}{7}\right) - \frac{3}{2}$

92. $\frac{2}{3} + \frac{5}{9} - \frac{4}{3} - \left(-\frac{1}{6}\right)$

93. $-\frac{9}{8} - \frac{1}{4} - \left(-\frac{5}{6}\right) + \frac{1}{8}$

For Exercises 94–101, evaluate the expressions for $a = -2$, $b = -6$, and $c = -1$.

94. $(a + b) - c$

95. $(a - b) + c$

96. $a - (b + c)$

97. $a + (b - c)$

98. $(a - b) - c$

99. $(a + b) + c$

100. $a - (b - c)$

101. $a + (b + c)$

For Exercises 102–107, evaluate the expression using the order of operations.

102. $\sqrt{29 + (-4)} - 7$

103. $8 - \sqrt{98 + (-3) + 5}$

104. $|10 + (-3)| - |-12 + (-6)|$

105. $|6 - 8| + |12 - 5|$

106. $\frac{3 - 4 + 5}{4 + (-2)}$

107. $\frac{12 - 14 + 6}{6 + (-2)}$

Chapter 1 Problem Recognition Exercises—Addition and Subtraction of Signed Numbers

This set of exercises allows you to practice recognizing the difference between a negative sign and a subtraction sign in the context of a problem.

1. State the rule for adding two negative numbers.

2. State the rule for adding a negative number to a positive number.

For Exercises 3–32, add or subtract as indicated.

3. $65 - 24$

4. $42 - 29$

5. $13 - (-18)$

6. $22 - (-24)$

7. $4.8 - 6.1$

8. $3.5 - 7.1$

9. $4 + (-20)$

10. $5 + (-12)$

11. $\frac{1}{3} - \frac{5}{12}$

12. $\frac{3}{8} - \frac{1}{12}$

13. $-32 - 4$

14. $-51 - 8$

15. $-6 + (-6)$

16. $-25 + (-25)$

17. $-4 - \left(-\frac{5}{6}\right)$

18. $-2 - \left(-\frac{2}{5}\right)$

19. $-60 + 55$

20. $-55 + 23$

21. $-18 - (-18)$

22. $-3 - (-3)$

23. $-3.5 - 4.2$

24. $-6.6 - 3.9$

25. $-90 + (-24)$

26. $-35 + (-21)$

27. $-14 + (-2) - 16$

28. $-25 + (-6) - 15$

29. $-42 + 12 + (-30)$

30. $-46 + 16 + (-40)$

31. $-10 - 8 - 6 - 4 - 2$

32. $-100 - 90 - 80 - 70 - 60$

For Exercises 33–34, evaluate the expression for $x = 3$ and $y = -5$.

33. $x - y$

34. $|x + y|$

For Exercises 35–36, write an algebraic expression for each English phrase and then simplify the result.

35. The sum of -8 and 20.

36. The difference of -11 and -2.

Multiplication and Division of Real Numbers

Section 1.5

Concepts

1. Multiplication of Real Numbers
2. Exponential Expressions
3. Division of Real Numbers
4. Applying the Order of Operations

1. Multiplication of Real Numbers

Multiplication of real numbers can be interpreted as repeated addition.

Example 1 Multiplying Real Numbers

Multiply the real numbers by writing the expressions as repeated addition.

a. $3(4)$ **b.** $3(-4)$

Solution:

a. $3(4) = 4 + 4 + 4 = 12$ Add 3 groups of 4.

b. $3(-4) = -4 + (-4) + (-4) = -12$ Add 3 groups of -4.

The results from Example 1 suggest that the product of a positive number and a negative number is *negative*. Consider the following pattern of products.

$$\begin{aligned} 4 \times 3 &= 12 \\ 4 \times 2 &= 8 \\ 4 \times 1 &= 4 \\ 4 \times 0 &= 0 \\ 4 \times -1 &= -4 \\ 4 \times -2 &= -8 \\ 4 \times -3 &= -12 \end{aligned}$$

The pattern decreases by 4 with each row.

Thus, the product of a positive number and a negative number must be *negative* for the pattern to continue.

Now suppose we have a product of two negative numbers. To determine the sign, consider the following pattern of products.

$$\begin{aligned} -4 \times 3 &= -12 \\ -4 \times 2 &= -8 \\ -4 \times 1 &= -4 \\ -4 \times 0 &= 0 \\ -4 \times -1 &= 4 \\ -4 \times -2 &= 8 \\ -4 \times -3 &= 12 \end{aligned}$$

The pattern increases by 4 with each row.

Thus, the product of two negative numbers must be *positive* for the pattern to continue.

We now summarize the rules for multiplying real numbers.

Multiplication of Real Numbers

1. The product of two real numbers with the *same* sign is positive.
2. The product of two real numbers with *different* signs is negative.
3. The product of any real number and zero is zero.

Example 2 **Multiplying Real Numbers**

Multiply the real numbers.

a. $-8(-4)$ **b.** $-2.5(-1.7)$ **c.** $-7(10)$

d. $\frac{1}{2}(-8)$ **e.** $0(-8.3)$ **f.** $-\frac{2}{7}\left(-\frac{7}{2}\right)$

Solution:

a. $-8(-4) = 32$
b. $-2.5(-1.7) = 4.25$ — *Same signs.* Product is positive.

c. $-7(10) = -70$
d. $\frac{1}{2}(-8) = -4$ — *Different signs.* Product is negative.

e. $0(-8.3)$
$= 0$ — The product of any real number and zero is zero.

f. $-\frac{2}{7}\left(-\frac{7}{2}\right)$
$= \frac{14}{14}$ — *Same signs.* Product is positive.
$= 1$ — Reduce to lowest terms.

Skill Practice Multiply.

1. $-9(3)$ **2.** $-6(-6)$ **3.** $0(-7)$

4. $1.5(-8)$ **5.** $-\frac{4}{3}\left(-\frac{3}{4}\right)$ **6.** $(-1.2)(-4.8)$

Observe the pattern for repeated multiplications.

$$(-1)(-1) = 1$$

$$\underbrace{(-1)(-1)}(-1) = (1)(-1) = -1$$

$$\underbrace{(-1)(-1)}(-1)(-1) = \underbrace{(1)(-1)}(-1) = (-1)(-1) = 1$$

$$\underbrace{(-1)(-1)}(-1)(-1)(-1) = \underbrace{(1)(-1)}(-1)(-1) = \underbrace{(-1)(-1)}(-1) = (1)(-1) = -1$$

The pattern demonstrated in these examples indicates that

- The product of an even number of negative factors is positive.
- The product of an odd number of negative factors is negative.

2. Exponential Expressions

Recall that for any real number b and any positive integer, n:

$$b^n = \underbrace{b \cdot b \cdot b \cdot b \ldots \cdot b}_{n \text{ factors of } b}$$

Be particularly careful when evaluating exponential expressions involving negative numbers. An exponential expression with a negative base is written with parentheses around the base, such as $(-2)^4$.

Skill Practice Answers

1. -27 **2.** 36 **3.** 0
4. -12 **5.** 1 **6.** 5.76

To evaluate $(-2)^4$, the base -2 is multiplied four times:

$$(-2)^4 = (-2)(-2)(-2)(-2) = 16$$

If parentheses are *not* used, the expression -2^4 has a different meaning:

- The expression -2^4 has a base of 2 (not -2) and can be interpreted as $-1 \cdot 2^4$. Hence,

$$-2^4 = -1(2)(2)(2)(2) = -16$$

- The expression -2^4 can also be interpreted as the opposite of 2^4. Hence,

$$-2^4 = -(2 \cdot 2 \cdot 2 \cdot 2) = -16$$

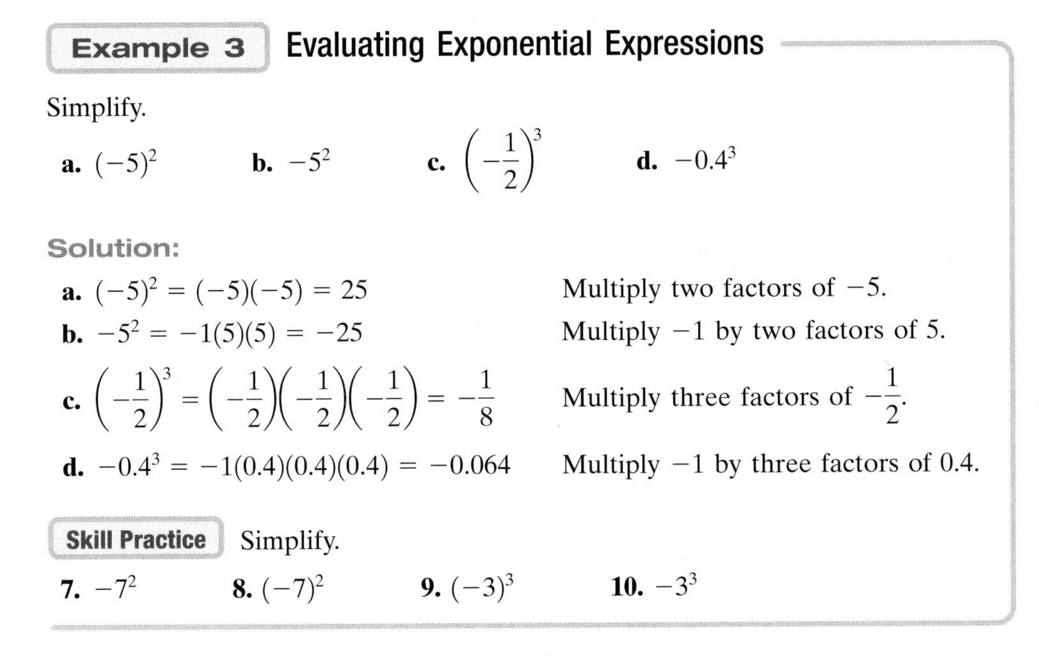

Example 3 Evaluating Exponential Expressions

Simplify.

a. $(-5)^2$ **b.** -5^2 **c.** $\left(-\frac{1}{2}\right)^3$ **d.** -0.4^3

Solution:

a. $(-5)^2 = (-5)(-5) = 25$ Multiply two factors of -5.

b. $-5^2 = -1(5)(5) = -25$ Multiply -1 by two factors of 5.

c. $\left(-\frac{1}{2}\right)^3 = \left(-\frac{1}{2}\right)\left(-\frac{1}{2}\right)\left(-\frac{1}{2}\right) = -\frac{1}{8}$ Multiply three factors of $-\frac{1}{2}$.

d. $-0.4^3 = -1(0.4)(0.4)(0.4) = -0.064$ Multiply -1 by three factors of 0.4.

Skill Practice Simplify.

7. -7^2 **8.** $(-7)^2$ **9.** $(-3)^3$ **10.** -3^3

3. Division of Real Numbers

Two numbers are *reciprocals* if their product is 1. For example, $-\frac{2}{7}$ and $-\frac{7}{2}$ are reciprocals because $-\frac{2}{7}(-\frac{7}{2}) = 1$. Symbolically, if a is a nonzero real number, then the reciprocal of a is $\frac{1}{a}$ because $a \cdot \frac{1}{a} = 1$. This definition also implies that a number and its reciprocal have the same sign.

The Reciprocal of a Real Number

Let a be a nonzero real number. Then, the **reciprocal** of a is $\frac{1}{a}$.

Recall that to subtract two real numbers, we add the opposite of the second number to the first number. In a similar way, division of real numbers is defined in terms of multiplication. To divide two real numbers, we multiply the first number by the reciprocal of the second number.

Division of Real Numbers

Let a and b be real numbers such that $b \neq 0$. Then, $a \div b = a \cdot \frac{1}{b}$.

Skill Practice Answers

7. -49 **8.** 49
9. -27 **10.** -27

Consider the quotient $10 \div 5$. The reciprocal of 5 is $\frac{1}{5}$, so we have

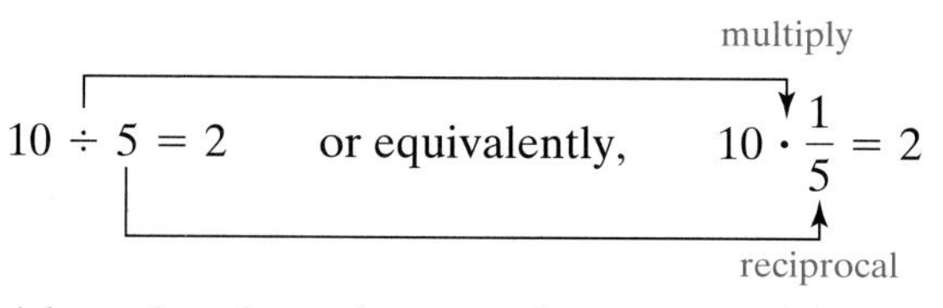

Because division of real numbers can be expressed in terms of multiplication, then the sign rules that apply to multiplication also apply to division.

Division of Real Numbers

1. The quotient of two real numbers with the *same* sign is positive.
2. The quotient of two real numbers with *different* signs is negative.

Example 4 Dividing Real Numbers

Divide the real numbers.

a. $200 \div (-10)$ **b.** $\frac{-48}{16}$ **c.** $\frac{-6.25}{-1.25}$ **d.** $\frac{-9}{-5}$

e. $15 \div -25$ **f.** $-\frac{3}{14} \div \frac{9}{7}$ **g.** $\frac{\frac{2}{5}}{-\frac{2}{5}}$

Solution:

a. $200 \div (-10) = -20$ *Different signs*. Quotient is negative.

b. $\frac{-48}{16} = -3$ *Different signs*. Quotient is negative.

c. $\frac{-6.25}{-1.25} = 5$ *Same signs*. Quotient is positive.

d. $\frac{-9}{-5} = \frac{9}{5}$ *Same signs*. Quotient is positive.

Because 5 does not divide into 9 evenly, the answer can be left as a fraction.

TIP: If the numerator and denominator of a fraction are both negative, then the quotient is positive. Therefore, $\frac{-9}{-5}$ can be simplified to $\frac{9}{5}$.

e. $15 \div -25$ *Different signs.* Quotient is negative.

$= \frac{15}{-25}$

$= -\frac{3}{5}$

TIP: If the numerator and denominator of a fraction have opposite signs, then the quotient will be negative. Therefore, a fraction has the same value whether the negative sign is written in the numerator, in the denominator, or in front of the fraction.

$$\frac{-3}{5} = \frac{3}{-5} = -\frac{3}{5}$$

f. $-\frac{3}{14} \div \frac{9}{7}$ *Different signs.* Quotient is negative.

$= -\frac{3}{14} \cdot \frac{7}{9}$ Multiply by the reciprocal of $\frac{9}{7}$ which is $\frac{7}{9}$.

$= -\frac{21}{126}$ Multiply fractions.

$= -\frac{1}{6}$ Reduce to lowest terms.

g. $\frac{\frac{2}{5}}{-\frac{2}{5}}$ This is equivalent to $\frac{2}{5} \div (-\frac{2}{5})$.

$= \frac{2}{5}\left(-\frac{5}{2}\right)$ Multiply by the reciprocal of $-\frac{2}{5}$, which is $-\frac{5}{2}$.

$= -1$ Multiply fractions.

Skill Practice Simplify.

11. $-14 \div 7$ **12.** $\frac{-20}{-5}$ **13.** $\frac{18}{-3}$ **14.** $\frac{-7}{-3}$

15. $\frac{-4}{-8}$ **16.** $\frac{3}{4} \div \left(\frac{-9}{16}\right)$ **17.** $\frac{-1}{1}$ **18.** $4.2 \div -0.2$

Multiplication can be used to check any division problem. If $\frac{a}{b} = c$, then $bc = a$ (provided that $b \neq 0$). For example,

$$\frac{8}{-4} = -2 \rightarrow \underline{\text{Check}}: \quad (-4)(-2) = 8 \checkmark$$

This relationship between multiplication and division can be used to investigate division problems involving the number zero.

1. The quotient of 0 and any nonzero number is 0. For example:

$$\frac{0}{6} = 0 \qquad \text{because } 6 \cdot 0 = 0 \checkmark$$

2. The quotient of any nonzero number and 0 is undefined. For example,

$$\frac{6}{0} = ?$$

Finding the quotient $\frac{6}{0}$ is equivalent to asking, "What number times zero will equal 6?" That is, $(0)(?) = 6$. No real number satisfies this condition. Therefore, we say that division by zero is undefined.

3. The quotient of 0 and 0 cannot be determined. Evaluating an expression of the form $\frac{0}{0} = ?$ is equivalent to asking, "What number times zero will equal 0?" That is, $(0)(?) = 0$. Any real number will satisfy this requirement; however, expressions involving $\frac{0}{0}$ are usually discussed in advanced mathematics courses.

Division Involving Zero

Let a represent a nonzero real number. Then,

1. $\frac{0}{a} = 0$ 2. $\frac{a}{0}$ is undefined

Skill Practice Answers

11. -2 **12.** 4 **13.** -6
14. $\frac{7}{3}$ **15.** $\frac{1}{2}$ **16.** $-\frac{4}{3}$
17. -1 **18.** -21

4. Applying the Order of Operations

Example 5 Applying the Order of Operations

Simplify the expressions.

a. $-36 \div (-27) \div (-9)$ **b.** $-8 + 8 \div (-2) \div (-6)$

c. $\dfrac{4 + \sqrt{30 - 5}}{-5 - 1}$ **d.** $\dfrac{24 - 2[-3 + (5 - 8)]^2}{2|-12 + 3|}$

Solution:

a. $\underbrace{-36 \div (-27)} \div (-9)$

$= \dfrac{-36}{-27} \div -9$ Divide from left to right.

$= \dfrac{4}{3} \div -9$ Reduce to lowest terms.

$= \dfrac{4}{3}\left(-\dfrac{1}{9}\right)$ Multiply by the reciprocal of -9, which is $-\frac{1}{9}$.

$= -\dfrac{4}{27}$ The product of two numbers with opposite signs is negative.

b. $-8 + 8 \div (-2) \div (-6)$

$= -8 + (-4) \div (-6)$ Perform division before addition.

$= -8 + \dfrac{4}{6}$ The quotient of -4 and -6 is positive $\frac{4}{6}$ or $\frac{2}{3}$.

$= -\dfrac{8}{1} + \dfrac{2}{3}$ Write -8 as a fraction.

$= -\dfrac{24}{3} + \dfrac{2}{3}$ Get a common denominator.

$= -\dfrac{22}{3}$ Add.

c. $\dfrac{4 + \sqrt{30 - 5}}{-5 - 1}$ Simplify numerator and denominator separately.

$= \dfrac{4 + \sqrt{25}}{-6}$ Simplify within the radical and simplify the denominator.

$= \dfrac{4 + 5}{-6}$ Simplify the radical.

$= \dfrac{9}{-6}$

$= \dfrac{3}{-2}$ or $-\dfrac{3}{2}$ Reduce to lowest terms. The quotient of two numbers with opposite signs is negative.

d. $\dfrac{24 - 2[-3 + (5 - 8)]^2}{2|-12 + 3|}$ Simplify numerator and denominator separately.

$= \dfrac{24 - 2[-3 + (-3)]^2}{2|-9|}$ Simplify within the inner parentheses and absolute value.

$= \dfrac{24 - 2[-6]^2}{2(9)}$ Simplify within brackets, [ ]. Simplify the absolute value.

$= \dfrac{24 - 2(36)}{2(9)}$ Simplify exponents.

$= \dfrac{24 - 72}{18}$ Perform multiplication before subtraction.

$= \dfrac{-48}{18}$ or $-\dfrac{8}{3}$ Reduce to lowest terms.

Skill Practice Simplify.

19. $-12 \div (-6) \div 2$

20. $-32 \div 4 \cdot 2 - 10$

21. $\dfrac{-4 - \sqrt{12 - 8}}{-6 - (-4)}$

22. $\dfrac{(3 + 15) - 1 - 2 + 11}{12 + 3(-4)}$

Example 6 Evaluating an Algebraic Expression

Given $y = -6$, evaluate the expressions.

a. y^2 **b.** $-y^2$

Solution:

a. y^2

$= (\quad)^2$ When substituting a number for a variable, use parentheses.

$= (-6)^2$ Substitute $y = -6$.

$= 36$ Square -6, that is, $(-6)(-6) = 36$.

b. $-y^2$

$= -(\quad)^2$ When substituting a number for a variable, use parentheses.

$= -(-6)^2$ Substitute $y = -6$.

$= -(36)$ Square -6.

$= -36$ Multiply by -1.

Skill Practice Given $a = -7$, evaluate the expressions.

23a. a^2 **b.** $-a^2$

Skill Practice Answers

19. 1 **20.** -26
21. 3 **22.** Undefined
23a. 49 **b.** -49

Calculator Connections

Be particularly careful when raising a negative number to an even power on a calculator. For example, the expressions $(-4)^2$ and -4^2 have different values. That is, $(-4)^2 = 16$ and $-4^2 = -16$. Verify these expressions on a calculator.

Scientific Calculator

To evaluate $(-4)^2$

Enter: (4 +/−) x^2 **Result**: 16

To evaluate -4^2 on a scientific calculator, it is important to square 4 first and then take its opposite.

Enter: 4 x^2 +/− **Result**: −16

Graphing Calculator

```
(-4)²
                16
-4²
               -16
```

The graphing calculator allows for several methods of denoting the multiplication of two real numbers. For example, consider the product of −8 and 4.

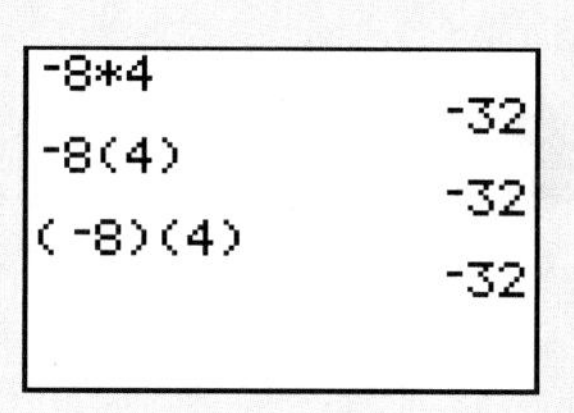

Calculator Exercises

Simplify the expression without the use of a calculator. Then use the calculator to verify your answer.

1. $-6(5)$ **2.** $\dfrac{-5.2}{2.6}$ **3.** $(-5)(-5)(-5)(-5)$ **4.** $(-5)^4$

5. -5^4 **6.** -2.4^2 **7.** $(-2.4)^2$

8. $(-1)(-1)(-1)$ **9.** $\dfrac{-8.4}{-2.1}$ **10.** $90 \div (-5)(2)$

Section 1.5 Practice Exercises

Boost *your* GRADE at mathzone.com!

MathZone

- Practice Problems
- Self-Tests
- NetTutor
- e-Professors
- Videos

Study Skills Exercises

1. Look through the text, and write down the page number that contains:

a. Avoiding Mistakes ________________

b. Tip box ________________

c. A key term (shown in bold) ________________

2. Define the key term **reciprocal of a real number**.

Review Exercises

For Exercises 3–7, determine if the expression is true or false.

3. $6 + (-2) > -5 + 6$

4. $|-6| + |-14| \leq |-3| + |-17|$

5. $16 - |-2| \leq |-3| - (-11)$

6. $\sqrt{36} - |-6| > 0$

7. $\sqrt{9} + |-3| \leq 0$

Concept 1: Multiplication of Real Numbers

For each product in Exercises 8–11, write an equivalent addition problem.

8. $5(4)$

9. $2(6)$

10. $3(-2)$

11. $5(-6)$

For Exercises 12–19, multiply the real numbers.

12. $8(-7)$

13. $(-3) \cdot 4$

14. $(-6) \cdot 7$

15. $9(-5)$

16. $(-11)(-13)$

17. $(-5)(-26)$

18. $(-30)(-8)$

19. $(-16)(-8)$

Concept 2: Exponential Expressions

For Exercises 20–27, simplify the exponential expression.

20. $(-6)^2$

21. $(-10)^2$

22. -6^2

23. -10^2

24. $\left(-\frac{2}{3}\right)^3$

25. $\left(-\frac{5}{2}\right)^3$

26. $(-0.2)^4$

27. $(-0.1)^4$

Concept 3: Division of Real Numbers

For Exercises 28–33, show how multiplication can be used to check the division problems.

28. $\frac{14}{-2} = -7$

29. $\frac{-18}{-6} = 3$

30. $\frac{0}{-5} = 0$

31. $\frac{0}{-4} = 0$

32. $\frac{6}{0}$ is undefined

33. $\frac{-4}{0}$ is undefined

For Exercises 34–41, divide the real numbers.

34. $\frac{54}{-9}$

35. $\frac{-27}{3}$

36. $\frac{-100}{-10}$

37. $\frac{-120}{-40}$

38. $\frac{-14}{-7}$

39. $\frac{-21}{-3}$

40. $\frac{13}{-65}$

41. $\frac{7}{-77}$

Mixed Exercises

For Exercises 42–117, multiply or divide as indicated.

42. $2 \cdot 3$

43. $8 \cdot 6$

44. $2(-3)$

45. $8(-6)$

46. $(-2)3$

47. $(-8)6$

48. $(-2)(-3)$

49. $(-8)(-6)$

50. $24 \div 3$

51. $52 \div 2$

52. $24 \div (-3)$

53. $52 \div (-2)$

54. $(-24) \div 3$

55. $(-52) \div 2$

56. $(-24) \div (-3)$

57. $(-52) \div (-2)$

58. $-6 \cdot 0$
59. $-8 \cdot 0$
60. $-18 \div 0$
61. $-42 \div 0$

62. $0\left(-\frac{2}{5}\right)$
63. $0\left(-\frac{1}{8}\right)$
64. $0 \div \left(-\frac{1}{10}\right)$
65. $0 \div \left(\frac{4}{9}\right)$

66. $\frac{-9}{6}$
67. $\frac{-15}{10}$
68. $\frac{-30}{-100}$
69. $\frac{-250}{-1000}$

70. $\frac{26}{-13}$
71. $\frac{52}{-4}$
72. $1.72(-4.6)$
73. $361.3(-14.9)$

74. $-0.02(-4.6)$
75. $-0.06(-2.15)$
76. $\frac{14.4}{-2.4}$
77. $\frac{50.4}{-6.3}$

78. $\frac{-5.25}{-2.5}$
79. $\frac{-8.5}{-27.2}$
80. $(-3)^2$
81. $(-7)^2$

82. -3^2
83. -7^2
84. $\left(-\frac{4}{3}\right)^3$
85. $\left(-\frac{1}{5}\right)^3$

86. $(-0.2)^3$
87. $(-0.1)^6$
88. -0.2^4
89. -0.1^4

90. $87 \div (-3)$
91. $96 \div (-6)$
92. $-4(-12)$
93. $(-5)(-11)$

94. $2.8(-5.1)$
95. $(7.21)(-0.3)$
96. $(-6.8) \div (-0.02)$
97. $(-12.3) \div (-0.03)$

98. $\left(-\frac{2}{15}\right)\left(\frac{25}{3}\right)$
99. $\left(-\frac{5}{16}\right)\left(\frac{4}{9}\right)$
100. $\left(-\frac{7}{8}\right) \div \left(-\frac{9}{16}\right)$
101. $\left(-\frac{22}{23}\right) \div \left(-\frac{11}{3}\right)$

102. $(-2)(-5)(-3)$
103. $(-6)(-1)(-10)$
104. $(-8)(-4)(-1)(-3)$
105. $(-6)(-3)(-1)(-5)$

106. $100 \div (-10) \div (-5)$
107. $150 \div (-15) \div (-2)$
108. $-12 \div (-6) \div (-2)$
109. $-36 \div (-2) \div 6$

110. $\frac{2}{5} \cdot \frac{1}{3} \cdot \left(-\frac{10}{11}\right)$
111. $\left(-\frac{9}{8}\right) \cdot \left(-\frac{2}{3}\right) \cdot \left(1\frac{5}{12}\right)$
112. $\left(1\frac{1}{3}\right) \div 3 \div \left(-\frac{7}{9}\right)$
113. $-\frac{7}{8} \div \left(3\frac{1}{4}\right) \div (-2)$

114. $12 \div (-2)(4)$
115. $(-6) \cdot 7 \div (-2)$
116. $\left(-\frac{12}{5}\right) \div (-6) \cdot \left(-\frac{1}{8}\right)$
117. $10 \cdot \frac{1}{3} \div \frac{25}{6}$

Concept 4: Applying the Order of Operations

For Exercises 118–131, perform the indicated operations.

118. $8 - 2^3 \cdot 5 + 3 - (-6)$
119. $-14 \div (-7) - 8 \cdot 2 + 3^3$
120. $-(2 - 8)^2 \div (-6) \cdot 2$

121. $-(3 - 5)^2 \cdot 6 \div (-4)$
122. $\frac{6(-4) - 2(5 - 8)}{-6 - 3 - 5}$
123. $\frac{3(-4) - 5(9 - 11)}{-9 - 2 - 3}$

124. $\frac{-4 + 5}{(-2) \cdot 5 + 10}$
125. $\frac{-3 + 10}{2(-4) + 8}$
126. $|-5| - |-7|$

127. $|-8| - |-2|$

128. $-|-1| - |5|$

129. $-|-10| - |6|$

130. $\dfrac{|2 - 9| - |5 - 7|}{10 - 15}$

131. $\dfrac{|-2 + 6| - |3 - 5|}{13 - 11}$

For Exercises 132–139, evaluate the expression for $x = -2$, $y = -4$, and $z = 6$.

132. $x^2 - 2y$

133. $3y^2 - z$

134. $4(2x - z)$

135. $6(3x + y)$

136. $\dfrac{3x + 2y}{y}$

137. $\dfrac{2z - y}{x}$

138. $\dfrac{x + 2y}{x - 2y}$

139. $\dfrac{x - z}{x^2 - z^2}$

140. Is the expression $\dfrac{10}{5x}$ equal to $10/5x$? Explain.

141. Is the expression $10/(5x)$ equal to $\dfrac{10}{5x}$? Explain.

For Exercises 142–149, translate the English phrase into an algebraic expression. Then evaluate the expression.

142. The product of -3.75 and 0.3

143. The product of -0.4 and -1.258

144. The quotient of $\frac{16}{5}$ and $\left(-\frac{8}{9}\right)$

145. The quotient of $\left(-\frac{3}{14}\right)$ and $\frac{1}{7}$

146. The number -0.4 plus the quantity 6 times -0.42

147. The number 0.5 plus the quantity -2 times 0.125

148. The number $-\frac{1}{4}$ minus the quantity 6 times $-\frac{1}{3}$

149. Negative five minus the quantity $\left(-\frac{5}{6}\right)$ times $\frac{3}{8}$

150. For 3 weeks, Jim pays \$2 a week for lottery tickets. Jim has one winning ticket for \$3. Write an expression that describes his net gain or loss. How much money has Jim won or lost?

151. Stephanie pays \$2 a week for 6 weeks for lottery tickets. Stephanie has one winning ticket for \$5. Write an expression that describes her net gain or loss. How much money has Stephanie won or lost?

152. Evaluate the expressions in parts (a) and (b).

a. $-4 - 3 - 2 - 1$

b. $-4(-3)(-2)(-1)$

c. Explain the difference between the operations in parts (a) and (b).

153. Evaluate the expressions in parts (a) and (b).

a. $-10 - 9 - 8 - 7$

b. $-10(-9)(-8)(-7)$

c. Explain the difference between the operations in parts (a) and (b).

Section 1.6 Properties of Real Numbers and Simplifying Expressions

Concepts

1. Commutative Properties of Real Numbers
2. Associative Properties of Real Numbers
3. Identity and Inverse Properties of Real Numbers
4. Distributive Property of Multiplication over Addition
5. Simplifying Algebraic Expressions
6. Clearing Parentheses and Combining *Like* Terms

1. Commutative Properties of Real Numbers

When getting dressed in the morning, it makes no difference whether you put on your left shoe first and then your right shoe, or vice versa. This example illustrates a process in which the order does not affect the outcome. Such a process or operation is said to be *commutative.*

In algebra, the operations of addition and multiplication are commutative because the order in which we add or multiply two real numbers does not affect the result. For example,

$$10 + 5 = 5 + 10 \quad \text{and} \quad 10 \cdot 5 = 5 \cdot 10$$

Commutative Properties of Real Numbers

If a and b are real numbers, then

1. $a + b = b + a$ **commutative property of addition**
2. $ab = ba$ **commutative property of multiplication**

It is important to note that although the operations of addition and multiplication are commutative, subtraction and division are *not* commutative. For example,

$$\underbrace{10 - 5}_{5} \neq \underbrace{5 - 10}_{-5} \quad \text{and} \quad \underbrace{10 \div 5}_{2} \neq \underbrace{5 \div 10}_{\frac{1}{2}}$$

Example 1 Applying the Commutative Property of Addition

Use the commutative property of addition to rewrite each expression.

a. $-3 + (-7)$ **b.** $3x^3 + 5x^4$

Solution:

a. $-3 + (-7) = -7 + (-3)$

b. $3x^3 + 5x^4 = 5x^4 + 3x^3$

Skill Practice Use the commutative property of addition to rewrite each expression.

1. $-5 + 9$ **2.** $7y + x$

Recall that subtraction is not a commutative operation. However, if we rewrite the difference of two numbers, $a - b$, as $a + (-b)$, we can apply the commutative property of addition. This is demonstrated in Example 2.

Skill Practice Answers

1. $9 + (-5)$ **2.** $x + 7y$

Example 2 Applying the Commutative Property of Addition

Rewrite the expression in terms of addition. Then apply the commutative property of addition.

a. $5a - 3b$ **b.** $z^2 - \frac{1}{4}$

Solution:

a. $5a - 3b$

$= 5a + (-3b)$ Rewrite subtraction as addition of $-3b$.

$= -3b + 5a$ Apply the commutative property of addition.

b. $z^2 - \frac{1}{4}$

$= z^2 + \left(-\frac{1}{4}\right)$ Rewrite subtraction as addition of $-\frac{1}{4}$.

$= -\frac{1}{4} + z^2$ Apply the commutative property of addition.

Skill Practice Rewrite each expression in terms of addition. Then apply the commutative property of addition.

3. $8m - 2n$ **4.** $\frac{1}{3}x - \frac{3}{4}$

In most cases, a detailed application of the commutative property will not be shown. Instead the process will be rewritten in one step. For example,

$$5a - 3b = -3b + 5a$$

Example 3 Applying the Commutative Property of Multiplication

Use the commutative property of multiplication to rewrite each expression.

a. $12(-6)$ **b.** $x \cdot 4$

Solution:

a. $12(-6) = -6(12)$

b. $x \cdot 4 = 4 \cdot x$ (or simply $4x$)

Skill Practice Use the commutative property of multiplication to rewrite each expression.

5. $-2(5)$ **6.** $y \cdot 6$

Skill Practice Answers

3. $8m + (-2n)$; $-2n + 8m$

4. $\frac{1}{3}x + \left(-\frac{3}{4}\right)$; $-\frac{3}{4} + \frac{1}{3}x$

5. $5(-2)$ **6.** $6y$

2. Associative Properties of Real Numbers

The associative property of real numbers states that the manner in which three or more real numbers are grouped under addition or multiplication will not affect the outcome. For example,

$$(5 + 10) + 2 = 5 + (10 + 2) \quad \text{and} \quad (5 \cdot 10)2 = 5(10 \cdot 2)$$
$$15 + 2 = 5 + 12 \qquad\qquad (50)2 = 5(20)$$
$$17 = 17 \qquad\qquad 100 = 100$$

Associative Properties of Real Numbers

If a, b, and c represent real numbers, then

1. $(a + b) + c = a + (b + c)$ **associative property of addition**
2. $(ab)c = a(bc)$ **associative property of multiplication**

Example 4 Applying the Associative Property of Multiplication

Use the associative property of multiplication to rewrite each expression. Then simplify the expression if possible.

a. $(5y)y$ **b.** $4(5z)$ **c.** $-\frac{3}{2}\left(-\frac{2}{3}w\right)$

Solution:

a. $(5y)y$

$= 5(y \cdot y)$ Apply the associative property of multiplication.

$= 5y^2$ Simplify.

b. $4(5z)$

$= (4 \cdot 5)z$ Apply the associative property of multiplication.

$= 20z$ Simplify.

c. $-\frac{3}{2}\left(-\frac{2}{3}w\right)$

$= \left[-\frac{3}{2}\left(-\frac{2}{3}\right)\right]w$ Apply the associative property of multiplication.

$= 1w$ Simplify.

$= w$

Note: In most cases, a detailed application of the associative property will not be shown when multiplying two expressions. Instead, the process will be written in one step, such as

$$(5y)y = 5y^2, \quad 4(5z) = 20z, \quad \text{and} \quad -\frac{3}{2}\left(-\frac{2}{3}w\right) = w$$

Skill Practice Use the associative property of multiplication to rewrite each expression. Simplify if possible.

7. $(-3z)z$ **8.** $-2(4x)$ **9.** $\frac{5}{4}\left(\frac{4}{5}t\right)$

Skill Practice Answers

7. $-3(z \cdot z); -3z^2$
8. $(-2 \cdot 4)x; -8x$
9. $\left(\frac{5}{4} \cdot \frac{4}{5}\right)t; t$

3. Identity and Inverse Properties of Real Numbers

The number 0 has a special role under the operation of addition. Zero added to any real number does not change the number. Therefore, the number 0 is said to be the *additive identity* (also called the *identity element of addition*). For example,

$$-4 + 0 = -4 \qquad 0 + 5.7 = 5.7 \qquad 0 + \frac{3}{4} = \frac{3}{4}$$

The number 1 has a special role under the operation of multiplication. Any real number multiplied by 1 does not change the number. Therefore, the number 1 is said to be the *multiplicative identity* (also called the *identity element of multiplication*). For example,

$$(-8)1 = -8 \qquad 1(-2.85) = -2.85 \qquad 1\left(\frac{1}{5}\right) = \frac{1}{5}$$

Identity Properties of Real Numbers

If a is a real number, then

1. $a + 0 = 0 + a = a$ **identity property of addition**

2. $a \cdot 1 = 1 \cdot a = a$ **identity property of multiplication**

The sum of a number and its opposite equals 0. For example, $-12 + 12 = 0$. For any real number, a, the opposite of a (also called the *additive inverse* of a) is $-a$ and $a + (-a) = -a + a = 0$. The inverse property of addition states that the sum of any number and its additive inverse is the identity element of addition, 0. For example,

Number	Additive Inverse (Opposite)	Sum
9	-9	$9 + (-9) = 0$
-21.6	21.6	$-21.6 + 21.6 = 0$
$\frac{2}{7}$	$-\frac{2}{7}$	$\frac{2}{7} + \left(-\frac{2}{7}\right) = 0$

If b is a nonzero real number, then the reciprocal of b (also called the *multiplicative inverse* of b) is $\frac{1}{b}$. The inverse property of multiplication states that the product of b and its multiplicative inverse is the identity element of multiplication, 1. Symbolically, we have $b \cdot \frac{1}{b} = \frac{1}{b} \cdot b = 1$. For example,

Number	Multiplicative Inverse (Reciprocal)	Product
7	$\frac{1}{7}$	$7 \cdot \frac{1}{7} = 1$
3.14	$\frac{1}{3.14}$	$3.14\left(\frac{1}{3.14}\right) = 1$
$-\frac{3}{5}$	$-\frac{5}{3}$	$-\frac{3}{5}\left(-\frac{5}{3}\right) = 1$

Inverse Properties of Real Numbers

If a is a real number and b is a nonzero real number, then

1. $a + (-a) = -a + a = 0$ **inverse property of addition**
2. $b \cdot \frac{1}{b} = \frac{1}{b} \cdot b = 1$ **inverse property of multiplication**

4. Distributive Property of Multiplication over Addition

The operations of addition and multiplication are related by an important property called the **distributive property of multiplication over addition**. Consider the expression $6(2 + 3)$. The order of operations indicates that the sum $2 + 3$ is evaluated first, and then the result is multiplied by 6:

$$\begin{aligned} &6(2 + 3) \\ &= 6(5) \\ &= 30 \end{aligned}$$

TIP: The mathematical definition of the distributive property is consistent with the everyday meaning of the word *distribute*. To distribute means to "spread out from one to many." In the mathematical context, the factor a is distributed to both b and c in the parentheses.

Notice that the same result is obtained if the factor of 6 is multiplied by each of the numbers 2 and 3, and then their products are added:

$6(2 + 3)$ The factor of 6 is *distributed* to the numbers 2 and 3.

$$\begin{aligned} &= 6(2) + 6(3) \\ &= 12 + 18 \\ &= 30 \end{aligned}$$

The distributive property of multiplication over addition states that this is true in general.

Distributive Property of Multiplication over Addition

If a, b, and c are real numbers, then

$$a(b + c) = ab + ac \quad \text{and} \quad (b + c)a = ab + ac$$

Example 5 Applying the Distributive Property

Apply the distributive property: $2(a + 6b + 7)$

Solution:

$2(a + 6b + 7)$

$= 2(a + 6b + 7)$

$= 2(a) + 2(6b) + 2(7)$ Apply the distributive property.

$= 2a + 12b + 14$ Simplify.

TIP: Notice that the parentheses are removed after the distributive property is applied. Sometimes this is referred to as *clearing parentheses*.

Skill Practice

10. Apply the distributive property.

$7(x + 4y + z)$

Because the difference of two expressions $a - b$ can be written in terms of addition as $a + (-b)$, the distributive property can be applied when the operation of subtraction is present within the parentheses. For example,

$5(y - 7)$

$= 5[y + (-7)]$ Rewrite subtraction as addition of -7.

$= 5[y + (-7)]$ Apply the distributive property.

$= 5(y) + 5(-7)$

$= 5y + (-35)$, or $5y - 35$ Simplify.

Example 6 Applying the Distributive Property

Use the distributive property to rewrite each expression.

a. $-(-3a + 2b + 5c)$ **b.** $-6(2 - 4x)$

Solution:

a. $-(-3a + 2b + 5c)$

$= -1(-3a + 2b + 5c)$ The negative sign preceding the parentheses can be interpreted as taking the opposite of the quantity that follows or as $-1(-3a + 2b + 5c)$

$= -1(-3a + 2b + 5c)$

$= -1(-3a) + (-1)(2b) + (-1)(5c)$ Apply the distributive property.

$= 3a + (-2b) + (-5c)$ Simplify.

$= 3a - 2b - 5c$

TIP: Notice that a negative factor preceding the parentheses changes the signs of all the terms to which it is multiplied.

$-1(-3a + 2b + 5c)$

$= +3a - 2b - 5c$

b. $-6(2 - 4x)$

$= -6[2 + (-4x)]$ Change subtraction to addition of $-4x$.

$= -6[2 + (-4x)]$

$= -6(2) + (-6)(-4x)$ Apply the distributive property. Notice that multiplying by -6 changes the signs of all terms to which it is applied.

$= -12 + 24x$ Simplify.

Skill Practice Use the distributive property to rewrite each expression.

11. $-6(-3a + 7b)$ **12.** $-(12x + 8y - 3z)$

Skill Practice Answers

10. $7x + 28y + 7z$
11. $18a - 42b$
12. $-12x - 8y + 3z$

Note: In most cases, the distributive property will be applied without as much detail as shown in Examples 5 and 6. Instead, the distributive property will be applied in one step.

$2(a + 6b + 7)$	$-(3a + 2b + 5c)$	$-6(2 - 4x)$
1 step $= 2a + 12b + 14$	1 step $= -3a - 2b - 5c$	1 step $= -12 + 24x$

5. Simplifying Algebraic Expressions

An algebraic expression is the sum of one or more terms. A term is a constant or the product of a constant and one or more variables. For example, the expression

$$-7x^2 + xy - 100 \quad \text{or} \quad -7x^2 + xy + (-100)$$

consists of the terms $-7x^2$, xy, and -100.

The terms $-7x^2$ and xy are **variable terms** and the term -100 is called a **constant term**. It is important to distinguish between a term and the factors within a term. For example, the quantity xy is one term, and the values x and y are factors within the term. The constant factor in a term is called the *numerical coefficient* (or simply **coefficient**) of the term. In the terms $-7x^2$, xy, and -100, the coefficients are -7, 1, and -100, respectively.

Terms are said to be *like* terms if they each have the same variables and the corresponding variables are raised to the same powers. For example,

***Like* Terms**			**Unlike Terms**			
$-3b$	and	$5b$	$-5c$	and	$7d$	(different variables)
$17xy$	and	$-4xy$	$6xy$	and	$3x$	(different variables)
$9p^2q^3$	and	p^2q^3	$4p^2q^3$	and	$8p^3q^2$	(different powers)
$5w$	and	$2w$	$5w$	and	2	(different variables)
7	and	10	7	and	$10a$	(different variables)

Example 7 **Identifying Terms, Factors, Coefficients, and *Like* Terms**

a. List the terms of the expression $5x^2 - 3x + 2$.

b. Identify the coefficient of the term $6yz^3$.

c. Which of the pairs are *like* terms: $8b, 3b^2$ or $4c^2d, -6c^2d$?

Solution:

a. The terms of the expression $5x^2 - 3x + 2$ are $5x^2$, $-3x$, and 2.

b. The coefficient of $6yz^3$ is 6.

c. $4c^2d$ and $-6c^2d$ are *like* terms.

Skill Practice

13. List the terms in the expression. $4xy - 9x^2 + 15$

14. Identify the coefficients of each term in the expression. $2a - 5b + c - 80$

15. Which of the pairs are *like* terms? $5x^3, 5x$ or $-7x^2, 11x^2$

Skill Practice Answers

13. $4xy, -9x^2, 15$

14. $2, -5, 1, -80$

15. $-7x^2$ and $11x^2$ are *like* terms.

6. Clearing Parentheses and Combining *Like* Terms

Two terms can be added or subtracted only if they are *like* terms. To add or subtract *like* terms, we use the distributive property as shown in the next example.

Example 8 **Using the Distributive Property to Add and Subtract *Like* Terms**

Add or subtract as indicated.

a. $7x + 2x$ **b.** $-2p + 3p - p$

Solution:

a. $7x + 2x$

$= (7 + 2)x$ Apply the distributive property.

$= 9x$ Simplify.

b. $-2p + 3p - p$

$= -2p + 3p - 1p$ Note that $-p$ equals $-1p$.

$= (-2 + 3 - 1)p$ Apply the distributive property.

$= (0)p$ Simplify.

$= 0$

Skill Practice Simplify by adding *like* terms.

16. $8x + 3x$ **17.** $-6a + 4a + a$

Although the distributive property is used to add and subtract *like* terms, it is tedious to write each step. Observe that adding or subtracting *like* terms is a matter of combining the coefficients and leaving the variable factors unchanged. This can be shown in one step, a shortcut that we will use throughout the text. For example,

$$7x + 2x = 9x \qquad -2p + 3p - 1p = 0p = 0 \qquad -3a - 6a = -9a$$

Example 9 **Adding and Subtracting *Like* Terms**

a. $3yz + 5 - 2yz + 9$ **b.** $1.2w^3 + 5.7w^3$

Solution:

a. $3yz + 5 - 2yz + 9$

$= 3yz - 2yz + 5 + 9$ Arrange *like* terms together. Notice that constants such as 5 and 9 are *like* terms.

$= 1yz + 14$ Combine *like* terms.

$= yz + 14$

Skill Practice Answers

16. $11x$ **17.** $-a$

b. $1.2w^3 + 5.7w^3$

$= 6.9w^3$ Combine *like* terms.

Skill Practice Simplify by adding *like* terms.

18. $4q + 3w - 2q - 3w + 2q$ **19.** $8x^2 + x + 5 + 4x - 3 + 2x^2$

Examples 10 and 11 illustrate how the distributive property is used to clear parentheses.

Example 10 **Clearing Parentheses and Combining *Like* Terms**

Simplify by clearing parentheses and combining *like* terms: $5 - 2(3x + 7)$

Solution:

$5 - 2(3x + 7)$ The order of operations indicates that we must perform multiplication before subtraction.

It is important to understand that a factor of -2 (not 2) will be multiplied to all terms within the parentheses. To see why this is so, we can rewrite the subtraction in terms of addition.

$= 5 + (-2)(3x + 7)$ Change subtraction to addition.

$= 5 + (-2)(3x + 7)$ A factor of -2 is to be distributed to terms in the parentheses.

$= 5 + (-2)(3x) + (-2)(7)$ Apply the distributive property.

$= 5 + (-6x) + (-14)$ Simplify.

$= 5 + (-14) + (-6x)$ Arrange *like* terms together.

$= -9 + (-6x)$ Combine *like* terms.

$= -9 - 6x$ Simplify by changing addition of the opposite to subtraction.

Skill Practice Clear the parentheses and combine *like* terms.

20. $3(2x - 1) + 4(x + 3)$

Example 11 **Clearing Parentheses and Combining *Like* Terms**

Simplify by clearing parentheses and combining *like* terms.

a. $10(5y + 2) - 6(y - 1)$ **b.** $\frac{1}{4}(4k + 2) - \frac{1}{2}(6k + 1)$

c. $-(4s - 6t) - (3t + 5s) - 2s$

Skill Practice Answers

18. $4q$

19. $10x^2 + 5x + 2$

20. $10x + 9$

Solution:

a. $10(5y + 2) - 6(y - 1)$

$= 50y + 20 - 6y + 6$ — Apply the distributive property. Notice that a factor of -6 is distributed through the second parentheses and changes the signs.

$= 50y - 6y + 20 + 6$ — Arrange *like* terms together.

$= 44y + 26$ — Combine *like* terms.

b. $\frac{1}{4}(4k + 2) - \frac{1}{2}(6k + 1)$

$= \frac{4}{4}k + \frac{2}{4} - \frac{6}{2}k - \frac{1}{2}$ — Apply the distributive property. Notice that a factor of $-\frac{1}{2}$ is distributed through the second parentheses and changes the signs.

$= k + \frac{1}{2} - 3k - \frac{1}{2}$ — Simplify fractions.

$= k - 3k + \frac{1}{2} - \frac{1}{2}$ — Arrange *like* terms together.

$= -2k + 0$ — Combine *like* terms.

$= -2k$

c. $-(4s - 6t) - (3t + 5s) - 2s$

$= -4s + 6t - 3t - 5s - 2s$ — Apply the distributive property.

$= -4s - 5s - 2s + 6t - 3t$ — Arrange *like* terms together.

$= -11s + 3t$ — Combine *like* terms.

Skill Practice Clear the parentheses and combine *like* terms.

21. $5(2y + 3) - 2(3y + 1)$

22. $\frac{1}{2}(8x + 1) + \frac{1}{3}(3x - 2)$

23. $-4(x + 2y) - (2x - y) - 5x$

Skill Practice Answers

21. $4y + 13$ **22.** $5x - \frac{1}{6}$

23. $-11x - 7y$

Section 1.6 Practice Exercises

Boost *your* GRADE at mathzone.com!

- Practice Problems
- Self-Tests
- NetTutor
- e-Professors
- Videos

Study Skills Exercises

1. Write down the page number(s) for the Chapter Summary for this chapter. Describe one way in which you can use the Summary found at the end of each chapter.

2. Define the key terms:

a. commutative properties **b. associative properties** **c. identity properties**

d. inverse properties **e. distributive property of multiplication over addition**

f. variable term **g. constant term** **h. coefficient** **i. *like* terms**

Review Exercises

For Exercises 3–18, perform the indicated operations.

3. $(-6) + 14$ **4.** $(-2) + 9$ **5.** $-13 - (-5)$ **6.** $-1 - (-19)$

7. $18 \div (-4)$ **8.** $-27 \div 5$ **9.** $-3 \cdot 0$ **10.** $0(-15)$

11. $\frac{1}{2} + \frac{3}{8}$ **12.** $\frac{7}{2} + \frac{5}{9}$ **13.** $\frac{25}{21} - \frac{6}{7}$ **14.** $\frac{8}{9} - \frac{1}{3}$

15. $\left(-\frac{3}{5}\right)\left(\frac{4}{27}\right)$ **16.** $\left(\frac{1}{6}\right)\left(-\frac{8}{3}\right)$ **17.** $\left(-\frac{11}{12}\right) \div \left(-\frac{5}{4}\right)$ **18.** $\left(-\frac{14}{15}\right) \div \left(-\frac{7}{5}\right)$

Concept 1: Commutative Properties of Real Numbers

For Exercises 19–26, rewrite each expression using the commutative property of addition or the commutative property of multiplication.

19. $5 + (-8)$ **20.** $7 + (-2)$ **21.** $8 + x$ **22.** $p + 11$

23. $5(4)$ **24.** $10(8)$ **25.** $x(-12)$ **26.** $y(-23)$

For Exercises 27–30, rewrite each expression using addition. Then apply the commutative property of addition.

27. $x - 3$ **28.** $y - 7$ **29.** $4p - 9$ **30.** $3m - 12$

Concept 2: Associative Properties of Real Numbers

For Exercises 31–38, use the associative property of multiplication to rewrite each expression. Then simplify the expression if possible.

31. $(4p)p$ **32.** $(-6y)y$ **33.** $-5(3x)$ **34.** $-12(4z)$

35. $\frac{6}{11}\left(\frac{11}{6}x\right)$ **36.** $\frac{3}{5}\left(\frac{5}{3}x\right)$ **37.** $-4\left(-\frac{1}{4}t\right)$ **38.** $-5\left(-\frac{1}{5}w\right)$

Concept 3: Identity and Inverse Properties of Real Numbers

39. What is another name for multiplicative inverse? **40.** What is another name for additive inverse?

41. What is the additive identity? **42.** What is the multiplicative identity?

Concept 4: The Distributive Property of Multiplication over Addition

For Exercises 43–64, use the distributive property to clear parentheses.

43. $6(5x + 1)$ **44.** $2(x + 7)$ **45.** $-2(a + 8)$ **46.** $-3(2z + 9)$

47. $3(5c - d)$ **48.** $4(w - 13z)$ **49.** $-7(y - 2)$ **50.** $-2(4x - 1)$

51. $-\frac{2}{3}(x - 6)$ **52.** $-\frac{1}{4}(2b - 8)$ **53.** $\frac{1}{3}(m - 3)$ **54.** $\frac{2}{5}(n - 5)$

55. $\frac{3}{8}(4 + 8s)$ **56.** $\frac{4}{9}(3 - 9t)$ **57.** $-(2p + 10)$ **58.** $-(7q + 1)$

59. $-(-3w - 5z)$ **60.** $-(-7a - b)$ **61.** $4(x + 2y - z)$ **62.** $-6(2a - b + c)$

63. $-(-6w + x - 3y)$ **64.** $-(-p - 5q - 10r)$

For Exercises 65–68, use the associative property or distributive property to clear parentheses.

65. $2(3 + x)$ **66.** $5(4 + y)$ **67.** $4(6z)$ **68.** $8(2p)$

For Exercises 69–77, match the statements with the properties of multiplication and addition.

69. $6 \cdot \frac{1}{6} = 1$

70. $7(4 \cdot 9) = (7 \cdot 4)9$

71. $2(3 + k) = 6 + 2k$

72. $3 \cdot 7 = 7 \cdot 3$

73. $5 + (-5) = 0$

74. $18 \cdot 1 = 18$

75. $(3 + 7) + 19 = 3 + (7 + 19)$

76. $23 + 6 = 6 + 23$

77. $3 + 0 = 3$

a. Commutative property of addition
b. Inverse property of multiplication
c. Commutative property of multiplication
d. Associative property of addition
e. Identity property of multiplication
f. Associative property of multiplication
g. Inverse property of addition
h. Identity property of addition
i. Distributive property of multiplication over addition

Concept 5: Simplifying Algebraic Expressions

For Exercises 78–81, list the terms and their coefficients for each expression.

78. $3xy - 6x^2 + y - 17$

Term	Coefficient

79. $2x - y + 18xy + 5$

Term	Coefficient

80. $x^4 - 10xy + 12 - y$

Term	Coefficient

81. $-x + 8y - 9x^2y - 3$

Term	Coefficient

82. Explain why $12x$ and $12x^2$ are not *like* terms.

83. Explain why $3x$ and $3xy$ are not *like* terms.

84. Explain why $7z$ and $\sqrt{13}z$ are *like* terms.

85. Explain why πx and $8x$ are *like* terms.

86. Write three different *like* terms.

87. Write three terms that are not *like*.

Concept 6: Clearing Parentheses and Combining *Like* Terms

For Exercises 88–97, simplify by combining *like* terms.

88. $5k - 10k - 12k + 16 + 7$

89. $-4p - 2p + 8p - 15 + 3$

90. $9x - 7x^2 + 12x + 14x^2$

91. $2y^2 - 8y + y - 5y^2 - 3y^2$

92. $4a + 2a^2 - 6a + 5 + 3a^2 - 2$

93. $8x^2 - 5x + 3 - 7 + 6x - x^2$

94. $\frac{1}{4}a + b - \frac{3}{4}a - 5b$

95. $\frac{2}{5} + 2t - \frac{3}{5} + t - \frac{6}{5}$

96. $2.8z - 8.1z + 6 - 15.2$

97. $2.4 - 8.4w - 2w + 0.9$

For Exercises 98–121, simplify by clearing parentheses and combining *like* terms.

98. $-3(2x - 4) + 10$

99. $-2(4a + 3) - 14$

100. $4(w + 3) - 12$

101. $5(2r + 6) - 30$

102. $5 - 3(x - 4)$

103. $4 - 2(3x + 8)$

104. $-3(2t + 4) + 8(2t - 4)$

105. $-5(5y + 9) + 3(3y + 6)$

106. $2(w - 5) - (2w + 8)$

107. $6(x + 3) - (6x - 5)$

108. $-\frac{1}{3}(6t + 9) + 10$

109. $-\frac{3}{4}(8 + 4q) + 7$

110. $10(5.1a - 3.1) + 4$

111. $100(-3.14p - 1.05) + 212$

112. $-4m + 2(m - 3) + 2m$

113. $-3b + 4(b + 2) - 8b$

114. $\frac{1}{2}(10q - 2) + \frac{1}{3}(2 - 3q)$

115. $\frac{1}{5}(15 - 4p) - \frac{1}{10}(10p + 5)$

116. $7n - 2(n - 3) - 6 + n$

117. $8k - 4(k - 1) + 7 - k$

118. $6(x + 3) - 12 - 4(x - 3)$

119. $5(y - 4) + 3 - 6(y - 7)$

120. $6.1(5.3z - 4.1) - 5.8$

121. $-3.6(1.7q - 4.2) + 14.6$

Expanding Your Skills

For Exercises 122–131, determine if the expressions are equivalent. If two expressions are not equivalent, state why.

122. $3a + b, b + 3a$

123. $4y + 1, 1 + 4y$

124. $2c + 7, 9c$

125. $5z + 4, 9z$

126. $5x - 3, 3 - 5x$

127. $6d - 7, 7 - 6d$

128. $5x - 3, -3 + 5x$

129. $8 - 2x, -2x + 8$

130. $5y + 6, 5 + 6y$

131. $7z - 2; 7 - 2z$

132. Which grouping of terms is easier computationally, $(14\frac{2}{7} + 2\frac{1}{3}) + \frac{2}{3}$ or $14\frac{2}{7} + (2\frac{1}{3} + \frac{2}{3})$?

133. Which grouping of terms is easier computationally, $(5\frac{1}{8} + 18\frac{2}{5}) + 1\frac{3}{5}$ or $5\frac{1}{8} + (18\frac{2}{5} + 1\frac{3}{5})$?

134. As a small child in school, the great mathematician Karl Friedrich Gauss (1777–1855) was said to have found the sum of the integers from 1 to 100 mentally:

$$1 + 2 + 3 + 4 + \cdots + 99 + 100$$

Rather than adding the numbers sequentially, he added the numbers in pairs:

$$(1 + 99) + (2 + 98) + (3 + 97) + \cdots$$

a. Use this technique to add the integers from 1 to 10.

$$1 + 2 + 3 + 4 + 5 + 6 + 7 + 8 + 9 + 10$$

b. Use this technique to add the integers from 1 to 20.

Chapter 1 SUMMARY

Section 1.1 Sets of Numbers and the Real Number Line

Key Concepts

Natural numbers: $\{1, 2, 3, \ldots\}$

Whole numbers: $\{0, 1, 2, 3, \ldots\}$

Integers: $\{\ldots -3, -2, -1, 0, 1, 2, 3, \ldots\}$

Rational numbers: The set of numbers that can be expressed in the form $\frac{p}{q}$, where p and q are integers and q does not equal 0.

Irrational numbers: The set of real numbers that are not rational.

Real numbers: The set of both the rational numbers and the irrational numbers.

$a < b$	"a is less than b."
$a > b$	"a is greater than b."
$a \leq b$	"a is less than or equal to b."
$a \geq b$	"a is greater than or equal to b."

Two numbers that are the same distance from zero but on opposite sides of zero on the number line are called **opposites**. The opposite of a is denoted $-a$.

The **absolute value** of a real number, a, denoted $|a|$, is the distance between a and 0 on the number line.

If $a \geq 0$, $|a| = a$

If $a < 0$, $|a| = -a$

Examples

Example 1

-5, 0, and 4 are integers.

$-\frac{5}{2}$, -0.5, and $0.\overline{3}$ are rational numbers.

$\sqrt{7}$, $-\sqrt{2}$, and π are irrational numbers.

Example 2

All real numbers can be located on a real number line.

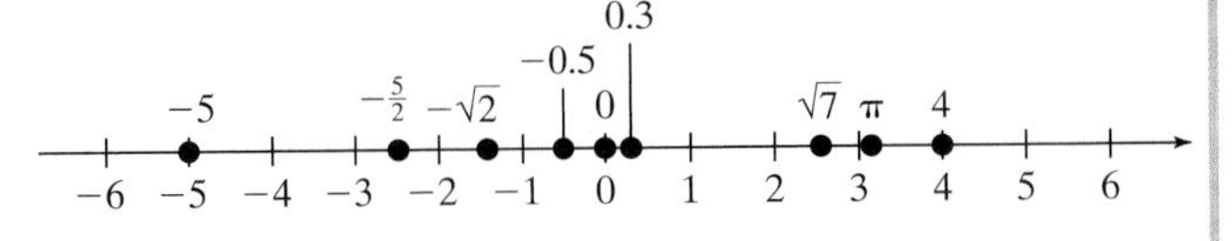

Example 3

$5 < 7$	"5 is less than 7."
$-2 > -10$	"-2 is greater than -10."
$y \leq 3.4$	"y is less than or equal to 3.4."
$x \geq \frac{1}{2}$	"x is greater than or equal to $\frac{1}{2}$."

Example 4

5 and -5 are opposites.

Example 5

$|7| = 7$

$|-7| = 7$

Section 1.2 Order of Operations

Key Concepts

A **variable** is a symbol or letter used to represent an unknown number.

A **constant** is a value that is not variable.

An algebraic **expression** is a collection of variables and constants under algebraic operations.

$b^n = \underbrace{b \cdot b \cdot b \cdot b \ldots \cdot b}_{n \text{ factors of } b}$ b is the **base**, n is the **exponent**

$\sqrt{x}$ is the positive **square root** of x.

The Order of Operations:

1. Simplify expressions within parentheses and other grouping symbols first.
2. Evaluate expressions involving exponents and radicals.
3. Do multiplication or division in the order that they occur from left to right.
4. Do addition or subtraction in the order that they occur from left to right.

Examples

Example 1

Variables: x, y, z, a, b

Constants: $2, -3, \pi$

Expressions: $2x + 5, 3a + b^2$

Example 2

$5^3 = 5 \cdot 5 \cdot 5 = 125$

Example 3

$\sqrt{49} = 7$

Example 4

$10 + 5(3 - 1)^2 - \sqrt{5 - 1}$

$= 10 + 5(2)^2 - \sqrt{4}$ Work within grouping symbols.

$= 10 + 5(4) - 2$ Simplify exponents.

$= 10 + 20 - 2$ Perform multiplication.

$= 30 - 2$ Add and subtract, left to right.

$= 28$

Section 1.3 Addition of Real Numbers

Key Concepts

Addition of Two Real Numbers:

Same Signs: Add the absolute values of the numbers and apply the common sign to the sum.

Different Signs: Subtract the smaller absolute value from the larger absolute value. Then apply the sign of the number having the larger absolute value.

Examples

Example 1

$-3 + (-4) = -7$

$-1.3 + (-9.1) = -10.4$

Example 2

$-5 + 7 = 2$

$\frac{2}{3} + \left(-\frac{7}{3}\right) = -\frac{5}{3}$

Section 1.4 Subtraction of Real Numbers

Key Concepts

Subtraction of Two Real Numbers:

Add the opposite of the second number to the first number. That is,

$a - b = a + (-b)$

Examples

Example 1

$7 - (-5) = 7 + (5) = 12$

$-3 - 5 = -3 + (-5) = -8$

$-11 - (-2) = -11 + (2) = -9$

Section 1.5 Multiplication and Division of Real Numbers

Key Concepts

Multiplication and Division of Two Real Numbers:

Same Signs:
Product is positive.
Quotient is positive.

Different Signs:
Product is negative.
Quotient is negative.

The **reciprocal** of a number a is $\frac{1}{a}$.

Multiplication and Division Involving Zero:

The product of any real number and 0 is 0.

The quotient of 0 and any nonzero real number is 0.

The quotient of any nonzero real number and 0 is undefined.

Examples

Example 1

$(-5)(-2) = 10$

$\frac{-20}{-4} = 5$

Example 2

$(-3)(7) = -21$

$\frac{-4}{8} = -\frac{1}{2}$

Example 3

The reciprocal of -6 is $-\frac{1}{6}$.

Example 4

$4 \cdot 0 = 0$

$0 \div 4 = 0$

$4 \div 0$ is undefined.

Section 1.6 Properties of Real Numbers and Simplifying Expressions

Key Concepts

The Properties of Real Numbers:

Commutative Property of Addition:

$a + b = b + a$

Associative Property of Addition:

$(a + b) + c = a + (b + c)$

Examples

Example 1

$-5 + (-7) = (-7) + (-5)$

Example 2

$(2 + 3) + 10 = 2 + (3 + 10)$

Identity Property of Addition: The number 0 is said to be the identity element for addition because:

$0 + a = a$ and $a + 0 = a$

Example 3

$0 + \frac{3}{4} = \frac{3}{4}$ and $\frac{3}{4} + 0 = \frac{3}{4}$

Inverse Property of Addition:

$a + (-a) = 0$ and $-a + a = 0$

Example 4

$1.5 + (-1.5) = 0$ and $-1.5 + 1.5 = 0$

Commutative Property of Multiplication:

$ab = ba$

Example 5

$(-3)(-4) = (-4)(-3)$

Associative Property of Multiplication:

$(ab)c = a(bc)$

Example 6

$(2 \cdot 3)6 = 2(3 \cdot 6)$

Identity Property of Multiplication: The number 1 is said to be the identity element for multiplication because:

$1 \cdot a = a$ and $a \cdot 1 = a$

Example 7

$1 \cdot 5 = 5$ and $5 \cdot 1 = 5$

Inverse Property of Multiplication:

$a \cdot \frac{1}{a} = 1$ and $\frac{1}{a} \cdot a = 1$

Example 8

$6 \cdot \frac{1}{6} = 1$ and $\frac{1}{6} \cdot 6 = 1$

The Distributive Property of Multiplication over Addition:

$a(b + c) = ab + ac$

Example 9

$2(x + 4y) = 2x + 8y$

$-(a + 6b - 5c) = -a - 6b + 5c$

A **term** is a constant or the product of a constant and one or more variables.

The **coefficient** of a term is the numerical factor of the term.

Example 10

$-2x$ is a term with coefficient -2.

yz^2 is a term with coefficient 1.

Like terms have the same variables, and the corresponding variables have the same powers.

Example 11

$3x$ and $-5x$ are *like* terms.

$4a^2b$ and $4ab$ are not *like* terms.

Two terms can be added or subtracted if they are *like* terms. Sometimes it is necessary to clear parentheses before adding or subtracting *like* terms.

Example 12

$-4d + 12d + d = 9d$

Example 13

$-2w - 4(w - 2) + 3$ Use the distributive property.

$= -2w - 4w + 8 + 3$ Combine *like* terms.

$= -6w + 11$

Chapter 1 Review Exercises

Section 1.1

1. Given the set $\{7, \frac{1}{3}, -4, 0, -\sqrt{3}, -0.\overline{2}, \pi, 1\}$,
   a. List the natural numbers.
   b. List the integers.
   c. List the whole numbers.
   d. List the rational numbers.
   e. List the irrational numbers.
   f. List the real numbers.

For Exercises 2–5, determine the absolute values.

2. $\left|\frac{1}{2}\right|$ 3. $|-6|$ 4. $|-\sqrt{7}|$ 5. $|0|$

For Exercises 6–13, identify whether the inequality is true or false.

6. $-6 > -1$ 7. $0 < -5$ 8. $-10 \le 0$

9. $5 \ne -5$ 10. $7 \ge 7$ 11. $7 \ge -7$

12. $0 \le -3$ 13. $-\frac{2}{3} \le -\frac{2}{3}$

Section 1.2

For Exercises 14–23, translate the English phrases into algebraic expressions.

14. The product of x and $\frac{2}{3}$
15. The quotient of 7 and y
16. The sum of 2 and $3b$
17. The difference of a and 5
18. Two more than $5k$
19. Seven less than $13z$
20. The quotient of 6 and x, decreased by 18
21. The product of y and 3, increased by 12
22. Three-eighths subtracted from z
23. Five subtracted from two times p

For Exercises 24–29, simplify the expressions.

24. 6^3 25. 15^2 26. $\sqrt{36}$

27. $\left(\frac{1}{4}\right)^2$ 28. $\frac{1}{\sqrt{100}}$ 29. $\left(\frac{3}{2}\right)^3$

For Exercises 30–33, perform the indicated operations.

30. $15 - 7 \cdot 2 + 12$ 31. $|-11| + |5| - (7 - 2)$

32. $4^2 - (5 - 2)^2$ 33. $22 - 3(8 \div 4)^2$

Section 1.3

For Exercises 34–46, add the rational numbers.

34. $-6 + 8$ 35. $14 + (-10)$

36. $21 + (-6)$ 37. $-12 + (-5)$

38. $\frac{2}{7} + \left(-\frac{1}{9}\right)$ 39. $\left(-\frac{8}{11}\right) + \left(\frac{1}{2}\right)$

40. $\left(-\frac{1}{10}\right) + \left(-\frac{5}{6}\right)$ 41. $\left(-\frac{5}{2}\right) + \left(-\frac{1}{5}\right)$

42. $-8.17 + 6.02$ 43. $2.9 + (-7.18)$

44. $13 + (-2) + (-8)$ 45. $-5 + (-7) + 20$

46. $2 + 5 + (-8) + (-7) + 0 + 13 + (-1)$

47. Under what conditions will the expression $a + b$ be negative?

48. The high temperatures (in degrees Celsius) for the province of Alberta, Canada, during a week in January were $-8, -1, -4, -3, -4, 0$, and 7. What was the average high temperature for that week? Round to the nearest tenth of a degree.

Section 1.4

For Exercises 49–61, subtract the rational numbers.

49. $13 - 25$

50. $31 - (-2)$

51. $-8 - (-7)$

52. $-2 - 15$

53. $\left(-\frac{7}{9}\right) - \frac{5}{6}$

54. $\frac{1}{3} - \frac{9}{8}$

55. $7 - 8.2$

56. $-1.05 - 3.2$

57. $-16.1 - (-5.9)$

58. $7.09 - (-5)$

59. $\frac{11}{2} - \left(-\frac{1}{6}\right) - \frac{7}{3}$

60. $-\frac{4}{5} - \frac{7}{10} - \left(-\frac{13}{20}\right)$

61. $6 - 14 - (-1) - 10 - (-21) - 5$

62. Under what conditions will the expression $a - b$ be negative?

For Exercises 63–67, write an algebraic expression and simplify.

63. -18 subtracted from -7

64. The difference of -6 and 41

65. Seven decreased by 13

66. Five subtracted from the difference of 20 and -7

67. The sum of 6 and -12, decreased by 21

68. In Nevada, the highest temperature ever recorded was 125°F and the lowest was -50°F. Find the difference between the highest and lowest temperatures. (*Source: Information Please Almanac*)

Section 1.5

For Exercises 69–88, multiply or divide as indicated.

69. $10(-17)$

70. $(-7)13$

71. $(-52) \div 26$

72. $(-48) \div (-16)$

73. $\frac{7}{4} \div \left(-\frac{21}{2}\right)$

74. $\frac{2}{3}\left(-\frac{12}{11}\right)$

75. $-\frac{21}{5} \cdot 0$

76. $\frac{3}{4} \div 0$

77. $0 \div (-14)$

78. $\frac{0}{3} \cdot \frac{1}{8}$

79. $(-0.45)(-5)$

80. $(-2.1) \div (-0.07)$

81. $\frac{-21}{14}$

82. $\frac{-13}{-52}$

83. $(5)(-2)(3)$

84. $(-6)(-5)(15)$

85. $\left(-\frac{1}{2}\right)\left(\frac{7}{8}\right)\left(-\frac{4}{7}\right)$

86. $\left(\frac{12}{13}\right)\left(-\frac{1}{6}\right)\left(\frac{13}{14}\right)$

87. $40 \div 4 \div (-5)$

88. $\frac{10}{11} \div \frac{7}{11} \div \frac{5}{9}$

For Exercises 89–92, perform the indicated operations.

89. $9 - 4[-2(4 - 8) - 5(3 - 1)]$

90. $\frac{8(-3) - 6}{-7 - (-2)}$

91. $\frac{2}{3} - \left(\frac{3}{8} + \frac{5}{6}\right) \div \frac{5}{3}$

92. $5.4 - (0.3)^2 \div 0.09$

For Exercises 93–96, evaluate the expressions with the given substitutions.

93. $3(x + 2) \div y$ for $x = 4$ and $y = -9$

94. $a^2 - bc$ for $a = -6$, $b = 5$, and $c = 2$

95. $w + xy - \sqrt{z}$ for $w = 12$, $x = 6$, $y = -5$, and $z = 25$

96. $(u - v)^2 + (u^2 - v^2)$ for $u = 5$ and $v = -3$

97. In statistics, the formula $x = \mu + z\sigma$ is used to find cutoff values for data that follow a bell-shaped curve. Find x if $\mu = 100$, $z = -1.96$, and $\sigma = 15$.

For Exercises 98–104, answer true or false. If a statement is false, explain why.

98. If n is positive, then $-n$ is negative.

99. If m is negative, then m^4 is negative.

100. If m is negative, then m^3 is negative.

101. If $m > 0$ and $n > 0$, then $mn > 0$.

102. If $p < 0$ and $q < 0$, then $pq < 0$.

103. A number and its reciprocal have the same signs.

104. A nonzero number and its opposite have different signs.

Section 1.6

For Exercises 105–112, answers may vary.

105. Give an example of the commutative property of addition.

106. Give an example of the associative property of addition.

107. Give an example of the inverse property of addition.

108. Give an example of the identity property of addition.

109. Give an example of the commutative property of multiplication.

110. Give an example of the associative property of multiplication.

111. Give an example of the inverse property of multiplication.

112. Give an example of the identity property of multiplication.

113. Explain why $5x - 2y$ is the same as $-2y + 5x$.

114. Explain why $3a - 9y$ is the same as $-9y + 3a$.

115. List the terms of the expression: $3y + 10x - 12 + xy$

116. Identify the coefficients for the terms listed in Exercise 115.

117. Simplify each expression by combining *like* terms.

a. $3a + 3b - 4b + 5a - 10$

b. $-6p + 2q + 9 - 13q - p + 7$

118. Use the distributive property to clear the parentheses.

a. $-2(4z + 9)$ **b.** $5(4w - 8y + 1)$

For Exercises 119–124, simplify the expressions.

119. $2p - (p + 5) + 3$

120. $6(h + 3) - 7h - 4$

121. $\frac{1}{2}(-6q) + q - 4\left(3q + \frac{1}{4}\right)$

122. $0.3b + 12(0.2 - 0.5b)$

123. $-4[2(x + 1) - (3x + 8)]$

124. $5[(7y - 3) + 3(y + 8)]$

Chapter 1 Test

1. Is $0.\overline{315}$ a rational number or irrational number? Explain your reasoning.

2. Plot the points on a number line: $|3|, 0, -2, 0.5, |-\frac{3}{2}|, \sqrt{16}$.

0

3. Use the number line in Exercise 2 to identify whether the statements are true or false.

a. $|3| < -2$ **b.** $0 \leq \left|-\frac{3}{2}\right|$

c. $-2 < 0.5$ **d.** $|3| \geq \left|-\frac{3}{2}\right|$

4. Use the definition of exponents to expand the expressions:

a. $(4x)^3$ **b.** $4x^3$

5. a. Translate the expression into an English phrase: $2(a - b)$. (Answers may vary.)

b. Translate the expression into an English phrase: $2a - b$. (Answers may vary.)

6. Translate the phrase into an algebraic expression: "The quotient of the principal square root of c and the square of d."

For Exercises 7–23, perform the indicated operations.

7. $18 + (-12)$

8. $-10 + (-9)$

9. $-15 - (-3)$

10. $21 - (-7)$

11. $-\frac{1}{8} + \left(-\frac{3}{4}\right)$

12. $-10.06 - (-14.72)$

13. $-14 + (-2) - 16$

14. $-84 \div 7$

15. $38 \div 0$

16. $7(-4)$

17. $-22 \cdot 0$

18. $(-16)(-2)(-1)(-3)$

19. $\frac{2}{5} \div \left(-\frac{7}{10}\right) \cdot \left(-\frac{7}{6}\right)$

20. $(8 - 10)\frac{3}{2} + (-5)$

21. $8 - [(2 - 4) - (8 - 9)]$

22. $\dfrac{\sqrt{5^2 - 4^2}}{|-12 + 3|}$

23. $\dfrac{|4 - 10|}{2 - 3(5 - 1)}$

24. Identify the property that justifies each statement.

a. $6(-8) = (-8)6$

b. $5 + 0 = 5$

c. $(2 + 3) + 4 = 2 + (3 + 4)$

d. $\frac{1}{7} \cdot 7 = 1$

e. $8[7(-3)] = (8 \cdot 7)(-3)$

For Exercises 25–29, simplify the expression.

25. $-5x - 4y + 3 - 7x + 6y - 7$

26. $-3(4m + 8p - 7)$

27. $-4[2(5 - x) - 8x] - 9$

28. $4(p - 5) - (8p + 3)$

29. $\frac{1}{2}(12p - 4) + \frac{1}{3}(2 - 6p)$

For Exercises 30–33, evaluate the expressions given the values $x = 4$, $y = -3$, and $z = -7$.

30. $y^2 - x$

31. $3x - 2y$

32. $y(x - 2)$

33. $-y^2 - 4x + z$

For Exercises 34–35, translate the English statement to an algebraic expression. Then simplify the expression.

34. Subtract -4 from 12.

35. Find the difference of 6 and 8.

Linear Equations and Inequalities

In this chapter we learn how to solve linear equations and inequalities in one variable. This important category of equations and inequalities is used in a variety of applications.

The list of words contains key terms used in this chapter. Search for them in the puzzle and in the text throughout the chapter. By the end of this chapter, you should be familiar with all of these terms.

Key Terms

compound	identity	notation
conditional	inequality	percent
consecutive	integer	property
contradiction	interest	set-builder
decimal	interval	solution
equation	linear	
fractions	literal	

```
            Z C
            E L
            Q A
    U M P   R N   I E L
  C D K I W E O P N V M X
C O N T R A D I C T I O N D
L M S S O C L T K E T K O E
I P L N N Y I I Z G U I I C
N O A O O T U D V E C N T I
E U R I I I B N Y R E T A M
A N E T T T T O A Q S E U A
R D T U A N E C U H N R Q L
  N I L T E S A A N O E E
  B L O O D L Z Y R C S A
    N S N I L Q G E F T
    I N T E R V A L R Q
      Y T R E P O R P
      T N E C R E P A
        P L M Q
```

Section 2.1 Addition, Subtraction, Multiplication, and Division Properties of Equality

Concepts

1. Definition of a Linear Equation in One Variable
2. Addition and Subtraction Properties of Equality
3. Multiplication and Division Properties of Equality
4. Translations

1. Definition of a Linear Equation in One Variable

An *equation* is a statement that indicates that two quantities are equal. The following are equations.

$$x = 5 \qquad y + 2 = 12 \qquad -4z = 28$$

All equations have an equal sign. Furthermore, notice that the equal sign separates the equation into two parts, the left-hand side and the right-hand side. A **solution to an equation** is a value of the variable that makes the equation a true statement. Substituting a solution to an equation for the variable makes the right-hand side equal to the left-hand side.

Equation	Solution	Check	
$x = 5$	5	$x = 5$ → $5 = 5$ ✔	Substitute 5 for x. Right-hand side equals left-hand side.
$y + 2 = 12$	10	$y + 2 = 12$ → $10 + 2 = 12$ ✔	Substitute 10 for y. Right-hand side equals left-hand side.
$-4z = 28$	-7	$-4z = 28$ → $-4(-7) = 28$ ✔	Substitute -7 for z. Right-hand side equals left-hand side.

Example 1 Determining Whether a Number is a Solution to an Equation

Determine whether the given number is a solution to the equation.

a. $-6w + 14 = 4;\quad 3$ **b.** $4x + 7 = 5;\quad -\frac{1}{2}$

Solution:

a. $-6w + 14 = 4;$

$-6(3) + 14 \stackrel{?}{=} 4$ Substitute 3 for w.

$-18 + 14 \stackrel{?}{=} 4$ Simplify.

$-4 \neq 4$ Right-hand side does not equal left-hand side. Thus, 3 *is not a solution* to the equation $-6w + 14 = 4$.

b. $4x + 7 = 5$

$4(-\frac{1}{2}) + 7 \stackrel{?}{=} 5$ Substitute $-\frac{1}{2}$ for x.

$-2 + 7 \stackrel{?}{=} 5$ Simplify.

$5 = 5$ ✔ Right-hand side equals the left-hand side. Thus, $-\frac{1}{2}$ *is a solution* to the equation $4x + 7 = 5$.

Skill Practice Determine if the number given is a solution to the equation.

1. $-2y + 5 = 9;\quad -2$ **2.** $4x - 1 = 7;\quad \frac{1}{4}$

In the study of algebra, you will encounter a variety of equations. In this chapter, we will focus on a specific type of equation called a linear equation in one variable.

Definition of a Linear Equation in One Variable

Let a and b be real numbers such that $a \neq 0$. A **linear equation in one variable** is an equation that can be written in the form

$$ax + b = 0$$

Notice that a linear equation in one variable has only one variable. Furthermore, because the variable has an implied exponent of 1, a linear equation is sometimes called a *first-degree equation*.

linear equation in one variable	*not* a linear equation in one variable
$2x + 3 = 0$	$4x^2 + 8 = 0$ (exponent on x is not 1)
$\frac{1}{5}a + \frac{2}{7} = 0$	$\frac{3}{4}a + \frac{5}{8}b = 0$ (more than one variable)

2. Addition and Subtraction Properties of Equality

If two equations have the same solution then the equations are said to be equivalent. For example, the following equations are equivalent because the solution for each equation is 6.

Equivalent Equations:	Check the Solution 6:
$2x - 5 = 7$	$2(6) - 5 = 7 \Rightarrow 12 - 5 = 7$ ✔
$2x = 12$	$2(6) = 12 \Rightarrow 12 = 12$ ✔
$x = 6$	$6 = 6 \Rightarrow 6 = 6$ ✔

To solve a linear equation, $ax + b = 0$, the goal is to find *all* values of x that make the equation true. One general strategy for solving an equation is to rewrite it as an equivalent but simpler equation. This process is repeated until the equation can be written in the form $x =$ number. The addition and subtraction properties of equality help us do this.

Addition and Subtraction Properties of Equality

Let a, b, and c represent algebraic expressions.

1. **Addition property of equality:** If $a = b$, then $a + c = b + c$
2. **Subtraction property of equality:** If $a = b$, then $a - c = b - c$

Skill Practice Answers

1. Yes
2. No

The addition and subtraction properties of equality indicate that adding or subtracting the same quantity to each side of an equation results in an equivalent equation. This is true because if two quantities are increased or decreased by the same amount, then the resulting quantities will also be equal (Figure 2-1).

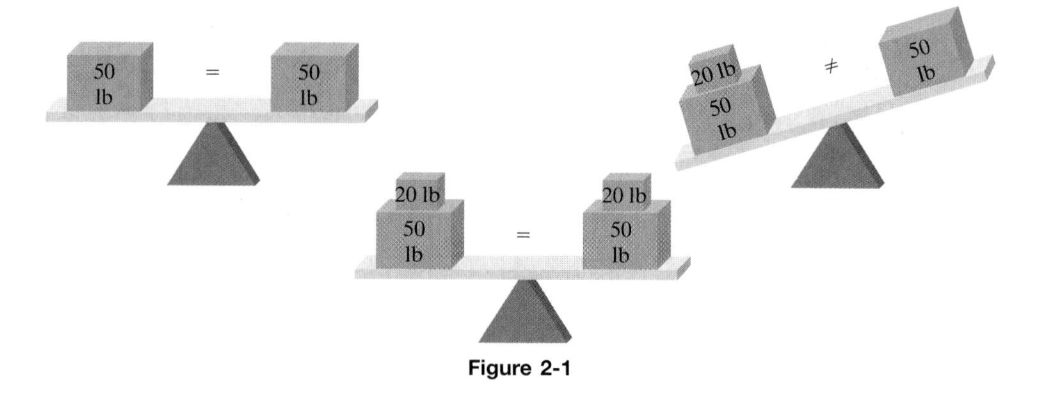

Figure 2-1

Example 2 Applying the Addition and Subtraction Properties of Equality

Solve the equations.

a. $p - 4 = 11$ **b.** $w + 5 = -2$ **c.** $\frac{9}{4} = q - \frac{3}{4}$ **d.** $-1.2 + z = 4.6$

Solution:

In each equation, the goal is to isolate the variable on one side of the equation. To accomplish this, we use the fact that the sum of a number and its opposite is zero and the difference of a number and itself is zero.

a. $p - 4 = 11$

$p - 4 + 4 = 11 + 4$ To isolate p, add 4 to both sides $(-4 + 4 = 0)$.

$p + 0 = 15$ Simplify.

$p = 15$

Check: $p - 4 = 11$ Check the solution by substituting $p = 15$ back in the original equation.

$15 - 4 \stackrel{?}{=} 11$

$11 = 11$ ✔ True

b. $w + 5 = -2$

$w + 5 - 5 = -2 - 5$ To isolate w, subtract 5 from both sides. $(5 - 5 = 0)$.

$w + 0 = -7$ Simplify.

$w = -7$

Check: $w + 5 = -2$ Check the solution by substituting $w = -7$ back in the original equation.

$-7 + 5 \stackrel{?}{=} -2$

$-2 = -2$ ✔ True

c. $\frac{9}{4} = q - \frac{3}{4}$

$\frac{9}{4} + \frac{3}{4} = q - \frac{3}{4} + \frac{3}{4}$ — To isolate q, add $\frac{3}{4}$ to both sides $(-\frac{3}{4} + \frac{3}{4} = 0)$.

$\frac{12}{4} = q + 0$ — Simplify.

$3 = q$ or equivalently, $q = 3$

Check: $\frac{9}{4} = q - \frac{3}{4}$ — Check the solution.

$\frac{9}{4} \stackrel{?}{=} 3 - \frac{3}{4}$ — Substitute $q = 3$ in the original equation.

$\frac{9}{4} \stackrel{?}{=} \frac{12}{4} - \frac{3}{4}$ — Get a common denominator.

$\frac{9}{4} = \frac{9}{4}$ ✔ — True

d. $-1.2 + z = 4.6$

$-1.2 + 1.2 + z = 4.6 + 1.2$ — To isolate z, add 1.2 to both sides.

$0 + z = 5.8$

$z = 5.8$

Check: $-1.2 + z = 4.6$ — Check the equation.

$-1.2 + 5.8 \stackrel{?}{=} 4.6$ — Substitute $z = 5.8$ in the original equation.

$4.6 = 4.6$ ✔ — True

Skill Practice Solve the equations.

3. $v - 7 = 2$ **4.** $14 + c = -2$

5. $\frac{1}{4} = a - \frac{2}{3}$ **6.** $-8.1 + w = 11.5$

TIP: The variable may be isolated on either side of the equation.

3. Multiplication and Division Properties of Equality

Adding or subtracting the same quantity to both sides of an equation results in an equivalent equation. In a similar way, multiplying or dividing both sides of an equation by the same nonzero quantity also results in an equivalent equation. This is stated formally as the multiplication and division properties of equality.

Multiplication and Division Properties of Equality

Let a, b, and c represent algebraic expressions.

1. **Multiplication property of equality:** If $a = b$, then $ac = bc$
2. **Division property of equality:** If $a = b$ then $\frac{a}{c} = \frac{b}{c}$ (provided $c \neq 0$)

Skill Practice Answers

3. $v = 9$ **4.** $c = -16$

5. $a = \frac{11}{12}$ **6.** $w = 19.6$

To understand the multiplication property of equality, suppose we start with a true equation such as $10 = 10$. If both sides of the equation are multiplied by a constant such as 3, the result is also a true statement (Figure 2-2).

$$10 = 10$$
$$3 \cdot 10 = 3 \cdot 10$$
$$30 = 30$$

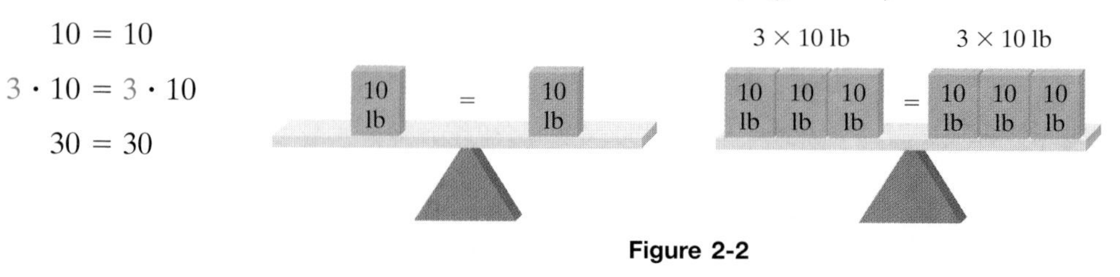

Figure 2-2

Similarly, if both sides of the equation are divided by a nonzero real number such as 2, the result is also a true statement (Figure 2-3).

$$10 = 10$$
$$\frac{10}{2} = \frac{10}{2}$$
$$5 = 5$$

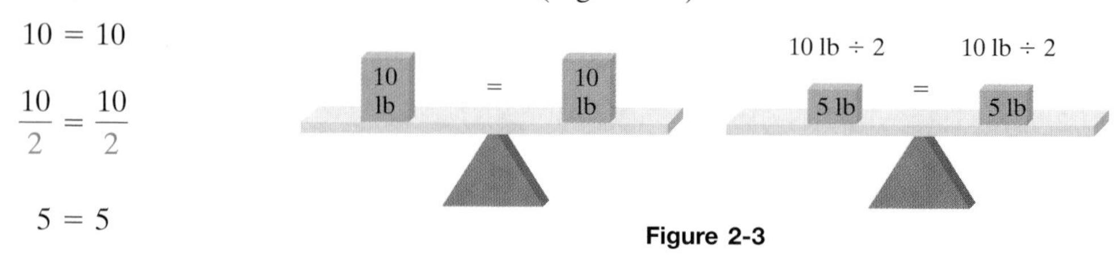

Figure 2-3

TIP: Recall that the product of a number and its reciprocal is 1. For example:

$$\frac{1}{5}(5) = 1$$
$$-\frac{7}{2}\left(-\frac{2}{7}\right) = 1$$

To solve an equation in the variable x, the goal is to write the equation in the form x = number. In particular, notice that we desire the coefficient of x to be 1. That is, we want to write the equation as $1x$ = number. Therefore, to solve an equation such as $5x = 15$, we can multiply both sides of the equation by the reciprocal of the x-term coefficient. In this case, multiply both sides by the reciprocal of 5, which is $\frac{1}{5}$.

$$5x = 15$$
$$\frac{1}{5}(5x) = \frac{1}{5}(15) \qquad \text{Multiply by } \tfrac{1}{5}.$$
$$1x = 3 \qquad \text{The coefficient of the } x\text{-term is now 1.}$$
$$x = 3$$

TIP: Recall that the quotient of a nonzero real number and itself is 1. For example:

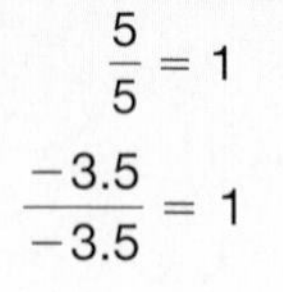

$$\frac{5}{5} = 1$$
$$\frac{-3.5}{-3.5} = 1$$

The division property of equality can also be used to solve the equation $5x = 15$ by dividing both sides by the coefficient of the x-term. In this case, divide both sides by 5 to make the coefficient of x equal to 1.

$$5x = 15$$
$$\frac{5x}{5} = \frac{15}{5} \qquad \text{Divide by 5.}$$
$$1x = 3 \qquad \text{The coefficient on the } x\text{-term is now 1.}$$
$$x = 3$$

Example 3 Applying the Multiplication and Division Properties of Equality

Solve the equations using the multiplication or division properties of equality.

a. $12x = 60$ **b.** $48 = -8w$ **c.** $-\frac{2}{9}q = \frac{1}{3}$

d. $-3.43 = -0.7z$ **e.** $\frac{d}{6} = -4$ **f.** $-x = 8$

Solution:

a. $12x = 60$

$\frac{12x}{12} = \frac{60}{12}$ To obtain a coefficient of 1 for the x-term, divide both sides by 12.

$1x = 5$ Simplify.

$x = 5$ Check: $12x = 60$

$12(5) \stackrel{?}{=} 60$

$60 = 60$ ✔ True

b. $48 = -8w$

$\frac{48}{-8} = \frac{-8w}{-8}$ To obtain a coefficient of 1 for the w-term, divide both sides by -8.

$-6 = 1w$ Simplify.

$-6 = w$ Check: $48w = -8w$

$48 \stackrel{?}{=} -8(-6)$

$48 = 48$ ✔ True

c. $-\frac{2}{9}q = \frac{1}{3}$

$\left(-\frac{9}{2}\right)\left(-\frac{2}{9}q\right) = \frac{1}{3}\left(-\frac{9}{2}\right)$ To obtain a coefficient of 1 for the q-term, multiply by the reciprocal of $-\frac{2}{9}$, which is $-\frac{9}{2}$.

$1q = -\frac{3}{2}$ Simplify. The product of a number and its reciprocal is 1.

$q = -\frac{3}{2}$ Check: $-\frac{2}{9}q = \frac{1}{3}$

$-\frac{2}{9}\left(-\frac{3}{2}\right) \stackrel{?}{=} \frac{1}{3}$

$\frac{1}{3} = \frac{1}{3}$ ✔ True

TIP: When applying the multiplication or division properties of equality to obtain a coefficient of 1 for the variable term, we will generally use the following convention:

- If the coefficient of the variable term is expressed as a fraction, we usually multiply both sides by its reciprocal.
- If the coefficient of the variable term is an integer or decimal, we divide both sides by the coefficient itself.

d. $-3.43 = -0.7z$

$\dfrac{-3.43}{-0.7} = \dfrac{-0.7z}{-0.7}$ To obtain a coefficient of 1 for the z-term, divide by -0.7.

$4.9 = 1z$ Simplify.

$4.9 = z$

$z = 4.9$

Check: $-3.43 = -0.7z$

$-3.43 \stackrel{?}{=} -0.7(4.9)$

$-3.43 = -3.43$ ✔ True

e. $\dfrac{d}{6} = -4$

$\dfrac{1}{6}d = -4$ $\frac{d}{6}$ is equivalent to $\frac{1}{6}d$.

$\dfrac{6}{1} \cdot \dfrac{1}{6}d = -4 \cdot \dfrac{6}{1}$ To obtain a coefficient of 1 for the d-term, multiply by the reciprocal of $\frac{1}{6}$, which is $\frac{6}{1}$.

$1d = -24$ Simplify.

$d = -24$

Check: $\dfrac{d}{6} = -4$

$\dfrac{-24}{6} \stackrel{?}{=} -4$

$-4 = -4$ ✔ True

f. $-x = 8$ Note that $-x$ is equivalent to $-1 \cdot x$.

$-1x = 8$

$\dfrac{-1x}{-1} = \dfrac{8}{-1}$ To obtain a coefficient of 1 for the x-term, divide by -1.

$x = -8$

Check: $-x = 8$

$-(-8) \stackrel{?}{=} 8$

$8 = 8$ ✔ True

TIP: In Example 3(f), we could also have *multiplied* both sides by -1 to create a coefficient of 1 on the x-term.

$$-x = 8$$
$$(-1)(-x) = (-1)8$$
$$x = -8$$

Skill Practice Solve the equations.

7. $4x = -20$ **8.** $1005 = -5p$ **9.** $\dfrac{2}{3}a = \dfrac{1}{4}$

10. $6.82 = 2.2w$ **11.** $\dfrac{x}{5} = -8$ **12.** $-y = -11$

It is important to distinguish between cases where the addition or subtraction properties of equality should be used to isolate a variable versus those in which the multiplication or division properties of equality should be used. Remember the goal is to isolate the variable term and obtain a coefficient of 1. Compare the equations:

$$5 + x = 20 \qquad \text{and} \qquad 5x = 20$$

Skill Practice Answers

7. $x = -5$ **8.** $p = -201$

9. $a = \dfrac{3}{8}$ **10.** $w = 3.1$

11. $x = -40$ **12.** $y = 11$

In the first equation, the relationship between 5 and x is addition. Therefore, we want to reverse the process by subtracting 5 from both sides. In the second equation, the relationship between 5 and x is multiplication. To isolate x, we reverse the process by dividing by 5 or equivalently, multiplying by the reciprocal, $\frac{1}{5}$.

$$5 + x = 20 \qquad \text{and} \qquad 5x = 20$$

$$5 - 5 + x = 20 - 5 \qquad\qquad \frac{5x}{5} = \frac{20}{5}$$

$$x = 15 \qquad\qquad x = 4$$

4. Translations

Example 4 Translating to a Linear Equation

Write an algebraic equation to represent each English sentence. Then solve the equation.

a. The quotient of a number and 4 is 6.

b. The product of a number and 4 is 6.

c. Negative twelve is equal to the sum of -5 and a number.

d. The value 1.4 subtracted from a number is 5.7.

Solution:

For each case we will let x represent the unknown number.

a. The quotient of a number and 4 is 6.

$$\frac{x}{4} = 6$$

$$4 \cdot \frac{x}{4} = 4 \cdot 6 \qquad \text{Multiply both sides by 4.}$$

$$\frac{4}{1} \cdot \frac{x}{4} = 4 \cdot 6$$

$$x = 24$$

b. The product of a number and 4 is 6.

$$4x = 6$$

$$\frac{4x}{4} = \frac{6}{4} \qquad \text{Divide both sides by 4.}$$

$$x = \frac{3}{2}$$

c. Negative twelve is equal to the sum of -5 and a number.

$$-12 = -5 + x$$

$$-12 + 5 = -5 + 5 + x \qquad \text{Add 5 to both sides.}$$

$$-7 = x$$

TIP: To help translate an English sentence to a mathematical equation, refer to page 61 for a table of equivalent expressions.

d. The value 1.4 subtracted from a number is 5.7.

$$x - 1.4 = 5.7$$

$$x - 1.4 + 1.4 = 5.7 + 1.4 \qquad \text{Add 1.4 to both sides.}$$

$$x = 7.1$$

Skill Practice Write an algebraic equation. Then solve the equation.

13. The sum of a number and 6 is -20.

14. The product of a number and -3 is -24.

15. 13 is equal to 5 subtracted from a number.

16. The quotient of a number and 6.3 is 1.2.

Skill Practice Answers

13. $x + 6 = -20; x = -26$
14. $-3x = -24; x = 8$
15. $13 = x - 5; x = 18$
16. $\frac{x}{6.3} = 1.2; x = 7.56$

Section 2.1 Practice Exercises

Boost *your* GRADE at mathzone.com!

- Practice Problems
- Self-Tests
- NetTutor
- e-Professors
- Videos

Study Skills Exercises

1. After getting a test back, it is a good idea to correct the test so that you do not make the same errors again. One recommended approach is to use a clean sheet of paper, and divide the paper down the middle vertically as shown. For each problem that you missed on the test, rework the problem correctly on the left-hand side of the paper. Then give a written explanation on the right-hand side of the paper. To reinforce the correct procedure, do four more problems of that type.

Take the time this week to make corrections from your last test.

Perform the correct math here.	Explain the process here.
↓	↓
$2 + 4(5)$ $= 2 + 20$ $= 22$	Do multiplication before addition.

2. Define the key terms:

a. linear equation in one variable **b. solution to an equation**

c. addition property of equality **d. subtraction property of equality**

e. multiplication property of equality **f. division property of equality**

Concept 1: Definition of a Linear Equation in One Variable

For Exercises 3–6, identify the following as either an expression or an equation.

3. $x - 4 + 5x$ **4.** $8x + 2 = 7$ **5.** $9 = 2x - 4$ **6.** $3x^2 + x = -3$

7. Explain how to determine if a number is a solution to an equation.

For Exercises 8–13, determine whether the given number is a solution to the equation.

8. $x - 1 = 5;\quad 4$ **9.** $x - 2 = 1;\quad -1$ **10.** $5x = -10;\quad -2$

11. $3x = 21;\quad 7$ **12.** $3x + 9 = 3;\quad -2$ **13.** $2x - 1 = -3;\quad -1$

Concept 2: Addition and Subtraction Properties of Equality

For Exercises 14–33, solve the equations using the addition or subtraction property of equality. Be sure to check your answers.

14. $x + 6 = 5$

15. $x - 2 = 10$

16. $q - 14 = 6$

17. $w + 3 = -5$

18. $2 + m = -15$

19. $-6 + n = 10$

20. $-23 = y - 7$

21. $-9 = -21 + b$

22. $4 + c = 4$

23. $-13 + b = -13$

24. $4.1 = 2.8 + a$

25. $5.1 = -2.5 + y$

26. $5 = z - \frac{1}{2}$

27. $-7 = p + \frac{2}{3}$

28. $x + \frac{5}{2} = \frac{1}{2}$

29. $x - \frac{2}{3} = \frac{7}{3}$

30. $-6.02 + c = -8.15$

31. $p + 0.035 = -1.12$

32. $3.245 + t = -0.0225$

33. $-1.004 + k = 3.0589$

Concept 3: Multiplication and Division Properties of Equality

For Exercises 34–53, solve the equations using the multiplication or division property of equality. Be sure to check your answers.

34. $6x = 54$

35. $2w = 8$

36. $12 = -3p$

37. $6 = -2q$

38. $-5y = 0$

39. $-3k = 0$

40. $-\frac{y}{5} = 3$

41. $-\frac{z}{7} = 1$

42. $\frac{4}{5} = -t$

43. $-\frac{3}{7} = -h$

44. $\frac{2}{5}a = -4$

45. $\frac{3}{8}b = -9$

46. $-\frac{1}{5}b = -\frac{4}{5}$

47. $-\frac{3}{10}w = \frac{2}{5}$

48. $-41 = -x$

49. $32 = -y$

50. $3.81 = -0.03p$

51. $2.75 = -0.5q$

52. $5.82y = -15.132$

53. $-32.3x = -0.4522$

Concept 4: Translations

For Exercises 54–63, write an algebraic equation to represent each English sentence. (Let x represent the unknown number.) Then solve the equation.

54. The sum of negative eight and a number is forty-two.

55. The sum of thirty-one and a number is thirteen.

56. The difference of a number and negative six is eighteen.

57. The sum of negative twelve and a number is negative fifteen.

58. The product of a number and seven is the same as negative sixty-three.

59. The product of negative three and a number is the same as twenty-four.

60. The quotient of a number and twelve is one-third.

61. Eighteen is equal to the quotient of a number and two.

62. The sum of a number and $\frac{5}{8}$ is $\frac{13}{8}$.

63. The difference of a number and $\frac{2}{3}$ is $\frac{1}{3}$.

Mixed Exercises

For Exercises 64–91, solve the equation using the appropriate property of equality.

64. $a - 9 = 1$

65. $b - 2 = -4$

66. $-9x = 1$

67. $-2k = -4$

68. $-\frac{2}{3}h = 8$

69. $\frac{3}{4}p = 15$

70. $\frac{2}{3} + t = 8$

71. $\frac{3}{4} + y = 15$

72. $\frac{r}{3} = -12$

73. $\frac{d}{-4} = 5$

74. $k + 16 = 32$

75. $-18 = -9 + t$

76. $16k = 32$

77. $-18 = -9t$

78. $7 = -4q$

79. $-3s = 10$

80. $-4 + q = 7$

81. $s - 3 = 10$

82. $-\frac{1}{3}d = 12$

83. $-\frac{2}{5}m = 10$

84. $4 = \frac{1}{2} + z$

85. $3 = \frac{1}{4} + p$

86. $1.2y = 4.8$

87. $4.3w = 8.6$

88. $4.8 = 1.2 + y$

89. $8.6 = w - 4.3$

90. $0.0034 = y - 0.405$

91. $-0.98 = m + 1.0034$

For Exercises 92–99, determine if the equation is a linear equation in one variable. Answer yes or no.

92. $4p + 5 = 0$

93. $3x - 5y = 0$

94. $4 + 2a^2 = 5$

95. $-8t = 7$

96. $x - 4 = 9$

97. $2x^3 + y = 0$

98. $19b = -3$

99. $13 + x = 19$

Expanding Your Skills

For Exercises 100–105, construct an equation with the given solution. Answers will vary.

100. $y = 6$

101. $x = 2$

102. $p = -4$

103. $t = -10$

104. $a = 0$

105. $k = 1$

For Exercises 106–109, simplify by collecting the *like* terms. Then solve the equation.

106. $5x - 4x + 7 = 8 - 2$

107. $2 + 3 = 2y + 1 - y$

108. $6p - 3p = 15 + 6$

109. $12 - 20 = 2t + 2t$

Solving Linear Equations

Section 2.2

Concepts

1. Solving Linear Equations Involving Multiple Steps
2. Steps to Solve a Linear Equation in One Variable
3. Conditional Equations, Identities, and Contradictions

1. Solving Linear Equations Involving Multiple Steps

In Section 2.1, we studied a one-step process to solve linear equations by using the addition, subtraction, multiplication, and division properties of equality. In the next example, we solve the equation $-2w - 7 = 11$. Solving this equation will require multiple steps. To understand the proper steps, always remember the ultimate goal—to isolate the variable. Therefore, we will first isolate the *term* containing the variable before dividing both sides by -2.

Example 1 Solving a Linear Equation

Solve the equation: $-2w - 7 = 11$

Solution:

$-2w - 7 = 11$

$-2w - 7 + 7 = 11 + 7$ Add 7 to both sides of the equation. This isolates the w-term.

$-2w = 18$

$\frac{-2w}{-2} = \frac{18}{-2}$ Next, apply the division property of equality to obtain a coefficient of 1 for w. Divide by -2 on both sides.

$1w = -9$

$w = -9$

Check:

$-2w - 7 = 11$

$-2(-9) - 7 \stackrel{?}{=} 11$ Substitute $w = -9$ in the original equation.

$18 - 7 \stackrel{?}{=} 11$

$11 = 11$ ✔ True

Skill Practice Solve.

1. $-5y + 25 = 10$

Example 2 Solving a Linear Equation

Solve the equation: $2 = \frac{1}{5}x + 3$

Skill Practice Answers

1. $y = 3$

Solution:

$$2 = \frac{1}{5}x + 3$$

$$2 - 3 = \frac{1}{5}x + 3 - 3$$ Subtract 3 from both sides. This isolates the x-term.

$$-1 = \frac{1}{5}x$$ Simplify.

$$5(-1) = 5 \cdot \left(\frac{1}{5}x\right)$$ Next, apply the multiplication property of equality to obtain a coefficient of 1 for x.

$$-5 = 1x$$

$$-5 = x$$ Simplify. The answer checks in the original equation.

Skill Practice Solve.

2. $2 = \frac{1}{2}a - 7$

In Example 3, the variable x appears on both sides of the equation. In this case, apply the addition or subtraction properties of equality to collect the variable terms on one side of the equation and the constant terms on the other side. Then use multiplication or division properties of equality to get a coefficient equal to 1.

Example 3 **Solving a Linear Equation**

Solve the equation: $6x - 4 = 2x - 8$

Solution:

$$6x - 4 = 2x - 8$$

$$6x - 2x - 4 = 2x - 2x - 8$$ Subtract $2x$ from both sides leaving $0x$ on the right-hand side.

$$4x - 4 = 0x - 8$$ Simplify.

$$4x - 4 = -8$$ The x-terms have now been combined on one side of the equation.

$$4x - 4 + 4 = -8 + 4$$ Add 4 to both sides of the equation. This combines the constant terms on the *other* side of the equation.

$$4x = -4$$

$$\frac{4x}{4} = \frac{-4}{4}$$ To obtain a coefficient of 1 for x, divide both sides of the equation by 4.

$$x = -1$$ Check:

$$6x - 4 = 2x - 8$$

$$6(-1) - 4 \stackrel{?}{=} 2(-1) - 8$$

$$-6 - 4 \stackrel{?}{=} -2 - 8$$

$$-10 = -10 \checkmark$$ True

Skill Practice Answers

2. $a = 18$

Skill Practice Solve.

3. $10x - 3 = 4x + 9$

TIP: It is important to note that the variable may be isolated on either side of the equation. We will solve the equation from Example 3 again, this time isolating the variable on the right-hand side.

$$\begin{aligned} 6x - 4 &= 2x - 8 \\ 6x - 6x - 4 &= 2x - 6x - 8 && \text{Subtract } 6x \text{ on both sides.} \\ 0x - 4 &= -4x - 8 \\ -4 &= -4x - 8 \\ -4 + 8 &= -4x - 8 + 8 && \text{Add } 8 \text{ to both sides.} \\ 4 &= -4x \\ \frac{4}{-4} &= \frac{-4x}{-4} && \text{Divide both sides by } -4. \\ -1 &= x \quad \text{or equivalently } x = -1 \end{aligned}$$

2. Steps to Solve a Linear Equation in One Variable

In some cases it is necessary to simplify both sides of a linear equation before applying the properties of equality. Therefore, we offer the following steps to solve a linear equation in one variable.

Steps to Solve a Linear Equation in One Variable

1. Simplify both sides of the equation.
   - Clear parentheses
   - Combine *like* terms
2. Use the addition or subtraction property of equality to collect the variable terms on one side of the equation.
3. Use the addition or subtraction property of equality to collect the constant terms on the *other* side of the equation.
4. Use the multiplication or division property of equality to make the coefficient of the variable term equal to 1.
5. Check your answer.

Example 4 Solving Linear Equations

Solve the equations:

a. $7 + 3 = 2(p - 3)$ **b.** $2.2y - 8.3 = 6.2y + 12.1$

Skill Practice Answers

3. $x = 2$

Solution:

a. $7 + 3 = 2(p - 3)$

$10 = 2p - 6$ **Step 1:** Simplify both sides of the equation by clearing parentheses and combining *like* terms.

Step 2: The variable terms are already on one side.

$10 + 6 = 2p - 6 + 6$ **Step 3:** Add 6 to both sides to collect the constant terms on the other side.

$16 = 2p$

$\frac{16}{2} = \frac{2p}{2}$ **Step 4:** Divide both sides by 2 to obtain a coefficient of 1 for p.

$8 = p$ **Step 5:** Check:

$7 + 3 = 2(p - 3)$

$10 \stackrel{?}{=} 2(8 - 3)$

$10 \stackrel{?}{=} 2(5)$

$10 = 10$ ✔ True

b. $2.2y - 8.3 = 6.2y + 12.1$ **Step 1:** The right- and left-hand sides are already simplified.

$2.2y - 2.2y - 8.3 = 6.2y - 2.2y + 12.1$ **Step 2:** Subtract $2.2y$ from both sides to collect the variable terms on one side of the equation.

$-8.3 = 4.0y + 12.1$

$-8.3 - 12.1 = 4.0y + 12.1 - 12.1$ **Step 3:** Subtract 12.1 from both sides to collect the constant terms on the other side.

$-20.4 = 4.0y$

$\frac{-20.4}{4.0} = \frac{4.0y}{4.0}$ **Step 4:** To obtain a coefficient of 1 for the y-term, divide both sides of the equation by 4.0.

$-5.1 = y$

$y = -5.1$ **Step 5:** Check:

$2.2y - 8.3 = 6.2y + 12.1$

$2.2(-5.1) - 8.3 \stackrel{?}{=} 6.2(-5.1) + 12.1$

$-11.22 - 8.3 \stackrel{?}{=} -31.62 + 12.1$

$-19.52 = -19.52$ ✔ True

Skill Practice Solve the equations.

4. $5 - 8 = -3(2x + 3)$ **5.** $1.5p + 2.3 = 3.5p - 1.9$

Skill Practice Answers

4. $x = -1$ **5.** $p = 2.1$

Example 5 **Solving Linear Equations**

Solve the equations:

a. $2 + 7x - 5 = 6(x + 3) + 2x$

b. $9 - (z - 3) + 4z = 4z - 5(z + 2) - 6$

Solution:

a. $2 + 7x - 5 = 6(x + 3) + 2x$

$-3 + 7x = 6x + 18 + 2x$ **Step 1:** Add *like* terms on the left. Clear parentheses on the right.

$-3 + 7x = 8x + 18$ Combine *like* terms.

$-3 + 7x - 7x = 8x - 7x + 18$ **Step 2:** Subtract $7x$ from both sides.

$-3 = x + 18$ Simplify.

$-3 - 18 = x + 18 - 18$ **Step 3:** Subtract 18 from both sides.

$-21 = x$ **Step 4:** Because the coefficient of the x term is already 1, there is no need to apply the multiplication or division property of equality.

$x = -21$

Step 5: The check is left to the reader.

b. $9 - (z - 3) + 4z = 4z - 5(z + 2) - 6$

$9 - z + 3 + 4z = 4z - 5z - 10 - 6$ **Step 1:** Clear parentheses.

$12 + 3z = -z - 16$ Combine *like* terms.

$12 + 3z + z = -z + z - 16$ **Step 2:** Add z to both sides.

$12 + 4z = -16$

$12 - 12 + 4z = -16 - 12$ **Step 3:** Subtract 12 from both sides.

$4z = -28$

$\frac{4z}{4} = \frac{-28}{4}$ **Step 4:** Divide both sides by 4.

$z = -7$ **Step 5:** The check is left to the reader.

Skill Practice Solve the equations.

6. $6y + 3 - y = 4(2y - 1)$ **7.** $3(3a - 4) - 4(5a + 2) = 7a - 2$

3. Conditional Equations, Identities, and Contradictions

The solutions to a linear equation are the values of x that make the equation a true statement. A linear equation has one unique solution. Some types of equations, however, have no solution while others have infinitely many solutions.

Skill Practice Answers

6. $y = \frac{7}{3}$ **7.** $a = -1$

I. Conditional Equations

An equation that is true for some values of the variable but false for other values is called a **conditional equation**. The equation $x + 4 = 6$, for example, is true on the condition that $x = 2$. For other values of x, the statement $x + 4 = 6$ is false.

II. Contradictions

Some equations have no solution, such as $x + 1 = x + 2$. There is no value of x, that when increased by 1 will equal the same value increased by 2. If we tried to solve the equation by subtracting x from both sides, we get the contradiction $1 = 2$. This indicates that the equation has no solution. An equation that has no solution is called a **contradiction**.

$$x + 1 = x + 2$$
$$x - x + 1 = x - x + 2$$
$$1 = 2 \quad \text{(contradiction)} \qquad \text{No solution}$$

III. Identities

An equation that has all real numbers as its solution set is called an **identity**. For example, consider the equation, $x + 4 = x + 4$. Because the left- and right-hand sides are identically equal, any real number substituted for x will result in equal quantities on both sides. If we subtract x from both sides of the equation, we get the identity $4 = 4$. In such a case, the solution is the set of all real numbers.

$$x + 4 = x + 4$$
$$x - x + 4 = x - x + 4$$
$$4 = 4 \quad \text{(identity)} \qquad \text{The solution is all real numbers.}$$

Example 6 Identifying Conditional Equations, Contradictions, and Identities

Identify each equation as a conditional equation, a contradiction, or an identity. Then describe the solution.

a. $4k - 5 = 2(2k - 3) + 1$ **b.** $2(b - 4) = 2b - 7$ **c.** $3x + 7 = 2x - 5$

Solution:

a.
$$4k - 5 = 2(2k - 3) + 1$$
$$4k - 5 = 4k - 6 + 1 \qquad \text{Clear parentheses.}$$
$$4k - 5 = 4k - 5 \qquad \text{Combine } \textit{like} \text{ terms.}$$
$$4k - 4k - 5 = 4k - 4k - 5 \qquad \text{Subtract } 4k \text{ from both sides.}$$
$$-5 = -5$$

This is an identity. The solution is all real numbers.

b.
$$2(b - 4) = 2b - 7$$
$$2b - 8 = 2b - 7 \qquad \text{Clear parentheses.}$$
$$2b - 2b - 8 = 2b - 2b - 7 \qquad \text{Subtract } 2b \text{ from both sides.}$$
$$-8 = -7 \quad \text{(Contradiction)}$$

This is a contradiction. There is no solution.

c.

$$3x + 7 = 2x - 5$$

$$3x - 2x + 7 = 2x - 2x - 5 \qquad \text{Subtract } 2x \text{ from both sides.}$$

$$x + 7 = -5 \qquad \text{Simplify.}$$

$$x + 7 - 7 = -5 - 7 \qquad \text{Subtract } 7 \text{ from both sides.}$$

$$x = -12$$

This is a conditional equation. The solution is $x = -12$. (The equation is true only on the condition that $x = -12$.)

Skill Practice Solve. Then describe the solution.

8. $5x - 1 = 2x + 3(x - 3)$ **9.** $4(2t - 1) + t = 3(3t - 1) - 1$

10. $6(v - 2) = 4v - 2(v + 1)$

Skill Practice Answers

8. No solution—the equation is a contradiction.
9. All real numbers—the equation is an identity.
10. $v = \frac{5}{2}$; conditional equation

Section 2.2 Practice Exercises

Study Skills Exercises

1. Several topics are given here about taking notes. Which would you do first to help make the most of note-taking? Put them in order of importance to you by labeling them with the numbers 1–6.

______ Read your notes after class to complete any abbreviations or incomplete sentences.

______ Highlight important terms and definitions.

______ Review your notes from the previous class.

______ Bring pencils (more than one) and paper to class.

______ Sit in class where you can clearly read the board and hear your instructor.

______ Keep your focus on the instructor looking for phrases such as, "The most important point is . . ." and "Here is where the problem usually occurs."

2. Define the key terms:

a. conditional equation **b. contradiction** **c. identity**

Review Exercises

For Exercises 3–6, simplify the expressions by clearing parentheses and combining *like* terms.

3. $5z + 2 - 7z - 3z$ **4.** $10 - 4w + 7w - 2 + w$

5. $-(-7p + 9) + (3p - 1)$ **6.** $8y - (2y + 3) - 19$

7. Explain the difference between simplifying an expression and solving an equation.

For Exercises 8–12, solve the equations using the addition, subtraction, multiplication, or division property of equality.

8. $5w = -30$

9. $-7y = 21$

10. $x + 8 = -15$

11. $z - 23 = -28$

12. $-\frac{9}{8} = -\frac{3}{4}k$

Concept 1: Solving Linear Equations Involving Multiple Steps

For Exercises 13–40, solve the equations using the steps outlined in the text.

13. $6z + 1 = 13$

14. $5x + 2 = -13$

15. $3y - 4 = 14$

16. $-7w - 5 = -19$

17. $-2p + 8 = 3$

18. $4q + 5 = 2$

19. $6 = 7m - 1$

20. $-9 = 4n - 1$

21. $-\frac{1}{2} - 4x = 8$

22. $2b - \frac{1}{4} = 5$

23. $0.2x + 3.1 = -5.3$

24. $-1.8 + 2.4a = -6.6$

25. $\frac{5}{8} = \frac{1}{4} - \frac{1}{2}p$

26. $\frac{6}{7} = \frac{1}{7} + \frac{5}{3}r$

27. $7w - 6w + 1 = 10 - 4$

28. $5v - 3 - 4v = 13$

29. $11h - 8 - 9h = -16$

30. $6u - 5 - 8u = -7$

31. $3a + 7 = 2a - 19$

32. $6b - 20 = 14 + 5b$

33. $-4r - 28 = -78 - r$

34. $-6x - 7 = -3 - 8x$

35. $-2z - 8 = -z$

36. $-7t + 4 = -6t$

37. $\frac{5}{6}x + \frac{2}{3} = -\frac{1}{6}x - \frac{5}{3}$

38. $\frac{3}{7}x - \frac{1}{4} = -\frac{4}{7}x - \frac{5}{4}$

39. $3y - 2 = 5y - 2$

40. $4 + 10t = -8t + 4$

Concept 2: Steps to Solve a Linear Equation in One Variable

For Exercises 41–58, solve the equations using the steps outlined in the text.

41. $3(2p - 4) = 15$

42. $4(t + 15) = 20$

43. $6(3x + 2) - 10 = -4$

44. $4(2k + 1) - 1 = 5$

45. $2(y - 3) - y = 6$

46. $4(w - 5) - 3w = 2$

47. $17(s + 3) = 4(s - 10) + 13$

48. $5(4 + p) = 3(3p - 1) - 9$

49. $6(3t - 4) + 10 = 5(t - 2) - (3t + 4)$

50. $-5y + 2(2y + 1) = 2(5y - 1) - 7$

51. $5 - 3(x + 2) = 5$

52. $1 - 6(2 - h) = 7$

53. $-2[(4p + 1) - (3p - 1)] = 5(3 - p) - 9$

54. $5 - (6k + 1) = 2[(5k - 3) - (k - 2)]$

55. $0.2w - 0.47 = 0.53 - 0.2(2w - 13)$

56. $0.4z - 0.15 = 0.65 - 0.3(6 - 2z)$

57. $3(-0.9n + 0.5) = -3.5n + 1.3$

58. $7(0.4m - 0.1) = 5.2m + 0.86$

Concept 3: Conditional Equations, Identities, and Contradictions

For Exercises 59–64, identify the equations as a conditional equation, a contradiction, or an identity. Then describe the solution.

59. $2(k - 7) = 2k - 13$

60. $5h + 4 = 5(h + 1) - 1$

61. $7x + 3 = 6(x - 2)$

62. $3y - 1 = 1 + 3y$

63. $3 - 5.2p = -5.2p + 3$

64. $2(q + 3) = 4q + q - 9$

65. A conditional linear equation has (choose one): One solution, no solution, or infinitely many solutions.

66. An equation that is a contradiction has (choose one): One solution, no solution, or infinitely many solutions.

67. An equation that is an identity has (choose one): One solution, no solution, or infinitely many solutions.

Mixed Exercises

For Exercises 68–91, find the solution, if possible.

68. $4p - 6 = 8 + 2p$

69. $\frac{1}{2}t - 2 = 3$

70. $2k + 9 = -8$

71. $3(y - 2) + 5 = 5$

72. $7(w - 2) = -14 - 3w$

73. $0.24 = 0.4m$

74. $2(x + 2) - 3 = 2x + 1$

75. $n + \frac{1}{4} = -\frac{1}{2}$

76. $0.5b = -23$

77. $3(2r + 1) = 6(r + 2) - 6$

78. $8 - 2q = 4$

79. $\frac{x}{7} - 3 = 1$

80. $2 - 4(y - 5) = -4$

81. $4 - 3(4p - 1) = -8$

82. $0.4(a + 20) = 6$

83. $2.2r - 12 = 3.4$

84. $10(2n + 1) - 6 = 20(n - 1) + 12$

85. $\frac{2}{5}y + 5 = -3$

86. $c + 0.123 = 2.328$

87. $4(2z + 3) = 8(z - 3) + 36$

88. $\frac{4}{5}t - 1 = \frac{1}{5}t + 5$

89. $6g - 8 = 4 - 3g$

90. $8 - (3q + 4) = 6 - q$

91. $6w - (8 + 2w) = 2(w - 4)$

Expanding Your Skills

92. Suppose $x = -5$ is a solution to the equation $x + a = 10$. Find the value of a.

93. Suppose $x = 6$ is a solution to the equation $x + a = -12$. Find the value of a.

94. Suppose $x = 3$ is a solution to the equation $ax = 12$. Find the value of a.

95. Suppose $x = 11$ is a solution to the equation $ax = 49.5$. Find the value of a.

96. Write an equation that is an identity. Answers may vary.

97. Write an equation that is a contradiction. Answers may vary.

Section 2.3 Linear Equations: Clearing Fractions and Decimals

Concepts

1. Clearing Fractions and Decimals
2. Solving Linear Equations with Fractions
3. Solving Linear Equations with Decimals

1. Clearing Fractions and Decimals

Linear equations that contain fractions can be solved in different ways. The first procedure, illustrated here, uses the method outlined in Section 2.2.

$$\frac{5}{6}x - \frac{3}{4} = \frac{1}{3}$$

$$\frac{5}{6}x - \frac{3}{4} + \frac{3}{4} = \frac{1}{3} + \frac{3}{4}$$ To isolate the variable term, add $\frac{3}{4}$ to both sides.

$$\frac{5}{6}x = \frac{4}{12} + \frac{9}{12}$$ Find the common denominator on the right-hand side.

$$\frac{5}{6}x = \frac{13}{12}$$ Simplify.

$$\frac{6}{5}\left(\frac{5}{6}x\right) = \frac{\overset{1}{\cancel{6}}}{5}\left(\frac{13}{\underset{2}{\cancel{12}}}\right)$$ Multiply by the reciprocal of $\frac{5}{6}$, which is $\frac{6}{5}$.

$$x = \frac{13}{10}$$

Sometimes it is simpler to solve an equation with fractions by eliminating the fractions first using a process called **clearing fractions**. To clear fractions in the equation $\frac{5}{6}x - \frac{3}{4} = \frac{1}{3}$, we can multiply both sides of the equation by the least common denominator (LCD) of all terms in the equation. In this case, the LCD of $\frac{5}{6}x$, $-\frac{3}{4}$, and $\frac{1}{3}$ is 12. Because each denominator in the equation is a factor of 12, we can simplify common factors to leave integer coefficients for each term.

Example 1 Solving a Linear Equation by Clearing Fractions

Solve the equation $\frac{5}{6}x - \frac{3}{4} = \frac{1}{3}$ by clearing fractions first.

Solution:

$$\frac{5}{6}x - \frac{3}{4} = \frac{1}{3}$$

$$12\left(\frac{5}{6}x - \frac{3}{4}\right) = 12\left(\frac{1}{3}\right)$$ Multiply both sides of the equation by the LCD, 12.

$$\frac{\overset{2}{\cancel{12}}}{1}\left(\frac{5}{\cancel{6}}x\right) - \frac{\overset{3}{\cancel{12}}}{1}\left(\frac{3}{\cancel{4}}\right) = \frac{\overset{4}{\cancel{12}}}{1}\left(\frac{1}{\cancel{3}}\right)$$ Apply the distributive property (recall that $12 = \frac{12}{1}$).

$$2(5x) - 3(3) = 4(1)$$ Simplify common factors to clear the fractions.

$$10x - 9 = 4$$

$$10x - 9 + 9 = 4 + 9$$ Add 9 to both sides.

$$10x = 13$$

$$\frac{10x}{10} = \frac{13}{10}$$ Divide both sides by 10.

$$x = \frac{13}{10}$$ Simplify.

TIP: Recall that the multiplication property of equality indicates that multiplying both sides of an equation by a nonzero constant results in an equivalent equation.

TIP: The fractions in this equation can be eliminated by multiplying both sides of the equation by *any* common multiple of the denominators. For example, try multiplying both sides of the equation by 24:

$$24\left(\frac{5}{6}x - \frac{3}{4}\right) = 24\left(\frac{1}{3}\right)$$

$$\frac{\overset{4}{\cancel{24}}}{1}\left(\frac{5}{\cancel{6}}x\right) - \frac{\overset{6}{\cancel{24}}}{1}\left(\frac{3}{\cancel{4}}\right) = \frac{\overset{8}{\cancel{24}}}{1}\left(\frac{1}{\cancel{3}}\right)$$

$$20x - 18 = 8$$

$$20x = 26$$

$$\frac{20x}{20} = \frac{26}{20}$$

$$x = \frac{13}{10}$$

Skill Practice Solve the equation by clearing fractions.

1. $\frac{2}{5}y + \frac{1}{2} = -\frac{7}{10}$

The same procedure used to clear fractions in an equation can be used to clear decimals. For example, consider the equation $0.05x + 0.25 = 0.2$. Because any terminating decimal can be written as a fraction, the equation can be interpreted as $\frac{5}{100}x + \frac{25}{100} = \frac{2}{10}$. A convenient common denominator for all terms in this equation is 100. Therefore, we can multiply the original equation by 100 to clear decimals.

Example 2 Solving a Linear Equation by Clearing Decimals

Solve the equation $0.05x + 0.25 = 0.2$ by clearing decimals first.

Solution:

$0.05x + 0.25 = 0.2$	
$100(0.05x + 0.25) = 100(0.2)$	Multiply both sides of the equation by 100.
$100(0.05x) + 100(0.25) = 100(0.2)$	Apply the distributive property.
$5x + 25 = 20$	Simplify (decimals have been cleared).
$5x + 25 - 25 = 20 - 25$	Subtract 25 from both sides.
$5x = -5$	
$\frac{5x}{5} = \frac{-5}{5}$	Divide both sides by 5.
$x = -1$	Simplify.

This equation can be checked by hand or by using a calculator.

$$0.05x + 0.25 = 0.2$$

$$0.05(-1) + 0.25 \stackrel{?}{=} 0.2$$

$$-0.05 + 0.25 \stackrel{?}{=} 0.2$$

$$0.2 = 0.2 \checkmark \quad \text{True}$$

TIP: Notice that multiplying a decimal number by 100 has the effect of moving the decimal point two places to the right. Similarly, multiplying by 10 moves the decimal point one place to the right, multiplying by 1000 moves the decimal point three places to the right, and so on.

Skill Practice Answers

1. $y = -3$

Skill Practice Solve the equation by clearing fractions.

2. $-0.12z - 0.5 = 1.3$

In this section, we combine the process for clearing fractions and decimals with the general strategies for solving linear equations. To solve a linear equation, it is important to follow the steps listed below.

Steps for Solving a Linear Equation in One Variable

1. Simplify both sides of the equation.
   - Clear parentheses
   - Consider clearing fractions or decimals (if any are present) by multiplying both sides of the equation by a common denominator of all terms.
   - Combine *like* terms
2. Use the addition or subtraction property of equality to collect the variable terms on one side of the equation.
3. Use the addition or subtraction property of equality to collect the constant terms on the other side of the equation.
4. Use the multiplication or division property of equality to make the coefficient of the variable term equal to 1.
5. Check your answer.

2. Solving Linear Equations with Fractions

Example 3 Solving Linear Equations with Fractions

a. $\frac{1}{6}x - \frac{2}{3} = \frac{1}{5}x - 1$ **b.** $\frac{1}{3}(x + 7) - \frac{1}{2}(x + 1) = 4$

Solution:

a.

$$\frac{1}{6}x - \frac{2}{3} = \frac{1}{5}x - 1$$ The LCD of $\frac{1}{6}x$, $-\frac{2}{3}$, and $\frac{1}{5}x$ is 30.

$$30\left(\frac{1}{6}x - \frac{2}{3}\right) = 30\left(\frac{1}{5}x - 1\right)$$ Multiply by the LCD, 30.

$$\frac{\overset{5}{\cancel{30}}}{1} \cdot \frac{1}{\cancel{6}}x - \frac{\overset{10}{\cancel{30}}}{1} \cdot \frac{2}{\cancel{3}} = \frac{\overset{6}{\cancel{30}}}{1} \cdot \frac{1}{\cancel{5}}x - 30(1)$$ Apply the distributive property (recall $30 = \frac{30}{1}$).

$$5x - 20 = 6x - 30$$ Clear fractions.

$$5x - 6x - 20 = 6x - 6x - 30$$ Subtract $6x$ from both sides.

$$-x - 20 = -30$$

$$-x - 20 + 20 = -30 + 20$$ Add 20 to both sides.

$$-x = -10$$

$$\frac{-x}{-1} = \frac{-10}{-1}$$ Divide both sides by -1.

$$x = 10$$ The check is left to the reader.

TIP: After clearing the parentheses, answer these two questions:
1. What is the LCD?
2. How many terms are in the equation? Then multiply each term by the LCD.

Skill Practice Answers

2. $z = -15$

b.

$$\frac{1}{3}(x + 7) - \frac{1}{2}(x + 1) = 4$$

$$\frac{1}{3}x + \frac{7}{3} - \frac{1}{2}x - \frac{1}{2} = 4$$ Clear parentheses.

$$6\left(\frac{1}{3}x + \frac{7}{3} - \frac{1}{2}x - \frac{1}{2}\right) = 6(4)$$ The LCD of $\frac{1}{3}x, \frac{7}{3}, -\frac{1}{2}x$, and $-\frac{1}{2}$ is 6.

$$\frac{\overset{2}{\cancel{6}}}{1} \cdot \frac{1}{\cancel{3}}x + \frac{\overset{2}{\cancel{6}}}{1} \cdot \frac{7}{\cancel{3}} + \frac{\overset{3}{\cancel{6}}}{1}\left(-\frac{1}{\cancel{2}}x\right) + \frac{\overset{3}{\cancel{6}}}{1}\left(-\frac{1}{\cancel{2}}\right) = 6(4)$$ Apply the distributive property.

$$2x + 14 - 3x - 3 = 24$$ Multiply fractions.

$$-x + 11 = 24$$ Combine *like* terms.

$$-x + 11 - 11 = 24 - 11$$ Subtract 11.

$$-x = 13$$

$$\frac{-x}{-1} = \frac{13}{-1}$$ Divide by -1.

$$x = -13$$ The check is left to the reader.

Skill Practice Solve the equations.

3. $\frac{2}{5}x - \frac{1}{2} = \frac{7}{4} + \frac{3}{10}x$ **4.** $\frac{1}{5}(z + 1) + \frac{1}{4}(z + 3) = 2$

TIP: In Example 3(b) both parentheses and fractions are present within the equation. In such a case, we recommend that you clear parentheses first. Then clear the fractions.

Example 4 Solving a Linear Equation with Fractions

Solve. $\frac{x - 2}{5} - \frac{x - 4}{2} = 2$

Solution:

$$\frac{x - 2}{5} - \frac{x - 4}{2} = \frac{2}{1}$$ The LCD of $\frac{x-2}{5}, \frac{x-4}{2}$, and $\frac{2}{1}$ is 10.

$$10\left(\frac{x - 2}{5} - \frac{x - 4}{2}\right) = 10\left(\frac{2}{1}\right)$$ Multiply both sides by 10.

$$\frac{\overset{2}{\cancel{10}}}{1} \cdot \frac{(x - 2)}{\cancel{5}} - \frac{\overset{5}{\cancel{10}}}{1} \cdot \frac{(x - 4)}{\cancel{2}} = \frac{10}{1} \cdot \left(\frac{2}{1}\right)$$ Apply the distributive property.

$$2(x - 2) - 5(x - 4) = 20$$ Clear fractions.

$$2x - 4 - 5x + 20 = 20$$ Apply the distributive property.

$$-3x + 16 = 20$$ Simplify both sides of the equation.

$$-3x + 16 - 16 = 20 - 16$$ Subtract 16 from both sides.

$$-3x = 4$$

$$\frac{-3x}{-3} = \frac{4}{-3}$$ Divide both sides by -3.

$$x = -\frac{4}{3}$$ The check is left to the reader.

Avoiding Mistakes:

In Example 4, several of the fractions in the equation have two terms in the numerator. It is important to enclose these fractions in parentheses when clearing fractions. In this way, we will remember to use the distributive property to multiply the factors shown in blue with both terms from the numerator of the fractions.

Skill Practice Answers

3. $x = \frac{45}{2}$ **4.** $z = \frac{7}{3}$

Skill Practice Solve the equation.

5. $\dfrac{x+1}{4} + \dfrac{x+2}{6} = 1$

3. Solving Linear Equations with Decimals

The process of **clearing decimals** is similar to clearing fractions in a linear equation. In this case, we multiply both sides of the equation by a convenient power of ten. We find the coefficient of the term within the equation that has the greatest number of digits following the decimal point. If the last nonzero digit in that coefficient is in the tenths place, we multiply by 10. If the coefficient is represented to the hundredths place, we multiply by 100, and so on.

Example 5 Solving Linear Equations Containing Decimals

Solve the equations by clearing decimals.

a. $2.5x + 3 = 1.7x - 6.6$ **b.** $0.2(x + 4) - 0.45(x + 9) = 12$

Solution:

a.

$2.5x + 3 = 1.7x - 6.6$

$10(2.5x + 3) = 10(1.7x - 6.6)$ Multiply both sides of the equation by 10.

$25x + 30 = 17x - 66$ Apply the distributive property.

$25x - 17x + 30 = 17x - 17x - 66$ Subtract $17x$ from both sides.

$8x + 30 = -66$

$8x + 30 - 30 = -66 - 30$ Subtract 30 from both sides.

$8x = -96$

$\dfrac{8x}{8} = \dfrac{-96}{8}$ Divide both sides by 8.

$x = -12$

b.

$0.2(x + 4) - 0.45(x + 9) = 12$

$0.2x + 0.8 - 0.45x - 4.05 = 12$ Clear parentheses first.

$100(0.2x + 0.8 - 0.45x - 4.05) = 100(12)$ Multiply both sides by 100.

$20x + 80 - 45x - 405 = 1200$ Apply the distributive property.

$-25x - 325 = 1200$ Simplify both sides.

$-25x - 325 + 325 = 1200 + 325$ Add 325 to both sides.

$-25x = 1525$

$\dfrac{-25x}{-25} = \dfrac{1525}{-25}$ Divide both sides by -25.

$x = -61$ The check is left to the reader.

TIP: The terms with the most digits following the decimal point are $-0.45x$ and -4.05. Each of these is written to the hundredths place. Therefore, we multiply both sides by 100.

Skill Practice Answers

5. $x = 1$

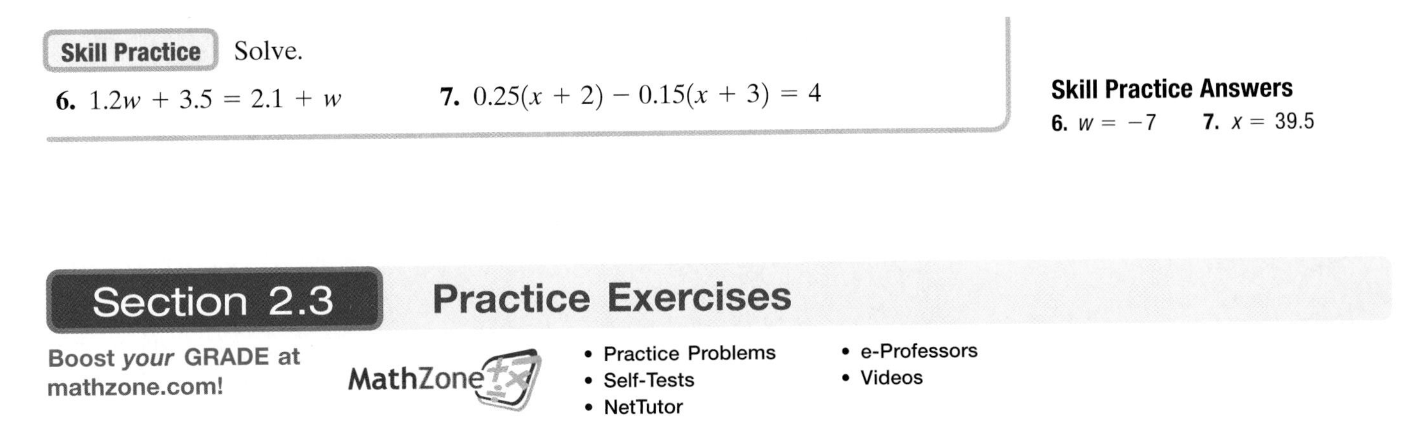

Skill Practice Solve.

6. $1.2w + 3.5 = 2.1 + w$ **7.** $0.25(x + 2) - 0.15(x + 3) = 4$

Skill Practice Answers

6. $w = -7$ **7.** $x = 39.5$

Section 2.3 Practice Exercises

Boost *your* GRADE at mathzone.com!

MathZone

- Practice Problems
- Self-Tests
- NetTutor
- e-Professors
- Videos

Study Skills Exercises

1. Instructors vary in what they emphasize on tests. For example, test material may come from the textbook, notes, handouts, or homework. What does your instructor emphasize?

2. Define the key terms:

   a. clearing fractions **b. clearing decimals**

Review Exercises

For Exercises 3–6, solve the equation.

3. $5(x + 2) - 3 = 4x + 5$

4. $-2(2x - 4x) = 6 + 18$

5. $3(2y + 3) - 4(-y + 1) = 7y - 10$

6. $-(3w + 4) + 5(w - 2) - 3(6w - 8) = 10$

7. Solve the equation and describe the solution set: $7x + 2 = 7(x - 12)$

8. Solve the equation and describe the solution set: $2(3x - 6) = 3(2x - 4)$

Concept 1: Clearing Fractions and Decimals

For Exercises 9–14, determine which of the values could be used to clear fractions or decimals in the given equation.

9. $\frac{2}{3}x - \frac{1}{6} = \frac{x}{9}$

Values: 6, 9, 12, 18, 24, 36

10. $\frac{1}{4}x - \frac{2}{7} = \frac{1}{2}x + 2$

Values: 4, 7, 14, 21, 28, 42

11. $0.02x + 0.5 = 0.35x + 1.2$

Values: 10; 100; 1000; 10,000

12. $0.003 - 0.002x = 0.1x$

Values: 10; 100; 1000; 10,000

13. $\frac{1}{6}x + \frac{7}{10} = x$

Values: 3, 6, 10, 30, 60

14. $2x - \frac{5}{2} = \frac{x}{3} - \frac{1}{4}$

Values: 2, 3, 4, 6, 12, 24

Concept 2: Solving Linear Equations with Fractions

For Exercises 15–36, solve the equation.

15. $\frac{1}{2}x + 3 = 5$

16. $\frac{1}{3}y - 4 = 9$

17. $\frac{1}{6}y + 2 = \frac{5}{12}$

18. $\frac{2}{15}z + 3 = \frac{7}{5}$

19. $\frac{1}{3}q + \frac{3}{5} = \frac{1}{15}q - \frac{2}{5}$

20. $\frac{3}{7}x - 5 = \frac{24}{7}x + 7$

21. $\frac{12}{5}w + 7 = 31 - \frac{3}{5}w$

22. $-\frac{1}{9}p - \frac{5}{18} = -\frac{1}{6}p + \frac{1}{3}$

23. $\frac{1}{4}(3m - 4) - \frac{1}{5} = \frac{1}{4}m + \frac{3}{10}$

24. $\frac{1}{25}(20 - t) = \frac{4}{25}t - \frac{3}{5}$

25. $\frac{1}{6}(5s + 3) = \frac{1}{2}(s + 11)$

26. $\frac{1}{12}(4n - 3) = \frac{1}{12}n - \frac{3}{4}$

27. $\frac{2}{3}x + 4 = \frac{2}{3}x - 6$

28. $-\frac{1}{9}a + \frac{2}{9} = \frac{1}{3} - \frac{1}{9}a$

29. $\frac{1}{6}(2c - 1) = \frac{1}{3}c - \frac{1}{6}$

30. $\frac{3}{2}b - 1 = \frac{1}{8}(12b - 8)$

31. $\frac{2x + 1}{3} + \frac{x - 1}{3} = 5$

32. $\frac{4y - 2}{5} - \frac{y + 4}{5} = -3$

33. $\frac{3w - 2}{6} = 1 - \frac{w - 1}{3}$

34. $\frac{z - 7}{4} = \frac{6z - 1}{8} - 2$

35. $\frac{x + 3}{3} - \frac{x - 1}{2} = 4$

36. $\frac{5y - 1}{2} - \frac{y + 4}{5} = 1$

Concept 3: Solving Linear Equations with Decimals

For Exercises 37–54, solve the equation.

37. $9.2y - 4.3 = 50.9$

38. $-6.3x + 1.5 = -4.8$

39. $21.1w + 4.6 = 10.9w + 35.2$

40. $0.05z + 0.2 = 0.15z - 10.5$

41. $0.2p - 1.4 = 0.2(p - 7)$

42. $0.5(3q + 87) = 1.5q + 43.5$

43. $0.20x + 53.60 = x$

44. $z + 0.06z = 3816$

45. $0.15(90) + 0.05p = 0.10(90 + p)$

46. $0.25(60) + 0.10x = 0.15(60 + x)$

47. $0.40(y + 10) + 0.60y = 2$

48. $0.75(x - 2) + 0.25x = 0.5$

49. $0.4x + 0.2 = -3.6 - 0.6x$

50. $0.12x + 3 - 0.8x = 0.22x - 0.6$

51. $0.06(x - 0.5) = 0.06x + 0.01$

52. $0.125x = 0.025(5x + 1)$

53. $-3.5x + 1.3 = -0.3(9x - 5)$

54. $x + 4 = 2(0.4x + 1.3)$

Mixed Exercises

For Exercises 55–81, solve the equation.

55. $2b + 23 = 6b - 5$

56. $-x = 7$

57. $\frac{y}{4} = -2$

58. $10p - 9 + 2p - 3 = 8p - 18$

59. $0.5(2a - 3) - 0.1 = 0.4(6 + 2a)$

60. $-\frac{5}{9}w + \frac{11}{12} = \frac{23}{36}$

61. $-6x = 0$

62. $15.2q = -2.4q - 176$

63. $9.8h + 2 = 3.8h + 20$

64. $-k - 41 = 3 - k$

65. $\frac{1}{4}(x + 4) = \frac{1}{5}(2x + 3)$

66. $7y + 3(2y + 5) = 10y + 17$

67. $2z - 7 = 2(z - 13)$

68. $x - 17.8 = -21.3$

69. $\frac{4}{5}w = 10$

70. $5c + 25 = 20$

71. $4b - 8 - b = -3b + 2(3b - 4)$

72. $36 = 6z + 9$

73. $-3a + 1 = 19$

74. $-5(1 - x) + x = -(6 - 2x) + 6$

75. $3(4h - 2) - (5h - 8) = 8 - (2h + 3)$

76. $1.72w - 0.04w = 0.42$

77. $\frac{3}{8}t - \frac{5}{8} = \frac{1}{2}t + \frac{1}{8}$

78. $3(8x - 1) + 10 = 6(5 + 4x) - 23$

79. $\frac{2x - 1}{4} + \frac{3x + 2}{6} = 2$

80. $\frac{w - 4}{6} - \frac{3w - 1}{2} = -1$

81. $\frac{2k + 5}{4} = 2 - \frac{k + 2}{3}$

82. The sum of $\frac{2}{5}$ and twice a number is the same as the sum of $\frac{11}{5}$ and the number. Find the number.

83. The difference of three times a number and $\frac{5}{9}$ is the same as the sum of twice the number and $\frac{1}{9}$. Find the number.

84. The sum of twice a number and $\frac{3}{4}$ is the same as the difference of four times the number and $\frac{1}{8}$. Find the number.

85. The difference of a number and $-\frac{11}{12}$ is the same as the difference of three times the number and $\frac{1}{6}$. Find the number.

Expanding Your Skills

For Exercises 86–89, solve the equation.

86. $\frac{1}{2}a + 0.4 = -0.7 - \frac{3}{5}a$

87. $\frac{3}{4}c - 0.11 = 0.23(c - 5)$

88. $0.8 + \frac{7}{10}b = \frac{3}{2}b - 0.8$

89. $0.78 - \frac{1}{25}h = \frac{3}{5}h - 0.5$

Applications of Linear Equations: Introduction to Problem Solving

Section 2.4

Concepts

1. Problem-Solving Strategies
2. Translations Involving Linear Equations
3. Consecutive Integer Problems
4. Applications of Linear Equations
5. Applications Involving Uniform Motion

1. Problem-Solving Strategies

Linear equations can be used to solve many real-world applications. However, with "word problems," students often do not know where to start. To help organize the problem-solving process, we offer the following guidelines.

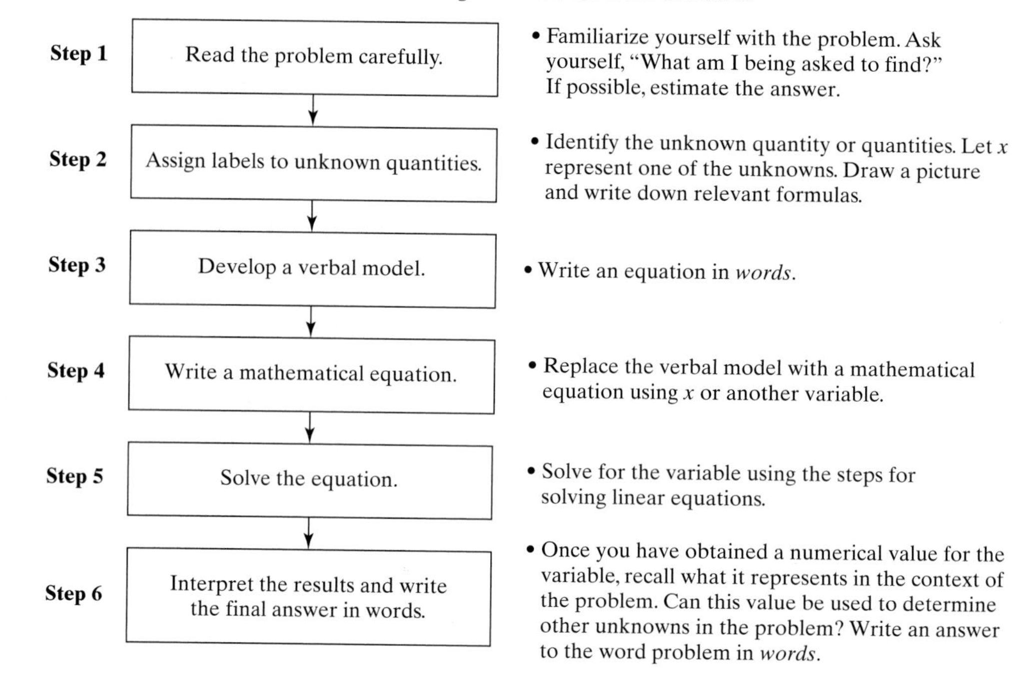

2. Translations Involving Linear Equations

We have already practiced translating an English sentence to a mathematical equation. Recall from Section 1.2 that several key words translate to the algebraic operations of addition, subtraction, multiplication, and division.

Addition: $a + b$	**Subtraction: $a - b$**
The sum of a and b	The difference of a and b
a plus b	a minus b
b added to a	b subtracted from a
b more than a	a decreased by b
a increased by b	b less than a
The total of a and b	

Multiplication: $a \cdot b$	**Division: $a \div b$**
The product of a and b	The quotient of a and b
a times b	a divided by b
a multiplied by b	b divided into a
	The ratio of a and b
	a over b
	a per b

Example 1 Translating to a Linear Equation

The sum of a number and negative eleven is negative fifteen. Find the number.

Solution:

Step 1: Read the problem.

Let x represent the unknown number. **Step 2:** Label the unknown.

$$\overset{\text{the sum of}}{\downarrow} \quad \overset{\text{is}}{\downarrow}$$

$$(\text{a number}) + (-11) = (-15)$$ **Step 3:** Develop a verbal model.

$$x + (-11) = -15$$ **Step 4:** Write an equation.

$$x + (-11) + 11 = -15 + 11$$ **Step 5:** Solve the equation.

$$x = -4$$

The number is -4. **Step 6:** Write the final answer in words.

Skill Practice

1. The sum of a number and negative seven is twelve. Find the number.

Example 2 Translating to a Linear Equation

Forty less than five times a number is fifty-two less than the number. Find the number.

Solution:

Step 1: Read the problem.

Let x represent the unknown number. **Step 2:** Label the unknown.

$$\begin{pmatrix}\text{5 times}\\ \text{a number}\end{pmatrix} - (40) = \begin{pmatrix}\text{the}\\ \text{number}\end{pmatrix} - (52)$$ **Step 3:** Develop a verbal model.

$$\downarrow \quad \downarrow \quad \downarrow \quad \downarrow$$

$$5x \quad - 40 \quad = \quad x \quad - 52$$ **Step 4:** Write an equation.

$$5x - 40 = x - 52$$ **Step 5:** Solve the equation.

$$5x - x - 40 = x - x - 52$$

$$4x - 40 = -52$$

$$4x - 40 + 40 = -52 + 40$$

$$4x = -12$$

$$\frac{4x}{4} = \frac{-12}{4}$$

$$x = -3$$

The number is -3. **Step 6:** Write the final answer in words.

Skill Practice

2. Thirteen more than twice a number is five more than the number. Find the number.

Avoiding Mistakes:

It is important to remember that subtraction is not a commutative operation. Therefore, the order in which two real numbers are subtracted affects the outcome. The expression "forty less than five times a number" must be translated as: $5x - 40$ (not $40 - 5x$). Similarly, "fifty-two less than the number" must be translated as: $x - 52$ (not $52 - x$).

Skill Practice Answers

1. The number is 19.
2. The number is -8.

Example 3 Translating to a Linear Equation

Twice the sum of a number and six is two more than three times the number. Find the number.

Solution:

Step 1: Read the problem.

Let x represent the unknown number. **Step 2:** Label the unknown.

twice, the sum, is, 2 more than, three times a number

$$2\ (x + 6) \ = \ 3x + 2$$

Step 3: Develop a verbal model.

Step 4: Write an equation.

Avoiding Mistakes:

It is important to enclose "the sum of a number and six" within parentheses so that the entire quantity is multiplied by 2.

Correct: $2(x + 6)$

Forgetting the parentheses would imply that the x-term only is multiplied by 2.

$$2(x + 6) = 3x + 2$$

Step 5: Solve the equation.

$$2x + 12 = 3x + 2$$

$$2x - 2x + 12 = 3x - 2x + 2$$

$$12 = x + 2$$

$$12 - 2 = x + 2 - 2$$

$$10 = x$$

The number is 10. **Step 6:** Write the final answer in words.

Skill Practice

3. Three times the sum of a number and eight is four more than the number. Find the number.

3. Consecutive Integer Problems

The word *consecutive* means "following one after the other in order without gaps." The numbers 6, 7, 8 are examples of three **consecutive integers**. The numbers $-4, -2, 0, 2$ are examples of **consecutive even integers**. The numbers 23, 25, 27 are examples of **consecutive odd integers**.

Notice that any two consecutive integers differ by 1. Therefore, if x represents an integer, then $(x + 1)$ represents the next larger consecutive integer (Figure 2-4).

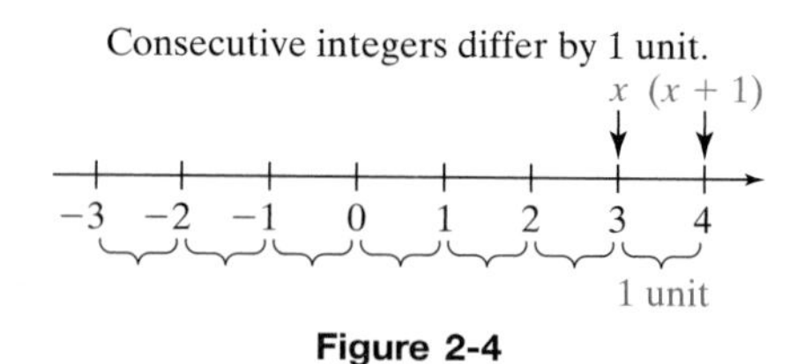

Figure 2-4

Skill Practice Answers

3. The number is −10.

Any two consecutive even integers differ by 2. Therefore, if x represents an even integer, then $(x + 2)$ represents the next consecutive larger even integer (Figure 2-5).

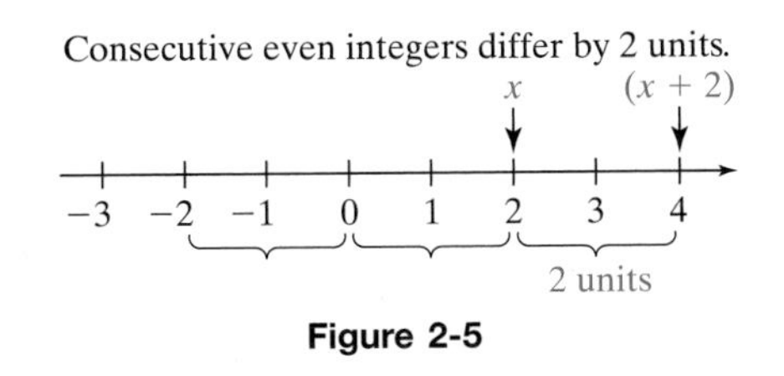

Figure 2-5

Likewise, any two consecutive odd integers differ by 2. If x represents an odd integer, then $(x + 2)$ is the next larger odd integer (Figure 2-6).

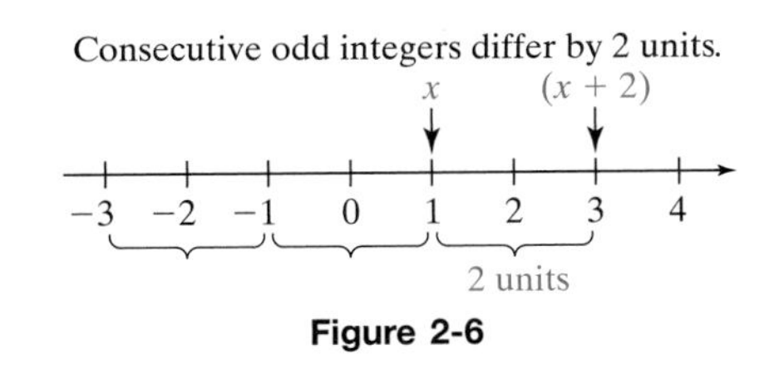

Figure 2-6

Example 4 Solving an Application Involving Consecutive Integers

The sum of two consecutive odd integers is −188. Find the integers.

Solution:

In this example we have two unknown integers. We can let x represent either of the unknowns.

Step 1: Read the problem.

Suppose x represents the first odd integer. **Step 2:** Label the variables.

Then $(x + 2)$ represents the second odd integer.

$$\begin{pmatrix}\text{first}\\\text{integer}\end{pmatrix} + \begin{pmatrix}\text{second}\\\text{integer}\end{pmatrix} = (\text{total})$$

Step 3: Write an equation in words.

$$x + (x + 2) = -188$$

Step 4: Write a mathematical equation.

$$x + (x + 2) = -188$$
$$2x + 2 = -188$$
$$2x + 2 - 2 = -188 - 2$$
$$2x = -190$$
$$\frac{2x}{2} = \frac{-190}{2}$$
$$x = -95$$

Step 5: Solve for x.

TIP: With word problems, it is advisable to check that the answer is reasonable.

The numbers -95 and -93 are consecutive odd integers. Furthermore, their sum is -188 as desired.

The first integer is $x = -95$.

The second integer is $x + 2 = -95 + 2 = -93$.

The two integers are -95 and -93.

Step 6: Interpret the results and write the answer in words.

Skill Practice

4. The sum of two consecutive even integers is 66. Find the integers.

Example 5 Solving an Application Involving Consecutive Integers

Ten times the smallest of three consecutive integers is twenty-two more than three times the sum of the integers. Find the integers.

Solution:

Step 1: Read the problem.

Let x represent the first integer.
$x + 1$ represents the second consecutive integer.
$x + 2$ represents the third consecutive integer.

Step 2: Label the variables.

$$\begin{pmatrix}\text{10 times}\\ \text{the first}\\ \text{integer}\end{pmatrix} = \begin{pmatrix}\text{3 times}\\ \text{the sum of}\\ \text{the integers}\end{pmatrix} + 22$$

Step 3: Write an equation in words.

(10 times the first integer → $10x$; is → $=$; 3 times → 3; 22 more than → $+ 22$)

$$10x = 3\underbrace{[(x) + (x + 1) + (x + 2)]}_{\text{the sum of the integers}} + 22$$

Step 4: Write a mathematical equation.

$$10x = 3(x + x + 1 + x + 2) + 22$$

Step 5: Solve the equation.

$$10x = 3(3x + 3) + 22$$

Clear parentheses.

$$10x = 9x + 9 + 22$$

Combine *like* terms.

$$10x = 9x + 31$$

$$10x - 9x = 9x - 9x + 31$$

$$x = 31$$

Isolate the x-terms on one side.

The first integer is $x = 31$.

The second integer is $x + 1 = 31 + 1 = 32$.

The third integer is $x + 2 = 31 + 2 = 33$.

Step 6: Interpret the results and write the answer in words.

The three integers are 31, 32, and 33.

Skill Practice Answers

4. The integers are 32 and 34.

Skill Practice

5. Five times the smallest of three consecutive integers is 117 more than the sum of the integers. Find the integers.

4. Applications of Linear Equations

Example 6 Using a Linear Equation in an Application

A carpenter cuts a 6-ft board in two pieces. One piece must be three times as long as the other. Find the length of each piece.

Solution:

In this problem, one piece must be three times as long as the other. Thus, if x represents the length of one piece, then $3x$ can represent the length of the other.

Step 1: Read the problem completely.

x represents the length of the smaller piece.
$3x$ represents the length of the longer piece.

Step 2: Label the unknowns. Draw a figure.

$3x$ x

$$\begin{pmatrix}\text{length of}\\ \text{one piece}\end{pmatrix} + \begin{pmatrix}\text{length of}\\ \text{other piece}\end{pmatrix} = \begin{pmatrix}\text{total length}\\ \text{of the board}\end{pmatrix}$$

Step 3: Set up a verbal equation.

$$x + 3x = 6$$

Step 4: Write an equation.

$$4x = 6$$

Step 5: Solve the equation.

$$\frac{4x}{4} = \frac{6}{4}$$

$$x = 1.5$$

The smaller piece is $x = 1.5$ ft.

The longer piece is $3x$ or $3(1.5 \text{ ft}) = 4.5$ ft.

Step 6: Interpret the results.

Skill Practice

6. A plumber cuts a 96-in. piece of pipe into two pieces. One piece is five times longer than the other piece. How long is each piece?

Example 7 Using a Linear Equation in an Application

The hit movies *Spider-Man* and *X-Men* together brought in \$169.3 million during their opening weekends. *Spider-Man* earned \$5.8 million more than twice what *X-Men* earned. How much revenue did each movie bring in during its opening weekend?

Skill Practice Answers

5. The integers are 60, 61, and 62.
6. One piece is 80 in. and the other is 16 in.

Solution:

In this example, we have two unknowns. The variable x can represent *either* quantity. However, the revenue from *Spider-Man* is given in terms of the revenue for *X-Men*. — **Step 1:** Read the problem.

Let x represent the revenue for *X-Men*. — **Step 2:** Label the variables.

Then $2x + 5.8$ represents the revenue for *Spider-Man*.

$$\begin{pmatrix}\text{Revenue from} \\ \textit{X-Men}\end{pmatrix} + \begin{pmatrix}\text{Revenue from} \\ \textit{Spider-Man}\end{pmatrix} = \begin{pmatrix}\text{Total} \\ \text{Revenue}\end{pmatrix}$$

Step 3: Set up a verbal equation.

$$x \quad + \quad 2x + 5.8 \quad = \quad 169.3$$

Step 4: Write an equation.

$$3x + 5.8 = 169.3$$

Step 5: Solve the equation.

$$3x + 5.8 - 5.8 = 169.3 - 5.8$$

$$3x = 163.5$$

$$\frac{3x}{3} = \frac{163.5}{3}$$

$$x = 54.5$$

Revenue from *X-Men*: $x = 54.5$

Step 6: Interpret the results.

Revenue from *Spider-Man*: $2x + 5.8 = 2(54.5) + 5.8 = 114.8$

The revenue from *X-Men* was $54.5 million for its opening weekend. The revenue from *Spider-Man* was $114.8 million.

Skill Practice

7. There are 40 students in an algebra class. There are four more women than men. How many women and men are in the class?

5. Applications Involving Uniform Motion

The formula: (distance) = (rate)(time) or simply, $d = rt$, relates the distance traveled to the rate of travel and the time of travel.

For example, if a car travels at 60 mph for 3 hours, then

$$d = (60 \text{ mph})(3 \text{ hours})$$
$$= 180 \text{ miles}$$

If a car travels at 60 mph for x hours, then

$$d = (60 \text{ mph})(x \text{ hours})$$
$$= 60x \text{ miles}$$

Skill Practice Answers

7. There are 22 women and 18 men.

Example 8 Solving an Application Involving Distance, Rate, and Time

One bicyclist rides 4 mph faster than another bicyclist. The faster rider takes 3 hr to complete a race, while the slower rider takes 4 hr. Find the speed for each rider.

Solution:

Step 1: Read the problem.

The problem is asking us to find the speed of each rider.

Let x represent the speed of the slower rider. Then $(x + 4)$ is the speed of the faster rider.

Step 2: Label the variables and organize the information given in the problem. A distance-rate-time chart may be helpful.

	Distance	Rate	Time
Faster rider	$3(x + 4)$	$x + 4$	3
Slower rider	$4(x)$	x	4

To complete the first column, we can use the relationship, $d = rt$.

faster rider's distance = (faster rate)(faster rider's time) = $(x + 4)(3)$

slower rider's distance = (slower rate)(slower rider's time) = $(x)(4)$

Because the riders are riding in the same race, their distances are equal.

$$\begin{pmatrix}\text{distance}\\ \text{by faster rider}\end{pmatrix} = \begin{pmatrix}\text{distance}\\ \text{by slower rider}\end{pmatrix}$$

Step 3: Set up a verbal model.

$$3(x + 4) = 4(x)$$

Step 4: Write a mathematical equation.

$$3x + 12 = 4x$$

Step 5: Solve the equation.

$$12 = x$$

Subtract $3x$ from both sides.

The variable x represents the slower rider's rate. The quantity $x + 4$ is the faster rider's rate. Thus, if $x = 12$, then $x + 4 = 16$.

The slower rider travels 12 mph and the faster rider travels 16 mph.

TIP: Check that the answer is reasonable. If the slower rider rides at 12 mph for 4 hr, he travels 48 mi. If the faster rider rides at 16 mph for 3 hr, he also travels 48 mi as expected.

Skill Practice

8. An express train travels 25 mph faster than a cargo train. It takes the express train 6 hr to travel a route, and it takes 9 hr for the cargo train to travel the same route. Find the speed of each train.

Example 9 Solving an Application Involving Distance, Rate, and Time

Two families that live 270 miles apart plan to meet for an afternoon picnic. To share the driving, they want to meet somewhere between their two homes. Both families leave at 9.00 A.M., but one family averages 12 mph faster than the other family. If the families meet at the designated spot $2\frac{1}{2}$ hours later, determine

a. The average rate of speed for each family.

b. The distance each family traveled to the picnic.

Skill Practice Answers

8. The express train travels 75 mph, and the cargo train travels 50 mph.

Solution:

For simplicity, we will call the two families, Family A and Family B. Let Family A be the family that travels at the slower rate (Figure 2-7).

Step 1: Read the problem and draw a sketch.

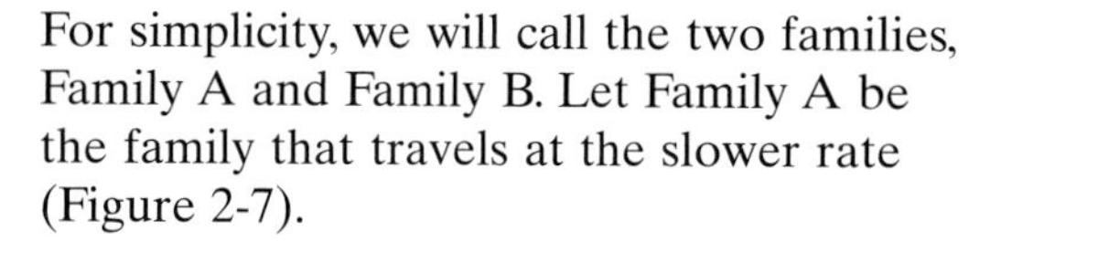

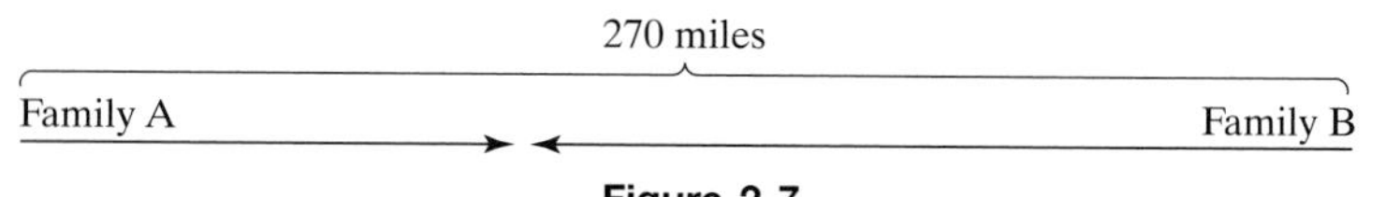

Figure 2-7

Let x represent the rate of Family A.

Then $(x + 12)$ is the rate of Family B.

Step 2: Label the variables.

	Distance	Rate	Time
Family A	$2.5x$	x	2.5
Family B	$2.5(x + 12)$	$x + 12$	2.5

To complete the first column, we can use the relationship $d = rt$.

The distance traveled by Family A is $(x)(2.5)$.

The distance traveled by Family B is $(x + 12)(2.5)$.

To set up an equation, recall that the total distance between the two families is given as 270 miles.

$$\begin{pmatrix}\text{distance}\\ \text{traveled by}\\ \text{Family A}\end{pmatrix} + \begin{pmatrix}\text{distance}\\ \text{traveled by}\\ \text{Family B}\end{pmatrix} = \begin{pmatrix}\text{total}\\ \text{distance}\end{pmatrix}$$

Step 3: Create a verbal equation.

$$2.5x \quad + \quad 2.5(x + 12) \quad = \quad 270$$

Step 4: Write a mathematical equation.

$$2.5x + 2.5(x + 12) = 270$$

$$2.5x + 2.5x + 30 = 270$$

$$5.0x + 30 = 270$$

$$5x = 240$$

$$\frac{5x}{5} = \frac{240}{5}$$

$$x = 48$$

Step 5: Solve for x.

a. Family A traveled 48 (mph).

Family B traveled $x + 12 = 48 + 12 = 60$ (mph).

Step 6: Interpret the results and write the answer in words.

b. To compute the distance each family traveled, use $d = rt$:

Family A traveled: (48 mph)(2.5 hr) = 120 miles

Family B traveled: (60 mph)(2.5 hr) = 150 miles

Skill Practice

9. A Piper Cub airplane has an average air speed that is 10 mph faster than a Cessna 150 airplane. If the combined distance traveled by these two small planes is 690 miles after 3 hr, what is the average speed of each plane?

Skill Practice Answers

9. The Cessna's speed is 110 mph, and the Piper Cub's speed is 120 mph.

Section 2.4 Practice Exercises

Study Skills Exercises

1. After doing a section of homework, check the odd-numbered answers in the back of the text. Choose a method to identify the exercises that you got wrong or had trouble with (i.e., circle the number or put a star by the number). List some reasons why it is important to label these problems.

2. Define the key terms:

a. consecutive integers **b. consecutive even integers** **c. consecutive odd integers**

Concept 2: Translations Involving Linear Equations

For Exercises 3–8, write an algebraic equation to represent the English sentence. Then solve the equation.

3. The sum of a number and sixteen is negative thirty-one. Find the number.

4. The sum of a number and negative twenty-one is fourteen. Find the number.

5. The difference of a number and six is negative three. Find the number.

6. The difference of a number and negative four is negative twelve. Find the number.

7. Sixteen less than a number is negative one. Find the number.

8. Ten less than a number is negative thirteen. Find the number.

For Exercises 9–18, use the problem-solving flowchart on page 150.

9. Six less than a number is -10. Find the number.

10. Fifteen less than a number is 41. Find the number.

11. Twice the sum of a number and seven is eight. Find the number.

12. Twice the sum of a number and negative two is sixteen. Find the number.

13. A number added to five is the same as twice the number. Find the number.

14. Three times a number is the same as the difference of twice the number and seven. Find the number.

15. The sum of six times a number and ten is equal to the difference of the number and fifteen. Find the number.

16. The difference of fourteen and three times a number is the same as the sum of the number and negative ten. Find the number.

17. If the difference of a number and four is tripled, the result is six more than the number. Find the number.

18. Twice the sum of a number and eleven is twenty-two less than three times the number. Find the number.

Concept 3: Consecutive Integer Problems

19. a. If x represents the smallest of three consecutive integers, write an expression to represent each of the next two consecutive integers.

b. If x represents the largest of three consecutive integers, write an expression to represent each of the previous two consecutive integers.

20. a. If x represents the smallest of three consecutive odd integers, write an expression to represent each of the next two consecutive odd integers.

b. If x represents the largest of three consecutive odd integers, write an expression to represent each of the previous two consecutive odd integers.

For Exercises 21–28, use the problem-solving flowchart from page 150.

21. The sum of two consecutive integers is -67. Find the integers.

22. The sum of two consecutive odd integers is 52. Find the integers.

23. The sum of two consecutive odd integers is 28. Find the integers.

24. The sum of three consecutive even integers is 66. Find the integers.

25. The sum of the page numbers on two facing pages in a book is 941. What are the page numbers?

x $x + 1$

26. Three raffle tickets are represented by three consecutive integers. If the sum of the three integers is 2,666,031, find the numbers.

x $x + 1$ $x + 2$

27. The perimeter of a pentagon (a five-sided polygon) is 80 in. The five sides are represented by consecutive integers. Find the measures of the sides.

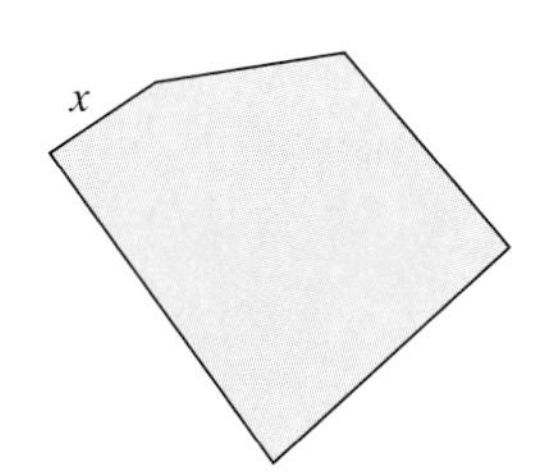

28. The perimeter of a pentagon (a five-sided polygon) is 95 in. The five sides are represented by consecutive integers. Find the measures of the sides.

Concept 4: Applications of Linear Equations

For Exercises 29–40, use the problem-solving flowchart (page 150) to solve the problems.

29. Karen's age is 12 years more than Clarann's age. The sum of their ages is 58. Find their ages.

30. Maria's age is 15 years less than Orlando's age. The sum of their ages is 29. Find their ages.

31. For a recent year, 104 more Democrats than Republicans were in the U.S. House of Representatives. If the total number of representatives in the House from these two parties was 434, find the number of representatives from each party.

32. For a recent year, 12 more Republicans than Democrats were in the U.S. House of Representatives. If the House had a total of 434 representatives from these two parties, find the number of Democrats and the number of Republicans.

33. A board is 86 cm in length and must be cut so that one piece is 20 cm longer than the other piece. Find the length of each piece.

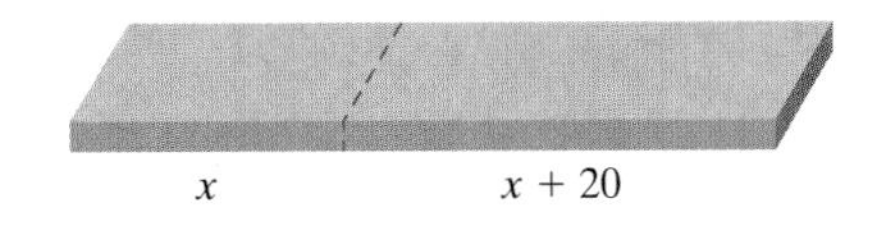

34. A rope is 54 in. in length and must be cut into two pieces. If one piece must be twice as long as the other, find the length of each piece.

35. Approximately 5.816 million people watch *The Oprah Winfrey Show*. This is 1.118 million more than watch *The Dr. Phil Show*. How many watch *The Dr. Phil Show*? (*Source: Neilson Media Research*)

36. Two of the largest Internet retailers are eBay and Amazon.com. Recently, the estimated U.S. sales of eBay were $0.1 billion less than twice the sales of Amazon.com. Given the total sales of $5.6 billion, determine the sales of eBay and Amazon.com.

37. The longest river in Africa is the Nile. It is 2455 km longer than the Congo River, also in Africa. The sum of the lengths of these rivers is 11,195 km. What is the length of each river?

38. The average depth of the Gulf of Mexico is three times the depth of the Red Sea. The difference between the average depths is 1078 m. What is the average depth of the Gulf of Mexico and the average depth of the Red Sea?

39. Asia and Africa are the two largest continents in the world. The land area of Asia is approximately 14,514,000 km^2 larger than the land area of Africa. Together their total area is 74,644,000 km^2. Find the land area of Asia and the land area of Africa.

40. Mt. Everest, the highest mountain in the world, is 2654 m higher than Mt. McKinley, the highest mountain in the United States. If the sum of their heights is 15,042 m, find the height of each mountain.

Concept 5: Applications Involving Uniform Motion

41. A woman can hike 1 mph faster down a trail to Cochita Lake than she can on the return trip uphill. It takes her 2 hr to get to the lake and 3 hr to return. What is her speed hiking down to the lake?

	Distance	Rate	Time
Downhill to the lake			
Uphill from the lake			

42. A car travels 20 mph slower in a bad rain storm than in sunny weather. The car travels the same distance in 2 hr in sunny weather as it does in 3 hr in rainy weather. Find the speed of the car in sunny weather.

43. Hazel and Emilie fly from Atlanta to San Diego. The flight from Atlanta to San Diego is against the wind and takes 4 hr. The return flight with the wind takes 3.5 hr. If the wind speed is 40 mph, find the speed of the plane in still air.

44. A boat on the Potomac River travels the same distance downstream in $\frac{2}{3}$ hr as it does going upstream in 1 hr. If the speed of the current is 3 mph, find the speed of the boat in still water.

45. Two cars are 200 miles apart and traveling toward each other on the same road. They meet in 2 hr. One car is traveling 4 mph faster than the other. What is the speed of each car?

46. Two cars are 238 miles apart and traveling toward each other along the same road. They meet in 2 hr. One car is traveling 5 mph slower than the other. What is the speed of each car?

47. After Hurricane Katrina, a rescue vehicle leaves a station at noon and heads for New Orleans. An hour later a second vehicle traveling 10 mph faster leaves the same station. By 4:00 P.M., the first vehicle reaches its destination, and the second is still 10 miles away. How fast is each vehicle?

48. A truck leaves a truck stop at 9:00 A.M. and travels toward Sturgis, Wyoming. At 10:00 A.M., a motorcycle leaves the same truck stop and travels the same route. The motorcycle travels 15 mph faster than the truck. By noon, the truck has traveled 20 miles further than the motorcycle. How fast is each vehicle?

49. Two boats traveling the same direction leave a harbor at noon. After 2 hr, they are 40 miles apart. If one boat travels twice as fast as the other, find the rate of each boat.

50. Two canoes travel down a river, starting at 9:00 A.M. One canoe travels twice as fast as the other. After 3.5 hr, the canoes are 5.25 miles apart. Find the speed of each canoe.

Mixed Exercises

51. A number increased by 58 is -22. Find the number.

52. The sum of a number and -14 is -32. Find the number.

53. Three consecutive integers are such that three times the largest exceeds the sum of the two smaller integers by 47. Find the integers.

54. Four times the smallest of three consecutive odd integers is 236 more than the sum of the other two integers. Find the integers.

55. A boat on the Hudson River travels the same distance in $\frac{1}{2}$ hour downstream as it can going upstream for an hour. If the boat travels 12 mph in still water, what is the speed of the current?

56. A flight from Orlando to Phoenix takes 4 hr with the wind. The return flight against the wind takes $4\frac{1}{2}$ hr. If the plane flies at 500 mph in still air, find the speed of the wind. (Round to the nearest mile per hour.)

57. Five times the difference of a number and three is four less than four times the number. Find the number.

58. Three times the difference of a number and seven is one less than twice the number. Find the number.

59. In a recent year, the estimated earnings for Jennifer Lopez was \$2.5 million more than half of the earnings for the band, U2. If the total earnings were \$106 million, what were the earnings for Jennifer Lopez and U2? (*Source: Forbes*)

60. In a recent year, the best selling DVD was *The Lord of the Rings: The Fellowship of the Rings*. Another big seller was *Harry Potter and the Sorcerer's Stone*. There were 90 million fewer Harry Potter DVDs sold than Lord of the Rings. If the two movies totaled 424 million, how many DVDs of each were sold? (*Source: VSDA VidTrac*)

61. A boat in distress, 21 nautical miles from a marina, travels toward the marina at 3 knots (nautical miles per hour). A coast guard cruiser leaves the marina and travels toward the boat at 25 knots. How long will it take for the boats to reach each other?

62. An air traffic controller observes a plane heading from New York to San Francisco traveling at 450 mph. At the same time, another plane leaves San Francisco and travels 500 mph to New York. If the distance between the airports is 2850 miles, how long will it take for the planes to pass each other?

63. If three is added to five times a number, the result is forty-three more than the number. Find the number.

64. If seven is added to three times a number, the result is thirty-one more than the number.

65. The deepest point in the Pacific Ocean is 676 m more than twice the deepest point in the Arctic Ocean. If the deepest point in the Pacific is 10,920 m, how many meters is the deepest point in the Arctic Ocean?

66. The area of Greenland is 201,900 km^2 less than three times the area of New Guinea. What is the area of New Guinea if the area of Greenland is 2,175,600 km^2?

Section 2.5 Applications Involving Percents

Concepts

1. Solving Basic Percent Equations
2. Applications Involving Sales Tax
3. Applications Involving Simple Interest

1. Solving Basic Percent Equations

Recall from Section R.3 that the word *percent* means "per hundred." For example:

Percent	Interpretation
63% of homes have a computer	63 out of 100 homes have a computer.
5% sales tax	5¢ in tax is charged for every 100¢ in merchandise.
15% commission	$15 is earned in commission for every $100 sold.

Percents come up in a variety of applications in day-to-day life. Many such applications follow the basic percent equation:

$$\text{amount} = (\text{percent})(\text{base}) \qquad \text{Basic percent equation}$$

In Example 1, we translate an English sentence into a percent equation.

Example 1 Solving Basic Percent Equations

a. What percent of 60 is 25.2?

b. 8.2 is 125% of what number?

c. 2% of 1500 is what number?

Solution:

a. Let x represent the unknown percent.

Step 1: Read the problem.
Step 2: Label the variables.

What percent of 60 is 25.2? — **Step 3:** Create a verbal model.

$$x \cdot 60 = 25.2$$ **Step 4:** Write a mathematical equation.

$$60x = 25.2$$ **Step 5:** Solve the equation.

$$\frac{60x}{60} = \frac{25.2}{60}$$

$$x = 0.42, \text{ or } 42\%$$ **Step 6:** Interpret the results and write the answer in words.

25.2 is 42% of 60.

Step 1: Read the problem.

b. Let x represent the unknown number. **Step 2:** Label the variables.

8.2 is 125% of what number? — **Step 3:** Create a verbal model.

$$8.2 = 1.25 \cdot x$$ **Step 4:** Write a mathematical equation.

$$8.2 = 1.25x$$ **Step 5:** Solve the equation.

$$\frac{8.2}{1.25} = \frac{1.25x}{1.25}$$

$$6.56 = x$$ **Step 6:** Interpret the results and write the answer in words.

8.2 is 125% of 6.56.

Avoiding Mistakes:

Be sure to use the decimal form of a percentage within an equation.

125% = 1.25

	Step 1: Read the problem.
c. Let x represent the unknown number.	**Step 2:** Label the variable.
2% of 1500 is what number?	**Step 3:** Create a verbal model.
$0.02 \cdot 1500 = x$	**Step 4:** Write a mathematical equation.
$30 = x$	**Step 5:** Solve the equation.
30 is 2% of 1500.	**Step 6:** Interpret the results.

Skill Practice

1. What percent of 85 is 11.9?

2. 279 is 90% of what number?

3. 115% of 82 is what number?

2. Applications Involving Sales Tax

One common use of percents is in computing **sales tax**.

$$\text{sales tax} = (\text{tax rate}) \cdot (\text{price of merchandise})$$

Example 2 Computing Sales Tax

A new digital camera costs \$429.95.

a. Compute the sales tax if the tax rate is 4%.

b. Determine the total cost, including tax.

Solution:

	Step 1: Read the problem.
a. Let x represent the amount of tax.	**Step 2:** Label the variable.
sales tax = (tax rate)(price of merchandise)	**Step 3:** Write a verbal equation.
$x = (0.04)(\$429.95)$	**Step 4:** Write a mathematical equation.
$x = \$17.198$	**Step 5:** Solve the equation.
$x = \$17.20$	Round to the nearest cent.
The tax on the merchandise is \$17.20.	**Step 6:** Interpret the results.

b. The total cost is found by:

total cost = cost of merchandise + amount of tax

Therefore the total cost is: \$429.95 + \$17.20 = \$447.15.

Skill Practice

4. Determine the total cost, including tax, of a portable CD player that sells for \$89. Assume that the tax rate is 6%.

Skill Practice Answers

1. 14% **2.** 310
3. 94.3 **4.** \$94.34

Example 3 **Applying Percents**

A video game is purchased for a total of $48.15, including sales tax. If the tax rate is 7%, find the original price of the video game before the sales tax.

Solution:

Step 1: Read the problem.

Let x represent the price of the video game. **Step 2:** Label variables.

$0.07x$ represents the amount of sales tax.

$$\begin{pmatrix}\text{Original}\\ \text{price}\end{pmatrix} + \begin{pmatrix}\text{sales}\\ \text{tax}\end{pmatrix} = \begin{pmatrix}\text{total}\\ \text{cost}\end{pmatrix}$$

Step 3: Write a verbal equation.

$$x + 0.07x = \$48.15$$

Step 4: Write a mathematical equation.

$$1.07x = 48.15$$

Step 5: Solve for x.

$$\frac{1.07x}{1.07} = \frac{48.15}{1.07}$$

Divide both sides by 1.07.

$$x = 45$$

Step 6: Interpret the results and write the answer in words.

The original price was $45.

Avoiding Mistakes:

The sales tax is computed on the *original* price of the merchandise, not the final price. That is, the sales tax is $0.07x$, not 0.07($48.15).

Skill Practice

5. The total price of a pair of jeans, including a 5% sales tax, is $27.30. Find the original price of the jeans.

3. Applications Involving Simple Interest

One important application of percents is in computing simple interest on a loan or on an investment.

Banks hold large quantities of money for their customers. However, because all bank customers are unlikely to withdraw all their money on a single day, a bank does not keep all the money in cash. Instead, it keeps some cash for day-to-day transactions but invests the remaining portion of the money. Because a bank uses its customer's money to make investments and because it wants to attract more customers, the bank pays interest on the money.

Simple interest is interest that is earned on principal (the original amount of money invested in an account). The following formula is used to compute simple interest:

$$\begin{pmatrix}\text{simple}\\ \text{interest}\end{pmatrix} = \begin{pmatrix}\text{principal}\\ \text{invested}\end{pmatrix}\begin{pmatrix}\text{annual}\\ \text{interest rate}\end{pmatrix}\begin{pmatrix}\text{time}\\ \text{in years}\end{pmatrix}$$

This formula is often written symbolically as $I = Prt$.

For example, to find the simple interest earned on $2000 invested at 7.5% interest for 3 years, we have

$$I = Prt$$

$$\text{Interest} = (\$2000)(0.075)(3)$$

$$= \$450$$

Skill Practice Answers

5. The original price was $26.00.

Example 4 Computing Interest

Wade borrows \$12,000 for a new car.

a. If he pays 7.5% simple interest over a 4-year period, how much total interest will he pay?

b. Determine the total amount to be paid back.

Solution:

a. Let I represent the amount of interest. — **Step 1:** Read the problem. **Step 2:** Label the variables.

Interest = (Principal)(interest rate)(time in years) — **Step 3:** Write an equation in words.

$I = (\$12{,}000)(0.075)(4)$ — **Step 4:** Write a mathematical equation.

$I = \$3600$ — **Step 5:** Solve the equation.

Wade will pay \$3600 in interest. — **Step 6:** Interpret the results.

b. The total amount to be paid back includes the amount borrowed plus interest.

$$\$12{,}000 + \$3600 = \$15{,}600$$

Skill Practice

6. How much interest must be paid if \$8000 is invested at a 5.5% simple interest rate for 2 years?

Example 5 Applying Simple Interest

Jorge wants to save money for his daughter's college education. If Jorge needs to have \$4340 at the end of 4 years, how much money would he need to invest at a 6% simple interest rate?

Solution:

Let P represent the original amount invested. — **Step 1:** Read the problem. **Step 2:** Label the variables.

$$\begin{pmatrix}\text{original}\\ \text{principal}\end{pmatrix} + (\text{interest}) = (\text{total})$$

Step 3: Write an equation in words.

Recall that interest is computed by the formula $I = Prt$.

$$(P) \quad + \quad (Prt) \quad = (\text{total})$$

$$P + P(0.06)(4) = 4340$$

Step 4: Write a mathematical equation.

$$P + 0.24P = 4340$$

$$1.24P = 4340$$

Step 5: Solve the equation.

$$\frac{1.24P}{1.24} = \frac{4340}{1.24}$$

$$P = 3500$$

Skill Practice Answers

6. The interest is \$880.

The original investment should be $3500. **Step 6:** Interpret the results and write the answer in words.

Skill Practice

7. Cassandra invested some money in her bank account, and after 10 years at 4% simple interest, it has grown to $7700. What was the initial amount invested?

Skill Practice Answers

7. The initial investment was $5500.

Section 2.5 Practice Exercises

Boost *your* GRADE at mathzone.com! MathZone

- Practice Problems
- Self-Tests
- NetTutor
- e-Professors
- Videos

Study Skills Exercises

1. Go to the online service called MathZone that accompanies this text (www.mathzone.com). Name two features that this online service offers that can help you in this course.

2. Define the key terms:

a. sales tax **b. simple interest**

Review Exercises

3. List the six steps to solve an application.

For Exercises 4–5, use the steps for problem solving to solve these applications.

4. Find two consecutive integers such that 3 times the larger is the same as 45 more than the smaller. Find the numbers.

5. The height of the Great Pyramid of Giza is 17 m more than twice the height of the pyramid found in Saqqara. If the difference in their heights is 77 m, find the height of each pyramid.

Concept 1: Solving Basic Percent Equations

For Exercises 6–17, find the missing values.

6. 45 is what percent of 360?

7. 338 is what percent of 520?

8. 544 is what percent of 640?

9. 576 is what percent of 800?

10. What is 0.5% of 150?

11. What is 9.5% of 616?

12. What is 142% of 740?

13. What is 156% of 280?

14. 177 is 20% of what number?

15. 126 is 15% of what number?

16. 275 is 12.5% of what number?

17. 594 is 45% of what number?

For Exercises 18–21, use the graph showing the distribution for leading forms of cancer in men. (*Source: Centers for Disease Control*)

Percent of Cancer Cases by Type (Men)

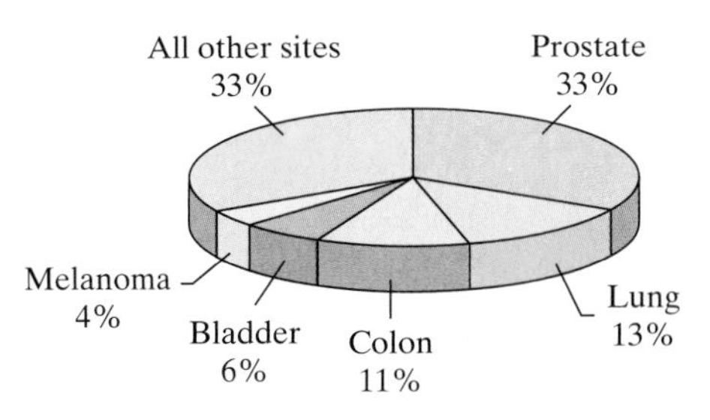

18. If there are 700,000 cases of cancer in men in the United States, approximately how many are prostate cancer?

19. Approximately how many cases of lung cancer would be expected in 700,000 cancer cases among men in the United States.

20. There were 14,000 cases of cancer of the pancreas diagnosed out of 700,000 cancer cases. What percent is this?

21. There were 21,000 cases of leukemia diagnosed out of 700,000 cancer cases. What percent is this?

22. A Pioneer car CD/MP3 player costs \$170. Circuit City has it on sale for 12% off with free installation.

a. What is the discount on the CD/MP3 player?

b. What is the sale price?

23. A laptop computer, originally selling for \$899.00, is on sale for 10% off.

a. What is the discount on the laptop?

b. What is the sale price?

24. A Sony digital camera is on sale for \$400.00. This price is 15% off the original price. What was the original price? Round to the nearest cent.

25. The *Star Wars III* DVD is on sale for \$18. If this represents an 18% sale price, what was the original price of the DVD?

26. The original price of an Audio Jukebox was \$250. It is on sale for \$220. What percent discount does this represent?

27. During the holiday season, the Xbox 360 sold for \$425.00 in stores. This product was in such demand that it sold for \$800 online. What percent markup does this represent? (Round to the nearest whole percent.)

Concept 2: Applications Involving Sales Tax

For Exercises 28–37, solve for the unknown quantity.

28. A Craftsman drill is on sale for \$99.99. If the sales tax rate is 7%, how much will Molly have to pay for the drill?

29. Patrick purchased four new tires that were regularly priced at \$94.99 each, but are on sale for \$20 off per tire. If the sales tax rate is 6%, how much will be charged to Patrick's VISA card?

30. The sales tax for a screwdriver set came to \$1.04. If the sales tax rate is 6.5%, what was the price of the screwdriver?

31. The sales tax for a picture frame came to \$1.32. If the sales tax rate is 5.5%, what was the price of the picture frame?

32. Sun Lei bought a laptop computer over the Internet for \$1800. The total cost, including tax, came to \$1890. What is the sales tax rate?

33. Jamie purchased a compact disc and paid \$18.26. If the disc price is \$16.99, what is the sales tax rate (round to the nearest tenth of a percent)?

34. When the Hendersons went to dinner, their total bill, including tax, was \$43.74. If the sales tax rate is 8%, what was the original price of the dinner?

35. The admission to Walt Disney World costs \$74.37 including taxes of 11%. What is the original price of a ticket?

36. A hotel room rented for five nights costs \$706.25 including 13% in taxes. Find the original price of the room rental for the five nights. Then find the price per night.

37. The price of four CDs is \$74.88, including a 4% sales tax. Find the original cost of a single CD assuming all CDs are the same price.

Concept 3: Applications Involving Simple Interest

For Exercises 38–47, solve these equations involving simple interest.

38. How much interest will Pam earn in 4 years if she invests \$3000 in an account that pays 3.5% simple interest?

39. How much interest will Roxanne have to pay if she borrows \$2000 for 2 years at a simple interest rate of 4%?

40. Bob borrowed some money for 1 year at 5% simple interest. If he had to pay back a total of \$1260, how much did he originally borrow?

41. Mike borrowed some money for 2 years at 6% simple interest. If he had to pay back a total of \$3640, how much did he originally borrow?

42. If \$1500 grows to \$1950 after 5 years, find the simple interest rate.

43. If \$9000 grows to \$10,440 in 2 years, find the simple interest rate.

44. A new bank offered simple interest loans at 11% for new customers. If a customer took out a loan for \$2000 to be paid back in 18 months ($\frac{18}{12} = \frac{3}{2}$ years), find

a. the interest on the loan

b. the total amount that the customer owes (principal + interest).

45. Rafael has \$3000 saved for a future trip to Europe. If he invests in an account that pays 4% simple interest, how much will he have after $2\frac{1}{2}$ years?

46. Perry is planning a vacation to Europe in 2 years. How much should he invest in a certificate of deposit that pays 3% simple interest to get the \$3500 that he needs for the trip? Round to the nearest dollar.

47. Sherica invested in a mutual fund and at the end of 20 years she has \$14,300 in her account. If the mutual fund returned an average yield of 8%, how much did she originally invest?

Expanding Your Skills

Percents are often used to represent rates. For example, salespeople may receive all or part of their salary as a percentage of their sales. This is called a **commission**. The **commission rate** is the percent of the sales that the salesperson receives in income.

$$\text{commission} = (\text{commission rate}) \cdot (\text{\$ in merchandise sold})$$

48. The local car dealership pays its sales personnel a commission of 25% of the dealer profit on each car sold. The dealer made a profit of \$18,250 on the cars Joëlle sold last month. What was her commission last month?

49. Dan sold a beachfront home for \$650,000. If his commission rate is 4%, what did he earn on the sale of that home?

50. A salesperson at You Bought It discount store earns 3% commission on all appliances that he sells. If Geoff's commission for the month was \$116.37, how much did he sell?

51. Anna makes a commission at an appliance store. In addition to her base salary, she earns a 2.5% commission on her sales. If Anna's commission for a month was \$260, how much did she sell?

52. For selling software, Tom received a bonus commission based on sales over \$500. If he received \$180 in commission for selling a total of \$2300 worth of software, what is his commission rate?

53. In addition to an hourly salary, Jessica earns a commission for selling ice cream bars at the beach. If she sells \$708 worth of ice cream and receives a commission of \$56.64, what is her commission rate?

54. Diane sells women's sportswear at a department store. She earns a regular salary and, as a bonus, she receives a commission of 4% on all sales over \$200. If Diane earned an extra \$25.80 last week in commission, how much merchandise did she sell over \$200?

Formulas and Applications of Geometry

Section 2.6

Concepts

1. Formulas
2. Geometry Applications

1. Formulas

Literal equations are equations that contain several variables. A formula is a literal equation with a specific application. For example, the perimeter of a triangle (distance around the triangle) can be found by the formula $P = a + b + c$, where a, b, and c are the lengths of the sides (Figure 2-8).

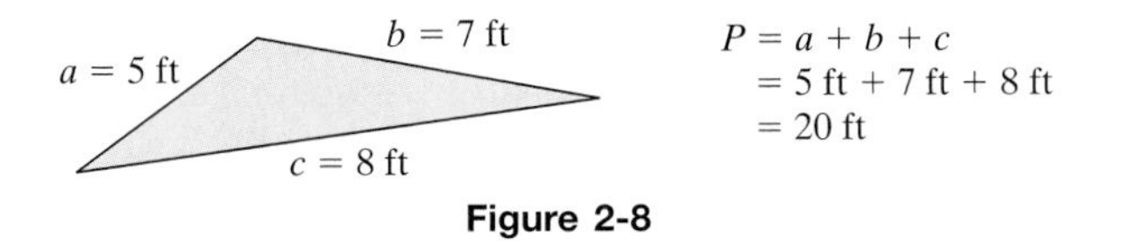

Figure 2-8

In this section, we will learn how to rewrite formulas to solve for a different variable within the formula. Suppose, for example, that the perimeter of a triangle is known and two of the sides are known (say, sides a and b). Then the third side, c, can be found by subtracting the lengths of the known sides from the perimeter (Figure 2-9).

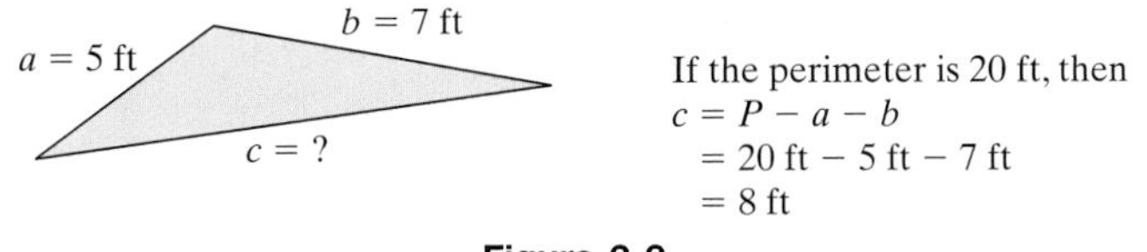

Figure 2-9

To solve a formula for a different variable, we use the same properties of equality outlined in the earlier sections of this chapter. For example, consider the two equations $2x + 3 = 11$ and $wx + y = z$. Suppose we want to solve for x in each case:

$2x + 3 = 11$		$wx + y = z$	
$2x + 3 - 3 = 11 - 3$	Subtract 3.	$wx + y - y = z - y$	Subtract y.
$2x = 8$		$wx = z - y$	
$\frac{2x}{2} = \frac{8}{2}$	Divide by 2.	$\frac{wx}{w} = \frac{z - y}{w}$	Divide by w.
$x = 4$		$x = \frac{z - y}{w}$	

The equation on the left has only one variable and we are able to simplify the equation to find a numerical value for x. The equation on the right has multiple variables. Because we do not know the values of w, y, and z, we are not able to simplify further. The value of x is left as a formula in terms of w, y, and z.

Example 1 Solving Formulas for an Indicated Variable

Solve the formulas for the indicated variables.

a. $d = rt$ for t **b.** $5x + 2y = 12$ for y

Solution:

a. $d = rt$ for t The goal is to isolate the variable t.

$\frac{d}{r} = \frac{rt}{r}$ Because the relationship between r and t is multiplication, we reverse the process by dividing both sides by r.

$\frac{d}{r} = t$, or equivalently $t = \frac{d}{r}$

TIP: The original equation $d = rt$ represents the distance traveled, d, in terms of the rate of speed, r, and the time of travel, t.

The equation $t = \frac{d}{r}$ represents the same relationship among the variables, however the time of travel is expressed in terms of the distance and rate.

b. $5x + 2y = 12$ for y The goal is to solve for y.

$5x - 5x + 2y = 12 - 5x$ Subtract $5x$ from both sides to isolate the y-term.

$2y = -5x + 12$ $-5x + 12$ is the same as $12 - 5x$.

$\frac{2y}{2} = \frac{-5x + 12}{2}$ Divide both sides by 2 to isolate y.

$y = \frac{-5x + 12}{2}$

TIP: In the expression $\frac{-5x + 12}{2}$ do not try to divide the 2 into the 12. The divisor of 2 is dividing the entire quantity, $-5x + 12$ (not just the 12).

We may, however, apply the divisor to each term individually in the numerator. That is, $\frac{-5x + 12}{2}$ can be written in several different forms. Each is correct.

$$y = \frac{-5x + 12}{2} \quad \text{or} \quad y = \frac{-5x}{2} + \frac{12}{2} \Rightarrow y = -\frac{5x}{2} + 6$$

Skill Practice Solve for the indicated variable.

1. $A = lw$ for l **2.** $-2a + 4b = 7$ for a

Example 2 **Solving Formulas for an Indicated Variable**

The formula $C = \frac{5}{9}(F - 32)$ is used to find the temperature, C, in degrees Celsius for a given temperature expressed in degrees Fahrenheit, F. Solve the formula $C = \frac{5}{9}(F - 32)$ for F.

Solution:

$$C = \frac{5}{9}(F - 32)$$

$$C = \frac{5}{9}F - \frac{5}{9} \cdot 32 \qquad \text{Clear parentheses.}$$

$$C = \frac{5}{9}F - \frac{160}{9} \qquad \text{Multiply: } \frac{5}{9} \cdot \frac{32}{1} = \frac{160}{9}.$$

$$9(C) = 9\left(\frac{5}{9}F - \frac{160}{9}\right) \qquad \text{Multiply by the LCD to clear fractions.}$$

$$9C = \frac{9}{1} \cdot \frac{5}{9}F - \frac{9}{1} \cdot \frac{160}{9} \qquad \text{Apply the distributive property.}$$

$$9C = 5F - 160 \qquad \text{Simplify.}$$

$$9C + 160 = 5F - 160 + 160 \qquad \text{Add 160 to both sides.}$$

$$9C + 160 = 5F$$

$$\frac{9C + 160}{5} = \frac{5F}{5} \qquad \text{Divide both sides by 5.}$$

$$\frac{9C + 160}{5} = F$$

The answer may be written in several forms:

$$F = \frac{9C + 160}{5} \quad \text{or} \quad F = \frac{9C}{5} + \frac{160}{5} \quad \Rightarrow \quad F = \frac{9}{5}C + 32$$

Skill Practice Solve the formula.

3. $M = \frac{1}{2}(a + b)$ for a

2. Geometry Applications

In Section R.4, we presented numerous facts and formulas related to geometry. Sometimes these are needed to solve applications in geometry.

Example 3 **Solving a Geometry Application Involving Perimeter**

The length of a rectangular lot is 1 m less than twice the width. If the perimeter is 190 m, find the length and width.

Skill Practice Answers

1. $l = \frac{A}{w}$

2. $a = \frac{4b - 7}{2}$ or $a = 2b - \frac{7}{2}$

3. $a = 2M - b$

Solution:

Step 1: Read the problem.

Let x represent the width of the rectangle. **Step 2:** Label the variables.

Then $2x - 1$ represents the length.

x

$2x - 1$

$P = 2l + 2w$ **Step 3:** Perimeter formula

$190 = 2(2x - 1) + 2(x)$ **Step 4:** Write an equation in terms of x.

$190 = 4x - 2 + 2x$ **Step 5:** Solve for x.

$190 = 6x - 2$

$192 = 6x$

$$\frac{192}{6} = \frac{6x}{6}$$

$32 = x$

The width is $x = 32$.

The length is $2x - 1 = 2(32) - 1 = 63$. **Step 6:** Interpret the results and write the answer in words.

The width of the rectangular lot is 32 m and the length is 63 m.

Skill Practice

4. The length of a rectangle is 10 ft less than twice the width. If the perimeter is 178 ft, find the length and width.

Example 4 Solving a Geometry Application Involving Complementary Angles

Two complementary angles are drawn such that one angle is 4° more than seven times the other angle. Find the measure of each angle.

Solution: **Step 1:** Read the problem.

Let x represent the measure of one angle. **Step 2:** Label the variables.

Then $7x + 4$ represents the measure of the other angle.
The angles are complementary, so their sum must be 90°.

$(7x + 4)°$

$x°$

Skill Practice Answers

4. length is 56 ft, width is 33 ft

$$\begin{pmatrix}\text{Measure of}\\ \text{first angle}\end{pmatrix} + \begin{pmatrix}\text{measure of}\\ \text{second angle}\end{pmatrix} = 90°$$

Step 3: Create a verbal equation.

$$x \quad + \quad 7x + 4 \quad = 90$$

Step 4: Write a mathematical equation.

$$8x + 4 = 90$$

Step 5: Solve for x.

$$8x = 86$$

$$\frac{8x}{8} = \frac{86}{8}$$

$$x = 10.75$$

Step 6: Interpret the results and write the answer in words.

One angle is $x = 10.75$.

The other angle is $7x + 4 = 7(10.75) + 4 = 79.25$.

The angles are 10.75° and 79.25°.

Skill Practice

5. Two supplementary angles are constructed so that the measure of one is 120° more than twice the other. Find the measures of the angles.

Example 5 Solving a Geometry Application

One angle in a triangle is twice as large as the smallest angle. The third angle is 10° more than seven times the smallest angle. Find the measure of each angle.

Solution:

Step 1: Read the problem.

Let x represent the measure of the smallest angle.

Then $2x$ and $7x + 10$ represent the measures of the other two angles.

Step 2: Label the variables.

$x°$, $(7x + 10)°$, $(2x)°$

The sum of the angles must be 180°.

Step 3: Create a verbal equation.

$$x + 2x + (7x + 10) = 180$$

Step 4: Write a mathematical equation.

$$x + 2x + 7x + 10 = 180$$

Step 5: Solve for x.

$$10x + 10 = 180$$

$$10x = 170$$

$$x = 17$$

Skill Practice Answers

5. 20° and 160°

The smallest angle is $x = 17$.

The other angles are $2x = 2(17) = 34$

$$7x + 10 = 7(17) + 10 = 129$$

The angles are 17°, 34°, and 129°.

Step 6: Interpret the results and write the answer in words.

Skill Practice

6. In a triangle, the measure of the largest angle is 80° greater than the measure of the smallest angle. The measure of the middle angle is twice that of the smallest. Find the measures of the angles.

Example 6 Solving a Geometry Application Involving Circumference

The circumference of a circle is 188.4 ft. Find the radius to the nearest tenth of a foot (Figure 2-10). Use 3.14 for π.

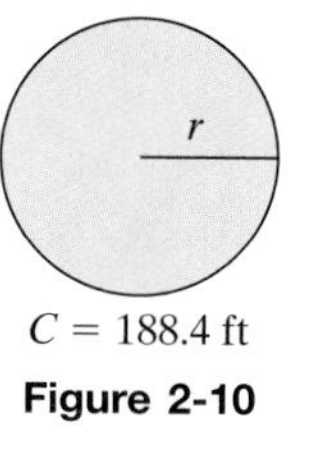

$C = 188.4$ ft

Figure 2-10

Solution:

$C = 2\pi r$ Use the formula for the circumference of a circle.

$188.4 = 2\pi r$ Substitute 188.4 for C.

$\frac{188.4}{2\pi} = \frac{2\pi r}{2\pi}$ Divide both sides by 2π.

$$r = \frac{188.4}{2\pi}$$

$$r \approx \frac{188.4}{2(3.14)}$$

$$= 30.0$$

The radius is approximately 30.0 ft.

Skill Practice

7. The area of a triangle is 52 cm^2 and the height is 10 cm. Find the measure of the base of the triangle.

Skill Practice Answers

6. 25°, 50°, and 105°

7. The base measures 10.4 cm.

TIP: In Example 6, we could have solved the equation $C = 2\pi r$ for the variable r first before substituting the value of C.

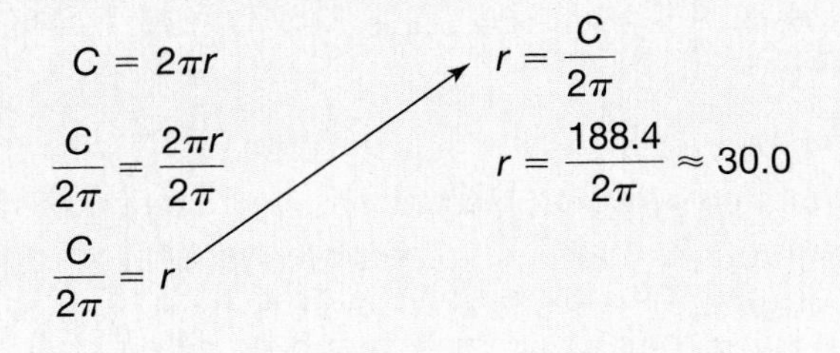

$$C = 2\pi r$$

$$\frac{C}{2\pi} = \frac{2\pi r}{2\pi}$$

$$\frac{C}{2\pi} = r$$

$$r = \frac{C}{2\pi}$$

$$r = \frac{188.4}{2\pi} \approx 30.0$$

Calculator Connections

In Example 6, we could obtain a more accurate result if we use the π key on the calculator.

Note that parentheses are required to divide 188.4 by the quantity 2π. This guarantees that the calculator follows the implied order of operations. Without parentheses, the calculator would divide 188.4 by 2 and then multiply the result by π.

Scientific Calculator

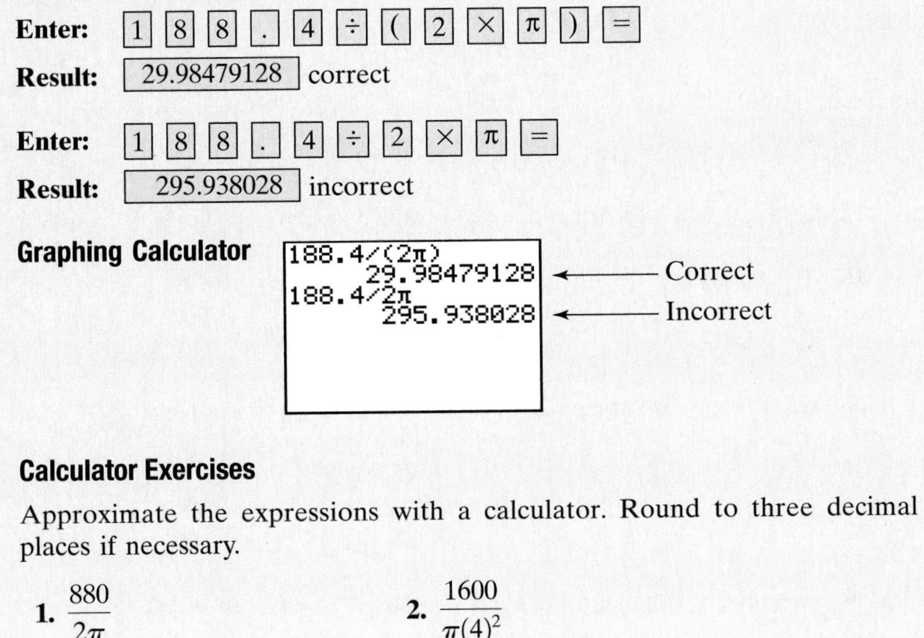

Enter: 1 8 8 . 4 ÷ (2 × π) =

Result: 29.98479128 correct

Enter: 1 8 8 . 4 ÷ 2 × π =

Result: 295.938028 incorrect

Graphing Calculator

Calculator Exercises

Approximate the expressions with a calculator. Round to three decimal places if necessary.

1. $\frac{880}{2\pi}$

2. $\frac{1600}{\pi(4)^2}$

3. $\frac{20}{(-0.05)(5)}$

4. $\frac{10}{0.5(6 + 4)}$

Section 2.6 Practice Exercises

Boost *your* GRADE at mathzone.com!

- Practice Problems
- Self-Tests
- NetTutor
- e-Professors
- Videos

Study Skills Exercises

1. A good technique for studying for a test is to choose four problems from each section of the chapter and write each of them along with the directions on a 3×5 card. On the back, put the page number where you found that problem. Then shuffle the cards and test yourself on the procedure to solve each problem. If you find one that you do not know how to solve, look at the page number and do several of that type. Write four problems you would choose for this section.

2. Define the key term: **literal equation**

Review Exercises

For Exercises 3–7, solve the equation.

3. $3(2y + 3) - 4(-y + 1) = 7y - 10$

4. $-(3w + 4) + 5(w - 2) - 3(6w - 8) = 10$

5. $\frac{1}{2}(x - 3) + \frac{3}{4} = 3x - \frac{3}{4}$

6. $\frac{5}{6}x + \frac{1}{2} = \frac{1}{4}(x - 4)$

7. $0.5(y + 2) - 0.3 = 0.4y + 0.5$

Concept 1: Formulas

For Exercises 8–39, solve for the indicated variable.

8. $P = a + b + c$ for a

9. $P = a + b + c$ for b

10. $x = y - z$ for y

11. $c + d = e$ for d

12. $p = 250 + q$ for q

13. $y = 35 + x$ for x

14. $d = rt$ for t

15. $d = rt$ for r

16. $PV = nrt$ for t

17. $P_1V_1 = P_2V_2$ for V_1

18. $x - y = 5$ for x

19. $x + y = -2$ for y

20. $3x + y = -19$ for y

21. $x - 6y = -10$ for x

22. $2x + 3y = 6$ for y

23. $5x + 2y = 10$ for y

24. $-2x - y = 9$ for x

25. $3x - y = -13$ for x

26. $4x - 3y = 12$ for y

27. $6x - 3y = 4$ for y

28. $ax + by = c$ for y

29. $ax + by = c$ for x

30. $A = P(1 + rt)$ for t

31. $P = 2(L + w)$ for L

32. $a = 2(b + c)$ for c

33. $3(x + y) = z$ for x

34. $Q = \frac{x + y}{2}$ for y

35. $Q = \frac{a - b}{2}$ for a

36. $M = \frac{a}{S}$ for a

37. $A = \frac{1}{3}(a + b + c)$ for c

38. $P = I^2R$ for R

39. $F = \frac{GMm}{d^2}$ for m

Concept 2: Geometry Applications

For Exercises 40–60, use the problem-solving flowchart (page 150) from Section 2.4.

40. The perimeter of a rectangular garden is 24 ft. The length is 2 ft more than the width. Find the length and the width of the garden.

41. In a small rectangular wallet photo, the width is 7 cm less than the length. If the border (perimeter) of the photo is 34 cm, find the length and width.

42. A builder buys a rectangular lot of land such that the length is 5 m less than two times the width. If the perimeter is 590 m, find the length and the width.

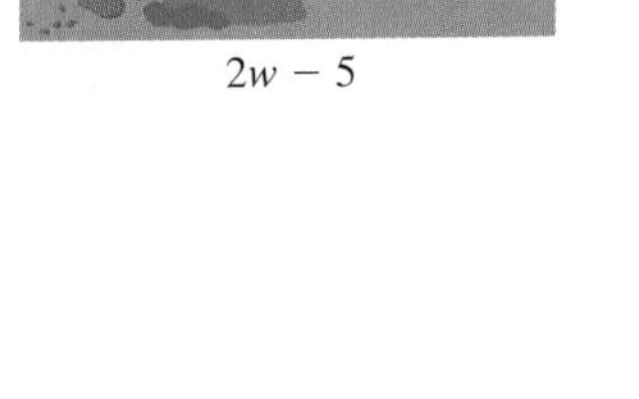

43. The perimeter of a rectangular pool is 140 yd. If the length is 10 yd more than the width, find the length and the width.

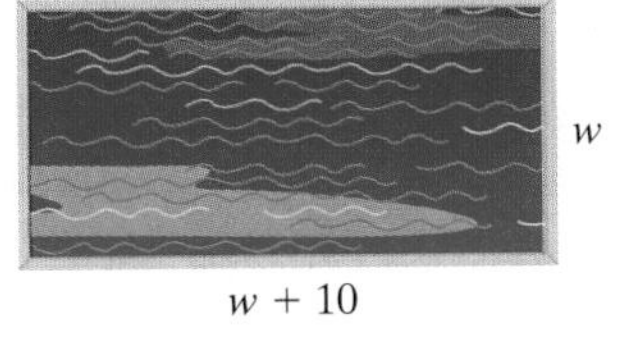

44. A triangular parking lot has two sides that are the same length and the third side is 5 m longer. If the perimeter is 71 m, find the lengths of the sides.

45. The perimeter of a triangle is 16 ft. One side is 3 ft longer than the shortest side. The third side is 1 ft longer than the shortest side. Find the lengths of all the sides.

46. The largest angle in a triangle is three times the smallest angle. The middle angle is two times the smallest angle. Given that the sum of the angles in a triangle is 180°, find the measure of each angle.

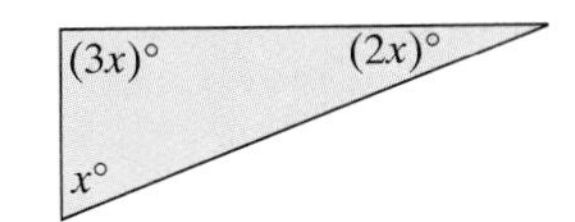

47. The smallest angle in a triangle measures 90° less than the largest angle. The middle angle measures 60° less than the largest angle. Find the measure of each angle.

48. The smallest angle in a triangle is half the largest angle. The middle angle measures 30° less than the largest angle. Find the measure of each angle.

49. The largest angle of a triangle is three times the middle angle. The smallest angle measures 10° less than the middle angle. Find the measure of each angle.

50. Find the value of x and the measure of each angle labeled in the figure.

51. Find the value of y and the measure of each angle labeled in the figure.

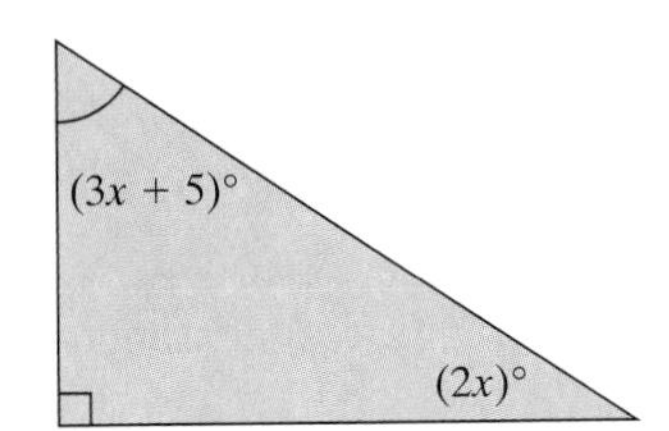

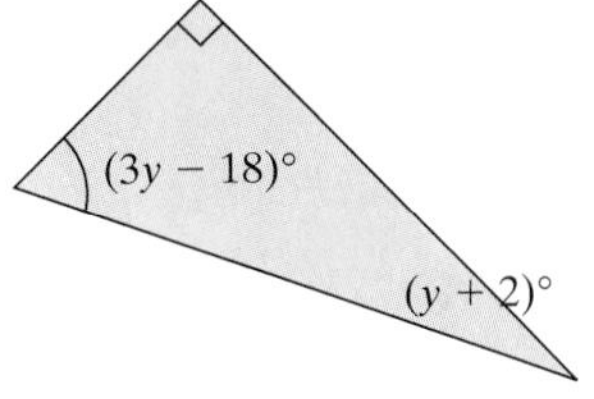

52. Sometimes memory devices are helpful for remembering mathematical facts. Recall that the sum of two complementary angles is 90°. That is, two complementary angles when added together form a right triangle or "corner." The words ***Complementary*** and ***Corner*** both start with the letter "***C***." Derive your own memory device for remembering that the sum of two supplementary angles is 180°.

Complementary angles form a "Corner" Supplementary angles . . .

53. Two angles are complementary. One angle is 20° less than the other angle. Find the measures of the angles.

54. Two angles are complementary. One angle is twice as large as the other angle. Find the measures of the angles.

55. Two angles are complementary. One angle is 4° less than three times the other angle. Find the measures of the angles.

56. Two angles are supplementary. One angle is three times as large as the other angle. Find the measures of the angles.

57. Two angles are supplementary. One angle is twice as large as the other angle. Find the measures of the angles.

58. Two angles are supplementary. One angle is 6° more than four times the other. Find the measures of the two angles.

59. Find the measures of the vertical angles labeled in the figure by first solving for x.

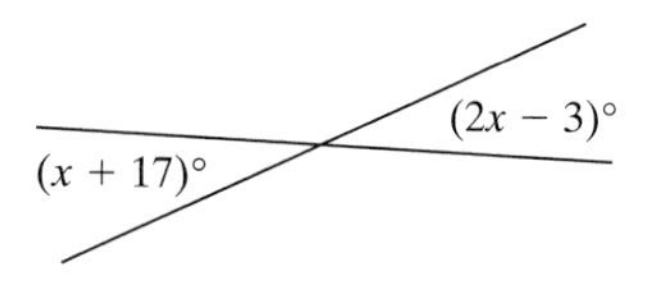

60. Find the measures of the vertical angles labeled in the figure by first solving for y.

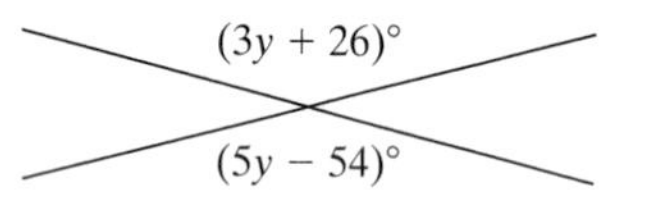

61. **a.** A rectangle has length l and width w. Write a formula for the area.

b. Solve the formula for the width, w.

c. The area of a rectangular volleyball court is 1740.5 ft^2 and the length is 59 ft. Find the width.

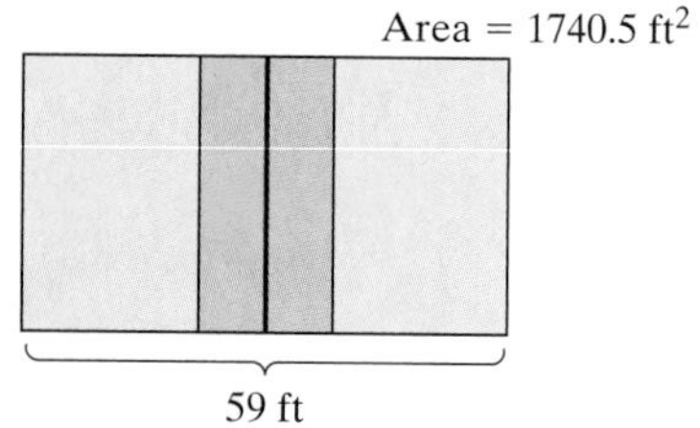

62. **a.** A parallelogram has height h and base b. Write a formula for the area.

b. Solve the formula for the base, b.

c. Find the base of the parallelogram pictured if the area is 40 m^2.

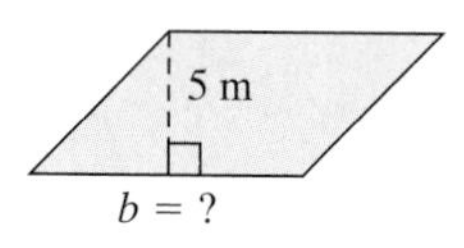

63. a. A rectangle has length l and width w. Write a formula for the perimeter.

b. Solve the formula for the length, l.

c. The perimeter of the soccer field at Giants Stadium is 338 m. If the width is 66 m, find the length.

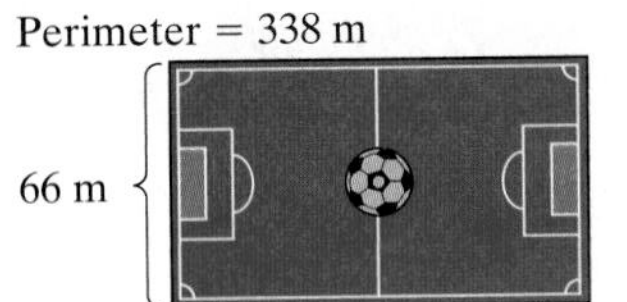

64. a. The length of each side of a square is s. Write a formula for the perimeter of the square.

b. Solve the formula for the length of a side, s.

c. The Pyramid of Khufu (known as the Great Pyramid) at Giza has a square base. If the distance around the bottom is 921.6 m, find the length of the sides at the bottom of the pyramid.

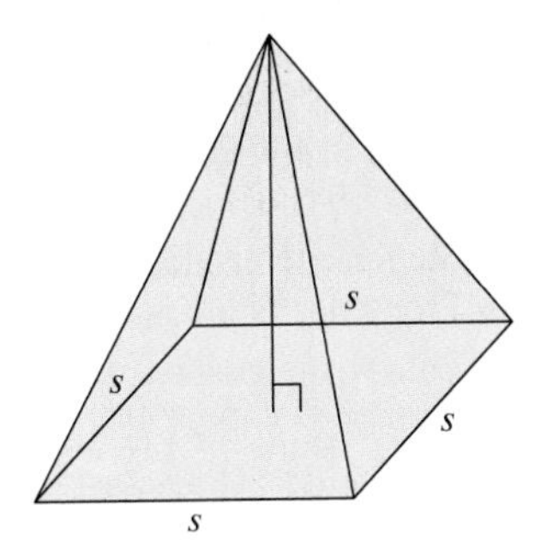

65. a. A triangle has height h and base b. Write a formula for the area.

b. Solve the formula for the height, h.

c. Find the height of the triangle pictured if the area is 12 km^2.

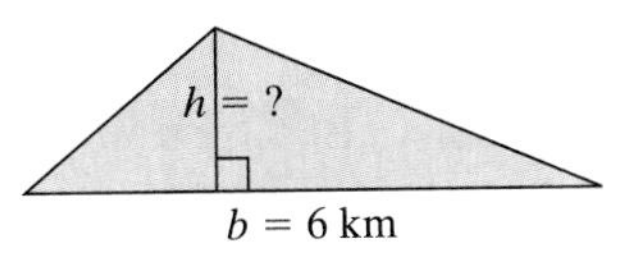

66. a. A circle has a radius of r. Write a formula for the circumference.

b. Solve the formula for the radius, r.

c. The circumference of the circular Buckingham Fountain in Chicago is approximately 880 ft. Find the radius. Round to the nearest foot.

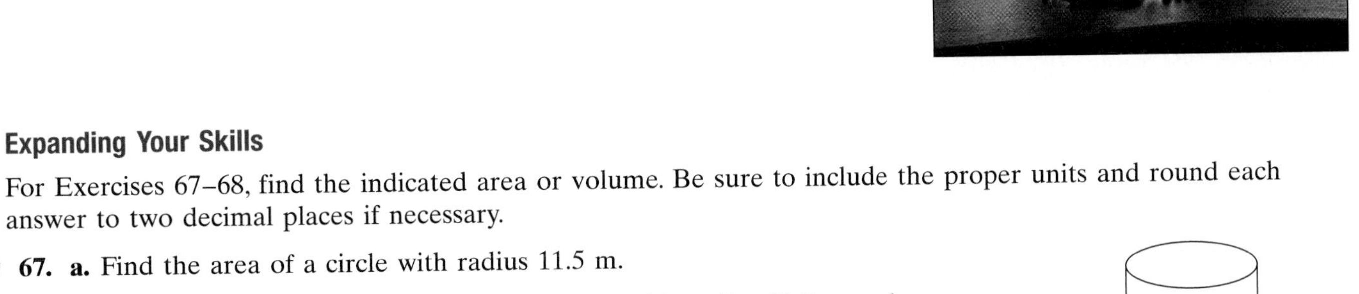

Expanding Your Skills

For Exercises 67–68, find the indicated area or volume. Be sure to include the proper units and round each answer to two decimal places if necessary.

67. a. Find the area of a circle with radius 11.5 m.

b. Find the volume of a right circular cylinder with radius 11.5 m and height 25 m.

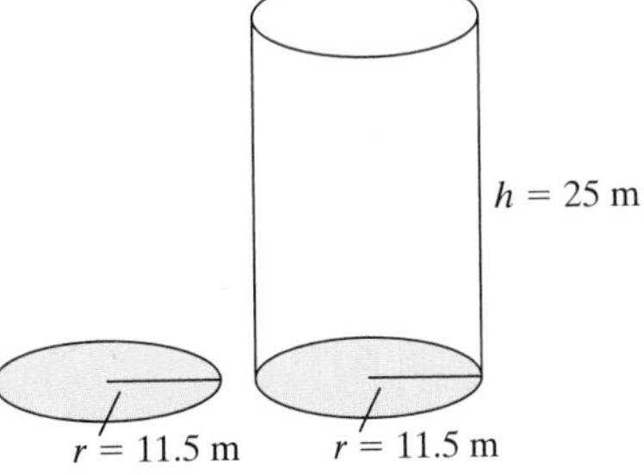

68. a. Find the area of a parallelogram with base 30 in. and height 12 in.

b. Find the area of a triangle with base 30 in. and height 12 in.

c. Compare the areas found in parts (a) and (b).

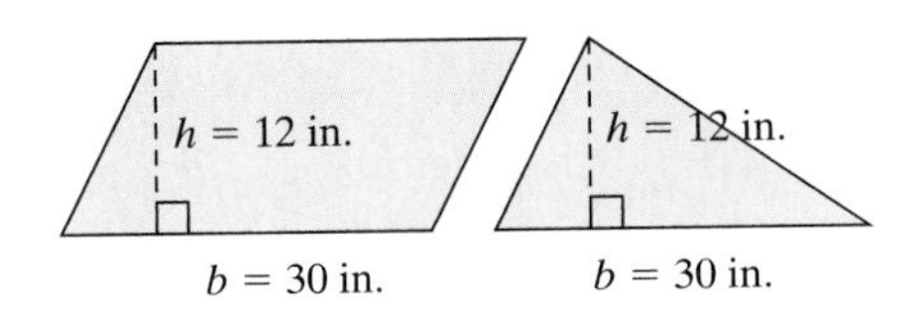

Section 2.7 Linear Inequalities

Concepts

1. Graphing Linear Inequalities
2. Set-Builder Notation and Interval Notation
3. Addition and Subtraction Properties of Inequality
4. Multiplication and Division Properties of Inequality
5. Solving Inequalities of the Form $a < x < b$
6. Applications of Linear Inequalities

1. Graphing Linear Inequalities

Recall that $a < b$ (equivalently $b > a$) means that a lies to the left of b on the number line. The statement $a > b$ (equivalently $b < a$) means that a lies to the right of b on the number line. If $a = b$, then a and b are represented by the same point on the number line.

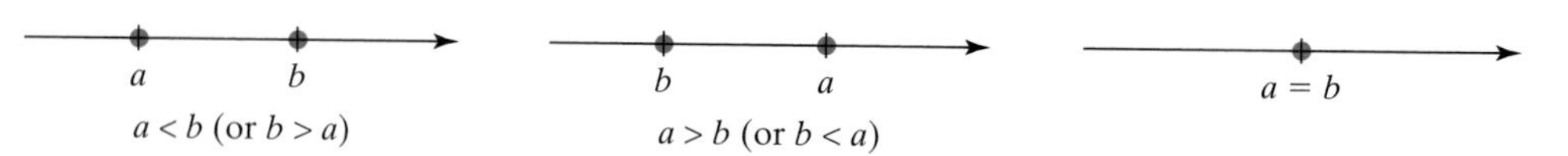

A Linear Inequality in One Variable

A **linear inequality in one variable**, x, is defined as any relationship that can be written in one of the following forms:

$$ax + b < 0, \quad ax + b \leq 0, \quad ax + b > 0, \quad \text{or} \quad ax + b \geq 0, \text{ where } a \neq 0.$$

The following inequalities are linear equalities in one variable.

$$2x - 3 < 0 \qquad -4z - 3 > 0 \qquad a \leq 4 \qquad 5.2y \geq 10.4$$

The number line is a tool used to visualize the solution set of an equation or inequality. For example, the solution set to the equation $x = 2$ is $\{2\}$ and may be graphed as a single point on the number line.

$x = 2$ −6 −5 −4 −3 −2 −1 0 1 2 3 4 5 6

The solution set to an inequality is the set of real numbers that make the inequality a true statement. For example, the solution set to the inequality $x \geq 2$ is all real numbers 2 or greater. Because the solution set has an infinite number of values, we cannot list all of the individual solutions. However, we can graph the solution set on the number line.

$x \geq 2$ −6 −5 −4 −3 −2 −1 0 1 2 3 4 5 6

The square bracket symbol, [, is used on the graph to indicate that the point $x = 2$ is included in the solution set. By convention, square brackets, either [or], are used to *include* a point on a graph. Parentheses, (or), are used to *exclude* a point on a graph.

The solution set of the inequality $x > 2$ includes the real numbers greater than 2 but not including 2. Therefore, a (symbol is used on the graph to indicate that $x = 2$ is not included.

$x > 2$ −6 −5 −4 −3 −2 −1 0 1 2 3 4 5 6

Example 1 Graphing Linear Inequalities

Graph the solution sets.

a. $x > -1$ **b.** $c \leq \frac{7}{3}$ **c.** $3 > y$

Solution:

a. $x > -1$

−6 −5 −4 −3 −2 −1 0 1 2 3 4 5 6

The solution set is the set of real numbers strictly greater than −1. Therefore, we graph the region on the number line to the right of −1. Because $x = -1$ is not included in the solution set, we use the (symbol at $x = -1$.

b. $c \leq \frac{7}{3}$ is equivalent to $c \leq 2\frac{1}{3}$.

$2\frac{1}{3}$

−6 −5 −4 −3 −2 −1 0 1 2 3 4 5 6

The solution set is the set of real numbers less than or equal to $2\frac{1}{3}$. Therefore, graph the region on the number line to the left of and including $2\frac{1}{3}$. Use the symbol] to indicate that $c = 2\frac{1}{3}$ is included in the solution set.

c. $3 > y$ This inequality reads "3 is greater than y." This is equivalent to saying, "y is less than 3."

$y < 3$

−6 −5 −4 −3 −2 −1 0 1 2 3 4 5 6

The solution set is the set of real numbers less than 3. Therefore, graph the region on the number line to the left of 3. Use the symbol) to denote that the endpoint, 3, is not included in the solution.

Skill Practice Graph the solution sets.

1. $y < 0$ **2.** $x \geq -\frac{5}{4}$ **3.** $5 \geq a$

TIP: Some textbooks use a closed circle or an open circle (● or ○) rather than a bracket or parenthesis to denote inclusion or exclusion of a value on the real number line. For example, the solution sets for the inequalities $x > -1$ and $c \leq \frac{7}{3}$ are graphed here.

$x > -1$

−6 −5 −4 −3 −2 −1 0 1 2 3 4 5 6

$c \leq \frac{7}{3}$

$\frac{7}{3}$

−6 −5 −4 −3 −2 −1 0 1 2 3 4 5 6

A statement that involves more than one inequality is called a **compound inequality**. One type of compound inequality is used to indicate that one number is between two others. For example, the inequality $-2 < x < 5$ means that $-2 < x$ and $x < 5$. In words, this is easiest to understand if we read the variable first: x is greater than −2 and x is less than 5. The numbers satisfied by these two conditions are those between −2 and 5.

Skill Practice Answers

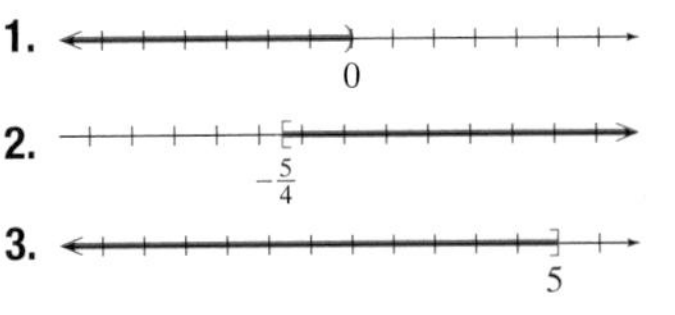

Example 2 Graphing a Compound Inequality

Graph the solution set of the inequality: $-4.1 < y \leq -1.7$

Solution:

$-4.1 < y \leq -1.7$ means that

$-4.1 < y$ and $y \leq -1.7$

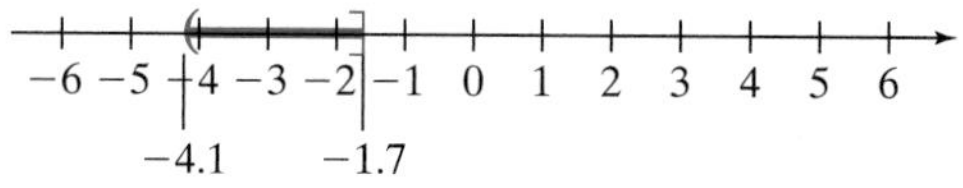

Shade the region of the number line greater than -4.1 and less than or equal to -1.7.

Skill Practice Graph the solution set.

4. $0 \leq y \leq 8.5$

2. Set-Builder Notation and Interval Notation

Graphing the solution set to an inequality is one way to define the set. Two other methods are to use **set-builder notation** or **interval notation**.

Set-Builder Notation

The solution to the inequality $x \geq 2$ can be expressed in set-builder notation as follows:

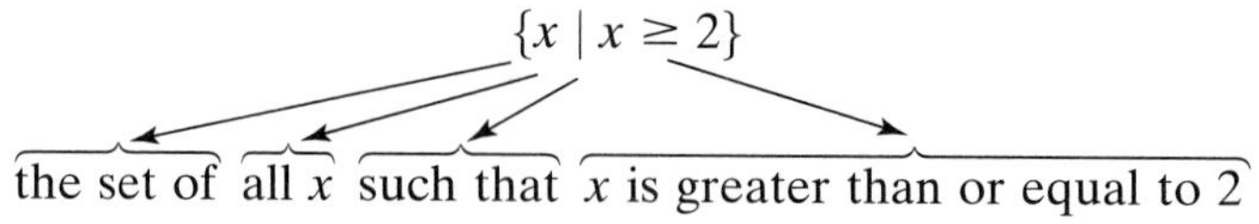

Interval Notation

To understand interval notation, first think of a number line extending infinitely far to the right and infinitely far to the left. Sometimes we use the infinity symbol, ∞, or negative infinity symbol, $-\infty$, to label the far right and far left ends of the number line (Figure 2-11).

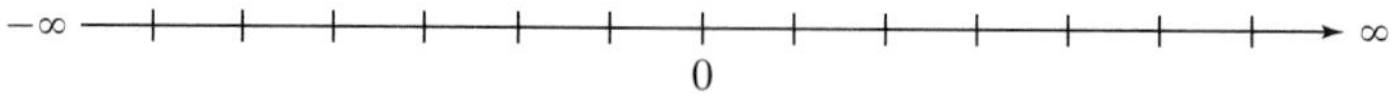

Figure 2-11

To express the solution set of an inequality in interval notation, sketch the graph first. Then use the endpoints to define the interval.

Inequality	Graph	Interval Notation
$x \geq 2$	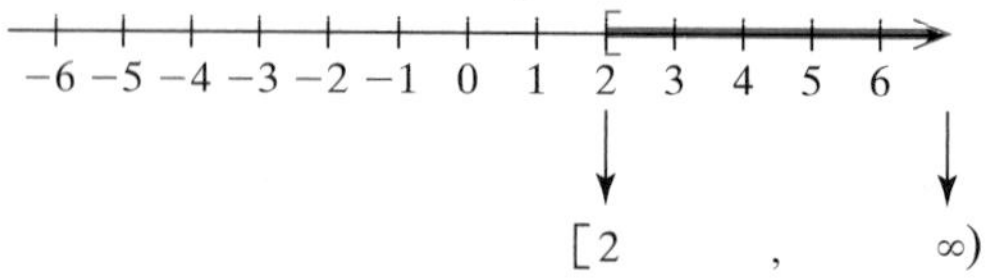	$[2, \infty)$

The graph of the solution set $x \geq 2$ begins at 2 and extends infinitely far to the right. The corresponding interval notation begins at 2 and extends to ∞. Notice that a square bracket [is used at 2 for both the graph and the interval notation. A parenthesis is always used at ∞ and for $-\infty$, because there is no endpoint.

Skill Practice Answers

4. 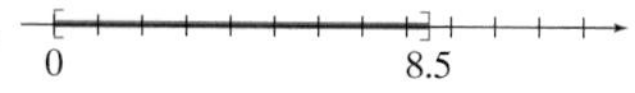

Using Interval Notation

- The endpoints used in interval notation are always written from left to right. That is, the smaller number is written first, followed by a comma, followed by the larger number.
- A parenthesis, (or), indicates that an endpoint is excluded from the set.
- A square bracket, [or], indicates that an endpoint is included in the set.
- Parentheses, (and), are always used with $-\infty$ and ∞, respectively.

Example 3 Using Set-Builder Notation and Interval Notation

Complete the chart.

Set-Builder Notation	Graph	Interval Notation
	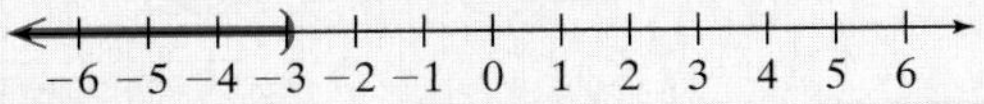	
		$[-\frac{1}{2}, \infty)$
$\{y \mid -2 \le y < 4\}$		

Solution:

Set-Builder Notation	Graph	Interval Notation
$\{x \mid x < -3\}$	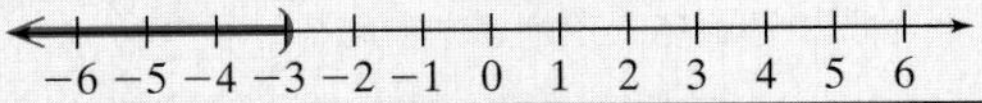	$(-\infty, -3)$
$\{x \mid x \ge -\frac{1}{2}\}$	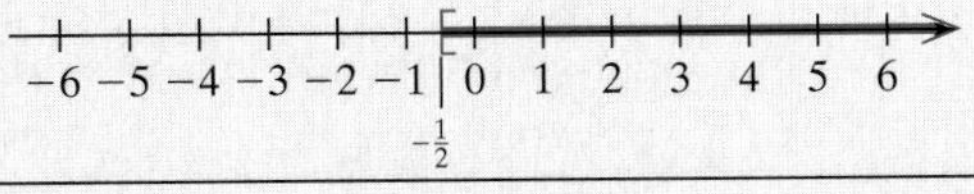	$[-\frac{1}{2}, \infty)$
$\{y \mid -2 \le y < 4\}$	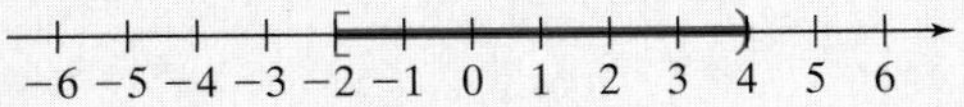	$[-2, 4)$

Skill Practice Graph the sets defined in Exercises 5–7, and give the set-builder notation and interval notation.

5. $\{x \mid x \ge -2\}$ **6.** $(-3, 1]$ **7.**

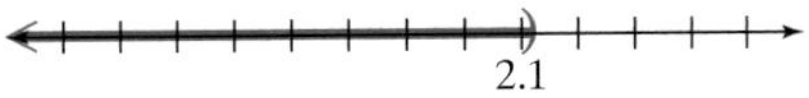

3. Addition and Subtraction Properties of Inequality

Solving linear inequalities is similar to solving linear equations. Recall that adding or subtracting the same quantity to both sides of an equation results in an equivalent equation. The addition and subtraction properties of inequality state that the same is true for an inequality.

Skill Practice Answers

5. $[-2, \infty)$;

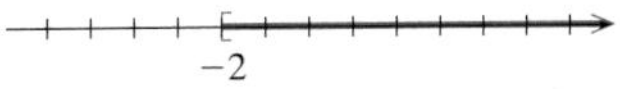

6. $\{x \mid -3 < x \le 1\}$;

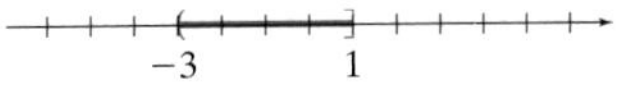

7. $\{x \mid x < 2.1\}$; $(-\infty, 2.1)$

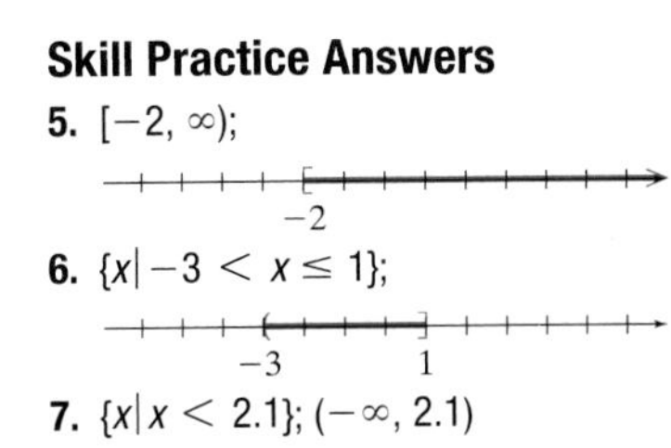

Addition and Subtraction Properties of Inequality

Let a, b, and c represent real numbers.

1. ***Addition Property of Inequality:** If $a < b$, then $a + c < b + c$
2. ***Subtraction Property of Inequality:** If $a < b$, then $a - c < b - c$

*These properties may also be stated for $a \leq b$, $a > b$, and $a \geq b$.

To illustrate the addition property of inequality, consider the inequality $5 > 3$. If we add a real number such as 4 to both sides, the left-hand side will still be greater than the right-hand side (Figure 2-12).

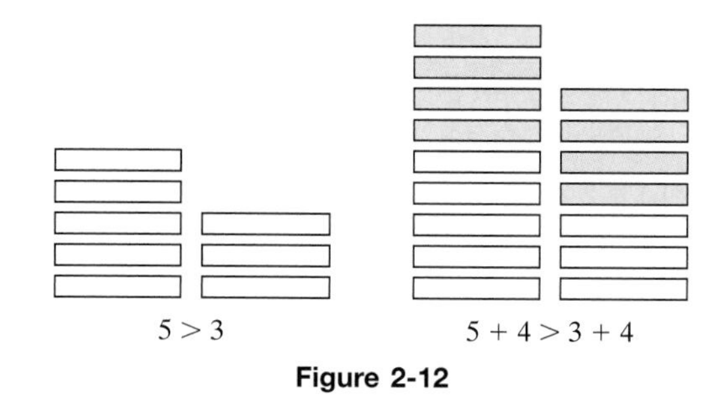

Figure 2-12

Example 4 Solving a Linear Inequality

Solve the inequality and graph the solution set. Express the solution set in set-builder notation and in interval notation.

$$-2p + 5 < -3p + 6$$

Solution:

$-2p + 5 < -3p + 6$	
$-2p + 3p + 5 < -3p + 3p + 6$	Addition property of inequality (add $3p$ to both sides)
$p + 5 < 6$	Simplify.
$p + 5 - 5 < 6 - 5$	Subtraction property of inequality
$p < 1$	

Graph: −6 −5 −4 −3 −2 −1 0 1 2 3 4 5 6

Set-builder notation: $\{p \mid p < 1\}$

Interval notation: $(-\infty, 1)$

Skill Practice Solve the inequality. Graph the solution set and express in interval notation.

8. $2y - 5 < y - 11$

Skill Practice Answers

8. −6

$(-\infty, -6)$

TIP: The solution to an inequality gives a set of values that make the original inequality true. Therefore, you can test your final answer by using *test points*. That is, pick a value in the proposed solution set and verify that it makes the original inequality false. Furthermore, any test point picked outside the solution set should make the original inequality false. For example,

−6 −5 −4 −3 −2 −1 0 1 2 3 4 5 6

Pick $p = -4$ as an arbitrary test point within the proposed solution set.

$$-2p + 5 < -3p + 6$$
$$-2(-4) + 5 \stackrel{?}{<} -3(-4) + 6$$
$$8 + 5 \stackrel{?}{<} 12 + 6$$
$$13 < 18 \checkmark \quad \text{True}$$

Pick $p = 3$ as an arbitrary test point outside the proposed solution set.

$$-2p + 5 < -3p + 6$$
$$-2(3) + 5 \stackrel{?}{<} -3(3) + 6$$
$$-6 + 5 \stackrel{?}{<} -9 + 6$$
$$-1 \stackrel{?}{<} -3 \quad \text{False}$$

4. Multiplication and Division Properties of Inequality

Multiplying both sides of an equation by the same quantity results in an equivalent equation. However, the same is not always true for an inequality. If you multiply or divide an inequality by a negative quantity, the direction of the inequality symbol must be reversed.

For example, consider multiplying or dividing the inequality, $4 < 5$ by -1.

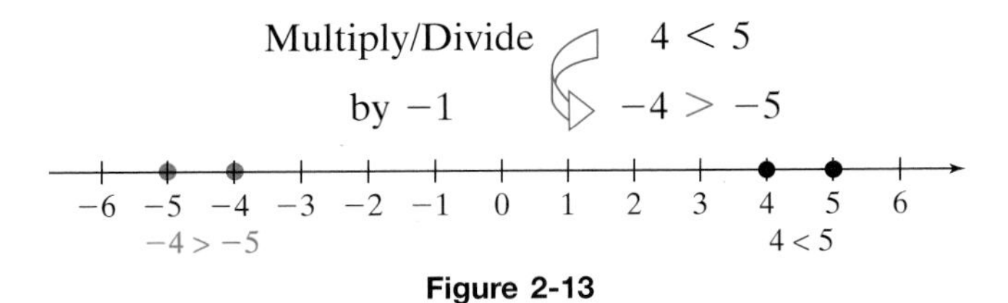

Figure 2-13

The number 4 lies to the left of 5 on the number line. However, −4 lies to the right of −5 (Figure 2-13). Changing the sign of two numbers changes their relative position on the number line. This is stated formally in the multiplication and division properties of inequality.

Multiplication and Division Properties of Inequality

Let a, b, and c represent real numbers. Then

*If c is positive and $a < b$, then $ac < bc$ and $\frac{a}{c} < \frac{b}{c}$.

*If c is negative and $a < b$, then $ac > bc$ and $\frac{a}{c} > \frac{b}{c}$.

The second statement indicates that if both sides of an inequality are multiplied or divided by a negative quantity, the inequality sign must be reversed.

*These properties may also be stated for $a \le b$, $a > b$, and $a \ge b$.

Example 5 **Solving a Linear Inequality**

Solve the inequality $-5x - 3 \leq 12$. Graph the solution set and write the answer in interval notation.

Solution:

$$-5x - 3 \leq 12$$

$$-5x - 3 + 3 \leq 12 + 3 \qquad \text{Add 3 to both sides.}$$

$$-5x \leq 15$$

$$\frac{-5x}{-5} \geq \frac{15}{-5} \qquad \text{Divide by } -5. \text{ Reverse the direction of the inequality sign.}$$

$$x \geq -3$$

Interval notation: $[-3, \infty)$

−6 −5 −4 −3 −2 −1 0 1 2 3 4 5 6

TIP: The inequality $-5x - 3 \leq 12$, could have been solved by isolating x on the right-hand side of the inequality. This would create a positive coefficient on the variable term and eliminate the need to divide by a negative number.

$$-5x - 3 \leq 12$$

$$-3 \leq 5x + 12$$

$$-15 \leq 5x \qquad \text{Notice that the coefficient of } x \text{ is positive.}$$

$$\frac{-15}{5} \leq \frac{5x}{5} \qquad \text{Do not reverse the inequality sign because we are dividing by a positive number.}$$

$-3 \leq x$, or equivalently, $x \geq -3$

Skill Practice Solve.

9. $-5p + 2 > 22$

Example 6 **Solving a Linear Inequality**

Solve the inequality. Graph the solution set and write the answer in interval notation.

$$1.4x + 4.5 - 0.2x < -0.3$$

Solution:

$$1.4x + 4.5 - 0.2x < -0.3$$

$$1.2x + 4.5 < -0.3 \qquad \text{Combine } \textit{like} \text{ terms.}$$

$$1.2x + 4.5 - 4.5 < -0.3 - 4.5 \qquad \text{Subtract 4.5 from both sides.}$$

$$1.2x < -4.8 \qquad \text{Simplify.}$$

$$\frac{1.2x}{1.2} < \frac{-4.8}{1.2} \qquad \text{Divide by 1.2. The direction of the inequality sign is } \textit{not} \text{ reversed because we divided by a positive number.}$$

$$x < -4$$

Interval notation: $(-\infty, -4)$

−6 −5 −4 −3 −2 −1 0 1 2 3 4 5 6

Skill Practice Answers

9. −4

$(-\infty, -4)$

Skill Practice Solve.

10. $0.1x + 4.2 < 1.1x - 0.8$

Example 7 Solving Linear Inequalities

Solve the inequality $-\frac{1}{4}k + \frac{1}{6} \le 2 + \frac{2}{3}k$. Graph the solution set and write the answer in interval notation.

Solution:

$$-\frac{1}{4}k + \frac{1}{6} \le 2 + \frac{2}{3}k$$

$$12\left(-\frac{1}{4}k + \frac{1}{6}\right) \le 12\left(2 + \frac{2}{3}k\right)$$ Multiply both sides by 12 to clear fractions. (Because we multiplied by a positive number, the inequality sign is *not* reversed.)

$$\frac{12}{1}\left(-\frac{1}{4}k\right) + \frac{12}{1}\left(\frac{1}{6}\right) \le 12(2) + \frac{12}{1}\left(\frac{2}{3}k\right)$$ Apply the distributive property.

$$-3k + 2 \le 24 + 8k$$ Simplify.

$$-3k - 8k + 2 \le 24 + 8k - 8k$$ Subtract $8k$ from both sides.

$$-11k + 2 \le 24$$

$$-11k + 2 - 2 \le 24 - 2$$ Subtract 2 from both sides.

$$-11k \le 22$$

$$\frac{-11k}{-11} \ge \frac{22}{-11}$$ Divide both sides by -11. Reverse the inequality sign.

$$k \ge -2$$

Graph: −4 −3 −2 −1 0 1 2 3 4

Interval notation: $[-2, \infty)$

Skill Practice Solve.

11. $\frac{1}{5}t + 7 \le \frac{1}{2}t - 2$

5. Solving Inequalities of the Form $a < x < b$

To solve a compound inequality of the form $a < x < b$ we can work with the inequality as a three-part inequality and isolate the variable, x, as demonstrated in the next example.

Skill Practice Answers

10. 5 $(5, \infty)$

11. 30 $[30, \infty)$

Example 8 **Solving a Compound Inequality of the Form $a < x < b$**

Solve the inequality: $-3 \le 2x + 1 < 7$. Graph the solution and write the answer in interval notation.

Solution:

To solve the compound inequality $-3 \le 2x + 1 < 7$ isolate the variable x in the middle. The operations performed on the middle portion of the inequality must also be performed on the left-hand side and right-hand side.

$$-3 \le 2x + 1 < 7$$

$$-3 - 1 \le 2x + 1 - 1 < 7 - 1$$ Subtract 1 from all three parts of the inequality.

$$-4 \le 2x < 6$$ Simplify.

$$\frac{-4}{2} \le \frac{2x}{2} < \frac{6}{2}$$ Divide by 2 in all three parts of the inequality.

$$-2 \le x < 3$$

Graph: −6 −5 −4 −3 −2 −1 0 1 2 3 4 5 6

Interval notation: $[-2, 3)$

Skill Practice Solve.

12. $-3 \le -5 + 2y < 11$

6. Applications of Linear Inequalities

Table 2-1 provides several commonly used translations to express inequalities.

Table 2-1

English Phrase	Mathematical Inequality
a is less than b	$a < b$
a is greater than b a exceeds b	$a > b$
a is less than or equal to b a is at most b a is no more than b	$a \le b$
a is greater than or equal to b a is at least b a is no less than b	$a \ge b$

Example 9 **Translating Expressions Involving Inequalities**

Translate the English phrases into mathematical inequalities:

a. Claude's annual salary, s, is no more than \$40,000.

b. A citizen must be at least 18 years old to vote. (Let a represent a citizen's age.)

c. An amusement park ride has a height requirement between 48 in. and 70 in. (Let h represent height in inches.)

Skill Practice Answers

12.

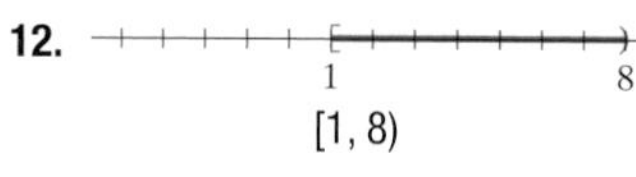

Solution:

a. $s \le 40{,}000$ Claude's annual salary, s, is no more than \$40,000.

b. $a \ge 18$ A citizen must be at least 18 years old to vote.

c. $48 < h < 70$ An amusement park ride has a height requirement between 48 in. and 70 in.

Skill Practice Translate the English phrase into a mathematical inequality.

13. Bill needs a score of at least 92 on the final exam. Let x represent Bill's score.

14. Fewer than 19 cars are in the parking lot. Let c represent the number of cars.

15. The heights, h, of women who wear petite size clothing are typically between 58 in. and 63 in., inclusive.

Linear inequalities are found in a variety of applications. See how the next example can help you determine the minimum grade you need on an exam to get an A in your math course.

Example 10 Solving an Application with Linear Inequalities

To earn an A in a math class, Alsha must average at least 90 on all of her tests. Suppose Alsha has scored 79, 86, 93, 90, and 95 on her first five math tests. Determine the minimum score she needs on her sixth test to get an A in the class.

Solution:

Let x represent the score on the sixth exam. Label the variable.

$$\left(\begin{array}{c}\text{Average of}\\ \text{all tests}\end{array}\right) \ge 90$$ Create a verbal model.

$$\frac{79 + 86 + 93 + 90 + 95 + x}{6} \ge 90$$ The average score is found by taking the sum of the test scores and dividing by the number of scores.

$$\frac{443 + x}{6} \ge 90$$ Simplify.

$$6\left(\frac{443 + x}{6}\right) \ge (90)6$$ Multiply both sides by 6 to clear fractions.

$$443 + x \ge 540$$ Solve the inequality.

$$x \ge 97$$ Interpret the results.

Alsha must score at least 97 on her sixth exam to receive an A in the course.

Skill Practice

16. To get at least a B in math, Simon must average 80 on all tests. Suppose Simon has scored 60, 72, 98, and 85 on the first four tests. What score does he need on the fifth test to receive a B?

Skill Practice Answers

13. $x \ge 92$
14. $c < 19$
15. $58 \le h \le 63$
16. Simon needs at least 85.

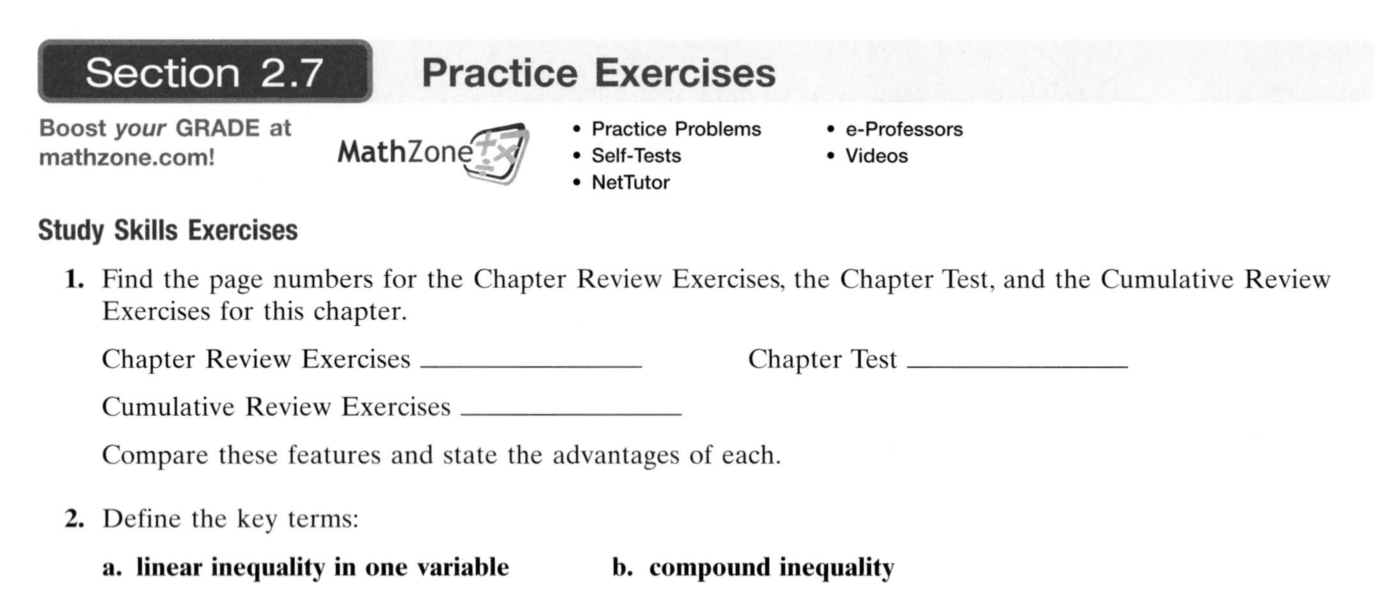

Section 2.7 Practice Exercises

Study Skills Exercises

1. Find the page numbers for the Chapter Review Exercises, the Chapter Test, and the Cumulative Review Exercises for this chapter.

   Chapter Review Exercises ______________ Chapter Test ______________

   Cumulative Review Exercises ______________

   Compare these features and state the advantages of each.

2. Define the key terms:

   a. linear inequality in one variable **b. compound inequality**

   c. set-builder notation **d. interval notation**

Review Problems

3. Solve the equation: $3(x + 2) - (2x - 7) = -(5x - 1) - 2(x + 6)$

4. Solve the equation: $6 - 8(x + 3) + 5x = 5x - (2x - 5) + 13$

Concepts 1–2: Graphing Linear Inequalities; Set-Builder Notation and Interval Notation

For Exercises 5–10, graph each inequality and write the solution set in interval notation.

Set-Builder Notation	Graph	Interval Notation
5. $\{x \mid x \geq 6\}$		
6. $\left\{x \mid \frac{1}{2} < x \leq 4\right\}$		
7. $\{x \mid x \leq 2.1\}$		
8. $\left\{x \mid x > \frac{7}{3}\right\}$		
9. $\{x \mid -2 < x \leq 7\}$		
10. $\{x \mid x < -5\}$		

For Exercises 11–16, write each set in set-builder notation and in interval notation.

Set-Builder Notation	Graph	Interval Notation
11.	$\frac{3}{4}$	
12.	-0.3	
13.	-1 8	
14.	0	
15.	-14	
16.	0 9	

For Exercises 17–22, graph each set and write the set in set-builder notation.

Set-Builder Notation	Graph	Interval Notation
17.		$[18, \infty)$
18.		$[-10, -2]$
19.		$(-\infty, -0.6)$
20.		$\left(-\infty, \frac{5}{3}\right)$
21.		$[-3.5, 7.1)$
22.		$[-10, \infty)$

Concepts 3–4: Properties of Inequality

For Exercises 23–30, solve the equation in part (a). For part (b), solve the inequality and graph the solution set.

23. a. $x + 3 = 6$
b. $x + 3 > 6$

24. a. $y - 6 = 12$
b. $y - 6 \geq 12$

25. a. $p - 4 = 9$
b. $p - 4 \leq 9$

26. a. $k + 8 = 10$
b. $k + 8 < 10$

27. a. $4c = -12$
b. $4c < -12$

28. a. $5d = -35$
b. $5d > -35$

29. a. $-10z = 15$
b. $-10z \leq 15$

30. a. $-2w = 14$
b. $-2w < 14$

Concept 5: Solving Inequalities of the Form $a < x < b$

For Exercises 31–36, graph the solution.

31. $-1 < y \leq 4$

32. $2.5 \leq t < 5.7$

33. $0 < x + 3 < 8$

34. $-2 \leq x - 4 \leq 3$

35. $8 \leq 4x \leq 24$

36. $-9 < 3x < 12$

Mixed Exercises

For Exercises 37–82, solve the inequality. Graph the solution set and write the set in interval notation.

37. $x + 5 \leq 6$

38. $y - 7 < 6$

39. $q - 7 > 3$

40. $r + 4 \geq -1$

41. $4 < 1 + z$

42. $3 > z - 6$

43. $2 \geq a - 6$

44. $7 \leq b + 12$

45. $3c > 6$

46. $4d \le 12$

47. $-3c > 6$

48. $-4d \le 12$

49. $-h \le -14$

50. $-q > -7$

51. $12 \ge -\frac{x}{2}$

52. $6 < -\frac{m}{3}$

53. $-2 \le p + 1 < 4$

54. $0 < k + 7 < 6$

55. $-3 < 6h - 3 < 12$

56. $-6 \le 4a - 2 \le 12$

57. $5 < \frac{1}{2}x < 6$

58. $-6 \le 3x \le 12$

59. $-5 \le 4x - 1 < 15$

60. $-2 < \frac{1}{3}x - 2 \le 2$

61. $0.6z \ge 54$

62. $-0.7w > 28$

63. $-\frac{2}{3}y < 6$

64. $\frac{3}{4}x \le -12$

65. $-2x - 4 \le 11$

66. $-3x + 1 > 0$

67. $-12 > 7x + 9$

68. $8 < 2x - 10$

69. $-7b - 3 \le 2b$

70. $3t \ge 7t - 35$

71. $4n + 2 < 6n + 8$

72. $2w - 1 \le 5w + 8$

73. $8 - 6(x - 3) > -4x + 12$

74. $3 - 4(h - 2) > -5h + 6$

75. $3(x + 1) - 2 \le \frac{1}{2}(4x - 8)$

76. $8 - (2x - 5) \ge \frac{1}{3}(9x - 6)$

77. $\frac{7}{6}p + \frac{4}{3} \ge \frac{11}{6}p - \frac{7}{6}$

78. $\frac{1}{3}w - \frac{1}{2} \le \frac{5}{6}w + \frac{1}{2}$

79. $\frac{1}{2}y - \frac{1}{4} > \frac{3}{4}y + 2$

80. $\frac{2}{5}t + \frac{1}{2} < \frac{1}{10}t - 1$

81. $-1.2a - 0.4 < -0.4a + 2$

82. $-0.4c + 1.2 > -2c - 0.4$

For Exercises 83–86, determine whether the given number is a solution to the inequality.

83. $-2x + 5 < 4 \quad x = -2$

84. $-3y - 7 > 5 \quad y = 6$

85. $4(p + 7) - 1 > 2 + p \quad p = 1$

86. $3 - k < 2(-1 + k) \quad k = 4$

Concept 6: Applications of Linear Inequalities

87. Let x represent a student's average in a math class. The grading scale is given here.

A	$93 \le x \le 100$
B+	$89 \le x < 93$
B	$84 \le x < 89$
C+	$80 \le x < 84$
C	$75 \le x < 80$
F	$0 \le x < 75$

a. Write the range of scores corresponding to each letter grade in interval notation.

b. If Stephan's average is 84.01, what grade will he receive?

c. If Estella's average is 79.89, what grade will she receive?

88. Let x represent a student's average in a science class. The grading scale is given here.

A	$90 \le x \le 100$
B+	$86 \le x < 90$
B	$80 \le x < 86$
C+	$76 \le x < 80$
C	$70 \le x < 76$
D+	$66 \le x < 70$
D	$60 \le x < 66$
F	$0 \le x < 60$

a. Write the range of scores corresponding to each letter grade in interval notation.

b. If Jacque's average is 89.99, what is her grade?

c. If Marc's average is 66.01, what is his grade?

For Exercises 89–99, translate the English phrase into a mathematical inequality.

89. The length of a fish, L, was at least 10 in.

90. Tasha's average test score, t, exceeded 90.

91. The wind speed, w, exceeded 75 mph.

92. The height of a cave, h, was no more than 2 ft.

93. The temperature of the water in Blue Spring, t, is no more than 72°F.

94. The temperature on the tennis court, t, was no less than 100°F.

95. The length of the hike, L, was no less than 8 km.

96. The depth, d, of a certain pool was at most 10 ft.

97. The amount of rain, a, in a recent storm was at most 2 in.

98. The snowfall, h, in Monroe County is between 2 inches and 5 inches.

99. The cost, c, of carpeting a room is between \$300 and \$400.

100. The average summer rainfall for Miami, Florida, for June, July, and August is 7.4 in. per month. If Miami receives 5.9 in. of rain in June and 6.1 in. in July, how much rain is required in August to exceed the 3-month summer average?

101. The average winter snowfall for Burlington, Vermont, for December, January, and February is 18.7 in. per month. If Burlington receives 22 in. of snow in December and 24 in. in January, how much snow is required in February to exceed the 3-month winter average?

102. An artist paints wooden birdhouses. She buys the birdhouses for \$9 each. However, for large orders, the price per birdhouse is discounted by a percentage off the original price. Let x represent the number of birdhouses ordered. The corresponding discount is given in the table.

Size of Order	Discount
$x \le 49$	0%
$50 \le x \le 99$	5%
$100 \le x \le 199$	10%
$x \ge 200$	20%

a. If the artist places an order for 190 birdhouses, compute the total cost.

b. Which costs more: 190 birdhouses or 200 birdhouses? Explain your answer.

103. A wholesaler sells T-shirts to a surf shop at \$8 per shirt. However, for large orders, the price per shirt is discounted by a percentage off the original price. Let x represent the number of shirts ordered. The corresponding discount is given in the table.

Number of Shirts Ordered	Discount
$x \le 24$	0%
$25 \le x \le 49$	2%
$50 \le x \le 99$	4%
$100 \le x \le 149$	6%
$x \ge 150$	8%

a. If the surf shop orders 50 shirts, compute the total cost.

b. Which costs more: 148 shirts or 150 shirts? Explain your answer.

104. Maggie sells lemonade at an art show. She has a fixed cost of \$75 to cover the registration fee for the art show. In addition, her cost to produce each lemonade is \$0.17. If x represents the number of lemonades, then the total cost to produce x lemonades is given by:

$$\text{Cost} = 75 + 0.17x$$

If Maggie sells each lemonade for \$2, then her revenue (the amount she brings in) for selling x lemonades is given by:

$$\text{Revenue} = 2.00x$$

a. Write an inequality that expresses the number of lemonades, x, that Maggie must sell to make a profit. Profit is realized when the revenue is greater than the cost (Revenue > Cost).

b. Solve the inequality in part (a).

105. Two rental car companies rent subcompact cars at a discount. Company A rents for \$14.95 per day plus 22 cents per mile. Company B rents for \$18.95 a day plus 18 cents per mile. Let x represent the number of miles driven in one day.

The cost to rent a subcompact car for one day from Company A is:

$$\text{Cost}_A = 14.95 + 0.22x$$

The cost to rent a subcompact car for one day from Company B is:

$$\text{Cost}_B = 18.95 + 0.18x$$

a. Write an inequality that expresses the number of miles, x, for which the daily cost to rent from Company A is less than the daily cost to rent from Company B.

b. Solve the inequality in part (a).

Expanding Your Skills

For Exercises 106–111, solve the inequality. Graph the solution set and write the set in interval notation.

106. $3(x + 2) - (2x - 7) \le (5x - 1) - 2(x + 6)$

107. $6 - 8(y + 3) + 5y > 5y - (2y - 5) + 13$

108. $-2 - \frac{w}{4} \le \frac{1 + w}{3}$

109. $\frac{z - 3}{4} - 1 > \frac{z}{2}$

110. $-0.703 < 0.122p - 2.472$

111. $3.88 - 1.335t \ge 5.66$

Chapter 2 SUMMARY

Section 2.1 Addition, Subtraction, Multiplication, and Division Properties of Equality

Key Concepts

An equation is an algebraic statement that indicates two expressions are equal. A **solution to an equation** is a value of the variable that makes the equation a true statement. The set of all solutions to an equation is the solution set of the equation.

A **linear equation in one variable** can be written in the form $ax + b = 0$, where $a \neq 0$.

Examples

Example 1

$2x + 1 = 9$ is an equation with solution $x = 4$.

Check: $2(4) + 1 = 9$

$8 + 1 = 9$

$9 = 9$ ✔ True

Addition Property of Equality:

If $a = b$, then $a + c = b + c$

Example 2

$$x - 5 = 12$$
$$x - 5 + 5 = 12 + 5$$
$$x = 17$$

Subtraction Property of Equality:

If $a = b$, then $a - c = b - c$

Example 3

$$z + 1.44 = 2.33$$
$$z + 1.44 - 1.44 = 2.33 - 1.44$$
$$z = 0.89$$

Multiplication Property of Equality:

If $a = b$, then $ac = bc$

Example 4

$$\frac{3}{4}x = 12$$
$$\frac{4}{3} \cdot \frac{3}{4}x = 12 \cdot \frac{4}{3}$$
$$x = 16$$

Division Property of Equality:

If $a = b$, then $\frac{a}{c} = \frac{b}{c}$ $(c \neq 0)$

Example 5

$$16 = 8y$$
$$\frac{16}{8} = \frac{8y}{8}$$
$$2 = y$$

Section 2.2 Solving Linear Equations

Key Concepts

Steps for Solving a Linear Equation in One Variable:

1. Simplify both sides of the equation.
   - Clear parentheses
   - Combine *like* terms
2. Use the addition or subtraction property of equality to collect the variable terms on one side of the equation.
3. Use the addition or subtraction property of equality to collect the constant terms on the other side of the equation.
4. Use the multiplication or division property of equality to make the coefficient of the variable term equal to 1.
5. Check your answer.

Examples

Example 1

$5y + 7 = 3(y - 1) + 2$

$5y + 7 = 3y - 3 + 2$ Clear parentheses.

$5y + 7 = 3y - 1$ Combine *like* terms.

$2y + 7 = -1$ Isolate variable term.

$2y = -8$ Isolate constant term.

$y = -4$ Divide both sides by 2.

Check:

$5(-4) + 7 \stackrel{?}{=} 3[(-4) - 1] + 2$

$-20 + 7 \stackrel{?}{=} 3(-5) + 2$

$-13 \stackrel{?}{=} -15 + 2$

$-13 = -13$ ✔ True

A **conditional equation** is true for some values of the variable but is false for other values.

Example 2

$x + 5 = 7$ is a conditional equation because it is true only on the condition that $x = 2$.

An equation that has all real numbers as its solution set is an **identity**.

Example 3

$x + 4 = 2(x + 2) - x$

$x + 4 = 2x + 4 - x$

$x + 4 = x + 4$

$4 = 4$ is an identity.

The solution is all real numbers.

An equation that has no solution is a **contradiction**.

Example 4

$y - 5 = 2(y + 3) - y$

$y - 5 = 2y + 6 - y$

$y - 5 = y + 6$

$-5 = 6$ is a contradiction.

There is no solution.

Section 2.3 Linear Equations: Clearing Fractions and Decimals

Key Concepts

To Clear Fractions or Decimals in an Equation:

Multiply both sides of an equation by an appropriate power of 10 to clear decimals.

Examples

Example 1

$$-1.2x - 5.1 = 16.5$$

$$10(-1.2x - 5.1) = 10(16.5) \qquad \text{Multiply both sides by 10.}$$

$$-12x - 51 = 165$$

$$-12x = 216$$

$$\frac{-12x}{-12} = \frac{216}{-12}$$

$$x = -18$$

Multiply both sides of the equation by a common denominator of all the fractions.

Example 2

$$\frac{1}{2}x - 2 - \frac{3}{4}x = \frac{7}{4}$$

$$\frac{4}{1}\left(\frac{1}{2}x - 2 - \frac{3}{4}x\right) = \frac{4}{1}\left(\frac{7}{4}\right) \qquad \text{Multiply by the LCD.}$$

$$2x - 8 - 3x = 7 \qquad \text{Apply distributive property.}$$

$$-x - 8 = 7 \qquad \text{Combine } like \text{ terms.}$$

$$-x = 15 \qquad \text{Add 8 to both sides.}$$

$$x = -15 \qquad \text{Divide by } -1.$$

Section 2.4 Applications of Linear Equations: Introduction to Problem Solving

Key Concepts

Problem-Solving Steps for Word Problems:

1. Read the problem carefully.
2. Assign labels to unknown quantities.
3. Develop a verbal model.
4. Write a mathematical equation.
5. Solve the equation.
6. Interpret the results and write the answer in words.

Examples

Example 1

The perimeter of a triangle is 54 m. The lengths of the sides are represented by three consecutive even integers. Find the lengths of the three sides.

1. Read the problem.
2. Let x represent one side, $x + 2$ represent the second side, and $x + 4$ represent the third side.

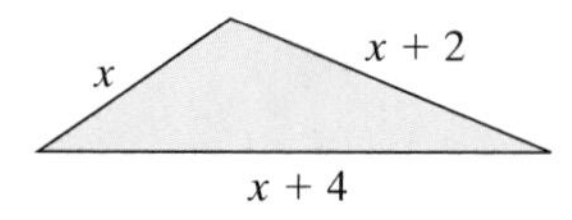

3. (First side) + (second side) + (third side) = perimeter
4. $x + (x + 2) + (x + 4) = 54$
5. $3x + 6 = 54$
   $3x = 48$
   $x = 16$
6. $x = 16$ represents the length of the shortest side. The lengths of the other sides are given by $x + 2 = 18$ and $x + 4 = 20$. The lengths of the three sides are 16 m, 18 m, and 20 m.

Section 2.5 Applications Involving Percents

Key Concepts

The following formula will help solve basic percent problems.

$$\text{amount} = (\text{percent})(\text{base})$$

One common use of percents is in computing **sales tax**.

Examples

Example 1

A dinette set costs \$1260.00 after a 5% sales tax is included. What was the price before tax?

$$\begin{pmatrix}\text{price}\\ \text{before tax}\end{pmatrix} + (\text{tax}) = \begin{pmatrix}\text{total}\\ \text{price}\end{pmatrix}$$

$$\begin{aligned} x + 0.05x &= 1260 \\ 1.05x &= 1260 \\ x &= 1200 \end{aligned}$$

The dinette set cost \$1200 before tax.

Another use of percent is in computing **simple interest** using the formula:

$$\begin{pmatrix}\text{simple}\\ \text{interest}\end{pmatrix} = \begin{pmatrix}\text{principal}\\ \text{invested}\end{pmatrix}\begin{pmatrix}\text{annual}\\ \text{interest}\\ \text{rate}\end{pmatrix}\begin{pmatrix}\text{time in}\\ \text{years}\end{pmatrix}$$

or $I = Prt$.

Example 2

John Li invests \$5400 at 2.5% simple interest. How much interest does he make after 5 years?

$I = Prt$

$I = (\$5400)(0.025)(5)$

$I = \$675$

Section 2.6 Formulas and Applications of Geometry

Key Concepts

A **literal equation** is an equation that has more than one variable. Often such an equation can be manipulated to solve for different variables.

Examples

Example 1

$P = 2a + b$, solve for a.

$$P - b = 2a + b - b$$

$$P - b = 2a$$

$$\frac{P - b}{2} = \frac{2a}{2}$$

$$\frac{P - b}{2} = a \quad \text{or} \quad a = \frac{P - b}{2}$$

Formulas from Section R.4 can be used in applications involving geometry.

Example 2

Find the length of a side of a square whose perimeter is 28 ft.

Use the formula $P = 4s$. Substitute 28 for P and solve:

$$P = 4s$$

$$28 = 4s$$

$$7 = s$$

The length of a side of the square is 7 ft.

Section 2.7 Linear Inequalities

Key Concepts

A **linear inequality in one variable**, x, is any relationship in the form: $ax + b < 0$, $ax + b > 0$, $ax + b \leq 0$, or $ax + b \geq 0$, where $a \neq 0$.

The solution set to an inequality can be expressed as a graph or in **set-builder notation** or in **interval notation**.

When graphing an inequality or when writing interval notation, a parenthesis, (or), is used to denote that an endpoint is *not included* in a solution set. A square bracket, [or], is used to show that an endpoint *is included* in a solution set. Parenthesis (or) are always used with $-\infty$ and ∞, respectively.

The inequality $a < x < b$ is used to show that x is greater than a and less than b. That is, x is *between* a and b.

Multiplying or dividing an inequality by a negative quantity requires the direction of the inequality sign to be reversed.

Examples

Example 1

$$-2x + 6 \geq 14$$

$$-2x + 6 - 6 \geq 14 - 6 \qquad \text{Subtract 6.}$$

$$-2x \geq 8 \qquad \text{Simplify.}$$

$$\frac{-2x}{-2} \leq \frac{8}{-2} \qquad \text{Divide by } -2. \text{ Reverse the inequality sign.}$$

$$x \leq -4$$

Set-builder notation: $\{x \mid x \leq -4\}$

Graph: 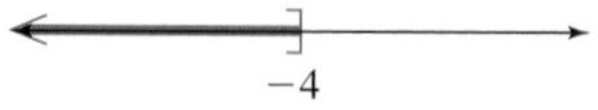

Interval notation: $(-\infty, -4]$

Chapter 2 Review Exercises

Section 2.1

1. Label the following as either an expression or an equation:

   a. $3x + y = 10$ b. $9x + 10y - 2xy$

   c. $4(x + 3) = 12$ d. $-5x = 7$

2. Explain how to determine whether an equation is linear in one variable.

3. Identify which equations are linear.

   a. $4x^2 + 8 = -10$ b. $x + 18 = 72$

   c. $-3 + 2y^2 = 0$ d. $-4p - 5 = 6p$

4. For the equation, $4y + 9 = -3$, determine if the given numbers are solutions.

   a. $y = 3$ b. $y = 0$

   c. $y = -3$ d. $y = -2$

For Exercises 5–12, solve the equation using the addition property, subtraction property, multiplication property, or division property of equality.

5. $a + 6 = -2$
6. $6 = z - 9$
7. $-\frac{3}{4} + k = \frac{9}{2}$
8. $0.1r = 7$
9. $-5x = 21$
10. $\frac{t}{3} = -20$
11. $-\frac{2}{5}k = \frac{4}{7}$
12. $-m = -27$
13. The quotient of a number and negative six is equal to negative ten. Find the number.
14. The difference of a number and $-\frac{1}{8}$ is $\frac{5}{12}$. Find the number.
15. Four subtracted from a number is negative twelve. Find the number.
16. Six subtracted from a number is negative eight. Find the number.

Section 2.2

For Exercises 17–28, solve the equation.

17. $4d + 2 = 6$
18. $5c - 6 = -9$
19. $-7c = -3c - 9$
20. $-28 = 5w + 2$
21. $\frac{b}{3} + 1 = 0$
22. $\frac{2}{3}h - 5 = 7$
23. $-3p + 7 = 5p + 1$
24. $4t - 6 = -12t + 16$
25. $4a - 9 = 3(a - 3)$
26. $3(2c + 5) = -2(c - 8)$
27. $7b + 3(b - 1) + 16 = 2(b + 8)$
28. $2 + (17 - x) + 2(x - 1) = 4(x + 2) - 8$
29. Explain the difference between an equation that is a contradiction and an equation that is an identity.
30. Label each equation as a conditional equation, a contradiction, or an identity.

    a. $x + 3 = 3 + x$ b. $3x - 19 = 2x + 1$

    c. $5x + 6 = 5x - 28$ d. $2x - 8 = 2(x - 4)$

    e. $-8x - 9 = -8(x - 9)$

Section 2.3

For Exercises 31–48, solve the equation.

31. $\frac{x}{8} - \frac{1}{4} = \frac{1}{2}$
32. $\frac{y}{15} - \frac{2}{3} = \frac{4}{5}$
33. $\frac{4z + 7}{5} = z + 2$
34. $\frac{5y - 3}{6} = 1 + y$
35. $\frac{1}{10}p - 3 = \frac{2}{5}p$
36. $\frac{1}{4}y - \frac{3}{4} = \frac{1}{2}y + 1$
37. $-\frac{1}{4}(2 - 3t) = \frac{3}{4}$
38. $\frac{2}{7}(w + 4) = \frac{1}{2}$
39. $17.3 - 2.7q = 10.55$

40. $4.9z + 4.6 = 3.2z - 2.2$

41. $5.74a + 9.28 = 2.24a - 5.42$

42. $62.84t - 123.66 = 4(2.36 + 2.4t)$

43. $0.05x + 0.10(24 - x) = 0.75(24)$

44. $0.20(x + 4) + 0.65x = 0.20(854)$

45. $100 - (t - 6) = -(t - 1)$

46. $3 - (x + 4) + 5 = 3x + 10 - 4x$

47. $5t - (2t + 14) = 3t - 14$

48. $9 - 6(2z + 1) = -3(4z - 1)$

Section 2.4

49. Twelve added to the sum of a number and two is forty-four. Find the number.

50. Twenty added to the sum of a number and six is thirty-seven. Find the number.

51. Three times a number is the same as the difference of twice the number and seven. Find the number.

52. Eight less than five times a number is forty-eight less than the number. Find the number.

53. Three times the largest of three consecutive even integers is 76 more than the sum of the other two integers. Find the integers.

54. Ten times the smallest of three consecutive integers is 213 more than the sum of the other two integers. Find the integers.

55. The perimeter of a triangle is 78 in. The lengths of the sides are represented by three consecutive integers. Find the lengths of the sides of the triangle.

56. The perimeter of a pentagon (a five-sided polygon) is 190 cm. The five sides are represented by consecutive integers. Find the measures of the sides.

57. The minimum salary for a major league baseball player in 1985 was $60,000. This was twice the minimum salary in 1980. What was the minimum salary in 1980?

58. The state of Indiana has approximately 2.1 million more people than Kentucky. Together their population totals 10.3 million. Approximately how many people are in each state?

59. A Cessna 182 airplane has an average air speed that is 50 mph slower than a Mooney airplane. If the combined distance traveled by these two small airplanes is 660 mi after 2 hr, what is the speed of each plane?

60. A bicyclist and a jogger leave a gym at the same time. The bicyclist rides 6 mph faster than the jogger can run. It takes the jogger 2 hr to complete the same distance completed by the bicyclist in 1 hr. Find the speed of each person.

Section 2.5

For Exercises 61–70, solve the problems involving percents.

61. What is 35% of 68?

62. What is 4% of 720?

63. 53.5 is what percent of 428?

64. 68.4 is what percent of 72?

65. 24 is 15% of what number?

66. 8.75 is 0.5% of what number?

67. A novel originally selling at $29.99 is on sale for 12% off. What is the sale price of the book? Round to the nearest cent.

68. What would be the total price (including tax) of a novel that sells for $26.39, if the sales tax rate is 7%?

69. Anna Tsao invested $3000 in an account paying 8% simple interest.

   a. How much interest will she earn in $3\frac{1}{2}$ years?

   b. What will her balance be at that time?

70. Eduardo invested money in an account earning 4% simple interest. At the end of 5 years, he had a total of $14,400. How much money did he originally invest?

Section 2.6

For Exercises 71–78, solve for the indicated variable.

71. $C = K - 273$ for K

72. $K = C + 273$ for C

73. $P = 4s$ for s

74. $P = 3s$ for s

75. $y = mx + b$ for x

76. $a + bx = c$ for x

77. $2x + 5y = -2$ for y

78. $4(a + b) = Q$ for b

For Exercises 79–85, use the appropriate geometry formula to solve the problem.

79. The volume of a cone is given by the formula $V = \frac{1}{3}\pi r^2 h$.

a. Solve the formula for h.

b. Find the height of a right circular cone whose volume is 47.8 in.^{3} and whose radius is 3 in. Round to the nearest tenth of an inch.

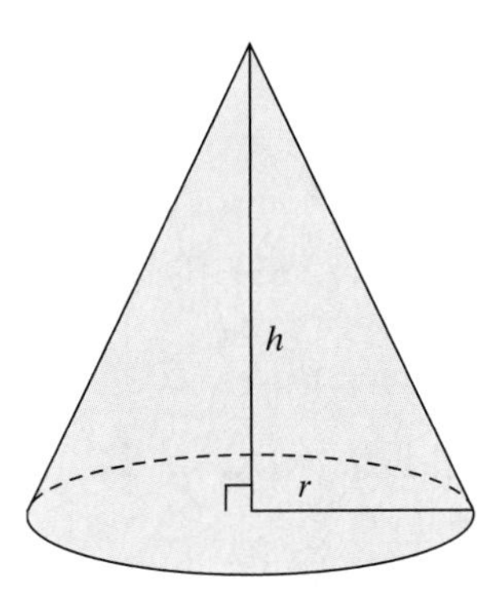

80. Find the height of a parallelogram whose area is 42 m^2 and whose base is 6 m.

81. The smallest angle of a triangle is 2° more than $\frac{1}{4}$ of the largest angle. The middle angle is 2° less than the largest angle. Find the measure of each angle.

82. One angle is 6° less than twice a second angle. If the two angles are complementary, what are their measures?

83. A rectangular window has a width 1 ft less than its length. The perimeter is 18 ft. Find the length and the width of the window.

84. Find the measure of the vertical angles by first solving for x.

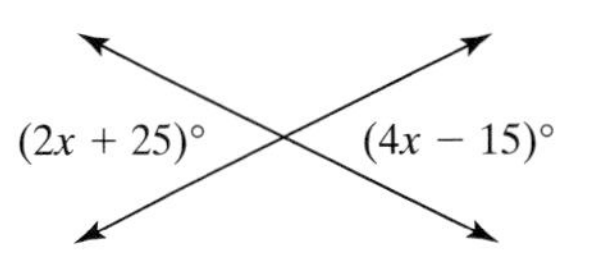

85. Find the measure of angle y.

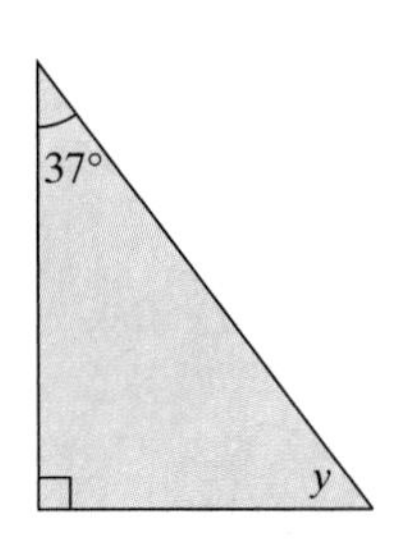

Section 2.7

86. Graph the inequalities and write the sets in interval notation.

a. $\{x | x > -2\}$

b. $\left\{x | x \leq \frac{1}{2}\right\}$

c. $\{x | -1 < x \leq 4\}$

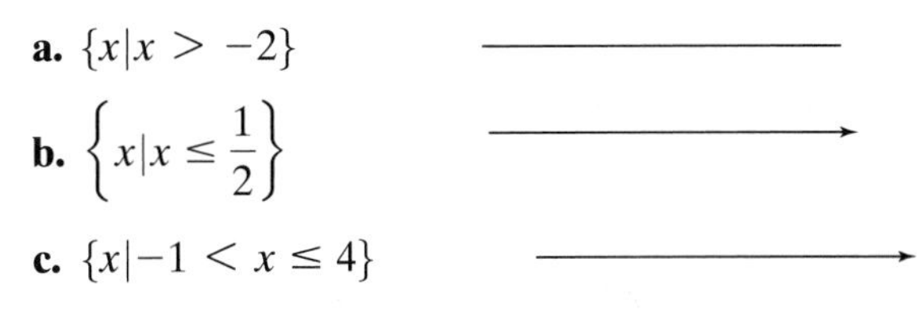

87. A landscaper buys potted geraniums from a nursery at a price of \$5 per plant. However, for large orders, the price per plant is discounted by a percentage off the original price. Let x represent the number of potted plants ordered. The corresponding discount is given in the table.

Number of Plants	Discount
$x \leq 99$	0%
$100 \leq x \leq 199$	2%
$200 \leq x \leq 299$	4%
$x \geq 300$	6%

a. Find the cost to purchase 130 plants.

b. Which costs more, 300 plants or 295 plants? Explain your answer.

For Exercises 88–97, solve the inequality. Graph the solution set and express the answer in interval notation.

88. $c + 6 < 23$

89. $3w - 4 > -5$

90. $-2x - 7 \geq 5$

91. $5(y + 2) \leq -4$

92. $-\frac{3}{7}a \leq -21$

93. $1.3 > 0.4t - 12.5$

94. $4k + 23 < 7k - 31$

95. $\frac{6}{5}h - \frac{1}{5} \leq \frac{3}{10} + h$

96. $-6 < 2b \leq 14$

97. $-2 \leq z + 4 \leq 9$

98. The summer average rainfall for Bermuda for June, July, and August is 5.3 in. per month. If Bermuda receives 6.3 in. of rain in June and 7.1 in. in July, how much rain is required in August to exceed the 3-month summer average?

99. Reggie sells hot dogs at a ballpark. He has a fixed cost of $33 to use the concession stand at the park. In addition, the cost for each hot dog is $0.40. If x represents the number of hot dogs sold, then the total cost is given by

$$\text{Cost} = 33 + 0.40x$$

If Reggie sells each hot dog for $1.50, then his revenue (the amount he brings in) for selling x hot dogs is given by

$$\text{Revenue} = 1.50x$$

a. Write an inequality that expresses the number of hot dogs, x, that Reggie must sell to make a profit. Profit is realized when the revenue is greater than the cost (revenue > cost).

b. Solve the inequality in part (a).

Chapter 2 Test

1. Which of the equations have $x = -3$ as a solution?

a. $4x + 1 = 10$

b. $6(x - 1) = x - 21$

c. $5x - 2 = 2x + 1$

d. $\frac{1}{3}x + 1 = 0$

2. a. Simplify: $3x - 1 + 2x + 8$

b. Solve: $3x - 1 = 2x + 8$

For Exercises 3 – 13, solve the equation.

3. $t + 3 = -13$

4. $8 = p - 4$

5. $\frac{t}{8} = -\frac{2}{9}$

6. $-3x + 5 = -2$

7. $2(p - 4) = p + 7$

8. $2 + d = 2 - 3(d - 5) - 2$

9. $\frac{3}{7} + \frac{2}{5}x = -\frac{1}{5}x + 1$

10. $3h + 1 = 3(h + 1)$

11. $\frac{3x + 1}{2} - \frac{4x - 3}{3} = 1$

12. $0.5c - 1.9 = 2.8 + 0.6c$

13. $-5(x + 2) + 8x = -2 + 3x - 8$

14. Solve the equation for y: $3x + y = -4$

15. Solve $C = 2\pi r$ for r.

16. 13% of what is 11.7?

17. One number is four plus one-half of another. The sum of the numbers is 31. Find the numbers.

18. The perimeter of a pentagon (a five-sided polygon) is 315 in. The five sides are represented by consecutive integers. Find the measures of the sides.

19. In the 1997–1998 season, a couple purchased two NHL hockey tickets and two NBA basketball tickets for $153.92. A hockey ticket cost $4.32 more than a basketball ticket. What were the prices of the individual tickets?

20. The total bill for a pair of basketball shoes (including sales tax) is \$87.74. If the tax rate is 7%, find the cost of the shoes before tax.

21. Clarita borrowed money at a 6% simple interest rate. If she paid back a total of \$8000 at the end of 10 years, how much did she originally borrow?

22. The length of a soccer field for international matches is 40 m less than twice its width. If the perimeter is 370 m, what are the dimensions of the field?

23. Given the triangle, find the measures of each angle by first solving for y.

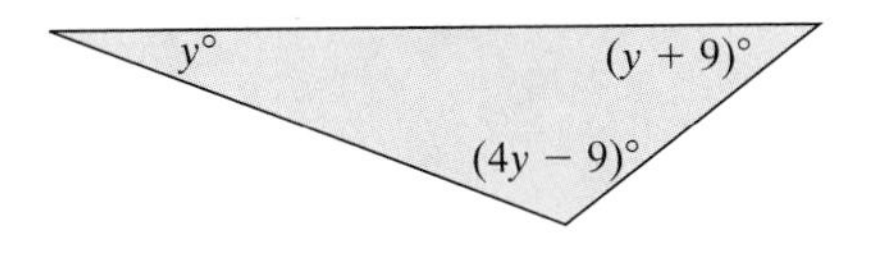

24. Two families leave their homes at the same time to meet for lunch. The families live 210 miles apart, and one family drives 5 mph slower than the other. If it takes them 2 hr to meet at a point between their homes, how fast does each family travel?

25. Two angles are complementary. One angle is 26° more than the other angle. What are the measures of the angles?

26. Graph the inequalities and write the sets in interval notation.

a. $\{x \mid x < 0\}$

b. $\{x \mid -2 \le x < 5\}$

For Exercises 27–29, solve the inequality. Graph the solution and write the solution set in interval notation.

27. $5x + 14 > -2x$

28. $2(3 - x) \ge 14$

29. $-13 \le 3p + 2 \le 5$

30. The average winter snowfall for Syracuse, New York, for December, January, and February is 27.5 in. per month. If Syracuse receives 24 in. of snow in December and 32 in. in January, how much snow is required in February to exceed the 3-month average?

Chapters 1–2 Cumulative Review Exercises

For Exercises 1–5, perform the indicated operations.

1. $\left| -\frac{1}{5} + \frac{7}{10} \right|$

2. $5 - 2[3 - (4 - 7)]$

3. $-\frac{2}{3} + \left(\frac{1}{2}\right)^2$

4. $-3^2 + (-5)^2$

5. $\sqrt{5 - (-20) - 3^2}$

For Exercises 6–9, translate the mathematical expressions and simplify the results.

6. The square root of the difference of five squared and nine

7. The sum of -14 and 12

8. List the terms of the expression: $-7x^2y + 4xy - 6$

9. Simplify: $-4[2x - 3(x + 4)] + 5(x - 7)$

For Exercises 10–15, solve the equations.

10. $8t - 8 = 24$

11. $-2.5x - 5.2 = 12.8$

12. $-5(p - 3) + 2p = 3(5 - p)$

13. $\frac{x + 3}{5} + \frac{x - 2}{2} = 2$

14. $\frac{2}{9}x - \frac{1}{3} = x + \frac{1}{9}$

15. $-0.6w = 48$

16. The sum of two consecutive odd integers is 156. Find the integers.

17. The total bill for a man's three-piece suit (including sales tax) is $374.50. If the tax rate is 7%, find the cost of the suit before tax.

18. The area of a triangle is 41 cm^2. Find the height of the triangle if the base is 12 cm.

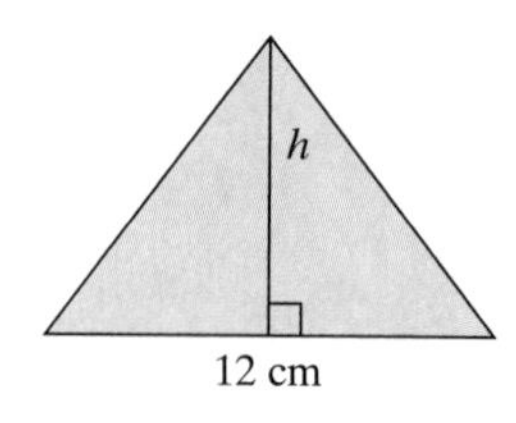

For Exercises 19–20, solve the inequality. Graph the solution set on a number line and express the solution in interval notation.

19. $-3x - 3(x + 1) < 9$

20. $-6 \le 2x - 4 \le 14$

Graphing Linear Equations in Two Variables

3

In this chapter we study graphing and focus on the graphs of lines.

As you work through the chapter, pay attention to key terms. Then work through this puzzle. When you fill in the blanks, notice that hyphens and spaces also go in the boxes.

Across

1. (0, b)
3. one of four regions in the *xy*-plane
4. The point (0, 0)
5. horizontal axis
7. lines with the same slope and different *y*-intercepts
8. (a, 0)

Down

1. vertical axis
2. lines that intersect at a right angle
6. "slant" of a line

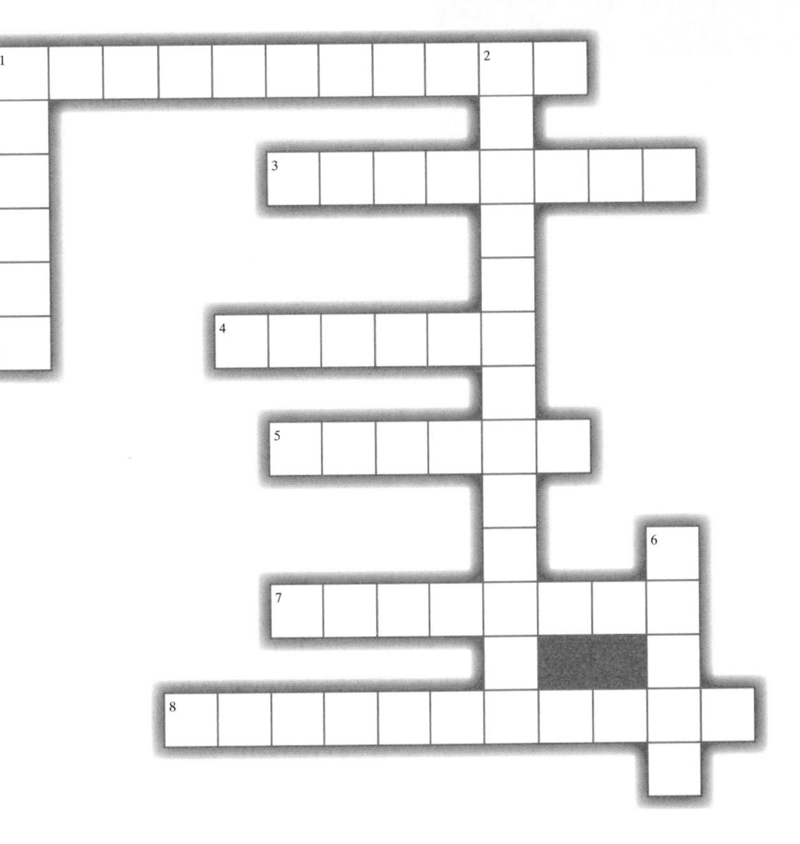

Section 3.1 Rectangular Coordinate System

Concepts

1. Interpreting Graphs
2. Plotting Points in a Rectangular Coordinate System
3. Applications of Plotting and Identifying Points

1. Interpreting Graphs

Mathematics is a powerful tool used by scientists and has directly contributed to the highly technical world we live in. Applications of mathematics have led to advances in the sciences, business, computer technology, and medicine.

One fundamental application of mathematics is the graphical representation of numerical information (or **data**). For example, Table 3-1 represents the number of clients admitted to a drug and alcohol rehabilitation program over a 12-month period.

Table 3-1

	Month	Number of Clients
Jan.	1	55
Feb.	2	62
March	3	64
April	4	60
May	5	70
June	6	73
July	7	77
Aug.	8	80
Sept.	9	80
Oct.	10	74
Nov.	11	85
Dec.	12	90

In table form, the information is difficult to picture and interpret. It appears that on a monthly basis, the number of clients fluctuates. However, when the data are represented in a graph, an upward trend is clear (Figure 3-1).

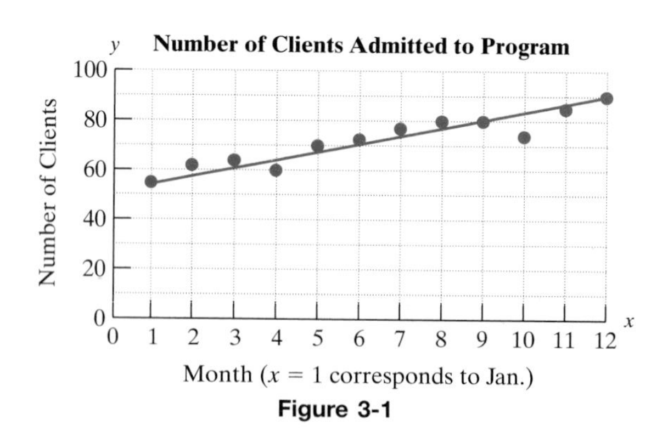

Figure 3-1

From the increase in clients shown in this graph, management for the rehabilitation center might make plans for the future. If the trend continues, management might consider expanding its facilities and increasing its staff to accommodate the expected increase in clients.

Example 1 Interpreting a Graph

Refer to Figure 3-1 and Table 3-1.

a. For which month was the number of clients the greatest?

b. How many clients were served in the first month (January)?

c. Which month corresponds to 60 clients served?

d. Between which two months did the number of clients decrease?

e. Between which two months did the number of clients remain the same?

Solution:

a. Month 12 (December) corresponds to the highest point on the graph, which represents the most clients.

b. In month 1 (January), there were 55 clients served.

c. Month 4 (April).

d. The number of clients decreased between months 3 and 4 and between months 9 and 10.

e. The number of clients remained the same between months 8 and 9.

Skill Practice

1. Refer to Figure 3-1.

 a. How many clients were served in October?

 b. Which month corresponds to 70 clients served?

 c. What is the difference between the number of clients in month 12 and month 1?

 d. For which month was the number of clients served the least?

2. Plotting Points in a Rectangular Coordinate System

In Example 1, two variables are represented, time and the number of clients. To picture two variables, we use a graph with two number lines drawn at right angles to each other (Figure 3-2). This forms a **rectangular coordinate system**. The horizontal line is called the ***x*-axis**, and the vertical line is called the ***y*-axis**. The point where the lines intersect is called the **origin**. On the x-axis, the numbers to the right of the origin are positive and the numbers to the left are negative. On the y-axis, the numbers above the origin are positive and the numbers below are negative. The x- and y-axes divide the graphing area into four regions called **quadrants**.

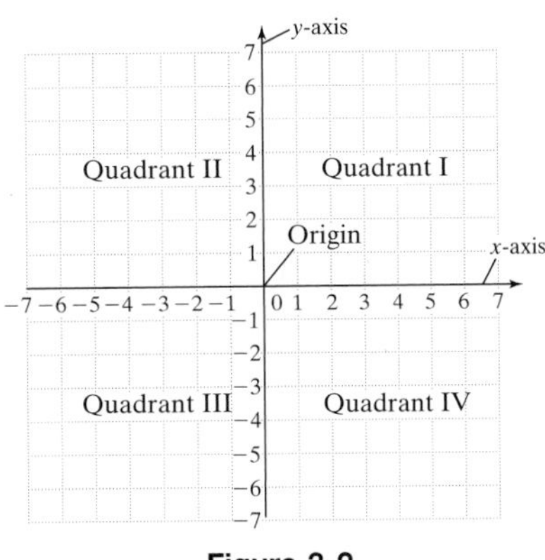

Figure 3-2

Skill Practice Answers

1a. 74
b. Month 5, May
c. 35
d. Month 1, January

Points graphed in a rectangular coordinate system are defined by two numbers as an **ordered pair**, (x, y). The first number (called the first coordinate, or the abscissa) is the horizontal position from the origin. The second number (called the second coordinate, or the ordinate) is the vertical position from the origin. Example 2 shows how points are plotted in a rectangular coordinate system.

Example 2 Plotting Points in a Rectangular Coordinate System

Plot the points.

a. $(4, 5)$ **b.** $(-4, -5)$ **c.** $(-1, 3)$ **d.** $(3, -1)$

e. $\left(\frac{1}{2}, -\frac{7}{3}\right)$ **f.** $(-2, 0)$ **g.** $(0, 0)$ **h.** $(\pi, 1.1)$

Solution:

See Figure 3-3.

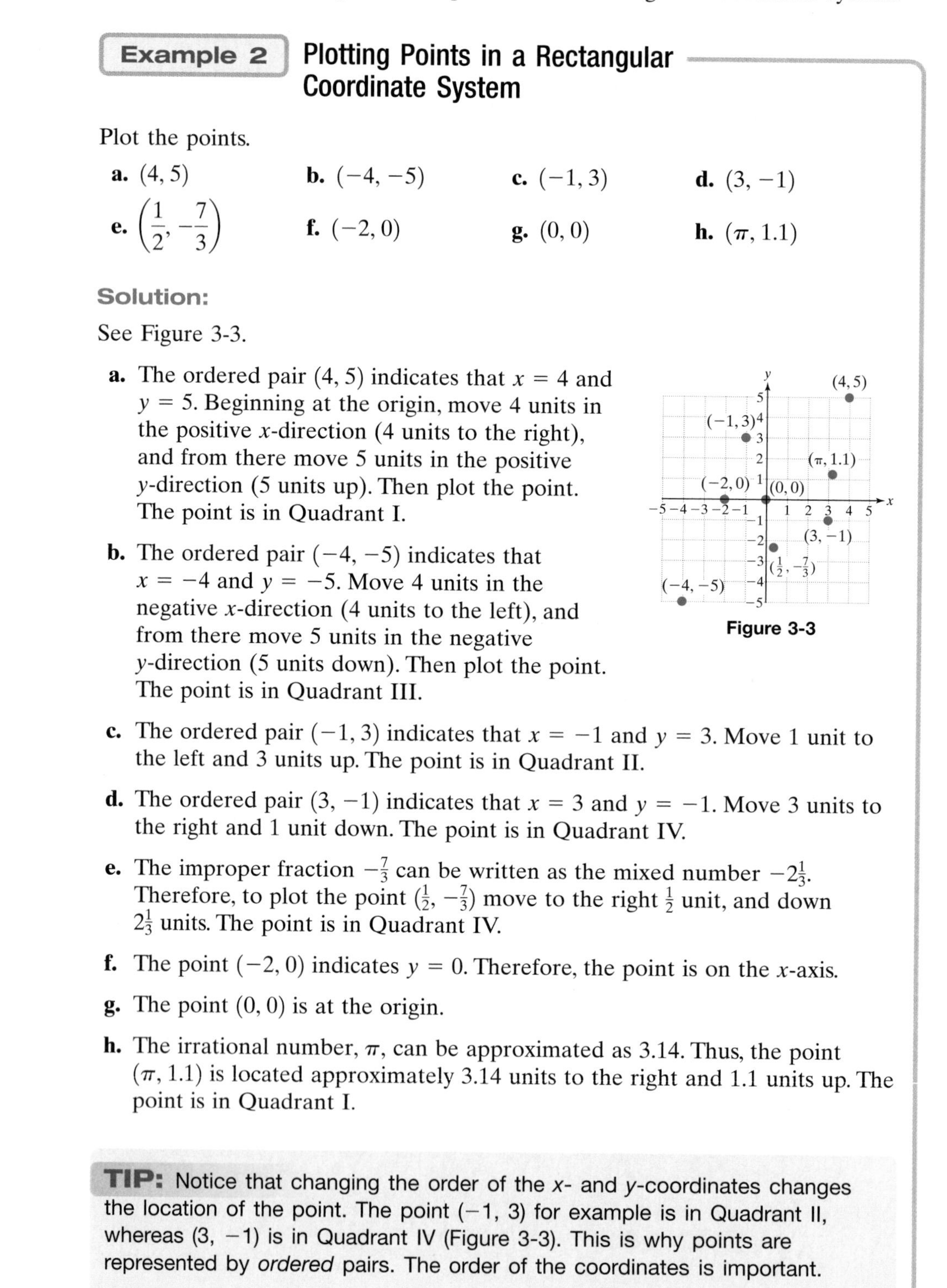

Figure 3-3

a. The ordered pair $(4, 5)$ indicates that $x = 4$ and $y = 5$. Beginning at the origin, move 4 units in the positive x-direction (4 units to the right), and from there move 5 units in the positive y-direction (5 units up). Then plot the point. The point is in Quadrant I.

b. The ordered pair $(-4, -5)$ indicates that $x = -4$ and $y = -5$. Move 4 units in the negative x-direction (4 units to the left), and from there move 5 units in the negative y-direction (5 units down). Then plot the point. The point is in Quadrant III.

c. The ordered pair $(-1, 3)$ indicates that $x = -1$ and $y = 3$. Move 1 unit to the left and 3 units up. The point is in Quadrant II.

d. The ordered pair $(3, -1)$ indicates that $x = 3$ and $y = -1$. Move 3 units to the right and 1 unit down. The point is in Quadrant IV.

e. The improper fraction $-\frac{7}{3}$ can be written as the mixed number $-2\frac{1}{3}$. Therefore, to plot the point $(\frac{1}{2}, -\frac{7}{3})$ move to the right $\frac{1}{2}$ unit, and down $2\frac{1}{3}$ units. The point is in Quadrant IV.

f. The point $(-2, 0)$ indicates $y = 0$. Therefore, the point is on the x-axis.

g. The point $(0, 0)$ is at the origin.

h. The irrational number, π, can be approximated as 3.14. Thus, the point $(\pi, 1.1)$ is located approximately 3.14 units to the right and 1.1 units up. The point is in Quadrant I.

TIP: Notice that changing the order of the x- and y-coordinates changes the location of the point. The point $(-1, 3)$ for example is in Quadrant II, whereas $(3, -1)$ is in Quadrant IV (Figure 3-3). This is why points are represented by *ordered* pairs. The order of the coordinates is important.

Skill Practice

2. Plot the points.

$A(3, 4)$ $B(-2, 2)$ $C(4, 0)$ $D\left(\frac{5}{2}, -\frac{1}{3}\right)$ $E(-5, -2)$

The effective use of graphs for mathematical models requires skill in identifying points and interpreting graphs.

Example 3 Identifying Points

Refer to the figure, and give the coordinates of each point and the quadrant or axis where it is located.

a. P

b. Q

c. R

d. S

e. T

f. U

Solution:

a. P The coordinates are $(1, 0)$, and the point is on the x-axis.

b. Q The coordinates are $(-3, -2)$, and the point is in Quadrant III.

c. R The coordinates are $(2, 5)$, and the point is in Quadrant I.

d. S The coordinates are $(4, -4)$, and the point is in Quadrant IV.

e. T The coordinates are $(-1, 1)$, and the point is in Quadrant II.

f. U The coordinates are $(0, -3)$, and the point is on the y-axis.

Skill Practice

3. Give the coordinates of the labeled points, and state the quadrant or axis where the point is located.

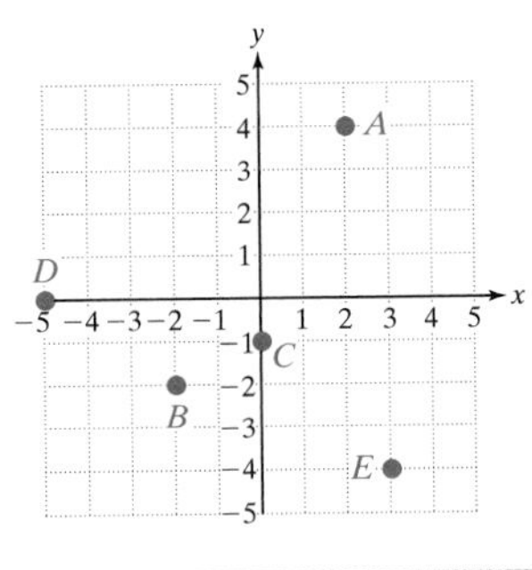

Skill Practice Answers

2.

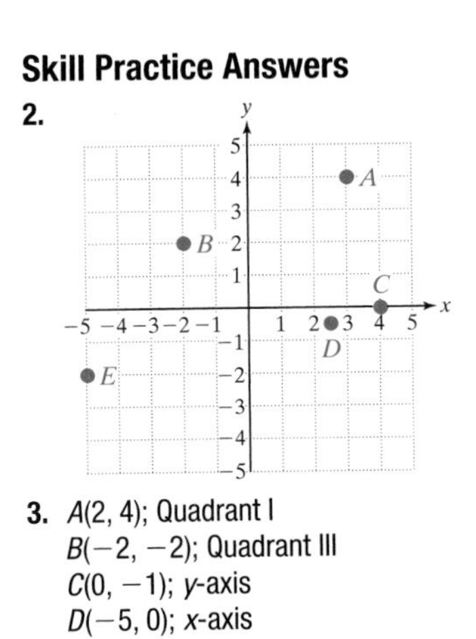

3. $A(2, 4)$; Quadrant I
$B(-2, -2)$; Quadrant III
$C(0, -1)$; y-axis
$D(-5, 0)$; x-axis
$E(3, -4)$; Quadrant IV

3. Applications of Plotting and Identifying Points

Example 4 Plotting Points in an Application

The daily low temperatures (in degrees Fahrenheit) for one week in January for Sudbury, Ontario, Canada, are given in Table 3-2.

a. Write an ordered pair for each row in the table using the day number as the x-coordinate and the temperature as the y-coordinate.

b. Plot the ordered pairs from part (a) on a rectangular coordinate system.

Table 3-2

Day Number, x	Temperature (°F), y
1	−3
2	−5
3	1
4	6
5	5
6	0
7	−4

Solution:

a. (1, −3) Each ordered pair represents the day number and the corresponding low temperature for that day.

(2, −5)

(3, 1)

(4, 6)

(5, 5)

(6, 0)

(7, −4)

b.

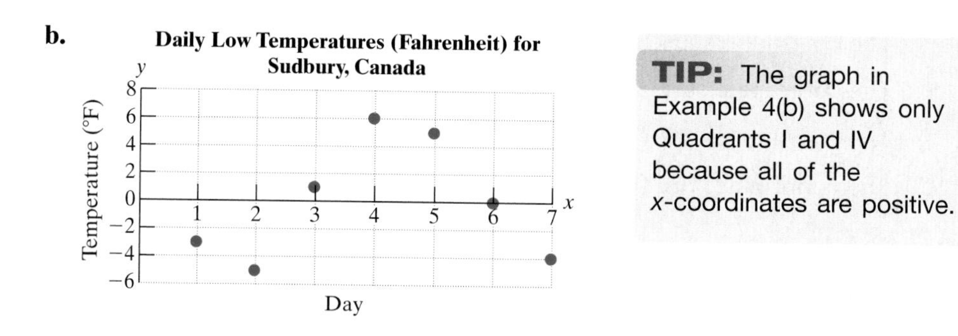

TIP: The graph in Example 4(b) shows only Quadrants I and IV because all of the x-coordinates are positive.

Skill Practice

4. The table shows the number of homes sold in one town for a 6-month period. Plot the ordered pairs.

Month, x	Number Sold, y
1	20
2	25
3	28
4	40
5	45
6	30

Skill Practice Answers

4. y 45 40 35 30 25 20 15 10 5 0 0 1 2 3 4 5 6 x Number Sold Month

Example 5 Determining Points from a Graph

A map of a national park is drawn so that the origin is placed at the ranger station (Figure 3-4). Four fire observation towers are located at points A, B, C, and D. Estimate the coordinates of the fire towers relative to the ranger station (all distances are in miles).

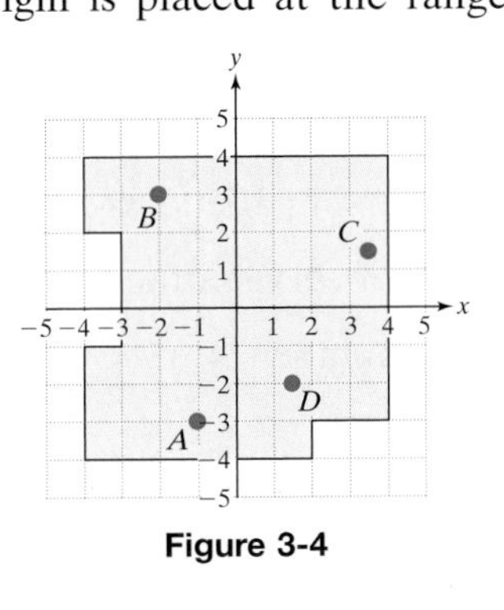

Figure 3-4

Solution:

Point A: $(-1, -3)$

Point B: $(-2, 3)$

Point C: $(3\frac{1}{2}, 1\frac{1}{2})$ or $(\frac{7}{2}, \frac{3}{2})$ or $(3.5, 1.5)$

Point D: $(1\frac{1}{2}, -2)$ or $(\frac{3}{2}, -2)$ or $(1.5, -2)$

Skill Practice

5. A map of a city is drawn so that the main office of a cell phone company is located at the origin. Cell towers are located at points A, B, C, and D. Estimate the coordinates of the towers.

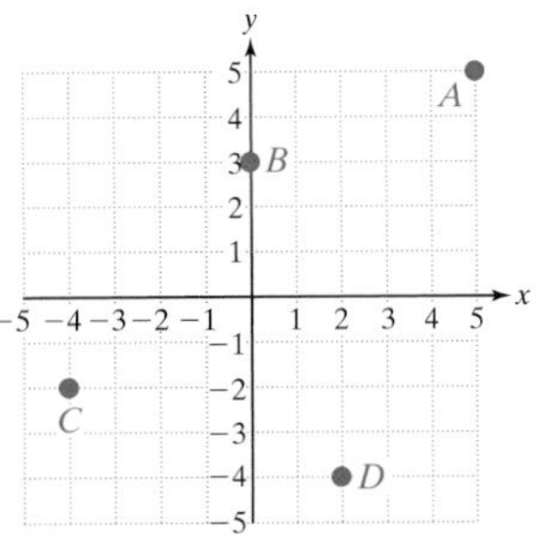

Skill Practice Answers

5. $A(5, 5)$
$B(0, 3)$
$C(-4, -2)$
$D(2, -4)$

Section 3.1 Practice Exercises

Boost *your* GRADE at mathzone.com!

- Practice Problems
- Self-Tests
- NetTutor
- e-Professors
- Videos

Study Skills Exercises

1. Before you proceed too much farther in Chapter 3, make your test corrections for the Chapter 2 test. See Exercise 1 of Section 2.1 for instructions.

2. Define the key terms:

a. **data** **b.** **ordered pair** **c.** **origin** **d.** **quadrant**

e. **rectangular coordinate system** **f.** ***x*-axis** **g.** ***y*-axis**

Concept 1: Interpreting Graphs

For Exercises 3–6, refer to the graphs to answer the questions.

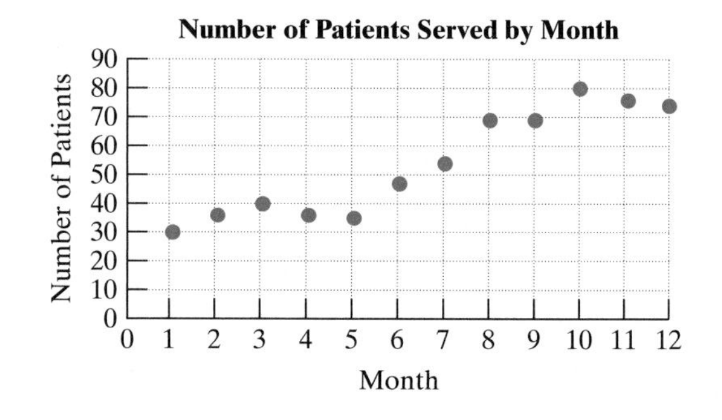

3. The number of patients served by a certain hospice care center is shown in the graph for the first 12 months after it opened.

   a. For which month was the number of patients greatest?

   b. How many patients did the center serve in the first month?

   c. Between which months did the number of patients decrease?

   d. Between which two months did the number of patients remain the same?

   e. Which month corresponds to 40 patients served?

   f. Approximately how many patients were served during the 10th month?

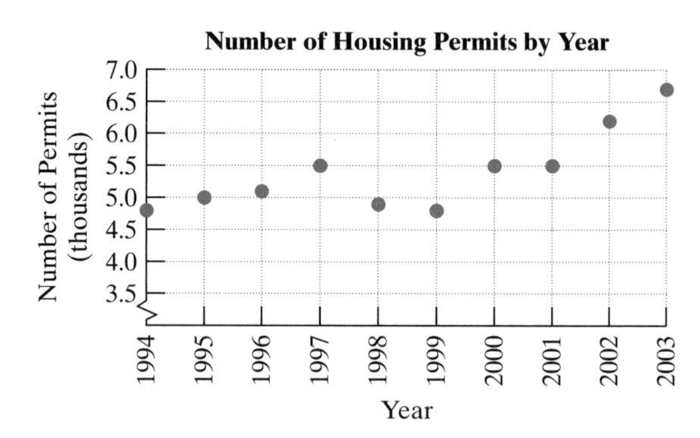

4. The number of housing permits (in thousands) issued by a certain county in Texas between 1994 and 2003 is shown in the graph.

   a. For which year was the number of permits greatest?

   b. How many permits did the county issue in 1997?

   c. Between which years did the number of permits decrease?

   d. Between which two consecutive years did the number of permits remain the same?

   e. Which year corresponds to 5000 permits issued?

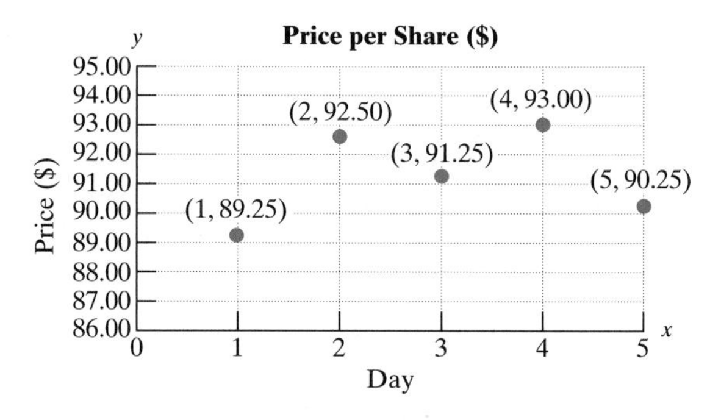

5. The price per share of a stock (in dollars) over a period of 5 days is shown in the graph.

   a. Interpret the meaning of the ordered pair (1, 89.25).

   b. What was the gain in price between day 3 and day 4?

   c. What was the loss in price between day 4 and day 5?

6. The price per share of a stock (in dollars) over a period of 5 days is shown in the graph.

a. Interpret the meaning of the ordered pair (1, 10.125).

b. What was the loss between day 4 and day 5?

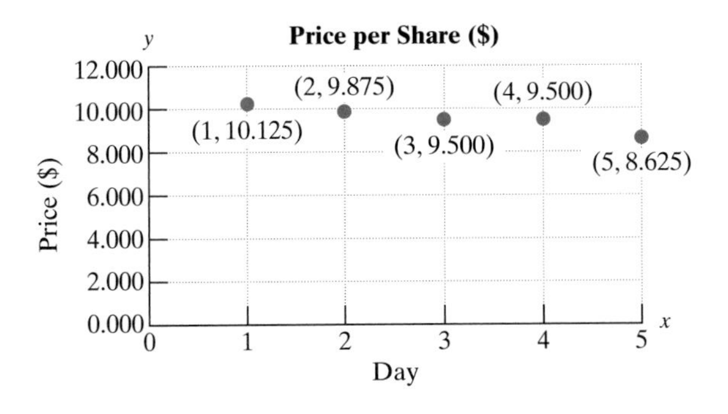

Concept 2: Plotting Points in a Rectangular Coordinate System

7. Plot the points on a rectangular coordinate system.

a. (2, 6) **b.** (6, 2)

c. (0, −3) **d.** (−3, 0)

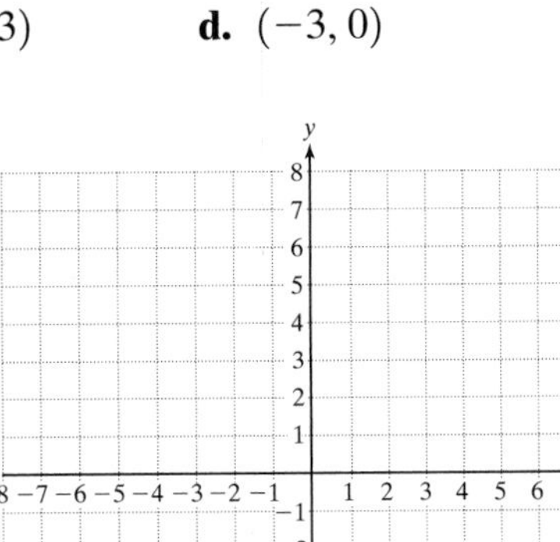

8. Plot the points on a rectangular coordinate system.

a. (−1, 2) **b.** (2, −1)

c. (0, 7) **d.** (7, 0)

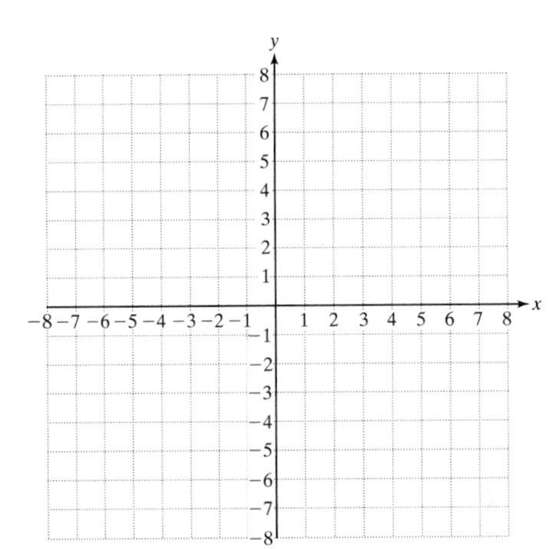

9. Plot the points on a rectangular coordinate system.

a. (4, 5) **b.** (−4, 5)

c. (4, −5) **d.** (−4, −5)

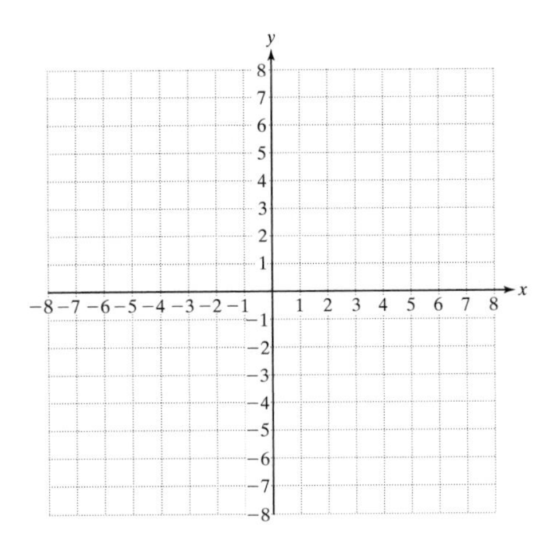

10. Plot the points on a rectangular coordinate system.

a. (2, 3) **b.** (−2, 3)

c. (2, −3) **d.** (−2, −3)

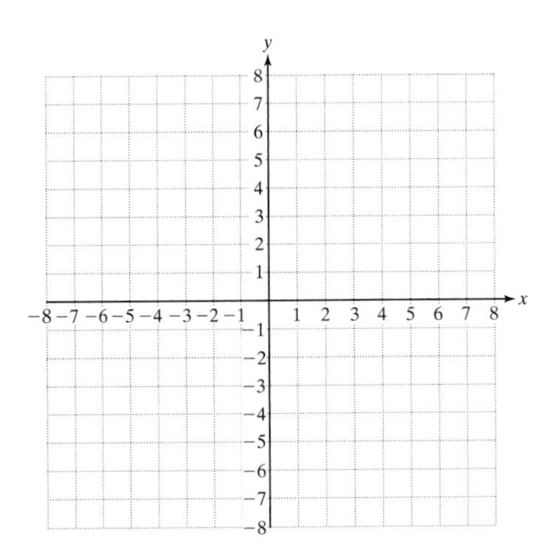

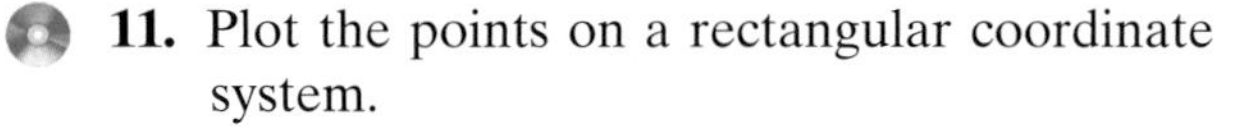

11. Plot the points on a rectangular coordinate system.

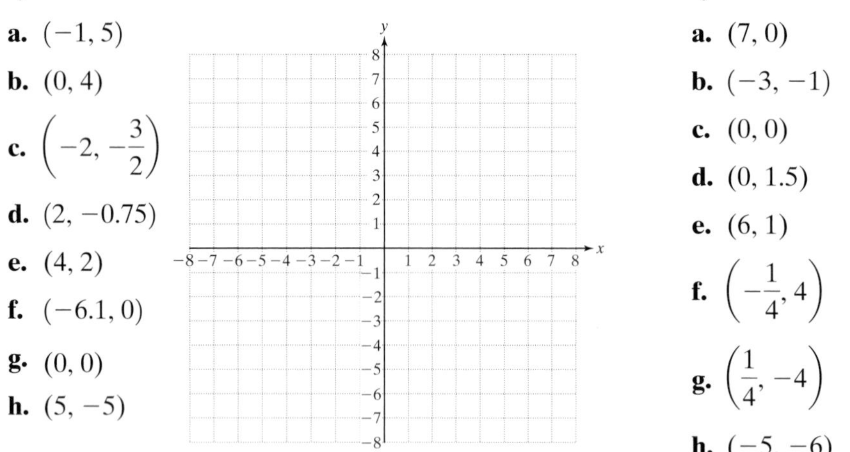

a. $(-1, 5)$

b. $(0, 4)$

c. $\left(-2, -\frac{3}{2}\right)$

d. $(2, -0.75)$

e. $(4, 2)$

f. $(-6.1, 0)$

g. $(0, 0)$

h. $(5, -5)$

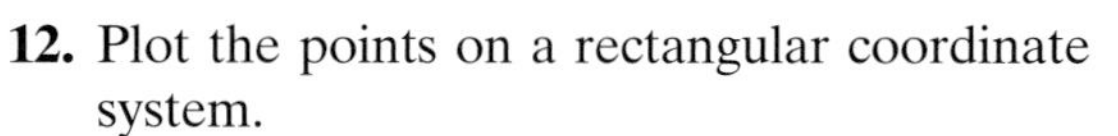

12. Plot the points on a rectangular coordinate system.

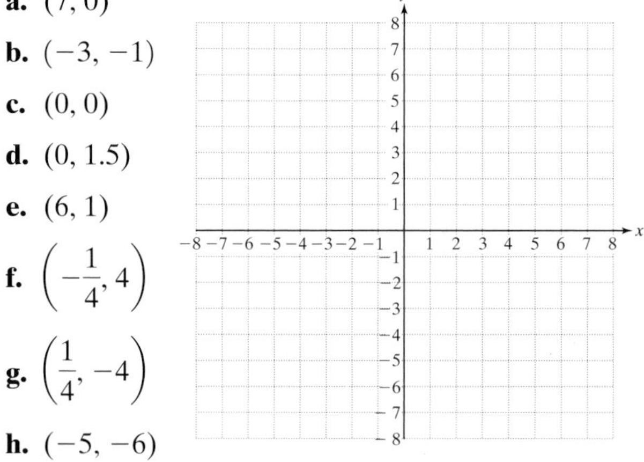

a. $(7, 0)$

b. $(-3, -1)$

c. $(0, 0)$

d. $(0, 1.5)$

e. $(6, 1)$

f. $\left(-\frac{1}{4}, 4\right)$

g. $\left(\frac{1}{4}, -4\right)$

h. $(-5, -6)$

For Exercises 13–20, identify the quadrant in which the given point is found.

13. $(13, -2)$ **14.** $(25, 16)$ **15.** $(-8, 14)$ **16.** $(-82, -71)$

17. $(-5, -19)$ **18.** $(-31, 6)$ **19.** $\left(\frac{5}{2}, \frac{7}{4}\right)$ **20.** $(9, -40)$

21. What is the x-coordinate of a point on the y-axis?

22. What is the y-coordinate of a point on the x-axis?

23. Where is the point $(\frac{7}{8}, 0)$ located?

24. Where is the point $(0, \frac{6}{5})$ located?

For Exercises 25–26, refer to the graph.

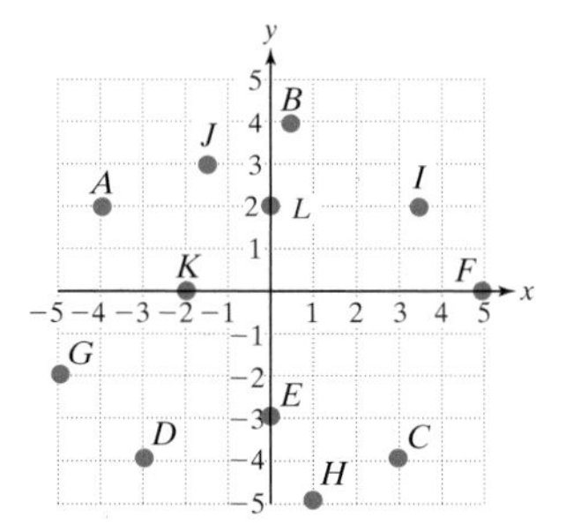

25. Estimate the coordinates of the points A, B, C, D, E, and F.

26. Estimate the coordinates of the points G, H, I, J, K, and L.

Concept 3: Applications of Plotting and Identifying Points

27. A movie theater has kept records of popcorn sales versus movie attendance.

a. Write the corresponding ordered pairs using the movie attendance as the x-variable and sales of popcorn as the y-variable. Interpret the meaning of the first ordered pair.

Movie Attendance (Number of People)	Sales of Popcorn ($)
250	225
175	193
315	330
220	209
450	570
400	480
190	185

b. Plot the data points on a rectangular coordinate system.

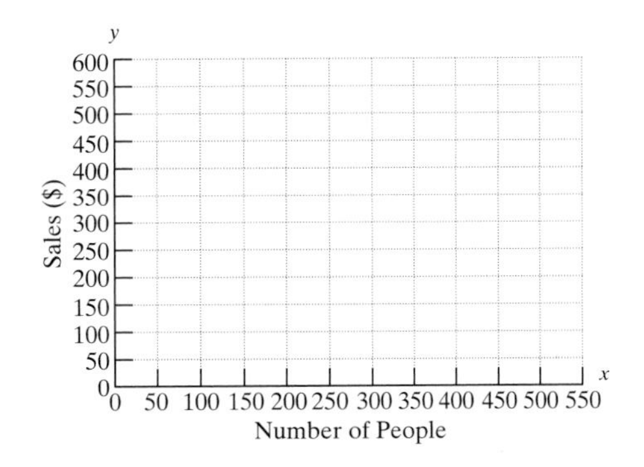

28. The age and systolic blood pressure (in millimeters of mercury, mm Hg) for eight different women are given in the table.

a. Write the corresponding ordered pairs using each woman's age as the x-variable and the systolic blood pressure as the y-variable. Interpret the meaning of the first ordered pair.

b. Plot the data points on a rectangular coordinate system.

Age (Years)	Systolic Blood Pressure (mm Hg)
57	149
41	120
71	158
36	115
64	151
25	110
40	118
77	165

29. The income level defining the poverty line for an individual is given for selected years between 1980 and 2005. Let x represent the number of years since 1980. Let y represent the income defining the poverty level.

(0, 4300) (5, 5600) (10, 6800)

(15, 7900) (20, 9000) (25, 10,500)

a. Interpret the meaning of the ordered pair (10, 6800).

b. Plot the points on a rectangular coordinate system.

(*Source: U.S. Department of the Census*)

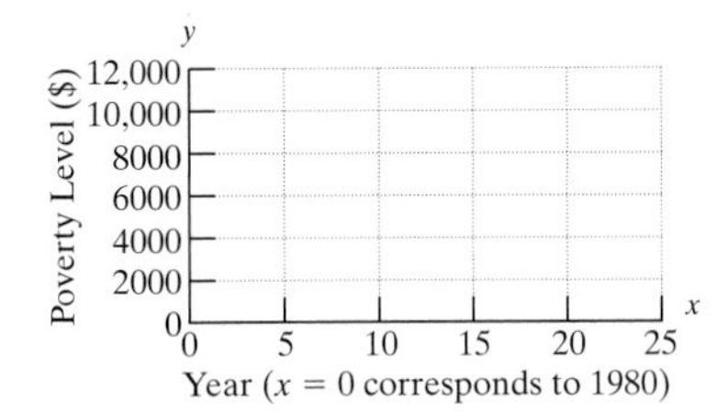

30. The following ordered pairs give the population of the U.S. colonies from 1700 to 1770. Let x represent the year, where $x = 0$ corresponds to 1700, $x = 10$ corresponds to 1710, and so on. Let y represent the population of the colonies.

(0, 251000) (10, 332000) (20, 466000)

(30, 629000) (40, 906000) (50, 1171000)

(60, 1594000) (70, 2148000)

a. Interpret the meaning of the ordered pair (10, 332000).

b. Plot the points on a rectangular coordinate system.

(*Source: Information Please Almanac*)

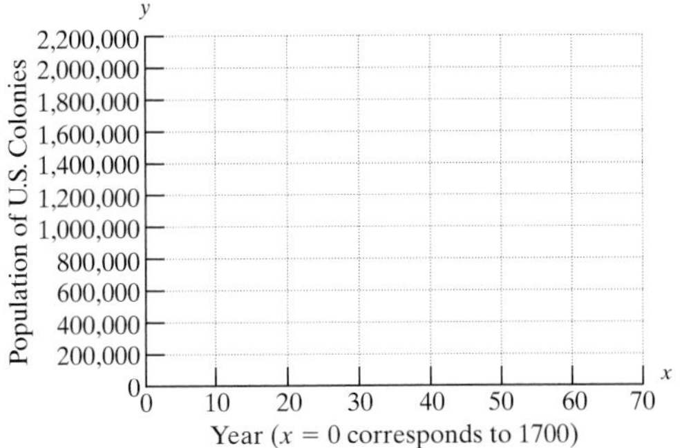

31. The following table shows the average temperature in degrees Celsius for Montreal, Quebec, Canada, by month.

a. Write the corresponding ordered pairs, letting $x = 1$ correspond to the month of January.

b. Plot the ordered pairs on a rectangular coordinate system.

Month, x		Temperature (°C), y
Jan.	1	−10.2
Feb.	2	−9.0
March	3	−2.5
April	4	5.7
May	5	13.0
June	6	18.3
July	7	20.9
Aug.	8	19.6
Sept.	9	14.8
Oct.	10	8.7
Nov.	11	2.0
Dec.	12	−6.9

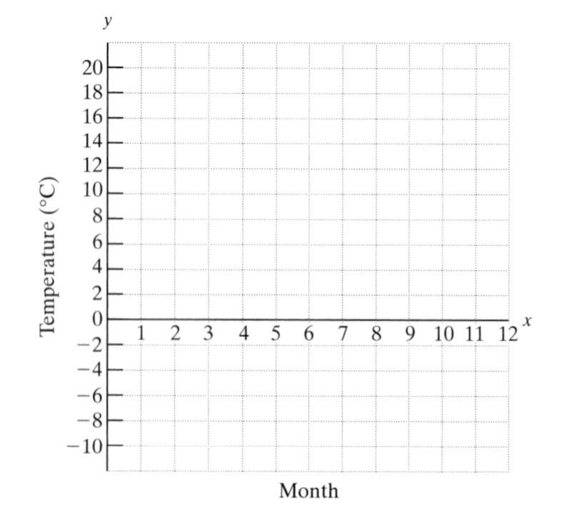

32. The table shows the average temperature in degrees Fahrenheit for Fairbanks, Alaska, by month.

a. Write the corresponding ordered pairs, letting $x = 1$ correspond to the month of January.

b. Plot the ordered pairs on a rectangular coordinate system.

Month, x		Temperature (°F), y
Jan.	1	−12.8
Feb.	2	−4.0
March	3	8.4
April	4	30.2
May	5	48.2
June	6	59.4
July	7	61.5
Aug.	8	56.7
Sept.	9	45.0
Oct.	10	25.0
Nov.	11	6.1
Dec.	12	−10.1

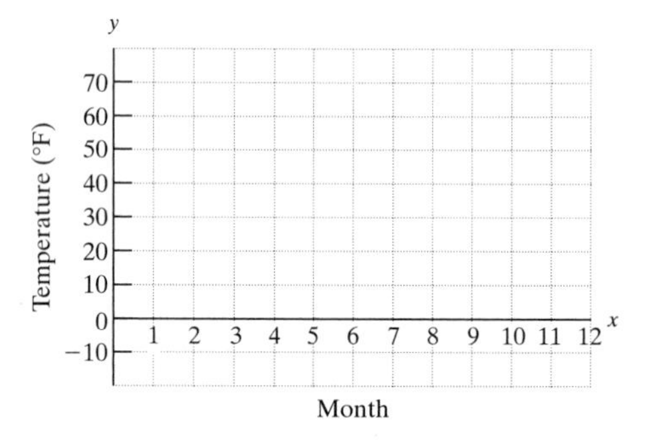

33. A map of a park is laid out with the Visitor Center located at the origin. Five visitors are in the park located at points A, B, C, D, and E. All distances are in meters.

a. Estimate the coordinates of each hiker.

b. How far apart are visitors C and D?

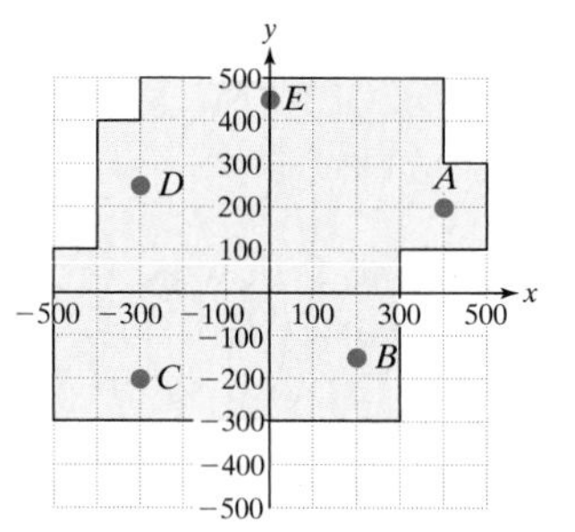

34. A townhouse has a sprinkler system in the backyard. With the water source at the origin, the sprinkler heads are located at points A, B, C, D, and E. All distances are in feet.

a. Estimate the coordinates of each sprinkler head.

b. How far is the distance from sprinkler head B to C?

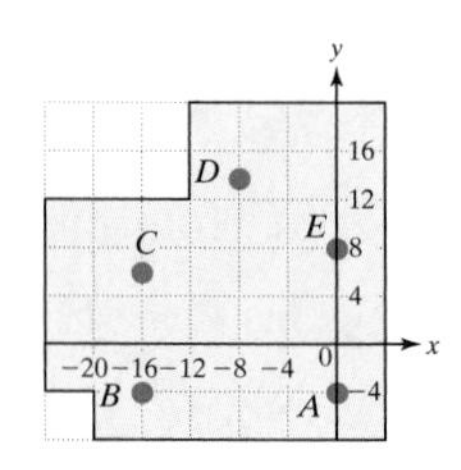

Linear Equations in Two Variables

Section 3.2

Concepts

1. Solutions to Linear Equations in Two Variables
2. Graphing Linear Equations in Two Variables by Plotting Points
3. x- and y-Intercepts
4. Horizontal and Vertical Lines

1. Solutions to Linear Equations in Two Variables

Recall that an equation in the form $ax + b = 0$, where $a \neq 0$, is called a linear equation in one variable. A solution to such an equation is a value of x that makes the equation a true statement. For example, $3x + 6 = 0$ has a solution of $x = -2$.

In this section, we will look at linear equations in *two* variables.

Definition of a Linear Equation in Two Variables

Let A, B, and C be real numbers such that A and B are not both zero. Then, an equation that can be written in the form:

$$Ax + By = C$$

is called a **linear equation in two variables.**

The equation $x + y = 4$ is a linear equation in two variables. A solution to such an equation is an ordered pair (x, y) that makes the equation a true statement. Several solutions to the equation $x + y = 4$ are listed here:

Solution:	Check:
(x, y)	$x + y = 4$
$(2, 2)$	$(2) + (2) = 4$ ✔
$(1, 3)$	$(1) + (3) = 4$ ✔
$(4, 0)$	$(4) + (0) = 4$ ✔
$(-1, 5)$	$(-1) + (5) = 4$ ✔

By graphing these ordered pairs, we see that the solution points line up (Figure 3-5).

Notice that there are infinitely many solutions to the equation $x + y = 4$ so they cannot all be listed. Therefore, to visualize all solutions to the equation $x + y = 4$, we draw the line through the points in the graph. Every point on the line represents an ordered pair solution to the equation $x + y = 4$, and the line represents the set of *all* solutions to the equation.

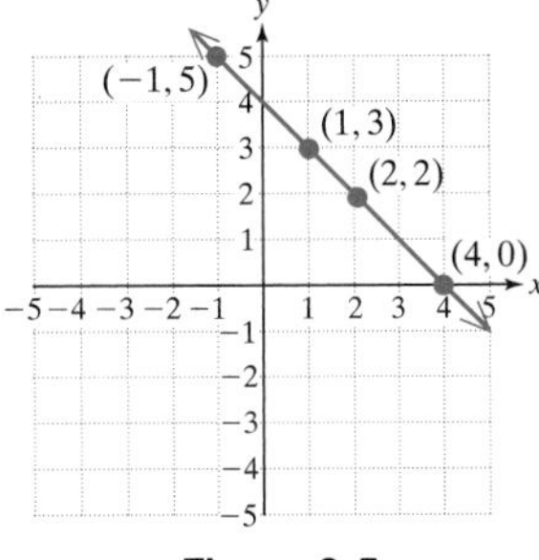

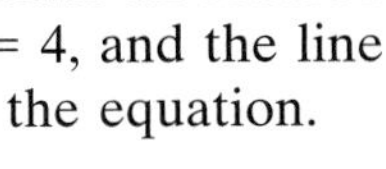

Figure 3-5

Example 1 Determining Solutions to a Linear Equation

For the linear equation, $4x - 5y = 8$, determine whether the given ordered pair is a solution.

a. $(2, 0)$ **b.** $(3, 1)$ **c.** $\left(1, -\frac{4}{5}\right)$

Solution:

a. $4x - 5y = 8$

$4(2) - 5(0) \stackrel{?}{=} 8$ Substitute $x = 2$ and $y = 0$.

$8 - 0 = 8$ ✔ (true) The ordered pair $(2, 0)$ is a solution.

b. $4x - 5y = 8$

$4(3) - 5(1) \stackrel{?}{=} 8$ Substitute $x = 3$ and $y = 1$.

$12 - 5 \neq 8$ The ordered pair (3, 1) is *not* a solution.

c. $4x - 5y = 8$

$4(1) - 5\left(-\frac{4}{5}\right) \stackrel{?}{=} 8$ Substitute $x = 1$ and $y = -\frac{4}{5}$.

$4 + 4 = 8$ ✔ (true) The ordered pair $\left(1, -\frac{4}{5}\right)$ is a solution.

Skill Practice

1. Given the equation $3x - 2y = -12$, determine whether the given ordered pair is a solution.

a. $(-2, 3)$ **b.** $(4, 0)$ **c.** $\left(1, \frac{15}{2}\right)$

2. Graphing Linear Equations in Two Variables by Plotting Points

The word *linear* means "relating to or resembling a line." It is not surprising then that the solution set for any linear equation in two variables forms a line in a rectangular coordinate system. Because two points determine a line, to graph a linear equation it is sufficient to find two solution points and draw the line between them. We will find three solution points and use the third point as a check point. This process is demonstrated in Example 2.

Example 2 Graphing a Linear Equation

Graph the equation $x - 2y = 8$.

Solution:

We will find three ordered pairs that are solutions to $x - 2y = 8$. To find the ordered pairs, choose arbitrary values of x or y, such as those shown in the table. Then complete the table to find the corresponding ordered pairs.

x	y	
2		→ (2,)
	−1	→ (, −1)
0		→ (0,)

From the first row, substitute $x = 2$:

$x - 2y = 8$

$(2) - 2y = 8$

$-2y = 6$

$y = -3$

From the second row, substitute $y = -1$:

$x - 2y = 8$

$x - 2(-1) = 8$

$x + 2 = 8$

$x = 6$

From the third row, substitute $x = 0$:

$x - 2y = 8$

$(0) - 2y = 8$

$-2y = 8$

$y = -4$

Skill Practice Answers

1a. Yes **b.** No **c.** Yes

The completed table is shown below with the corresponding ordered pairs.

x	y	
2	−3	→ (2, −3)
6	−1	→ (6, −1)
0	−4	→ (0, −4)

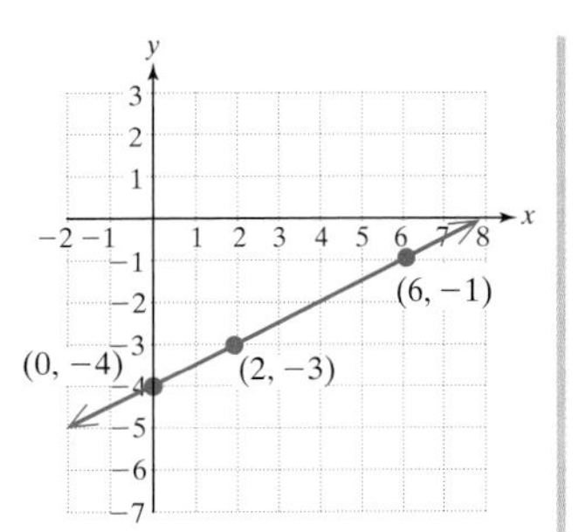

Figure 3-6

To graph the equation, plot the three solutions and draw the line through the points (Figure 3-6).

Skill Practice

2. Graph the equation $2x + y = 6$.

TIP: Only two points are needed to graph a line. However, in Example 2, we found a third ordered pair, (0, −4). Notice that this point "lines up" with the other two points. If the three points do not line up, then we know that a mistake was made in solving for at least one of the ordered pairs.

In Example 2, the original values for x and y given in the table were picked arbitrarily by the authors. It is important to note, however, that once you pick an arbitrary value for x, the corresponding y-value is determined by the equation. Similarly, once you pick an arbitrary value for y, the x-value is determined by the equation.

Example 3 Graphing a Linear Equation

Graph the equation $4x + 3y = 15$.

Solution:

We will find three ordered pairs that are solutions to the equation $4x + 3y = 15$. In the table, we have selected arbitrary values for x and y and must complete the ordered pairs. Notice that in this case, we are choosing zero for x and zero for y to illustrate that the resulting equation is often easy to solve.

x	y	
0		→ (0,)
	0	→ (, 0)
3		→ (3,)

From the first row, substitute $x = 0$:

$$4x + 3y = 15$$
$$4(0) + 3y = 15$$
$$3y = 15$$
$$y = 5$$

From the second row, substitute $y = 0$:

$$4x + 3y = 15$$
$$4x + 3(0) = 15$$
$$4x = 15$$
$$x = \frac{15}{4} \text{ or } 3\frac{3}{4}$$

From the third row, substitute $x = 3$:

$$4x + 3y = 15$$
$$4(3) + 3y = 15$$
$$12 + 3y = 15$$
$$3y = 3$$
$$y = 1$$

Skill Practice Answers

2. $2x + y = 6$

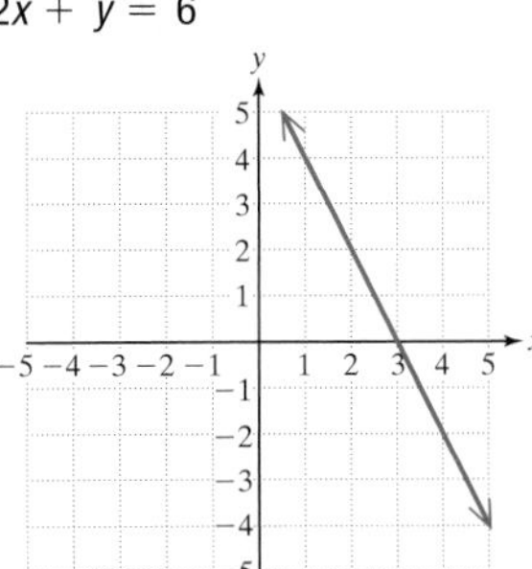

The completed table is shown with the corresponding ordered pairs.

x	y	
0	5	$\rightarrow (0, 5)$
$3\frac{3}{4}$	0	$\rightarrow (3\frac{3}{4}, 0)$
3	1	$\rightarrow (3, 1)$

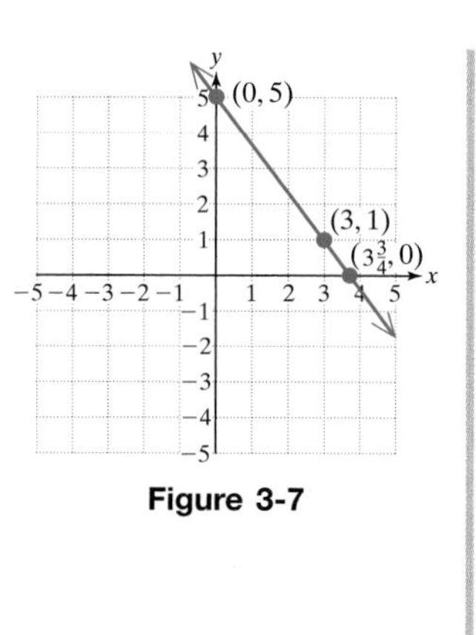

Figure 3-7

To graph the equation, plot the three solutions and draw the line through the points (Figure 3-7).

Skill Practice

3. Graph the equation $2x + 3y = 12$.

Example 4 Graphing a Linear Equation

Graph the line $y = -\frac{1}{3}x + 1$.

Solution:

Because the y-variable is isolated in the equation, it is easy to substitute a value for x and simplify the right-hand side to find y. Since any number for x can be picked, choose numbers that are multiples of 3 that will simplify easily when multiplied by $-\frac{1}{3}$.

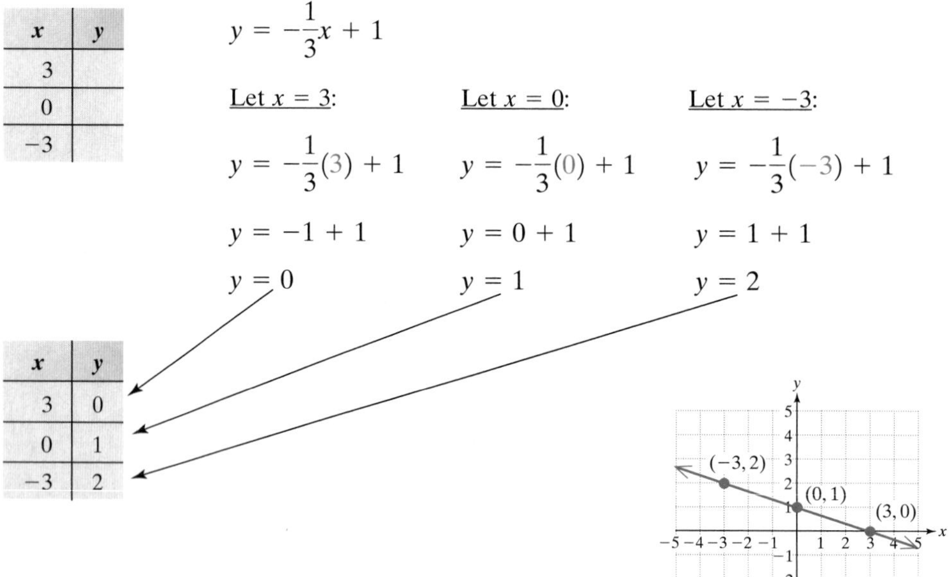

x	y
3	
0	
−3	

$$y = -\frac{1}{3}x + 1$$

Let $x = 3$:	Let $x = 0$:	Let $x = -3$:
$y = -\frac{1}{3}(3) + 1$	$y = -\frac{1}{3}(0) + 1$	$y = -\frac{1}{3}(-3) + 1$
$y = -1 + 1$	$y = 0 + 1$	$y = 1 + 1$
$y = 0$	$y = 1$	$y = 2$

x	y
3	0
0	1
−3	2

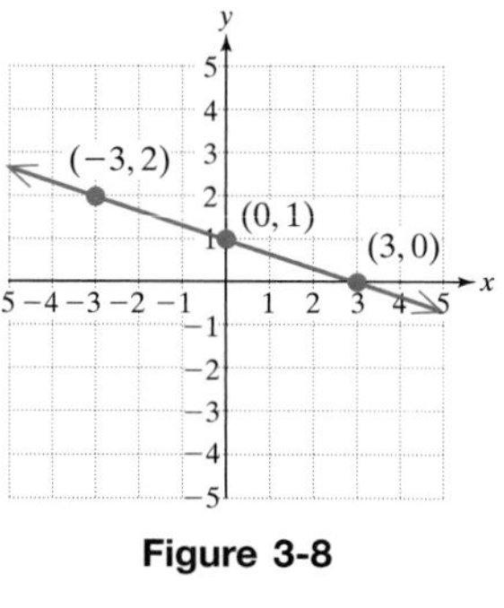

Figure 3-8

The line through the three ordered pairs $(3, 0)$, $(0, 1)$, and $(-3, 2)$ is shown in Figure 3-8. The line represents the set of all solutions to the equation $y = -\frac{1}{3}x + 1$.

Skill Practice

4. Graph the line $y = \frac{1}{2}x + 3$.

Skill Practice Answers

3. $2x + 3y = 12$

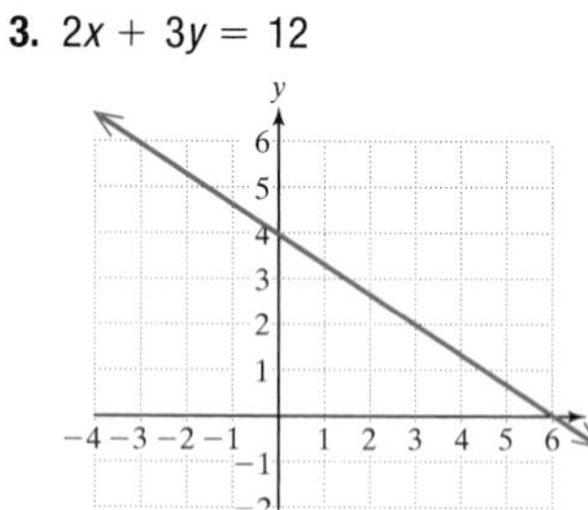
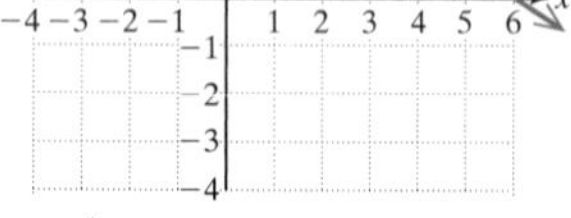

4. $y = \frac{1}{2}x + 3$

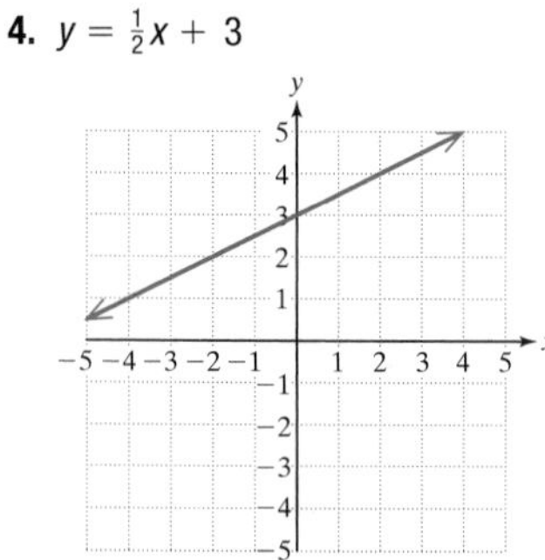

3. *x*- and *y*-Intercepts

The x- and y-intercepts are the points where the graph intersects the x- and y-axes, respectively. From Example 4, we see that the x-intercept is at the point $(3, 0)$ and the y-intercept is at the point $(0, 1)$. Notice that an x-intercept is a point on the x-axis and must have a y-coordinate of 0. Likewise, a y-intercept is a point on the y-axis and has an x-coordinate of 0.

***Definition of *x*- and *y*-Intercepts**

An **x-intercept** is a point $(a, 0)$ where a graph intersects the x-axis.

A **y-intercept** is a point $(0, b)$ where a graph intersects the y-axis.

Although any two points may be used to graph a line, in some cases it is convenient to use the x- and y-intercepts of the line. To find the x- and y-intercepts of any two-variable equation in x and y, follow these steps:

Finding *x*- and *y*-Intercepts

Step 1. Find the x-intercept(s) by substituting $y = 0$ into the equation and solving for x.

Step 2. Find the y-intercept(s) by substituting $x = 0$ into the equation and solving for y.

Example 5 **Finding the *x*- and *y*-Intercepts of a Line**

Given the equation $-3x + 2y = 8$,

a. Find the x-intercept. **b.** Find the y-intercept.

c. Graph the equation.

Solution:

a. To find the x-intercept, substitute $y = 0$.

$$-3x + 2y = 8$$
$$-3x + 2(0) = 8$$
$$-3x = 8$$
$$\frac{-3x}{-3} = \frac{8}{-3}$$
$$x = -\frac{8}{3}$$

The x-intercept is $(-\frac{8}{3}, 0)$.

b. To find the y-intercept, substitute $x = 0$.

$$-3x + 2y = 8$$
$$-3(0) + 2y = 8$$
$$2y = 8$$
$$y = 4$$

The y-intercept is $(0, 4)$.

*In some applications, an x-intercept is defined as the x-coordinate of a point of intersection that a graph makes with the x-axis. For example, if an x-intercept is at the point $(3, 0)$, it is sometimes stated simply as 3 (the y-coordinate is assumed to be 0). Similarly, a y-intercept is sometimes defined as the y-coordinate of a point of intersection that a graph makes with the y-axis. For example, if a y-intercept is at the point $(0, 7)$, it may be stated simply as 7 (the x-coordinate is assumed to be 0).

TIP: A third point such as $(-2, 1)$, can be plotted to check the line.

c. The line through the ordered pairs $(-\frac{8}{3}, 0)$ and $(0, 4)$ is shown in Figure 3-9. Note that the point $(-\frac{8}{3}, 0)$ can be written as $(-2\frac{2}{3}, 0)$.

The line represents the set of all solutions to the equation $-3x + 2y = 8$.

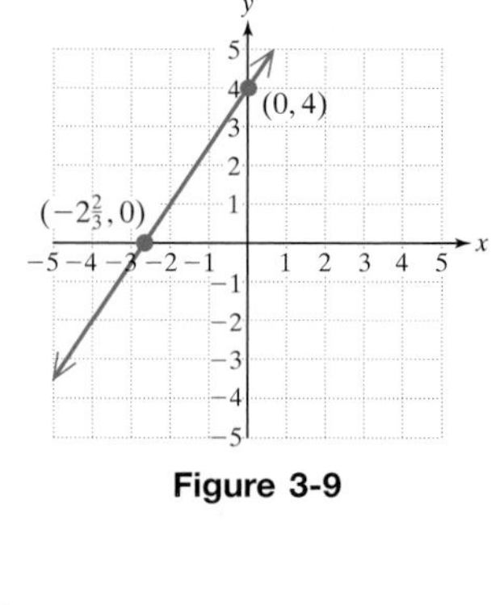

Figure 3-9

Skill Practice

5. Given the equation $4x - 3y = -9$,

a. Find the x-intercept.

b. Find the y-intercept.

c. Graph the equation.

Example 6 Finding the x- and y-Intercepts of a Line

Given the equation $4x + 5y = 0$,

a. Find the x-intercept.

b. Find the y-intercept.

c. Graph the line.

Solution:

a. To find the x-intercept, substitute $y = 0$.

$$4x + 5y = 0$$
$$4x + 5(0) = 0$$
$$4x = 0$$
$$x = 0$$

The x-intercept is $(0, 0)$.

b. To find the y-intercept, substitute $x = 0$.

$$4x + 5y = 0$$
$$4(0) + 5y = 0$$
$$5y = 0$$
$$y = 0$$

The y-intercept is $(0, 0)$.

c. Because the x-intercept and the y-intercept are the same point (the origin), one or more additional points are needed to graph the line. In the table, we have arbitrarily selected additional values for x and y to find two more points on the line.

x	y
-5	
	2

Let $x = -5$:

$$4x + 5y = 0$$
$$4(-5) + 5y = 0$$
$$-20 + 5y = 0$$
$$5y = 20$$
$$y = 4$$

Let $y = 2$:

$$4x + 5y = 0$$
$$4x + 5(2) = 0$$
$$4x + 10 = 0$$
$$4x = -10$$
$$x = -\frac{10}{4}$$
$$x = -\frac{5}{2}$$

Skill Practice Answers

5a. $\left(-\frac{9}{4}, 0\right)$ **b.** $(0, 3)$

c. $4x - 3y = -9$

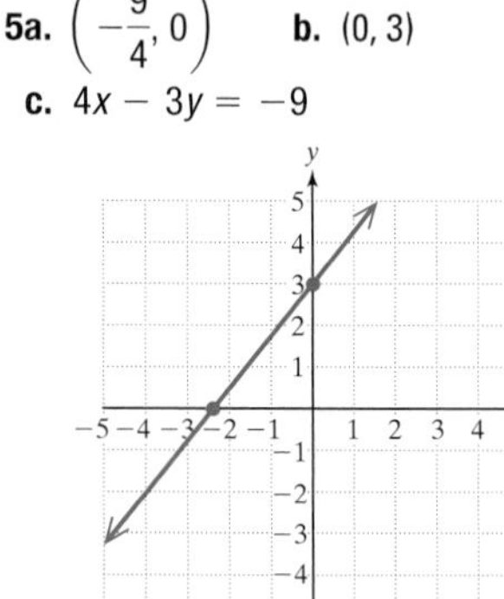

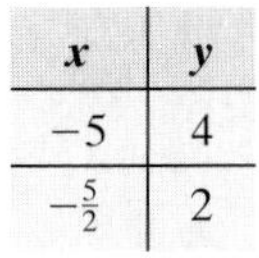

x	y
-5	4
$-\frac{5}{2}$	2

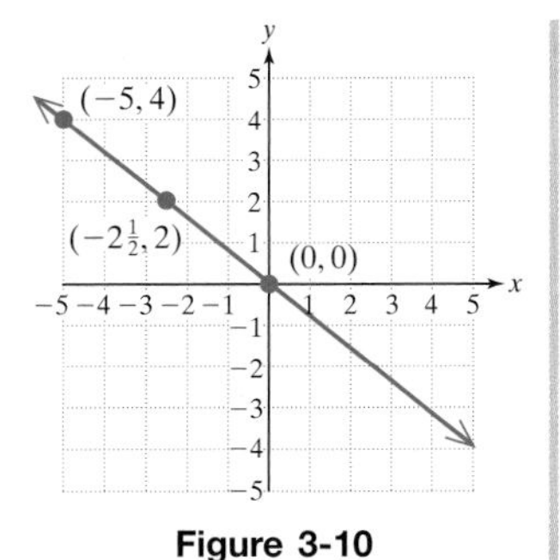

Figure 3-10

The line through the ordered pairs (0, 0), (−5, 4), and $(-\frac{5}{2}, 2)$ is shown in Figure 3-10. Note that the point $(-\frac{5}{2}, 2)$ can be written as $(-2\frac{1}{2}, 2)$.

The line represents the set of all solutions to the equation $4x + 5y = 0$.

Skill Practice

6. Given the equation $2x - 3y = 0$,

a. Find the x-intercept.

b. Find the y-intercept.

c. Graph the line. (*Hint:* You may need to find an additional point.)

4. Horizontal and Vertical Lines

Recall that a linear equation can be written in the form $Ax + By = C$, where A and B are not both zero. However, if A or B is 0, then the line is either parallel to the x-axis (horizontal) or parallel to the y-axis (vertical), respectively.

Definitions of Vertical and Horizontal Lines

1. A **vertical line** is a line that can be written in the form, $x = k$, where k is a constant.
2. A **horizontal line** is a line that can be written in the form, $y = k$, where k is a constant.

Example 7 Graphing a Horizontal Line

Graph the line $y = 3$.

Solution:

Because this equation is in the form $y = k$, the line is horizontal and must cross the y-axis at $y = 3$ (Figure 3-11).

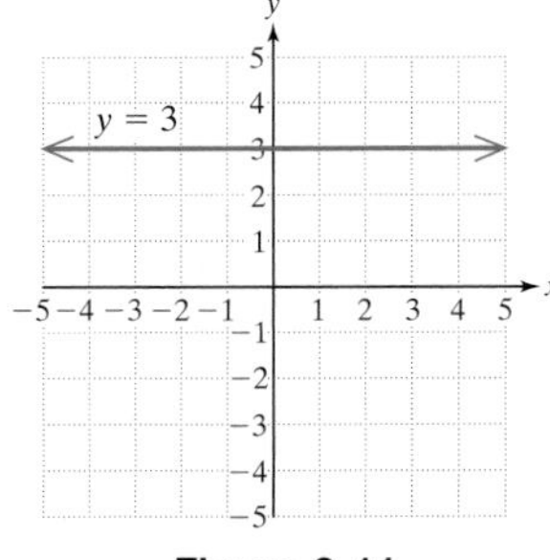

Figure 3-11

Skill Practice Answers

6a. (0, 0) **b.** (0, 0)

c. $2x - 3y = 0$

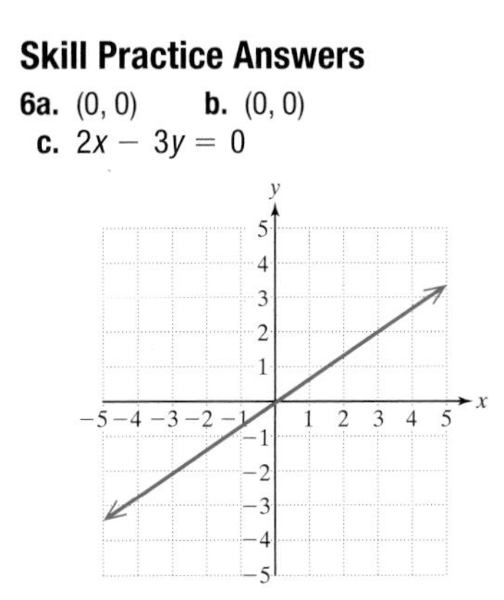

Alternative Solution:

Create a table of values for the equation $y = 3$. The choice for the y-coordinate must be 3, but x can be any real number.

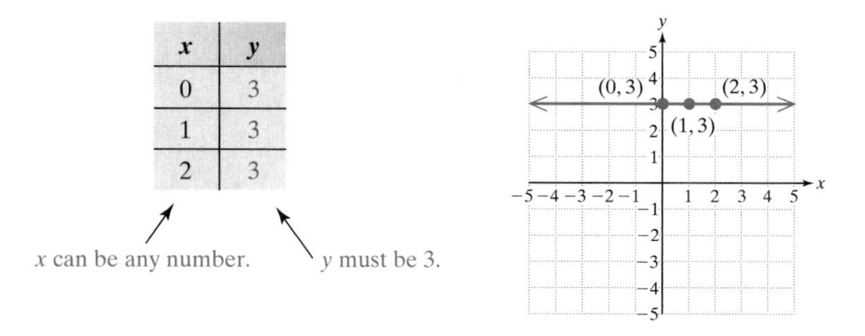

x	y
0	3
1	3
2	3

TIP: Notice that a horizontal line has a y-intercept, but does not have an x-intercept (unless the horizontal line is the x-axis itself).

Skill Practice

7. Graph the line $y = -2$.

Example 8 Graphing a Vertical Line

Graph the line $4x = -8$.

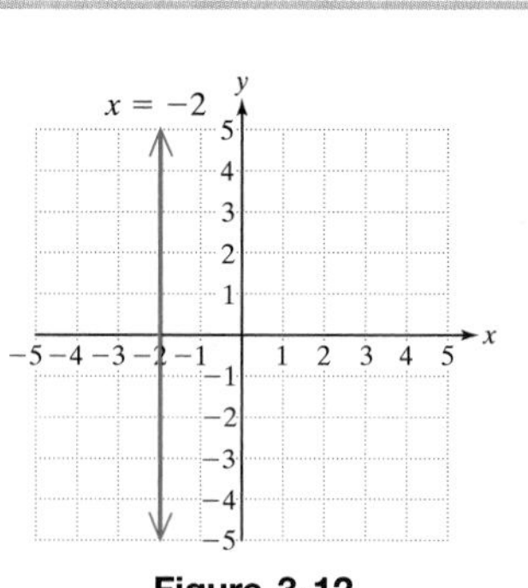

Figure 3-12

Solution:

Because the equation does not have a y-variable, we can solve the equation for x.

$$4x = -8 \quad \text{is equivalent to} \quad x = -2$$

This equation is in the form $x = k$, indicating that the line is vertical and must cross the x-axis at $x = -2$ (Figure 3-12).

Alternative Solution:

Create a table of values for the equation $x = -2$. The choice for the x-coordinate must be -2, but y can be any real number.

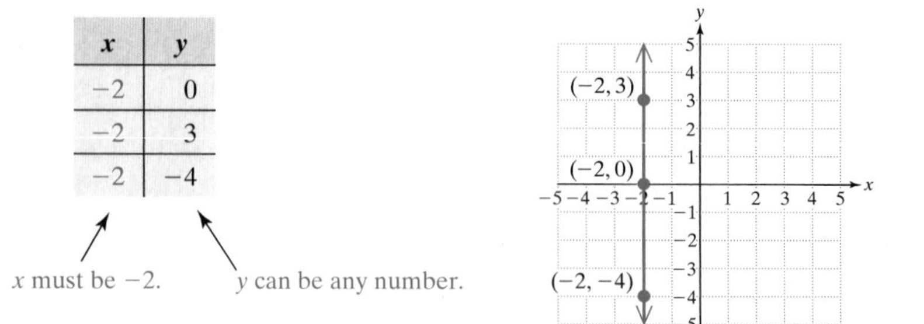

x	y
−2	0
−2	3
−2	−4

TIP: Notice that a vertical line has an x-intercept but does not have a y-intercept (unless the vertical line is the y-axis itself).

Skill Practice

8. Graph the line $x = 4$.

Skill Practice Answers

7. $y = -2$

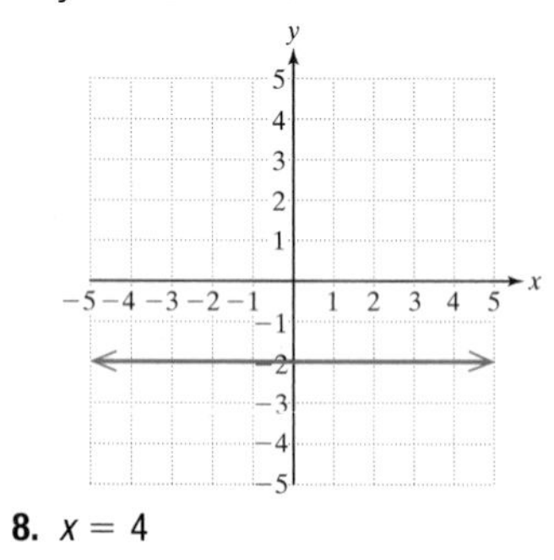

8. $x = 4$

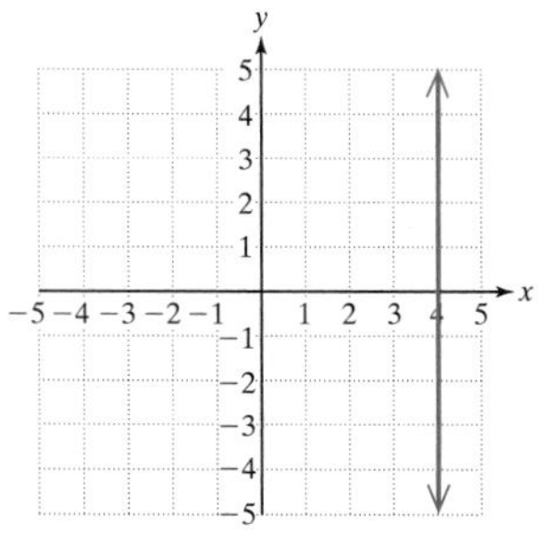

Calculator Connections

A viewing window of a graphing calculator shows a portion of a rectangular coordinate system. The standard viewing window for many calculators shows the x-axis between -10 and 10 and the y-axis between -10 and 10 (Figure 3-13). Furthermore, the scale defined by the "tick" marks on both the x- and y-axes is usually set to 1.

The "Standard Viewing Window"

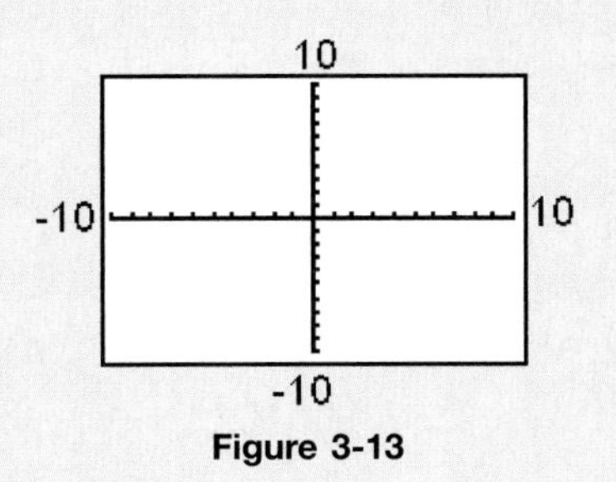

Figure 3-13

To graph an equation in x and y on a graphing calculator, the equation must be written with the y-variable isolated. Therefore, the equation $x + 3y = 3$ must first be written as $y = -\frac{1}{3}x + 1$ before it can be entered into a graphing calculator. To enter the equation $y = -\frac{1}{3}x + 1$, use parentheses around the fraction $\frac{1}{3}$. The *Graph* option displays the graph of the line.

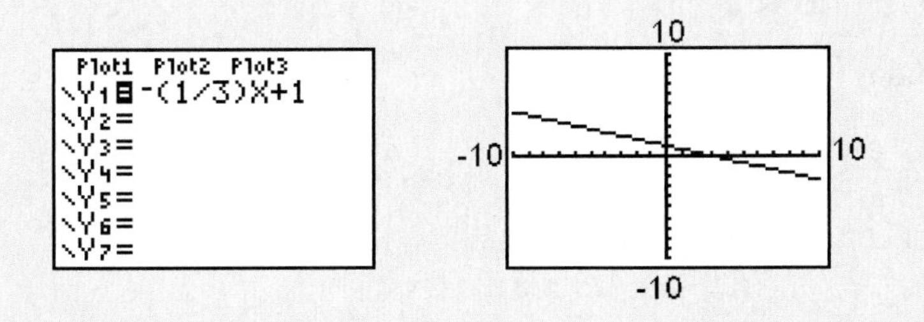

Sometimes the standard viewing window does not provide an adequate display for the graph of an equation. For example, the graph of $y = -x + 15$ is visible only in a small portion of the upper right corner of the standard viewing window.

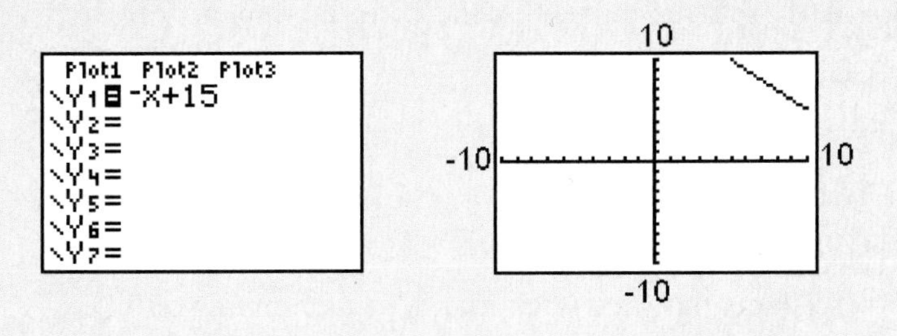

To see where this line crosses the x- and y-axes, we can change the viewing window to accommodate larger values of x and y. Most calculators have a *Range* feature or *Window* feature that allows the user to change the minimum and maximum x- and y-values.

To get a better picture of the equation $y = -x + 15$, change the minimum x-value to -10 and the maximum x-value to 20. Similarly, use a minimum y-value of -10 and a maximum y-value of 20.

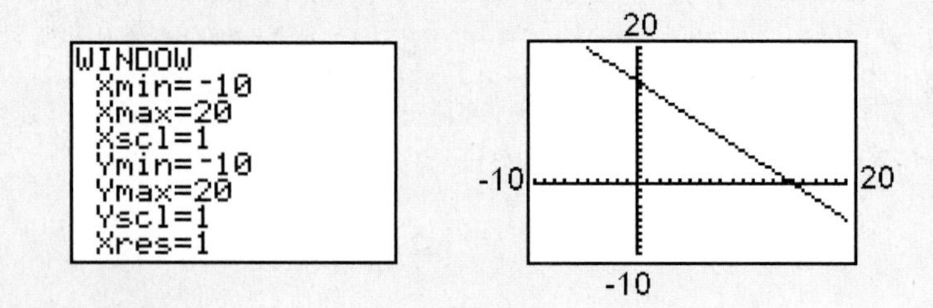

Calculator Exercises

Graph the equations on the standard viewing window.

1. $y = -2x + 5$

2. $y = 3x - 1$

3. $y = \frac{1}{2}x - \frac{7}{2}$

4. $y = -\frac{3}{4}x + \frac{5}{3}$

5. $4x - 7y = 21$

6. $2x + 3y = 12$

Graph the equations on the given viewing window.

7. $y = 3x + 15$

Window: $-10 \le x \le 10$
$-5 \le y \le 20$

8. $y = -2x - 25$

Window: $-30 \le x \le 30$
$-30 \le y \le 30$

Xscl = 3 (sets the x-axis tick marks to increments of 3)

Yscl = 3 (sets the y-axis tick marks to increments of 3)

9. $y = -0.2x + 0.04$

Window: $-0.1 \le x \le 0.3$
$-0.1 \le y \le 0.1$

Xscl = 0.01 (sets the x-axis tick marks to increments of 0.01)

Yscl = 0.01 (sets the y-axis tick marks to increments of 0.01)

10. $y = 0.3x - 0.5$

Window: $-1 \le x \le 3$
$-1 \le y \le 1$

Xscl = 0.1 (sets the x-axis tick marks to increments of 0.1)

Yscl = 0.1 (sets the y-axis tick marks to increments of 0.1)

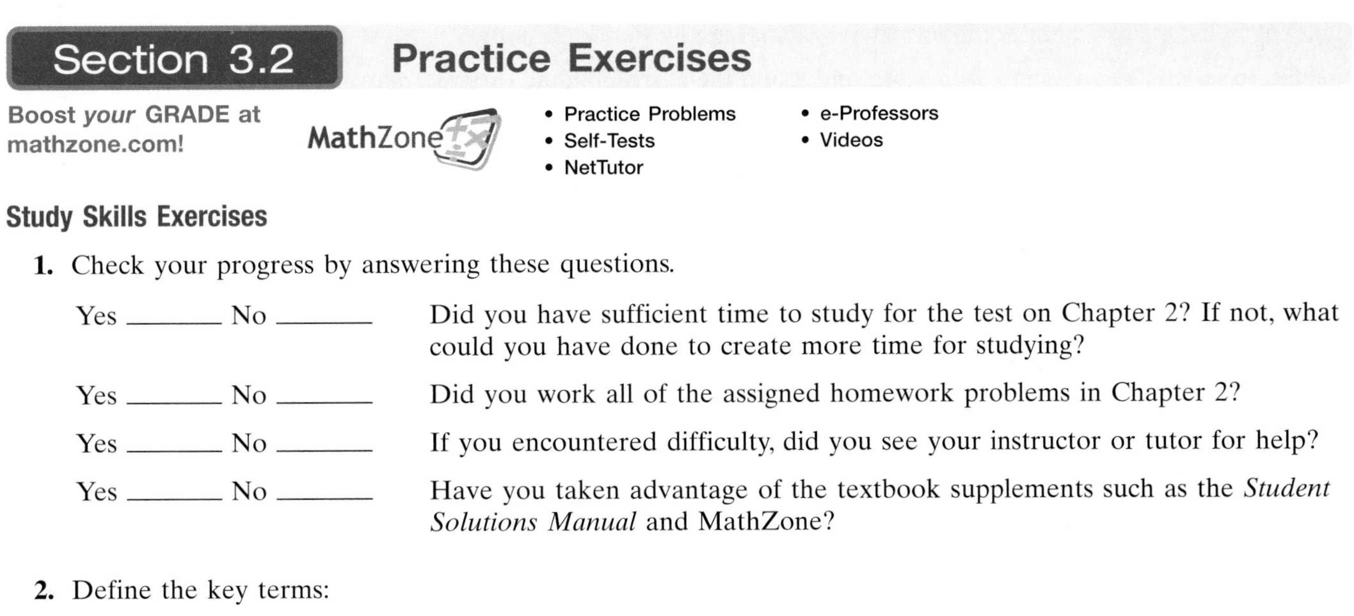

Section 3.2 Practice Exercises

Boost *your* GRADE at mathzone.com!

MathZone

- Practice Problems
- Self-Tests
- NetTutor
- e-Professors
- Videos

Study Skills Exercises

1. Check your progress by answering these questions.

Yes _______ No _______ Did you have sufficient time to study for the test on Chapter 2? If not, what could you have done to create more time for studying?

Yes _______ No _______ Did you work all of the assigned homework problems in Chapter 2?

Yes _______ No _______ If you encountered difficulty, did you see your instructor or tutor for help?

Yes _______ No _______ Have you taken advantage of the textbook supplements such as the *Student Solutions Manual* and MathZone?

2. Define the key terms:

a. horizontal line **b. linear equation in two variables** **c. vertical line**

d. *x*-intercept **e. *y*-intercept**

Review Exercises

For Exercises 3–8, refer to the figure to give the coordinates of the labeled points, and state the quadrant or axis where the point is located.

3. A **4.** B **5.** C

6. D **7.** E **8.** F

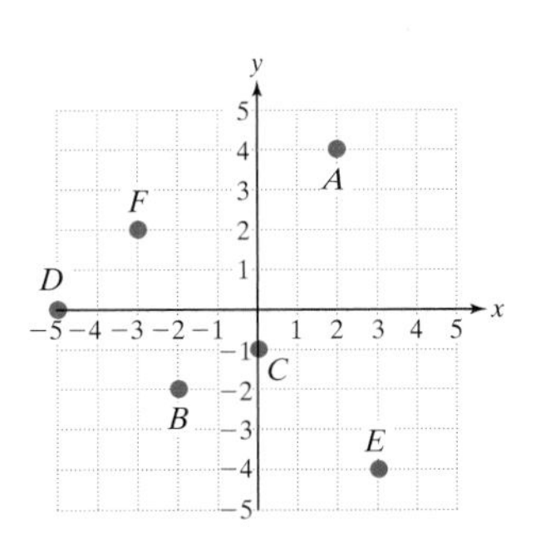

Concept 1: Solutions to Linear Equations in Two Variables

For Exercises 9–17, determine if the given ordered pair is a solution to the equation.

9. $x - y = 6;\quad (8, 2)$

10. $y = 3x - 2;\quad (1, 1)$

11. $y = -\frac{1}{3}x + 3;\quad (-3, 4)$

12. $y = -\frac{5}{2}x + 5;\quad (-2, 0)$

13. $4x + 5y = 20;\quad (-5, -4)$

14. $y = 7;\quad (0, 7)$

15. $y = -2;\quad (-2, 6)$

16. $x = 1;\quad (0, 1)$

17. $x = -5;\quad (-5, 6)$

Concept 2: Graphing Linear Equations in Two Variables by Plotting Points

For Exercises 18–33, complete each table, and graph the corresponding ordered pairs. Draw the line defined by the points to represent all solutions to the equation.

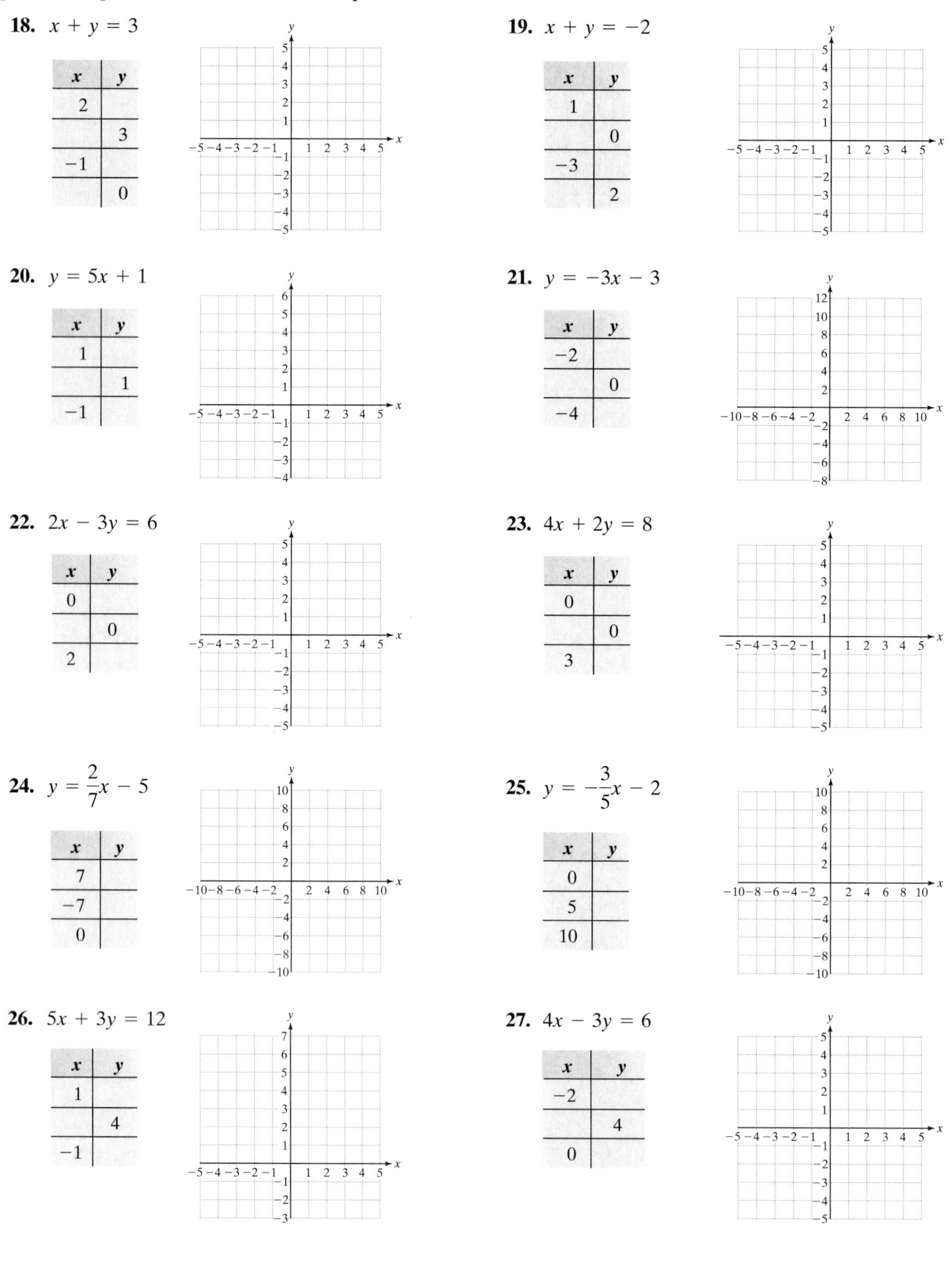

18. $x + y = 3$

x	y
2	
	3
−1	
	0

19. $x + y = -2$

x	y
1	
	0
−3	
	2

20. $y = 5x + 1$

x	y
1	
	1
−1	

21. $y = -3x - 3$

x	y
−2	
	0
−4	

22. $2x - 3y = 6$

x	y
0	
	0
2	

23. $4x + 2y = 8$

x	y
0	
	0
3	

24. $y = \frac{2}{7}x - 5$

x	y
7	
−7	
0	

25. $y = -\frac{3}{5}x - 2$

x	y
0	
5	
10	

26. $5x + 3y = 12$

x	y
1	
	4
−1	

27. $4x - 3y = 6$

x	y
−2	
	4
0	

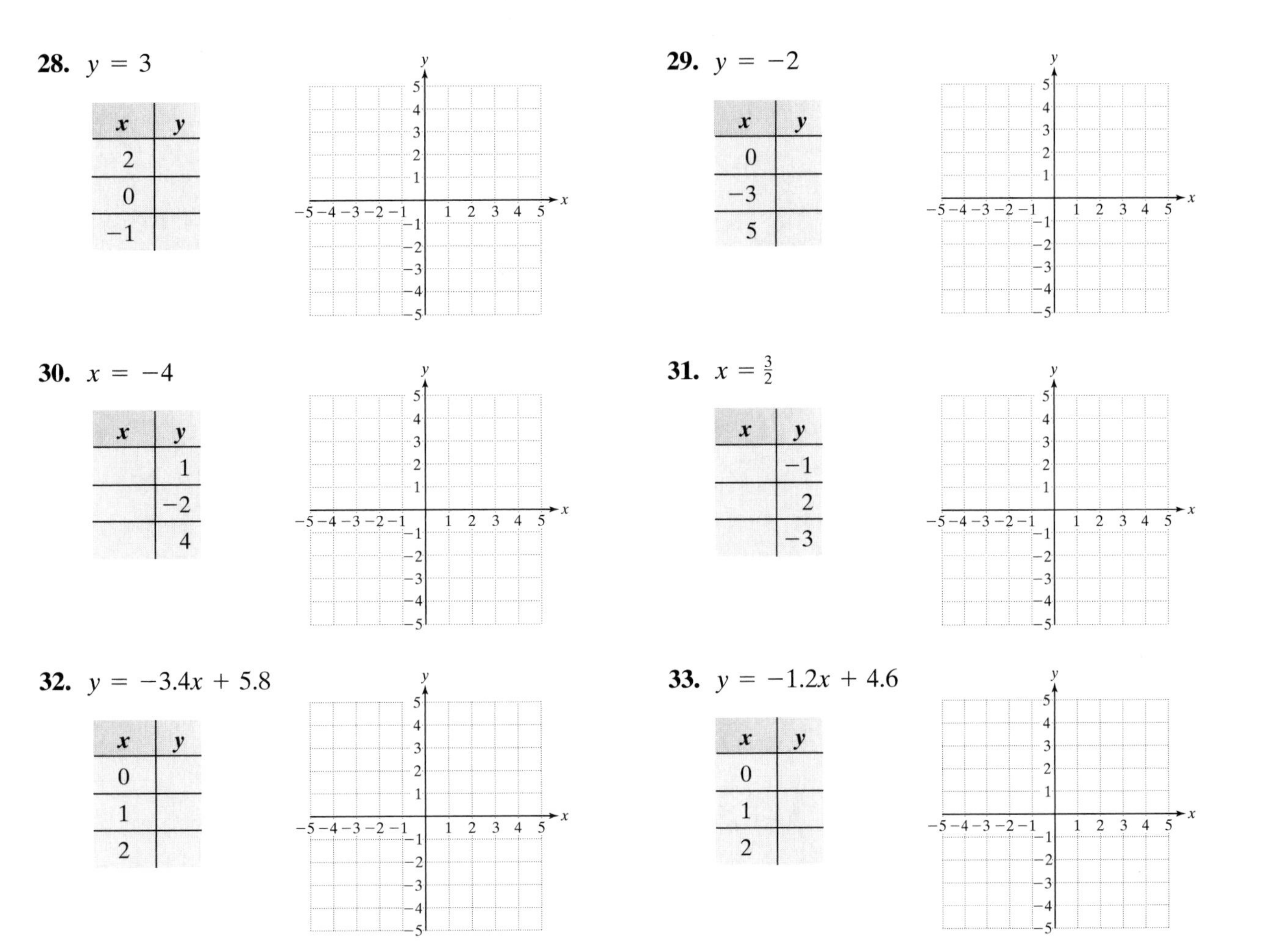

28. $y = 3$

x	y
2	
0	
−1	

29. $y = -2$

x	y
0	
−3	
5	

30. $x = -4$

x	y
	1
	−2
	4

31. $x = \frac{3}{2}$

x	y
	−1
	2
	−3

32. $y = -3.4x + 5.8$

x	y
0	
1	
2	

33. $y = -1.2x + 4.6$

x	y
0	
1	
2	

For Exercises 34–47, graph the lines by making a table of at least three ordered pairs and plotting the points.

34. $x = y + 2$

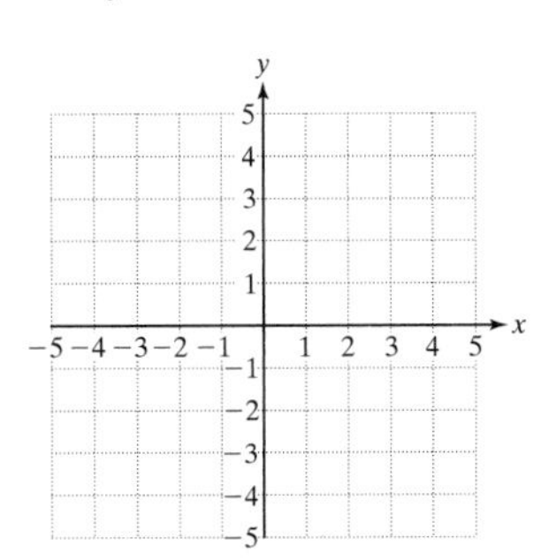

35. $x - y = 4$

36. $-3x + y = -6$

37. $2x - 5y = 10$

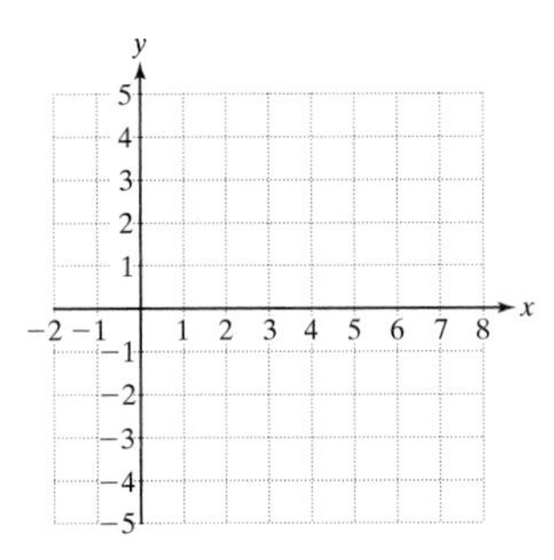

38. $y = 4x$

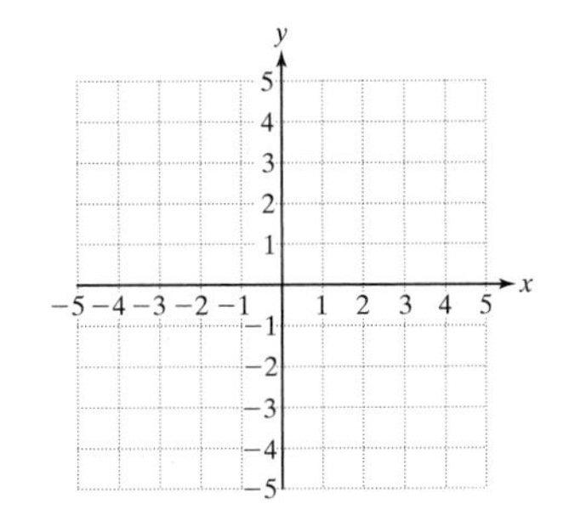

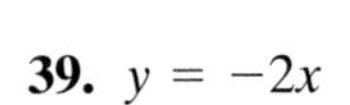

39. $y = -2x$

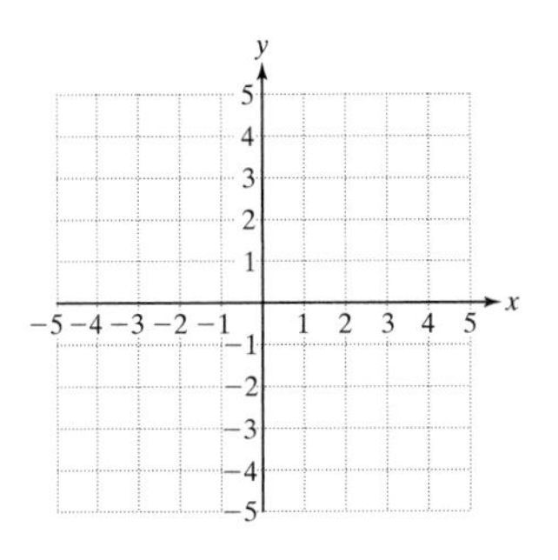

40. $y = -\frac{1}{2}x + 3$

41. $y = \frac{1}{4}x - 2$

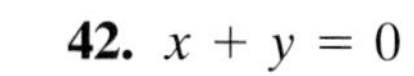

42. $x + y = 0$

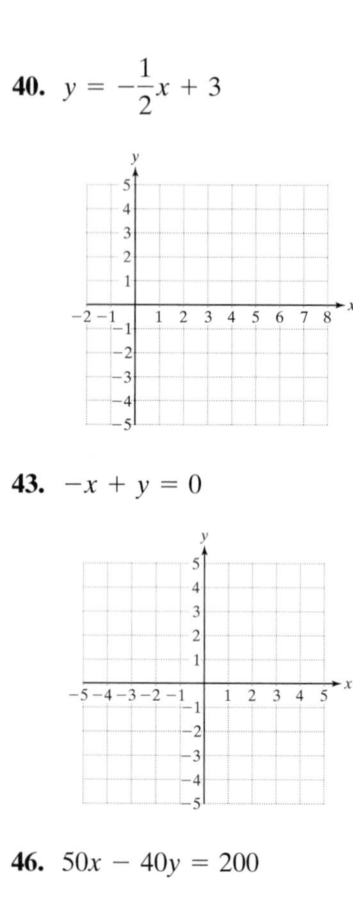

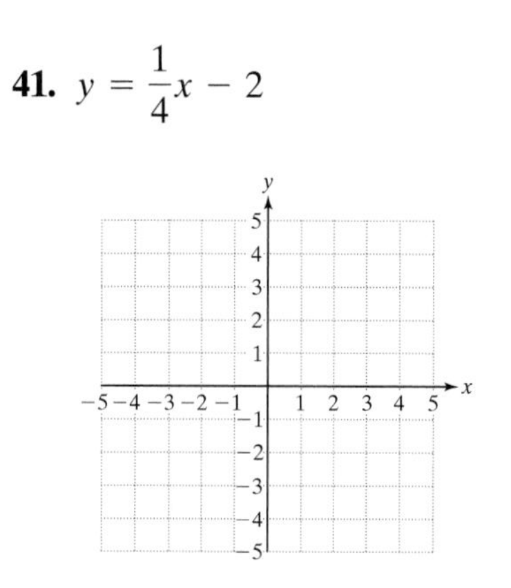

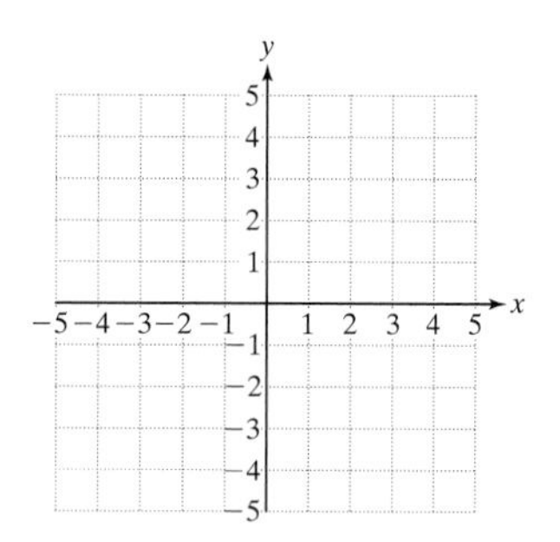

43. $-x + y = 0$

44. $2x + 3y = 8$

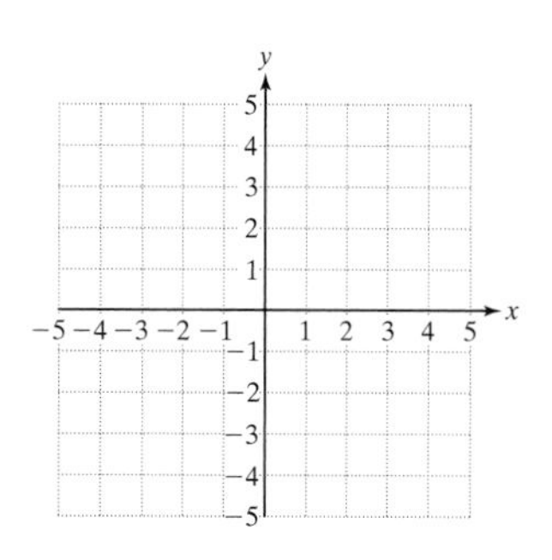

45. $4x - 5y = 15$

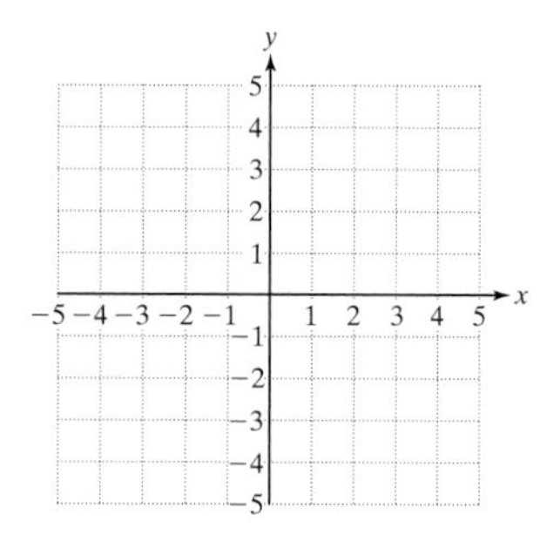

46. $50x - 40y = 200$

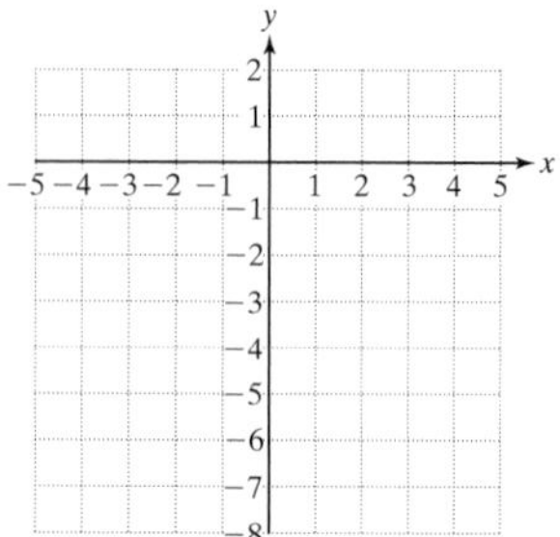

47. $-30x - 20y = 60$

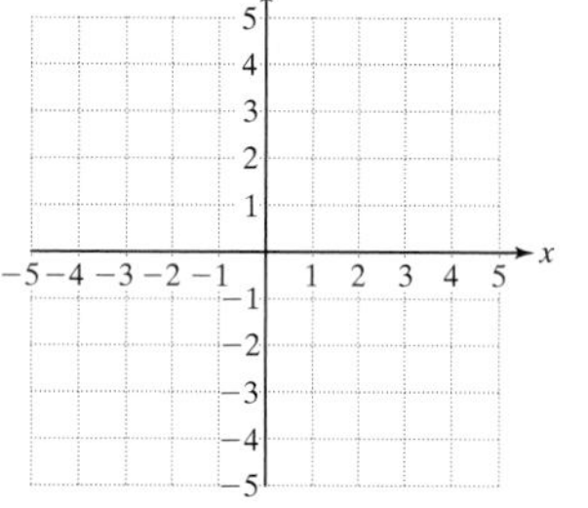

Concept 3: *x*- and *y*-Intercepts

48. The x-intercept is on which axis?

49. The y-intercept is on which axis?

For Exercises 50–53, estimate the coordinates of the x- and y-intercepts.

50.

51.

52. **53.**

For Exercises 54–67, find the x- and y-intercepts (if they exist), and graph the line.

54. $5x + y = 5$

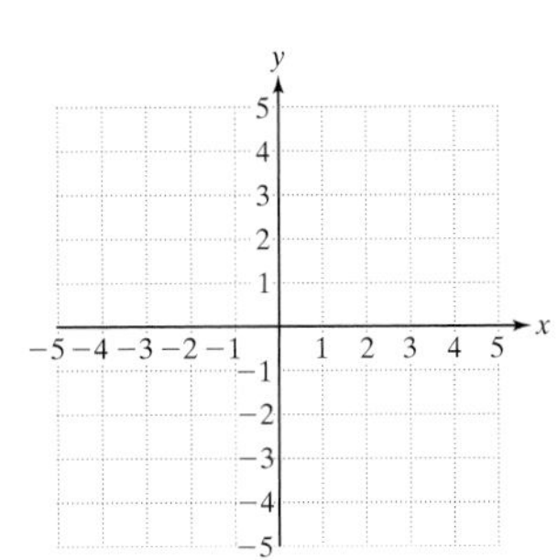

55. $x - 3y = -9$

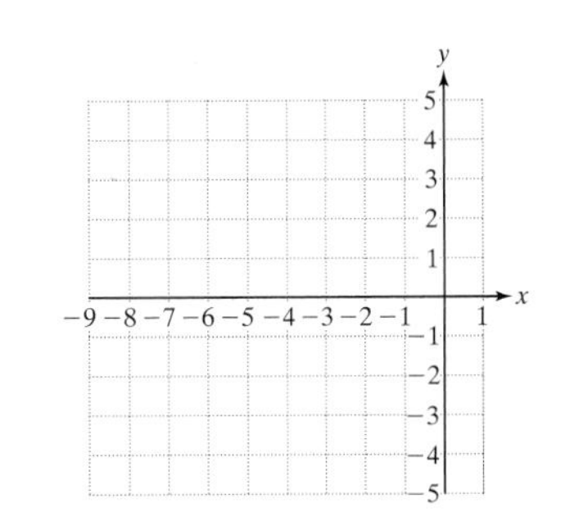

56. $y = \frac{2}{3}x - 1$

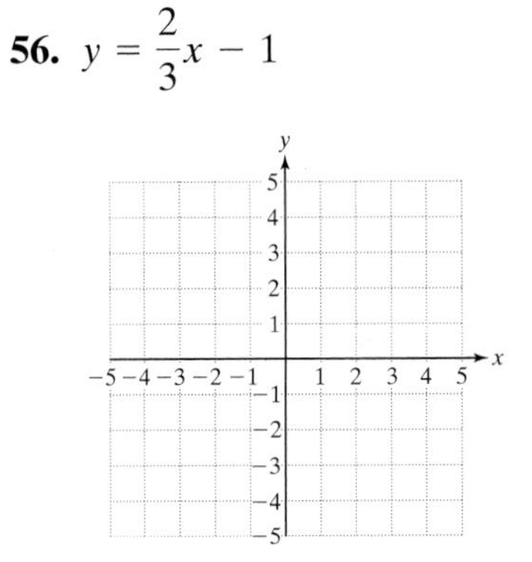

57. $y = -\frac{3}{4}x + 2$

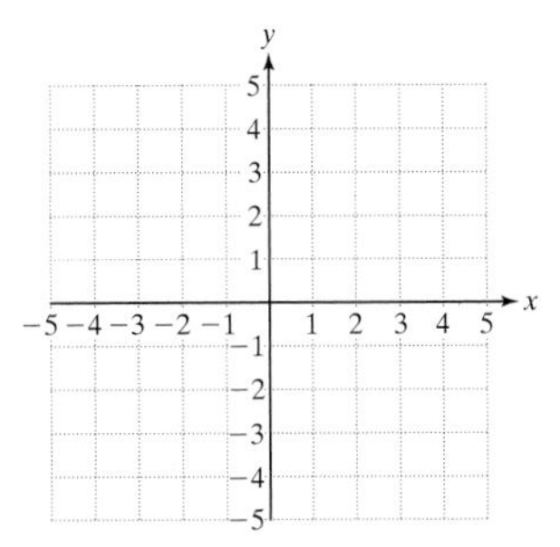

58. $x - 3 = y$

59. $2x + 8 = y$

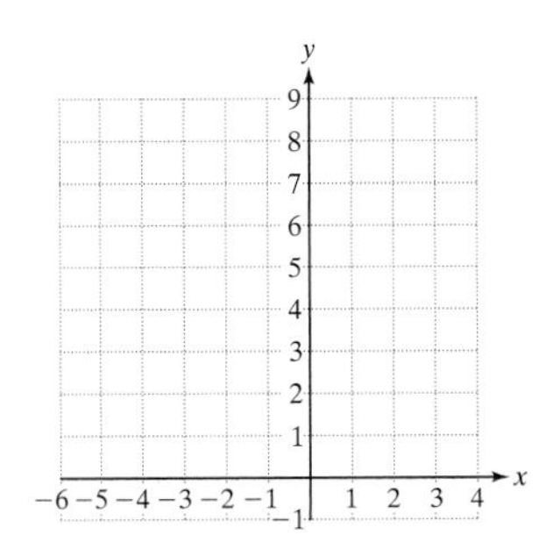

60. $-3x + y = 0$

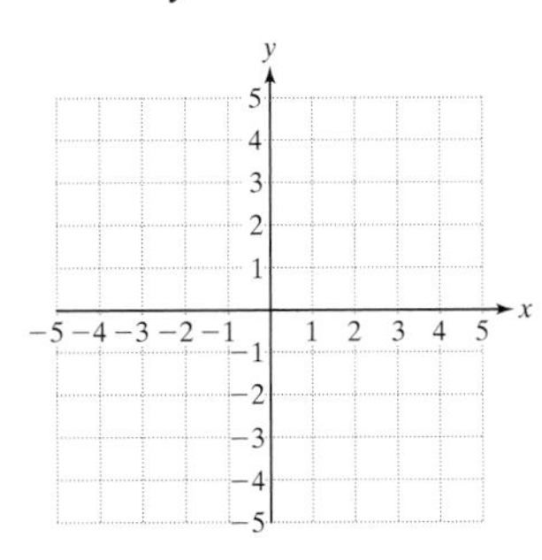

61. $2x - 2y = 0$

62. $25y = 10x + 100$

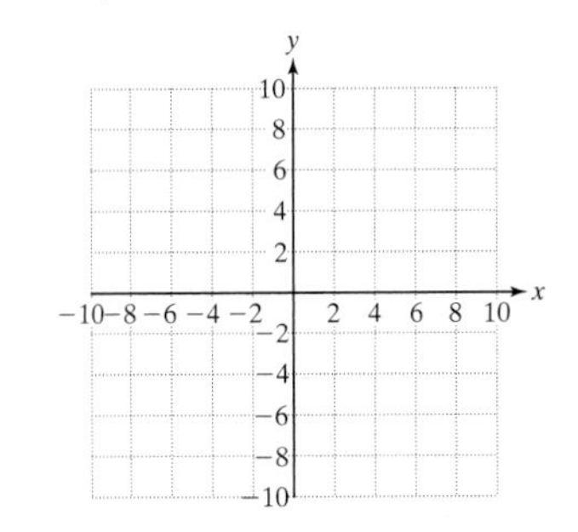

63. $20x = -40y + 200$

64. $1.2x - 2.4y = 3.6$

65. $-8.1x - 10.8y = 16.2$

66. $x = 2y$

67. $x = -5y$

Concept 4: Horizontal and Vertical Lines

68. Explain why not every line has both an x- and a y-intercept.

69. Which of the lines will have only one intercept?

a. $2x - 3y = 6$ **b.** $x = 5$ **c.** $2y = 8$ **d.** $-x + y = 0$

70. Which of the lines will have only one intercept?

a. $y = 2$ **b.** $x + y = 0$ **c.** $2x - 10 = 2$ **d.** $x + 4y = 8$

For Exercises 71–74, answer true or false. If the statement is false, rewrite it to be true.

71. The line $x = 3$ is horizontal.

72. The line $y = -4$ is horizontal.

73. A line parallel to the y-axis is vertical.

74. A line perpendicular to the x-axis is vertical.

For Exercises 75–86,

a. Identify as representing a horizontal or vertical line.

b. Graph the line.

c. Identify the x- and y-intercepts if they exist.

75. $x = 3$

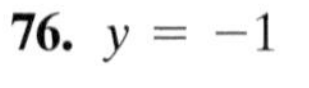

76. $y = -1$

77. $-2y = 8$

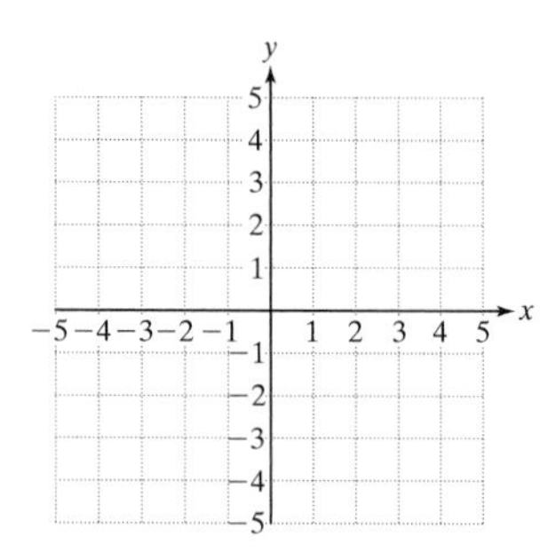

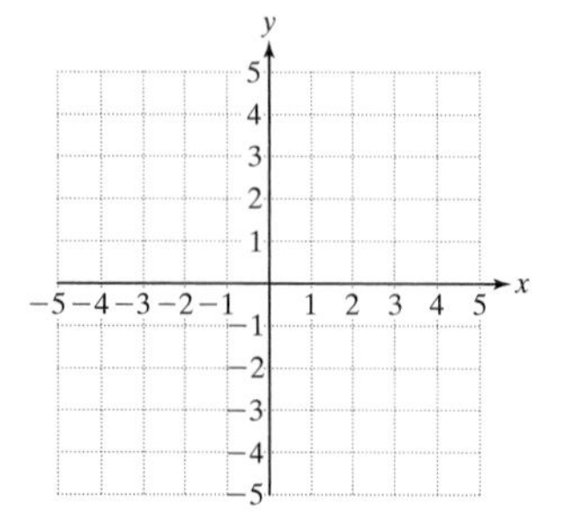

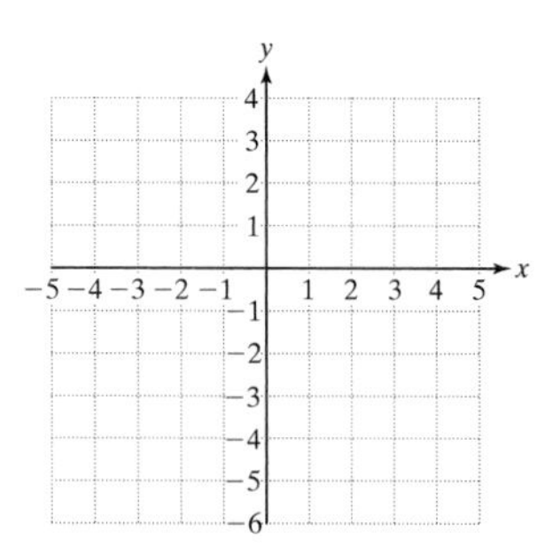

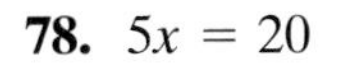

78. $5x = 20$

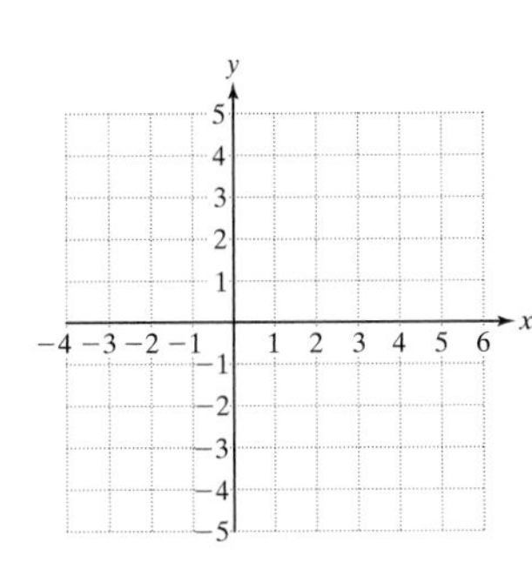

79. $x + 3 = 7$

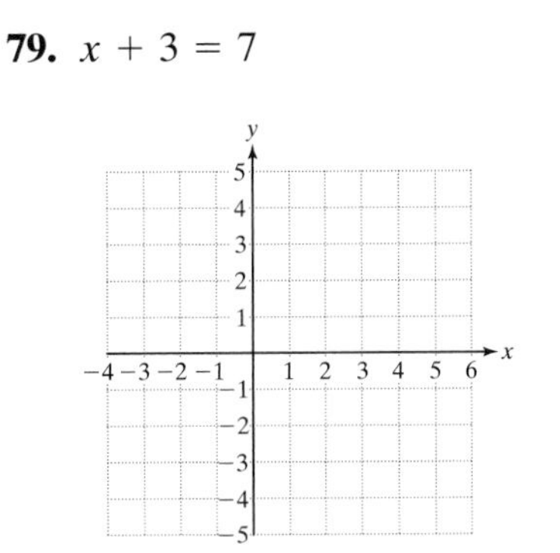

80. $y - 8 = -13$

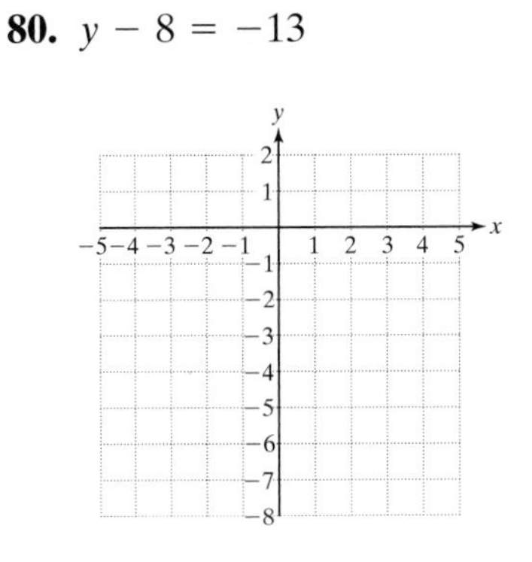

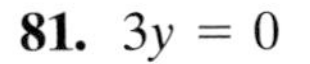

81. $3y = 0$

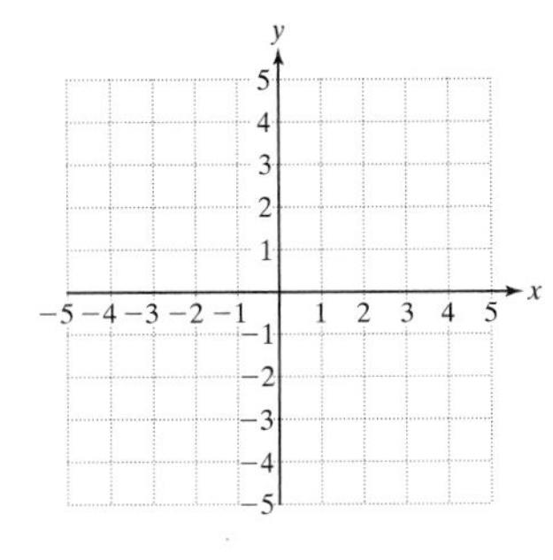

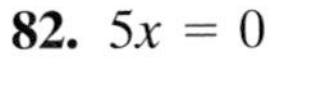

82. $5x = 0$

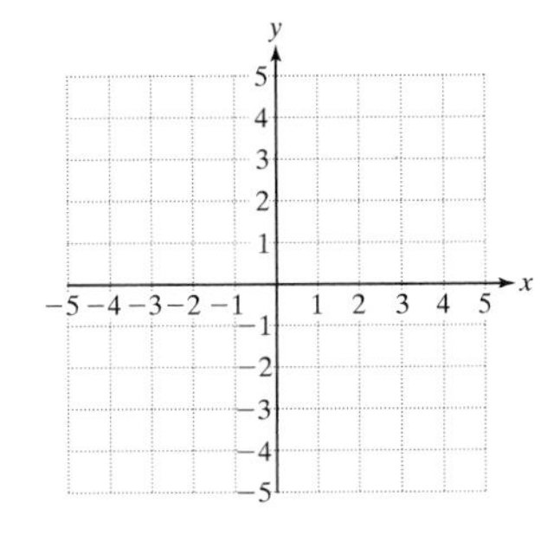

83. $2x + 7 = 10$

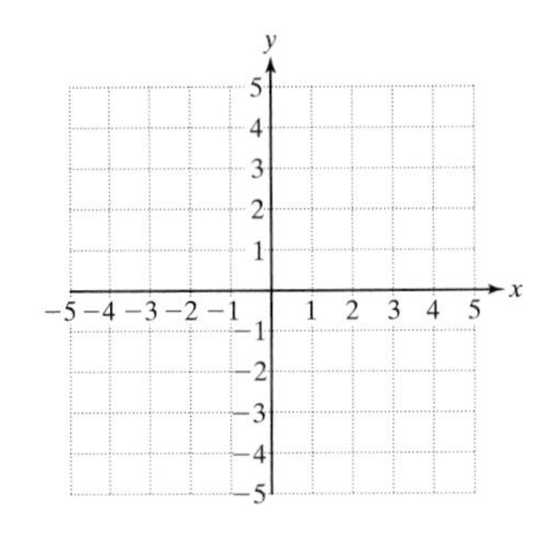

84. $-3y + 2 = 9$

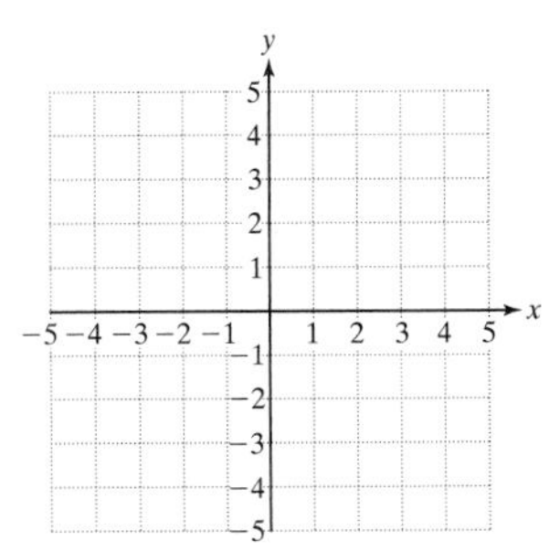

85. $9 = 3 + 4y$

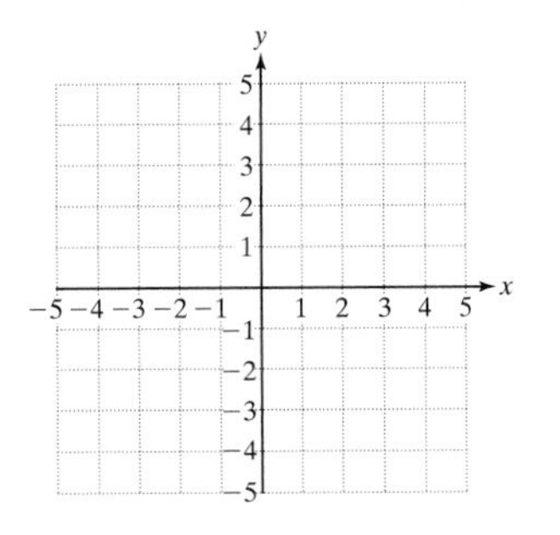

86. $7 = -2x - 5$

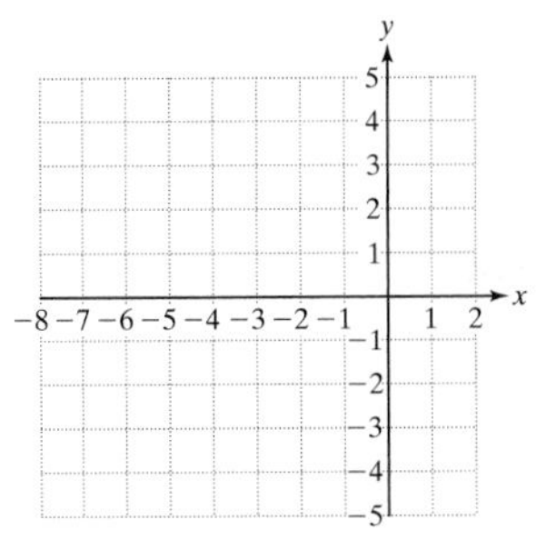

Expanding Your Skills

87. The students in the ninth grade at Atlantic High School pick up aluminum cans to be recycled. The current value of aluminum is \$0.69 per pound. If the students pay \$20 to rent a truck to haul the cans, then the following equation expresses the amount of money that they earn, y, given the number of pounds of aluminum, x.

$$y = 0.69x - 20 \quad (x \geq 0)$$

a. Let $x = 55$ and solve for y.

b. Let $y = 80.05$ and solve for x.

c. Write the ordered pairs from parts (a) and (b), and interpret their meaning in the context of the problem.

d. Graph the ordered pairs and the line defined by the points.

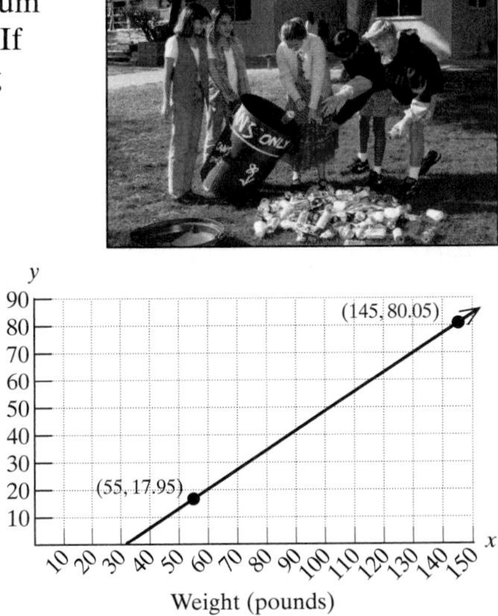

88. The store "CDs R US" sells all compact disks for \$13.99. The following equation represents the revenue, y, (in dollars) generated by selling x CDs.

$$y = 13.99x \quad (x \geq 0)$$

a. Find y when $x = 13$.

b. Find x when $y = 279.80$.

c. Write the ordered pairs from parts (a) and (b), and interpret their meaning in the context of the problem.

d. Graph the ordered pairs and the line defined by the points.

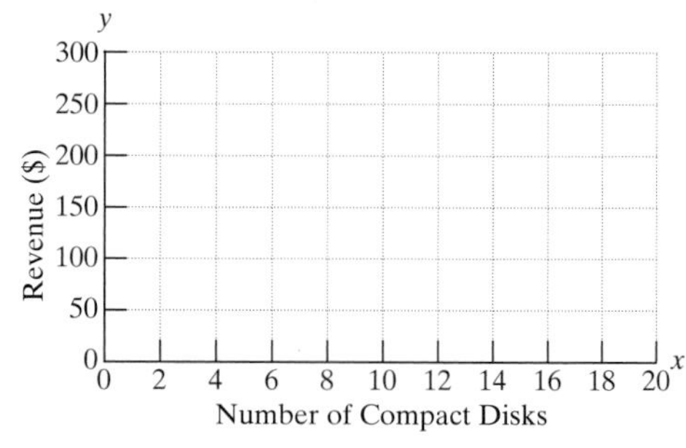

89. The value of a car depreciates once it is driven off of the dealer's lot. For a Hyundai Accent, the value of the car is given by the equation $y = -1531x + 11{,}599$ $(x \geq 0)$ where y is the value of the car in dollars x years after its purchase. (*Source: Kelly Blue Book*)

a. Find y when $x = 1$.

b. Find x when $y = 7006$.

c. Write the ordered pairs from parts (a) and (b), and interpret their meaning in the context of the problem.

Section 3.3 Slope of a Line

Concepts

1. Introduction to Slope
2. Slope Formula
3. Parallel and Perpendicular Lines
4. Applications of Slope

1. Introduction to Slope

The x- and y-intercepts represent the points where a line crosses the x- and y-axes. Another important feature of a line is its slope. Geometrically, the slope of a line measures the "steepness" of the line. For example, two ski runs are depicted by the lines in Figure 3-14.

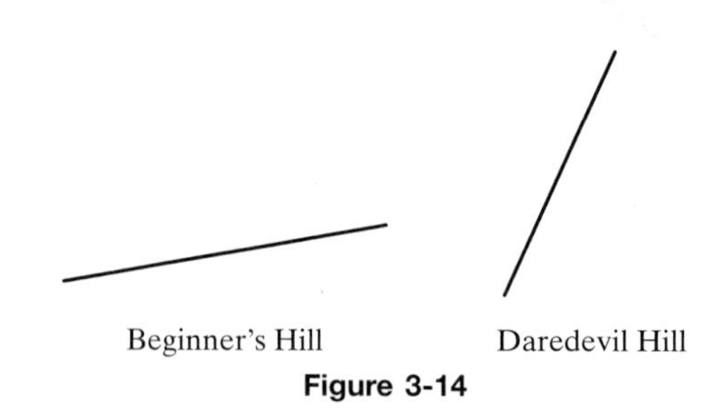

Figure 3-14

By visual inspection, Daredevil Hill is "steeper" than Beginner's Hill. To measure the slope of a line quantitatively, consider two points on the line. The slope of the line is the ratio of the vertical change (change in y) between the two points and the horizontal change (change in x). As a memory device, we might think of the slope of a line as "rise over run." See Figure 3-15.

$$\text{Slope} = \frac{\text{change in } y}{\text{change in } x} = \frac{\text{rise}}{\text{run}}$$

Change in x (run)

Change in y (rise)

Figure 3-15

To move from point A to point B on Beginner's Hill, rise 2 ft and move to the right 6 ft (Figure 3-16).

To move from point A to point B on Daredevil Hill, rise 12 ft and move to the right 6 ft (Figure 3-17).

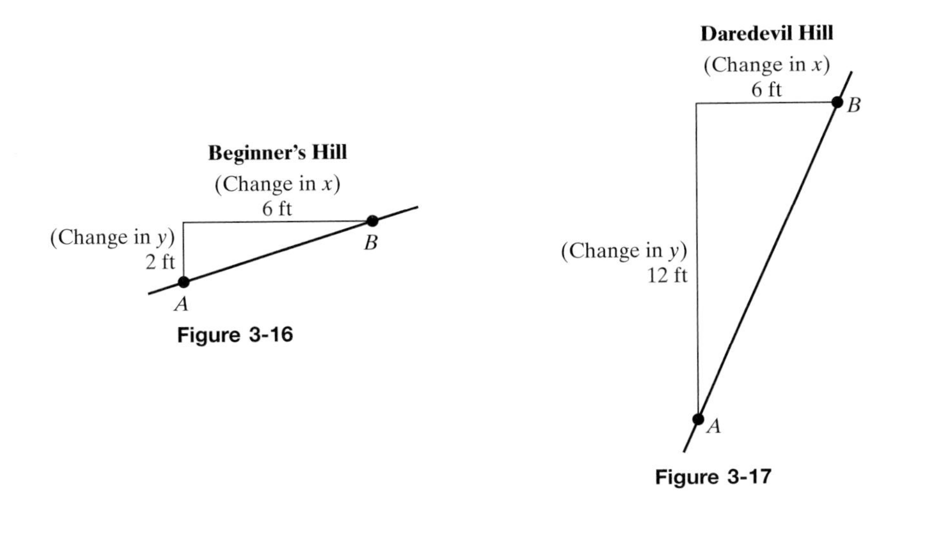

Figure 3-16

Figure 3-17

$$\text{Slope} = \frac{\text{change in } y}{\text{change in } x} = \frac{2 \text{ ft}}{6 \text{ ft}} = \frac{1}{3}$$

$$\text{Slope} = \frac{\text{change in } y}{\text{change in } x} = \frac{12 \text{ ft}}{6 \text{ ft}} = \frac{2}{1} = 2$$

The slope of Daredevil Hill is greater than the slope of Beginner's Hill, confirming the observation that Daredevil Hill is steeper. On Daredevil Hill, there is a 12-ft change in elevation for every 6 ft of horizontal distance (a 2:1 ratio). On Beginner's Hill there is only a 2-ft change in elevation for every 6 ft of horizontal distance (a 1:3 ratio).

Example 1 Finding Slope in an Application

Find the slope of the ramp up the stairs.

Solution:

$$\text{Slope} = \frac{\text{change in } y}{\text{change in } x} = \frac{7\frac{1}{2} \text{ ft}}{15 \text{ ft}}$$

$$\frac{7\frac{1}{2}}{15} = \frac{\frac{15}{2}}{\frac{15}{1}}$$ Write the mixed number as an improper fraction.

$$\frac{15}{2} \cdot \frac{1}{15} = \frac{1}{2}$$ Multiply by the reciprocal and simplify.

The slope is $\frac{1}{2}$.

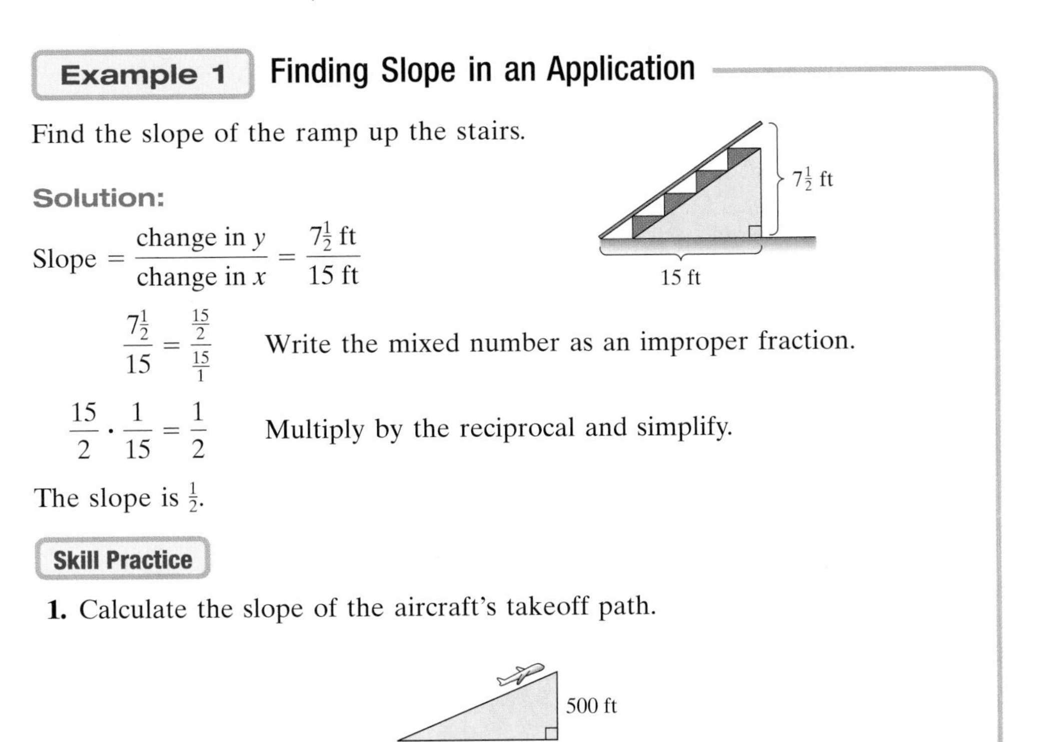

Skill Practice

1. Calculate the slope of the aircraft's takeoff path.

Skill Practice Answers

1. $\frac{500}{6000} = \frac{1}{12}$

2. Slope Formula

The slope of a line may be found using any two points on the line—call these points (x_1, y_1) and (x_2, y_2). The change in y between the points can be found by taking the difference of the y-values: $y_2 - y_1$. The change in x can be found by taking the difference of the x-values in the same order: $x_2 - x_1$ (Figure 3-18).

The slope of a line is often symbolized by the letter m and is given by the following formula.

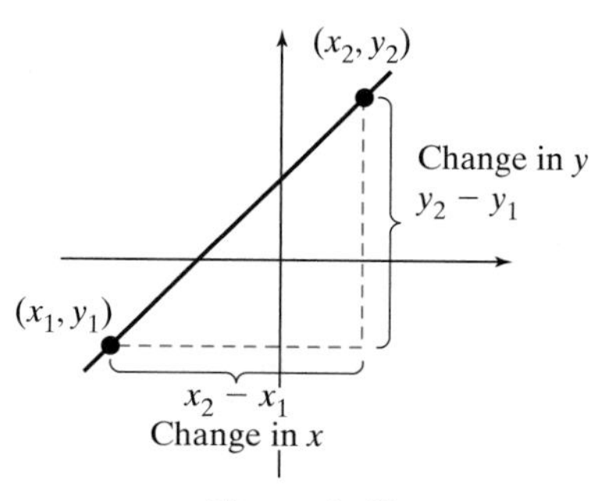

Figure 3-18

Definition of the Slope of a Line

The **slope** of a line passing through the distinct points (x_1, y_1) and (x_2, y_2) is

$$m = \frac{y_2 - y_1}{x_2 - x_1} \quad \text{provided } x_2 - x_1 \neq 0$$

Example 2 Finding the Slope of a Line Given Two Points

Find the slope of the line through the points $(-1, 3)$ and $(-4, -2)$.

Solution:

To use the slope formula, first label the coordinates of each point and then substitute the coordinates into the slope formula.

$(-1, 3)$ and $(-4, -2)$
(x_1, y_1) $\quad$ (x_2, y_2) — Label the points.

$$m = \frac{y_2 - y_1}{x_2 - x_1} = \frac{(-2) - (3)}{(-4) - (-1)}$$ Apply the slope formula.

$$= \frac{-5}{-3}; \quad \text{hence, } m = \frac{5}{3}$$ Simplify to lowest terms.

The slope of the line can be verified from the graph (Figure 3-19).

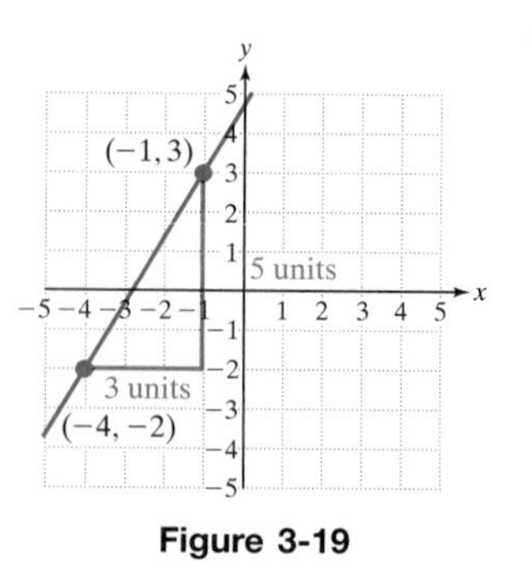

Figure 3-19

Skill Practice Find the slope of the line through the given points.

2. $(-5, 2)$ and $(1, 3)$

Skill Practice Answers

2. $\frac{1}{6}$

TIP: The slope formula is not dependent on which point is labeled (x_1, y_1) and which point is labeled (x_2, y_2). In Example 2, reversing the order in which the points are labeled results in the same slope.

$(-1, 3)$ and $(-4, -2)$
(x_2, y_2) $\quad$ (x_1, y_1) $\quad$ Label the points.

$$m = \frac{(3) - (-2)}{(-1) - (-4)} = \frac{5}{3} \qquad \text{Apply the slope formula.}$$

When you apply the slope formula, you will see that the slope of a line may be positive, negative, zero, or undefined.

- Lines that increase, or rise, from left to right have a positive slope.
- Lines that decrease, or fall, from left to right have a negative slope.
- Horizontal lines have a slope of zero.
- Vertical lines have an undefined slope.

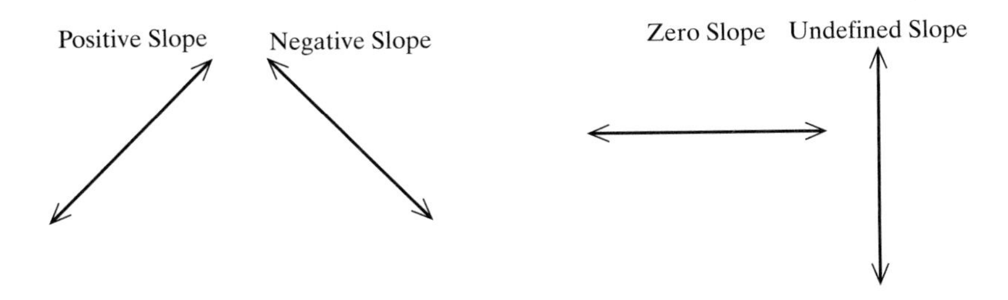

Example 3 Finding the Slope of a Line Given Two Points

Find the slope of the line passing through the points $(-5, 0)$ and $(2, -3)$.

Solution:

$(-5, 0)$ and $(2, -3)$
(x_1, y_1) $\quad$ (x_2, y_2) $\quad$ Label the points.

$$m = \frac{y_2 - y_1}{x_2 - x_1} = \frac{(-3) - (0)}{(2) - (-5)} \qquad \text{Apply the slope formula.}$$

$$= \frac{-3}{7} \quad \text{or} \quad -\frac{3}{7} \qquad \text{Simplify.}$$

By graphing the points $(-5, 0)$ and $(2, -3)$, we can verify that the slope is $-\frac{3}{7}$ (Figure 3-20). Notice that the line slopes downward from left to right.

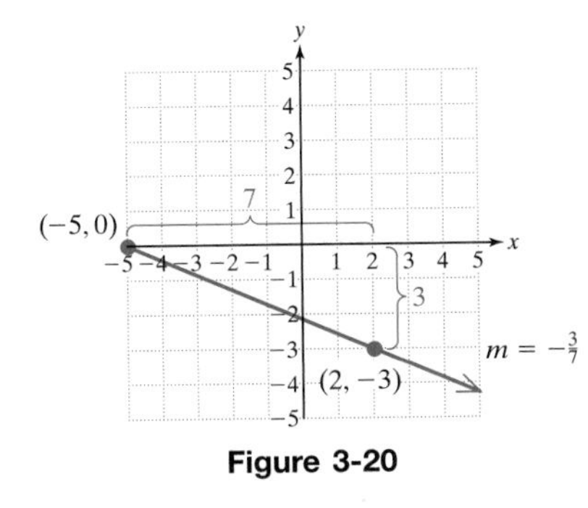

Figure 3-20

Skill Practice Find the slope of the line through the given points.

3. $(0, -8)$ and $(-2, -2)$

Skill Practice Answers

3. -3

Example 4 Determining the Slope of a Horizontal and Vertical Line

a. Find the slope of the line passing through the points (2, −1) and (2, 4).

b. Find the slope of the line passing through the points (3, −2) and (−4, −2).

Solution:

a. (2, −1) and (2, 4)

(x_1, y_1) (x_2, y_2) Label the points.

$$m = \frac{y_2 - y_1}{x_2 - x_1} = \frac{(4) - (-1)}{(2) - (2)}$$ Apply the slope formula.

$$m = \frac{5}{0}$$ Undefined

Because the slope, m, is undefined, we expect the points to form a vertical line as shown in Figure 3-21.

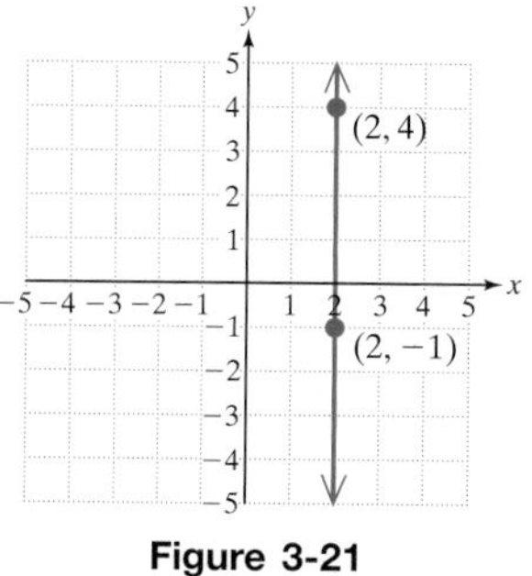

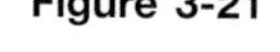
Figure 3-21

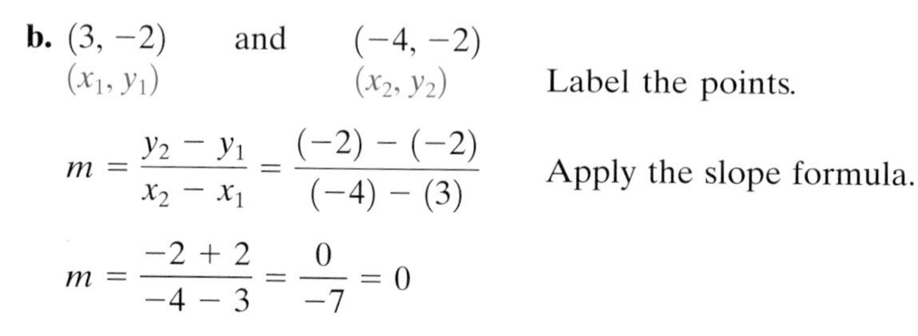

b. (3, −2) and (−4, −2)

(x_1, y_1) (x_2, y_2) Label the points.

$$m = \frac{y_2 - y_1}{x_2 - x_1} = \frac{(-2) - (-2)}{(-4) - (3)}$$ Apply the slope formula.

$$m = \frac{-2 + 2}{-4 - 3} = \frac{0}{-7} = 0$$

Because the slope is 0, we expect the points to form a horizontal line, as shown in Figure 3-22.

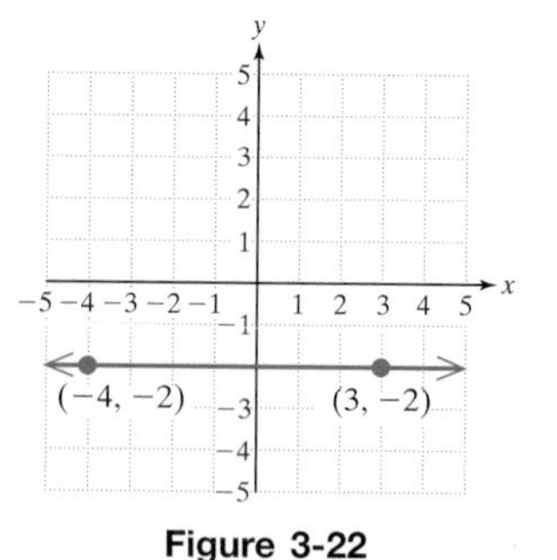

Figure 3-22

Skill Practice

4. Find the slope of the line through the given points.

a. (5, 6) and (5, −2) **b.** (3, 8) and (−5, 8)

Skill Practice Answers

4a. Undefined **b.** 0

3. Parallel and Perpendicular Lines

Lines in the same plane that do not intersect are called **parallel lines**. Parallel lines have the same slope and different y-intercepts (Figure 3-23).

Lines that intersect at a right angle are **perpendicular lines**. If two lines are perpendicular, then the slope of one line is the opposite of the reciprocal of the slope of the other line (provided neither line is vertical) (Figure 3-24).

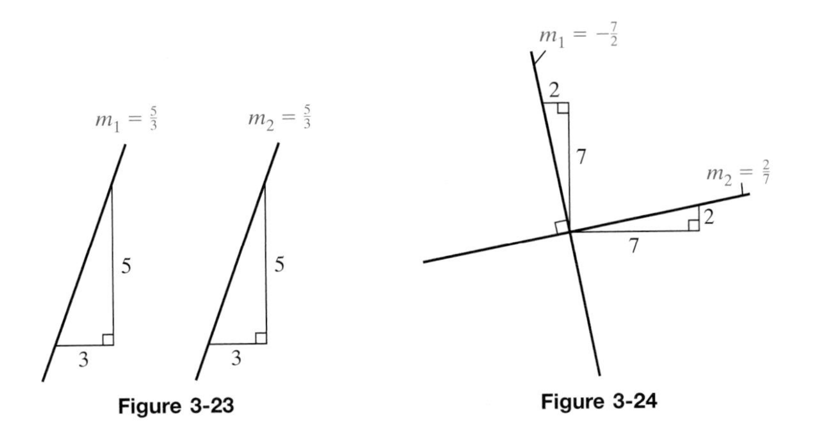

Figure 3-23

Figure 3-24

Slopes of Parallel Lines

If m_1 and m_2 represent the slopes of two parallel (nonvertical) lines, then

$$m_1 = m_2.$$

See Figure 3-23.

Slopes of Perpendicular Lines

If $m_1 \neq 0$ and $m_2 \neq 0$ represent the slopes of two perpendicular lines, then

$$m_1 = -\frac{1}{m_2} \text{ or equivalently, } m_1 m_2 = -1. \text{ See Figure 3-24.}$$

Example 5 Determining the Slope of Parallel and Perpendicular Lines

Suppose a given line has a slope of $-\frac{1}{4}$.

a. Find the slope of a line parallel to the given line.

b. Find the slope of a line perpendicular to the given line.

Solution:

a. Parallel lines must have the same slope. The slope of a line parallel to the given line is $m = -\frac{1}{4}$.

b. Perpendicular lines must have opposite and reciprocal slopes. The slope of a line perpendicular to the given line is $m = +\frac{4}{1}$ or simply, $m = 4$.

Skill Practice

5. A given line has a slope of $\frac{5}{3}$.

a. Find the slope of a line parallel to the given line.

b. Find the slope of a line perpendicular to the given line.

4. Applications of Slope

In many applications, the interpretation of slope refers to the *rate of change* of the y-variable to the x-variable.

Example 6 Interpreting Slope in an Application

Mario earns \$10.00/hr working for a landscaping company. Shannelle earns \$15.00/hr working for an in-home nursing agency. Figure 3-25 shows their total earnings versus the number of hours they work.

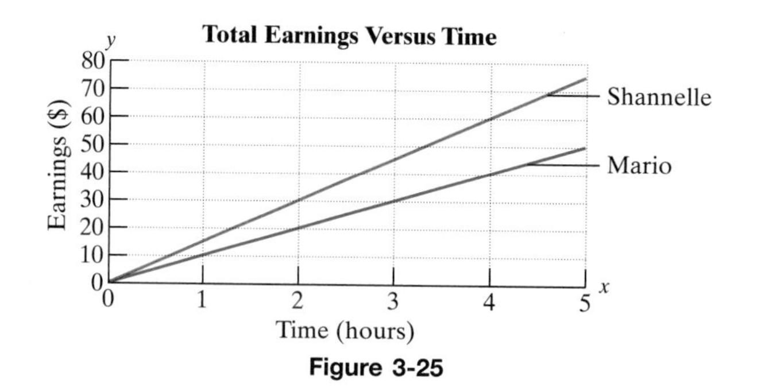

Figure 3-25

a. Find the slope of the line representing Mario's earnings.

b. Find the slope of the line representing Shannelle's earnings.

TIP: To find the slope for Example 6(a) we can pick points from the graph, such as (1, 10) and (2, 20). Then use the slope formula to calculate the slope.

$m = \dfrac{y_2 - y_1}{x_2 - x_1}$ becomes

$m = \dfrac{20 - 10}{2 - 1} = \dfrac{10}{1} = 10$

Solution:

a. After 1 hr, Mario earns \$10. After 2 hr, he earns \$20, and so on. For each 1-hr change in time, there is a \$10 increase in wages. The slope of the line representing Mario's earnings is \$10/hr.

b. After 1 hr, Shannelle earns \$15. After 2 hr, she earns \$30, and so on. For each 1-hr change in time, there is a \$15 increase in wages. The slope of the line representing Shannelle's earnings is \$15/hr.

Skill Practice

6. Anita invested in a stock that went up \$50 in 4 months. David invested in a stock that went up \$15 in 2 months. Let the variable y represent the amount that a stock is worth. Let x represent the time in months that a stock is invested.

a. Find the slope of the line represented by Anita's earnings.

b. Find the slope of the line representing David's earnings.

Skill Practice Answers

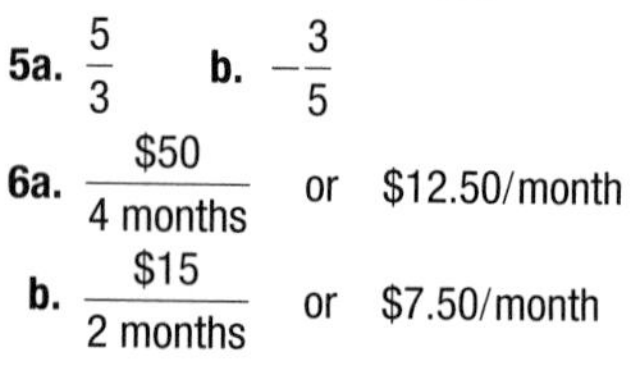

5a. $\frac{5}{3}$ **b.** $-\frac{3}{5}$

6a. $\frac{\$50}{4 \text{ months}}$ or \$12.50/month

b. $\frac{\$15}{2 \text{ months}}$ or \$7.50/month

Example 7 Interpreting Slope in an Application

Figure 3-26 shows the annual median income for males in the United States between 2000 and 2005. The trend is approximately linear. Find the slope of the line and interpret the meaning of the slope in the context of this problem.

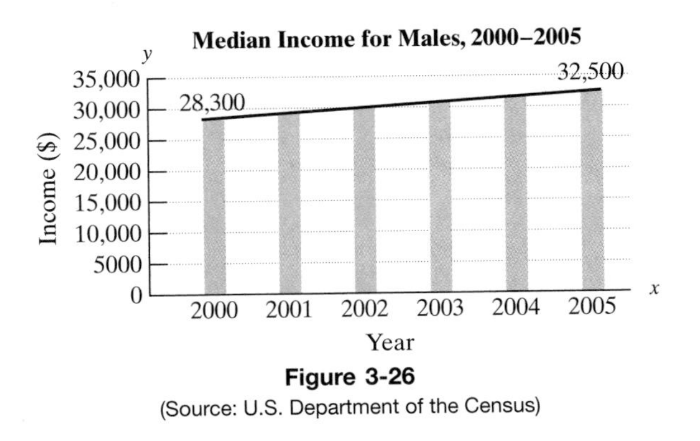

Figure 3-26
(Source: U.S. Department of the Census)

Solution:

To determine the slope, we need to know two points on the line. From the graph, the median income for males in 2000 was $28,300. This corresponds to the ordered pair (2000, 28300). In 2005, the median income was $32,500. This corresponds to the ordered pair (2005, 32500).

$(2000, 28300)$ and $(2005, 32500)$
(x_1, y_1) $\quad(x_2, y_2)$ Label the points.

$$m = \frac{y_2 - y_1}{x_2 - x_1} = \frac{32{,}500 - 28{,}300}{2005 - 2000}$$ Apply the slope formula.

$$= \frac{4200}{5} \text{ or } 840$$ Simplify.

The slope indicates that the median income for males in the United States increased at a rate of approximately $840 per year between 2000 and 2005.

Skill Practice

7. In the year 2000, the population of Alaska was approximately 630,000. By 2005, it had grown to 670,000. Use the ordered pairs (2000, 630000) and (2005, 670000) to determine the slope of the line through the points. Then interpret the meaning in the context of this problem.

Skill Practice Answers

7. $m = 8000$; The population of Alaska is increasing by 8000 people per year.

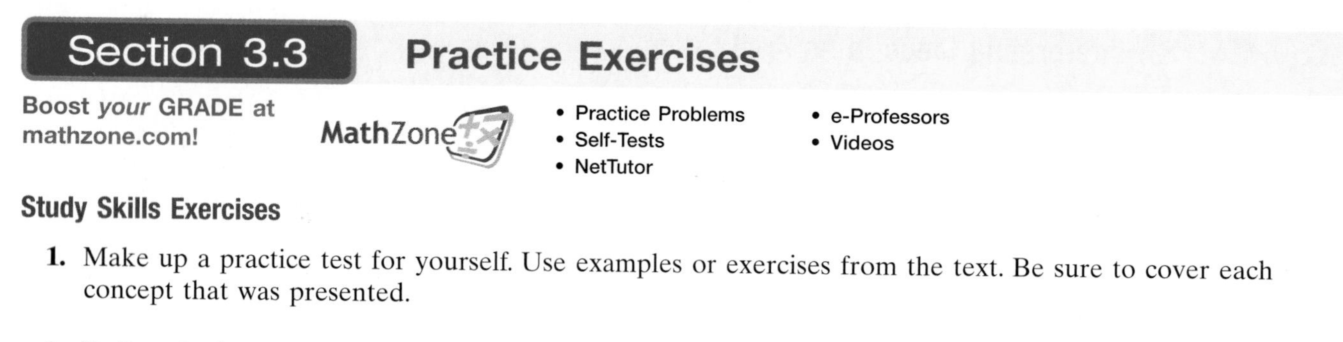

Section 3.3 Practice Exercises

Boost *your* GRADE at mathzone.com! MathZone

- Practice Problems
- Self-Tests
- NetTutor
- e-Professors
- Videos

Study Skills Exercises

1. Make up a practice test for yourself. Use examples or exercises from the text. Be sure to cover each concept that was presented.

2. Define the key terms:

a. **parallel lines** **b.** **perpendicular lines** **c.** **slope**

Review Exercises

For Exercises 3–8, find the x- and y-intercepts (if they exist). Then graph the lines.

3. $x - 3y = 6$

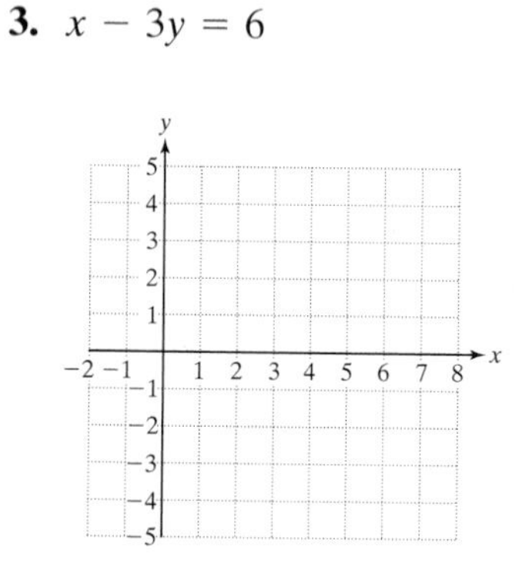

4. $x - 5 = 2$

5. $y = \frac{2}{3}x$

6. $2y - 3 = 0$

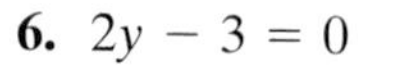

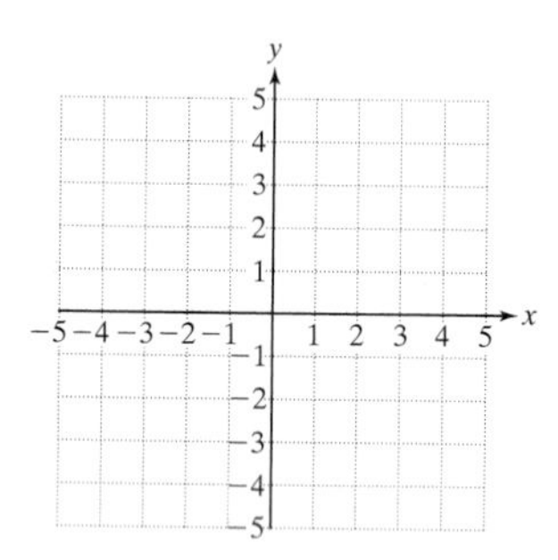

7. $4x + y = 8$

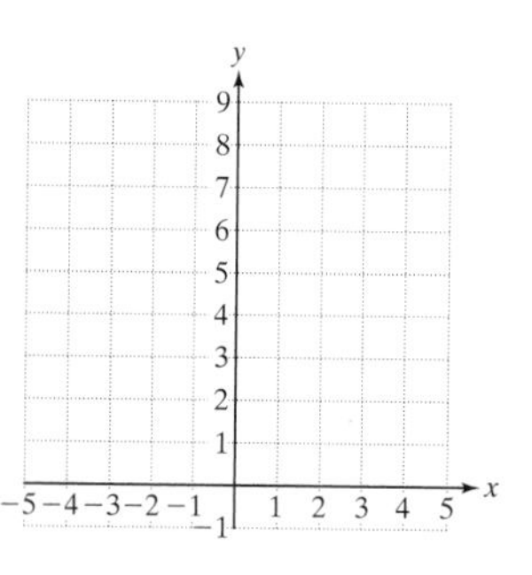

8. $2x = 4y$

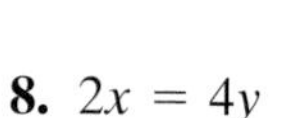

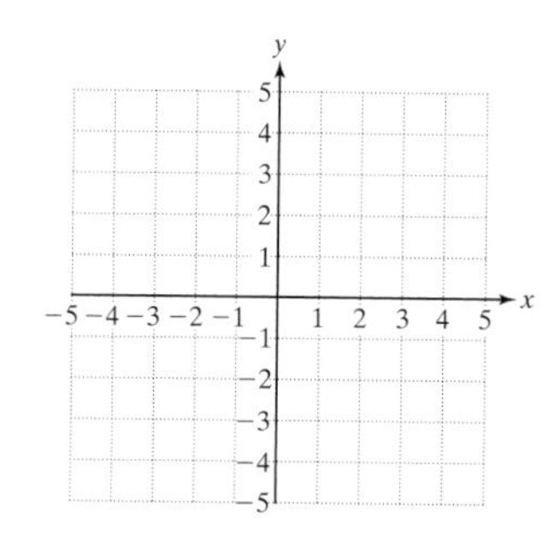

Concept 1: Introduction to Slope

9. Determine the pitch (slope) of the roof.

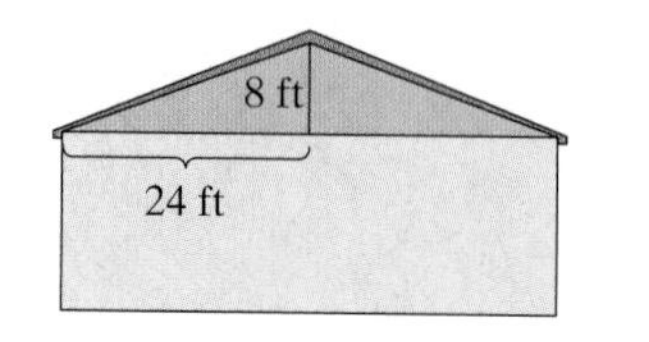

10. Determine the slope of the stairs.

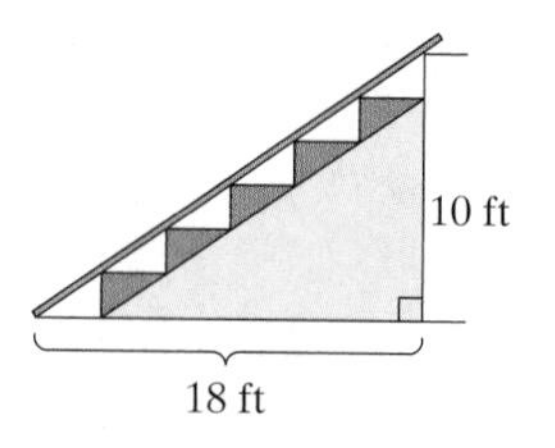

11. Determine the slope of the ramp.

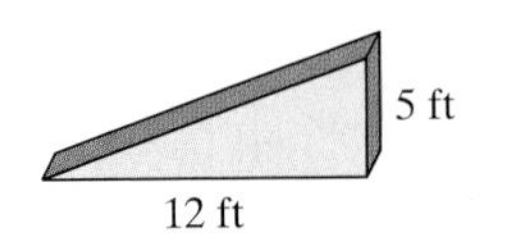

12. Determine the slope of the treadmill.

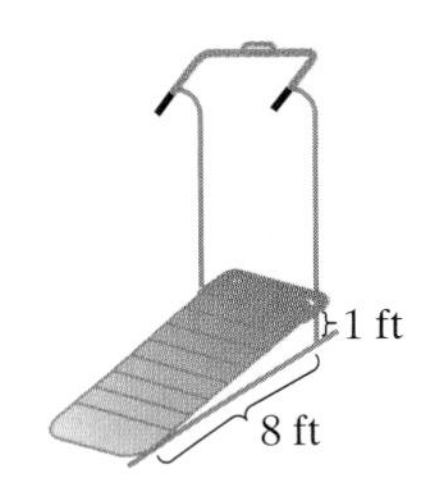

Concept 2: Slope Formula

For Exercises 13–16, fill in the blank with the appropriate term: zero, negative, positive, or undefined.

13. The slope of a line parallel to the y-axis is ________.

14. The slope of a horizontal line is ____.

15. The slope of a line that rises from left to right is ______.

16. The slope of a line that falls from left to right is ________.

For Exercises 17–24, label the lines as having a positive, negative, zero, or undefined slope.

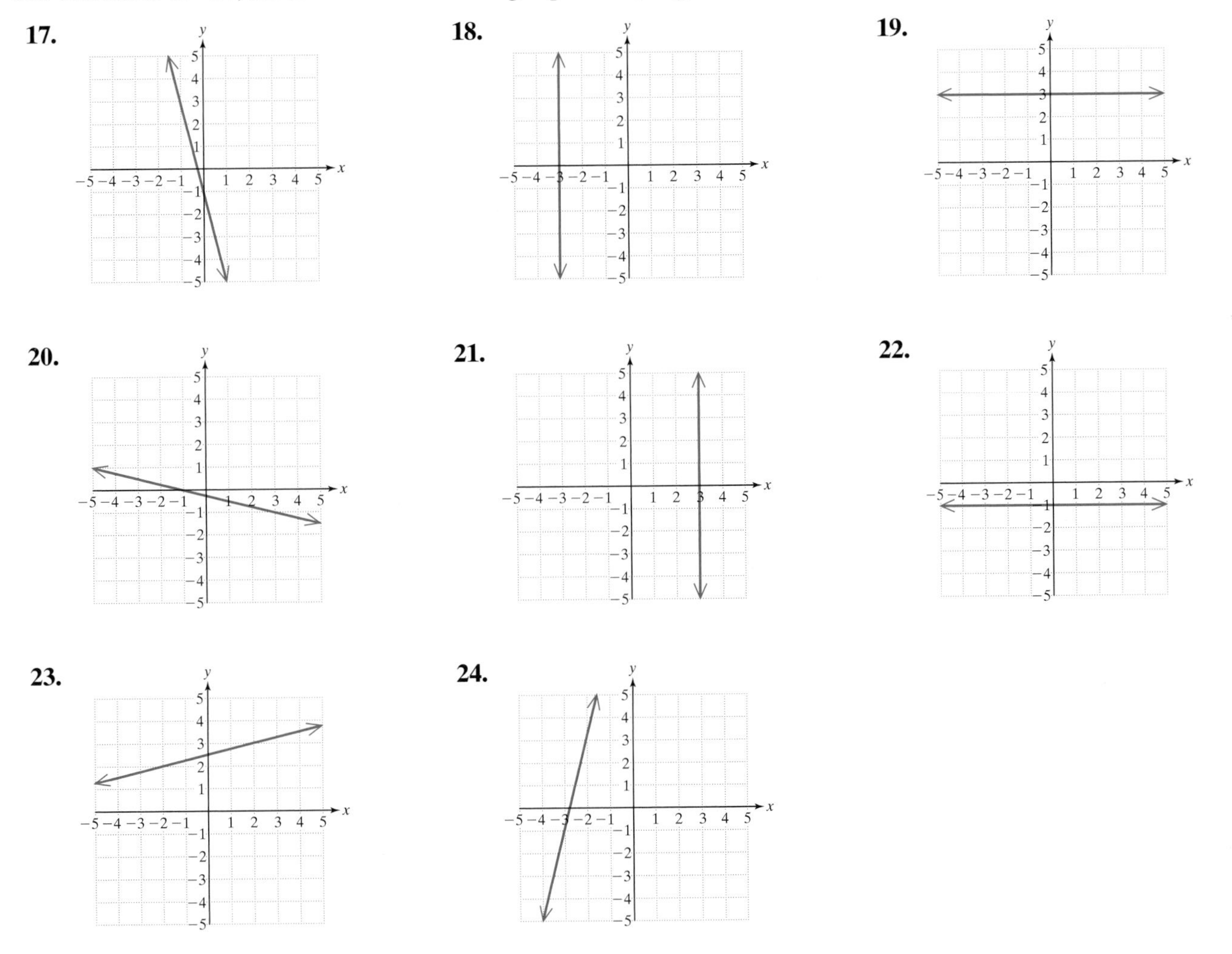

For Exercises 25–32, determine the slope by using the slope formula and any two points on the line. Check your answer by drawing a right triangle and labeling the "rise" and "run."

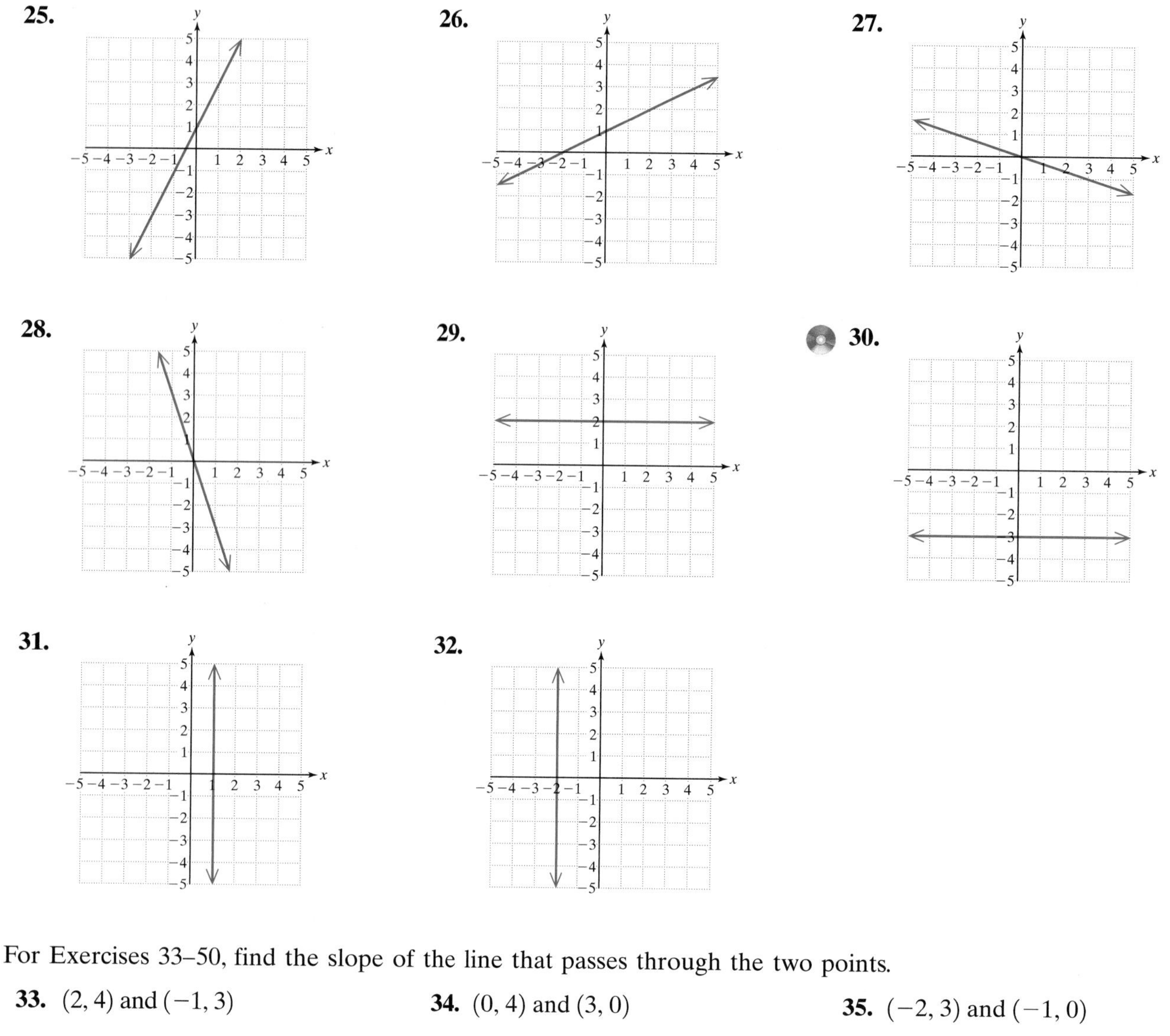

For Exercises 33–50, find the slope of the line that passes through the two points.

33. $(2, 4)$ and $(-1, 3)$

34. $(0, 4)$ and $(3, 0)$

35. $(-2, 3)$ and $(-1, 0)$

36. $(-3, -4)$ and $(1, -5)$

37. $(1, 5)$ and $(-4, 2)$

38. $(-6, -1)$ and $(-2, -3)$

39. $(5, 3)$ and $(-2, 3)$

40. $(0, -1)$ and $(-4, -1)$

41. $(2, -7)$ and $(2, 5)$

42. $(-4, 3)$ and $(-4, -4)$

43. $\left(\frac{1}{2}, \frac{3}{5}\right)$ and $\left(\frac{1}{4}, -\frac{4}{5}\right)$

44. $\left(-\frac{2}{7}, \frac{1}{3}\right)$ and $\left(\frac{8}{7}, -\frac{5}{6}\right)$

45. $(3, -1)$ and $(-5, 6)$

46. $(-6, 5)$ and $(-10, 4)$

47. $(6.8, -3.4)$ and $(-3.2, 1.1)$

48. $(-3.15, 8.25)$ and $(6.85, -4.25)$

49. $(1994, 3.5)$ and $(2000, 2.6)$

50. $(1988, 4.65)$ and $(1998, 9.25)$

Concept 3: Parallel and Perpendicular Lines

For Exercises 51–58, information regarding the slope of a line is given.

a. Determine the slope of a line parallel to the given line.

b. Determine the slope of a line perpendicular to the given line.

51. $m = -2$

52. $m = \frac{2}{3}$

53. $m = 0$

54. The slope is undefined.

55. $m = \frac{4}{5}$

56. $m = -4$

57. The slope is undefined.

58. $m = 0$

For Exercises 59–66, find the slopes of the lines l_1 and l_2 determined by the two given points. Then identify whether l_1 and l_2 are parallel, perpendicular, or neither.

59. l_1: $(2, 4)$ and $(-1, -2)$
l_2: $(1, 7)$ and $(0, 5)$

60. l_1: $(0, 0)$ and $(-2, 4)$
l_2: $(1, -5)$ and $(-1, -1)$

61. l_1: $(1, 9)$ and $(0, 4)$
l_2: $(5, 2)$ and $(10, 1)$

62. l_1: $(3, -4)$ and $(-1, -8)$
l_2: $(5, -5)$ and $(-2, 2)$

63. l_1: $(4, 4)$ and $(0, 3)$
l_2: $(1, 7)$ and $(-1, -1)$

64. l_1: $(3, 5)$ and $(-2, -5)$
l_2: $(2, 0)$ and $(-4, -3)$

65. l_1: $(3.1, 6.3)$ and $(3.1, -5.7)$
l_2: $(1.2, 4.7)$ and $(1.2, -5.3)$

66. l_1: $(4.5, -6.7)$ and $(-2.3, -6.7)$
l_2: $(-2.2, 6.7)$ and $(-1.4, 6.7)$

Concept 4: Applications of Slope

67. In 1980, there were 304 thousand male inmates in federal and state prisons. By 2005, the number increased to 1479 thousand.

Let x represent the year, and let y represent the number of prisons (in thousands).

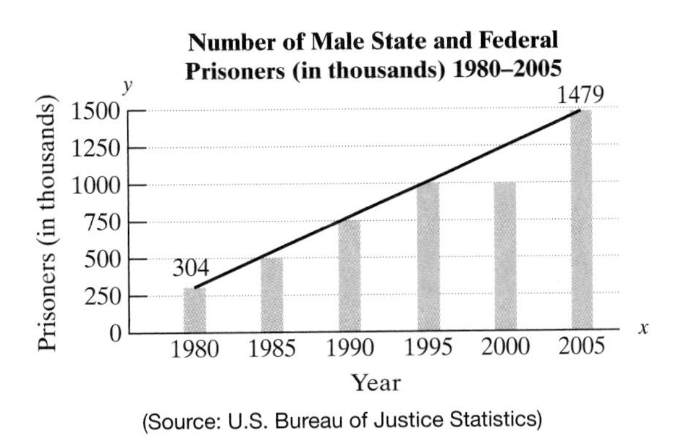

(Source: U.S. Bureau of Justice Statistics)

a. Using the ordered pairs (1980, 304) and (2005, 1479), find the slope of the line.

b. Interpret the slope in the context of this problem.

68. In the year 1980, there were 12 thousand female inmates in federal and state prisons. By 2005, the number increased to 102 thousand.

Let x represent the year, and let y represent the number of prisoners (in thousands).

a. Using the ordered pairs (1980, 12) and (2005, 102), find the slope of the line.

b. Interpret the slope in the context of this problem.

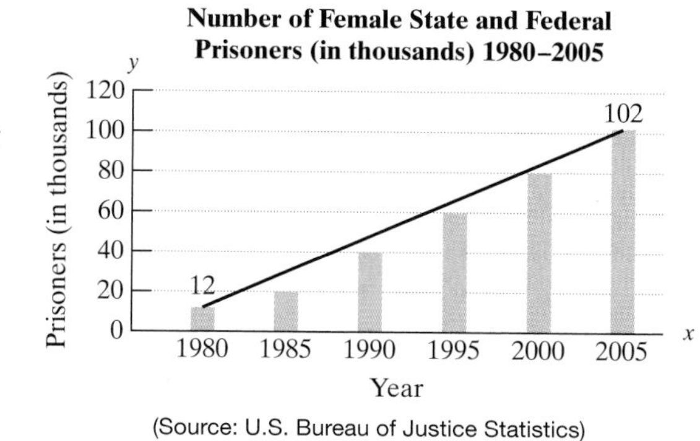

(Source: U.S. Bureau of Justice Statistics)

69. The following graph shows the median income for females in the United States between 2000 and 2005.

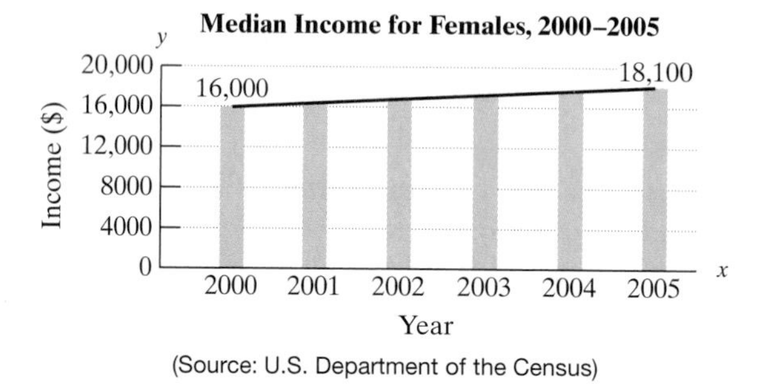

(Source: U.S. Department of the Census)

a. Find the slope of the line and interpret the meaning in the context of this problem.

b. Compare the slopes for the rise in median income per year for women and for men (see Example 7). Based on the slopes of the two lines, will the median income for women ever catch up to the median income for men? Explain.

70. Jorge is paid by the hour according to the equation

$P = 11.50x$ $\quad$ P is his total pay (in dollars) and x is the number of hours worked.

a. How much money will Jorge earn if he works 20 hr?

b. How much money will Jorge earn if he works 21 hr?

c. How much money will Jorge earn if he works 22 hr?

d. What is the slope of the line? Interpret the meaning of the slope in the context of this problem.

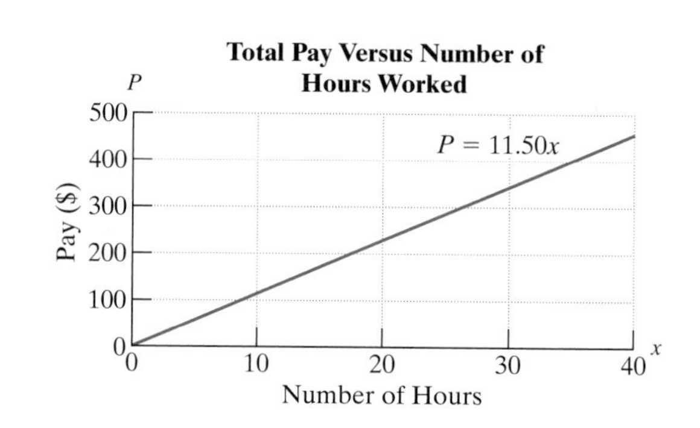

71. The distance, d (in miles), between a lightning strike and an observer is given by the equation $d = 0.2t$, where t is the time (in seconds) between seeing lightning and hearing thunder.

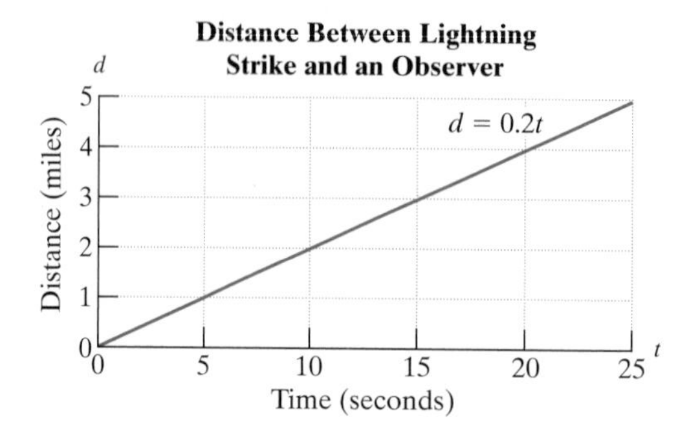

a. If an observer counts 5 sec between seeing lightning and hearing thunder, how far away was the lightning strike?

b. If an observer counts 10 sec between seeing lightning and hearing thunder, how far away was the lightning strike?

c. If an observer counts 15 sec between seeing lightning and hearing thunder, how far away was the lightning strike?

d. What is the slope of the line? Interpret the meaning of the slope in the context of this problem.

Mixed Exercises

For Exercises 72–77, determine the slope of the line passing through points A and B.

72. Point A is located 3 units up and 4 units to the right of point B.

73. Point A is located 2 units up and 5 units to the left of point B.

74. Point A is located 3 units up and 3 units to the left of point B.

75. Point A is located 2 units down and 2 units to the left of point B.

76. Point A is located 5 units to the right of point B.

77. Point A is located 3 units down from point B.

78. Graph the line through the point $(1, -2)$ having slope $\frac{2}{3}$. Then give two other points on the line.

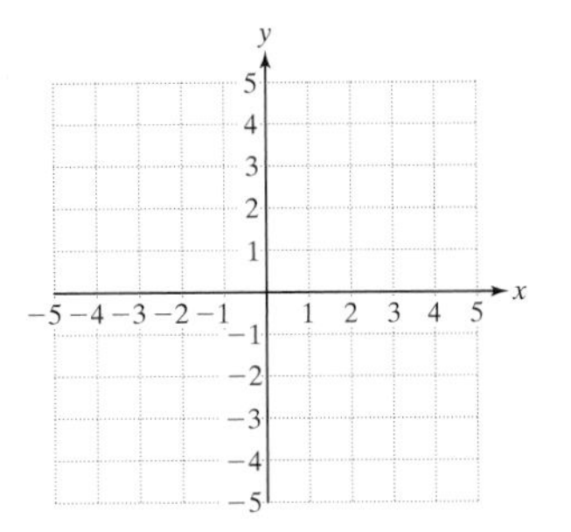

79. Graph the line through the point $(-2, -3)$ having slope $\frac{3}{4}$. Then give two other points on the line.

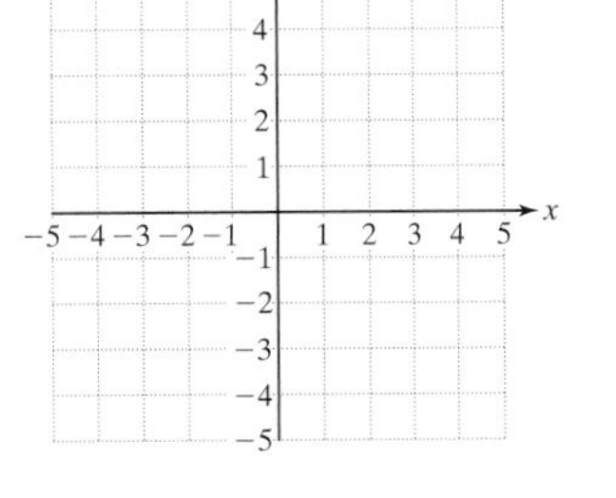

80. Graph the line through the point $(2, 2)$ having slope -3. Then give two other points on the line.

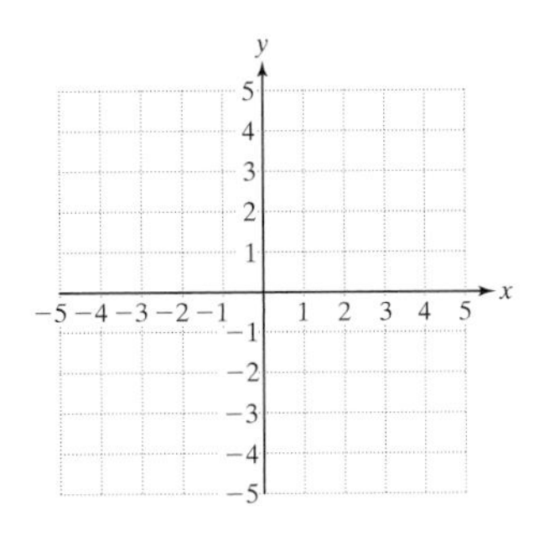

81. Graph the line through the point $(-1, 3)$ having slope -2. Then give two other points on the line.

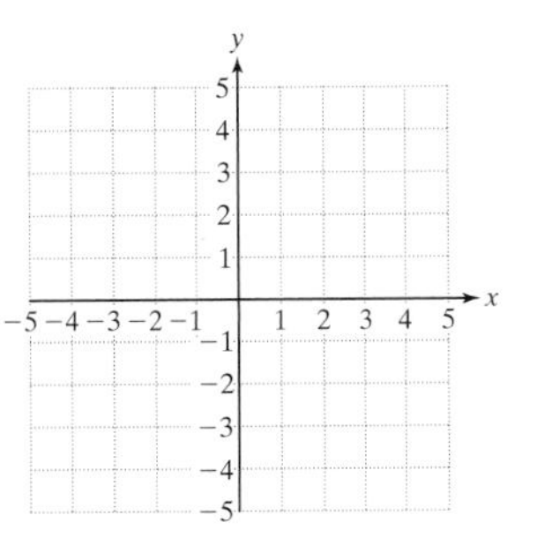

For Exercises 82–89, draw a line as indicated. Answers may vary.

82. Draw a line with a positive slope and a positive y-intercept.

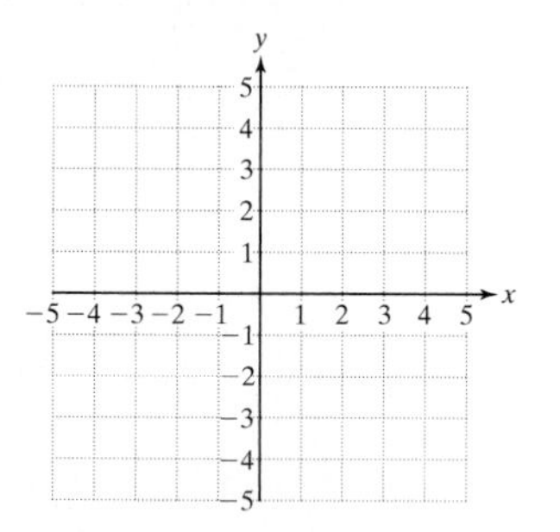

83. Draw a line with a positive slope and a negative y-intercept.

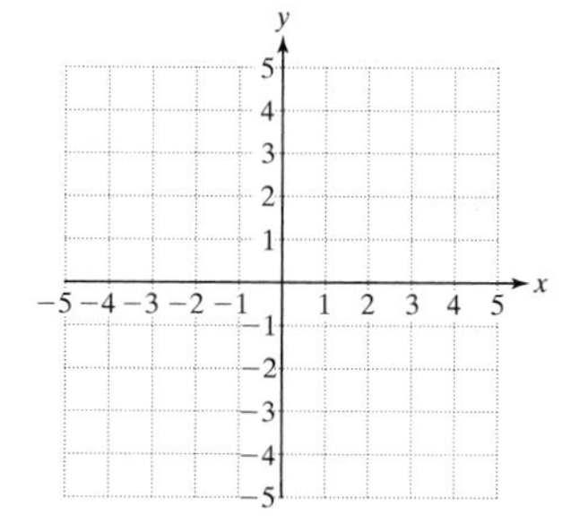

84. Draw a line with a negative slope and a negative y-intercept.

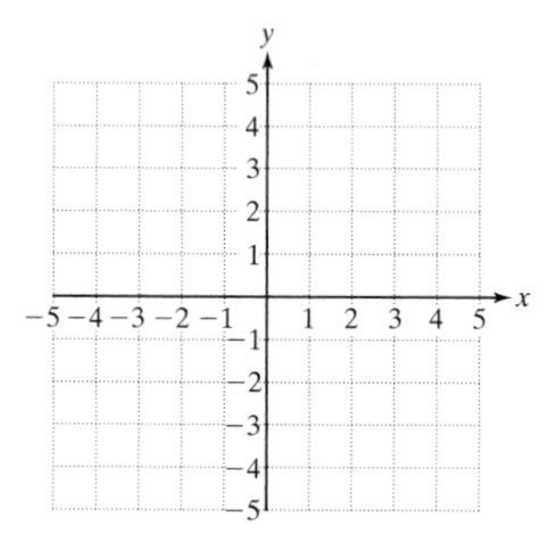

85. Draw a line with a negative slope and positive y-intercept.

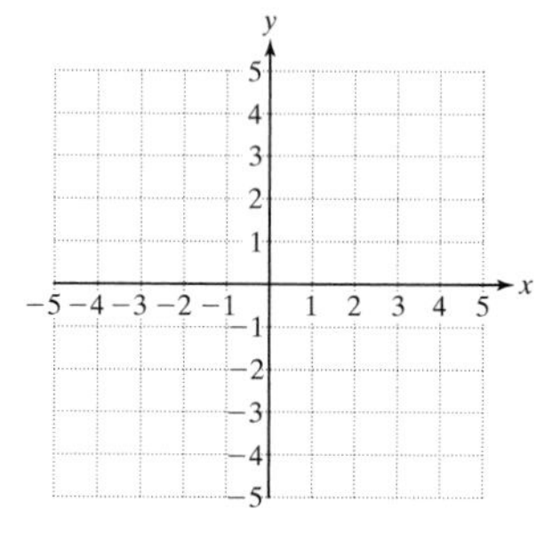

86. Draw a line with a zero slope and a positive y-intercept.

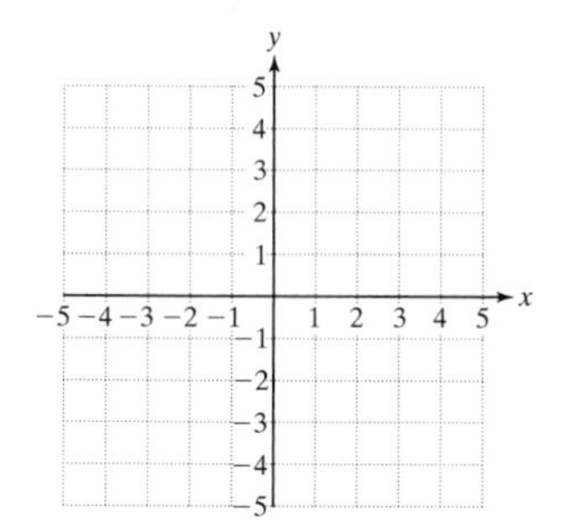

87. Draw a line with a zero slope and a negative y-intercept.

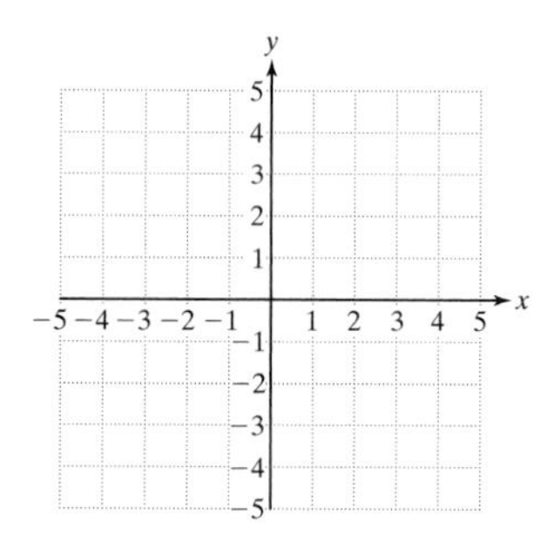

88. Draw a line with undefined slope and a negative x-intercept.

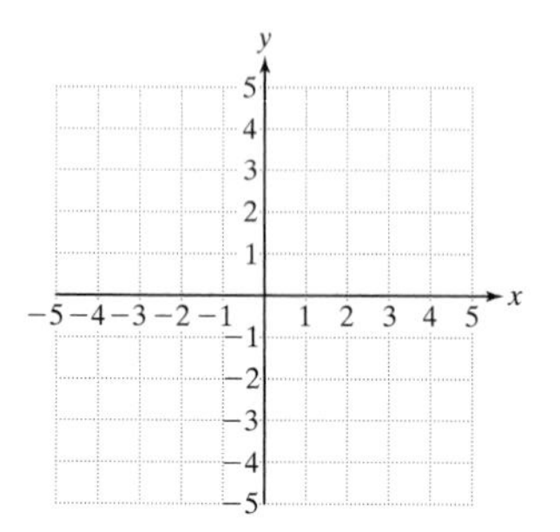

89. Draw a line with undefined slope and a positive x-intercept.

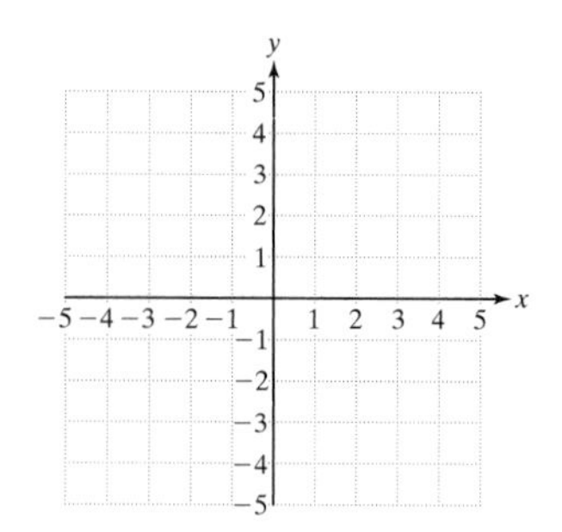

Expanding Your Skills

90. Find the slope between the points $(a + b, 4m - n)$ and $(a - b, m + 2n)$.

91. Find the slope between the points $(3c - d, s + t)$ and $(c - 2d, s - t)$.

92. Find the x-intercept of the line $ax + by = c$.

93. Find the y-intercept of the line $ax + by = c$.

94. Find another point on the line that contains the point $(2, -1)$ and has a slope of $\frac{2}{5}$.

95. Find another point on the line that contains the point $(-3, 4)$ and has a slope of $\frac{1}{4}$.

Slope-Intercept Form of a Line

Section 3.4

Concepts

1. Slope-Intercept Form of a Line
2. Graphing a Line from Its Slope and y-Intercept
3. Determining Whether Two Lines Are Parallel, Perpendicular, or Neither
4. Writing an Equation of a Line Given Its Slope and y-Intercept

1. Slope-Intercept Form of a Line

In Section 3.2, we learned that an equation of the form $Ax + By = C$ (where A and B are not both zero) represents a line in a rectangular coordinate system. An equation of a line written in this way is in **standard form**. In this section, we will learn a new form, called **slope-intercept form**, which is useful in determining the slope and y-intercept of a line.

Let $(0, b)$ represent the y-intercept of a line. Let (x, y) represent any other point on the line. Then the slope of the line can be found as follows:

Let $(0, b)$ represent (x_1, y_1), and let (x, y) represent (x_2, y_2). Apply the slope formula.

$$m = \frac{(y_2 - y_1)}{(x_2 - x_1)} \rightarrow m = \frac{y - b}{x - 0}$$ Apply the slope formula.

$$m = \frac{y - b}{x}$$ Simplify.

$$mx = \left(\frac{y - b}{x}\right)x$$ Multiply by x to clear fractions.

$$mx = y - b$$

$$mx + b = y - b + b$$ To isolate y, add b to both sides.

$$mx + b = y \quad \text{or} \quad y = mx + b$$ The equation is in slope-intercept form.

Slope-Intercept Form of a Line

$y = mx + b$ is the slope-intercept form of a line.

m is the slope and the point $(0, b)$ is the y-intercept.

Example 1 Identifying the Slope and y-Intercept of a Line

For each equation, identify the slope and y-intercept.

a. $y = 3x - 1$ **b.** $y = 4x$ **c.** $y = -2.7x + 5$ **d.** $y = 5$

Solution:

Each equation is written in slope-intercept form, $y = mx + b$. The slope is the coefficient of x, and the y-intercept is determined by the constant term.

a. $y = 3x - 1$ — The slope is 3. — The y-intercept is $(0, -1)$.

b. $y = 4x$ can be written as $y = 4x + 0$. — The slope is 4. — The y-intercept is $(0, 0)$.

c. $y = -2.7x + 5$ — The slope is -2.7. — The y-intercept is $(0, 5)$.

d. $y = 5$ can be written as $y = 0x + 5$. — The slope is 0. — The y-intercept is $(0, 5)$.

Skill Practice Identify the slope and the y-intercept.

1. $y = 4x + 6$ **2.** $y = -\frac{3}{4}x$ **3.** $y = 3.56x - 4.27$ **4.** $y = -7$

Given the equation of a line, we can write the equation in slope-intercept form by solving the equation for the y-variable. This is demonstrated in Example 2.

Example 2 Identifying the Slope and y-Intercept of a Line

Given the line $-5x - 2y = 6$,

a. Write the slope-intercept form of the line.

b. Identify the slope and y-intercept.

Solution:

a. Write the equation in slope-intercept form, $y = mx + b$, by solving for y.

$$-5x - 2y = 6$$

$$-2y = 5x + 6 \qquad \text{Add } 5x \text{ to both sides.}$$

$$\frac{-2y}{-2} = \frac{5x}{-2} + \frac{6}{-2} \qquad \text{Divide both sides by } -2.$$

$$y = -\frac{5}{2}x - 3 \qquad \text{Slope-intercept form}$$

b. The slope is $-\frac{5}{2}$, and the y-intercept is $(0, -3)$.

Skill Practice

5. Given the equation of the line $2x - 6y = -3$,

a. Write the slope-intercept form of the line.

b. Identify the slope and the y-intercept.

Skill Practice Answers

1. slope: 4; y-intercept: $(0, 6)$

2. slope: $-\frac{3}{4}$; y-intercept: $(0, 0)$

3. slope: 3.56; y-intercept: $(0, -4.27)$

4. slope: 0; y-intercept: $(0, -7)$

5a. $y = \frac{1}{3}x + \frac{1}{2}$

b. slope is $\frac{1}{3}$; y-intercept is: $\left(0, \frac{1}{2}\right)$

2. Graphing a Line from Its Slope and *y*-Intercept

Slope-intercept form is a useful tool to graph a line. The *y*-intercept is a known point on the line. The slope indicates the direction of the line and can be used to find a second point. Using slope-intercept form to graph a line is demonstrated in Example 3.

Example 3 **Graphing a Line Using the Slope and *y*-Intercept**

Graph the line $y = -\frac{5}{2}x - 3$ by using the slope and *y*-intercept.

Solution:

First plot the *y*-intercept, $(0, -3)$.

The slope, $m = -\frac{5}{2}$ can be written as

$$m = \frac{-5}{2}$$ ← The change in *y* is −5. ← The change in *x* is 2.

To find a second point on the line, start at the *y*-intercept and move down 5 units and to the right 2 units. Then draw the line through the two points (Figure 3-27).

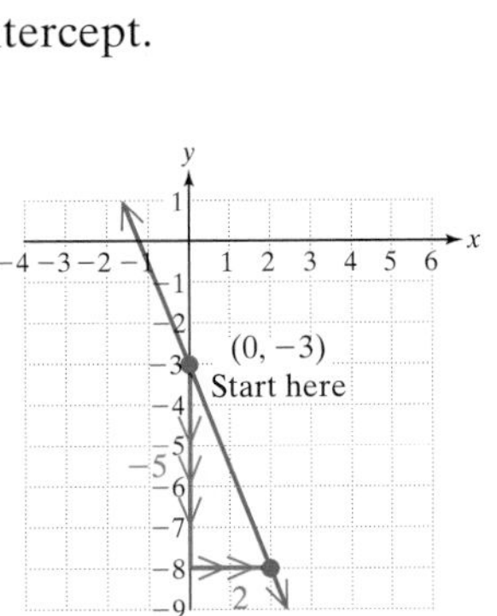

Figure 3-27

Similarly, the slope can be written as

$$m = \frac{5}{-2}$$ ← The change in *y* is 5. ← The change in *x* is −2.

To find a second point, start at the *y*-intercept and move up 5 units and to the left 2 units. Then draw the line through the two points (Figure 3-28).

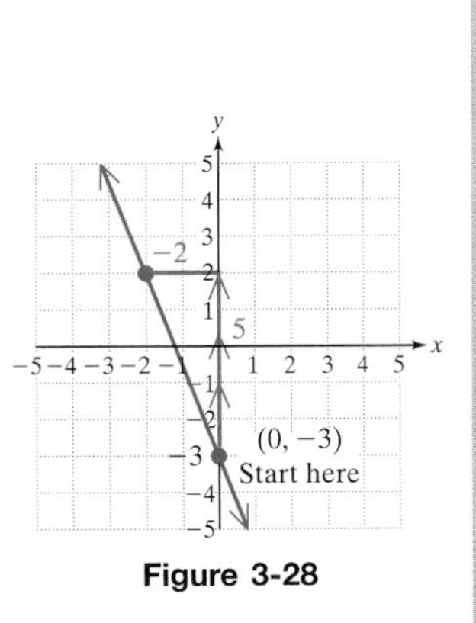

Figure 3-28

Skill Practice

6. Graph the line $y = 2x - 3$ by using the slope and the *y*-intercept.

Example 4 **Graphing a Line from Its Slope and *y*-Intercept**

Graph the line $y = 4x$ by using the slope and *y*-intercept.

Solution:

The line can be written as $y = 4x + 0$. Therefore, we can plot the *y*-intercept at $(0, 0)$. The slope $m = 4$ can be written as

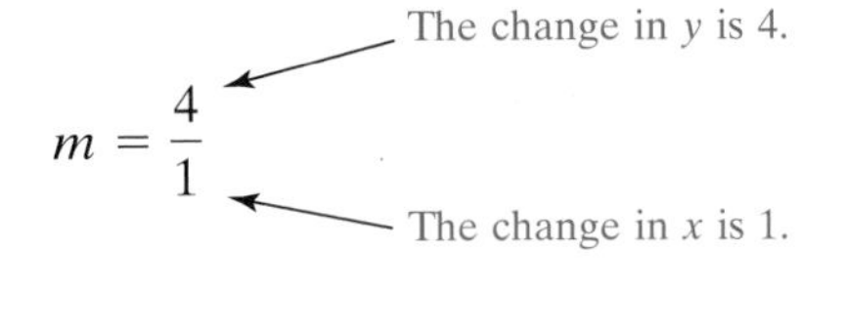

Skill Practice Answers

6. $y = 2x - 3$

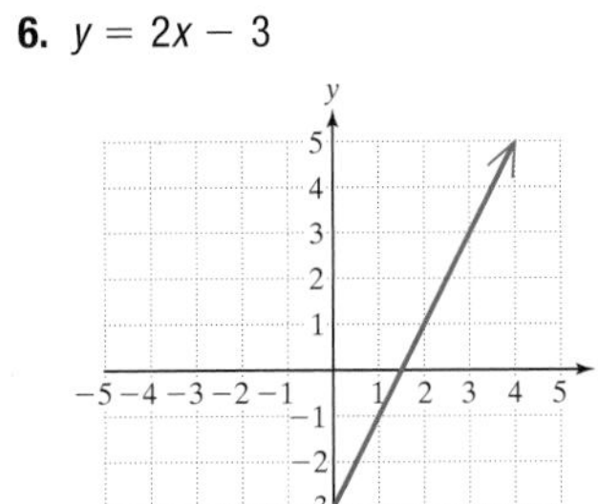

To find a second point on the line, start at the y-intercept and move up 4 units and to the right 1 unit. Then draw the line through the two points (Figure 3-29).

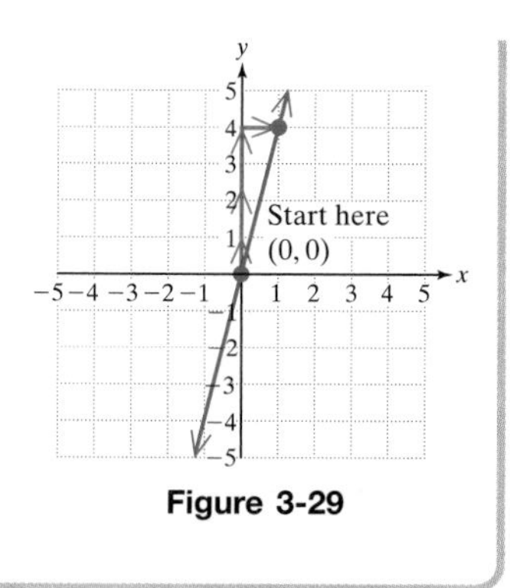

Figure 3-29

Skill Practice

7. Graph the line by using the slope and the y-intercept.

$$y = -\frac{1}{4}x$$

3. Determining Whether Two Lines Are Parallel, Perpendicular, or Neither

The slope-intercept form provides a means to find the slope of a line by inspection. Furthermore, if the slopes of two lines are known, then we can compare the slopes to determine if the lines are parallel, perpendicular, or neither parallel nor perpendicular. (Recall that two distinct nonvertical lines are parallel if their slopes are equal. Two lines are perpendicular if the slope of one line is the opposite of the reciprocal of the slope of the other line.)

Example 5 **Determining If Two Lines Are Parallel, Perpendicular, or Neither**

For each pair of lines, determine if they are parallel, perpendicular, or neither.

a. l_1: $y = 3x - 5$
l_2: $y = 3x + 1$

b. l_1: $x - 3y = -9$
l_2: $3x = -y + 4$

c. l_1: $y = \frac{3}{2}x + 2$
l_2: $y = \frac{2}{3}x + 1$

d. l_1: $x = 2$
l_2: $2y = 8$

Solution:

a. l_1: $y = 3x - 5$ The slope of l_1 is 3.
l_2: $y = 3x + 1$ The slope of l_2 is 3.

Because the slopes are the same, the lines are parallel.

b. First write the equation of each line in slope-intercept form.

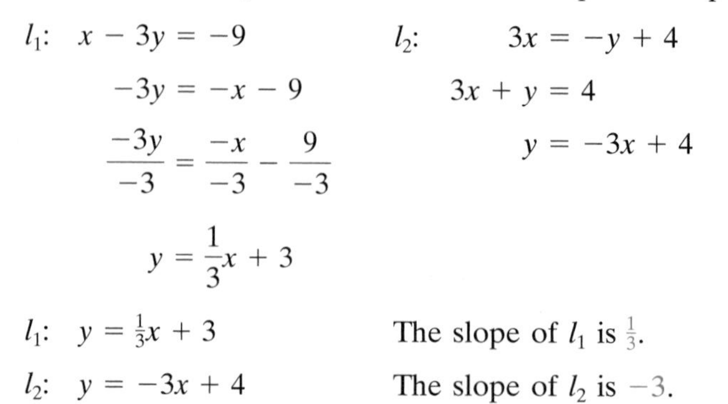

l_1: $x - 3y = -9$

$$-3y = -x - 9$$

$$\frac{-3y}{-3} = \frac{-x}{-3} - \frac{9}{-3}$$

$$y = \frac{1}{3}x + 3$$

l_2: $3x = -y + 4$

$$3x + y = 4$$

$$y = -3x + 4$$

l_1: $y = \frac{1}{3}x + 3$ The slope of l_1 is $\frac{1}{3}$.
l_2: $y = -3x + 4$ The slope of l_2 is -3.

The slope of $\frac{1}{3}$ is the opposite of the reciprocal of -3. Therefore, the lines are perpendicular.

Skill Practice Answers

7. $y = -\frac{1}{4}x$

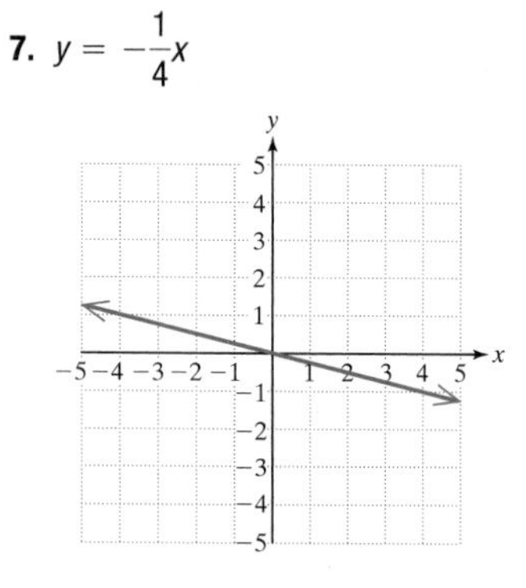

c. l_1: $y = \frac{3}{2}x + 2$ The slope of l_1 is $\frac{3}{2}$.

l_2: $y = \frac{2}{3}x + 1$ The slope of l_2 is $\frac{2}{3}$.

The slopes are not the same. Therefore, the lines are not parallel. The values of the slopes are reciprocals, but they are not opposite in sign. Therefore, the lines are not perpendicular. The lines are neither parallel nor perpendicular.

d. The equation $x = 2$ represents a vertical line because the equation is in the form $x = k$.

The equation $2y = 8$ can be simplified to $y = 4$, which represents a horizontal line.

In this example, we do not need to analyze the slopes because vertical lines and horizontal lines are perpendicular.

Skill Practice For each pair of lines determine if they are parallel, perpendicular, or neither.

8. $y = 3x - 5$

$y = -3x - 15$

9. $x - 5y = 10$

$5x - 1 = -y$

10. $y = \frac{5}{6}x - \frac{1}{2}$

$y = \frac{5}{6}x + \frac{1}{2}$

11. $y = -5$

$x = 6$

4. Writing an Equation of a Line Given Its Slope and y-Intercept

The slope-intercept form of a line can be used to write an equation of a line when the slope is known and the y-intercept is known.

Example 6 **Writing an Equation of a Line Using Slope-Intercept Form**

Write an equation of the line with a slope of $\frac{2}{3}$ and y-intercept $(0, 8)$.

Solution:

The slope is given as $m = \frac{2}{3}$, and the y-intercept $(0, b)$ is given as $(0, 8)$. Substitute the values $m = \frac{2}{3}$ and $b = 8$, into the slope-intercept form of a line.

$$y = mx + b$$

$$y = \frac{2}{3}x + 8$$

Skill Practice

12. Write an equation of the line with slope of -4 and y-intercept $(0, -10)$.

Skill Practice Answers

8. Neither **9.** Perpendicular
10. Parallel **11.** Perpendicular
12. $y = -4x - 10$

Example 7 Writing an Equation of a Line Given Its Slope and *y*-Intercept

Write an equation of the line with a slope of -2 that passes through the origin.

Solution:

The slope is given as $m = -2$. Furthermore, the line passes through the origin. Therefore, the y-intercept is $(0, 0)$ and the corresponding value of b is 0. The slope-intercept form of the line becomes:

$$y = mx + b$$

$$y = -2x + 0 \text{ or equivalently } y = -2x.$$

Skill Practice

13. Write an equation of the line with a slope of $\frac{6}{5}$ that passes through the origin.

Skill Practice Answers

13. $y = \frac{6}{5}x$

Calculator Connections

In Example 5(b) we found that the lines $y = \frac{1}{3}x + 3$ and $y = -3x + 4$ are perpendicular. We can verify our results by graphing the lines on a graphing calculator.

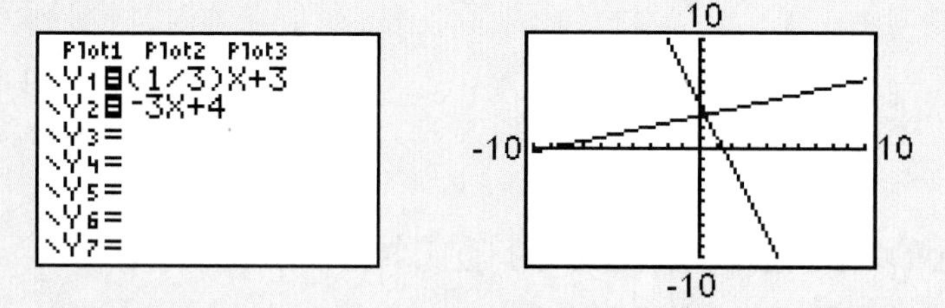

Notice that the lines do not appear perpendicular in the calculator display. That is, they do not appear to form a right angle at the point of intersection. Because many calculators have a rectangular screen, the standard viewing window is elongated in the horizontal direction. To eliminate this distortion, try using a *ZSquare* option. This feature will set the viewing window so that equal distances on the display denote an equal number of units on the graph.

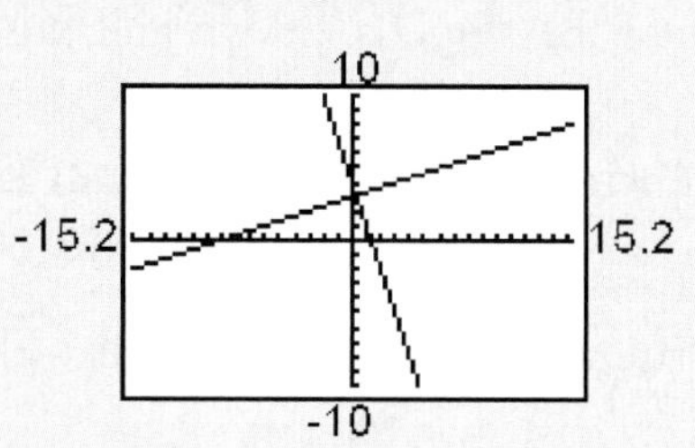

Calculator Exercises

For each pair of lines, determine if the lines are parallel, perpendicular, or neither. Then use a square viewing window to graph the lines on a graphing calculator to verify your results.

1. $x + y = 1$, $x - y = -3$

2. $3x + y = -2$, $6x + 2y = 6$

3. $2x - y = 4$, $3x + 2y = 4$

4. Graph the lines: $y = x + 1$ and $y = 0.99x + 3$. Are these lines parallel? Explain.

5. Graph the lines $y = -2x - 1$ and $y = -2x - 0.99$. Are these lines the same? Explain.

Study Skills Exercises

1. When taking a test, go through the test and do all the problems that you know first. Then go back and work on the problems that were more difficult. Give yourself a time limit for how much time you spend on each problem (maybe 3 to 5 minutes the first time through). Circle the importance of each statement.

	not important	somewhat important	very important
a. Read through the entire test first.	1	2	3
b. If time allows, go back and check each problem.	1	2	3
c. Write out all steps instead of doing the work in your head.	1	2	3

2. Define the key terms:

a. slope-intercept form of a line **b. standard form of a line**

Review Exercises

For Exercises 3–9, determine the x- and y-intercepts, if they exist.

3. $x - 5y = 10$ **4.** $3x + y = -12$ **5.** $3y = -9$ **6.** $2 + y = 5$

7. $-4x = 6y$ **8.** $-x + 3 = 8$ **9.** $5x = 20$

Concept 1: Slope-Intercept Form of a Line

For Exercises 10–29, identify the slope and y-intercept, if they exist.

10. $y = -2x + 3$ **11.** $y = \frac{2}{3}x + 5$ **12.** $y = x - 2$

13. $y = -x + 6$ **14.** $y = -x$ **15.** $y = -4x$

16. $y = \frac{3}{4}x - 1$ **17.** $y = x - \frac{5}{3}$ **18.** $2x - 5y = 4$

19. $3x + 2y = 9$ **20.** $3x - y = 5$ **21.** $7x - 3y = -6$

22. $x + y = 6$ **23.** $x - y = 1$ **24.** $x + 6 = 8$

25. $-4 + x = 1$ **26.** $-8y = 2$ **27.** $1 - y = 9$

28. $3y - 2x = 0$ **29.** $5x = 6y$

Concept 2: Graphing a Line from Its Slope and y-Intercept

For Exercises 30–33, graph the line using the slope and y-intercept.

30. Graph the line through the point (0, 2), having a slope of -4.

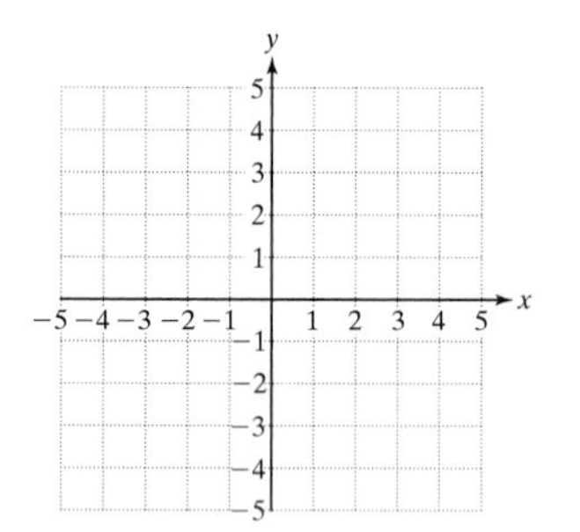

31. Graph the line through the point (0, -1), having a slope of -3.

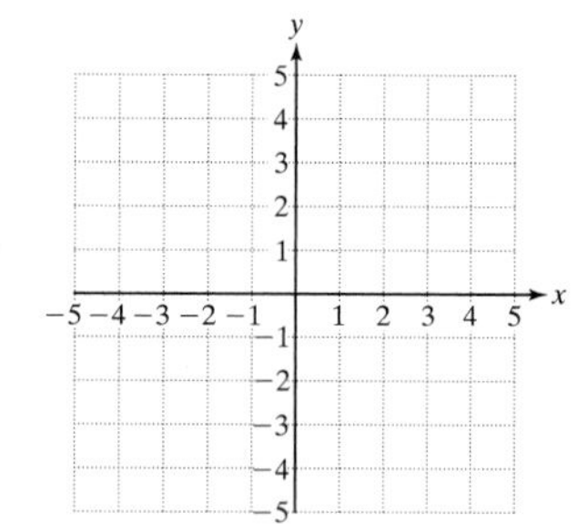

32. Graph the line through the point (0, -5), having a slope of $\frac{3}{2}$.

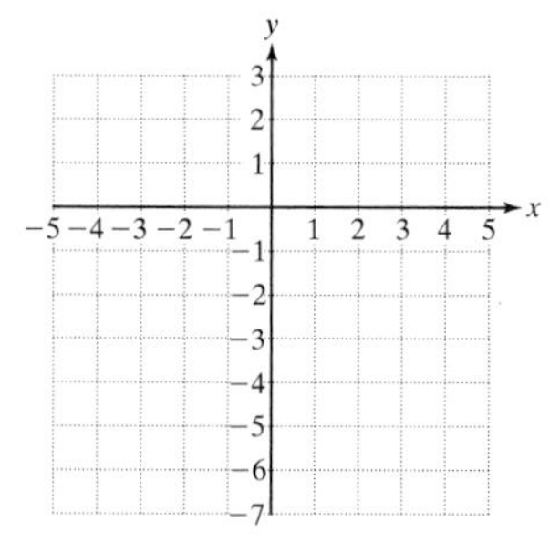

33. Graph the line through the point (0, 3), having a slope of $-\frac{1}{4}$.

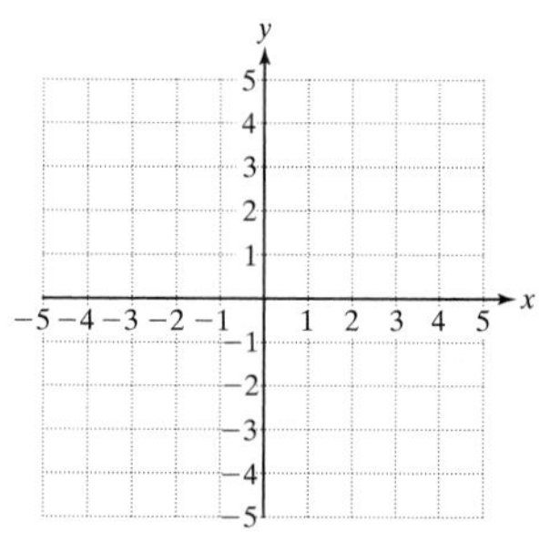

For Exercises 34–39, match the equation with the graph (a–f) by identifying if the slope is positive or negative and if the y-intercept is positive, negative, or zero.

34. $y = 2x + 3$

35. $y = -3x - 2$

36. $y = -\frac{1}{3}x + 3$

37. $y = \frac{1}{2}x - 2$

38. $y = x$

39. $y = -2x$

a.

b.

c.

d.

e.

f.

For Exercises 40–55, write each equation in slope-intercept form (if possible) and graph the line.

40. $x - 2y = 6$

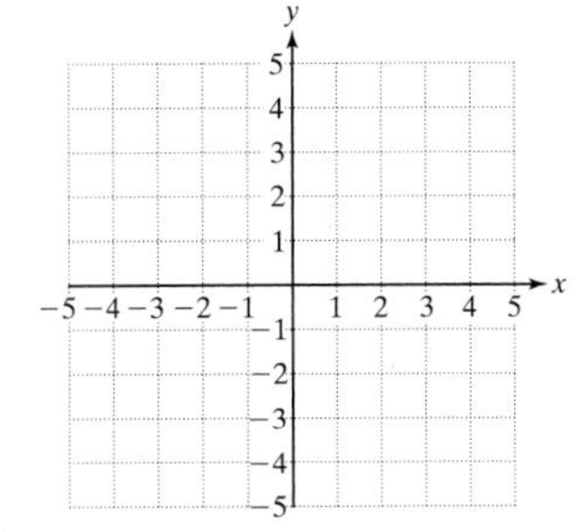

41. $5x - 2y = 2$

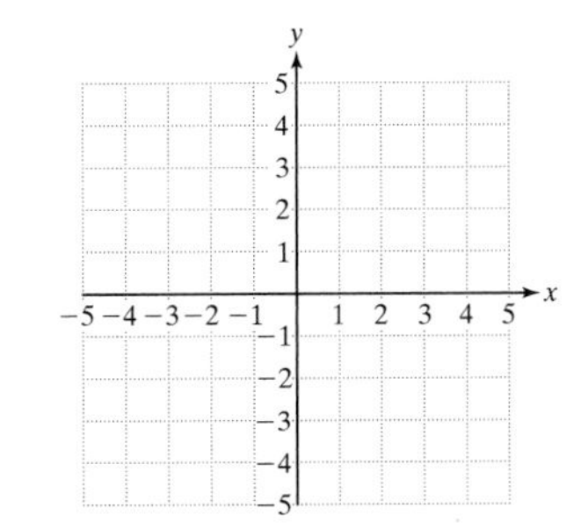

42. $2x + y = 9$

43. $-6x + y = 8$

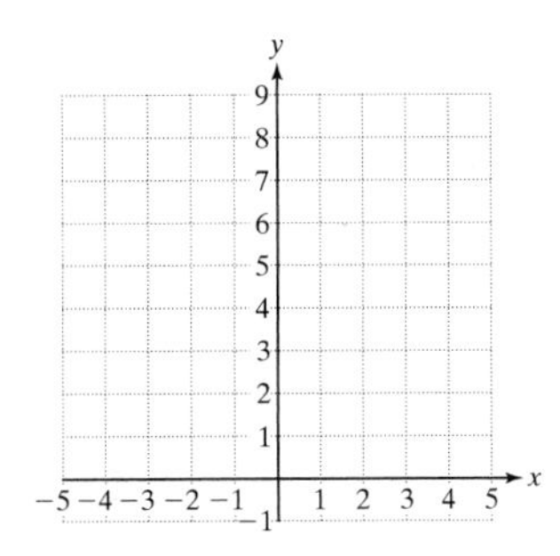

44. $2x = -4y + 6$

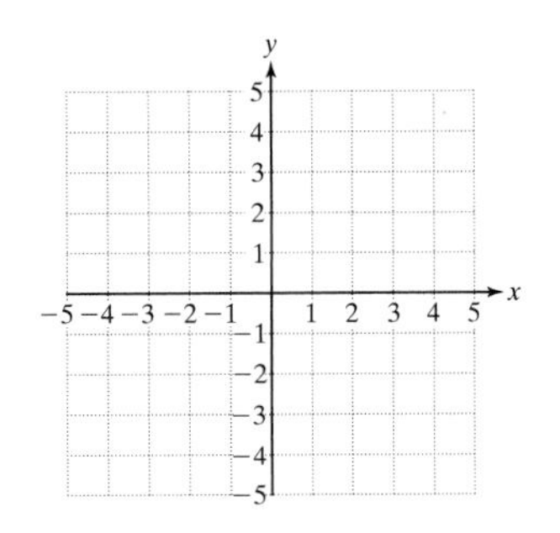

45. $3x = y - 7$

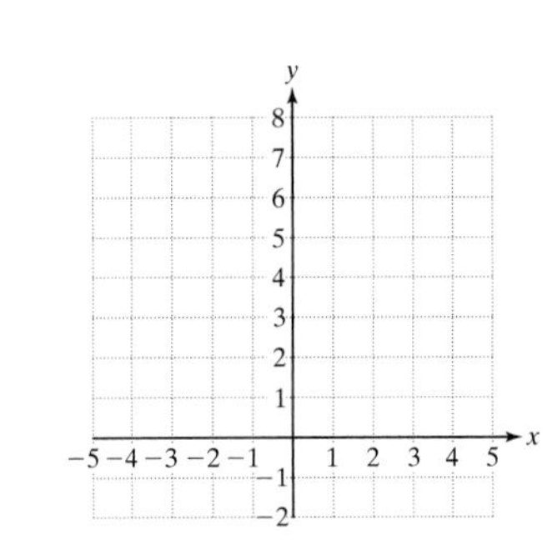

46. $x + y = 0$

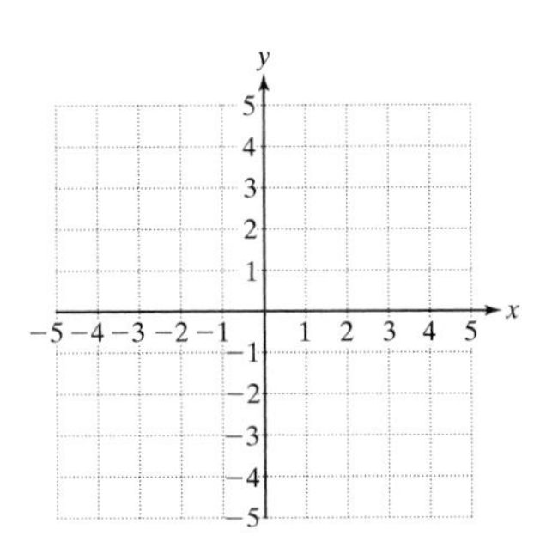

47. $x - y = 0$

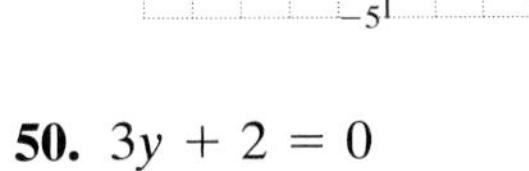

48. $5y = 4x$

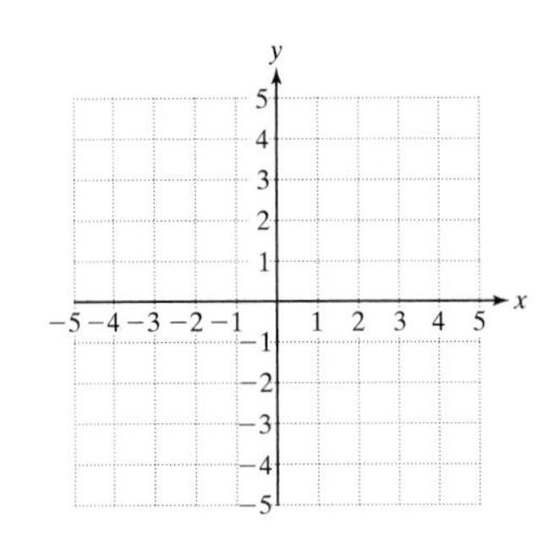

49. $-2x = 5y$

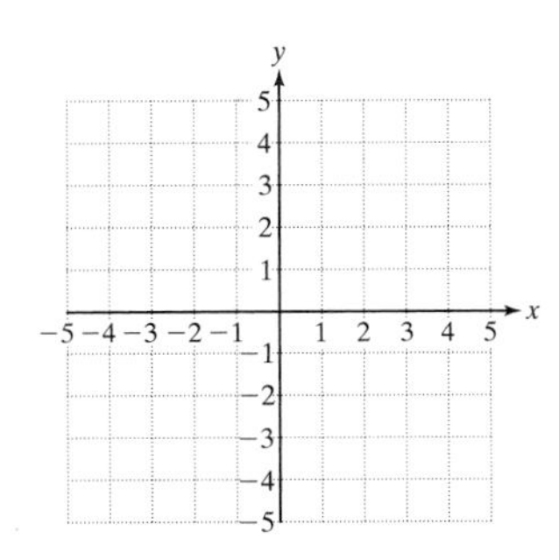

50. $3y + 2 = 0$

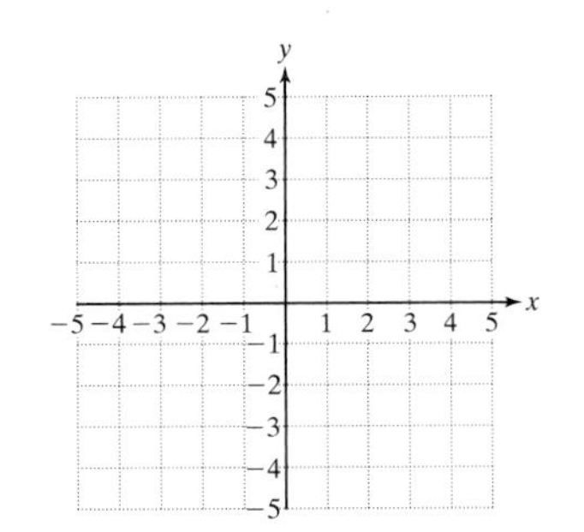

51. $1 + 5y = 6$

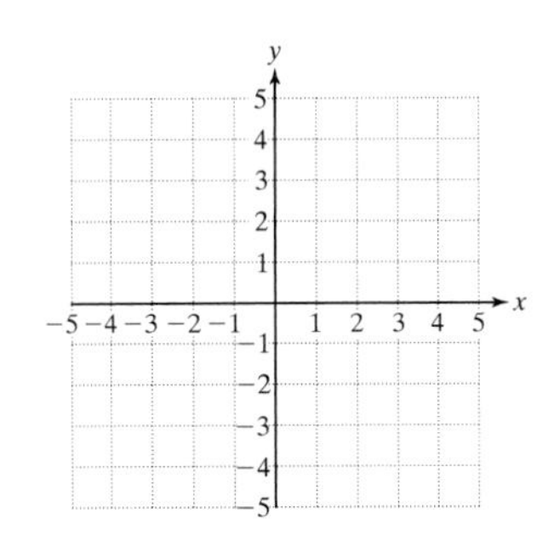

52. $3x + 1 = 7$

53. $-2x - 5 = 1$

54. $\frac{1}{2}x + \frac{1}{4}y = \frac{1}{2}$

55. $\frac{1}{3}x - \frac{1}{6}y = \frac{1}{2}$

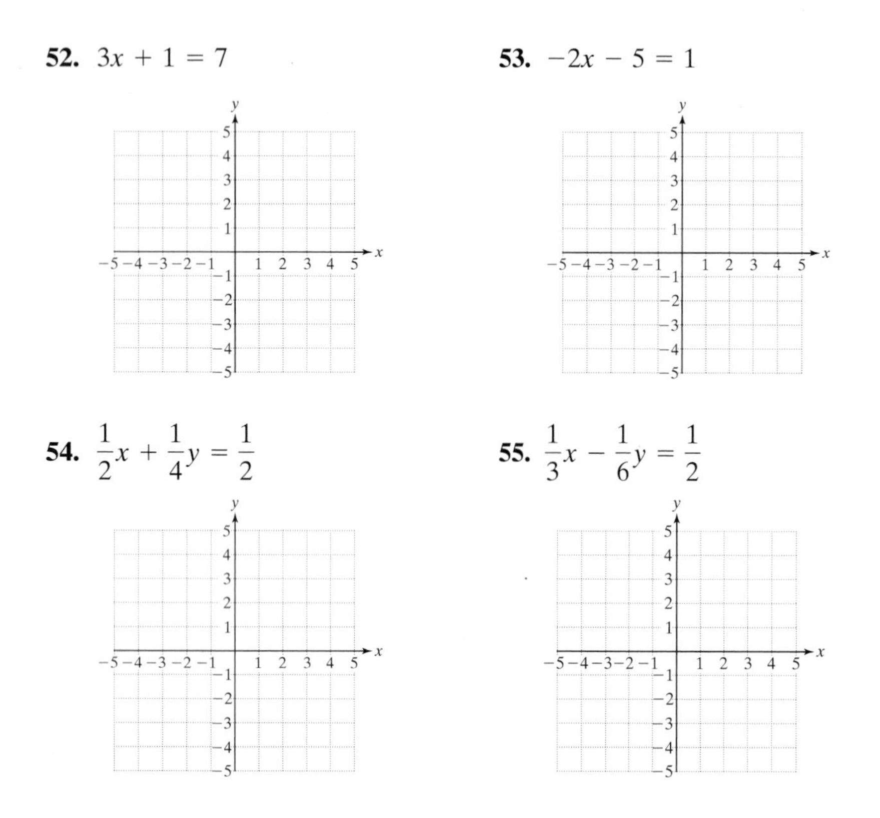

Concept 3: Determining Whether Two Lines Are Parallel, Perpendicular, or Neither

For Exercises 56–61, let m_1 and m_2 represent the slopes of two lines. Determine if the lines are parallel, perpendicular, or neither.

56. $m_1 = -2,\ m_2 = \frac{1}{2}$

57. $m_1 = \frac{2}{3},\ m_2 = \frac{3}{2}$

58. $m_1 = 1,\ m_2 = \frac{4}{4}$

59. $m_1 = \frac{3}{4},\ m_2 = -\frac{8}{6}$

60. $m_1 = \frac{2}{7},\ m_2 = -\frac{2}{7}$

61. $m_1 = 5,\ m_2 = 5$

For Exercises 62–77, determine if the lines l_1 and l_2 are parallel, perpendicular, or neither.

62. l_1: $y = -2x - 3$
l_2: $y = \frac{1}{2}x + 4$

63. l_1: $y = \frac{4}{3}x - 2$
l_2: $y = -\frac{3}{4}x + 6$

64. l_1: $y = \frac{4}{5}x - \frac{1}{2}$
l_2: $y = \frac{5}{4}x - \frac{2}{3}$

65. l_1: $y = \frac{1}{5}x + 1$
l_2: $y = 5x - 3$

66. l_1: $y = -9x + 6$
l_2: $y = -9x - 1$

67. l_1: $y = 4x - 1$
l_2: $y = 4x + \frac{1}{2}$

68. l_1: $x = 3$
l_2: $y = \frac{7}{4}$

69. l_1: $y = \frac{2}{3}$
l_2: $x = 6$

70. l_1: $2x = 4$
l_2: $6 = x$

71. l_1: $2y = 7$
l_2: $y = 4$

72. l_1: $2x + 3y = 6$
l_2: $3x - 2y = 12$

73. l_1: $4x + 5y = 20$
l_2: $5x - 4y = 60$

74. l_1: $4x + 2y = 6$
l_2: $4x + 8y = 16$

75. l_1: $3x + y = 5$
l_2: $x + 3y = 18$

76. l_1: $y = \frac{1}{5}x - 3$
l_2: $2x - 10y = 20$

77. l_1: $y = \frac{1}{3}x + 2$
l_2: $-x + 3y = 12$

Concept 4: Writing an Equation of a Line Given Its Slope and *y*-Intercept

For Exercises 78–87, write an equation of the line given the following information. Write the answer in slope-intercept form if possible.

78. The slope is $-\frac{1}{3}$, and the y-intercept is $(0, 2)$.

79. The slope is $\frac{2}{3}$, and the y-intercept is $(0, -1)$.

80. The slope is 10, and the y-intercept is $(0, -19)$.

81. The slope is -14, and the y-intercept is $(0, 2)$.

82. The slope is 0, and the y-intercept is -11.

83. The slope is 0, and the y-intercept is $\frac{6}{7}$.

84. The slope is 5, and the line passes through the origin.

85. The slope is -3, and the line passes through the origin.

86. The slope is 6, and the line passes through the point $(0, -2)$.

87. The slope is -4, and the line passes through the point $(0, -3)$.

Expanding Your Skills

88. The cost for a rental car is \$49.95 per day plus a flat fee of \$31.95 for insurance. The equation, $C = 49.95x + 31.95$ represents the total cost, C (in dollars), to rent the car for x days.

a. Identify the slope. Interpret the meaning of the slope in the context of this problem.

b. Identify the C-intercept. Interpret the meaning of the C-intercept in the context of this problem.

c. Use the equation to determine how much it would cost to rent the car for 1 week.

89. A phone bill is determined each month by a \$16.95 flat fee plus \$0.10/min of long distance. The equation $C = 0.10x + 16.95$ represents the total monthly cost, C, for x minutes of long distance.

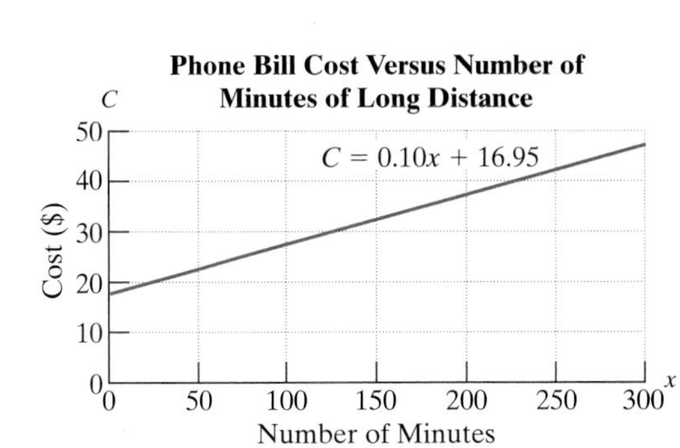

a. Identify the slope. Interpret the meaning of the slope in the context of this problem.

b. Identify the C-intercept. Interpret the meaning of the C-intercept in the context of this problem.

c. Use the equation to determine the total cost of 234 min of long distance.

90. A linear equation is said to be written in standard form if it can be written as $Ax + By = C$, where A and B are not both zero. Write the equation $Ax + By = C$ in slope-intercept form to show that the slope is given by the ratio, $-\frac{A}{B}$. $(B \neq 0.)$

For Exercises 91–94, use the result of Exercise 90 to find the slope of the line.

91. $2x + 5y = 8$

92. $6x + 7y = -9$

93. $4x - 3y = -5$

94. $11x - 8y = 4$

Section 3.5 Point-Slope Formula

Concepts

1. Writing an Equation of a Line Using the Point-Slope Formula
2. Writing an Equation of a Line Through Two Points
3. Writing an Equation of a Line Parallel or Perpendicular to Another Line
4. Different Forms of Linear Equations: A Summary

1. Writing an Equation of a Line Using the Point-Slope Formula

In Section 3.4, the slope-intercept form of a line was used as a tool to construct an equation of a line. Another useful tool to determine an equation of a line is the point-slope formula. The point-slope formula can be derived from the slope formula as follows:

Suppose a line passes through a given point (x_1, y_1) and has slope m. If (x, y) is any other point on the line, then:

$$m = \frac{y - y_1}{x - x_1} \qquad \text{Slope formula}$$

$$m(x - x_1) = \frac{y - y_1}{\cancel{x - x_1}}\cancel{(x - x_1)} \qquad \text{Clear fractions}$$

$$m(x - x_1) = y - y_1$$

or

$$y - y_1 = m(x - x_1) \qquad \text{Point-slope formula}$$

Point-Slope Formula

The **point-slope formula** is given by

$$y - y_1 = m(x - x_1)$$

where m is the slope of the line and (x_1, y_1) is a known point on the line.

Example 1 demonstrates how to use the point-slope formula to find an equation of a line when a point on the line and slope are given.

Example 1 Writing an Equation of a Line Using the Point-Slope Formula

Use the point-slope formula to find an equation of the line having a slope of 3 and passing through the point $(-2, -4)$. Write the answer in slope-intercept form.

Solution:

The slope of the line is given: $m = 3$.

A point on the line is given: $(x_1, y_1) = (-2, -4)$.

The point-slope formula:

$$y - y_1 = m(x - x_1)$$

$$y - (-4) = 3[x - (-2)] \qquad \text{Substitute } m = 3, x_1 = -2, \text{ and } y_1 = -4.$$

$$y + 4 = 3(x + 2)$$ Simplify. Because the final answer is required in slope-intercept form, simplify the equation and solve for y.

$$y + 4 = 3x + 6 \qquad \text{Apply the distributive property.}$$

$$y = 3x + 6 - 4 \qquad \text{Subtract 4 from both sides.}$$

$$y = 3x + 2 \qquad \text{Slope-intercept form}$$

Skill Practice

1. Write an equation of the line passing through the point $(-1, 5)$ and having slope -4.

The equation $y = 3x + 2$ from Example 1 is graphed in Figure 3-30. Notice that the line does indeed pass through the point $(-2, -4)$.

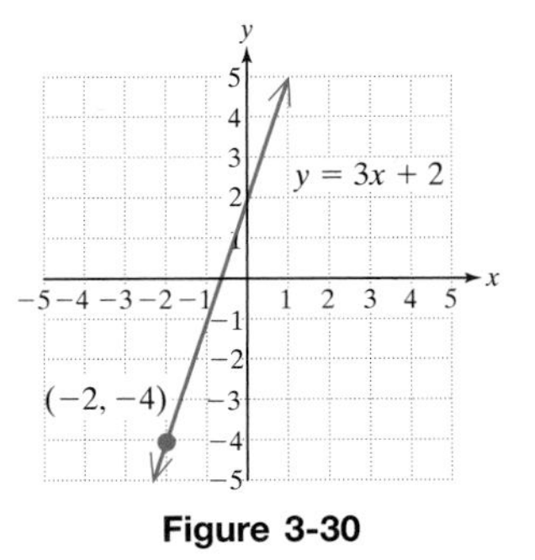

Figure 3-30

2. Writing an Equation of a Line Through Two Points

Example 2 is similar to Example 1; however, the slope must first be found from two given points.

Example 2 Writing an Equation of a Line Through Two Points

Use the point-slope formula to find an equation of the line passing through the points $(-2, 5)$ and $(4, -1)$. Write the final answer in slope-intercept form.

Solution:

Given two points on a line, the slope can be found with the slope formula.

$(-2, 5)$ and $(4, -1)$
(x_1, y_1) (x_2, y_2) Label the points.

$$m = \frac{y_2 - y_1}{x_2 - x_1} = \frac{(-1) - (5)}{(4) - (-2)} = \frac{-6}{6} = -1$$

To apply the point-slope formula, use the slope, $m = -1$ and either given point. We will choose the point $(-2, 5)$ as (x_1, y_1).

$y - y_1 = m(x - x_1)$

$y - 5 = -1[x - (-2)]$ Substitute $m = -1$, $x_1 = -2$, and $y_1 = 5$.

$y - 5 = -1(x + 2)$ Simplify.

$y - 5 = -x - 2$

$y = -x + 3$ Slope-intercept form

TIP: The point-slope formula can be applied using either given point for (x_1, y_1). In Example 2, using the point $(4, -1)$ for (x_1, y_1) produces the same result.

$y - y_1 = m(x - x_1)$
$y - (-1) = -1(x - 4)$
$y + 1 = -x + 4$
$y = -x + 3$

Skill Practice

2. Use the point-slope formula to write an equation of the line passing through the two points $(1, -1)$ and $(-1, -5)$. Write the final answer in slope-intercept form.

Skill Practice Answers

1. $y = -4x + 1$
2. $y = 2x - 3$

The solution to Example 2 can be checked by graphing the line $y = -x + 3$ using the slope and y-intercept. Notice that the line passes through the points $(-2, 5)$ and $(4, -1)$ as expected. See Figure 3-31.

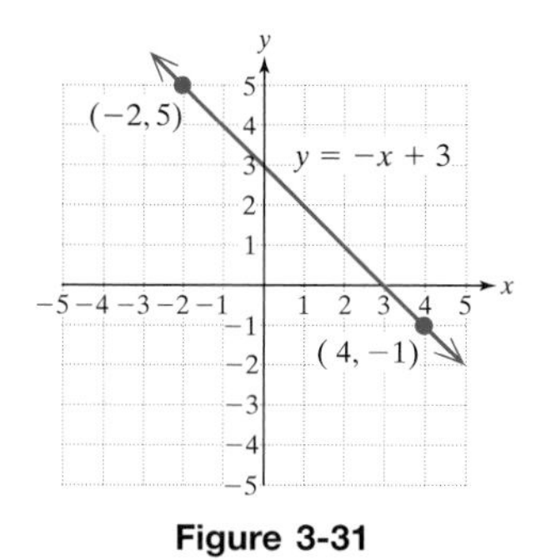

Figure 3-31

3. Writing an Equation of a Line Parallel or Perpendicular to Another Line

Example 3 Writing an Equation of a Line Parallel to Another Line

Use the point-slope formula to find an equation of the line passing through the point $(-1, 0)$ and parallel to the line $y = -4x + 3$. Write the final answer in slope-intercept form.

Solution:

Figure 3-32 shows the line $y = -4x + 3$ (pictured in black) and a line parallel to it (pictured in blue) that passes through the point $(-1, 0)$. The equation of the given line, $y = -4x + 3$, is written in slope-intercept form, and its slope is easily identified as -4. The line parallel to the given line must also have a slope of -4.

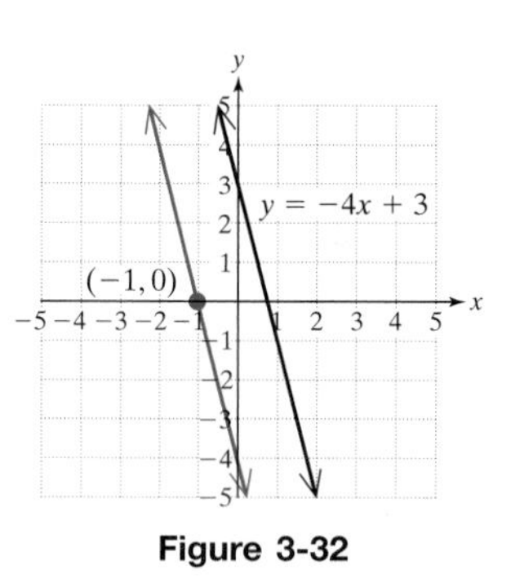

Figure 3-32

Apply the point-slope formula using $m = -4$ and the point $(x_1, y_1) = (-1, 0)$.

$$y - y_1 = m(x - x_1)$$
$$y - 0 = -4[x - (-1)]$$
$$y = -4(x + 1)$$
$$y = -4x - 4$$

Skill Practice

3. Use the point-slope formula to write an equation of the line passing through the point (8, 2) and parallel to the line $y = \frac{3}{4}x - \frac{1}{2}$.

Example 4 Writing an Equation of a Line Perpendicular to Another Line

Use the point-slope formula to find an equation of the line passing through the point $(-3, 1)$ and perpendicular to the line $3x + y = -2$. Write the final answer in slope-intercept form.

Solution:

The given line can be written in slope-intercept form as $y = -3x - 2$. The slope of this line is -3. Therefore, the slope of a line perpendicular to the given line is $\frac{1}{3}$.

Apply the point-slope formula with $m = \frac{1}{3}$, and $(x_1, y_1) = (-3, 1)$.

$y - y_1 = m(x - x_1)$ Point-slope formula

$y - (1) = \frac{1}{3}[x - (-3)]$ Substitute $m = \frac{1}{3}$, $x_1 = -3$, and $y_1 = 1$.

$y - 1 = \frac{1}{3}(x + 3)$ To write the final answer in slope-intercept form, simplify the equation and solve for y.

$y - 1 = \frac{1}{3}x + 1$ Apply the distributive property.

$y = \frac{1}{3}x + 2$ Add 1 to both sides.

A sketch of the perpendicular lines $y = \frac{1}{3}x + 2$ and $y = -3x - 2$ is shown in Figure 3-33. Notice that the line $y = \frac{1}{3}x + 2$ passes through the point $(-3, 1)$.

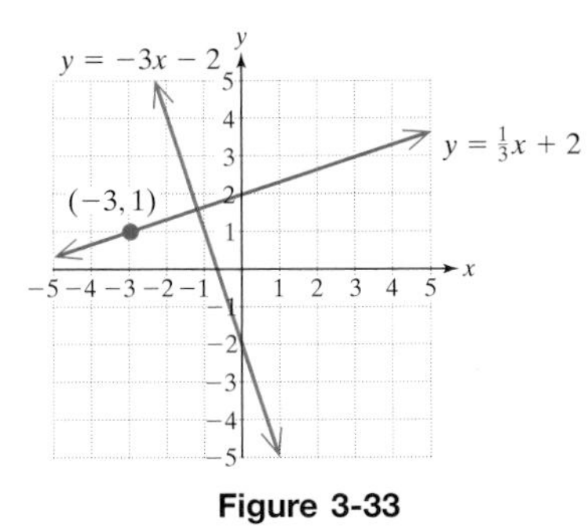

Figure 3-33

Skill Practice

4. Write an equation of the line passing through the point (10, 4) and perpendicular to the line $x + 2y = 1$.

4. Different Forms of Linear Equations: A Summary

A linear equation can be written in several different forms as summarized in Table 3-3.

Skill Practice Answers

3. $y = \frac{3}{4}x - 4$

4. $y = 2x - 16$

Table 3-3

Form	Example	Comments
Standard Form $Ax + By = C$	$4x + 2y = 8$	A and B must not both be zero.
Horizontal Line $y = k$ (k is constant)	$y = 4$	The slope is zero, and the y-intercept is $(0, k)$.
Vertical Line $x = k$ (k is constant)	$x = -1$	The slope is undefined, and the x-intercept is $(k, 0)$.
Slope-Intercept Form $y = mx + b$ the slope is m y-intercept is $(0, b)$	$y = -3x + 7$ Slope $= -3$ y-intercept is $(0, 7)$	Solving a linear equation for y results in slope-intercept form. The coefficient of the x-term is the slope, and the constant defines the location of the y-intercept.
Point-Slope Formula $y - y_1 = m(x - x_1)$	$m = -3$ $(x_1, y_1) = (4, 2)$ $y - 2 = -3(x - 4)$	This formula is typically used to build an equation of a line when a point on the line is known and the slope of the line is known.

Although standard form and slope-intercept form can be used to express an equation of a line, often the slope-intercept form is used to give a *unique* representation of the line. For example, the following linear equations are all written in standard form, yet they each define the same line.

$$2x + 5y = 10$$
$$-4x - 10y = -20$$
$$6x + 15y = 30$$
$$\frac{2}{5}x + y = 2$$

The line can be written uniquely in slope-intercept form as: $y = -\frac{2}{5}x + 2$.

Although it is important to understand and apply slope-intercept form and the point-slope formula, they are not necessarily applicable to all problems, particularly when dealing with a horizontal or vertical line.

Example 5 Writing an Equation of a Line

Find an equation of the line passing through the point $(2, -4)$ and perpendicular to the x-axis.

Solution:

Because the line is perpendicular to the x-axis, the line must be vertical. Recall that all vertical lines can be written in the form $x = k$, where k is a constant. A quick sketch can help find the value of the constant. See Figure 3-34.

Because the line must pass through a point whose x-coordinate is 2, then the equation of the line must be $x = 2$.

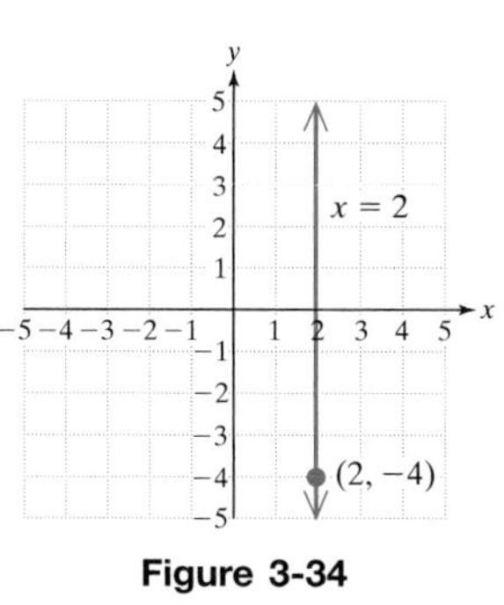

Figure 3-34

Skill Practice

5. Write an equation for the vertical line that passes through the point $(-7, 2)$.

Skill Practice Answers

5. $x = -7$

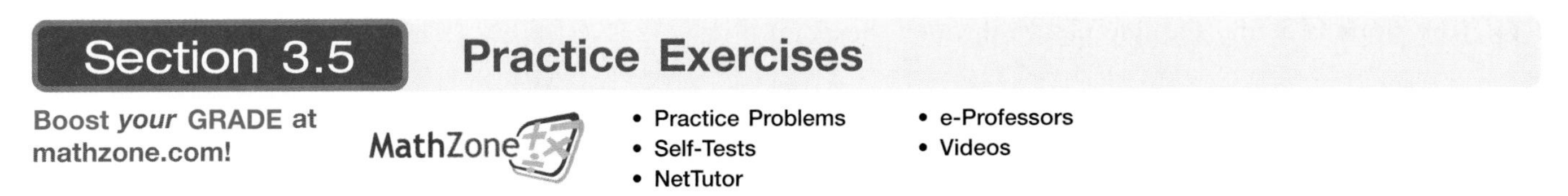

Section 3.5 Practice Exercises

Study Skills Exercises

1. Prepare a one-page summary sheet with the most important information that you need for the test. On the day of the test, look at this sheet several times to refresh your memory instead of trying to memorize new information.

2. Define the key term: **point-slope formula**

Review Exercises

For Exercises 3–6, graph the equations.

3. $2x - 3y = -3$

4. $y = -2x$

5. $3 - y = 9$

6. $y = \frac{4}{5}x$

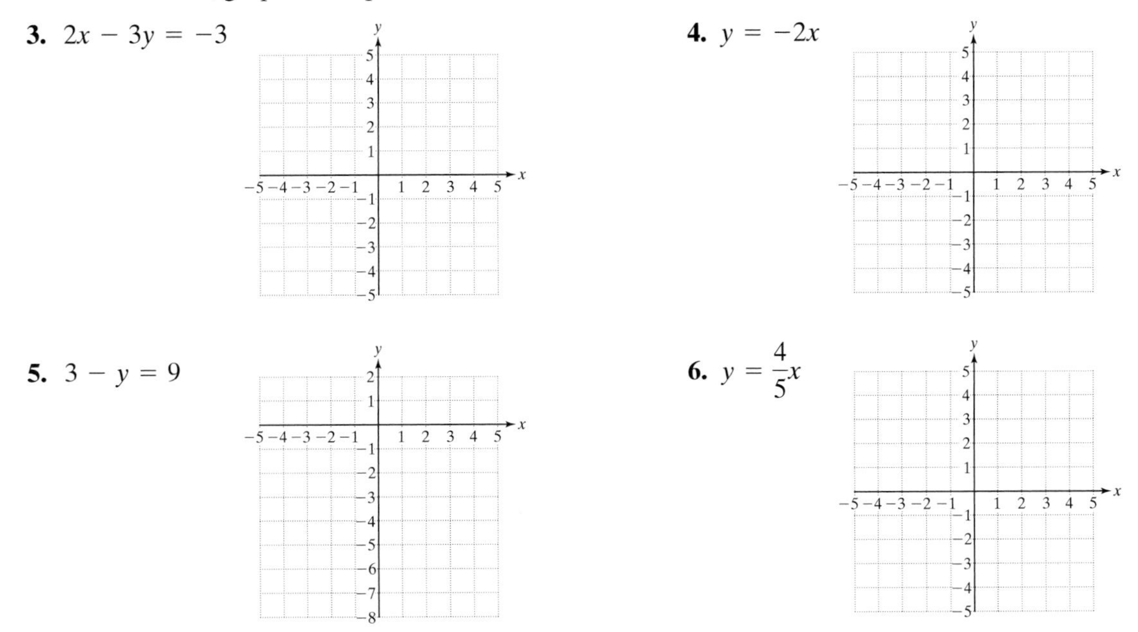

For Exercises 7–10, find the slope of the line that passes through the given points.

7. $(1, -3)$ and $(2, 6)$

8. $(2, -4)$ and $(-2, 4)$

9. $(-2, 5)$ and $(5, 5)$

10. $(6.1, 2.5)$ and $(6.1, -1.5)$

Concept 1: Writing an Equation of a Line Using the Point-Slope Formula

For Exercises 11–22, use the point-slope formula (if possible) to write an equation of the line given the following information.

11. The slope is 3, and the line passes through the point $(-2, 1)$.

12. The slope is -2, and the line passes through the point $(1, -5)$.

13. The slope is -4, and the line passes through the point $(-3, -2)$.

14. The slope is 5, and the line passes through the point $(-1, -3)$.

15. The slope is $-\frac{1}{2}$, and the line passes through $(-1, 0)$.

16. The slope is $-\frac{3}{4}$, and the line passes through $(2, 0)$.

17. The slope is $\frac{1}{4}$, and the line passes through the point $(-8, 6)$.

18. The slope is $\frac{2}{5}$, and the line passes through the point $(-5, 4)$.

19. The slope is 4.5, and the line passes through the point $(5.2, -2.2)$.

20. The slope is -3.6, and the line passes through the point $(10.0, 8.2)$.

21. The slope is 0, and the line passes through the point $(3, -2)$.

22. The slope is 0, and the line passes through the point $(0, 5)$.

Concept 2: Writing an Equation of a Line through Two Points

For Exercises 23–26, find an equation of the line through the given points. Write the final answer in slope-intercept form.

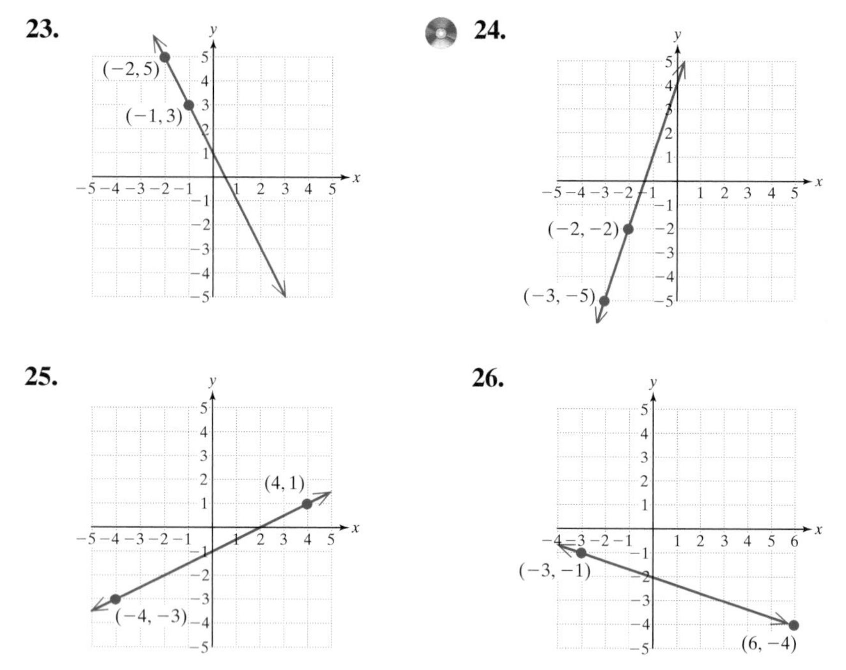

For Exercises 27–32, use the point-slope formula to write an equation of the line given the following information.

27. The line passes through the points $(-2, -6)$ and $(1, 0)$.

28. The line passes through the points $(-2, 5)$ and $(0, 1)$.

29. The line passes through the points $(1, -3)$ and $(-7, 2)$.

30. The line passes through the points $(0, -4)$ and $(-1, -3)$.

31. The line passes through the points $(2.2, -3.3)$ and $(12.2, -5.3)$.

32. The line passes through the points $(4.7, -2.2)$ and $(-0.3, 6.8)$.

Concept 3: Writing an Equation of a Line Parallel or Perpendicular to Another Line

For Exercises 33–44, use the point-slope formula to write an equation of the line given the following information.

33. The line passes through the point $(-3, 1)$ and is parallel to the line $y = 4x + 3$.

34. The line passes through the point $(4, -1)$ and is parallel to the line $y = 3x + 1$.

35. The line passes through the point $(4, 0)$ and is parallel to the line $3x + 2y = 8$.

36. The line passes through the point $(2, 0)$ and is parallel to the line $5x + 3y = 6$.

37. The line passes through the point $(-5, 2)$ and is perpendicular to the line $y = \frac{1}{2}x + 3$.

38. The line passes through the point $(-2, -2)$ and is perpendicular to the line $y = \frac{1}{3}x - 5$.

39. The line passes through the point $(0, -6)$ and is perpendicular to the line $-5x + y = 4$.

40. The line passes through the point $(0, -8)$ and is perpendicular to the line $2x - y = 5$.

41. The line passes through the point $(4, 4)$ and is parallel to the line $3x - y = 6$.

42. The line passes through the point $(-1, -7)$ and is parallel to the line $5x + y = -5$.

43. The line passes through the point $(-2, -1)$ and is perpendicular to $4x - y = 3$.

44. The line passes through the point $(-4, 3)$ and is perpendicular to $3x - 2y = 9$.

Concept 4: Different Forms of Linear Equations: A Summary

For Exercises 45–50, match the form or formula on the left with its name on the right.

45. $x = k$	i. Standard form
46. $y = mx + b$	ii. Point-slope formula
47. $m = \dfrac{y_2 - y_1}{x_2 - x_1}$	iii. Horizontal line
48. $y - y_1 = m(x - x_1)$	iv. Vertical line
49. $y = k$	v. Slope-intercept form
50. $Ax + By = C$	vi. Slope formula

For Exercises 51–62, find an equation for the line given the following information.

51. The line passes through the point (3, 1) and is parallel to the line $y = -4$. See the figure.

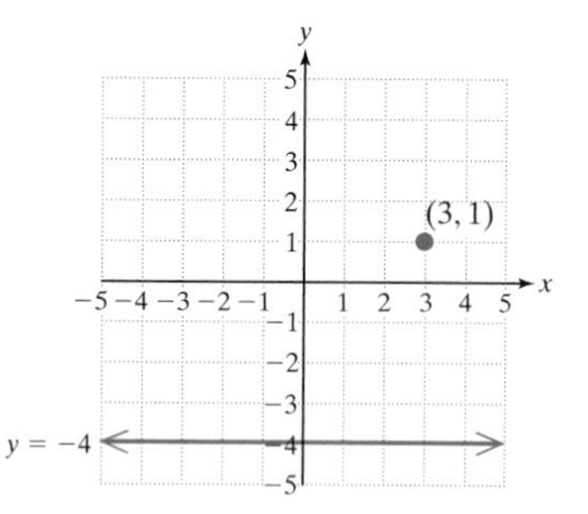

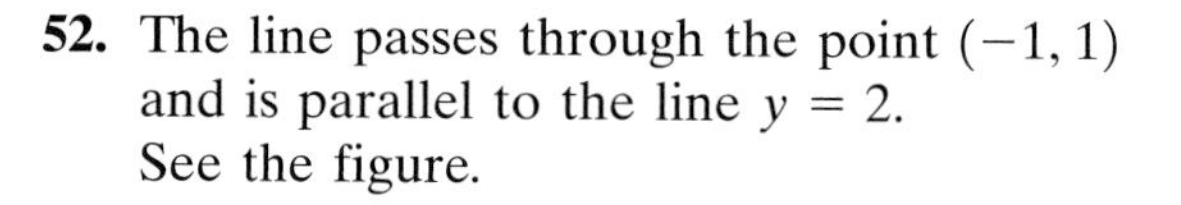

52. The line passes through the point (−1, 1) and is parallel to the line $y = 2$. See the figure.

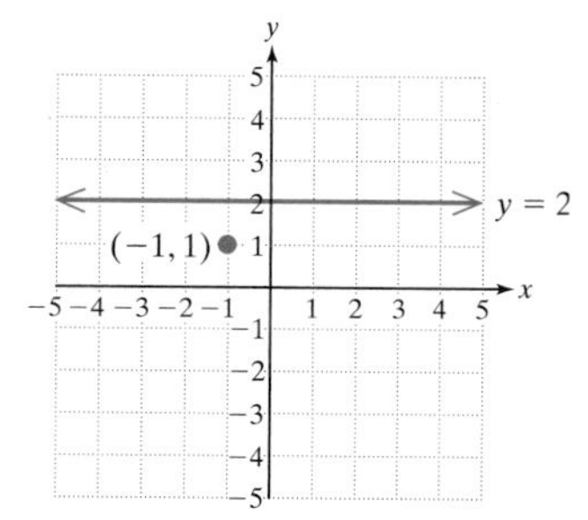

53. The line passes through the point (2, 6) and is perpendicular to the line $y = 1$. (*Hint:* Sketch the line first.)

54. The line passes through the point (0, 3) and is perpendicular to the line $y = -5$. (*Hint:* Sketch the line first.)

55. The line passes through the point $(\frac{5}{2}, \frac{1}{2})$ and is parallel to the line $x = 4$.

56. The line passes through the point $(-6, \frac{2}{3})$ and is parallel to the line $x = -2$.

57. The line passes through the point (2, 2) and is perpendicular to the line $x = 0$.

58. The line passes through the point (5, −2) and is perpendicular to the line $x = 0$.

59. The slope is undefined, and the line passes through the point (−6, −3).

60. The slope is undefined, and the line passes through the point (2, −1).

61. The line passes through the points (−4, 0) and (−4, 3).

62. The line passes through the points (1, 3) and (1, −4).

Expanding Your Skills

63. The following table represents the percentage of females, y, who smoked for selected years. Let x represent the number of years after 1965. Let y represent percentage of women who smoked.

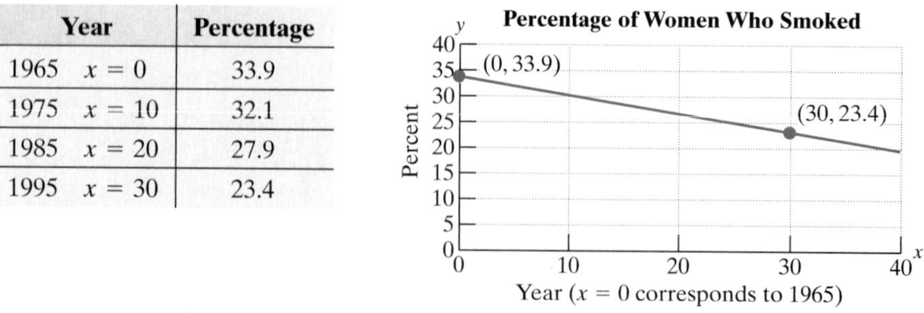

Year		Percentage
1965	$x = 0$	33.9
1975	$x = 10$	32.1
1985	$x = 20$	27.9
1995	$x = 30$	23.4

(Source: U.S. National Center for Health Statistics.)

a. Find the slope of the line between the points $(0, 33.9)$ and $(30, 23.4)$.

b. Find an equation of the line between the points $(0, 33.9)$ and $(30, 23.4)$. Write the answer in slope-intercept form.

c. Use the equation from part (b) to estimate the percentage of women who smoked in the year 2000.

64. The following table represents the median selling price, y, of new privately owned one-family houses sold in the Midwest from 1980 to 2005. Let x represent the number of years after 1980. Let y represent price in thousands of dollars. (Source: U.S. Bureau of Census.)

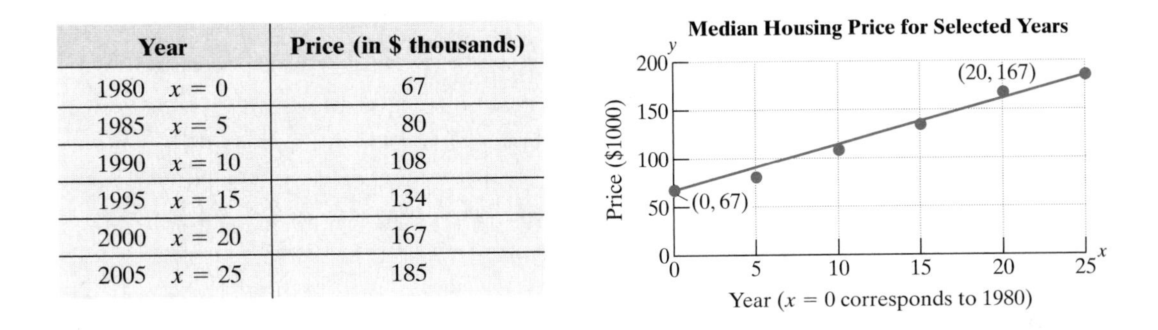

Year	Price (in $ thousands)
1980 $x = 0$	67
1985 $x = 5$	80
1990 $x = 10$	108
1995 $x = 15$	134
2000 $x = 20$	167
2005 $x = 25$	185

a. Find the slope of the line between the points $(0, 67)$ and $(20, 167)$.

b. Find an equation of the line between the points $(0, 67)$ and $(20, 167)$. Write the answer in slope-intercept form.

c. Use the equation from part (b) to estimate the median price of a one-family house sold in the Midwest in the year 2008.

Applications of Linear Equations

Section 3.6

Concepts

1. Interpreting a Linear Equation in Two Variables
2. Writing a Linear Equation Using Observed Data Points
3. Writing a Linear Equation Given a Fixed Value and a Rate of Change

1. Interpreting a Linear Equation in Two Variables

Linear equations can often be used to describe (or model) the relationship between two variables in a real-world event. In an xy-coordinate system, the variable being predicted by the mathematical equation is called the **dependent variable** (or response variable) and is represented by y. The variable used to make the prediction is called the **independent variable** (or predictor variable) and is represented by x.

Example 1 Interpreting a Linear Equation

The cost, y, of a speeding ticket (in dollars) is given by $y = 10x + 100$, where $x > 0$ is the number of miles per hour over the speed limit.

a. Which is the independent variable?

b. Which variable is the dependent variable?

c. What is the slope of the line?

d. Interpret the meaning of the slope in terms of cost and the number of miles per hour over the speed limit.

e. Graph the line.

Solution:

a. The independent variable is the number of miles over the speed limit and is represented by x.

b. The dependent variable is the cost of the speeding ticket and is represented by y. The cost of the ticket *depends* on the number of miles per hour over the speed limit.

c. The equation is written in slope-intercept form where $m = 10$.

d. The slope $m = 10$ or $\frac{10}{1}$ indicates that there is a \$10 increase in the cost of the speeding ticket for every 1 mph over the speed limit.

e.

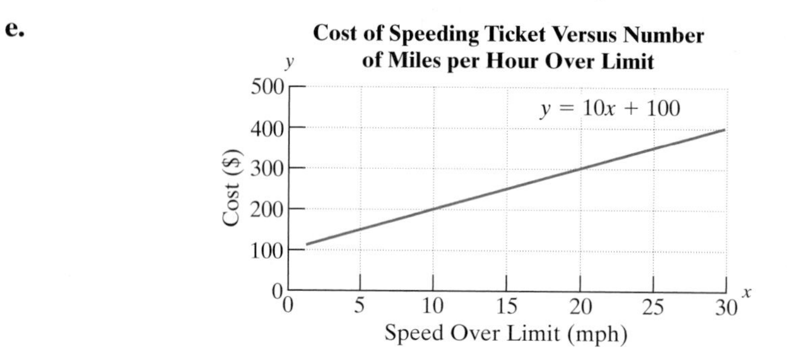

Skill Practice

1. The cost, y, for a local move by a small moving company is given by $y = 60x + 100$, where $x > 0$ is the number of hours required for the move.
a. Which variable is the independent variable?
b. Which variable is the dependent variable?
c. What is the slope of the line?
d. Interpret the meaning of the slope in terms of cost and the number of hours it takes to move.
e. Graph the line.

Example 2 Interpreting a Linear Equation

The total number of crimes in the United States decreased from the year 1990 to 2006 (Figure 3-35). The decrease followed a trend that is approximately linear and can be represented by the linear equation:

$N = -0.28x + 14.8$ where N is the number of crimes (in millions) and x is the number of years since 1990.

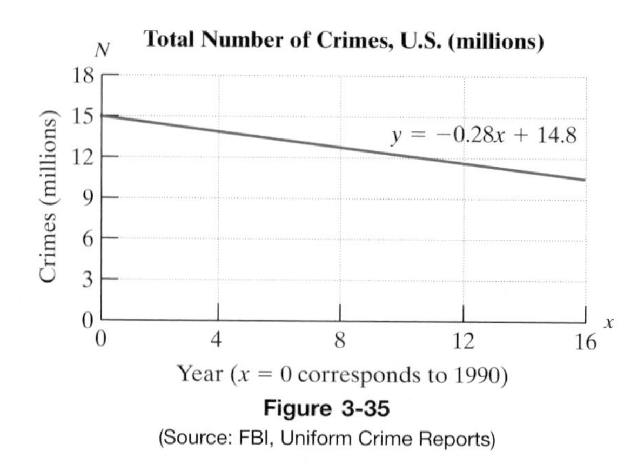

Figure 3-35
(Source: FBI, Uniform Crime Reports)

Skill Practice Answers

1a. Number of hours, x
b. Cost, y **c.** 60
d. There is an increase in cost of \$60 for each hour of the move.
e.

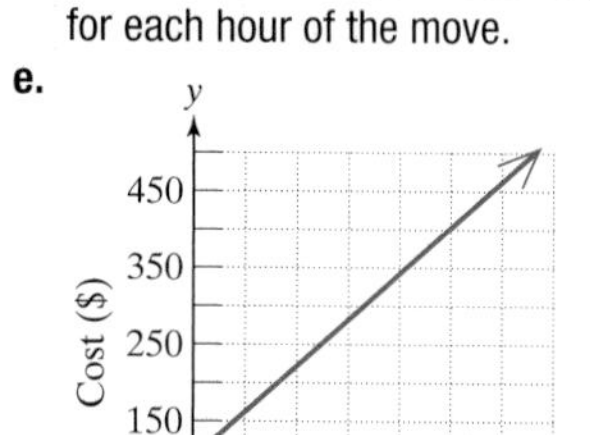

a. Which is the independent variable?

b. Which is the dependent variable?

c. Use the equation to predict the number of crimes in 2000.

d. What is the N-intercept of the line? Interpret the meaning of the N-intercept in terms of the number of crimes and the year.

e. What is the slope of the line? Interpret the meaning of the slope in terms of the number of crimes and the year.

f. Is it possible for this linear trend to continue indefinitely?

g. From the equation, determine the value of the x-intercept. Round to the nearest whole unit. Interpret the meaning of the x-intercept in terms of the number of crimes and the year. Realistically, is it possible for this linear trend to continue to its x-intercept?

Solution:

a. The number of years since 1990 is the independent variable. It is represented by x.

b. The number of reported crimes is the dependent variable. It is represented by N.

c. The year 2000 is 10 years after 1990. Therefore, substitute $x = 10$ into the linear equation.

$$N = -0.28x + 14.8$$

$$N = -0.28(10) + 14.8 \qquad \text{Substitute } x = 10.$$

$$= 12$$

The number of reported crimes in the U.S. in 2000 was approximately 12 million.

d. Notice that the variable N is playing the role of y (the dependent variable). The equation is written in slope-intercept form, $N = -0.28x + 14.8$. The N-intercept is $(0, 14.8)$ and indicates that in the year $x = 0$ (1990), the number of reported crimes in the United States was approximately 14.8 million.

e. From the slope intercept form of the line, $N = -0.28x + 14.8$, the slope is -0.28 or equivalently, $\frac{-0.28}{1}$. The slope indicates that the number of crimes decreased by 0.28 million per year during this time period.

f. It is not possible for the linear trend to continue indefinitely because eventually the number of crimes would reduce to a negative number.

g. To find the x-intercept, substitute $N = 0$.

$$N = -0.28x + 14.8$$

$$0 = -0.28x + 14.8 \qquad \text{Substitute } N = 0.$$

$$-14.8 = -0.28x \qquad \text{Subtract 14.8 from both sides.}$$

$$\frac{-14.8}{-0.28} = \frac{-0.28x}{-0.28} \qquad \text{Divide both sides by } -0.28.$$

$$53 \approx x$$

The x-intercept is $(53, 0)$ and indicates that approximately 53 years after 1990 (the year 2043), the number of crimes will be 0. Although the concept of having zero reported crimes is appealing, it is not realistic. This shows that the linear trend will not continue indefinitely.

Skill Practice

2. The number of tigers, y, in India decreased between the years 1900 to 1980 according to the linear equation $y = -350x + 42{,}000$, where x is the number of years since 1900. (*Source:* Environmental Investigation Agency)

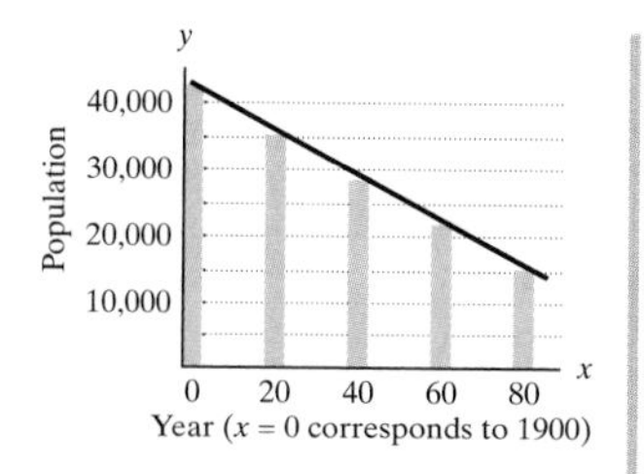

a. Which is the independent variable?

b. Which is the dependent variable?

c. Use the equation to predict the number of tigers in 1960.

d. What is the slope of the line? Interpret the meaning of the slope in terms of the number of tigers and the year.

e. Find the x-intercept. Interpret the meaning of the x-intercept in terms of the number of tigers.

2. Writing a Linear Equation Using Observed Data Points

Example 3 Writing a Linear Equation from Observed Data Points

In the 1990s sales for SUVs increased in the United States until higher gas prices eventually slowed the trend. Let x represent the number of years since 1994, and let y represent the yearly number of SUVs sold (in millions) in the United States. From Figure 3-36, we see that SUV sales followed a trend that was approximately linear.

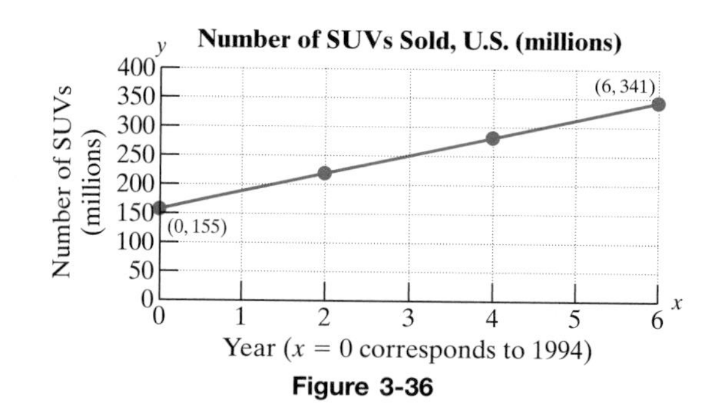

Figure 3-36

a. Use the data points from Figure 3-36 to find a linear equation that represents the number of SUVs sold in the U.S. x years after 1994.

b. Use the linear equation found in part (a) to estimate the number of SUVs sold for the year 1998.

Solution:

a. From the graph, the two given data values are (0, 155) and (6, 341). From these points we can find the slope of the line.

$(0, 155)$ $(6, 341)$
(x_1, y_1) (x_2, y_2) Label the points.

Skill Practice Answers

2a. The number of years since 1900, x
b. The number of tigers, y
c. 21,000 tigers
d. -350; The number of tigers decreases by 350 per year.
e. (0, 120) If the trend continues, there will be 0 tigers in the year 2020.

$$m = \frac{y_2 - y_1}{x_2 - x_1} = \frac{341 - 155}{6 - 0}$$

$$= \frac{186}{6} = \frac{31}{1} = 31$$

The slope is 31 and indicates that there has been an increase of 31 million SUVs sold per year in the United States.

With $m = 31$ and the y-intercept given as (0, 155), we have the following linear equation in slope-intercept form:

$$y = 31x + 155$$

b. The year 1998 is 4 years after the base year of 1994. Therefore, substitute $x = 4$ into the linear equation to approximate the number of SUVs sold in 1998.

$$y = 31x + 155$$

$$y = 31(4) + 155 \qquad \text{Substitute } x = 4.$$

$$= 279$$

The number of SUVs sold in the U.S. in 1998 was approximately 279 million.

Skill Practice

3. Soft drink sales at a concession stand at a softball stadium have increased linearly over the course of the summer softball season.
   a. Use the given data points to find a linear equation that relates the sales, y, to week number, x.
   b. Use the equation to predict the number of soft drinks sold in week 10.

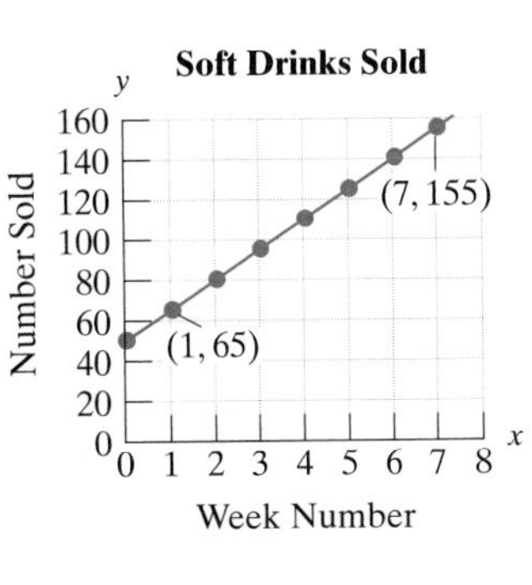

3. Writing a Linear Equation Given a Fixed Value and a Rate of Change

Another way to look at the equation $y = mx + b$ is to identify the term mx as the variable term and the term b as the constant term. The value of the term mx will change with the value of x (this is why the slope, m, is called a *rate of change*). However, the term b will remain constant regardless of the value of x. With these ideas in mind, we can write a linear equation if the rate of change and the constant are known.

Example 4 Finding a Linear Equation

A small word-processing business has a fixed monthly cost of \$6000 (this includes rent, utilities, and salaries of its employees). The business has a variable cost of \$5 per project (this includes mostly paper and computer supplies).

a. Write a linear equation to compute the total cost, y, for 1 month if x projects are completed.

b. Use the equation to compute the cost to run the business for 1 month if 800 projects are completed.

Skill Practice Answers

3a. $y = 15x + 50$
b. 200 soft drinks

Solution:

a. \$6000 is the constant (fixed) monthly cost. The variable cost is \$5 per project. If the slope, m, is replaced with 5, and b is replaced with 6000, then the equation $y = mx + b$ becomes

$y = 5x + 6000$ where y represents the total cost of completing x projects in 1 month

b. Because x represents the number of projects, substitute $x = 800$.

$$y = 5(800) + 6000$$
$$= 4000 + 6000$$
$$= 10{,}000$$

The total cost of completing 800 projects is \$10,000.

Skill Practice

4. The commission for buying stock at a discount brokerage firm is \$12.95 plus \$.02 per share of stock.

a. Write a linear equation to compute the commission, y, on a purchase of x shares of stock.

b. Find the commission that would be charged to purchase 250 shares of stock.

Calculator Connections

In Example 2, the equation $N = -0.28x + 14.8$ was used to represent the number of crimes, N (in millions), in the United States versus the number of years, x, since 1990. The equation is based on data between 1990 and 2006. This corresponds to x-values between 0 and 16. To graph the equation on a graphing calculator, the viewing window can be set for x between 0 and 16. The calculator interprets the values of N as the y-variable. We set the viewing window to accommodate N-values up to 20.

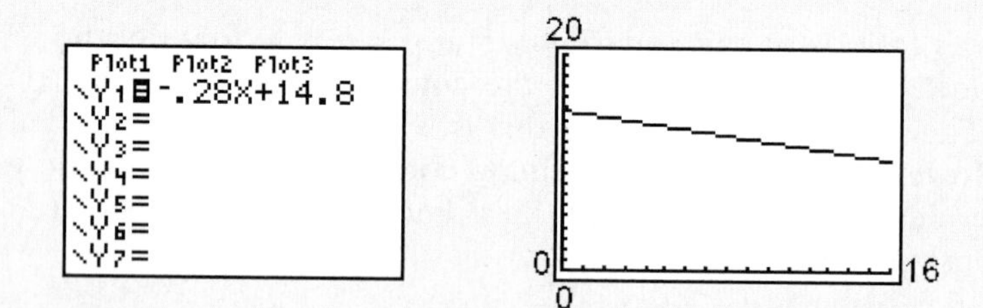

An *Eval* feature can be used to find solutions to an equation by evaluating the value of y for user-defined values of x. For example, entering a value of 10 for x results in 12 for y. This indicates that for $x = 10$ (the year 2000) there were approximately 12 million reported crimes in the United States.

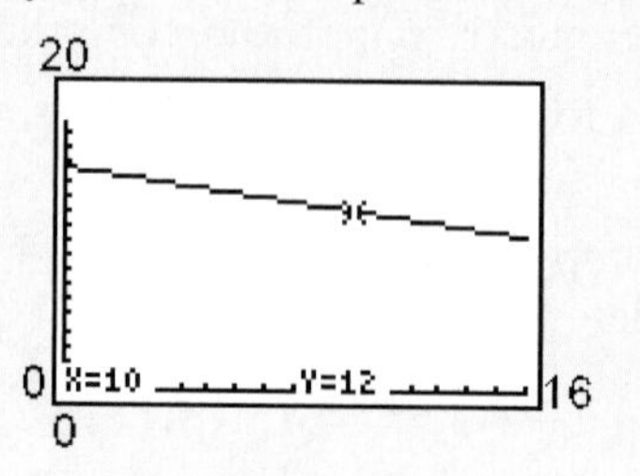

Skill Practice Answers

4a. $y = 0.02x + 12.95$

b. \$17.95

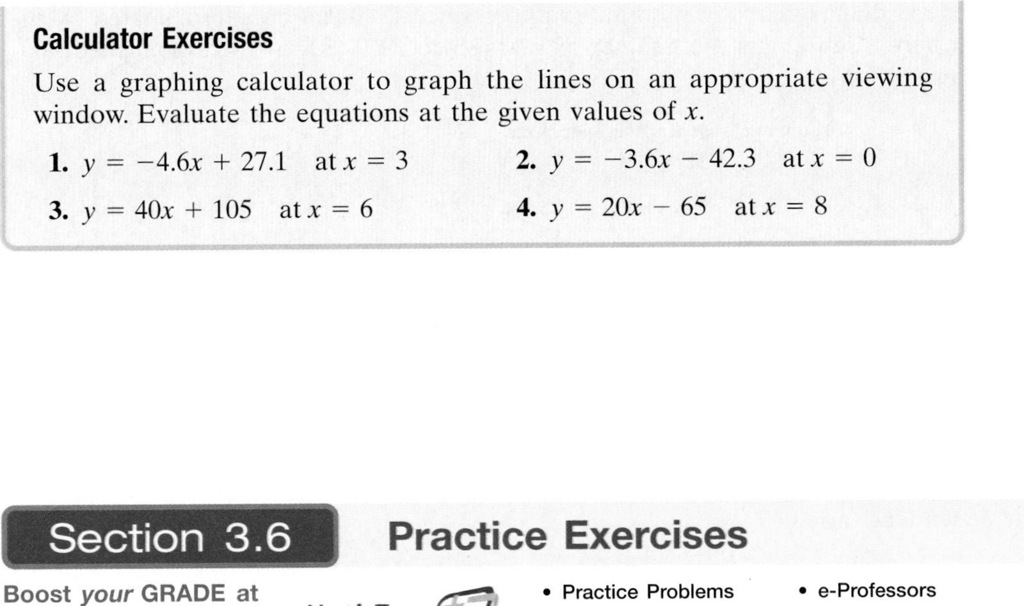

Calculator Exercises

Use a graphing calculator to graph the lines on an appropriate viewing window. Evaluate the equations at the given values of x.

1. $y = -4.6x + 27.1$ at $x = 3$

2. $y = -3.6x - 42.3$ at $x = 0$

3. $y = 40x + 105$ at $x = 6$

4. $y = 20x - 65$ at $x = 8$

Section 3.6 Practice Exercises

Study Skills Exercises

1. On test day, take a look at any formulas or important points that you had to memorize before you enter the classroom. Then when you sit down to take your test, write these formulas on the test or scrap paper. This is called a memory dump. Write down the formulas from Chapter 3.

2. Define the key terms:

a. dependent variable **b. independent variable**

Review Exercises

For Exercises 3–7, find the x- and y-intercepts of the lines, if possible.

3. $5x + 6y = 30$

4. $3x + 4y = 1$

5. $y = -2x - 4$

6. $y = 5x$

7. $y = -9$

Concept 1: Interpreting a Linear Equation in Two Variables

8. The electric bill charge for a certain utility company is \$0.095 per kilowatt-hour. The total cost, y, depends on the number of kilowatt-hours, x, according to the equation

$$y = 0.095x$$

a. Which variable is the independent variable?

b. Which variable is the dependent variable?

c. Determine the cost of using 1000 kilowatt-hours.

d. Determine the cost of using 2000 kilowatt-hours.

e. Determine the y-intercept. Interpret the meaning of the y-intercept in the context of this problem.

f. Determine the slope. Interpret the meaning of the slope in the context of this problem.

9. The minimum hourly wage, y (in dollars/hour), in the United States since 1960 can be approximated by the equation $y = 0.10x + 0.82$, where x represents the number of years since 1960 ($x = 0$ corresponds to 1960, $x = 1$ corresponds to 1961, and so on).

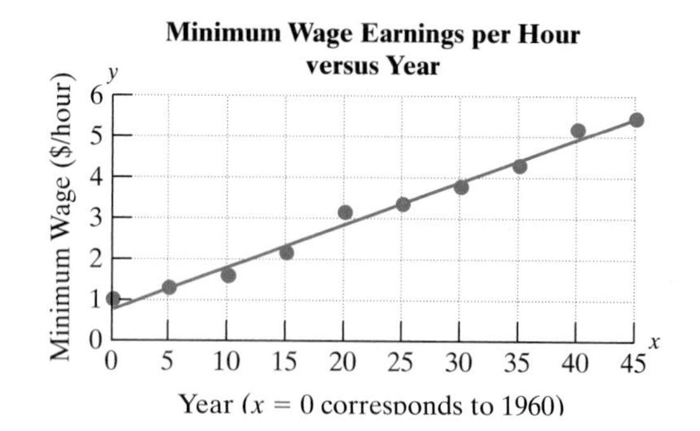

a. Which variable is the independent variable?

b. Which variable is the dependent variable?

c. Approximate the minimum wage in 1985.

d. Use the equation to predict the minimum wage in 2010.

e. Find the y-intercept. Interpret the meaning of the y-intercept in the context of this problem.

f. Find the slope. Interpret the meaning of the slope in the context of this problem.

10. The graph depicts the rise in the number of jail inmates in the United States since 1995. Two linear equations are given: one to describe the number of female inmates and one to describe the number of male inmates by year.

Let y represent the number of inmates (in thousands). Let x represent the number of years since 1995.

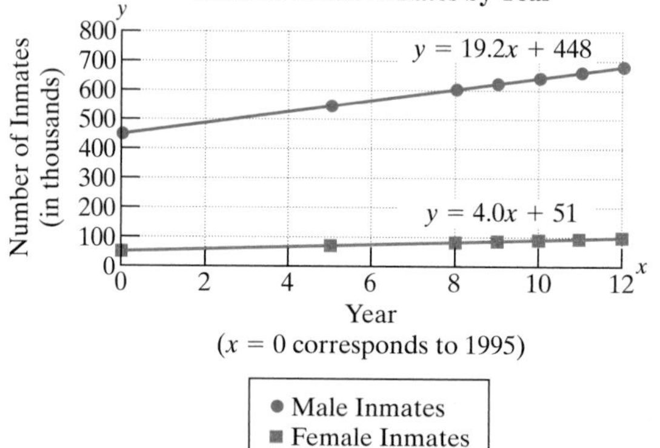

(Source: U.S. Bureau of Justice Statistics)

a. What is the slope of the line representing the number of female inmates? Interpret the meaning of the slope in the context of this problem.

b. What is the slope of the line representing the number of male inmates? Interpret the meaning of the slope in the context of this problem.

c. Which group, males or females, has the larger slope? What does this imply about the rise in the number of male and female prisoners?

11. The following graph shows the number of points scored by Shaquille O'Neal and by Allen Iverson according to the number of minutes played for several games. Two linear equations are given: one to describe the number of points scored by O'Neal and one to describe the number of points scored by Iverson. In both equations, y represents the number of points scored and x represents the number of minutes played.

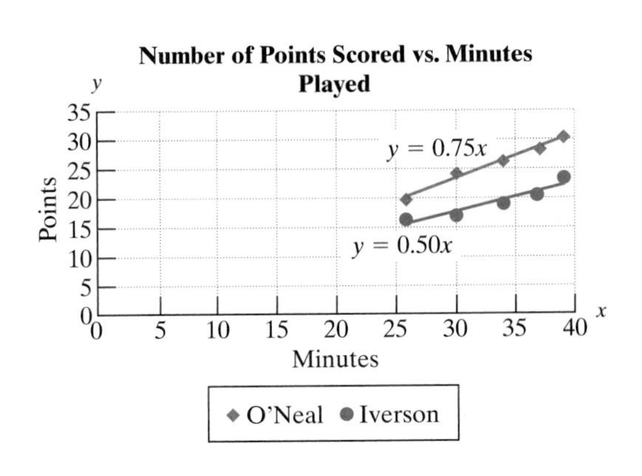

a. What is the slope of the line representing the number of points scored by O'Neal? Interpret the meaning of the slope in the context of this problem.

b. What is the slope of the line representing the number of points scored by Iverson? Interpret the meaning of the slope in the context of this problem.

c. According to these linear equations, approximately how many points would each player expect to score if he played for 36 minutes? Round to the nearest point.

12. The water bill charge for a certain utility company is \$4.20 per 1000 gallons used. The total cost, y, depends on the number of thousands of gallons of water used, x, according to the equation

$$y = 4.20x \quad x \geq 0$$

a. Determine the cost of using 3000 gallons. (*Hint:* $x = 3$.)

b. Determine the cost of using 5000 gallons.

c. Determine the y-intercept. Interpret the meaning of the y-intercept in the context of this problem.

d. Determine the slope. Interpret the meaning of the slope in the context of this problem.

13. The average daily temperature in January for cities along the eastern seaboard of the United States and Canada generally decreases for cities farther north. A city's latitude in the northern hemisphere is a measure of how far north it is on the globe.

The average temperature, y, (measured in degrees Fahrenheit) can be described by the equation

$y = -2.333x + 124.0$ where x is the latitude of the city.

City	x Latitude (°N)	y Average Daily Temperature (°F)
Jacksonville, FL	30.3	52.4
Miami, FL	25.8	67.2
Atlanta, GA	33.8	41.0
Baltimore, MD	39.3	31.8
Boston, MA	42.3	28.6
Atlantic City, NJ	39.4	30.9
New York, NY	40.7	31.5
Portland, ME	43.7	20.8
Charlotte, NC	35.2	39.3
Norfolk, VA	36.9	39.1

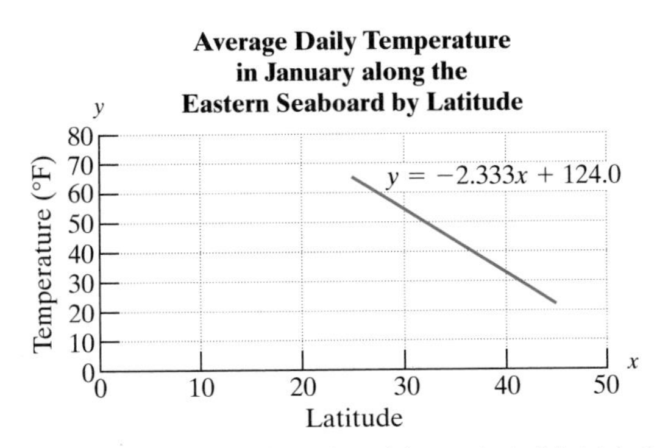

(Source: U.S. National Oceanic and Atmospheric Administration.)

a. Which variable is the dependent variable?

b. Which variable is the independent variable?

c. Use the equation to predict the average daily temperature in January for Philadelphia, Pennsylvania, whose latitude is 40.0°N. Round to one decimal place.

d. Use the equation to predict the average daily temperature in January for Edmundston, New Brunswick, Canada, whose latitude is 47.4°N. Round to one decimal place.

e. What is the slope of the line? Interpret the meaning of the slope in terms of the latitude and temperature.

f. From the equation, determine the value of the x-intercept. Round to one decimal place. Interpret the meaning of the x-intercept in terms of latitude and temperature.

Concept 2: Writing a Linear Equation Using Observed Data Points

14. The average length of stay for community hospitals has been decreasing in the United States from 1980 to 2005. Let x represent the number of years since 1980. Let y represent the average length of a hospital stay in days.

a. Find a linear equation that relates the average length of hospital stays versus the year.

b. Use the linear equation found in part (a) to predict the average length of stay in community hospitals in the year 2010. Round to the nearest day.

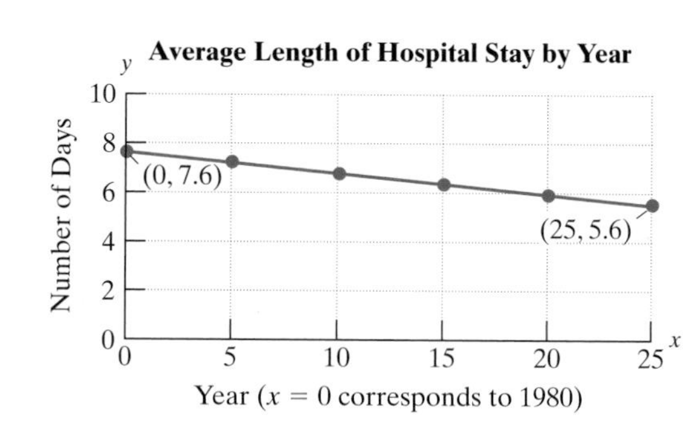

15. The graph shows the average height for boys based on age. Let x represent a boy's age, and let y represent his height (in inches).

a. Find a linear equation that represents the height of a boy versus his age.

b. Use the linear equation found in part (a) to predict the average height of a 5-year-old boy.

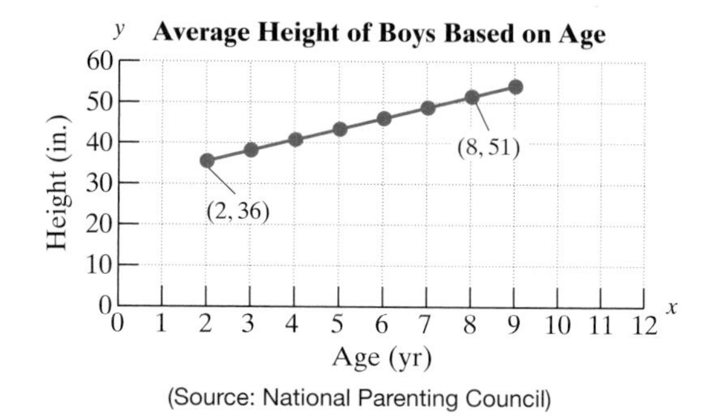

(Source: National Parenting Council)

16. The figure depicts a relationship between a person's height, y (in inches), and the length of the person's arm, x (measured in inches from shoulder to wrist).

a. Use the points (17, 57.75) and (24, 82.25) to find a linear equation relating height to arm length.

b. What is the slope of the line? Interpret the slope in the context of this problem.

c. Use the equation from part (a) to estimate the height of a person whose arm length is 21.5 in.

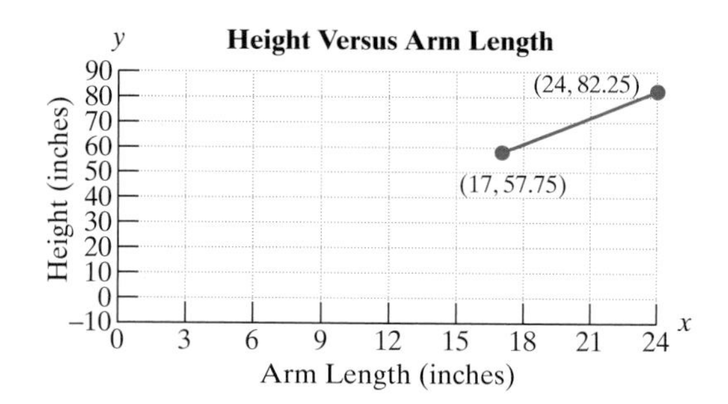

17. In a certain city, the time required to commute to work, y (in minutes), by car is related linearly to the distance traveled, x (in miles).

a. Use the points (5, 12) and (16, 34) to find a linear equation relating the commute time to work to the distance traveled.

b. What is the slope of the line? Interpret the slope in the context of this problem.

c. Use the equation from part (a) to find the time required to commute to work for a motorist who lives 18 miles away.

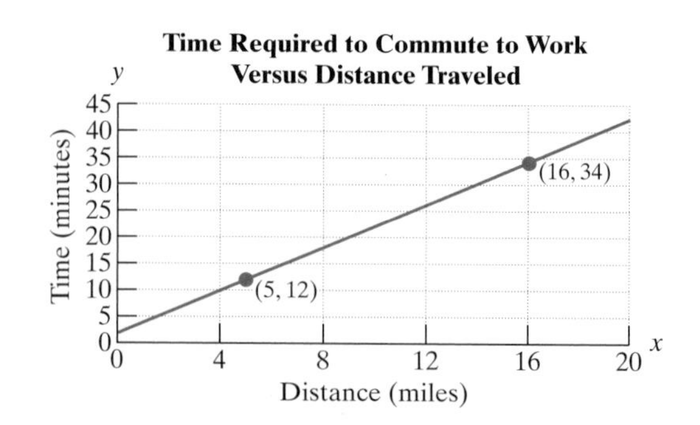

Concept 3: Writing a Linear Equation Given a Fixed Value and a Rate of Change

18. The cost to rent a car, y, for 1 day is \$20 plus \$0.25 per mile.

a. Write a linear equation to compute the cost, y, of driving a car x miles for 1 day.

b. Use the equation to compute the cost of driving 258 miles in the rental car for 1 day.

19. A phone bill is determined each month by a \$18.95 flat fee plus \$0.08 per minute of long distance.

a. Write a linear equation to compute the monthly cost of a phone bill, y, if x minutes of long distance are used.

b. Use the equation to compute the phone bill for a month in which 1 hr and 27 min of long distance was used.

20. A tennis instructor charges a student \$25 per lesson plus a one-time court fee of \$20.

a. Write a linear equation to compute the total cost, y, for x tennis lessons.

b. What is the total cost to a student who takes 20 tennis lessons?

21. The cost to rent a 10 ft by 10 ft storage space is \$90 per month plus a nonrefundable deposit of \$105.

a. Write a linear equation to compute the cost, y, of renting a 10 ft by 10 ft space for x months.

b. What is the cost of renting such a storage space for 1 year (12 months)?

22. A business has a fixed monthly cost of \$1200. In addition, the business has a variable cost of \$35 for each item produced.

a. Write a linear equation to compute the total cost, y, for 1 month if x items are produced.

b. Use the equation to compute the cost for 1 month if 100 items are produced.

23. An air-conditioning and heating company has a fixed monthly cost of \$5000. Furthermore, each service call costs the company \$25.

a. Write a linear equation to compute the total cost, y, for 1 month if x service calls are made.

b. Use the equation to compute the cost for 1 month if 150 service calls are made.

24. A bakery that specializes in bread rents a booth at a flea market. The daily cost to rent the booth is \$100. Each loaf of bread costs the bakery \$0.80 to produce.

a. Write a linear equation to compute the total cost, y, for 1 day if x loaves of bread are produced.

b. Use the equation to compute the cost for 1 day if 200 loaves of bread are produced.

25. A beverage company rents a booth at an art show to sell lemonade. The daily cost to rent a booth is \$35. Each lemonade costs \$0.50 to produce.

a. Write a linear equation to compute the total cost, y, for 1 day if x lemonades are produced.

b. Use the equation to compute the cost for 1 day if 350 lemonades are produced.

Chapter 3 SUMMARY

Section 3.1 Rectangular Coordinate System

Key Concepts

Graphical representation of numerical **data** is often helpful to study problems in real-world applications.

A **rectangular coordinate system** is made up of a horizontal line called the ***x*-axis** and a vertical line called the ***y*-axis**. The point where the lines meet is the **origin**. The four regions of the plane are called **quadrants**.

The point (x, y) is an **ordered pair**. The first element in the ordered pair is the point's horizontal position from the origin. The second element in the ordered pair is the point's vertical position from the origin.

Example

Example 1

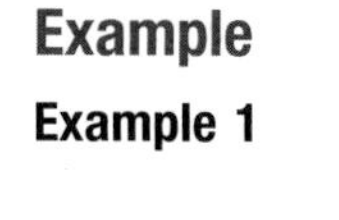

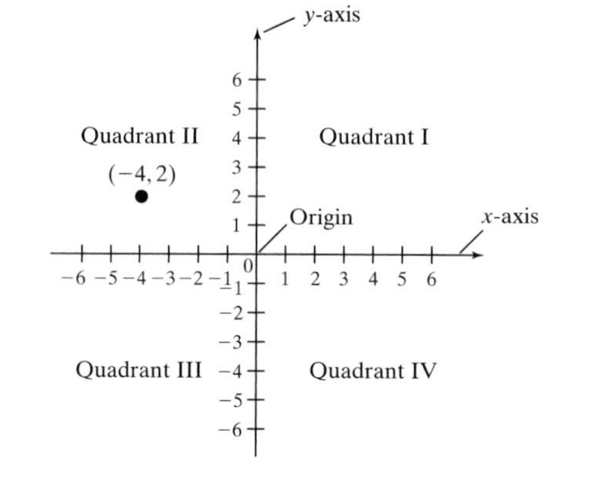

Section 3.2 Linear Equations in Two Variables

Key Concepts

An equation written in the form $Ax + By = C$ (where A and B are not both zero) is a **linear equation in two variables**.

A solution to a linear equation in x and y is an ordered pair (x, y) that makes the equation a true statement. The graph of the set of all solutions of a linear equation in two variables is a line in a rectangular coordinate system.

A linear equation can be graphed by finding at least two solutions and graphing the line through the points.

Examples

Example 1

Graph the equation $2x + y = 2$.

Select arbitrary values of x or y such as those shown in the table. Then complete the table to find the corresponding ordered pairs.

x	y	
0	2	⟶ $(0, 2)$
−1	4	⟶ $(-1, 4)$
1	0	⟶ $(1, 0)$

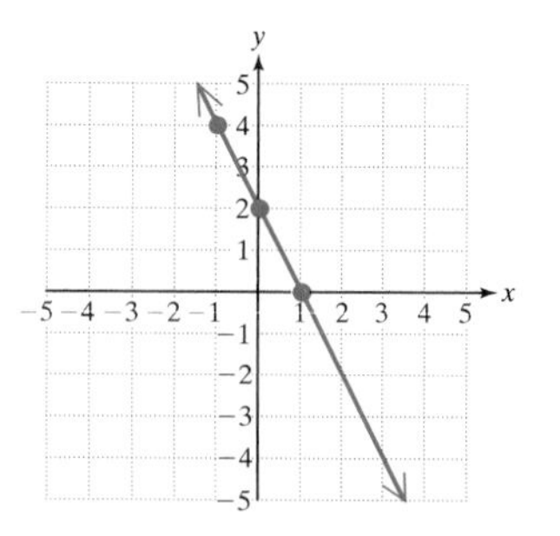

An ***x*-intercept** of a graph is a point $(a, 0)$ where the graph intersects the x-axis.

A ***y*-intercept** of a graph is a point $(0, b)$ where the graph intersects the y-axis.

Example 2

For the line $2x + y = 2$, the x-intercept is $(1, 0)$ and the y-intercept is $(0, 2)$.

A **vertical line** can be written in the form $x = k$.
A **horizontal line** can be written in the form $y = k$.

Example 3

$x = 3$ is a vertical line

$y = 3$ is a horizontal line

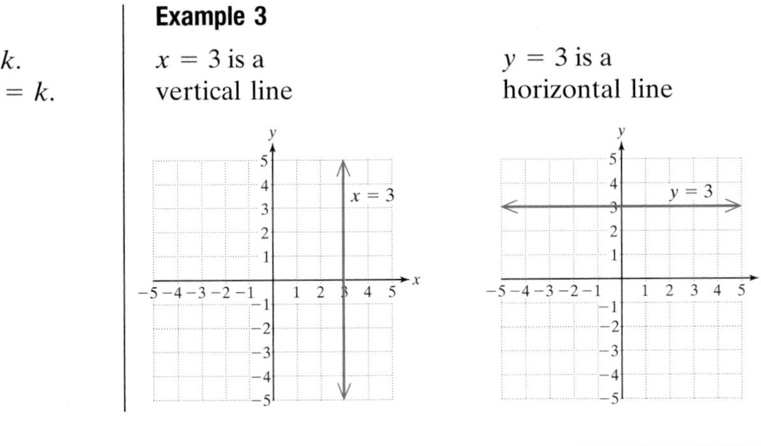

Section 3.3 Slope of a Line

Key Concepts

The **slope**, m, of a line between two points (x_1, y_1) and (x_2, y_2) is given by

$$m = \frac{y_2 - y_1}{x_2 - x_1} \quad \text{or} \quad \frac{\text{change in } y}{\text{change in } x}$$

The slope of a line may be positive, negative, zero, or undefined.

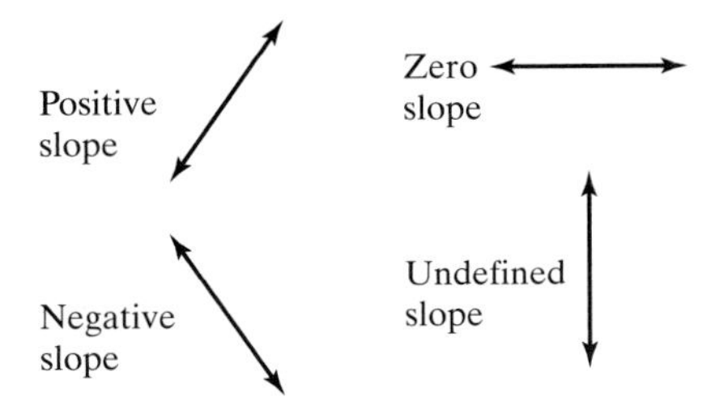

If m_1 and m_2 represent the slopes of two **parallel lines** (nonvertical), then $m_1 = m_2$.

If $m_1 \neq 0$ and $m_2 \neq 0$ represent the slopes of two nonvertical **perpendicular lines**, then

$$m_1 = -\frac{1}{m_2} \text{ or equivalently, } m_1 m_2 = -1.$$

Examples

Example 1

Find the slope of the line between $(1, -5)$ and $(-3, 7)$.

$$m = \frac{7 - (-5)}{-3 - 1} = \frac{12}{-4} = -3$$

Example 2

The slope of the line $y = -2$ is 0 because the line is horizontal.

Example 3

The slope of the line $x = 4$ is undefined because the line is vertical.

Example 4

The slopes of two distinct lines are given. Determine whether the lines are parallel, perpendicular, or neither.

a. $m_1 = -7$ and $m_2 = -7$ Parallel

b. $m_1 = -\frac{1}{5}$ and $m_2 = 5$ Perpendicular

c. $m_1 = -\frac{3}{2}$ and $m_2 = -\frac{2}{3}$ Neither

Section 3.4 Slope-Intercept Form of a Line

Key Concepts

The **slope-intercept form** of a line is

$$y = mx + b$$

where m is the slope of the line and $(0, b)$ is the y-intercept.

Slope-intercept form is used to identify the slope and y-intercept of a line when the equation is given.

Slope-intercept form can also be used to graph a line.

Examples

Example 1

Find the slope and y-intercept.

$7x - 2y = 4$

$$-2y = -7x + 4 \qquad \text{Solve for } y.$$

$$\frac{-2y}{-2} = \frac{-7x}{-2} + \frac{4}{-2}$$

$$y = \frac{7}{2}x - 2$$

The slope is $\frac{7}{2}$. The y-intercept is $(0, -2)$.

Example 2

Graph the line.

$$y = \frac{7}{2}x - 2$$

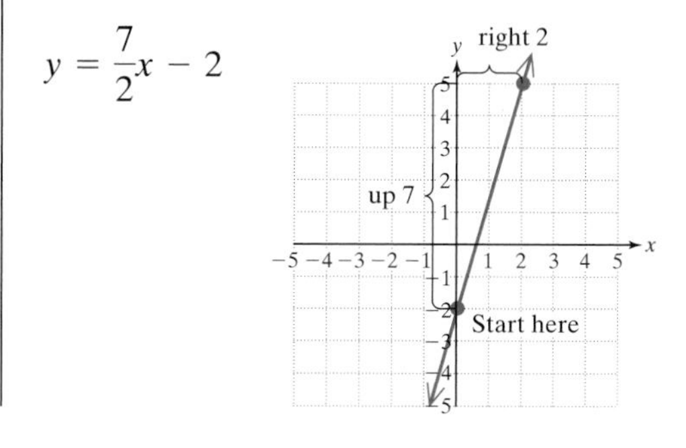

Section 3.5 Point-Slope Formula

Key Concepts

The **point-slope formula** is used primarily to construct an equation of a line given a point and the slope.

Equations of Lines—A Summary:

Standard form: $Ax + By = C$
Horizontal line: $y = k$
Vertical line: $x = k$
Slope-intercept form: $y = mx + b$
Point-slope formula: $y - y_1 = m(x - x_1)$

Examples

Example 1

Find an equation of the line passing through the point $(6, -4)$ and having a slope of $-\frac{1}{2}$.

Label the given information:
$m = -\frac{1}{2}$ and $(x_1, y_1) = (6, -4)$

$$y - y_1 = m(x - x_1)$$

$$y - (-4) = -\frac{1}{2}(x - 6)$$

$$y + 4 = -\frac{1}{2}x + 3$$

$$y = -\frac{1}{2}x - 1$$

Section 3.6 Applications of Linear Equations

Key Concepts

Linear equations can often be used to describe or model the relationship between variables in a real-world event. In such applications, the slope may be interpreted as a rate of change.

The two variables used in an application are called the independent and the dependent variables. The **independent variable** is used to make a prediction and is represented on the horizontal axis. The **dependent variable** is the response to the value of the independent variable and is represented on the vertical axis.

Examples

Example 1

The number of drug-related arrests for a small city has been growing approximately linearly since 1980.

Let y represent the number of drug arrests, and let x represent the number of years after 1980.

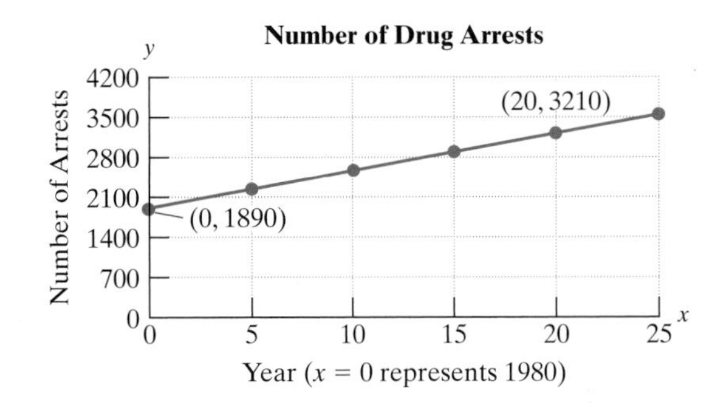

a. Use the ordered pairs (0, 1890) and (20, 3210) to find an equation of the line shown in the graph.

$$m = \frac{y_2 - y_1}{x_2 - x_1} = \frac{3210 - 1890}{20 - 0}$$

$$= \frac{1320}{20} = 66$$

The slope is 66, indicating that the number of drug arrests is increasing at a rate of 66 per year. $m = 66$, and the y-intercept is (0, 1890). Hence:

$y = mx + b \Rightarrow y = 66x + 1890$

b. Use the equation in part (a) to predict the number of drug-related arrests in the year 2010. (The year 2010 is 30 years after 1980. Hence, $x = 30$.)

$y = 66(30) + 1890$

$y = 3870$

The number of drug arrests is predicted to be 3870 by the year 2010.

Chapter 3 Review Exercises

Section 3.1

1. Graph the points on a rectangular coordinate system.

a. $\left(\frac{1}{2}, 5\right)$ **b.** $(-1, 4)$ **c.** $(2, -1)$

d. $(0, 3)$ **e.** $(0, 0)$ **f.** $\left(-\frac{8}{5}, 0\right)$

g. $(-2, -5)$ **h.** $(3, 1)$

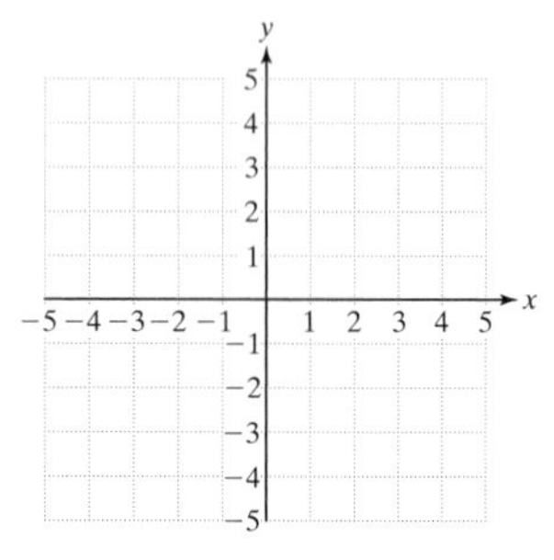

2. Estimate the coordinates of the points A, B, C, D, E, and F.

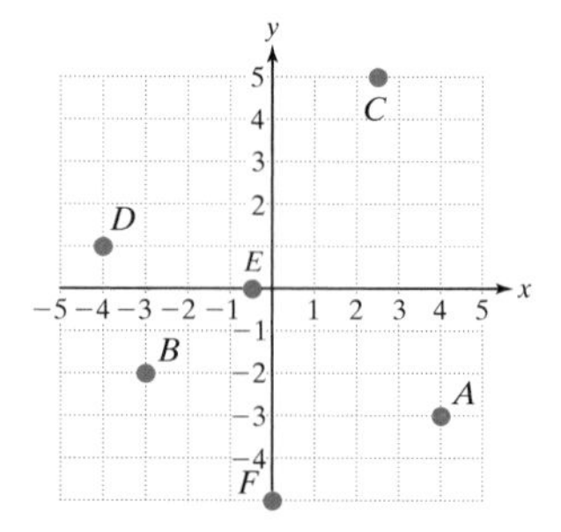

For Exercises 3–8, determine the quadrant in which the given point is found.

3. $(-2, -10)$ **4.** $(-4, 6)$

5. $(3, -5)$ **6.** $\left(\frac{1}{2}, \frac{7}{5}\right)$

7. $(\pi, -2.7)$ **8.** $(-1.2, -6.8)$

9. On which axis is the point $(2, 0)$ found?

10. On which axis is the point $(0, -3)$ found?

11. The price per share of a stock (in dollars) over a period of 5 days is shown in the graph.

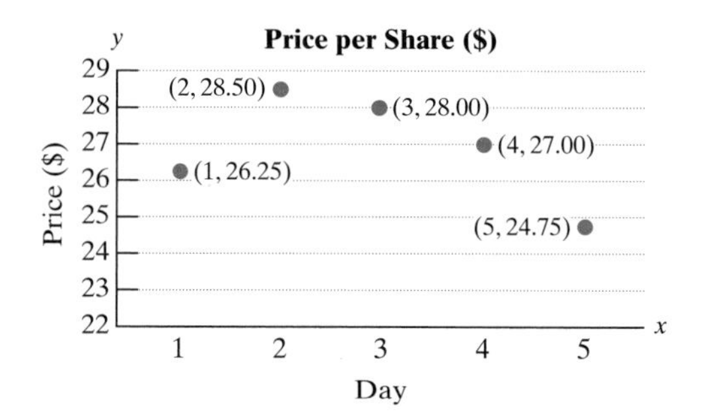

a. Interpret the meaning of the ordered pair $(1, 26.25)$.

b. On which day was the price the highest?

c. What was the increase in price between day 1 and day 2?

12. The number of space shuttle launches for selected years is given by the ordered pairs. Let x represent the number of years since 1995. Let y represent the number of launches.

$(1, 7)$ $(2, 8)$ $(3, 5)$ $(4, 3)$

$(5, 5)$ $(6, 6)$ $(7, 5)$ $(8, 1)$

a. Interpret the meaning of the ordered pair $(8, 1)$.

b. Plot the points on a rectangular coordinate system.

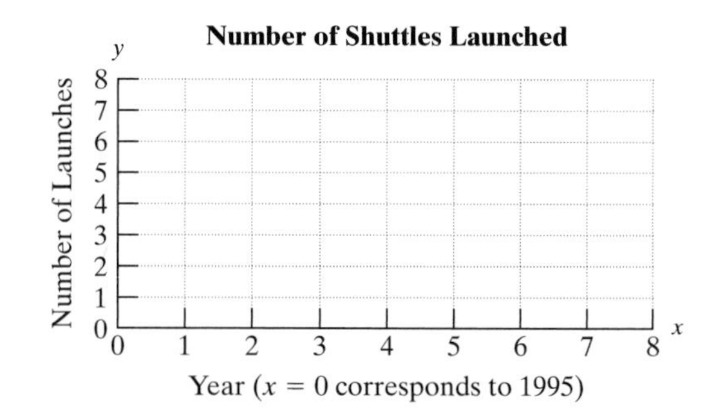

Section 3.2

For Exercises 13–16, determine if the given ordered pair is a solution to the equation.

13. $5x - 3y = 12$; $(0, 4)$

14. $2x - 4y = -6$; $(3, 0)$

15. $y = \frac{1}{3}x - 2$; $(9, 1)$

16. $y = -\frac{2}{5}x + 1$; $(-10, 5)$

For Exercises 17–20, complete the table and graph the corresponding ordered pairs. Graph the line through the points to represent all solutions to the equation.

17. $3x - y = 5$

x	y
2	
	4
1	

18. $\frac{1}{2}x + 3y = 6$

x	y
	2
−2	
	3

19. $y = \frac{2}{3}x - 1$

x	y
0	
3	
−6	

20. $y = -2x - 3$

x	y
0	
−3	
1	

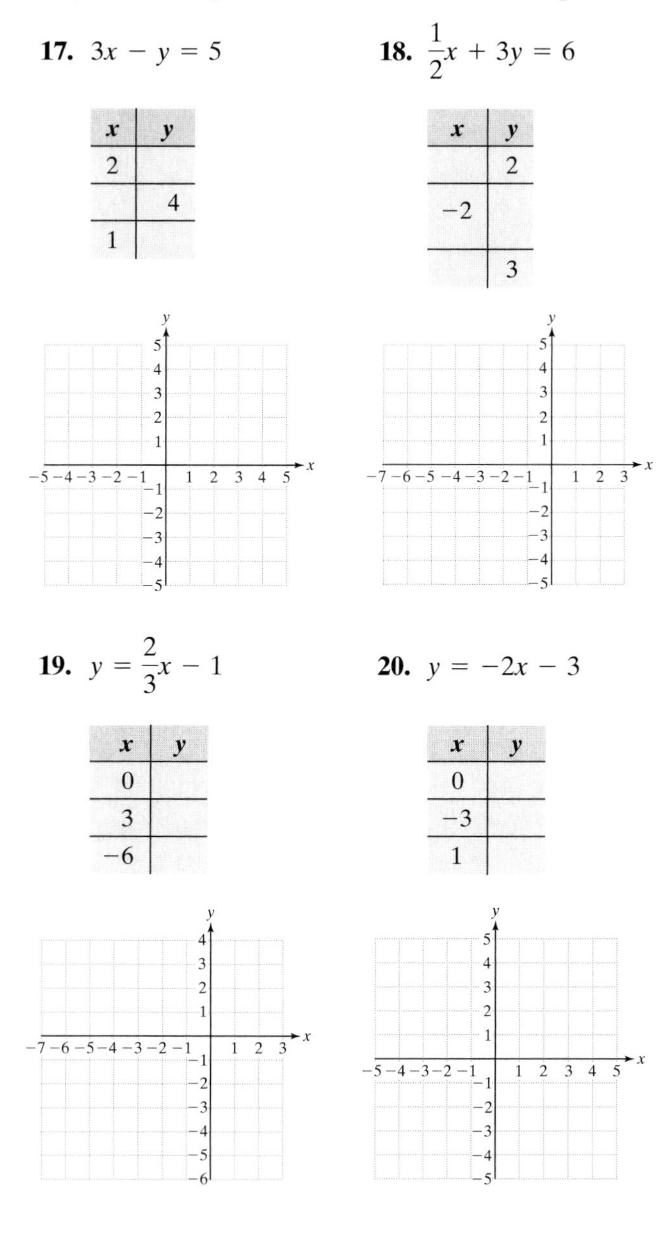

For Exercises 21–24, graph the line.

21. $x + 2y = 4$

22. $x - y = 5$

23. $y = 3x - 2$

24. $y = \frac{1}{4}x$

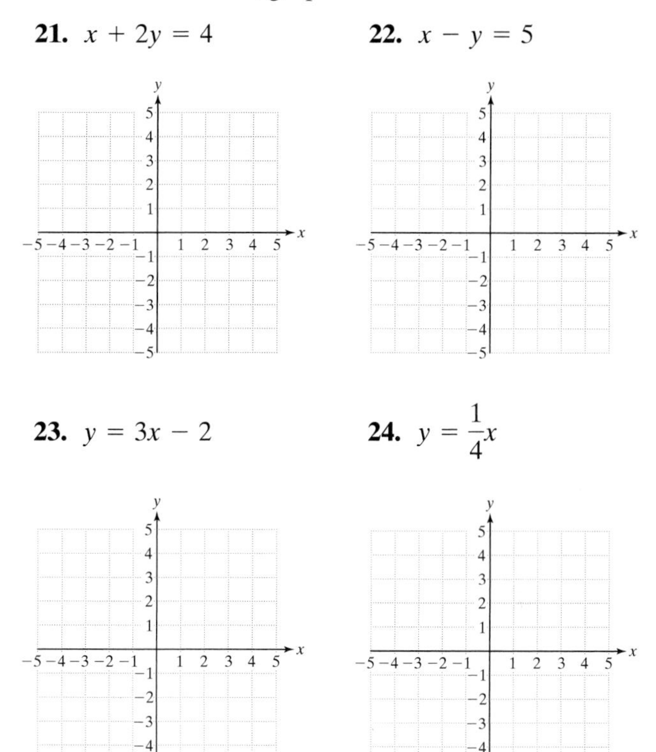

For Exercises 25–28, identify the line as horizontal or vertical. Then graph the line.

25. $3x - 2 = 10$

26. $2x + 1 = -2$

27. $6y + 1 = 13$

28. $5y - 1 = 14$

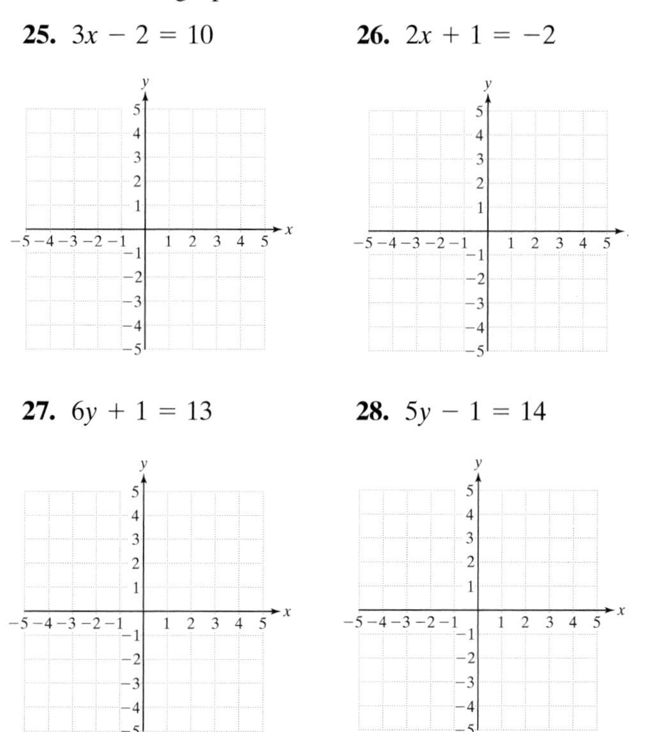

For Exercises 29–36, find the x- and y-intercepts if they exist.

29. $-4x + 8y = 12$

30. $2x + y = 6$

31. $y = 8x$

32. $5x - y = 0$

33. $6y = -24$

34. $2y - 3 = 1$

35. $2x + 5 = 0$

36. $-3x + 1 = 0$

Section 3.3

37. What is the slope of the ladder leaning up against the wall?

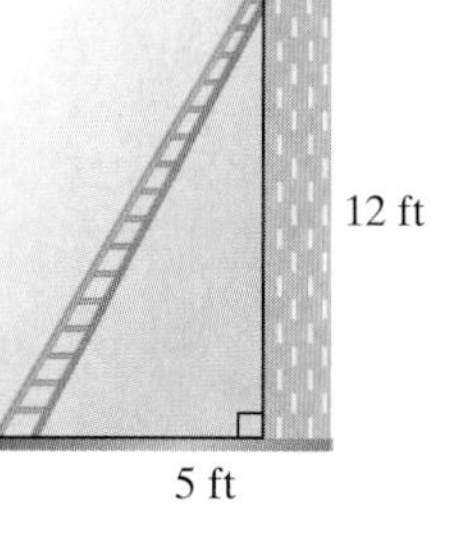

38. Point A is located 4 units down and 2 units to the right of point B. What is the slope of the line through points A and B?

39. Determine the slope of the line that passes through the points $(7, -9)$ and $(-5, -1)$.

40. Determine the slope of the line that has x- and y-intercepts of $(-1, 0)$ and $(0, 8)$.

41. Determine the slope of the line that passes through the points $(3, 0)$ and $(3, -7)$.

42. Determine the slope of the horizontal line $y = -1$.

43. A given line has a slope of -5.

a. What is the slope of a line parallel to the given line?

b. What is the slope of a line perpendicular to the given line?

44. A given line has a slope of 0.

a. What is the slope of a line parallel to the given line?

b. What is the slope of a line perpendicular to the given line?

For Exercises 45–48, find the slopes of the lines l_1 and l_2 from the two given points. Then determine whether l_1 and l_2 are parallel, perpendicular, or neither.

45. l_1: $(3, 7)$ and $(0, 5)$
l_2: $(6, 3)$ and $(-3, -3)$

46. l_1: $(-2, 1)$ and $(-1, 9)$
l_2: $(0, -6)$ and $(2, 10)$

47. l_1: $(0, \frac{5}{6})$ and $(2, 0)$
l_2: $(0, \frac{6}{5})$ and $(-\frac{1}{2}, 0)$

48. l_1: $(1, 1)$ and $(1, -8)$
l_2: $(4, -5)$ and $(7, -5)$

Section 3.4

For Exercises 49–54, write each equation in slope-intercept form. Identify the slope and the y-intercept.

49. $5x - 2y = 10$

50. $3x + 4y = 12$

51. $x - 3y = 0$

52. $5y - 8 = 4$

53. $2y = -5$

54. $y - x = 0$

For Exercises 55–59, determine whether the lines l_1 and l_2 are parallel, perpendicular, or neither.

55. l_1: $y = \frac{3}{5}x + 3$
l_2: $y = \frac{5}{3}x + 1$

56. l_1: $2x - 5y = 10$
l_2: $5x + 2y = 20$

57. l_1: $3x + 2y = 6$
l_2: $-6x - 4y = 4$

58. l_1: $y = \frac{1}{4}x - 3$
l_2: $-x + 4y = 8$

59. l_1: $2x = 4$
l_2: $y = 6$

60. Write an equation of the line whose slope is $-\frac{4}{3}$ and whose y-intercept is $(0, -1)$.

61. Write an equation of the line that passes through the origin and has a slope of 5.

Section 3.5

62. Write a linear equation in two variables in slope-intercept form. (Answers may vary.)

63. Write a linear equation in two variables in standard form. (Answers may vary.)

64. Write the slope formula to find the slope of the line between the points (x_1, y_1) and (x_2, y_2).

65. Write the point-slope formula.

66. Write an equation of a vertical line (answers may vary).

67. Write an equation of a horizontal line (answers may vary).

For Exercises 68–73, use the point-slope formula to write an equation of a line given the following information.

68. The slope is -6, and the line passes through the point $(-1, 8)$.

69. The slope is $\frac{2}{3}$, and the line passes through the point $(5, 5)$.

70. The line passes through the points $(0, -4)$ and $(8, -2)$.

71. The line passes through the points $(2, -5)$ and $(8, -5)$.

72. The line passes through the point $(5, 12)$ and is perpendicular to the line $y = -\frac{5}{6}x - 3$.

73. The line passes through the point $(-6, 7)$ and is parallel to the line $4x - y = 0$.

Section 3.6

74. The graph shows the average height for girls based on age (*Source:* National Parenting Council). Let x represent a girl's age, and let y represent her height (in inches).

$$y = 2.4x + 31$$

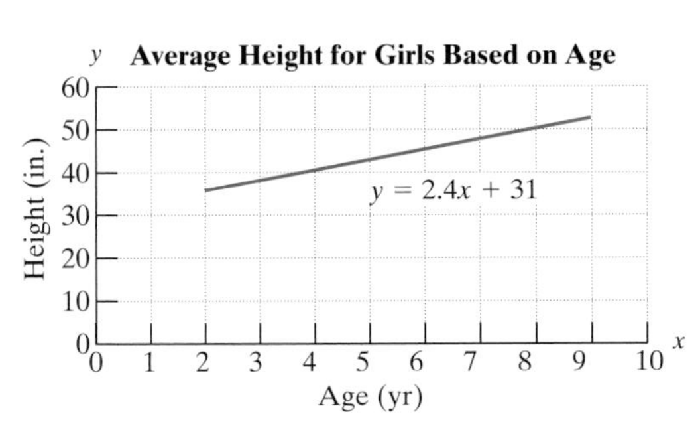

a. Which variable is the independent variable?

b. Which variable is the dependent variable?

c. Use the equation to estimate the average height of a 7-year-old girl.

d. What is the slope of the line? Interpret the meaning of the slope in the context of the problem.

75. The number of drug prescriptions has increased between 1995 and 2007 (see graph). Let x represent the number of years since 1995. Let y represent the number of prescriptions (in millions).

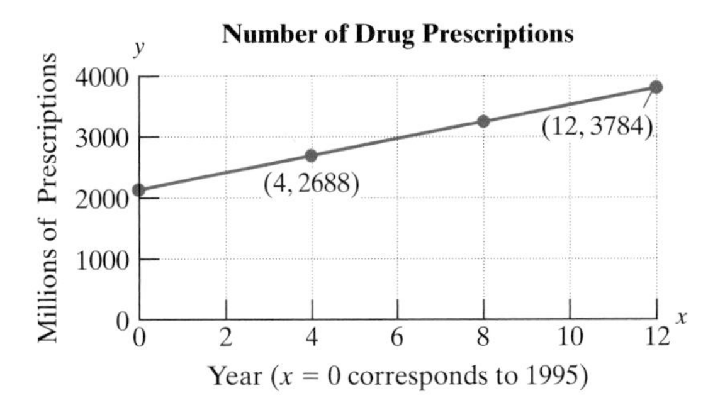

a. Using the ordered pairs (4, 2688) and (12, 3784) find the slope of the line.

b. Interpret the meaning of the slope in the context of this problem.

c. Find a linear equation that represents the number of prescriptions, y, versus the year, x.

d. Predict the number of prescriptions for the year 2008.

76. The amount of money that U.S. consumers had in outstanding automobile loans beginning at year 1992 is shown in the graph. Let x represent the number of years since 1992. Let y represent the total debt in auto loans (in billions of dollars).

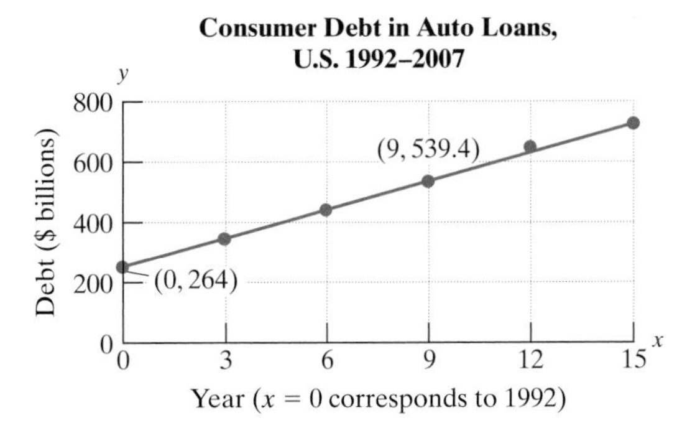

a. Find a linear equation that represents the total debt in auto loans, y, versus the year, x.

b. Use the linear equation found in part (a) to predict the debt in the year 2010.

77. A water purification company charges \$20 per month and a \$55 installation fee.

a. Write a linear equation to compute the total cost, y, of renting this system for x months.

b. Use the equation from part (a) to determine the total cost to rent the system for 9 months.

78. A small cleaning company has a fixed monthly cost of \$700 and a variable cost of \$8 per service call.

a. Write a linear equation to compute the total cost, y, of making x service calls in one month.

b. Use the equation from part (a) to determine the total cost of making 80 service calls.

Chapter 3 Test

1. In which quadrant is the given point found?

a. $\left(-\frac{7}{2}, 4\right)$ **b.** $(4.6, -2)$ **c.** $(-37, -45)$

2. What is the y-coordinate for a point on the x-axis?

3. What is the x-coordinate for a point on the y-axis?

4. The following table depicts a boy's height versus his age. Let x represent the boy's age and y represent his height.

Age (years), x	Height (inches), y
5	46
7	50
9	55
11	60

a. Write the data as ordered pairs and interpret the meaning of the first ordered pair.

b. Graph the ordered pairs on a rectangular coordinate system.

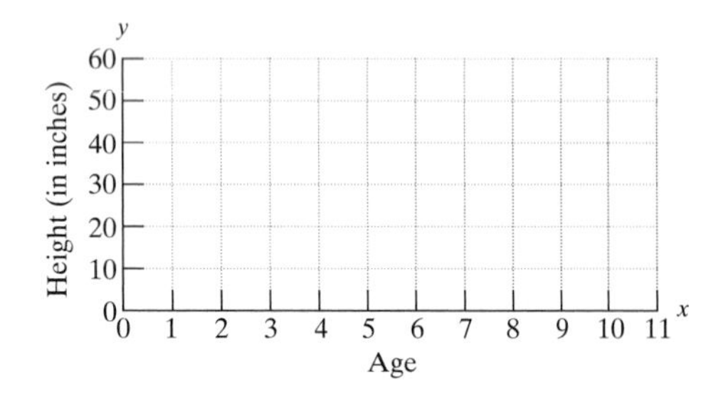

c. From the graph and table estimate the boy's height at age 10.

d. The data appear to follow an upward trend up to the boy's teenage years. Do you think this trend will continue? Would it be reasonable to use these data to predict the boy's height at age 25?

5. Determine whether the ordered pair is a solution to the equation $2x - y = 6$

a. $(0, 6)$ **b.** $(4, 2)$

c. $(3, 0)$ **d.** $\left(\frac{9}{2}, 3\right)$

6. Given the equation $y = \frac{1}{4}x - 2$, complete the table. Plot the ordered pairs and graph the line through the points to represent the set of all solutions to the equation.

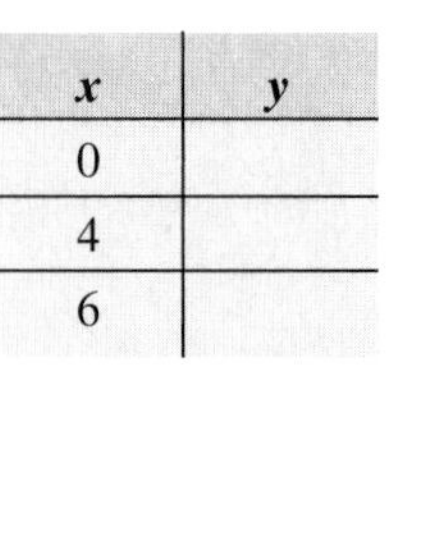

x	y
0	
4	
6	

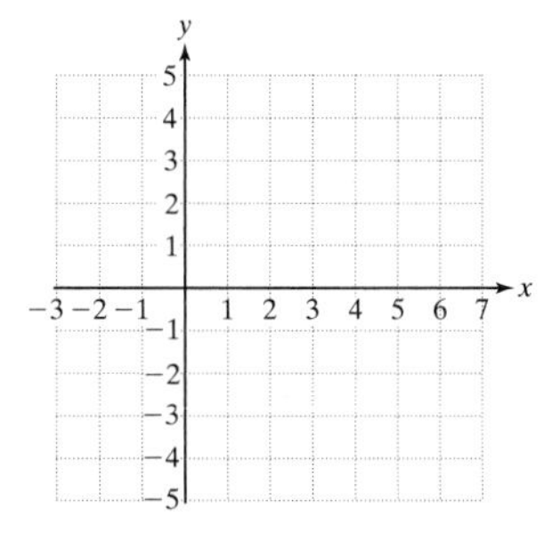

7. If x represents an adult's age, then the person's maximum recommended heart rate, y, during exercise is approximated by the equation

$$y = 220 - x \quad (x \geq 18)$$

a. Use the equation to find the maximum recommended heart rate for a person who is 18 years old.

b. Use the equation to complete the following ordered pairs: (20,), (30,), (40,), (50,), (60,).

For Exercises 8–9, determine whether the equation represents a horizontal or vertical line. Then graph the line.

8. $-6y = 18$ **9.** $5x + 1 = 8$

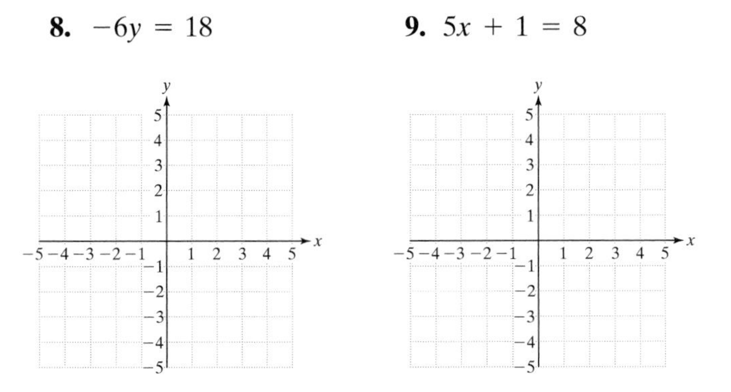

10. Find the x-intercept and the y-intercept of the line $-4x + 3y = 6$.

11. What is the slope of the hill?

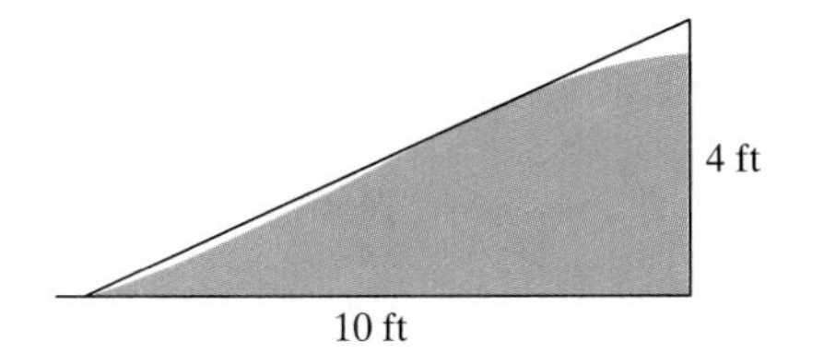

12. a. Find the slope of the line that passes through the points $(-2, 0)$ and $(-5, -1)$.

b. Find the slope of the line $4x - 3y = 9$.

13. a. What is the slope of a line parallel to the line $x + 4y = -16$?

b. What is the slope of a line perpendicular to the line $x + 4y = -16$?

14. a. What is the slope of the line $x = 5$?

b. What is the slope of the line $y = -3$?

For Exercises 15–18, find the x- and y-intercepts if they exist, and graph the lines.

15. $y = 8x + 2$

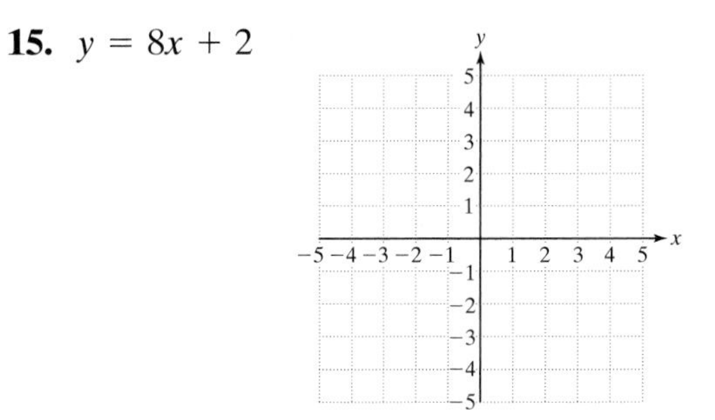

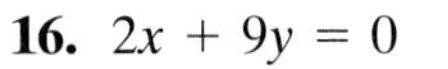

16. $2x + 9y = 0$

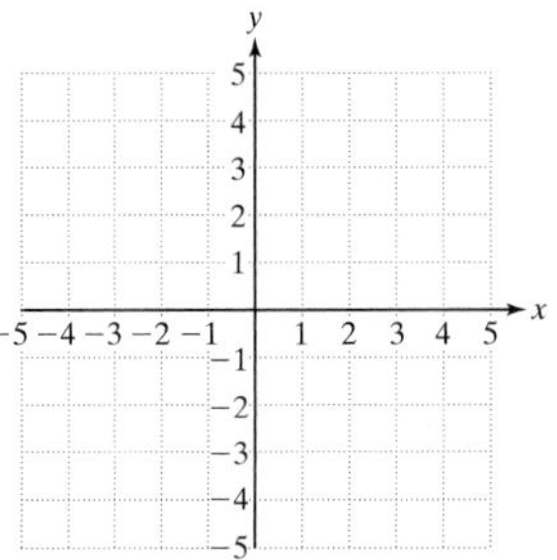

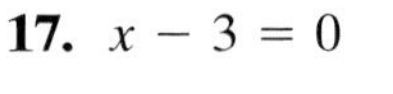

17. $x - 3 = 0$

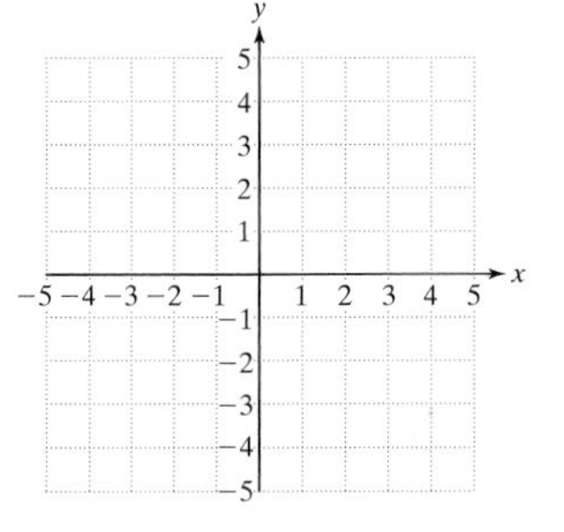

18. $-4y = 12$

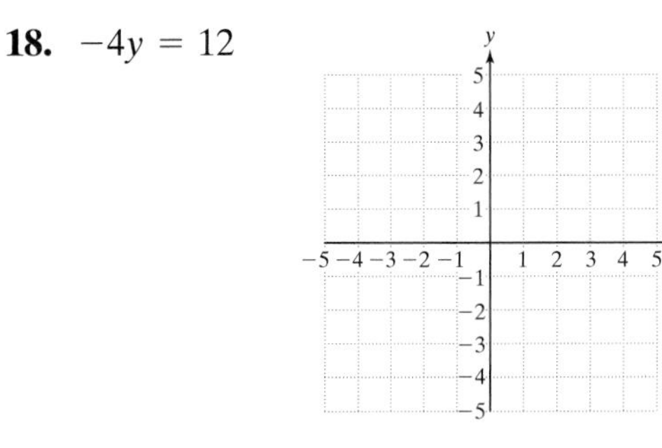

19. Determine whether the lines l_1 and l_2 are parallel, perpendicular, or neither.

l_1: $2y = 3x - 3$ $\quad$ l_2: $4x = -6y + 1$

20. Write an equation of the line that has y-intercept $(0, \frac{1}{2})$ and slope $\frac{1}{4}$.

21. Write an equation of the line that passes through the points (2, 8), and (4, 1).

22. Write an equation of the line that passes through the point $(2, -6)$ and is parallel to the x-axis.

23. Write an equation of the line that passes through the point (3, 0) and is parallel to the line $2x + 6y = -5$.

24. Write an equation of the line that passes through the point $(-3, -1)$ and is perpendicular to the line $x + 3y = 9$.

25. Hurricane Floyd dumped rain at an average rate of $\frac{3}{4}$ in./hr on Southport, NC. Further inland, in Lumberton, NC, the storm dropped $\frac{1}{2}$ in. of rain per hour. The following graph depicts the total amount of rainfall (in inches) versus the time (in hours) for both locations in North Carolina.

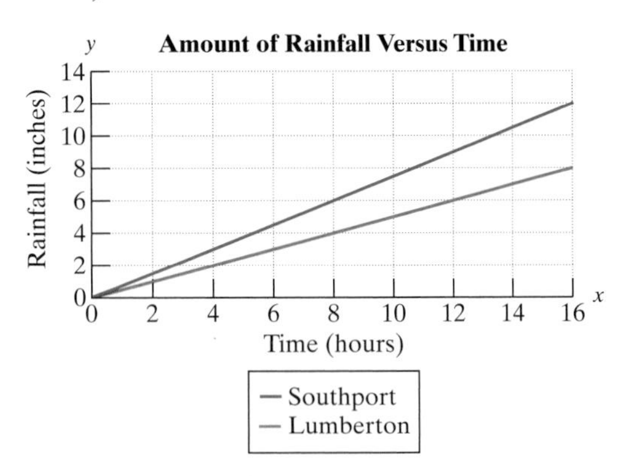

a. What is the slope of the line representing the rainfall for Southport?

b. What is the slope of the line representing the rainfall for Lumberton?

26. To attend a state fair, the cost is \$10 per person to cover exhibits and musical entertainment. There is an additional cost of \$1.50 per ride.

a. Write an equation that gives the total cost, y, of visiting the state fair and going on x rides.

b. Use the equation from part (a) to determine the cost of going to the state fair and going on 10 rides.

27. The number of medical doctors for selected years is shown in the graph. Let x represent the number of years since 1980, and let y represent the number of medical doctors (in thousands) in the United States.

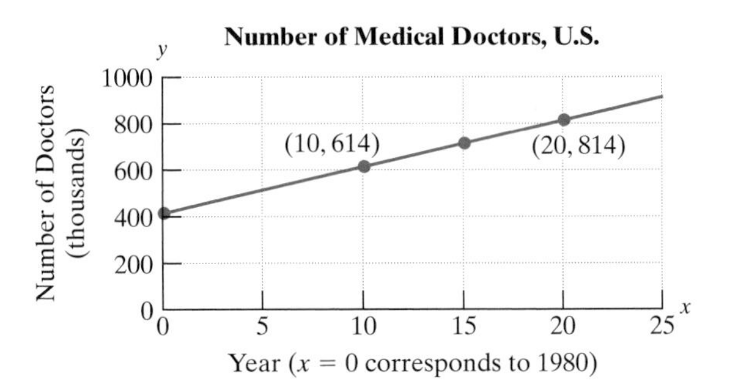

a. Find the slope of the line shown in the graph. Interpret the meaning of the slope in the context of this problem.

b. Find an equation of the line.

c. Use the equation from part (b) to predict the number of medical doctors in the United States for the year 2010.

Chapters 1–3 Cumulative Review Exercises

1. Identify the numbers as rational or irrational.

a. -3 **b.** $\frac{5}{4}$ **c.** $\sqrt{10}$ **d.** 0

2. Write the opposite and the absolute value for each number.

a. $-\frac{2}{3}$ **b.** 5.3

3. Simplify the expression using the order of operations: $32 \div 2 \cdot 4 + 5$

4. Add: $3 + (-8) + 2 + (-10)$

5. Subtract: $16 - 5 - (-7)$

For Exercises 6–7, translate the English phrase into an algebraic expression. Then evaluate the expression.

6. The quotient of $\frac{3}{4}$ and $-\frac{7}{8}$.

7. The product of -2.1 and -6.

8. Name the property that is illustrated by the following statement. $6 + (8 + 2) = (6 + 8) + 2$

For Exercises 9–12, solve the equation.

9. $6x - 10 = 14$

10. $3(m + 2) - 3 = 2m + 8$

11. $\frac{2}{3}y - \frac{1}{6} = y + \frac{4}{3}$

12. $1.7z + 2 = -2(0.3z + 1.3)$

13. The area of Texas is 267,277 mi^2. If this is 712 mi^2 less than 29 times the area of Maine, find the area of Maine.

14. For the formula $3a + b = c$, solve for a.

15. Graph the line $-6x + 2y = 0$.

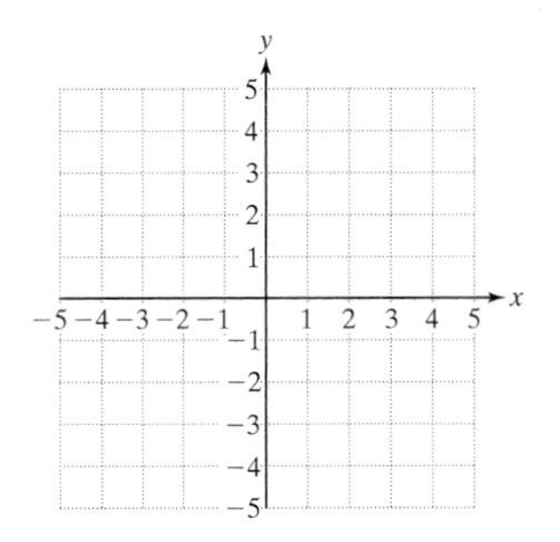

16. Find the x- and y-intercepts of $-2x + 4y = 4$.

17. Write the equation in slope-intercept form. Then identify the slope and the y-intercept.
$3x + 2y = -12$

18. Explain why the line $2x + 3 = 5$ has only one intercept.

19. Find an equation of a line passing through $(2, -5)$ with slope -3.

20. Find an equation of the line passing through $(0, 6)$ and $(-3, 4)$.

Systems of Linear Equations and Inequalities in Two Variables

4

This chapter is devoted to solving systems of linear equations and inequalities. Applications of systems of equations involve two or more variables subject to two or more constraints. For example:

At a movie theater, one group of students bought three drinks and two small popcorns for a total of \$13.00 (excluding tax). Another group bought five drinks and three popcorns for \$20.50. There are two unknown quantities in this scenario: the cost per drink, and the cost per popcorn.

Fill in the blanks below using trial and error to determine the cost per drink and the cost per popcorn. You have the correct answer if both equations are true.

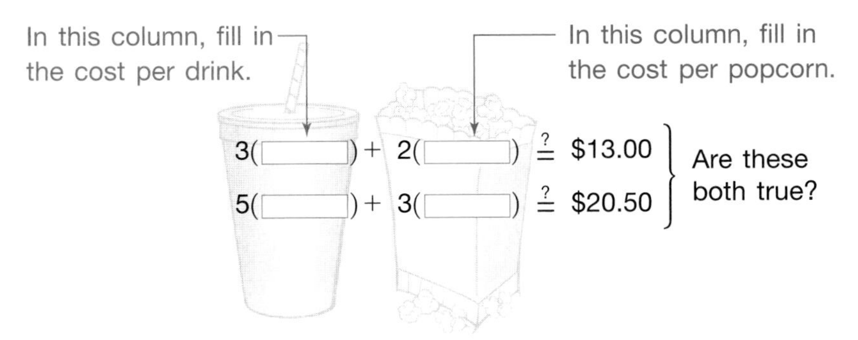

If you have trouble with this puzzle, don't fret. Later in the chapter, we'll use the power of algebra to set up and solve a system of equations that will take the guess work away!

Section 4.1 Solving Systems of Equations by the Graphing Method

Concepts

1. Determining Solutions to a System of Linear Equations
2. Dependent and Inconsistent Systems of Linear Equations
3. Solving Systems of Linear Equations by Graphing

1. Determining Solutions to a System of Linear Equations

Recall from Section 3.2 that a linear equation in two variables has an infinite number of solutions. The set of all solutions to a linear equation forms a line in a rectangular coordinate system. Two or more linear equations form a **system of linear equations**. For example, here are three systems of equations:

$$x - 3y = -5 \qquad y = \tfrac{1}{4}x - \tfrac{3}{4} \qquad 5a + b = 4$$

$$2x + 4y = 10 \qquad -2x + 8y = -6 \qquad -10a - 2b = 8$$

A **solution to a system of linear equations** is an ordered pair that is a solution to both individual equations in the system.

Example 1 Determining Solutions to a System of Linear Equations

Determine whether the ordered pairs are solutions to the system.

$$x + y = 4$$

$$-2x + y = -5$$

a. (3, 1) **b.** (0, 4)

Solution:

a. Substitute the ordered pair (3, 1) into both equations:

$$x + y = 4 \longrightarrow (3) + (1) \stackrel{?}{=} 4 \checkmark \quad \text{True}$$

$$-2x + y = -5 \longrightarrow -2(3) + (1) \stackrel{?}{=} -5 \checkmark \quad \text{True}$$

Because the ordered pair (3, 1) is a solution to both equations, it is a solution to the *system* of equations.

b. Substitute the ordered pair (0, 4) into both equations.

$$x + y = 4 \longrightarrow (0) + (4) \stackrel{?}{=} 4 \checkmark \quad \text{True}$$

$$-2x + y = -5 \longrightarrow -2(0) + (4) \stackrel{?}{=} -5 \quad \text{False}$$

Because the ordered pair (0, 4) is not a solution to the second equation, it is *not* a solution to the system of equations.

Avoiding Mistakes:

It is important to test an ordered pair in *both* equations to determine if the ordered pair is a solution.

Skill Practice

1. Determine whether the ordered pairs are solutions to the system.

$$5x - 2y = 24$$

$$2x + y = 6$$

a. (6, 3) **b.** (4, −2)

Skill Practice Answers

1a. No
b. Yes

A solution to a system of two linear equations may be interpreted graphically as a point of intersection between the two lines. Using slope-intercept form to graph the lines from Example 1, we have

$$x + y = 4 \longrightarrow y = -x + 4$$

$$-2x + y = -5 \longrightarrow y = 2x - 5$$

Notice that the lines intersect at (3, 1) (Figure 4-1).

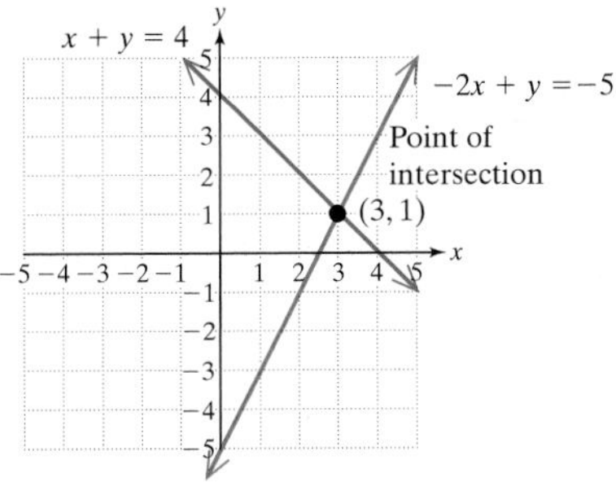

Figure 4-1

2. Dependent and Inconsistent Systems of Linear Equations

When two lines are drawn in a rectangular coordinate system, three geometric relationships are possible:

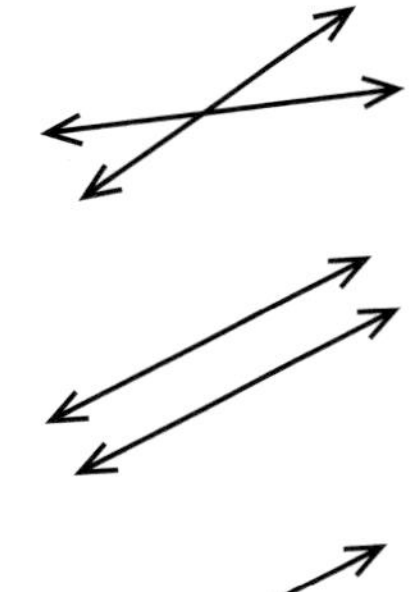

1. Two lines may intersect at *exactly one point.*
2. Two lines may intersect at *no point*. This occurs if the lines are parallel.
3. Two lines may intersect at *infinitely many points* along the line. This occurs if the equations represent the same line (the lines coincide).

If a system of linear equations has one or more solutions, the system is **consistent**. If a system of linear equations has no solution, it is **inconsistent**.

If two equations represent the same line, then all points along the line are solutions to the system of equations. In such a case, the system is characterized as a **dependent system**. An **independent system** is one in which the two equations represent different lines.

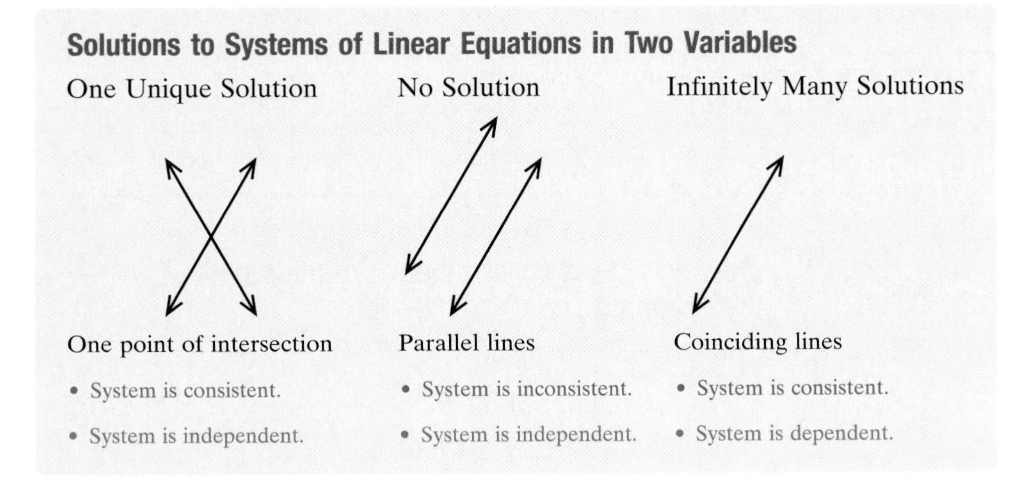

3. Solving Systems of Linear Equations by Graphing

One way to find a solution to a system of equations is to graph the equations and find the point (or points) of intersection. This is called the *graphing method* to solve a system of equations.

Example 2 **Solving a System of Linear Equations by Graphing**

Solve the system by the graphing method.

$$y = 2x$$
$$y = 2$$

Solution:

The equation $y = 2x$ is written in slope-intercept form as $y = 2x + 0$. The line passes through the origin, with a slope of 2.

The line $y = 2$ is a horizontal line and has a slope of 0.

Because the lines have different slopes, the lines must be different and nonparallel. From this, we know that the lines must intersect at exactly one point. Graph the lines to find the point of intersection (Figure 4-2).

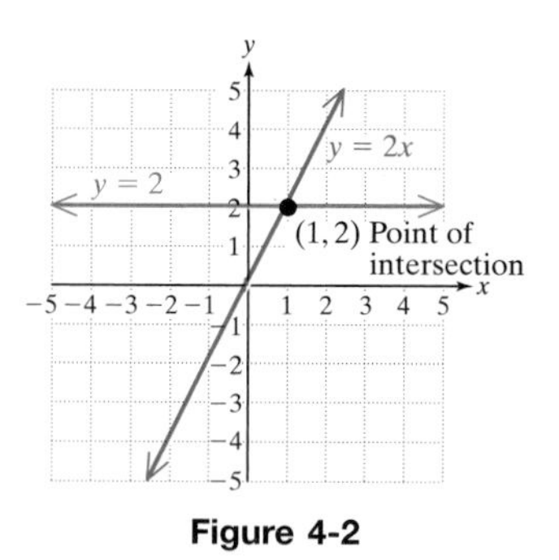

Figure 4-2

The point (1, 2) appears to be the point of intersection. This can be confirmed by substituting $x = 1$ and $y = 2$ into both original equations.

$$y = 2x \longrightarrow (2) \stackrel{?}{=} 2(1) \checkmark \quad \text{True}$$
$$y = 2 \longrightarrow (2) \stackrel{?}{=} 2 \checkmark \quad \text{True}$$

The solution is (1, 2).

Skill Practice Solve the system by graphing.

2. $y = -3x$
$x = -1$

Skill Practice Answers

2. $(-1, 3)$; $x = -1$; $y = -3x$

Example 3 **Solving a System of Linear Equations by Graphing**

Solve the system by the graphing method.

$$x - 2y = -2$$
$$-3x + 2y = 6$$

Solution:

To graph each equation, write the equation in slope-intercept form: $y = mx + b$.

Equation 1	**Equation 2**
$x - 2y = -2$	$-3x + 2y = 6$
$-2y = -x - 2$	$2y = 3x + 6$
$\frac{-2y}{-2} = \frac{-x}{-2} - \frac{2}{-2}$	$\frac{2y}{2} = \frac{3x}{2} + \frac{6}{2}$
$y = \frac{1}{2}x + 1$	$y = \frac{3}{2}x + 3$

From their slope-intercept forms, we see that the lines have different slopes, indicating that the lines are different and nonparallel. Therefore, the lines must intersect at exactly one point. Graph the lines to find that point (Figure 4-3).

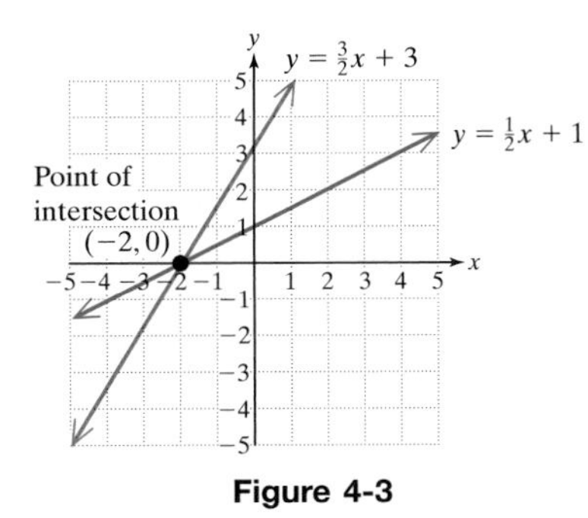

Figure 4-3

The point $(-2, 0)$ appears to be the point of intersection. This can be confirmed by substituting $x = -2$ and $y = 0$ into both equations.

$$x - 2y = -2 \longrightarrow (-2) - 2(0) \stackrel{?}{=} -2 \checkmark \quad \text{True}$$

$$-3x + 2y = 6 \longrightarrow -3(-2) + 2(0) \stackrel{?}{=} 6 \checkmark \quad \text{True}$$

The solution is $(-2, 0)$.

Skill Practice Solve the system by graphing.

3. $y = 2x - 3$

$6x + 2y = 4$

TIP: In Examples 2 and 3, the lines could also have been graphed by using the x- and y-intercepts or by using a table of points. However, the advantage of writing the equations in slope-intercept form is that we can compare the slopes and y-intercepts of each line.

1. If the slopes differ, the lines are different and nonparallel and must cross in exactly one point.
2. If the slopes are the same and the y-intercepts are different, the lines are parallel and will not intersect.
3. If the slopes are the same and the y-intercepts are the same, the two equations represent the same line.

Skill Practice Answers

3.

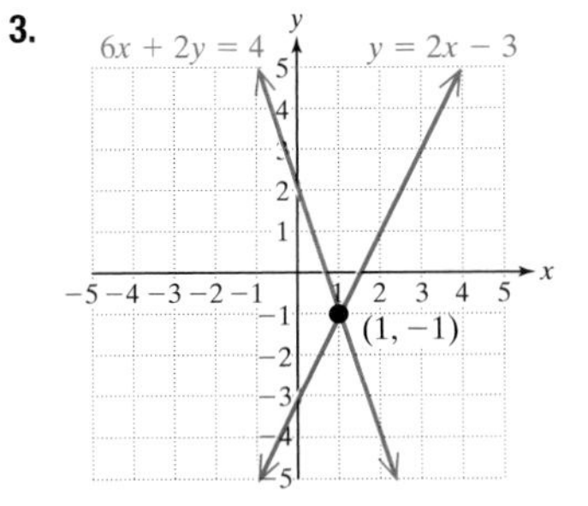

Example 4 Graphing an Inconsistent System

Solve the system by graphing.

$$-x + 3y = -6$$
$$6y = 2x + 6$$

Solution:

To graph the lines, write each equation in slope-intercept form.

Equation 1

$$-x + 3y = -6$$
$$3y = x - 6$$
$$\frac{3y}{3} = \frac{x}{3} - \frac{6}{3}$$
$$y = \frac{1}{3}x - 2$$

Equation 2

$$6y = 2x + 6$$
$$\frac{6y}{6} = \frac{2x}{6} + \frac{6}{6}$$
$$y = \frac{1}{3}x + 1$$

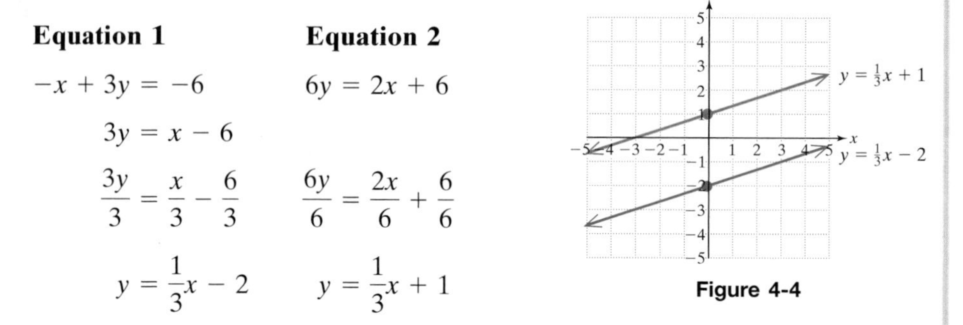

Figure 4-4

Because the lines have the same slope but different y-intercepts, they are parallel (Figure 4-4). Two parallel lines do not intersect, which implies that the system has no solution. The system is inconsistent.

Skill Practice Solve the system by graphing.

4. $4x + y = 8$

$y = -4x + 3$

Example 5 Graphing a Dependent System

Solve the system by graphing.

$$x + 4y = 8$$
$$y = -\frac{1}{4}x + 2$$

Solution:

Write the first equation in slope-intercept form. The second equation is already in slope-intercept form.

Equation 1

$$x + 4y = 8$$
$$4y = -x + 8$$
$$\frac{4y}{4} = \frac{-x}{4} + \frac{8}{4}$$
$$y = -\frac{1}{4}x + 2$$

Equation 2

$$y = -\frac{1}{4}x + 2$$

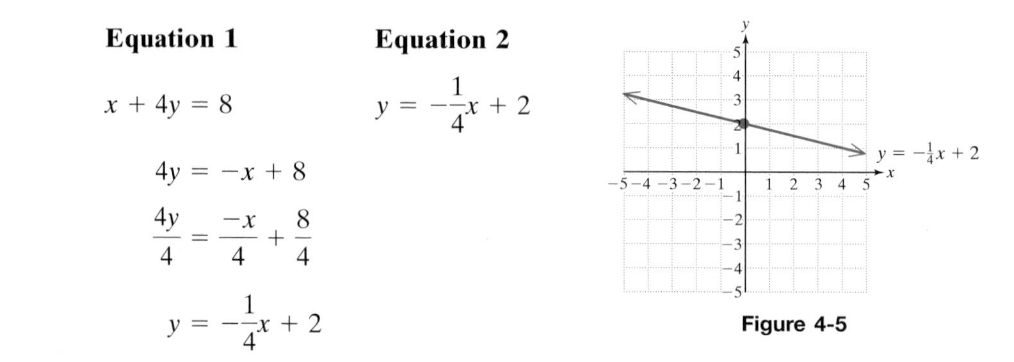

Figure 4-5

Skill Practice Answers

4.

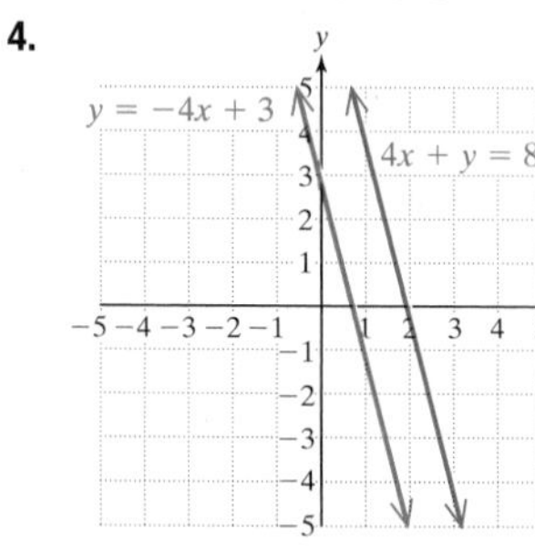

No solution, the lines are parallel. The system is inconsistent.

Notice that the slope-intercept forms of the two lines are identical. Therefore, the equations represent the same line (Figure 4-5). The system is dependent, and the solution to the system of equations is the set of all points on the line.

Because the ordered pairs in the solution set cannot all be listed, we can write the solution in set-builder notation. Furthermore, the equations $x + 4y = 8$ and $y = -\frac{1}{4}x + 2$ represent the same line. Therefore, the solution set may be written as $\{(x, y) \mid x + 4y = 8\}$ or as $\{(x, y) \mid y = -\frac{1}{4}x + 2\}$.

Skill Practice Solve the system by graphing.

5. $x - 3y = 4$

$y = \frac{1}{3}x - \frac{4}{3}$

Calculator Connections

The solution to a system of equations can be found by using either a *Trace* feature or an *Intersect* feature on a graphing calculator to find the point of intersection.

For example, consider the system:

$$-2x + y = 6$$
$$5x + y = -1$$

First graph the equations together on the same viewing window. Recall that to enter the equations into the calculator, the equations must be written with the y-variable isolated.

$$-2x + y = 6 \xrightarrow{\text{isolate } y} y = 2x + 6$$
$$5x + y = -1 \longrightarrow y = -5x - 1$$

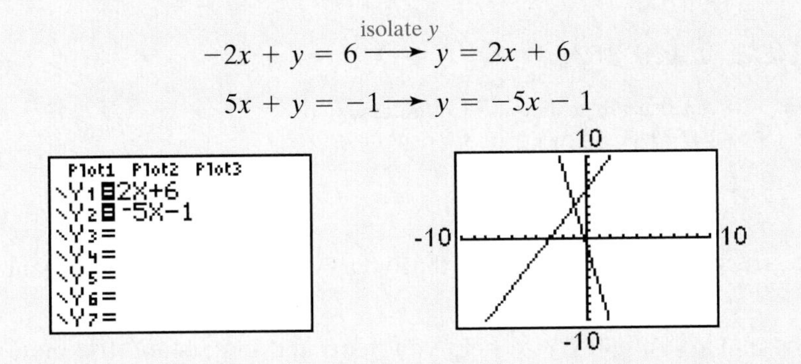

By inspection of the graph, it appears that the solution is $(-1, 4)$. The *Trace* option on the calculator may come close to $(-1, 4)$ but may not show the exact solution (Figure 4-6). However, an *Intersect* feature on a graphing calculator may provide the exact solution (Figure 4-7). See your user's manual for further details.

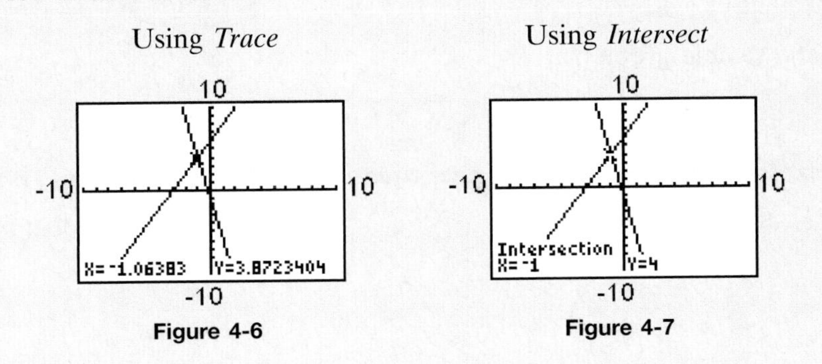

Figure 4-6 **Figure 4-7**

Skill Practice Answers

5.

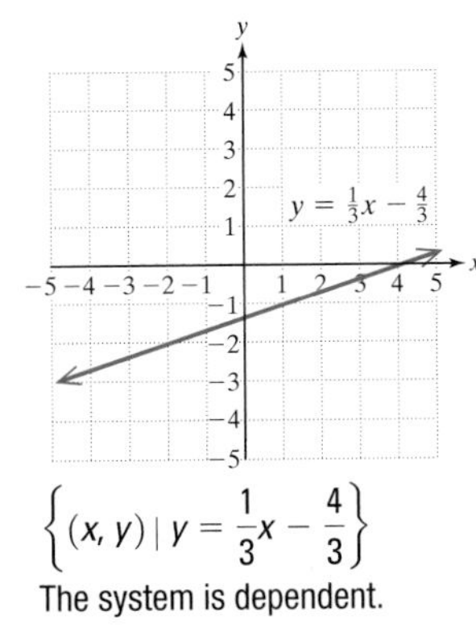

$\left\{(x, y) \mid y = \frac{1}{3}x - \frac{4}{3}\right\}$

The system is dependent.

Calculator Exercises

Use a graphing calculator to graph each linear equation on the same viewing window. Use a *Trace* or *Intersect* feature to find the point(s) of intersection.

1. $y = 2x - 3$
$y = -4x + 9$

2. $y = -\frac{1}{2}x + 2$
$y = \frac{1}{3}x - 3$

3. $x + y = 4$ (Example 1)
$-2x + y = -5$

4. $x - 2y = -2$ (Example 3)
$-3x + 2y = 6$

5. $-x + 3y = -6$ (Example 4)
$6y = 2x + 6$

6. $x + 4y = 8$ (Example 5)
$y = -\frac{1}{4}x + 2$

Section 4.1 Practice Exercises

Study Skills Exercises

1. Figure out your grade at this point. Are you earning the grade that you want? If not, maybe organizing a study group would help.

In a study group, check the activities that you might try to help you learn and understand the material.

_____ Quiz each other by asking each other questions.

_____ Practice teaching each other.

_____ Share and compare class notes.

_____ Support and encourage each other.

_____ Work together on exercises and sample problems.

2. Define the key terms:

a. system of linear equations **b. solution to a system of linear equations** **c. consistent system**

d. inconsistent system **e. dependent system** **f. independent system**

Concept 1: Determining Solutions to a System of Linear Equations

For Exercises 3–10, determine if the given point is a solution to the system.

3. $3x - y = 7$ $(2, -1)$
$x - 2y = 4$

4. $x - y = 3$ $(4, 1)$
$x + y = 5$

5. $4y = -3x + 12$ $(0, 4)$
$y = \frac{2}{3}x - 4$

6. $y = -\frac{1}{3}x + 2$ $(9, -1)$
$x = 2y + 6$

7. $3x - 6y = 9$ $\left(4, \frac{1}{2}\right)$
$x - 2y = 3$

8. $x - y = 4$ $(6, 2)$
$3x - 3y = 12$

9. $\frac{1}{3}x = \frac{2}{5}y - \frac{4}{5}$ $(0, 2)$
$\frac{3}{4}x + \frac{1}{2}y = 2$

10. $\frac{1}{4}x + \frac{1}{2}y = \frac{3}{2}$ $(4, 1)$
$y = \frac{3}{2}x - 6$

For Exercises 11–14, match the graph of the system of equations with the appropriate description of the solution.

11.

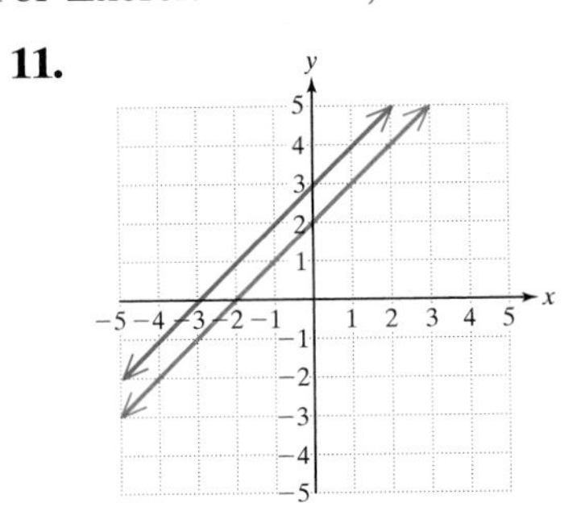

a. The solution is (1, 3).

b. No solution.

c. There are infinitely many solutions.

d. The solution is (0, 0).

12.

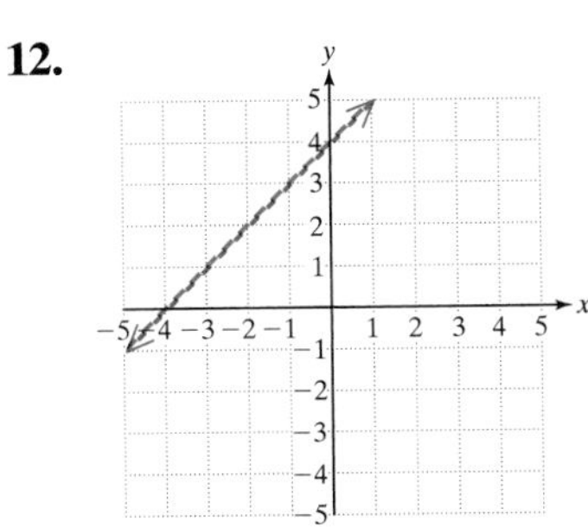

13.

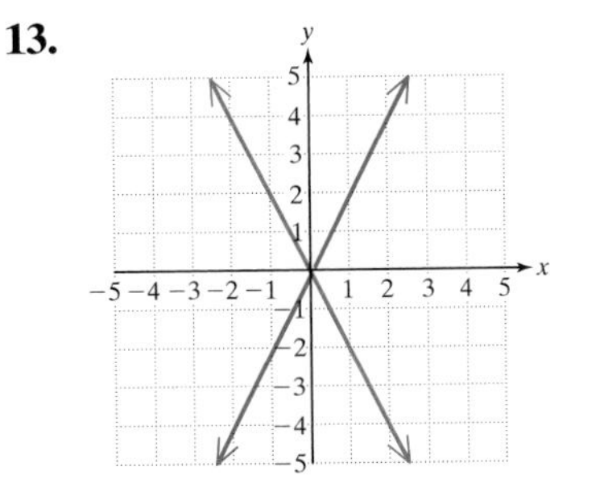

14.

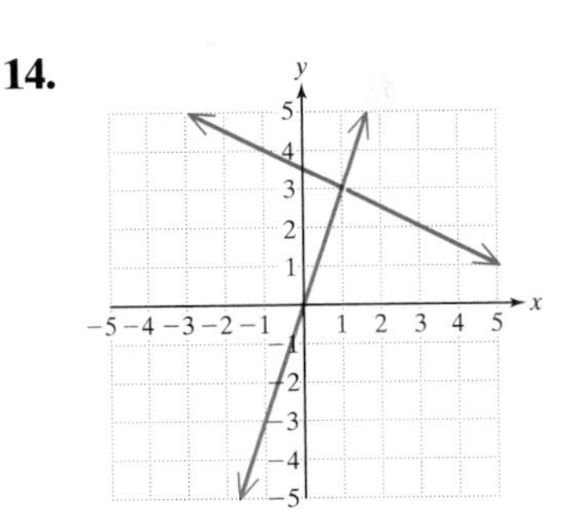

Concept 2: Dependent and Inconsistent Systems of Linear Equations

15. Graph each system of equations.

a. $y = 2x - 3$
$y = 2x + 5$

b. $y = 2x + 1$
$y = 4x - 5$

c. $y = 3x - 5$
$y = 3x - 5$

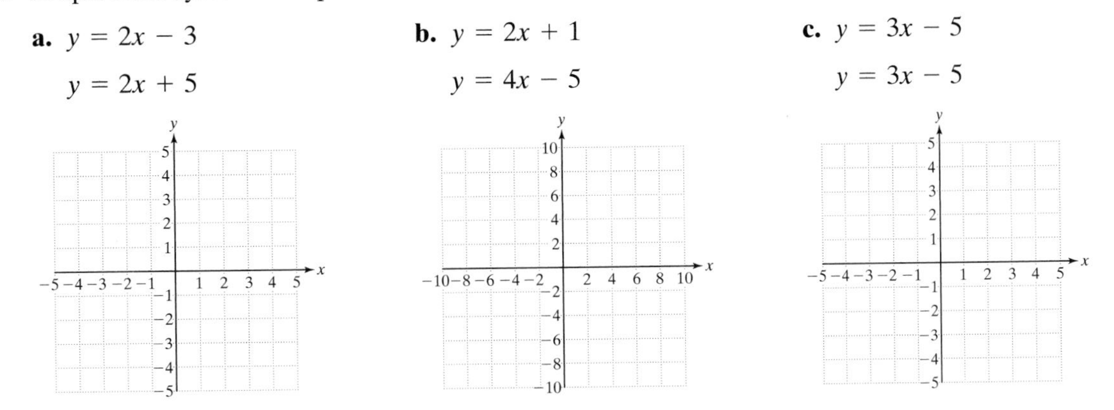

For Exercises 16–26, determine which system of equations (a, b, or c) makes the statement true. (*Hint:* Refer to the graphs from Exercise 15.)

16. The lines are parallel.

17. The lines coincide.

18. The lines intersect at exactly one point.

19. The system is inconsistent.

20. The system is dependent.

21. The lines have the same slope but different y-intercepts.

22. The lines have the same slope and same y-intercept.

23. The lines have different slopes.

24. The system has exactly one solution.

25. The system has infinitely many solutions.

26. The system has no solution.

a. $y = 2x - 3$
$y = 2x + 5$

b. $y = 2x + 1$
$y = 4x - 5$

c. $y = 3x - 5$
$y = 3x - 5$

Concept 3: Solving Systems of Linear Equations by Graphing

For Exercises 27–52, solve the systems by graphing. If a system does not have a unique solution, identify the system as inconsistent or dependent.

27. $y = -x + 4$
$y = x - 2$

28. $y = 3x + 2$
$y = 2x$

29. $2x + y = 0$
$3x + y = 1$

30. $x + y = -1$
$2x - y = -5$

31. $2x + y = 6$
$x = 1$

32. $4x + 3y = 9$
$x = 3$

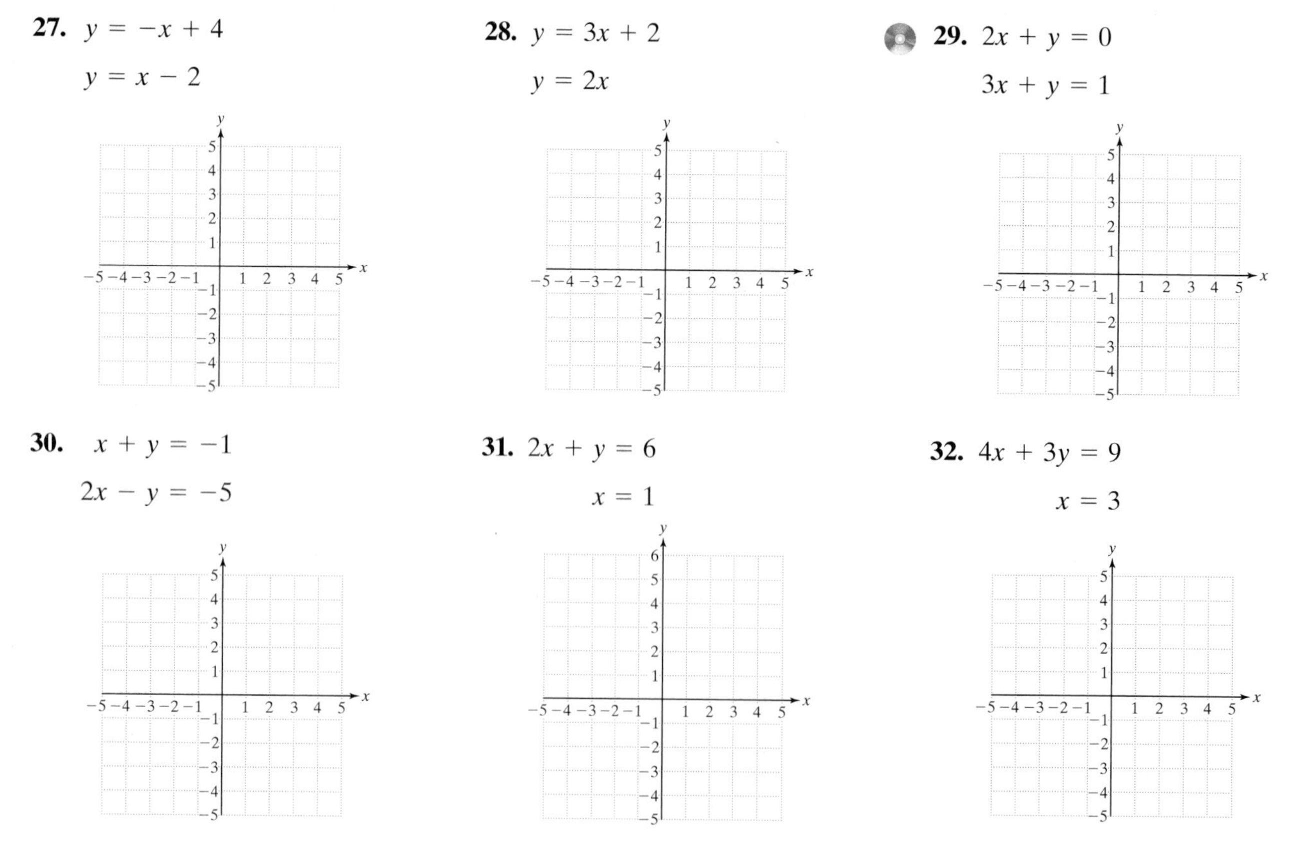

33. $-6x - 3y = 0$
$4x + 2y = 4$

34. $2x - 6y = 12$
$-3x + 9y = 12$

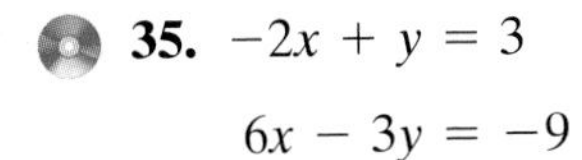

35. $-2x + y = 3$
$6x - 3y = -9$

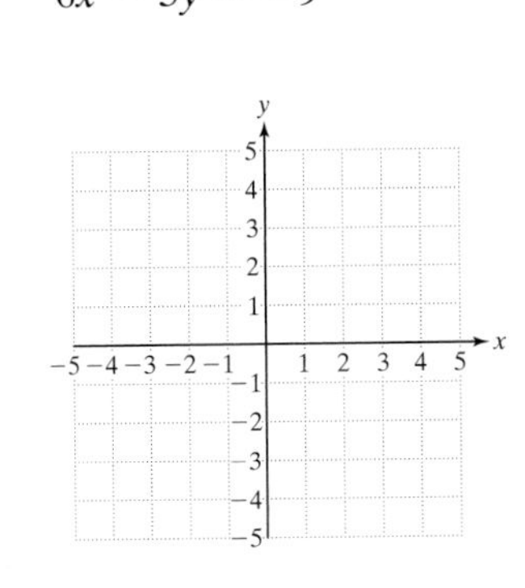

36. $x + 3y = 0$
$-2x - 6y = 0$

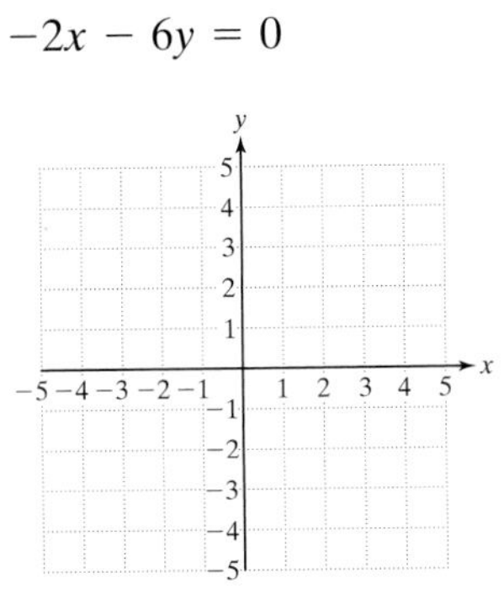

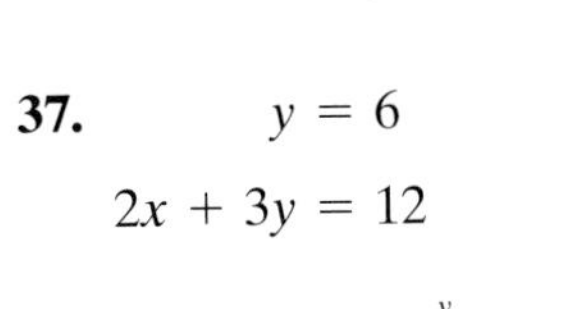

37. $y = 6$
$2x + 3y = 12$

38. $y = -2$
$x - 2y = 10$

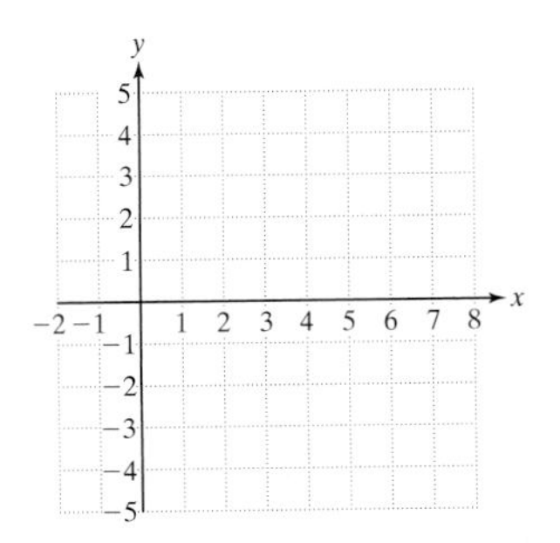

39. $-5x + 3y = -9$
$y = \frac{5}{3}x - 3$

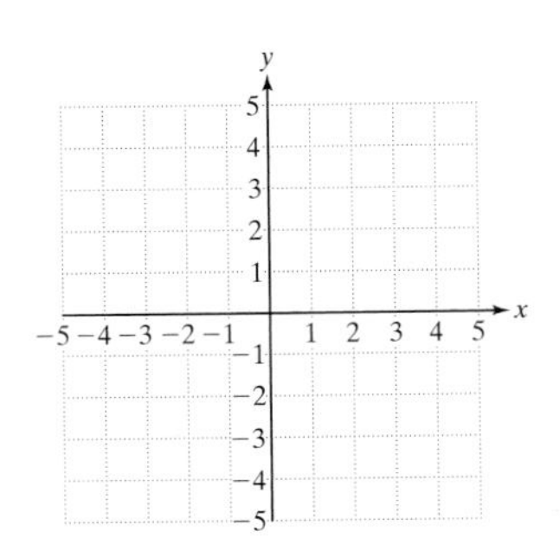

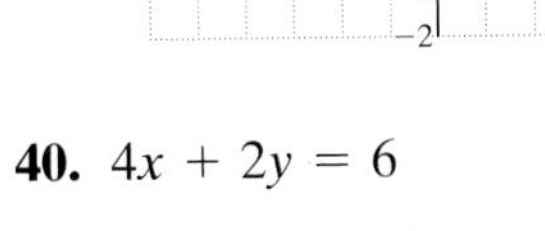

40. $4x + 2y = 6$
$y = -2x + 3$

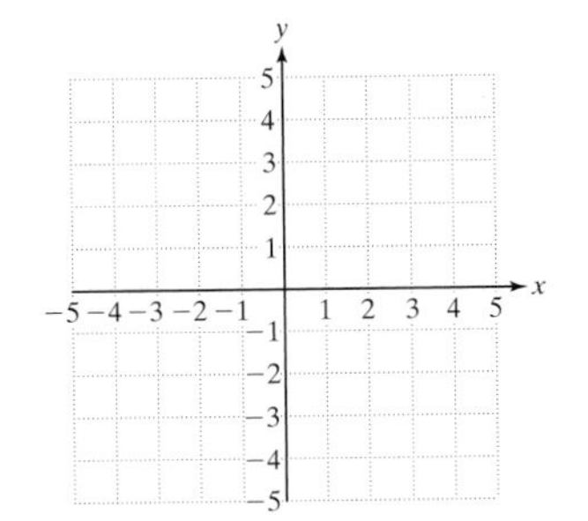

41. $x = 4 + y$
$3y = -3x$

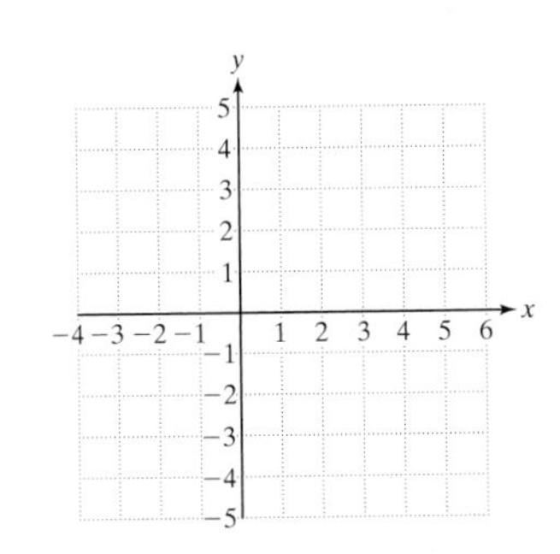

42. $3y = 4x$
$x - y = -1$

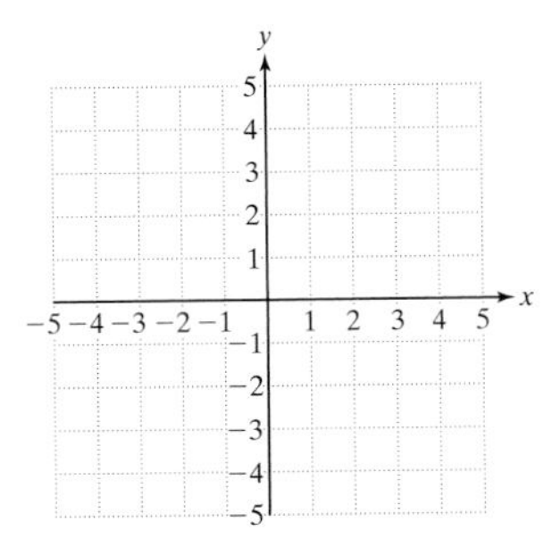

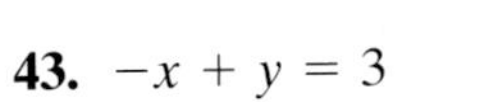

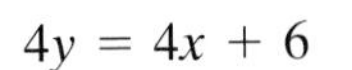

43. $-x + y = 3$
$4y = 4x + 6$

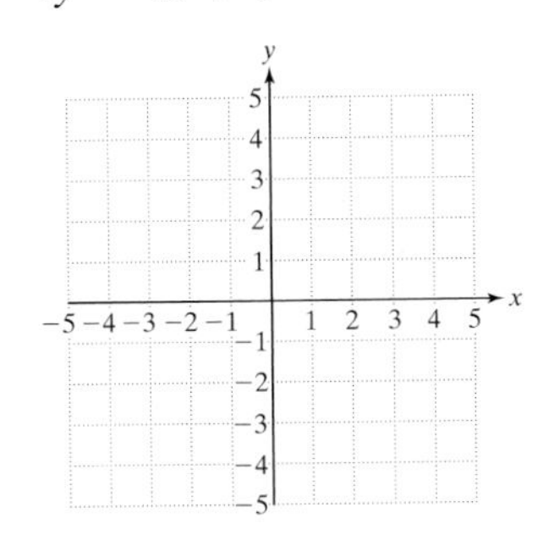

44. $x - y = 4$
$3y = 3x + 6$

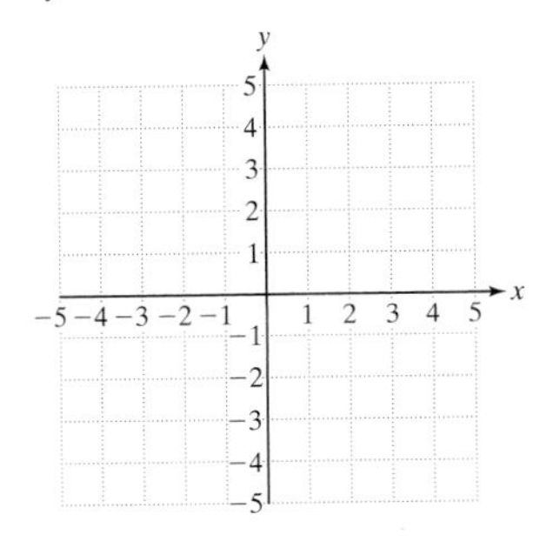

45. $x = 4$
$2y = 4$

46. $-3x = 6$
$y = 2$

47. $2x + 3y = 8$
$-4x - 6y = 6$

48. $4x + 4y = 8$
$5x + 5y = 5$

49. $2x + y = 4$
$4x - 2y = -4$

50. $6x + 6y = 3$
$2x - y = 4$

51. $y = 0.5x + 2$
$-x + 2y = 4$

52. $3x - 4y = 6$
$-6x + 8y = -12$

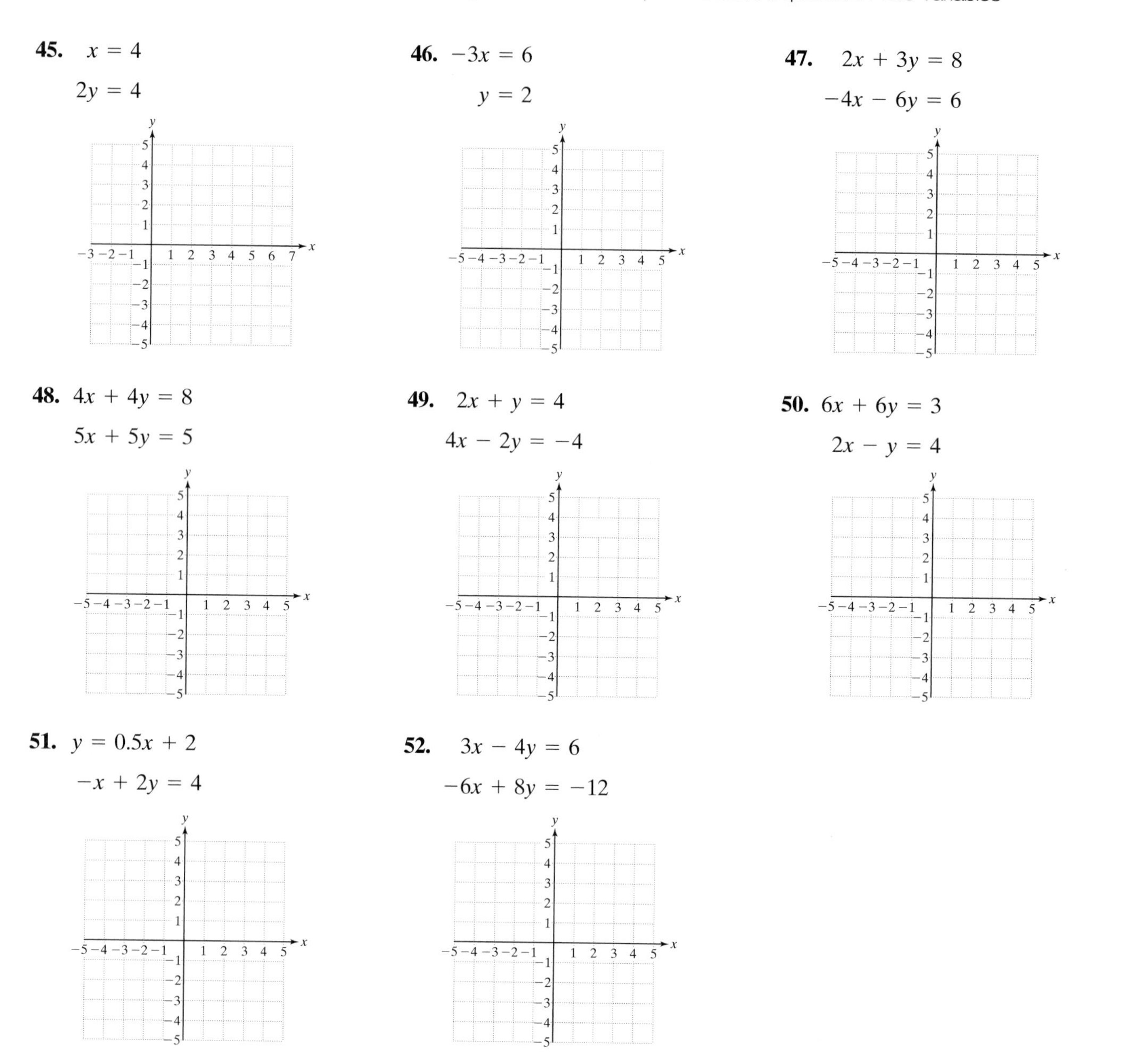

53. Two tennis instructors have two different fee schedules. Owen charges \$25 per lesson plus a one-time court fee of \$20 at the tennis club. Joan charges \$30 per lesson but does not require a court fee. The total cost, y, depends on the number of lessons, x, according to the equations

Owen: $y = 25x + 20$

Joan: $y = 30x$

From the graph, determine the number of lessons for which the total cost is the same for both instructors.

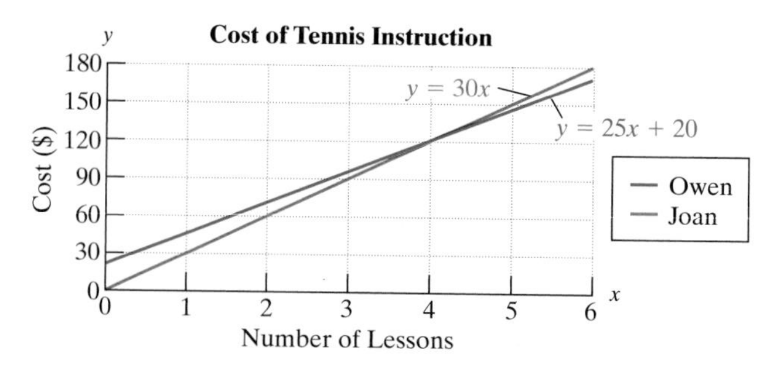

54. The cost to rent a 10 ft by 10 ft storage space is different for two different storage companies. The Storage Bin charges \$90 per month plus a nonrefundable deposit of \$120. AAA Storage charges \$110 per month with no deposit. The total cost, y, to rent a 10 ft by 10 ft space depends on the number of months, x, according to the equations

The Storage Bin: $y = 90x + 120$

AAA Storage: $y = 110x$

From the graph, determine the number of months required for which the cost to rent space is equal for both companies.

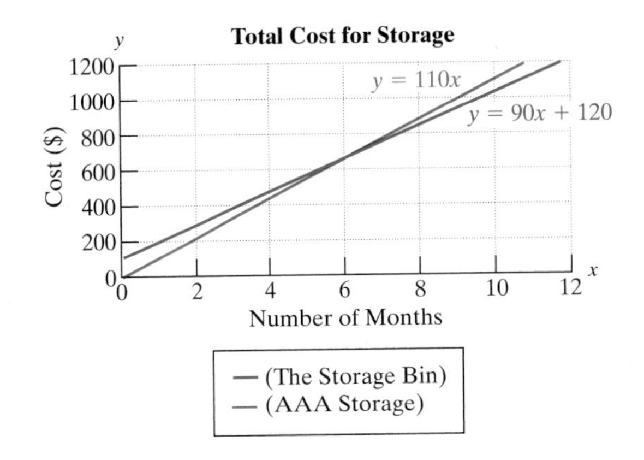

For the systems graphed in Exercises 55–56, explain why the ordered pair cannot be a solution to the system of equations.

55. $(-3, 1)$

56. $(-1, -4)$

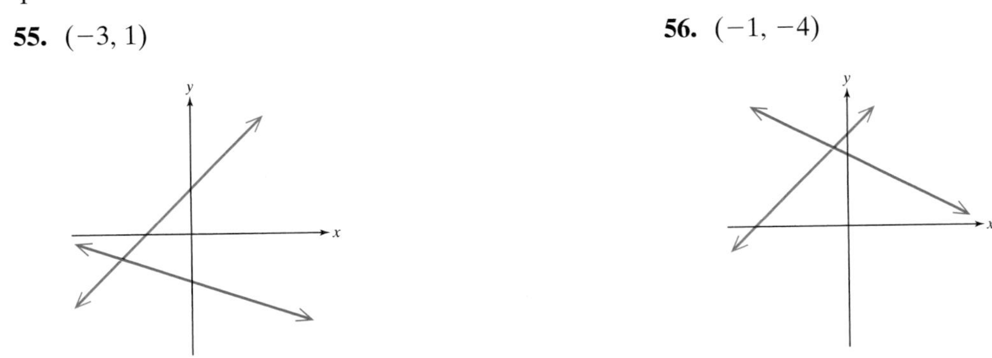

Expanding Your Skills

57. Write a system of linear equations whose solution is $(2, 1)$.

58. Write a system of linear equations whose solution is $(1, 4)$.

59. One equation in a system of linear equations is $x + y = 4$. Write a second equation such that the system will have no solution. (Answers may vary.)

60. One equation in a system of linear equations is $x - y = 3$. Write a second equation such that the system will have infinitely many solutions. (Answers may vary.)

Section 4.2 Solving Systems of Equations by the Substitution Method

Concepts

1. Solving Systems of Linear Equations by the Substitution Method
2. Solutions to Systems of Linear Equations: A Summary
3. Applications of the Substitution Method

1. Solving Systems of Linear Equations by the Substitution Method

In Section 4.1, we used the graphing method to find the solution set to a system of equations. However, sometimes it is difficult to determine the solution using this method because of limitations in the accuracy of the graph. This is particularly true when the coordinates of a solution are not integer values or when the solution is a point not sufficiently close to the origin. Identifying the coordinates of the point $(\frac{3}{17}, -\frac{23}{9})$ or $(-251, 8349)$ for example, might be difficult from a graph.

In this section and the next, we will cover two algebraic methods to solve a system of equations that do not require graphing. The first method, called the *substitution method*, is demonstrated in Examples 1–5.

Example 1 Solving a System of Linear Equations Using the Substitution Method

Solve by using the substitution method.

$$x = 2y - 3$$
$$-4x + 3y = 2$$

Solution:

The variable x has been isolated in the first equation. The quantity $2y - 3$ is equal to x and therefore can be substituted for x in the second equation. This leaves the second equation in terms of y only.

First equation: $x = 2y - 3$

Second equation: $-4x + 3y = 2$

$-4(2y - 3) + 3y = 2$ This equation now contains only one variable.

$-8y + 12 + 3y = 2$ Solve the resulting equation.

$$-5y + 12 = 2$$
$$-5y = -10$$
$$y = 2$$

To find x, substitute $y = 2$ back into the first equation.

$$x = 2y - 3$$
$$x = 2(2) - 3$$
$$x = 1$$

Check the ordered pair $(1, 2)$ in both original equations.

$x = 2y - 3 \longrightarrow 1 \stackrel{?}{=} 2(2) - 3$ ✔ True

$-4x + 3y = 2 \longrightarrow -4(1) + 3(2) \stackrel{?}{=} 2$ ✔ True

The solution is $(1, 2)$ because it checks in both original equations.

Skill Practice

1. Solve by using the substitution method.

$$y = -2x + 4$$
$$3x - y = -4$$

In Example 1, we eliminated the x-variable from the second equation by substituting an equivalent expression for x. The resulting equation was relatively simple to solve because it had only one variable. This is the premise of the substitution method.

The substitution method can be summarized as follows.

Solving a System of Equations by the Substitution Method

1. Isolate one of the variables from one equation.
2. Substitute the quantity found in step 1 into the other equation.
3. Solve the resulting equation.
4. Substitute the value found in step 3 back into the equation in step 1 to find the value of the remaining variable.
5. Check the solution in both original equations and write the answer as an ordered pair.

Example 2 Solving a System of Linear Equations Using the Substitution Method

Solve the system using the substitution method.

$$x + y = 4$$
$$-5x + 3y = -12$$

Solution:

The x- or y-variable in the first equation is easy to isolate because the coefficients are both 1. While either variable can be isolated, we arbitrarily choose the x-variable.

$x + y = 4 \longrightarrow x = 4 - y$ **Step 1:** Solve the first equation for x.

$-5(4 - y) + 3y = -12$ **Step 2:** Substitute $4 - y$ for x in the other equation.

$-20 + 5y + 3y = -12$ **Step 3:** Solve for y.

$-20 + 8y = -12$

$8y = 8$

$y = 1$

$x = 4 - y$ **Step 4:** Substitute $y = 1$ into the equation $x = 4 - y$.

$x = 4 - 1$

$x = 3$

Skill Practice Answers

1. (0, 4)

Step 5: Check the ordered pair (3, 1) in both original equations.

$x + y = 4$ $\quad (3) + (1) \stackrel{?}{=} 4$ ✔ True

$-5x + 3y = -12$ $\quad -5(3) + 3(1) \stackrel{?}{=} -12$ ✔ True

The solution is (3, 1) because it checks in both original equations.

Skill Practice Solve the system by the substitution method.

2. $2x + 3y = -2$

$-x + y = 1$

TIP: The solution to a system of linear equations can be confirmed by graphing. The system from Example 2 is graphed here.

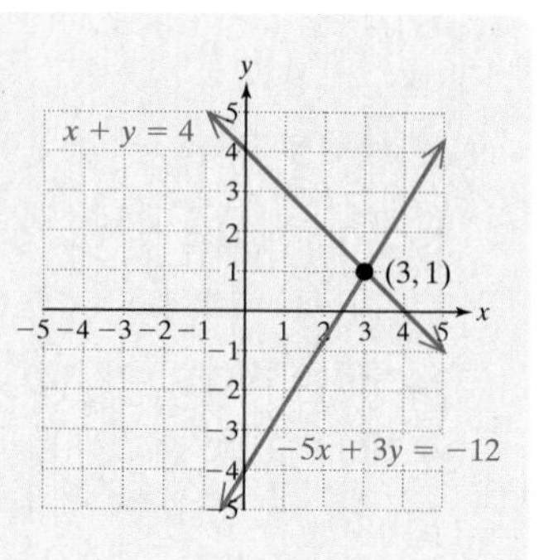

Example 3 Solving a System of Linear Equations Using the Substitution Method

Solve the system by using the substitution method.

$$3x + 5y = 17$$
$$2x - y = -6$$

Solution:

The y-variable in the second equation is the easiest variable to isolate because its coefficient is -1.

$$3x + 5y = 17$$
$$2x - y = -6 \longrightarrow -y = -2x - 6$$
$$y = 2x + 6$$

Step 1: Solve the second equation for y.

$$3x + 5(2x + 6) = 17$$

Step 2: Substitute the quantity $2x + 6$ for y in the other equation.

$$3x + 10x + 30 = 17$$

Step 3: Solve for x.

$$13x + 30 = 17$$
$$13x = 17 - 30$$
$$13x = -13$$
$$x = -1$$

Avoiding Mistakes:

Do not substitute $y = 2x + 6$ into the same equation from which it came. This mistake will result in an identity:

$$2x - y = -6$$
$$2x - (2x + 6) = -6$$
$$2x - 2x - 6 = -6$$
$$-6 = -6$$

Skill Practice Answers

2. $(-1, 0)$

$y = 2x + 6$ $y = 2(-1) + 6$ $y = -2 + 6$ $y = 4$	**Step 4:**	Substitute $x = -1$ into the equation $y = 2x + 6$.
	Step 5:	The ordered pair $(-1, 4)$ can be checked in the original equations to verify the answer.

$$3x + 5y = 17 \longrightarrow 3(-1) + 5(4) \stackrel{?}{=} 17 \longrightarrow -3 + 20 = 17 \checkmark \quad \text{True}$$
$$2x - y = -6 \longrightarrow 2(-1) - (4) \stackrel{?}{=} -6 \longrightarrow -2 - 4 = -6 \checkmark \quad \text{True}$$

The solution is $(-1, 4)$.

Skill Practice Solve the system by the substitution method.

3. $x + 4y = 11$

$2x - 5y = -4$

Recall from Section 4.1, that a system of linear equations may represent two parallel lines. In such a case, there is no solution to the system.

Example 4 Solving an Inconsistent System Using Substitution

Solve the system by using the substitution method.

$$2x + 3y = 6$$
$$y = -\tfrac{2}{3}x + 4$$

Solution:

$2x + 3y = 6$ $y = -\frac{2}{3}x + 4$	**Step 1:**	The variable y is already isolated in the second equation.
$2x + 3(-\frac{2}{3}x + 4) = 6$	**Step 2:**	Substitute $y = -\frac{2}{3}x + 4$ from the second equation into the first equation.
$2x - 2x + 12 = 6$	**Step 3:**	Solve the resulting equation.
$12 = 6$ (contradiction)		

The equation results in a contradiction. There are no values of x and y that will make 12 equal to 6. Therefore, there is no solution, and the system is inconsistent.

Skill Practice Solve the system by the substitution method.

4. $y = -\dfrac{1}{2}x + 3$

$2x + 4y = 5$

Skill Practice Answers

3. (3, 2) **4.** No solution.

TIP: The answer to Example 3 can be verified by writing each equation in slope-intercept form and graphing the lines.

Equation 1

$$2x + 3y = 6$$

$$3y = -2x + 6$$

$$\frac{3y}{3} = \frac{-2x}{3} + \frac{6}{3}$$

$$y = -\frac{2}{3}x + 2$$

Equation 2

$$y = -\frac{2}{3}x + 4$$

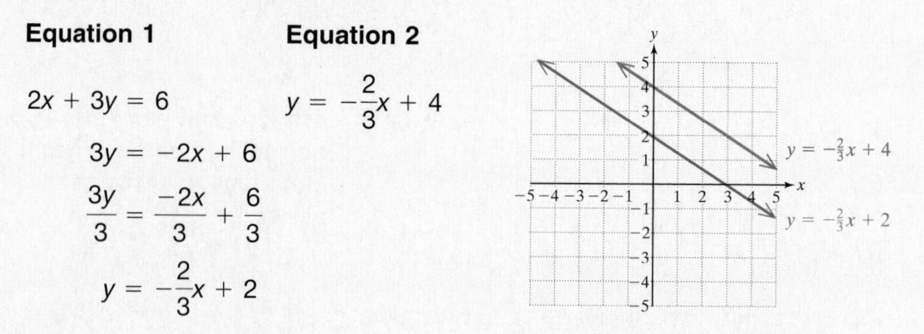

The equations indicate that the lines have the same slope but different *y*-intercepts. Therefore, the lines must be parallel. There is no point of intersection, indicating that the system has no solution.

Recall that a system of two linear equations may represent the same line. In such a case, the solution is the set of all points on the line.

Example 5 Solving a Dependent System Using Substitution

Solve the system by using the substitution method.

$$\frac{1}{2}x - \frac{1}{4}y = 1$$

$$6x - 3y = 12$$

Solution:

$$\frac{1}{2}x - \frac{1}{4}y = 1$$

$$6x - 3y = 12$$

To make the first equation easier to work with, we have the option of clearing fractions.

$$\frac{1}{2}x - \frac{1}{4}y = 1 \xrightarrow{\text{multiply by 4}} 4\left(\frac{1}{2}x\right) - 4\left(\frac{1}{4}y\right) = 4(1) \rightarrow 2x - y = 4$$

Now the system becomes:

$$2x - y = 4$$

$$6x - 3y = 12$$

The *y*-variable in the first equation is the easiest to isolate because its coefficient is −1.

$$2x - y = 4 \xrightarrow{\text{solve for } y} -y = -2x + 4 \rightarrow y = 2x - 4$$

$$6x - 3y = 12$$

Step 1: Isolate one of the variables.

$$6x - 3(2x - 4) = 12$$

Step 2: Substitute $y = 2x - 4$ from the first equation into the second equation.

$6x - 6x + 12 = 12$ **Step 3:** Solve the resulting equation.

$12 = 12$ (identity)

Because the equation produces an identity, all values of x make this equation true. Thus, x can be any real number. Substituting any real number, x, into the equation $y = 2x - 4$ produces an ordered pair on the line $y = 2x - 4$. Hence, the solution set to the system of equations is the set of all ordered pairs on the line $y = 2x - 4$. This can be written as $\{(x, y) \mid y = 2x - 4\}$. The system is dependent.

Skill Practice Solve the system by the substitution method.

5. $2x + \frac{1}{3}y = -\frac{1}{3}$

$12x + 2y = -2$

TIP: The solution to Example 5 can be verified by writing each equation in slope-intercept form and graphing the lines.

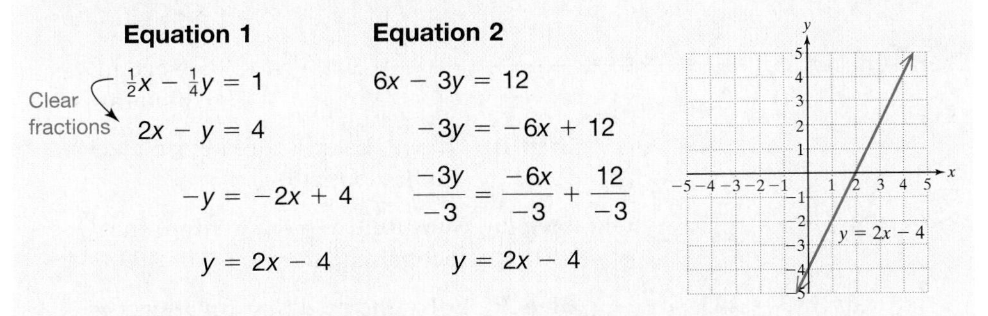

Equation 1	Equation 2
$\frac{1}{2}x - \frac{1}{4}y = 1$	$6x - 3y = 12$
Clear fractions $2x - y = 4$	$-3y = -6x + 12$
$-y = -2x + 4$	$\frac{-3y}{-3} = \frac{-6x}{-3} + \frac{12}{-3}$
$y = 2x - 4$	$y = 2x - 4$

Notice that the slope-intercept forms for both equations are identical. The equations represent the same line, indicating that the system is dependent. Each point on the line is a solution to the system of equations.

2. Solutions to Systems of Linear Equations: A Summary

The following summary reviews the three different geometric relationships between two lines and the solutions to the corresponding systems of equations.

Solutions to a System of Two Linear Equations

1. The lines may intersect at one point (yielding one unique solution).
2. The lines may be parallel and have no point of intersection (yielding no solution). This is detected algebraically when a contradiction (false statement) is obtained (for example, $0 = -3$ and $12 = 6$).
3. The lines may be the same and intersect at all points on the line (yielding an infinite number of solutions). This is detected algebraically when an identity is obtained (for example, $0 = 0$ and $12 = 12$).

Skill Practice Answers

5. Infinitely many solutions: $\{(x, y) \mid 12x + 2y = -2\}$

3. Applications of the Substitution Method

In Chapter 2, we solved word problems using one linear equation and one variable. In this chapter we investigate application problems with two unknowns. In such a case, we can use two variables to represent the unknown quantities. However, if two variables are used, we must write a system of *two* distinct equations.

Example 6 Applying the Substitution Method

One number is 3 more than 4 times another. Their sum is 133. Find the numbers.

Solution:

We can use two variables to represent the two unknown numbers.

Let x represent one number.
Let y represent the other number. Label the variables.

We must now write two equations. Each of the first two sentences gives a relationship between x and y:

One number is 3 more than 4 times another. ⟶ $x = 4y + 3$ (first equation)

Their sum is 133. ⟶ $x + y = 133$ (second equation)

Step 1: Notice that x is already isolated in the first equation.

$(4y + 3) + y = 133$ **Step 2:** Substitute $x = 4y + 3$ into the second equation, $x + y = 133$.

$5y + 3 = 133$ **Step 3:** Solve the resulting equation.

$5y = 130$

$y = 26$

$x = 4y + 3$

$x = 4(26) + 3$ **Step 4:** To solve for x, substitute $y = 26$ into the equation $x = 4y + 3$.

$x = 104 + 3$

$x = 107$

One number is 26, and the other is 107.

TIP: Check that the numbers 26 and 107 meet the conditions of Example 6.

- 4 times 26 is 104. Three more than 104 is 107. ✔
- The sum of the numbers should be 133: $26 + 107 = 133$. ✔

Skill Practice

6. One number is 16 more than another. Their sum is 92. Use a system of equations to find the numbers.

Skill Practice Answers

6. One number is 38, and the other number is 54.

Example 7 **Using the Substitution Method in a Geometry Application**

Two angles are supplementary. The measure of one angle is 15° more than twice the measure of the other angle. Find the measures of the two angles.

Solution:

Let x represent the measure of one angle.
Let y represent the measure of the other angle.

The sum of the measures of supplementary angles is 180° $\rightarrow$ $x + y = 180$

The measure of one angle is 15° more than twice the other angle $\rightarrow$ $x = 2y + 15$

$$x + y = 180$$

$$x = 2y + 15$$

Step 1: The x-variable in the second equation is already isolated.

$$(2y + 15) + y = 180$$

Step 2: Substitute $x = 2y + 15$ from the second equation into the first equation.

$$2y + 15 + y = 180$$

Step 3: Solve the resulting equation.

$$3y + 15 = 180$$

$$3y = 165$$

$$y = 55$$

Step 4: Substitute $y = 55$ into the equation $x = 2y + 15$.

$$x = 2y + 15$$

$$x = 2(55) + 15$$

$$x = 110 + 15$$

$$x = 125$$

One angle is 55°, and the other is 125°.

TIP: Check that the angles 55° and 125° meet the conditions of Example 7.

- Because $55° + 125° = 180°$, the angles are supplementary. ✔
- The angle 125° is 15° more than twice 55°: $125° = 2(55°) + 15°$. ✔

Skill Practice

7. The measure of one angle is 2° less than 3 times the measure of another angle. The angles are complementary. Use a system of equations to find the measures of the two angles.

Skill Practice Answers

7. The measures of the angles are 23° and 67°.

Section 4.2 Practice Exercises

Review Exercises

For Exercises 1–6, write each pair of lines in slope-intercept form. Then identify whether the lines intersect in exactly one point or if the lines are parallel or coinciding.

1. $2x - y = 4$
$-2y = -4x + 8$

2. $x - 2y = 5$
$3x = 6y + 15$

3. $2x + 3y = 6$
$x - y = 5$

4. $x - y = -1$
$x + 2y = 4$

5. $2x = \frac{1}{2}y + 2$
$4x - y = 13$

6. $4y = 3x$
$3x - 4y = 15$

Concept 1: Solving Systems of Linear Equations by the Substitution Method

For Exercises 7–10, solve each system using the substitution method.

7. $3x + 2y = -3$
$y = 2x - 12$

8. $4x - 3y = -19$
$y = -2x + 13$

9. $x = -4y + 16$
$3x + 5y = 20$

10. $x = -y + 3$
$-2x + y = 6$

11. Given the system: $4x - 2y = -6$
$3x + y = 8$

a. Which variable from which equation is easiest to isolate and why?

b. Solve the system using the substitution method.

12. Given the system: $x - 5y = 2$
$11x + 13y = 22$

a. Which variable from which equation is easiest to isolate and why?

b. Solve the system using the substitution method.

For Exercises 13–48, solve each system using the substitution method.

13. $x - 3y = -1$
$2x = 4y + 2$

14. $3x + y = 1$
$2y = x + 9$

15. $-2x + 5y = 5$
$x - 4y = -10$

16. $3x - 7y = -2$
$2x + y = 27$

17. $4x - y = -1$
$2x + 4y = 13$

18. $5x - 3y = -2$
$10x - y = 1$

19. $4x - 3y = 11$
$x = 5$

20. $y = -3x - 9$
$y = 12$

21. $4x = 8y + 4$
$5x - 3y = 5$

22. $3y = 6x - 6$
$-3x + y = -4$

23. $x - 3y = -11$
$6x - y = 2$

24. $-2x - y = 9$
$x + 7y = 15$

25. $3x + 2y = -1$
$\frac{3}{2}x + y = 4$

26. $5x - 2y = 6$
$-\frac{5}{2}x + y = 5$

27. $10x - 30y = -10$
$2x - 6y = -2$

28. $3x + 6y = 6$
$-6x - 12y = -12$

29. $2x + y = 3$
$y = -7$

30. $-3x = 2y + 23$
$x = -1$

31. $x + 2y = -2$
$4x = -2y - 17$

32. $x + y = 1$
$2x - y = -2$

33. $y = -\frac{1}{2}x - 4$
$y = 4x - 13$

34. $y = \frac{2}{3}x - 3$
$y = 6x - 19$

35. $y = \frac{1}{2}x + 8$
$y - 4 = -2(x + 3)$

36. $x = -2y + 7$
$x - 3 = 6(y + 2)$

37. $3x + 2y = 4$
$2x - 3y = -6$

38. $4x + 3y = 4$
$-2x + 5y = -2$

39. $y = 0.25x + 1$
$-x + 4y = 4$

40. $y = 0.75x - 3$
$-3x + 4y = -12$

41. $11x + 6y = 17$
$5x - 4y = 1$

42. $3x - 8y = 7$
$10x - 5y = 45$

43. $x + 2y = 4$
$4y = -2x - 8$

44. $-y = x - 6$
$2x + 2y = 4$

45. $\frac{1}{3}(2x + y) = 1$
$x + y = 4$

46. $2(x - y) = 4$
$3x + y = 10$

47. $\frac{x}{3} + \frac{y}{2} = -4$
$x - 3y = 6$

48. $x - 2y = -5$
$\frac{2x}{3} + \frac{y}{3} = 0$

Concept 3: Applications of the Substitution Method

For Exercises 49–58, set up a system of linear equations and solve for the indicated quantities.

49. Two numbers have a sum of 106. One number is 10 less than the other. Find the numbers.

50. Two positive numbers have a difference of 8. The larger number is 2 less than 3 times the smaller number. Find the numbers.

51. The difference between two positive numbers is 26. The larger number is three times the smaller. Find the numbers.

52. The sum of two numbers is 956. One number is 94 less than 6 times the other. Find the numbers.

53. Two angles are supplementary. One angle is 15° more than 10 times the other angle. Find the measure of each angle.

54. Two angles are complementary. One angle is 1° less than 6 times the other angle. Find the measure of each angle.

55. Two angles are complementary. One angle is 10° more than 3 times the other angle. Find the measure of each angle.

56. Two angles are supplementary. One angle is 5° less than twice the other angle. Find the measure of each angle.

57. In a right triangle, one of the acute angles is 6° less than the other acute angle. Find the measure of each acute angle.

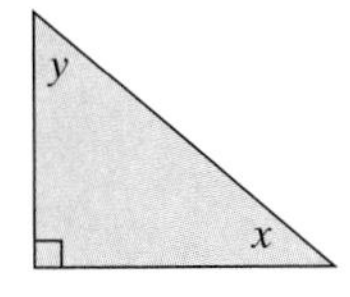

58. In a right triangle, one of the acute angles is 9° less than twice the other acute angle. Find the measure of each acute angle.

Expanding Your Skills

59. The following system of equations is dependent and has infinitely many solutions. Find three ordered pairs that are solutions to the system of equations.

$$y = 2x + 3$$
$$-4x + 2y = 6$$

60. The following system of equations is dependent and has infinitely many solutions. Find three ordered pairs that are solutions to the system of equations.

$$y = -x + 1$$
$$2x + 2y = 2$$

Section 4.3 Solving Systems of Equations by the Addition Method

Concepts

1. Solving Systems of Linear Equations by the Addition Method
2. Summary of Methods for Solving Linear Equations in Two Variables

1. Solving Systems of Linear Equations by the Addition Method

Thus far in Chapter 4 we have used the graphing method and the substitution method to solve a system of linear equations in two variables. In this section, we present another algebraic method to solve a system of linear equations, called the *addition method* (sometimes called the *elimination method*). The purpose of the addition method is to eliminate one variable.

Example 1 Solving a System of Linear Equations Using the Addition Method

Solve the system using the addition method.

$$x + y = -2$$
$$x - y = -6$$

Solution:

Notice that the coefficients of the y-variables are opposites:

$$x + 1y = -2 \quad \text{Coefficient is 1.}$$
$$x - 1y = -6 \quad \text{Coefficient is } -1.$$

Because the coefficients of the y-variables are opposites, we can add the two equations to eliminate the y-variable.

$$x + y = -2$$
$$\underline{x - y = -6}$$
$$2x \quad = -8 \quad \leftarrow \text{After adding the equations, we have one equation and one variable.}$$

$$2x = -8 \quad \text{Solve the resulting equation.}$$
$$x = -4$$

To find the value of y, substitute $x = -4$ into *either* of the original equations.

$$x + y = -2 \quad \text{First equation}$$
$$(-4) + y = -2$$
$$y = -2 + 4$$
$$y = 2$$

The solution is $(-4, 2)$.

Check:

$$x + y = -2 \longrightarrow (-4) + (2) \stackrel{?}{=} -2 \longrightarrow -2 = -2 \checkmark \quad \text{True}$$
$$x - y = -6 \longrightarrow (-4) - (2) \stackrel{?}{=} -6 \longrightarrow -6 = -6 \checkmark \quad \text{True}$$

TIP: Notice that the value $x = -4$ could have been substituted into the second equation to obtain the same value for y.

$$x - y = -6$$
$$(-4) - y = -6$$
$$-y = -6 + 4$$
$$-y = -2$$
$$y = 2$$

Skill Practice Solve the system using the addition method.

1. $x + y = 13$

   $2x - y = 2$

It is important to note that the addition method works on the premise that the two equations have *opposite* values for the coefficients of one of the variables. Sometimes it is necessary to manipulate the original equations to create two coefficients that are opposites. This is accomplished by multiplying one or both equations by an appropriate constant. The process is outlined as follows.

Solving a System of Equations by the Addition Method

1. Write both equations in standard form: $Ax + By = C$.
2. Clear fractions or decimals (optional).
3. Multiply one or both equations by nonzero constants to create opposite coefficients for one of the variables.
4. Add the equations from step 3 to eliminate one variable.
5. Solve for the remaining variable.
6. Substitute the known value from step 5 into one of the original equations to solve for the other variable.
7. Check the solution in both equations.

Example 2 Solving a System of Linear Equations Using the Addition Method

Solve the system using the addition method.

$$3x + 5y = 17$$
$$2x - y = -6$$

Solution:

$3x + 5y = 17$ **Step 1:** Both equations are already written in standard form.

$2x - y = -6$ **Step 2:** There are no fractions or decimals.

Notice that neither the coefficients of x nor the coefficients of y are opposites. However, multiplying the second equation by 5 creates the term $-5y$ in the second equation. This is the opposite of the term $+5y$ in the first equation.

$$\begin{aligned} 3x + 5y &= 17 \\ 2x - y &= -6 \end{aligned} \xrightarrow{\text{Multiply by 5}} \begin{aligned} 3x + 5y &= 17 \\ 10x - 5y &= -30 \\ \hline 13x \quad &= -13 \end{aligned}$$

Step 3: Multiply the second equation by 5.

Step 4: Add the equations.

Step 5: Solve the equation.

$$13x = -13$$
$$x = -1$$

Step 6: Substitute $x = -1$ into one of the original equations.

$3x + 5y = 17$ First equation

$$3(-1) + 5y = 17$$
$$5y = 20$$
$$y = 4$$

Skill Practice Answers

1. (5, 8)

The solution is $(-1, 4)$.

Step 7: Check the solution in both original equations.

Check:

$3x + 5y = 17 \longrightarrow 3(-1) + 5(4) \stackrel{?}{=} 17 \longrightarrow -3 + 20 = 17$ ✔ True

$2x - y = -6 \longrightarrow 2(-1) - (4) \stackrel{?}{=} -6 \longrightarrow -2 - 4 = -6$ ✔ True

Skill Practice Solve the system using the addition method.

2. $4x + 3y = 3$

$x - 2y = 9$

In Example 3, the system of equations uses the variables a and b instead of x and y. In such a case, we will write the solution as an ordered pair with the variables written in alphabetical order, such as (a, b).

Example 3 Solving a System of Linear Equations Using the Addition Method

Solve the system using the addition method.

$$5b = 7a + 8$$
$$-4a - 2b = -10$$

Solution:

Step 1: Write the equations in standard form.

The first equation becomes: $5b = 7a + 8 \longrightarrow -7a + 5b = 8$

The system becomes:

$$-7a + 5b = 8$$
$$-4a - 2b = -10$$

Step 2: There are no fractions or decimals.

Step 3: We need to obtain opposite coefficients on either the a or b term.

Notice that neither the coefficients of a nor the coefficients of b are opposites. However, it is possible to change the coefficients of b to 10 and -10 (this is because the LCM of 5 and 2 is 10). This is accomplished by multiplying the first equation by 2 and the second equation by 5.

$$\begin{aligned} -7a + 5b &= 8 \xrightarrow{\text{Multiply by 2}} & -14a + 10b &= 16 \\ -4a - 2b &= -10 \xrightarrow[\text{Multiply by 5}]{} & -20a - 10b &= -50 \\ & & \hline -34a \quad &= -34 \end{aligned}$$

Step 4: Add the equations.

$$-34a = -34$$

$$\frac{-34a}{-34} = \frac{-34}{-34}$$

$$a = 1$$

Step 5: Solve the resulting equation.

Skill Practice Answers

2. $(3, -3)$

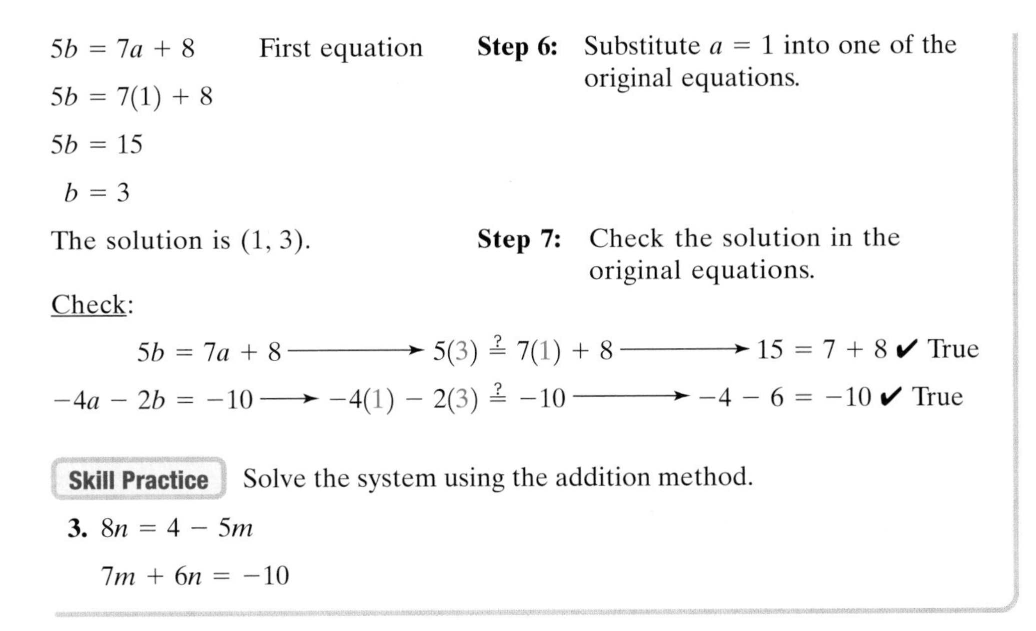

$5b = 7a + 8$ First equation

$5b = 7(1) + 8$

$5b = 15$

$b = 3$

Step 6: Substitute $a = 1$ into one of the original equations.

The solution is (1, 3).

Step 7: Check the solution in the original equations.

Check:

$5b = 7a + 8 \longrightarrow 5(3) \stackrel{?}{=} 7(1) + 8 \longrightarrow 15 = 7 + 8$ ✔ True

$-4a - 2b = -10 \longrightarrow -4(1) - 2(3) \stackrel{?}{=} -10 \longrightarrow -4 - 6 = -10$ ✔ True

Skill Practice Solve the system using the addition method.

3. $8n = 4 - 5m$

$7m + 6n = -10$

Example 4 Solving a System of Linear Equations Using the Addition Method

Solve the system using the addition method.

$$3(x - 10) = 7y + 11$$
$$-2(x - y) = 2x - 18$$

Solution:

Step 1: Write the equations in standard form.

$$3(x - 10) = 7y + 11 \xrightarrow{\text{Clear parentheses}} 3x - 30 = 7y + 11 \xrightarrow{\text{Write as } Ax + By = C} 3x - 7y = 41$$

$$-2(x - y) = 2x - 18 \longrightarrow -2x + 2y = 2x - 18 \longrightarrow -4x + 2y = -18$$

Step 2: There are no fractions or decimals.

Notice that neither the coefficients of x nor the coefficients of y are opposites. However, it is possible to change the coefficients of x to 12 and -12 (notice that 12 is the LCM of 3 and 4). This is accomplished by multiplying the first equation by 4 and the second equation by 3.

$$3x - 7y = 41 \xrightarrow{\text{Multiply by 4}} 12x - 28y = 164$$
$$-4x + 2y = -18 \xrightarrow[\text{Multiply by 3}]{} \underline{-12x + 6y = -54}$$
$$-22y = 110$$

Step 3: Create opposite coefficients of x.

Step 4: Add the equations.

$$-22y = 110$$
$$\frac{-22y}{-22} = \frac{110}{-22}$$
$$y = -5$$

Step 5: Solve the resulting equation.

Skill Practice Answers

3. $(-4, 3)$

$$3x - 7y = 41 \quad \text{First equation}$$
$$3x - 7(-5) = 41$$
$$3x + 35 = 41$$
$$3x = 6$$
$$x = 2$$

Step 6: Substitute $y = -5$ into one of the equations.

The solution is $(2, -5)$.

Step 7: Check the solution in the original equations.

Check:

$3(x - 10) = 7y + 11 \longrightarrow 3(2 - 10) \stackrel{?}{=} 7(-5) + 11 \longrightarrow -24 = -24$ ✔ True

$-2(x - y) = 2x - 18 \longrightarrow -2[2 - (-5)] \stackrel{?}{=} 2(2) - 18 \longrightarrow -14 = -14$ ✔ True

Skill Practice Solve the system using the addition method.

4. $x = 5(y + 2) - x$

$7y = 3(x - 5)$

TIP: When using the addition method, it makes no difference which variable is eliminated. In Example 4 we eliminated x. However, we could easily have eliminated y by changing the coefficients of y to -14 and 14. This would be accomplished by multiplying the first equation by 2 and the second equation by 7.

$$3x - 7y = 41 \xrightarrow{\text{Multiply by 2}} 6x - 14y = 82$$
$$-4x + 2y = -18 \xrightarrow{\text{Multiply by 7}} \underline{-28x + 14y = -126}$$
$$-22x \qquad = -44$$

Because $-22x = -44$, then $x = 2$. Substituting $x = 2$ into either original equation yields $y = -5$.

Example 5 Solving a System of Linear Equations Using the Addition Method

Solve the system using the addition method.

$$34x - 22y = 4$$
$$17x - 88y = -19$$

Solution:

The equations are already in standard form. There are no fractions or decimals to clear.

$$34x - 22y = 4 \longrightarrow 34x - 22y = 4$$
$$17x - 88y = -19 \xrightarrow{\text{Multiply by } -2} \underline{-34x + 176y = 38}$$
$$154y = 42$$

$$154y = 42$$
$$\frac{154y}{154} = \frac{42}{154}$$
$$y = \frac{3}{11}$$

Skill Practice Answers

4. (5, 0)

To find the value of x, we normally substitute y into one of the original equations and solve for x. In this example, we will show an alternative approach for finding x. By repeating the addition method, this time eliminating y, we can solve for x. This approach enables us to avoid substitution of the fractional value for y.

$$\begin{aligned} 34x - 22y &= 4 \xrightarrow{\text{Multiply by } -4} & -136x + 88y &= -16 \\ 17x - 88y &= -19 \longrightarrow & 17x - 88y &= -19 \\ & & \hline -119x \quad &= -35 \end{aligned}$$

$-119x = -35$ Solve for x.

$$\frac{-119x}{-119} = \frac{-35}{-119}$$

$x = \frac{5}{17}$ Simplify.

The solution is $(\frac{5}{17}, \frac{3}{11})$. These values can be checked in the original equations:

$$34x - 22y = 4 \qquad\qquad 17x - 88y = -19$$

$$34\left(\frac{5}{17}\right) - 22\left(\frac{3}{11}\right) \stackrel{?}{=} 4 \qquad\qquad 17\left(\frac{5}{17}\right) - 88\left(\frac{3}{11}\right) \stackrel{?}{=} -19$$

$10 - 6 = 4$ ✔ True $\qquad\qquad 5 - 24 = -19$ ✔ True

Skill Practice Solve the system using the addition method.

5. $15x - 16y = 1$

$45x + 4y = 16$

Example 6 Solving an Inconsistent System of Linear Equations

Solve the system using the addition method.

$$2x - 5y = 10$$

$$\frac{1}{2}x = 1 + \frac{5}{4}y$$

Solution:

$$2x - 5y = 10 \longrightarrow 2x - 5y = 10$$

$$\frac{1}{2}x = 1 + \frac{5}{4}y \longrightarrow \frac{1}{2}x - \frac{5}{4}y = 1$$

Step 1: Write the equations in standard form.

Step 2: Multiply both sides of the second equation by 4 to clear fractions.

$$\frac{1}{2}x - \frac{5}{4}y = 1 \longrightarrow 4\left(\frac{1}{2}x - \frac{5}{4}y\right) = 4(1) \longrightarrow 2x - 5y = 4$$

Skill Practice Answers

5. $\left(\frac{1}{3}, \frac{1}{4}\right)$

Now the system becomes
$$2x - 5y = 10$$
$$2x - 5y = 4$$

To make either the x-coefficients or y-coefficients opposites, multiply either equation by -1.

$$2x - 5y = 10 \xrightarrow{\text{Multiply by } -1} -2x + 5y = -10$$
$$2x - 5y = 4 \longrightarrow \underline{2x - 5y = 4}$$
$$0 = -6$$

Step 3: Create opposite coefficients.

Step 4: Add the equations.

Because the equation results in a contradiction, there is no solution, and the system of equations is inconsistent. Writing each line in slope-intercept form verifies that the lines are parallel (Figure 4-8).

$$2x - 5y = 10 \xrightarrow{\text{Slope-intercept form}} y = \frac{2}{5}x - 2$$

$$\frac{1}{2}x = 1 + \frac{5}{4}y \xrightarrow{\text{Slope-intercept form}} y = \frac{2}{5}x - \frac{4}{5}$$

There is no solution.

Figure 4-8

Skill Practice Solve the system using the addition method.

6. $\frac{2}{3}x = 2 + \frac{3}{4}y$

$8x - 9y = 6$

Example 7 Solving a Dependent System of Linear Equations

Solve the system by the addition method.

$$3x - y = 4$$
$$2y = 6x - 8$$

Solution:

$$3x - y = 4 \longrightarrow 3x - y = 4$$

Step 1: Write the equations in standard form.

$$2y = 6x - 8 \longrightarrow -6x + 2y = -8$$

Step 2: There are no fractions or decimals.

Notice that the equations differ exactly by a factor of -2, which indicates that these two equations represent the same line. Multiply the first equation by 2 to create opposite coefficients for the variables.

$$3x - y = 4 \xrightarrow{\text{Multiply by } 2} 6x - 2y = 8$$
$$-6x + 2y = -8 \qquad \underline{-6x + 2y = -8}$$
$$0 = 0$$

Step 3: Create opposite coefficients.

Step 4: Add the equations.

Skill Practice Answers

6. No solution.

Because the resulting equation is an identity, the original equations represent the same line. This can be confirmed by writing each equation in slope-intercept form.

$$3x - y = 4 \longrightarrow -y = -3x + 4 \longrightarrow y = 3x - 4$$
$$-6x + 2y = -8 \longrightarrow 2y = 6x - 8 \longrightarrow y = 3x - 4$$

The solution is the set of all points on the line, or equivalently, $\{(x, y) \mid 3x - y = 4\}$.

Skill Practice Solve the system by using the addition method.

7. $3x = 3y + 15$
$2x = 10 + 2y$

2. Summary of Methods for Solving Linear Equations in Two Variables

If no method of solving a system of linear equations is specified, you may use the method of your choice. However, we recommend the following guidelines:

1. If one of the equations is written with a variable isolated, the substitution method is a good choice. For example:

$$2x + 5y = 2 \qquad \text{or} \qquad y = \frac{1}{3}x - 2$$
$$x = y - 6 \qquad\qquad x - 6y = 9$$

2. If both equations are written in standard form, $Ax + By = C$, where none of the variables has coefficients of 1 or -1, then the addition method is a good choice.

$$4x + 5y = 12$$
$$5x + 3y = 15$$

3. If both equations are written in standard form, $Ax + By = C$, and at least one variable has a coefficient of 1 or -1, then either the substitution method or the addition method is a good choice.

Skill Practice Answers

7. $\{(x, y) \mid 3x = 3y + 15\}$

Section 4.3 Practice Exercises

Study Skills Exercise

1. Now that you have learned three methods of solving a system of linear equations with two variables, choose a system and solve it all three ways. There are two advantages to this. One is to check your answer (you should get the same answer using all three methods). The second advantage is to show you which method is the easiest for you to use.

Solve the system by using the graphing method, the substitution method, and the addition method.

$$2x + y = -7$$
$$x - 10 = 4y$$

Review Exercises

For Exercises 2–5, check to see if the given ordered pair is a solution to the system.

2. $x + y = 8$ $(5, 3)$
$y = x - 2$

3. $x = y + 1$ $(3, 2)$
$-x + 2y = 0$

4. $3x + 2y = 14$ $(5, -2)$
$5x - 2y = 29$

5. $x = 2y - 11$ $(-3, 4)$
$-x + 5y = 23$

Concept 1: Solving Systems of Linear Equations by the Addition Method

For Exercises 6–7, answer as true or false.

6. Given the system

$$5x - 4y = 1$$
$$7x - 2y = 5$$

a. To eliminate the y-variable using the addition method, multiply the second equation by 2.

b. To eliminate the x-variable, multiply the first equation by 7 and the second equation by -5.

7. Given the system

$$3x + 5y = -1$$
$$9x - 8y = -26$$

a. To eliminate the x-variable using the addition method, multiply the first equation by -3.

b. To eliminate the y-variable, multiply the first equation by 8 and the second equation by -5.

8. Given the system

$$3x - 4y = 2$$
$$17x + y = 35$$

a. Which variable, x or y, is easier to eliminate using the addition method?

b. Solve the system using the addition method.

9. Given the system

$$-2x + 5y = -15$$
$$6x - 7y = 21$$

a. Which variable, x or y, is easier to eliminate using the addition method?

b. Solve the system using the addition method.

For Exercises 10–21, solve the systems using the addition method.

10. $x + 2y = 8$
$5x - 2y = 4$

11. $2x - 3y = 11$
$-4x + 3y = -19$

12. $a + b = 3$
$3a + b = 13$

13. $-2u + 6v = 10$
$-2u + v = -5$

14. $-3x + y = 1$
$-6x - 2y = -2$

15. $5m - 2n = 4$
$3m + n = 9$

16. $3x - 5y = 13$
$x - 2y = 5$

17. $7a + 2b = -1$
$3a - 4b = 19$

18. $6c - 2d = -2$
$5c + 3d = 17$

19. $2s + 3t = -1$
$5s - 2t = 7$

20. $6y - 4z = -2$
$4y + 6z = 42$

21. $4k - 2r = -4$
$2k + 4r = -32$

22. In solving a system of equations, suppose you get the statement $0 = 5$. How many solutions will the system have? What can you say about the graphs of these equations?

23. In solving a system of equations, suppose you get the statement $0 = 0$. How many solutions will the system have? What can you say about the graphs of these equations?

24. In solving a system of equations, suppose you get the statement $3 = 3$. How many solutions will the system have? What can you say about the graphs of these equations?

25. In solving a system of equations, suppose you get the statement $2 = -5$. How many solutions will the system have? What can you say about the graphs of these equations?

For Exercises 26–37, solve the system of equations using the addition method.

26. $-2x + y = -5$
$8x - 4y = 12$

27. $x - 3y = 2$
$-5x + 15y = 10$

28. $x + 2y = 2$
$-3x - 6y = -6$

29. $4x - 3y = 6$
$-12x + 9y = -18$

30. $3a + 2b = 11$
$7a - 3b = -5$

31. $4y + 5z = -2$
$5y - 3z = 16$

32. $3x - 5y = 7$
$5x - 2y = -1$

33. $4s + 3t = 9$
$3s + 4t = 12$

34. $2(x + 1) = -3y + 9$
$3x - 10 = -4y$

35. $-3(x - 2) + 7y = 5$
$5y = 2x$

36. $4x - 5y = 0$
$8(x - 1) = 10y$

37. $y = 2x + 1$
$-3(2x - y) = 0$

Concept 2: Summary of Methods for Solving Linear Equations in Two Variables

For Exercises 38–57, solve the system of equations by either the addition method or the substitution method.

38. $5x - 2y = 4$
$y = -3x + 9$

39. $-x = 8y + 5$
$4x - 3y = -20$

40. $0.1x + 0.1y = 0.6$
$0.1x - 0.1y = 0.1$

41. $0.1x + 0.1y = 0.2$
$0.1x - 0.1y = 0.3$

42. $3x = 5y - 9$
$2y = 3x + 3$

43. $10x - 5 = 3y$
$2x + 3y = 1$

44. $y = -5x - 5$
$6x - 3 = -3y$

45. $4x + 5y = -2$
$6x = -10y + 2$

46. $x = -\frac{1}{2}$
$6x - 5y = -8$

47. $4x - 2y = 1$
$y = 3$

48. $0.02x + 0.04y = 0.12$
$0.03x - 0.05y = -0.15$

49. $-0.03x + 0.01y = 0.01$
$-0.06x - 0.02y = -0.02$

50. $8x - 16y = 24$
$2x - 4y = 0$

51. $y = -\frac{1}{2}x - 5$
$2x + 4y = -8$

52. $\frac{m}{2} + \frac{n}{5} = \frac{13}{10}$
$3(m - n) = m - 10$

53. $\frac{a}{4} - \frac{3b}{2} = \frac{15}{2}$
$\frac{1}{5}(a + 2b) = -2$

54. $2(m - 3n) = m + 4$
$3m + 8 = 5m - n$

55. $m - 3n = 10$
$3(m + 4n) = -12$

56. $9a - 2b = 8$
$6(3a + 1) = 4b + 22$

57. $a = 5 + 2b$
$3(a - 2b) = 15$

For Exercises 58–61, set up a system of linear equations, and solve for the indicated quantities.

58. The sum of two positive numbers is 26. Their difference is 14. Find the numbers.

59. The difference of two positive numbers is 2. The sum of the numbers is 36. Find the numbers.

60. Eight times the smaller of two numbers plus 2 times the larger number is 44. Three times the smaller number minus 2 times the larger number is zero. Find the numbers.

61. Six times the smaller of two numbers minus the larger number is -9. Ten times the smaller number plus five times the larger number is 5. Find the numbers.

For Exercises 62–65, use the addition method to eliminate the x-variable to solve for y. Then use the addition method to eliminate the y-variable to solve for x. Write the solution as an ordered pair.

62. $2x + 3y = 6$
$x - y = 5$

63. $6x + 6y = 8$
$9x - 18y = -3$

64. $2x - 5y = 4$
$3x - 3y = 4$

65. $6x - 5y = 7$
$4x - 6y = 7$

For Exercises 66–68, solve the system by using each of the three methods: (a) the graphing method, (b) the substitution method, and (c) the addition method.

66. $2x + y = 1$
$-4x - 2y = -2$

67. $3x + y = 6$
$-2x + 2y = 4$

68. $2x - 2y = 6$
$5y = 5x + 5$

Expanding Your Skills

69. Explain why a system of linear equations cannot have exactly two solutions.

70. The solution to the following system of linear equations is (1, 2). Find A and B.

$$Ax + 3y = 8$$
$$x + By = -7$$

71. The solution to the following system of linear equations is $(-3, 4)$. Find A and B.

$$4x + Ay = -32$$
$$Bx + 6y = 18$$

Applications of Linear Equations in Two Variables

Section 4.4

Concepts

1. Applications Involving Cost

In Sections 2.4–2.6, we solved several applied problems by setting up a linear equation in one variable. When solving an application that involves two unknowns, sometimes it is convenient to use a system of linear equations in two variables.

Example 1 **Using a System of Linear Equations Involving Cost**

At a movie theater a couple buys one large popcorn and two drinks for \$5.75. A group of teenagers buys two large popcorns and five drinks for \$13.00. Find the cost of one large popcorn and the cost of one drink.

Solution:

In this application we have two unknowns, which we can represent by x and y.

Let x represent the cost of one large popcorn.
Let y represent the cost of one drink.

We must now write two equations. Each of the first two sentences in the problem gives a relationship between x and y:

$$\begin{pmatrix}\text{Cost of 1}\\ \text{large popcorn}\end{pmatrix} + \begin{pmatrix}\text{cost of 2}\\ \text{drinks}\end{pmatrix} = \begin{pmatrix}\text{total}\\ \text{cost}\end{pmatrix} \longrightarrow x + 2y = 5.75$$

$$\begin{pmatrix}\text{Cost of 2}\\ \text{large popcorns}\end{pmatrix} + \begin{pmatrix}\text{cost of 5}\\ \text{drinks}\end{pmatrix} = \begin{pmatrix}\text{total}\\ \text{cost}\end{pmatrix} \longrightarrow 2x + 5y = 13.00$$

To solve this system, we may either use the substitution method or the addition method. We will use the substitution method by solving for x in the first equation.

$x + 2y = 5.75 \longrightarrow x = -2y + 5.75$ Isolate x in the first equation.

$$2x + 5y = 13.00$$

$$2(-2y + 5.75) + 5y = 13.00$$ Substitute $x = -2y + 5.75$ into the other equation.

$$-4y + 11.50 + 5y = 13.00$$ Solve for y.

$$y + 11.50 = 13.00$$

$$y = 1.50$$

$$x = -2y + 5.75$$

$$x = -2(1.50) + 5.75$$ Substitute $y = 1.50$ into the equation $x = -2y + 5.75$.

$$x = -3.00 + 5.75$$

$$x = 2.75$$

The cost of one large popcorn is \$2.75 and the cost of one drink is \$1.50.

Check by verifying that the solutions meet the specified conditions.

1 popcorn + 2 drinks = 1(\$2.75) + 2(\$1.50) = \$5.75 ✔ True

2 popcorns + 5 drinks = 2(\$2.75) + 5(\$1.50) = \$13.00 ✔ True

Skill Practice

1. Lynn went to a fast-food restaurant and spent \$9.00. She purchased 4 hamburgers and 5 orders of fries. The next day, Ricardo went to the same restaurant and purchased 10 hamburgers and 7 orders of fries. He spent \$18.10. Use a system of equations to determine the cost of a burger and the cost of an order of fries.

2. Applications Involving Principal and Interest

In Section 2.5, we learned that simple interest is interest computed on the principal amount of money invested (or borrowed). Simple interest, I, is found by using the formula

$I = Prt$ where P is the principal,
r is the annual interest rate, and
t is the time in years.

If the amount of time is taken to be 1 year, we have: $I = Pr(1)$ or simply $I = Pr$.

In Example 2, we apply the concept of simple interest to two accounts to produce a desired amount of interest after 1 year.

Example 2 Using a System of Linear Equations Involving Investments

Joanne has a total of \$6000 to deposit in two accounts. One account earns 3.5% simple interest and the other earns 2.5% simple interest. If the total amount of interest at the end of 1 year is \$195, find the amount she deposited in each account.

Solution:

Let x represent the principal deposited in the 2.5% account.
Let y represent the principal deposited in the 3.5% account.

	2.5% Account	3.5% Account	Total
Principal	x	y	6000
Interest $(I = Pr)$	$0.025x$	$0.035y$	195

Each row of the table yields an equation in x and y:

$$\begin{pmatrix}\text{Principal}\\ \text{invested}\\ \text{at } 2.5\%\end{pmatrix} + \begin{pmatrix}\text{principal}\\ \text{invested}\\ \text{at } 3.5\%\end{pmatrix} = \begin{pmatrix}\text{total}\\ \text{principal}\end{pmatrix} \longrightarrow x + y = 6000$$

$$\begin{pmatrix}\text{Interest}\\ \text{earned}\\ \text{at } 2.5\%\end{pmatrix} + \begin{pmatrix}\text{interest}\\ \text{earned}\\ \text{at } 3.5\%\end{pmatrix} = \begin{pmatrix}\text{total}\\ \text{interest}\end{pmatrix} \longrightarrow 0.025x + 0.035y = 195$$

Skill Practice Answers

1. The cost of a burger is \$1.25 and the cost of an order of fries is \$0.80.

We will choose the addition method to solve the system of equations. First multiply the second equation by 1000 to clear decimals.

$$x + y = 6000 \longrightarrow x + y = 6000$$
$$0.025x + 0.035y = 195 \xrightarrow[\text{Multiply by 1000}]{} 25x + 35y = 195{,}000$$

$$\begin{aligned} x + y = 6000 &\xrightarrow{\text{Multiply by } -25} & -25x - 25y &= -150{,}000 \\ 25x + 35y = 195{,}000 &\longrightarrow & 25x + 35y &= 195{,}000 \\ & & \hline 10y &= 45{,}000 \end{aligned}$$

$$\frac{10y}{10} = \frac{45{,}000}{10}$$

$y = 4500$ The amount invested in the 3.5% account is \$4500.

$x + y = 6000$ Substitute $y = 4500$ into the equation $x + y = 6000$.

$x + 4500 = 6000$

$x = 1500$ The amount invested in the 2.5% account is \$1500.

Joanne deposited \$1500 in the 2.5% account and \$4500 in the 3.5% account.

To check the solution, verify that the conditions of the problem have been met.

1. The sum of \$1500 and \$4500 is \$6000 as desired. ✔
2. The interest earned on \$1500 at 2.5% is: 0.025(\$1500) = \$37.5
   The interest earned on \$4500 at 3.5% is: 0.035(\$4500) = \$157.5
   Total interest: \$195.00 ✔

Skill Practice

2. Addie has a total of \$8000 in two accounts. One pays 5% interest, and the other pays 6.5% interest. At the end of one year, she earned \$475 interest. Use a system of equations to determine the amount invested in each account.

3. Applications Involving Mixtures

Example 3 Using a System of Linear Equations in a Mixture Application

A 10% alcohol solution is mixed with a 40% alcohol solution to produce 30 L of a 20% alcohol solution. Find the number of liters of 10% solution and the number of liters of 40% solution required for this mixture.

Solution:

Each solution contains a percentage of alcohol plus some other mixing agent such as water. Before we set up a system of equations to model this situation,

Skill Practice Answers

2. \$3000 is invested at 5%, and \$5000 is invested at 6.5%.

it is helpful to have background understanding of the problem. In Figure 4-9, the liquid depicted in blue is pure alcohol and the liquid shown in gray is the mixing agent (such as water). Together these liquids form a solution. (Realistically the mixture may not separate as shown, but this image may be helpful for your understanding.)

Let x represent the number of liters of 10% solution.
Let y represent the number of liters of 40% solution.

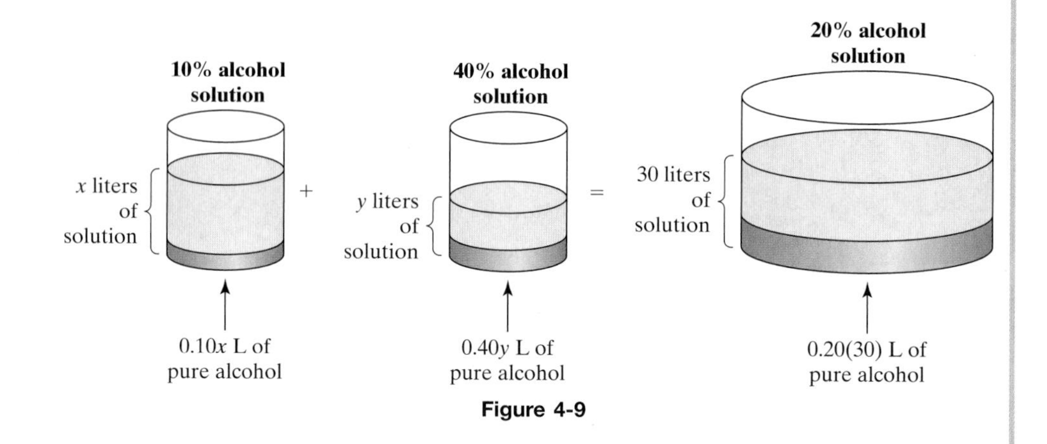

Figure 4-9

The information given in the statement of the problem can be organized in a chart.

	10% Alcohol	40% Alcohol	20% Alcohol
Number of liters of solution	x	y	30
Number of liters of pure alcohol	$0.10x$	$0.40y$	$0.20(30) = 6$

From the first row, we have

$$\begin{pmatrix}\text{Amount of}\\ \text{10\% solution}\end{pmatrix} + \begin{pmatrix}\text{amount of}\\ \text{40\% solution}\end{pmatrix} = \begin{pmatrix}\text{total amount}\\ \text{of 20\% solution}\end{pmatrix} \rightarrow x + y = 30$$

From the second row, we have

$$\begin{pmatrix}\text{Amount of}\\ \text{alcohol in}\\ \text{10\% solution}\end{pmatrix} + \begin{pmatrix}\text{amount of}\\ \text{alcohol in}\\ \text{40\% solution}\end{pmatrix} = \begin{pmatrix}\text{total amount of}\\ \text{alcohol in}\\ \text{20\% solution}\end{pmatrix} \rightarrow 0.10x + 0.40y = 6$$

We will solve the system with the addition method by first clearing decimals.

$$\begin{aligned} x + y &= 30 \xrightarrow{\phantom{\text{Multiply by 10}}} x + y = 30 \xrightarrow{\text{Multiply by } -1} -x - y &= -30 \\ 0.10x + 0.40y &= 6 \xrightarrow[\text{Multiply by 10}]{} x + 4y = 60 \xrightarrow{\phantom{\text{Multiply by } -1}} \underline{x + 4y} &= \underline{60} \\ & & 3y &= 30 \end{aligned}$$

$3y = 30$ After eliminating the x-variable, solve for y.

$y = 10$ 10 L of 40% solution is needed.

$x + y = 30$ Substitute $y = 10$ into either of the original equations.

$x + (10) = 30$

$x = 20$ 20 L of 10% solution is needed.

10 L of 40% solution must be mixed with 20 L of 10% solution.

Skill Practice

3. How many ounces of 20% and 35% acid solution should be mixed together to obtain 15 ounces of 30% acid solution?

4. Applications Involving Distance, Rate, and Time

The following formula relates the distance traveled to the rate and time of travel.

$d = rt$ distance = rate · time

For example, if a car travels at 60 mph for 3 hours, then

$$d = (60 \text{ mph})(3 \text{ hours})$$
$$= 180 \text{ miles}$$

If a car travels at 60 mph for x hours, then

$$d = (60 \text{ mph})(x \text{ hours})$$
$$= 60x \text{ miles}$$

The relationship $d = rt$ is used in Example 4.

Example 4 **Using a System of Linear Equations in a Distance, Rate, Time Application**

A plane travels with a tail wind from Kansas City, Missouri, to Denver, Colorado, a distance of 600 miles in 2 hours. The return trip against a head wind takes 3 hours. Find the speed of the plane in still air, and find the speed of the wind.

Solution:

Let p represent the speed of the plane in still air.
Let w represent the speed of the wind.

Notice that when the plane travels with the wind, the net speed is $p + w$. When the plane travels against the wind, the net speed is $p - w$.

The information given in the problem can be organized in a chart.

	Distance	Rate	Time
With a tail wind	600	$p + w$	2
Against a head wind	600	$p - w$	3

Skill Practice Answers

3. 10 ounces of the 35% solution, and 5 ounces of the 20% solution.

To set up two equations in p and w, recall that $d = rt$.

From the first row, we have

$$\begin{pmatrix}\text{Distance}\\\text{with the wind}\end{pmatrix} = \begin{pmatrix}\text{rate with}\\\text{the wind}\end{pmatrix}\begin{pmatrix}\text{time traveled}\\\text{with the wind}\end{pmatrix} \longrightarrow 600 = (p + w) \cdot 2$$

From the second row, we have

$$\begin{pmatrix}\text{Distance}\\\text{against the wind}\end{pmatrix} = \begin{pmatrix}\text{rate against}\\\text{the wind}\end{pmatrix}\begin{pmatrix}\text{time traveled}\\\text{against the wind}\end{pmatrix} \longrightarrow 600 = (p - w) \cdot 3$$

Using the distributive property to clear parentheses produces the following system:

$$2p + 2w = 600$$
$$3p - 3w = 600$$

The coefficients on the w-variable can be changed to 6 and -6 by multiplying the first equation by 3 and the second equation by 2.

$$\begin{aligned} 2p + 2w = 600 &\xrightarrow{\text{Multiply by 3}} 6p + 6w = 1800 \\ 3p - 3w = 600 &\xrightarrow[\text{Multiply by 2}]{} \underline{6p - 6w = 1200} \\ & \qquad\quad 12p \qquad = 3000 \end{aligned}$$

$$12p = 3000$$

$$\frac{12p}{12} = \frac{3000}{12}$$

$p = 250$ The speed of the plane in still air is 250 mph.

TIP: To create opposite coefficients on the w-variables, we could have divided the first equation by 2 and divided the second equation by 3:

$$\begin{aligned} 2p + 2w = 600 &\xrightarrow{\text{Divide by 2}} p + w = 300 \\ 3p - 3w = 600 &\xrightarrow[\text{Divide by 3}]{} \underline{p - w = 200} \\ & \qquad\quad 2p \qquad = 500 \\ & \qquad\qquad\quad p = 250 \end{aligned}$$

$2p + 2w = 600$ Substitute $p = 250$ into the first equation.

$2(250) + 2w = 600$

$500 + 2w = 600$

$2w = 100$

$w = 50$ The speed of the wind is 50 mph.

The speed of the plane in still air is 250 mph. The speed of the wind is 50 mph.

Skill Practice

4. Dan and Cheryl paddled their canoe 40 miles in 5 hours with the current and 16 miles in 8 hours against the current. Find the speed of the current and the speed of the canoe in still water.

Skill Practice Answers

4. The speed of the current is 3 mph. The speed of the canoe in still water is 5 mph.

5. Miscellaneous Mixture Applications

Example 5 **Solving a Miscellaneous Mixture Application**

At the start of business each day, Petersen's Bakery in St. Augustine has twice the number of $1 bills as $5 bills in the cash register. If the register holds a total of $175 in $1 and $5 bills at the start of the day, how many of each type of bill does it have?

Solution:

First label the unknown quantities.

Let x represent the number of $1 bills.
Let y represent the number of $5 bills.

We must now write two equations that relate x and y.

Since the number of $1 bills is twice the number of $5 bills, we have: $\longrightarrow$ $x = 2y$

Since the total value in the register is $175, we have:

$$\begin{pmatrix}\text{Value of}\\ \text{the \$1 bills}\end{pmatrix} + \begin{pmatrix}\text{value of}\\ \text{the \$5 bills}\end{pmatrix} = \begin{pmatrix}\text{total}\\ \text{value}\end{pmatrix} \longrightarrow 1x + 5y = 175$$

The system can be written as:

$$\begin{aligned} x &= 2y \\ x + 5y &= 175 \end{aligned}$$

In the first equation, the value of x is isolated. Therefore, the substitution method is a good choice to solve the system of equations.

$x + 5y = 175$	Second equation
$(2y) + 5y = 175$	Substitute $x = 2y$ into the second equation.
$7y = 175$	Solve the resulting equation.
$\frac{7y}{7} = \frac{175}{7}$	
$y = 25$	There are twenty-five $5 bills.

$x = 2y$	Now find x by substituting the known value of y.
$= 2(25)$	Substitute $y = 25$.
$= 50$	There are fifty $1 bills.

The cash register holds twenty-five $5 bills and fifty $1 bills.

Skill Practice

5. A postal worker has four times as many 39¢ stamps as 3¢ stamps. If the total value of the stamps is $15.90, how many of each type of stamp are there?

Skill Practice Answers

5. There are ten 3¢ stamps and forty 39¢ stamps.

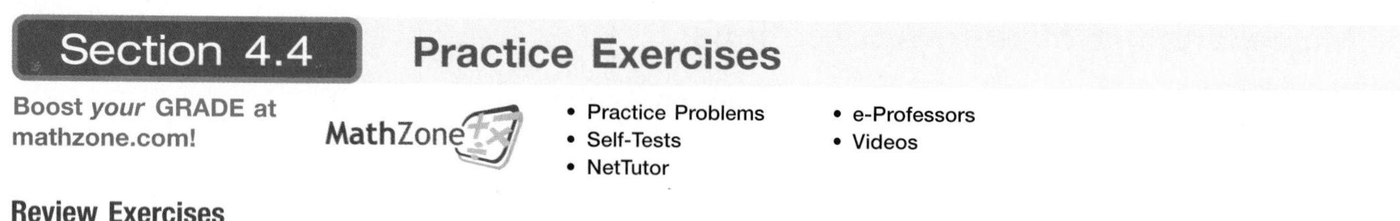

Review Exercises

For Exercises 1–4, solve each system of equations by three different methods:

a. Graphing method **b.** Substitution method **c.** Addition method

1. $-2x + y = 6$
$2x + y = 2$

2. $x - y = 2$
$x + y = 6$

3. $y = -2x + 6$
$4x - 2y = 8$

4. $2x = y + 4$
$4x = 2y + 8$

For Exercises 5–8, set up a system of linear equations in two variables to solve for the unknown quantities.

5. One number is eight more than twice another. Their sum is 20. Find the numbers.

6. The difference of two positive numbers is 264. The larger number is three times the smaller number. Find the numbers.

7. Two angles are complementary. The measure of one angle is 10° less than nine times the measure of the other. Find the measure of each angle.

8. Two angles are supplementary. The measure of one angle is 9° more than twice the measure of the other angle. Find the measure of each angle.

Concept 1: Applications Involving Cost

9. Kent bought three tapes and two CDs for \$62.50. Demond bought one tape and four CDs for \$72.50. Find the cost of one tape and the cost of one CD.

10. Tanya bought three adult tickets and one child's ticket to a movie for \$23.00. Li bought two adult tickets and five children's tickets for \$30.50. Find the cost of one adult ticket and the cost of one children's ticket.

11. Linda bought 100 shares of a technology stock and 200 shares of a mutual fund for \$3800. Her sister, Sandy, bought 300 shares of technology stock and 50 shares of a mutual fund for \$5350. Find the cost per share of the technology stock, and the cost per share of the mutual fund.

12. Two videos and three DVDs can be rented for \$19.15. Four videos and one DVD can be rented for \$17.35. Find the cost to rent one video and the cost to rent one DVD.

Concept 2: Applications Involving Principal and Interest

13. Shanelle invested \$10,000, and at the end of 1 year, she received \$805 in interest. She invested part of the money in an account earning 10% simple interest and the remaining money in an account earning 7% simple interest. How much did she invest in each account?

	10% Account	7% Account	Total
Principal invested			
Interest earned			

14. $12,000 was invested in two accounts, one earning 12% simple interest and the other earning 8% simple interest. If the total interest at the end of 1 year was $1240, how much was invested in each account?

	12% Account	8% Account	Total
Principal invested			
Interest earned			

15. Troy borrowed a total of $12,000 in two different loans to help pay for his new Chevy Silverado. One loan charges 9% simple interest, and the other charges 6% simple interest. If he is charged $810 in interest after 1 year, find the amount borrowed at each rate.

16. Blake has a total of $4000 to invest in two accounts. One account earns 2% simple interest, and the other earns 5% simple interest. How much should be invested in both accounts to earn exactly $155 at the end of 1 year?

Concept 3: Applications Involving Mixtures

17. How much 50% disinfectant solution must be mixed with a 40% disinfectant solution to produce 25 gal of a 46% disinfectant solution?

	50% Mixture	40% Mixture	46% Mixture
Amount of solution			
Amount of disinfectant			

18. How many gallons of 20% antifreeze solution and a 10% antifreeze solution must be mixed to obtain 40 gal of a 16% antifreeze solution?

19 How much 45% disinfectant solution must be mixed with a 30% disinfectant solution to produce 20 gal of a 39% disinfectant solution?

20. How many gallons of a 25% antifreeze solution and a 15% antifreeze solution must be mixed to obtain 15 gal of a 23% antifreeze solution?

Concept 4: Applications Involving Distance, Rate, and Time

21. It takes a boat 2 hr to go 16 miles downstream with the current and 4 hr to return against the current. Find the speed of the boat in still water and the speed of the current.

	Distance	Rate	Time
Downstream			
Return			

22. A boat takes 1.5 hr to go 12 miles upstream against the current. It can go 24 miles downstream with the current in the same amount of time. Find the speed of the current and the speed of the boat in still water.

23. A plane can fly 960 miles with the wind in 3 hr. It takes the same amount of time to fly 840 miles against the wind. What is the speed of the plane in still air and the speed of the wind?

24. A plane flies 720 miles with the wind in 3 hr. The return trip takes 4 hr. What is the speed of the wind and the speed of the plane in still air?

Concept 5: Miscellaneous Mixture Applications

25. Debi has \$2.80 in a collection of dimes and nickels. The number of nickels is five more than the number of dimes. Find the number of each type of coin.

26. A child is collecting state quarters and new \$1 coins. If she has a total of 25 coins, and the number of quarters is nine more than the number of dollar coins, how many of each type of coin does she have?

27. In the 1961–1962 NBA basketball season, Wilt Chamberlain of the Philadelphia Warriors made 2432 baskets. Some of the baskets were free throws (worth 1 point each) and some were field goals (worth 2 points each). The number of field goals was 762 more than the number of free throws.

a. How many field goals did he make and how many free throws did he make?

b. What was the total number of points scored?

c. If Wilt Chamberlain played 80 games during this season, what was the average number of points per game?

28. In the 1971–1972 NBA basketball season, Kareem Abdul-Jabbar of the Milwaukee Bucks made 1663 baskets. Some of the baskets were free throws (worth 1 point each) and some were field goals (worth 2 points each). The number of field goals he scored was 151 more than twice the number of free throws.

a. How many field goals did he make and how many free throws did he make?

b. What was the total number of points scored?

c. If Kareem Abdul-Jabbar played 81 games during this season, what was the average number of points per game?

29. A small plane can fly 350 miles with a tailwind in $1\frac{3}{4}$ hours. In the same amount of time, the same plane can travel only 210 miles with a headwind. What is the speed of the plane in still air and the speed of the wind?

30. A plane takes 2 hr to travel 1000 miles with the wind. It can travel only 880 miles against the wind in the same time. Find the speed of the wind and the speed of the plane in still air.

31. At the holidays, Erica likes to sell a candy/nut mixture to her neighbors. She wants to combine candy that costs \$1.80 per pound with nuts that cost \$1.20 per pound. If Erica needs 20 lb of mixture that will sell for \$1.56 per pound, how many pounds of candy and how many pounds of nuts should she use?

32. Mary Lee's natural food store sells a combination of teas. The most popular is a mixture of a tea that sells for \$3.00 per pound with one that sells for \$4.00 per pound. If she needs 40 lb of tea that will sell for \$3.65 per pound, how many pounds of each tea should she use?

33. A total of \$60,000 is invested in two accounts, one that earns 5.5% simple interest, and one that earns 6.5% simple interest. If the total interest at the end of 1 year is \$3750, find the amount invested in each account.

34. Jacques borrows a total of \$15,000. Part of the money is borrowed from a bank that charges 12% simple interest per year. Jacques borrows the remaining part of the money from his sister and promises to pay her 7% simple interest per year. If Jacques' total interest for the year is \$1475, find the amount he borrowed from each source.

35. Miracle-Gro All-Purpose Plant Food contains 15% nitrogen. Green Light Super Bloom contains 12% nitrogen. How much Miracle-Gro and how much Green Light fertilizer must be mixed to obtain 60 oz of a mixture that is 13% nitrogen?

36. A textile manufacturer wants to combine a mixture of 20% dye with a mixture that is 50% dye to form 200 gal of a mixture that is 42.5% dye. How much of the 20% and 50% dye mixtures should he use?

37. In the 1994 Super Bowl, the Dallas Cowboys scored four more points than twice the number of points scored by the Buffalo Bills. If the total number of points scored by both teams was 43, find the number of points scored by each team.

38. In the 1973 Super Bowl, the Miami Dolphins scored twice as many points as the Washington Redskins. If the total number of points scored by both teams was 21, find the number of points scored by each team.

Expanding Your Skills

39. In a survey conducted among 500 college students, 340 said that the campus lacked adequate lighting. If $\frac{4}{5}$ of the women and $\frac{1}{2}$ of the men said that they thought the campus lacked adequate lighting, how many men and how many women were in the survey?

40. A thousand people were surveyed in southern California, and 445 said that they worked out at least three times a week. If $\frac{1}{2}$ of the women and $\frac{3}{8}$ of the men said that they worked out at least three times a week, how many men and how many women were in the survey?

41. During a 1-hour television program, there were 22 commercials. Some commercials were 15 sec and some were 30 sec long. Find the number of 15-sec commercials and the number of 30-sec commercials if the total playing time for commercials was 9.5 min.

Linear Inequalities in Two Variables

Section 4.5

Concept

1. Graphing Linear Inequalities in Two Variables

1. Graphing Linear Inequalities in Two Variables

A **linear inequality in two variables** x and y is an inequality that can be written in one of the following forms: $ax + by < c$, $ax + by > c$, $ax + by \leq c$, or $ax + by \geq c$.

A solution to a linear inequality in two variables is an ordered pair that makes the inequality true. For example, solutions to the inequality $x + y < 3$ are ordered pairs (x, y) such that the sum of the x- and y-coordinates is less than 3. Several such examples are $(0, 0)$, $(-2, -2)$, $(3, -2)$, and $(-4, 1)$. There are actually infinitely many solutions to this inequality, and therefore it is convenient to express the solution set as a graph. The shaded area in Figure 4-10 represents all solutions (x, y), whose coordinates total less than 3.

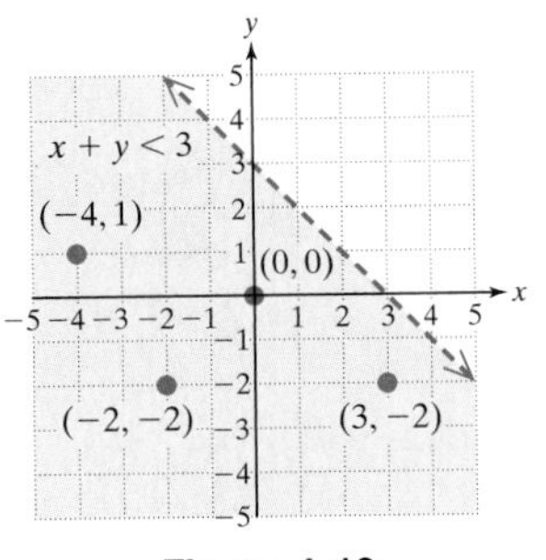

Figure 4-10

To graph a linear inequality in two variables we will use a process called the **test point method**. To use the test point method, first graph the related equation. In this case, the related equation represents a line in the xy-plane. Then choose a test point *not* on the line to determine which side of the line to shade. This process is demonstrated in Example 1.

Example 1 Graphing a Linear Inequality in Two Variables

Graph the solution set. $2x + y \leq 3$

Solution:

$2x + y \leq 3 \longrightarrow 2x + y = 3$ **Step 1:** Set up the related equation.

Step 2: Graph the related equation.

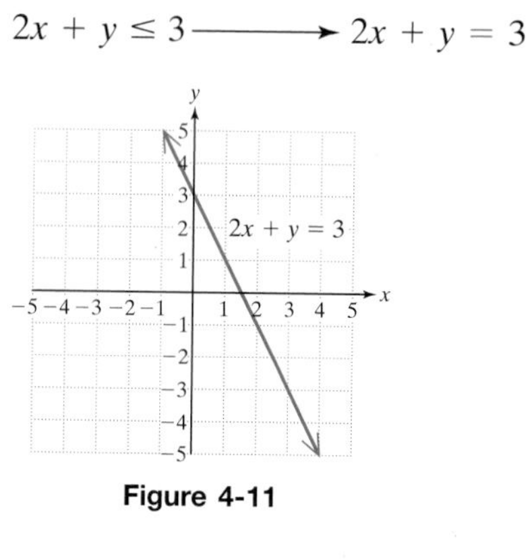

Figure 4-11

Graph the line by either setting up a table of points, or by using the slope-intercept form (Figure 4-11).

Table:

x	y
1	1
0	3
$\frac{3}{2}$	0

Slope-intercept form:

$2x + y = 3$

$y = -2x + 3$

Step 3: The solution to $2x + y \leq 3$ includes points for which $2x + y$ is less than *or equal to 3*. Because equality is included, points on the line $2x + y = 3$ are included.

Now we must determine which side of the line to shade. To do so, we choose an arbitrary test point *not* on the line. The point (0, 0) is a convenient choice.

Test point: (0, 0)

$$2x + y \leq 3$$

$$2(0) + (0) \overset{?}{\leq} 3$$

$$0 \leq 3 \checkmark \quad \text{True}$$

The test point (0, 0) is true in the original inequality. This means that the region from which the test point was taken is part of the solution set. Therefore, shade below the line (Figure 4-12).

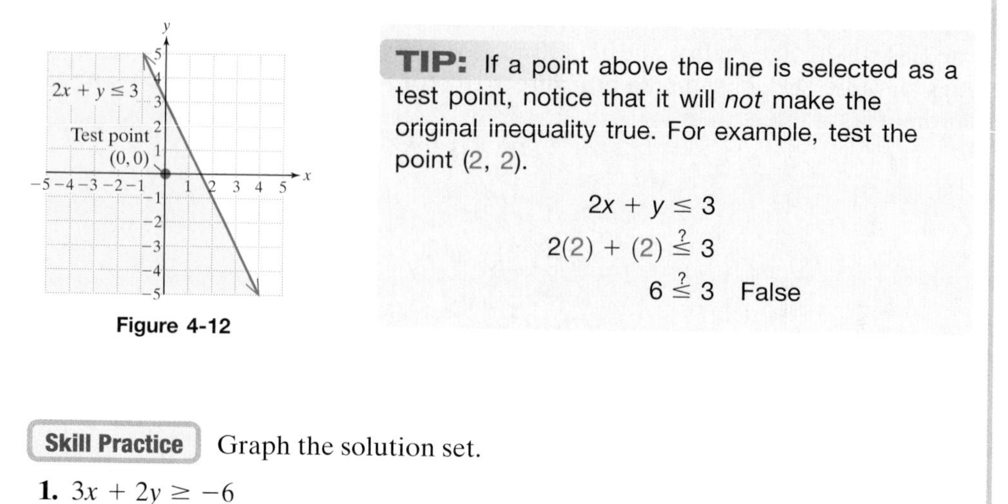

Figure 4-12

TIP: If a point above the line is selected as a test point, notice that it will *not* make the original inequality true. For example, test the point (2, 2).

$$2x + y \leq 3$$

$$2(2) + (2) \overset{?}{\leq} 3$$

$$6 \overset{?}{\leq} 3 \quad \text{False}$$

Skill Practice Graph the solution set.

1. $3x + 2y \geq -6$

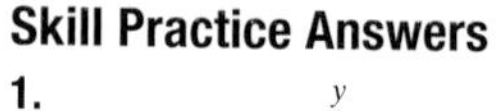

Skill Practice Answers

1.

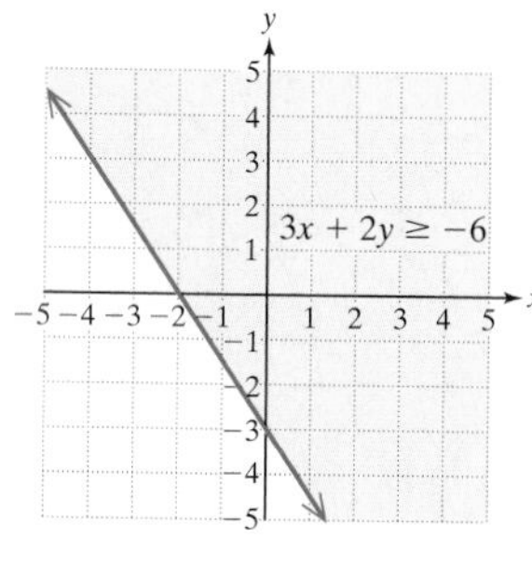

Now suppose the inequality from Example 1 had the strict inequality symbol, <. That is, consider the inequality $2x + y < 3$. The boundary line $2x + y = 3$ is *not* included in the solution set, because the expression $2x + y$ must be *strictly less than 3* (not equal to 3). To show that the boundary line is not included in the solution set, we draw a dashed line (Figure 4-13).

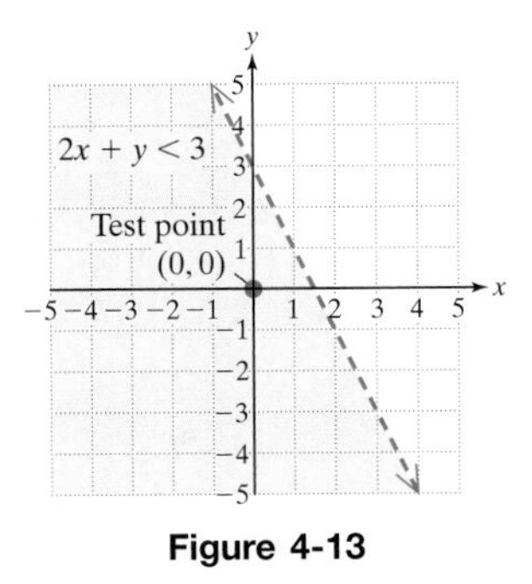

Figure 4-13

The test point method to graph linear inequalities in two variables is summarized as follows:

Test Point Method: Summary

1. Set up the related equation.
2. Graph the related equation from step 1. The equation will be a boundary line in the xy-plane.
   - If the original inequality is a strict inequality, < or >, then the line is *not* part of the solution set. Graph the line as a *dashed line.*
   - If the original inequality is not strict, ≤ or ≥, then the line *is* part of the solution set. Graph the line as a *solid line.*
3. Choose a point not on the line and substitute its coordinates into the original inequality.
   - If the test point makes the inequality true, then the region it represents is part of the solution set. Shade that region.
   - If the test point makes the inequality false, then the other region is part of the solution set and should be shaded.

Example 2 Graphing a Linear Inequality in Two Variables

Graph the solution set. $4x - 2y > 6$

Solution:

$4x - 2y > 6 \longrightarrow 4x - 2y = 6$ **Step 1:** Set up the related equation.

Step 2: Graph the equation. Draw a dashed line because the inequality is strict, > (Figure 4-14).

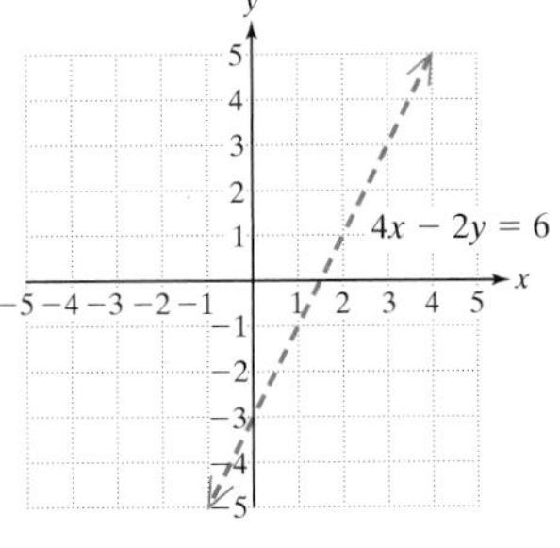

Figure 4-14

Table:

x	y
0	−3
$\frac{3}{2}$	0
2	1

Slope-intercept form:

$$4x - 2y = 6$$
$$-2y = -4x + 6$$
$$y = 2x - 3$$

Step 3: Choose a test point. Again (0, 0) is a good choice because, when substituted into the original inequality, the arithmetic will be minimal.

$$4x - 2y > 6$$
$$4(0) - 2(0) \overset{?}{>} 6$$
$$0 \overset{?}{>} 6 \quad \text{False}$$

The test point from above the line does not check in the original inequality. Therefore, shade below the line (Figure 4-15).

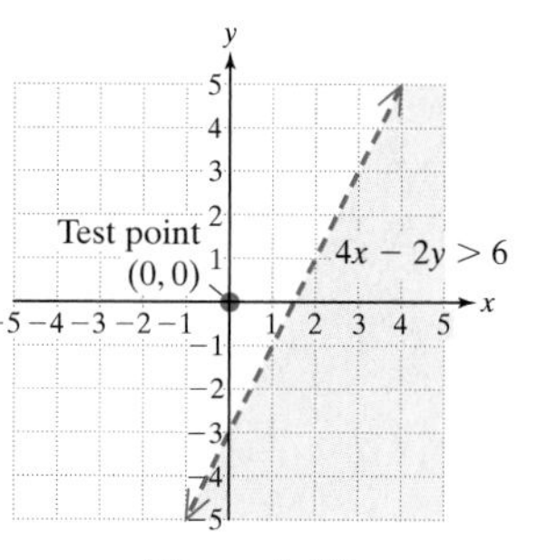

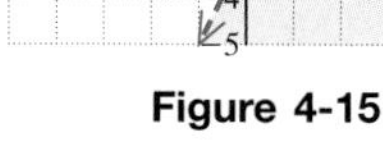
Figure 4-15

Skill Practice Graph the solution set.

2. $6x - 2y < -6$

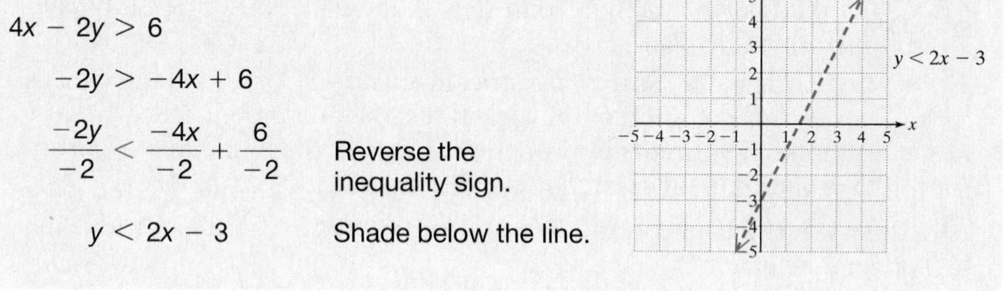

TIP: An inequality can also be graphed by first solving the inequality for *y*. Then,

- Shade *below* the line if the inequality is of the form $y < mx + b$ or $y \leq mx + b$.
- Shade *above* the line if the inequality is of the form $y > mx + b$ or $y \geq mx + b$.

From Example 2, we have

$$4x - 2y > 6$$
$$-2y > -4x + 6$$
$$\frac{-2y}{-2} < \frac{-4x}{-2} + \frac{6}{-2} \quad \text{Reverse the inequality sign.}$$
$$y < 2x - 3 \quad \text{Shade below the line.}$$

Example 3 Graphing a Linear Inequality in Two Variables

Graph the solution set. $2y \geq 5x$

Solution:

$2y \geq 5x \longrightarrow 2y = 5x$

Step 1: Set up the related equation.

Step 2: Graph the equation. Draw a solid line because the symbol $\geq$ is used (Figure 4-16).

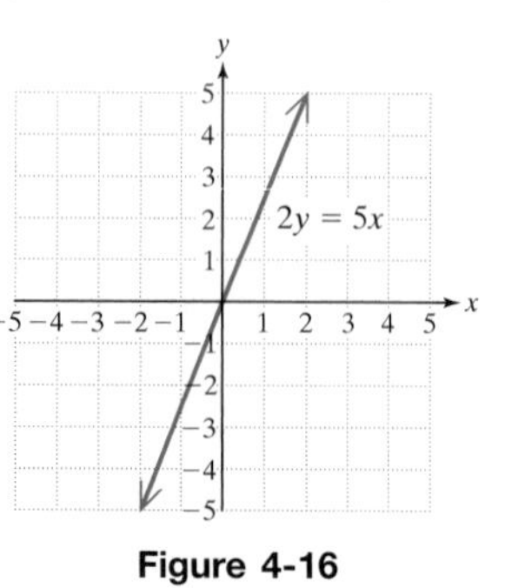

Figure 4-16

Table:

x	y
0	0
1	$\frac{5}{2}$
−1	$-\frac{5}{2}$

Slope-intercept form:

$$2y = 5x$$
$$y = \frac{5}{2}x$$

Skill Practice Answers

2.

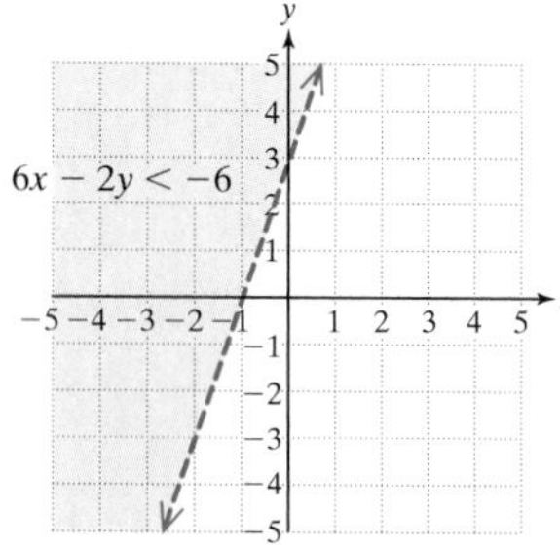

Step 3: The point (0, 0) cannot be used as a test point because it is on the boundary line. Choose a different point such as (1, 1).

$$2y \geq 5x$$

$$2(1) \stackrel{?}{\geq} 5(1)$$

$$2 \stackrel{?}{\geq} 5 \quad \text{False}$$

The test point from below the line does not check in the original inequality. Therefore, shade above the line (Figure 4-17).

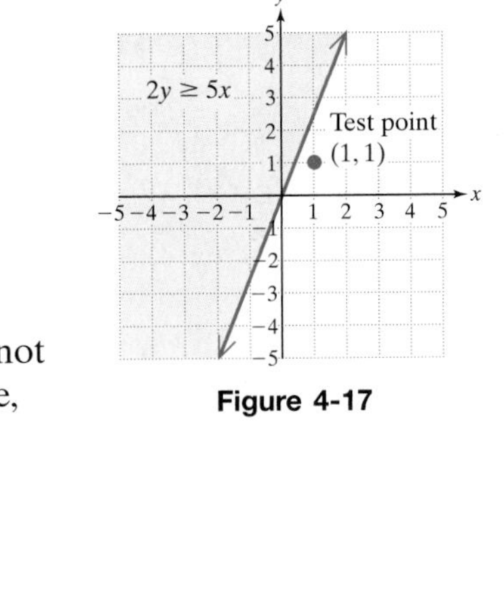

Figure 4-17

Skill Practice Graph the solution set.

3. $-2y < x$

Example 4 Graphing a Linear Inequality in Two Variables

Graph the solution set. $2x > -4$

Solution:

$2x > -4 \longrightarrow 2x = -4$

Step 1: Set up the related equation.

Step 2: Graph the equation. The equation represents a vertical line.

$$2x = -4$$

$$x = -2$$

Draw a dashed vertical line (Figure 4-18).

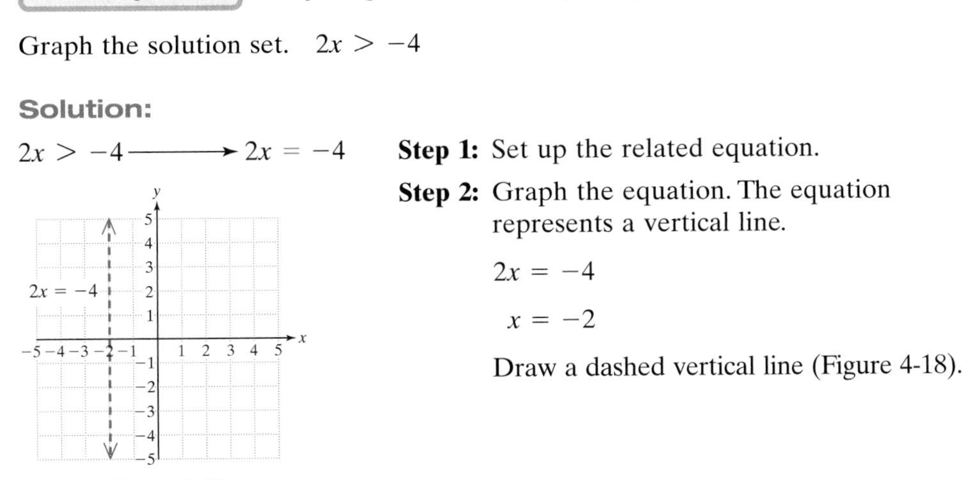

Figure 4-18

Step 3: Choose a test point such as (0, 0).

$$2x > -4$$

$$2(0) \stackrel{?}{>} -4$$

$$0 > -4 \checkmark \quad \text{True}$$

The test point from the right of the line checks in the original inequality. Therefore, shade to the right of the line (Figure 4-19).

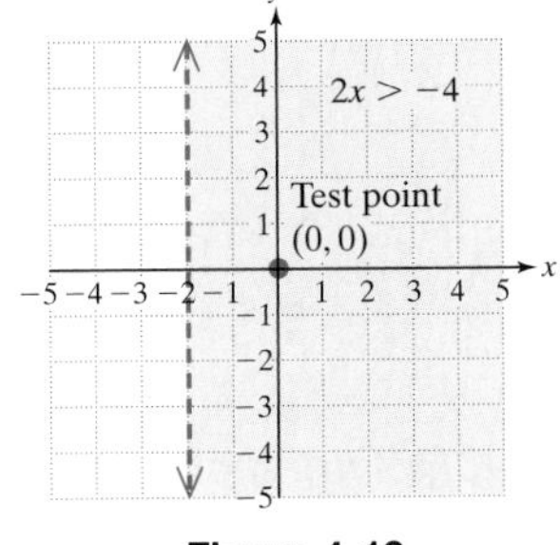

Figure 4-19

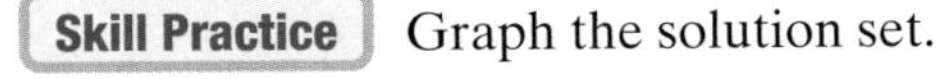

Skill Practice Graph the solution set.

4. $4x \geq 12$

Skill Practice Answers

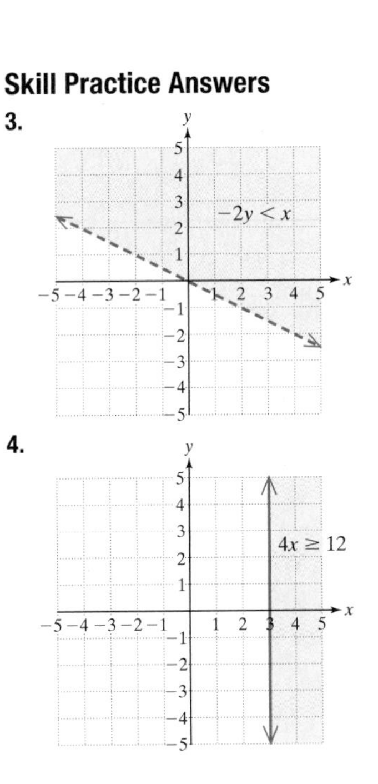

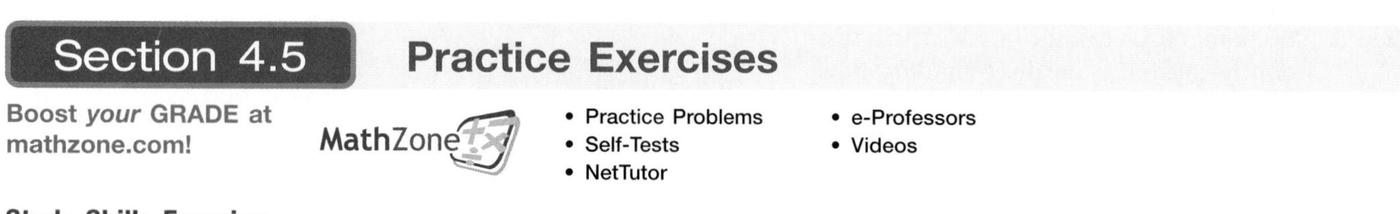

Section 4.5 Practice Exercises

Study Skills Exercise

1. Define the key terms:

a. linear inequality in two variables **b. test point method**

Review Exercises

For Exercises 2–4, graph the equations using the slope and the y-intercept.

2. $y = 5x + 1$ **3.** $y = \frac{3}{5}x + 2$ **4.** $y = -\frac{4}{3}x$

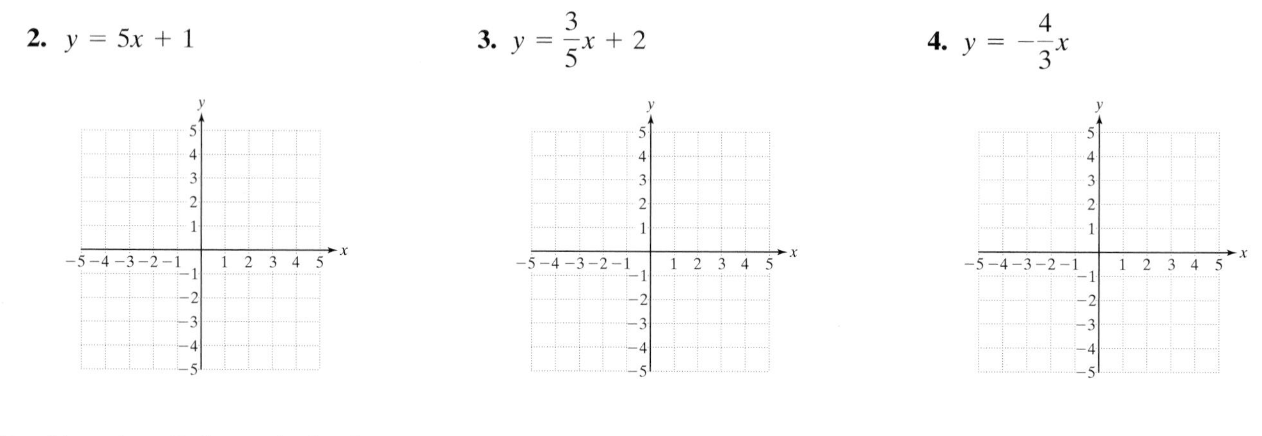

For Exercises 5–6, graph the lines.

5. $x = -3$ **6.** $y = -\frac{5}{2}$

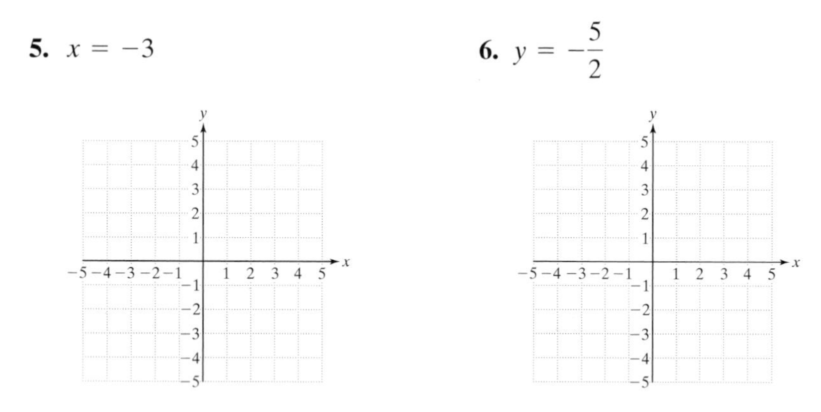

Concept 1: Graphing Linear Inequalities in Two Variables

7. When is a solid line used in the graph of a linear inequality in two variables?

8. When is a dashed line used in the graph of a linear inequality in two variables?

9. What does the shaded region represent in the graph of a linear inequality in two variables?

10. When graphing a linear inequality in two variables, how do you determine which side of the boundary line to shade?

11. Which is the graph of $-2x - y \leq 2$?

a. **b.** **c.**

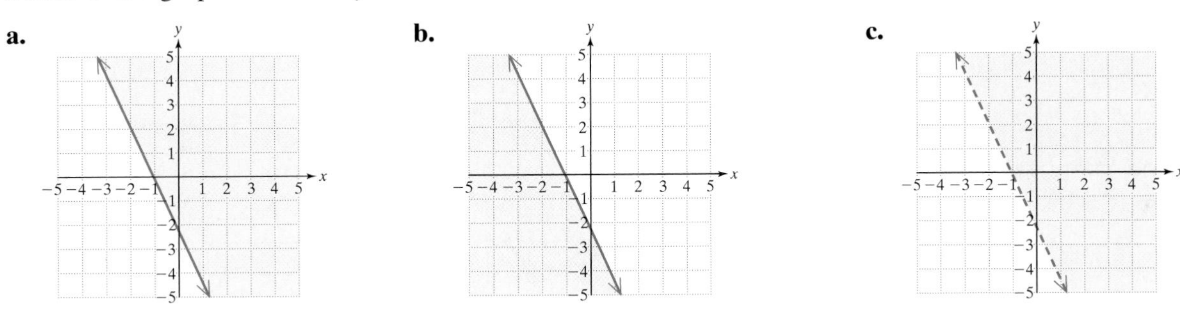

12. Which is the graph of $-3x + y > -1$?

a. **b.** **c.**

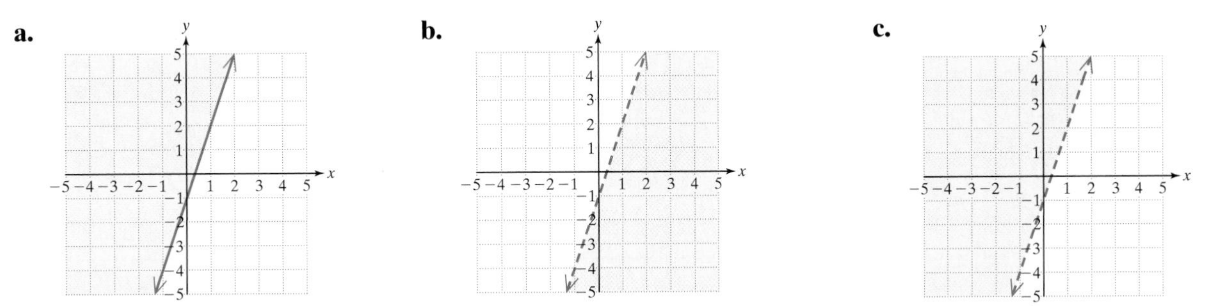

For Exercises 13–18, graph the linear inequalities. Then write three ordered pairs that are solutions to each inequality.

13. $y \geq -x + 5$

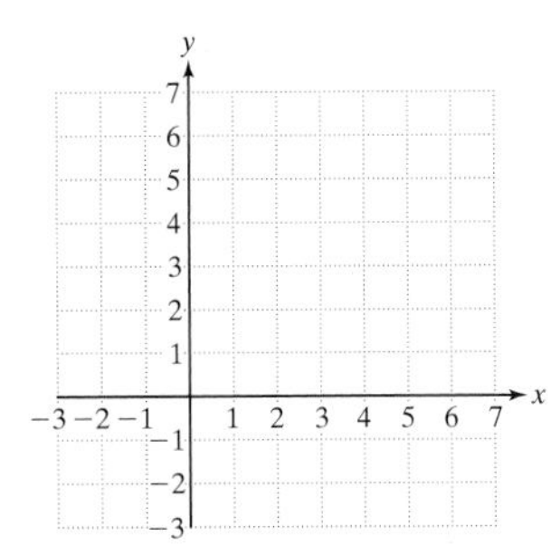

14. $y \leq 2x - 1$

15. $y < 4x$

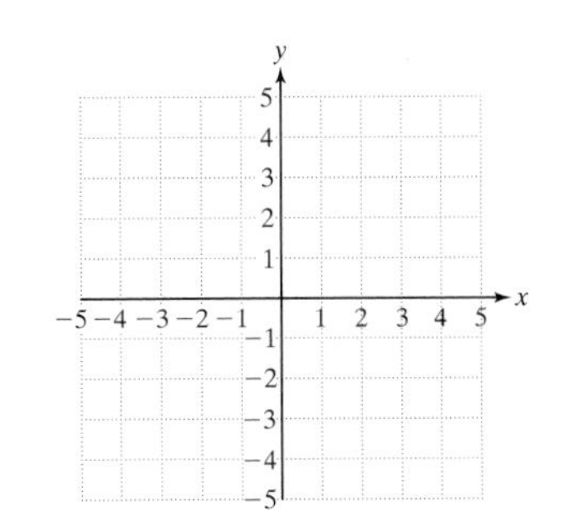

16. $y > -5x$

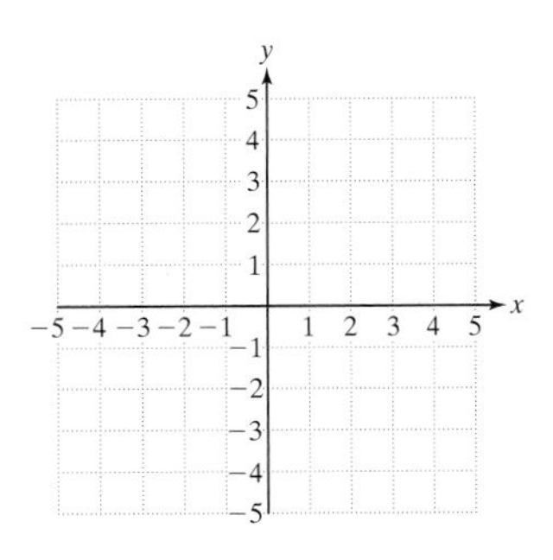

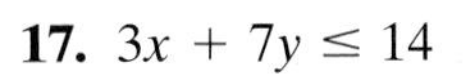

17. $3x + 7y \leq 14$

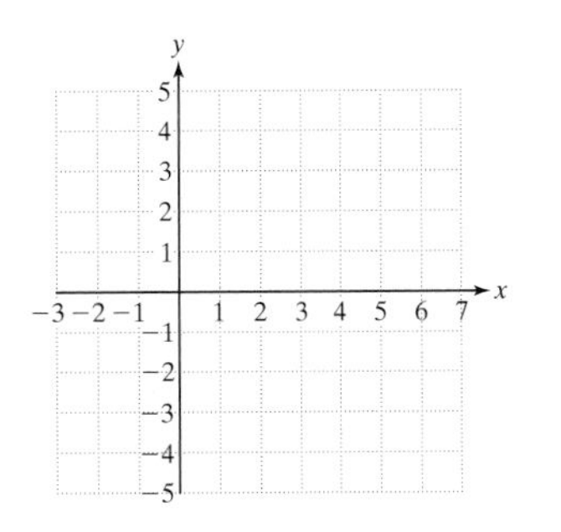

18. $5x - 6y \geq 18$

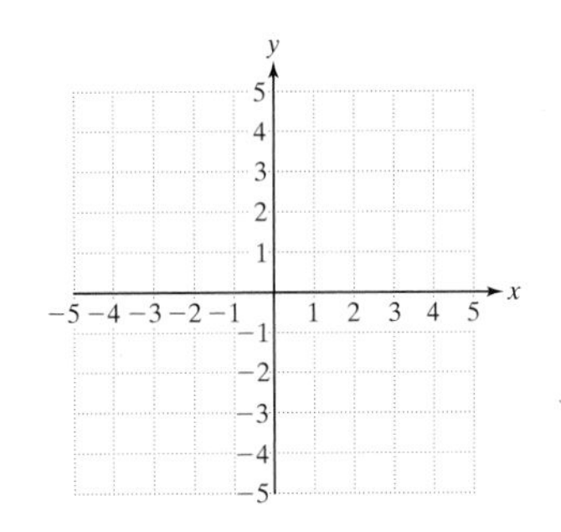

For Exercises 19–36, graph the linear inequalities.

19. $x - y > 6$

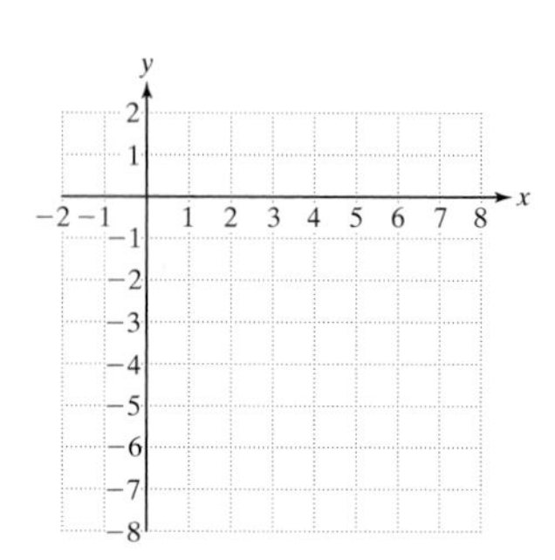

20. $x + y < 5$

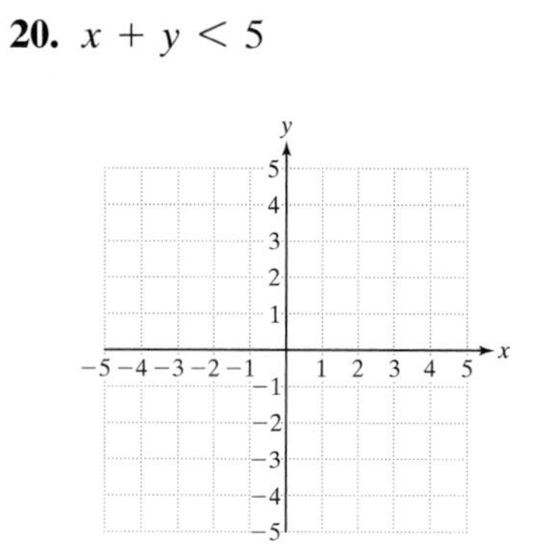

21. $x \geq -1$

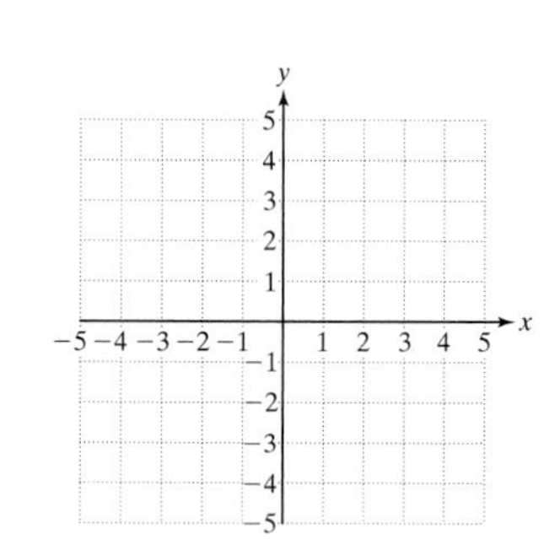

22. $x \leq 6$

23. $y < 3$

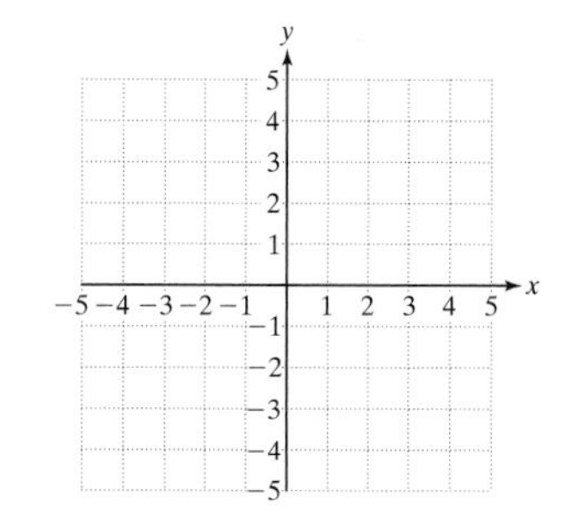

24. $y > -3$

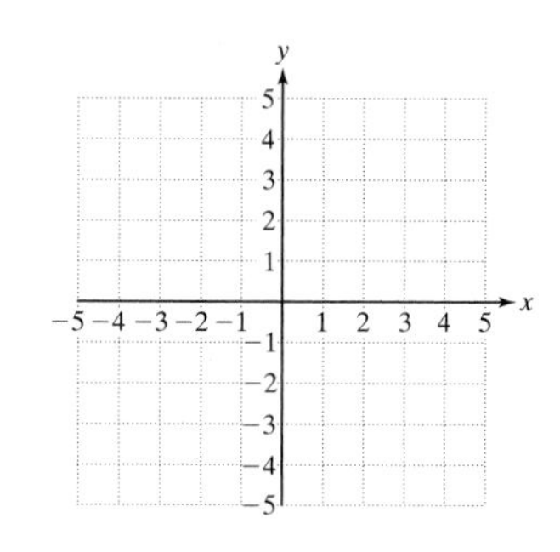

25. $y \leq -\frac{3}{4}x + 2$

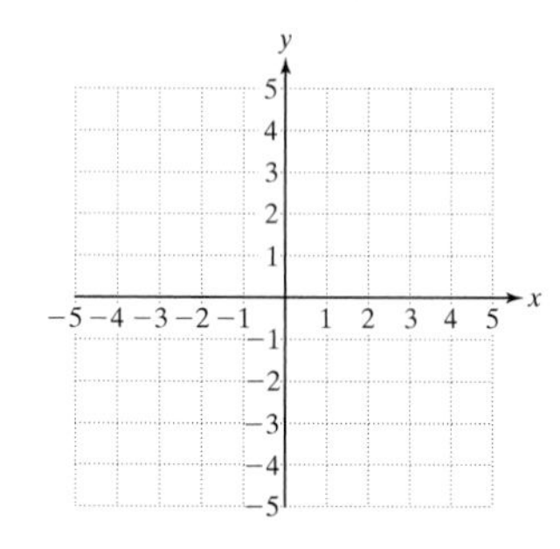

26. $y \geq \frac{2}{3}x + 1$

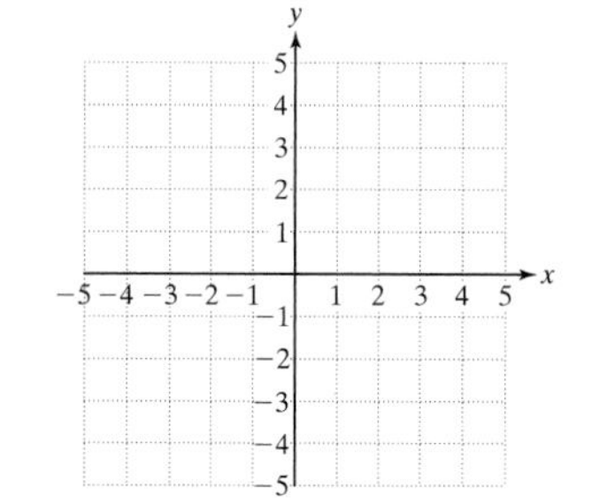

27. $y - 2x > 0$

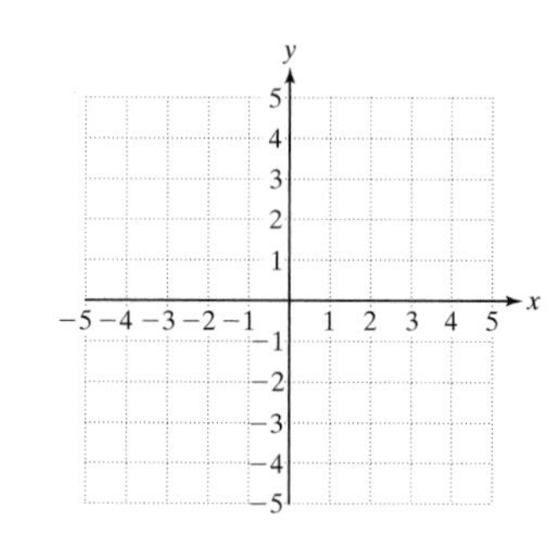

28. $y + 3x < 0$

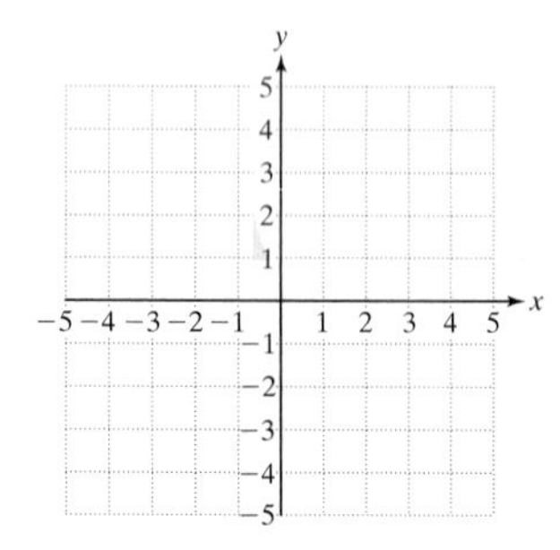

29. $x \leq 0$

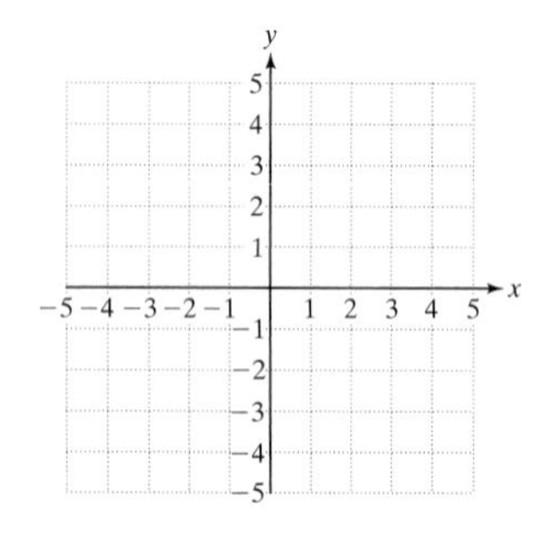

30. $y \leq 0$

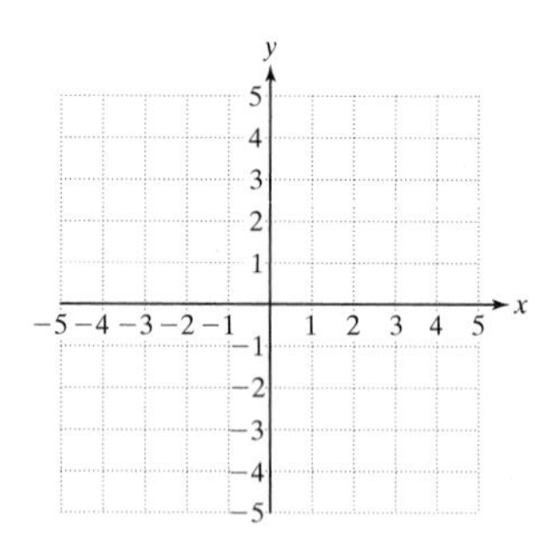

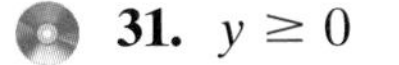

31. $y \geq 0$

32. $x \geq 0$

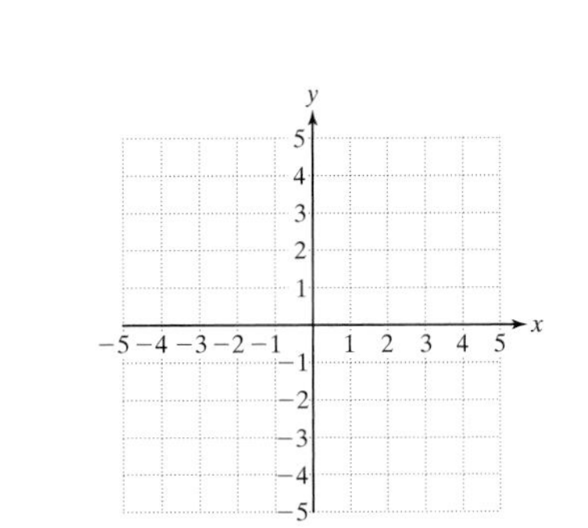

33. $-x \leq \frac{1}{2}y - 2$

34. $-3 + 2x \leq -y$

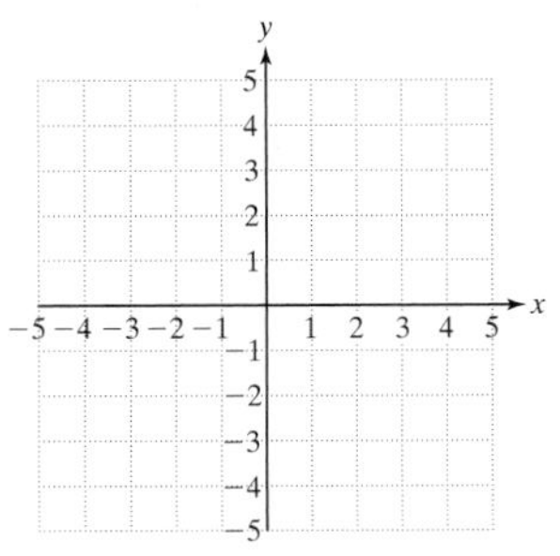

35. $2x > 3y$

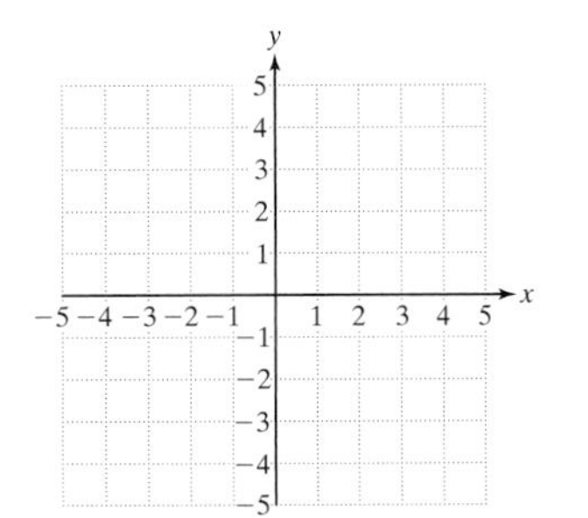

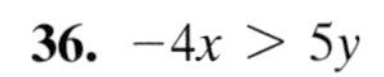

36. $-4x > 5y$

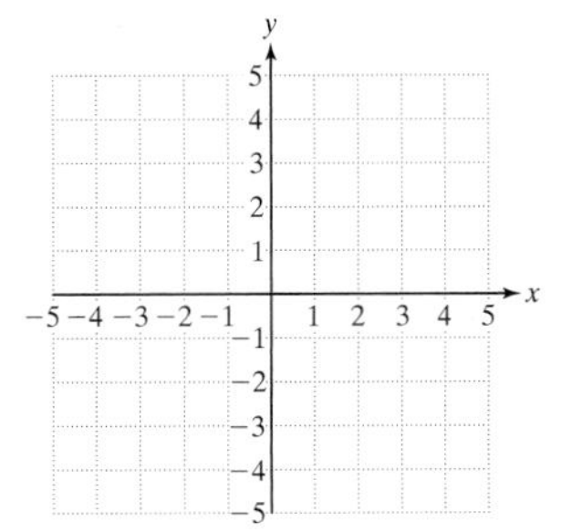

37. **a.** Describe the graph of the inequality $x + y > 4$. Find three solutions to the inequality (answers will vary).

    b. Describe the graph of the equation $x + y = 4$. Find three solutions to the equation (answers will vary).

    c. Describe the graph of the inequality $x + y < 4$. Find three solutions to the inequality (answers will vary).

38. **a.** Describe the graph of the inequality $x + y < 3$. Find three solutions to the inequality (answers will vary).

    b. Describe the graph of the equation $x + y = 3$. Find three solutions to the equation (answers will vary).

    c. Describe the graph of the inequality $x + y > 3$. Find three solutions to the inequality (answers will vary).

Systems of Linear Inequalities in Two Variables

Section 4.6

Concept

1. Graphing Systems of Linear Inequalities in Two Variables

1. Graphing Systems of Linear Inequalities in Two Variables

In Sections 4.1–4.4, we studied systems of linear equations in two variables. Graphically, a solution to such a system is a point of intersection between two lines. In this section, we will study systems of linear *inequalities* in two variables. Graphically, the solution set to such a system is the intersection (or "overlap") of the shaded regions of each individual inequality.

Example 1 Graphing a System of Linear Inequalities

Graph the solution set. $x \geq 2$

$y < 3$

Solution:

We begin by graphing each inequality individually.

$x \geq 2 \xrightarrow{\text{Related equation}} x = 2$ (vertical line)

The solution to $x \geq 2$ is represented by the points on and to the right of the line $x = 2$ (Figure 4-20).

$y < 3 \xrightarrow{\text{Related equation}} y = 3$ (horizontal line)

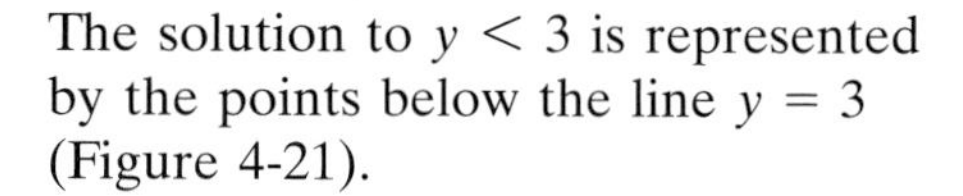

The solution to $y < 3$ is represented by the points below the line $y = 3$ (Figure 4-21).

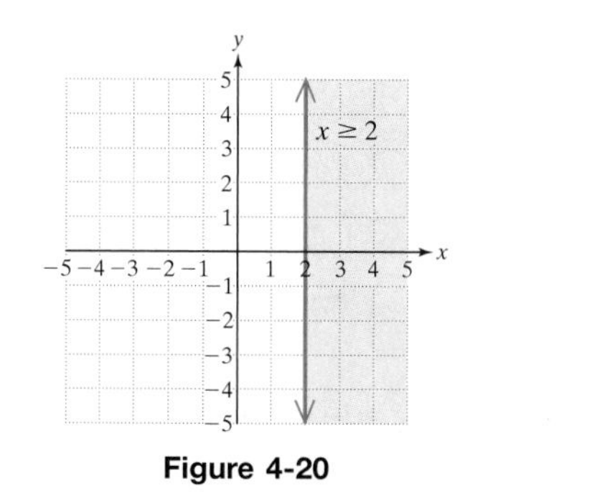

Figure 4-20

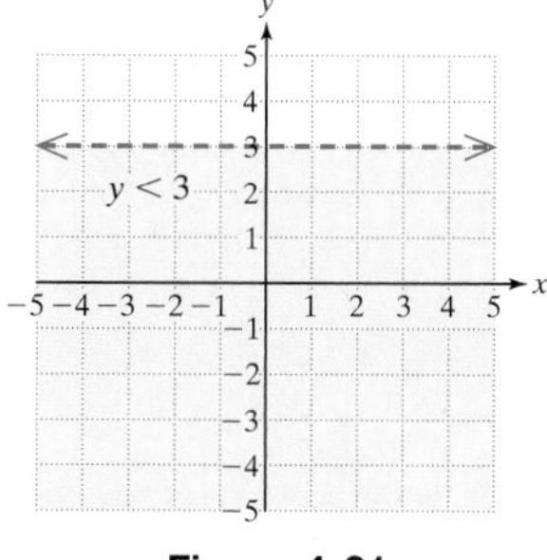

Figure 4-21

Next, we draw these regions on the same graph. The intersection ("overlap") is shown in purple (Figure 4-22).

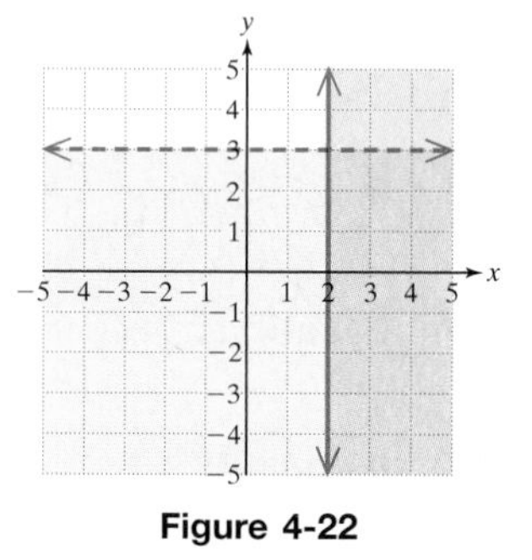

Figure 4-22

In Figure 4-23, we show the solution to the system of inequalities. Notice that any portions of the lines not bounding the solution should be dashed.

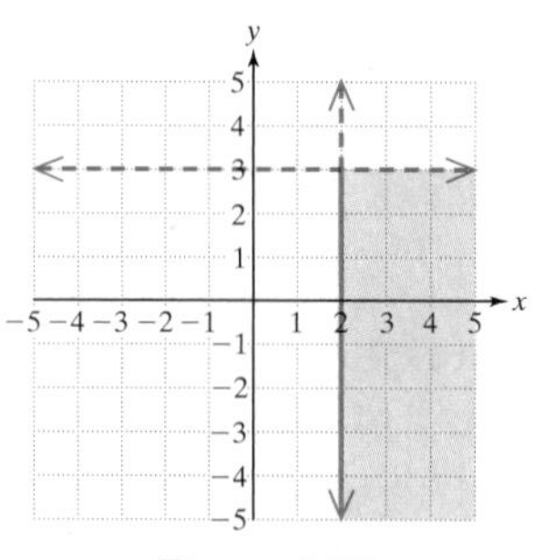

Figure 4-23

Skill Practice

1. Graph the solution set.

$$x < 1$$
$$y \geq -3$$

Example 2 **Graphing a System of Linear Inequalities**

Graph the solution set. $y > \frac{1}{2}x - 2$

$x + y \leq 1$

Solution:

Sketch each inequality.

$y > \frac{1}{2}x - 2 \xrightarrow{\text{Related equation}} y = \frac{1}{2}x - 2$

The line $y = \frac{1}{2}x - 2$ is drawn in red in Figure 4-24. Substituting the test point (0, 0) into the inequality results in a true statement. Therefore, we shade above the line.

$x + y \leq 1 \xrightarrow{\text{Related equation}} x + y = 1$

The line $x + y = 1$ is drawn in blue in Figure 4-25. Substituting the test point (0, 0) into the inequality results in a true statement. Therefore, we shade below the line.

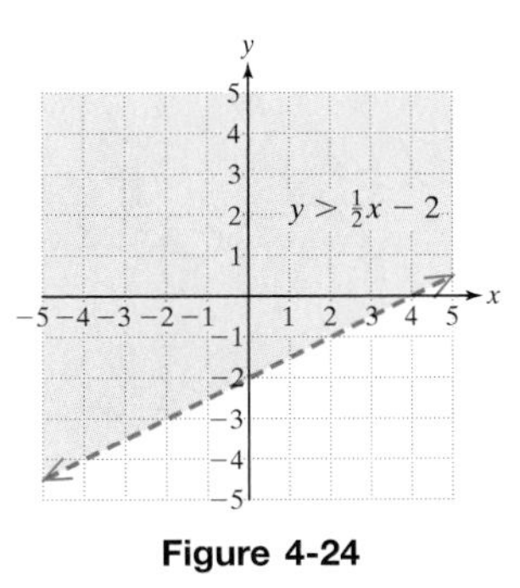

Figure 4-24

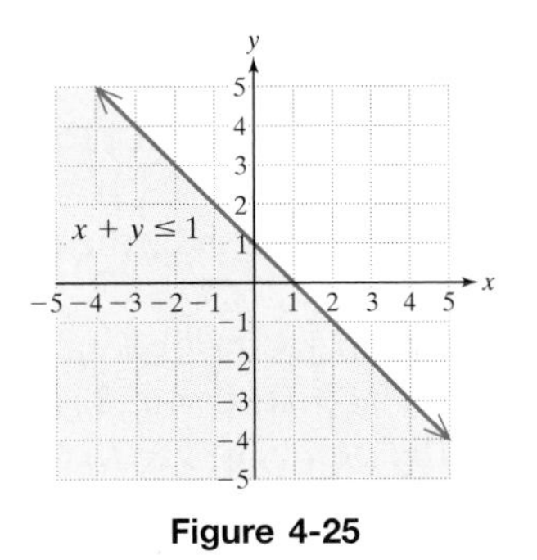

Figure 4-25

Next, we draw these regions on the same graph. The intersection ("overlap") is shown in purple (Figure 4-26).

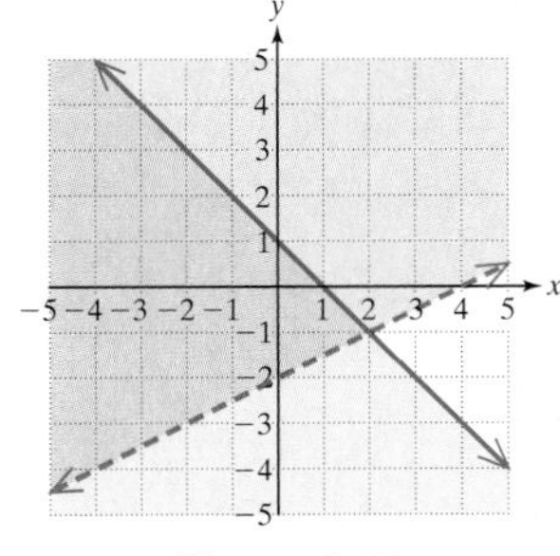
Figure 4-26

Skill Practice Answers

1.

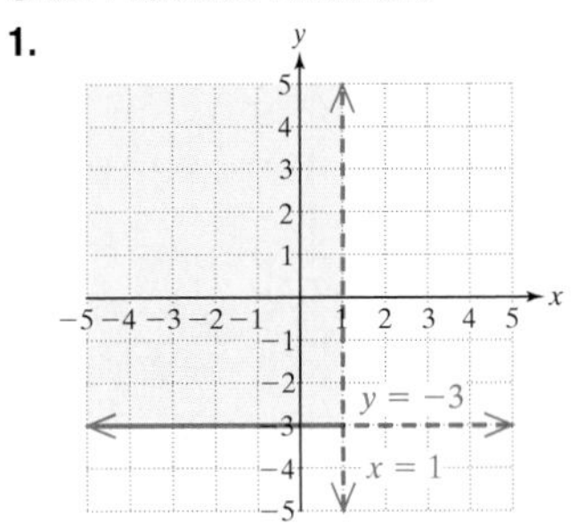

In Figure 4-27, we show the solution to the system of inequalities. Notice that the portions of the lines not bounding the solution are dashed.

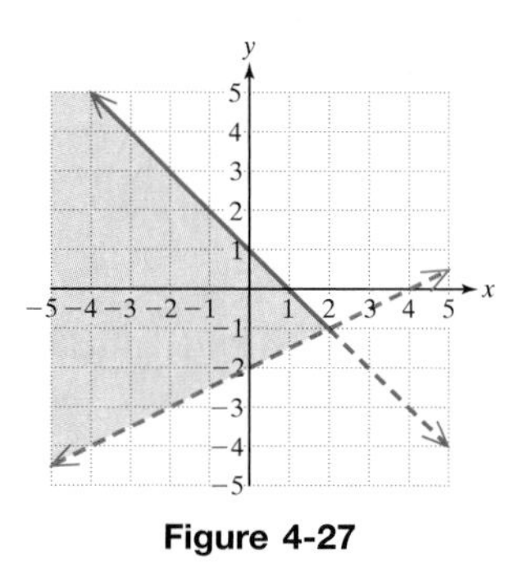

Figure 4-27

Skill Practice

2. Graph the solution set.

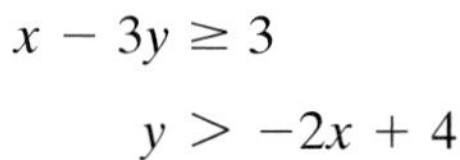

$$x - 3y \geq 3$$
$$y > -2x + 4$$

Example 3 Graphing a System of Linear Inequalities

Graph the solution set. $3y \leq 6x$

$$x \geq 0$$

Solution:

The solution to the inequality $3y \leq 6x$ is bounded by the line $3y = 6x$ (shown in red in Figure 4-28). Substituting the test point (2, 0) results in a true statement. Therefore, we shade the region below the line.

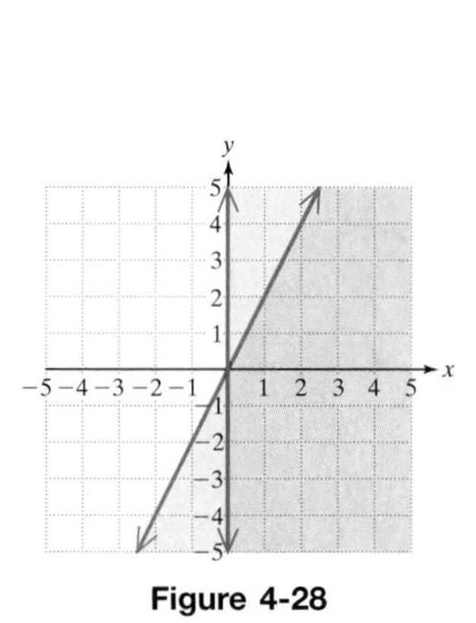

Figure 4-28

The solution to the inequality $x \geq 0$ is bounded by the line $x = 0$ (the y-axis). The inequality $x \geq 0$ represents the points to the *right* of and on the y-axis.

The intersection of the two regions is shown in purple in Figure 4-28.

In Figure 4-29, we show the solution to the system of inequalities. Notice that the portions of the lines outside the region of intersection are dashed.

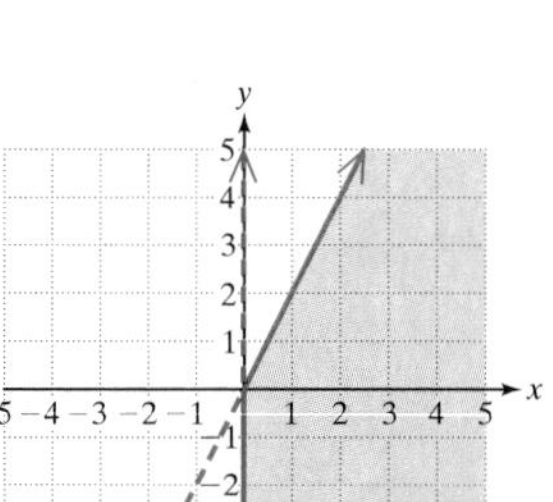

Figure 4-29

Skill Practice

3. Graph the solution set.

$$2y \geq -x$$
$$y \leq 0$$

Skill Practice Answers

2. $y = -2x + 4$, $x - 3y = 3$

3.

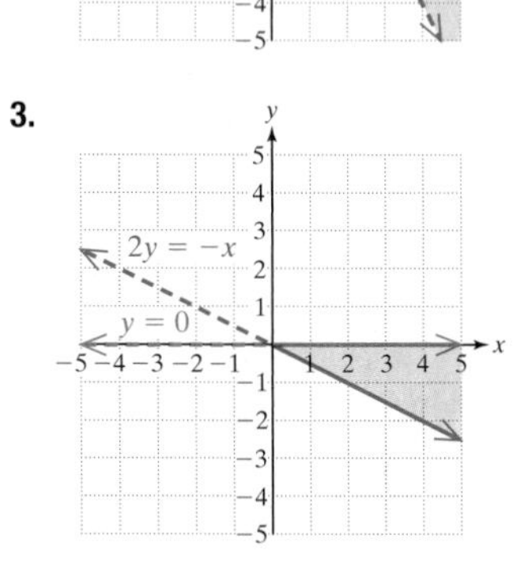

Section 4.6 Practice Exercises

Boost *your* GRADE at mathzone.com!

MathZone

- Practice Problems
- Self-Tests
- NetTutor
- e-Professors
- Videos

Study Skills Exercise

1. It is not too early to think about your final exam. Write the page number of the cumulative review for Chapters 1–4. Make this exercise set part of your homework this week.

Review Exercises

For Exercises 2–3, solve the systems of equations by graphing. Check by substituting the ordered pairs into both original equations.

2. $2x + y = 3$
$-3x - y = -5$

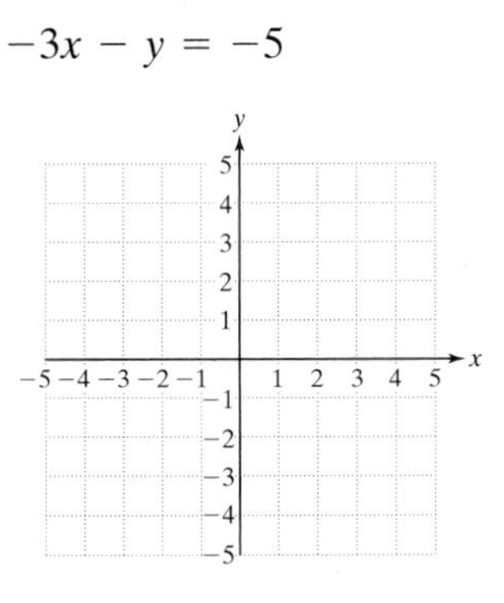

3. $2x + 2y = 6$
$y = x + 3$

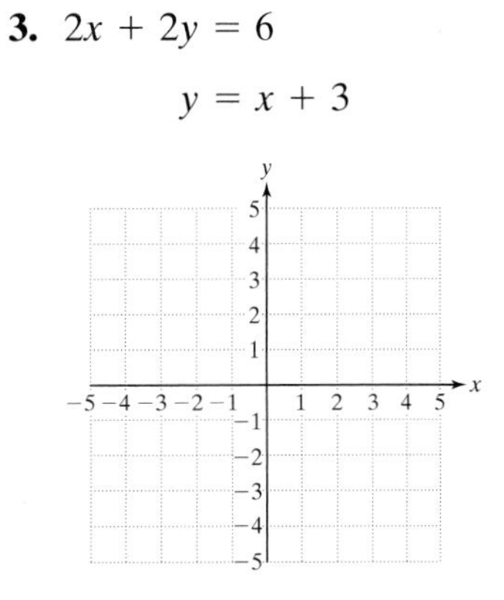

For Exercises 4–5, graph the equations and inequalities.

4. a. $-3x = y + 1$

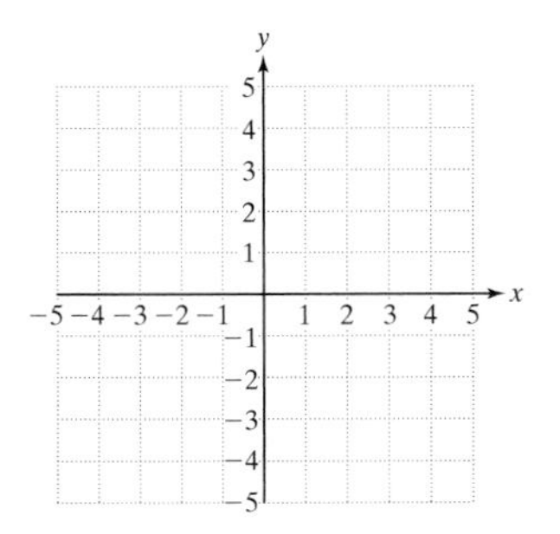

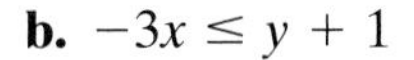

b. $-3x \leq y + 1$

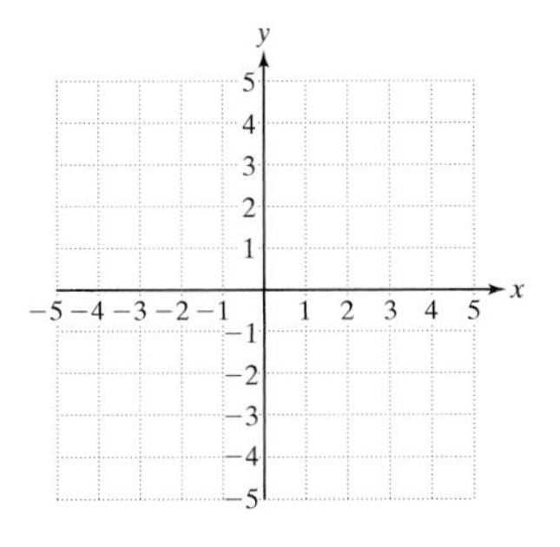

5. a. $2y = 4x - 2$

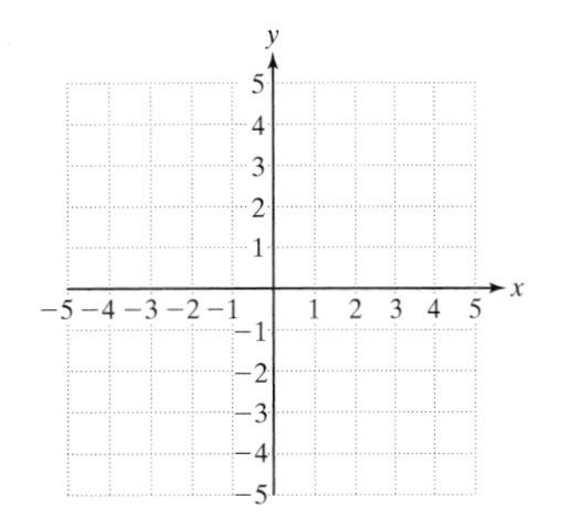

b. $2y < 4x - 2$

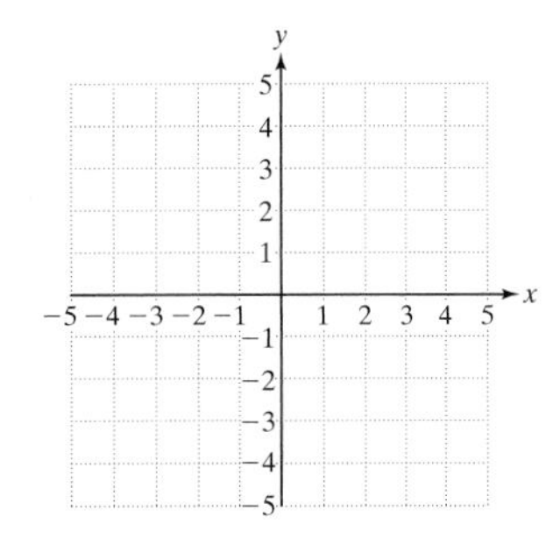

Concept 1: Graphing Systems of Linear Inequalities in Two Variables

For Exercises 6–27, graph the solution set.

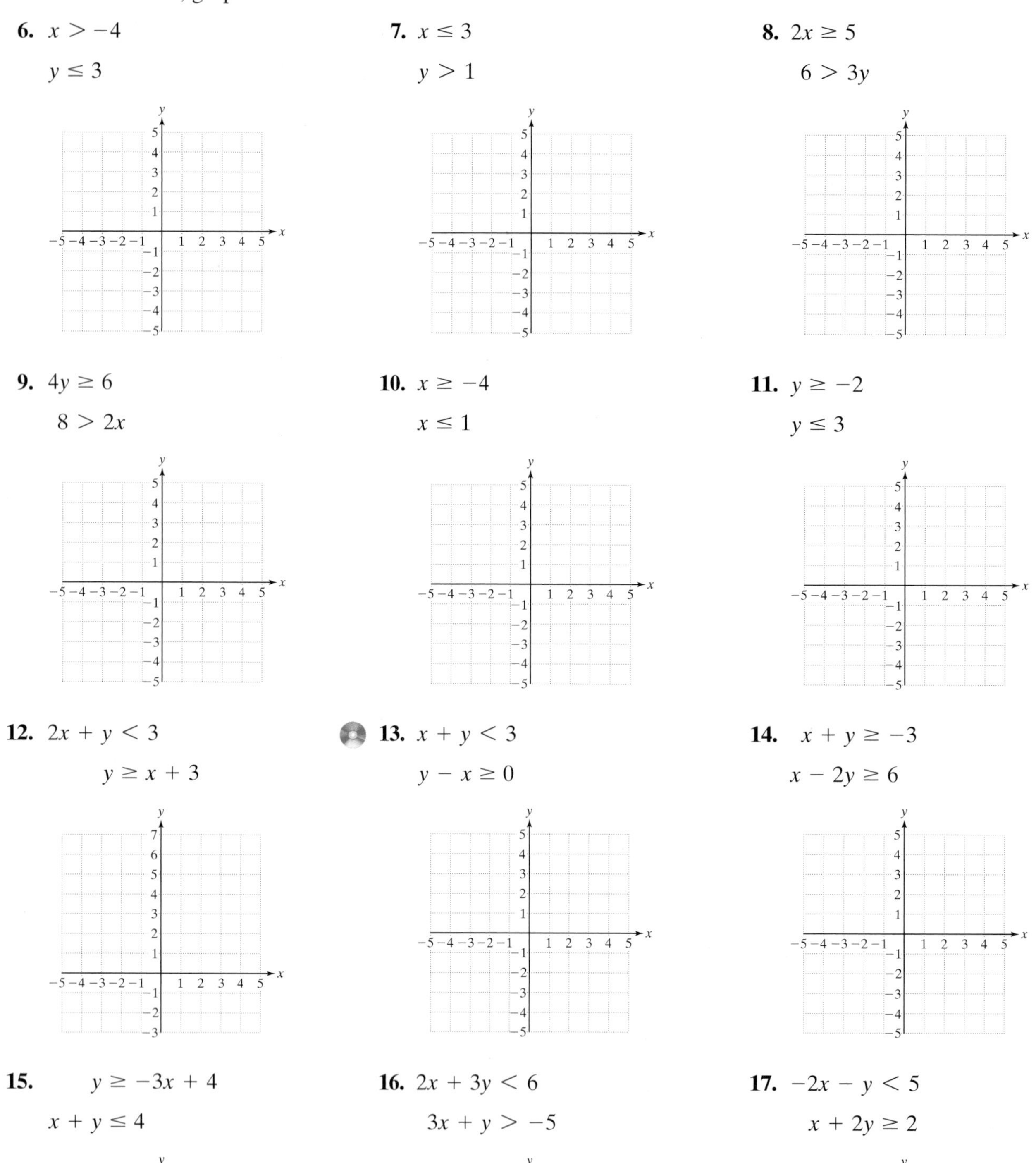

6. $x > -4$
$y \leq 3$

7. $x \leq 3$
$y > 1$

8. $2x \geq 5$
$6 > 3y$

9. $4y \geq 6$
$8 > 2x$

10. $x \geq -4$
$x \leq 1$

11. $y \geq -2$
$y \leq 3$

12. $2x + y < 3$
$y \geq x + 3$

13. $x + y < 3$
$y - x \geq 0$

14. $x + y \geq -3$
$x - 2y \geq 6$

15. $y \geq -3x + 4$
$x + y \leq 4$

16. $2x + 3y < 6$
$3x + y > -5$

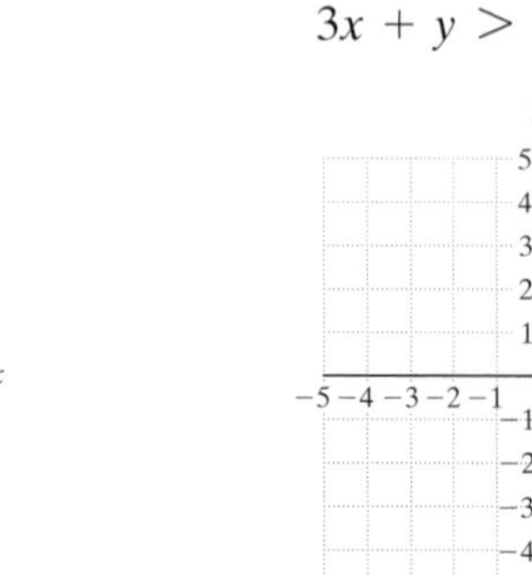

17. $-2x - y < 5$
$x + 2y \geq 2$

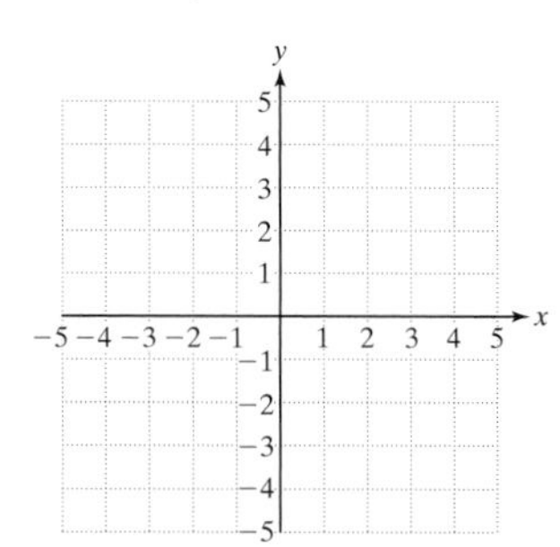

18. $y > 2x$

$y > -4x$

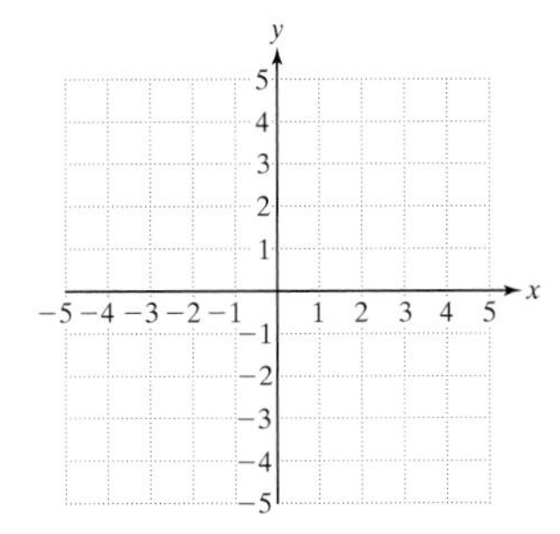

19. $2y \geq 6x$

$y \leq x$

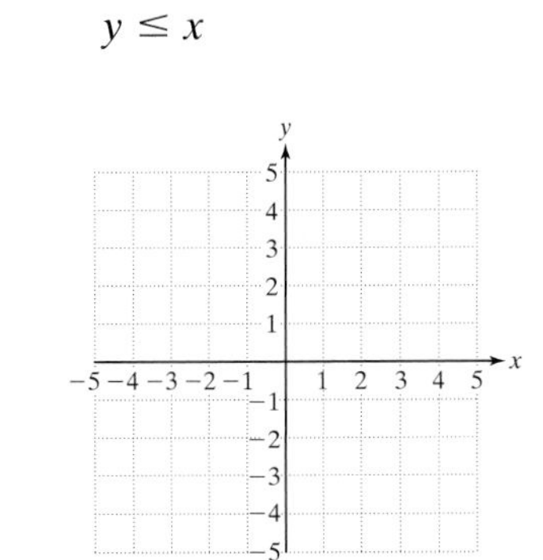

20. $y < \frac{1}{2}x - 1$

$5x + y \leq -12$

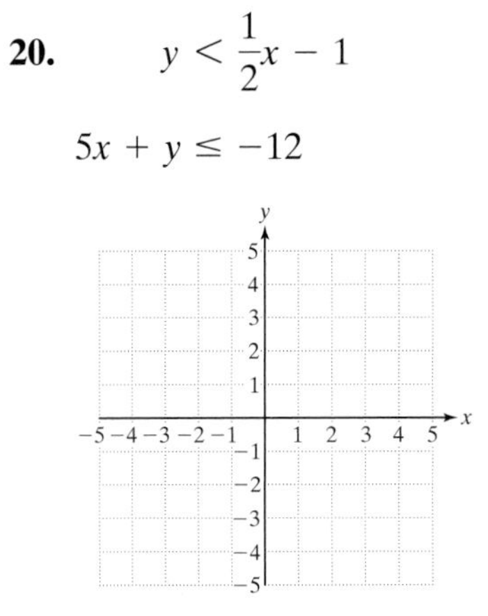

21. $y \geq \frac{1}{3}x + 2$

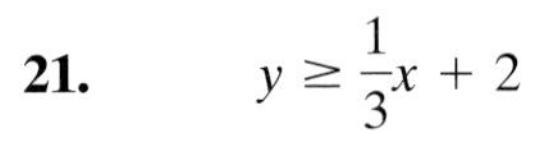

$4x + y < -2$

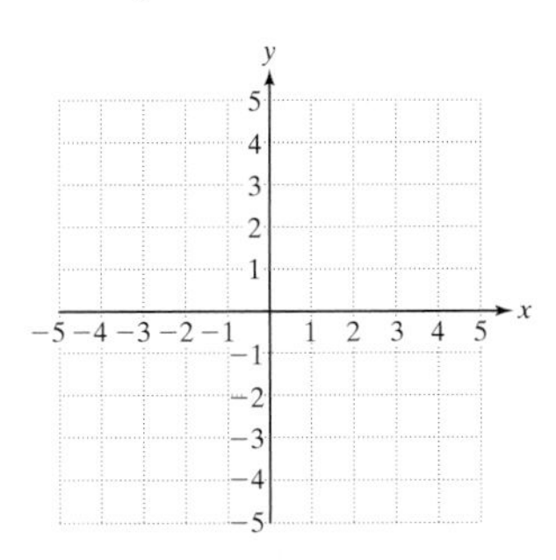

22. $y < 4$

$4x + 3y \geq 12$

23. $x \geq -3$

$2x + 4y < 4$

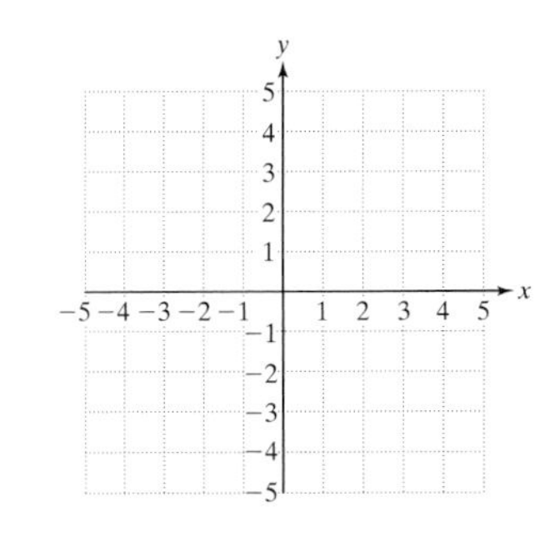

24. $y \leq x$

$y \geq x - 3$

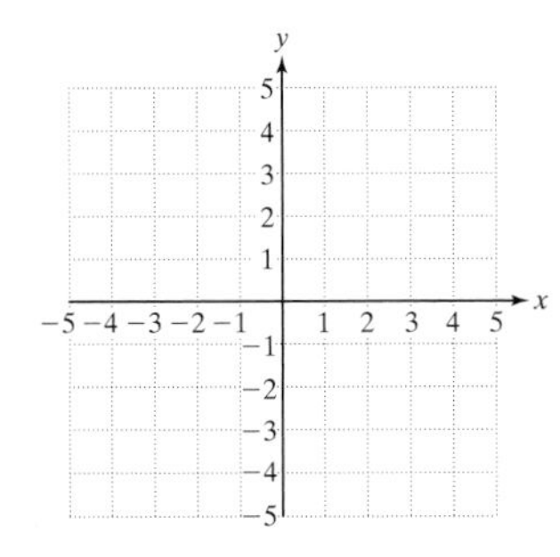

25. $x + y < 0$

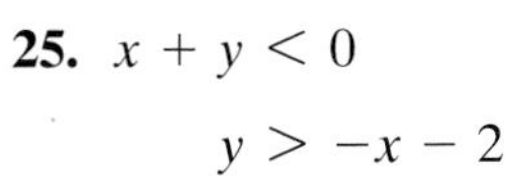

$y > -x - 2$

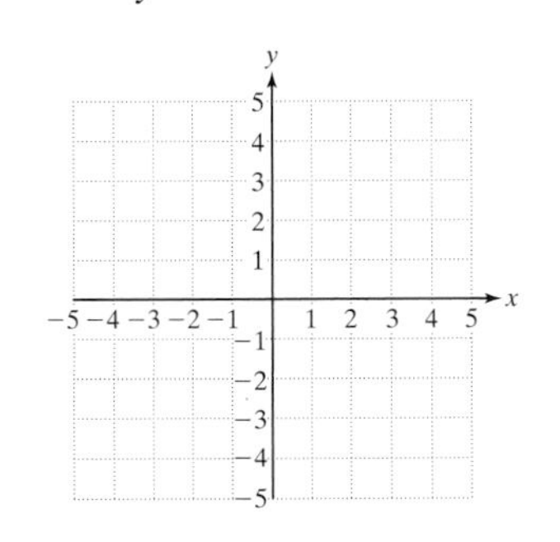

26. $y < 2x - 4$

$y > 2x + 1$

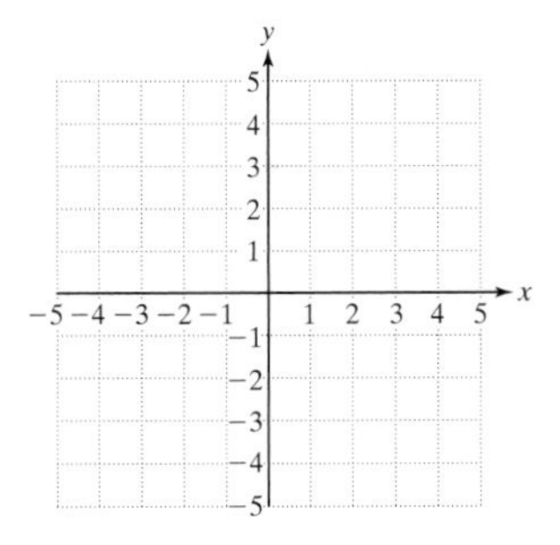

27. $y > -3x - 1$

$y < -3x - 3$

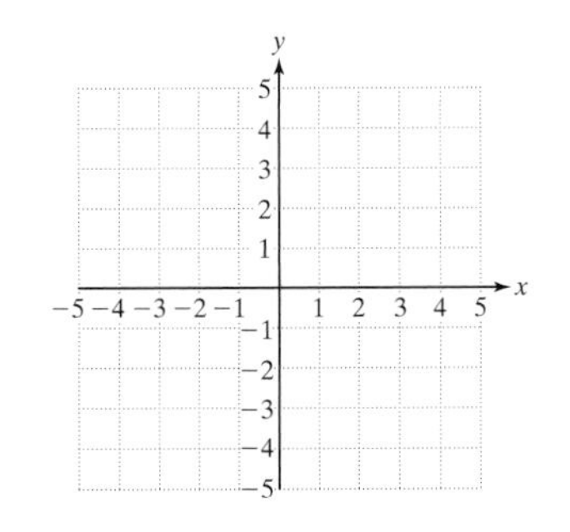

Expanding Your Skills

28. Describe the region bounded by the inequalities $x > 0$ and $y > 0$.

29. Describe the region bounded by the inequalities $x < 0$ and $y > 0$.

For Exercises 30–31, graph the solution set.

30. $x > -1$

$y > -1$

$x < 1$

$y < 1$

31. $x < 2$

$y < 2$

$x > -2$

$y > -2$

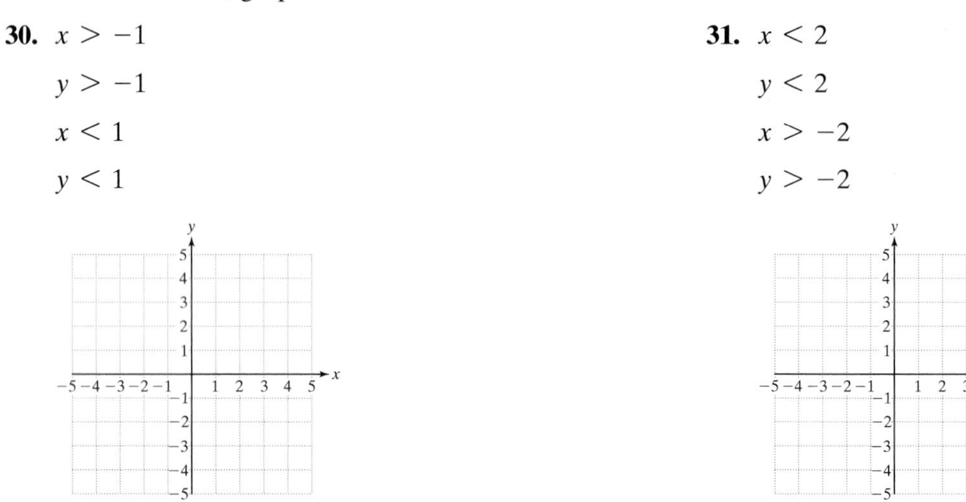

SUMMARY

Section 4.1 Solving Systems of Equations by the Graphing Method

Key Concepts

A **system of two linear equations** can be solved by graphing.

A **solution to a system of linear equations** is an ordered pair that satisfies each equation in the system. Graphically, this represents a point of intersection of the lines.

There may be one solution, infinitely many solutions, or no solution.

One solution
Consistent
Independent

Infinitely many solutions
Consistent
Dependent

No solution
Inconsistent
Independent

A system of equations is **consistent** if there is at least one solution. A system is **inconsistent** if there is no solution.

A linear system in x and y is **dependent** if two equations represent the same line. The solution set is the set of all points on the line.

If two linear equations represent different lines, then the system of equations is **independent**.

Examples

Example 1

Solve by graphing. $x + y = 3$

$$2x - y = 0$$

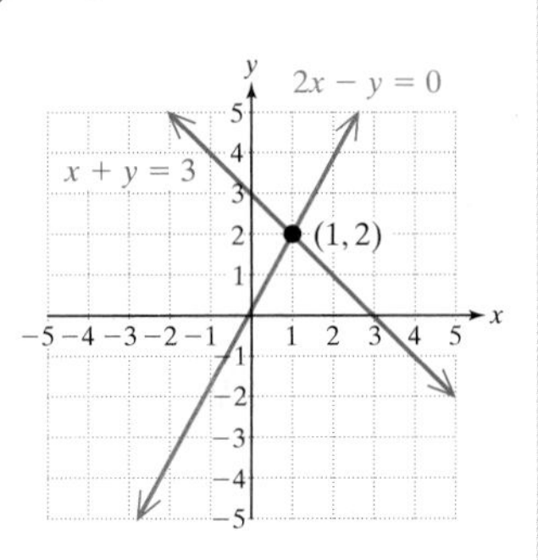

The solution is (1, 2).

Example 2

Solve by graphing. $3x - 2y = 2$

$$-6x + 4y = 4$$

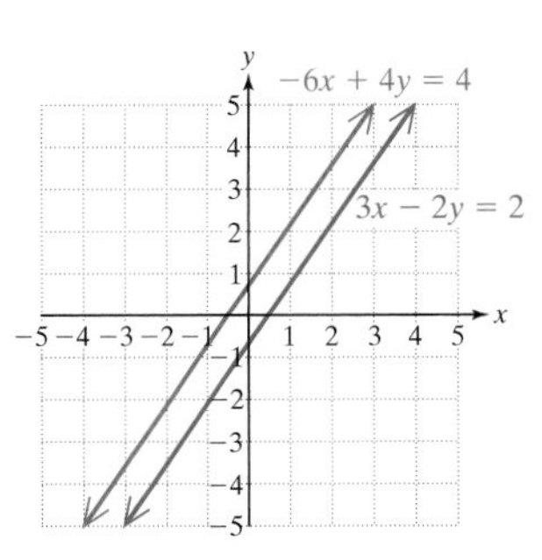

There is no solution. The system is inconsistent.

Example 3

Solve by graphing. $x + 2y = 2$

$$-3x - 6y = -6$$

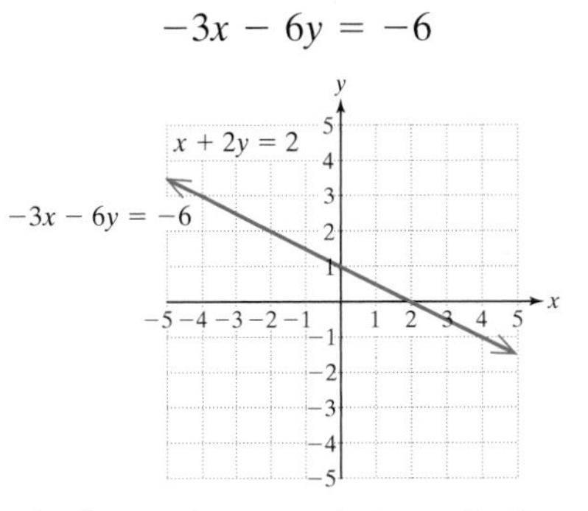

The system is dependent, and the solution set consists of all points on the line, given by $\{(x, y) | x + 2y = 2\}$

Section 4.2 Solving Systems of Equations by the Substitution Method

Key Concepts

Steps to Solve a System of Equations by the Substitution Method:

1. Isolate one of the variables from one equation.
2. Substitute the quantity found in step 1 into the other equation.
3. Solve the resulting equation.
4. Substitute the value found in step 3 back into the equation in step 1 to find the remaining variable.
5. Check the solution in both original equations and write the answer as an ordered pair.

An inconsistent system has no solution and is detected algebraically by a contradiction (such as $0 = 3$). When graphed, the lines are parallel.

If two linear equations represent the same line, the system is dependent. This is detected algebraically by an identity (such as $0 = 0$).

Examples

Example 1

Solve by the substitution method.

$$x + 4y = -11$$
$$3x - 2y = -5$$

Isolate x in the first equation: $x = -4y - 11$
Substitute into the second equation.

$3(-4y - 11) - 2y = -5$ Solve the equation.

$$-12y - 33 - 2y = -5$$
$$-14y = 28$$
$$y = -2$$

$x = -4y - 11$ Substitute $y = -2$.

$x = -4(-2) - 11$ Solve for x.

$$x = -3$$

The solution is $(-3, -2)$ and checks in both original equations.

Example 2

Solve by the substitution method.

$$3x + y = 4$$
$$-6x - 2y = 2$$

Isolate y in the first equation: $y = -3x + 4$.
Substitute into the second equation.

$$-6x - 2(-3x + 4) = 2$$
$$-6x + 6x - 8 = 2$$

$-8 = 2$ Contradiction

The system is inconsistent and has no solution.

Example 3

Solve by the substitution method.

$y = x + 2$ y is already isolated.

$$x - y = -2$$

$x - (x + 2) = -2$ Substitute $y = x + 2$ into the second equation.

$$x - x - 2 = -2$$

$-2 = -2$ Identity

The system is dependent. The solution set is all points on the line $y = x + 2$ or $\{(x, y) \mid y = x + 2\}$.

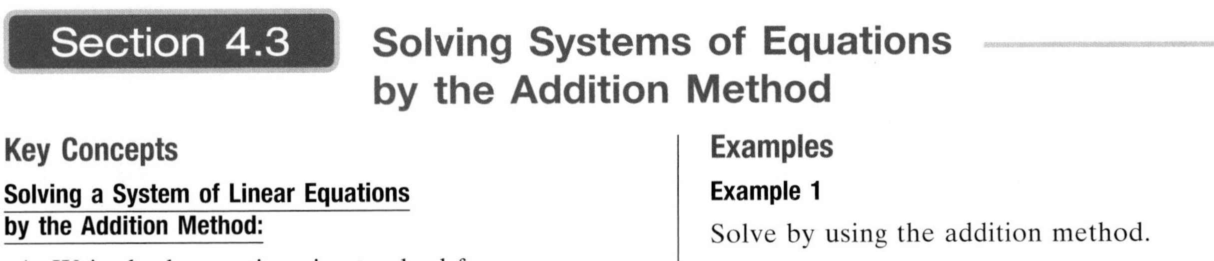

Section 4.3 Solving Systems of Equations by the Addition Method

Key Concepts

Solving a System of Linear Equations by the Addition Method:

1. Write both equations in standard form: $Ax + By = C$.
2. Clear fractions or decimals (optional).
3. Multiply one or both equations by a nonzero constant to create opposite coefficients for one of the variables.
4. Add the equations to eliminate one variable.
5. Solve for the remaining variable.
6. Substitute the known value into one of the original equations to solve for the other variable.
7. Check the solution in both equations.

Examples

Example 1

Solve by using the addition method.

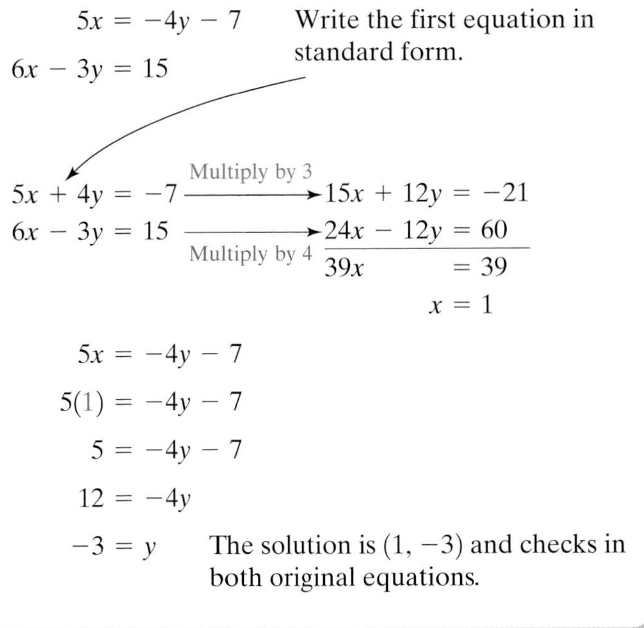

$$5x = -4y - 7 \qquad \text{Write the first equation in standard form.}$$
$$6x - 3y = 15$$

$$5x + 4y = -7 \xrightarrow{\text{Multiply by 3}} 15x + 12y = -21$$
$$6x - 3y = 15 \xrightarrow{\text{Multiply by 4}} 24x - 12y = 60$$
$$39x = 39$$
$$x = 1$$

$$5x = -4y - 7$$
$$5(1) = -4y - 7$$
$$5 = -4y - 7$$
$$12 = -4y$$
$$-3 = y$$

The solution is $(1, -3)$ and checks in both original equations.

Section 4.4 Applications of Linear Equations in Two Variables

Examples

Example 1

A riverboat travels 36 miles with the current to a marina in 2 hr. The return trip takes 3 hr against the current. Find the speed of the current and the speed of the boat in still water.

Let x represent the speed of the boat in still water. Let y represent the speed of the current.

	Distance	Rate	Time
Against current	36	$x - y$	3
With current	36	$x + y$	2

Distance = (rate)(time)

$$36 = (x - y) \cdot 3 \longrightarrow 36 = 3x - 3y$$
$$36 = (x + y) \cdot 2 \longrightarrow 36 = 2x - 2y$$

$$36 = 3x - 3y \xrightarrow{\text{Multiply by 2}} 72 = 6x - 6y$$
$$36 = 2x + 2y \xrightarrow{\text{Multiply by 3}} 108 = 6x + 6y$$
$$180 = 12x$$
$$15 = x$$

$$36 = 2(15) + 2y$$
$$36 = 30 + 2y$$
$$6 = 2y$$
$$3 = y$$

The speed of the boat in still water is 15 mph, and the speed of the current is 3 mph.

Example 2

Diane invests $15,000 more in an account earning 8% simple interest than in an account earning 5% simple interest. If the total interest after 1 year is $1850, how much was invested in each account?

	8%	5%	Total
Principal	x	y	
Interest	$0.08x$	$0.05y$	1850

$$x = y + 15{,}000$$

$$0.08x + 0.05y = 1850$$

Substitute $x = y + 15{,}000$ into the second equation:

$$0.08(y + 15{,}000) + 0.05y = 1850$$

$$0.08y + 1200 + 0.05y = 1850$$

$$0.13y + 1200 = 1850$$

$$0.13y = 650$$

$$\frac{0.13y}{0.13} = \frac{650}{0.13}$$

$$y = 5000$$

$$x = y + 15{,}000$$

$$x = 5000 + 15{,}000$$

$$x = 20{,}000$$

The amount invested in the 8% account is $20,000, and the amount invested at 5% is $5000.

Section 4.5 Linear Inequalities in Two Variables

Key Concepts

A **linear inequality in two variables** can be written in one of the forms: $ax + by < c$, $ax + by > c$, $ax + by \le c$, or $ax + by \ge c$.

Steps for Using the Test Point Method to Solve a Linear Inequality in Two Variables:

1. Set up the related *equation*.
2. Graph the related equation. This will be a line in the xy-plane.
   - If the original inequality is a strict inequality, $<$ or $>$, then the line is *not* part of the solution set. Therefore, graph the boundary as a dashed line.
   - If the original inequality is not strict, $\le$ or $\ge$, then the line *is* part of the solution set. Therefore, graph the boundary as a solid line.

Example

Example 1

Graph the inequality. $2x - y < 4$

1. The related equation is $2x - y = 4$.
2. Graph the equation $2x - y = 4$ (dashed line).

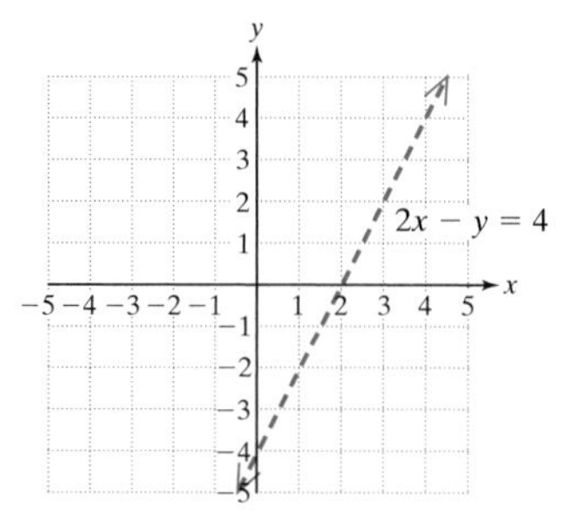

3. Choose a point not on the line and substitute its coordinates into the original inequality.
   - If the test point makes the inequality true, then the region it represents is part of the solution set. Shade that region.
   - If the test point makes the inequality false, then the other region is part of the solution set and should be shaded.

3. Choose an arbitrary test point not on the line such as $(0, 0)$.

$$2x - y < 4$$
$$2(0) - (0) \stackrel{?}{<} 4$$
$$0 < 4 \checkmark \quad \text{True}$$

Shade the region represented by the test point (in this case, above the line).

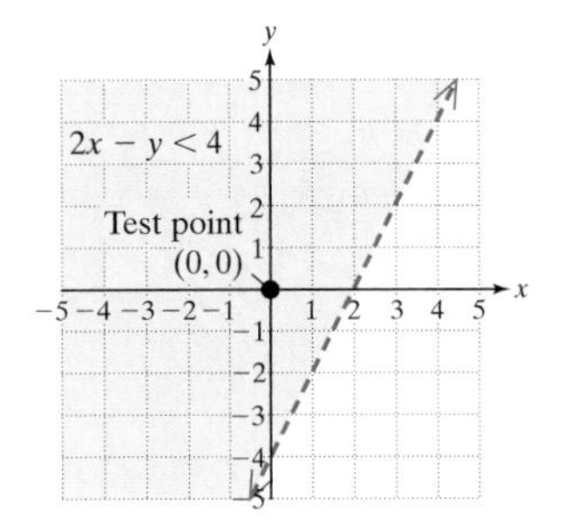

Section 4.6 Systems of Linear Inequalities in Two Variables

Key Concepts

The solution set for a system of linear inequalities in two variables is the intersection ("overlap") of the solution sets of the individual inequalities.

Graphing the Solution Set for a System of Linear Inequalities

1. First sketch the solutions to the individual inequalities.
2. Then, sketch the two shaded regions on the same graph.

Example

Example 1

Graph the solution set. $x + y < 2$
$x \geq 1$

The solution to the inequality $x + y < 2$ is the set of points below the line $x + y = 2$ (shown in red).

The solution to the inequality $x \geq 1$ is the set of points on and to the right of the vertical line $x = 1$ (shown in blue).

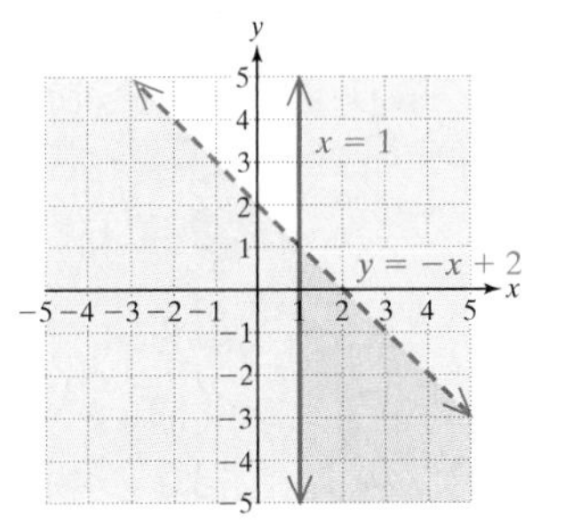

3. The solution is the intersection of the shaded regions. However, be sure that any portion of the lines not bounding the solution set are dashed.

The solution to the system of inequalities is the region shown in purple. Note that the portions of the lines not adjacent to the shaded region are dashed.

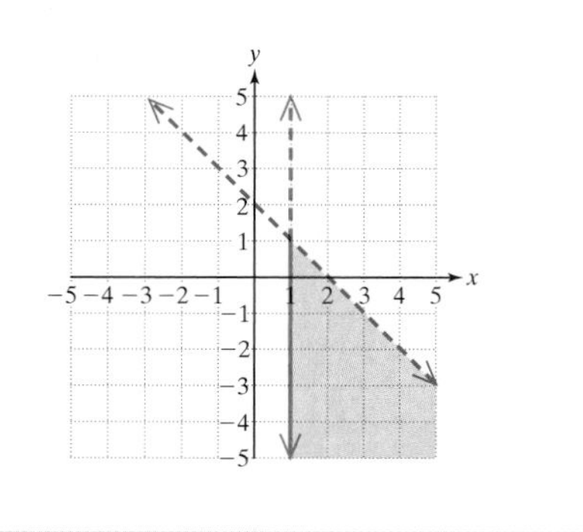

Chapter 4 Review Exercises

Section 4.1

For Exercises 1–4, determine if the ordered pair is a solution to the system.

1. $x - 4y = -4$ $(4, 2)$
$x + 2y = 8$

2. $x - 6y = 6$ $(12, 1)$
$-x + y = 4$

3. $3x + y = 9$ $(1, 3)$
$y = 3$

4. $2x - y = 8$ $(2, -4)$
$x = 2$

For Exercises 5–10, identify whether the system represents intersecting lines, parallel lines, or coinciding lines by comparing their slopes and y-intercepts.

5. $y = -\frac{1}{2}x + 4$
$y = x - 1$

6. $y = -3x + 4$
$y = 3x + 4$

7. $y = -\frac{4}{7}x + 3$
$y = -\frac{4}{7}x - 5$

8. $y = 5x - 3$
$y = \frac{1}{5}x - 3$

9. $y = 9x - 2$
$9x - y = 2$

10. $x = -5$
$y = 2$

For Exercises 11–18, solve the systems by graphing. If a system does not have a unique solution, identify the system as inconsistent or dependent.

11. $y = -\frac{2}{3}x - 2$
$-x + 3y = -6$

12. $y = -2x - 1$
$2x + 3y = 5$

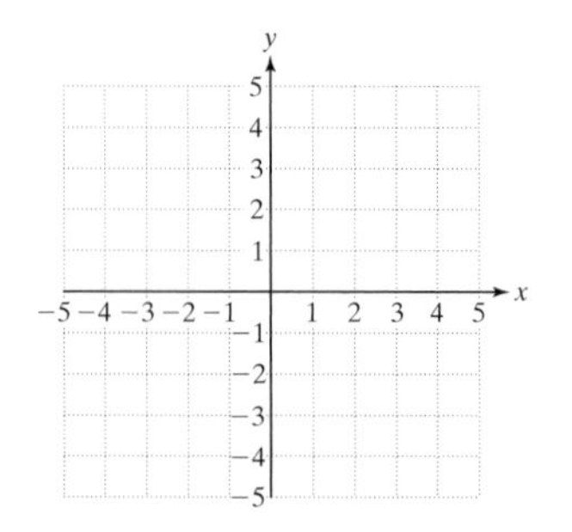

13. $4x = -2y + 10$

$2x + y = 5$

14. $10y = 2x - 10$

$-x + 5y = -5$

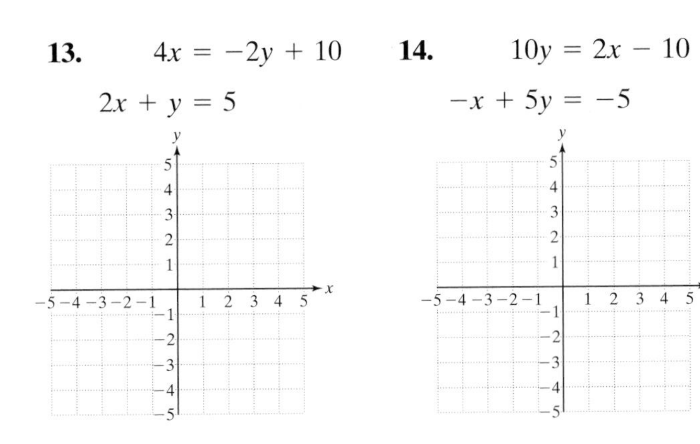

15. $6x - 3y = 9$

$y = -1$

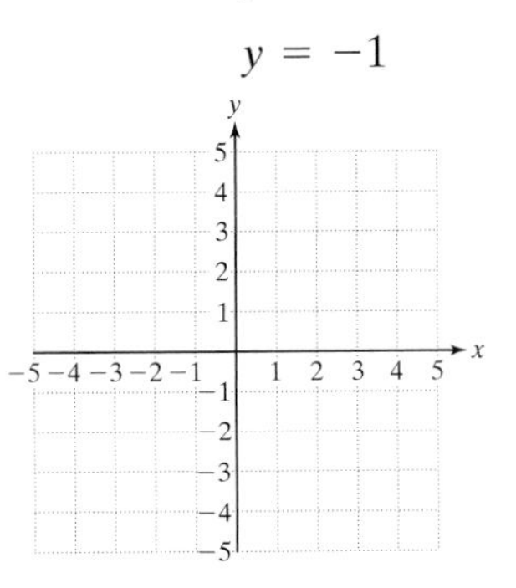

16. $5x + y = -11$

$x = -1$

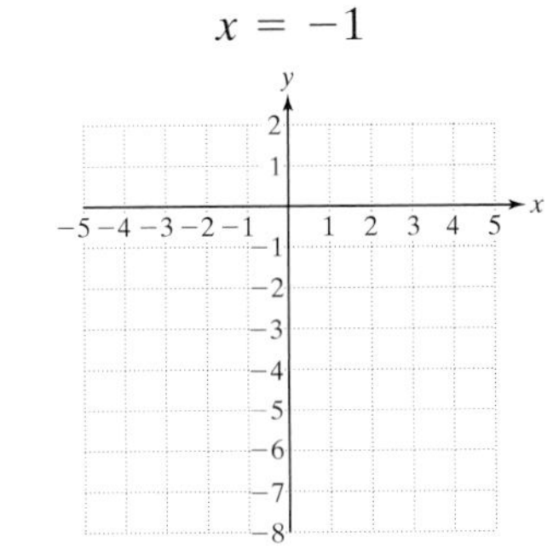

17. $x - 7y = 14$

$-2x + 14y = 14$

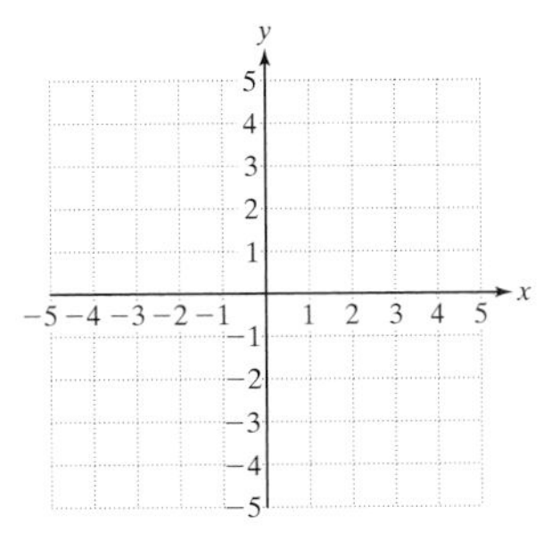

18. 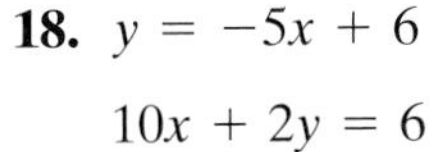

$y = -5x + 6$

$10x + 2y = 6$

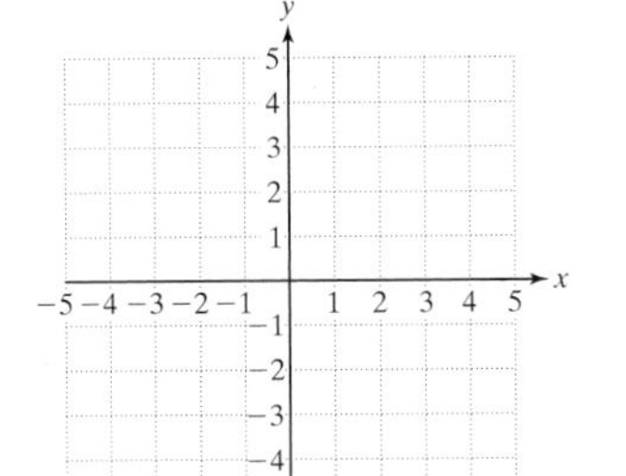

19. A rental car company rents a compact car for \$20 a day plus \$0.25 per mile. A midsize car rents for \$30 a day plus \$0.20 per mile.

The cost, y_c, to rent a compact car for one day is given by the equation:

$y_c = 20 + 0.25x$ where x is the number of miles driven

The cost, y_m, to rent a midsize car for one day is given by the equation:

$y_m = 30 + 0.20x$ where x is the number of miles driven

Find the number of miles at which the cost to rent either car would be the same, and confirm your answer with the graph.

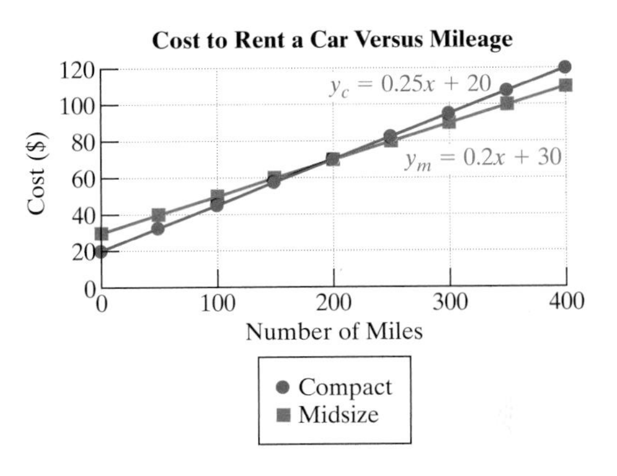

Section 4.2

For Exercises 20–23, solve the systems using the substitution method.

20. $6x + y = 2$

$y = 3x - 4$

21. $2x + 3y = -5$

$x = y - 5$

22. $2x + 6y = 10$

$x = -3y + 6$

23. $4x + 2y = 4$

$y = -2x + 2$

24. Given the system:

$$x + 2y = 11$$
$$5x + 4y = 40$$

a. Which variable from which equation is easiest to isolate and why?

b. Solve the system using the substitution method.

25. Given the system:

$$4x - 3y = 9$$
$$2x + y = 12$$

a. Which variable from which equation is easiest to isolate and why?

b. Solve the system using the substitution method.

For Exercises 26–29, solve the systems using the substitution method.

26. $3x - 2y = 23$
$x + 5y = -15$

27. $x + 5y = 20$
$3x + 2y = 8$

28. $x - 3y = 9$
$5x - 15y = 45$

29. $-3x + y = 15$
$6x - 2y = 12$

30. The difference of two positive numbers is 42. The larger number is 2 more than 6 times the smaller number. Find the numbers.

31. In a right triangle, one of the acute angles is 6° less than the other acute angle. Find the measure of each acute angle.

32. Two angles are supplementary. One angle measures 14° less than two times the other angle. Find the measure of each angle.

Section 4.3

33. Explain the process for solving a system of two equations using the addition method.

34. Given the system:

$$3x - 5y = 1$$
$$2x - y = -4$$

a. Which variable, x or y, is easier to eliminate using the addition method? (Answers may vary.)

b. Solve the system using the addition method.

35. Given the system:

$$9x - 2y = 14$$
$$4x + 3y = 14$$

a. Which variable, x or y, is easier to eliminate using the addition method? (Answers may vary.)

b. Solve the system using the addition method.

For Exercises 36–43, solve the systems using the addition method.

36. $2x + 3y = 1$
$x - 2y = 4$

37. $x + 3y = 0$
$-3x - 10y = -2$

38. $8(x + 1) = -6y + 6$
$10x = 9y - 8$

39. $12x = 5(y + 1)$
$5y = -1 - 4x$

40. $-4x - 6y = -2$
$6x + 9y = 3$

41. $-8x - 4y = 16$
$10x + 5y = 5$

42. $\frac{1}{2}x - \frac{3}{4}y = -\frac{1}{2}$
$\frac{1}{3}x + y = -\frac{10}{3}$

43. $0.5x - 0.2y = 0.5$
$0.4x + 0.7y = 0.4$

44. Given the system:

$$4x + 9y = -7$$
$$y = 2x - 13$$

a. Which method would you choose to solve the system, the substitution method or the addition method? Explain your choice. (Answers may vary.)

b. Solve the system.

45. Given the system:

$$5x - 8y = -2$$
$$3x - y = -5$$

a. Which method would you choose to solve the system, the substitution method or the addition method? Explain your choice. (Answers may vary.)

b. Solve the system.

Section 4.4

46. Miami Metrozoo charges $11.50 for adult admission and $6.75 for children under 12. The total bill before tax for a school group of 60 people is $443. How many adults and how many children were admitted?

47. Emillo invested $20,000, and at the end of 1 year he received $1525 in interest. If he invested part of the money at 5% simple interest and the remaining money at 8% simple interest, how much did he invest in each account?

48. To produce a 16% alcohol solution, a chemist mixes a 20% alcohol solution and a 14% alcohol solution. How much 20% solution and how much 14% solution must be used to produce 15 L of a 16% alcohol solution?

49. A boat travels 80 miles downstream with the current in 4 hr and 80 miles upstream against the current in 5 hr. Find the speed of the current and the speed of the boat in still water.

50. Suzanne has a collection of new quarters and new \$1 coins. She has four more quarters than dollar coins and the total value of the coins is \$4.75. How many of each coin does Suzanne have?

51. At Conseco Fieldhouse, home of the Indiana Pacers, the total cost of a soft drink and a hot dog is \$8.00. The price of the hot dog is \$1.00 more than the cost of the soft drink. Find the cost of a soft drink and the cost of a hot dog.

52. In a recent election 5700 votes were cast and 3675 voters voted for the winning candidate. If $\frac{5}{8}$ of the women and $\frac{2}{3}$ of the men voted for the winning candidate, how many men and how many women voted?

53. Ray played two rounds of golf at Pebble Beach for a total score of 154. If his score in the second round is 10 more than his score in the first round, find the scores for each round.

Section 4.5

For Exercises 54–57, graph the inequalities. Then write three ordered pairs that are in the solution set (answers may vary).

54. $y < 3x - 1$

55. $y > -2x + 6$

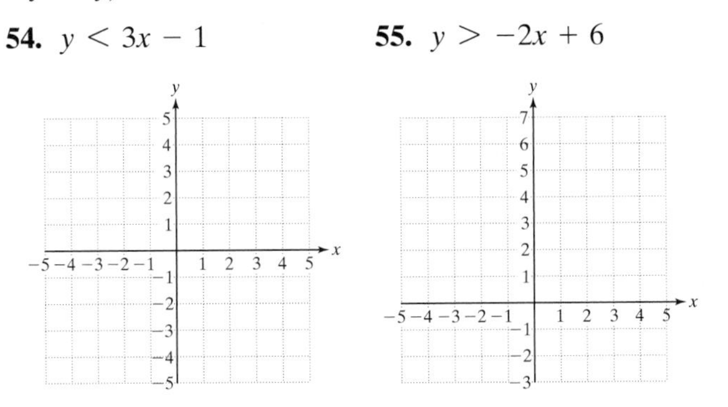

56. $-2x - 3y \geq 8$

57. $4x - 2y \leq 10$

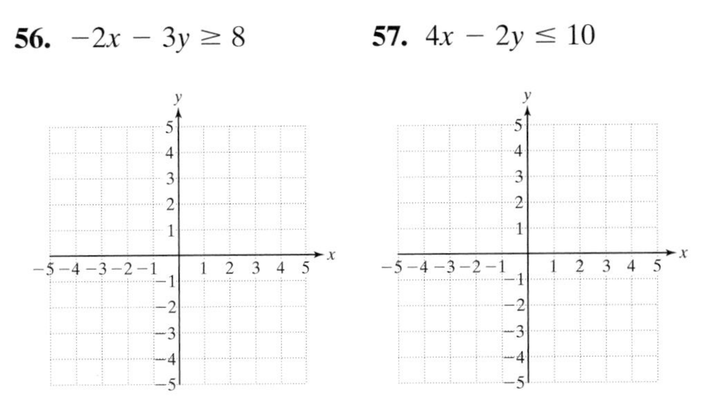

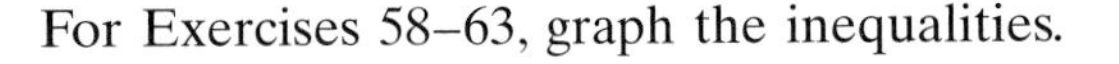

For Exercises 58–63, graph the inequalities.

58. $x - 5y \geq 0$

59. $7x - y \leq 0$

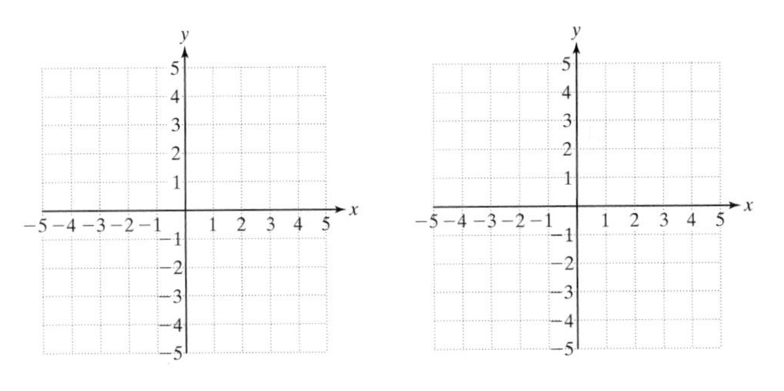

60. $x > 5$

61. $y < -4$

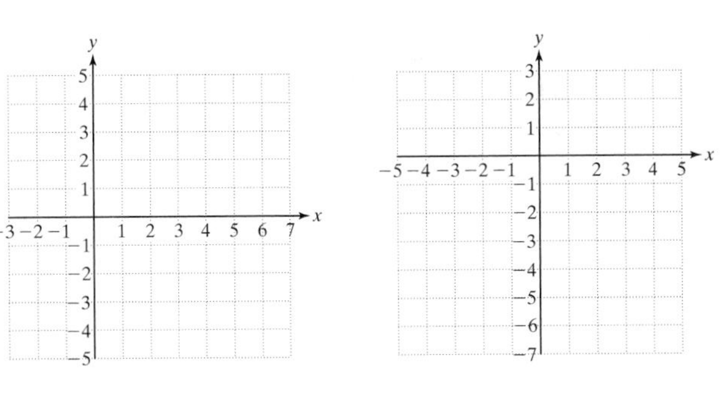

62. $x \geq 0$

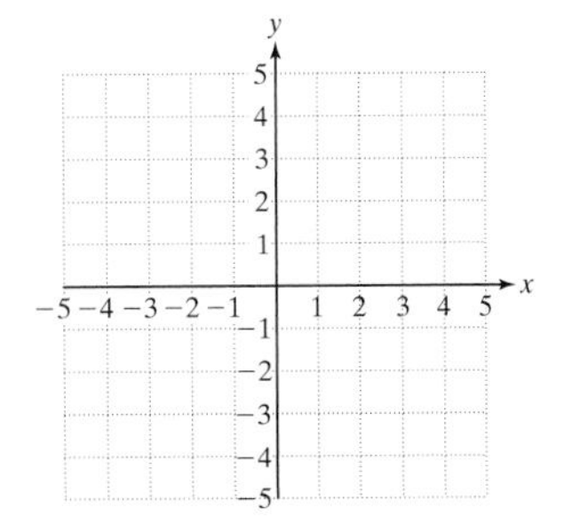

63. $y \geq 0$

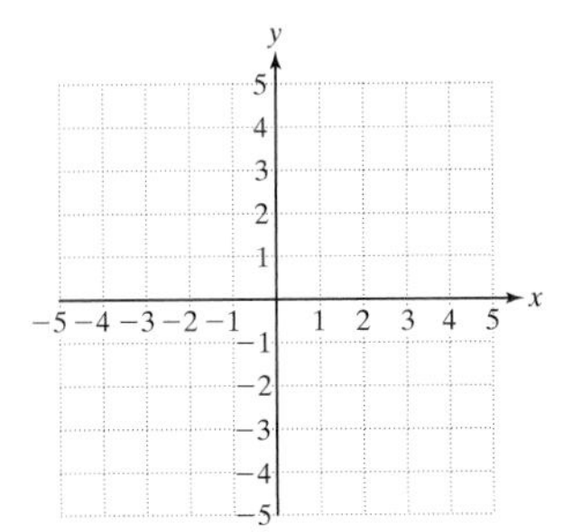

Section 4.6

For Exercises 64–69, sketch the systems of inequalities.

64. $2x - y \geq 8$

$x + y \leq 3$

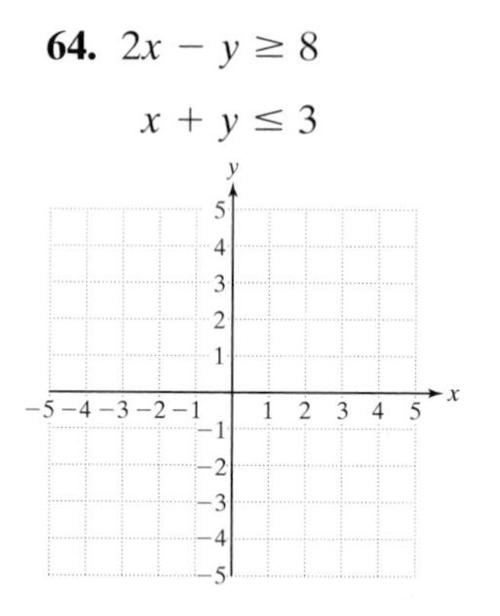

65. $y \leq x - 1$

$x + 2y \geq 4$

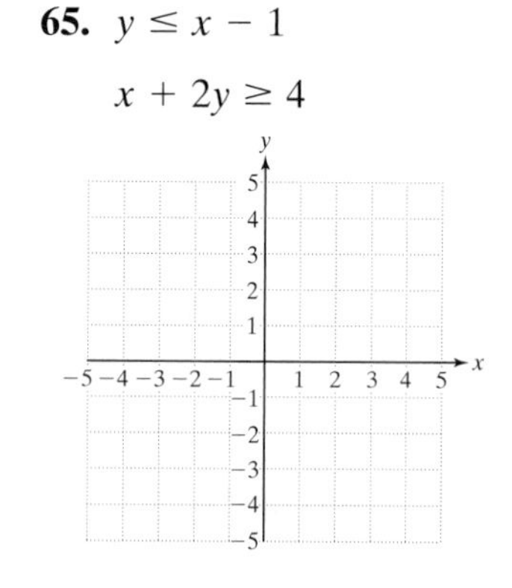

66. $y \leq 2x$

$-2x - y > -3$

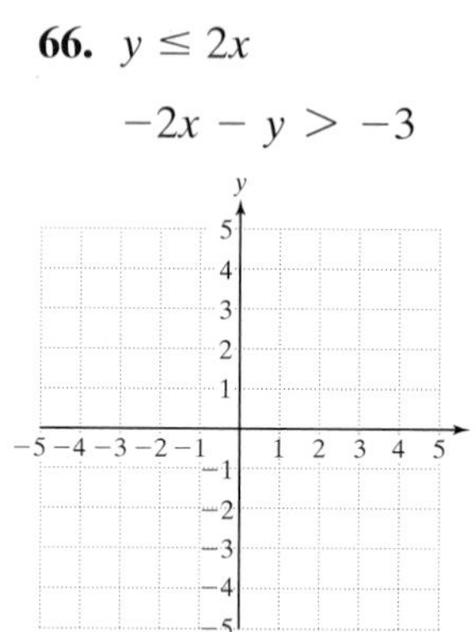

67. $y \leq 4$

$2x - y < 1$

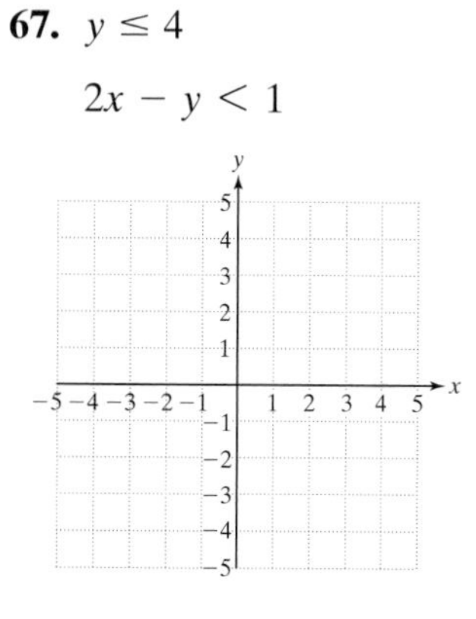

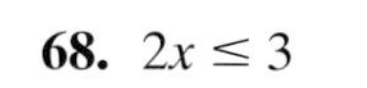

68. $2x \leq 3$

$x \geq -2$

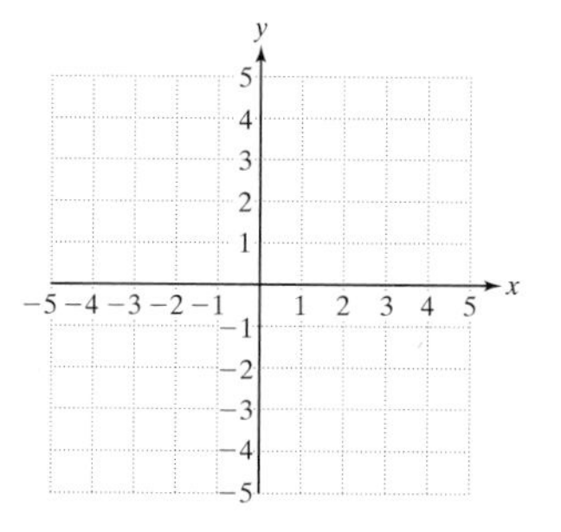

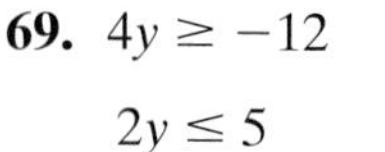

69. $4y \geq -12$

$2y \leq 5$

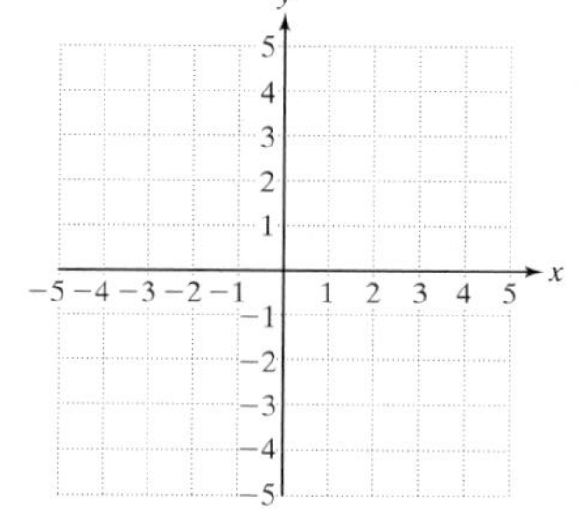

Chapter 4 Test

1. Write each line in slope-intercept form. Then determine if the lines represent intersecting lines, parallel lines, or coinciding lines.

$$5x + 2y = -6$$

$$-\frac{5}{2}x - y = -3$$

For Exercises 2–3 solve the system by graphing.

2. $y = 2x - 4$

$-2x + 3y = 0$

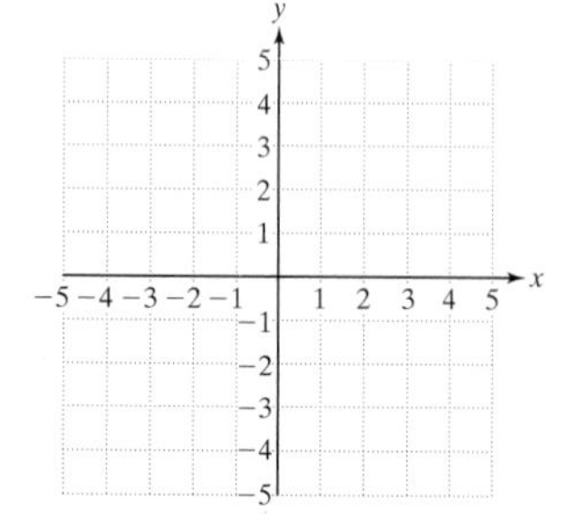

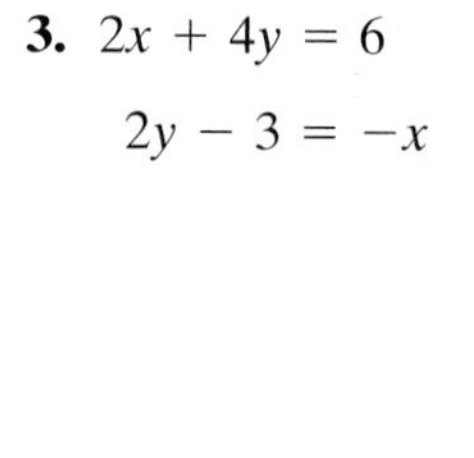

3. $2x + 4y = 6$

$2y - 3 = -x$

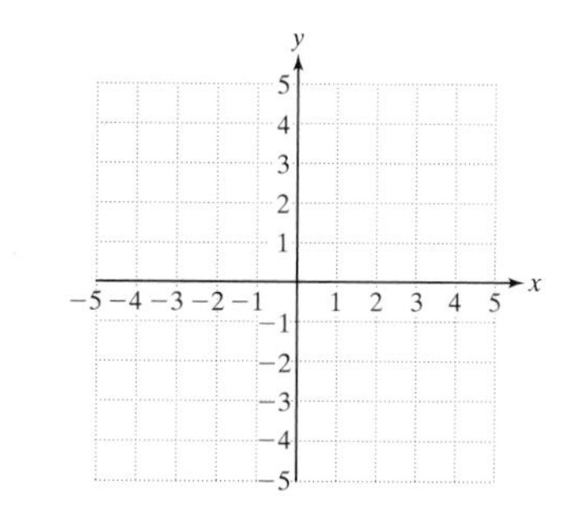

4. Solve the system using the substitution method.

$$x = 5y - 2$$

$$2x + y = -4$$

5. In the 2005 WNBA (basketball) season, the league's leading scorer was Sheryl Swoopes from the Houston Comets. Swoopes scored 17 points more than the second leading scorer, Lauren Jackson from the Seattle Storm. Together they scored a total of 1211 points. How many points did each player score?

6. Solve the system using the addition method.

$$3x - 6y = 8$$
$$2x + 3y = 3$$

7. How many milliliters of a 50% acid solution and how many milliliters of a 20% acid solution must be mixed to produce 36 mL of a 30% acid solution?

8. a. How many solutions does a system of two linear equations have if the equations represent parallel lines?

   b. How many solutions does a system of two linear equations have if the equations represent coinciding lines?

   c. How many solutions does a system of two linear equations have if the equations represent intersecting lines?

For Exercises 9–14, solve the systems using any method.

9. $\frac{1}{3}x + y = \frac{7}{3}$

   $x = \frac{3}{2}y - 11$

10. $2(x - 6) = y$

    $2x - \frac{1}{2}y = x + 5$

11. $3x - 4y = 29$

    $2x + 5y = -19$

12. $2x = 6y - 14$

    $2y = 3 - x$

13. $-0.25x - 0.05y = 0.2$

    $10x + 2y = -8$

14. $3(x + y) = -2y - 7$

    $-3y = 10 - 4x$

15. At *Best Buy*, Latrell buys four CDs and two DVDs for $54 from the sale rack. Kendra buys two CDs and three DVDs from the same rack for $49. What is the price per CD and the price per DVD?

16. The cost to ride the trolley one-way in San Diego is $2.25. Kelly and Hazel had to buy eight tickets for their group.

    a. What was the total amount of money required?

    b. Kelly and Hazel had only quarters and $1 bills. They also determined that they used twice as many quarters as $1 bills. How many quarters and how many $1 bills did they use?

17. Suppose a total of $5000 is borrowed from two different loans. One loan charges 10% simple interest, and the other charges 8% simple interest. How much was borrowed at each rate if $424 in interest is charged at the end of 1 year?

18. During the first 13 years of his football career, Jerry Rice scored a total of 166 touchdowns. One touchdown was scored on a kickoff return, and the remaining 165 were scored rushing or receiving. The number of receiving touchdowns he scored was 5 more than 15 times the number of rushing touchdowns he scored. How many receiving touchdowns and how many rushing touchdowns did he score?

19. A plane travels 880 miles in 2 hr against the wind and 1000 miles in 2 hr with the same wind. Find the speed of the plane in still air and the speed of the wind.

20. The number of calories in a piece of cake is 20 less than 3 times the number of calories in a scoop of ice cream. Together, the cake and ice cream have 460 calories. How many calories are in each?

21. A police force has 240 officers. If there are 116 more men than women, find the number of men and the number of women on the force.

22. Graph the inequality $5x - y \geq -6$.

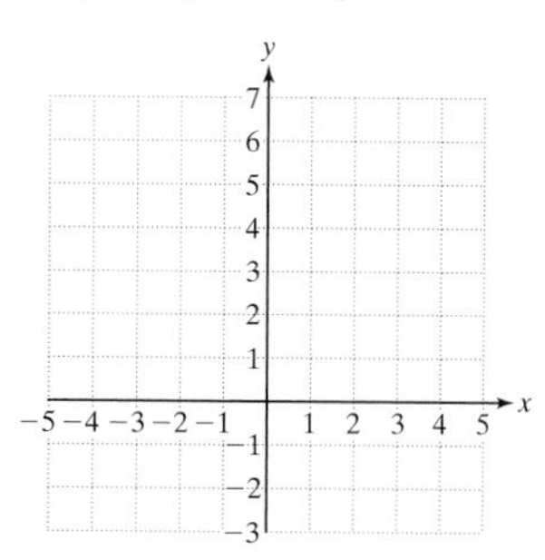

23. Graph the solution set.

$$2x + y > 1$$
$$x + y < 2$$

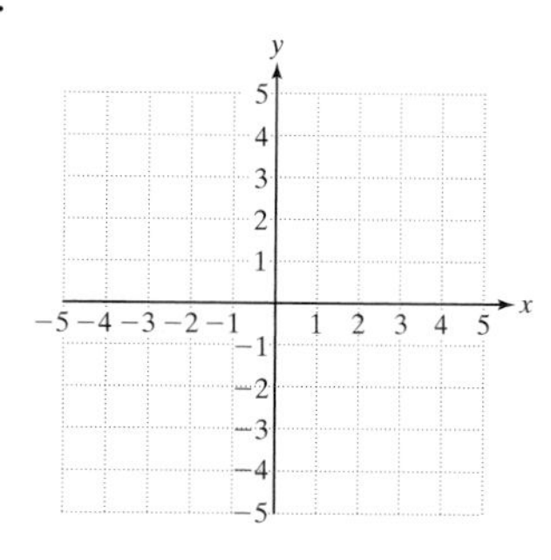

Chapters 1–4 Cumulative Review Exercises

1. Simplify.

$$\frac{|2 - 5| + 10 \div 2 + 3}{\sqrt{10^2 - 8^2}}$$

2. Solve for x: $\frac{1}{3}x - \frac{3}{4} = \frac{1}{2}(x + 2)$

3. Solve for a: $-4(a + 3) + 2 = -5(a + 1) + a$

4. Solve for y: $3x - 2y = 6$

5. Solve for z. Graph the solution set on a number line and write the solution in interval notation:

$$-2(3z + 1) \leq 5(z - 3) + 10$$

6. The largest angle in a triangle is 110°. Of the remaining two angles, one is 4° less than the other angle. Find the measure of the three angles.

7. Two hikers start at opposite ends of an 18-mile trail and walk toward each other. One hiker walks predominately down hill and averages 2 mph faster than the other hiker. Find the average rate of each hiker if they meet in 3 hr.

8. Jesse Ventura became the 38th governor of Minnesota by receiving 37% of the votes. If approximately 2,060,000 votes were cast, how many did Mr. Ventura get?

9. The YMCA wants to raise \$2500 for its summer program for disadvantaged children. If the YMCA has already raised \$900, what percent of its goal has been achieved?

10. Two angles are complementary. One angle measures 17° more than the other angle. Find the measure of each angle.

11. Solve for x: $z = \dfrac{x - m}{5}$

12. Solve for y: $2x - 3y = 6$

13. The slope of a given line is $-\frac{2}{3}$.

a. What is the slope of a line parallel to the given line?

b. What is the slope of a line perpendicular to the given line?

14. Find an equation of the line passing through the point $(2, -3)$ and having a slope of -3. Write the final answer in slope-intercept form.

15. Sketch the following equations on the same graph.

a. $2x + 5y = 10$

b. $2y = 4$

c. Find the point of intersection and check the solution in each equation.

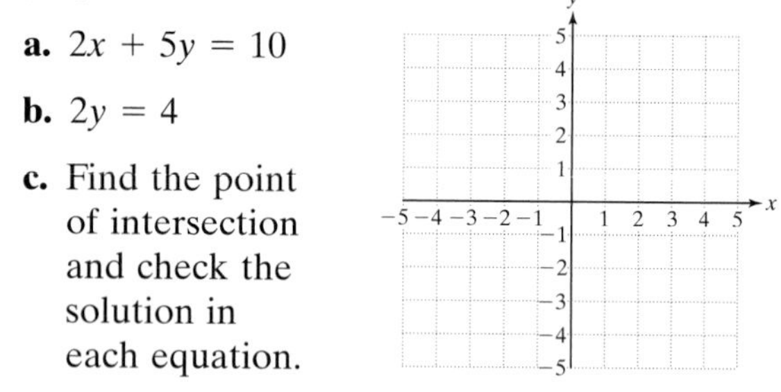

16. Solve the system of equations by using the substitution method.

$$2x + 5y = 10$$
$$2y = 4$$

17. a. Graph the line $2x + y = 3$.

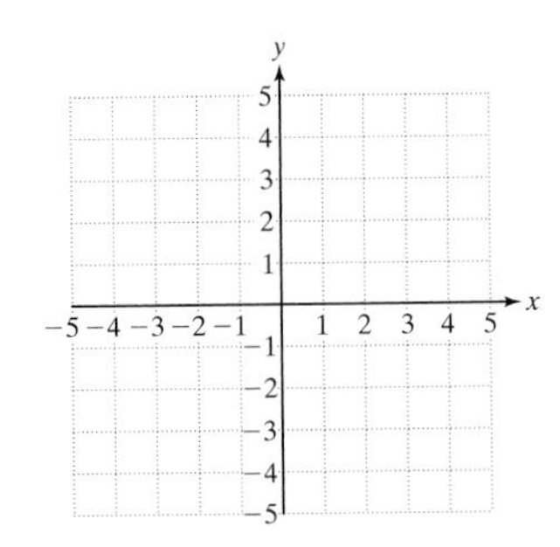

b. Graph the inequality $2x + y < 3$.

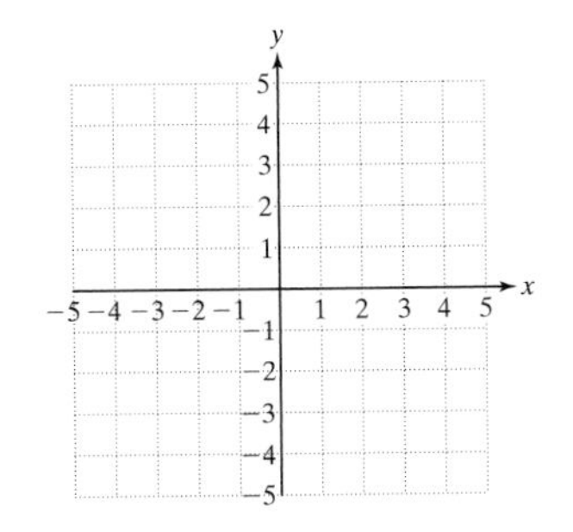

c. Explain the difference between the graphs in parts (a) and (b).

18. How many gallons of a 15% antifreeze solution should be mixed with a 60% antifreeze solution to produce 60 gal of a 45% antifreeze solution?

19. Use a system of linear equations to solve for x and y.

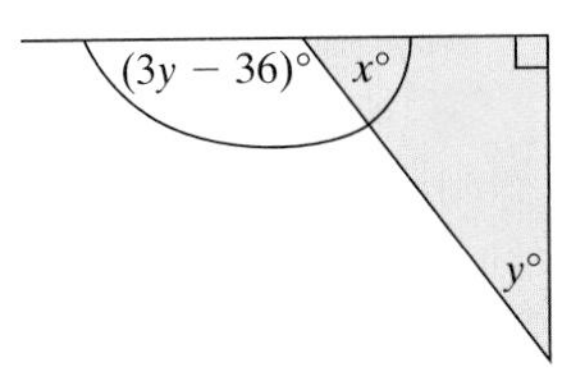

20. In 1920, the average speed for the winner of the Indianapolis 500 car race was 88.6 mph. By 1990, the average speed of the winner was 186.0 mph.

a. Find the slope of the line shown in the figure. Round to one decimal place.

b. Interpret the meaning of the slope in the context of this problem.

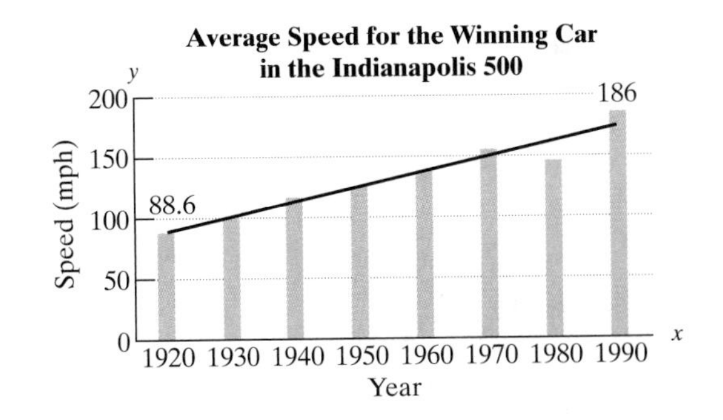

Polynomials and Properties of Exponents

In this chapter we introduce the concept of a polynomial and learn how to add, subtract, multiply, and divide polynomials. In addition, we cover the fundamental properties and applications of expressions involving exponents, including scientific notation.

Complete the fill-it-in puzzle using terms from this chapter. Write the words left to right and down.

Key Terms

base
binomial
coefficient
conjugates
degree
exponent
leading
magnitude
monomial
polynomial
power
reciprocal
trinomial

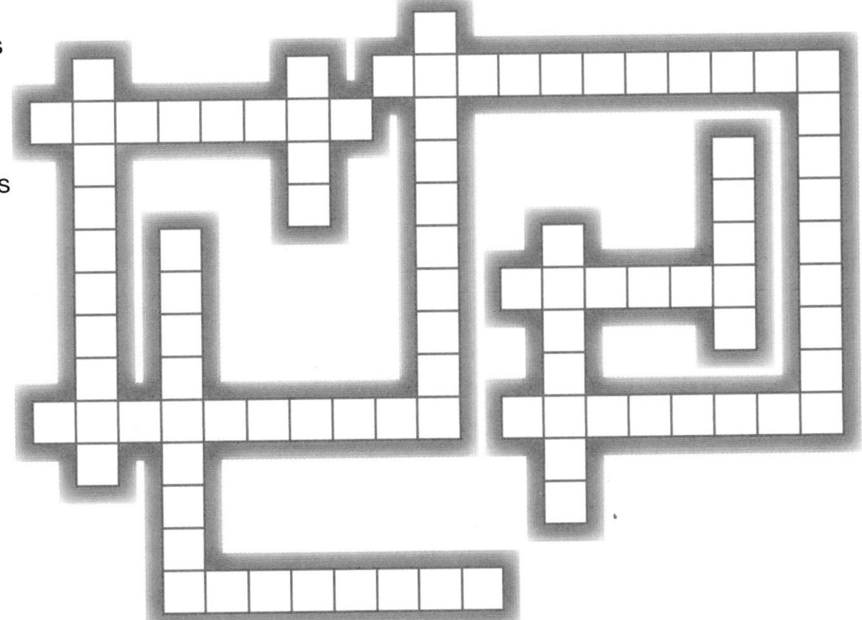

Section 5.1

Exponents: Multiplying and Dividing Common Bases

Concepts

1. Review of Exponential Notation
2. Evaluating Expressions with Exponents
3. Multiplying and Dividing Common Bases
4. Simplifying Expressions with Exponents
5. Applications of Exponents

1. Review of Exponential Notation

Recall that an **exponent** is used to show repeated multiplication of the **base**.

Definition of b^n

Let b represent any real number and n represent a positive integer. Then,

$$b^n = \underbrace{b \cdot b \cdot b \cdot b \ldots \cdot b}_{n \text{ factors of } b}$$

Example 1 Evaluating Expressions with Exponents

For each expression, identify the exponent and base. Then evaluate the expression.

a. 6^2 **b.** $\left(-\frac{1}{2}\right)^3$ **c.** 0.8^4

Solution:

Expression	Base	Exponent	Result
a. 6^2	6	2	$(6)(6) = 36$
b. $\left(-\frac{1}{2}\right)^3$	$-\frac{1}{2}$	3	$\left(-\frac{1}{2}\right)\left(-\frac{1}{2}\right)\left(-\frac{1}{2}\right) = -\frac{1}{8}$
c. 0.8^4	0.8	4	$(0.8)(0.8)(0.8)(0.8) = 0.4096$

Skill Practice For each expression, identify the base and exponent.

1. 8^3 **2.** $(-4)^2$ **3.** 0.02^4

Note that if no exponent is explicitly written for an expression, then the expression has an implied exponent of 1. For example,

$$x = x^1$$
$$y = y^1$$
$$5 = 5^1$$

2. Evaluating Expressions with Exponents

Recall from Section 1.2 that particular care must be taken when evaluating exponential expressions involving negative numbers. An exponential expression with a negative base is written with parentheses around the base, such as $(-3)^2$.

To evaluate $(-3)^2$, we have: $(-3)^2 = (-3)(-3) = 9$

If no parentheses are present, the expression -3^2, is the *opposite* of 3^2, or equivalently, $-1 \cdot 3^2$.

Hence: $-3^2 = -1(3^2) = -1(3)(3) = -9$

Skill Practice Answers

1. Base 8; exponent 3
2. Base -4; exponent 2
3. Base 0.02; exponent 4

Example 2 Evaluating Expressions with Exponents

Evaluate each expression.

a. -5^4 **b.** $(-5)^4$ **c.** $(-0.2)^3$ **d.** -0.2^3

Solution:

a. -5^4

$= -1 \cdot 5^4$	5 is the base to the power of 4.
$= -1 \cdot 5 \cdot 5 \cdot 5 \cdot 5$	Multiply -1 times four factors of 5.
$= -625$	

b. $(-5)^4$

$= (-5)(-5)(-5)(-5)$	Parentheses indicate that -5 is the base to the power of 4.
$= 625$	Multiply four factors of -5.

c. $(-0.2)^3$ Parentheses indicate that -0.2 is the base to the power of 3.

$= (-0.2)(-0.2)(-0.2)$	Multiply three factors of -0.2.
$= -0.008$	

d. -0.2^3

$= -1 \cdot 0.2^3$	0.2 is the base to the power of 3.
$= -1 \cdot 0.2 \cdot 0.2 \cdot 0.2$	Multiply -1 times three factors of 0.2.
$= -0.008$	

Skill Practice Evaluate.

4. -2^4 **5.** $(-2)^4$ **6.** $(-0.1)^3$ **7.** -0.1^3

Example 3 Evaluating Expressions with Exponents

Evaluate each expression for $a = 2$ and $b = -3$.

a. $5a^2$ **b.** $(5a)^2$ **c.** $5ab^2$ **d.** $(b + a)^2$

Solution:

a. $5a^2$

$= 5(\ )^2$	Use parentheses to substitute a number for a variable.
$= 5(2)^2$	Substitute $a = 2$.
$= 5(4)$	Simplify.
$= 20$	

b. $(5a)^2$

$= [5(\ )]^2$	Use parentheses to substitute a number for a variable.
$= [5(2)]^2$	Substitute $a = 2$.
$= (10)^2$	Simplify inside the parentheses first.
$= 100$	

Skill Practice Answers

4. -16 **5.** 16
6. -0.001 **7.** -0.001

c. $5ab^2$

$= 5(2)(-3)^2$ Substitute $a = 2, b = -3$.

$= 5(2)(9)$ Simplify exponents first.

$= 90$ Multiply.

TIP: In the expression $5ab^2$, the exponent, 2, applies only to the variable b. The constant 5 and the variable a both have an implied exponent of 1.

d. $(b + a)^2$

$= [(-3) + (2)]^2$ Substitute $b = -3$ and $a = 2$.

$= (-1)^2$ Simplify within the parentheses first.

$= 1$

Avoiding Mistakes:

Be sure to follow the order of operations. In Example 3(d), it would be incorrect to square the terms within the parentheses before adding.

Skill Practice Evaluate each expression for $x = -2$ and $y = 5$.

8. $6x^2$ **9.** $(6x)^2$ **10.** $-2xy^2$ **11.** $(x - y)^2$

3. Multiplying and Dividing Common Bases

In this section, we investigate the effect of multiplying or dividing two quantities with the same base. For example, consider the expressions: x^5x^2 and $\frac{x^5}{x^2}$. Simplifying each expression, we have:

$$x^5x^2 = (x \cdot x \cdot x \cdot x \cdot x)(x \cdot x) = \overbrace{x \cdot x \cdot x \cdot x \cdot x \cdot x \cdot x}^{\text{7 factors of } x} = x^7$$

$$\frac{x^5}{x^2} = \frac{x \cdot x \cdot x \cdot \cancel{x} \cdot \cancel{x}}{\cancel{x} \cdot \cancel{x}} = \frac{x \cdot x \cdot x}{1} = x^3$$

These examples suggest that to multiply two quantities with the same base, we add the exponents. To divide two quantities with the same base, we subtract the exponent in the denominator from the exponent in the numerator. These rules are stated formally as Properties 1 and 2 of exponents, respectively.

Multiplication of Like Bases

Assume that $a \neq 0$ is a real number and that m and n represent positive integers. Then,

$$\text{Property 1:} \quad a^m a^n = a^{m+n}$$

Division of Like Bases

Assume that $a \neq 0$ is a real number and that m and n represent positive integers such that $m > n$. Then,

$$\text{Property 2:} \quad \frac{a^m}{a^n} = a^{m-n}$$

Skill Practice Answers

8. 24 **9.** 144
10. 100 **11.** 49

Example 4 Simplifying Expressions with Exponents

Simplify the expressions.

a. w^3w^4 **b.** 2^32^4 **c.** $\frac{t^6}{t^4}$ **d.** $\frac{5^6}{5^4}$ **e.** $\frac{z^4z^5}{z^3}$ **f.** $\frac{10^7}{10^2 \cdot 10}$

Solution:

a. w^3w^4 $\quad (w \cdot w \cdot w)(w \cdot w \cdot w \cdot w)$

$= w^{3+4}$ $\quad$ Add the exponents.

$= w^7$

b. 2^32^4 $\quad (2 \cdot 2 \cdot 2)(2 \cdot 2 \cdot 2 \cdot 2)$

$= 2^{3+4}$ $\quad$ Add the exponents (the base is unchanged).

$= 2^7$ or 128

c. $\frac{t^6}{t^4}$ $\quad \frac{\not{t} \cdot \not{t} \cdot \not{t} \cdot \not{t} \cdot t \cdot t}{\not{t} \cdot \not{t} \cdot \not{t} \cdot \not{t}}$

$= t^{6-4}$ $\quad$ Subtract the exponents.

$= t^2$

d. $\frac{5^6}{5^4}$ $\quad \frac{\not{5} \cdot \not{5} \cdot \not{5} \cdot \not{5} \cdot 5 \cdot 5}{\not{5} \cdot \not{5} \cdot \not{5} \cdot \not{5}}$

$= 5^{6-4}$ $\quad$ Subtract the exponents (the base is unchanged).

$= 5^2$ or 25

e. $\frac{z^4z^5}{z^3}$

$= \frac{z^{4+5}}{z^3}$ $\quad$ Add the exponents in the numerator (the base is unchanged).

$= \frac{z^9}{z^3}$

$= z^{9-3}$ $\quad$ Subtract the exponents.

$= z^6$

f. $\frac{10^7}{10^2 \cdot 10}$

$= \frac{10^7}{10^2 \cdot 10^1}$ $\quad$ Note that 10 is equivalent to 10^1.

$= \frac{10^7}{10^{2+1}}$ $\quad$ Add the exponents in the denominator (the base is unchanged).

$= \frac{10^7}{10^3}$

$= 10^{7-3}$ $\quad$ Subtract the exponents.

$= 10^4$ or 10,000 $\quad$ Simplify.

Avoiding Mistakes:

When we use Property 1 to add exponents, the base does not change. In Example 4(b), we have $2^32^4 = 2^7$.

Skill Practice Simplify the expressions.

12. $q^4 \cdot q^8$ **13.** $8^5 \cdot 8^{10}$ **14.** $\dfrac{y^{15}}{y^8}$ **15.** $\dfrac{2^{15}}{2^8}$

16. $\dfrac{a^3a^8}{a^7}$ **17.** $\dfrac{3^3 \cdot 3^8}{3^7}$

4. Simplifying Expressions with Exponents

Example 5 Simplifying Expressions with Exponents

Use the commutative and associative properties of real numbers and the properties of exponents to simplify the expressions.

a. $(3p^2q^4)(2pq^5)$ **b.** $\dfrac{16w^9z^3}{3w^8z}$

Solution:

a. $(3p^2q^4)(2pq^5)$

$= (3 \cdot 2)(p^2p)(q^4q^5)$ Apply the associative and commutative properties of multiplication to group coefficients and common bases.

$= (3 \cdot 2)p^{2+1}q^{4+5}$ Add the exponents when multiplying common bases.

$= 6p^3q^9$ Simplify.

b. $\dfrac{16w^9z^3}{3w^8z}$

$= \left(\dfrac{16}{3}\right)\left(\dfrac{w^9}{w^8}\right)\left(\dfrac{z^3}{z}\right)$ Group like coefficients and factors.

$= \left(\dfrac{16}{3}\right)w^{9-8}z^{3-1}$ Subtract the exponents when dividing common bases.

$= \left(\dfrac{16}{3}\right)wz^2$ or $\dfrac{16wz^2}{3}$ Simplify.

Skill Practice Simplify the expressions.

18. $(4x^2y^3)(3x^5y^7)$ **19.** $\dfrac{3^5x^4y^7}{3^2xy^3}$

5. Applications of Exponents

Recall that **simple interest** on an investment or loan is computed by the formula $I = Prt$, where P is the amount of principal, r is the interest rate, and t is the time in years. Simple interest is based only on the original principal. However, in most day-to-day applications, the interest computed on money invested or

Skill Practice Answers

12. q^{12} **13.** 8^{15}
14. y^7 **15.** 2^7 or 128
16. a^4 **17.** 3^4 or 81
18. $12x^7y^{10}$
19. $3^3x^3y^4$ or $27x^3y^4$

borrowed is compound interest. **Compound interest** is computed on the original principal and on the interest already accrued.

Suppose \$1000 is invested at 8% interest for 3 years. Compare the total amount in the account if the money earns simple interest versus if the interest is compounded annually.

Simple Interest

The simple interest earned is given by $I = Prt$

$$= (1000)(0.08)(3)$$

$$= \$240$$

Thus, the total amount in the account after 3 years is \$1240.

Compound Interest (Annual)

To compute interest compounded annually over a period of 3 years, compute the interest earned in the first year. Then add the principal plus the interest earned in the first year. This value then becomes the principal on which to base the interest earned in the second year. We repeat this process, finding the interest for the second and third years based on the principal and interest earned in the preceding years. This process is outlined using a table.

Year	Interest Earned $I = Prt$	Total Amount in the Account
First year	$I = (\$1000)(0.08)(1) = \$80$	$\$1000 + \$80 = \$1080$
Second year	$I = (\$1080)(0.08)(1) = \$86.40$	$\$1080 + \$86.40 = \$1166.40$
Third year	$I = (\$1166.40)(0.08)(1) \approx \$93.31$	$\$1166.40 + 93.31 =$ **\$1259.71**

The total amount in the account found by compounding interest annually is \$1259.71.

The difference in the account balance for interest compounded annually versus for simple interest is $\$1259.71 - \$1240 = \$19.71$.

The total amount, A, in an account earning compound annual interest can be computed quickly using the following formula:

$A = P(1 + r)^t$ where P is the amount of principal, r is the annual interest rate (expressed in decimal form), and t is the number of years.

For example, for \$1000 invested at 8% interest compounded annually for 3 years, we have $P = 1000$, $r = 0.08$, and $t = 3$.

$$A = P(1 + r)^t$$

$$A = 1000(1 + 0.08)^3$$

$$= 1000(1.08)^3$$

$$= 1000(1.259712)$$

$$= 1259.712$$

Rounding to the nearest cent, we have $A = \$1259.71$, as expected.

Example 6 Using Exponents in an Application

Find the amount in an account after 8 years if the initial investment is \$7000, invested at 2.25% interest compounded annually.

Solution:

Identify the values for each variable.

$P = 7000$

$r = 0.0225$ Note that the decimal form of a percent is used for calculations.

$t = 8$

$A = P(1 + r)^t$

$= 7000(1 + 0.0225)^8$ Substitute.

$= 7000(1.0225)^8$ Simplify inside the parentheses.

$\approx 7000(1.194831142)$ Approximate $(1.0225)^8$.

$= 8363.82$ Multiply (round to the nearest cent).

The amount in the account after 8 years is \$8363.82.

Skill Practice

20. Find the amount in an account after 3 years if the initial investment is \$4000 invested at 5% interest compounded annually.

Calculator Connections

In Example 6, it was necessary to evaluate the expression $(1.0225)^8$. Recall that the [^] or [y^x] key can be used to enter expressions with exponents.

Scientific Calculator

Enter: [1] [.] [0] [2] [2] [5] [y^x] [8] [=] **Result:** 1.194831142

Graphing Calculator

```
1.0225^8
        1.194831142
```

Calculator Exercises

Use a calculator to evaluate the expressions.

1. $(1.06)^5$ **2.** $(1.02)^{40}$ **3.** $5000(1.06)^5$

4. $2000(1.02)^{40}$ **5.** $3000(1 + 0.06)^2$ **6.** $1000(1 + 0.05)^3$

Skill Practice Answers

20. \$4630.50

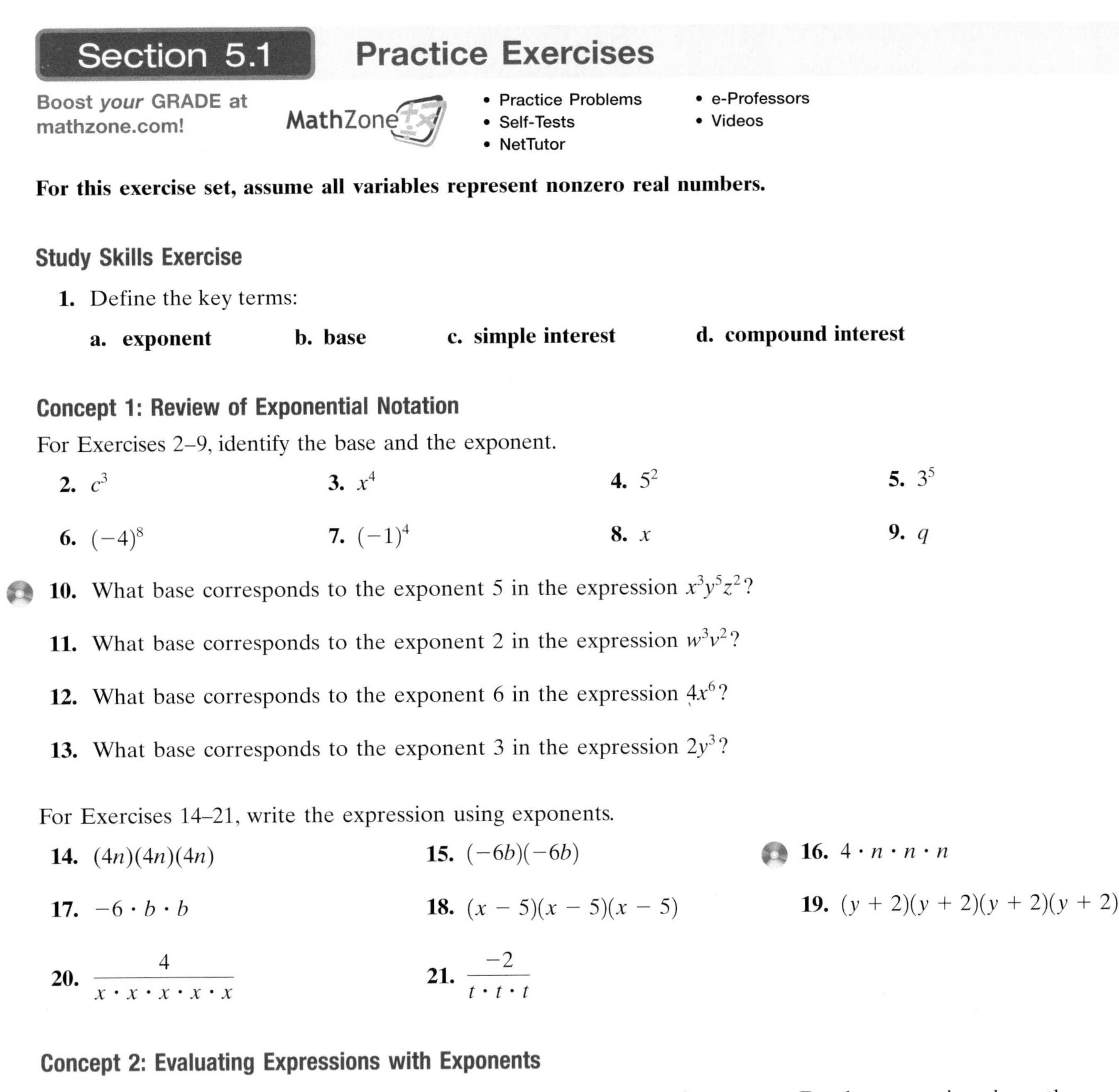

Section 5.1 Practice Exercises

Boost *your* GRADE at mathzone.com!

MathZone

- Practice Problems
- Self-Tests
- NetTutor
- e-Professors
- Videos

For this exercise set, assume all variables represent nonzero real numbers.

Study Skills Exercise

1. Define the key terms:

a. **exponent** **b.** **base** **c.** **simple interest** **d.** **compound interest**

Concept 1: Review of Exponential Notation

For Exercises 2–9, identify the base and the exponent.

2. c^3 **3.** x^4 **4.** 5^2 **5.** 3^5

6. $(-4)^8$ **7.** $(-1)^4$ **8.** x **9.** q

10. What base corresponds to the exponent 5 in the expression $x^3y^5z^2$?

11. What base corresponds to the exponent 2 in the expression w^3v^2?

12. What base corresponds to the exponent 6 in the expression $4x^6$?

13. What base corresponds to the exponent 3 in the expression $2y^3$?

For Exercises 14–21, write the expression using exponents.

14. $(4n)(4n)(4n)$ **15.** $(-6b)(-6b)$ **16.** $4 \cdot n \cdot n \cdot n$

17. $-6 \cdot b \cdot b$ **18.** $(x - 5)(x - 5)(x - 5)$ **19.** $(y + 2)(y + 2)(y + 2)(y + 2)$

20. $\dfrac{4}{x \cdot x \cdot x \cdot x \cdot x}$ **21.** $\dfrac{-2}{t \cdot t \cdot t}$

Concept 2: Evaluating Expressions with Exponents

For Exercises 22–29, evaluate the two expressions and compare the answers. Do the expressions have the same value?

22. -5^2 and $(-5)^2$ **23.** -3^4 and $(-3)^4$ **24.** -2^5 and $(-2)^5$ **25.** -5^3 and $(-5)^3$

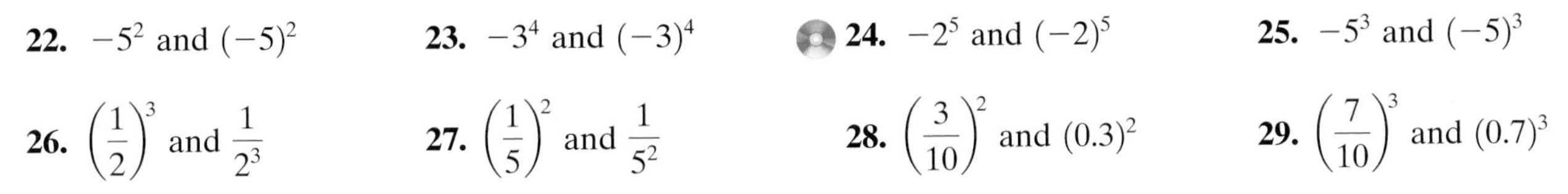

26. $\left(\dfrac{1}{2}\right)^3$ and $\dfrac{1}{2^3}$ **27.** $\left(\dfrac{1}{5}\right)^2$ and $\dfrac{1}{5^2}$ **28.** $\left(\dfrac{3}{10}\right)^2$ and $(0.3)^2$ **29.** $\left(\dfrac{7}{10}\right)^3$ and $(0.7)^3$

For Exercises 30–39, evaluate the expressions.

30. 16^1 **31.** 20^1 **32.** $(-1)^{21}$ **33.** $(-1)^{30}$

34. 0^6 **35.** 0^4 **36.** $\left(-\frac{1}{3}\right)^2$ **37.** $\left(-\frac{1}{4}\right)^3$

38. $-\left(\frac{2}{5}\right)^2$ **39.** $-\left(\frac{3}{5}\right)^2$

For Exercises 40–47, simplify using the order of operations.

40. $3 \cdot 2^4$ **41.** $2 \cdot 0^5$ **42.** $-4(-1)^7$ **43.** $-3(-1)^4$

44. $6^2 - 3^3$ **45.** $4^3 + 2^3$ **46.** $2 \cdot 3^2 + 4 \cdot 2^3$ **47.** $6^2 - 3 \cdot 1^3$

For Exercises 48–59, evaluate each expression for $a = -4$ and $b = 5$.

48. $-4b^2$ **49.** $5a^2$ **50.** $(-4b)^2$ **51.** $(5a)^2$

52. $(a + b)^2$ **53.** $(a - b)^2$ **54.** $a^2 + 2ab + b^2$ **55.** $a^2 - 2ab + b^2$

56. $-10ab^2$ **57.** $-6a^3b$ **58.** $-10a^2b$ **59.** $-a^2b$

Concept 3: Multiplying and Dividing Common Bases

60. Expand the following expressions first. Then simplify using exponents.

a. $x^4 \cdot x^3$ **b.** $5^4 \cdot 5^3$

61. Expand the following expressions first. Then simplify using exponents.

a. $y^2 \cdot y^4$ **b.** $3^2 \cdot 3^4$

For Exercises 62–73, simplify the expressions. Write the answers in exponent form.

62. z^5z^3 **63.** w^4w^7 **64.** $a \cdot a^8$ **65.** p^4p

66. $4^5 \cdot 4^9$ **67.** $6^7 \cdot 6^5$ **68.** $\left(\frac{2}{3}\right)^3\left(\frac{2}{3}\right)$ **69.** $\left(\frac{1}{x}\right)\left(\frac{1}{x}\right)^2$

70. $c^5c^2c^7$ **71.** $b^7b^2b^8$ **72.** $x \cdot x^4 \cdot x^{10} \cdot x^3$ **73.** $z^7 \cdot z^{11} \cdot z^{60} \cdot z$

74. Expand the following expressions. Then simplify.

a. $\frac{p^8}{p^3}$ **b.** $\frac{8^8}{8^3}$

75. Expand the following expressions. Then simplify.

a. $\frac{w^5}{w^2}$ **b.** $\frac{4^5}{4^2}$

For Exercises 76–93, simplify the expressions. Write the answers in exponent form.

76. $\frac{x^8}{x^6}$ **77.** $\frac{z^5}{z^4}$ **78.** $\frac{a^{10}}{a}$ **79.** $\frac{b^{12}}{b}$

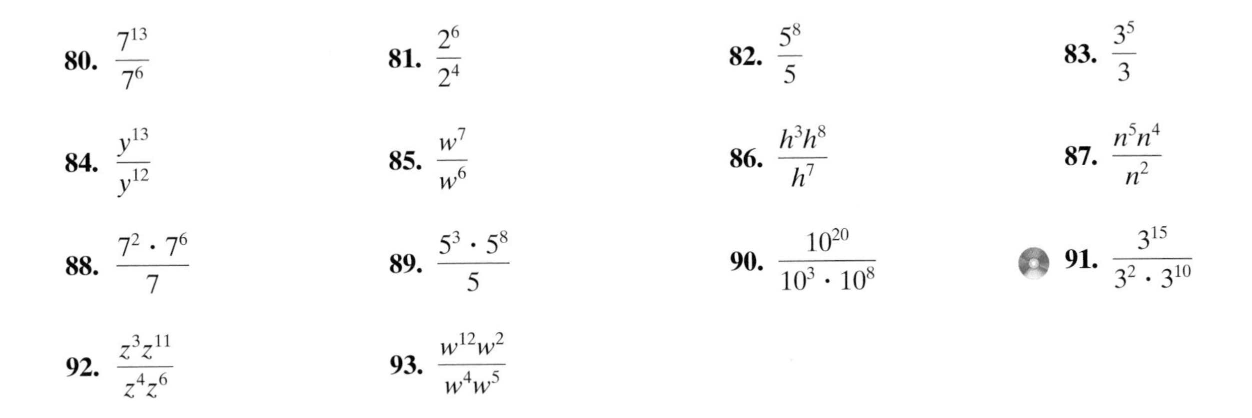

80. $\frac{7^{13}}{7^6}$ **81.** $\frac{2^6}{2^4}$ **82.** $\frac{5^8}{5}$ **83.** $\frac{3^5}{3}$

84. $\frac{y^{13}}{y^{12}}$ **85.** $\frac{w^7}{w^6}$ **86.** $\frac{h^3h^8}{h^7}$ **87.** $\frac{n^5n^4}{n^2}$

88. $\frac{7^2 \cdot 7^6}{7}$ **89.** $\frac{5^3 \cdot 5^8}{5}$ **90.** $\frac{10^{20}}{10^3 \cdot 10^8}$ **91.** $\frac{3^{15}}{3^2 \cdot 3^{10}}$

92. $\frac{z^3z^{11}}{z^4z^6}$ **93.** $\frac{w^{12}w^2}{w^4w^5}$

Concept 4: Simplifying Expressions with Exponents

For Exercises 94–109, use the commutative and associative properties of real numbers and the properties of exponents to simplify the expressions.

94. $(5a^2b)(8a^3b^4)$ **95.** $(10xy^3)(3x^4y)$ **96.** $(r^6s^4)(13r^2s)$ **97.** $(6p^2q^8)(7p^5q^3)$

98. $\left(\frac{2}{3}m^{13}n^8\right)(24m^7n^2)$ **99.** $\left(\frac{1}{4}c^6d^6\right)(28c^2d^7)$ **100.** $\frac{14c^4d^5}{7c^3d}$ **101.** $\frac{36h^5k^2}{9h^3k}$

102. $\frac{2x^3y^5}{8xy^3}$ **103.** $\frac{13w^8z^3}{26w^2z}$ **104.** $\frac{25h^3jk^5}{12h^2k}$ **105.** $\frac{15m^5np^{12}}{4mp^9}$

106. $(-4p^6q^8r^4)(2pqr^2)$ **107.** $(-5a^4bc)(-10a^2b)$ **108.** $\frac{-12s^2tu^3}{4su^2}$ **109.** $\frac{15w^5x^{10}y^3}{-15w^4x}$

Concept 5: Applications of Exponents

Use the formula $A = P(1 + r)^t$ for Exercises 110–113.

110. Find the amount in an account after 2 years if the initial investment is \$5000, invested at 7% interest compounded annually.

111. Find the amount in an account after 5 years if the initial investment is \$2000, invested at 4% interest compounded annually.

112. Find the amount in an account after 3 years if the initial investment is \$4000, invested at 6% interest compounded annually.

113. Find the amount in an account after 4 years if the initial investment is \$10,000, invested at 5% interest compounded annually.

For Exercises 114–117, use the geometry formulas found in Section R.4.

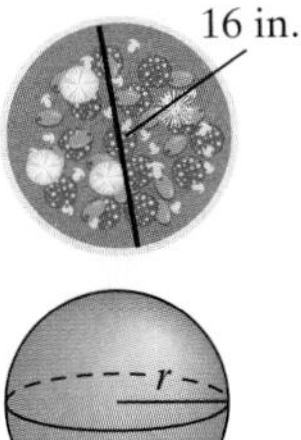

114. Find the area of the pizza shown in the figure. Round to the nearest square inch.

115. Find the area of a circular pool 50 ft in diameter. Round to the nearest square foot.

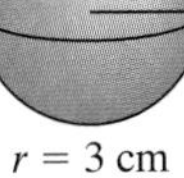

$r = 3$ cm

116. Find the volume of the sphere shown in the figure. Round to the nearest cubic centimeter.

117. Find the volume of a spherical balloon that is 8 in. in diameter. Round to the nearest cubic inch.

Expanding Your Skills

For Exercises 118–125, simplify the expressions using the addition or subtraction rules of exponents. Assume that a, b, m, and n represent positive integers.

118. $x^n x^{n+1}$ **119.** $y^a y^{2a}$ **120.** $p^{3m+5}p^{-m-2}$ **121.** $q^{4b-3}q^{-4b+4}$

122. $\dfrac{z^{b+1}}{z^b}$ **123.** $\dfrac{w^{5n+3}}{w^{2n}}$ **124.** $\dfrac{r^{3a+3}}{r^{3a}}$ **125.** $\dfrac{t^{3+2m}}{t^{2m}}$

Section 5.2 More Properties of Exponents

Concepts

1. Power Rule for Exponents
2. The Properties $(ab)^m = a^m b^m$ and $\left(\dfrac{a}{b}\right)^m = \dfrac{a^m}{b^m}$
3. Simplifying Expressions with Exponents

1. Power Rule for Exponents

The expression $(x^2)^3$ indicates that the quantity x^2 is cubed.

$$(x^2)^3 = (x^2)(x^2)(x^2) = (x \cdot x)(x \cdot x)(x \cdot x) = x^6$$

From this example, it appears that to raise a base to successive powers, we multiply the exponents and leave the base unchanged. This is stated formally as the power rule for exponents.

Power Rule for Exponents

Assume that $a \neq 0$ is a real number and that m and n represent positive integers. Then,

$$\text{Property 3:} \quad (a^m)^n = a^{m \cdot n}$$

Example 1 Simplifying Expressions with Exponents

Simplify the expressions.

a. $(s^4)^2$ **b.** $(3^4)^2$ **c.** $(x^2x^5)^4$

Solution:

a. $(s^4)^2$

$= s^{4 \cdot 2}$ Multiply exponents (the base is unchanged).

$= s^8$

b. $(3^4)^2$

$= 3^{4 \cdot 2}$ Multiply exponents (the base is unchanged).

$= 3^8$ or 6561

c. $(x^2x^5)^4$

$= (x^7)^4$ Simplify inside the parentheses by adding exponents.

$= x^{7 \cdot 4}$ Multiply exponents (the base is unchanged).

$= x^{28}$

Skill Practice Simplify the expressions.

1. $(y^3)^5$ **2.** $(2^8)^{10}$ **3.** $(q^5q^4)^3$

Skill Practice Answers

1. y^{15} **2.** 2^{80} **3.** q^{27}

2. The Properties $(ab)^m = a^m b^m$ and $\left(\frac{a}{b}\right)^m = \frac{a^m}{b^m}$

Consider the following expressions and their simplified forms:

$$(xy)^3 = (xy)(xy)(xy) = (x \cdot x \cdot x)(y \cdot y \cdot y) = x^3y^3$$

$$\left(\frac{x}{y}\right)^3 = \left(\frac{x}{y}\right)\left(\frac{x}{y}\right)\left(\frac{x}{y}\right) = \left(\frac{x \cdot x \cdot x}{y \cdot y \cdot y}\right) = \frac{x^3}{y^3}$$

The expressions were simplified using the commutative and associative properties of multiplication. The simplified forms for each expression could have been reached in one step by applying the exponent to each factor inside the parentheses.

Power of a Product and Power of a Quotient

Assume that a and b are real numbers such that $b \neq 0$. Let m represent a positive integer. Then,

Property 4: $(ab)^m = a^m b^m$

Property 5: $\left(\frac{a}{b}\right)^m = \frac{a^m}{b^m}$

Avoiding Mistakes:

The power rule of exponents can be applied to a product of bases but in general cannot be applied to a sum or difference of bases.

$$(ab)^n = a^n b^n$$

but $(a + b)^n \neq a^n + b^n$

Applying these properties of exponents, we have

$$(xy)^3 = x^3y^3 \quad \text{and} \quad \left(\frac{x}{y}\right)^3 = \frac{x^3}{y^3}$$

3. Simplifying Expressions with Exponents

Example 2 Simplifying Expressions with Exponents

Simplify the expressions.

a. $(-2xyz)^4$ **b.** $(5x^2y^7)^3$ **c.** $\left(\frac{2}{5}\right)^3$ **d.** $\left(\frac{1}{3xy^4}\right)^2$

Solution:

a. $(-2xyz)^4$

$= (-2)^4x^4y^4z^4$ or $16x^4y^4z^4$ Raise each factor within parentheses to the fourth power.

b. $(5x^2y^7)^3$

$= 5^3(x^2)^3(y^7)^3$ Raise each factor within parentheses to the third power.

$= 125x^6y^{21}$ Multiply exponents and simplify.

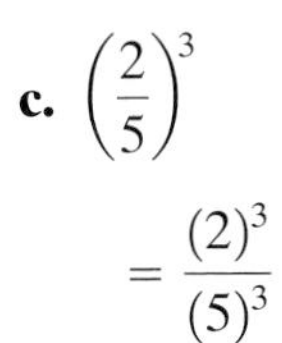

c. $\left(\frac{2}{5}\right)^3$

$= \frac{(2)^3}{(5)^3}$ Apply the exponent to each factor in parentheses.

$= \frac{8}{125}$ Simplify.

d. $\left(\dfrac{1}{3xy^4}\right)^2$

$= \dfrac{1^2}{3^2x^2(y^4)^2}$ Square each factor within parentheses.

$= \dfrac{1}{9x^2y^8}$ Multiply exponents and simplify.

Skill Practice Simplify the expressions.

4. $(3x^4y^{10})^3$ **5.** $(-5ab)^3$ **6.** $\left(\dfrac{4}{3}\right)^4$ **7.** $\left(\dfrac{2x^3}{y^5}\right)^2$

The properties of exponents can be used along with the properties of real numbers to simplify complicated expressions.

Example 3 Simplifying Expressions with Exponents

Simplify the expressions.

a. $\dfrac{(x^2)^6(x^3)}{(x^7)^2}$ **b.** $(3cd^2)(2cd^3)^3$ **c.** $\left(\dfrac{x^7yz^4}{8xz^3}\right)^2$

Solution:

a. $\dfrac{(x^2)^6(x^3)}{(x^7)^2}$ Clear parentheses by applying the power rule.

$= \dfrac{x^{2\cdot 6}x^3}{x^{7\cdot 2}}$ Multiply exponents.

$= \dfrac{x^{12}x^3}{x^{14}}$

$= \dfrac{x^{12+3}}{x^{14}}$ Add exponents in the numerator.

$= \dfrac{x^{15}}{x^{14}}$

$= x^{15-14}$ Subtract exponents.

$= x$ Simplify.

b. $(3cd^2)(2cd^3)^3$ Clear parentheses by applying the power rule.

$= 3cd^2 \cdot 2^3c^3d^9$ Raise each factor in the second parentheses to the third power.

$= 3 \cdot 2^3cc^3d^2d^9$ Group like factors.

$= 3 \cdot 8c^{1+3}d^{2+9}$ Add exponents from like factors.

$= 24c^4d^{11}$ Simplify.

Skill Practice Answers

4. $27x^{12}y^{30}$ **5.** $-125a^3b^3$
6. $\dfrac{256}{81}$ **7.** $\dfrac{4x^6}{y^{10}}$

c. $\left(\dfrac{x^7yz^4}{8xz^3}\right)^2$

$= \left(\dfrac{x^{7-1}yz^{4-3}}{8}\right)^2$ Simplify inside the parentheses by subtracting exponents from like factors.

$= \left(\dfrac{x^6yz}{8}\right)^2$

$= \dfrac{(x^6)^2y^2z^2}{8^2}$ Apply the power rule of exponents.

$= \dfrac{x^{12}y^2z^2}{64}$

Skill Practice Simplify the expressions.

8. $\dfrac{k^5 \cdot k^8}{(k^2)^4}$ **9.** $(x^4y^8z^{10})^4(xyz^3)^3$ **10.** $\left(\dfrac{2w^2xy^4}{6xy^3}\right)^2$

Skill Practice Answers

8. k^5 **9.** $x^{19}y^{35}z^{49}$

10. $\dfrac{w^4y^2}{9}$

Section 5.2 Practice Exercises

Boost *your* GRADE at mathzone.com!

MathZone

- Practice Problems
- Self-Tests
- NetTutor
- e-Professors
- Videos

For this exercise set assume all variables represent nonzero real numbers.

Review Exercises

For Exercises 1–8, simplify.

1. $4^2 \cdot 4^7$ **2.** $5^8 \cdot 5^3 \cdot 5$ **3.** $a^{13} \cdot a \cdot a^6$ **4.** $y^{14}y^3$

5. $\dfrac{d^{13}d}{d^5}$ **6.** $\dfrac{3^8 \cdot 3}{3^2}$ **7.** $\dfrac{7^{11}}{7^5}$ **8.** $\dfrac{z^4}{z^3}$

9. Explain when to add exponents versus when to subtract exponents.

10. Explain when to add exponents versus when to multiply exponents.

Concept 1: Power Rule for Exponents

For Exercises 11–22, simplify and write answers in exponent form.

11. $(5^3)^4$ **12.** $(2^8)^7$ **13.** $(12^3)^2$ **14.** $(6^4)^4$

15. $(y^7)^2$ **16.** $(z^6)^4$ **17.** $(w^5)^5$ **18.** $(t^3)^6$

19. $(a^2a^4)^6$ **20.** $(z \cdot z^3)^2$ **21.** $(y^3y^4)^2$ **22.** $(w^5w)^4$

23. Evaluate the two expressions and compare the answers: $(2^2)^3$ and $(2^3)^2$.

24. Evaluate the two expressions and compare the answers: $(4^4)^2$ and $(4^2)^4$.

25. Evaluate the two expressions and compare the answers. Which expression is greater? Why?

$$2^{(2^4)} \quad \text{and} \quad (2^2)^4$$

26. Evaluate the two expressions and compare the answers. Which expression is greater? Why?

$$3^{(2^4)} \quad \text{and} \quad (3^2)^4$$

Concept 3: Simplifying Expressions with Exponents

For Exercises 27–42, use the appropriate property to clear the parentheses.

27. $(5w)^2$ **28.** $(4y)^3$ **29.** $(srt)^4$ **30.** $(wxy)^6$

31. $\left(\frac{2}{r}\right)^4$ **32.** $\left(\frac{1}{t}\right)^8$ **33.** $\left(\frac{x}{y}\right)^5$ **34.** $\left(\frac{w}{z}\right)^7$

35. $(-3a)^4$ **36.** $(2x)^5$ **37.** $(-3abc)^3$ **38.** $(-5xyz)^2$

39. $\left(-\frac{4}{x}\right)^3$ **40.** $\left(-\frac{1}{w}\right)^4$ **41.** $\left(-\frac{a}{b}\right)^2$ **42.** $\left(-\frac{r}{s}\right)^3$

For Exercises 43–76, simplify the expressions.

43. $(6u^2v^4)^3$ **44.** $(3a^5b^2)^6$ **45.** $5(x^2y)^4$ **46.** $18(u^3v^4)^2$

47. $(-h^4)^7$ **48.** $(-k^6)^3$ **49.** $(-m^2)^6$ **50.** $(-n^3)^8$

51. $\left(\frac{4}{rs^4}\right)^5$ **52.** $\left(\frac{2}{h^7k}\right)^3$ **53.** $\left(\frac{3p}{q^3}\right)^5$ **54.** $\left(\frac{5x^2}{y^3}\right)^4$

55. $\frac{y^8(y^3)^4}{(y^2)^3}$ **56.** $\frac{(w^3)^2(w^4)^5}{(w^4)^2}$ **57.** $(x^2)^5(x^3)^7$ **58.** $(y^3)^4(y^2)^5$

59. $(a^2b)^3(a^4b^3)^5$ **60.** $(c^3d^5)^2(cd^3)^3$ **61.** $(-2p^2q^4)^4$ **62.** $(-7x^4y^5)^2$

63. $(-m^7n^3)^5$ **64.** $(-a^3b^6)^7$ **65.** $\frac{(5a^3b)^4(a^2b)^4}{(5ab)^2}$ **66.** $\frac{(6s^3)^2(s^4t^5)^2}{(3s^4t^2)^2}$

67. $\frac{(21x^5y)(2x^8y^4)}{14xy}$ **68.** $\frac{(4u^3v^3)(9u^4v)}{12u^5v^2}$ **69.** $\left(\frac{2c^3d^4}{3c^2d}\right)^2$ **70.** $\left(\frac{x^3y^5z}{5xy^2}\right)^2$

71. $(2c^3d^2)^5\left(\frac{c^6d^8}{4c^2d}\right)^3$ **72.** $\left(\frac{s^5t^6}{2s^2t}\right)^2(10s^3t^3)^2$ **73.** $\left(\frac{-3a^3b}{c^2}\right)^3$ **74.** $\left(\frac{-4x^2}{y^4z}\right)^3$

75. $\frac{(-8b^6)^2(b^3)^5}{4b}$ **76.** $\frac{(-6a^2)^2(a^3)^4}{9a}$

Expanding Your Skills

For Exercises 77–84, simplify the expressions using the addition or subtraction properties of exponents. Assume that a, b, m, and n represent positive integers.

77. $(x^m)^2$ **78.** $(y^3)^n$ **79.** $(5a^{2n})^3$ **80.** $(3b^4)^m$

81. $\left(\frac{m^2}{n^3}\right)^b$ **82.** $\left(\frac{x^5}{y^3}\right)^m$ **83.** $\left(\frac{3a^3}{5b^4}\right)^n$ **84.** $\left(\frac{4m^6}{3n^2}\right)^b$

Definitions of b^0 and b^{-n}

Section 5.3

Concepts

1. Definition of b^0
2. Definition of b^{-n}
3. Properties of Integer Exponents: A Summary
4. Simplifying Expressions with Exponents

In Sections 5.1 and 5.2, we learned several rules that allow us to manipulate expressions containing *positive* integer exponents. In this section, we present definitions that can be used to simplify expressions with negative exponents or with an exponent of zero.

1. Definition of b^0

To begin, consider the following pattern.

$3^3 = 27$ (divide by 3)
$3^2 = 9$ (divide by 3)
$3^1 = 3$ (divide by 3)
$3^0 = 1$

As the exponents decrease by 1, the resulting expressions are divided by 3.

For the pattern to continue, we define $3^0 = 1$.

This pattern suggests that we should define an expression with a zero exponent as follows.

Avoiding Mistakes:

$b^0 = 1$ provided that b is not zero. Therefore, the expression 0^0 cannot be simplified by this rule.

Definition of b^0

Let b be a nonzero real number. Then, $b^0 = 1$.

Example 1 Simplifying Expressions with a Zero Exponent

Simplify.

a. 4^0 **b.** $(-4)^0$ **c.** -4^0

d. z^0 **e.** $-4z^0$ **f.** $(4z)^0$

Solution:

a. $4^0 = 1$ By definition

b. $(-4)^0 = 1$ By definition

c. $-4^0 = -1 \cdot 4^0 = -1 \cdot 1 = -1$ The exponent 0 applies only to 4.

d. $z^0 = 1$ By definition

e. $-4z^0 = -4 \cdot z^0 = -4 \cdot 1 = -4$ The exponent 0 applies only to z.

f. $(4z)^0 = 1$ The parentheses indicate that the exponent, 0, applies to both factors 4 and z.

Skill Practice Evaluate the expressions. Assume all variables represent nonzero real numbers.

1. 7^0 **2.** $(-7)^0$ **3.** -5^0

4. $(4 + -8)^0$ **5.** $2x^0$ **6.** $(2x)^0$

Skill Practice Answers

1. 1 **2.** 1 **3.** −1
4. 1 **5.** 2 **6.** 1

The definition of b^0 is consistent with the other properties of exponents learned thus far. For example, we know that $1 = \frac{5^3}{5^3}$. If we subtract exponents, the result is 5^0.

subtract exponents

$$1 = \frac{5^3}{5^3} = 5^{3-3} = 5^0.$$ Therefore, 5^0 must be defined as 1.

2. Definition of b^{-n}

To understand the concept of a *negative* exponent, consider the following pattern.

$3^3 = 27$
divide by 3
$3^2 = 9$
divide by 3
$3^1 = 3$
divide by 3
$3^0 = 1$
divide by 3

As the exponents decrease by 1, the resulting expressions are divided by 3.

$3^{-1} = \frac{1}{3}$ ← For the pattern to continue, we define $3^{-1} = \frac{1}{3^1} = \frac{1}{3}$.

$3^{-2} = \frac{1}{9}$ ← For the pattern to continue, we define $3^{-2} = \frac{1}{3^2} = \frac{1}{9}$.

$3^{-3} = \frac{1}{27}$ ← For the pattern to continue, we define $3^{-3} = \frac{1}{3^3} = \frac{1}{27}$.

This pattern suggests that $3^{-n} = \frac{1}{3^n}$ for all integers, n. In general, we have the following definition involving negative exponents.

Definition of b^{-n}

Let n be an integer and b be a nonzero real number. Then,

$$b^{-n} = \left(\frac{1}{b}\right)^n \quad \text{or} \quad \frac{1}{b^n}$$

The definition of b^{-n} implies that to evaluate b^{-n}, take the reciprocal of the base and change the sign of the exponent.

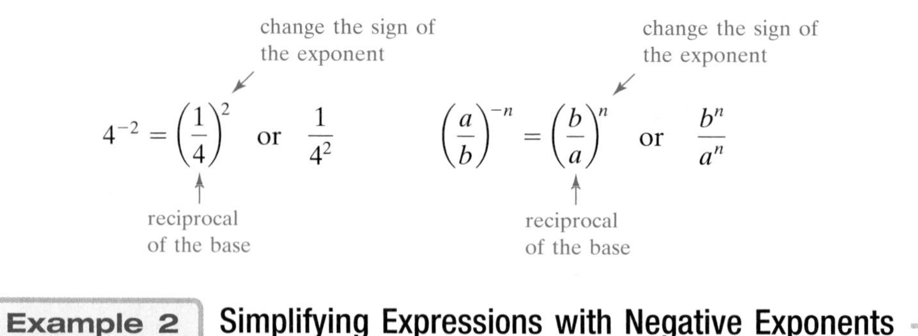

Example 2 **Simplifying Expressions with Negative Exponents**

Simplify.

a. c^{-3} **b.** 5^{-1} **c.** $(-3)^{-4}$ **d.** $\left(\frac{1}{4}\right)^{-2}$

e. $\left(-\frac{3}{5}\right)^{-3}$ **f.** $\frac{1}{y^{-5}}$ **g.** $(5x)^{-3}$ **h.** $5x^{-3}$

Solution:

a. $c^{-3} = \frac{1}{c^3}$ — By definition

b. $5^{-1} = \frac{1}{5^1}$ — By definition

$= \frac{1}{5}$ — Simplify.

c. $(-3)^{-4} = \frac{1}{(-3)^4}$ — The base is -3 and must be enclosed in parentheses.

$= \frac{1}{81}$ — Simplify. Note that $(-3)^4 = (-3)(-3)(-3)(-3) = 81$.

d. $\left(\frac{1}{4}\right)^{-2} = 4^2$ — Take the reciprocal of the base, and change the sign of the exponent.

$= 16$ — Simplify.

e. $\left(-\frac{3}{5}\right)^{-3} = \left(-\frac{5}{3}\right)^3$ — Take the reciprocal of the base, and change the sign of the exponent.

$= -\frac{125}{27}$ — Simplify.

f. $\frac{1}{y^{-5}} = \left(\frac{1}{y}\right)^{-5}$ — Apply the power of a quotient rule from Section 5.2.

$= (y)^5$ — Take the reciprocal of the base, and change the sign of the exponent.

$= y^5$

g. $(5x)^{-3} = \left(\frac{1}{5x}\right)^3$ — Take the reciprocal of the base, and change the sign of the exponent.

$= \frac{(1)^3}{(5x)^3}$ — Apply the exponent of 3 to each factor within parentheses.

$= \frac{1}{125x^3}$ — Simplify.

h. $5x^{-3} = 5 \cdot x^{-3}$ — Note that the exponent, -3, applies only to x.

$= 5 \cdot \frac{1}{x^3}$ — Rewrite x^{-3} as $\frac{1}{x^3}$.

$= \frac{5}{x^3}$ — Multiply.

Skill Practice Evaluate.

7. 3^{-3} **8.** x^{-1} **9.** $(-5)^{-2}$ **10.** $\left(\frac{1}{5}\right)^{-3}$

11. $\left(-\frac{5}{3}\right)^{-3}$ **12.** $\frac{2}{z^{-4}}$ **13.** $(2w)^{-4}$ **14.** $2w^{-4}$

Skill Practice Answers

7. $\frac{1}{27}$ **8.** $\frac{1}{x}$
9. $\frac{1}{25}$ **10.** 125
11. $-\frac{27}{125}$ **12.** $2z^4$
13. $\frac{1}{16w^4}$ **14.** $\frac{2}{w^4}$

It is important to note that the definition of b^{-n} is consistent with the other properties of exponents learned thus far. For example, consider the expression

$$\frac{x^4}{x^7} = \frac{\cancel{x} \cdot \cancel{x} \cdot \cancel{x} \cdot \cancel{x}}{\cancel{x} \cdot \cancel{x} \cdot \cancel{x} \cdot \cancel{x} \cdot x \cdot x \cdot x} = \frac{1}{x^3}$$

By subtracting exponents, we have $\frac{x^4}{x^7} = x^{4-7} = x^{-3}$ (subtract exponents)

Hence, $x^{-3} = \frac{1}{x^3}$

3. Properties of Integer Exponents: A Summary

The definitions of b^0 and b^{-n} allow us to extend the properties of exponents learned in Sections 5.1 and 5.2 to include integer exponents. These are summarized in Table 5-1.

Table 5-1

Properties of Integer Exponents

Assume that a and b are real numbers ($b \neq 0$) and that m and n represent integers.

Property	Example	Details/Notes
Multiplication of Like Bases 1. $b^m b^n = b^{m+n}$	$b^2 b^4 = b^{2+4} = b^6$	$b^2 b^4 = (b \cdot b)(b \cdot b \cdot b \cdot b) = b^6$
Division of Like Bases 2. $\frac{b^m}{b^n} = b^{m-n}$	$\frac{b^5}{b^2} = b^{5-2} = b^3$	$\frac{b^5}{b^2} = \frac{\cancel{b} \cdot \cancel{b} \cdot b \cdot b \cdot b}{\cancel{b} \cdot \cancel{b}} = b^3$
The Power Rule 3. $(b^m)^n = b^{m \cdot n}$	$(b^4)^2 = b^{4 \cdot 2} = b^8$	$(b^4)^2 = (b \cdot b \cdot b \cdot b)(b \cdot b \cdot b \cdot b) = b^8$
Power of a Product 4. $(ab)^m = a^m b^m$	$(ab)^3 = a^3 b^3$	$(ab)^3 = (ab)(ab)(ab)$ $= (a \cdot a \cdot a)(b \cdot b \cdot b) = a^3 b^3$
Power of a Quotient 5. $\left(\frac{a}{b}\right)^m = \frac{a^m}{b^m}$	$\left(\frac{a}{b}\right)^3 = \frac{a^3}{b^3}$	$\left(\frac{a}{b}\right)^3 = \left(\frac{a}{b}\right)\left(\frac{a}{b}\right)\left(\frac{a}{b}\right) = \frac{a \cdot a \cdot a}{b \cdot b \cdot b} = \frac{a^3}{b^3}$

Definitions

Assume that b is a real number ($b \neq 0$) and that n represents an integer.

Definition	Example	Details/Notes
$b^0 = 1$	$(4)^0 = 1$	Any nonzero quantity raised to the zero power equals 1.
$b^{-n} = \left(\frac{1}{b}\right)^n = \frac{1}{b^n}$	$b^{-5} = \left(\frac{1}{b}\right)^5 = \frac{1}{b^5}$	To simplify a negative exponent, take the reciprocal of the base and make the exponent positive.

4. Simplifying Expressions with Exponents

Example 3 Simplifying Expressions with Exponents

Simplify the following expressions. Write the answers with positive exponents only. Assume all variables are nonzero.

a. $\dfrac{a^3b^{-2}}{c^{-5}}$ **b.** $\dfrac{x^2x^{-7}}{x^3}$ **c.** $\dfrac{z^2}{w^{-4}w^4z^{-8}}$

d. $(-4ab^{-2})^{-3}$ **e.** $\left(\dfrac{2p^{-4}q^3}{5p^2q}\right)^{-1}$

Solution:

a. $\dfrac{a^3b^{-2}}{c^{-5}}$

$= \dfrac{a^3}{1} \cdot \dfrac{b^{-2}}{1} \cdot \dfrac{1}{c^{-5}}$

$= \dfrac{a^3}{1} \cdot \dfrac{1}{b^2} \cdot \dfrac{c^5}{1}$ Simplify negative exponents.

$= \dfrac{a^3c^5}{b^2}$ Multiply.

b. $\dfrac{x^2x^{-7}}{x^3}$

$= \dfrac{x^{2+(-7)}}{x^3}$ Add the exponents in the numerator.

$= \dfrac{x^{-5}}{x^3}$ Simplify.

$= x^{-5-3}$ Subtract the exponents.

$= x^{-8}$

$= \dfrac{1}{x^8}$ Simplify the negative exponent.

c. $\dfrac{z^2}{w^{-4}w^4z^{-8}}$

$= \dfrac{z^2}{w^{-4+4}z^{-8}}$ Add the exponents in the denominator.

$= \dfrac{z^2}{w^0z^{-8}}$

$= \dfrac{z^2}{(1)z^{-8}}$ Recall that $w^0 = 1$.

$= z^{2-(-8)}$ Subtract the exponents.

$= z^{10}$ Simplify.

d. $(-4ab^{-2})^{-3}$

$= (-4)^{-3}a^{-3}(b^{-2})^{-3}$ Apply the power rule of exponents.

$= (-4)^{-3}a^{-3}b^{6}$

$= \dfrac{1}{(-4)^3} \cdot \dfrac{1}{a^3} \cdot b^6$ Simplify the negative exponents.

$= \dfrac{1}{-64} \cdot \dfrac{1}{a^3} \cdot b^6$ Simplify.

$= -\dfrac{b^6}{64a^3}$ Multiply fractions.

TIP: Example 3(e) can also be simplified by clearing parentheses first:

$$\frac{2^{-1}p^4q^{-3}}{5^{-1}p^{-2}q^{-1}}$$

e. $\left(\dfrac{2p^{-4}q^3}{5p^2q}\right)^{-1}$ The negative exponent outside the parentheses can be eliminated by taking the reciprocal of the quantity within the parentheses.

$= \left(\dfrac{5p^2q}{2p^{-4}q^3}\right)^1$ Take the reciprocal of the base and make the outer exponent positive.

$= \dfrac{5p^2q}{2p^{-4}q^3}$

$= \dfrac{5p^{2-(-4)}q^{1-3}}{2}$ Subtract the exponents.

$= \dfrac{5p^6q^{-2}}{2}$ Simplify.

$= \dfrac{5p^6}{2} \cdot \dfrac{1}{q^2}$ Simplify the negative exponent.

$= \dfrac{5p^6}{2q^2}$ Simplify.

Skill Practice Simplify the expressions. Use positive exponents only. Assume all variables are nonzero.

15. $\dfrac{x^{-6}}{y^4z^{-8}}$ **16.** $\dfrac{x^5 \cdot x^{-8}}{x^2}$ **17.** $\dfrac{a^7 \cdot a^{-4}}{a^{-3}}$

18. $(x^{-2} \cdot y^3)^{-2}$ **19.** $\left(\dfrac{3x^{-3}y^{-2}}{4xy^{-3}}\right)^{-2}$

Example 4 Simplifying an Expression with Exponents

Simplify the expression $2^{-1} + 3^{-1} + 5^0$. Write the answer with positive exponents only.

Skill Practice Answers

15. $\dfrac{z^8}{y^4x^6}$ **16.** $\dfrac{1}{x^5}$ **17.** a^6

18. $\dfrac{x^4}{y^6}$ **19.** $\dfrac{16x^8}{9y^2}$

Solution:

$2^{-1} + 3^{-1} + 5^0$

$= \frac{1}{2} + \frac{1}{3} + 1$ Simplify negative exponents. Simplify $5^0 = 1$.

$= \frac{3}{6} + \frac{2}{6} + \frac{6}{6}$ Get a common denominator.

$= \frac{11}{6}$ Simplify.

Skill Practice Simplify the expression.

20. $2^{-1} + 4^{-2} + 3^0$

Skill Practice Answers

20. $\frac{25}{16}$

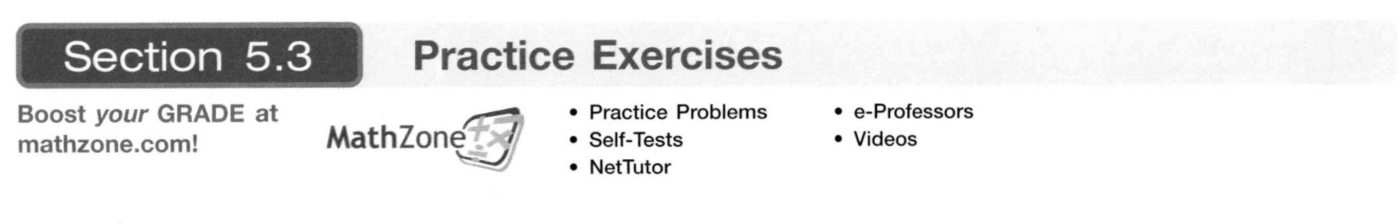

Section 5.3 Practice Exercises

Boost *your* GRADE at mathzone.com!

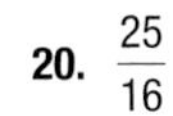

- Practice Problems
- Self-Tests
- NetTutor
- e-Professors
- Videos

For this set of exercises, assume all variables represent nonzero real numbers.

Study Skills Exercise

1. To help you remember the properties of exponents, write them on 3 × 5 cards. On each card, write a property on one side and an example using that property on the other side. Keep these cards with you, and when you have a spare moment (such as waiting at the doctor's office), pull out these cards and go over the properties.

Review Exercises

For Exercises 2–11, simplify the expressions.

2. b^3b^8

3. c^7c^2

4. $\frac{x^6}{x^2}$

5. $\frac{y^9}{y^8}$

6. $\frac{9^4 \cdot 9^8}{9}$

7. $\frac{3^{14}}{3^3 \cdot 3^5}$

8. $(6ab^3c^2)^5$

9. $(7w^7z^2)^4$

10. $\left(\frac{s^2t^5}{4}\right)^3$

11. $\left(\frac{5k^3}{h^7}\right)^2$

Concept 1: Definition of b^0

12. Simplify.

a. 8^0

b. $\frac{8^4}{8^4}$

13. Simplify.

a. d^0

b. $\frac{d^3}{d^3}$

For Exercises 14–27, simplify the expression.

14. p^0

15. k^0

16. 5^0

17. 2^0

18. -4^0

19. -1^0

20. $(-6)^0$

21. $(-2)^0$

22. $(8x)^0$

23. $(-3y^3)^0$

24. $-7x^0$

25. $6y^0$

26. ab^0

27. $-pq^0$

Concept 2: Definition of b^{-n}

28. Simplify and write the answers with positive exponents.

a. t^{-5} b. $\dfrac{t^3}{t^8}$

29. Simplify and write the answers with positive exponents.

a. 4^{-3} b. $\dfrac{4^2}{4^5}$

For Exercises 30–49, simplify.

30. $\left(\dfrac{2}{7}\right)^{-3}$

31. $\left(\dfrac{5}{4}\right)^{-1}$

32. $\left(-\dfrac{1}{5}\right)^{-2}$

33. $\left(-\dfrac{1}{3}\right)^{-3}$

34. a^{-3}

35. c^{-5}

36. 12^{-1}

37. 4^{-2}

38. $(4b)^{-2}$

39. $(3z)^{-1}$

40. $6x^{-2}$

41. $7y^{-1}$

42. $(-8)^{-2}$

43. -8^{-2}

44. -5^{-3}

45. $(-w)^{-2}$

46. $(-t)^{-3}$

47. $(-r)^{-5}$

48. $\dfrac{1}{a^{-5}}$

49. $\dfrac{1}{b^{-6}}$

Concept 3: Properties of Integer Exponents: A Summary

50. Explain what is wrong with the following logic. $\dfrac{x^4}{x^{-6}} = x^{4-6} = x^{-2}$

51. Explain what is wrong with the following logic. $\dfrac{y^5}{y^{-3}} = y^{5-3} = y^2$

52. Explain what is wrong with the following logic. $2a^{-3} = \dfrac{1}{2a^3}$

53. Explain what is wrong with the following logic. $5b^{-2} = \dfrac{1}{5b^2}$

Concept 4: Simplifying Expressions with Exponents

For Exercises 54–95, simplify the expression. Write the answer with positive exponents only.

54. $x^{-8}x^4$

55. s^5s^{-6}

56. $a^{-8}a^8$

57. q^3q^{-3}

58. $y^{17}y^{-13}$

59. $b^{20}b^{-14}$

60. $(m^{-6}n^9)^3$

61. $(c^4d^{-5})^{-2}$

62. $(-3j^{-5}k^6)^4$

63. $(6xy^{-11})^{-3}$

64. $\dfrac{p^3}{p^9}$

65. $\dfrac{q^2}{q^{10}}$

66. $\dfrac{r^{-5}}{r^{-2}}$

67. $\dfrac{u^{-2}}{u^{-6}}$

68. $\dfrac{a^2}{a^{-6}}$

69. $\dfrac{p^3}{p^{-5}}$

70. $\dfrac{y^{-2}}{y^6}$

71. $\dfrac{s^{-4}}{s^3}$

72. $\dfrac{7^3}{7^2 \cdot 7^8}$

73. $\dfrac{3^4 \cdot 3}{3^7}$

74. $\dfrac{a^2a}{a^3}$

75. $\dfrac{t^5}{t^2t^3}$

76. $\dfrac{a^{-1}b^2}{a^3b^8}$

77. $\dfrac{k^{-4}h^{-1}}{k^6h}$

78. $\dfrac{w^{-8}(w^2)^{-5}}{w^3}$

79. $\dfrac{p^2p^{-7}}{(p^2)^3}$

80. $\dfrac{3^{-2}}{3}$

81. $\dfrac{5^{-1}}{5}$

82. $\left(\dfrac{p^{-1}q^5}{p^{-6}}\right)^0$

83. $\left(\dfrac{ab^{-4}}{a^{-5}}\right)^0$

84. $(8x^3y^0)^{-2}$

85. $(3u^2v^0)^{-3}$

86. $(-8y^{-12})(2y^{16}z^{-2})$

87. $(5p^{-2}q^5)(-2p^{-4}q^{-1})$

88. $\dfrac{-18a^{10}b^6}{108a^{-2}b^6}$

89. $\dfrac{-35x^{-4}y^{-3}}{-21x^2y^{-3}}$

90. $\dfrac{(-4c^{12}d^7)^2}{(5c^{-3}d^{10})^{-1}}$

91. $\dfrac{(s^3t^{-2})^4}{(3s^{-4}t^6)^{-2}}$

92. $\left(\dfrac{2}{p^6p^3}\right)^{-3}$

93. $\left(\dfrac{5x}{x^7}\right)^{-2}$

94. $\left(\dfrac{5cd^{-3}}{10d^5}\right)^{-1}$

95. $\left(\dfrac{4m^{10}n^4}{2m^{12}n^{-2}}\right)^{-1}$

For Exercises 96–105, simplify the expression.

96. $5^{-1} + 2^{-2}$

97. $4^{-2} + 8^{-1}$

98. $10^0 - 10^{-1}$

99. $3^0 - 3^{-2}$

100. $2^{-2} + 1^{-2}$

101. $4^{-1} + 8^{-1}$

102. $4 \cdot 5^0 - 2 \cdot 3^{-1}$

103. $2 \cdot 4^0 - 3 \cdot 4^{-1}$

104. $5 \cdot 2^{-3} + 2 \cdot 4^{-1}$

105. $2 \cdot 3^{-2} + 3 \cdot 9^{-1}$

Scientific Notation

Section 5.4

Concepts

1. Introduction to Scientific Notation
2. Writing Numbers in Scientific Notation
3. Writing Numbers without Scientific Notation
4. Multiplying and Dividing Numbers in Scientific Notation
5. Applications of Scientific Notation

1. Introduction to Scientific Notation

In many applications in mathematics, it is necessary to work with very large or very small numbers. For example, the number of movie tickets sold in the United States in 2004 is estimated to be 1,500,000,000. The weight of a flea is approximately 0.00066 lb. To avoid writing numerous zeros in very large or small numbers, scientific notation was devised as a shortcut. Scientific notation is useful when performing calculations and when comparing the relative sizes of very large or very small numbers.

The principle behind scientific notation is to use a power of 10 to express the magnitude of the number. Consider the following powers of 10:

$$10^0 = 1$$

$$10^1 = 10 \qquad 10^{-1} = \frac{1}{10^1} = \frac{1}{10} = 0.1$$

$$10^2 = 100 \qquad 10^{-2} = \frac{1}{10^2} = \frac{1}{100} = 0.01$$

$$10^3 = 1000 \qquad 10^{-3} = \frac{1}{10^3} = \frac{1}{1000} = 0.001$$

$$10^4 = 10{,}000 \qquad 10^{-4} = \frac{1}{10^4} = \frac{1}{10{,}000} = 0.0001$$

In the base-10 numbering system, each place value to the left and right of the decimal point represents a different power of 10 (Figure 5-1).

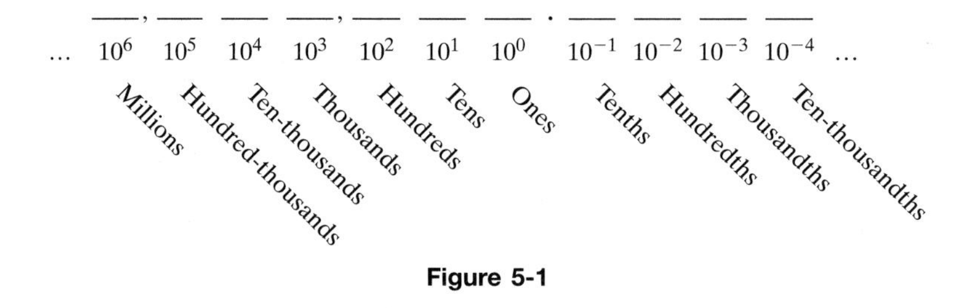

Figure 5-1

Therefore, a number such as 4000 can be written as 4.0×1000, or equivalently, 4.0×10^3. Similarly, the number 0.07 can be written as 7.0×0.01, or equivalently, 7.0×10^{-2}.

Definition of a Number Written in Scientific Notation

A number expressed in the form: $a \times 10^n$, where $1 \le |a| < 10$ and n is an integer is said to be written in **scientific notation**.

The numbers 4.0×10^3 and 7.0×10^{-2} are both expressed in scientific notation. To write a positive number in scientific notation, we apply the following guidelines:

1. Move the decimal point so that its new location is to the right of the first nonzero digit. The number should now be greater than or equal to 1, but less than 10. Count the number of places that the decimal point is moved.
2. If the original number is *large* (greater than or equal to 10), use the number of places the decimal point was moved as a *positive* power of 10.

$$450{,}000 = 4.5 \times 100{,}000 = 4.5 \times 10^5$$

5 places

3. If the original number is *small* (between 0 and 1), use the number of places the decimal point was moved as a *negative* power of 10.

$$0.0002 = 2.0 \times 0.0001 = 2.0 \times 10^{-4}$$

4 places

4. If the original number is greater than or equal to 1 but less than 10, use 0 as the power of 10.

$7.592 = 7.592 \times 10^0$ *Note:* A number between 1 and 10 is seldom written in scientific notation.

2. Writing Numbers in Scientific Notation

Example 1 Writing Numbers in Scientific Notation

Write the numbers in scientific notation.

a. 53,000 **b.** 0.00053

Solution:

a. $53{,}000. = 5.3 \times 10^4$

To write 53,000 in scientific notation, the decimal point must be moved four places to the left. Because 53,000 is larger than 10, a *positive* power of 10 is used.

b. $0.00053 = 5.3 \times 10^{-4}$

To write 0.00053 in scientific notation, the decimal point must be moved four places to the right. Because 0.00053 is less than 1, a *negative* power of 10 is used.

Skill Practice Write the numbers in scientific notation.

1. 175,000,000 **2.** 0.000005

Example 2 Writing Numbers in Scientific Notation

Write the numerical values in scientific notation.

a. The number of movie tickets sold in the United States in 2004 is estimated to be 1,500,000,000.

b. The weight of a flea is approximately 0.00066 lb.

c. The temperature on a January day in Fargo dropped to −43°F.

d. A bench is 8.2 ft long.

Solution:

a. $1{,}500{,}000{,}000 = 1.5 \times 10^9$

b. $0.00066 \text{ lb} = 6.6 \times 10^{-4} \text{ lb}$

c. $-43°\text{F} = -4.3 \times 10^1 \text{ °F}$

d. $8.2 \text{ ft} = 8.2 \times 10^0 \text{ ft}$

Skill Practice Write the numbers in scientific notation.

3. The population of the Earth is approximately 6,400,000,000.

4. The weight of a grain of salt is approximately 0.000002 ounces.

5. The lowest recorded temperature is −128.6°F.

6. A gallon of water weighs 8.33 lb.

Skill Practice Answers

1. 1.75×10^8
2. 5.0×10^{-6}
3. 6.4×10^9
4. 2.0×10^{-6} oz
5. -1.286×10^2 °F
6. 8.33×10^0 lb

TIP: For a number written in scientific notation, the power of 10 is sometimes called the **order of magnitude** (or simply the magnitude) of the number.

- The order of magnitude of the number of movie tickets sold in 2004 in the United States is $\$10^9$ (billions of dollars).
- The mass of a flea is on the order of 10^{-4} lb (ten-thousandths of a pound).

3. Writing Numbers without Scientific Notation

Example 3 Writing Numbers Without Scientific Notation

Write the numerical values without scientific notation.

a. The mass of a proton is approximately 1.67×10^{-24} g.

b. The "nearby" star Vega is approximately 1.552×10^{14} miles from Earth.

Solution:

a. 1.67×10^{-24} g $=$ 0.000 000 000 000 000 000 000 001 67 g

Because the power of 10 is negative, the value of 1.67×10^{-24} is a decimal number between 0 and 1. Move the decimal point 24 places to the *left.*

b. 1.552×10^{14} miles $=$ 155,200,000,000,000 miles

Because the power of 10 is a positive integer, the value of 1.552×10^{14} is a large number greater than 10. Move the decimal point 14 places to the *right.*

Skill Practice Write the numerical values without scientific notation.

7. The probability of winning the California Super Lotto Jackpot is 5.5×10^{-8}.

8. The Sun's mass is 2×10^{30} kilograms.

4. Multiplying and Dividing Numbers in Scientific Notation

To multiply or divide two numbers in scientific notation, use the commutative and associative properties of multiplication to group the powers of 10. For example,

$$400 \times 2000 = (4 \times 10^2)(2 \times 10^3) = (4 \cdot 2) \times (10^2 \cdot 10^3) = 8 \times 10^5$$

$$\frac{0.00054}{150} = \frac{5.4 \times 10^{-4}}{1.5 \times 10^2} = \left(\frac{5.4}{1.5}\right) \times \left(\frac{10^{-4}}{10^2}\right) = 3.6 \times 10^{-6}$$

Example 4 Multiplying and Dividing Numbers in Scientific Notation

a. $(8.7 \times 10^4)(2.5 \times 10^{-12})$

b. $\dfrac{4.25 \times 10^{13}}{8.5 \times 10^{-2}}$

Solution:

a. $(8.7 \times 10^4)(2.5 \times 10^{-12})$

$= (8.7 \cdot 2.5) \times (10^4 \cdot 10^{-12})$ Commutative and associative properties of multiplication

Skill Practice Answers

7. 0.000 000 055

8. 2,000,000,000,000,000,000,000,000,000,000

$= 21.75 \times 10^{-8}$ The number 21.75 is not in proper scientific notation because 21.75 is not between 1 and 10.

$= (2.175 \times 10^{1}) \times 10^{-8}$ Rewrite 21.75 as 2.175×10^{1}.

$= 2.175 \times (10^{1} \times 10^{-8})$ Associative property of multiplication

$= 2.175 \times 10^{-7}$ Simplify.

b. $\dfrac{4.25 \times 10^{13}}{8.5 \times 10^{-2}}$

$= \left(\dfrac{4.25}{8.5}\right) \times \left(\dfrac{10^{13}}{10^{-2}}\right)$ Commutative and associative properties

$= 0.5 \times 10^{15}$ The number 0.5×10^{15} is not in proper scientific notation because 0.5 is not between 1 and 10.

$= (5.0 \times 10^{-1}) \times 10^{15}$ Rewrite 0.5 as 5.0×10^{-1}.

$= 5.0 \times (10^{-1} \times 10^{15})$ Associative property of multiplication

$= 5.0 \times 10^{14}$ Simplify.

Skill Practice Multiply.

9. $(7 \times 10^{5})(5 \times 10^{3})$

10. $\dfrac{1 \times 10^{-2}}{4 \times 10^{-7}}$

5. Applications of Scientific Notation

Example 5 Applying Scientific Notation

If a spacecraft travels at 1.6×10^{4} mph, how long will it take the craft to travel to Mars if the distance is approximately 8.0×10^{7} miles?

Solution:

Since $d = rt$, then $t = \dfrac{d}{r}$

$$t = \frac{8.0 \times 10^{7} \text{ miles}}{1.6 \times 10^{4} \text{ miles/hour}}$$

$$= \left(\frac{8.0}{1.6}\right) \times \left(\frac{10^{7}}{10^{4}}\right) \text{hr}$$

$$= 5.0 \times 10^{3} \text{ hr}$$

The time required to travel to Mars is approximately $5.0 \times 10^{3} = 5000$ hr, or 208 days.

Skill Practice

11. The Earth's orbit is approximately 9.49×10^{11} meters. The Earth travels through its orbit at a speed of approximately 2.6×10^{9} meters per day. How many days does it take for the Earth to complete the orbit?

Skill Practice Answers

9. 3.5×10^{9}
10. 2.5×10^{4}
11. Approximately 3.65×10^{2} or 365 days.

Calculator Connections

Both scientific and graphing calculators can perform calculations involving numbers written in scientific notation. Most calculators use an [EE] key or an [EXP] key to enter the power of 10.

Scientific Calculator

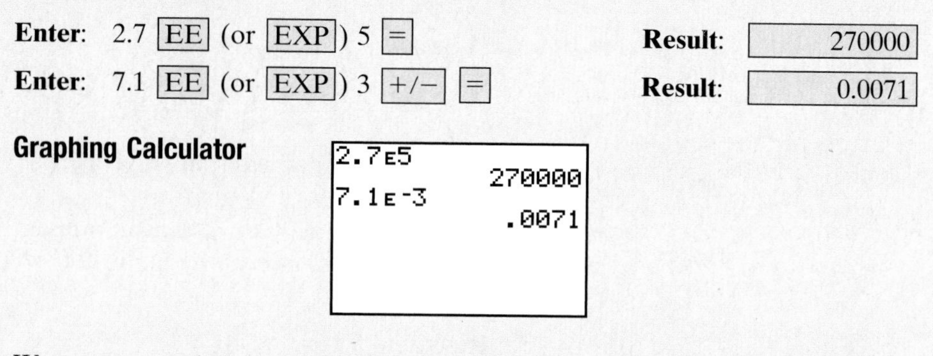

TIP: Note that the symbol E on the calculator screen means × 10. Therefore, 5 E 14 means 5×10^{14}.

We recommend that you use parentheses to enclose each number written in scientific notation when performing calculations. Try using your calculator to perform the calculations from Example 4.

a. $(8.7 \times 10^4)(2.5 \times 10^{-12})$ **b.** $\dfrac{4.25 \times 10^{13}}{8.5 \times 10^{-2}}$

Scientific Calculator

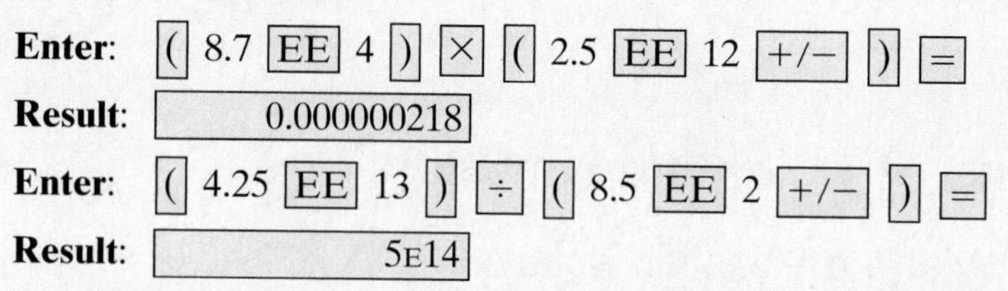

Notice that the answer to part (b) is shown on the calculator in scientific notation. The calculator does not have enough room to display 14 zeros.

Graphing Calculator

```
(8.7E4)*(2.5E-12
)
          2.175E-7
(4.25E13)/(8.5E-
2)
              5E14
```

Calculator Exercises

Use a calculator to perform the indicated operations.

1. $(5.2 \times 10^6)(4.6 \times 10^{-3})$
2. $(2.19 \times 10^{-8})(7.84 \times 10^{-4})$
3. $\dfrac{4.76 \times 10^{-5}}{2.38 \times 10^9}$
4. $\dfrac{8.5 \times 10^4}{4.0 \times 10^{-1}}$
5. $\dfrac{(9.6 \times 10^7)(4.0 \times 10^{-3})}{2.0 \times 10^{-2}}$
6. $\dfrac{(5.0 \times 10^{-12})(6.4 \times 10^{-5})}{(1.6 \times 10^{-8})(4.0 \times 10^2)}$

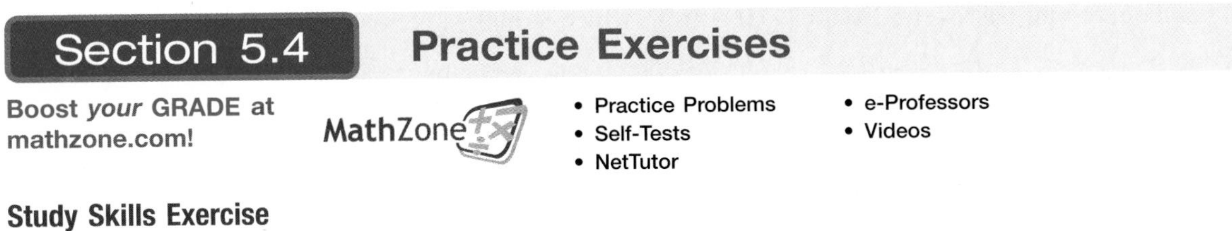

Study Skills Exercise

1. Define the key terms:

a. scientific notation **b. order of magnitude**

Review Exercises

For Exercises 2–13, simplify the expression. Assume all variables represent nonzero real numbers.

2. a^3a^{-4} **3.** b^5b^8 **4.** $10^3 \cdot 10^{-4}$ **5.** $10^5 \cdot 10^8$

6. $\dfrac{x^3}{x^6}$ **7.** $\dfrac{y^2}{y^7}$ **8.** $\dfrac{10^3}{10^6}$ **9.** $\dfrac{10^2}{10^7}$

10. $\dfrac{z^9z^4}{z^3}$ **11.** $\dfrac{w^{-2}w^5}{w^{-1}}$ **12.** $\dfrac{10^9 \cdot 10^4}{10^3}$ **13.** $\dfrac{10^{-2} \cdot 10^5}{10^{-1}}$

Concept 1: Introduction to Scientific Notation

For Exercises 14–21, simplify the expressions.

14. 10^3 **15.** 10^5 **16.** 10^{-2} **17.** 10^{-4}

18. 10^0 **19.** 10^1 **20.** 10^{-1} **21.** 10^2

Concept 2: Writing Numbers in Scientific Notation

22. Explain how scientific notation might be valuable in studying astronomy. Answers may vary.

23. Explain how you would write the number 0.000 000 000 23 in scientific notation.

24. Explain how you would write the number 23,000,000,000,000 in scientific notation.

For Exercises 25–36, write the number in scientific notation.

25. 50,000 **26.** 900,000 **27.** 208,000

28. 420,000,000 **29.** 6,010,000 **30.** 75,000

31. 0.000008 **32.** 0.003 **33.** 0.000125

34. 0.00000025 **35.** 0.006708 **36.** 0.02004

For Exercises 37–42, write the numbers in scientific notation.

37. The mass of a proton is approximately 0.000 000 000 000 000 000 000 0017 g.

38. The total combined salaries of the president, vice president, senators, and representatives of the United States federal government is approximately $85,000,000.

39. The Bill Gates Foundation has over $27,000,000,000 from which it makes contributions to global charities.

40. One gram is equivalent to 0.0035 oz.

41. In the world's largest tanker disaster, *Amoco Cadiz* spilled 68,000,000 gal of oil off Portsall, France, causing widespread environmental damage over 100 miles of Brittany coast.

42. The human heart pumps about 1400 L of blood per day. That would mean that it pumps approximately 10,000,000 L per year.

Concept 3: Writing Numbers without Scientific Notation

43. Explain how you would write the number 3.1×10^{-9} without scientific notation.

44. Explain how you would write the number 3.1×10^{9} without scientific notation.

For Exercises 45–60, write the numbers without scientific notation.

45. 5×10^{-5}

46. 2×10^{-7}

47. 2.8×10^{3}

48. 9.1×10^{6}

49. 6.03×10^{-4}

50. 7.01×10^{-3}

51. 2.4×10^{6}

52. 3.1×10^{4}

53. 1.9×10^{-2}

54. 2.8×10^{-6}

55. 7.032×10^{3}

56. 8.205×10^{2}

57. One picogram (pg) is equal to 1×10^{-12} g.

58. A nanometer (nm) is approximately 3.94×10^{-8} in.

59. A normal diet contains between 1.6×10^{3} Cal and 2.8×10^{3} Cal per day.

60. The total land area of Texas is approximately 2.62×10^{5} square miles.

Concept 4: Multiplying and Dividing Numbers in Scientific Notation

For Exercises 61–80, multiply or divide as indicated. Write the answers in scientific notation.

61. $(2.5 \times 10^{6})(2.0 \times 10^{-2})$

62. $(2.0 \times 10^{-7})(3.0 \times 10^{13})$

63. $(1.2 \times 10^{4})(3 \times 10^{7})$

64. $(3.2 \times 10^{-3})(2.5 \times 10^{8})$

65. $\dfrac{7.7 \times 10^{6}}{3.5 \times 10^{2}}$

66. $\dfrac{9.5 \times 10^{11}}{1.9 \times 10^{3}}$

67. $\dfrac{9.0 \times 10^{-6}}{4.0 \times 10^{7}}$

68. $\dfrac{7.0 \times 10^{-2}}{5.0 \times 10^{9}}$

69. $(8.0 \times 10^{10})(4.0 \times 10^{3})$

70. $(6.0 \times 10^{-4})(3.0 \times 10^{-2})$

71. $(3.2 \times 10^{-4})(7.6 \times 10^{-7})$

72. $(5.9 \times 10^{12})(3.6 \times 10^{9})$

73. $\dfrac{2.1 \times 10^{11}}{7.0 \times 10^{-3}}$

74. $\dfrac{1.6 \times 10^{14}}{8.0 \times 10^{-5}}$

75. $\dfrac{5.7 \times 10^{-2}}{9.5 \times 10^{-8}}$

76. $\dfrac{2.72 \times 10^{-6}}{6.8 \times 10^{-4}}$

77. $6{,}000{,}000{,}000 \times 0.0000000023$

78. $0.000055 \times 40{,}000$

79. $\dfrac{0.0000000003}{6000}$

80. $\dfrac{420{,}000}{0.0000021}$

Concept 5: Applications of Scientific Notation

81. If a piece of paper is 3.0×10^{-3} in. thick, how thick is a stack of 1.25×10^{3} pieces of paper?

82. A box of staples contains 5.0×10^{3} staples and weighs 15 oz. How much does one staple weigh? Write your answer in scientific notation.

83. Bill Gates owned approximately 1,100,000,000 shares of Microsoft stock. If the stock price was $27 per share, how much was Bill Gates' stock worth?

84. A state lottery had a jackpot of $\$5.2 \times 10^{7}$. This week the winner was a group of office employees that included 13 people. How much would each person receive?

85. Dinosaurs became extinct about 65 million years ago.

   a. Write the number 65 million in scientific notation.

   b. How many days is 65 million years?

   c. How many hours is 65 million years?

   d. How many seconds is 65 million years?

86. The Earth is 111,600,000 km from the Sun.

   a. Write the number 111,600,000 in scientific notation.

   b. If there are 1000 m in a kilometer, how many meters is the Earth from the Sun?

   c. If there are 100 cm in a meter, how many centimeters is the Earth from the Sun?

87. For a recent year, the U.S. national debt was about $7,500,000,000,000. If the U.S. population was approximately 300,000,000, how much would each person pay to relieve the country of its debt?

88. The area of Japan is approximately 380,000 km^2. If the population is about 129,200,000, what is the density (people per square kilometer) of Japan?

89. The longest paper clip chain is 2.961×10^{3} ft. If the chain took 2.3688×10^{5} clips, what is the length of each paper clip in feet?

90. A sports figure had a 5-year (60-month) contract. If he makes $\$1.5 \times 10^{6}$ per month, how much will he make in the 5-year period?

2. Applications of Polynomials

Example 2 Using Polynomials in an Application

A child throws a ball upward and the height of the ball, h (in feet), can be computed by the following equation:

$h = -16t^2 + 64t + 2$ where t is the time (in seconds) after the ball is released.

a. Find the height of the ball after 0.5 sec, 1 sec, and 1.5 sec.

b. Find the height of the ball at the time of release.

Solution:

a. $h = -16t^2 + 64t + 2$

$= -16(0.5)^2 + 64(0.5) + 2$ Substitute $t = 0.5$.

$= -16(0.25) + 32 + 2$

$= -4 + 32 + 2$

$= 30$ The height of the ball after 0.5 sec is 30 ft.

$h = -16t^2 + 64t + 2$

$= -16(1)^2 + 64(1) + 2$ Substitute $t = 1$.

$= -16(1) + 64 + 2$

$= -16 + 64 + 2$

$= 50$ The height of the ball after 1 sec is 50 ft.

$h = -16t^2 + 64t + 2$

$= -16(1.5)^2 + 64(1.5) + 2$ Substitute $t = 1.5$.

$= -16(2.25) + 96 + 2$

$= -36 + 96 + 2$

$= 62$ The height of the ball after 1.5 sec is 62 ft.

b. $h = -16t^2 + 64t + 2$

$= -16(0)^2 + 64(0) + 2$ At the time of release, $t = 0$.

$= 0 + 0 + 2$

$= 2$ The height of the ball at the time of release is 2 ft.

Skill Practice

2. The number of cases, C (in millions), of soft drinks sold by a company the first 10 years it is in business is given by

$$C = -3t^2 + 25t + 150 \quad (0 \le t \le 10)$$

where t is the number of years the company has been in business.

a. Find the number of cases sold after 2 years.

b. Find the number of cases sold after 5 years.

Skill Practice Answers

2a. 188 million
b. 200 million

3. Addition of Polynomials

Recall that two terms are said to be *like* terms if they each have the same variables, and the corresponding variables are raised to the same powers.

Like Terms: $3x^2, -7x^2$ (same exponents, same variables) $\quad -5yz^3, yz^3$ (same exponents, same variables)

Unlike Terms: $9z^2, 12z^6$ (different exponents) $\quad \frac{1}{3}w^6, \frac{2}{5}p^6$ (different variables) $\quad 4y, 7$ (different variables)

Recall that the distributive property is used to add or subtract *like* terms. For example,

$3x^2 + 9x^2 - 2x^2$

$= (3 + 9 - 2)x^2$ Apply the distributive property.

$= (10)x^2$ Simplify.

$= 10x^2$

Example 3 **Adding Polynomials**

Add the polynomials.

a. $3x^2y + 5x^2y$ **b.** $(-3c^3 + 5c^2 - 7c) + (11c^3 + 6c^2 + 3)$

c. $\left(\frac{1}{4}w^2 - \frac{2}{3}w\right) + \left(\frac{3}{4}w^2 + \frac{1}{6}w - \frac{1}{2}\right)$

Solution:

a. $3x^2y + 5x^2y$

$= (3 + 5)x^2y$ Apply the distributive property.

$= (8)x^2y$

$= 8x^2y$ Simplify.

b. $(-3c^3 + 5c^2 - 7c) + (11c^3 + 6c^2 + 3)$

$= -3c^3 + 11c^3 + 5c^2 + 6c^2 - 7c + 3$ Clear parentheses, and group *like* terms.

$= 8c^3 + 11c^2 - 7c + 3$ Combine *like* terms.

TIP: Polynomials can also be added by combining *like* terms in columns. The sum of the polynomials from Example 3(b) is shown here.

$$\begin{array}{r} -3c^3 + 5c^2 - 7c + 0 \\ +\,11c^3 + 6c^2 + 0c + 3 \\ \hline 8c^3 + 11c^2 - 7c + 3 \end{array}$$

Place holders such as 0 and 0c may be used to help line up *like* terms.

TIP: Although the distributive property is used to combine *like* terms, the process is simplified by combining the coefficients of *like* terms.

Avoiding Mistakes:

Example 3(c) is an expression, not an equation. Therefore, we cannot clear fractions.

c. $\left(\frac{1}{4}w^2 - \frac{2}{3}w\right) + \left(\frac{3}{4}w^2 + \frac{1}{6}w - \frac{1}{2}\right)$

$= \frac{1}{4}w^2 + \frac{3}{4}w^2 - \frac{2}{3}w + \frac{1}{6}w - \frac{1}{2}$ Clear parentheses, and group *like* terms.

$= \frac{1}{4}w^2 + \frac{3}{4}w^2 - \frac{4}{6}w + \frac{1}{6}w - \frac{1}{2}$ Get common denominators for *like* terms.

$= \frac{4}{4}w^2 - \frac{3}{6}w - \frac{1}{2}$ Add *like* terms.

$= w^2 - \frac{1}{2}w - \frac{1}{2}$ Simplify.

Skill Practice Add the polynomials.

3. $(7q^2 - 2q + 4) + (5q^2 + 6q - 9)$

4. $\left(\frac{1}{2}t^3 - 1\right) + \left(\frac{2}{3}t^3 + \frac{1}{5}\right)$

5. $(5x^2 + x - 2) + (x + 4) + (3x^2 + 4x)$

4. Subtraction of Polynomials

The *opposite* (or additive inverse) of a real number a is $-a$. Similarly, if A is a polynomial, then $-A$ is its opposite.

Example 4 Finding the Opposite of a Polynomial

Find the opposite of the polynomials.

a. $5x$ **b.** $3a - 4b - c$ **c.** $5.5y^4 - 2.4y^3 + 1.1y - 3$

Solution:

TIP: Notice that the sign of each term is changed when finding the opposite of a polynomial.

a. The opposite of $5x$ is $-(5x)$, or $-5x$.

b. The opposite of $3a - 4b - c$ is $-(3a - 4b - c)$ or equivalently, $-3a + 4b + c$.

c. The opposite of $5.5y^4 - 2.4y^3 + 1.1y - 3$ is $-(5.5y^4 - 2.4y^3 + 1.1y - 3)$, or equivalently, $-5.5y^4 + 2.4y^3 - 1.1y + 3$.

Skill Practice Find the opposite of the polynomials.

6. $x - 3$ **7.** $3y^2 - 2xy + 6x + 2$ **8.** $\frac{1}{3}x^3 - \frac{1}{2}x + \frac{5}{3}$

Subtraction of two polynomials is similar to subtracting real numbers. Add the opposite of the second polynomial to the first polynomial.

Skill Practice Answers

3. $12q^2 + 4q - 5$
4. $\frac{7}{6}t^3 - \frac{4}{5}$
5. $8x^2 + 6x + 2$
6. $-x + 3$
7. $-3y^2 + 2xy - 6x - 2$
8. $-\frac{1}{3}x^3 + \frac{1}{2}x - \frac{5}{3}$

Definition of Subtraction of Polynomials

If A and B are polynomials, then $A - B = A + (-B)$.

Example 5 Subtracting Polynomials

Subtract the polynomials.

a. $(-4p^4 + 5p^2 - 3) - (11p^2 + 4p - 6)$

b. $(a^2 - 2ab + 7b^2) - (-8a^2 - 6ab + 2b^2)$

Solution:

a. $(-4p^4 + 5p^2 - 3) - (11p^2 + 4p - 6)$

$= (-4p^4 + 5p^2 - 3) + (-11p^2 - 4p + 6)$ Add the opposite of the second polynomial.

$= -4p^4 + 5p^2 - 11p^2 - 4p - 3 + 6$ Group *like* terms.

$= -4p^4 - 6p^2 - 4p + 3$ Combine *like* terms.

TIP: Two polynomials can also be subtracted in columns by adding the opposite of the second polynomial to the first polynomial. Place holders (shown in red) may be used to help line up *like* terms.

$$\begin{array}{r} -4p^4 + 0p^3 + 5p^2 + 0p - 3 \\ -(0p^4 + 0p^3 + 11p^2 + 4p - 6) \\ \hline \end{array} \xrightarrow{\text{add the opposite}} \begin{array}{r} -4p^4 + 0p^3 + 5p^2 + 0p - 3 \\ + \; -0p^4 - 0p^3 - 11p^2 - 4p + 6 \\ \hline -4p^4 \qquad\quad - 6p^2 - 4p + 3 \end{array}$$

b. $(a^2 - 2ab + 7b^2) - (-8a^2 - 6ab + 2b^2)$

$= (a^2 - 2ab + 7b^2) + (8a^2 + 6ab - 2b^2)$ Add the opposite of the second polynomial.

$= a^2 + 8a^2 - 2ab + 6ab + 7b^2 - 2b^2$ Group *like* terms.

$= 9a^2 + 4ab + 5b^2$ Combine *like* terms.

TIP: Recall that $a - b = a + (-b)$, or equivalently, $a + -1b$. Therefore, subtraction of polynomials can be simplified by applying the distributive property to clear parentheses.

$(a^2 - 2ab + 7b^2) - (-8a^2 - 6ab + 2b^2)$

$= a^2 - 2ab + 7b^2 - 1(-8a^2 - 6ab + 2b^2)$

$= a^2 - 2ab + 7b^2 + 8a^2 + 6ab - 2b^2$ Apply the distributive property.

$= a^2 + 8a^2 - 2ab + 6ab + 7b^2 - 2b^2$ Group *like* terms.

$= 9a^2 + 4ab + 5b^2$ Combine *like* terms.

Skill Practice Subtract the polynomials.

9. $(x^2 + 3x - 2) - (4x^2 + 6x + 1)$

10. $(-3y^2 + xy + 2x^2) - (-2y^2 - 3xy - 8x^2)$

Example 6 Subtracting Polynomials

Subtract $\frac{1}{3}t^4 + \frac{1}{2}t^2$ from $t^2 - 4$, and simplify the result.

Solution:

To subtract a from b, we write $b - a$. Thus, to subtract $\overbrace{\frac{1}{3}t^4 + \frac{1}{2}t^2}^{a}$ from $\overbrace{t^2 - 4}^{b}$, we have

$$\overbrace{(t^2 - 4)}^{b} \overbrace{-}^{-} \overbrace{\left(\frac{1}{3}t^4 + \frac{1}{2}t^2\right)}^{a}$$

$= t^2 - 4 - \frac{1}{3}t^4 - \frac{1}{2}t^2$ Apply the distributive property.

$= -\frac{1}{3}t^4 + t^2 - \frac{1}{2}t^2 - 4$ Group *like* terms in descending order.

$= -\frac{1}{3}t^4 + \frac{2}{2}t^2 - \frac{1}{2}t^2 - 4$ The t^2-terms are the only *like* terms. Get a common denominator for the t^2-terms.

$= -\frac{1}{3}t^4 + \frac{1}{2}t^2 - 4$ Add *like* terms.

Skill Practice

11. Subtract $\frac{1}{2}y^2 - \frac{1}{3}y$ from $y^3 + 3y^2 - y$.

5. Polynomials and Applications to Geometry

Example 7 Application of Polynomials in Geometry

If the perimeter of the triangle in Figure 5-2 can be represented by the polynomial $2x^2 + 5x + 6$, find a polynomial that represents the length of the missing side.

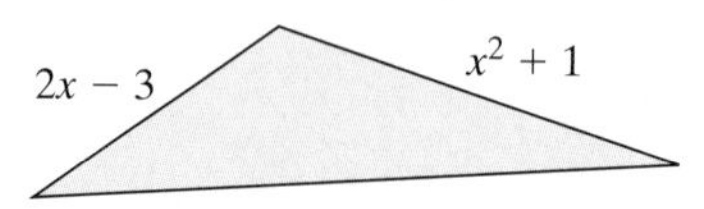

Figure 5-2

Skill Practice Answers

9. $-3x^2 - 3x - 3$

10. $-y^2 + 4xy + 10x^2$

11. $y^3 + \frac{5}{2}y^2 - \frac{2}{3}y$

Solution:

The missing side of the triangle can be found by subtracting the sum of the two known sides from the perimeter.

$$\begin{pmatrix}\text{Length}\\\text{of missing}\\\text{side}\end{pmatrix} = (\text{perimeter}) - \begin{pmatrix}\text{sum of the}\\\text{two known sides}\end{pmatrix}$$

$$\begin{pmatrix}\text{Length}\\\text{of missing}\\\text{side}\end{pmatrix} = (2x^2 + 5x + 6) - [(2x - 3) + (x^2 + 1)]$$

$= 2x^2 + 5x + 6 - [2x - 3 + x^2 + 1]$ Clear inner parentheses.

$= 2x^2 + 5x + 6 - (x^2 + 2x - 2)$ Combine *like* terms within [ ].

$= 2x^2 + 5x + 6 - x^2 - 2x + 2$ Apply the distributive property.

$= 2x^2 - x^2 + 5x - 2x + 6 + 2$ Group *like* terms.

$= x^2 + 3x + 8$ Combine *like* terms.

The polynomial $x^2 + 3x + 8$ represents the length of the missing side.

Skill Practice

12. If the perimeter of the triangle is represented by the polynomial $6x - 9$, find the polynomial that represents the missing side.

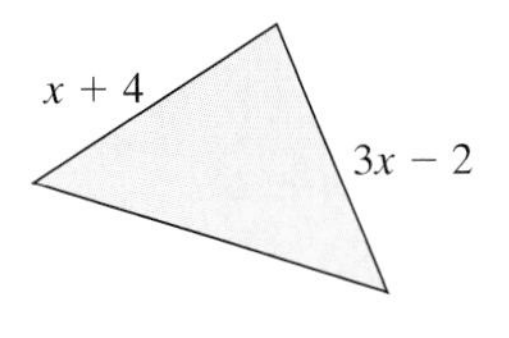

Skill Practice Answers

12. $2x - 11$

Section 5.5 Practice Exercises

Study Skills Exercise

1. Define the key terms:

a. polynomial **b. coefficient** **c. degree of term**

d. monomial **e. binomial** **f. trinomial**

g. leading term **h. leading coefficient** **i. degree of polynomial**

Review Exercises

For Exercises 2–7, simplify the expression. Assume all variables represent nonzero real numbers.

2. $\dfrac{p^3 \cdot 4p}{p^2}$

3. $(3x)(5x^{-4})$

4. $(6y^{-3})(2y^9)$

5. $\dfrac{8t^{-6}}{4t^{-2}}$

6. $\dfrac{8^3 \cdot 8^{-4}}{8^{-2} \cdot 8^6}$

7. $\dfrac{3^4 \cdot 3^{-8}}{3^{12} \cdot 3^{-4}}$

8. Explain the difference between 3.0×10^7 and 3^7.

9. Explain the difference between 4.0×10^{-2} and 4^{-2}.

10. Write the polynomial in descending order. $10 - 8a - a^3 + 2a^2 + a^5$

Concept 1: Introduction to Polynomials

11. Write the polynomial in descending order:

$$6 + 7x^2 - 7x^4 + 9x$$

12. Write the polynomial in descending order:

$$\frac{1}{2}y + y^2 - 12y^4 + y^3 - 6$$

For Exercises 13–24, categorize the expression as a monomial, a binomial, or a trinomial. Then identify the coefficient and degree of the leading term.

13. $10a^2 + 5a$

14. $7z + 13z^2 - 15$

15. $6x^2$

16. 9

17. $2t - t^4 - 5t$

18. $7x + 2$

19. $12y^4 - 3y + 1$

20. $5bc^2$

21. 23

22. $4 - 2c$

23. $-32xyz$

24. $w^4 - w^2$

Concept 2: Applications of Polynomials

25. A ball is dropped off a building and the height of the ball, h (in feet), can be computed by the equation:

$h = -16t^2 + 150$ where t is the time (in seconds) after the ball is released.

a. Find the height of the ball after 1 sec, 1.5 sec, and 2 sec.

b. Find the height of the building by determining the height at the time of release.

26. An object is dropped off a building and the height of the object, h (in meters), can be computed by the equation:

$h = -4.9t^2 + 45$ where t is the time (in seconds) after the object is released.

a. Find the height of the object after 1 sec, 1.5 sec, and 2 sec.

b. Find the height of the building by determining the height at the time of release.

27. A small business produces candles. The equation $P = -0.02x^2 + 10x - 60$ gives the profit, P in dollars, for x boxes of candles produced.

a. Find the profit when $x = 100$, 250, and 300.

b. What is P when $x = 0$. What does this number mean in the context of the problem?

28. An engineering student has a radio controlled model airplane. The student flies the plane over an open field and has a programmed flight path for landing. Four hundred horizontal feet from the landing point, the plane begins its descent (see figure). The flight path for landing can be modeled by:

$h = -\frac{1}{320{,}000}x^3 + \frac{3}{1600}x^2$ where h is the altitude of the plane in feet and x is the horizontal distance from the landing point.

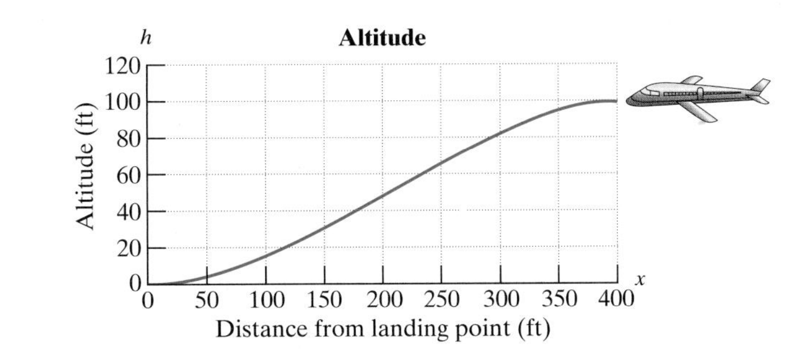

a. Find the altitude when the plane is 400 horizontal feet from the landing point.

b. Find the altitude when the plane is 100 horizontal feet from the landing point.

Concept 3: Addition of Polynomials

29. Explain why the terms $3x$ and $3x^2$ are not *like* terms.

30. Explain why the terms $4w^3$ and $4z^3$ are not *like* terms.

For Exercises 31–46, add the polynomials.

31. $23x^2y + 12x^2y$

32. $-5ab^3 + 17ab^3$

33. $(6y + 3x) + (4y - 3x)$

34. $(2z - 5h) + (-3z + h)$

35. $3b^2 + (5b^2 - 9)$

36. $4c + (3 - 10c)$

37. $(7y^2 + 2y - 9) + (-3y^2 - y)$

38. $(-3w^2 + 4w - 6) + (5w^2 + 2)$

39. $(6a + 2b - 5c) + (-2a - 2b - 3c)$

40. $(-13x + 5y + 10z) + (-3x - 3y + 2z)$

41. $\left(\frac{2}{5}a + \frac{1}{4}b - \frac{5}{6}\right) + \left(\frac{3}{5}a - \frac{3}{4}b - \frac{7}{6}\right)$

42. $\left(\frac{5}{9}x + \frac{1}{10}y\right) + \left(-\frac{4}{9}x + \frac{3}{10}y\right)$

43. $\left(z - \frac{8}{3}\right) + \left(\frac{4}{3}z^2 - z + 1\right)$

44. $\left(-\frac{7}{5}r + 1\right) + \left(-\frac{3}{5}r^2 + \frac{7}{5}r + 1\right)$

45. $(7.9t^3 + 2.6t - 1.1) + (-3.4t^2 + 3.4t - 3.1)$

46. $(0.34y^2 + 1.23) + (3.42y - 7.56)$

Concept 4: Subtraction of Polynomials

For Exercises 47–54, find the opposite of each polynomial.

47. $4h - 5$

48. $5k - 12$

49. $-2m^2 + 3m - 15$

50. $-n^2 - 6n + 9$

51. $3v^3 + 5v^2 + 10v + 22$

52. $7u^4 + 3v^2 + 17$

53. $-9t^4 - 8t - 39$

54. $-5r^5 - 3r^3 - r - 23$

For Exercises 55–74, subtract the polynomials.

55. $4a^3b^2 - 12a^3b^2$

56. $5yz^4 - 14yz^4$

57. $-32x^3 - 21x^3$

58. $-23c^5 - 12c^5$

59. $(7a - 7) - (12a - 4)$

60. $(4x + 3v) - (-3x + v)$

61. $(4k + 3) - (-12k - 6)$

62. $(3h - 15) - (8h - 13)$

63. $25s - (23s - 14)$

64. $3x^2 - (-x^2 - 12)$

65. $(5t^2 - 3t - 2) - (2t^2 + t + 1)$

66. $(k^2 + 2k + 1) - (3k^2 - 6k + 2)$

67. $(10r - 6s + 2t) - (12r - 3s - t)$

68. $(a - 14b + 7c) - (-3a - 8b + 2c)$

69. $\left(\frac{7}{8}x + \frac{2}{3}y - \frac{3}{10}\right) - \left(\frac{1}{8}x + \frac{1}{3}y\right)$

70. $\left(r - \frac{1}{12}s\right) - \left(\frac{1}{2}r - \frac{5}{12}s - \frac{4}{11}\right)$

71. $\left(\frac{2}{3}h^2 - \frac{1}{5}h - \frac{3}{4}\right) - \left(\frac{4}{3}h^2 - \frac{4}{5}h + \frac{7}{4}\right)$

72. $\left(\frac{3}{8}p^3 - \frac{5}{7}p^2 - \frac{2}{5}\right) - \left(\frac{5}{8}p^3 - \frac{2}{7}p^2 + \frac{7}{5}\right)$

73. $(4.5x^4 - 3.1x^2 - 6.7) - (2.1x^4 + 4.4x)$

74. $(1.3c^3 + 4.8) - (4.3c^2 - 2c - 2.2)$

75. Find the difference of $(4b^3 + 6b - 7)$ and $(-12b^2 + 11b + 5)$.

76. Find the difference of $(-5y^2 + 3y - 21)$ and $(-4y^2 - 5y + 23)$.

77. Subtract $(3x^3 - 5x + 10)$ from $(-2x^2 + 6x - 21)$.

78. Subtract $(7a^5 - 2a^3 - 5a)$ from $(3a^5 - 9a^2 + 3a - 8)$.

Concept 5: Polynomials and Applications to Geometry

79. Find a polynomial that represents the perimeter of the figure.

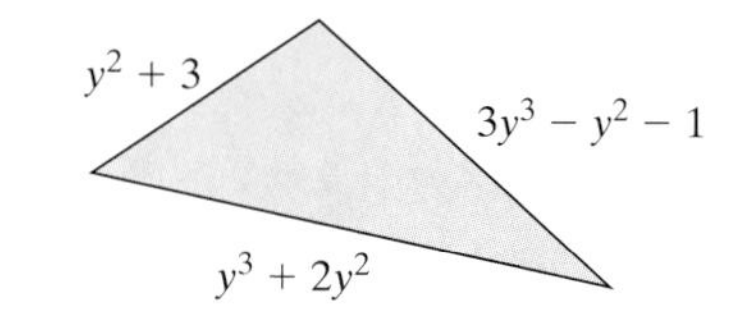

80. Find a polynomial that represents the perimeter of the figure.

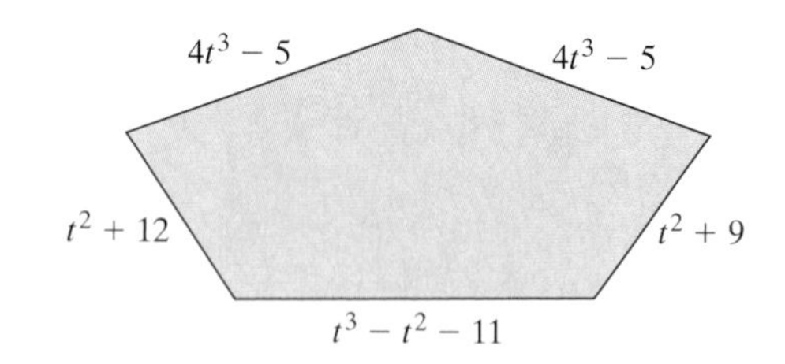

81. If the perimeter of the figure can be represented by the polynomial $5a^2 - 2a + 1$, find a polynomial that represents the length of the missing side.

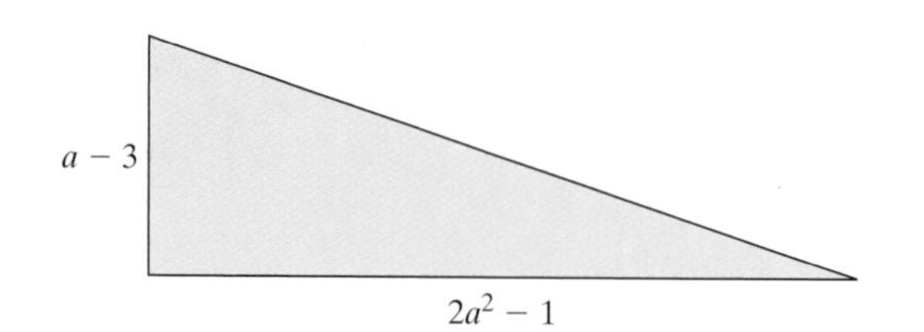

82. If the perimeter of the figure can be represented by the polynomial $6w^3 - 2w - 3$, find a polynomial that represents the length of the missing side.

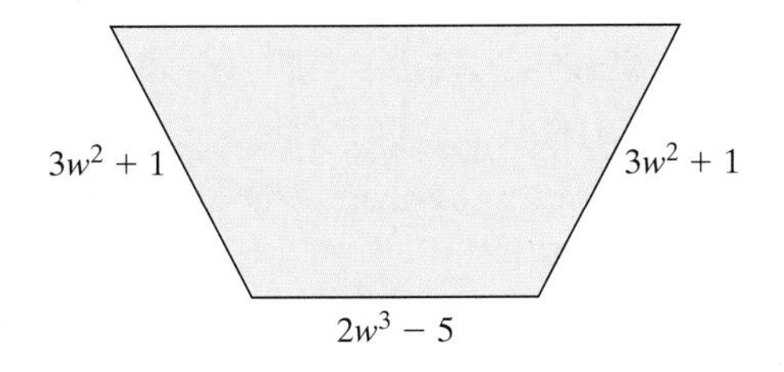

Mixed Exercises

For Exercises 83–98, perform the indicated operation.

83. $(2ab^2 + 9a^2b) + (7ab^2 - 3ab + 7a^2b)$

84. $(8x^2y - 3xy - 6xy^2) + (3x^2y - 12xy)$

85. $(4z^5 + z^3 - 3z + 13) - (-z^4 - 8z^3 + 15)$

86. $(-15t^4 - 23t^2 + 16t) - (21t^3 + 18t^2 + t)$

87. $(9x^4 + 2x^3 - x + 5) + (9x^3 - 3x^2 + 8x + 3) - (7x^4 - x + 12)$

88. $(-6y^3 - 9y^2 + 23) - (7y^2 + 2y - 11) + (3y^3 - 25)$

89. $(5w^2 - 3w + 2) + (-4w + 6) - (7w^2 - 10)$

90. $(10u^3 - 5u^2 + 4) - (2u^3 + 5u^2 + u) - (u^3 - 3u + 9)$

91. $(7p^2q - 3pq^2) - (8p^2q + pq) + (4pq - pq^2)$

92. $(12c^2d - 2cd + 8cd^2) - (-c^2d + 4cd) - (5cd - 2cd^2)$

93. $(5x - 2x^3) + (2x^3 - 5x)$

94. $(p^2 - 4p + 2) - (2 + p^2 - 4p)$

95. $(2a^2b - 4ab + ab^2) - (-5ab^2 + 2a^2b + ab)$

96. $-3xy + 7xy^2 + 5x^2y + (3x^2y - 8xy - 11xy^2)$

97. $[(3y^2 - 5y) - (2y^2 + y - 1)] + (10y^2 - 4y - 5)$

98. $(12c^3 - 5c^2 - 2c) + [(7c^3 - 2c^2 + c) - (4c^3 + 4c)]$

Expanding Your Skills

99. Write a binomial of degree 3. (Answers may vary.)

100. Write a trinomial of degree 6. (Answers may vary.)

101. Write a monomial of degree 5. (Answers may vary.)

102. Write a monomial of degree 1. (Answers may vary.)

103. Write a trinomial with the leading coefficient -6. (Answers may vary.)

104. Write a binomial with the leading coefficient 13. (Answers may vary.)

Section 5.6 Multiplication of Polynomials

Concepts

1. Multiplication of Polynomials
2. Special Case Products: Difference of Squares and Perfect Square Trinomials
3. Applications to Geometry

1. Multiplication of Polynomials

The properties of exponents covered in Sections 5.1–5.3 can be used to simplify many algebraic expressions including the multiplication of monomials. To multiply monomials, first use the associative and commutative properties of multiplication to group coefficients and like bases. Then simplify the result by using the properties of exponents.

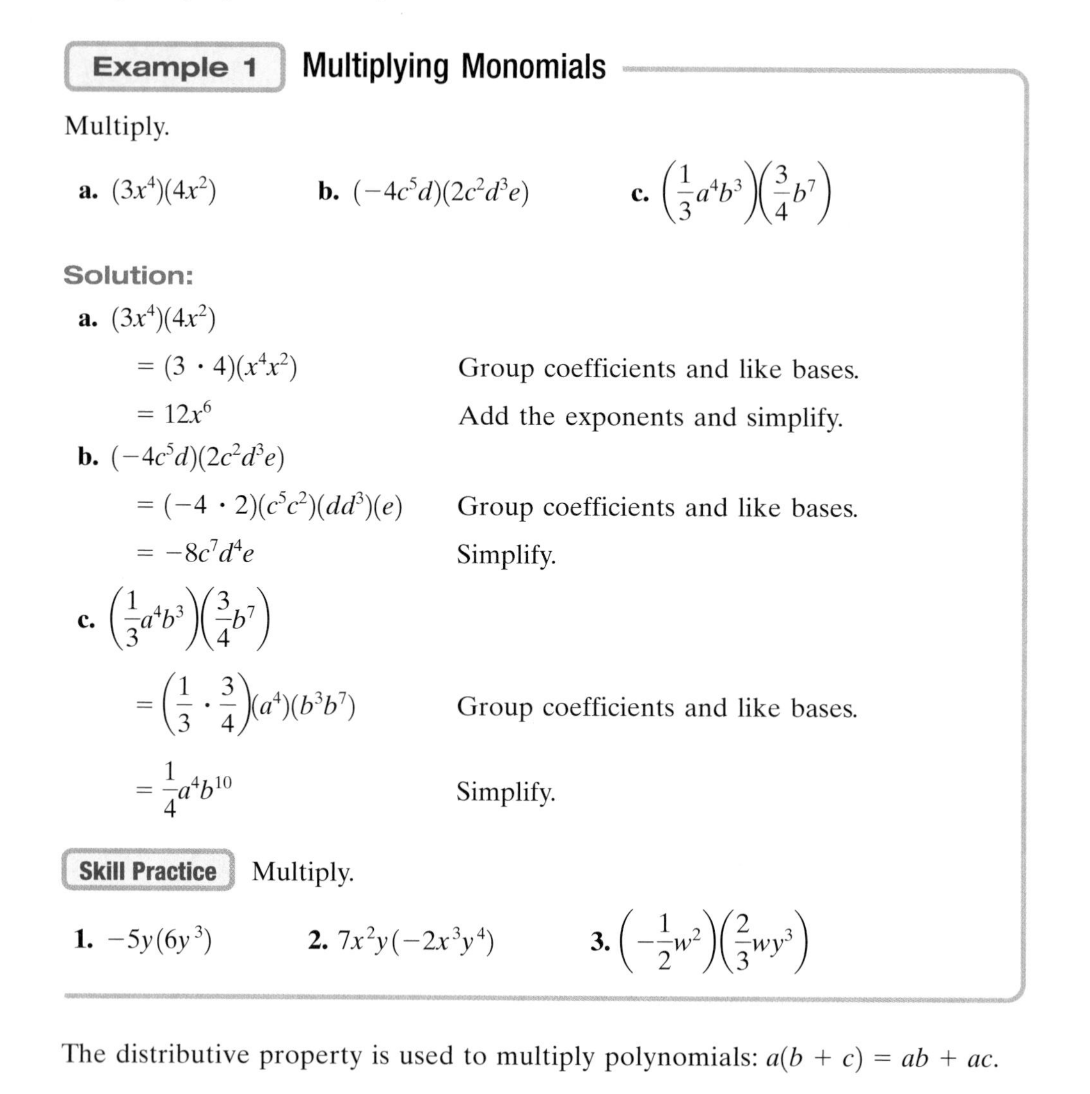

Example 1 **Multiplying Monomials**

Multiply.

a. $(3x^4)(4x^2)$ **b.** $(-4c^5d)(2c^2d^3e)$ **c.** $\left(\frac{1}{3}a^4b^3\right)\left(\frac{3}{4}b^7\right)$

Solution:

a. $(3x^4)(4x^2)$

$= (3 \cdot 4)(x^4x^2)$ Group coefficients and like bases.

$= 12x^6$ Add the exponents and simplify.

b. $(-4c^5d)(2c^2d^3e)$

$= (-4 \cdot 2)(c^5c^2)(dd^3)(e)$ Group coefficients and like bases.

$= -8c^7d^4e$ Simplify.

c. $\left(\frac{1}{3}a^4b^3\right)\left(\frac{3}{4}b^7\right)$

$= \left(\frac{1}{3} \cdot \frac{3}{4}\right)(a^4)(b^3b^7)$ Group coefficients and like bases.

$= \frac{1}{4}a^4b^{10}$ Simplify.

Skill Practice Multiply.

1. $-5y(6y^3)$ **2.** $7x^2y(-2x^3y^4)$ **3.** $\left(-\frac{1}{2}w^2\right)\left(\frac{2}{3}wy^3\right)$

The distributive property is used to multiply polynomials: $a(b + c) = ab + ac$.

Example 2 **Multiplying a Polynomial by a Monomial**

Multiply.

a. $2t(4t - 3)$ **b.** $-3a^2\left(-4a^2 + 2a - \frac{1}{3}\right)$

Solution:

a. $2t(4t - 3)$ Multiply each term of the polynomial by $2t$.

$= (2t)(4t) + 2t(-3)$ Apply the distributive property.

$= 8t^2 - 6t$ Simplify each term.

Skill Practice Answers

1. $-30y^4$
2. $-14x^5y^5$
3. $-\frac{1}{3}w^3y^3$

b. $-3a^2\left(-4a^2 + 2a - \frac{1}{3}\right)$ Multiply each term of the polynomial by $-3a^2$.

$= (-3a^2)(-4a^2) + (-3a^2)(2a) + (-3a^2)\left(-\frac{1}{3}\right)$ Apply the distributive property.

$= 12a^4 - 6a^3 + a^2$ Simplify each term.

Skill Practice Multiply.

4. $-4p(2p^2 - 6p + \frac{1}{4})$ **5.** $5a^2b^3(-3ab^2 + 4ab - 6b^3)$

Thus far, we have shown polynomial multiplication involving monomials. Next, the distributive property will be used to multiply polynomials with more than one term.

$(x + 3)(x + 5) = (x + 3)x + (x + 3)5$ Apply the distributive property.

$= (x + 3)x + (x + 3)5$ Apply the distributive property again.

$= (x)(x) + (3)(x) + (x)(5) + (3)(5)$

$= x^2 + 3x + 5x + 15$

$= x^2 + 8x + 15$ Combine *like* terms.

Note: Using the distributive property results in multiplying each term of the first polynomial by each term of the second polynomial.

$$(x + 3)(x + 5) = (x)(x) + (x)(5) + (3)(x) + (3)(5)$$
$$= x^2 + 5x + 3x + 15$$
$$= x^2 + 8x + 15$$

Example 3 Multiplying a Polynomial by a Polynomial

Multiply the polynomials.

a. $(c - 7)(c + 2)$ **b.** $(10x + 3y)(2x - 4y)$

c. $(y - 2)(3y^2 + y - 5)$

Solution:

a. $(c - 7)(c + 2)$ Multiply each term in the first polynomial by each term in the second.

$= (c)(c) + (c)(2) + (-7)(c) + (-7)(2)$ Apply the distributive property.

$= c^2 + 2c - 7c - 14$ Simplify.

$= c^2 - 5c - 14$ Combine *like* terms.

Skill Practice Answers

4. $-8p^3 + 24p^2 - p$

5. $-15a^3b^5 + 20a^3b^4 - 30a^2b^6$

TIP: Notice that the product of two *binomials* equals the sum of the products of the **F**irst terms, the **O**uter terms, the **I**nner terms, and the **L**ast terms. The acronym, **FOIL** (First Outer Inner Last) can be used as a memory device to multiply two binomials.

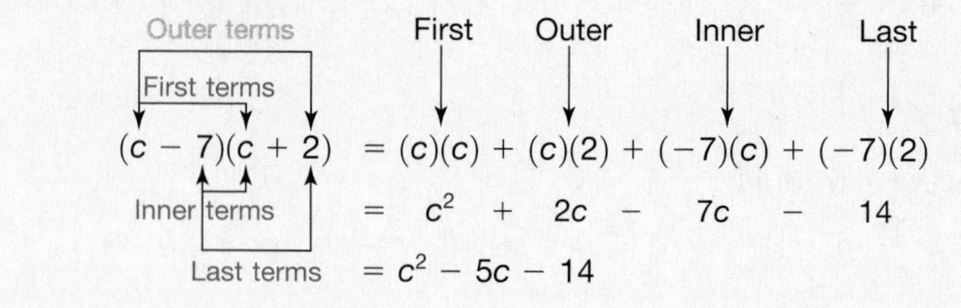

$$(c - 7)(c + 2) = (c)(c) + (c)(2) + (-7)(c) + (-7)(2)$$
$$= c^2 + 2c - 7c - 14$$
$$= c^2 - 5c - 14$$

b. $(10x + 3y)(2x - 4y)$ — Multiply each term in the first polynomial by each term in the second.

$= (10x)(2x) + (10x)(-4y) + (3y)(2x) + (3y)(-4y)$ — Apply the distributive property.

$= 20x^2 - 40xy + 6xy - 12y^2$ — Simplify each term.

$= 20x^2 - 34xy - 12y^2$ — Combine *like* terms.

c. $(y - 2)(3y^2 + y - 5)$ — Multiply each term in the first polynomial by each term in the second.

$= (y)(3y^2) + (y)(y) + (y)(-5) + (-2)(3y^2) + (-2)(y) + (-2)(-5)$

$= 3y^3 + y^2 - 5y - 6y^2 - 2y + 10$ — Simplify each term.

$= 3y^3 - 5y^2 - 7y + 10$ — Combine *like* terms.

Avoiding Mistakes:

It is important to note that the acronym **FOIL** does not apply to Example 3(c) because the product does not involve two binomials.

TIP: Multiplication of polynomials can be performed vertically by a process similar to column multiplication of real numbers. For example,

$$\begin{array}{r} 235 \\ \times\ 21 \\ \hline 235 \\ 4700 \\ \hline 4935 \end{array} \qquad \begin{array}{r} 3y^2 + y - 5 \\ \times \qquad y - 2 \\ \hline -6y^2 - 2y + 10 \\ 3y^3 + y^2 - 5y + 0 \\ \hline 3y^3 - 5y^2 - 7y + 10 \end{array}$$

Note: When multiplying by the column method, it is important to *align like* terms vertically before adding terms.

Skill Practice Multiply and simplify.

6. $(x + 2)(x + 8)$ **7.** $(4a - 3c)(5a - 2c)$ **8.** $(2y + 4)(3y^2 - 5y + 2)$

2. Special Case Products: Difference of Squares and Perfect Square Trinomials

In some cases the product of two binomials takes on a special pattern.

I. The first special case occurs when multiplying the sum and difference of the same two terms. For example,

$$(2x + 3)(2x - 3)$$
$$= 4x^2 - 6x + 6x - 9$$
$$= 4x^2 - 9$$

Notice that the middle terms are opposites. This leaves only the difference between the square of the first term and the square of the second term. For this reason, the product is called a *difference of squares.*

Note: The sum and difference of the same two terms are called **conjugates**. Thus, the expressions $2x + 3$ and $2x - 3$ are conjugates of each other.

II. The second special case involves the square of a binomial. For example,

$$(3x + 7)^2$$
$$= (3x + 7)(3x + 7)$$
$$= 9x^2 + 21x + 21x + 49$$
$$= 9x^2 + 42x + 49$$
$$= (3x)^2 + 2(3x)(7) + (7)^2$$

When squaring a binomial, the product will be a trinomial called a *perfect square trinomial.* The first and third terms are formed by squaring each term of the binomial. The middle term equals twice the product of the terms in the binomial.

Note: The expression $(3x - 7)^2$ also expands to a perfect square trinomial, but the middle term will be negative:

$$(3x - 7)(3x - 7) = 9x^2 - 21x - 21x + 49 = 9x^2 - 42x + 49$$

Special Case Product Formulas

1. $(a + b)(a - b) = a^2 - b^2$ The product is called a **difference of squares.**

2. $\left.\begin{array}{l}(a + b)^2 = a^2 + 2ab + b^2\\(a - b)^2 = a^2 - 2ab + b^2\end{array}\right\}$ The product is called a **perfect square trinomial.**

You should become familiar with these special case products because they will be used again in the next chapter to factor polynomials.

Example 4 Finding Special Products

Multiply the conjugates.

a. $(x - 9)(x + 9)$ **b.** $\left(\frac{1}{2}p - 6\right)\left(\frac{1}{2}p + 6\right)$

Skill Practice Answers

6. $x^2 + 10x + 16$
7. $20a^2 - 23ac + 6c^2$
8. $6y^3 + 2y^2 - 16y + 8$

TIP: The product of two conjugates can be checked by applying the distributive property:

$(x - 9)(x + 9)$

$= x^2 + 9x - 9x - 81$

$= x^2 - 81$

Solution:

a. $(x - 9)(x + 9)$ Apply the formula: $(a + b)(a - b) = a^2 - b^2$.

$a^2 - b^2$

$= (x)^2 - (9)^2$ Substitute $a = x$ and $b = 9$.

$= x^2 - 81$

b. $\left(\frac{1}{2}p - 6\right)\left(\frac{1}{2}p + 6\right)$ Apply the formula: $(a + b)(a - b) = a^2 - b^2$.

$a^2 - b^2$

$= \left(\frac{1}{2}p\right)^2 - (6)^2$ Substitute $a = \frac{1}{2}p$ and $b = 6$.

$= \frac{1}{4}p^2 - 36$ Simplify each term.

Skill Practice Multiply the conjugates.

9. $(a + 7)(a - 7)$ **10.** $(\frac{4}{5}x - 10)(\frac{4}{5}x + 10)$

Example 5 Finding Special Products

Square the binomials.

a. $(3w - 4)^2$ **b.** $(5x^2 + 2)^2$

Solution:

a. $(3w - 4)^2$ Apply the formula: $(a - b)^2 = a^2 - 2ab + b^2$.

$a^2 - 2ab + b^2$

$= (3w)^2 - 2(3w)(4) + (4)^2$ Substitute $a = 3w$ and $b = 4$.

$= 9w^2 - 24w + 16$ Simplify each term.

TIP: The square of a binomial can be checked by explicitly writing the product of the two binomials and applying the distributive property:

$(3w - 4)^2 = (3w - 4)(3w - 4) = 9w^2 - 12w - 12w + 16$
$= 9w^2 - 24w + 16$

b. $(5x^2 + 2)^2$ Apply the formula: $(a + b)^2 = a^2 + 2ab + b^2$.

$a^2 + 2ab + b^2$

$= (5x^2)^2 + 2(5x^2)(2) + (2)^2$ Substitute $a = 5x^2$ and $b = 2$.

$= 25x^4 + 20x^2 + 4$ Simplify each term.

Avoiding Mistakes:

The property for squaring two factors is different than the property of squaring two terms: $(ab)^2 = a^2b^2$ but $(a + b)^2 = a^2 + 2ab + b^2$

Skill Practice Square the binomials.

11. $(x + 3)^2$ **12.** $(3c^2 - 4)^2$

Skill Practice Answers

9. $a^2 - 49$ **10.** $\frac{16}{25}x^2 - 100$
11. $x^2 + 6x + 9$
12. $9c^4 - 24c^2 + 16$

3. Applications to Geometry

Example 6 **Using Special Case Products in an Application of Geometry**

Find a polynomial that represents the volume of the cube (Figure 5-3).

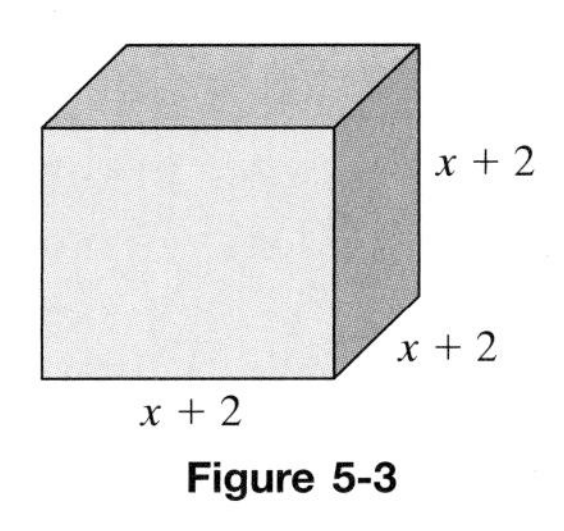

Figure 5-3

Solution:

Volume = (length)(width)(height)

$$V = (x + 2)(x + 2)(x + 2) \quad \text{or} \quad V = (x + 2)^3$$

To expand $(x + 2)(x + 2)(x + 2)$, multiply the first two factors. Then multiply the result by the last factor.

$$V = \underbrace{(x + 2)(x + 2)}(x + 2)$$

$$= (x^2 + 4x + 4)(x + 2)$$

TIP: $(x + 2)(x + 2) = (x + 2)^2$ and results in a perfect square trinomial.

$$(x + 2)^2 = (x)^2 + 2(x)(2) + (2)^2$$
$$= x^2 + 4x + 4$$

$$= (x^2)(x) + (x^2)(2) + (4x)(x) + (4x)(2) + (4)(x) + (4)(2)$$ Apply the distributive property.

$$= x^3 + 2x^2 + 4x^2 + 8x + 4x + 8$$ Group *like* terms.

$$= x^3 + 6x^2 + 12x + 8$$ Combine *like* terms.

The volume of the cube can be represented by

$$V = (x + 2)^3 = x^3 + 6x^2 + 12x + 8.$$

Skill Practice

13. Find the polynomial that represents the area of the square.

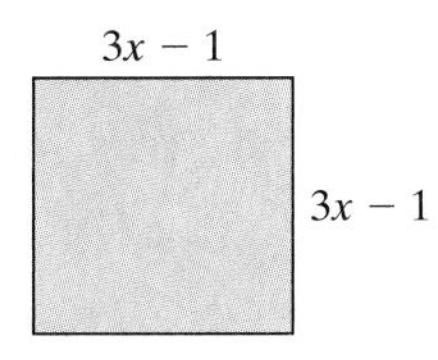

Skill Practice Answers

13. The area of the square can be represented by $9x^2 - 6x + 1$.

Example 7 Using the Product of Polynomials in Geometry

Find a polynomial that represents the area of the triangle (Figure 5-4).

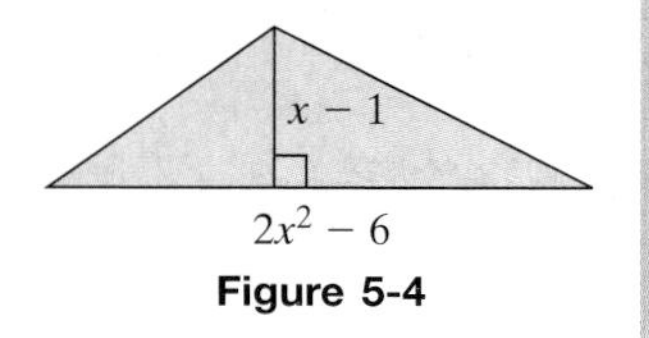

Figure 5-4

Solution:

The area of a triangle is found by $A = \frac{1}{2}bh$. In this case, $b = 2x^2 - 6$ and $h = x - 1$; therefore, $A = \frac{1}{2}(2x^2 - 6)(x - 1)$.

$A = \frac{1}{2}(2x^2 - 6)(x - 1)$ Apply the distributive property.

$= (x^2 - 3)(x - 1)$

$= (x^2 - 3)(x - 1)$ Multiply each term in the first polynomial by each term in the second.

$= (x^2)(x) + (x^2)(-1) + (-3)(x) + (-3)(-1)$ Apply the distributive property.

$= x^3 - x^2 - 3x + 3$

The area of the triangle can be represented by the polynomial $x^3 - x^2 - 3x + 3$.

Skill Practice

14. Find the polynomial that represents the area of the triangle.

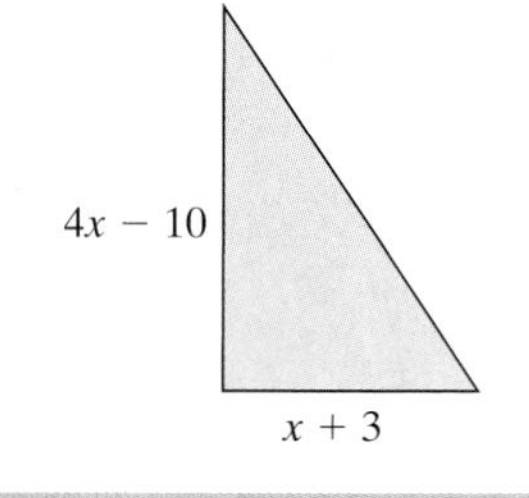

Skill Practice Answers

14. The area of the triangle can be represented by $2x^2 + x - 15$

Section 5.6 Practice Exercises

Boost *your* GRADE at mathzone.com!

MathZone

- Practice Problems
- Self-Tests
- NetTutor
- e-Professors
- Videos

Study Skills Exercise

1. Define the key terms:

a. conjugates **b. difference of squares** **c. perfect square trinomial**

Review Exercises

For Exercises 2–13, simplify the expressions (if possible).

2. $4x + 5x$ **3.** $2y^2 - 4y^2$ **4.** $(4x)(5x)$ **5.** $(2y^2)(-4y^2)$

6. $-5a^3b - 2a^3b$ **7.** $7uvw^2 + uvw^2$ **8.** $(-5a^3b)(-2a^3b)$ **9.** $(7uvw^2)(uvw^2)$

10. $-c + 4c^2$ **11.** $3t + 3t^3$ **12.** $(-c)(4c^2)$ **13.** $(3t)(3t^3)$

Concept 1: Multiplication of Polynomials

For Exercises 14–22, multiply the expressions.

14. $8(4x)$

15. $-2(6y)$

16. $-10(5z)$

17. $7(3p)$

18. $(x^{10})(4x^3)$

19. $(a^{13}b^4)(12ab^4)$

20. $(4m^3n^7)(-3m^6n)$

21. $(2c^7d)(-c^3d^{11})$

22. $(-5u^2v)(-8u^3v^2)$

For Exercises 23–52, multiply the polynomials.

23. $8pq(2pq - 3p + 5q)$

24. $5ab(2ab + 6a - 3b)$

25. $(k^2 - 13k - 6)(-4k)$

26. $(h^2 + 5h - 12)(-2h)$

27. $-15pq(3p^2 + p^3q^2 - 2q)$

28. $-4u^2v(2u - 5uv^3 + v)$

29. $(y - 10)(y + 9)$

30. $(x + 5)(x - 6)$

31. $(m - 12)(m - 2)$

32. $(n - 7)(n - 2)$

33. $(p - 2)(p + 1)$

34. $(q + 11)(q - 5)$

35. $(w + 8)(w + 3)$

36. $(z + 10)(z + 4)$

37. $(p - 3)(p - 11)$

38. $(y - 7)(y - 10)$

39. $(6x - 1)(2x + 5)$

40. $(3x + 7)(x - 8)$

41. $(4a - 9)(2a - 1)$

42. $(3b + 5)(b - 5)$

43. $(3t - 7)(3t + 1)$

44. $(5w - 2)(2w - 5)$

45. $(3x + 4)(x + 8)$

46. $(7y + 1)(3y + 5)$

47. $(5s + 3)(s^2 + s - 2)$

48. $(t - 4)(2t^2 - t + 6)$

49. $(3w - 2)(9w^2 + 6w + 4)$

50. $(z + 5)(z^2 - 5z + 25)$

51. $(p^2 + p - 5)(p^2 + 4p - 1)$

52. $(-x^2 - 2x + 4)(x^2 + 2x - 6)$

Concept 2: Special Case Products: Difference of Squares and Perfect Square Trinomials

For Exercises 53–64, multiply the conjugates.

53. $(3a - 4b)(3a + 4b)$

54. $(5y + 7x)(5y - 7x)$

55. $(9k + 6)(9k - 6)$

56. $(2h - 5)(2h + 5)$

57. $\left(\frac{1}{2} - t\right)\left(\frac{1}{2} + t\right)$

58. $\left(r + \frac{1}{4}\right)\left(r - \frac{1}{4}\right)$

59. $(u^3 + 5v)(u^3 - 5v)$

60. $(8w^2 - x)(8w^2 + x)$

61. $(2 - 3a)(2 + 3a)$

62. $(1 - 4x^2)(1 + 4x^2)$

63. $\left(\frac{2}{3} - p\right)\left(\frac{2}{3} + p\right)$

64. $\left(\frac{1}{8} - q\right)\left(\frac{1}{8} + q\right)$

For Exercises 65–76, square the binomials.

65. $(a + b)^2$

66. $(a - b)^2$

67. $(x - y)^2$

68. $(x + y)^2$

69. $(2c + 5)^2$

70. $(5d - 9)^2$

71. $(3t^2 - 4s)^2$

72. $(u^2 + 4v)^2$

73. $(7 - t)^2$

74. $(4 + w)^2$

75. $(3 + 4q)^2$

76. $(2 - 3b)^2$

77. **a.** Evaluate $(2 + 4)^2$ by working within the parentheses first.

b. Evaluate $2^2 + 4^2$.

c. Compare the answers to parts (a) and (b) and make a conjecture about $(a + b)^2$ and $a^2 + b^2$.

78. **a.** Evaluate $(6 - 5)^2$ by working within the parentheses first.

b. Evaluate $6^2 - 5^2$.

c. Compare the answers to parts (a) and (b) and make a conjecture about $(a - b)^2$ and $a^2 - b^2$.

79. **a.** Simplify $(3x + y)^2$.

b. Simplify $(3xy)^2$.

c. Compare the answers to parts (a) and (b) to make a conjecture about $(a + b)^2$ and $(ab)^2$.

Concept 3: Applications to Geometry

80. Find a polynomial expression that represents the area of the rectangle shown in the figure.

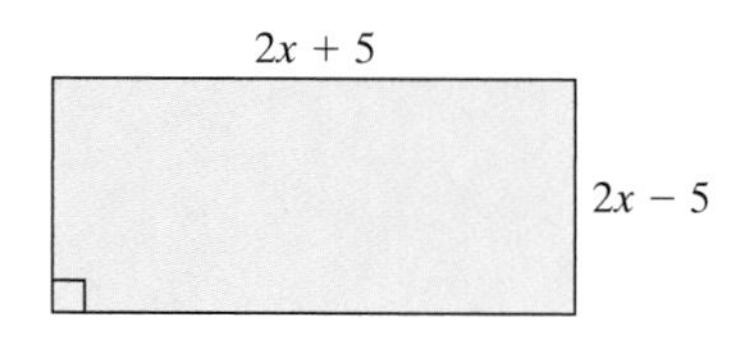

81. Find a polynomial expression that represents the area of the rectangle shown in the figure.

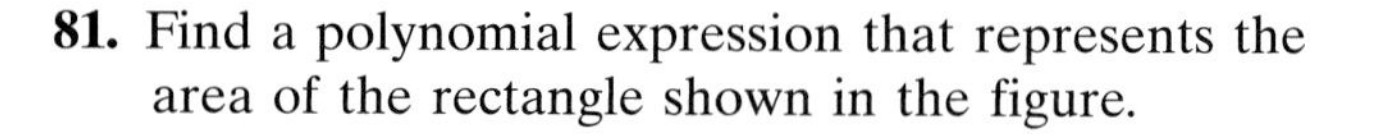

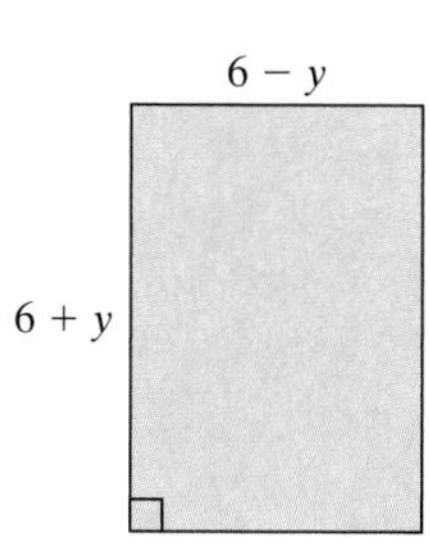

82. Find a polynomial expression that represents the area of the square shown in the figure.

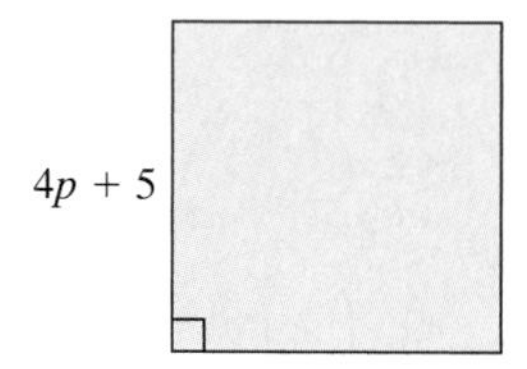

83. Find a polynomial expression that represents the area of the square shown in the figure.

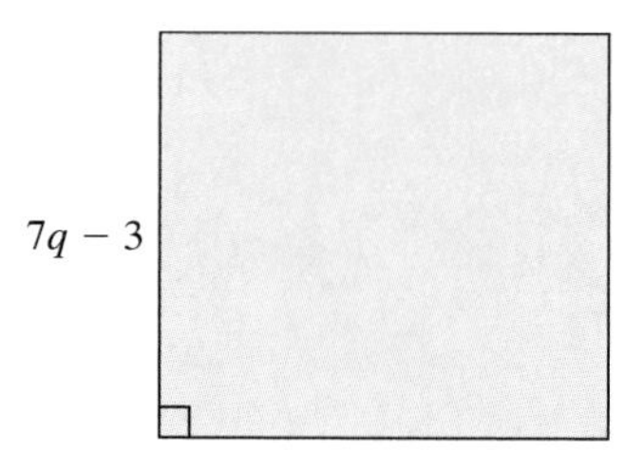

84. Find a polynomial that represents the area of the triangle shown in the figure.
(Recall: $A = \frac{1}{2}bh$)

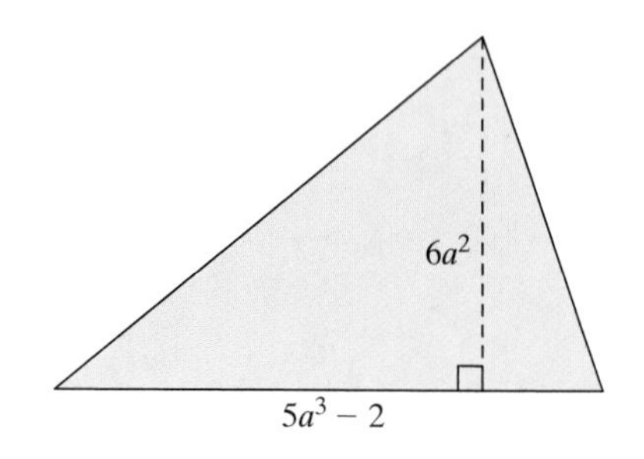

85. Find a polynomial that represents the area of the triangle shown in the figure.

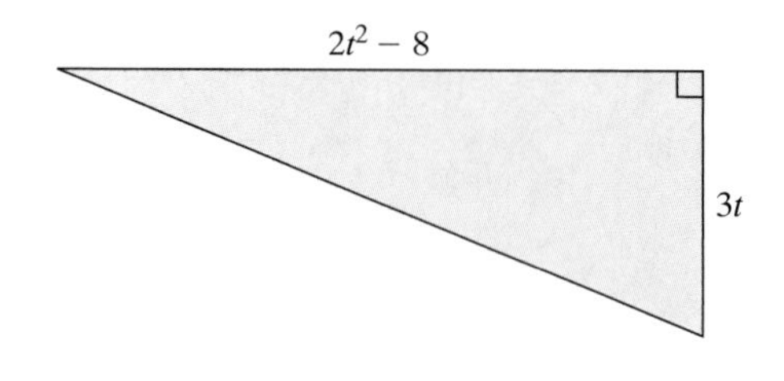

86. Find a polynomial that represents the volume of the cube shown in the figure.
(Recall: $V = s^3$)

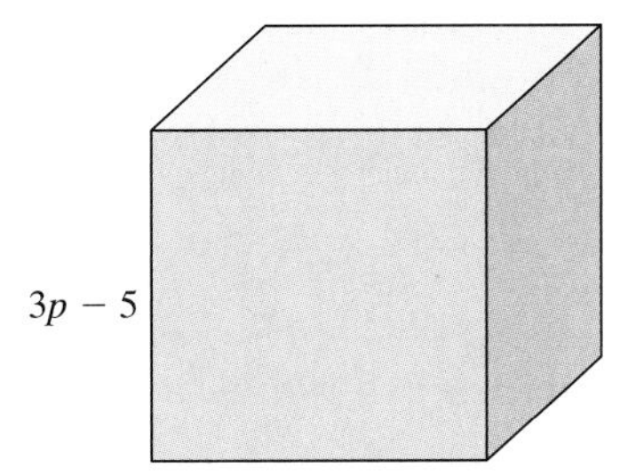

87. Find a polynomial that represents the volume of the rectangular solid shown in the figure.
(Recall: $V = lwh$)

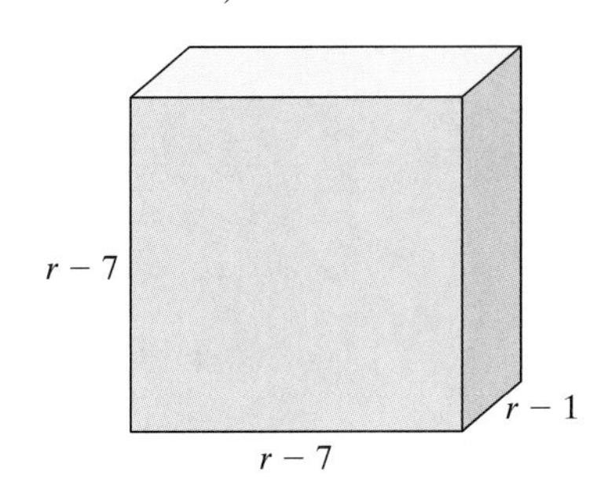

Mixed Exercises

For Exercises 88–117, multiply the expressions.

88. $(7x + y)(7x - y)$
89. $(9w - 4z)(9w + 4z)$
90. $(5s + 3t)^2$
91. $(5s - 3t)^2$
92. $(7x - 3y)(3x - 8y)$
93. $(5a - 4b)(2a - b)$
94. $\left(\frac{2}{3}t + 2\right)(3t + 4)$
95. $\left(\frac{1}{5}s + 6\right)(5s - 3)$
96. $(5z + 3)(z^2 + 4z - 1)$
97. $(2k - 5)(2k^2 + 3k + 5)$
98. $(3a - 2)(5a + 1 + 2a^2)$
99. $(u + 4)(2 - 3u + u^2)$
100. $(y^2 + 2y + 4)(y - 5)$
101. $(w^2 - w + 6)(w + 2)$
102. $\left(\frac{1}{3}m - n\right)^2$
103. $\left(\frac{2}{5}p - q\right)^2$
104. $6w^2(7w - 14)$
105. $4v^3(v + 12)$
106. $(4y - 8.1)(4y + 8.1)$
107. $(2h + 2.7)(2h - 2.7)$
108. $(3c^2 + 4)(7c^2 - 8)$
109. $(5k^3 - 9)(k^3 - 2)$
110. $(3.1x + 4.5)^2$
111. $(2.5y + 1.1)^2$
112. $(k - 4)^3$
113. $(h + 3)^3$
114. $(5x + 3)^3$
115. $(2a - 4)^3$
116. $(y^2 + 2y + 1)(2y^2 - y + 3)$
117. $(2w^2 - w - 5)(3w^2 + 2w + 1)$

Expanding Your Skills

For Exercises 118–121, multiply the expressions containing more than two factors.

118. $2a(3a - 4)(a + 5)$
119. $5x(x + 2)(6x - 1)$
120. $(x - 3)(2x + 1)(x - 4)$
121. $(y - 2)(2y - 3)(y + 3)$

122. What binomial when multiplied by $(3x + 5)$ will produce a product of $6x^2 - 11x - 35$? [*Hint:* Let the quantity $(a + b)$ represent the unknown binomial.] Then find a and b such that $(3x + 5)(a + b) = 6x^2 - 11x - 35$.

123. What binomial when multiplied by $(2x - 4)$ will produce a product of $2x^2 + 8x - 24$?

For Exercises 124–126, determine what values of k would create a perfect square trinomial.

124. $x^2 + kx + 25$
125. $w^2 + kw + 9$
126. $a^2 + ka + 16$

Section 5.7 Division of Polynomials

Concepts

1. Division by a Monomial
2. Long Division

Division of polynomials will be presented in this section as two separate cases. The first case illustrates division by a monomial divisor. The second case illustrates long division by a polynomial with two or more terms.

1. Division by a Monomial

To divide a polynomial by a monomial, divide each individual term in the polynomial by the divisor and simplify the result.

Dividing a Polynomial by a Monomial

If a, b, and c are polynomials such that $c \neq 0$, then

$$\frac{a+b}{c} = \frac{a}{c} + \frac{b}{c} \qquad \text{Similarly,} \qquad \frac{a-b}{c} = \frac{a}{c} - \frac{b}{c}$$

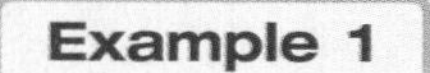

Example 1 Dividing a Polynomial by a Monomial

Divide the polynomials.

a. $\dfrac{5a^3 - 10a^2 + 20a}{5a}$ **b.** $(12y^2z^3 - 15yz^2 + 6y^2z) \div (-6y^2z)$

Solution:

a. $\dfrac{5a^3 - 10a^2 + 20a}{5a}$

$= \dfrac{5a^3}{5a} - \dfrac{10a^2}{5a} + \dfrac{20a}{5a}$ Divide each term in the numerator by $5a$.

$= a^2 - 2a + 4$ Simplify each term using the properties of exponents.

b. $(12y^2z^3 - 15yz^2 + 6y^2z) \div (-6y^2z)$

$= \dfrac{12y^2z^3 - 15yz^2 + 6y^2z}{-6y^2z}$

$= \dfrac{12y^2z^3}{-6y^2z} - \dfrac{15yz^2}{-6y^2z} + \dfrac{6y^2z}{-6y^2z}$ Divide each term by $-6y^2z$.

$= -2z^2 + \dfrac{5z}{2y} - 1$ Simplify each term.

Skill Practice Divide the polynomials.

1. $(36a^4 - 48a^3 + 12a^2) \div (6a^3)$ **2.** $\dfrac{-15x^3y^4 + 25x^2y^3 - 5xy^2}{-5xy^2}$

Skill Practice Answers

1. $6a - 8 + \dfrac{2}{a}$

2. $3x^2y^2 - 5xy + 1$

2. Long Division

If the divisor has two or more terms, a *long division* process similar to the division of real numbers is used. Take a minute to review the long division process for real numbers by dividing 5074 by 31.

$$31\overline{)5074} \qquad \begin{array}{r} 1 \\ 31\overline{)5074} \\ \underline{-31} \\ 19 \end{array} \quad \text{Subtract.}$$

$$\begin{array}{r} 16 \\ 31\overline{)5074} \\ \underline{-31}\downarrow \\ 197 \\ \underline{-186} \\ 11 \end{array} \quad \begin{array}{l} \\ \\ \\ \text{Bring down next column and repeat the process.} \\ \\ \text{Subtract.} \end{array}$$

$$\begin{array}{r} 163 \\ 31\overline{)5074} \\ \underline{-31} \\ 197 \\ \underline{-186} \\ 114 \\ \underline{-93} \\ 21 \end{array} \quad \begin{array}{l} \leftarrow \text{Quotient} \\ \\ \\ \\ \\ \\ \\ \leftarrow \text{Remainder} \end{array}$$

Therefore, $5074 \div 31 = 163\frac{21}{31}$.

A similar procedure is used for long division of polynomials, as shown in Example 2.

Example 2 Using Long Division to Divide Polynomials

Divide the polynomials using long division: $(2x^2 - x + 3) \div (x - 3)$

Solution:

$x - 3\overline{)2x^2 - x + 3}$

Divide the leading term in the dividend by the leading term in the divisor.

$$\frac{2x^2}{x} = 2x$$

This is the first term in the quotient.

$$\begin{array}{r} 2x\phantom{{}- x + 3} \\ x - 3\overline{)2x^2 - x + 3} \\ -(2x^2 - 6x)\phantom{{}+3} \end{array}$$

Multiply $2x$ by the divisor $2x(x - 3) = 2x^2 - 6x$ and subtract the result.

$$\begin{array}{r} 2x\phantom{{}- x + 3} \\ x - 3\overline{)2x^2 - x + 3} \\ \underline{-2x^2 + 6x}\phantom{{}+3} \\ 5x\phantom{{}+3} \end{array}$$

Subtract the quantity $2x^2 - 6x$. To do this, add the opposite.

$$\begin{array}{r} 2x + 5 \\ x - 3\overline{)2x^2 - x + 3} \\ \underline{-2x^2 + 6x}\downarrow \\ 5x + 3 \end{array}$$

Bring down the next column, and repeat the process.

Divide the leading term by x: $(5x)/x = 5$. Place 5 in the quotient.

$$\begin{array}{r} 2x + 5 \\ x - 3\overline{)2x^2 - x + 3} \\ -2x^2 + 6x \\ \hline 5x + 3 \\ -(5x - 15) \end{array}$$

Multiply the divisor by 5: $5(x - 3) = 5x - 15$ and subtract the result.

$$\begin{array}{r} 2x + 5 \\ x - 3\overline{)2x^2 - x + 3} \\ -2x^2 + 6x \\ \hline 5x + 3 \\ -5x + 15 \\ \hline 18 \end{array}$$

Subtract the quantity $5x - 15$ by adding the opposite.

The remainder is 18.

Summary:

The quotient is	$2x + 5$
The remainder is	18
The divisor is	$x - 3$
The dividend is	$2x^2 - x + 3$

The solution to a long division problem is usually written in the form:

$$\text{quotient} + \frac{\text{remainder}}{\text{divisor}}$$

Hence,

$$(2x^2 - x + 3) \div (x - 3) = 2x + 5 + \frac{18}{x - 3}$$

Skill Practice Divide the polynomials using long division.

3. $(3x^2 + 2x - 5) \div (x + 2)$

The division of polynomials can be checked in the same fashion as the division of real numbers. To check Example 2, we have

$$\text{Dividend} = (\text{divisor})(\text{quotient}) + \text{remainder}$$
$$2x^2 - x + 3 \stackrel{?}{=} (x - 3)(2x + 5) + (18)$$
$$\stackrel{?}{=} 2x^2 + 5x - 6x - 15 + (18)$$
$$= 2x^2 - x + 3 ✔$$

Example 3 Using Long Division to Divide Polynomials

Divide the polynomials using long division: $(2w^3 + 8w^2 - 16) \div (2w + 4)$

Solution:

First note that the dividend has a missing power of w and can be written as $2w^3 + 8w^2 + 0w - 16$. The term $0w$ is a place holder for the missing term. It is helpful to use the place holder to keep the powers of w lined up.

$$\begin{array}{r} w^2 \\ 2w + 4\overline{)2w^3 + 8w^2 + 0w - 16} \\ -(2w^3 + 4w^2) \end{array}$$

Divide $2w^3 \div 2w = w^2$. This is the first term of the quotient.

Then multiply $w^2(2w + 4) = 2w^3 + 4w^2$.

Skill Practice Answers

3. $3x - 4 + \dfrac{3}{x + 2}$

$$\begin{array}{r}
w^2 \\
2w + 4\overline{)2w^3 + 8w^2 + 0w - 16} \\
-2w^3 - 4w^2 \\
\hline
4w^2 + 0w
\end{array}$$

Subtract by adding the opposite.

Bring down the next column, and repeat the process.

$$\begin{array}{r}
w^2 + 2w \\
2w + 4\overline{)2w^3 + 8w^2 + 0w - 16} \\
-2w^3 - 4w^2 \\
\hline
4w^2 + 0w \\
-(4w^2 + 8w)
\end{array}$$

Divide $4w^2$ by the leading term in the divisor. $4w^2 \div 2w = 2w$. Place $2w$ in the quotient.

Multiply $2w(2w + 4) = 4w^2 + 8w$.

$$\begin{array}{r}
w^2 + 2w \\
2w + 4\overline{)2w^3 + 8w^2 + 0w - 16} \\
-2w^3 - 4w^2 \\
\hline
4w^2 + 0w \\
-4w^2 - 8w \\
\hline
-8w - 16
\end{array}$$

Subtract by adding the opposite.

Bring down the next column and repeat.

$$\begin{array}{r}
w^2 + 2w - 4 \\
2w + 4\overline{)2w^3 + 8w^2 + 0w - 16} \\
-2w^3 - 4w^2 \\
\hline
4w^2 + 0w \\
-4w^2 - 8w \\
\hline
-8w - 16 \\
-(-8w - 16)
\end{array}$$

Divide $-8w$ by the leading term in the divisor. $-8w \div 2w = -4$. Place -4 in the quotient.

Multiply $-4(2w + 4) = -8w - 16$.

$$\begin{array}{r}
w^2 + 2w - 4 \\
2w + 4\overline{)2w^3 + 8w^2 + 0w - 16} \\
-2w^3 - 4w^2 \\
\hline
4w^2 + 0w \\
-4w^2 - 8w \\
\hline
-8w - 16 \\
8w + 16 \\
\hline
0
\end{array}$$

Subtract by adding the opposite.

The remainder is 0.

The quotient is $w^2 + 2w - 4$, and the remainder is 0.

Skill Practice Divide the polynomials using long division.

4. $\dfrac{4x^3 - 11x - 5}{2x + 1}$

In Example 3, the remainder is zero. Therefore, we say that $2w + 4$ divides *evenly* into $2w^3 + 8w^2 - 16$. For this reason, the divisor and quotient are factors of $2w^3 + 8w^2 - 16$. To check, we have

$$\begin{aligned}
\text{Dividend} &= (\text{divisor})(\text{quotient}) + \text{remainder} \\
2w^3 + 8w^2 - 16 &\stackrel{?}{=} (2w + 4)(w^2 + 2w - 4) + 0 \\
&\stackrel{?}{=} 2w^3 + 4w^2 - 8w + 4w^2 + 8w - 16 \\
&= 2w^3 + 8w^2 - 16 \checkmark
\end{aligned}$$

Skill Practice Answers

4. $2x^2 - x - 5$

Example 4 Using Long Division to Divide Polynomials

Divide the polynomials using long division.

$$\frac{2y + y^4 - 5}{1 + y^2}$$

Solution:

First note that both the dividend and divisor should be written in descending order:

$$\frac{y^4 + 2y - 5}{y^2 + 1}$$

Also note that the dividend and the divisor have missing powers of y. Leave place holders.

$$y^2 + 0y + 1\overline{)y^4 + 0y^3 + 0y^2 + 2y - 5}$$

$$\begin{array}{r} y^2 \\ y^2 + 0y + 1\overline{)y^4 + 0y^3 + 0y^2 + 2y - 5} \\ -(\underline{y^4 + 0y^3 + y^2}) \end{array}$$

Divide $y^4 \div y^2 = y^2$. This is the first term of the quotient.

Multiply $y^2(y^2 + 0y + 1) = y^4 + 0y^3 + y^2$

$$\begin{array}{r} y^2 \\ y^2 + 0y + 1\overline{)y^4 + 0y^3 + 0y^2 + 2y - 5} \\ \underline{-y^4 - 0y^3 - y^2} \\ -y^2 + 2y - 5 \end{array}$$

Subtract by adding the opposite.

Bring down the next columns.

$$\begin{array}{r} y^2 - 1 \\ y^2 + 0y + 1\overline{)y^4 + 0y^3 + 0y^2 + 2y - 5} \\ \underline{-y^4 - 0y^3 - y^2} \\ -y^2 + 2y - 5 \\ -(\underline{-y^2 - 0y - 1}) \end{array}$$

Divide $-y^2 \div y^2 = -1$.

Multiply $-1(y^2 + 0y + 1) = -y^2 - 0y - 1$.

$$\begin{array}{r} y^2 - 1 \\ y^2 + 0y + 1\overline{)y^4 + 0y^3 + 0y^2 + 2y - 5} \\ \underline{-y^4 - 0y^3 - y^2} \\ -y^2 + 2y - 5 \\ \underline{y^2 + 0y + 1} \\ 2y - 4 \end{array}$$

Subtract by adding the opposite.

Remainder

Therefore, $\dfrac{y^4 + 2y - 5}{y^2 + 1} = y^2 - 1 + \dfrac{2y - 4}{y^2 + 1}$

Skill Practice Divide the polynomials using long division.

5. $(4 - x^2 + x^3) \div (2 + x^2)$

Skill Practice Answers

5. $x - 1 + \dfrac{-2x + 6}{x^2 + 2}$

Example 5 **Determining Whether Long Division Is Necessary**

Determine whether long division is necessary for each division of polynomials.

a. $\dfrac{2p^5 - 8p^4 + 4p - 16}{p^2 - 2p + 1}$ b. $\dfrac{2p^5 - 8p^4 + 4p - 16}{2p^2}$

c. $(3z^3 - 5z^2 + 10) \div (15z^3)$ d. $(3z^3 - 5z^2 + 10) \div (3z + 1)$

Solution:

a. $\dfrac{2p^5 - 8p^4 + 4p - 16}{p^2 - 2p + 1}$ The divisor has three terms. Use long division.

b. $\dfrac{2p^5 - 8p^4 + 4p - 16}{2p^2}$ The divisor has one term. No long division.

c. $(3z^3 - 5z^2 + 10) \div (15z^3)$ The divisor has one term. No long division.

d. $(3z^3 - 5z^2 + 10) \div (3z + 1)$ The divisor has two terms. Use long division.

Skill Practice Determine whether long division is necessary for each division of polynomials.

6. $\dfrac{6x^3 - x^2 + 3x - 5}{2x + 3}$ 7. $\dfrac{4x^4 - 3x^2}{2x^2}$

8. $(4y^3 - 3y^2 + 1) \div (7y)$ 9. $(p^2 - p - 12) \div (p - 4)$

TIP: Recall that

- Long division is used when the divisor has *two or more terms*.
- If the divisor has *one term*, then divide each term in the dividend by the monomial divisor.

Skill Practice Answers

6. Long division
7. No long division
8. No long division
9. Long division

Section 5.7 Practice Exercises

Boost *your* GRADE at mathzone.com!

- Practice Problems
- Self-Tests
- NetTutor
- e-Professors
- Videos

Review Exercises

For Exercises 1–10, perform the indicated operations.

1. $(6z^5 - 2z^3 + z - 6) - (10z^4 + 2z^3 + z^2 + z)$

2. $(7a^2 + a - 6) + (2a^2 + 5a + 11)$

3. $(10x + y)(x - 3y)$

4. $8b^2(2b^2 - 5b + 12)$

5. $(10x + y) + (x - 3y)$

6. $(2w^3 + 5)^2$

7. $\left(\frac{4}{3}y^2 - \frac{1}{2}y + \frac{3}{8}\right) - \left(\frac{1}{3}y^2 + \frac{1}{4}y - \frac{1}{8}\right)$

8. $\left(\frac{7}{8}w - 1\right)\left(\frac{7}{8}w + 1\right)$

9. $(a + 3)(a^2 - 3a + 9)$

10. $(2x + 1)(5x - 3)$

Concept 1: Division by a Monomial

11. There are two methods for dividing polynomials. Explain when long division is used.

12. Explain how to check a polynomial division problem.

13. a. Divide $\dfrac{15t^3 + 18t^2}{3t}$

b. Check by multiplying the quotient by the divisor.

14. a. Divide $(-9y^4 + 6y^2 - y) \div (3y)$

b. Check by multiplying the quotient by the divisor.

For Exercises 15–30, divide the polynomials.

15. $(6a^2 + 4a - 14) \div (2)$

16. $\dfrac{4b^2 + 16b - 12}{4}$

17. $\dfrac{-5x^2 - 20x + 5}{-5}$

18. $\dfrac{-3y^3 + 12y - 6}{-3}$

19. $\dfrac{3p^3 - p^2}{p}$

20. $(7q^4 + 5q^2) \div q$

21. $(4m^2 + 8m) \div 4m^2$

22. $\dfrac{n^2 - 8}{n}$

23. $\dfrac{14y^4 - 7y^3 + 21y^2}{-7y^2}$

24. $(25a^5 - 5a^4 + 15a^3 - 5a) \div (-5a)$

25. $(4x^3 - 24x^2 - x + 8) \div (4x)$

26. $\dfrac{20w^3 + 15w^2 - w + 5}{10w}$

27. $\dfrac{-a^3b^2 + a^2b^2 - ab^3}{-a^2b^2}$

28. $(3x^4y^3 - x^2y^2 - xy^3) \div (-x^2y^2)$

29. $(6t^4 - 2t^3 + 3t^2 - t + 4) \div (2t^3)$

30. $\dfrac{2y^3 - 2y^2 + 3y - 9}{2y^2}$

Concept 2: Long Division

31. a. Divide $(z^2 + 7z + 11) \div (z + 5)$

b. Check by multiplying the quotient by the divisor and adding the remainder.

32. a. Divide $\dfrac{2w^2 - 7w + 3}{w - 4}$

b. Check by multiplying the quotient by the divisor and adding the remainder.

For Exercises 33–54, divide the polynomials.

33. $\dfrac{t^2 + 4t + 3}{t + 1}$

34. $(3x^2 + 8x + 4) \div (x + 2)$

35. $(7b^2 - 3b - 4) \div (b - 1)$

36. $\dfrac{w^2 - w - 2}{w - 2}$

37. $\dfrac{5k^2 - 29k - 6}{5k + 1}$

38. $(4y^2 + 25y - 21) \div (4y - 3)$

39. $(4p^3 + 12p^2 + p - 12) \div (2p + 3)$

40. $\dfrac{12a^3 - 2a^2 - 17a - 5}{3a + 1}$

41. $\dfrac{-k - 6 + k^2}{1 + k}$

42. $(1 + h^2 + 3h) \div (2 + h)$

43. $(4x^3 - 8x^2 + 15x - 16) \div (2x - 3)$

44. $\dfrac{3b^3 + b^2 + 17b - 49}{3b - 5}$

45. $\dfrac{9 + a^2}{a + 3}$

46. $(3 + m^2) \div (m + 3)$

47. $(4x^3 - 3x - 26) \div (x - 2)$

48. $(4y^3 + y + 1) \div (2y + 1)$

49. $(w^4 + 5w^3 - 5w^2 - 15w + 7) \div (w^2 - 3)$

50. $\dfrac{p^4 - p^3 - 4p^2 - 2p - 15}{p^2 + 2}$

51. $\dfrac{2n^4 + 5n^3 - 11n^2 - 20n + 12}{2n^2 + 3n - 2}$

52. $(6y^4 - 5y^3 - 8y^2 + 16y - 8) \div (2y^2 - 3y + 2)$

53. $(5x^3 - 4x - 9) \div (5x^2 + 5x + 1)$

54. $\dfrac{3a^3 - 5a + 16}{3a^2 - 6a + 7}$

55. Show that $(x^3 - 8) \div (x - 2)$ is *not* $(x^2 + 4)$.

56. Explain why $(y^3 + 27) \div (y + 3)$ is *not* $(y^2 + 9)$.

For Exercises 57–68, determine which method to use to divide the polynomials: monomial division or long division. Then use that method to divide the polynomials.

57. $\dfrac{9a^3 + 12a^2}{3a}$

58. $\dfrac{3y^2 + 17y - 12}{y + 6}$

59. $(p^3 + p^2 - 4p - 4) \div (p^2 - p - 2)$

60. $(q^3 + 1) \div (q + 1)$

61. $\dfrac{t^4 + t^2 - 16}{t + 2}$

62. $\dfrac{-8m^5 - 4m^3 + 4m^2}{-2m^2}$

63. $(w^4 + w^2 - 5) \div (w^2 - 2)$

64. $(2k^2 + 9k + 7) \div (k + 1)$

65. $\dfrac{n^3 - 64}{n - 4}$

66. $\dfrac{15s^2 + 34s + 28}{5s + 3}$

67. $(9r^3 - 12r^2 + 9) \div (-3r^2)$

68. $(6x^4 - 16x^3 + 15x^2 - 5x + 10) \div (3x + 1)$

Expanding Your Skills

For Exercises 69–76, divide the polynomials and note any patterns.

69. $(x^2 - 1) \div (x - 1)$

70. $(x^3 - 1) \div (x - 1)$

71. $(x^4 - 1) \div (x - 1)$

72. $(x^5 - 1) \div (x - 1)$

73. $x^2 \div (x - 1)$

74. $x^3 \div (x - 1)$

75. $x^4 \div (x - 1)$

76. $x^5 \div (x - 1)$

Chapter 5 Problem Recognition Exercises—Operations on Polynomials

Perform the indicated operations and simplify.

1. $(2x - 4)(x^2 - 2x + 3)$

2. $(3y^2 + 8)(-y^2 - 4)$

3. $(2x - 4) + (x^2 - 2x + 3)$

4. $(3y^2 + 8) - (-y^2 - 4)$

5. $(6y - 7)^2$

6. $(3z + 2)^2$

7. $(6y - 7)(6y + 7)$

8. $(3z + 2)(3z - 2)$

9. $(-2x^4 - 6x^3 + 8x^2) \div (2x^2)$

10. $(-15m^3 + 12m^2 - 3m) \div (-3m)$

11. $(4x + y)^2$

12. $(2a + b)^2$

13. $(4xy)^2$

14. $(2ab)^2$

15. $(m^3 - 4m^2 - 6) - (3m^2 + 7m) + (-m^3 - 9m + 6)$

16. $(n^4 + 2n^2 - 3n) + (4n^2 + 2n - 1) - (4n^5 + 6n - 3)$

17. $(8x^3 + 2x + 6) \div (x - 2)$

18. $(-4x^3 + 2x^2 - 5) \div (x - 3)$

19. $(2x - y)(3x^2 + 4xy - y^2)$

20. $(3a + b)(2a^2 - ab + 2b^2)$

21. $(x + y^2)(x^2 - xy^2 + y^4)$

22. $(m^2 + 1)(m^4 - m^2 + 1)$

23. $(a^2 + 2b) - (a^2 - 2b)$

24. $(y^3 - 6z) - (y^3 + 6z)$

25. $(a^2 + 2b)(a^2 - 2b)$

26. $(y^3 - 6z)(y^3 + 6z)$

27. $(8u + 3v)^2$

28. $(2p - t)^2$

29. $\dfrac{8p^2 + 4p - 6}{2p - 1}$

30. $\dfrac{4v^2 - 8v + 8}{2v + 3}$

31. $\dfrac{12x^3y^7}{3xy^5}$

32. $\dfrac{-18p^2q^4}{2pq^3}$

33. $(2a - 9)(5a - 6)$

34. $(7a + 1)(4a - 3)$

35. $\left(\frac{3}{7}x - \frac{1}{2}\right)\left(\frac{3}{7}x + \frac{1}{2}\right)$

36. $\left(\frac{2}{5}y + \frac{4}{3}\right)\left(\frac{2}{5}y - \frac{4}{3}\right)$

37. $\left(\frac{1}{9}x^3 + \frac{2}{3}x^2 + \frac{1}{6}x - 3\right) - \left(\frac{4}{3}x^3 + \frac{1}{9}x^2 + \frac{2}{3}x + 1\right)$

38. $\left(\frac{1}{10}y^2 - \frac{3}{5}y - \frac{1}{15}\right) - \left(\frac{7}{5}y^2 + \frac{3}{10}y - \frac{1}{3}\right)$

39. $(0.05x^2 - 0.16x - 0.75) + (1.25x^2 - 0.14x + 0.25)$

40. $(1.6w^3 + 2.8w + 6.1) + (3.4w^3 - 4.1w^2 - 7.3)$

Chapter 5 SUMMARY

Section 5.1 Exponents: Multiplying and Dividing Common Bases

Key Concepts

Definition

$b^n = \underbrace{b \cdot b \cdot b \cdot b \cdots b}_{n \text{ factors of } b}$ b is the base, n is the exponent

Multiplying Common Bases

$a^m a^n = a^{m+n}$

Dividing Common Bases

$\dfrac{a^m}{a^n} = a^{m-n} \quad (a \neq 0)$

Examples

Example 1

$3^4 = 3 \cdot 3 \cdot 3 \cdot 3 = 81$ 3 is the base, 4 is the exponent

Example 2

Compare: $(-5)^2$ versus -5^2

$(-5)^2 = (-5)(-5) = 25$

versus

$-5^2 = -1(5^2) = -1(5)(5) = -25$

Example 3

Simplify: $x^3 \cdot x^4 \cdot x^2 \cdot x = x^{10}$

Example 4

Simplify: $\dfrac{c^4 d^{10}}{cd^5} = c^{4-1}d^{10-5} = c^3 d^5$

Section 5.2 More Properties of Exponents

Key Concepts

Power Rule for Exponents

$(a^m)^n = a^{mn} \quad (a \neq 0,\ m,\ n \text{ positive integers})$

Power of a Product and Power of a Quotient

Assume m and n are positive integers and a and b are real numbers where $b \neq 0$.

$(ab)^m = a^m b^m$ and $\left(\dfrac{a}{b}\right)^m = \dfrac{a^m}{b^m}$

Examples

Example 1

Simplify: $(x^4)^5 = x^{20}$

Example 2

Simplify: $(4uv^2)^3 = 4^3u^3(v^2)^3 = 64u^3v^6$

Example 3

Simplify: $\left(\dfrac{p^5q^3}{5pq^2}\right)^2 = \left(\dfrac{p^{5-1}q^{3-2}}{5}\right)^2 = \left(\dfrac{p^4q}{5}\right)^2$

$= \dfrac{p^8q^2}{25}$

Section 5.3 Definitions of b^0 and b^{-n}

Key Concepts

Definitions

If b is a real number such that $b \neq 0$ and n is an integer, then:

1. $b^0 = 1$
2. $b^{-n} = \left(\frac{1}{b}\right)^n = \frac{1}{b^n}$

Examples

Example 1

Simplify: $4^0 = 1$

Example 2

Simplify: $y^{-7} = \frac{1}{y^7}$

Example 3

Simplify: $\left(\frac{2a^3b}{a^{-2}c^{-4}}\right)^{-2}$

$$= \left(\frac{2a^{3-(-2)}b}{c^{-4}}\right)^{-2}$$

$$= \left(\frac{2a^5b}{c^{-4}}\right)^{-2} = \frac{2^{-2}a^{-10}b^{-2}}{c^8}$$

$$= \frac{1}{2^2a^{10}b^2c^8}$$

$$= \frac{1}{4a^{10}b^2c^8}$$

Section 5.4 Scientific Notation

Key Concepts

A number written in **scientific notation** is expressed in the form:

$a \times 10^n$ where $1 \leq |a| < 10$ and n is an integer. The value 10^n is sometimes called the **order of magnitude** or simply the magnitude of the number.

Examples

Example 1

Write the numbers in scientific notation:

$35{,}000 = 3.5 \times 10^4$

$0.000\,000\,548 = 5.48 \times 10^{-7}$

Example 2

Multiply: $(3.5 \times 10^4)(2.0 \times 10^{-6})$

$= 7.0 \times 10^{-2}$

Example 3

Divide: $\frac{8.4 \times 10^{-9}}{2.1 \times 10^3} = 4.0 \times 10^{-9-3} = 4.0 \times 10^{-12}$

Section 5.5 Addition and Subtraction of Polynomials

Key Concepts

A **polynomial** in one variable is a finite sum of terms of the form ax^n, where a is a real number and the exponent, n, is a nonnegative integer. For each term, a is called the **coefficient** of the term and n is the **degree of the term**. The term with highest degree is the **leading term**, and its coefficient is called the **leading coefficient**. The **degree of the polynomial** is the largest degree of all its terms.

To add or subtract polynomials, add or subtract *like* terms.

Examples

Example 1

Given: $4x^5 - 8x^3 + 9x - 5$

Coefficients of each term: $4, -8, 9, -5$

Degree of each term: $5, 3, 1, 0$

Leading term: $4x^5$

Leading coefficient: 4

Degree of polynomial: 5

Example 2

Perform the indicated operations:

$(2x^4 - 5x^3 + 1) - (x^4 + 3) + (x^3 - 4x - 7)$

$= 2x^4 - 5x^3 + 1 - x^4 - 3 + x^3 - 4x - 7$

$= 2x^4 - x^4 - 5x^3 + x^3 - 4x + 1 - 3 - 7$

$= x^4 - 4x^3 - 4x - 9$

Section 5.6 Multiplication of Polynomials

Key Concepts

Multiplying Monomials

Use the commutative and associative properties of multiplication to group coefficients and like bases.

Multiplying Polynomials

Multiply each term in the first polynomial by each term in the second polynomial.

Product of Conjugates

Results in a **difference of squares**

$(a + b)(a - b) = a^2 - b^2$

Square of a Binomial

Results in a **perfect square trinomial**

$(a + b)^2 = a^2 + 2ab + b^2$

$(a - b)^2 = a^2 - 2ab + b^2$

Examples

Example 1

Multiply: $(5a^2b)(-2ab^3)$

$= (5 \cdot -2)(a^2a)(bb^3)$

$= -10a^3b^4$

Example 2

Multiply: $(x - 2)(3x^2 - 4x + 11)$

$= 3x^3 - 4x^2 + 11x - 6x^2 + 8x - 22$

$= 3x^3 - 10x^2 + 19x - 22$

Example 3

Multiply: $(3w - 4v)(3w + 4v)$

$= (3w)^2 - (4v)^2$

$= 9w^2 - 16v^2$

Example 4

Multiply: $(5c - 8d)^2$

$= (5c)^2 - 2(5c)(8d) + (8d)^2$

$= 25c^2 - 80cd + 64d^2$

Section 5.7 Division of Polynomials

Key Concepts

Division of Polynomials

1. Division by a monomial, use the properties:

$$\frac{a+b}{c} = \frac{a}{c} + \frac{b}{c} \quad \text{and} \quad \frac{a-b}{c} = \frac{a}{c} - \frac{b}{c}$$

2. If the divisor has more than one term, use long division.

Examples

Example 1

Divide: $\dfrac{-3x^2 - 6x + 9}{-3x}$

$$= \frac{-3x^2}{-3x} - \frac{6x}{-3x} + \frac{9}{-3x}$$

$$= x + 2 - \frac{3}{x}$$

Example 2

Divide: $(3x^2 - 5x + 1) \div (x + 2)$

$$\begin{array}{r} 3x - 11 \\ x + 2 \overline{)3x^2 - 5x + 1} \\ -(3x^2 + 6x) \\ \hline -11x + 1 \\ -(-11x - 22) \\ \hline 23 \end{array}$$

$$3x - 11 + \frac{23}{x+2}$$

Chapter 5 Review Exercises

Section 5.1

For Exercises 1–4, identify the base and the exponent.

1. 5^3 **2.** x^4 **3.** $(-2)^0$ **4.** y

5. Evaluate the expressions.

a. 6^2 **b.** $(-6)^2$ **c.** -6^2

6. Evaluate the expressions.

a. 4^3 **b.** $(-4)^3$ **c.** -4^3

For Exercises 7–18, simplify and write the answers in exponent form. Assume that all variables represent nonzero real numbers.

7. $5^3 \cdot 5^{10}$ **8.** a^7a^4

9. $x \cdot x^6 \cdot x^2$ **10.** $6^3 \cdot 6 \cdot 6^5$

11. $\dfrac{10^7}{10^4}$ **12.** $\dfrac{y^{14}}{y^8}$

13. $\dfrac{b^9}{b}$ **14.** $\dfrac{7^8}{7}$

15. $\dfrac{k^2k^3}{k^4}$ **16.** $\dfrac{8^4 \cdot 8^7}{8^{11}}$

17. $\dfrac{2^8 \cdot 2^{10}}{2^3 \cdot 2^7}$ **18.** $\dfrac{q^3q^{12}}{qq^8}$

19. Explain why $2^2 \cdot 4^4$ does *not* equal 8^6.

20. Explain why $\frac{10^5}{5^2}$ does *not* equal 2^3.

For Exercises 21–22, use the formula

$$A = P(1 + r)^t$$

21. Find the amount in an account after 3 years if the initial investment is \$6000, invested at 6% interest compounded annually.

22. Find the amount in an account after 2 years if the initial investment is \$20,000, invested at 5% interest compounded annually.

Section 5.2

For Exercises 23–40, simplify the expressions. Write the answers in exponent form. Assume all variables represent nonzero real numbers.

23. $(7^3)^4$

24. $(c^2)^6$

25. $(p^4p^2)^3$

26. $(9^5 \cdot 9^2)^4$

27. $\left(\frac{a}{b}\right)^2$

28. $\left(\frac{1}{3}\right)^4$

29. $\left(\frac{5}{c^2d^5}\right)^2$

30. $\left(-\frac{m^2}{4n^6}\right)^5$

31. $(2ab^2)^4$

32. $(-x^7y)^2$

33. $\left(\frac{-3x^3}{5y^2z}\right)^3$

34. $\left(\frac{r^3}{s^2t^6}\right)^5$

35. $\frac{a^4(a^2)^8}{(a^3)^3}$

36. $\frac{(8^3)^4 \cdot 8^{10}}{(8^4)^5}$

37. $\frac{(4h^2k)^2(h^3k)^4}{(2hk^3)^2}$

38. $\frac{(p^3q)^3(2p^2q^4)^4}{(8p)(pq^3)^2}$

39. $\left(\frac{2x^4y^3}{4xy^2}\right)^2$

40. $\left(\frac{a^4b^6}{ab^4}\right)^3$

Section 5.3

For Exercises 41–62, simplify the expressions. Assume all variables represent nonzero real numbers.

41. 8^0

42. $(-b)^0$

43. 1^0

44. $-x^0$

45. $2y^0$

46. $(2y)^0$

47. z^{-5}

48. 10^{-4}

49. $(6a)^{-2}$

50. $6a^{-2}$

51. $4^0 + 4^{-2}$

52. $9^{-1} + 9^0$

53. $t^{-6}t^{-2}$

54. r^8r^{-9}

55. $\frac{12x^{-2}y^3}{6x^4y^{-4}}$

56. $\frac{8ab^{-3}c^0}{10a^{-5}b^{-4}c^{-1}}$

57. $(-2m^2n^{-4})^{-4}$

58. $(3u^{-5}v^2)^{-3}$

59. $\frac{(k^{-6})^{-2}(k^3)}{5k^{-6}k^0}$

60. $\frac{(3h)^{-2}(h^{-5})^{-3}}{h^{-4}h^8}$

61. $2 \cdot 3^{-1} - 6^{-1}$

62. $2^{-1} - 2^{-2} + 2^0$

Section 5.4

63. Write the numbers in scientific notation.

a. In a recent year there were 97,000,000 packages of M&Ms sold in the United States.

b. The thickness of a piece of paper is 0.0042 in.

c. The area of the Pacific Ocean is 166,241,000 km^2.

64. Write the numbers without scientific notation.

a. A pH of 10 means the hydrogen ion concentration is 1×10^{-10} units.

b. When it was released, *The Lord of the Rings: The Fellowship of the Ring* sold 2.573×10^8 DVDs.

c. A fund-raising event for neurospinal research raised \$$2.56 \times 10^5$.

For Exercises 65–68, perform the indicated operations. Write the answers in scientific notation.

65. $(4.1 \times 10^{-6})(2.3 \times 10^{11})$

66. $\frac{9.3 \times 10^3}{6.0 \times 10^{-7}}$

67. $\frac{2000}{0.000008}$

68. $(0.000078)(21,000,000)$

69. Use your calculator to evaluate 5^{20}. Why is scientific notation necessary on your calculator to express the answer?

70. Use your calculator to evaluate $(0.4)^{30}$. Why is scientific notation necessary on your calculator to express the answer?

71. The average distance between the Earth and Sun is 9.3×10^7 miles.

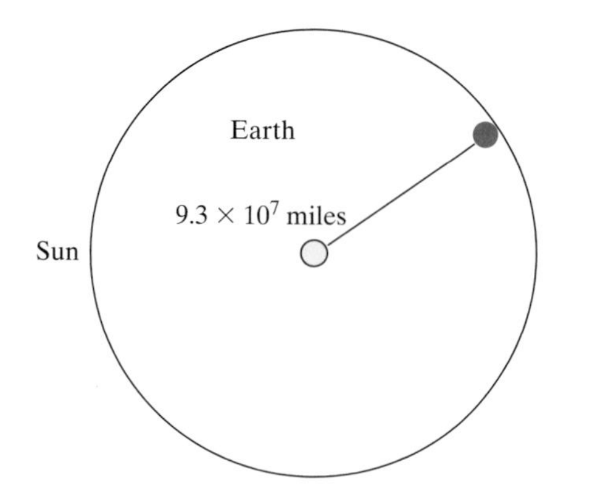

a. If the Earth's orbit is approximated by a circle, find the total distance the Earth travels around the Sun in one orbit. (*Hint:* The circumference of a circle is given by $C = 2\pi r$.) Express the answer in scientific notation.

b. If the Earth makes one complete trip around the Sun in 1 year (365 days $= 8.76 \times 10^3$ hr), find the average speed that the Earth travels around the Sun in miles per hour. Express the answer in scientific notation.

72. The average distance between the planet Mercury and the Sun is 3.6×10^7 miles.

a. If Mercury's orbit is approximated by a circle, find the total distance Mercury travels around the Sun in one orbit. (*Hint:* The circumference of a circle is given by $C = 2\pi r$.) Express the answer in scientific notation.

b. If Mercury makes one complete trip around the Sun in 88 days (2.112×10^3 hr), find the average speed that Mercury travels around the Sun in miles per hour. Express the answer in scientific notation.

Section 5.5

73. For the polynomial $7x^4 - x + 6$

a. Classify as a monomial, a binomial, or a trinomial.

b. Identify the degree of the polynomial.

c. Identify the leading coefficient.

74. For the polynomial $2y^3 - 5y^7$

a. Classify as a monomial, a binomial, or a trinomial.

b. Identify the degree of the polynomial.

c. Identify the leading coefficient.

For Exercises 75–80, add or subtract as indicated.

75. $(4x + 2) + (3x - 5)$

76. $(7y^2 - 11y - 6) - (8y^2 + 3y - 4)$

77. $(9a^2 - 6) - (-5a^2 + 2a)$

78. $(8w^4 - 6w + 3) + (2w^4 + 2w^3 - w + 1)$

79. $\left(5x^3 - \frac{1}{4}x^2 + \frac{5}{8}x + 2\right) + \left(\frac{5}{2}x^3 + \frac{1}{2}x^2 - \frac{1}{8}x\right)$

80. $(-0.02b^5 + b^4 - 0.7b + 0.3) + (0.03b^5 - 0.1b^3 + b + 0.03)$

81. Subtract $(9x^2 + 4x + 6)$ from $(7x^2 - 5x)$.

82. Find the difference of $(x^2 - 5x - 3)$ and $(6x^2 + 4x + 9)$.

83. Write a trinomial of degree 2 with a leading coefficient of -5. (Answers may vary.)

84. Write a binomial of degree 6 with leading coefficient 6. (Answers may vary.)

85. Find a polynomial that represents the perimeter of the given rectangle.

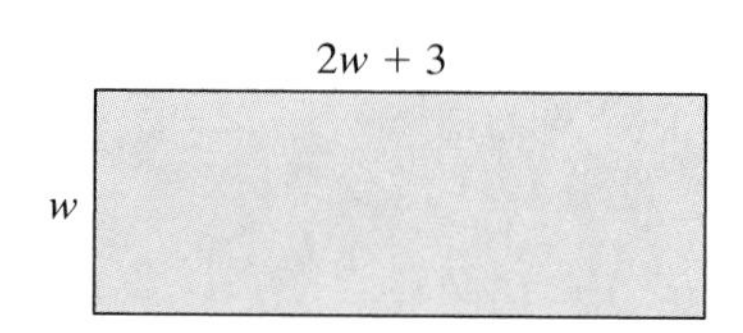

Section 5.6

For Exercises 86–101, multiply the expressions.

86. $(25x^4y^3)(-3x^2y)$

87. $(9a^6)(2a^2b^4)$

88. $5c(3c^3 - 7c + 5)$

89. $(x^2 + 5x - 3)(-2x)$

90. $(5k - 4)(k + 1)$

91. $(4t - 1)(5t + 2)$

92. $(q + 8)(6q - 1)$

93. $(2a - 6)(a + 5)$

94. $\left(7a + \frac{1}{2}\right)^2$

95. $(b - 4)^2$

96. $(4p^2 + 6p + 9)(2p - 3)$

97. $(2w - 1)(-w^2 - 3w - 4)$

98. $(b - 4)(b + 4)$

99. $\left(\frac{1}{3}r^4 - s^2\right)\left(\frac{1}{3}r^4 + s^2\right)$

100. $(-7z^2 + 6)^2$

101. $(2h + 3)(h^4 - h^3 + h^2 - h + 1)$

102. Find a polynomial that represents the area of the given rectangle.

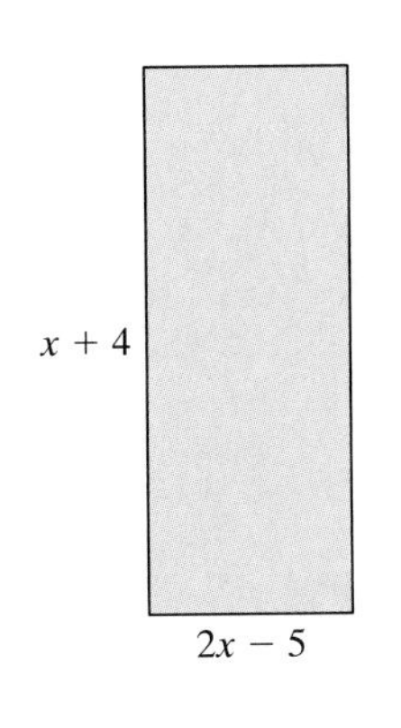

Section 5.7

For Exercises 103–117, divide the polynomials.

103. $\dfrac{20y^3 - 10y^2}{5y}$

104. $(18a^3b^2 - 9a^2b - 27ab^2) \div 9ab$

105. $(12x^4 - 8x^3 + 4x^2) \div (-4x^2)$

106. $\dfrac{10z^7w^4 - 15z^3w^2 - 20zw}{-20z^2w}$

107. $\dfrac{6m^3 + 5m^2 - 6m}{6m}$

108. $\dfrac{18n^4 - 6n^3 + 12n^2}{3n^2}$

109. $\dfrac{x^2 + 7x + 10}{x + 5}$

110. $(2t^2 + t - 10) \div (t - 2)$

111. $(2p^2 + p - 16) \div (2p + 7)$

112. $\dfrac{5a^2 + 27a - 22}{5a - 3}$

113. $\dfrac{b^3 - 125}{b - 5}$

114. $(z^3 + 4z^2 + 5z + 20) \div (5 + z^2)$

115. $(-3y - 4y^3 + 5y^2 + y^4 + 2) \div (y^2 + 3)$

116. $(3t^4 - 8t^3 + t^2 - 4t - 5) \div (3t^2 + t + 1)$

117. $\dfrac{2w^4 + w^3 + 4w - 3}{2w^2 - w + 3}$

Chapter 5 Test

Assume all variables represent nonzero real numbers.

1. Expand the expression using the definition of exponents, then simplify: $\dfrac{3^4 \cdot 3^3}{3^6}$

For Exercises 2–11, simplify the expression. Write the answer with positive exponents only.

2. $9^5 \cdot 9$

3. $\dfrac{q^{10}}{q^2}$

4. $(3a^2b)^3$

5. $\left(\dfrac{2x}{y^3}\right)^4$

6. $(-7)^0$

7. c^{-3}

8. $\dfrac{14^3 \cdot 14^9}{14^{10} \cdot 14}$

9. $\dfrac{(s^2t)^3(7s^4t)^4}{(7s^2t^3)^2}$

10. $(2a^0b^{-6})^2$

11. $\left(\dfrac{6a^{-5}b}{8ab^{-2}}\right)^{-2}$

12. a. Write the number in scientific notation: 43,000,000,000

b. Write the number without scientific notation: 5.6×10^{-6}

13. The average amount of water flowing over Niagara Falls is 1.68×10^5 m^3/min.

 a. How many cubic meters of water flow over the falls in one day?

 b. How many cubic meters of water flow over the falls in one year?

14. Write the polynomial in descending order: $4x + 5x^3 - 7x^2 + 11$.

 a. Identify the degree of the polynomial.

 b. Identify the leading coefficient of the polynomial.

15. Perform the indicated operations.

 $(7w^2 - 11w - 6) + (8w^2 + 3w + 4) - (-9w^2 - 5w + 2)$

16. Subtract $(3x^2 - 5x^3 + 2x)$ from $(10x^3 - 4x^2 + 1)$.

For Exercises 17–23, multiply the polynomials.

17. $-2x^3(5x^2 + x - 15)$

18. $(4a - 3)(2a - 1)$

19. $(4y - 5)(y^2 - 5y + 3)$

20. $(2 + 3b)(2 - 3b)$

21. $(5z - 6)^2$

22. $(10 - 3w)(10 + 3w)$

23. $(y^2 - 5y + 2)(y - 6)$

24. Find the perimeter and the area of the rectangle shown in the figure.

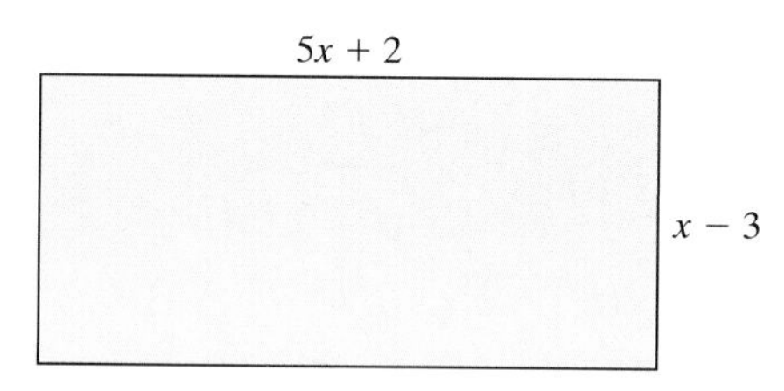

For Exercises 25–27, divide:

25. $(-12x^8 + x^6 - 8x^3) \div (4x^2)$

26. $\dfrac{2y^2 - 13y + 21}{y - 3}$

27. $(2w^3 - 2w - 5w^2 + 5) \div (2w + 3)$

Chapters 1–5 Cumulative Review Exercises

For Exercises 1–2, simplify completely.

1. $-5 - \frac{1}{2}[4 - 3(-7)]$

2. $|-3^2 + 5|$

3. Translate the phrase into a mathematical expression and simplify:

 The difference of the square of five and the square root of four.

4. Solve for x: $\frac{1}{2}(x - 6) + \frac{2}{3} = \frac{1}{4}x$

5. Solve for y: $-2y - 3 = -5(y - 1) + 3y$

6. For a point in a rectangular coordinate system, in which quadrant are both the x- and y-coordinates negative?

7. For a point in a rectangular coordinate system, on which axis is the x-coordinate zero and the y-coordinate nonzero?

8. In a triangle, one angle measures 23° more than the smallest angle. The third angle measures 10° more than the sum of the other two angles. Find the measure of each angle.

9. A snow storm lasts for 9 hr and dumps snow at a rate of $1\frac{1}{2}$ in./hr. If there was already 6 in. of snow on the ground before the storm, the snow depth is given by the equation:

 $y = \frac{3}{2}x + 6$ where y is the snow depth in inches and $x \geq 0$ is the time in hours.

Time (h) x	Snow Depth (in.) y
0	
2	
4	
6	
8	
9	

a. Find the snow depth after 4 hr.

b. Find the snow depth at the end of the storm.

c. How long had it snowed when the total depth of snow was $14\frac{1}{4}$ in.?

d. Complete the table and graph the corresponding ordered pairs.

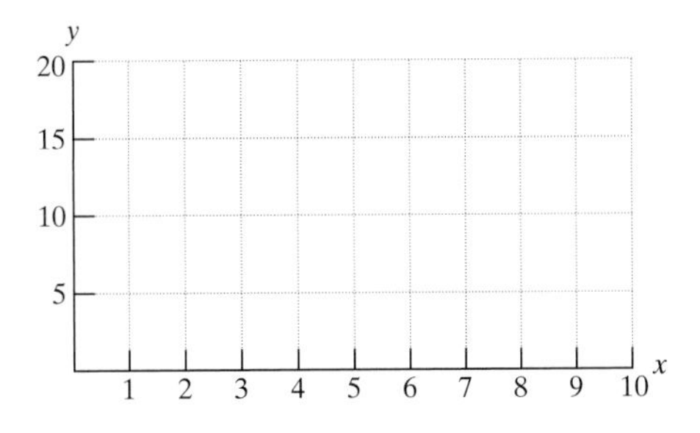

10. Solve the system of equations.

$$5x + 3y = -3$$
$$3x + 2y = -1$$

11. Solve the inequality. Graph the solution set on the real number line and express the solution in interval notation. $2 - 3(2x + 4) \leq -2x - (x - 5)$

For Exercises 12–15, perform the indicated operations.

12. $(2x^2 + 3x - 7) - (-3x^2 + 12x + 8)$

13. $(2y + 3z)(-y - 5z)$

14. $(4t - 3)^2$

15. $\left(\frac{2}{5}a + \frac{1}{3}\right)\left(\frac{2}{5}a - \frac{1}{3}\right)$

For Exercises 16–17, divide the polynomials.

16. $(12a^4b^3 - 6a^2b^2 + 3ab) \div (-3ab)$

17. $\dfrac{4m^3 - 5m + 2}{m - 2}$

For Exercises 18–19, use the properties of exponents to simplify the expressions. Write the answers with positive exponents only. Assume all variables represent nonzero real numbers.

18. $\left(\dfrac{2c^2d^4}{8cd^6}\right)^2$

19. $\dfrac{10a^{-2}b^{-3}}{5a^0b^{-6}}$

20. Perform the indicated operations, and write the final answer in scientific notation.

$$\frac{(8.2 \times 10^{-2})(6.8 \times 10^{-6})}{2.0 \times 10^{-5}}$$

Factoring Polynomials

Chapter 6 is devoted to a mathematical operation called factoring. The applications of factoring are far-reaching, and in this chapter, we use factoring as a tool to solve a type of equation called a quadratic equation.

As you work through the chapter, pay attention to key terms. Then work through this puzzle.

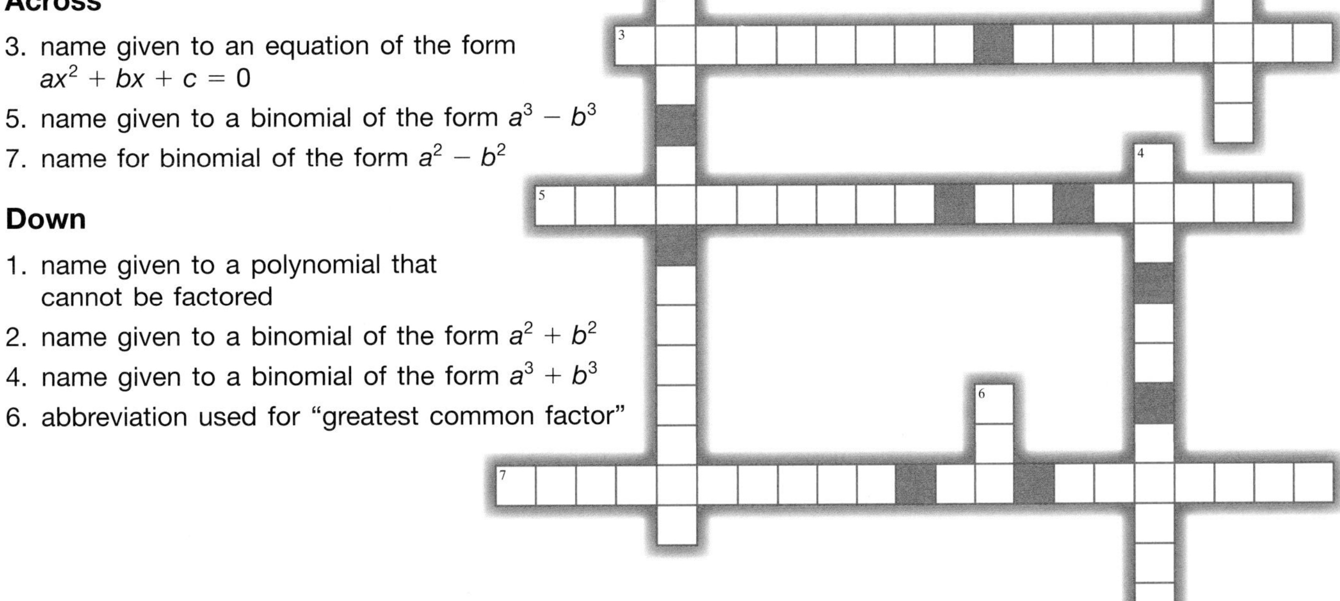

Across

3. name given to an equation of the form $ax^2 + bx + c = 0$
5. name given to a binomial of the form $a^3 - b^3$
7. name for binomial of the form $a^2 - b^2$

Down

1. name given to a polynomial that cannot be factored
2. name given to a binomial of the form $a^2 + b^2$
4. name given to a binomial of the form $a^3 + b^3$
6. abbreviation used for "greatest common factor"

Section 6.1 Greatest Common Factor and Factoring by Grouping

Concepts

1. Identifying the Greatest Common Factor
2. Factoring out the Greatest Common Factor
3. Factoring out a Negative Factor
4. Factoring out a Binomial Factor
5. Factoring by Grouping

1. Identifying the Greatest Common Factor

Chapter 6 is devoted to a mathematical operation called **factoring**. To factor an integer means to write the integer as a product of two or more integers. To factor a polynomial means to express the polynomial as a product of two or more polynomials.

In the product $2 \cdot 5 = 10$, for example, 2 and 5 are factors of 10.

In the product $(3x + 4)(2x - 1) = 6x^2 + 5x - 4$, the quantities $(3x + 4)$ and $(2x - 1)$ are factors of $6x^2 + 5x - 4$.

We begin our study of factoring by factoring integers. The number 20, for example, can be factored as $1 \cdot 20$, $2 \cdot 10$, $4 \cdot 5$, or $2 \cdot 2 \cdot 5$. The product $2 \cdot 2 \cdot 5$ (or equivalently $2^2 \cdot 5$) consists only of prime numbers and is called the **prime factorization**.

The **greatest common factor** (denoted **GCF**) of two or more integers is the greatest factor common to each integer. To find the greatest common factor of two integers, it is often helpful to express the numbers as a product of prime factors as shown in the next example.

Example 1 Identifying the GCF

Find the greatest common factor.

a. 24 and 36 **b.** 105, 40, and 60

Solution:

First find the prime factorization of each number. Then find the product of common factors.

a.

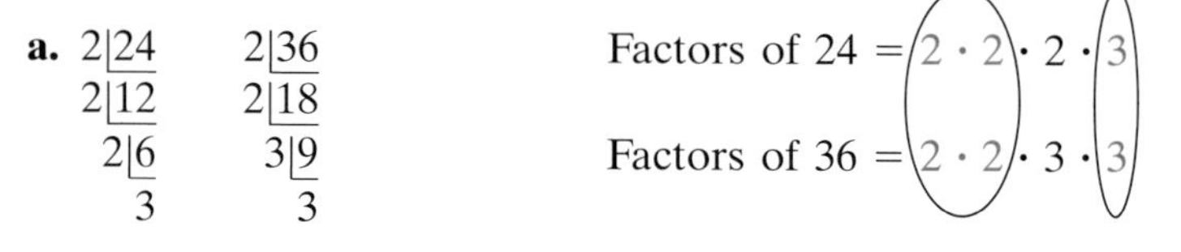

The numbers 24 and 36 share two factors of 2 and one factor of 3. Therefore, the greatest common factor is $2 \cdot 2 \cdot 3 = 12$.

b. 5|105, 3|21, 7   5|40, 2|8, 2|4, 2   5|60, 3|12, 2|4, 2

Factors of $105 = 3 \cdot 7 \cdot 5$

Factors of $40 = 2 \cdot 2 \cdot 2 \cdot 5$

Factors of $60 = 2 \cdot 2 \cdot 3 \cdot 5$

The greatest common factor is 5.

Skill Practice

1. Find the GCF of 12 and 20. **2.** Find the GCF of 45, 75, and 30.

Skill Practice Answers

1. 4 **2.** 15

In Example 2, we find the greatest common factor of two or more variable terms.

Example 2 Identifying the Greatest Common Factor

Find the GCF among each group of terms.

a. $7x^3, 14x^2, 21x^4$ **b.** $15a^4b, 25a^3b^2$ **c.** $8c^2d^7e, 6c^3d^4$

Solution:

List the factors of each term.

a.
$7x^3 = 7 \cdot x \cdot x \cdot x$
$14x^2 = 2 \cdot 7 \cdot x \cdot x$
$21x^4 = 3 \cdot 7 \cdot x \cdot x \cdot x \cdot x$

The GCF is $7x^2$.

b.
$15a^4b = 3 \cdot 5 \cdot a \cdot a \cdot a \cdot a \cdot b$
$25a^3b^2 = 5 \cdot 5 \cdot a \cdot a \cdot a \cdot b \cdot b$

The GCF is $5a^3b$.

TIP: Notice that the expressions $15a^4b$ and $25a^3b^2$ share factors of 5, a, and b. The GCF is the product of the common factors, where each factor is raised to the lowest power to which it occurs in all the original expressions.

$15a^4b = 3 \cdot 5a^4b$
$25a^3b^2 = 5^2a^3b^2$

Lowest power of 5 is 1: 5^1
Lowest power of a is 3: a^3
Lowest power of b is 1: b^1

The GCF is $5a^3b$.

c.
$8c^2d^7e = 2^3c^2d^7e$
$6c^3d^4 = 2 \cdot 3c^3d^4$

The common factors are 2, c, and d.

The lowest power of 2 is 1: 2^1
The lowest power of c is 2: c^2
The lowest power of d is 4: d^4

The GCF is $2c^2d^4$.

Skill Practice Find the GCF.

3. $10z^3, 15z^5, 40z$ **4.** $6w^3y^5, 21w^4y^2$ **5.** $9m^2np^8, 15n^4p^5$

Sometimes polynomials share a common binomial factor as shown in Example 3.

Example 3 Finding the Greatest Common Binomial Factor

Find the greatest common factor between the terms: $3x(a + b)$ and $2y(a + b)$

Solution:

$3x(a + b)$
$2y(a + b)$

The only common factor is the binomial $(a + b)$. The GCF is $(a + b)$.

Skill Practice Answers

3. $5z$ **4.** $3w^3y^2$ **5.** $3np^5$

Skill Practice

6. Find the greatest common binomial factor of the terms:

$a(x + 2)$ and $b(x + 2)$

2. Factoring out the Greatest Common Factor

The process of factoring a polynomial is the reverse process of multiplying polynomials. Both operations use the distributive property: $ab + ac = a(b + c)$.

Multiply

$$5y(y^2 + 3y + 1) = 5y(y^2) + 5y(3y) + 5y(1)$$
$$= 5y^3 + 15y^2 + 5y$$

Factor

$$5y^3 + 15y^2 + 15y = 5y(y^2) + 5y(3y) + 5y(1)$$
$$= 5y(y^2 + 3y + 1)$$

Steps to Factor out the Greatest Common Factor

1. Identify the GCF of all terms of the polynomial.
2. Write each term as the product of the GCF and another factor.
3. Use the distributive property to remove the GCF.

Note: To check the factorization, multiply the polynomials to remove parentheses.

Example 4 Factoring out the Greatest Common Factor

Factor out the GCF.

a. $4x - 20$ **b.** $6w^2 + 3w$

c. $15y^3 + 12y^4$ **d.** $9a^4b - 18a^5b + 27a^6b$

Solution:

a. $4x - 20$ — The GCF is 4.

$= 4(x) - 4(5)$ — Write each term as the product of the GCF and another factor.

$= 4(x - 5)$ — Use the distributive property to factor out the GCF.

TIP: Any factoring problem can be checked by multiplying the factors:

Check: $4(x - 5) = 4x - 20$ ✔

Skill Practice Answers

6. $(x + 2)$

b. $6w^2 + 3w$ — The GCF is $3w$.

$= 3w(2w) + 3w(1)$ — Write each term as the product of $3w$ and another factor.

$= 3w(2w + 1)$ — Use the distributive property to factor out the GCF.

Check: $3w(2w + 1) = 6w^2 + 3w$ ✔

c. $15y^3 + 12y^4$ — The GCF is $3y^3$.

$= 3y^3(5) + 3y^3(4y)$ — Write each term as the product of $3y^3$ and another factor.

$= 3y^3(5 + 4y)$ — Use the distributive property to factor out the GCF.

Check: $3y^3(5 + 4y) = 15y^3 + 12y^4$ ✔

d. $9a^4b - 18a^5b + 27a^6b$ — The GCF is $9a^4b$.

$= 9a^4b(1) - 9a^4b(2a) + 9a^4b(3a^2)$ — Write each term as the product of $9a^4b$ and another factor.

$= 9a^4b(1 - 2a + 3a^2)$ — Use the distributive property to factor out the GCF.

Check: $9a^4b(1 - 2a + 3a^2) = 9a^4b - 18a^5b + 27a^6b$ ✔

Skill Practice Factor out the GCF.

7. $6w + 18$ **8.** $21m^3 - 7m^2$ **9.** $9y^2 - 6y^5$ **10.** $50s^3t - 40st^2 + 10st$

Avoiding Mistakes:

In Example 4(b), the GCF, $3w$, is equal to one of the terms of the polynomial. In such a case, you must leave a 1 in place of that term after the GCF is factored out.

The greatest common factor of the polynomial $2x + 5y$ is 1. If we factor out the GCF, we have $1(2x + 5y)$. A polynomial whose only factors are itself and 1 is called a **prime polynomial**. A prime polynomial cannot be factored further.

3. Factoring out a Negative Factor

Sometimes it is advantageous to factor out the *opposite* of the GCF when the leading coefficient of the polynomial is negative. This is demonstrated in the next example. Notice that this *changes the signs* of the remaining terms inside the parentheses.

Example 5 Factoring out a Negative Factor

a. Factor out -3 from the polynomial $-3x^2 + 6x - 33$.

b. Factor out the quantity $-4pq$ from the polynomial $-12p^3q - 8p^2q^2 + 4pq^3$.

Skill Practice Answers

7. $6(w + 3)$
8. $7m^2(3m - 1)$
9. $3y^2(3 - 2y^3)$
10. $10st(5s^2 - 4t + 1)$

Solution:

a. $-3x^2 + 6x - 33$ — The GCF is 3. However, in this case, we will factor out the *opposite* of the GCF, -3.

$= -3(x^2) + (-3)(-2x) + (-3)(11)$ — Write each term as the product of -3 and another factor.

$= -3[x^2 + (-2x) + 11]$ — Factor out -3.

$= -3(x^2 - 2x + 11)$ — Simplify. Notice that each sign within the trinomial has changed.

<u>Check</u>: $-3(x^2 - 2x + 11) = -3x^2 + 6x - 33$ ✔ — Check by multiplying.

b. $-12p^3q - 8p^2q^2 + 4pq^3$ — The GCF is $4pq$. However, in this case, we will factor out the *opposite* of the GCF, $-4pq$.

$= -4pq(3p^2) + (-4pq)(2pq) + (-4pq)(-q^2)$ — Write each term as the product of $-4pq$ and another factor.

$= -4pq[3p^2 + 2pq + (-q^2)]$ — Factor out $-4pq$. Notice that each sign within the trinomial has changed.

$= -4pq(3p^2 + 2pq - q^2)$ — To verify that this is the correct factorization and that the signs are correct, multiply the factors.

<u>Check</u>: $-4pq(3p^2 + 2pq - q^2) = -12p^3q - 8p^2q^2 + 4pq^3$ ✔

Skill Practice

11. Factor out the opposite of the GCF: $-2x^2 - 10x + 16$

12. Factor out $-5xy$ from the polynomial: $-10x^2y + 5xy - 15xy^2$

4. Factoring out a Binomial Factor

The distributive property can also be used to factor out a common factor that consists of more than one term as shown in Example 6.

Example 6 **Factoring out a Binomial Factor**

Factor out the GCF: $2w(x + 3) - 5(x + 3)$

Skill Practice Answers

11. $-2(x^2 + 5x - 8)$
12. $-5xy(2x - 1 + 3y)$

Solution:

$2w(x + 3) - 5(x + 3)$	The greatest common factor is the quantity $(x + 3)$.
$= (x + 3)(2w) - (x + 3)(5)$	Write each term as the product of $(x + 3)$ and another factor.
$= (x + 3)(2w - 5)$	Use the distributive property to factor out the GCF.

Skill Practice Factor out the GCF.

13. $8y(a + b) + 9(a + b)$

5. Factoring by Grouping

When two binomials are multiplied, the product before simplifying contains four terms. For example:

$$(x + 4)(3a + 2b) = (x + 4)(3a) + (x + 4)(2b)$$

$$= (x + 4)(3a) + (x + 4)(2b)$$

$$= 3ax + 12a + 2bx + 8b$$

In Example 7, we learn how to reverse this process. That is, given a four-term polynomial, we will factor it as a product of two binomials. The process is called *factoring by grouping.*

Example 7 **Factoring by Grouping**

Factor by grouping: $3ax + 12a + 2bx + 8b$

Solution:

$3ax + 12a + 2bx + 8b$	**Step 1:** Identify and factor out the GCF from all four terms. In this case, the GCF is 1.
$= 3ax + 12a \mid + 2bx + 8b$	Group the first pair of terms and the second pair of terms.
$= 3a(x + 4) + 2b(x + 4)$	**Step 2:** Factor out the GCF from each pair of terms. *Note:* The two terms now share a common binomial factor of $(x + 4)$.
$= (x + 4)(3a + 2b)$	**Step 3:** Factor out the common binomial factor.

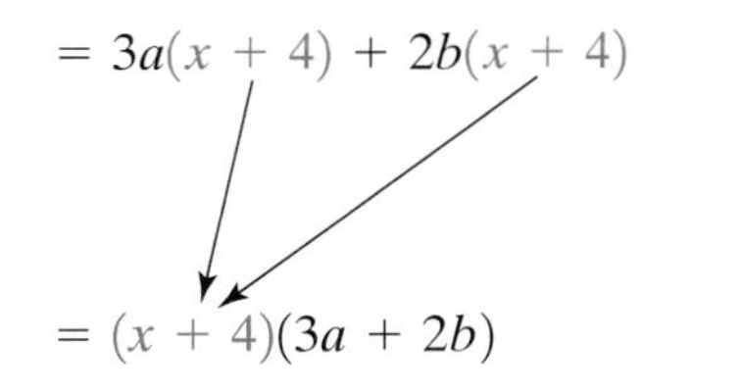

Check: $(x + 4)(3a + 2b) = 3ax + 2bx + 12a + 8b$ ✔

Note: Step 2 results in two terms with a common binomial factor. If the two binomials are different, step 3 cannot be performed. In such a case, the original polynomial may not be factorable by grouping, or different pairs of terms may need to be grouped and inspected.

Skill Practice Answers

13. $(a + b)(8y + 9)$

Skill Practice Factor by grouping.

14. $5x + 10y + ax + 2ay$

TIP: One frequently asked question when factoring is whether the order can be switched between the factors. The answer is yes. Because multiplication is commutative, the order in which the factors are written does not matter.

$$(x + 4)(3a + 2b) = (3a + 2b)(x + 4)$$

Steps to Factoring by Grouping

To factor a four-term polynomial by grouping:

1. Identify and factor out the GCF from all four terms.
2. Factor out the GCF from the first pair of terms. Factor out the GCF from the second pair of terms. (Sometimes it is necessary to factor out the opposite of the GCF.)
3. If the two terms share a common binomial factor, factor out the binomial factor.

Example 8 Factoring by Grouping

Factor the polynomials by grouping.

a. $ax + ay - bx - by$ **b.** $16w^4 - 40w^3 - 12w^2 + 30w$

Solution:

a. $ax + ay - bx - by$ **Step 1:** Identify and factor out the GCF from all four terms. In this case, the GCF is 1.

$= ax + ay \mid - bx - by$ Group the first pair of terms and the second pair of terms.

$= a(x + y) - b(x + y)$ **Step 2:** Factor out a from the first pair of terms.

Factor out $-b$ from the second pair of terms. (This causes sign changes within the second parentheses. The terms in parentheses now match.)

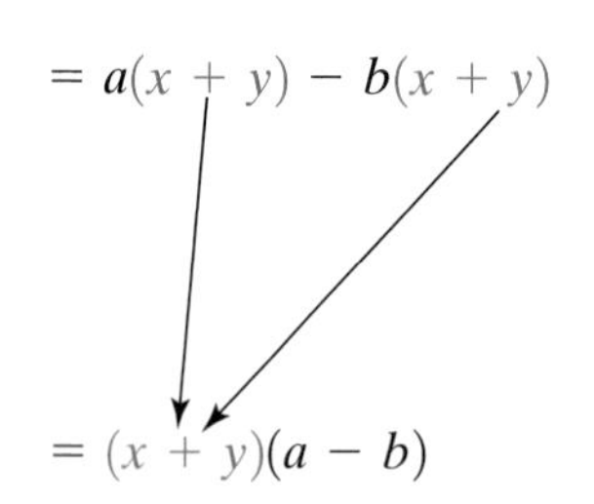

$= (x + y)(a - b)$ **Step 3:** Factor out the common binomial factor.

Check: $(x + y)(a - b) = x(a) + x(-b) + y(a) + y(-b)$

$= ax - bx + ay - by$ ✔

Avoiding Mistakes:

In step 2, the expression $a(x + y) - b(x + y)$ is not yet factored because it is a *difference*, not a product. To factor the expression, you must carry it one step further.

$$a(x + y) - b(x + y)$$
$$= (x + y)(a - b)$$

The factored form must be represented as a product.

Skill Practice Answers

14. $(x + 2y)(5 + a)$

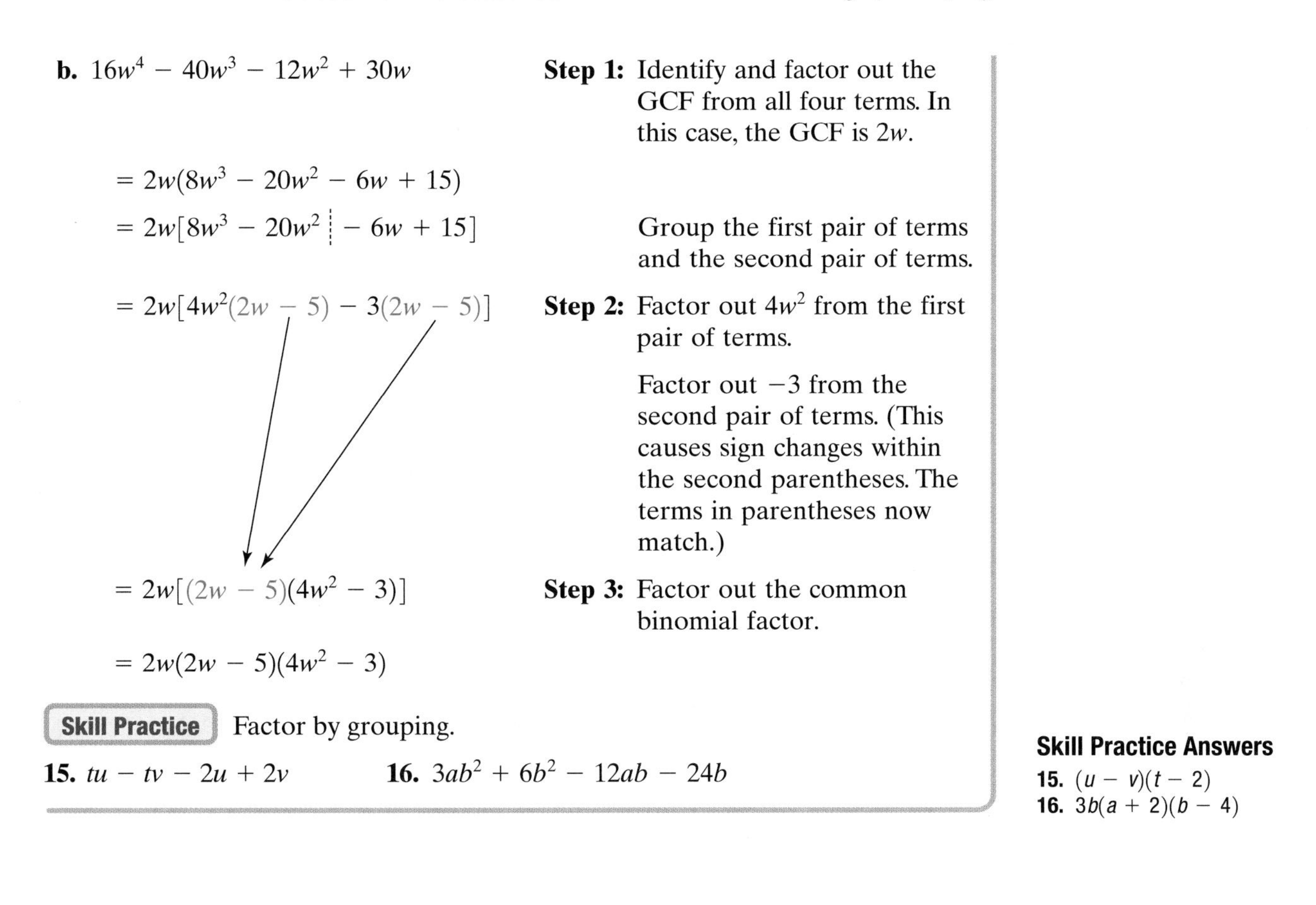

b. $16w^4 - 40w^3 - 12w^2 + 30w$

Step 1: Identify and factor out the GCF from all four terms. In this case, the GCF is $2w$.

$= 2w(8w^3 - 20w^2 - 6w + 15)$

$= 2w[8w^3 - 20w^2 \mid - 6w + 15]$

Group the first pair of terms and the second pair of terms.

$= 2w[4w^2(2w - 5) - 3(2w - 5)]$

Step 2: Factor out $4w^2$ from the first pair of terms.

Factor out -3 from the second pair of terms. (This causes sign changes within the second parentheses. The terms in parentheses now match.)

$= 2w[(2w - 5)(4w^2 - 3)]$

Step 3: Factor out the common binomial factor.

$= 2w(2w - 5)(4w^2 - 3)$

Skill Practice Factor by grouping.

15. $tu - tv - 2u + 2v$ **16.** $3ab^2 + 6b^2 - 12ab - 24b$

Skill Practice Answers

15. $(u - v)(t - 2)$
16. $3b(a + 2)(b - 4)$

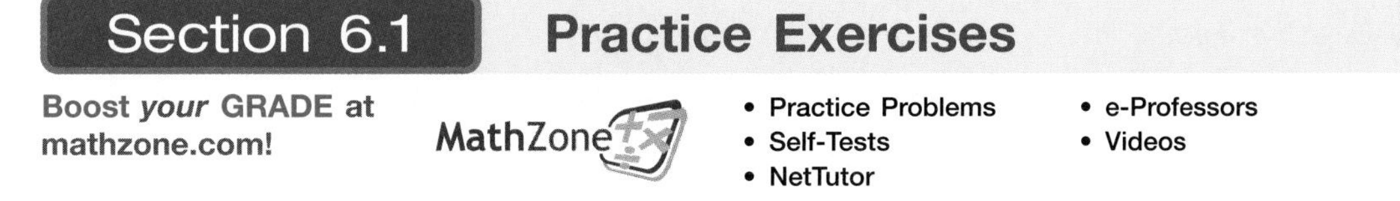

Section 6.1 Practice Exercises

Boost *your* GRADE at mathzone.com!

MathZone

- Practice Problems
- Self-Tests
- NetTutor
- e-Professors
- Videos

Study Skills Exercises

1. The final exam is just around the corner. Your old tests and quizzes provide good material to study for the final exam. Use your old tests to make a list of the chapters on which you need to concentrate. Ask your professor for help if there are still concepts that you do not understand.

2. Define the key terms:

a. factoring **b. greatest common factor (GCF)**

c. prime factorization **d. prime polynomial**

Concept 1: Identifying the Greatest Common Factor

For Exercises 3–18, identify the greatest common factor for the group of integers.

3. 28, 63 **4.** 24, 40 **5.** 42, 30, 60

6. 20, 52, 32 **7.** $3xy, 7y$ **8.** $10mn, 11n$

9. $8w^2, 4x^3$ **10.** $15t^2, 10u^2$ **11.** $2a^2b, 3ab^2$

12. $3x^3y^2, 5xy^4$

13. $12w^3z, 16w^2z$

14. $20cd, 15c^3d$

15. $8x^3y^4z^2, 12xy^5z^4, 6x^2y^8z^3$

16. $15r^2s^2t^5, 5r^3s^4t^3, 30r^4s^3t^2$

17. $7(x - y), 9(x - y)$

18. $(2a - b), 3(2a - b)$

Concept 2: Factoring out the Greatest Common Factor

19. **a.** Use the distributive property to multiply $3(x - 2y)$.

b. Use the distributive property to factor $3x - 6y$.

20. **a.** Use the distributive property to multiply $a^2(5a + b)$.

b. Use the distributive property to factor $5a^3 + a^2b$.

For Exercises 21–40, factor out the GCF.

21. $4p + 12$

22. $3q - 15$

23. $5c^2 - 10c + 15$

24. $16d^3 + 24d^2 + 32d$

25. $x^5 + x^3$

26. $y^2 - y^3$

27. $t^4 - 4t + 8t^2$

28. $7r^3 - r^5 + r^4$

29. $2ab + 4a^3b$

30. $5u^3v^2 - 5uv$

31. $38x^2y - 19x^2y^4$

32. $100a^5b^3 + 16a^2b$

33. $6x^3y^5 - 18xy^9z$

34. $15mp^7q^4 + 12m^4q^3$

35. $5 + 7y^3$

36. $w^3 - 5u^3v^2$

37. $42p^3q^2 + 14pq^2 - 7p^4q^4$

38. $8m^2n^3 - 24m^2n^2 + 4m^3n$

39. $t^5 + 2rt^3 - 3t^4 + 4r^2t^2$

40. $u^2v + 5u^3v^2 - 2u^2 + 8uv$

Concept 3: Factoring out a Negative Factor

41. For the polynomial $-2x^3 - 4x^2 + 8x$

a. Factor out $-2x$.

b. Factor out $2x$.

42. For the polynomial $-9y^5 + 3y^3 - 12y$

a. Factor out $-3y$.

b. Factor out $3y$.

43. Factor out -1 from the polynomial $-8t^2 - 9t - 2$.

44. Factor out -1 from the polynomial $-6x^3 - 2x - 5$.

For Exercises 45–50, factor out the opposite of the greatest common factor.

45. $-15p^3 - 30p^2$

46. $-24m^3 - 12m^4$

47. $-q^4 + 2q^2 - 9q$

48. $-r^3 + 9r^2 - 5r$

49. $-7x - 6y - 2z$

50. $-4a + 5b - c$

Concept 4: Factoring out a Binomial Factor

For Exercises 51–56, factor out the GCF.

51. $13(a + 6) - 4b(a + 6)$

52. $7(x^2 + 1) - y(x^2 + 1)$

53. $8v(w^2 - 2) + (w^2 - 2)$

54. $t(r + 2) + (r + 2)$

55. $21x(x + 3) + 7x^2(x + 3)$

56. $5y^3(y - 2) - 15y(y - 2)$

Concept 5: Factoring by Grouping

For Exercises 57–76, factor by grouping.

57. $8a^2 - 4ab + 6ac - 3bc$

58. $4x^3 + 3x^2y + 4xy^2 + 3y^3$

59. $3q + 3p + qr + pr$

60. $xy - xz + 7y - 7z$

61. $6x^2 + 3x + 4x + 2$

62. $4y^2 + 8y + 7y + 14$

63. $2t^2 + 6t - 5t - 15$

64. $2p^2 - p - 6p + 3$

65. $6y^2 - 2y - 9y + 3$

66. $5a^2 + 30a - 2a - 12$

67. $b^4 + b^3 - 4b - 4$

68. $8w^5 + 12w^2 - 10w^3 - 15$

69. $3j^2k + 15k + j^2 + 5$

70. $2ab^2 - 6ac + b^2 - 3c$

71. $14w^6x^6 + 7w^6 - 2x^6 - 1$

72. $18p^4q - 9p^5 - 2q + p$

73. $ay + bx + by + ax$
(*Hint:* Rearrange the terms.)

74. $2c + 3ay + ac + 6y$

75. $vw^2 - 3 + w - 3wv$

76. $2x^2 + 6m + 12 + x^2m$

For Exercises 77–82, factor out the GCF first. Then factor by grouping.

77. $15x^4 + 15x^2y^2 + 10x^3y + 10xy^3$

78. $2a^3b - 4a^2b + 32ab - 64b$

79. $4abx - 4b^2x - 4ab + 4b^2$

80. $p^2q - pq^2 - rp^2q + rpq^2$

81. $6st^2 - 18st - 6t^4 + 18t^3$

82. $15j^3 - 10j^2k - 15j^2k^2 + 10jk^3$

83. The formula $P = 2l + 2w$ represents the perimeter, P, of a rectangle given the length, l, and the width, w. Factor out the GCF, and write an equivalent formula in factored form.

84. The formula $P = 2a + 2b$ represents the perimeter, P, of a parallelogram given the base, b, and an adjacent side, a. Factor out the GCF, and write an equivalent formula in factored form.

85. The formula $S = 2\pi r^2 + 2\pi rh$ represents the surface area, S, of a cylinder with radius, r, and height, h. Factor out the GCF, and write an equivalent formula in factored form.

86. The formula $A = P + Prt$ represents the total amount of money, A, in an account that earns simple interest at a rate, r, for t years. Factor out the GCF, and write an equivalent formula in factored form.

Expanding Your Skills

87. Factor out $\frac{1}{7}$ from $\frac{1}{7}x^2 + \frac{3}{7}x - \frac{5}{7}$.

88. Factor out $\frac{1}{5}$ from $\frac{6}{5}y^2 - \frac{4}{5}y + \frac{1}{5}$.

89. Factor out $\frac{1}{4}$ from $\frac{5}{4}w^2 + \frac{3}{4}w + \frac{9}{4}$.

90. Factor out $\frac{1}{6}$ from $\frac{1}{6}p^2 - \frac{3}{6}p + \frac{5}{6}$.

91. Write a polynomial that has a GCF of $3x$. (Answers may vary.)

92. Write a polynomial that has a GCF of $7y$. (Answers may vary.)

93. Write a polynomial that has a GCF of $4p^2q$. (Answers may vary.)

94. Write a polynomial that has a GCF of $2ab^2$. (Answers may vary.)

Section 6.2

Factoring Trinomials of the Form $x^2 + bx + c$ (Optional)

Concept

1. Factoring Trinomials with a Leading Coefficient of 1

1. Factoring Trinomials with a Leading Coefficient of 1

In Section 5.6, we learned how to multiply two binomials. We also saw that such a product often results in a trinomial. For example,

$$(x + 3)(x + 7) = \underbrace{x^2}_{\text{Product of first terms}} + \underbrace{7x + 3x}_{\text{Sum of products of inner terms and outer terms}} + \underbrace{21}_{\text{Product of last terms}} = x^2 + 10x + 21$$

In this section, we want to reverse the process. That is, given a trinomial, we want to *factor* it as a product of two binomials. In particular, we begin our study with the case in which a trinomial has a leading coefficient of 1.

Consider the trinomial $x^2 + bx + c$. To produce a leading term of x^2, we can construct binomials of the form $(x + \quad)(x + \quad)$. The remaining terms can be satisfied by two integers, p and q, whose product is c and whose sum is b.

Factors of c

$$x^2 + bx + c = (x + p)(x + q) = x^2 + qx + px + pq$$

$$= x^2 + \underbrace{(q + p)}_{\text{Sum} = b}x + \underbrace{pq}_{\text{Product} = c}$$

This process is demonstrated in Example 1.

Example 1 **Factoring a Trinomial of the Form $x^2 + bx + c$**

Factor: $x^2 + 4x - 45$

Solution:

$x^2 + 4x - 45 = (x + \square)(x + \square)$ The product $x \cdot x = x^2$.

We must fill in the blanks with two integers whose product is -45 and whose sum is 4. The factors must have opposite signs to produce a negative product. The possible factorizations of -45 are:

Product = -45	Sum
$-1 \cdot 45$	44
$-3 \cdot 15$	12
$-5 \cdot 9$	4
$-9 \cdot 5$	-4
$-15 \cdot 3$	-12
$-45 \cdot 1$	-44

$$x^2 + 4x - 45 = (x + \square)(x + \square)$$

$= (x + (-5))(x + 9)$ Fill in the blanks with -5 and 9,

$= (x - 5)(x + 9)$ Factored form

Check:
$(x - 5)(x + 9) = x^2 + 9x - 5x - 45$
$= x^2 + 4x - 45$ ✔

Skill Practice Factor.

1. $x^2 - 5x - 14$

One frequently asked question is whether the order of factors can be reversed. The answer is yes because multiplication of polynomials is a commutative operation. Therefore, in Example 1, we can express the factorization as $(x - 5)(x + 9)$ or as $(x + 9)(x - 5)$.

Example 2 Factoring a Trinomial of the Form $x^2 + bx + c$

Factor: $w^2 - 15w + 50$

Solution:

$w^2 - 15w + 50 = (w + \square)(w + \square)$ The product $w \cdot w = w^2$.

Find two integers whose product is 50 and whose sum is -15. To form a positive product, the factors must be either both positive or both negative. The sum must be negative, so we will choose negative factors of 50.

Product = 50	Sum
$(-1)(-50)$	-51
$(-2)(-25)$	-27
$(-5)(-10)$	-15

$$w^2 - 15w + 50 = (w + \square)(w + \square)$$

$= (w + (-5))(w + (-10))$

$= (w - 5)(w - 10)$ Factored form

Skill Practice Factor.

2. $z^2 - 16z + 48$

Practice will help you become proficient in factoring polynomials. As you do your homework, keep these important guidelines in mind:

- To factor a trinomial, write the trinomial in descending order such as $x^2 + bx + c$.
- For all factoring problems, always factor out the GCF from all terms first.

Furthermore, we offer the following rules for determining the signs within the binomial factors.

Skill Practice Answers

1. $(x - 7)(x + 2)$
2. $(z - 4)(z - 12)$

Sign Rules for Factoring Trinomials

Given the trinomial $x^2 + bx + c$, the signs within the binomial factors are determined as follows:

1. If c is *positive*, then the signs in the binomials must be the same (either both positive or both negative). The correct choice is determined by the middle term. If the middle term is positive, then both signs must be positive. If the middle term is negative, then both signs must be negative.

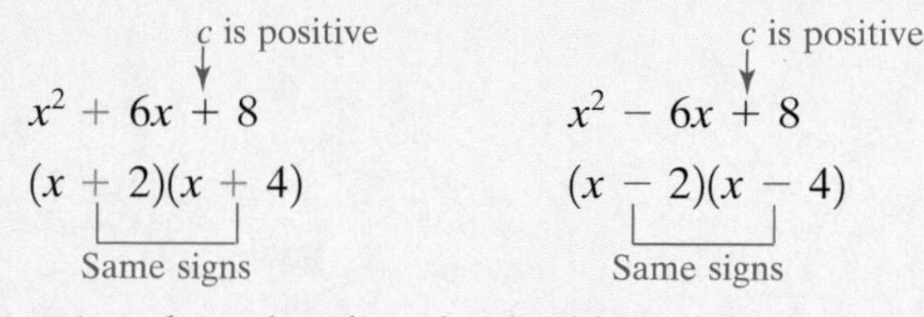

2. If c is *negative*, then the signs in the binomials must be different.

c is negative → $x^2 + 2x - 35$ → $(x + 7)(x - 5)$ Different signs

c is negative → $x^2 - 2x - 35$ → $(x - 7)(x + 5)$ Different signs

Example 3 Factoring Trinomials

Factor.

a. $-8p - 48 + p^2$ **b.** $-40t - 30t^2 + 10t^3$

c. $-a^2 + 6a - 8$ **d.** $2c^2 + 22cd + 60d^2$

Solution:

a. $-8p - 48 + p^2$

$= p^2 - 8p - 48$ Write in descending order.

$= (p \quad \square)(p \quad \square)$ Find two integers whose product is -48 and whose sum is -8. The numbers are -12 and 4.

$= (p - 12)(p + 4)$ Factored form

b. $-40t - 30t^2 + 10t^3$

$= 10t^3 - 30t^2 - 40t$ Write in descending order.

$= 10t(t^2 - 3t - 4)$ Factor out the GCF.

$= 10t(t \quad \square)(t \quad \square)$ Find two integers whose product is -4 and whose sum is -3. The numbers are -4 and 1.

$= 10t(t - 4)(t + 1)$ Factored form

c. $-a^2 + 6a - 8$ It is generally easier to factor a trinomial with a *positive* leading coefficient. Therefore, we will factor out -1 from all terms.

$= -1(a^2 - 6a + 8)$

$= -1(a \quad \square)(a \quad \square)$ Find two integers whose product is 8 and whose sum is -6. The numbers are -4 and -2.

$= -1(a - 4)(a - 2)$

TIP: Recall that factoring out -1 from a polynomial changes the signs of the terms within parentheses.

d. $2c^2 + 22cd + 60d^2$

$= 2(c^2 + 11cd + 30d^2)$	Factor out the GCF.
$= 2(c \quad \square d)(c \quad \square d)$	Notice that the second pair of terms has a factor of d. This will produce a product of d^2.
$= 2(c + 5d)(c + 6d)$	Find two integers whose product is 30 and whose sum is 11. The numbers are 5 and 6.

Skill Practice Factor.

3. $-5w + w^2 - 6$ **4.** $30y^3 + 2y^4 + 112y^2$

5. $-x^2 + x + 12$ **6.** $3a^2 - 15ab + 12b^2$

To factor a trinomial of the form $x^2 + bx + c$, we must find two integers whose product is c and whose sum is b. If no such integers exist, then the trinomial is not factorable and is called a **prime polynomial**.

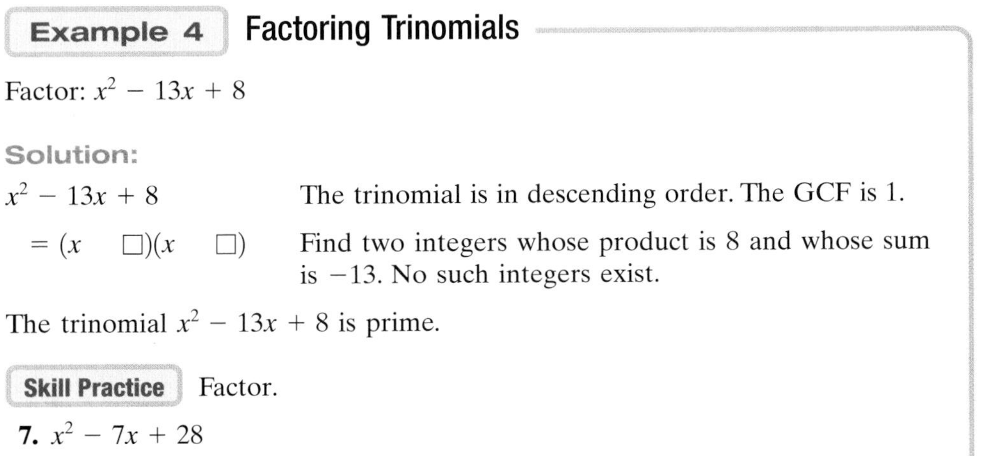

Example 4 Factoring Trinomials

Factor: $x^2 - 13x + 8$

Solution:

$x^2 - 13x + 8$	The trinomial is in descending order. The GCF is 1.
$= (x \quad \square)(x \quad \square)$	Find two integers whose product is 8 and whose sum is -13. No such integers exist.

The trinomial $x^2 - 13x + 8$ is prime.

Skill Practice Factor.

7. $x^2 - 7x + 28$

Skill Practice Answers

3. $(w - 6)(w + 1)$
4. $2y^2(y + 8)(y + 7)$
5. $-(x - 4)(x + 3)$
6. $3(a - b)(a - 4b)$
7. Prime

Section 6.2 Practice Exercises

Study Skills Exercises

1. Sometimes the problems on a test do not appear in the same order as the concepts appear in the text. In order to better prepare for a test, try to practice on problems taken from the book and placed in random order. Choose 30 problems from various chapters, randomize the order, and use them to review for the test. Repeat the process several times for additional practice.

2. Define the key term **prime polynomial**.

Review Exercises

For Exercises 3–6, factor completely.

3. $4x^3y^7 - 12x^4y^5 + 8xy^8$

4. $9a^6b^3 - 27a^3b^6 - 3a^2b^2$

5. $ax + 2bx - 5a - 10b$

6. $m^2 - mx - 3pm + 3px$

Concept 1: Factoring Trinomials with a Leading Coefficient of 1

For Exercises 7–20, factor completely.

7. $x^2 + 10x + 16$

8. $y^2 + 18y + 80$

9. $z^2 - 11z + 18$

10. $w^2 - 7w + 12$

11. $z^2 - 9z + 18$

12. $w^2 + 4w - 12$

13. $p^2 + 3p - 40$

14. $a^2 - 10a + 9$

15. $t^2 + 6t - 40$

16. $m^2 - 12m + 11$

17. $x^2 - 3x + 20$

18. $y^2 + 6y + 18$

19. $n^2 + 8n + 16$

20. $v^2 + 10v + 25$

For Exercises 21–24, assume that b and c represent positive integers.

21. When factoring a polynomial of the form $x^2 + bx + c$, pick an appropriate combination of signs.

a. (+)(+) **b.** (−)(−) **c.** (+)(−)

22. When factoring a polynomial of the form $x^2 + bx - c$, pick an appropriate combination of signs.

a. (+)(+) **b.** (−)(−) **c.** (+)(−)

23. When factoring a polynomial of the form $x^2 - bx - c$, pick an appropriate combination of signs.

a. (+)(+) **b.** (−)(−) **c.** (+)(−)

24. When factoring a polynomial of the form $x^2 - bx + c$, pick an appropriate combination of signs.

a. (+)(+) **b.** (−)(−) **c.** (+)(−)

25. Which is the correct factorization of $y^2 - y - 12$?

$(y - 4)(y + 3)$ or $(y + 3)(y - 4)$

26. Which is the correct factorization of $x^2 + 14x + 13$?

$(x + 13)(x + 1)$ or $(x + 1)(x + 13)$

27. Which is the correct factorization of $w^2 + 2w + 1$?

$(w + 1)(w + 1)$ or $(w + 1)^2$

28. Which is the correct factorization of $z^2 - 4z + 4$?

$(z - 2)(z - 2)$ or $(z - 2)^2$

29. In what order should a trinomial be written before attempting to factor it?

For Exercises 30–35, factor completely.

30. $-13x + x^2 - 30$

31. $12y - 160 + y^2$

32. $-18w + 65 + w^2$

33. $17t + t^2 + 72$

34. $22t + t^2 + 72$

35. $10q - 1200 + q^2$

36. Referring to page 457, write two important guidelines to follow when factoring trinomials.

For Exercises 37–48, factor completely. Be sure to factor out the GCF.

37. $3x^2 - 30x - 72$

38. $2z^2 + 4z - 198$

39. $8p^3 - 40p^2 + 32p$

40. $5w^4 - 35w^3 + 50w^2$

41. $y^4z^2 - 12y^3z^2 + 36y^2z^2$

42. $t^4u^2 + 6t^3u^2 + 9t^2u^2$

43. $-x^2 + 10x - 24$

44. $-y^2 - 12y - 35$

45. $-m^2 + m + 6$

46. $-n^2 + 5n + 6$

47. $-4 - 2c^2 - 6c$

48. $-40d - 30 - 10d^2$

Mixed Exercises

For Exercises 49–66, factor completely.

49. $x^3y^3 - 19x^2y^3 + 60xy^3$

50. $y^2z^5 + 17yz^5 + 60z^5$

51. $12p^2 - 96p + 84$

52. $5w^2 - 40w - 45$

53. $-2m^2 + 22m - 20$

54. $-3x^2 - 36x - 81$

55. $c^2 + 6cd + 5d^2$

56. $x^2 + 8xy + 12y^2$

57. $a^2 - 9ab + 14b^2$

58. $m^2 - 15mn + 44n^2$

59. $a^2 + 4a + 18$

60. $b^2 - 6a + 15$

61. $2q + q^2 - 63$

62. $-32 - 4t + t^2$

63. $x^2 + 20x + 100$

64. $z^2 - 24z + 144$

65. $t^2 + 18t - 40$

66. $d^2 + 2d - 99$

67. A student factored a trinomial as $(2x - 4)(x - 3)$. The instructor did not give full credit. Why?

68. A student factored a trinomial as $(y + 2)(5y - 15)$. The instructor did not give full credit. Why?

69. What polynomial factors as $(x - 4)(x + 13)$?

70. What polynomial factors as $(q - 7)(q + 10)$?

Expanding Your Skills

71. Find all integers, b, that make the trinomial $x^2 + bx + 6$ factorable.

72. Find all integers, b, that make the trinomial $x^2 + bx + 10$ factorable.

73. Find a value of c that makes the trinomial $x^2 + 6x + c$ factorable.

74. Find a value of c that makes the trinomial $x^2 + 8x + c$ factorable.

Section 6.3 Factoring Trinomials: Trial-and-Error Method

Concepts

1. Factoring Trinomials by the Trial-and-Error Method
2. Identifying GCF and Factoring Trinomials
3. Factoring Perfect Square Trinomials

In Section 6.2, we learned how to factor trinomials of the form $x^2 + bx + c$. These trinomials have a leading coefficient of 1. In this section and the next, we will consider the more general case in which the leading coefficient may be *any* integer. That is, we will factor trinomials of the form $ax^2 + bx + c$ (where $a \neq 0$). The method presented in this section is called the trial-and-error method.

1. Factoring Trinomials by the Trial-and-Error Method

To understand the basis of factoring trinomials of the form $ax^2 + bx + c$, first consider the multiplication of two binomials:

$$(2x + 3)(1x + 2) = \overset{\text{Product of } 2 \cdot 1}{2x^2} + \underbrace{\mathbf{4x + 3x}}_{\text{Sum of products of inner terms and outer terms}} + \overset{\text{Product of } 3 \cdot 2}{6} = 2x^2 + 7x + 6$$

To factor the trinomial, $2x^2 + 7x + 6$, this operation is reversed. Hence,

$$2x^2 + 7x + 6 = (\square x \quad \square)(\square x \quad \square)$$

Factors of 2 (first blanks); Factors of 6 (last blanks)

We need to fill in the blanks so that the product of the first terms in the binomials is $2x^2$ and the product of the last terms in the binomials is 6. Furthermore, the factors of $2x^2$ and 6 must be chosen so that the sum of the products of the inner terms and outer terms equals $7x$.

To produce the product $2x^2$, we might try the factors $2x$ and x within the binomials:

$$(2x \quad \square)(x \quad \square)$$

To produce a product of 6, the remaining terms in the binomials must either both be positive or both be negative. To produce a positive middle term, we will try positive factors of 6 in the remaining blanks until the correct product is found. The possibilities are $1 \cdot 6, 2 \cdot 3, 3 \cdot 2$, and $6 \cdot 1$.

$(2x + 1)(x + 6) = 2x^2 + 12x + 1x + 6 = 2x^2 + 13x + 6$ Wrong middle term

$(2x + 2)(x + 3) = 2x^2 + 6x + 2x + 6 = 2x^2 + 8x + 6$ Wrong middle term

$(2x + 3)(x + 2) = 2x^2 + 4x + 3x + 6 = 2x^2 + 7x + 6$ Correct!

$(2x + 6)(x + 1) = 2x^2 + 2x + 6x + 6 = 2x^2 + 8x + 6$ Wrong middle term

The correct factorization of $2x^2 + 7x + 6$ is $(2x + 3)(x + 2)$. ✔

As this example shows, we factor a trinomial of the form $ax^2 + bx + c$ by shuffling the factors of a and c within the binomials until the correct product is obtained. However, sometimes it is not necessary to test all the possible combinations of factors.

In the previous example, the GCF of the original trinomial is 1. Therefore, any binomial factor that shares a common factor *greater than 1* does not need to be considered. In this case, the possibilities $(2x + 2)(x + 3)$ and $(2x + 6)(x + 1)$ cannot work.

$$\underbrace{(2x + 2)}_{\text{Common factor of 2}}(x + 3) \qquad \underbrace{(2x + 6)}_{\text{Common factor of 2}}(x + 1)$$

The steps to factor a trinomial by the trial-and-error method are outlined in the following box.

Trial-and-Error Method to Factor $ax^2 + bx + c$

1. Factor out the GCF.
2. List all pairs of positive factors of a and pairs of positive factors of c. Consider the reverse order for one of the lists of factors.
3. Construct two binomials of the form:

Factors of a

$$(\square x \quad \square)(\square x \quad \square)$$

Factors of c

Test each combination of factors and signs until the correct product is found. If no combination of factors produces the correct product, the trinomial cannot be factored further and is a **prime polynomial**.

Before we begin our next example, keep these two important guidelines in mind:

- For any factoring problem you encounter, always factor out the GCF from all terms first.
- To factor a trinomial, write the trinomial in the form $ax^2 + bx + c$.

Example 1 Factoring a Trinomial by the Trial-and-Error Method

Factor the trinomial by the trial-and-error method: $10x^2 - 9x - 1$

Solution:

$10x^2 - 9x - 1$ **Step 1:** Factor out the GCF from all terms. In this case, the GCF is 1.

The trinomial is written in the form $ax^2 + bx + c$.

To factor $10x^2 - 9x - 1$, two binomials must be constructed in the form:

Factors of 10

$(\square x \quad \square)(\square x \quad \square)$

Factors of -1

Step 2: To produce the product $10x^2$, we might try $5x$ and $2x$, or $10x$ and $1x$. To produce a product of -1, we will try the factors $(1)(-1)$ and $(-1)(1)$.

Step 3: Construct all possible binomial factors using different combinations of the factors of $10x^2$ and -1.

$(5x + 1)(2x - 1) = 10x^2 - 5x + 2x - 1 = 10x^2 - 3x - 1$ Wrong middle term

$(5x - 1)(2x + 1) = 10x^2 + 5x - 2x - 1 = 10x^2 + 3x - 1$ Wrong middle term

Because the numbers 1 and -1 did not produce the correct trinomial when coupled with $5x$ and $2x$, try using $10x$ and $1x$.

$(10x - 1)(1x + 1) = 10x^2 + 10x - 1x - 1 = 10x^2 + 9x - 1$ Wrong middle term

$(10x + 1)(1x - 1) = 10x^2 - 10x + 1x - 1 = 10x^2 - 9x - 1$ Correct!

Hence, $10x^2 - 9x - 1 = (10x + 1)(x - 1)$.

Skill Practice Factor using the trial-and-error method.

1. $3b^2 + 8b + 4$

In Example 1, the factors of -1 must have opposite signs to produce a negative product. Therefore, one binomial factor is a sum and one is a difference. Determining the correct signs is an important aspect of factoring trinomials. We suggest the following guidelines:

Sign Rules for the Trial-and-Error Method

Given the trinomial $ax^2 + bx + c$, $(a > 0)$, the signs can be determined as follows:

1. If *c is positive*, then the signs in the binomials must be the same (either both positive or both negative). The correct choice is determined by the middle term. If the middle term is positive, then both signs must be positive. If the middle term is negative, then both signs must be negative.

c is positive	c is positive
$20x^2 + 43x + 21$	$20x^2 - 43x + 21$
$(4x + 3)(5x + 7)$	$(4x - 3)(5x - 7)$
Same signs	Same signs

2. If *c is negative*, then the signs in the binomial must be different. The middle term in the trinomial determines which factor gets the positive sign and which gets the negative sign.

c is negative	c is negative
$x^2 + 3x - 28$	$x^2 - 3x - 28$
$(x + 7)(x - 4)$	$(x - 7)(x + 4)$
Different signs	Different signs

Example 2 Factoring a Trinomial

Factor the trinomial: $13y - 6 + 8y^2$

Solution:

$13y - 6 + 8y^2$

$= 8y^2 + 13y - 6$ Write the polynomial in descending order.

$(\square y \quad \square)(\square y \quad \square)$ **Step 1:** The GCF is 1.

Skill Practice Answers

1. $(3b + 2)(b + 2)$

Factors of 8	Factors of 6
$1 \cdot 8$	$1 \cdot 6$
$2 \cdot 4$	$2 \cdot 3$
	$3 \cdot 2$ (reverse order)
	$6 \cdot 1$ (reverse order)

Step 2: List the positive factors of 8 and positive factors of 6. Consider the reverse order in only one list of factors.

Step 3: Construct all possible binomial factors using different combinations of the factors of 8 and 6.

$(2y \quad 1)(4y \quad 6)$
$(2y \quad 2)(4y \quad 3)$
$(2y \quad 3)(4y \quad 2)$
$(2y \quad 6)(4y \quad 1)$
$(1y \quad 1)(8y \quad 6)$
$(1y \quad 3)(8y \quad 2)$

Without regard to signs, these factorizations cannot work because the terms in the binomials share a common factor greater than 1.

Test the remaining factorizations. Keep in mind that to produce a product of -6, the signs within the parentheses must be opposite (one positive and one negative). Also, the sum of the products of the inner terms and outer terms must be combined to form $13y$.

$(1y \quad 6)(8y \quad 1)$ *Incorrect.* Wrong middle term.
Regardless of the signs, the product of inner terms, $48y$, and the product of outer terms, $1y$, cannot be combined to form the middle term $13y$.

$(1y \quad 2)(8y \quad 3)$ *Correct.* The terms $16y$ and $3y$ can be combined to form the middle term $13y$, provided the signs are applied correctly. We require $+16y$ and $-3y$.

Hence, the correct factorization of $8y^2 + 13y - 6$ is $(y + 2)(8y - 3)$.

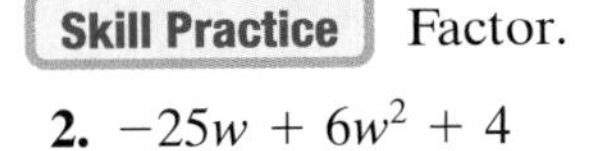

Skill Practice Factor.

2. $-25w + 6w^2 + 4$

2. Identifying GCF and Factoring Trinomials

Remember that the first step in any factoring problem is to remove the GCF. By removing the GCF, the remaining terms of the trinomial will be simpler and may have smaller coefficients.

Example 3 Factoring a Trinomial by the Trial-and-Error Method

Factor the trinomial by the trial-and-error method: $40x^3 - 104x^2 + 10x$

Solution:

$40x^3 - 104x^2 + 10x$

$= 2x(20x^2 - 52x + 5)$ **Step 1:** The GCF is $2x$.

Skill Practice Answers

2. $(6w - 1)(w - 4)$

TIP: Notice that when the GCF, $2x$, is removed from the original trinomial, the new trinomial has smaller coefficients. This makes the factoring process simpler. It is easier to list the factors of 20 and 5 rather than the factors of 40 and 10.

$= 2x(20x^2 - 52x + 5)$

$= 2x(\square x \quad \square)(\square x \quad \square)$

Step 2: List the factors of 20 and factors of 5. Consider the reverse order in one list of factors.

Factors of 20	Factors of 5
$1 \cdot 20$	$1 \cdot 5$
$2 \cdot 10$	$5 \cdot 1$
$4 \cdot 5$	

Step 3: Construct all possible binomial factors using different combinations of the factors of 20 and factors of 5. The signs in the parentheses must both be negative.

$= 2x(1x - 1)(20x - 5)$
$= 2x(2x - 1)(10x - 5)$
$= 2x(4x - 1)(5x - 5)$

Incorrect. The binomials contain a GCF greater than 1.

$= 2x(1x - 5)(20x - 1)$ *Incorrect.* Wrong middle term.

$$2x(x - 5)(20x - 1) = 2x(20x^2 - 1x - 100x + 5) = 2x(20x^2 - 101x + 5)$$

$= 2x(4x - 5)(5x - 1)$ *Incorrect.* Wrong middle term.

$$2x(4x - 5)(5x - 1) = 2x(20x^2 - 4x - 25x + 5) = 2x(20x^2 - 29x + 5)$$

$= 2x(2x - 5)(10x - 1)$ *Correct.*

$$2x(2x - 5)(10x - 1) = 2x(20x^2 - 2x - 50x + 5) = 2x(20x^2 - 52x + 5) = 40x^3 - 104x^2 + 10x$$

The correct factorization is $2x(2x - 5)(10x - 1)$.

Skill Practice Factor.

3. $8t^3 + 38t^2 + 24t$

Often it is easier to factor a trinomial when the leading coefficient is positive. If the leading coefficient is negative, consider factoring out the opposite of the GCF.

Example 4 Factoring a Trinomial by the Trial-and-Error Method

Factor: $-45x^2 - 3xy + 18y^2$

Solution:

$-45x^2 - 3xy + 18y^2$

$= -3(15x^2 + xy - 6y^2)$ **Step 1:** Factor out -3 to make the leading term positive.

$= -3(\square x \quad \square y)(\square x \quad \square y)$ **Step 2:** List the factors of 15 and 6.

Skill Practice Answers

3. $2t(4t + 3)(t + 4)$

Factors of 15	Factors of 6	$-3(15x^2 + xy - 6y^2)$
$1 \cdot 15$	$1 \cdot 6$	
$3 \cdot 5$	$2 \cdot 3$	
	$3 \cdot 2$	
	$6 \cdot 1$	**Step 3:** We will construct all binomial combinations, without regard to signs first.

$-3(x \quad y)(15x \quad 6y)$
$-3(x \quad 2y)(15x \quad 3y)$
$-3(3x \quad 3y)(5x \quad 2y)$
$-3(3x \quad 6y)(5x \quad y)$

Incorrect. The binomials contain a common factor.

Test the remaining factorizations. The signs within parentheses must be opposite to produce a product of $-6y^2$. Also, the sum of the products of the inner terms and outer terms must be combined to form $1xy$.

$-3(x \quad 3y)(15x \quad 2y)$ *Incorrect.* Regardless of signs, $45xy$ and $2xy$ cannot be combined to equal xy.

$-3(x \quad 6y)(15x \quad y)$ *Incorrect.* Regardless of signs, $90xy$ and xy cannot be combined to equal xy.

$-3(3x \quad y)(5x \quad 6y)$ *Incorrect.* Regardless of signs, $5xy$ and $18xy$ cannot be combined to equal xy.

$-3(3x \quad 2y)(5x \quad 3y)$ *Correct.* The terms $10xy$ and $9xy$ can be combined to form xy provided that the signs are applied correctly. We require $10xy$ and $-9xy$.

$-3(3x + 2y)(5x - 3y)$ Factored form

TIP: Do not forget to write the GCF in the final answer.

Skill Practice Factor.

4. $-4x^2 + 26xy - 40y^2$

3. Factoring Perfect Square Trinomials

Recall from Section 5.6 that the square of a binomial always results in a **perfect square trinomial**.

$$(a + b)^2 = (a + b)(a + b) \xrightarrow{\text{Multiply}} = a^2 + 2ab + b^2$$
$$(a - b)^2 = (a - b)(a - b) \xrightarrow{\text{Multiply}} = a^2 - 2ab + b^2$$

For example, $(3x + 5)^2 = (3x)^2 + 2(3x)(5) + (5)^2$
$= 9x^2 + 30x + 25$ (perfect square trinomial)

We now want to reverse this process by factoring a perfect square trinomial. The trial-and-error method can always be used; however, if we recognize the pattern for a perfect square trinomial, we can use one of the following formulas to reach a quick solution.

Skill Practice Answers

4. $-2(2x - 5y)(x - 4y)$

Factored Form of a Perfect Square Trinomial

$$a^2 + 2ab + b^2 = (a + b)^2$$
$$a^2 - 2ab + b^2 = (a - b)^2$$

For example, $9x^2 + 30x + 25$ is a perfect square trinomial with $a = 3x$ and $b = 5$. Therefore, it factors as

$$9x^2 + 30x + 25 = (3x)^2 + 2(3x)(5) + (5)^2 = (3x + 5)^2$$
$$\uparrow \quad \uparrow\uparrow\uparrow \quad \uparrow \quad \uparrow\uparrow$$
$$a^2 + 2\,(a)\,(b) + (b)^2 = (a + b)^2$$

To apply the factored form of a perfect square trinomial, we must first be sure that the trinomial is indeed a perfect square trinomial.

Checking for a Perfect Square Trinomial

1. Determine whether the first and third terms are both perfect squares and have positive coefficients.
2. If this is the case, identify a and b, and determine if the middle term equals $2ab$.

Example 5 Factoring Perfect Square Trinomials

Factor the trinomials completely.

a. $x^2 + 14x + 49$

b. $25y^2 - 20y + 4$

c. $18c^3 - 48c^2d + 32cd^2$

d. $5w^2 + 50w + 45$

Solution:

a. $x^2 + 14x + 49$

The GCF is 1.

- The first and third terms are positive.
- The first term is a perfect square: $x^2 = (x)^2$.
- The third term is a perfect square: $49 = (7)^2$.
- The middle term is twice the product of x and 7: $14x = 2(x)(7)$

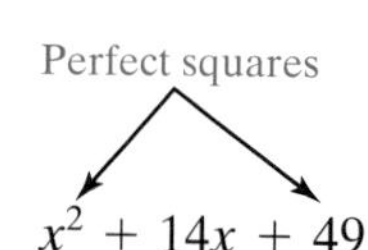

$x^2 + 14x + 49$

$= (x)^2 + 2(x)(7) + (7)^2$

Hence, the trinomial is in the form $a^2 + 2ab + b^2$, where $a = x$ and $b = 7$.

$= (x + 7)^2$

Factor as $(a + b)^2$.

b. $25y^2 - 20y + 4$

The GCF is 1.

Perfect squares

$25y^2 - 20y + 4$

- The first and third terms are positive.
- The first term is a perfect square: $25y^2 = (5y)^2$.
- The third term is a perfect square: $4 = (2)^2$.

$= (5y)^2 - 2(5y)(2) + (2)^2$

- In the middle: $20y = 2(5y)(2)$

$= (5y - 2)^2$

Factor as $(a - b)^2$.

c. $18c^3 - 48c^2d + 32cd^2$

$= 2c(9c^2 - 24cd + 16d^2)$

The GCF is $2c$.

Perfect squares

$= 2c(9c^2 - 24cd + 16d^2)$

- The first and third terms are positive.
- The first term is a perfect square: $9c^2 = (3c)^2$.
- The third term is a perfect square: $16d^2 = (4d)^2$.

$= 2c[(3c)^2 - 2(3c)(4d) + (4d)^2]$

- In the middle: $24cd = 2(3c)(4d)$

$= 2c(3c - 4d)^2$

Factor as $(a - b)^2$.

d. $5w^2 + 50w + 45$

$= 5(w^2 + 10w + 9)$

The GCF is 5.

Perfect squares

$= 5(w^2 + 10w + 9)$

The first and third terms are perfect squares.

$w^2 = (w)^2$ and $9 = (3)^2$

However, the middle term is not 2 times the product of w and 3. Therefore, this is not a perfect square trinomial.

$$10w \neq 2(w)(3)$$

$= 5(w + 9)(w + 1)$

To factor, use the trial-and-error method.

Skill Practice Factor completely.

5. $x^2 - 6x + 9$

6. $81w^2 + 72w + 16$

7. $5z^2 + 20z + 20$

8. $4x^2 + 13x + 9$

Skill Practice Answers

5. $(x - 3)^2$ **6.** $(9w + 4)^2$

7. $5(z + 2)^2$ **8.** $(4x + 9)(x + 1)$

TIP: To help you identify a perfect square trinomial, we recommend that you familiarize yourself with the first several perfect squares.

$(1)^2 = 1$	$(6)^2 = 36$	$(11)^2 = 121$
$(2)^2 = 4$	$(7)^2 = 49$	$(12)^2 = 144$
$(3)^2 = 9$	$(8)^2 = 64$	$(13)^2 = 169$
$(4)^2 = 16$	$(9)^2 = 81$	$(14)^2 = 196$
$(5)^2 = 25$	$(10)^2 = 100$	$(15)^2 = 225$

If you do not recognize that a trinomial is a perfect square trinomial, you may still use the trial-and-error method to factor it.

Recall that a prime polynomial is a polynomial whose only factors are itself and 1. Not every trinomial is factorable by the methods presented in this text.

Example 6 Factoring a Trinomial by the Trial-and-Error Method

Factor the trinomial by the trial-and-error method: $14p^2 - 56p + 21$

Solution:

$14p^2 - 56p + 21$

$= 7(2p^2 - 8p + 3)$

Step 1: The GCF is 7. Now factor the trinomial $2p^2 - 8p + 3$

$= 7(2p^2 - 8p + 3)$

$= 7(1p \;\square)(2p \;\square)$

Step 2: List the factors of 2 and the factors of 3.

Factors of 2	Factors of 3
$1 \cdot 2$	$1 \cdot 3$
	$3 \cdot 1$

Step 3: Construct all possible binomial factors using different combinations of the factors of 2 and 3. Because the third term in the trinomial is positive, both signs in the binomial must be the same. Because the middle term coefficient is negative, both signs will be negative.

$7(p - 1)(2p - 3) = 7(2p^2 - 3p - 2p + 3)$

$= 7(2p^2 - 5p + 3)$ *Incorrect.* Wrong middle term.

$7(p - 3)(2p - 1) = 7(2p^2 - p - 6p + 3)$

$= 7(2p^2 - 7p + 3)$ *Incorrect.* Wrong middle term.

None of the combinations of factors results in the correct product. Therefore, the polynomial $2p^2 - 8p + 3$ is prime and cannot be factored further.

The factored form of $14p^2 - 56p + 21$ is $7(2p^2 - 8p + 3)$.

Skill Practice Factor.

9. $3a^2 + a + 4$

Skill Practice Answers

9. Prime

Section 6.3 Practice Exercises

Study Skills Exercises

1. In addition to studying the material for a test, here are some other activities that people use when preparing for a test. Circle the importance of each statement.

	not important	somewhat important	very important
a. Get a good night's sleep the night before the test.	1	2	3
b. Eat a good meal before the test.	1	2	3
c. Wear comfortable clothes on the day of the test.	1	2	3
d. Arrive early to class on the day of the test.	1	2	3

2. Define the key terms:

a. prime polynomial **b. perfect square trinomial**

Review Exercises

For Exercises 3–8, factor completely.

3. $21a^2b^2 + 12ab^2 - 15a^2b$ **4.** $5uv^2 - 10u^2v + 25u^2v^2$ **5.** $mn - m - 2n + 2$

6. $5x - 10 - xy + 2y$ **7.** $6a^2 - 30a - 84$ **8.** $10b^2 + 20b - 240$

Concept 1: Factoring Trinomials by the Trial-and-Error Method

For Exercises 9–11, assume a, b, and c represent positive integers.

9. When factoring a polynomial of the form $ax^2 + bx + c$, pick an appropriate combination of signs.

a. (+)(+) **b.** (−)(−) **c.** (+)(−)

10. When factoring a polynomial of the form $ax^2 - bx - c$, pick an appropriate combination of signs.

a. (+)(+) **b.** (−)(−) **c.** (+)(−)

11. When factoring a polynomial of the form $ax^2 - bx + c$, pick an appropriate combination of signs.

a. (+)(+) **b.** (−)(−) **c.** (+)(−)

For Exercises 12–29, factor completely by using the trial-and-error method.

12. $2y^2 - 3y - 2$ **13.** $2w^2 + 5w - 3$ **14.** $3n^2 + 13n + 4$

15. $2a^2 + 7a + 6$ **16.** $5x^2 - 14x - 3$ **17.** $7y^2 + 9y - 10$

18. $12c^2 - 5c - 2$ **19.** $6z^2 + z - 12$ **20.** $-12 + 10w^2 + 37w$

21. $-10 + 10p^2 + 21p$ **22.** $-5q - 6 + 6q^2$ **23.** $17a - 2 + 3a^2$

24. $6b - 23 + 4b^2$

25. $8 + 7x^2 - 18x$

26. $-8 + 25m^2 - 10m$

27. $8q^2 + 31q - 4$

28. $6y^2 - 19xy - 20x^2$

29. $12y^2 - 73yz + 6z^2$

Concept 2: Identifying GCF and Factoring Trinomials

For Exercises 30–39, factor completely. Be sure to factor out the GCF first.

30. $2m^2 - 12m - 80$

31. $3c^2 - 33c + 72$

32. $2y^5 + 13y^4 + 6y^3$

33. $3u^8 - 13u^7 + 4u^6$

34. $-a^2 - 15a + 34$

35. $-x^2 - 7x - 10$

36. $-12u^3 - 22u^2 + 20u$

37. $-18z^4 + 15z^3 + 12z^2$

38. $80m^2 - 100mp - 30p^2$

39. $60w^2 + 550wz - 500z^2$

Concept 3: Factoring Perfect Square Trinomials

40. Multiply. $(3x + 5)^2$

41. Multiply. $(2y - 7)^2$

42. **a.** Which trinomial is a perfect square trinomial? $x^2 + 4x + 4$ or $x^2 + 5x + 4$

b. Factor the trinomials from part (a).

43. **a.** Which trinomial is a perfect square trinomial? $x^2 + 13x + 36$ or $x^2 + 12x + 36$

b. Factor the trinomials from part (a).

For Exercises 44–55, factor completely. (*Hint:* Look for the pattern of a perfect square trinomial.)

44. $x^2 + 18x + 81$

45. $y^2 - 8y + 16$

46. $25z^2 - 20z + 4$

47. $36p^2 + 60p + 25$

48. $49a^2 + 42ab + 9b^2$

49. $25m^2 - 30mn + 9n^2$

50. $-2y + y^2 + 1$

51. $4 + w^2 - 4w$

52. $80z^2 + 120z + 45$

53. $36p^2 - 24p + 4$

54. $9y^2 + 78x + 25$

55. $4y^2 + 20y + 9$

Mixed Exercises

For Exercises 56–97, factor the trinomial completely.

56. $20z - 18 - 2z^2$

57. $25t - 5t^2 - 30$

58. $42 - 13q + q^2$

59. $-5w - 24 + w^2$

60. $6t^2 + 7t - 3$

61. $4p^2 - 9p + 2$

62. $4m^2 - 20m + 25$

63. $16r^2 + 24r + 9$

64. $5c^2 - c + 2$

65. $7s^2 + 2s + 9$

66. $6x^2 - 19xy + 10y^2$

67. $15p^2 + pq - 2q^2$

68. $12m^2 + 11mn - 5n^2$

69. $4a^2 + 5ab - 6b^2$

70. $6r^2 + rs - 2s^2$

71. $18x^2 - 9xy - 2y^2$

72. $4s^2 - 8st + t^2$

73. $6u^2 - 10uv + 5v^2$

74. $10t^2 - 23t - 5$

75. $16n^2 + 14n + 3$

76. $14w^2 + 13w - 12$

77. $12x^2 - 16x + 5$

78. $x^2 + 7x - 18$

79. $y^2 - 6y - 40$

80. $a^2 - 10a - 24$

81. $b^2 + 6b - 7$

82. $r^2 + 5r - 24$

83. $t^2 + 20t + 100$

84. $x^2 + 9xy + 20y^2$

85. $p^2 - 13pq + 36q^2$

86. $v^2 + 2v + 15$

87. $x^2 - x - 1$

88. $a^2 + 21ab + 20b^2$ **89.** $x^2 - 17xy - 18y^2$ **90.** $t^2 - 10t + 21$ **91.** $z^2 - 15z + 36$

92. $5d^3 + 3d^2 - 10d$ **93.** $3y^3 - y^2 + 12y$ **94.** $4b^3 - 4b^2 - 80b$ **95.** $2w^2 + 20w + 42$

96. $x^2y^2 - 13xy^2 + 30y^2$ **97.** $p^2q^2 - 14pq^2 + 33q^2$

Expanding Your Skills

Each pair of trinomials looks similar but differs by one sign. Factor each trinomial, and see how their factored forms differ.

98. **a.** $x^2 - 10x - 24$
b. $x^2 - 10x + 24$

99. **a.** $x^2 - 13x - 30$
b. $x^2 - 13x + 30$

100. **a.** $x^2 - 5x - 6$
b. $x^2 - 5x + 6$

101. **a.** $x^2 - 10x + 9$
b. $x^2 + 10x + 9$

For Exercises 102–105, factor completely.

102. $x^4 + 10x^2 + 9$ **103.** $y^4 + 4y^2 - 21$

104. $w^4 + 2w^2 - 15$ **105.** $p^4 - 13p^2 + 40$

Factoring Trinomials: AC-Method

Section 6.4

Concepts

1. Factoring Trinomials by the AC-Method
2. Factoring Perfect Square Trinomials

In Section 6.2, we factored trinomials with a leading coefficient of 1. In Section 6.3, we learned the trial-and-error method to factor the more general case in which the leading coefficient is any integer. In this section, we provide an alternative method to factor trinomials, called the ac-method.

1. Factoring Trinomials by the AC-Method

The product of two binomials results in a four-term expression that can sometimes be simplified to a trinomial. To factor the trinomial, we want to reverse the process.

Multiply:

Multiply the binomials. Add the middle terms.

$$(2x + 3)(x + 2) = \longrightarrow 2x^2 + 4x + 3x + 6 = \longrightarrow 2x^2 + 7x + 6$$

Factor:

$$2x^2 + 7x + 6 = \longrightarrow 2x^2 + 4x + 3x + 6 = \longrightarrow (2x + 3)(x + 2)$$

Rewrite the middle term as a sum or difference of terms. Factor by grouping.

To factor a trinomial, $ax^2 + bx + c$, by the ac-method, we rewrite the middle term, bx, as a sum or difference of terms. The goal is to produce a four-term polynomial that can be factored by grouping. The process is outlined as follows.

AC-Method Factor $ax^2 + bx + c$ ($a \neq 0$)

1. Multiply the coefficients of the first and last terms (ac).
2. Find two integers whose product is ac and whose sum is b. (If no pair of integers can be found, then the trinomial cannot be factored further and is a **prime polynomial**.)
3. Rewrite the middle term, bx, as the sum of two terms whose coefficients are the integers found in step 2.
4. Factor by grouping.

The ac-method for factoring trinomials is illustrated in Example 1. However, before we begin, keep these two important guidelines in mind:

- For any factoring problem you encounter, always factor out the GCF from all terms first.
- To factor a trinomial, write the trinomial in the form $ax^2 + bx + c$.

Example 1 Factoring a Trinomial by the AC-Method

Factor the trinomial by the ac-method: $2x^2 + 7x + 6$

Solution:

$2x^2 + 7x + 6$ — Factor out the GCF from all terms. In this case, the GCF is 1.

$2x^2 + 7x + 6$ — **Step 1:** The trinomial is written in the form $ax^2 + bx + c$.

$a = 2, b = 7, c = 6$ — Find the product $ac = (2)(6) = 12$.

12	12
$1 \cdot 12$	$(-1)(-12)$
$2 \cdot 6$	$(-2)(-6)$
$3 \cdot 4$	$(-3)(-4)$

Step 2: List all factors of ac and search for the pair whose sum equals the value of b. That is, list the factors of 12 and find the pair whose sum equals 7.

The numbers 3 and 4 satisfy both conditions: $3 \cdot 4 = 12$ and $3 + 4 = 7$.

$2x^2 + 7x + 6$

$= 2x^2 + 3x + 4x + 6$ — **Step 3:** Write the middle term of the trinomial as the sum of two terms whose coefficients are the selected pair of numbers: 3 and 4.

$= 2x^2 + 3x \mid + 4x + 6$ — **Step 4:** Factor by grouping.

$= x(2x + 3) + 2(2x + 3)$

$= (2x + 3)(x + 2)$

Check: $(2x + 3)(x + 2) = 2x^2 + 4x + 3x + 6$
$= 2x^2 + 7x + 6$ ✔

Skill Practice Factor by the ac-method.

1. $2x^2 + 5x + 3$

Skill Practice Answers

1. $(x + 1)(2x + 3)$

TIP: One frequently asked question is whether the order matters when we rewrite the middle term of the trinomial as two terms (step 3). The answer is no. From the previous example, the two middle terms in step 3 could have been reversed to obtain the same result:

$$\begin{aligned} &2x^2 + 7x + 6 \\ &= 2x^2 + 4x + 3x + 6 \\ &= 2x(x + 2) + 3(x + 2) \\ &= (x + 2)(2x + 3) \end{aligned}$$

This example also points out that the order in which two factors are written does not matter. The expression $(x + 2)(2x + 3)$ is equivalent to $(2x + 3)(x + 2)$ because multiplication is a commutative operation.

Example 2 Factoring Trinomials by the AC-Method

Factor the trinomial by the ac-method: $-2x + 8x^2 - 3$

Solution:

$-2x + 8x^2 - 3$ — First rewrite the polynomial in the form $ax^2 + bx + c$.

$= 8x^2 - 2x - 3$ — The GCF is 1.

$a = 8, b = -2, c = -3$ — **Step 1:** Find the product $ac = (8)(-3) = -24$.

Step 2: List all the factors of -24 and find the pair of factors whose sum equals -2.

-24	-24
$-1 \cdot 24$	$-24 \cdot 1$
$-2 \cdot 12$	$-12 \cdot 2$
$-3 \cdot 8$	$-8 \cdot 3$
$-4 \cdot 6$	$-6 \cdot 4$

The numbers -6 and 4 satisfy both conditions: $(-6)(4) = -24$ and $-6 + 4 = -2$.

$= 8x^2 - 2x - 3$ — **Step 3:** Write the middle term of the trinomial as two terms whose coefficients are the selected pair of numbers, -6 and 4.

$= 8x^2 - 6x + 4x - 3$

$= 8x^2 - 6x \;|\; + 4x - 3$ — **Step 4:** Factor by grouping.

$= 2x(4x - 3) + 1(4x - 3)$

$= (4x - 3)(2x + 1)$

Check: $(4x - 3)(2x + 1) = 8x^2 + 4x - 6x - 3$
$= 8x^2 - 2x - 3$ ✔

Skill Practice Factor by the ac-method.

2. $13w + 6w^2 + 6$

Avoiding Mistakes:

Before factoring a trinomial, be sure to write the trinomial in descending order. That is, write it in the form $ax^2 + bx + c$.

Skill Practice Answers

2. $(2w + 3)(3w + 2)$

Example 3 Factoring a Trinomial by the AC-Method

Factor the trinomial by the ac-method: $10x^3 - 85x^2 + 105x$

Solution:

$10x^3 - 85x^2 + 105x$ The GCF is $5x$.

$= 5x(2x^2 - 17x + 21)$ The trinomial is in the form $ax^2 + bx + c$.

$a = 2, b = -17, c = 21$ **Step 1:** Find the product $ac = (2)(21) = 42$.

42	42
$1 \cdot 42$	$(-1)(-42)$
$2 \cdot 21$	$(-2)(-21)$
$3 \cdot 14$	$(-3)(-14)$
$6 \cdot 7$	$(-6)(-7)$

Step 2: List all the factors of 42 and find the pair whose sum equals -17.

The numbers -3 and -14 satisfy both conditions: $(-3)(-14) = 42$ and $-3 + (-14) = -17$.

$= 5x(2x^2 - 17x + 21)$ **Step 3:** Write the middle term of the trinomial as two terms whose coefficients are the selected pair of numbers, -3 and -14.

$= 5x(2x^2 - 3x - 14x + 21)$

$= 5x(2x^2 - 3x \mid - 14x + 21)$ **Step 4:** Factor by grouping.

$= 5x[x(2x - 3) - 7(2x - 3)]$

$= 5x[(2x - 3)(x - 7)]$

$= 5x(2x - 3)(x - 7)$

Avoiding Mistakes:

Be sure to bring down the GCF in each successive step as you factor.

TIP: Notice when the GCF is removed from the original trinomial, the new trinomial has smaller coefficients. This makes the factoring process simpler because the product *ac* is smaller. It is much easier to list the factors of 42 than the factors of 1050.

Original trinomial	**With the GCF factored out**
$10x^3 - 85x^2 + 105x$	$5x(2x^2 - 17x + 21)$
$ac = (10)(105) = 1050$	$ac = (2)(21) = 42$

Skill Practice Factor by the ac-method.

3. $9y^3 - 30y^2 + 24y$

In most cases, it is easier to factor a trinomial with a positive leading coefficient.

Example 4 Factoring a Trinomial by the AC-Method

Factor: $-18x^2 + 21xy + 15y^2$

Skill Practice Answers

3. $3y(3y - 4)(y - 2)$

Solution:

$-18x^2 + 21xy + 15y^2$

$= -3(6x^2 - 7xy - 5y^2)$ — Factor out -3 to make the leading term positive.

Step 1: The product $ac = (6)(-5) = -30$.

Step 2: The numbers -10 and 3 have a product of -30 and a sum of -7.

$= -3[6x^2 - 10xy + 3xy - 5y^2]$ — **Step 3:** Rewrite the middle term, $-7xy$ as $-10xy + 3xy$.

$= -3[6x^2 - 10xy \mid + 3xy - 5y^2]$ — **Step 4:** Factor by grouping.

$= -3[2x(3x - 5y) + y(3x - 5y)]$

$= -3(3x - 5y)(2x + y)$ — Factored form

Skill Practice Factor.

4. $-8x^2 - 8xy + 30y^2$

TIP: Do not forget to write the GCF in the final answer.

2. Factoring Perfect Square Trinomials

Recall from Section 5.6 that the square of a binomial always results in a **perfect square trinomial**.

$$(a + b)^2 = (a + b)(a + b) \xrightarrow{\text{Multiply}} = a^2 + 2ab + b^2$$

$$(a - b)^2 = (a - b)(a - b) \xrightarrow{\text{Multiply}} = a^2 - 2ab + b^2$$

For example, $(3x + 5)^2 = (3x)^2 + 2(3x)(5) + (5)^2$

$= 9x^2 + 30x + 25$ (perfect square trinomial)

We now want to reverse this process by factoring a perfect square trinomial. The ac-method or the trial-and-error method can always be used; however, if we recognize the pattern for a perfect square trinomial, we can use one of the following formulas to reach a quick solution.

Factored Form of a Perfect Square Trinomial

$$a^2 + 2ab + b^2 = (a + b)^2$$

$$a^2 - 2ab + b^2 = (a - b)^2$$

For example, $9x^2 + 30x + 25$ is a perfect square trinomial with $a = 3x$ and $b = 5$. Therefore, it factors as

$$9x^2 + 30x + 25 = (3x)^2 + 2(3x)(5) + (5)^2 = (3x + 5)^2$$

$$a^2 + 2(a)(b) + (b)^2 = (a + b)^2$$

To apply the factored form of a perfect square trinomial, we must first be sure that the trinomial is indeed a perfect square trinomial.

Skill Practice Answers

4. $-2(2x - 3y)(2x + 5y)$

Checking for a Perfect Square Trinomial

1. Determine whether the first and third terms are both perfect squares and have positive coefficients.
2. If this is the case, identify a and b, and determine if the middle term equals $2ab$.

Example 5 Factoring Perfect Square Trinomials

Factor the trinomials completely.

a. $x^2 + 10x + 25$ **b.** $49y^2 - 28y + 4$

c. $8w^3 - 24w^2z + 18wz^2$ **d.** $2x^2 + 52x + 50$

Solution:

a. $x^2 + 10x + 25$

The GCF is 1.

- The first and third terms are positive.
- The first term is a perfect square: $x^2 = (x)^2$
- The third term is a perfect square: $25 = (5)^2$
- The middle term is twice the product of x and 5: $10x = 2(x)(5)$.

Perfect squares

$= x^2 + 10x + 25$

$= (x)^2 + 2(x)(5) + (5)^2$

$= (x + 5)^2$

This is a perfect square trinomial with $a = x$ and $b = 5$. Factor as $a^2 + 2ab + b^2 = (a + b)^2$.

b. $49y^2 - 28y + 4$

The GCF is 1.

- The first and third terms are positive.
- The first term is a perfect square: $49y^2 = (7y)^2$
- The third term is a perfect square: $4 = (2)^2$
- In the middle: $28y = 2(7y)(2)$.

Perfect squares

$= 49y^2 - 28y + 4$

$= (7y)^2 - 2(7y)(2) + (2)^2$

$= (7y - 2)^2$

This is a perfect square trinomial with $a = 7y$ and $b = 2$. Factor as $a^2 - 2ab + b^2 = (a - b)^2$.

c. $8w^3 - 24w^2z + 18wz^2$

$= 2w(4w^2 - 12wz + 9z^2)$

Perfect squares

The GCF is $2w$.

- $4w^2$ is a perfect square: $4w^2 = (2w)^2$
- $9z^2$ is a perfect square: $9z^2 = (3z)^2$

$= 2w[(2w)^2 - 2(2w)(3z) + (3z)^2]$

- In the middle: $12wz = 2(2w)(3z)$

$= 2w(2w - 3z)^2$

This is a perfect square trinomial with $a = 2w$ and $b = 3z$. Factor as $a^2 - 2ab + b^2 = (a - b)^2$.

d. $2x^2 + 52x + 50$

$= 2(x^2 + 26x + 25)$

The GCF is 2.

Middle term does not fit the pattern for a perfect square trinomial.

- The first and third terms are perfect squares. $x^2 = (x)^2$ and $25 = (5)^2$.
- However, the middle term is *not* 2 times the product of x and 5. Therefore, this is not a perfect square trinomial. $26x \neq 2(x)(5)$

$= 2(x + 25)(x + 1)$

Factor by using either the ac-method or the trial-and-error method.

Skill Practice Factor.

5. $y^2 + 20y + 100$

6. $16x^2 - 24x + 9$

7. $27w^3 + 18w^2 + 3w$

8. $9x^2 + 15x + 4$

TIP: To help you identify a perfect square trinomial, we recommend that you familiarize yourself with the first several perfect squares.

$(1)^2 = 1$	$(6)^2 = 36$	$(11)^2 = 121$
$(2)^2 = 4$	$(7)^2 = 49$	$(12)^2 = 144$
$(3)^2 = 9$	$(8)^2 = 64$	$(13)^2 = 169$
$(4)^2 = 16$	$(9)^2 = 81$	$(14)^2 = 196$
$(5)^2 = 25$	$(10)^2 = 100$	$(15)^2 = 225$

If you do not recognize that a trinomial is a perfect square trinomial, you may still use the trial-and-error method or ac-method to factor it.

Recall that a prime polynomial is a polynomial whose only factors are itself and 1. It also should be noted that not every trinomial is factorable by the methods presented in this text.

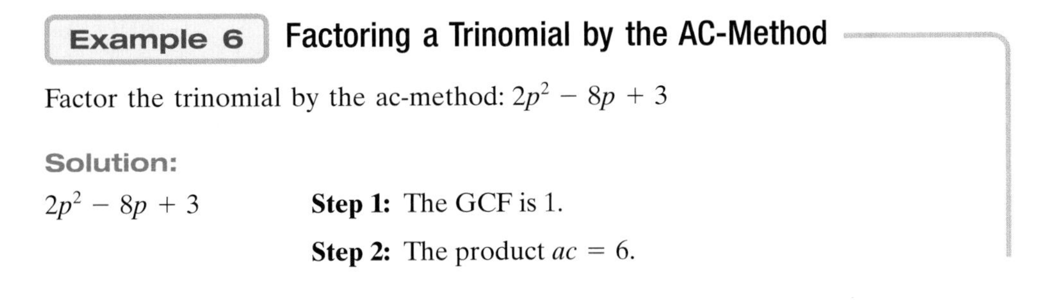

Example 6 Factoring a Trinomial by the AC-Method

Factor the trinomial by the ac-method: $2p^2 - 8p + 3$

Solution:

$2p^2 - 8p + 3$

Step 1: The GCF is 1.

Step 2: The product $ac = 6$.

Skill Practice Answers

5. $(y + 10)^2$ **6.** $(4x - 3)^2$
7. $3w(3w + 1)^2$
8. $(3x + 1)(3x + 4)$

$\underline{6}$	$\underline{6}$
$1 \cdot 6$	$(-1)(-6)$
$2 \cdot 3$	$(-2)(-3)$

Step 3: List the factors of 6. Notice that no pair of factors has a sum of -8. Therefore, the trinomial cannot be factored.

The trinomial $2p^2 - 8p + 3$ is a prime polynomial.

Skill Practice Factor.

9. $4x^2 + 5x + 2$

Skill Practice Answer

9. The trinomial is prime.

Section 6.4 Practice Exercises

Boost *your* GRADE at mathzone.com!

MathZone

- Practice Problems
- Self-Tests
- NetTutor
- e-Professors
- Videos

Study Skills Exercise

1. Define the key terms:

a. prime polynomial **b. perfect square trinomial**

Review Exercises

For Exercises 2–4, factor completely.

2. $5x(x - 2) - 2(x - 2)$

3. $8(y + 5) + 9y(y + 5)$

4. $6ab + 24b - 12a - 48$

Concept 1: Factoring Trinomials by the AC-Method

For Exercises 5–12, find the pair of integers whose product and sum are given.

5. Product: 12 Sum: 13

6. Product: 12 Sum: 7

7. Product: 8 Sum: -9

8. Product: -4 Sum: -3

9. Product: -20 Sum: 1

10. Product: -6 Sum: -1

11. Product: -18 Sum: 7

12. Product: -72 Sum: -6

For Exercises 13–30, factor the trinomials using the ac-method.

13. $3x^2 + 13x + 4$

14. $2y^2 + 7y + 6$

15. $4w^2 - 9w + 2$

16. $2p^2 - 3p - 2$

17. $2m^2 + 5m - 3$

18. $6n^2 + 7n - 3$

19. $8k^2 - 6k - 9$

20. $9h^2 - 12h + 4$

21. $4k^2 - 20k + 25$

22. $16h^2 + 24h + 9$

23. $5x^2 + x + 7$

24. $4y^2 - y + 2$

25. $10 + 9z^2 - 21z$

26. $13x + 4x^2 - 12$

27. $50y + 24 + 14y^2$

28. $-24 + 10w + 4w^2$

29. $12y^2 + 8yz - 15z^2$

30. $20a^2 + 3ab - 9b^2$

Concept 2: Factoring Perfect Square Trinomials

31. Multiply. $(5y - 7)^2$

32. Multiply. $(3x + 4)^2$

33. **a.** Which trinomial is a perfect square trinomial? $4x^2 - 25x + 25$ or $4x^2 - 20x + 25$

b. Factor the trinomials from part (a).

34. **a.** Which trinomial is a perfect square trinomial? $9x^2 + 12x + 4$ or $9x^2 + 15x + 4$

b. Factor the trinomials from part (a).

For Exercises 35–46, factor completely. (*Hint:* Look for the pattern of a perfect square trinomial.)

35. $t^2 - 16t + 64$

36. $n^2 + 18n + 81$

37. $49q^2 - 28q + 4$

38. $64y^2 - 80y + 25$

39. $36x + 4x^2 + 81$

40. $42p + 9p^2 + 49$

41. $32x^2 + 80xy + 50y^2$

42. $12w^2 - 36wz + 27z^2$

43. $4c^2 + 8c + 4$

44. $36x^2 - 24x + 4$

45. $9y^2 + 15y + 4$

46. $4w^2 + 20w + 9$

Mixed Exercises

For Exercises 47–79, factor completely.

47. $20p^2 - 19p + 3$

48. $4p^2 + 5pq - 6q^2$

49. $6u^2 - 19uv + 10v^2$

50. $15m^2 + mn - 2n^2$

51. $12a^2 + 11ab - 5b^2$

52. $3r^2 - rs - 14s^2$

53. $3h^2 + 19hk - 14k^2$

54. $2x^2 - 13xy + y^2$

55. $3p^2 + 20pq - q^2$

56. $3 - 14z + 16z^2$

57. $10w + 1 + 16w^2$

58. $b^2 + 16 - 8b$

59. $1 + q^2 - 2q$

60. $25x - 5x^2 - 30$

61. $20a - 18 - 2a^2$

62. $-6 - t + t^2$

63. $-6 + m + m^2$

64. $72x^2 + 18x - 2$

65. $20y^2 - 78y - 8$

66. $p^3 - 6p^2 - 27p$

67. $w^5 - 11w^4 + 28w^3$

68. $3x^3 + 10x^2 + 7x$

69. $4r^3 + 3r^2 - 10r$

70. $2p^3 - 38p^2 + 120p$

71. $4q^3 - 4q^2 - 80q$

72. $x^2y^2 + 14x^2y + 33x^2$

73. $a^2b^2 + 13ab^2 + 30b^2$

74. $-k^2 - 7k - 10$

75. $-m^2 - 15m + 34$

76. $-3n^2 - 3n + 90$

77. $-2h^2 + 28h - 90$

78. $x^4 - 7x^2 + 10$

79. $m^4 + 10m^2 + 21$

80. Is the expression $(2x + 4)(x - 7)$ factored completely? Explain why or why not.

81. Is the expression $(3x + 1)(5x - 10)$ factored completely? Explain why or why not.

Section 6.5 Factoring Binomials

Concepts

1. Factoring a Difference of Squares
2. Factoring a Sum or Difference of Cubes
3. Factoring Binomials: A Summary

1. Factoring a Difference of Squares

Up to this point, we have learned several methods of factoring, including:

- Factoring out the greatest common factor from a polynomial
- Factoring a four-term polynomial by grouping
- Recognizing and factoring perfect square trinomials
- Factoring trinomials by the ac-method or by the trial-and-error method

In this section, we will learn how to factor binomials. We begin by factoring a difference of squares. Recall from Section 5.6 that the product of two conjugates results in a **difference of squares**:

$$(a + b)(a - b) = a^2 - b^2$$

Therefore, to factor a difference of squares, the process is reversed. Identify a and b and construct the conjugate factors.

Factored Form of a Difference of Squares

$$a^2 - b^2 = (a + b)(a - b)$$

In addition to recognizing numbers that are perfect squares, it is helpful to recognize that a variable expression is a perfect square if its exponent is a multiple of 2. For example:

Perfect Squares

$x^2 = (x)^2$

$x^4 = (x^2)^2$

$x^6 = (x^3)^2$

$x^8 = (x^4)^2$

$x^{10} = (x^5)^2$

Example 1 Factoring Differences of Squares

Factor the binomials.

a. $y^2 - 25$ **b.** $49s^2 - 4t^4$ **c.** $18w^2z - 2z$

Solution:

a. $y^2 - 25$ The binomial is a difference of squares.

$= (y)^2 - (5)^2$ Write in the form: $a^2 - b^2$, where $a = y, b = 5$.

$= (y + 5)(y - 5)$ Factor as $(a + b)(a - b)$.

b. $49s^2 - 4t^4$ The binomial is a difference of squares.

$= (7s)^2 - (2t^2)^2$ Write in the form $a^2 - b^2$, where $a = 7s$ and $b = 2t^2$.

$= (7s + 2t^2)(7s - 2t^2)$ Factor as $(a + b)(a - b)$.

c. $18w^2z - 2z$ The GCF is $2z$.

$= 2z(9w^2 - 1)$ $(9w^2 - 1)$ is a difference of squares.

$= 2z[(3w)^2 - (1)^2]$ Write in the form: $a^2 - b^2$, where $a = 3w, b = 1$.

$= 2z(3w + 1)(3w - 1)$ Factor as $(a + b)(a - b)$.

Skill Practice Factor completely.

1. $a^2 - 64$ **2.** $25q^2 - 49w^2$ **3.** $98m^3n - 50mn$

The difference of squares $a^2 - b^2$ factors as $(a - b)(a + b)$. However, the *sum* of squares is not factorable.

Sum of Squares

Suppose a and b have no common factors. Then the **sum of squares** $a^2 + b^2$ is *not* factorable over the real numbers.

That is, $a^2 + b^2$ is prime over the real numbers.

To see why $a^2 + b^2$ is not factorable, consider the product of binomials:

$(a + b)(a - b) = a^2 - b^2$ Wrong sign

$(a + b)(a + b) = a^2 + 2ab + b^2$ Wrong middle term

$(a - b)(a - b) = a^2 - 2ab + b^2$ Wrong middle term

After exhausting all possibilities, we see that if a and b share no common factors, then the sum of squares $a^2 + b^2$ is a prime polynomial.

Example 2 Factoring Binomials

Factor the binomials, if possible.

a. $p^2 - 9$ **b.** $p^2 + 9$

Solution:

a. $p^2 - 9$ Difference of squares

$= (p - 3)(p + 3)$ Factor as $a^2 - b^2 = (a - b)(a + b)$.

b. $p^2 + 9$ Sum of squares

Prime (cannot be factored)

Skill Practice Factor the binomials, if possible.

4. $t^2 - 144$ **5.** $t^2 + 144$

Skill Practice Answers

1. $(a + 8)(a - 8)$
2. $(5q + 7w)(5q - 7w)$
3. $2mn(7m + 5)(7m - 5)$
4. $(t - 12)(t + 12)$
5. Prime

2. Factoring a Sum or Difference of Cubes

A binomial $a^2 - b^2$ is a difference of squares and can be factored as $(a - b)(a + b)$. Furthermore, if a and b share no common factors, then a sum of squares $a^2 + b^2$ is not factorable over the real numbers. In this section, we will learn that both a difference of cubes, $a^3 - b^3$, and a sum of cubes, $a^3 + b^3$, are factorable.

Factoring a Sum or Difference of Cubes

Sum of Cubes: $a^3 + b^3 = (a + b)(a^2 - ab + b^2)$

Difference of Cubes: $a^3 - b^3 = (a - b)(a^2 + ab + b^2)$

Multiplication can be used to confirm the formulas for factoring a sum or difference of cubes:

$$(a + b)(a^2 - ab + b^2) = a^3 - \cancel{a^2b} + \cancel{ab^2} + \cancel{a^2b} - \cancel{ab^2} + b^3 = a^3 + b^3 \checkmark$$

$$(a - b)(a^2 + ab + b^2) = a^3 + \cancel{a^2b} + \cancel{ab^2} - \cancel{a^2b} - \cancel{ab^2} - b^3 = a^3 - b^3 \checkmark$$

To help you remember the formulas for factoring a sum or difference of cubes, keep the following guidelines in mind:

- The factored form is the product of a binomial and a trinomial.
- The first and third terms in the trinomial are the squares of the terms within the binomial factor.
- Without regard to signs, the middle term in the trinomial is the product of terms in the binomial factor.

Square the first term of the binomial. Product of terms in the binomial

$$x^3 + 8 = (x)^3 + (2)^3 = (x + 2)[(x)^2 - (x)(2) + (2)^2]$$

Square the last term of the binomial.

- The sign within the binomial factor is the same as the sign of the original binomial.
- The first and third terms in the trinomial are always positive.
- The sign of the middle term in the trinomial is opposite the sign within the binomial.

Same sign Positive

$$x^3 + 8 = (x)^3 + (2)^3 = (x + 2)[(x)^2 - (x)(2) + (2)^2]$$

Opposite signs

To help you recognize a sum or difference of cubes, we recommend that you familiarize yourself with the first several perfect cubes:

Perfect Cube	Perfect Cube
$1 = (1)^3$	$216 = (6)^3$
$8 = (2)^3$	$343 = (7)^3$
$27 = (3)^3$	$512 = (8)^3$
$64 = (4)^3$	$729 = (9)^3$
$125 = (5)^3$	$1000 = (10)^3$

It is also helpful to recognize that a variable expression is a perfect cube if its exponent is a multiple of 3. For example,

Perfect Cube

$$x^3 = (x)^3$$
$$x^6 = (x^2)^3$$
$$x^9 = (x^3)^3$$
$$x^{12} = (x^4)^3$$

Example 3 Factoring a Sum of Cubes

Factor: $w^3 + 64$

Solution:

$w^3 + 64$ — w^3 and 64 are perfect cubes.

$= (w)^3 + (4)^3$ — Write as $a^3 + b^3$, where $a = w$, $b = 4$.

$a^3 + b^3 = (a + b)(a^2 - ab + b^2)$ — Apply the formula for a sum of cubes.

$$(w)^3 + (4)^3 = (w + 4)[(w)^2 - (w)(4) + (4)^2]$$

$= (w + 4)(w^2 - 4w + 16)$ — Simplify.

Skill Practice Factor.

6. $p^3 + 125$

Example 4 Factoring a Difference of Cubes

Factor: $27p^3 - q^6$

Solution:

$27p^3 - q^6$ — $27p^3$ and q^6 are perfect cubes.

$(3p)^3 - (q^2)^3$ — Write as $a^3 - b^3$, where $a = 3p$, $b = q^2$.

$a^3 - b^3 = (a - b)(a^2 + ab + b^2)$ — Apply the formula for a difference of cubes.

$$(3p)^3 - (q^2)^3 = (3p - q^2)[(3p)^2 + (3p)(q^2) + (q^2)^2]$$

$= (3p - q^2)(9p^2 + 3pq^2 + q^4)$ — Simplify.

Skill Practice Factor.

7. $8y^3 - 27z^6$

Skill Practice Answers

6. $(p + 5)(p^2 - 5p + 25)$
7. $(2y - 3z^2)(4y^2 + 6yz^2 + 9z^4)$

3. Factoring Binomials: A Summary

After removing the GCF, the next step in any factoring problem is to recognize what type of pattern it follows. Exponents that are divisible by 2 are perfect squares and those divisible by 3 are perfect cubes. The formulas for factoring binomials are summarized in the following box:

Factoring Binomials

1. Difference of Squares: $a^2 - b^2 = (a + b)(a - b)$
2. Difference of Cubes: $a^3 - b^3 = (a - b)(a^2 + ab + b^2)$
3. Sum of Cubes: $a^3 + b^3 = (a + b)(a^2 - ab + b^2)$

Example 5 Factoring Binomials

Factor completely.

a. $27y^3 + 1$ **b.** $m^2 - \frac{1}{4}$ **c.** $3y^4 - 48$ **d.** $z^6 - 8w^3$

Solution:

a. $27y^3 + 1$ — Sum of cubes: $27y^3 = (3y)^3$ and $1 = (1)^3$.

$= (3y)^3 + (1)^3$ — Write as $a^3 + b^3$, where $a = 3y$ and $b = 1$.

$= (3y + 1)((3y)^2 - (3y)(1) + (1)^2)$ — Apply the formula $a^3 + b^3 = (a + b)(a^2 - ab + b^2)$.

$= (3y + 1)(9y^2 - 3y + 1)$ — Simplify.

b. $m^2 - \frac{1}{4}$ — Difference of squares

$= (m)^2 - \left(\frac{1}{2}\right)^2$ — Write as $a^2 - b^2$, where $a = m$ and $b = \frac{1}{2}$.

$= \left(m + \frac{1}{2}\right)\left(m - \frac{1}{2}\right)$ — Apply the formula $a^2 - b^2 = (a + b)(a - b)$.

c. $3y^4 - 48$

$= 3(y^4 - 16)$ — Factor out the GCF. The binomial is a difference of squares.

$= 3[(y^2)^2 - (4)^2]$ — Write as $a^2 - b^2$, where $a = y^2$ and $b = 4$.

$= 3(y^2 + 4)(y^2 - 4)$ — Apply the formula $a^2 - b^2 = (a + b)(a - b)$.

$y^2 + 4$ is a sum of squares and cannot be factored.

$= 3(y^2 + 4)(y + 2)(y - 2)$ — $y^2 - 4$ is a difference of squares and can be factored further.

d. $z^6 - 8w^3$ — Difference of cubes: $z^6 = (z^2)^3$ and $8w^3 = (2w)^3$

$= (z^2)^3 - (2w)^3$ — Write as $a^3 - b^3$, where $a = z^2$ and $b = 2w$.

$= (z^2 - 2w)[(z^2)^2 + (z^2)(2w) + (2w)^2]$ — Apply the formula $a^3 - b^3 = (a - b)(a^2 + ab + b^2)$.

$= (z^2 - 2w)(z^4 + 2z^2w + 4w^2)$ — Simplify.

Each of the factorizations in Example 3 can be checked by multiplying.

Skill Practice Factor completely.

8. $1000x^3 + 1$ **9.** $25p^2 - \frac{1}{9}$ **10.** $2x^4 - 2$ **11.** $27a^6 - b^3$

Some factoring problems require more than one method of factoring. In general, when factoring a polynomial, be sure to factor completely.

Example 6 Factoring Polynomials

Factor completely.

a. $w^4 - 81$ **b.** $4x^3 + 4x^2 - 25x - 25$

Solution:

a. $w^4 - 81$ — The GCF is 1. $w^4 - 81$ is a difference of squares.

$= (w^2)^2 - (9)^2$ — Write in the form: $a^2 - b^2$, where $a = w^2$, $b = 9$.

$= (w^2 + 9)(w^2 - 9)$ — Factor as $(a + b)(a - b)$.

$= (w^2 + 9)(w + 3)(w - 3)$ — Note that $w^2 - 9$ can be factored further as a difference of squares. (The binomial $w^2 + 9$ is a sum of squares and cannot be factored further.)

b. $4x^3 + 4x^2 - 25x - 25$ — The GCF is 1.

$= 4x^3 + 4x^2 \mid - 25x - 25$ — The polynomial has four terms. Factor by grouping.

$= 4x^2(x + 1) - 25(x + 1)$

$= (x + 1)(4x^2 - 25)$ — $4x^2 - 25$ is a difference of squares.

$= (x + 1)(2x + 5)(2x - 5)$

Skill Practice Factor completely, if possible.

12. $y^4 - 625$ **13.** $x^3 + 6x^2 - 4x - 24$

Skill Practice Answers

8. $(10x + 1)(100x^2 - 10x + 1)$
9. $\left(5p - \frac{1}{3}\right)\left(5p + \frac{1}{3}\right)$
10. $2(x^2 + 1)(x - 1)(x + 1)$
11. $(3a^2 - b)(9a^4 + 3a^2b + b^2)$
12. $(y^2 + 25)(y + 5)(y - 5)$
13. $(x + 6)(x + 2)(x - 2)$

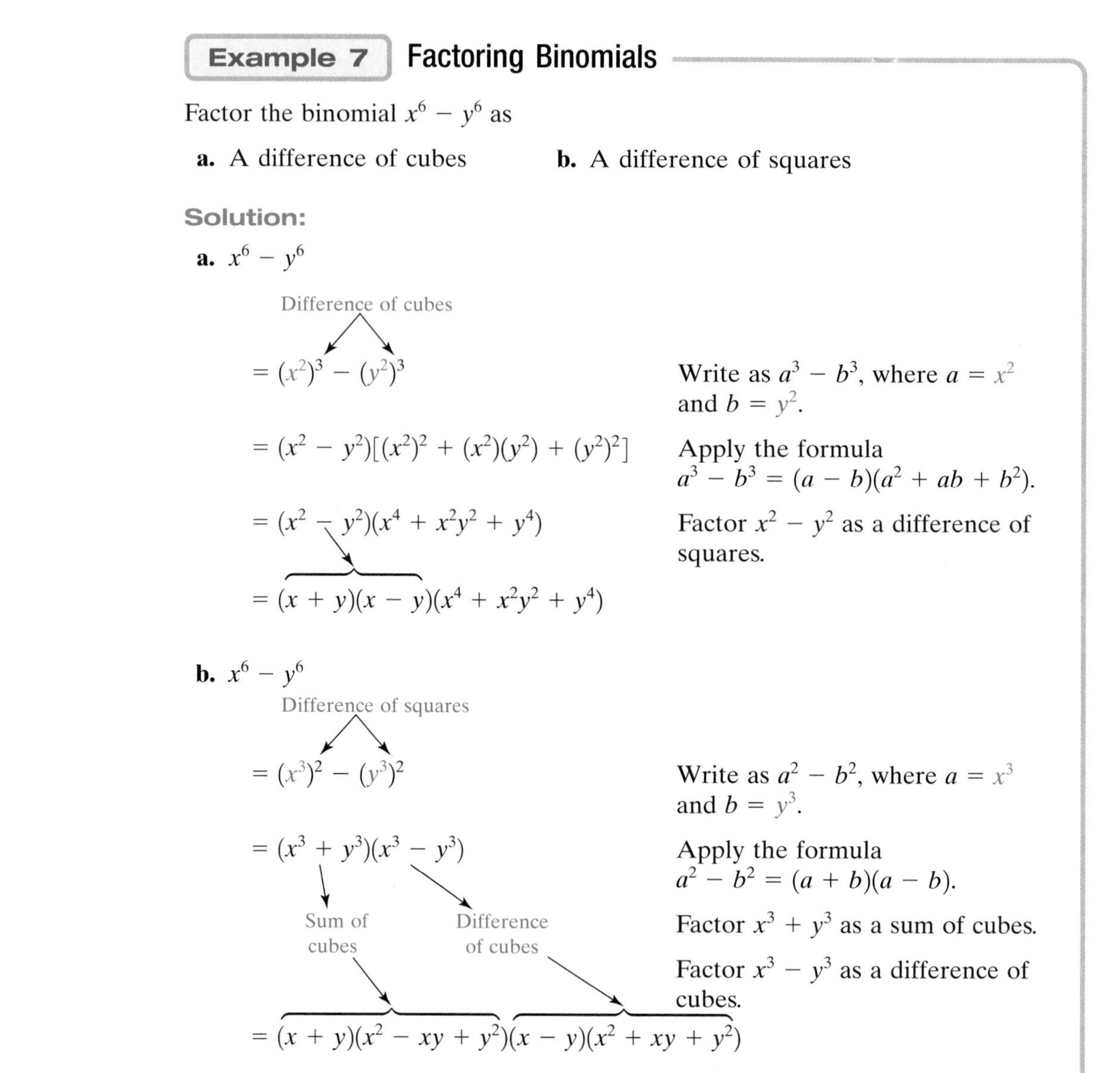

Example 7 Factoring Binomials

Factor the binomial $x^6 - y^6$ as

a. A difference of cubes **b.** A difference of squares

Solution:

a. $x^6 - y^6$

Difference of cubes

$= (x^2)^3 - (y^2)^3$ Write as $a^3 - b^3$, where $a = x^2$ and $b = y^2$.

$= (x^2 - y^2)[(x^2)^2 + (x^2)(y^2) + (y^2)^2]$ Apply the formula $a^3 - b^3 = (a - b)(a^2 + ab + b^2)$.

$= (x^2 - y^2)(x^4 + x^2y^2 + y^4)$ Factor $x^2 - y^2$ as a difference of squares.

$= (x + y)(x - y)(x^4 + x^2y^2 + y^4)$

b. $x^6 - y^6$

Difference of squares

$= (x^3)^2 - (y^3)^2$ Write as $a^2 - b^2$, where $a = x^3$ and $b = y^3$.

$= (x^3 + y^3)(x^3 - y^3)$ Apply the formula $a^2 - b^2 = (a + b)(a - b)$.

Sum of cubes; Difference of cubes

Factor $x^3 + y^3$ as a sum of cubes.

Factor $x^3 - y^3$ as a difference of cubes.

$= (x + y)(x^2 - xy + y^2)(x - y)(x^2 + xy + y^2)$

Notice that the expressions x^6 and y^6 are both perfect squares and perfect cubes because both exponents are multiples of 2 and of 3. Consequently, $x^6 - y^6$ can be factored initially as either the difference of squares or as the difference of cubes. In such a case, it is recommended that you factor the expression as a difference of squares first because it factors more completely into polynomials of lower degree. Hence,

$$x^6 - y^6 = (x + y)(x^2 - xy + y^2)(x - y)(x^2 + xy + y^2)$$

Skill Practice Factor completely.

14. $z^6 - 64$

Skill Practice Answers

14. $(z + 2)(z - 2)(z^2 + 2z + 4)(z^2 - 2z + 4)$

Section 6.5 Practice Exercises

Study Skills Exercise

1. Define the key terms:

a. difference of squares **b. sum of squares** **c. difference of cubes** **d. sum of cubes**

Review Exercises

For Exercises 2–10, factor completely.

2. $3x^2 + x - 10$ **3.** $6x^2 - 17x + 5$ **4.** $6a^2b + 3a^3b$

5. $15x^2y^5 - 10xy^6$ **6.** $5p^2q + 20p^2 - 3pq - 12p$ **7.** $ax + ab - 6x - 6b$

8. $2xy - 3x - 4y + 6$ **9.** $25x^2 + 30x + 9$ **10.** $81y^2 - 36y + 4$

Concept 1: Factoring a Difference of Squares

11. What binomial factors as $(x - 5)(x + 5)$?

12. What binomial factors as $(n - 3)(n + 3)$?

13. What binomial factors as $(2p - 3q)(2p + 3q)$?

14. What binomial factors as $(7x - 4y)(7x + 4y)$?

For Exercises 15–34, factor the binomials completely.

15. $x^2 - 36$ **16.** $r^2 - 81$ **17.** $w^2 - 100$ **18.** $t^2 - 49$

19. $4a^2 - 121b^2$ **20.** $9x^2 - y^2$ **21.** $49m^2 - 16n^2$ **22.** $100a^2 - 49b^2$

23. $9q^2 + 16$ **24.** $36 + s^2$ **25.** $y^2 - 4z^2$ **26.** $b^2 - 144c^2$

27. $a^2 - b^4$ **28.** $y^4 - x^2$ **29.** $25p^2q^2 - 1$ **30.** $81s^2t^2 - 1$

31. $c^6 - 25$ **32.** $z^6 - 4$ **33.** $25 - 16t^2$ **34.** $64 - h^2$

Concept 2: Factoring a Sum or Difference of Cubes

35. Identify the expressions that are perfect cubes:

$$\{x^3, 8, 9, y^6, a^4, b^2, 3p^3, 27q^3, w^{12}, r^3s^6\}$$

36. Identify the expressions that are perfect cubes:

$$\{z^9, -81, 30, 8, 6x^3, y^{15}, 27a^3, b^2, p^3q^2, -1\}$$

37. From memory, write the formula to factor a sum of cubes:

$a^3 + b^3 =$ ______________

38. From memory, write the formula to factor a difference of cubes:

$a^3 - b^3 =$ ______________

For Exercises 39–58, factor the sums or differences of cubes.

39. $y^3 - 8$ **40.** $x^3 + 27$ **41.** $1 - p^3$ **42.** $q^3 + 1$

43. $w^3 + 64$ **44.** $8 - t^3$ **45.** $x^3 - 1000$ **46.** $8y^3 - 27$

47. $64t^3 + 1$ **48.** $125r^3 + 1$ **49.** $1000a^3 + 27$ **50.** $216b^3 - 125$

51. $n^3 - \frac{1}{8}$ **52.** $\frac{8}{27} + m^6$ **53.** $a^3 + b^6$ **54.** $u^6 - v^3$

55. $x^9 + 64y^3$ **56.** $125w^3 - z^9$ **57.** $25m^{12} + 16$ **58.** $36p^6 + 49q^4$

Concept 3: Factoring Binomials: A Summary

For Exercises 59–90, factor completely.

59. $x^4 - 4$ **60.** $b^4 - 25$ **61.** $a^2 + 9$ **62.** $w^2 + 36$

63. $t^3 + 64$ **64.** $u^3 + 27$ **65.** $g^3 - 4$ **66.** $h^3 - 25$

67. $4b^3 + 108$ **68.** $3c^3 - 24$ **69.** $5p^2 - 125$ **70.** $2q^4 - 8$

71. $\frac{1}{64} - 8h^3$ **72.** $\frac{1}{125} + k^6$ **73.** $x^4 - 16$ **74.** $p^4 - 81$

75. $q^6 - 64$ **76.** $a^6 - 1$ **77.** $\frac{4x^2}{9} - w^2$ **78.** $\frac{16y^2}{25} - x^2$

79. $2x^3 + 3x^2 - 2x - 3$ **80.** $3x^3 + x^2 - 12x - 4$ **81.** $16x^4 - y^4$ **82.** $1 - t^4$

83. $81y^4 - 16$ **84.** $u^4 - 256$ **85.** $81k^2 + 30k + 1$ **86.** $9h^2 - 15h + 4$

87. $k^3 + 4k^2 - 9k - 36$ **88.** $w^3 - 2w^2 - 4w + 8$ **89.** $2t^3 - 10t^2 - 2t + 10$ **90.** $9a^3 + 27a^2 - 4a - 12$

Expanding Your Skills

For Exercises 91–98, factor the difference of squares.

91. $(y - 3)^2 - 9$ **92.** $(x - 2)^2 - 4$ **93.** $(2p + 1)^2 - 36$ **94.** $(4q + 3)^2 - 25$

95. $16 - (t + 2)^2$ **96.** $81 - (a + 5)^2$ **97.** $100 - (2b - 5)^2$ **98.** $49 - (3k - 7)^2$

Section 6.6 General Factoring Summary

Concepts

1. Factoring Strategy
2. Mixed Practice

1. Factoring Strategy

To factor a polynomial, remember always to look for the greatest common factor first. Then identify the number of terms and the type of factoring problem the polynomial represents.

Factoring Strategy

1. Factor out the GCF (Section 6.1).
2. Identify whether the polynomial has two terms, three terms, or more than three terms.
3. If the polynomial has more than three terms, try factoring by grouping (Section 6.1).
4. If the polynomial has three terms, check first for a perfect square trinomial. Otherwise, factor the trinomial with the ac-method or the trial-and-error method (Sections 6.3 or 6.4, respectively).
5. If the polynomial has two terms, determine if it fits the pattern for
   - A difference of squares: $a^2 - b^2 = (a - b)(a + b)$
   - A sum of squares: $a^2 + b^2$ prime
   - A difference of cubes: $a^3 - b^3 = (a - b)(a^2 + ab + b^2)$
   - A sum of cubes: $a^3 + b^3 = (a + b)(a^2 - ab + b^2)$ (Section 6.5)
6. Be sure to factor the polynomial completely.
7. Check by multiplying.

2. Mixed Practice

Example 1 Factoring Polynomials

Factor out the GCF and identify the number of terms and type of factoring pattern represented by the polynomial. Then factor the polynomial completely.

a. $abx^2 - 3ax + 5bx - 15$

b. $20y^2 - 110y - 210$

c. $4p^3 + 20p^2 + 25p$

d. $w^3 + 1000$

e. $t^4 - \frac{1}{16}$

f. $y^3 - 5y^2 - 4y + 20$

Solution:

a. $abx^2 - 3ax + 5bx - 15$

The GCF is 1. The polynomial has four terms. Therefore, factor by grouping.

$abx^2 - 3ax \;\vdots\; + 5bx - 15$

$= ax(bx - 3) + 5(bx - 3)$

$= (bx - 3)(ax + 5)$

Check: $(bx - 3)(ax + 5) = abx^2 + 5bx - 3ax - 15$ ✔

b. $20y^2 - 110y - 210$

$= 10(2y^2 - 11y - 21)$

$= 10(2y + 3)(y - 7)$

The GCF is 10. The polynomial has three terms. The trinomial is not a perfect square trinomial. Use either the ac-method or the trial-and-error method.

Check: $10(2y + 3)(y - 7) = 10(2y^2 - 14y + 3y - 21)$

$= 10(2y^2 - 11y - 21)$

$= 20y^2 - 110y - 210$ ✔

c. $4p^3 + 20p^2 + 25p$

$= p(4p^2 + 20p + 25)$ — The GCF is p. The polynomial has three terms and is a perfect square trinomial, $a^2 + 2ab + b^2$, where $a = 2p$ and $b = 5$.

$= p(2p + 5)^2$ — Apply the formula $a^2 + 2ab + b^2 = (a + b)^2$.

<u>Check</u>: $p(2p + 5)^2 = p[(2p)^2 + 2(2p)(5) + (5)^2]$

$= p(4p^2 + 20p + 25)$

$= 4p^3 + 20p^2 + 25p$ ✔

d. $w^3 + 1000$

$= (w)^3 + (10)^3$ — The GCF is 1. The polynomial has two terms. The binomial is a sum of cubes, $a^3 + b^3$, where $a = w$ and $b = 10$.

$= (w + 10)(w^2 - 10w + 100)$ — Apply the formula $a^3 + b^3 = (a + b)(a^2 - ab + b^2)$.

<u>Check</u>: $(w + 10)(w^2 - 10w + 100) = w^3 - \cancel{10w^2} + \cancel{100w} + \cancel{10w^2} - \cancel{100w} + 1000$

$= w^3 + 100$ ✔

e. $t^4 - \dfrac{1}{16}$ — The GCF is 1. The binomial is a difference of squares, $a^2 - b^2$.

$= \left(t^2 - \dfrac{1}{4}\right)\left(t^2 + \dfrac{1}{4}\right)$ — Apply the formula $a^2 - b^2 = (a - b)(a + b)$.

$= \left(t - \dfrac{1}{2}\right)\left(t + \dfrac{1}{2}\right)\left(t^2 + \dfrac{1}{4}\right)$ — The binomial $t^2 - \frac{1}{4}$ is also a difference of squares.

<u>Check</u>: $(t - \frac{1}{2})(t + \frac{1}{2})(t^2 + \frac{1}{4}) = (t^2 + \cancel{\frac{1}{2}t} - \cancel{\frac{1}{2}t} - \frac{1}{4})(t^2 + \frac{1}{4})$

$= (t^2 - \frac{1}{4})(t^2 + \frac{1}{4})$

$= t^4 + \cancel{\frac{1}{4}t^2} - \cancel{\frac{1}{4}t^2} - \frac{1}{16}$

$= t^4 - \frac{1}{16}$ ✔

f. $y^3 - 5y^2 - 4y + 20$ — The GCF is 1. The polynomial has four terms. Factor by grouping.

$= y^3 - 5y^2 \; | \; -4y + 20$

$= y^2(y - 5) - 4(y - 5)$

$= (y - 5)(y^2 - 4)$ — The expression $y^2 - 4$ is a difference of squares and can be factored further as $(y - 2)(y + 2)$.

$= (y - 5)(y - 2)(y + 2)$

<u>Check</u>: $(y - 5)(y - 2)(y + 2) = (y - 5)(y^2 - \cancel{2y} + \cancel{2y} - 4)$

$= (y - 5)(y^2 - 4)$

$= (y^3 - 4y - 5y^2 + 20)$

$= y^3 - 5y^2 - 4y + 20$ ✔

Skill Practice Factor completely.

1. $3ac + 6bc + ad + 2bd$ **2.** $18y^3 - 36y^2 + 10y$ **3.** $25t^2 - 20t + 4$

4. $4x^3 + 32y^3$ **5.** $w^4 - \dfrac{1}{81}$ **6.** $p^3 + 7p^2 - 9p - 63$

Skill Practice Answers

1. $(a + 2b)(3c + d)$
2. $2y(3y - 1)(3y - 5)$
3. $(5t - 2)^2$
4. $4(x + 2y)(x^2 - 2xy + 4y^2)$
5. $(w - \frac{1}{3})(w + \frac{1}{3})(w^2 + \frac{1}{9})$
6. $(p - 3)(p + 3)(p + 7)$

Section 6.6 Practice Exercises

Boost *your* GRADE at mathzone.com!

- Practice Problems
- Self-Tests
- NetTutor
- e-Professors
- Videos

Study Skills Exercise

1. This section summarizes the different methods of factoring that we have learned. In your own words, write a strategy for factoring any expression completely. Write it as if you were explaining the factoring process to another person.

Concept 1: Factoring Strategy

2. What is meant by a prime factor?

3. When factoring a trinomial, what pattern do you look for first before using the ac-method or trial-and-error method?

4. What is the first step in factoring any polynomial?

5. When factoring a binomial, what patterns can you look for?

6. What technique should be considered when factoring a four-term polynomial?

Concept 2: Mixed Practice

For Exercises 7–74,

a. Factor out the GCF from each polynomial. Then identify the category in which the polynomial best fits. Choose from
- difference of squares
- sum of squares
- difference of cubes
- sum of cubes
- trinomial (perfect square trinomial)
- trinomial (nonperfect square trinomial)
- four terms-grouping
- none of these

b. Factor the polynomial completely.

7. $2a^2 - 162$ **8.** $y^2 + 4y + 3$ **9.** $6w^2 - 6w$ **10.** $16z^4 - 81$

11. $3t^2 + 13t + 4$ **12.** $5r^3 + 5$ **13.** $3ac + ad - 3bc - bd$ **14.** $x^3 - 125$

15. $y^3 + 8$ **16.** $7p^2 - 29p + 4$ **17.** $3q^2 - 9q - 12$ **18.** $-2x^2 + 8x - 8$

19. $18a^2 + 12a$ **20.** $54 - 2y^3$ **21.** $4t^2 - 100$ **22.** $4t^2 - 31t - 8$

23. $10c^2 + 10c + 10$ **24.** $2xw - 10x + 3yw - 15y$ **25.** $x^3 + 0.001$ **26.** $4q^2 - 9$

27. $64 + 16k + k^2$ **28.** $s^2t + 5t + 6s^2 + 30$ **29.** $2x^2 + 2x - xy - y$ **30.** $w^3 + y^3$

31. $a^3 - c^3$ **32.** $3y^2 + y + 1$ **33.** $c^2 + 8c + 9$ **34.** $a^2 + 2a + 1$

35. $b^2 + 10b + 25$ **36.** $-t^2 - 4t + 32$ **37.** $-p^3 - 5p^2 - 4p$ **38.** $x^2y^2 - 49$

39. $6x^2 - 21x - 45$ **40.** $20y^2 - 14y + 2$ **41.** $5a^2bc^3 - 7abc^2$

42. $8a^2 - 50$ **43.** $t^2 + 2t - 63$ **44.** $b^2 + 2b - 80$

45. $ab + ay - b^2 - by$ **46.** $6x^3y^4 + 3x^2y^5$ **47.** $14u^2 - 11uv + 2v^2$

48. $9p^2 - 36pq + 4q^2$ **49.** $4q^2 - 8q - 6$ **50.** $9w^2 + 3w - 15$

51. $9m^2 + 16n^2$ **52.** $5b^2 - 30b + 45$ **53.** $6r^2 + 11r + 3$

54. $4s^2 + 4s - 15$ **55.** $16a^4 - 1$ **56.** $p^3 + p^2c - 9p - 9c$

57. $81u^2 - 90uv + 25v^2$ **58.** $4x^2 + 16$ **59.** $x^2 - 5x - 6$

60. $q^2 + q - 7$ **61.** $2ax - 6ay + 4bx - 12by$ **62.** $8m^3 - 10m^2 - 3m$

63. $21x^4y + 41x^3y + 10x^2y$ **64.** $2m^4 - 128$ **65.** $8uv - 6u + 12v - 9$

66. $4t^2 - 20t + st - 5s$ **67.** $12x^2 - 12x + 3$ **68.** $p^2 + 2pq + q^2$

69. $6n^3 + 5n^2 - 4n$ **70.** $4k^3 + 4k^2 - 3k$ **71.** $64 - y^2$

72. $36b - b^3$ **73.** $b^2 - 4b + 10$ **74.** $y^2 + 6y + 8$

Expanding Your Skills

For Exercises 75–78, factor completely.

75. $\frac{64}{125}p^3 - \frac{1}{8}q^3$ **76.** $\frac{1}{1000}r^3 + \frac{8}{27}s^3$ **77.** $a^{12} + b^{12}$ **78.** $a^9 - b^9$

Use Exercises 79–80 to investigate the relationship between division and factoring.

79. **a.** Use long division to divide $x^3 - 8$ by $(x - 2)$.

b. Factor $x^3 - 8$.

80. **a.** Use long division to divide $y^3 + 27$ by $(y + 3)$.

b. Factor $y^3 + 27$.

81. What trinomial multiplied by $(x - 2)$ gives a difference of cubes?

82. What trinomial multiplied by $(p + 3)$ gives a sum of cubes?

83. Write a binomial that when multiplied by $(4x^2 - 2x + 1)$ produces a sum of cubes.

84. Write a binomial that when multiplied by $(9y^2 + 15y + 25)$ produces a difference of cubes.

85. **a.** Write a polynomial that represents the area of the shaded region in the figure.

b. Factor the expression from part (a).

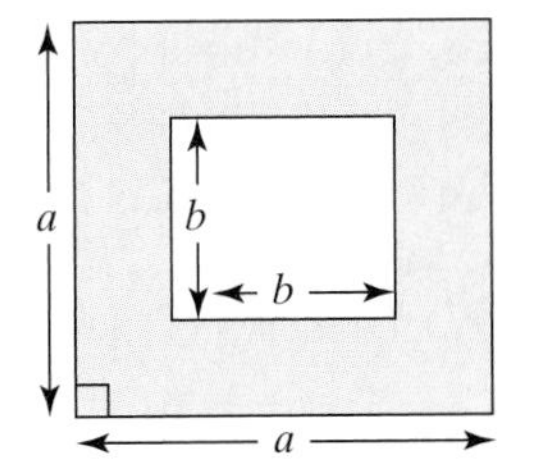

86. **a.** Write a polynomial that represents the area of the shaded region in the figure.

b. Factor the expression from part (a).

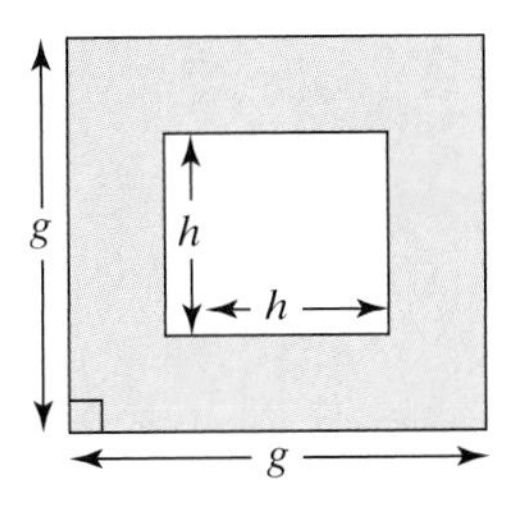

87. Use a difference of squares to find the product 67×73. [*Hint:* Write the product as $(70 - 3)(70 + 3)$.]

88. Use a difference of squares to find the product 85×75.

Solving Equations Using the Zero Product Rule

Section 6.7

Concepts

1. Definition of a Quadratic Equation
2. Zero Product Rule
3. Solving Equations Using the Zero Product Rule
4. Applications of Quadratic Equations
5. Pythagorean Theorem

1. Definition of a Quadratic Equation

In Section 2.1, we solved linear equations in one variable. These are equations of the form $ax + b = 0$ $(a \neq 0)$. A linear equation in one variable is sometimes called a first-degree polynomial equation because the highest degree of all its terms is 1. A second-degree polynomial equation in one variable is called a quadratic equation.

Definition of a Quadratic Equation in One Variable

If a, b, and c are real numbers such that $a \neq 0$, then a **quadratic equation** is an equation that can be written in the form

$$ax^2 + bx + c = 0$$

The following equations are quadratic because they can each be written in the form $ax^2 + bx + c = 0$, $(a \neq 0)$.

$-4x^2 + 4x = 1$	$x(x - 2) = 3$	$(x - 4)(x + 4) = 9$
$-4x^2 + 4x - 1 = 0$	$x^2 - 2x = 3$	$x^2 - 16 = 9$
	$x^2 - 2x - 3 = 0$	$x^2 - 25 = 0$
		$x^2 + 0x - 25 = 0$

2. Zero Product Rule

One method for solving a quadratic equation is to factor the equation and apply the zero product rule. The **zero product rule** states that if the product of two factors is zero, then one or both of its factors is zero.

Zero Product Rule

If $ab = 0$, then $a = 0$ or $b = 0$.

For example, the quadratic equation $x^2 - x - 12 = 0$ can be written in factored form as $(x - 4)(x + 3) = 0$. By the zero product rule, one or both factors must be zero. Hence, either $x - 4 = 0$ or $x + 3 = 0$. Therefore, to solve the quadratic equation, set each factor equal to zero and solve for x.

$(x - 4)(x + 3) = 0$			Apply the zero product rule.
$x - 4 = 0$	or	$x + 3 = 0$	Set each factor equal to zero.
$x = 4$	or	$x = -3$	Solve each equation for x.

3. Solving Equations Using the Zero Product Rule

Quadratic equations, like linear equations, arise in many applications in mathematics, science, and business. The following steps summarize the factoring method for solving a quadratic equation.

Steps for Solving a Quadratic Equation by Factoring

1. Write the equation in the form: $ax^2 + bx + c = 0$.
2. Factor the equation completely.
3. Apply the zero product rule. That is, set each factor equal to zero, and solve the resulting equations.

Note: The solution(s) found in step 3 may be checked by substitution in the original equation.

Example 1 Solving Quadratic Equations

Solve the quadratic equations.

a. $2x^2 - 9x = 5$ **b.** $4x^2 + 24x = 0$ **c.** $5x(5x + 2) = 10x + 9$

Solution:

a. $2x^2 - 9x = 5$

$2x^2 - 9x - 5 = 0$			Write the equation in the form $ax^2 + bx + c = 0$.
$(2x + 1)(x - 5) = 0$			Factor the polynomial completely.
$2x + 1 = 0$	or	$x - 5 = 0$	Set each factor equal to zero.
$2x = -1$	or	$x = 5$	Solve each equation.
$x = -\frac{1}{2}$	or	$x = 5$	

Check: $x = -\frac{1}{2}$

$$2x^2 - 9x = 5$$

$$2\left(-\frac{1}{2}\right)^2 - 9\left(-\frac{1}{2}\right) \stackrel{?}{=} 5$$

$$2\left(\frac{1}{4}\right) + \frac{9}{2} \stackrel{?}{=} 5$$

$$\frac{1}{2} + \frac{9}{2} \stackrel{?}{=} 5$$

$$\frac{10}{2} = 5 \checkmark$$

Check: $x = 5$

$$2x^2 - 9x = 5$$

$$2(5)^2 - 9(5) \stackrel{?}{=} 5$$

$$2(25) - 45 \stackrel{?}{=} 5$$

$$50 - 45 = 5 \checkmark$$

b. $4x^2 + 24x = 0$ The equation is already in the form $ax^2 + bx + c = 0$. (Note that $c = 0$.)

$4x(x + 6) = 0$ Factor completely.

$4x = 0$ or $x + 6 = 0$ Set each factor equal to zero.

$x = 0$ or $x = -6$ Each solution checks in the original equation.

c. $5x(5x + 2) = 10x + 9$

$25x^2 + 10x = 10x + 9$ Clear parentheses.

$25x^2 + 10x - 10x - 9 = 0$ Set the equation equal to zero.

$25x^2 - 9 = 0$ The equation is in the form $ax^2 + bx + c = 0$. (Note that $b = 0$.)

$(5x - 3)(5x + 3) = 0$ Factor completely.

$5x - 3 = 0$ or $5x + 3 = 0$ Set each factor equal to zero.

$5x = 3$ or $5x = -3$ Solve each equation.

$\frac{5x}{5} = \frac{3}{5}$ or $\frac{5x}{5} = \frac{-3}{5}$

$x = \frac{3}{5}$ or $x = -\frac{3}{5}$ Each solution checks in the original equation.

Skill Practice Solve the quadratic equations.

1. $2y^2 + 19y = -24$
2. $5s^2 = 45$
3. $4z(z + 3) = 4z + 5$

The zero product rule can be used to solve higher degree polynomial equations provided the equations can be set to zero and written in factored form.

Skill Practice Answers

1. $y = -8,\ y = -\frac{3}{2}$
2. $s = 3,\ s = -3$
3. $z = -\frac{5}{2},\ z = \frac{1}{2}$

Example 2 Solving Higher Degree Polynomial Equations

Solve the equations.

a. $-6(y+3)(y-5)(2y+7)=0$ **b.** $w^3+5w^2-9w-45=0$

Solution:

a. $-6(y+3)(y-5)(2y+7)=0$

The equation is already in factored form and equal to zero.

Set each factor equal to zero.

Solve each equation for y.

$-6 \neq 0$ or $y+3=0$ or $y-5=0$ or $2y+7=0$

No solution, $y=-3$ or $y=5$ or $y=-\frac{7}{2}$

Notice that when the constant factor is set equal to zero, the result is a contradiction $-6=0$. The constant factor does not produce a solution to the equation. Therefore, the only solutions are $y=-3$, $y=5$, and $y=-\frac{7}{2}$. Each solution can be checked in the original equation.

b. $w^3+5w^2-9w-45=0$

This is a higher degree polynomial equation.

$w^3+5w^2 \mid -9w-45=0$

The equation is already set equal to zero. Now factor.

$w^2(w+5)-9(w+5)=0$

$(w+5)(w^2-9)=0$

Because there are four terms, try factoring by grouping.

$(w+5)(w-3)(w+3)=0$

w^2-9 is a difference of squares and can be factored further.

$w+5=0$ or $w-3=0$ or $w+3=0$

Set each factor equal to zero.

$w=-5$ or $w=3$ or $w=-3$

Solve each equation.

Each solution checks in the original equation.

Skill Practice Solve the equations.

4. $5(p-4)(p+7)(2p-9)=0$
5. $x^3+x^2-6x=0$

4. Applications of Quadratic Equations

Example 3 Using a Quadratic Equation in a Geometry Application

A rectangular sign has an area of 40 ft^2. If the width is 3 feet shorter than the length, what are the dimensions of the sign?

Skill Practice Answers

4. $p=4$, $p=-7$, $p=\frac{9}{2}$
5. $x=0$, $x=-3$, $x=2$

Solution:

Let x represent the length of the sign. Then $x - 3$ represents the width (Figure 6-1).

Label the variables.

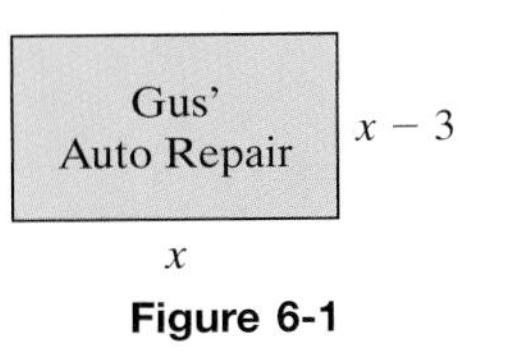

Figure 6-1

The problem gives information about the length of the sides and about the area. Therefore, we can form a relationship by using the formula for the area of a rectangle.

$A = l \cdot w$ — Area equals length times width.

$40 = x(x - 3)$ — Set up an algebraic equation.

$40 = x^2 - 3x$ — Clear parentheses.

$0 = x^2 - 3x - 40$ — Write the equation in the form, $ax^2 + bx + c = 0$.

$0 = (x - 8)(x + 5)$ — Factor the equation.

$0 = x - 8 \quad \text{or} \quad 0 = x + 5$ — Set each factor equal to zero.

$8 = x \quad \text{or} \quad -5 \not= x$ — Because x represents the length of a rectangle, reject the negative solution.

The variable x represents the length of the sign. Thus, the length is 8 ft.

The expression $x - 3$ represents the width. The width is 8 ft $-$ 3 ft, or 5 ft.

Skill Practice

6. The length of a rectangle is 5 ft more than the width. The area is 36 ft^2.

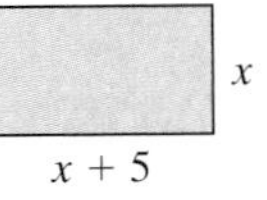

Find the length and width of the rectangle.

Example 4 Translating to a Quadratic Equation

The product of two consecutive integers is 48 more than the larger integer. Find the integers.

Solution:

Let x represent the first (smaller) integer.

Then $x + 1$ represents the second (larger) integer. — Label the variables.

(First integer)(second integer) = (second integer) + 48 — Verbal model

$x(x + 1) = (x + 1) + 48$ — Algebraic equation

$x^2 + x = x + 49$ — Simplify.

$x^2 + x - x - 49 = 0$ — Set the equation equal to zero.

Skill Practice Answers

6. The width is 4 ft, and the length is 9 ft.

$$x^2 - 49 = 0$$

$$(x - 7)(x + 7) = 0$$ Factor.

$x - 7 = 0$ or $x + 7 = 0$ Set each factor equal to zero.

$x = 7$ or $x = -7$ Solve for x.

Recall that x represents the smaller integer. Therefore, there are two possibilities for the pairs of consecutive integers.

If $x = 7$, then the larger integer is $x + 1$ or $7 + 1 = 8$.

If $x = -7$, then the larger integer is $x + 1$ or $-7 + 1 = -6$.

The integers are 7 and 8, or -7 and -6.

Skill Practice

7. The product of two consecutive odd integers is 9 more than ten times the smaller integer. Find the pairs of integers.

Example 5 Using a Quadratic Equation in an Application

A stone is dropped off a 64-ft cliff and falls into the ocean below. The height of the stone above sea level is given by the equation

$h = -16t^2 + 64$ where h is the stone's height in feet, and t is the time in seconds.

Find the time required for the stone to hit the water.

Solution:

When the stone hits the water, its height is zero. Therefore, substitute $h = 0$ into the equation.

$h = -16t^2 + 64$ The equation is quadratic.

$0 = -16t^2 + 64$ Substitute $h = 0$.

$0 = -16(t^2 - 4)$ Factor out the GCF.

$0 = -16(t - 2)(t + 2)$ Factor as a difference of squares.

$-16 \neq 0$ or $t - 2 = 0$ or $t + 2 = 0$ Set each factor to zero.

No solution, $t = 2$ or $t \neq -2$ Solve for t.

The negative value of t is rejected because the stone cannot fall for a negative time. Therefore, the stone hits the water after 2 seconds.

Skill Practice

8. An object is launched into the air from the ground and its height is given by $h = -16t^2 + 144t$, where h is the height in feet after t seconds. Find the time required for the object to hit the ground.

Skill Practice Answers

7. The pairs of consecutive odd integers are 9 and 11, or -1 and 1.
8. 9 seconds

In Example 5, we can analyze the path of the stone as it falls from the cliff. Compute the height values at various times between 0 and 2 seconds (Table 6-1 and Table 6-2). The ordered pairs can be graphed where t is used in place of x and h is used in place of y.

Table 6-1

Time, t (sec)	Height, h (ft)
0.0	
0.5	
1.0	
1.5	
2.0	

$\longrightarrow h = -16(0.0)^2 + 64 = 64 \longrightarrow$

$\longrightarrow h = -16(0.5)^2 + 64 = 60 \longrightarrow$

$\longrightarrow h = -16(1.0)^2 + 64 = 48 \longrightarrow$

$\longrightarrow h = -16(1.5)^2 + 64 = 28 \longrightarrow$

$\longrightarrow h = -16(2.0)^2 + 64 = 0 \longrightarrow$

Table 6-2

Time, t (sec)	Height, h (ft)
0.0	64
0.5	60
1.0	48
1.5	28
2.0	0

The graph of the height of the stone versus time is shown in Figure 6-2.

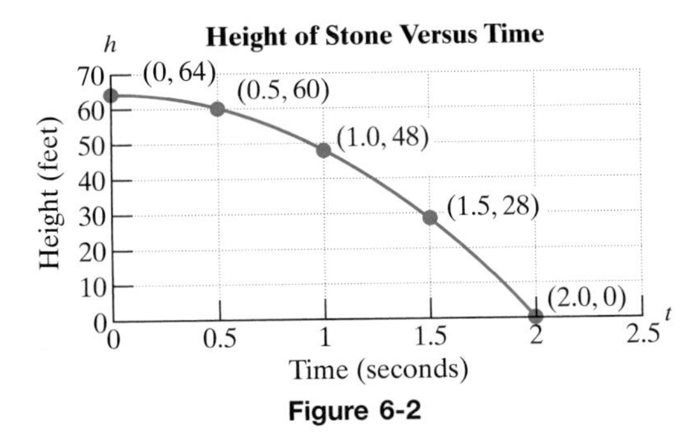

Figure 6-2

5. Pythagorean Theorem

Recall that a right triangle is a triangle that contains a 90° angle. Furthermore, the sum of the squares of the two legs (the shorter sides) of a right triangle equals the square of the hypotenuse (the longest side). This important fact is known as the Pythagorean theorem. The Pythagorean theorem is an enduring landmark of mathematical history from which many mathematical ideas have been built. Although the theorem is named after Pythagoras (sixth century B.C.E.), a Greek mathematician and philosopher, it is thought that the ancient Babylonians were familiar with the principle more than a thousand years earlier.

For the right triangle shown in Figure 6-3, the **Pythagorean theorem** is stated as:

$$a^2 + b^2 = c^2$$

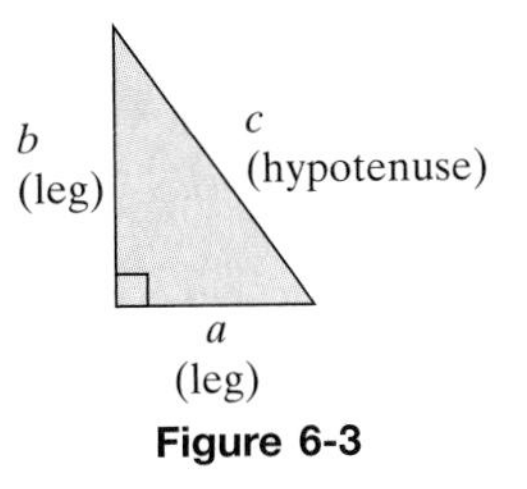

Figure 6-3

In this formula, a and b are the legs of the right triangle and c is the hypotenuse. Notice that the hypotenuse is the longest side of the right triangle and is opposite the 90° angle.

The triangle shown below is a right triangle. Notice that the lengths of the sides satisfy the Pythagorean theorem.

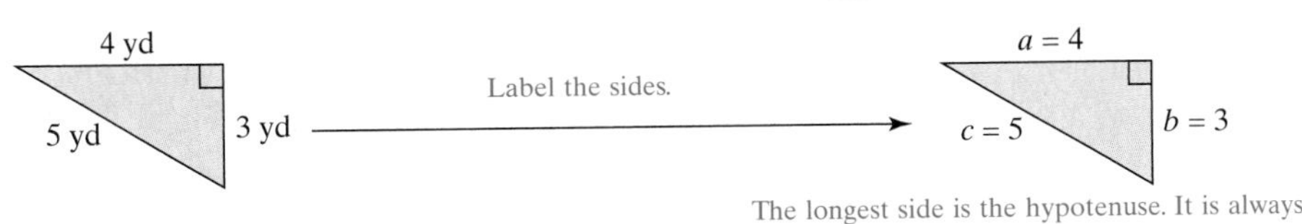

$a^2 + b^2 = c^2$	Apply the Pythagorean theorem.
$(4)^2 + (3)^2 = (5)^2$	Substitute $a = 4$, $b = 3$, and $c = 5$.
$16 + 9 = 25$	
$25 = 25$ ✔	

Example 6 Applying the Pythagorean Theorem

Find the length of the missing side of the right triangle.

? 10 ft 6 ft

Solution:

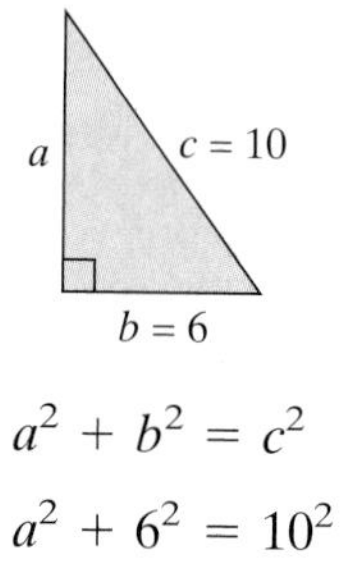

Label the triangle.

$a^2 + b^2 = c^2$	Apply the Pythagorean theorem.
$a^2 + 6^2 = 10^2$	Substitute $b = 6$ and $c = 10$.
$a^2 + 36 = 100$	Simplify.
	The equation is quadratic. Set the equation equal to zero.
$a^2 + 36 - 100 = 100 - 100$	Subtract 100 from both sides.
$a^2 - 64 = 0$	
$(a + 8)(a - 8) = 0$	Factor.
$a + 8 = 0$ or $a - 8 = 0$	Set each factor equal to zero.
$a \cancel{=} -8$ or $a = 8$	Because x represents the length of a side of a triangle, reject the negative solution.

The third side is 8 ft.

Skill Practice

9. Find the length of the missing side.

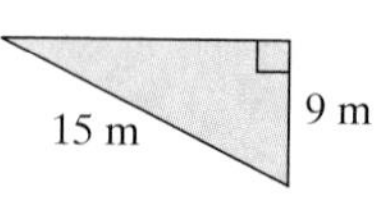

Example 7 Using a Quadratic Equation in an Application

A 13-ft board is used as a ramp to unload furniture off a loading platform. If the distance between the top of the board and the ground is 7 ft less than the distance between the bottom of the board and the base of the platform, find both distances.

Skill Practice Answers

9. The length of the third side is 12 m.

Solution:

Let x represent the distance between the bottom of the board and the base of the platform. Then $x - 7$ represents the distance between the top of the board and the ground (Figure 6-4).

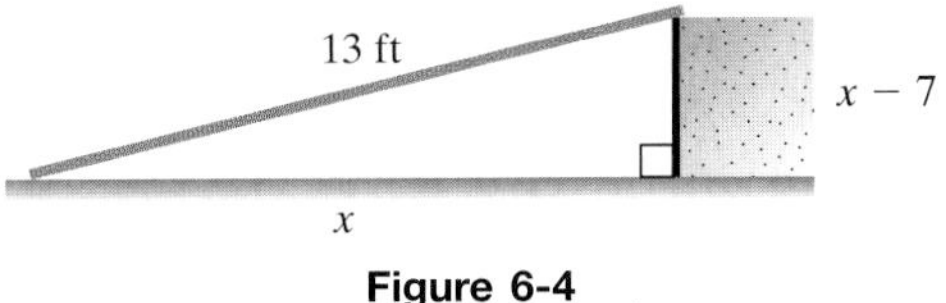

Figure 6-4

$a^2 + b^2 = c^2$	Pythagorean theorem
$x^2 + (x - 7)^2 = (13)^2$	
$x^2 + [(x)^2 - 2(x)(7) + (7)^2] = 169$	
$x^2 + x^2 - 14x + 49 = 169$	
$2x^2 - 14x + 49 = 169$	Combine *like* terms.
$2x^2 - 14x + 49 - 169 = 169 - 169$	Set the equation equal to zero.
$2x^2 - 14x - 120 = 0$	Write the equation in the form $ax^2 + bx + c = 0$.
$2(x^2 - 7x - 60) = 0$	Factor.
$2(x - 12)(x + 5) = 0$	
$2 \neq 0$ or $x - 12 = 0$ or $x + 5 = 0$	Set each factor equal to zero.
$x = 12$ or $x \neq -5$	Solve both equations for x.

Recall that x represents the distance between the bottom of the board and the base of the platform. We reject the negative value of x because a distance cannot be negative. Therefore, the distance between the bottom of the board and the base of the platform is 12 ft. The distance between the top of the board and the ground is $x - 7 = 5$ ft.

Avoiding Mistakes:

Recall that the square of a binomial results in a perfect square trinomial.

$(a - b)^2 = a^2 - 2ab + b^2$

Don't forget the middle term.

Skill Practice

10. A 5-yd ladder leans against a wall. The distance from the bottom of the wall to the top of the ladder is 1 yd more than the distance from the bottom of the wall to the bottom of the ladder. Find both distances.

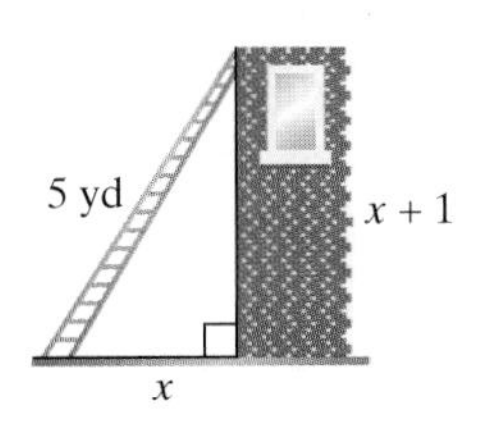

Skill Practice Answers

10. The distance along the wall to the top of the ladder is 4 yd. The distance on the ground from the ladder to the wall is 3 yd.

Section 6.7 Practice Exercises

Study Skills Exercise

1. Define the key terms:

a. quadratic equation **b. Pythagorean theorem** **c. zero product rule**

Review Exercises

For Exercises 2–7, factor completely.

2. $6a - 8 - 3ab + 4b$

3. $4b^2 - 44b + 120$

4. $8u^2v^2 - 4uv$

5. $3x^2 + 10x - 8$

6. $3h^2 - 75$

7. $4x^2 + 16y^2$

Concept 1: Definition of a Quadratic Equation

For Exercises 8–13, identify the equations as linear, quadratic, or neither.

8. $4 - 5x = 0$

9. $5x^3 + 2 = 0$

10. $3x - 6x^2 = 0$

11. $1 - x + 2x^2 = 0$

12. $7x^4 + 8 = 0$

13. $3x + 2 = 0$

Concept 2: Zero Product Rule

For Exercises 14–21, solve the equations using the zero product rule.

14. $(x - 5)(x + 1) = 0$

15. $(x + 3)(x - 1) = 0$

16. $(3x - 2)(3x + 2) = 0$

17. $(2x - 7)(2x + 7) = 0$

18. $2(x - 7)(x - 7) = 0$

19. $3(x + 5)(x + 5) = 0$

20. $x(x - 4)(2x + 3) = 0$

21. $x(3x + 1)(x + 1) = 0$

22. For a quadratic equation of the form $ax^2 + bx + c = 0$, what must be done before applying the zero product rule?

23. What are the requirements needed to use the zero product rule to solve a quadratic equation or higher degree polynomial equation?

Concept 3: Solving Equations Using the Zero Product Rule

For Exercises 24–67, solve the equations.

24. $p^2 - 2p - 15 = 0$

25. $y^2 - 7y - 8 = 0$

26. $z^2 + 10z - 24 = 0$

27. $w^2 - 10w + 16 = 0$

28. $2q^2 - 7q - 4 = 0$

29. $4x^2 - 11x - 3 = 0$

30. $0 = 9x^2 - 4$

31. $4a^2 - 49 = 0$

32. $2k^2 - 28k + 96 = 0$

33. $0 = 2t^2 + 20t + 50$

34. $0 = 2m^3 - 5m^2 - 12m$

35. $3n^3 + 4n^2 + n = 0$

36. $(3p + 1)(p - 3)(p + 6) = 0$

37. $(2x - 1)(x - 10)(x + 7) = 0$

38. $x^3 - 16x = 0$

39. $t^3 - 36t = 0$

40. $3x^2 - 14x - 5 = 0$

41. $2y^2 + 3y - 9 = 0$

42. $16m^2 = 9$

43. $9n^2 = 1$

44. $2y^3 + 14y^2 = -20y$

45. $3d^3 - 6d^2 = 24d$

46. $5t - 2(t - 7) = 0$

47. $8h = 5(h - 9) + 6$

48. $2c(c - 8) = -30$

49. $3q(q - 3) = 12$

50. $b^3 = -4b^2 - 4b$

51. $x^3 + 36x = 12x^2$

52. $3(a^2 + 2a) = 2a^2 - 9$

53. $9(k - 1) = -4k^2$

54. $2n(n + 2) = 6$

55. $3p(p - 1) = 18$

56. $x(2x + 5) - 1 = 2x^2 + 3x + 2$

57. $3z(z - 2) - z = 3z^2 + 4$

58. $27q^2 = 9q$

59. $21w^2 = 14w$

60. $3(c^2 - 2c) = 0$

61. $2(4d^2 + d) = 0$

62. $y^3 - 3y^2 - 4y + 12 = 0$

63. $t^3 + 2t^2 - 16t - 32 = 0$

64. $(x - 1)(x + 2) = 18$

65. $(w + 5)(w - 3) = 20$

66. $(p + 2)(p + 3) = 1 - p$

67. $(k - 6)(k - 1) = -k - 2$

Concept 4: Applications of Quadratic Equations

68. If eleven is added to the square of a number, the result is sixty. Find all such numbers.

69. If a number is added to two times its square, the result is thirty-six. Find all such numbers.

70. If twelve is added to six times a number, the result is twenty-eight less than the square of the number. Find all such numbers.

71. The square of a number is equal to twenty more than the number. Find all such numbers.

72. The product of two consecutive odd integers is sixty-three. Find all such integers.

73. The product of two consecutive even integers is forty-eight. Find all such integers.

74. The sum of the squares of two consecutive integers is one more than ten times the larger number. Find all such integers.

75. The sum of the squares of two consecutive integers is nine less than ten times the sum of the integers. Find all such integers.

76. The length of a rectangular room is 5 yd more than the width. If 300 yd² of carpeting cover the room, what are the dimensions of the room?

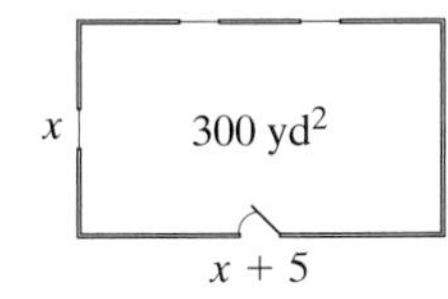

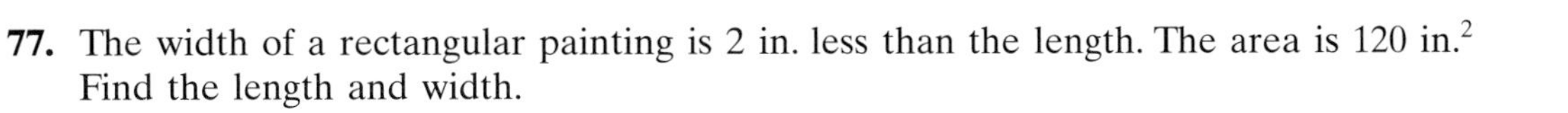

77. The width of a rectangular painting is 2 in. less than the length. The area is 120 in.² Find the length and width.

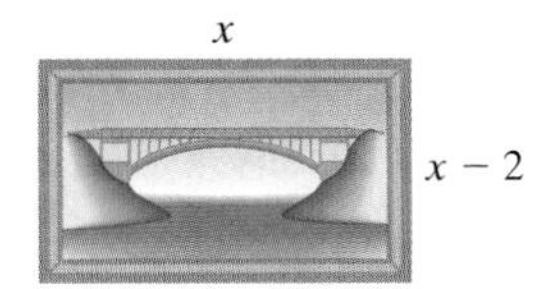

78. The width of a rectangular slab of concrete is 3 m less than the length. The area is 28 m^2.

a. What are the dimensions of the rectangle?

b. What is the perimeter of the rectangle?

79. The width of a rectangular picture is 7 in. less than the length. The area of the picture is 78 $in.^2$

a. What are the dimensions of the rectangle?

b. What is the perimeter of the rectangle?

80. The base of a triangle is 1 ft less than twice the height. The area is 14 ft^2. Find the base and height of the triangle.

81. The height of a triangle is 5 cm less than 3 times the base. The area is 125 cm^2. Find the base and height of the triangle.

82. In a physics experiment, a ball is dropped off a 144-ft platform. The height of the ball above the ground is given by the equation

$h = -16t^2 + 144$ where h is the ball's height in feet, and t is the time in seconds after the ball is dropped ($t \geq 0$).

Find the time required for the ball to hit the ground. (*Hint:* Let $h = 0$.)

83. A stone is dropped off a 256-ft cliff. The height of the stone above the ground is given by the equation

$h = -16t^2 + 256$ where h is the stone's height in feet, and t is the time in seconds after the stone is dropped ($t \geq 0$).

Find the time required for the stone to hit the ground.

84. An object is shot straight up into the air from ground level with an initial speed of 24 ft/sec. The height of the object (in feet) is given by the equation

$h = -16t^2 + 24t$ where t is the time in seconds after launch ($t \geq 0$).

Find the time(s) when the object is at ground level.

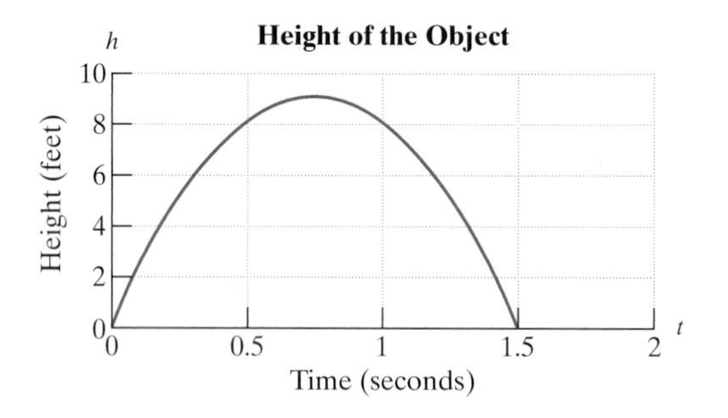

85. A rocket is launched straight up into the air from the ground with initial speed of 64 ft/sec. The height of the rocket (in feet) is given by the equation

$h = -16t^2 + 64t$ where t is the time in seconds after launch ($t \geq 0$).

Find the time(s) when the ball is at ground level.

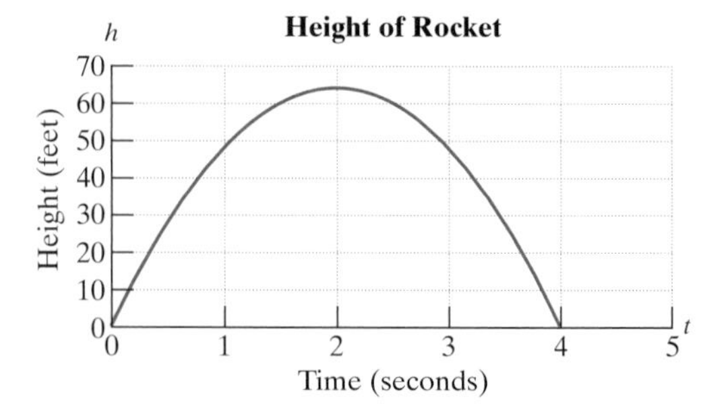

Concept 5: Pythagorean Theorem

86. Sketch a right triangle and label the sides with the words *leg* and *hypotenuse.*

87. State the Pythagorean theorem.

For Exercises 88–91, find the length of the missing side of the right triangle.

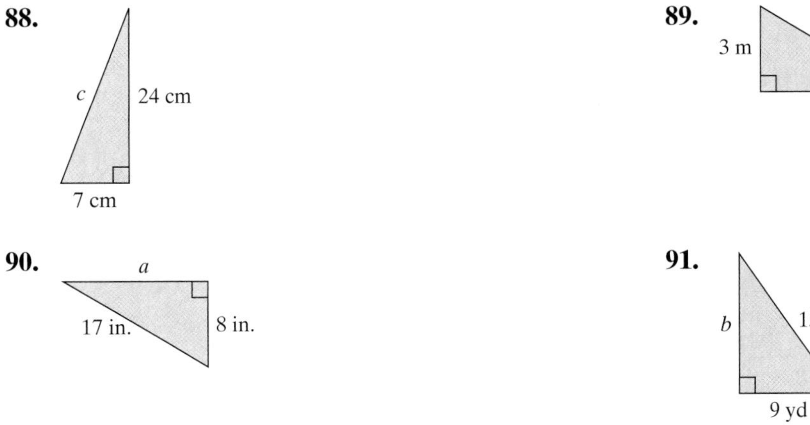

91.

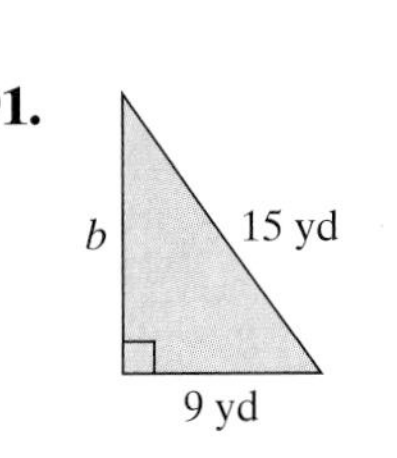

92. Find the length of the supporting brace.

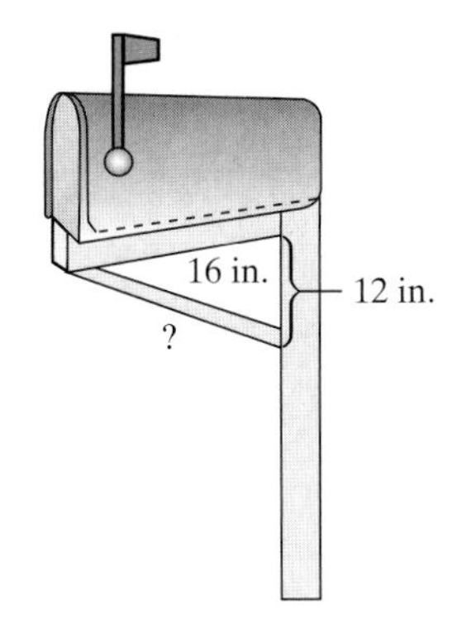

93. Find the height of the airplane above the ground.

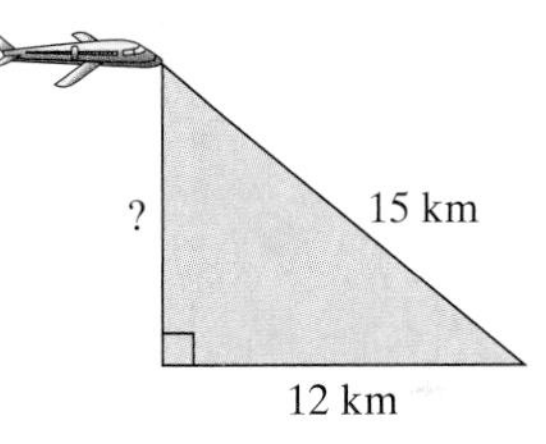

94. A 17-ft ladder rests against the side of a house. The distance between the top of the ladder and the ground is 7 ft more than the distance between the base of the ladder and the bottom of the house. Find both distances.

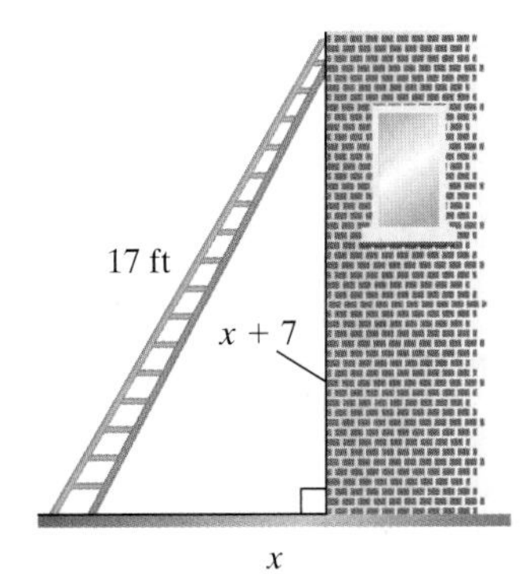

95. Darcy holds the end of a kite string 3 ft (1 yd) off the ground and wants to estimate the height of the kite. Her friend Jenna is 24 yd away from her, standing directly under the kite as shown in the figure. If Darcy has 30 yd of string out, find the height of the kite (ignore the sag in the string).

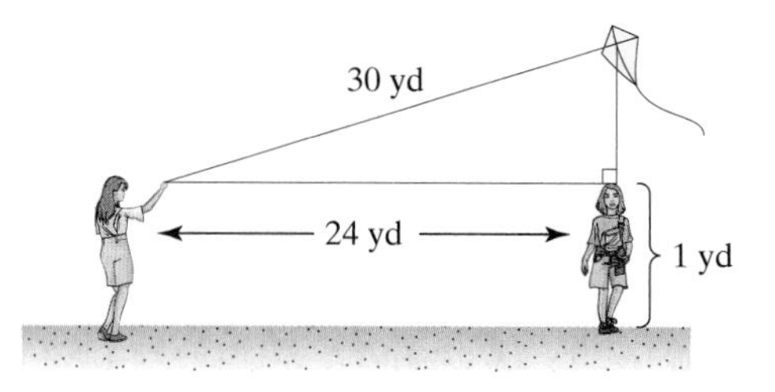

96. Two boats leave a marina. One travels east, and the other travels south. After 30 min, the second boat has traveled 1 mile farther than the first boat and the distance between the boats is 5 miles. Find the distance each boat traveled.

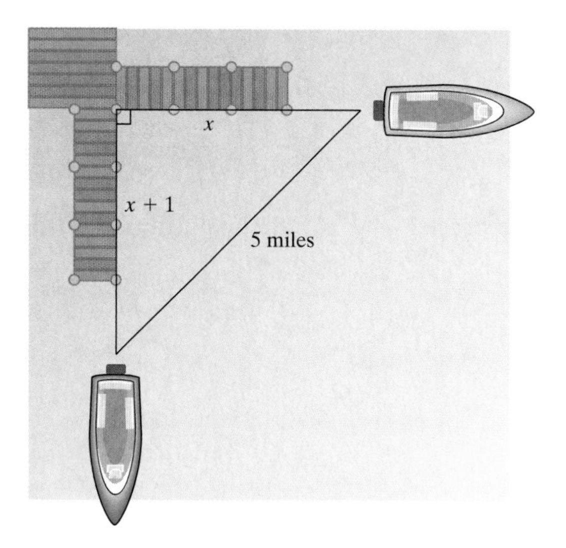

97. One leg of a right triangle is 4 m less than the hypotenuse. The other leg is 2 m less than the hypotenuse. Find the length of the hypotenuse.

98. The longer leg of a right triangle is 1 cm less than twice the shorter leg. The hypotenuse is 1 cm greater than twice the shorter leg. Find the length of the shorter leg.

Chapter 6 SUMMARY

Section 6.1 Greatest Common Factor and Factoring by Grouping

Key Concepts

The **greatest common factor** (GCF) is the greatest factor common to all terms of a polynomial. To factor out the GCF from a polynomial, use the distributive property.

A four-term polynomial may be factorable by grouping.

Steps to Factoring by Grouping

1. Identify and factor out the GCF from all four terms.
2. Factor out the GCF from the first pair of terms. Factor out the GCF or its opposite from the second pair of terms.
3. If the two terms share a common binomial factor, factor out the binomial factor.

Examples

Example 1

$3x(a + b) - 5(a + b)$ Greatest common factor is $(a + b)$.

$= (a + b)(3x - 5)$

Example 2

$60xa - 30xb - 80ya + 40yb$

$= 10[6xa - 3xb - 8ya + 4yb]$ Factor out GCF.

$= 10[3x(2a - b) - 4y(2a - b)]$ Factor by grouping.

$= 10(2a - b)(3x - 4y)$

Section 6.2 Factoring Trinomials of the Form $x^2 + bx + c$ (Optional)

Key Concepts

Factoring a Trinomial with a Leading Coefficient of 1:

A trinomial of the form $x^2 + bx + c$ factors as

$$x^2 + bx + c = (x \quad \square)(x \quad \square)$$

where the remaining terms are given by two integers whose product is c and whose sum is b.

Examples

Example 1

Factor: $x^2 - 14x + 45$

$x^2 - 14x + 45$ The integers -5 and -9 have a product of 45 and a sum of -14.

$= (x \quad \square)(x \quad \square)$

$= (x - 5)(x - 9)$

Section 6.3 Factoring Trinomials: Trial-and-Error Method

Key Concepts

Trial-and-Error Method for Factoring Trinomials in the Form $ax^2 + bx + c$ (where $a \neq 0$)

1. Factor out the GCF from all terms.
2. List the pairs of factors of a and the pairs of factors of c. Consider the reverse order in one of the lists.
3. Construct two binomials of the form

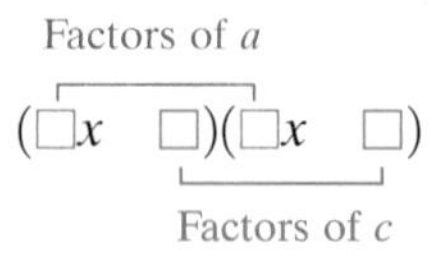

4. Test each combination of factors and signs until the product forms the correct trinomial.
5. If no combination of factors produces the correct product, then the trinomial is prime.

The factored form of a **perfect square trinomial** is the square of a binomial:

$a^2 + 2ab + b^2 = (a + b)^2$

$a^2 - 2ab + b^2 = (a - b)^2$

Examples

Example 1

$10y^2 + 35y - 20$

$= 5(2y^2 + 7y - 4)$

The pairs of factors of 2 are: $2 \cdot 1$
The pairs of factors of -4 are:

$-1(4)$ $1(-4)$

$-2(2)$ $2(-2)$

$-4(1)$ $4(-1)$

$(2y - 2)(y + 2) = 2y^2 + 2y - 4$ No

$(2y - 4)(y + 1) = 2y^2 - 2y - 4$ No

$(2y + 1)(y - 4) = 2y^2 - 7y - 4$ No

$(2y + 2)(y - 2) = 2y^2 - 2y - 4$ No

$(2y + 4)(y - 1) = 2y^2 + 2y - 4$ No

$(2y - 1)(y + 4) = 2y^2 + 7y - 4$ Yes

Example 2

$9w^2 - 30wz + 25z^2$

$= (3w)^2 - 2(3w)(5z) + (5z)^2$

$= (3w - 5z)^2$

Section 6.4 Factoring Trinomials: AC-Method

Key Concepts

AC-Method for Factoring Trinomials of the Form $ax^2 + bx + c$ (where $a \neq 0$)

1. Factor out the GCF from all terms.
2. Find the product ac.
3. Find two integers whose product is ac and whose sum is b. (If no pair of integers can be found, then the trinomial is prime.)
4. Rewrite the middle term (bx) as the sum of two terms whose coefficients are the numbers found in step 3.
5. Factor the polynomial by grouping.

The factored form of a **perfect square trinomial** is the square of a binomial:

$a^2 + 2ab + b^2 = (a + b)^2$

$a^2 - 2ab + b^2 = (a - b)^2$

Examples

Example 1

$10y^2 + 35y - 20$

$= 5(2y^2 + 7y - 4)$ First factor out GCF.

Identify the product $ac = (2)(-4) = -8$.

Find two integers whose product is -8 and whose sum is 7. The numbers are 8 and -1.

$5[2y^2 + 8y - 1y - 4]$

$= 5[2y(y + 4) - 1(y + 4)]$

$= 5(y + 4)(2y - 1)$

Example 2

Factor: $25y^2 + 10y + 1$

$= (5y)^2 + 2(5y)(1) + (1)^2$

$= (5y + 1)^2$

Section 6.5 Factoring Binomials

Key Concepts

Factoring a Difference of Squares

$a^2 - b^2 = (a - b)(a + b)$

Factoring a Sum or Difference of Cubes

$a^3 - b^3 = (a - b)(a^2 + ab + b^2)$

$a^3 + b^3 = (a + b)(a^2 - ab + b^2)$

Examples

Example 1

$25z^2 - 4y^2$

$= (5z - 2y)(5z + 2y)$

Example 2

$m^3 - 64$

$= (m)^3 - (4)^3$

$= (m - 4)(m^2 + 4m + 16)$

Example 3

$x^6 + 8y^3$

$= (x^2)^3 + (2y)^3$

$= (x^2 + 2y)(x^4 - 2x^2y + 4y^2)$

Section 6.6 General Factoring Summary

Key Concepts

Factoring Strategy

1. Factor out the greatest common factor, GCF.
2. Identify whether the polynomial has two terms, three terms, or more than three terms.
3. If the polynomial has more than three terms, try factoring by grouping (Section 6.1).
4. If the polynomial has three terms, check first for a perfect square trinomial. Otherwise, factor the trinomial with the trial-and-error method or ac-method (Sections 6.3 and 6.4, respectively).
5. If the polynomial has two terms, determine if it fits the pattern for a difference of squares, difference of cubes, or sum of cubes (Section 6.5).
6. Be sure to factor completely. Check that no factor can be factored further.
7. Check by multiplying.

Examples

Example 1

$y^3 + 2y^2 - 9y - 18$

$= y^3 + 2y^2 \mid - 9y - 18$

$= y^2(y + 2) - 9(y + 2)$

$= (y + 2)(y^2 - 9)$

$= (y + 2)(y - 3)(y + 3)$

Example 2

$9x^3 + 9x^2 - 4x$

$= x(9x^2 + 9x - 4)$ Factor out the GCF.

$= x(3x + 4)(3x - 1)$ Factor the trinomial.

Example 3

$5x^3 + 5$

$= 5(x^3 + 1)$ Factor out the GCF.

$= 5(x + 1)(x^2 - x + 1)$ Factor the sum of cubes.

Section 6.7 Solving Equations Using the Zero Product Rule

Key Concepts

An equation of the form $ax^2 + bx + c = 0$, where $a \neq 0$, is a **quadratic equation**.

The zero product rule states that if $ab = 0$, then either $a = 0$ or $b = 0$. The zero product rule can be used to solve a quadratic equation or a higher degree polynomial equation that is factored and set to zero.

Examples

Example 1

The equation $2x^2 - 17x + 30 = 0$ is a quadratic equation.

Example 2

$3w(w - 4)(2w + 1) = 0$

$3w = 0$ or $w - 4 = 0$ or $2w + 1 = 0$

$w = 0$ or $w = 4$ or $w = -\frac{1}{2}$

Example 3

$4x^2 = 34x - 60$

$4x^2 - 34x + 60 = 0$

$2(2x^2 - 17x + 30) = 0$

$2(2x - 5)(x - 6) = 0$

$2 \neq 0$ or $2x - 5 = 0$ or $x - 6 = 0$

$x = \frac{5}{2}$ or $x = 6$

Chapter 6 Review Exercises

Section 6.1

For Exercises 1–4, identify the greatest common factor between each group of terms.

1. $15a^2b^4, 30a^3b, 9a^5b^3$

2. $3(x + 5), x(x + 5)$

3. $2c^3(3c - 5), 4c(3c - 5)$

4. $-2wyz, -4xyz$

For Exercises 5–10, factor out the greatest common factor.

5. $6x^2 + 2x^4 - 8x$

6. $11w^3y^3 - 44w^2y^5$

7. $-t^2 + 5t$

8. $-6u^2 - u$

9. $3b(b + 2) - 7(b + 2)$

10. $2(5x + 9) + 8x(5x + 9)$

For Exercises 11–14, factor by grouping.

11. $7w^2 + 14w + wb + 2b$

12. $b^2 - 2b + yb - 2y$

13. $60y^2 - 45y - 12y + 9$

14. $6a - 3a^2 - 2ab + a^2b$

Section 6.2

For Exercises 15–24, factor completely.

15. $x^2 - 10x + 21$

16. $y^2 - 19y + 88$

17. $-6z + z^2 - 72$

18. $-39 + q^2 - 10q$

19. $3p^2w + 36pw + 60w$

20. $2m^4 + 26m^3 + 80m^2$

21. $-t^2 + 10t - 16$

22. $-w^2 - w + 20$

23. $a^2 + 12ab + 11b^2$

24. $c^2 - 3cd - 18d^2$

Section 6.3

For Exercises 25–28, let *a*, *b*, and *c* represent positive integers.

25. When factoring a polynomial of the form $ax^2 - bx - c$, should the signs of the binomials be both positive, both negative, or different?

26. When factoring a polynomial of the form $ax^2 - bx + c$, should the signs of the binomials be both positive, both negative, or different?

27. When factoring a polynomial of the form $ax^2 + bx + c$, should the signs of the binomials be both positive, both negative, or different?

28. When factoring a polynomial of the form $ax^2 + bx - c$, should the signs of the binomials be both positive, both negative, or different?

For Exercises 29–40, factor the trinomial using the trial-and-error method.

29. $2y^2 - 5y - 12$

30. $4w^2 - 5w - 6$

31. $10z^2 + 29z + 10$

32. $8z^2 + 6z - 9$

33. $2p^2 - 5p + 1$

34. $5r^2 - 3r + 7$

35. $10w^2 - 60w - 270$

36. $3y^2 - 18y - 48$

37. $9c^2 - 30cd + 25d^2$

38. $x^2 + 12x + 36$

39. $v^4 - 2v^2 - 3$

40. $x^4 + 7x^2 + 10$

41. In Exercises 29–40, which trinomials are perfect square trinomials?

Section 6.4

For Exercises 42–43, find a pair of integers whose product and sum are given.

42. Product: -5 sum: 4

43. Product: 15 sum: -8

For Exercises 44–57, factor the trinomial using the ac-method.

44. $3c^2 - 5c - 2$

45. $4y^2 + 13y + 3$

46. $t^2 + 13t + 12$

47. $4x^3 + 17x^2 - 15x$

48. $w^3 + 4w^2 - 5w$

49. $p^2 - 8pq + 15q^2$

50. $40v^2 + 22v - 6$

51. $40s^2 + 30s - 100$

52. $a^3b - 10a^2b^2 + 24ab^3$

53. $2z^6 + 8z^5 - 42z^4$

54. $3m + 9m^2 - 2$

55. $10 + 6p^2 + 19p$

56. $49x^2 + 140x + 100$

57. $9w^2 - 6wz + z^2$

58. In Exercises 42–57, which trinomials are perfect square trinomials?

Section 6.5

For Exercises 59–62, write the formula to factor each binomial, if possible.

59. $a^2 - b^2$

60. $a^2 + b^2$

61. $a^3 + b^3$

62. $a^3 - b^3$

For Exercises 63–78, factor completely.

63. $a^2 - 49$

64. $d^2 - 64$

65. $100 - 81t^2$

66. $4 - 25k^2$

67. $x^2 + 16$

68. $y^2 + 121$

69. $64 + a^3$

70. $125 - b^3$

71. $p^6 + 8$

72. $q^6 - \frac{1}{27}$

73. $6x^3 - 48$

74. $7y^3 + 7$

75. $2c^4 - 18$

76. $72x^2 - 2y^2$

77. $p^3 + 3p^2 - 16p - 48$

78. $4k - 8 - k^3 + 2k^2$

Section 6.6

For Exercises 79–94, factor completely using the factoring strategy found on page 491.

79. $6y^2 - 11y - 2$

80. $3p^2 - 6p + 3$

81. $x^3 - 36x$

82. $k^2 - 13k + 42$

83. $7ac - 14ad - bc + 2bd$

84. $q^4 - 64q$

85. $8h^2 + 20$

86. $2t^2 + t + 3$

87. $m^2 - 8m$

88. $x^3 + 4x^2 - x - 4$

89. $12s^3t - 45s^2t^2 - 12st^3$

90. $5p^4q - 20q^3$

91. $18a^2 + 39a - 15$

92. $w^4 + w^3 - 56w^2$

93. $8n + n^4$

94. $14m^3 - 14$

Section 6.7

95. For which of the following equations can the zero product rule be applied directly? Explain.

$(x - 3)(2x + 1) = 0$ or $(x - 3)(2x + 1) = 6$

For Exercises 96–111, solve the equation using the zero product rule.

96. $(4x - 1)(3x + 2) = 0$

97. $(a - 9)(2a - 1) = 0$

98. $3w(w + 3)(5w + 2) = 0$

99. $6u(u - 7)(4u - 9) = 0$

100. $7k^2 - 9k - 10 = 0$

101. $4h^2 - 23h - 6 = 0$

102. $q^2 - 144 = 0$

103. $r^2 = 25$

104. $5v^2 - v = 0$

105. $x(x - 6) = -8$

106. $36t^2 + 60t = -25$

107. $9s^2 + 12s = -4$

108. $3(y^2 + 4) = 20y$

109. $2(p^2 - 66) = -13p$

110. $2y^3 - 18y^2 = -28y$

111. $x^3 - 4x = 0$

112. The base of a parallelogram is 1 ft longer than twice the height. If the area is 78 ft^2, what are the base and height of the parallelogram?

113. A ball is tossed into the air from ground level with initial speed of 16 ft/sec. The height of the ball is given by the equation.

$h = -16x^2 + 16x \quad (x \geq 0)$ where h is the ball's height in feet, and x is the time in seconds

Find the time(s) when the ball is at ground level.

114. Find the length of the ramp.

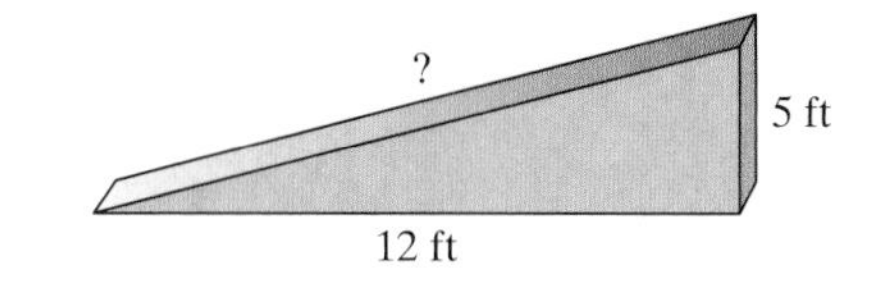

115. A right triangle has one leg that is 2 ft longer than the other leg. The hypotenuse is 2 ft less than twice the shorter leg. Find the length of all sides of the triangle.

116. If the square of a number is subtracted from 60, the result is -4. Find all such numbers.

117. The product of two consecutive integers is 44 more than 14 times their sum.

118. The base of a triangle is 1 m longer than twice the height. If the area of the triangle is 18 m^2, find the base and height.

Chapter 6 Test

1. Factor out the GCF: $15x^4 - 3x + 6x^3$

2. Factor by grouping: $7a - 35 - a^2 + 5a$

3. Factor the trinomial: $6w^2 - 43w + 7$

4. Factor the difference of squares: $169 - p^2$

5. Factor the perfect square trinomial: $q^2 - 16q + 64$

6. Factor the sum of cubes: $8 + t^3$

For Exercises 7–25, factor completely.

7. $a^2 + 12a + 32$

8. $x^2 + x - 42$

9. $2y^2 - 17y + 8$

10. $6z^2 + 19z + 8$

11. $9t^2 - 100$

12. $v^2 - 81$

13. $3a^2 + 27ab + 54b^2$

14. $c^4 - 1$

15. $xy - 7x + 3y - 21$

16. $49 + p^2$

17. $-10u^2 + 30u - 20$

18. $12t^2 - 75$

19. $5y^2 - 50y + 125$

20. $21q^2 + 14q$

21. $2x^3 + x^2 - 8x - 4$

22. $y^3 - 125$

23. $m^2n^2 - 81$

24. $16a^2 - 64b^2$

25. $64x^3 - 27y^6$

For Exercises 26–30, solve the equation.

26. $(2x - 3)(x + 5) = 0$

27. $x^2 - 7x = 0$

28. $x^2 - 6x = 16$

29. $x(5x + 4) = 1$

30. $y^3 + 10y^2 - 9y - 90 = 0$

31. A tennis court has an area of 312 yd^2. If the length is 2 yd more than twice the width, find the dimensions of the court.

32. The product of two consecutive odd integers is 35. Find the integers.

33. The height of a triangle is 5 in. less than the length of the base. The area is 42 in^2. Find the length of the base and the height of the triangle.

34. The hypotenuse of a right triangle is 2 ft less than three times the shorter leg. The longer leg is 3 ft less than three times the shorter leg. Find the length of the shorter leg.

Chapters 1–6 Cumulative Review Exercises

For Exercises 1–2, simplify completely.

1. $\dfrac{|4 - 25 \div (-5) \cdot 2|}{\sqrt{8^2 + 6^2}}$

2. Solve for t: $5 - 2(t + 4) = 3t + 12$

3. Solve for y: $3x - 2y = 8$

4. A child's piggy bank has \$3.80 in quarters, dimes, and nickels. The number of nickels is two more than the number of quarters. The number of dimes is three less than the number of quarters. Find the number of each type of coin in the bank.

5. Solve the inequality. Graph the solution on a number line and write the solution set in interval notation.

$-\dfrac{5}{12}x \leq \dfrac{5}{3}$

6. Given the equation $y = x + 4$

a. Is the equation linear?

b. Identify the slope.

c. Identify the y-intercept.

d. Identify the x-intercept.

e. Graph the line.

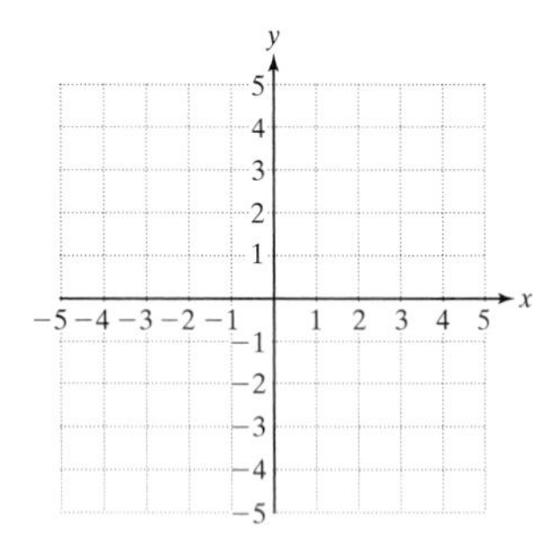

7. Consider the equation, $x = 5$.

a. Does the equation present a horizontal or vertical line?

b. Determine the slope of the line, if it exists.

c. Identify the x-intercept, if it exists.

d. Identify the y-intercept, if it exists.

8. Find an equation of the line passing through the point $(-3, 5)$ and having a slope of 3. Write the final answer in slope-intercept form.

9. Solve the system. $\begin{aligned} 2x - 3y &= 4 \\ 5x - 6y &= 13 \end{aligned}$

For Exercises 10–12, perform the indicated operations.

10. $2\left(\dfrac{1}{3}y^3 - \dfrac{3}{2}y^2 - 7\right) - \left(\dfrac{2}{3}y^3 + \dfrac{1}{2}y^2 + 5y\right)$

11. $(4p^2 - 5p - 1)(2p - 3)$

12. $(2w - 7)^2$

13. Divide using long division: $(r^4 + 2r^3 - 5r + 1) \div (r - 3)$

14. $\dfrac{c^{12}c^{-5}}{c^3}$

15. Divide. Write the final answer in scientific notation: $\dfrac{8.0 \times 10^{-3}}{5.0 \times 10^{-6}}$

For Exercises 16–19, factor completely.

16. $w^4 - 16$

17. $2ax + 10bx - 3ya - 15yb$

18. $4x^2 - 8x - 5$

19. $y^3 - 27$

For Exercise 20, solve the equation.

20. $4x(2x - 1)(x + 5) = 0$

Rational Expressions

In Chapter 7 we define a rational expression as the ratio of two polynomials. Then we will learn how to add, subtract, multiply, and divide rational expressions. Then we study equations and applications involving rational expressions.

Working with rational expressions is similar to working with fractions so it is worthwhile to review operations of fractions first.

Fill in the appropriate operations ($+$, $-$, $\cdot$, $\div$) into the equations to make them true. Use each of the symbols $+$, $-$, $\cdot$, $\div$ in each column only once.

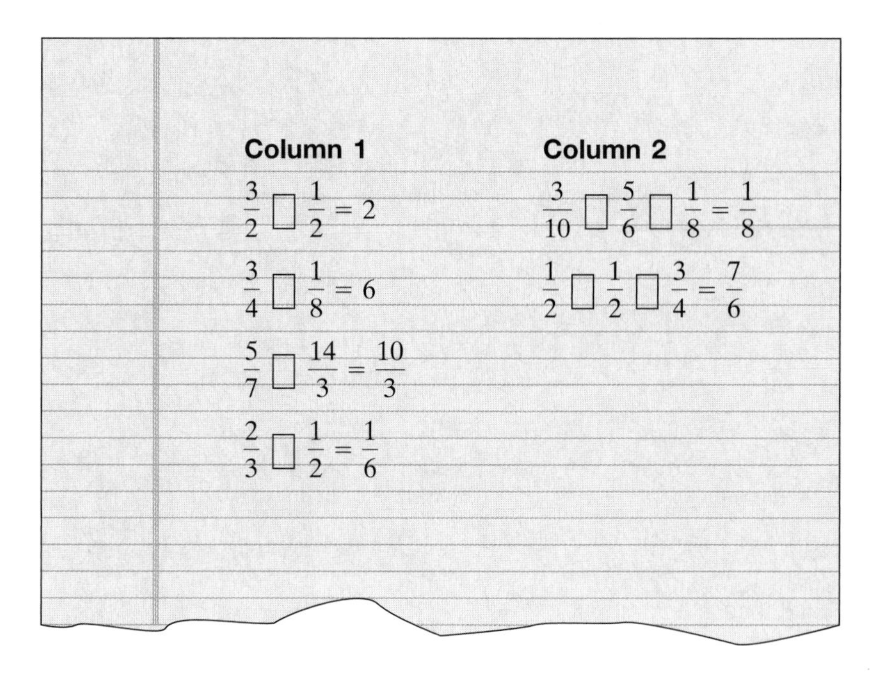

Section 7.1 Introduction to Rational Expressions

Concepts

1. Definition of a Rational Expression
2. Evaluating Rational Expressions
3. Domain of a Rational Expression
4. Simplifying Rational Expressions to Lowest Terms
5. Simplifying a Ratio of −1

1. Definition of a Rational Expression

In Section 1.1, we defined a rational number as the ratio of two integers, $\frac{p}{q}$, where $q \neq 0$.

Examples of rational numbers: $\frac{2}{3}, -\frac{1}{5}, 9$

In a similar way, we define a **rational expression** as the ratio of two polynomials, $\frac{p}{q}$, where $q \neq 0$.

Examples of rational expressions: $\frac{3x - 6}{x^2 - 4}, \quad \frac{3}{4}, \quad \frac{6r^5 + 2r}{7}$

2. Evaluating Rational Expressions

Example 1 Evaluating Rational Expressions

Evaluate the rational expression (if possible) for the given values of x: $\frac{12}{x - 3}$

a. $x = 0$ **b.** $x = 1$ **c.** $x = -3$ **d.** $x = 3$

Solution:

Substitute the given value for the variable. Then use the order of operations to simplify.

a. $\frac{12}{x - 3}$

$\frac{12}{(0) - 3}$ Substitute $x = 0$.

$= \frac{12}{-3}$

$= -4$

b. $\frac{12}{x - 3}$

$\frac{12}{(1) - 3}$ Substitute $x = 1$.

$= \frac{12}{-2}$

$= -6$

c. $\frac{12}{x - 3}$

$\frac{12}{(-3) - 3}$ Substitute $x = -3$.

$= \frac{12}{-6}$

$= -2$

d. $\frac{12}{x - 3}$

$\frac{12}{(3) - 3}$ Substitute $x = 3$.

$= \frac{12}{0}$ Undefined. Recall that division by zero is undefined.

Skill Practice

1. Evaluate the expression for the given values of x.

$$\frac{x - 3}{x + 5}$$

a. $x = 2$ **b.** $x = 0$ **c.** $x = 3$ **d.** $x = -5$

Skill Practice Answers

1a. $-\frac{1}{7}$ b. $-\frac{3}{5}$
c. 0 d. Undefined

3. Domain of a Rational Expression

In Example 1(d), the expression $12/(x - 3)$ is undefined for $x = 3$. The fact that a rational expression may be defined for some values of the variable but not for others leads us to an important concept called the domain of an expression.

Informal Definition of the Domain of an Algebraic Expression

Given an algebraic expression, the **domain** of the expression is the set of real numbers that when substituted for the variable makes the expression result in a real number.

For a rational expression, the domain will consist of the real numbers that when substituted into the expression does not make the denominator equal to zero. Therefore, in Example 1(d), because $12/(x - 3)$ is undefined for $x = 3$, the domain is all real numbers except 3. We write this in set-builder notation as

$$\{x \mid x \text{ is a real number and } x \neq 3\}$$

Steps to Find the Domain of a Rational Expression

1. Set the denominator equal to zero and solve the resulting equation.
2. The domain is the set of real numbers *excluding* the values found in step 1.

Example 2 **Finding the Domain of Rational Expressions**

Find the domain of the expressions.

a. $\dfrac{y - 3}{2y + 7}$ **b.** $\dfrac{-5}{x}$ **c.** $\dfrac{a + 10}{a^2 - 25}$ **d.** $\dfrac{2x^3 + 5}{x^2 + 9}$

Solution:

a. $\dfrac{y - 3}{2y + 7}$

$2y + 7 = 0$ Set the denominator equal to zero.

$2y = -7$ Solve the equation.

$\dfrac{2y}{2} = \dfrac{-7}{2}$

$y = -\dfrac{7}{2}$ The domain is the set of real numbers except $-\frac{7}{2}$.

Domain: $\{y \mid y \text{ is a real number and } y \neq -\frac{7}{2}\}$

b. $\dfrac{-5}{x}$

$x = 0$ Set the denominator equal to zero.

Domain: $\{x \mid x \text{ is a real number and } x \neq 0\}$ The domain is the set of real numbers except 0.

c. $\dfrac{a + 10}{a^2 - 25}$

$a^2 - 25 = 0$	Set the denominator equal to zero. The equation is quadratic.
$(a - 5)(a + 5) = 0$	Factor the equation.
$a - 5 = 0$ or $a + 5 = 0$	Set each factor equal to zero.
$a = 5$ or $a = -5$	The domain is the set of real numbers except 5 and -5.

Domain: $\{a \mid a \text{ is a real number and } a \neq 5, a \neq -5\}$

d. $\dfrac{2x^3 + 5}{x^2 + 9}$

The quantity x^2 cannot be negative for any real number, x, so the denominator $x^2 + 9$ cannot equal zero. Therefore, no numbers are excluded from the domain. The domain is the set of all real numbers.

Skill Practice Find the domain.

2. $\dfrac{a + 2}{2a - 8}$ **3.** $\dfrac{t}{t^2 + 10t}$ **4.** $\dfrac{w - 4}{w^2 - 9}$ **5.** $\dfrac{8}{z^4 + 1}$

4. Simplifying Rational Expressions to Lowest Terms

In many cases, it is advantageous to simplify or reduce a fraction to lowest terms. The same is true for rational expressions.

The method for simplifying rational expressions mirrors the process for simplifying fractions. In each case, factor the numerator and denominator. Common factors in the numerator and denominator form a ratio of 1 and can be reduced.

Simplifying a fraction: $\dfrac{21}{35} \xrightarrow{\text{factor}} \dfrac{3 \cdot \overset{1}{\cancel{7}}}{5 \cdot \cancel{7}} = \dfrac{3}{5} \cdot (1) = \dfrac{3}{5}$

Simplifying a rational expression: $\dfrac{2x - 6}{x^2 - 9} \xrightarrow{\text{factor}} \dfrac{2\overset{1}{\cancel{(x - 3)}}}{(x + 3)\cancel{(x - 3)}} = \dfrac{2}{(x + 3)}(1) = \dfrac{2}{x + 3}$

Informally, to simplify a rational expression to lowest terms, we reduce common factors whose ratio is 1. Formally, this is accomplished by applying the fundamental principle of rational expressions.

Fundamental Principle of Rational Expressions

Let p, q, and r represent polynomials. Then

$$\frac{pr}{qr} = \frac{p}{q} \text{ for } q \neq 0 \text{ and } r \neq 0$$

Skill Practice Answers

2. $\{a \mid a \text{ is a real number and } a \neq 4\}$
3. $\{t \mid t \text{ is a real number and } t \neq 0, t \neq -10\}$
4. $\{w \mid w \text{ is a real number and } w \neq 3, w \neq -3\}$
5. The set of all real numbers

Example 3 **Simplifying a Rational Expression to Lowest Terms**

Given the expression: $\dfrac{2p - 14}{p^2 - 49}$

a. Factor the numerator and denominator.

b. Determine the domain of the expression.

c. Simplify the expression to lowest terms.

Solution:

a. $\dfrac{2p - 14}{p^2 - 49}$

$= \dfrac{2(p - 7)}{(p + 7)(p - 7)}$ Factor out the GCF in the numerator. Factor the denominator as a difference of squares.

b. $(p + 7)(p - 7) = 0$ To find the domain restrictions, set the denominator equal to zero. The equation is quadratic.

$p + 7 = 0$ or $p - 7 = 0$ Set each factor equal to 0.

$p = -7$ or $p = 7$ The domain is all real numbers except -7 and 7.

Domain: $\{p \mid p \text{ is a real number and } p \neq -7, p \neq 7\}$

c. $\dfrac{2\overset{1}{\cancel{(p - 7)}}}{(p + 7)\cancel{(p - 7)}}$ Reduce common factors whose ratio is 1.

$= \dfrac{2}{p + 7}$ (provided $p \neq 7$ and $p \neq -7$)

Skill Practice

6. Given: $\dfrac{5z + 25}{z^2 + 3z - 10}$

a. Factor the numerator and the denominator.

b. Determine the domain of the expression.

c. Simplify the rational expression to lowest terms.

Avoiding Mistakes:

The domain of a rational expression is always determined *before* simplifying the expression to lowest terms.

In Example 3, it is important to note that the expressions

$$\frac{2p - 14}{p^2 - 49} \quad \text{and} \quad \frac{2}{p + 7}$$

are equal for all values of p that make each expression a real number. Therefore,

$$\frac{2p - 14}{p^2 - 49} = \frac{2}{p + 7}$$

for all values of p except $p = 7$ and $p = -7$. (At $p = 7$ and $p = -7$, the original expression is undefined.) This is why the domain of an expression is always determined before the expression is simplified.

Skill Practice Answers

6a. $\dfrac{5(z + 5)}{(z + 5)(z - 2)}$

b. $\{z \mid z \text{ is a real number and } z \neq -5, z \neq 2\}$

c. $\dfrac{5}{z - 2}$

From this point forward, we will write statements of equality between two rational expressions with the assumption that they are equal for all values of the variable for which each expression is defined.

Example 4 Simplifying Rational Expressions to Lowest Terms

Simplify the rational expressions to lowest terms.

a. $\dfrac{18a^4}{9a^5}$ **b.** $\dfrac{2c - 8}{10c^2 - 80c + 160}$

Solution:

a. $\dfrac{18a^4}{9a^5}$

$= \dfrac{2 \cdot 3 \cdot 3 \cdot a \cdot a \cdot a \cdot a}{3 \cdot 3 \cdot a \cdot a \cdot a \cdot a \cdot a}$ Factor the numerator and denominator.

$= \dfrac{2 \cdot (3 \cdot 3 \cdot a \cdot a \cdot a \cdot a)}{(3 \cdot 3 \cdot a \cdot a \cdot a \cdot a) \cdot a}$ Reduce common factors to lowest terms.

$= \dfrac{2}{a}$

b. $\dfrac{2c - 8}{10c^2 - 80c + 160}$

$= \dfrac{2(c - 4)}{10(c^2 - 8c + 16)}$ Factor out the GCF.

$= \dfrac{2(c - 4)}{10(c - 4)^2}$ The denominator is a perfect square trinomial.

$= \dfrac{\overset{1}{\cancel{2}}\overset{1}{\cancel{(c - 4)}}}{\cancel{2} \cdot 5\cancel{(c - 4)}(c - 4)}$ Reduce common factors whose ratio is 1.

$= \dfrac{1}{5(c - 4)}$

Avoiding Mistakes:

Given the expression

$$\frac{2c - 8}{10c^2 - 80c + 160}$$

do not be tempted to reduce before factoring. The terms $2c$ and $10c^2$ cannot be "canceled" because they are *terms*, not factors.

Similarly, the -8 in the numerator cannot be "canceled" with the $-80c$ or 160 in the denominator because they are terms, not factors.

Skill Practice Simplify to lowest terms.

7. $\dfrac{15q^3}{9q^2}$ **8.** $\dfrac{x^2 - 1}{2x^2 - x - 3}$

The process to simplify a rational expression to lowest terms is based on the identity property of multiplication. Therefore, this process applies only to factors (remember that factors are multiplied). For example,

$$\frac{3x}{3y} = \frac{\overset{1}{\cancel{3}} \cdot x}{\cancel{3} \cdot y} = 1 \cdot \frac{x}{y} = \frac{x}{y}$$

Simplify

Skill Practice Answers

7. $\dfrac{5q}{3}$ **8.** $\dfrac{x - 1}{2x - 3}$

Terms that are added or subtracted cannot be reduced to lowest terms. For example,

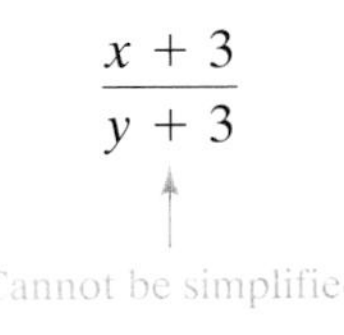

The objective of simplifying a rational expression to lowest terms is to create an equivalent expression that is simpler to use. Consider the rational expression from Example 4(b) in its original form and in its simplified form. If we choose an arbitrary value of c from the domain of the original expression and substitute that value into each expression, we see that the simplified form is easier to evaluate. For example, substitute $c = 3$:

	Original Expression	**Simplified Expression**
	$\dfrac{2c - 8}{10c^2 - 80c + 160}$	$\dfrac{1}{5(c - 4)}$
Substitute $c = 3$.	$= \dfrac{2(3) - 8}{10(3)^2 - 80(3) + 160}$	$= \dfrac{1}{5(3 - 4)}$
	$= \dfrac{6 - 8}{10(9) - 240 + 160}$	$= \dfrac{1}{5(-1)}$
	$= \dfrac{-2}{90 - 240 + 160}$	$= -\dfrac{1}{5}$
	$= \dfrac{-2}{10}$ or $-\dfrac{1}{5}$	

5. Simplifying a Ratio of −1

When two factors are identical in the numerator and denominator, they form a ratio of 1 and can be reduced. Sometimes we encounter two factors that are opposites and form a ratio of −1. For example,

Simplified Form	**Details/Notes**
$\dfrac{-5}{5} = -1$	The ratio of a number and its opposite is −1.
$\dfrac{100}{-100} = -1$	The ratio of a number and its opposite is −1.
$\dfrac{x + 7}{-x - 7} = -1$	$\dfrac{x + 7}{-x - 7} = \dfrac{x + 7}{-1(x + 7)} = \dfrac{\overset{1}{\cancel{x + 7}}}{-1\cancel{(x + 7)}} = \dfrac{1}{-1} = -1$ (Factor out −1)
$\dfrac{2 - x}{x - 2} = -1$	$\dfrac{2 - x}{x - 2} = \dfrac{-1(-2 + x)}{x - 2} = \dfrac{-1\overset{1}{\cancel{(x - 2)}}}{\cancel{x - 2}} = \dfrac{-1}{1} = -1$

Recognizing factors that are opposites is useful when simplifying rational expressions.

Example 5 Simplifying Rational Expressions to Lowest Terms

Simplify the rational expressions to lowest terms.

a. $\dfrac{3c - 3d}{d - c}$ **b.** $\dfrac{5 - y}{y^2 - 25}$

Solution:

a. $\dfrac{3c - 3d}{d - c}$

$= \dfrac{3(c - d)}{d - c}$ Factor the numerator and denominator.

Notice that $(c - d)$ and $(d - c)$ are opposites and form a ratio of -1.

$= \dfrac{3\overset{-1}{\cancel{(c - d)}}}{\cancel{d - c}}$ Details: $\dfrac{3(c - d)}{d - c} = \dfrac{3(c - d)}{-1(-d + c)} = \dfrac{3(c - d)}{-1(c - d)}$

$= 3(-1)$

$= -3$

b. $\dfrac{5 - y}{y^2 - 25}$

$= \dfrac{5 - y}{(y - 5)(y + 5)}$ Factor the numerator and denominator.

Notice that $5 - y$ and $y - 5$ are opposites and form a ratio of -1.

$= \dfrac{\overset{-1}{\cancel{5 - y}}}{\cancel{(y - 5)}(y + 5)}$ Details: $\dfrac{5 - y}{(y - 5)(y + 5)} = \dfrac{-1(-5 + y)}{(y - 5)(y + 5)} = \dfrac{-1(y - 5)}{(y - 5)(y + 5)}$

$= \dfrac{-1}{y + 5}$ or $\dfrac{1}{-(y + 5)}$ or $-\dfrac{1}{y + 5}$

TIP: It is important to recognize that a rational expression may be written in several equivalent forms, particularly when a negative factor is present. For example, since two numbers with opposite signs form a negative quotient, the number $-\frac{3}{4}$ can be written as $\frac{-3}{4}$ or as $\frac{3}{-4}$. The $-$ sign can be written in the numerator, in the denominator, or out in front of the fraction.

For this reason, the expression in Example 5(b) can be written in a variety of forms.

Skill Practice Simplify to lowest terms.

9. $\dfrac{b - a}{a^2 - b^2}$ **10.** $\dfrac{20 - 5y}{y^2 - 16}$

Skill Practice Answers

9. $\dfrac{-1}{a + b}$ or $\dfrac{1}{-(a + b)}$ or $-\dfrac{1}{a + b}$

10. $\dfrac{-5}{y + 4}$ or $\dfrac{5}{-(y + 4)}$ or $-\dfrac{5}{y + 4}$

Section 7.1 Practice Exercises

Boost *your* GRADE at mathzone.com!

MathZone

- Practice Problems
- Self-Tests
- NetTutor
- e-Professors
- Videos

Study Skills Exercises

1. Review Section R.2 in this text. Write an example of how to simplify (reduce) a fraction, multiply two fractions, divide two fractions, add two fractions, and subtract two fractions. Then as you learn about rational expressions, compare the operations on rational expressions with those on fractions. This is a great place to use 3×5 cards again. Write an example of an operation with fractions on one side and the same operation with rational expressions on the other side.

2. Define the key terms:

a. rational expression **b. domain**

Concept 1: Definition of a Rational Expression

3. a. What is a rational number?

b. What is a rational expression?

4. a. Write an example of a rational number. (Answers will vary.)

b. Write an example of a rational expression. (Answers will vary.)

Concept 2: Evaluating Rational Expressions

For Exercises 5–11, substitute the given number into the expression and simplify (if possible).

5. $\dfrac{1}{x - 6}$ let $x = -2$

6. $\dfrac{w - 10}{w + 6}$ let $w = 0$

7. $\dfrac{w - 4}{2w + 8}$ let $w = 0$

8. $\dfrac{y - 8}{2y^2 + y - 1}$ let $y = 8$

9. $\dfrac{y + 3}{3y^2 - 25y - 18}$ let $y = -3$

10. $\dfrac{(a - 7)(a + 1)}{(a - 2)(a + 5)}$ let $a = 2$

11. $\dfrac{(a + 4)(a + 1)}{(a - 4)(a - 1)}$ let $a = 1$

12. A bicyclist rides 24 miles against a wind and returns 24 miles with the same wind. His average speed for the return trip traveling with the wind is 8 mph faster than his speed going out against the wind. If x represents the bicyclist's speed going out against the wind, then the total time, t, required for the round trip is given by

$$t = \frac{24}{x} + \frac{24}{x + 8}$$ where $x > 0$ and t is measured in hours.

a. Find the time required for the round trip if the cyclist rides 12 mph against the wind.

b. Find the time required for the round trip if the cyclist rides 24 mph against the wind.

13. The manufacturer of mountain bikes has a fixed cost of $56,000, plus a variable cost of $140 per bike. The average cost per bike, y (in dollars), is given by the equation:

$y = \dfrac{56{,}000 + 140x}{x}$ where x represents the number of bikes produced and $x > 0$.

a. Find the average cost per bike if the manufacturer produces 1000 bikes.

b. Find the average cost per bike if the manufacturer produces 2000 bikes.

c. Find the average cost per bike if the manufacturer produces 10,000 bikes.

Concept 3: Domain of a Rational Expression

For Exercises 14–23, write the domain.

14. $\dfrac{5}{k + 2}$

15. $\dfrac{-3}{h - 4}$

16. $\dfrac{x + 5}{(2x - 5)(x + 8)}$

17. $\dfrac{4y + 1}{(3y + 7)(y + 3)}$

18. $\dfrac{b + 12}{b^2 + 5b + 6}$

19. $\dfrac{c - 11}{c^2 - 5c - 6}$

20. $\dfrac{x - 4}{x^2 + 9}$

21. $\dfrac{x + 1}{x^2 + 4}$

22. $\dfrac{y^2 - y - 12}{12}$

23. $\dfrac{z^2 + 10z + 9}{9}$

24. Construct a rational expression that is undefined for $x = 2$. (Answers will vary.)

25. Construct a rational expression that is undefined for $x = 5$. (Answers will vary.)

26. Construct a rational expression that is undefined for $x = -3$ and $x = 7$. (Answers will vary.)

27. Construct a rational expression that is undefined for $x = -1$ and $x = 4$. (Answers will vary.)

28. Evaluate the expressions for $x = 4$.

a. $\dfrac{5x + 5}{x^2 - 1}$ **b.** $\dfrac{5}{x - 1}$

29. Evaluate the expressions for $x = 3$.

a. $\dfrac{2x^2 - 4x - 6}{2x^2 - 18}$ **b.** $\dfrac{x + 1}{x + 3}$

30. Evaluate the expressions for $x = -1$.

a. $\dfrac{3x^2 - 2x - 1}{6x^2 - 7x - 3}$ **b.** $\dfrac{x - 1}{2x - 3}$

31. Evaluate the expressions for $x = 4$.

a. $\dfrac{(x + 5)^2}{x^2 + 6x + 5}$ **b.** $\dfrac{x + 5}{x + 1}$

Concept 4: Simplifying Rational Expressions to Lowest Terms

For Exercises 32–41,

a. Write the domain in set-builder notation.

b. Simplify the expression to lowest terms.

32. $\dfrac{3y + 6}{6y + 12}$

33. $\dfrac{8x - 8}{4x - 4}$

34. $\dfrac{t^2 - 1}{t + 1}$

35. $\dfrac{r^2 - 4}{r - 2}$

36. $\dfrac{7w}{21w^2 - 35w}$

37. $\dfrac{12a^2}{24a^2 - 18a}$

38. $\dfrac{9x^2 - 4}{6x + 4}$

39. $\dfrac{8b - 20}{4b^2 - 25}$

40. $\dfrac{a^2 + 3a - 10}{a^2 + a - 6}$

41. $\dfrac{t^2 + 3t - 10}{t^2 + t - 20}$

For Exercises 42–83, simplify the expression to lowest terms.

42. $\dfrac{7b^2}{21b}$

43. $\dfrac{15c^3}{3c^5}$

44. $\dfrac{18st^5}{12st^3}$

45. $\dfrac{20a^4b^2}{25ab^2}$

46. $\dfrac{-24x^2y^5z}{8xy^4z^3}$

47. $\dfrac{60rs^4t^2}{-12r^4s^2t^3}$

48. $\dfrac{3(y + 2)}{6(y + 2)}$

49. $\dfrac{8(x - 1)}{4(x - 1)}$

50. $\dfrac{(p - 3)(p + 5)}{(p + 5)(p + 4)}$

51. $\dfrac{(c + 4)(c - 1)}{(c + 4)(c + 2)}$

52. $\dfrac{(m + 11)}{4(m + 11)(m - 11)}$

53. $\dfrac{(n - 7)}{9(n + 2)(n - 7)}$

54. $\dfrac{x(2x + 1)^2}{4x^3(2x + 1)}$

55. $\dfrac{(p + 2)(p - 3)^4}{(p + 2)^2(p - 3)^2}$

56. $\dfrac{5}{20a - 25}$

57. $\dfrac{7}{14c - 21}$

58. $\dfrac{4w - 8}{w^2 - 4}$

59. $\dfrac{3x + 15}{x^2 - 25}$

60. $\dfrac{3x^2 - 6x}{9xy + 18x}$

61. $\dfrac{6p^2 + 12p}{2pq - 4p}$

62. $\dfrac{2x + 4}{x^2 - 3x - 10}$

63. $\dfrac{5z + 15}{z^2 - 4z - 21}$

64. $\dfrac{a^2 - 49}{a - 7}$

65. $\dfrac{b^2 - 64}{b - 8}$

66. $\dfrac{q^2 + 25}{q + 5}$

67. $\dfrac{r^2 + 36}{r + 6}$

68. $\dfrac{y^2 + 6y + 9}{2y^2 + y - 15}$

69. $\dfrac{h^2 + h - 6}{h^2 + 2h - 8}$

70. $\dfrac{3x^2 + 7x - 6}{x^2 + 7x + 12}$

71. $\dfrac{x^2 - 5x - 14}{2x^2 - x - 10}$

72. $\dfrac{5q^2 + 5}{q^4 - 1}$

73. $\dfrac{4t^2 + 16}{t^4 - 16}$

74. $\dfrac{ac - ad + 2bc - 2bd}{2ac + ad + 4bc + 2bd}$ (*Hint:* Factor by grouping.)

75. $\dfrac{3pr - ps - 3qr + qs}{3pr - ps + 3qr - qs}$ (*Hint:* Factor by grouping.)

76. $\dfrac{2t^2 - 3t}{2t^4 - 13t^3 + 15t^2}$

77. $\dfrac{4m^3 + 3m^2}{4m^3 + 7m^2 + 3m}$

78. $\dfrac{49p^2 - 28pq + 4q^2}{14p - 4q}$

79. $\dfrac{3x - 3y}{2x^2 - 4xy + 2y^2}$

80. $\dfrac{5x^3 + 4x^2 - 45x - 36}{x^2 - 9}$

81. $\dfrac{x^2 - 1}{ax^3 - bx^2 - ax + b}$

82. $\dfrac{2x^2 - xy - 3y^2}{2x^2 - 11xy + 12y^2}$

83. $\dfrac{2c^2 + cd - d^2}{5c^2 + 3cd - 2d^2}$

Concept 5: Simplifying a Ratio of −1

84. What is the relationship between $x - 2$ and $2 - x$?

85. What is the relationship between $w + p$ and $-w - p$?

For Exercises 86–99, simplify to lowest terms.

86. $\dfrac{x-5}{5-x}$ **87.** $\dfrac{8-p}{p-8}$ **88.** $\dfrac{-4-y}{4+y}$ **89.** $\dfrac{z+10}{-z-10}$

90. $\dfrac{3y-6}{12-6y}$ **91.** $\dfrac{4q-4}{12-12q}$ **92.** $\dfrac{2m-7n}{7n-2m}$ **93.** $\dfrac{3a^2-5}{5-3a^2}$

94. $\dfrac{k+5}{5-k}$ **95.** $\dfrac{2+n}{2-n}$ **96.** $\dfrac{10x-12}{10x+12}$ **97.** $\dfrac{4t-16}{16+4t}$

98. $\dfrac{x^2-x-12}{16-x^2}$ **99.** $\dfrac{49-b^2}{b^2-10b+21}$

Expanding Your Skills

For Exercises 100–103, factor and simplify to lowest terms.

100. $\dfrac{w^3-8}{w^2+2w+4}$ **101.** $\dfrac{y^3+27}{y^2-3y+9}$ **102.** $\dfrac{z^2-16}{z^3-64}$ **103.** $\dfrac{x^2-25}{x^3+125}$

Section 7.2 Multiplication and Division of Rational Expressions

Concepts

1. Multiplication of Rational Expressions
2. Division of Rational Expressions

1. Multiplication of Rational Expressions

Recall from Section R.2 that to multiply fractions, we multiply the numerators and multiply the denominators. The same is true for multiplying rational expressions.

Multiplication of Rational Expressions

Let p, q, r, and s represent polynomials, such that $q \neq 0, s \neq 0$. Then,

$$\frac{p}{q} \cdot \frac{r}{s} = \frac{pr}{qs}$$

For example:

Multiply the Fractions	**Multiply the Rational Expressions**
$\dfrac{2}{3} \cdot \dfrac{5}{7} = \dfrac{10}{21}$	$\dfrac{2x}{3y} \cdot \dfrac{5z}{7} = \dfrac{10xz}{21y}$

Sometimes it is possible to simplify a ratio of common factors to 1 *before* multiplying. To do so, we must first factor the numerators and denominators of each fraction.

$$\frac{15}{14} \cdot \frac{21}{10} = \frac{3 \cdot \overset{1}{\cancel{5}}}{2 \cdot \cancel{7}} \cdot \frac{3 \cdot \overset{1}{\cancel{7}}}{2 \cdot \cancel{5}} = \frac{9}{4}$$

The same process is also used to multiply rational expressions.

Steps to Multiply Rational Expressions

1. Factor the numerators and denominators of all rational expressions.
2. Simplify the ratios of common factors to 1 or -1.
3. Multiply the remaining factors in the numerator, and multiply the remaining factors in the denominator.

Example 1 Multiplying Rational Expressions

Multiply.

a. $\dfrac{5a^2b}{2} \cdot \dfrac{6a}{10b}$ **b.** $\dfrac{3c-3d}{6c} \cdot \dfrac{2}{c^2-d^2}$ **c.** $\dfrac{35-5x}{5x+5} \cdot \dfrac{x^2+5x+4}{x^2-49}$

Solution:

a. $\dfrac{5a^2b}{2} \cdot \dfrac{6a}{10b}$

$= \dfrac{5 \cdot a \cdot a \cdot b}{2} \cdot \dfrac{2 \cdot 3 \cdot a}{2 \cdot 5 \cdot b}$ Factor into prime factors.

$= \dfrac{\overset{1}{\cancel{5}} \cdot a \cdot a \cdot \overset{1}{\cancel{b}}}{2} \cdot \dfrac{\overset{1}{\cancel{2}} \cdot 3 \cdot a}{\cancel{2} \cdot \cancel{5} \cdot \cancel{b}}$ Simplify.

$= \dfrac{3a^3}{2}$ Multiply remaining factors.

b. $\dfrac{3c-3d}{6c} \cdot \dfrac{2}{c^2-d^2}$

$= \dfrac{3(c-d)}{2 \cdot 3 \cdot c} \cdot \dfrac{2}{(c-d)(c+d)}$ Factor into prime factors.

$= \dfrac{\overset{1}{\cancel{3}}\overset{1}{\cancel{(c-d)}}}{\cancel{2} \cdot \cancel{3} \cdot c} \cdot \dfrac{\overset{1}{\cancel{2}}}{\cancel{(c-d)}(c+d)}$ Simplify.

$= \dfrac{1}{c(c+d)}$

Avoiding Mistakes:

If all the factors in the numerator reduce to a ratio of 1, do not forget to write the factor of 1 in the numerator.

c. $\dfrac{35-5x}{5x+5} \cdot \dfrac{x^2+5x+4}{x^2-49}$

$= \dfrac{5(7-x)}{5(x+1)} \cdot \dfrac{(x+4)(x+1)}{(x-7)(x+7)}$ Factor the numerators and denominators completely.

$= \dfrac{\overset{1}{\cancel{5}}\overset{-1}{\cancel{(7-x)}}}{\cancel{5}\cancel{(x+1)}} \cdot \dfrac{(x+4)\overset{1}{\cancel{(x+1)}}}{\cancel{(x-7)}(x+7)}$ Simplify the ratios of common factors to 1 or -1.

$= \dfrac{-1(x+4)}{x+7}$

$= \dfrac{-(x+4)}{x+7}$ or $\dfrac{x+4}{-(x+7)}$ or $-\dfrac{x+4}{x+7}$

TIP: The ratio $\frac{7-x}{x-7} = -1$ because $7-x$ and $x-7$ are opposites.

Skill Practice Multiply.

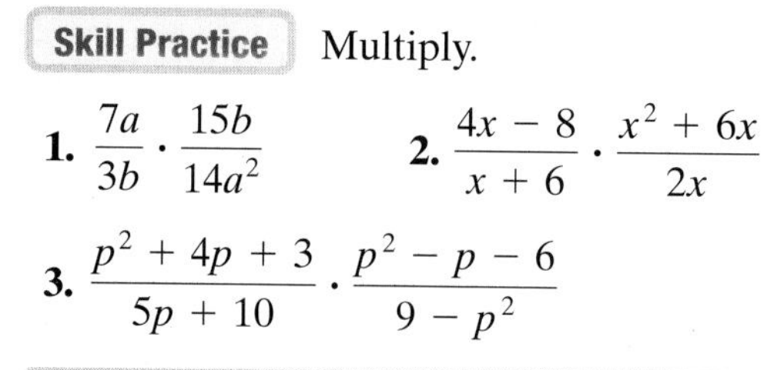

1. $\dfrac{7a}{3b} \cdot \dfrac{15b}{14a^2}$

2. $\dfrac{4x-8}{x+6} \cdot \dfrac{x^2+6x}{2x}$

3. $\dfrac{p^2+4p+3}{5p+10} \cdot \dfrac{p^2-p-6}{9-p^2}$

2. Division of Rational Expressions

Recall that to divide fractions, multiply the first fraction by the reciprocal of the second.

$$\frac{21}{10} \div \frac{49}{15} \xrightarrow[\text{of the second fraction}]{\text{Multiply by the reciprocal}} \frac{21}{10} \cdot \frac{15}{49} \xrightarrow{\text{Factor}} \frac{3 \cdot \overset{1}{\cancel{7}}}{2 \cdot \cancel{5}} \cdot \frac{3 \cdot \overset{1}{\cancel{5}}}{\cancel{7} \cdot 7} = \frac{9}{14}$$

The same process is used to divide rational expressions.

Division of Rational Expressions

Let p, q, r, and s represent polynomials, such that $q \neq 0, r \neq 0, s \neq 0$. Then,

$$\frac{p}{q} \div \frac{r}{s} = \frac{p}{q} \cdot \frac{s}{r} = \frac{ps}{qr}$$

Example 2 Dividing Rational Expressions

Divide.

a. $\dfrac{5t-15}{2} \div \dfrac{t^2-9}{10}$ b. $\dfrac{p^2-11p+30}{10p^2-250} \div \dfrac{30p-5p^2}{2p+4}$ c. $\dfrac{\frac{3x}{4y}}{\frac{5x}{6y}}$

Solution:

a. $\dfrac{5t-15}{2} \div \dfrac{t^2-9}{10}$

$= \dfrac{5t-15}{2} \cdot \dfrac{10}{t^2-9}$ Multiply the first fraction by the reciprocal of the second.

$= \dfrac{5(t-3)}{2} \cdot \dfrac{2 \cdot 5}{(t-3)(t+3)}$ Factor each polynomial.

$= \dfrac{5\overset{1}{\cancel{(t-3)}}}{\cancel{2}} \cdot \dfrac{\overset{1}{\cancel{2}} \cdot 5}{\cancel{(t-3)}(t+3)}$ Reduce common factors.

$= \dfrac{25}{t+3}$

Skill Practice Answers

1. $\dfrac{5}{2a}$ 2. $2(x-2)$

3. $\dfrac{-(p+1)}{5}$ or $\dfrac{p+1}{-5}$ or $-\dfrac{p+1}{5}$

b. $\dfrac{p^2 - 11p + 30}{10p^2 - 250} \div \dfrac{30p - 5p^2}{2p + 4}$

$= \dfrac{p^2 - 11p + 30}{10p^2 - 250} \cdot \dfrac{2p + 4}{30p - 5p^2}$ Multiply the first fraction by the reciprocal of the second.

Factor the trinomial.
$p^2 - 11p + 30 = (p - 5)(p - 6)$

Factor out the GCF.
$2p + 4 = 2(p + 2)$

$= \dfrac{(p - 5)(p - 6)}{2 \cdot 5(p - 5)(p + 5)} \cdot \dfrac{2(p + 2)}{5p(6 - p)}$ Factor out the GCF. Then factor the difference of squares.
$10p^2 - 250 = 10(p^2 - 25)$
$= 2 \cdot 5(p - 5)(p + 5)$

Factor out the GCF.
$30p - 5p^2 = 5p(6 - p)$

$= \dfrac{\overset{1}{\cancel{(p - 5)}}\overset{-1}{\cancel{(p - 6)}}}{\cancel{2} \cdot 5\cancel{(p - 5)}(p + 5)} \cdot \dfrac{\overset{1}{\cancel{2}}(p + 2)}{5p\cancel{(6 - p)}}$ Reduce common factors.

$= -\dfrac{(p + 2)}{25p(p + 5)}$

c. $\dfrac{\dfrac{3x}{4y}}{\dfrac{5x}{6y}}$ ← This fraction bar denotes division (÷).

$= \dfrac{3x}{4y} \div \dfrac{5x}{6y}$

$= \dfrac{3x}{4y} \cdot \dfrac{6y}{5x}$ Multiply by the reciprocal of the second fraction.

$= \dfrac{3 \cdot \overset{1}{\cancel{x}}}{\cancel{2} \cdot 2 \cdot \cancel{y}} \cdot \dfrac{\overset{1}{\cancel{2}} \cdot 3 \cdot \overset{1}{\cancel{y}}}{5 \cdot \cancel{x}}$ Reduce common factors.

$= \dfrac{9}{10}$

TIP: $(p - 6)$ and $(6 - p)$ are opposites and form a ratio of -1.

$$\frac{p - 6}{6 - p} = \frac{p - 6}{-1(-6 + p)} = \frac{p - 6}{-1(p - 6)} = -1$$

TIP: A fraction with one or more rational expressions in its numerator or denominator is called a *complex fraction*, for example,

$$\frac{\frac{3x}{4y}}{\frac{5x}{6y}}$$

Skill Practice Divide the rational expressions.

4. $\dfrac{7y - 14}{y + 1} \div \dfrac{y^2 + 2y - 8}{2y + 2}$ **5.** $\dfrac{4x^2 - 9}{2x^2 - x - 3} \div \dfrac{20x + 30}{x^2 + 7x + 6}$

6. $\dfrac{\dfrac{a^3b}{9c}}{\dfrac{4ab}{3c^3}}$

Skill Practice Answers

4. $\dfrac{14}{y + 4}$ **5.** $\dfrac{x + 6}{10}$ **6.** $\dfrac{a^2c^2}{12}$

Section 7.2 Practice Exercises

Review Exercises

1. Explain the difference between multiplying the fractions $\frac{2}{3} \cdot \frac{5}{9}$ and dividing the fractions $\frac{2}{3} \div \frac{5}{9}$.

For Exercises 2–9, multiply or divide the fractions.

2. $\frac{3}{5} \cdot \frac{1}{2}$

3. $\frac{6}{7} \cdot \frac{5}{12}$

4. $\frac{3}{4} \div \frac{3}{8}$

5. $\frac{18}{5} \div \frac{2}{5}$

6. $6 \cdot \frac{5}{12}$

7. $\frac{7}{25} \cdot 5$

8. $\frac{\frac{21}{4}}{\frac{7}{5}}$

9. $\frac{\frac{9}{2}}{\frac{3}{4}}$

Concept 1: Multiplication of Rational Expressions

For Exercises 10–21, multiply.

10. $\frac{2xy}{5x^2} \cdot \frac{15}{4y}$

11. $\frac{7s}{t^2} \cdot \frac{t^2}{14s^2}$

12. $\frac{6x^3}{9x^6y^2} \cdot \frac{18x^4y^7}{4y}$

13. $\frac{10a^2b}{15b^2} \cdot \frac{30b}{2a^3}$

14. $\frac{4x - 24}{20x} \cdot \frac{5x}{8}$

15. $\frac{5a + 20}{a} \cdot \frac{3a}{10}$

16. $\frac{3y + 18}{y^2} \cdot \frac{4y}{6y + 36}$

17. $\frac{2p - 4}{6p} \cdot \frac{4p^2}{8p - 16}$

18. $\frac{10}{2 - a} \cdot \frac{a - 2}{16}$

19. $\frac{b - 3}{6} \cdot \frac{20}{3 - b}$

20. $\frac{b^2 - a^2}{a - b} \cdot \frac{a}{a^2 - ab}$

21. $\frac{(x - y)^2}{x^2 + xy} \cdot \frac{x}{y - x}$

Concept 2: Division of Rational Expressions

For Exercises 22–35, divide.

22. $\frac{4x}{7y} \div \frac{2x^2}{21xy}$

23. $\frac{6cd}{5d^2} \div \frac{8c^3}{10d}$

24. $\frac{8m^4n^5}{5n^6} \div \frac{24mn}{15m^3}$

25. $\frac{10a^3b}{3a} \div \frac{5b}{9ab}$

26. $\frac{4a + 12}{6a - 18} \div \frac{3a + 9}{5a - 15}$

27. $\frac{8b - 16}{3b + 3} \div \frac{5b - 10}{2b + 2}$

28. $\frac{3x - 21}{6x^2 - 42x} \div \frac{7}{12x}$

29. $\frac{4a^2 - 4a}{9a - 9} \div \frac{5}{12a}$

30. $\frac{m^2 - n^2}{9} \div \frac{3n - 3m}{27m}$

31. $\frac{9 - b^2}{15b + 15} \div \frac{b - 3}{5b}$

32. $\frac{3p + 4q}{p^2 + 4pq + 4q^2} \div \frac{4}{p + 2q}$

33. $\frac{x^2 + 2xy + y^2}{2x - y} \div \frac{x + y}{5}$

34. $\frac{p^2 - 2p - 3}{p^2 - p - 6} \div \frac{p^2 - 1}{p^2 + 2p}$

35. $\frac{4t^2 - 1}{t^2 - 5t} \div \frac{2t^2 + 5t + 2}{t^2 - 3t - 10}$

Mixed Exercises

For Exercises 36–61, multiply or divide as indicated.

36. $(w + 3) \cdot \dfrac{w}{2w^2 + 5w - 3}$

37. $\dfrac{5t + 1}{5t^2 - 31t + 6} \cdot (t - 6)$

38. $\dfrac{\dfrac{5t - 10}{12}}{\dfrac{4t - 8}{8}}$

39. $\dfrac{\dfrac{6m + 6}{5}}{\dfrac{3m + 3}{10}}$

40. $\dfrac{q + 1}{5q^2 - 28q - 12} \cdot (5q + 2)$

41. $(r - 5) \cdot \dfrac{4r}{2r^2 - 7r - 15}$

42. $\dfrac{2a^2 + 13a - 24}{8a - 12} \div (a + 8)$

43. $\dfrac{3y^2 + 20y - 7}{5y + 35} \div (3y - 1)$

44. $\dfrac{y^2 + 5y - 36}{y^2 - 2y - 8} \cdot \dfrac{y + 2}{y - 6}$

45. $\dfrac{z^2 - 11z + 28}{z - 1} \cdot \dfrac{z + 1}{z^2 - 6z - 7}$

46. $\dfrac{t^2 + 4t - 5}{t^2 + 7t + 10} \cdot \dfrac{t + 4}{t - 1}$

47. $\dfrac{p^2 - 3p + 2}{p^2 - 4p + 3} \cdot \dfrac{p + 1}{p - 2}$

48. $(5t - 1) \div \dfrac{5t^2 + 9t - 2}{3t + 8}$

49. $(2q - 3) \div \dfrac{2q^2 + 5q - 12}{q - 7}$

50. $\dfrac{x^2 + 2x - 3}{x^2 - 3x + 2} \cdot \dfrac{x^2 + 2x - 8}{x^2 + 4x + 3}$

51. $\dfrac{y^2 + y - 12}{y^2 - y - 20} \cdot \dfrac{y^2 + y - 30}{y^2 - 2y - 3}$

52. $\dfrac{\dfrac{w^2 - 6w + 9}{8}}{\dfrac{9 - w^2}{4w + 12}}$

53. $\dfrac{\dfrac{p^2 - 6p + 8}{24}}{\dfrac{16 - p^2}{6p + 6}}$

54. $\dfrac{k^2 + 3k + 2}{k^2 + 5k + 4} \div \dfrac{k^2 + 5k + 6}{k^2 + 10k + 24}$

55. $\dfrac{4h^2 - 5h + 1}{h^2 + h - 2} \div \dfrac{6h^2 - 7h + 2}{2h^2 + 3h - 2}$

56. $\dfrac{ax + a + bx + b}{2x^2 + 4x + 2} \cdot \dfrac{4x + 4}{a^2 + ab}$

57. $\dfrac{3my + 9m + ny + 3n}{9m^2 + 6mn + n^2} \cdot \dfrac{30m + 10n}{5y^2 + 15y}$

58. $\dfrac{y^4 - 1}{2y^2 - 3y + 1} \div \dfrac{2y^2 + 2}{8y^2 - 4y}$

59. $\dfrac{x^4 - 16}{6x^2 + 24} \div \dfrac{x^2 - 2x}{3x}$

60. $\dfrac{x^2 - xy - 2y^2}{x + 2y} \div \dfrac{x^2 - 4xy + 4y^2}{x^2 - 4y^2}$

61. $\dfrac{4m^2 - 4mn - 3n^2}{8m^2 - 18n^2} \div \dfrac{3m + 3n}{6m^2 + 15mn + 9n^2}$

For Exercises 62–67, multiply or divide as indicated.

62. $\dfrac{b^3 - 3b^2 + 4b - 12}{b^4 - 16} \cdot \dfrac{3b^2 + 5b - 2}{3b^2 - 10b + 3} \div \dfrac{3}{6b - 12}$

63. $\dfrac{x^2 - 25}{3x^2 + 3xy} \cdot \dfrac{x^2 + 4x + xy + 4y}{x^2 + 9x + 20} \div \dfrac{x - 5}{x}$

64. $\dfrac{a^2 - 5a}{a^2 + 7a + 12} \div \dfrac{a^3 - 7a^2 + 10a}{a^2 + 9a + 18} \div \dfrac{a + 6}{a + 4}$

65. $\dfrac{t^2 + t - 2}{t^2 + 5t + 6} \div \dfrac{t - 1}{t} \div \dfrac{5t - 5}{t + 3}$

66. $\dfrac{p^3 - q^3}{p - q} \cdot \dfrac{p + q}{2p^2 + 2pq + 2q^2}$

67. $\dfrac{r^3 + s^3}{r - s} \div \dfrac{r^2 + 2rs + s^2}{r^2 - s^2}$

Section 7.3 Least Common Denominator

Concepts

1. Writing Equivalent Rational Expressions
2. Least Common Denominator
3. Writing Rational Expressions with the Least Common Denominator

1. Writing Equivalent Rational Expressions

In Sections 7.1 and 7.2, we learned how to simplify, multiply, and divide rational expressions. Our next goal is to add and subtract rational expressions. As with fractions, rational expressions can be added or subtracted only if they have the same denominator. Therefore, we must first learn how to identify a common denominator between two or more rational expressions. Then we must learn how to convert a rational expression into an equivalent rational expression with the indicated denominator.

Using the identity property of multiplication, we know that for $q \neq 0$ and $r \neq 0$,

$$\frac{p}{q} = \frac{p}{q} \cdot 1 = \frac{p}{q} \cdot \frac{r}{r} = \frac{pr}{qr}$$

This principle is used to convert a rational expression into an equivalent expression with a different denominator. For example, $\frac{1}{2}$ can be converted into an equivalent expression with a denominator of 12 as follows:

$$\frac{1}{2} = \frac{1}{2} \cdot \frac{6}{6} = \frac{1 \cdot 6}{2 \cdot 6} = \frac{6}{12}$$

In this example, we multiplied $\frac{1}{2}$ by a convenient form of 1. The ratio $\frac{6}{6}$ was chosen so that the product produced a new denominator of 12. Notice that multiplying $\frac{1}{2}$ by $\frac{6}{6}$ is equivalent to multiplying the numerator and denominator of the original expression by 6. In general, if the numerator and denominator of a rational expression are both multiplied by the same nonzero quantity, the value of the expression remains unchanged.

Example 1 Creating Equivalent Fractions

Convert each expression into an equivalent expression with the indicated denominator.

a. $\frac{5}{x} = \frac{\quad}{xyz}$

b. $\frac{7}{5p^2} = \frac{\quad}{20p^6}$

c. $\frac{w}{w + 5} = \frac{\quad}{(w + 5)(w - 2)}$

d. $\frac{1}{5} = \frac{\quad}{5x + 20}$

Solution:

a. $\frac{5}{x} = \frac{\quad}{xyz}$

$\frac{5 \cdot yz}{x \cdot yz} = \frac{5yz}{xyz}$

To convert $\frac{5}{x}$ to an equivalent fraction with a denominator of xyz, multiply both numerator and denominator by the missing factor of yz.

TIP: When multiplying both the numerator and denominator of the fraction $\frac{5}{x}$ by yz, we are actually multiplying the fraction by 1. This is because $\frac{yz}{yz} = 1$. Hence,

$$\frac{5}{x} = \frac{5}{x} \cdot 1 = \frac{5}{x} \cdot \left(\frac{yz}{yz}\right) = \frac{5 \cdot yz}{x \cdot yz} = \frac{5yz}{xyz}$$

b. $\dfrac{7}{5p^2} = \dfrac{\quad}{20p^6}$ Multiply the numerator and denominator of the fraction by the missing factor of $4p^4$.

$$\frac{7 \cdot 4p^4}{5p^2 \cdot 4p^4} = \frac{28p^4}{20p^6}$$

c. $\dfrac{w}{w+5} = \dfrac{\quad}{(w+5)(w-2)}$ Multiply numerator and denominator by the missing factor of $(w-2)$.

$$\frac{w}{w+5} = \frac{w \cdot (w-2)}{(w+5) \cdot (w-2)} \quad \text{or} \quad \frac{w^2 - 2w}{(w+5)(w-2)}$$

d. $\dfrac{1}{5} = \dfrac{\quad}{5x+20}$

$\dfrac{1}{5} = \dfrac{\quad}{5(x+4)}$ Factor the denominator of the second expression. The factor missing from the denominator of the first expression is $(x + 4)$.

Multiply numerator and denominator by $(x + 4)$.

$$\frac{1}{5} = \frac{1 \cdot (x+4)}{5 \cdot (x+4)} = \frac{x+4}{5(x+4)}$$

Skill Practice Convert each expression to an equivalent expression with the indicated denominator.

1. $\dfrac{12}{a} = \dfrac{\quad}{abc}$ **2.** $\dfrac{6}{7y} = \dfrac{\quad}{14y^3}$

3. $\dfrac{x}{x+4} = \dfrac{\quad}{(x+6)(x+4)}$ **4.** $\dfrac{1}{8} = \dfrac{\quad}{8y-16}$

TIP: Notice that in Example 1(c) we multiplied the polynomials in the numerator but left the denominator in factored form. This convention is followed because when we add and subtract rational expressions in the next section, the terms in the numerators must be combined.

2. Least Common Denominator

Recall from Section R.2 that to add or subtract fractions, the fractions must have a common denominator. The same is true for rational expressions. In this section, we present a method to find the least common denominator of two rational expressions.

The **least common denominator (LCD)** of two or more rational expressions is defined as the least common multiple of the denominators. For example, consider the fractions $\frac{1}{20}$ and $\frac{1}{8}$. By inspection, you can probably see that the least common denominator is 40. To understand why, find the prime factorization of both denominators:

$$20 = 2^2 \cdot 5 \quad \text{and} \quad 8 = 2^3$$

A common multiple of 20 and 8 must be a multiple of 5, a multiple of 2^2, and a multiple of 2^3. However, any number that is a multiple of $2^3 = 8$ is automatically a multiple of $2^2 = 4$. Therefore, it is sufficient to construct the least common denominator as the product of unique prime factors, in which each factor is raised to its highest power.

$$\text{The LCD of } \frac{1}{20} \text{ and } \frac{1}{8} \text{ is } 2^3 \cdot 5 = 40.$$

Skill Practice Answers

1. $\dfrac{12}{a} = \dfrac{12bc}{abc}$ **2.** $\dfrac{6}{7y} = \dfrac{12y^2}{14y^3}$

3. $\dfrac{x}{x+4} = \dfrac{x^2+6x}{(x+4)(x+6)}$

4. $\dfrac{1}{8} = \dfrac{y-2}{8(y-2)}$

Steps to Find the Least Common Denominator of Two or More Rational Expressions

1. Factor all denominators completely.
2. The LCD is the product of unique factors from the denominators, in which each factor is raised to the highest power to which it appears in any denominator.

Example 2 Finding the Least Common Denominator of Rational Expressions

Find the LCD of the following sets of rational expressions.

a. $\dfrac{5}{14}; \dfrac{3}{49}; \dfrac{1}{8}$ **b.** $\dfrac{5}{3x^2z}; \dfrac{7}{x^5y^3}$ **c.** $\dfrac{a+b}{a^2-25}; \dfrac{1}{2a-10}$

d. $\dfrac{x-5}{x^2-2x}; \dfrac{1}{x^2-4x+4}$

Solution:

a. $\dfrac{5}{14}; \dfrac{3}{49}; \dfrac{1}{8}$

$= \dfrac{5}{2 \cdot 7}; \dfrac{3}{7^2}; \dfrac{1}{2^3}$ **Step 1:** Factor the denominators.

The LCD is $2^3 \cdot 7^2 = 392$. **Step 2:** The LCD is the product of unique factors, each raised to its highest power.

b. $\dfrac{5}{3x^2z}; \dfrac{7}{x^5y^3}$

$= \dfrac{5}{3x^2z}; \dfrac{7}{x^5y^3}$ **Step 1:** The denominators are already factored.

The LCD is $3x^5y^3z$. **Step 2:** The LCD is the product of unique factors, each raised to its highest power.

c. $\dfrac{a+b}{a^2-25}; \dfrac{1}{2a-10}$

$= \dfrac{a+b}{(a-5)(a+5)}; \dfrac{1}{2(a-5)}$ **Step 1:** Factor the denominators.

The LCD is $2(a-5)(a+5)$. **Step 2:** The LCD is the product of unique factors, each raised to its highest power.

d. $\dfrac{x-5}{x^2-2x}; \dfrac{1}{x^2-4x+4}$

$= \dfrac{x-5}{x(x-2)}; \dfrac{1}{(x-2)^2}$ **Step 1:** Factor the denominators.

The LCD is $x(x-2)^2$. **Step 2:** The LCD is the product of unique factors, each raised to its highest power.

Skill Practice Find the LCD for each set of expressions.

5. $\frac{3}{8}; \frac{7}{10}; \frac{1}{15}$

6. $\frac{1}{5a^3b^2}; \frac{1}{10a^4b}$

7. $\frac{x}{x^2 - 16}; \frac{2}{3x + 12}$

8. $\frac{6}{t^2 + 5t - 14}; \frac{8}{t^2 - 3t + 2}$

3. Writing Rational Expressions with the Least Common Denominator

To add or subtract two rational expressions, the expressions must have the same denominator. Therefore, we must first practice the skill of converting each rational expression into an equivalent expression with the LCD as its denominator. The process is as follows: Identify the LCD for the two expressions. Then, multiply the numerator and denominator of each fraction by the factors from the LCD that are missing from the original denominators.

Example 3 Converting to the Least Common Denominator

Find the LCD of each pair of rational expressions. Then convert each expression to an equivalent fraction with the denominator equal to the LCD.

a. $\frac{3}{2ab}; \frac{6}{5a^2}$ **b.** $\frac{4}{x + 1}; \frac{7}{x - 4}$ **c.** $\frac{w + 2}{w^2 - w - 12}; \frac{1}{w^2 - 9}$

Solution:

a. $\frac{3}{2ab}; \frac{6}{5a^2}$

The LCD is $10a^2b$.

$$\frac{3}{2ab} = \frac{3 \cdot 5a}{2ab \cdot 5a} = \frac{15a}{10a^2b}$$

The first expression is missing the factor $5a$ from the denominator.

$$\frac{6}{5a^2} = \frac{6 \cdot 2b}{5a^2 \cdot 2b} = \frac{12b}{10a^2b}$$

The second expression is missing the factor $2b$ from the denominator.

b. $\frac{4}{x + 1}; \frac{7}{x - 4}$

The LCD is $(x + 1)(x - 4)$.

$$\frac{4}{x + 1} = \frac{4(x - 4)}{(x + 1)(x - 4)} = \frac{4x - 16}{(x + 1)(x - 4)}$$

The first expression is missing the factor $(x - 4)$ from the denominator.

$$\frac{7}{x - 4} = \frac{7(x + 1)}{(x - 4)(x + 1)} = \frac{7x + 7}{(x - 4)(x + 1)}$$

The second expression is missing the factor $(x + 1)$ from the denominator.

Skill Practice Answers

5. 120 **6.** $10a^4b^2$
7. $3(x - 4)(x + 4)$
8. $(t + 7)(t - 2)(t - 1)$

c. $\dfrac{w+2}{w^2-w-12}; \dfrac{1}{w^2-9}$

To find the LCD, factor each denominator.

$\dfrac{w+2}{(w-4)(w+3)}; \dfrac{1}{(w-3)(w+3)}$

The LCD is $(w-4)(w+3)(w-3)$.

$$\frac{w+2}{(w-4)(w+3)} = \frac{(w+2)(w-3)}{(w-4)(w+3)(w-3)} = \frac{w^2-w-6}{(w-4)(w+3)(w-3)}$$

The first expression is missing the factor $(w-3)$ from the denominator.

$$\frac{1}{(w-3)(w+3)} = \frac{1(w-4)}{(w-3)(w+3)(w-4)} = \frac{w-4}{(w-3)(w+3)(w-4)}$$

The second expression is missing the factor $(w-4)$ from the denominator.

Skill Practice For each pair of expressions, find the LCD, and then convert each expression to an equivalent fraction with the denominator equal to the LCD.

9. $\dfrac{2}{rs^2}; \dfrac{-1}{r^3s}$ **10.** $\dfrac{5}{x-3}; \dfrac{x}{x+1}$ **11.** $\dfrac{z}{z^2-4}; \dfrac{-3}{z^2-z-2}$

Example 4 Converting to the Least Common Denominator

Find the LCD of the expressions $\dfrac{3}{x-7}$ and $\dfrac{1}{7-x}$.

TIP: In Example 4, the expressions

$\dfrac{3}{x-7}$ and $\dfrac{1}{7-x}$

have opposite factors in the denominators. In such a case, you do not need to include *both* factors in the LCD.

Solution:

Notice that the expressions $x-7$ and $7-x$ are opposites and differ by a factor of -1. Therefore, we may use either $x-7$ or $7-x$ as a common denominator. Each case is detailed in the following conversions.

Converting to the Denominator $x - 7$

$\dfrac{3}{x-7}; \dfrac{1}{7-x}$

Leave the first fraction unchanged because it has the desired LCD.

$$\frac{1}{7-x} = \frac{(-1)1}{(-1)(7-x)}$$

Multiply the *second* rational expression by the ratio $\frac{-1}{-1}$ to change its denominator to $x-7$.

$$= \frac{-1}{-7+x}$$

Apply the distributive property.

$$= \frac{-1}{x-7}$$

Skill Practice Answers

9. $\dfrac{2r^2}{r^3s^2}; \dfrac{-s}{r^3s^2}$

10. $\dfrac{5x+5}{(x-3)(x+1)}; \dfrac{x^2-3x}{(x+1)(x-3)}$

11. $\dfrac{z^2+z}{(z-2)(z+2)(z+1)}; \dfrac{-3z-6}{(z-2)(z+2)(z+1)}$

Converting to the Denominator $7 - x$

$\frac{3}{x-7}; \frac{1}{7-x}$ Leave the second fraction unchanged because it has the desired LCD.

$$\frac{3}{x-7} = \frac{(-1)3}{(-1)(x-7)};$$ Multiply the first rational expression by the ratio $\frac{-1}{-1}$ to change its denominator to $7 - x$.

$$= \frac{-3}{-x+7}$$ Apply the distributive property.

$$= \frac{-3}{7-x}$$

Skill Practice

12a. Find the LCD of the expressions.

b. Then convert each expression to an equivalent fraction with denominator equal to the LCD.

$$\frac{9}{w-2}; \frac{11}{2-w}$$

Skill Practice Answers

12a. The LCD is $(w - 2)$ or $(2 - w)$.

b. $\frac{9}{w-2} = \frac{9}{w-2}$; $\frac{11}{2-w} = \frac{-11}{w-2}$ or $\frac{9}{w-2} = \frac{-9}{2-w}$; $\frac{11}{2-w} = \frac{11}{2-w}$

Section 7.3 Practice Exercises

Study Skills Exercise

1. Define the key term **least common denominator**.

Review Exercises

2. Evaluate the expression for the given values of x. $\frac{2x}{x+5}$

a. $x = 1$ **b.** $x = 5$ **c.** $x = -5$

For Exercises 3–4, write the domain in set-builder notation. Then reduce the expression to lowest terms.

3. $\frac{3x+3}{5x^2-5}$

4. $\frac{x+2}{x^2-3x-10}$

For Exercises 5–8, multiply or divide as indicated.

5. $\frac{a+3}{a+7} \cdot \frac{a^2+3a-10}{a^2+a-6}$

6. $\frac{6(a+2b)}{2(a-3b)} \cdot \frac{4(a+3b)(a-3b)}{9(a+2b)(a-2b)}$

7. $\frac{16y^2}{9y+36} \div \frac{8y^3}{3y+12}$

8. $\frac{5b^2+6b+1}{b^2+5b+6} \div (5b+1)$

Concept 1: Writing Equivalent Rational Expressions

For Exercises 9–24, convert the expressions into equivalent expressions with the indicated denominator.

9. $\dfrac{6}{7} = \dfrac{}{42}$

10. $\dfrac{4}{9} = \dfrac{}{72}$

11. $\dfrac{2}{13} = \dfrac{}{39}$

12. $\dfrac{1}{8} = \dfrac{}{64}$

13. $\dfrac{3}{p^2q} = \dfrac{}{5p^3q}$

14. $\dfrac{2}{3rs} = \dfrac{}{18rs^3}$

15. $\dfrac{2x}{yz} = \dfrac{}{6y^2z^4}$

16. $\dfrac{8a}{b^2c} = \dfrac{}{2b^4c^5}$

17. $\dfrac{w+6}{w-7} = \dfrac{}{(w-7)(w+2)}$

18. $\dfrac{z-1}{z+1} = \dfrac{}{(z+1)(z-3)}$

19. $\dfrac{-4}{z-3} = \dfrac{}{5z-15}$

20. $\dfrac{-8}{3a+2} = \dfrac{}{12a+8}$

21. $\dfrac{5}{x+2} = \dfrac{}{x^2-4}$

22. $\dfrac{3y}{4y-5} = \dfrac{}{16y^2-25}$

23. $\dfrac{6}{x-3} = \dfrac{}{3-x}$

24. $\dfrac{2}{a-9} = \dfrac{}{9-a}$

25. Which of the expressions are equivalent to $-\dfrac{5}{x-3}$? Circle all that apply.

a. $\dfrac{-5}{x-3}$ **b.** $\dfrac{5}{-x+3}$ **c.** $\dfrac{5}{3-x}$ **d.** $\dfrac{5}{-(x-3)}$

26. Which of the expressions are equivalent to $\dfrac{4-a}{6}$? Circle all that apply.

a. $\dfrac{a-4}{-6}$ **b.** $\dfrac{a-4}{6}$ **c.** $\dfrac{-(4-a)}{-6}$ **d.** $-\dfrac{a-4}{6}$

Concept 2: Least Common Denominator

27. Explain why the least common denominator of $\frac{1}{x^3}$, $\frac{1}{x^5}$, and $\frac{1}{x^4}$ is x^5.

28. Explain why the least common denominator of $\frac{2}{y^3}$, $\frac{9}{y^6}$, and $\frac{4}{y^5}$ is y^6.

29. Explain why the least common denominator of

$$\frac{1}{x+3} \quad \text{and} \quad \frac{3}{x-2}$$

is $(x+3)(x-2)$.

30. Explain why the least common denominator of

$$\frac{7}{y-8} \quad \text{and} \quad \frac{3}{y+1}$$

is $(y-8)(y+1)$.

31. Explain why a common denominator of

$$\frac{b+1}{b-1} \quad \text{and} \quad \frac{b}{1-b}$$

could be either $(b-1)$ or $(1-b)$.

32. Explain why a common denominator of

$$\frac{1}{6-t} \quad \text{and} \quad \frac{t}{t-6}$$

could be either $(6-t)$ or $(t-6)$.

For Exercises 33–50, identify the LCD.

33. $\frac{4}{15}; \frac{5}{9}$

34. $\frac{7}{12}; \frac{1}{18}$

35. $\frac{3}{16}; \frac{1}{4}$

36. $\frac{1}{2}; \frac{11}{12}$

37. $\frac{1}{7}; \frac{2}{9}$

38. $\frac{2}{3}; \frac{5}{8}$

39. $\frac{1}{3x^2y}; \frac{8}{9xy^3}$

40. $\frac{5}{2a^4b^2}; \frac{1}{8ab^3}$

41. $\frac{6}{w^2}; \frac{7}{y}$

42. $\frac{2}{r}; \frac{3}{s^2}$

43. $\frac{p}{(p+3)(p-1)}; \frac{2}{(p+3)(p+2)}$

44. $\frac{6}{(q+4)(q-4)}; \frac{q^2}{(q+1)(q+4)}$

45. $\frac{7}{3t(t+1)}; \frac{10t}{9(t+1)^2}$

46. $\frac{13x}{15(x-1)^2}; \frac{5}{3x(x-1)}$

47. $\frac{y}{y^2-4}; \frac{3y}{y^2+5y+6}$

48. $\frac{4}{w^2-3w+2}; \frac{w}{w^2-4}$

49. $\frac{5}{3-x}; \frac{7}{x-3}$

50. $\frac{4}{x-6}; \frac{9}{6-x}$

Concept 3: Writing Rational Expressions with the Least Common Denominator

For Exercises 51–74, find the LCD. Then convert each expression to an equivalent expression with the denominator equal to the LCD.

51. $\frac{6}{5x^2}; \frac{1}{x}$

52. $\frac{3}{y}; \frac{7}{9y^2}$

53. $\frac{4}{5x^2}; \frac{y}{6x^3}$

54. $\frac{3}{15b^2}; \frac{c}{3b^2}$

55. $\frac{5}{6a^2b}; \frac{a}{12b}$

56. $\frac{x}{15y^2}; \frac{y}{5xy}$

57. $\frac{6}{m+4}; \frac{3}{m-1}$

58. $\frac{3}{n-5}; \frac{7}{n+2}$

59. $\frac{6}{2x-5}; \frac{1}{x+3}$

60. $\frac{4}{m+3}; \frac{-3}{5m+1}$

61. $\frac{6}{(w+3)(w-8)}; \frac{w}{(w-8)(w+1)}$

62. $\frac{t}{(t+2)(t+12)}; \frac{18}{(t-2)(t+2)}$

63. $\frac{6p}{p^2-4}; \frac{3}{p^2+4p+4}$

64. $\frac{5}{q^2-6q+9}; \frac{q}{q^2-9}$

65. $\frac{1}{a-4}; \frac{a}{4-a}$

66. $\frac{3b}{2b-5}; \frac{2b}{5-2b}$

67. $\frac{4}{x-7}; \frac{y}{14-2x}$

68. $\frac{4}{3x-15}; \frac{z}{5-x}$

69. $\frac{1}{a+b}; \frac{6}{-a-b}$

70. $\frac{p}{-q-8}; \frac{1}{q+8}$

71. $\frac{-3}{24y+8}; \frac{5}{18y+6}$

72. $\frac{r}{10r+5}; \frac{2}{16r+8}$

73. $\frac{3}{5z}; \frac{1}{z+4}$

74. $\frac{-1}{4a-8}; \frac{5}{4a}$

Expanding Your Skills

For Exercises 75–78, find the LCD. Then convert each expression to an equivalent expression with the denominator equal to the LCD.

75. $\dfrac{z}{z^2 + 9z + 14}; \dfrac{-3z}{z^2 + 10z + 21}; \dfrac{5}{z^2 + 5z + 6}$

76. $\dfrac{6}{w^2 - 3w - 4}; \dfrac{1}{w^2 + 6w + 5}; \dfrac{-9w}{w^2 + w - 20}$

77. $\dfrac{3}{p^3 - 8}; \dfrac{p}{p^2 - 4}; \dfrac{5p}{p^2 + 2p + 4}$

78. $\dfrac{7}{q^3 + 125}; \dfrac{q}{q^2 - 25}; \dfrac{12}{q^2 - 5q + 25}$

Section 7.4 Addition and Subtraction of Rational Expressions

Concepts

1. Addition and Subtraction of Rational Expressions with the Same Denominator
2. Addition and Subtraction of Rational Expressions with Different Denominators
3. Using Rational Expressions in Translations

1. Addition and Subtraction of Rational Expressions with the Same Denominator

To add or subtract rational expressions, the expressions must have the same denominator. As with fractions, we add or subtract rational expressions with the same denominator by combining the terms in the numerator and then writing the result over the common denominator. Then, if possible, we simplify the expression to lowest terms.

Addition and Subtraction of Rational Expressions

Let p, q, and r represent polynomials where $q \neq 0$. Then,

1. $\dfrac{p}{q} + \dfrac{r}{q} = \dfrac{p + r}{q}$
2. $\dfrac{p}{q} - \dfrac{r}{q} = \dfrac{p - r}{q}$

Example 1 Adding and Subtracting Rational Expressions with a Common Denominator

Add or subtract as indicated.

a. $\dfrac{1}{12} + \dfrac{7}{12}$

b. $\dfrac{2}{5p} - \dfrac{7}{5p}$

c. $\dfrac{2}{3d + 5} + \dfrac{7d}{3d + 5}$

d. $\dfrac{x^2}{x - 3} - \dfrac{-5x + 24}{x - 3}$

Solution:

a. $\dfrac{1}{12} + \dfrac{7}{12}$ The fractions have the same denominator.

$= \dfrac{1 + 7}{12}$ Add the terms in the numerators, and write the result over the common denominator.

$= \dfrac{8}{12}$

$= \dfrac{2}{3}$ Simplify to lowest terms.

b. $\dfrac{2}{5p} - \dfrac{7}{5p}$ The rational expressions have the same denominator.

$= \dfrac{2 - 7}{5p}$ Subtract the terms in the numerators, and write the result over the common denominator.

$= \dfrac{-5}{5p}$

$= \dfrac{(-\overset{1}{\cancel{5}})}{\cancel{5}p}$ Simplify to lowest terms.

$= -\dfrac{1}{p}$

c. $\dfrac{2}{3d + 5} + \dfrac{7d}{3d + 5}$ The rational expressions have the same denominator.

$= \dfrac{2 + 7d}{3d + 5}$ Add the terms in the numerators, and write the result over the common denominator.

$= \dfrac{7d + 2}{3d + 5}$ Because the numerator and denominator share no common factors, the expression is in lowest terms.

d. $\dfrac{x^2}{x - 3} - \dfrac{-5x + 24}{x - 3}$ The rational expressions have the same denominator.

$= \dfrac{x^2 - (-5x + 24)}{x - 3}$ Subtract the terms in the numerators, and write the result over the common denominator.

$= \dfrac{x^2 + 5x - 24}{x - 3}$ Simplify the numerator.

$= \dfrac{(x + 8)(x - 3)}{(x - 3)}$ Factor the numerator and denominator to determine if the rational expression can be simplified.

$= \dfrac{(x + 8)\overset{1}{\cancel{(x - 3)}}}{\cancel{(x - 3)}}$ Simplify to lowest terms.

$= x + 8$

Avoiding Mistakes:

When subtracting rational expressions, use parentheses to group the terms in the numerator that follow the subtraction sign. This will help you remember to apply the distributive property.

Skill Practice Add or subtract as indicated.

1. $\dfrac{3}{14} + \dfrac{4}{14}$
2. $\dfrac{3}{7d} - \dfrac{6}{7d}$
3. $\dfrac{x^2 + 2}{x + 3} + \dfrac{4x + 1}{x + 3}$
4. $\dfrac{4t - 9}{2t + 1} - \dfrac{t - 5}{2t + 1}$

2. Addition and Subtraction of Rational Expressions with Different Denominators

To add or subtract two rational expressions with unlike denominators, we must convert the expressions to equivalent expressions with the same denominator. For example, consider adding

$$\frac{1}{10} + \frac{12}{5y}$$

The LCD is $10y$. For each expression, identify the factors from the LCD that are missing from the denominator. Then multiply the numerator and denominator of the expression by the missing factor(s).

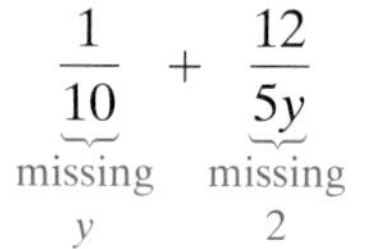

$$= \frac{1 \cdot y}{10 \cdot y} + \frac{12 \cdot 2}{5y \cdot 2}$$

$$= \frac{y}{10y} + \frac{24}{10y}$$ The rational expressions now have the same denominators.

$$= \frac{y + 24}{10y}$$ Add the numerators.

Avoiding Mistakes:

In the expression $\frac{y + 24}{10y}$, notice that you cannot reduce the 24 and 10 because 24 is not a factor in the numerator. It is a term. Only factors can be reduced.

After successfully adding or subtracting two rational expressions, always check to see if the final answer is simplified. If necessary, factor the numerator and denominator, and reduce common factors. The expression

$$\frac{y + 24}{10y}$$

is in lowest terms because the numerator and denominator do not share any common factors.

Steps to Add or Subtract Rational Expressions

1. Factor the denominators of each rational expression.
2. Identify the LCD.
3. Rewrite each rational expression as an equivalent expression with the LCD as its denominator.
4. Add or subtract the numerators, and write the result over the common denominator.
5. Simplify to lowest terms.

Skill Practice Answers

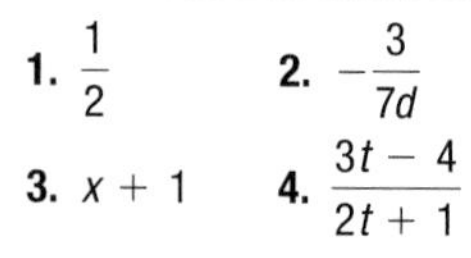

1. $\frac{1}{2}$ 2. $-\frac{3}{7d}$

3. $x + 1$ 4. $\frac{3t - 4}{2t + 1}$

Example 2 Adding and Subtracting Rational Expressions with Unlike Denominators

Add or subtract as indicated.

a. $\dfrac{4}{7k} - \dfrac{3}{k^2}$ **b.** $\dfrac{2q - 4}{3} - \dfrac{q + 1}{2}$ **c.** $\dfrac{1}{x - 5} + \dfrac{-10}{x^2 - 25}$

Solution:

a. $\dfrac{4}{7k} - \dfrac{3}{k^2}$

Step 1: The denominators are already factored.

Step 2: The LCD is $7k^2$.

$$= \frac{4 \cdot k}{7k \cdot k} - \frac{3 \cdot 7}{k^2 \cdot 7}$$

Step 3: Write each expression with the LCD.

$$= \frac{4k}{7k^2} - \frac{21}{7k^2}$$

$$= \frac{4k - 21}{7k^2}$$

Step 4: Subtract the numerators, and write the result over the LCD.

Step 5: The expression is in lowest terms because the numerator and denominator share no common factors.

b. $\dfrac{2q - 4}{3} - \dfrac{q + 1}{2}$

Step 1: The denominators are already factored.

Step 2: The LCD is 6.

$$= \frac{2(2q - 4)}{2 \cdot 3} - \frac{3(q + 1)}{3 \cdot 2}$$

Step 3: Write each expression with the LCD.

$$= \frac{2(2q - 4) - 3(q + 1)}{6}$$

Step 4: Subtract the numerators, and write the result over the LCD.

$$= \frac{4q - 8 - 3q - 3}{6}$$

$$= \frac{q - 11}{6}$$

Step 5: The expression is in lowest terms because the numerator and denominator share no common factors.

Avoiding Mistakes:

Do not reduce after rewriting the fractions with the LCD. You will revert back to the original expression.

c. $\dfrac{1}{x-5} + \dfrac{-10}{x^2-25}$

$= \dfrac{1}{x-5} + \dfrac{-10}{(x-5)(x+5)}$ **Step 1:** Factor the denominators.

Step 2: The LCD is $(x-5)(x+5)$.

$= \dfrac{1(x+5)}{(x-5)(x+5)} + \dfrac{-10}{(x-5)(x+5)}$ **Step 3:** Write each expression with the LCD.

$= \dfrac{1(x+5)+(-10)}{(x-5)(x+5)}$ **Step 4:** Add the numerators, and write the result over the LCD.

$= \dfrac{x+5-10}{(x-5)(x+5)}$

$= \dfrac{\overset{1}{\cancel{x-5}}}{\cancel{(x-5)}(x+5)}$ **Step 5:** Simplify.

$= \dfrac{1}{x+5}$

Skill Practice Add or subtract as indicated.

5. $\dfrac{4}{3x} + \dfrac{1}{2x^2}$ **6.** $\dfrac{q}{12} - \dfrac{q-2}{4}$ **7.** $\dfrac{1}{x-4} + \dfrac{-8}{x^2-16}$

Example 3 Subtracting Rational Expressions with Different Denominators

Subtract the rational expressions. $\dfrac{p+2}{p-1} - \dfrac{2}{p+6} - \dfrac{14}{p^2+5p-6}$

Solution:

$\dfrac{p+2}{p-1} - \dfrac{2}{p+6} - \dfrac{14}{p^2+5p-6}$

$= \dfrac{p+2}{p-1} - \dfrac{2}{p+6} - \dfrac{14}{(p-1)(p+6)}$ **Step 1:** Factor the denominators.

Step 2: The LCD is $(p-1)(p+6)$.

Step 3: Write each expression with the LCD.

$= \dfrac{(p+2)(p+6)}{(p-1)(p+6)} - \dfrac{2(p-1)}{(p+6)(p-1)} - \dfrac{14}{(p-1)(p+6)}$

$= \dfrac{(p+2)(p+6)-2(p-1)-14}{(p-1)(p+6)}$ **Step 4:** Combine the numerators, and write the result over the LCD.

Skill Practice Answers

5. $\dfrac{8x+3}{6x^2}$ **6.** $\dfrac{-q+3}{6}$

7. $\dfrac{1}{x+4}$

$$= \frac{p^2 + 6p + 2p + 12 - 2p + 2 - 14}{(p - 1)(p + 6)}$$

Step 5: Clear parentheses in the numerator.

$$= \frac{p^2 + 6p}{(p - 1)(p + 6)}$$

Combine *like* terms.

$$= \frac{p(p + 6)}{(p - 1)(p + 6)}$$

Factor the numerator to determine if the expression is in lowest terms.

$$= \frac{p\cancel{(p + 6)}}{(p - 1)\cancel{(p + 6)}}$$

Simplify to lowest terms.

$$= \frac{p}{p - 1}$$

Skill Practice Subtract.

8. $\frac{2y}{y - 1} - \frac{1}{y} - \frac{2y + 1}{y^2 - y}$

When the denominators of two rational expressions are opposites, we can produce identical denominators by multiplying one of the expressions by the ratio $\frac{-1}{-1}$. This is demonstrated in Example 4.

Example 4 Adding Rational Expressions with Different Denominators

Add the rational expressions. $\frac{1}{d - 7} + \frac{5}{7 - d}$

Solution:

$$\frac{1}{d - 7} + \frac{5}{7 - d}$$

The expressions $d - 7$ and $7 - d$ are opposites and differ by a factor of -1. Therefore, multiply the numerator and denominator of *either* expression by -1 to obtain a common denominator.

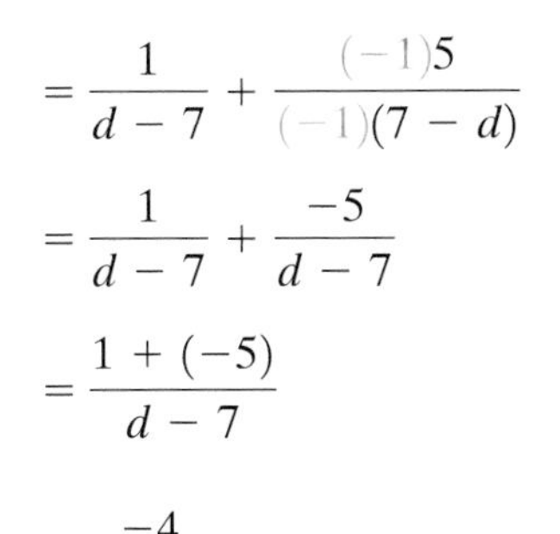

$$= \frac{1}{d - 7} + \frac{(-1)5}{(-1)(7 - d)}$$

Note that $-1(7 - d) = -7 + d$ or $d - 7$.

$$= \frac{1}{d - 7} + \frac{-5}{d - 7}$$

Simplify.

$$= \frac{1 + (-5)}{d - 7}$$

Add the terms in the numerators, and write the result over the common denominator.

$$= \frac{-4}{d - 7}$$

Skill Practice Add.

9. $\frac{3}{p - 8} + \frac{1}{8 - p}$

Skill Practice Answers

8. $\frac{2y - 3}{y - 1}$ **9.** $\frac{2}{p - 8}$ or $\frac{-2}{8 - p}$

3. Using Rational Expressions in Translations

Example 5 Using Rational Expressions in Translations

Translate the English phrase into a mathematical expression. Then simplify by combining the rational expressions.

The difference of the reciprocal of x and the quotient of x and 3

Solution:

The difference of the reciprocal of x and the quotient of x and 3

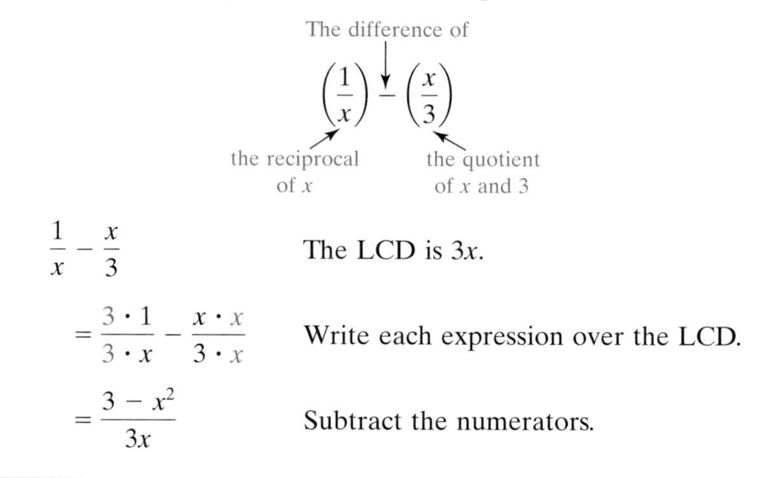

$$\left(\frac{1}{x}\right) - \left(\frac{x}{3}\right)$$

$\frac{1}{x} - \frac{x}{3}$ The LCD is $3x$.

$= \frac{3 \cdot 1}{3 \cdot x} - \frac{x \cdot x}{3 \cdot x}$ Write each expression over the LCD.

$= \frac{3 - x^2}{3x}$ Subtract the numerators.

Skill Practice Translate the English phrase into a mathematical expression. Then simplify by combining the rational expressions.

10. The sum of 1 and the quotient of 2 and a

Skill Practice Answers

10. $1 + \frac{2}{a}; \frac{a + 2}{a}$

Section 7.4 Practice Exercises

Boost *your* GRADE at mathzone.com!

MathZone

- Practice Problems
- Self-Tests
- NetTutor
- e-Professors
- Videos

Review Exercises

1. Write the domain of the expression. $\frac{x + 4}{x^2 - 36}$

2. For the rational expression $\frac{x^2 - 4x - 5}{x^2 - 7x + 10}$

a. Find the value of the expression (if possible) when $x = 0, 1, -1, 2$, and 5.

b. Factor the denominator and identify the domain. Write the domain in set-builder notation.

c. Reduce the expression to lowest terms.

3. For the rational expression $\dfrac{a^2 + a - 2}{a^2 - 4a - 12}$

a. Find the value of the expression (if possible) when $a = 0, 1, -2, 2$, and 6.

b. Factor the denominator, and identify the domain. Write the domain in set-builder notation.

c. Reduce the expression to lowest terms.

For Exercises 4–5, multiply or divide as indicated.

4. $\dfrac{2b^2 - b - 3}{2b^2 - 3b - 9} \div \dfrac{b^2 - 1}{4b + 6}$

5. $\dfrac{6t - 1}{5t - 30} \cdot \dfrac{10t - 25}{2t^2 - 3t - 5}$

Concept 1: Addition and Subtraction of Rational Expressions with the Same Denominator

For Exercises 6–27, add or subtract the expressions with like denominators as indicated.

6. $\dfrac{7}{8} + \dfrac{3}{8}$

7. $\dfrac{1}{3} + \dfrac{7}{3}$

8. $\dfrac{9}{16} - \dfrac{3}{16}$

9. $\dfrac{14}{15} - \dfrac{4}{15}$

10. $\dfrac{5a}{a + 2} - \dfrac{3a - 4}{a + 2}$

11. $\dfrac{2b}{b - 3} - \dfrac{b - 9}{b - 3}$

12. $\dfrac{5c}{c + 6} + \dfrac{30}{c + 6}$

13. $\dfrac{12}{2 + d} + \dfrac{6d}{2 + d}$

14. $\dfrac{5}{t - 8} - \dfrac{2t + 1}{t - 8}$

15. $\dfrac{7p + 1}{2p + 1} - \dfrac{p - 4}{2p + 1}$

16. $\dfrac{9x^2}{3x - 7} - \dfrac{49}{3x - 7}$

17. $\dfrac{4w^2}{2w - 1} - \dfrac{1}{2w - 1}$

18. $\dfrac{m^2}{m + 5} + \dfrac{10m + 25}{m + 5}$

19. $\dfrac{k^2}{k - 3} - \dfrac{6k - 9}{k - 3}$

20. $\dfrac{2a}{a + 2} + \dfrac{4}{a + 2}$

21. $\dfrac{5b}{b + 4} + \dfrac{20}{b + 4}$

22. $\dfrac{x^2}{x + 5} - \dfrac{25}{x + 5}$

23. $\dfrac{y^2}{y - 7} - \dfrac{49}{y - 7}$

24. $\dfrac{r}{r^2 + 3r + 2} + \dfrac{2}{r^2 + 3r + 2}$

25. $\dfrac{x}{x^2 - x - 12} - \dfrac{4}{x^2 - x - 12}$

26. $\dfrac{1}{3y^2 + 22y + 7} - \dfrac{-3y}{3y^2 + 22y + 7}$

27. $\dfrac{5}{2x^2 + 13x + 20} + \dfrac{2x}{2x^2 + 13x + 20}$

For Exercises 28–29, find an expression that represents the perimeter of the figure (assume that $x > 0$, $y > 0$, and $t > 0$).

28. **29.**

Concept 2: Addition and Subtraction of Rational Expressions with Different Denominators

For Exercises 30–71, add or subtract the expressions with unlike denominators as indicated.

30. $\dfrac{5}{4} + \dfrac{3}{2a}$

31. $\dfrac{11}{6p} + \dfrac{-7}{4p}$

32. $\dfrac{4}{5xy^3} + \dfrac{2x}{15y^2}$

33. $\dfrac{5}{3a^2b} + \dfrac{-7}{6b^2}$

34. $\dfrac{2}{s^3t^3} - \dfrac{3}{s^4t}$

35. $\dfrac{1}{p^2q} - \dfrac{2}{pq^3}$

36. $\dfrac{z}{3z - 9} - \dfrac{z - 2}{z - 3}$

37. $\dfrac{3w - 8}{2w - 4} - \dfrac{w - 3}{w - 2}$

38. $\dfrac{5}{a + 1} + \dfrac{4}{3a + 3}$

39. $\dfrac{2}{c - 4} + \dfrac{1}{5c - 20}$

40. $\dfrac{k}{k^2 - 9} - \dfrac{4}{k - 3}$

41. $\dfrac{7}{h + 5} - \dfrac{2h - 3}{h^2 - 25}$

42. $\dfrac{3a - 7}{6a + 10} - \dfrac{10}{3a^2 + 5a}$

43. $\dfrac{k + 2}{8k} - \dfrac{3 - k}{12k}$

44. $\dfrac{10}{3x - 7} + \dfrac{5}{7 - 3x}$

45. $\dfrac{8}{2w - 1} + \dfrac{4}{1 - 2w}$

46. $\dfrac{6a}{a^2 - b^2} + \dfrac{2a}{a^2 + ab}$

47. $\dfrac{7x}{x^2 + 2xy + y^2} + \dfrac{3x}{x^2 + xy}$

48. $\dfrac{p}{3} - \dfrac{4p - 1}{-3}$

49. $\dfrac{r}{7} - \dfrac{r - 5}{-7}$

50. $\dfrac{4n}{n - 8} - \dfrac{2n - 1}{8 - n}$

51. $\dfrac{m}{m - 2} - \dfrac{3m + 1}{2 - m}$

52. $\dfrac{5}{x} + \dfrac{3}{x + 2}$

53. $\dfrac{6}{y - 1} + \dfrac{9}{y}$

54. $\dfrac{5}{p - 3} - \dfrac{2}{p - 1}$

55. $\dfrac{1}{7x} + \dfrac{5}{2y^2}$

56. $\dfrac{y}{4y + 2} + \dfrac{3y}{6y + 3}$

57. $\dfrac{4}{q^2 - 2q} - \dfrac{5}{3q - 6}$

58. $\dfrac{4w}{w^2 + 2w - 3} + \dfrac{2}{1 - w}$

59. $\dfrac{z - 23}{z^2 - z - 20} - \dfrac{2}{5 - z}$

60. $\dfrac{3a - 8}{a^2 - 5a + 6} + \dfrac{a + 2}{a^2 - 6a + 8}$

61. $\dfrac{3b + 5}{b^2 + 4b + 3} + \dfrac{-b + 5}{b^2 + 2b - 3}$

62. $\dfrac{3x}{x^2 + x - 6} + \dfrac{x}{x^2 + 5x + 6}$

63. $\dfrac{x}{x^2 + 5x + 4} - \dfrac{2x}{x^2 - 2x - 3}$

64. $\dfrac{3y}{2y^2 - y - 1} - \dfrac{4y}{2y^2 - 7y - 4}$

65. $\dfrac{5}{6y^2 - 7y - 3} + \dfrac{4y}{3y^2 + 4y + 1}$

66. $\dfrac{3}{2p - 1} - \dfrac{4p + 4}{4p^2 - 1}$

67. $\dfrac{1}{3q - 2} - \dfrac{6q + 4}{9q^2 - 4}$

68. $\dfrac{m}{m + n} - \dfrac{m}{m - n} + \dfrac{1}{m^2 - n^2}$

69. $\dfrac{x}{x + y} - \dfrac{2xy}{x^2 - y^2} + \dfrac{y}{x - y}$

70. $\dfrac{2}{a + b} + \dfrac{2}{a - b} - \dfrac{4a}{a^2 - b^2}$

71. $\dfrac{-2x}{x^2 - y^2} + \dfrac{1}{x + y} - \dfrac{1}{x - y}$

For Exercises 72–73, find an expression that represents the perimeter of the figure (assume that $x > 0$ and $t > 0$).

72. **73.**

Concept 3: Using Rational Expressions in Translations

74. Let a number be represented by n. Write the reciprocal of n.

75. Write the reciprocal of the sum of a number and 6.

76. Write the quotient of 5 and the sum of a number and 2.

77. Let a number be represented by p. Write the quotient of 12 and p.

For Exercises 78–81, translate the English phrases into algebraic expressions. Then simplify by combining the rational expressions.

78. The sum of a number and the quantity seven times the reciprocal of the number.

79. The sum of a number and the quantity five times the reciprocal of the number.

80. The difference of the reciprocal of n and the quotient of 2 and n.

81. The difference of the reciprocal of m and the quotient of $3m$ and 7.

Expanding Your Skills

For Exercises 82–87, perform the indicated operations.

82. $\dfrac{-3}{w^3 + 27} - \dfrac{1}{w^2 - 9}$

83. $\dfrac{m}{m^3 - 1} + \dfrac{1}{(m - 1)^2}$

84. $\dfrac{2p}{p^2 + 5p + 6} - \dfrac{p + 1}{p^2 + 2p - 3} + \dfrac{3}{p^2 + p - 2}$

85. $\dfrac{3t}{8t^2 + 2t - 1} - \dfrac{5t}{2t^2 - 9t - 5} + \dfrac{2}{4t^2 - 21t + 5}$

86. $\dfrac{3m}{m^2 + 3m - 10} + \dfrac{5}{4 - 2m} - \dfrac{1}{m + 5}$

87. $\dfrac{2n}{3n^2 - 8n - 3} + \dfrac{1}{6 - 2n} - \dfrac{3}{3n + 1}$

For Exercises 88–91, simplify by applying the order of operations.

88. $\left(\dfrac{2}{k + 1} + 3\right)\left(\dfrac{k + 1}{4k + 7}\right)$

89. $\left(\dfrac{p + 1}{3p + 4}\right)\left(\dfrac{1}{p + 1} + 2\right)$

90. $\left(\dfrac{1}{10a} - \dfrac{b}{10a^2}\right) \div \left(\dfrac{1}{10} - \dfrac{b}{10a}\right)$

91. $\left(\dfrac{1}{2m} + \dfrac{n}{2m^2}\right) \div \left(\dfrac{1}{4} + \dfrac{n}{4m}\right)$

Problem Recognition Exercises—Operations on Rational Expressions

In Sections 7.1–7.4, we learned how to simplify, add, subtract, multiply, and divide rational expressions. The procedure for each operation is different, and it takes considerable practice to determine the correct method to apply for a given problem. The following review exercises give you the opportunity to practice the specific techniques for simplifying rational expressions.

1. Subtract. $\dfrac{5}{3x+1} - \dfrac{2x-4}{3x+1}$

2. Divide. $\dfrac{\dfrac{w+1}{w^2-16}}{\dfrac{w+1}{w+4}}$

3. Multiply. $\dfrac{3}{y} \cdot \dfrac{y^2-5y}{6y-9}$

4. Add. $\dfrac{-1}{x+3} + \dfrac{2}{2x-1}$

5. Simplify. $\dfrac{x-9}{9x-x^2}$

6. Add and subtract. $\dfrac{1}{p} - \dfrac{3}{p^2+3p} + \dfrac{p}{3p+9}$

7. Divide. $\dfrac{c^2+5c+6}{c^2+c-2} \div \dfrac{c}{c-1}$

8. Multiply. $\dfrac{2x^2-5x-3}{x^2-9} \cdot \dfrac{x^2+6x+9}{10x+5}$

9. Simplify. $\dfrac{6a^2b^3}{72ab^7c}$

10. Subtract. $\dfrac{2a}{a+b} - \dfrac{b}{a-b} - \dfrac{-4ab}{a^2-b^2}$

11. Divide. $\dfrac{p^2+10pq+25q^2}{p^2+6pq+5q^2} \div \dfrac{10p+50q}{2p^2-2q^2}$

12. Add. $\dfrac{3k-8}{k-5} + \dfrac{k-12}{k-5}$

13. Simplify. $\dfrac{20x^2+10x}{4x^3+4x^2+x}$

14. Multiply.

$$\frac{w^2-81}{w^2+10w+9} \cdot \frac{w^2+w+2zw+2z}{w^2-9w+zw-9z}$$

15. Divide. $\dfrac{8x^2-18x-5}{4x^2-25} \div \dfrac{4x^2-11x-3}{3x-9}$

16. Simplify. $\dfrac{xy+7x+5y+35}{x^2+ax+5x+5a}$

17. Subtract. $\dfrac{a}{a^2-9} - \dfrac{3}{6a-18}$

18. Add. $\dfrac{4}{y^2-36} + \dfrac{2}{y^2-4y-12}$

19. Multiply. $(t^2+5t-24)\left(\dfrac{t+8}{t-3}\right)$

20. Simplify. $\dfrac{6b^2-7b-10}{b-2}$

For Exercises 21–22, determine the domain.

21. $\dfrac{x-3}{x+1}$

22. $\dfrac{x-2}{x^2-9}$

Complex Fractions

Section 7.5

Concepts

1. Simplifying Complex Fractions (Method I)
2. Simplifying Complex Fractions (Method II)

1. Simplifying Complex Fractions (Method I)

A **complex fraction** is a fraction whose numerator or denominator contains one or more rational expressions. For example,

$$\frac{\frac{1}{ab}}{\frac{2}{b}} \quad \text{and} \quad \frac{1 + \frac{3}{4} - \frac{1}{6}}{\frac{1}{2} + \frac{1}{3}}$$

are complex fractions.

Two methods will be presented to simplify complex fractions. The first method (Method I) follows the order of operations to simplify the numerator and denominator separately before dividing. The process is summarized as follows.

Steps to Simplify a Complex Fraction (Method I)

1. Add or subtract expressions in the numerator to form a single fraction. Add or subtract expressions in the denominator to form a single fraction.
2. Divide the rational expressions from step 1 by multiplying the numerator of the complex fraction by the reciprocal of the denominator of the complex fraction.
3. Simplify to lowest terms if possible.

Example 1 Simplifying Complex Fractions (Method I)

Simplify the expression. $\dfrac{\frac{1}{ab}}{\frac{2}{b}}$

Solution:

Step 1: The numerator and denominator of the complex fraction are already single fractions.

$$\frac{\frac{1}{ab}}{\frac{2}{b}}$$ ← This fraction bar denotes division (÷).

$$= \frac{1}{ab} \div \frac{2}{b}$$

$$= \frac{1}{ab} \cdot \frac{b}{2}$$

Step 2: Multiply the numerator of the complex fraction by the reciprocal of $\frac{2}{b}$, which is $\frac{b}{2}$.

$$= \frac{1}{a\not{b}} \cdot \frac{\overset{1}{\not{b}}}{2}$$

Step 3: Reduce common factors and simplify.

$$= \frac{1}{2a}$$

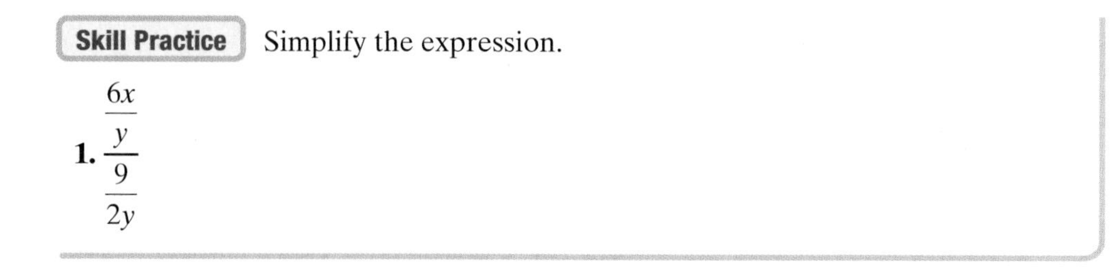

Skill Practice Simplify the expression.

1. $\dfrac{\dfrac{6x}{y}}{\dfrac{9}{2y}}$

Sometimes it is necessary to simplify the numerator and denominator of a complex fraction before the division can be performed. This is illustrated in the next example.

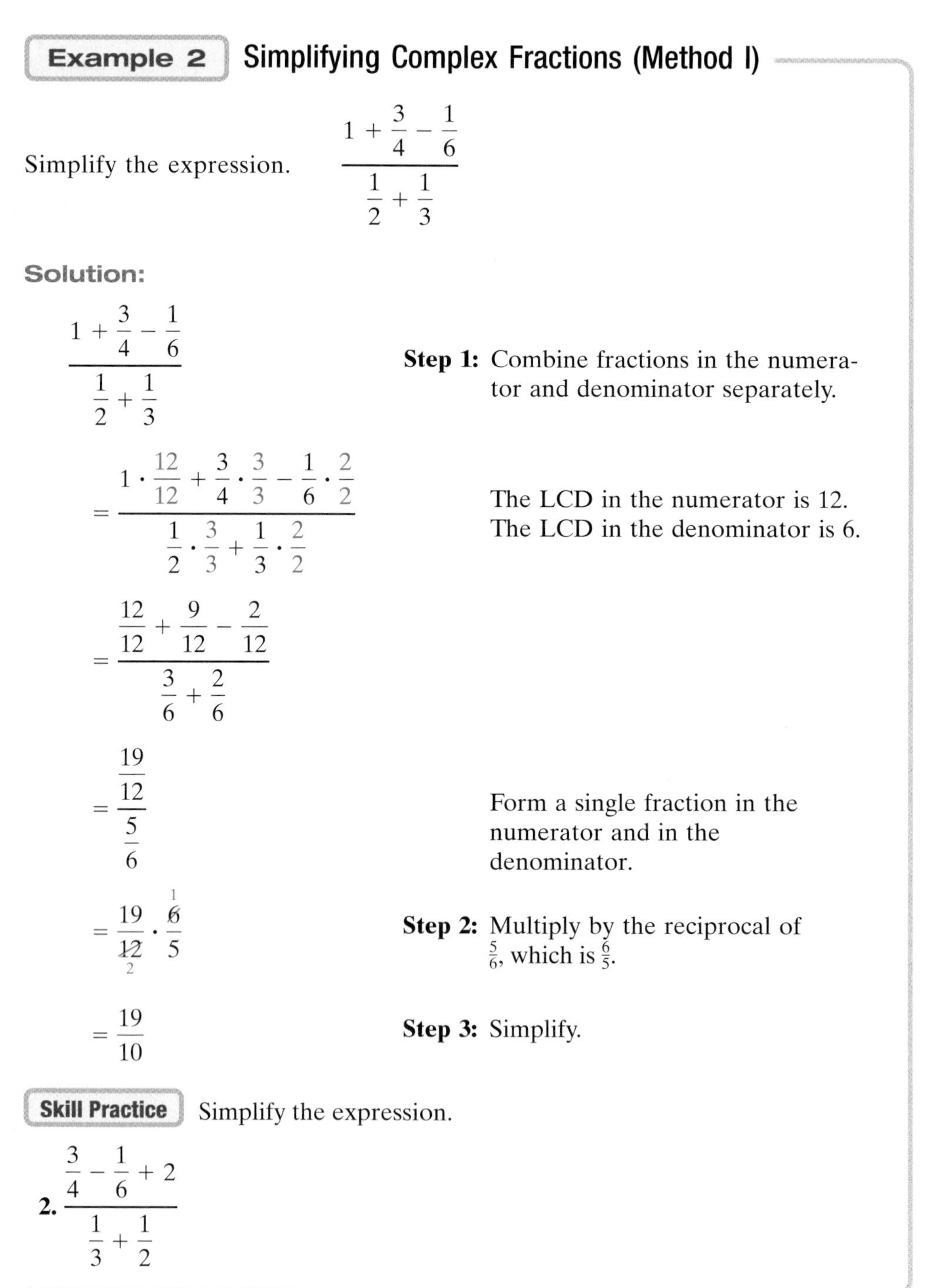

Example 2 Simplifying Complex Fractions (Method I)

Simplify the expression. $\dfrac{1 + \dfrac{3}{4} - \dfrac{1}{6}}{\dfrac{1}{2} + \dfrac{1}{3}}$

Solution:

$\dfrac{1 + \dfrac{3}{4} - \dfrac{1}{6}}{\dfrac{1}{2} + \dfrac{1}{3}}$ **Step 1:** Combine fractions in the numerator and denominator separately.

$= \dfrac{1 \cdot \dfrac{12}{12} + \dfrac{3}{4} \cdot \dfrac{3}{3} - \dfrac{1}{6} \cdot \dfrac{2}{2}}{\dfrac{1}{2} \cdot \dfrac{3}{3} + \dfrac{1}{3} \cdot \dfrac{2}{2}}$ The LCD in the numerator is 12. The LCD in the denominator is 6.

$= \dfrac{\dfrac{12}{12} + \dfrac{9}{12} - \dfrac{2}{12}}{\dfrac{3}{6} + \dfrac{2}{6}}$

$= \dfrac{\dfrac{19}{12}}{\dfrac{5}{6}}$ Form a single fraction in the numerator and in the denominator.

$= \dfrac{19}{\cancel{12}_{2}} \cdot \dfrac{\cancel{6}^{1}}{5}$ **Step 2:** Multiply by the reciprocal of $\frac{5}{6}$, which is $\frac{6}{5}$.

$= \dfrac{19}{10}$ **Step 3:** Simplify.

Skill Practice Simplify the expression.

2. $\dfrac{\dfrac{3}{4} - \dfrac{1}{6} + 2}{\dfrac{1}{3} + \dfrac{1}{2}}$

Skill Practice Answers

1. $\dfrac{4x}{3}$

2. $\dfrac{31}{10}$

Example 3 Simplifying Complex Fractions (Method I)

Simplify the expression. $\dfrac{\dfrac{1}{x} + \dfrac{1}{y}}{x - \dfrac{y^2}{x}}$

Solution:

$\dfrac{\dfrac{1}{x} + \dfrac{1}{y}}{x - \dfrac{y^2}{x}}$ The LCD in the numerator is xy. The LCD in the denominator is x.

$= \dfrac{\dfrac{1 \cdot y}{x \cdot y} + \dfrac{1 \cdot x}{y \cdot x}}{\dfrac{x \cdot x}{1 \cdot x} - \dfrac{y^2}{x}}$ Rewrite the expressions using common denominators.

$= \dfrac{\dfrac{y}{xy} + \dfrac{x}{xy}}{\dfrac{x^2}{x} - \dfrac{y^2}{x}}$

$= \dfrac{\dfrac{y + x}{xy}}{\dfrac{x^2 - y^2}{x}}$ Form single fractions in the numerator and denominator.

$= \dfrac{y + x}{xy} \cdot \dfrac{x}{x^2 - y^2}$ Multiply by the reciprocal of the denominator.

$= \dfrac{\overset{1}{\cancel{y + x}}}{x y} \cdot \dfrac{\overset{1}{\cancel{x}}}{(\cancel{x + y})(x - y)}$ Factor and reduce. Note that $(y + x) = (x + y)$.

$= \dfrac{1}{y(x - y)}$ Simplify.

Skill Practice Simplify the expression.

3. $\dfrac{1 - \dfrac{q}{p}}{\dfrac{p}{q} - \dfrac{q}{p}}$

2. Simplifying Complex Fractions (Method II)

We will now simplify the expressions from Examples 2 and 3 again using a second method to simplify complex fractions (Method II). Recall that multiplying the numerator and denominator of a rational expression by the same quantity does not change the value of the expression because we are multiplying by a number equivalent to 1. This is the basis for Method II.

Skill Practice Answers

3. $\dfrac{q}{p + q}$

Steps to Simplifying a Complex Fraction (Method II)

1. Multiply the numerator and denominator of the complex fraction by the LCD of *all* individual fractions within the expression.
2. Apply the distributive property, and simplify the numerator and denominator.
3. Simplify to lowest terms if possible.

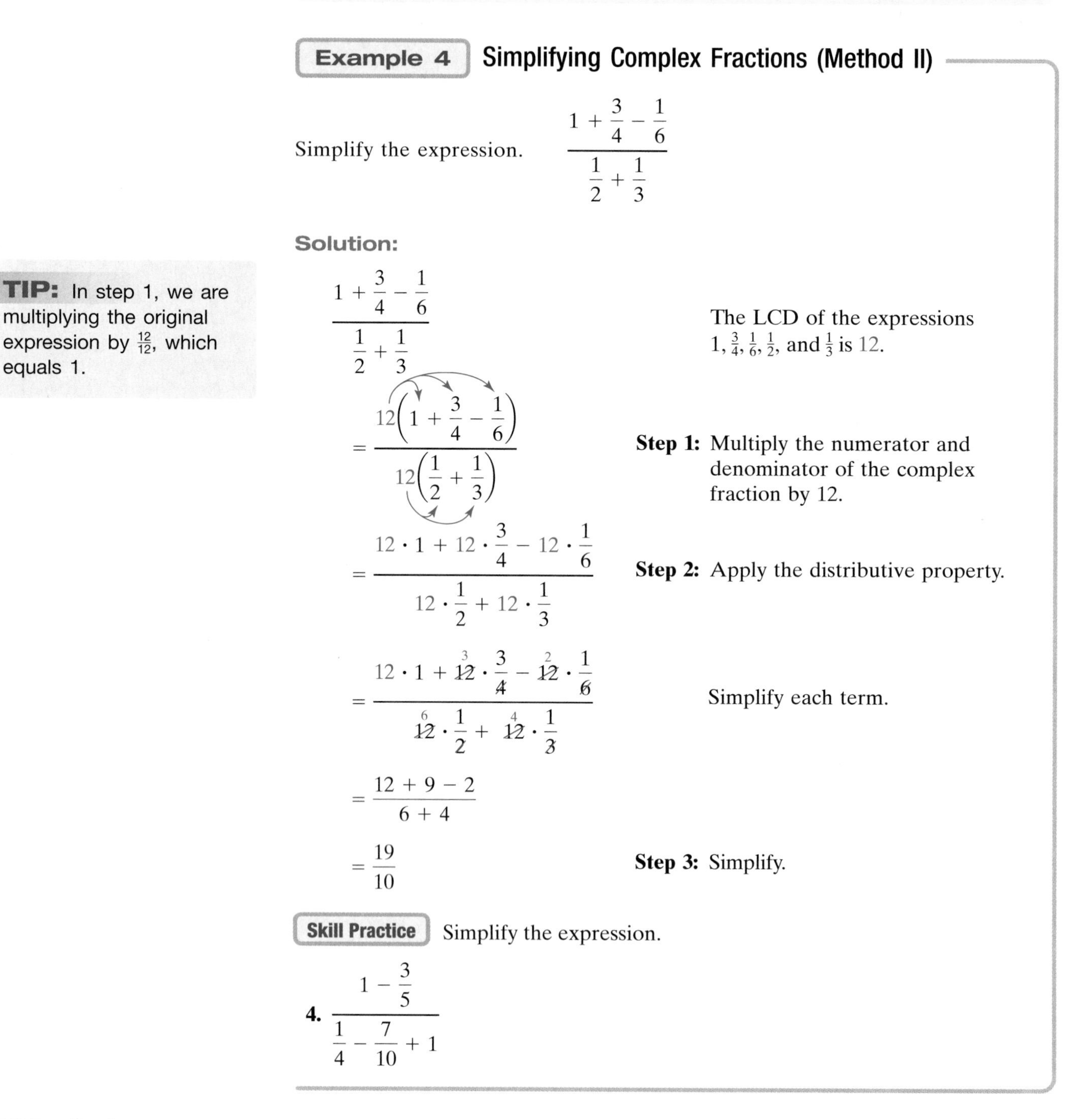

Example 4 Simplifying Complex Fractions (Method II)

Simplify the expression. $\dfrac{1+\frac{3}{4}-\frac{1}{6}}{\frac{1}{2}+\frac{1}{3}}$

Solution:

TIP: In step 1, we are multiplying the original expression by $\frac{12}{12}$, which equals 1.

$\dfrac{1+\frac{3}{4}-\frac{1}{6}}{\frac{1}{2}+\frac{1}{3}}$ The LCD of the expressions $1, \frac{3}{4}, \frac{1}{6}, \frac{1}{2}$, and $\frac{1}{3}$ is 12.

$= \dfrac{12\left(1+\frac{3}{4}-\frac{1}{6}\right)}{12\left(\frac{1}{2}+\frac{1}{3}\right)}$ **Step 1:** Multiply the numerator and denominator of the complex fraction by 12.

$= \dfrac{12\cdot 1 + 12\cdot\frac{3}{4} - 12\cdot\frac{1}{6}}{12\cdot\frac{1}{2} + 12\cdot\frac{1}{3}}$ **Step 2:** Apply the distributive property.

$= \dfrac{12\cdot 1 + \overset{3}{\cancel{12}}\cdot\frac{3}{\cancel{4}} - \overset{2}{\cancel{12}}\cdot\frac{1}{\cancel{6}}}{\overset{6}{\cancel{12}}\cdot\frac{1}{\cancel{2}} + \overset{4}{\cancel{12}}\cdot\frac{1}{\cancel{3}}}$ Simplify each term.

$= \dfrac{12+9-2}{6+4}$

$= \dfrac{19}{10}$ **Step 3:** Simplify.

Skill Practice Simplify the expression.

4. $\dfrac{1-\frac{3}{5}}{\frac{1}{4}-\frac{7}{10}+1}$

Skill Practice Answers

4. $\dfrac{8}{11}$

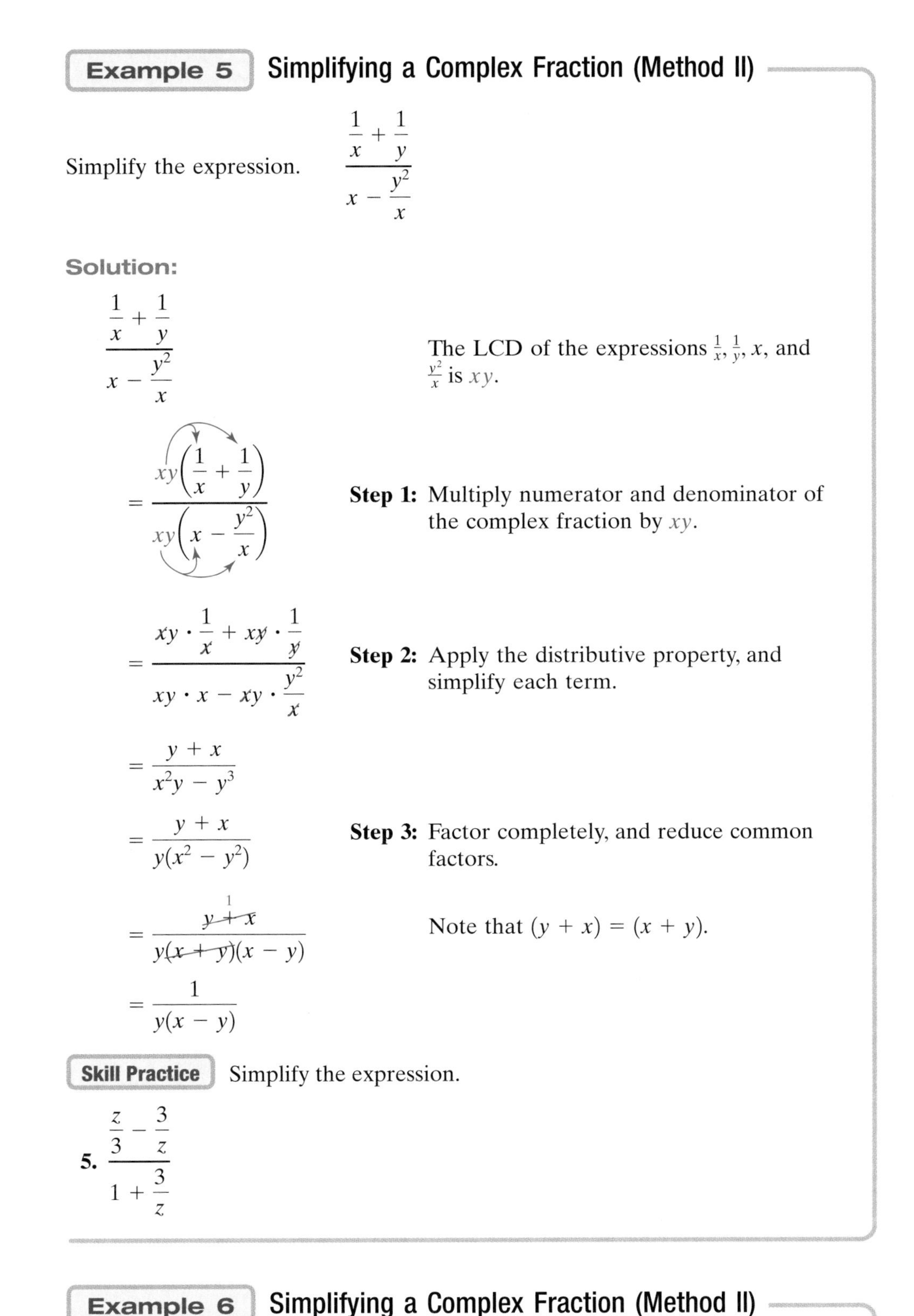

Example 5 Simplifying a Complex Fraction (Method II)

Simplify the expression. $\dfrac{\dfrac{1}{x} + \dfrac{1}{y}}{x - \dfrac{y^2}{x}}$

Solution:

$\dfrac{\dfrac{1}{x} + \dfrac{1}{y}}{x - \dfrac{y^2}{x}}$ The LCD of the expressions $\frac{1}{x}, \frac{1}{y}, x$, and $\frac{y^2}{x}$ is xy.

$= \dfrac{xy\left(\dfrac{1}{x} + \dfrac{1}{y}\right)}{xy\left(x - \dfrac{y^2}{x}\right)}$ **Step 1:** Multiply numerator and denominator of the complex fraction by xy.

$= \dfrac{xy \cdot \dfrac{1}{x} + xy \cdot \dfrac{1}{y}}{xy \cdot x - xy \cdot \dfrac{y^2}{x}}$ **Step 2:** Apply the distributive property, and simplify each term.

$= \dfrac{y + x}{x^2y - y^3}$

$= \dfrac{y + x}{y(x^2 - y^2)}$ **Step 3:** Factor completely, and reduce common factors.

$= \dfrac{\overset{1}{\cancel{y + x}}}{y\cancel{(x + y)}(x - y)}$ Note that $(y + x) = (x + y)$.

$= \dfrac{1}{y(x - y)}$

Skill Practice Simplify the expression.

5. $\dfrac{\dfrac{z}{3} - \dfrac{3}{z}}{1 + \dfrac{3}{z}}$

Example 6 Simplifying a Complex Fraction (Method II)

Simplify the expression. $\dfrac{\dfrac{1}{k + 1} - 1}{\dfrac{1}{k + 1} + 1}$

Skill Practice Answers

5. $\dfrac{z - 3}{3}$

Solution:

$$\frac{\dfrac{1}{k+1}-1}{\dfrac{1}{k+1}+1}$$

The LCD of $\dfrac{1}{k+1}$ and 1 is $(k+1)$.

$$=\frac{(k+1)\left(\dfrac{1}{k+1}-1\right)}{(k+1)\left(\dfrac{1}{k+1}+1\right)}$$

Step 1: Multiply numerator and denominator of the complex fraction by $(k+1)$.

$$=\frac{(k+1)\cdot\dfrac{1}{(k+1)}-(k+1)\cdot 1}{(k+1)\cdot\dfrac{1}{(k+1)}+(k+1)\cdot 1}$$

Step 2: Apply the distributive property.

$$=\frac{1-(k+1)}{1+(k+1)}$$

Simplify.

$$=\frac{1-k-1}{1+k+1}$$

$$=\frac{-k}{k+2}$$

Step 3: The expression is already in lowest terms.

Skill Practice Simplify the expression.

6. $\dfrac{\dfrac{4}{p-3}+1}{1+\dfrac{2}{p-3}}$

Skill Practice Answers

6. $\dfrac{p+1}{p-1}$

Section 7.5 Practice Exercises

Boost *your* GRADE at mathzone.com!

MathZone

- Practice Problems
- Self-Tests
- NetTutor
- e-Professors
- Videos

Study Skills Exercise

1. Define the key term **complex fraction**.

Review Exercises

For Exercises 2–3, write the domain in set-builder notation, and simplify the expression.

2. $\dfrac{y(2y+9)}{y^2(2y+9)}$

3. $\dfrac{a+5}{2a^2+7a-15}$

For Exercises 4–9, perform the indicated operations.

4. $\dfrac{2}{w-2}+\dfrac{3}{w}$

5. $\dfrac{6}{5}-\dfrac{3}{5k-10}$

6. $\dfrac{p^2+2p}{2p-1}\cdot\dfrac{10p^2-5p}{12p^3+24p^2}$

7. $\dfrac{x^2 - 2xy + y^2}{x^4 - y^4} \div \dfrac{3x^2y - 3xy^2}{x^2 + y^2}$

8. $\left(\dfrac{1}{z} - \dfrac{1}{2z}\right) \div \left(\dfrac{1}{2} + \dfrac{1}{2z}\right)$

9. $\left(\dfrac{2}{3a^2} - \dfrac{3}{b}\right) \div \left(\dfrac{5}{ab} - 4\right)$

Concepts 1–2: Simplifying Complex Fractions (Methods I and II)

For Exercises 10–35, simplify the complex fractions.

10. $\dfrac{\frac{7}{18y}}{\frac{2}{9}}$

11. $\dfrac{\frac{a^2}{2a-3}}{\frac{5a}{8a-12}}$

12. $\dfrac{\frac{3x+2y}{2y}}{\frac{6x+4y}{2}}$

13. $\dfrac{\frac{2x-10}{4}}{\frac{x^2-5x}{3x}}$

14. $\dfrac{\frac{8a^4b^3}{3c}}{\frac{a^7b^2}{9c}}$

15. $\dfrac{\frac{12x^2}{5y}}{\frac{8x^6}{9y^2}}$

16. $\dfrac{\frac{4r^3s}{t^5}}{\frac{2s^7}{r^2t^9}}$

17. $\dfrac{\frac{5p^4q}{w^4}}{\frac{10p^2}{qw^2}}$

18. $\dfrac{\frac{1}{8}+\frac{4}{3}}{\frac{1}{2}-\frac{5}{12}}$

19. $\dfrac{\frac{8}{9}-\frac{1}{3}}{\frac{7}{6}+\frac{1}{9}}$

20. $\dfrac{\frac{1}{h}+\frac{1}{k}}{\frac{1}{hk}}$

21. $\dfrac{\frac{1}{b}+1}{\frac{1}{b}}$

22. $\dfrac{\frac{n+1}{n^2-9}}{\frac{2}{n+3}}$

23. $\dfrac{\frac{5}{k-5}}{\frac{k+1}{k^2-25}}$

24. $\dfrac{2+\frac{1}{x}}{4+\frac{1}{x}}$

25. $\dfrac{6+\frac{6}{k}}{1+\frac{1}{k}}$

26. $\dfrac{\frac{m}{7}-\frac{7}{m}}{\frac{1}{7}+\frac{1}{m}}$

27. $\dfrac{\frac{2}{p}+\frac{p}{2}}{\frac{p}{3}-\frac{3}{p}}$

28. $\dfrac{\frac{1}{5}-\frac{1}{y}}{\frac{7}{10}+\frac{1}{y^2}}$

29. $\dfrac{\frac{1}{m^2}+\frac{2}{3}}{\frac{1}{m}-\frac{5}{6}}$

30. $\dfrac{\frac{8}{a+4}+2}{\frac{12}{a+4}-2}$

31. $\dfrac{\frac{2}{w+1}+3}{\frac{3}{w+1}+4}$

32. $\dfrac{1-\frac{4}{t^2}}{1-\frac{2}{t}-\frac{8}{t^2}}$

33. $\dfrac{1-\frac{9}{p^2}}{1-\frac{1}{p}-\frac{6}{p^2}}$

34. $\dfrac{t+4+\frac{3}{t}}{t-4-\frac{5}{t}}$

35. $\dfrac{\frac{9}{4m}+\frac{9}{2m^2}}{\frac{3}{2}+\frac{3}{m}}$

For Exercises 36–39, translate the English phrases into algebraic expressions. Then simplify the expressions.

36. The sum of one-half and two-thirds, divided by five.

37. The quotient of ten and the difference of two-fifths and one-fourth.

38. The quotient of three and the sum of two-thirds and three-fourths.

39. The difference of three-fifths and one-half, divided by four.

40. In electronics, resistors oppose the flow of current. For two resistors in parallel, the total resistance is given by

$$R = \frac{1}{\frac{1}{R_1} + \frac{1}{R_2}}$$

a. Find the total resistance if $R_1 = 2\ \Omega$ (ohms) and $R_2 = 3\ \Omega$.

b. Find the total resistance if $R_1 = 10\ \Omega$ and $R_2 = 15\ \Omega$.

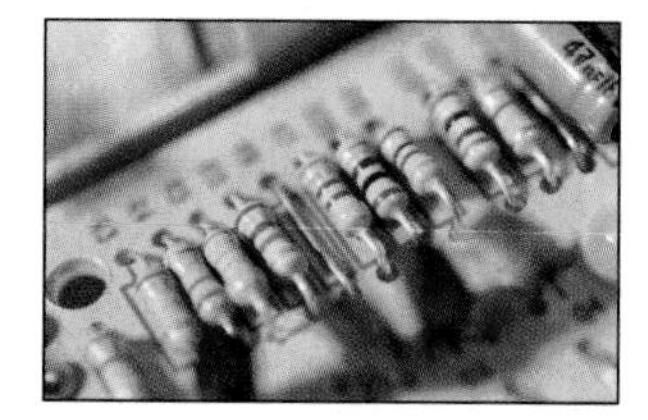

41. Suppose that Joëlle makes a round trip to a location that is d miles away. If the average rate going to the location is r_1 and the average rate on the return trip is given by r_2, the average rate of the entire trip, R, is given by

$$R = \frac{2d}{\frac{d}{r_1} + \frac{d}{r_2}}$$

a. Find the average rate of a trip to a destination 30 miles away when the average rate going there was 60 mph and the average rate returning home was 45 mph. (Round to the nearest tenth of a mile per hour.)

b. Find the average rate of a trip to a destination that is 50 miles away if the driver travels at the same rates as in part (a). (Round to the nearest tenth of a mile per hour.)

c. Compare your answers from parts (a) and (b) and explain the results in the context of the problem.

Expanding Your Skills

For Exercises 42–45, simplify the complex fractions using either method.

42. $\dfrac{\frac{1}{z^2 - 9} + \frac{2}{z + 3}}{\frac{3}{z - 3}}$

43. $\dfrac{\frac{5}{w^2 - 25} - \frac{3}{w + 5}}{\frac{4}{w - 5}}$

44. $\dfrac{\frac{2}{x - 1} + 2}{\frac{2}{x + 1} - 2}$

45. $\dfrac{\frac{1}{y - 3} + 1}{\frac{2}{y + 3} - 1}$

For Exercises 46–48, simplify the complex fractions. (*Hint:* Use the order of operations and begin with the fraction on the lower right.)

46. $1 + \dfrac{1}{1 + 1}$

47. $1 + \dfrac{1}{1 + \frac{1}{1 + 1}}$

48. $1 + \dfrac{1}{1 + \frac{1}{1 + \frac{1}{1 + 1}}}$

Rational Equations

Section 7.6

Concepts

1. Introduction to Rational Equations
2. Solving Rational Equations
3. Solving Formulas Involving Rational Equations

1. Introduction to Rational Equations

Thus far we have studied two specific types of equations in one variable: linear equations and quadratic equations. Recall,

$ax + b = 0$, where $a \neq 0$, is a **linear equation.**

$ax^2 + bx + c = 0$, where $a \neq 0$, is a **quadratic equation.**

We will now study another type of equation called a rational equation.

Definition of a Rational Equation

An equation with one or more rational expressions is called a **rational equation.**

The following equations are rational equations:

$$\frac{y}{2} + \frac{y}{4} = 6 \qquad \frac{1}{x} + \frac{1}{3} = \frac{5}{6} \qquad \frac{6}{t^2 - 7t + 12} + \frac{2t}{t - 3} = \frac{3t}{t - 4}$$

To understand the process of solving a rational equation, first review the process of clearing fractions from Section 2.3.

Example 1 Solving a Rational Equation

Solve. $\frac{y}{2} + \frac{y}{4} = 6$

Solution:

$\frac{y}{2} + \frac{y}{4} = 6$ The LCD of all terms in the equation is 4.

$4\left(\frac{y}{2} + \frac{y}{4}\right) = 4(6)$ Multiply both sides of the equation by 4 to clear fractions.

$4 \cdot \frac{y}{2} + 4 \cdot \frac{y}{4} = 4(6)$ Apply the distributive property.

$2y + y = 24$ Clear fractions.

$3y = 24$ Solve the resulting equation (linear).

$y = 8$

Check: $\frac{y}{2} + \frac{y}{4} = 6$

$$\frac{(8)}{2} + \frac{(8)}{4} \stackrel{?}{=} 6$$

$$4 + 2 \stackrel{?}{=} 6$$

$$6 = 6 \checkmark$$

Skill Practice Solve the equation.

1. $\dfrac{t}{5} - \dfrac{t}{4} = 2$

2. Solving Rational Equations

The same process of clearing fractions is used to solve rational equations when variables are present in the denominator.

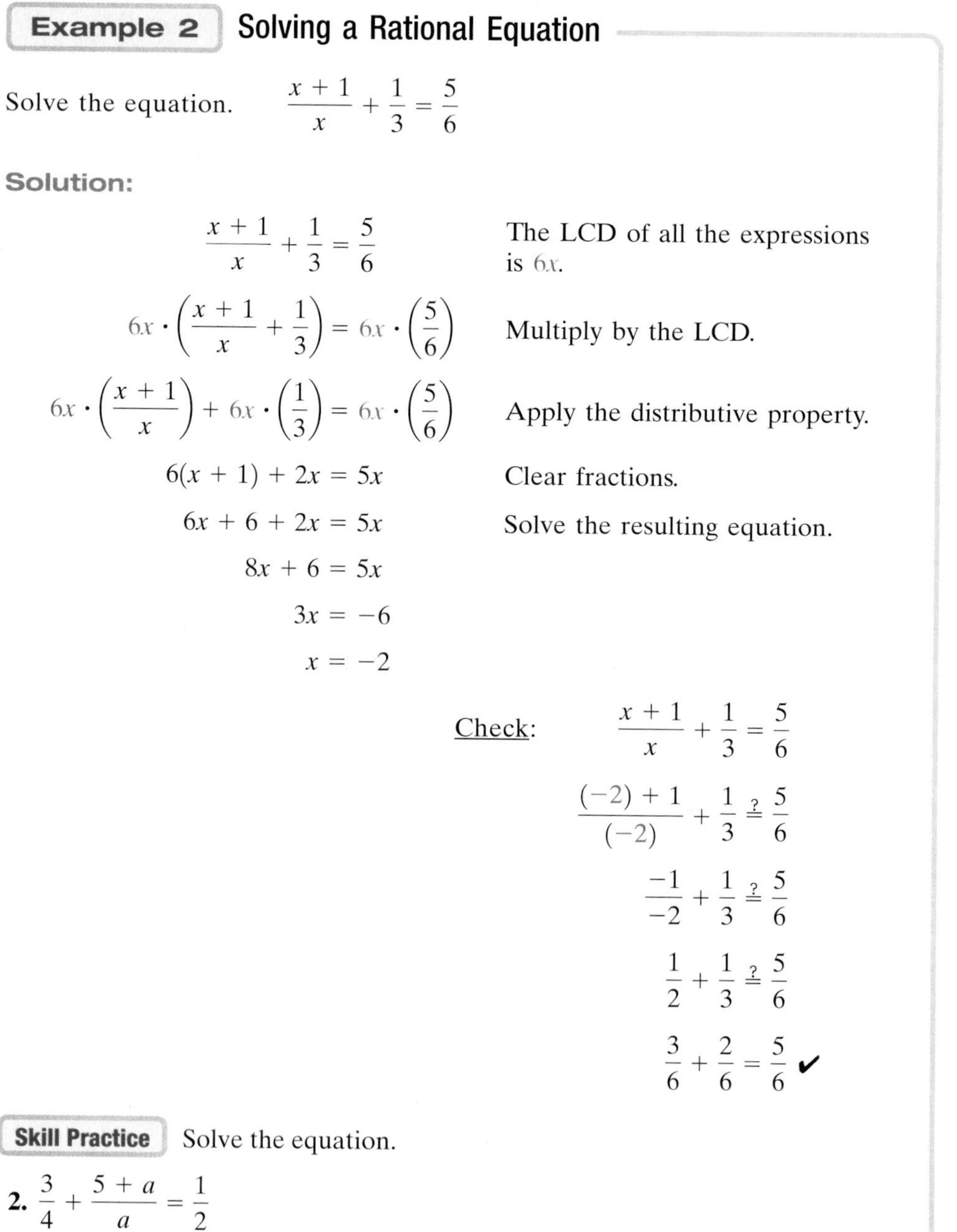

Example 2 Solving a Rational Equation

Solve the equation. $\dfrac{x+1}{x} + \dfrac{1}{3} = \dfrac{5}{6}$

Solution:

$\dfrac{x+1}{x} + \dfrac{1}{3} = \dfrac{5}{6}$ The LCD of all the expressions is $6x$.

$6x \cdot \left(\dfrac{x+1}{x} + \dfrac{1}{3}\right) = 6x \cdot \left(\dfrac{5}{6}\right)$ Multiply by the LCD.

$6x \cdot \left(\dfrac{x+1}{x}\right) + 6x \cdot \left(\dfrac{1}{3}\right) = 6x \cdot \left(\dfrac{5}{6}\right)$ Apply the distributive property.

$6(x + 1) + 2x = 5x$ Clear fractions.

$6x + 6 + 2x = 5x$ Solve the resulting equation.

$8x + 6 = 5x$

$3x = -6$

$x = -2$

Check: $\dfrac{x+1}{x} + \dfrac{1}{3} = \dfrac{5}{6}$

$\dfrac{(-2)+1}{(-2)} + \dfrac{1}{3} \stackrel{?}{=} \dfrac{5}{6}$

$\dfrac{-1}{-2} + \dfrac{1}{3} \stackrel{?}{=} \dfrac{5}{6}$

$\dfrac{1}{2} + \dfrac{1}{3} \stackrel{?}{=} \dfrac{5}{6}$

$\dfrac{3}{6} + \dfrac{2}{6} = \dfrac{5}{6}$ ✔

Skill Practice Solve the equation.

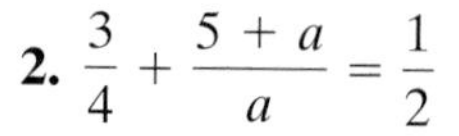

2. $\dfrac{3}{4} + \dfrac{5+a}{a} = \dfrac{1}{2}$

Skill Practice Answers

1. $t = -40$ **2.** $a = -4$

Example 3 Solving a Rational Equation

Solve the equation. $1 + \frac{3a}{a-2} = \frac{6}{a-2}$

Solution:

$$1 + \frac{3a}{a-2} = \frac{6}{a-2}$$ The LCD of all the expressions is $a - 2$.

$$(a-2)\left(1 + \frac{3a}{a-2}\right) = (a-2)\left(\frac{6}{a-2}\right)$$ Multiply by the LCD.

$$(a-2)1 + (a-2)\left(\frac{3a}{a-2}\right) = (a-2)\left(\frac{6}{a-2}\right)$$ Apply the distributive property.

$$a - 2 + 3a = 6$$ Solve the resulting equation (linear).

$$4a - 2 = 6$$

$$4a = 8$$

$$a = 2$$

Check: $1 + \frac{3a}{a-2} = \frac{6}{a-2}$

$$1 + \frac{3(2)}{(2)-2} \stackrel{?}{=} \frac{6}{(2)-2}$$

$$1 + \frac{6}{0} \stackrel{?}{=} \frac{6}{0}$$

The denominator is 0 when $a = 2$.

Because the value $a = 2$ makes the denominator zero in one (or more) of the rational expressions within the equation, the equation is undefined for $a = 2$. That is, $a = 2$ is not in the domain of the equation, therefore, it is an extraneous solution. No other potential solutions exist for the equation.

The equation $1 + \frac{3a}{a-2} = \frac{6}{a-2}$ has no solution.

Skill Practice Solve the equation.

3. $\frac{x}{x+1} - 2 = \frac{-1}{x+1}$

Examples 1–3 show that the steps to solve a rational equation mirror the process of clearing fractions from Section 2.3. However, there is one significant difference. The solutions of a rational equation must not make the denominator equal to zero for any expression within the equation. When $a = 2$ is substituted into the expression

$$\frac{3a}{a-2} \quad \text{or} \quad \frac{6}{a-2}$$

Skill Practice Answers

3. No solution; ($x = -1$ does not check.)

the denominator is zero and the expression is undefined. Hence, $a = 2$ cannot be a solution to the equation

$$1 + \frac{3a}{a-2} = \frac{6}{a-2}$$

The steps to solve a rational equation are summarized as follows.

Steps to Solve a Rational Equation

1. Factor the denominators of all rational expressions.
2. Identify the LCD of all expressions in the equation.
3. Multiply both sides of the equation by the LCD.
4. Solve the resulting equation.
5. Check potential solutions in the original equation.

Example 4 Solving Rational Equations

Solve the equations.

a. $1 - \frac{4}{p} = -\frac{3}{p^2}$ **b.** $\frac{6}{t^2 - 7t + 12} + \frac{2t}{t-3} = \frac{3t}{t-4}$

Solution:

a. $1 - \frac{4}{p} = -\frac{3}{p^2}$

Step 1: The denominators are already factored.

Step 2: The LCD of all expressions is p^2.

$$p^2\left(1 - \frac{4}{p}\right) = p^2\left(-\frac{3}{p^2}\right)$$

Step 3: Multiply by the LCD.

$$p^2(1) - p^2\left(\frac{4}{p}\right) = p^2\left(-\frac{3}{p^2}\right)$$

Apply the distributive property.

$$p^2 - 4p = -3$$

Step 4: Solve the resulting equation (quadratic).

$$p^2 - 4p + 3 = 0$$
$$(p-3)(p-1) = 0$$

Set the equation equal to zero and factor.

$$p - 3 = 0 \quad \text{or} \quad p - 1 = 0$$

Set each factor equal to zero.

$$p = 3 \quad \text{or} \quad p = 1$$

Step 5: Check: $p = 3$

$$1 - \frac{4}{p} = -\frac{3}{p^2}$$
$$1 - \frac{4}{(3)} \stackrel{?}{=} -\frac{3}{(3)^2}$$
$$\frac{3}{3} - \frac{4}{3} \stackrel{?}{=} -\frac{3}{9}$$
$$-\frac{1}{3} = -\frac{1}{3} \checkmark$$

Check: $p = 1$

$$1 - \frac{4}{p} = -\frac{3}{p^2}$$
$$1 - \frac{4}{(1)} \stackrel{?}{=} -\frac{3}{(1)^2}$$
$$1 - 4 \stackrel{?}{=} -3$$
$$-3 = -3 \checkmark$$

Both solutions $p = 3$ and $p = 1$ check.

b. $\dfrac{6}{t^2 - 7t + 12} + \dfrac{2t}{t - 3} = \dfrac{3t}{t - 4}$

$\dfrac{6}{(t - 3)(t - 4)} + \dfrac{2t}{t - 3} = \dfrac{3t}{t - 4}$

Step 1: Factor the denominators.

Step 2: The LCD is $(t - 3)(t - 4)$.

Step 3: Multiply by the LCD on both sides.

$$(t - 3)(t - 4)\left(\frac{6}{(t - 3)(t - 4)} + \frac{2t}{t - 3}\right) = (t - 3)(t - 4)\left(\frac{3t}{t - 4}\right)$$

$$\cancel{(t - 3)}\cancel{(t - 4)}\left(\frac{6}{\cancel{(t - 3)}\cancel{(t - 4)}}\right) + \cancel{(t - 3)}(t - 4)\left(\frac{2t}{\cancel{t - 3}}\right) = (t - 3)\cancel{(t - 4)}\left(\frac{3t}{\cancel{t - 4}}\right)$$

$$6 + 2t(t - 4) = 3t(t - 3)$$

$6 + 2t^2 - 8t = 3t^2 - 9t$

Step 4: Solve the resulting equation.

$0 = 3t^2 - 2t^2 - 9t + 8t - 6$

$0 = t^2 - t - 6$

$0 = (t - 3)(t + 2)$

Because the resulting equation is quadratic, set the equation equal to zero and factor.

$t - 3 = 0$ or $t + 2 = 0$

$t = 3$ or $t = -2$

Set each factor equal to zero.

Step 5: Check the potential solutions in the original equation.

TIP: Note that $t = 3$ and $t = 4$ are not defined in the original expressions. Therefore, they cannot be solutions to the original equation.

Check: $t = 3$

$t = 3$ cannot be a solution to the equation because it will make the denominator zero in the original equation.

$\dfrac{6}{t^2 - 7t + 12} + \dfrac{2t}{t - 3} = \dfrac{3t}{t - 4}$

$\dfrac{6}{(3)^2 - 7(3) + 12} + \dfrac{2(3)}{(3) - 3} \stackrel{?}{=} \dfrac{3(3)}{(3) - 4}$

$\dfrac{6}{0} + \dfrac{6}{0} \stackrel{?}{=} \dfrac{9}{-1}$

zero in the denominator

Check: $t = -2$

$\dfrac{6}{t^2 - 7t + 12} + \dfrac{2t}{t - 3} = \dfrac{3t}{t - 4}$

$\dfrac{6}{(-2)^2 - 7(-2) + 12} + \dfrac{2(-2)}{(-2) - 3} \stackrel{?}{=} \dfrac{3(-2)}{(-2) - 4}$

$\dfrac{6}{4 + 14 + 12} + \dfrac{-4}{-5} \stackrel{?}{=} \dfrac{-6}{-6}$

$\dfrac{6}{30} + \dfrac{4}{5} \stackrel{?}{=} 1$

$\dfrac{1}{5} + \dfrac{4}{5} = 1$ ✔

$t = -2$ is a solution.

TIP: $t = 3$ is not a solution because it is not in the domain of the equation.

The only solution is $t = -2$.

Skill Practice Solve the equations.

4. $\dfrac{z}{2} - \dfrac{1}{2z} = \dfrac{12}{z}$

5. $\dfrac{-8}{x^2 + 6x + 8} + \dfrac{x}{x + 4} = \dfrac{2}{x + 2}$

Skill Practice Answers

4. $z = 5$ or $z = -5$
5. $x = 4$; ($x = -4$ does not check.)

Example 5 Translating to a Rational Equation

Ten times the reciprocal of a number is added to four. The result is equal to the quotient of twenty-two and the number. Find the number.

Solution:

Let x represent the number.

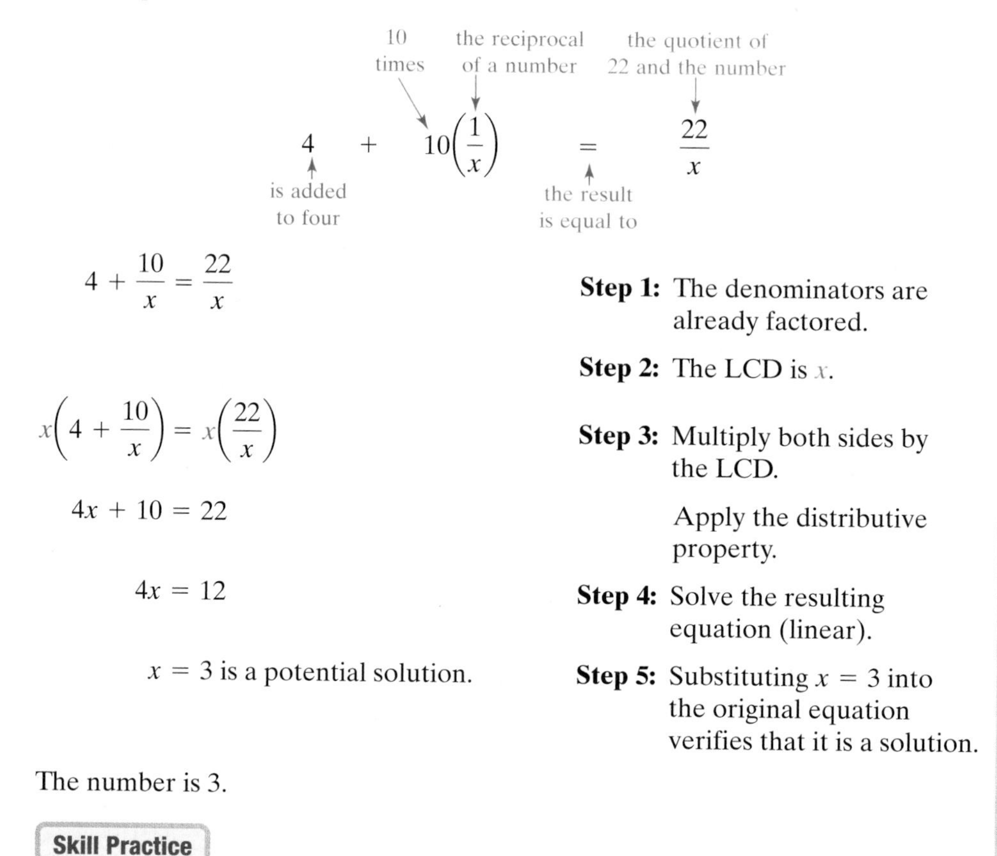

$$4 + \frac{10}{x} = \frac{22}{x}$$

Step 1: The denominators are already factored.

Step 2: The LCD is x.

$$x\left(4 + \frac{10}{x}\right) = x\left(\frac{22}{x}\right)$$

Step 3: Multiply both sides by the LCD.

$$4x + 10 = 22$$

Apply the distributive property.

$$4x = 12$$

Step 4: Solve the resulting equation (linear).

$x = 3$ is a potential solution.

Step 5: Substituting $x = 3$ into the original equation verifies that it is a solution.

The number is 3.

Skill Practice

6. The quotient of ten and a number is two less than four times the reciprocal of the number. Find the number.

3. Solving Formulas Involving Rational Equations

A rational equation may have more than one variable. To solve for a specific variable within a rational equation, we can still apply principles of clearing fractions.

Example 6 Solving a Formula Involving a Rational Equation

Solve for k. $\quad F = \dfrac{ma}{k}$

Skill Practice Answers

6. The number is -3.

Solution:

To solve for k, we must clear fractions so that k appears in the numerator.

$F = \dfrac{ma}{k}$	The LCD is k.
$k \cdot (F) = k \cdot \left(\dfrac{ma}{k}\right)$	Multiply both sides of the equation by the LCD.
$kF = ma$	Clear fractions.
$\dfrac{kF}{F} = \dfrac{ma}{F}$	Divide both sides by F.
$k = \dfrac{ma}{F}$	

Skill Practice

7. Solve for t. $C = \dfrac{rt}{d}$

Example 7 Solving a Formula Involving a Rational Equation

Solve for b. $h = \dfrac{2A}{B + b}$

Solution:

To solve for b, we must clear fractions so that b appears in the numerator.

$h = \dfrac{2A}{B + b}$	The LCD is $(B + b)$.
$h(B + b) = \left(\dfrac{2A}{B + b}\right) \cdot (B + b)$	Multiply both sides of the equation by the LCD.
$hB + hb = 2A$	Apply the distributive property.
$hb = 2A - hB$	Subtract hB from both sides to isolate the b term.
$\dfrac{hb}{h} = \dfrac{2A - hB}{h}$	Divide by h.
$b = \dfrac{2A - hB}{h}$	

Avoiding Mistakes:

Algebra is case-sensitive. The variables B and b represent different values.

Skill Practice

8. Solve the formula for x. $y = \dfrac{3}{x - 2}$

Skill Practice Answers

7. $t = \dfrac{Cd}{r}$

8. $x = \dfrac{3 + 2y}{y}$ or $x = \dfrac{3}{y} + 2$

TIP: The solution to Example 7 can be written in several forms. The quantity

$$\frac{2A - hB}{h}$$

can be left as a single rational expression or can be split into two fractions and simplified.

$$b = \frac{2A - hB}{h} = \frac{2A}{h} - \frac{hB}{h} = \frac{2A}{h} - B$$

Example 8 Solving a Formula Involving a Rational Expression

Solve for z. $\quad y = \dfrac{x - z}{x + z}$

Solution:

To solve for z, we must clear fractions so that z appears in the numerator only.

$y = \dfrac{x - z}{x + z}$	LCD is $(x + z)$.
$y(x + z) = \left(\dfrac{x - z}{\cancel{x + z}}\right)\overset{1}{\cancel{(x + z)}}$	Multiply both sides of the equation by the LCD.
$yx + yz = x - z$	Apply the distributive property.
$yz + z = x - yx$	Collect z terms on one side of the equation.
$z(y + 1) = x - yx$	Factor out a z.
$z = \dfrac{x - yx}{y + 1}$	Divide by $y + 1$ to solve for z.

Skill Practice

9. Solve for h. $\quad \dfrac{b}{x} = \dfrac{a}{h} + 1$

Skill Practice Answers

9. $h = \dfrac{xa}{b - x}$ or $\dfrac{-ax}{x - b}$

Section 7.6 Practice Exercises

Boost *your* GRADE at mathzone.com!

- Practice Problems
- Self-Tests
- NetTutor
- e-Professors
- Videos

Study Skills Exercise

1. Define the key terms:

a. linear equation **b. quadratic equation** **c. rational equation**

Review Exercises

For Exercises 2–7, perform the indicated operations.

2. $\dfrac{2}{x-3} - \dfrac{3}{x^2-x-6}$

3. $\dfrac{2x-6}{4x^2+7x-2} \div \dfrac{x^2-5x+6}{x^2-4}$

4. $\dfrac{2y}{y-3} + \dfrac{4}{y^2-9}$

5. $\dfrac{h-\dfrac{1}{h}}{\dfrac{1}{5}-\dfrac{1}{5h}}$

6. $\dfrac{w-4}{w^2-9} \cdot \dfrac{w-3}{w^2-8w+16}$

7. $1+\dfrac{1}{x}-\dfrac{12}{x^2}$

Concept 1: Introduction to Rational Equations

For Exercises 8–13, solve the equations by first clearing the fractions.

8. $\dfrac{1}{3}z + \dfrac{2}{3} = -2z + 10$

9. $\dfrac{5}{2} + \dfrac{1}{2}b = 5 - \dfrac{1}{3}b$

10. $\dfrac{3}{2}p + \dfrac{1}{3} = \dfrac{2p-3}{4}$

11. $\dfrac{5}{3} - \dfrac{1}{6}k = \dfrac{3k+5}{4}$

12. $\dfrac{2x-3}{4} + \dfrac{9}{10} = \dfrac{x}{5}$

13. $\dfrac{4y+2}{3} - \dfrac{7}{6} = -\dfrac{y}{6}$

14. For the equation

$$\frac{1}{w} - \frac{1}{2} = -\frac{1}{4}$$

a. Identify the domain of the equation.

b. Identify the LCD of all the denominators of the equation.

c. Solve the equation.

15. For the equation

$$\frac{3}{z} - \frac{4}{5} = -\frac{1}{5}$$

a. Identify the domain of the equation.

b. Identify the LCD of all the denominators of the equation.

c. Solve the equation.

16. For the equation

$$\frac{x+1}{x^2+2x-3} = \frac{1}{x+3} - \frac{1}{x-1}$$

a. Identify the domain of the equation.

b. Identify the LCD of all the denominators of the equation.

c. Solve the equation.

17. For the equation

$$\frac{10}{x-2} - \frac{40}{x^2+x-6} = \frac{12}{x+3}$$

a. Identify the domain of the equation.

b. Identify the LCD of all the denominators of the equation.

c. Solve the equation.

Concept 2: Solving Rational Equations

For Exercises 18–47, solve the equations.

18. $\dfrac{1}{8} = \dfrac{3}{5} + \dfrac{5}{y}$

19. $\dfrac{2}{7} - \dfrac{1}{x} = \dfrac{2}{3}$

20. $\dfrac{4}{t} = \dfrac{3}{t} + \dfrac{1}{8}$

21. $\dfrac{9}{b} - \dfrac{8}{b} = \dfrac{1}{4}$

22. $\dfrac{5}{6x} + \dfrac{7}{x} = 1$

23. $\dfrac{14}{3x} - \dfrac{5}{x} = 2$

24. $1 - \dfrac{2}{y} = \dfrac{3}{y^2}$

25. $1 - \dfrac{2}{m} = \dfrac{8}{m^2}$

26. $\dfrac{a+1}{a} = 1 + \dfrac{a-2}{2a}$

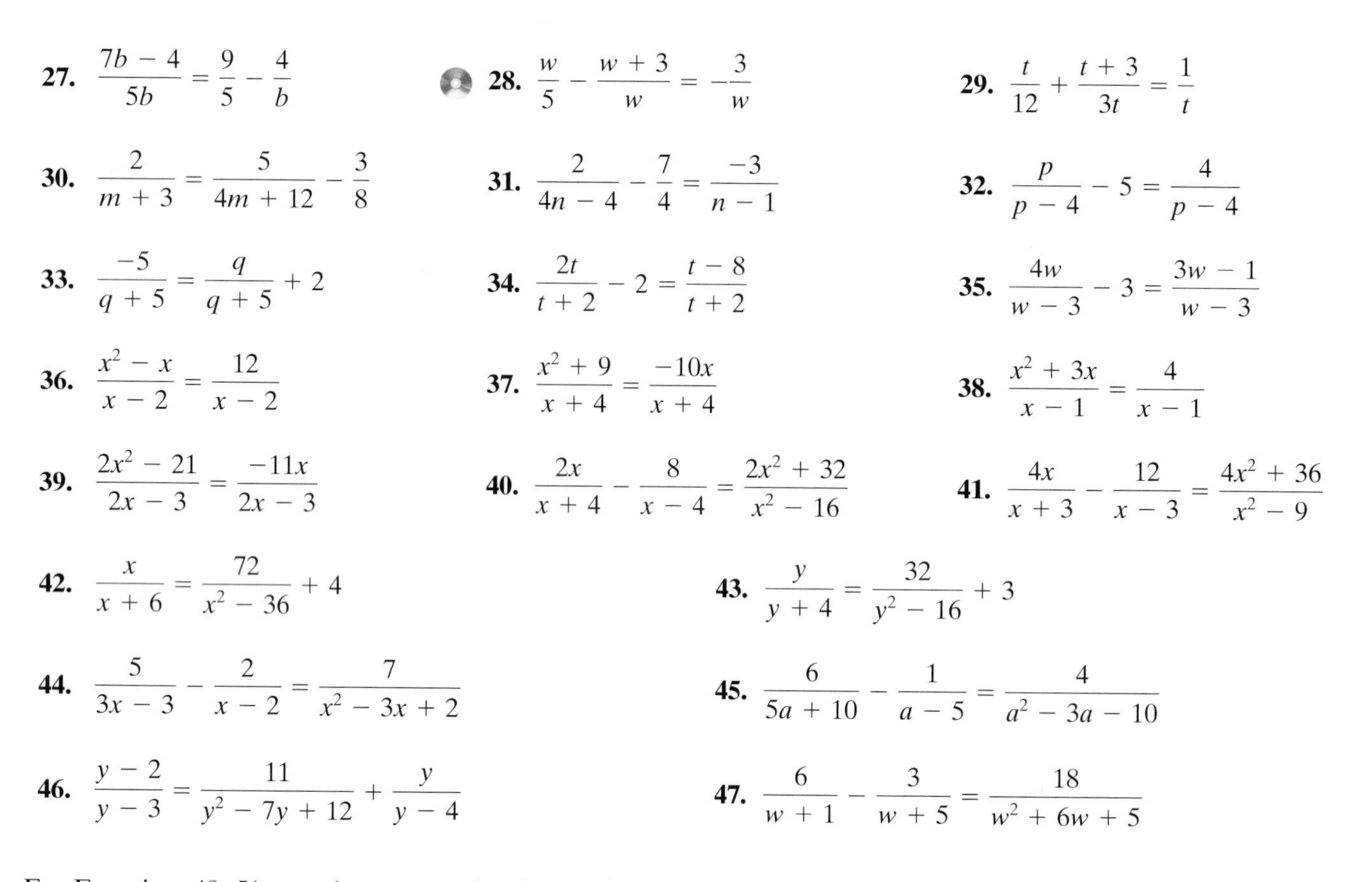

27. $\dfrac{7b - 4}{5b} = \dfrac{9}{5} - \dfrac{4}{b}$

28. $\dfrac{w}{5} - \dfrac{w + 3}{w} = -\dfrac{3}{w}$

29. $\dfrac{t}{12} + \dfrac{t + 3}{3t} = \dfrac{1}{t}$

30. $\dfrac{2}{m + 3} = \dfrac{5}{4m + 12} - \dfrac{3}{8}$

31. $\dfrac{2}{4n - 4} - \dfrac{7}{4} = \dfrac{-3}{n - 1}$

32. $\dfrac{p}{p - 4} - 5 = \dfrac{4}{p - 4}$

33. $\dfrac{-5}{q + 5} = \dfrac{q}{q + 5} + 2$

34. $\dfrac{2t}{t + 2} - 2 = \dfrac{t - 8}{t + 2}$

35. $\dfrac{4w}{w - 3} - 3 = \dfrac{3w - 1}{w - 3}$

36. $\dfrac{x^2 - x}{x - 2} = \dfrac{12}{x - 2}$

37. $\dfrac{x^2 + 9}{x + 4} = \dfrac{-10x}{x + 4}$

38. $\dfrac{x^2 + 3x}{x - 1} = \dfrac{4}{x - 1}$

39. $\dfrac{2x^2 - 21}{2x - 3} = \dfrac{-11x}{2x - 3}$

40. $\dfrac{2x}{x + 4} - \dfrac{8}{x - 4} = \dfrac{2x^2 + 32}{x^2 - 16}$

41. $\dfrac{4x}{x + 3} - \dfrac{12}{x - 3} = \dfrac{4x^2 + 36}{x^2 - 9}$

42. $\dfrac{x}{x + 6} = \dfrac{72}{x^2 - 36} + 4$

43. $\dfrac{y}{y + 4} = \dfrac{32}{y^2 - 16} + 3$

44. $\dfrac{5}{3x - 3} - \dfrac{2}{x - 2} = \dfrac{7}{x^2 - 3x + 2}$

45. $\dfrac{6}{5a + 10} - \dfrac{1}{a - 5} = \dfrac{4}{a^2 - 3a - 10}$

46. $\dfrac{y - 2}{y - 3} = \dfrac{11}{y^2 - 7y + 12} + \dfrac{y}{y - 4}$

47. $\dfrac{6}{w + 1} - \dfrac{3}{w + 5} = \dfrac{18}{w^2 + 6w + 5}$

For Exercises 48–51, translate to a rational equation and solve.

48. The reciprocal of a number is added to three. The result is the quotient of 25 and the number. Find the number.

49. The difference of three and the reciprocal of a number is equal to the quotient of 20 and the number. Find the number.

50. If a number added to five is divided by the difference of the number and two, the result is three-fourths. Find the number.

51. If twice a number added to three is divided by the number plus one, the result is three-halves. Find the number.

Concept 3: Solving Formulas Involving Rational Equations

For Exercises 52–69, solve for the indicated variable.

52. $K = \dfrac{ma}{F}$ for m

53. $K = \dfrac{ma}{F}$ for a

54. $K = \dfrac{IR}{E}$ for E

55. $K = \dfrac{IR}{E}$ for R

56. $I = \dfrac{E}{R + r}$ for R

57. $I = \dfrac{E}{R + r}$ for r

58. $h = \dfrac{2A}{B + b}$ for B

59. $\dfrac{C}{\pi r} = 2$ for r

60. $\dfrac{V}{\pi h} = r^2$ for h

61. $\dfrac{V}{lw} = h$ for w

62. $x = \dfrac{at + b}{t}$ for t

63. $\dfrac{T + mf}{m} = g$ for m

64. $\dfrac{x - y}{xy} = z$ for x

65. $\dfrac{w - n}{wn} = P$ for w

66. $a + b = \dfrac{2A}{h}$ for h

67. $1 + rt = \dfrac{A}{P}$ for P

68. $\dfrac{1}{R} = \dfrac{1}{R_1} + \dfrac{1}{R_2}$ for R

69. $\dfrac{b + a}{ab} = \dfrac{1}{f}$ for b

Chapter 7 Problem Recognition Exercises—Comparing Rational Equations and Rational Expressions

Often adding or subtracting rational expressions is confused with solving rational equations. When adding rational expressions, we combine the terms to simplify the expression. When solving an equation, we clear the fractions and find numerical solutions, if possible. Both processes begin with finding the LCD, but the LCD is used differently in each process. Compare these two examples.

Example 1:

Add. $\dfrac{4}{x} + \dfrac{x}{3}$ (The LCD is $3x$.)

$$= \frac{3}{3} \cdot \left(\frac{4}{x}\right) + \left(\frac{x}{3}\right) \cdot \frac{x}{x}$$

$$= \frac{12}{3x} + \frac{x^2}{3x}$$

$$= \frac{12 + x^2}{3x}$$ The final answer is a rational expression.

Example 2:

Solve. $\dfrac{4}{x} + \dfrac{x}{3} = -\dfrac{8}{3}$ (The LCD is $3x$.)

$$\frac{3x}{1}\left(\frac{4}{x} + \frac{x}{3}\right) = \frac{3x}{1}\left(-\frac{8}{3}\right)$$

$$12 + x^2 = -8x$$

$$x^2 + 8x + 12 = 0$$

$$(x + 2)(x + 6) = 0$$

$$x + 2 = 0 \text{ or } x - 2 = 0$$

$x = -2$ or $x = 2$ The final answers are numbers.

For Exercises 1–12, solve the equation or simplify the expression by combining the terms.

1. $\dfrac{y}{2y + 4} - \dfrac{2}{y^2 + 2y}$

2. $\dfrac{1}{x + 2} + 2 = \dfrac{x + 11}{x + 2}$

3. $\dfrac{5t}{2} - \dfrac{t - 2}{3} = 5$

4. $3 - \dfrac{2}{a - 5}$

5. $\dfrac{7}{6p^2} + \dfrac{2}{9p} + \dfrac{1}{3p^2}$

6. $\dfrac{3b}{b + 1} - \dfrac{2b}{b - 1}$

7. $4 + \dfrac{2}{h - 3} = 5$

8. $\dfrac{2}{w + 1} + \dfrac{3}{(w + 1)^2}$

9. $\dfrac{1}{x - 6} - \dfrac{3}{x^2 - 6x} = \dfrac{4}{x}$

10. $\dfrac{3}{m} - \dfrac{6}{5} = -\dfrac{3}{m}$

11. $\dfrac{7}{2x + 2} + \dfrac{3x}{4x + 4}$

12. $\dfrac{10}{2t - 1} - 1 = \dfrac{t}{2t - 1}$

Section 7.7 Applications of Rational Equations and Proportions

Concepts

1. Solving Proportions
2. Applications of Proportions and Similar Triangles
3. Distance, Rate, and Time Applications
4. Work Applications

1. Solving Proportions

In this section, we look at how rational equations can be used to solve a variety of applications. The first type of rational equation that will be applied is called a proportion.

Definition of a Proportion

An equation that equates two ratios or rates is called a **proportion**. Thus, for $b \neq 0$ and $d \neq 0$, $\frac{a}{b} = \frac{c}{d}$ is a proportion.

A proportion can be solved by multiplying both sides of the equation by the LCD and clearing fractions.

Example 1 Solving a Proportion

Solve the proportion. $\dfrac{3}{11} = \dfrac{123}{w}$

Solution:

$\dfrac{3}{11} = \dfrac{123}{w}$ The LCD is $11w$.

$11w\left(\dfrac{3}{11}\right) = 11w\left(\dfrac{123}{w}\right)$ Multiply by the LCD and clear fractions.

$3w = 11 \cdot 123$ Solve the resulting equation (linear).

$3w = 1353$

$\dfrac{3w}{3} = \dfrac{1353}{3}$

$w = 451$

Check: $w = 451$

$\dfrac{3}{11} = \dfrac{123}{w}$

$\dfrac{3}{11} \stackrel{?}{=} \dfrac{123}{(451)}$

$\dfrac{3}{11} = \dfrac{3}{11}$ ✔ Simplify to lowest terms.

Skill Practice Solve the proportion.

1. $\dfrac{10}{b} = \dfrac{2}{33}$

Skill Practice Answers

1. $b = 165$

TIP: The cross products of any proportion are equal. That is, for $b \neq 0$ and $d \neq 0$, the proportion $\frac{a}{b} = \frac{c}{d}$ is equivalent to $ad = bc$. Some rational equations are proportions and can be solved by equating the cross products. Consider the proportion from Example 1:

$$\frac{3}{11} \times \frac{123}{w}$$

$$3 \cdot w = 11 \cdot 123 \quad \text{Equate the cross products.}$$

$$3w = 1353 \quad \text{Solve the resulting equation.}$$

$$\frac{3w}{3} = \frac{1353}{3}$$

$$w = 451$$

2. Applications of Proportions and Similar Triangles

Example 2 Using a Proportion in an Application

For a recent year, the population of Alabama was approximately 4.2 million. At that time, Alabama had seven representatives in the U.S. House of Representatives. In the same year, North Carolina had a population of approximately 7.2 million. If representation in the House is based on population in equal proportions for each state, how many representatives did North Carolina have?

Solution:

Let x represent the number of representatives for North Carolina.

Set up a proportion by writing two equivalent ratios.

$$\frac{\text{Population of Alabama}}{\text{number of representatives}} \rightarrow \frac{4.2}{7} = \frac{7.2}{x} \leftarrow \frac{\text{Population of North Carolina}}{\text{number of representatives}}$$

$$\frac{4.2}{7} = \frac{7.2}{x}$$

$$7x \cdot \frac{4.2}{7} = 7x \cdot \frac{7.2}{x} \quad \text{Multiply by the LCD, } 7x.$$

$$4.2x = (7.2)(7) \quad \text{Solve the resulting equation (linear).}$$

$$4.2x = 50.4$$

$$\frac{4.2x}{4.2} = \frac{50.4}{4.2}$$

$$x = 12 \quad \text{North Carolina had 12 representatives.}$$

TIP: The equation from Example 2 could have been solved by first equating the cross products:

$$\frac{4.2}{7} \times \frac{7.2}{x}$$

$$4.2x = (7.2)(7)$$

$$4.2x = 50.4$$

$$x = 12$$

Skill Practice

2. A college keeps the ratio of students to faculty at 105 to 2. If the student population at the college is 1575, how many faculty members are needed?

Skill Practice Answers

2. 30 faculty members are needed.

Proportions are used in geometry with **similar triangles**. Two triangles are said to be similar if their corresponding angles have equal measures. In such a case, the lengths of the corresponding sides are proportional. The triangles in Figure 7-1 are similar. Therefore, the following ratios are equivalent.

$$\frac{a}{x} = \frac{b}{y} = \frac{c}{z}$$

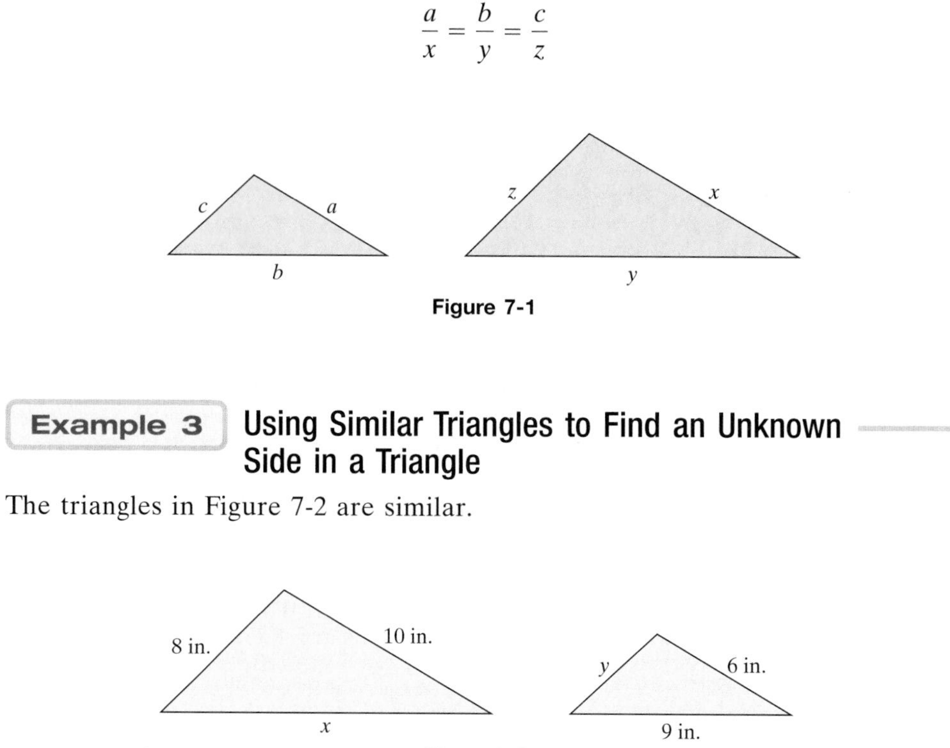

Figure 7-1

Example 3 Using Similar Triangles to Find an Unknown Side in a Triangle

The triangles in Figure 7-2 are similar.

Figure 7-2

a. Solve for x. **b.** Solve for y.

Solution:

a. The lengths of the upper right sides of the triangles are given. These form a known ratio of $\frac{10}{6}$. Because the triangles are similar, the ratio of the other corresponding sides must be equal to $\frac{10}{6}$. To solve for x, we have:

Bottom side from large triangle →	$\dfrac{x}{9 \text{ in.}} = \dfrac{10 \text{ in.}}{6 \text{ in.}}$	← Right side from large triangle
Bottom side from small triangle →		← Right side from small triangle

$$\frac{x}{9} = \frac{10}{6} \qquad \text{The LCD is 18.}$$

$$18 \cdot \left(\frac{x}{9}\right) = 18 \cdot \left(\frac{10}{6}\right) \qquad \text{Multiply by the LCD.}$$

$$2x = 30 \qquad \text{Clear fractions.}$$

$$x = 15 \qquad \text{Divide by 2.}$$

The length of side x is 15 in.

b. To solve for y, the ratio of the upper left sides of the triangles must equal $\frac{10}{6}$.

$$\frac{\text{Left side from large triangle}}{\text{Left side from small triangle}} \rightarrow \frac{8 \text{ in.}}{y} = \frac{10 \text{ in.}}{6 \text{ in.}} \leftarrow \frac{\text{Right side from large triangle}}{\text{Right side from small triangle}}$$

$\frac{8}{y} = \frac{10}{6}$ The LCD is $6y$.

$6y \cdot \left(\frac{8}{y}\right) = 6y \cdot \left(\frac{10}{6}\right)$ Multiply by the LCD.

$48 = 10y$ Clear fractions.

$\frac{48}{10} = \frac{10y}{10}$

$4.8 = y$

The length of side y is 4.8 in.

Skill Practice

3. The two triangles shown are similar triangles. Solve for the lengths of the missing sides.

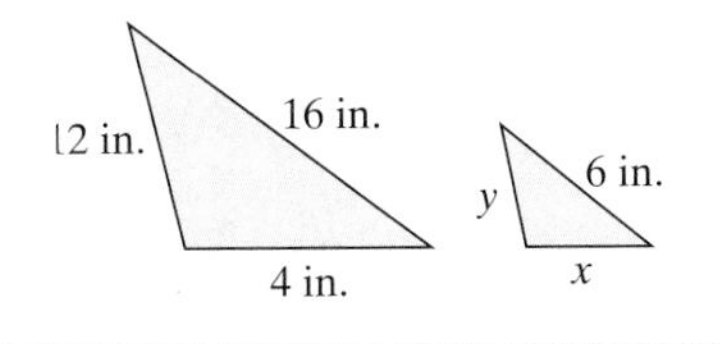

Example 4 Using Similar Triangles in an Application

The shadow cast by a yardstick is 2 ft long. The shadow cast by a tree is 11 ft long. Find the height of the tree.

Solution:

Let x represent the height of the tree. Label the variable.

We will assume that the measurements were taken at the same time of day. Therefore, the angle of the sun is the same on both objects, and we can set up similar triangles (Figure 7-3).

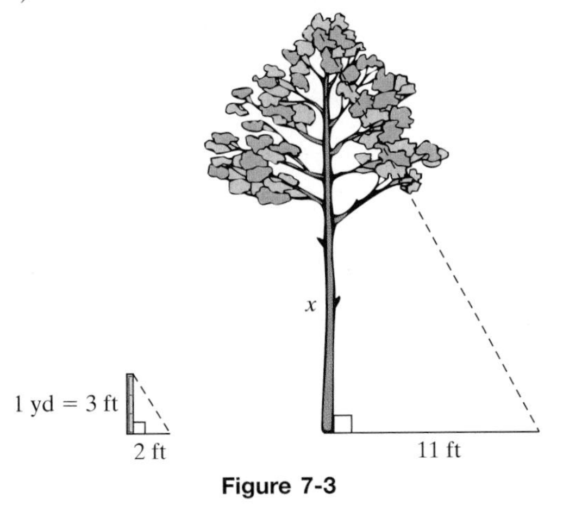

Figure 7-3

Skill Practice Answers

3. $x = 1.5$ in., and $y = 4.5$ in.

Create a verbal model.

$$\frac{\text{Height of yardstick}}{\text{Length of yardstick's shadow}} \longrightarrow \frac{3\text{ ft}}{2\text{ ft}} = \frac{x}{11\text{ ft}} \longleftarrow \frac{\text{Height of tree}}{\text{Length of tree's shadow}}$$

$$\frac{3}{2} = \frac{x}{11}$$ Write a mathematical equation.

$$\overset{11}{\cancel{22}} \cdot \left(\frac{3}{\cancel{2}}\right) = \left(\frac{x}{\cancel{11}}\right) \cdot \overset{2}{\cancel{22}}$$ Multiply by the LCD.

$$33 = 2x$$ Solve the equation.

$$\frac{33}{2} = \frac{2x}{2}$$

$$16.5 = x$$ Interpret the results and write the answer in words.

The tree is 16.5 ft high.

Skill Practice

4. The sun casts a 3.2-ft shadow of a 6-ft man. At the same time, the sun casts an 80-ft shadow of a building. How tall is the building?

3. Distance, Rate, and Time Applications

In Sections 2.4 and 4.4, we presented applications involving the relationship among the variables distance, rate, and time. Recall that $d = rt$.

Example 5 **Using a Rational Equation in a Distance, Rate, and Time Application**

A small plane flies 440 miles with the wind from Memphis, Tennessee, to Oklahoma City, Oklahoma. In the same amount of time, the plane flies 340 miles against the wind from Oklahoma City to Little Rock, Arkansas (see Figure 7-4). If the wind speed is 30 mph, find the speed of the plane in still air.

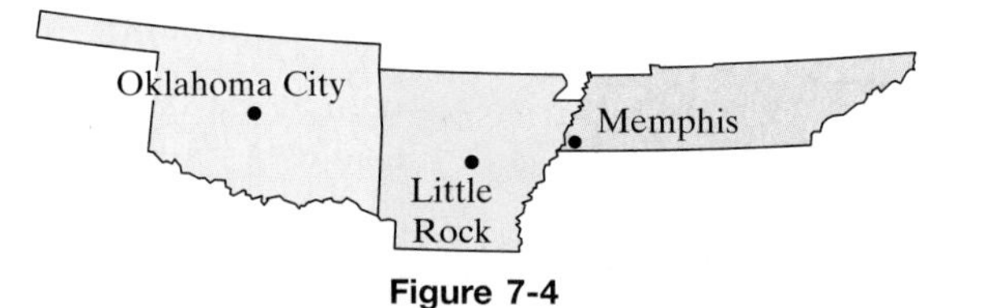

Figure 7-4

Solution:

Let x represent the speed of the plane in still air.

Organize the given information in a chart.

	Distance	Rate	Time
With the wind	440	$x + 30$	$\frac{440}{x + 30}$
Against the wind	340	$x - 30$	$\frac{340}{x - 30}$

Because $d = rt$, then $t = \frac{d}{r}$.

Skill Practice Answers

4. The building is 150 ft tall.

The plane travels with the wind for the same amount of time as it travels against the wind, so we can equate the two expressions for time.

$$\begin{pmatrix}\text{Time with}\\ \text{the wind}\end{pmatrix} = \begin{pmatrix}\text{time against}\\ \text{the wind}\end{pmatrix}$$

$$\frac{440}{x + 30} = \frac{340}{x - 30}$$ The LCD is $(x + 30)(x - 30)$.

$$(x + 30)(x - 30) \cdot \frac{440}{x + 30} = (x + 30)(x - 30) \cdot \frac{340}{x - 30}$$

$$440(x - 30) = 340(x + 30)$$

$$440x - 13{,}200 = 340x + 10{,}200$$ Solve the resulting linear equation.

$$100x = 23{,}400$$

$$x = 234$$

The plane's speed in still air is 234 mph.

TIP: The equation

$$\frac{440}{x + 30} = \frac{340}{x - 30}$$

is a proportion. The fractions can also be cleared by equating the cross products.

$$\frac{440}{x + 30} \times \frac{340}{x - 30}$$

$$440(x - 30) = 340(x + 30)$$

Skill Practice

5. Alison paddles her kayak in a river where the current of the water is 2 mph. She can paddle 20 miles with the current in the same time that she can paddle 10 miles against the current. Find the speed of the kayak in still water.

Example 6 Using a Rational Equation in a Distance, Rate, and Time Application

A motorist drives 100 miles between two cities in a bad rainstorm. For the return trip in sunny weather, she averages 10 mph faster and takes $\frac{1}{2}$ hour less time. Find the average speed of the motorist in the rainstorm and in sunny weather.

Solution:

Let x represent the motorist's speed during the rain.

Then $x + 10$ represents the speed in sunny weather.

	Distance	Rate	Time
Trip during rainstorm	100	x	$\frac{100}{x}$
Trip during sunny weather	100	$x + 10$	$\frac{100}{x + 10}$

Because $d = rt$, then $t = \frac{d}{r}$.

Skill Practice Answers

5. The speed is 6 mph.

Because the same distance is traveled in $\frac{1}{2}$ hr less time, the difference between the time of the trip during the rainstorm and the time during sunny weather is $\frac{1}{2}$ hr.

$$\begin{pmatrix}\text{Time during} \\ \text{the rainstorm}\end{pmatrix} - \begin{pmatrix}\text{time during} \\ \text{sunny weather}\end{pmatrix} = \left(\frac{1}{2}\text{ hr}\right)$$ Verbal model

$$\frac{100}{x} - \frac{100}{x + 10} = \frac{1}{2}$$ Mathematical equation

$$2x(x + 10)\left(\frac{100}{x} - \frac{100}{x + 10}\right) = 2x(x + 10)\left(\frac{1}{2}\right)$$ Multiply by the LCD.

$$2x(x + 10)\left(\frac{100}{x}\right) - 2x(x + 10)\left(\frac{100}{x + 10}\right) = 2x(x + 10)\left(\frac{1}{2}\right)$$ Apply the distributive property.

$$200(x + 10) - 200x = x(x + 10)$$ Clear fractions.

$$200x + 2000 - 200x = x^2 + 10x$$ Solve the resulting equation (quadratic).

$$2000 = x^2 + 10x$$

$$0 = x^2 + 10x - 2000$$ Set the equation equal to zero.

$$0 = (x - 40)(x + 50)$$ Factor.

$x = 40$ or $x \neq -50$

Because a rate of speed cannot be negative, reject $x = -50$. Therefore, the speed of the motorist in the rainstorm is 40 mph. Because $x + 10 = 40 + 10 = 50$, the average speed for the return trip in sunny weather is 50 mph.

Avoiding Mistakes:

The equation

$$\frac{100}{x} - \frac{100}{x + 10} = \frac{1}{2}$$

is not a proportion because the left-hand side has more than one fraction. Do not try to multiply the cross products. Instead, multiply by the LCD to clear fractions.

Skill Practice

6. Harley rode his mountain bike 12 miles to the top of the mountain and the same distance back down. His speed going up was 8 mph slower than coming down. The ride up took 2 hours longer than coming down. Find his speeds.

4. Work Applications

Suppose Winston can paint a room in 2 hours. Then he paints $\frac{1}{2}$ room per hour. Suppose Clyde can paint the room in 4 hours. Then he paints $\frac{1}{4}$ room per hour. In general, we can define a work rate as follows:

Work rate: $\frac{1}{t}$ jobs per hour, where t is the total time required to complete the job.

Furthermore, if we multiply a work rate by the time an individual works, we compute the portion of the job completed. That is,

$$\text{Portion of job completed} = (\text{Work rate})(\text{Time})$$

Therefore, in 3 hours, Clyde would paint $(\frac{1}{4}\text{ room/hr})(3\text{ hr}) = \frac{3}{4}$ room. We use this basic principle to solve equations involving "work."

Skill Practice Answers

6. Uphill speed was 4 mph; downhill speed was 12 mph.

Example 7 Using a Rational Equation in a "Work" Application

A new printing press can print the morning edition in 2 hours, whereas the old printer required 4 hours. How long would it take to print the morning edition if both printers were working together?

Solution:

Let x represent the time required for both printers working together to complete the job.

One method to approach this problem is to determine the portion of the job that each printer can complete in 1 hour and extend that rate to the portion of the job completed in x hours.

- The old printer can perform the job in 4 hours. Therefore, it completes $\frac{1}{4}$ of the job in 1 hour and $\frac{1}{4}x$ jobs in x hours.
- The new printer can perform the job in 2 hours. Therefore, it completes $\frac{1}{2}$ of the job in 1 hour and $\frac{1}{2}x$ jobs in x hours.

	Work Rate	Time	Portion of Job Completed
Old printer	$\frac{1 \text{ job}}{4 \text{ hr}}$	x hours	$\frac{1}{4}x$
New printer	$\frac{1 \text{ job}}{2 \text{ hr}}$	x hours	$\frac{1}{2}x$

The sum of the portions of the job completed by each printer must equal one whole job.

$$\begin{pmatrix}\text{Portion of job}\\ \text{completed by}\\ \text{old printer}\end{pmatrix} + \begin{pmatrix}\text{portion of job}\\ \text{completed by}\\ \text{new printer}\end{pmatrix} = \begin{pmatrix}1\\ \text{whole}\\ \text{job}\end{pmatrix}$$

$$\frac{1}{4}x + \frac{1}{2}x = 1 \qquad \text{The LCD is 4.}$$

$$4\left(\frac{1}{4}x + \frac{1}{2}x\right) = 4(1) \qquad \text{Multiply by the LCD.}$$

$$4 \cdot \frac{1}{4}x + 4 \cdot \frac{1}{2}x = 4 \cdot 1 \qquad \text{Apply the distributive property.}$$

$$x + 2x = 4 \qquad \text{Solve the resulting linear equation.}$$

$$3x = 4$$

$$x = \frac{4}{3} \quad \text{or} \quad x = 1\frac{1}{3}$$

The time required to print the morning edition using both printers is $1\frac{1}{3}$ hr.

Skill Practice

7. The computer at a bank can process and prepare the bank statements in 30 hours. A new faster computer can do the job in 20 hours. If the bank uses both computers together, how long will it take to process the statements?

Skill Practice Answers

7. 12 hours

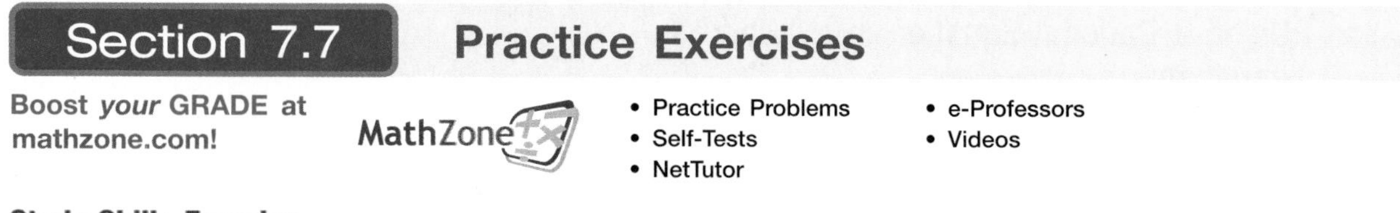

Section 7.7 Practice Exercises

Study Skills Exercise

1. Define the key terms:

a. proportion **b. similar triangles**

Review Exercises

For Exercises 2–8, determine whether each of the following is an equation or an expression. If it is an equation, solve it. If it is an expression, perform the indicated operation.

2. $\dfrac{b}{5} + 3 = 9$

3. $\dfrac{m}{m-1} - \dfrac{2}{m+3}$

4. $\dfrac{2}{a+5} + \dfrac{5}{a^2-25}$

5. $\dfrac{3y+6}{20} \div \dfrac{4y+8}{8}$

6. $\dfrac{z^2+z}{24} \cdot \dfrac{8}{z+1}$

7. $\dfrac{3}{p+3} = \dfrac{12p+19}{p^2+7p+12} - \dfrac{5}{p+4}$

8. $\dfrac{\dfrac{1}{t^2} + \dfrac{2}{3}}{\dfrac{1}{t} - \dfrac{5}{6}}$

Concept 1: Solving Proportions

For Exercises 9–22, solve the proportions.

9. $\dfrac{8}{5} = \dfrac{152}{p}$

10. $\dfrac{6}{7} = \dfrac{96}{y}$

11. $\dfrac{19}{76} = \dfrac{z}{4}$

12. $\dfrac{15}{135} = \dfrac{w}{9}$

13. $\dfrac{5}{3} = \dfrac{a}{8}$

14. $\dfrac{b}{14} = \dfrac{3}{8}$

15. $\dfrac{2}{1.9} = \dfrac{x}{38}$

16. $\dfrac{16}{1.3} = \dfrac{30}{p}$

17. $\dfrac{y+1}{2y} = \dfrac{2}{3}$

18. $\dfrac{w-2}{4w} = \dfrac{1}{6}$

19. $\dfrac{9}{2z-1} = \dfrac{3}{z}$

20. $\dfrac{1}{t} = \dfrac{1}{4-t}$

21. $\dfrac{8}{9a-1} = \dfrac{5}{3a+2}$

22. $\dfrac{4p+1}{3} = \dfrac{2p-5}{6}$

23. Charles' law describes the relationship between the initial and final temperature and volume of a gas held at a constant pressure.

$$\frac{V_i}{V_f} = \frac{T_i}{T_f}$$

a. Solve the equation for V_f.

b. Solve the equation for T_f.

24. The relationship between the area, height, and base of a triangle is given by the proportion

$$\frac{A}{b} = \frac{h}{2}, \text{ where } A \text{ is area, } b \text{ is the base, and } h \text{ is the height.}$$

a. Solve the equation for A. **b.** Solve the equation for b.

Concept 2: Applications of Proportions and Similar Triangles

For Exercises 25–32, solve using proportions.

25. Toni drives her Honda Civic 132 miles on the highway on 4 gallons of gas. At this rate how many miles can she drive on 9 gallons of gas?

26. Tim takes his pulse for 10 seconds and counts 12 beats. How many beats per minute is this?

27. Suppose a household of 4 people produces 128 lb of garbage in one week. At this rate, how many pounds will 48 people produce in one week?

28. Property tax on a $180,000 house is $4000. At this rate, how much tax would be paid on a $216,000 home?

29. Martin won an election by a ratio 5 to 4. If he received 5420 votes, how many votes did his opponent receive?

30. Cooking oatmeal requires 1 cup of water for every $\frac{1}{2}$ cup of oats. How many cups of water will be required for $\frac{3}{4}$ cup of oats?

31. A map has a scale of 75 miles/in. If two cities measure 3.5 in. apart, how many miles does this represent?

32. A map has a scale of 50 miles/in. If two cities measure 6.5 in. apart, how many miles does this represent?

33. $\triangle ABC$ is similar to $\triangle DEF$.

a. Find the length of $\overline{EF}$.

b. Find the length of $\overline{DF}$.

34. Figure $ABCD$ is similar to Figure $EFGH$.

a. Find the length of $\overline{EH}$.

b. Find the length of $\overline{AB}$.

c. Find the length of $\overline{BC}$.

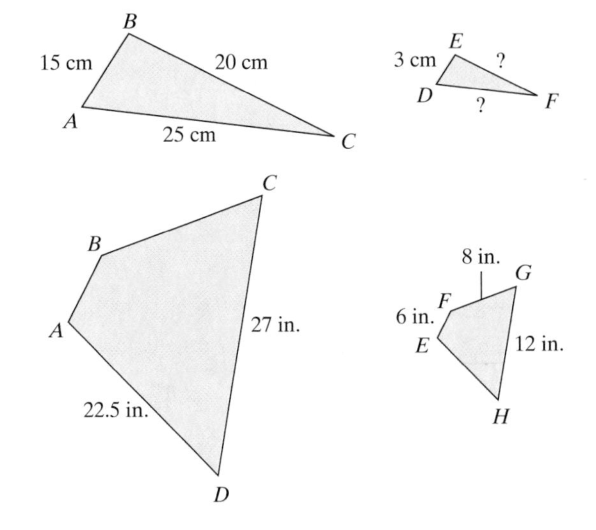

35. Solve for x and y.

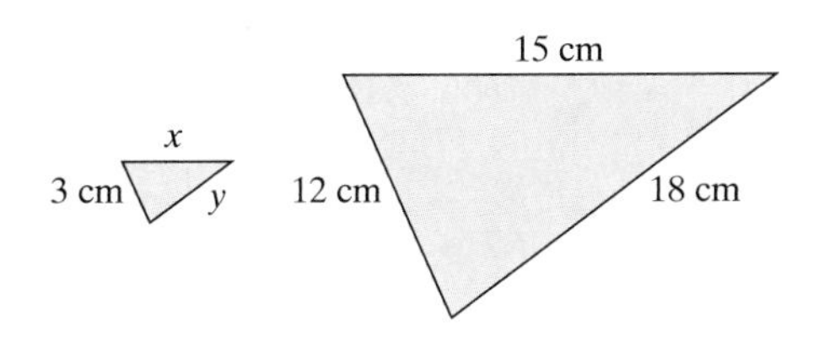

36. Solve for x and y.

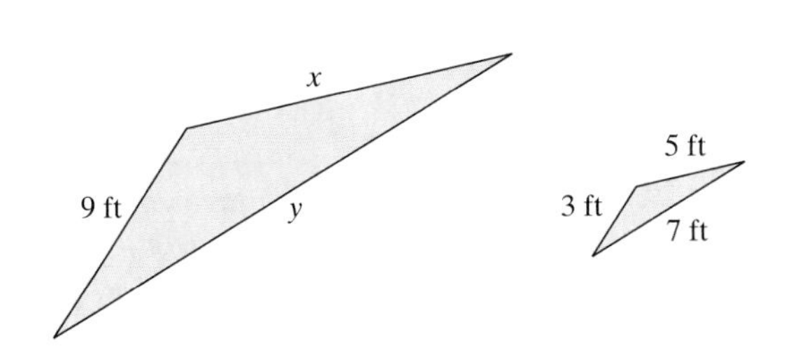

37. To estimate the height of a light pole, a mathematics student measures the length of a shadow cast by a meterstick and the length of the shadow cast by the light pole. Find the height of the light pole (see figure).

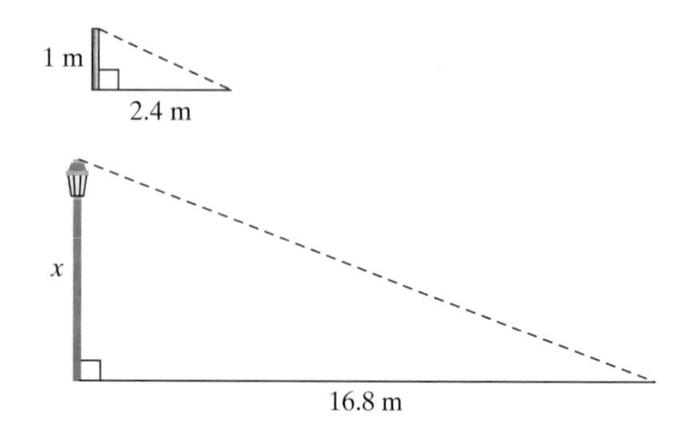

38. To estimate the height of a building, a student measures the length of a shadow cast by a yardstick and the length of the shadow cast by the building (see figure). Find the height of the building.

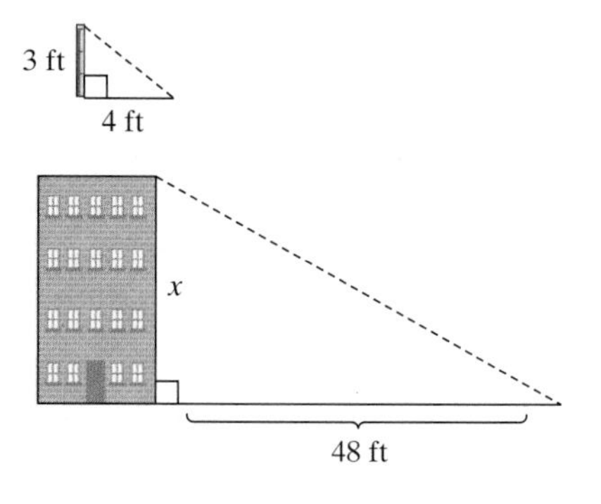

39. A 6-ft-tall man standing 54 ft from a light post casts an 18-ft shadow. What is the height of the light post?

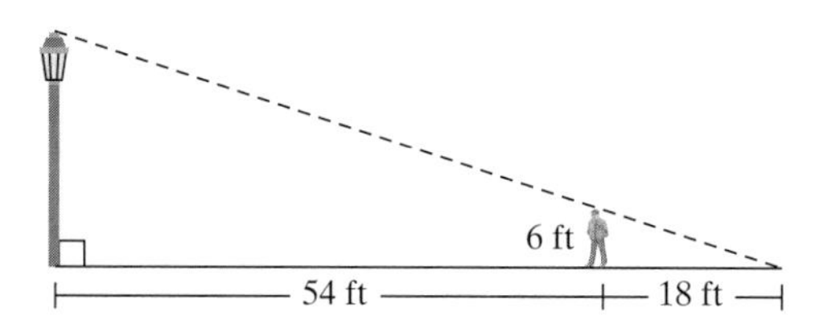

40. For a science project at school, a student must measure the height of a tree. The student measures the length of the shadow of the tree and then measures the length of the shadow cast by a yardstick. Use similar triangles to find the height of the tree.

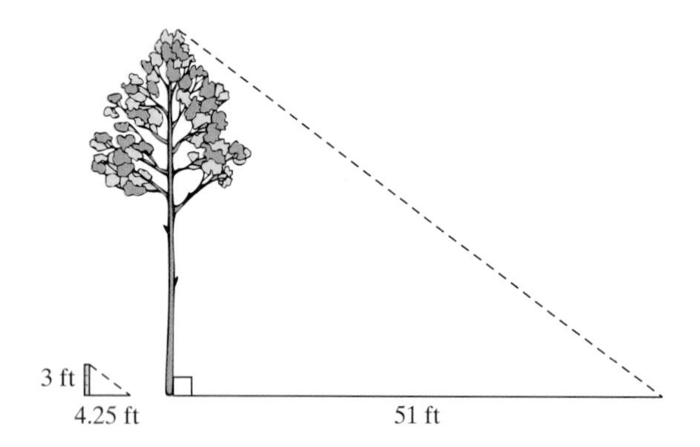

Concept 3: Distance, Rate, and Time Applications

41. A boat travels 54 miles upstream against the current in the same amount of time it takes to travel 66 miles downstream with the current. If the boat has the speed of 20 mph in still water, what is the speed of the current? (Use $t = \frac{d}{r}$ to complete the table.)

	Distance	Rate	Time
With the current (downstream)			
Against the current (upstream)			

42. A fisherman travels 9 miles downstream with the current in the same time that he travels 3 miles upstream against the current. If the speed of the current is 6 mph, what is the speed at which the fisherman travels in still water?

43. A plane flies 630 miles with the wind in the same time that it takes to fly 455 miles against the wind. If this plane flies at the rate of 217 mph in still air, what is the speed of the wind? (Use $t = \frac{d}{r}$ to complete the table.)

	Distance	Rate	Time
With the wind			
Against the wind			

44. A plane flies 370 miles with the wind in the same time that it takes to fly 290 miles against the wind. If the speed of the wind is 20 mph, what is the speed of the plane in still air?

45. Devon can cross-country ski 5 km/hr faster than his sister Shanelle. Devon skis 45 km in the same time Shanelle skis 30 km. Find their speeds.

46. Brooke walks 2 km/hr slower than her older sister Adrianna. Brooke can walk 12 km in the same amount of time that Adriana can walk 18 km. Find their speeds.

47. One motorist travels 15 mph faster than another. The slower driver takes 2 hours longer to travel 360 mi than the faster driver. What are the speeds of the two motorists?

48. A train travels 180 miles in 1 hour less time than a bus traveling the same distance. If the speed of the bus is 15 mph slower than the speed of the train, find the speed of the train and the bus.

49. Kendra flew 900 miles to Cincinnati, Ohio. When she returned she traveled 30 mph slower. If it took her 1 hour longer on her return flight, what were her speeds going to and returning from Cincinnati?

50. A plane flew 500 km from Atlanta, Georgia, to Louisville, Kentucky. When returning to Atlanta, the flight took $\frac{1}{2}$ hr less time. If the rate to Louisville was 50 km/hr slower than the rate returning, find the two rates in km/hr.

51. Sergio rode his bike 4 miles. Then he got a flat tire and had to walk back 4 miles. It took him 1 hr longer to walk than it did to ride. If his rate walking was 9 mph less than his rate riding, find the two rates.

52. Amber jogs 10 km in $\frac{3}{4}$ hr less than she can walk the same distance. If her walking rate is 3 km/hr less than her jogging rate, find her rates jogging and walking (in km/hr).

Concept 4: Work Applications

53. If it takes a person 2 hr to paint a room, what fraction of the room would be painted in 1 hr?

54. If it takes a copier 3 hr to complete a job, what fraction of the job would be completed in 1 hr?

55. If the cold-water faucet is left on, the sink will fill in 10 min. If the hot-water faucet is left on, the sink will fill in 12 min. How long would it take the sink to fill if both faucets are left on?

56. The CUT-IT-OUT lawn mowing company consists of two people: Tina and Bill. If Tina cuts a lawn by herself, she can do it in 4 hr. If Bill cuts the same lawn himself, it takes him an hour longer than Tina. How long would it take them if they worked together?

57. A manuscript needs to be printed. One printer can do the job in 50 min, and another printer can do the job in 40 min. How long would it take if both printers were used?

58. A pump can empty a small pond in 4 hr. Another more efficient pump can do the job in 3 hr. How long would it take to empty the pond if both pumps were used?

59. Tim and Al are bricklayers. Tim can construct an outdoor grill in 5 days. If Al helps Tim, they can build it in only 2 days. How long would it take Al to build the grill alone?

60. Norma is a new and inexperienced secretary. It takes her 3 hr to prepare a mailing. If her boss helps her, the mailing can be completed in 1 hr. How long would it take the boss to do the job by herself?

61. A pipe can fill a reservoir in 16 hr. A drainage pipe can drain the reservoir in 24 hr. How long would it take to fill the reservoir if the drainage pipe were left open by mistake? (*Hint:* The rate at which water drains should be negative.)

62. A hole in the bottom of a child's plastic swimming pool can drain the pool in 60 min. If the pool had no hole, a hose could fill the pool in 40 min. How long would it take the hose to fill the pool with the hole?

63. A new copy machine works three times faster than the older model. It takes 12 minutes to complete a job when both machines are working together. Find the time required for the new copy machine to complete the job by itself.

64. A cold-water faucet can fill a tub twice as fast as a hot-water faucet. Together it takes 12 min to fill the tub. How long will it take to fill the tub using only the cold-water faucet?

Expanding Your Skills

For Exercises 65–68, solve using proportions.

65. The ratio of smokers to nonsmokers in a restaurant is 2 to 7. There are 100 more nonsmokers than smokers. How many smokers and nonsmokers are in the restaurant?

66. The ratio of fiction to nonfiction books sold in a bookstore is 5 to 3. One week there are 180 more fiction books sold than nonfiction. Find the number of fiction and nonfiction books sold during that week.

67. There are 440 students attending a biology lecture. The ratio of male to female students at the lecture is 6 to 5. How many men and women are attending the lecture?

68. The ratio of dogs to cats at the humane society is 5 to 8. There are a total of 650 dogs and cats. How many dogs and how many cats are at the humane society?

Section 7.8 Variation (Optional)

Concepts

1. Definition of Direct and Inverse Variation
2. Translations Involving Variation
3. Applications of Variation

1. Definition of Direct and Inverse Variation

In this section, we introduce the concept of variation. Direct and inverse variation models can show how one quantity varies in proportion to another.

Definition of Direct and Inverse Variation

Let k be a nonzero constant. Then the following statements are equivalent:

1. y **varies directly** as x.
   y is directly proportional to x. } $y = kx$
2. y **varies inversely** as x.
   y is inversely proportional to x. } $y = \frac{k}{x}$

Note: The value of k is called the constant of variation.

For a car traveling at 30 mph, the equation $d = 30t$ indicates that the distance traveled is *directly proportional* to the time of travel. For positive values of k, when two variables are directly related, as one variable increases, the other variable will also increase. Likewise, if one variable decreases, the other will decrease. In the equation $d = 30t$, the longer the time of the trip, the greater the distance traveled; the shorter the time of the trip, the shorter the distance traveled.

For positive values of k, when two variables are *inversely proportional*, as one variable increases, the other will decrease, and vice versa. Consider a car traveling between Toronto and Montreal, a distance of 500 km. The time required to make the trip is inversely related to the speed of travel: $t = 500/r$. As the rate of speed, r, increases, the quotient $500/r$ will decrease. Hence the time will decrease. Similarly, as the rate of speed decreases, the trip will take longer.

2. Translations Involving Variation

The first step in using a variation model is to translate an English phrase into an equivalent mathematical equation.

Example 1 Translating to a Variation Model

Translate each expression into an equivalent mathematical model.

a. The circumference of a circle varies directly as the radius.

b. At a constant temperature, the volume of a gas varies inversely as the pressure.

c. The length of time of a meeting is directly proportional to the *square* of the number of people present.

Solution:

a. Let C represent circumference and r represent radius. The variables are directly related, so use the model $C = kr$.

b. Let V represent volume and P represent pressure. Because the variables are inversely related, use the model $V = \frac{k}{P}$.

c. Let t represent time, and let N be the number of people present at a meeting. Because t is directly related to N^2, use the model $t = kN^2$.

TIP: Do not forget to use the constant of variation, k, in the equation.

Skill Practice Translate each expression into an equivalent mathematical model.

1. The distance, d, driven in a particular time varies directly with the speed of the car, s.
2. The weight of an individual kitten, w, varies inversely with the number of kittens in the litter, n.
3. The value of v varies inversely as the square root of b.

Sometimes a variable varies directly as the product of two or more other variables. In this case, we have joint variation.

Definition of Joint Variation

Let k be a nonzero constant. Then the following statements are equivalent:

y **varies jointly** as w and z.
y is jointly proportional to w and z.
} $y = kwz$

Example 2 Translating to a Variation Model

Translate each expression into an equivalent mathematical model.

a. y varies jointly as u and the square root of v.

b. The gravitational force of attraction between two planets varies jointly as the product of their masses and inversely as the square of the distance between them.

Solution:

a. $y = ku\sqrt{v}$

b. Let m_1 and m_2 represent the masses of the two planets. Let F represent the gravitational force of attraction and d represent the distance between the planets.

The variation model is $$F = \frac{km_1m_2}{d^2}$$

Skill Practice Translate each expression into an equivalent mathematical model.

4. The value of q varies jointly as u and v.
5. The value of x varies directly as the square of y and inversely as z.

3. Applications of Variation

Consider the variation models $y = kx$ and $y = k/x$. In either case, if values for x and y are known, we can solve for k. Once k is known, we can use the variation equation to find y if x is known, or to find x if y is known. This concept is the basis for solving many problems involving variation.

Skill Practice Answers

1. $d = ks$
2. $w = \frac{k}{n}$
3. $v = \frac{k}{\sqrt{b}}$
4. $q = kuv$
5. $x = \frac{ky^2}{z}$

Steps to Find a Variation Model

1. Write a general variation model that relates the variables given in the problem. Let k represent the constant of variation.
2. Solve for k by substituting known values of the variables into the model from step 1.
3. Substitute the value of k into the original variation model from step 1.

Example 3 Solving an Application Involving Direct Variation

The variable z varies directly as w. When w is 16, z is 56.

a. Write a variation model for this situation. Use k as the constant of variation.

b. Solve for the constant of variation.

c. Find the value of z when w is 84.

Solution:

a. $z = kw$

b. $z = kw$

$56 = k(16)$ Substitute known values for z and w. Then solve for the unknown value of k.

$\frac{56}{16} = \frac{k(16)}{16}$ To isolate k, divide both sides by 16.

$\frac{7}{2} = k$ Simplify $\frac{56}{16}$ to $\frac{7}{2}$.

c. With the value of k known, the variation model can now be written as $z = \frac{7}{2}w$.

$z = \frac{7}{2}(84)$ To find z when $w = 84$, substitute $w = 84$ into the equation.

$z = 294$

Skill Practice

6. The variable t varies directly as the square of v. When v is 8, t is 32.

a. Write a variation model for this relationship.

b. Solve for the constant of variation.

c. Find t when $v = 10$.

Example 4 Solving an Application Involving Direct Variation

The speed of a racing canoe in still water varies directly as the square root of the length of the canoe.

a. If a 16-ft canoe can travel 6.2 mph in still water, find a variation model that relates the speed of a canoe to its length.

b. Find the speed of a 25-ft canoe.

Skill Practice Answers

6a. $t = kv^2$ **b.** $\frac{1}{2}$ **c.** 50

Solution:

a. Let s represent the speed of the canoe and L represent the length. The general variation model is $s = k\sqrt{L}$. To solve for k, substitute the known values for s and L.

$s = k\sqrt{L}$

$6.2 = k\sqrt{16}$ Substitute $s = 6.2$ mph and $L = 16$ ft.

$6.2 = k \cdot 4$

$\frac{6.2}{4} = \frac{4k}{4}$ Solve for k.

$k = 1.55$

$s = 1.55\sqrt{L}$ Substitute $k = 1.55$ into the model $s = k\sqrt{L}$.

b. $s = 1.55\sqrt{L}$

$= 1.55\sqrt{25}$ Find the speed when $L = 25$ ft.

$= 7.75$ mph

Skill Practice

7. The amount of water needed by a mountain hiker varies directly as the time spent hiking. The hiker needs 2.4 liters for a 4-hour hike. How much water will be needed for a 5-hour hike?

Example 5 Solving an Application Involving Inverse Variation

The loudness of sound measured in decibels varies inversely as the square of the distance between the listener and the source of the sound. If the loudness of sound is 17.92 decibels at a distance of 10 ft from a stereo speaker, what is the decibel level 20 ft from the speaker?

Solution:

Let L represent the loudness of sound in decibels and d represent the distance in feet. The inverse relationship between decibel level and the square of the distance is modeled by

$$L = \frac{k}{d^2}$$

$17.92 = \frac{k}{(10)^2}$ Substitute $L = 17.92$ decibels and $d = 10$ ft.

$17.92 = \frac{k}{100}$

$(17.92)100 = \frac{k}{100} \cdot 100$ Solve for k (clear fractions).

$k = 1792$

$L = \frac{1792}{d^2}$ Substitute $k = 1792$ into the original model $L = \frac{k}{d^2}$.

Skill Practice Answers

7. 3 liters

With the value of k known, we can find L for any value of d.

$$L = \frac{1792}{(20)^2} \qquad \text{Find the loudness when } d = 20 \text{ ft.}$$

$$= 4.48 \text{ decibels}$$

Notice that the loudness of sound is 17.92 decibels at a distance 10 ft from the speaker. When the distance from the speaker is increased to 20 ft, the decibel level decreases to 4.48 decibels. This is consistent with an inverse relationship. For $k > 0$, as one variable is increased, the other is decreased. It also seems reasonable that the further one moves away from the source of a sound, the softer the sound becomes.

Skill Practice

8. The yield on a bond varies inversely as the price. The yield on a particular bond is 5% when the price is \$100. Find the yield when the price is \$125.

Example 6 Solving an Application Involving Joint Variation

In the early morning hours of August 29, 2005, Hurricane Katrina plowed into the Gulf Coast of the United States, bringing unprecedented destruction to southern Louisiana, Mississippi, and Alabama.

The kinetic energy of an object varies jointly as the weight of the object at sea level and as the square of its velocity. During a hurricane, a $\frac{1}{2}$-lb stone traveling at 60 mph has 81 J (joules) of kinetic energy. Suppose the wind speed doubles to 120 mph. Find the kinetic energy.

Solution:

Let E represent the kinetic energy, let w represent the weight, and let v represent the velocity of the stone. The variation model is

$$E = kwv^2$$

$$81 = k(0.5)(60)^2 \qquad \text{Substitute } E = 81 \text{ J}, w = 0.5 \text{ lb, and } v = 60 \text{ mph.}$$

$$81 = k(0.5)(3600) \qquad \text{Simplify exponents.}$$

$$81 = k(1800)$$

Skill Practice Answers

8. 4%

$$\frac{81}{1800} = \frac{k(1800)}{1800} \quad \text{Divide by 1800.}$$

$$0.045 = k \quad \text{Solve for } k.$$

With the value of k known, the model $E = kwv^2$ can now be written as $E = 0.045wv^2$. We now find the kinetic energy of a $\frac{1}{2}$-lb stone traveling at 120 mph.

$$E = 0.045(0.5)(120)^2$$
$$= 324$$

The kinetic energy of a $\frac{1}{2}$-lb stone traveling at 120 mph is 324 J.

Skill Practice

9. The amount of simple interest earned in an account varies jointly as the interest rate and time of the investment. An account earns \$72 dollars in 4 years at 2% interest. How much interest would be earned in 3 years at a rate of 5%?

In Example 6, when the velocity increased by a factor of 2, the kinetic energy increased by a factor of 4 (note that 324 J = 4 · 81 J). This factor of 4 occurs because the kinetic energy is proportional to the *square* of the velocity. When the velocity increased by 2 times, the kinetic energy increased by 2^2 times.

Skill Practice Answers
9. \$135

Section 7.8 Practice Exercises

Boost *your* GRADE at mathzone.com!

- Practice Problems
- Self-Tests
- NetTutor
- e-Professors
- Videos

Study Skills Exercise

1. Define the key terms:

a. direct variation **b. inverse variation** **c. joint variation**

Review Exercises

For Exercises 2–7, perform the indicated operation, or solve the equation.

2. $\dfrac{5p}{p+2} + \dfrac{10}{p+2}$

3. $\dfrac{2y}{3} - \dfrac{3y-1}{5} = 1$

4. $\dfrac{3}{q-1} \cdot \dfrac{2q^2+3q-5}{6q+24}$

5. $\dfrac{a}{4} + \dfrac{3}{a} = 2$

6. $\dfrac{3}{b^2+5b-14} - \dfrac{2}{b^2-49}$

7. $\dfrac{a+\dfrac{a}{b}}{\dfrac{a}{b}-a}$

Concept 1: Definition of Direct and Inverse Variation

8. Suppose y varies directly as x, and $k > 0$.

a. If x increases, then y will (increase or decrease).

b. If x decreases, then y will (increase or decrease).

9. Suppose y varies inversely as x, and $k > 0$.

a. If x increases, then y will (increase or decrease).

b. If x decreases, then y will (increase or decrease).

Concept 2: Translations Involving Variation

For Exercises 10–17, write a variation model. Use k as the constant of variation.

10. T varies directly as q.

11. P varies inversely as r.

12. W varies inversely as the square of p.

13. Y varies directly as the square root of z.

14. Q is directly proportional to x and inversely proportional to the cube of y.

15. M is directly proportional to the square of p and inversely proportional to the cube of n.

16. L varies jointly as w and the square root of v.

17. X varies jointly as w and the square of y.

For Exercises 18–23, find the constant of variation, k.

18. y varies directly as x and when x is 4, y is 18.

19. m varies directly as x and when x is 8, m is 22.

20. p varies inversely as q and when q is 16, p is 32.

21. T varies inversely as x and when x is 40, T is 200.

22. y varies jointly as w and v. When w is 50 and v is 0.1, y is 8.75.

23. N varies jointly as t and p. When t is 1 and p is 7.5, N is 330.

Concept 3: Applications of Variation

Solve Exercises 24–29 using the steps found on page 587.

24. Z varies directly as the square of w. $Z = 14$ when $w = 4$. Find Z when $w = 8$.

25. Q varies inversely as the square of p. $Q = 4$ when $p = 3$. Find Q when $p = 2$.

26. L varies jointly as a and the square root of b. $L = 72$ when $a = 8$ and $b = 9$. Find L when $a = \frac{1}{2}$ and $b = 36$.

27. Y varies jointly as the cube of x and the square root of w. $Y = 128$ when $x = 2$ and $w = 16$. Find Y when $x = \frac{1}{2}$ and $w = 64$.

28. B varies directly as m and inversely as n. $B = 20$ when $m = 10$ and $n = 3$. Find B when $m = 15$ and $n = 12$.

29. R varies directly as s and inversely as t. $R = 14$ when $s = 2$ and $t = 9$. Find R when $s = 4$ and $t = 3$.

For Exercises 30–41, use a variation model to solve for the unknown value.

30. The amount of pollution entering the atmosphere varies directly as the number of people living in an area. If 80,000 people cause 56,800 tons of pollutants, how many tons enter the atmosphere in a city with a population of 500,000?

31. The area of a picture projected on a wall varies directly as the square of the distance from the projector to the wall. If a 10-ft distance produces a 16-ft^2 picture, what is the area of a picture produced when the projection unit is moved to a distance 20 ft from the wall?

32. The stopping distance of a car varies directly as the square of the speed of the car. If a car traveling at 40 mph has a stopping distance of 109 ft, find the stopping distance of a car that is traveling at 25 mph. (Round your answer to one decimal place.)

33. The intensity of a light source varies inversely as the square of the distance from the source. If the intensity is 48 lumens at a distance of 5 ft, what is the intensity when the distance is 8 ft?

34. The current in a wire varies directly as the voltage and inversely as the resistance. If the current is 9 A (amperes) when the voltage is 90 V (volts) and the resistance is 10 Ω (ohms), find the current when the voltage is 185 V and the resistance is 10 Ω.

35. The power in an electric circuit varies jointly as the current and the square of the resistance. If the power is 144 W (watts) when the current is 4 A and the resistance is 6 Ω, find the power when the current is 3 A and the resistance is 10 Ω.

36. The resistance of a wire varies directly as its length and inversely as the square of its diameter. A 40-ft wire 0.1 in. in diameter has a resistance of 4 Ω. What is the resistance of a 50-ft wire with a diameter of 0.2 in.?

37. The frequency of a vibrating string varies inversely as its length. A 24-in. piano string vibrates at 252 cycles/sec. What would be the frequency of an 18-in. piano string?

38. The weight of a medicine ball varies directly as the cube of its radius. A ball with a radius of 3 in. weighs 4.32 lb. How much would a medicine ball weigh if its radius is 5 in.?

39. The surface area of a cube varies directly as the square of the length of an edge. The surface area is 24 ft^2 when the length of an edge is 2 ft. Find the surface area of a cube with an edge that is 5 ft.

40. The strength of a wooden beam varies jointly as the width of the beam and the square of the thickness of the beam and inversely as the length of the beam. A beam that is 48 in. long, 6 in. wide, and 2 in. thick can support a load of 417 lb. Find the maximum load that can be safely supported by a board that is 12 in. wide, 72 in. long, and 4 in. thick.

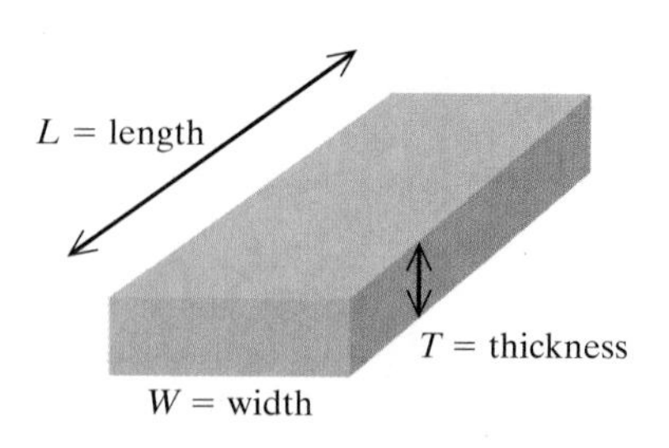

41. The period of a pendulum is the length of time required to complete one swing back and forth. The period varies directly as the square root of the length of the pendulum. If it takes 1.8 sec for a 0.81-m pendulum to complete one period, what is the period of a 1-m pendulum?

Expanding Your Skills

42. The area, A, of a square varies directly as the square of the length, l, of its sides.

a. Write a general variation model with k as the constant of variation.

b. If the length of the sides is doubled, what effect will that have on the area?

c. If the length of a side is tripled, what effect will that have on the area?

43. In a physics laboratory, a spring is fixed to the ceiling. With no weight attached to the end of the spring, the spring is said to be in its equilibrium position. As weights are applied to the end of the spring, the force stretches the spring a distance, d, from its equilibrium position. A student in the laboratory collects the following data:

Force, F (lb)	2	4	6	8	10
Distance, d (cm)	2.5	5.0	7.5	10.0	12.5

a. Based on the data, do you suspect a direct relationship between force and distance or an inverse relationship?

b. Find a variation model that describes the relationship between force and distance.

Chapter 7 SUMMARY

Section 7.1 Introduction to Rational Expressions

Key Concepts

A **rational expression** is a ratio of the form $\frac{p}{q}$ where p and q are polynomials and $q \neq 0$.

The **domain** of an algebraic expression is the set of real numbers that when substituted for the variable makes the expression result in a real number. For a rational expression, the domain is all real numbers except those that make the denominator zero.

Simplifying a Rational Expression to Lowest Terms

Factor the numerator and denominator completely, and reduce factors whose ratio is equal to 1 or to -1. A rational expression written in lowest terms will still have the same restrictions on the domain as the original expression.

Examples

Example 1

$\dfrac{x+2}{x^2-5x-14}$ is a rational expression.

Example 2

To find the domain of $\dfrac{x+2}{x^2-5x-14}$ factor the denominator: $\dfrac{x+2}{(x+2)(x-7)}$

The domain is $\{x \mid x \text{ is a real number and } x \neq -2, x \neq 7\}$.

Example 3

Simplify to lowest terms. $\dfrac{x+2}{x^2-5x-14}$

$\dfrac{x+2}{(x+2)(x-7)}$ Simplify.

$= \dfrac{1}{x-7}$ (provided $x \neq 7, x \neq -2$).

Section 7.2 Multiplication and Division of Rational Expressions

Key Concepts

Multiplying Rational Expressions

Factor the numerator and denominator completely. Then reduce factors whose ratio is 1 or -1.

Examples

Example 1

Multiply. $\dfrac{b^2-a^2}{a^2-2ab+b^2} \cdot \dfrac{a^2-3ab+2b^2}{2a+2b}$

$$= \frac{(b-a)(b+a)}{(a-b)(a-b)} \cdot \frac{(a-2b)(a-b)}{2(a+b)}$$

$$= -\frac{a-2b}{2} \quad \text{or} \quad \frac{2b-a}{2}$$

Dividing Rational Expressions

Multiply the first expression by the reciprocal of the second expression. That is, for $q \neq 0$, $r \neq 0$, and $s \neq 0$,

$$\frac{p}{q} \div \frac{r}{s} = \frac{p}{q} \cdot \frac{s}{r}$$

Example 2

Divide. $\dfrac{2c^2d^5}{15e^4} \div \dfrac{6c^4d^3}{20e}$

$$= \frac{2c^2d^5}{15e^4} \cdot \frac{20e}{6c^4d^3}$$

$$= \frac{40c^2d^5e}{90c^4d^3e^4}$$

$$= \frac{4d^2}{9c^2e^3}$$

Section 7.3 Least Common Denominator

Key Concepts

Converting a Rational Expression to an Equivalent Expression with a Different Denominator

Multiply numerator and denominator of the rational expression by the missing factors necessary to create the desired denominator.

Examples

Example 1

Convert $\dfrac{-3}{x-2}$ to an equivalent expression with the indicated denominator:

$$\frac{-3}{x-2} = \frac{}{5x^2 - 20}$$

$$\frac{-3}{x-2} = \frac{}{5(x^2 - 4)} \qquad \text{Factor.}$$

$$\frac{-3}{x-2} = \frac{}{5(x-2)(x+2)}$$

Multiply numerator and denominator by the missing factors from the denominator.

$$\frac{-3 \cdot 5(x+2)}{(x-2) \cdot 5(x+2)} = \frac{-15x - 30}{5(x-2)(x+2)}$$

Finding the Least Common Denominator (LCD) of Two or More Rational Expressions

1. Factor all denominators completely.
2. The LCD is the product of unique factors from the denominators, where each factor is raised to its highest power.

Example 2

Identify the LCD. $\dfrac{1}{8x^3y^2z}; \dfrac{5}{6xy^4}$

1. Write the denominators as a product of prime factors:

   $$\frac{1}{2^3x^3y^2z}; \frac{5}{2 \cdot 3xy^4}$$

2. The LCD is $2^3 3x^3y^4z$ or $24x^3y^4z$

Section 7.4 Addition and Subtraction of Rational Expressions

Key Concepts

To add or subtract rational expressions, the expressions must have the same denominator.

Steps to Add or Subtract Rational Expressions

1. Factor the denominators of each rational expression.
2. Identify the LCD.
3. Rewrite each rational expression as an equivalent expression with the LCD as its denominator.
4. Add or subtract the numerators, and write the result over the common denominator.
5. Simplify.

Examples

Example 1

Subtract. $\dfrac{c}{c^2 - c - 12} - \dfrac{1}{2c - 8}$

$$= \frac{c}{(c-4)(c+3)} - \frac{1}{2(c-4)}$$

The LCD is $2(c - 4)(c + 3)$.

$$= \frac{2c}{2(c-4)(c+3)} - \frac{1(c+3)}{2(c-4)(c+3)}$$

$$= \frac{2c - (c+3)}{2(c-4)(c+3)}$$

$$= \frac{2c - c - 3}{2(c-4)(c+3)} = \frac{c-3}{2(c-4)(c+3)}$$

Section 7.5 Complex Fractions

Key Concepts

Complex fractions can be simplified by using Method I or Method II.

Method I

1. Simplify the numerator and denominator of the complex fraction separately to form a single fraction in the numerator and a single fraction in the denominator.
2. Perform the division represented by the complex fraction. (Multiply the numerator of the complex fraction by the reciprocal of the denominator of the complex fraction.)
3. Simplify to lowest terms, if possible.

Examples

Example 1

Simplify. $\dfrac{1 - \dfrac{4}{w^2}}{1 - \dfrac{1}{w} - \dfrac{6}{w^2}} = \dfrac{\dfrac{w^2}{w^2} - \dfrac{4}{w^2}}{\dfrac{w^2}{w^2} - \dfrac{w}{w^2} - \dfrac{6}{w^2}}$

$$= \frac{\dfrac{w^2 - 4}{w^2}}{\dfrac{w^2 - w - 6}{w^2}} = \frac{w^2 - 4}{w^2} \cdot \frac{w^2}{w^2 - w - 6}$$

$$= \frac{(w-2)\cancel{(w+2)}}{\cancel{w^2}} \cdot \frac{\cancel{w^2}}{(w-3)\cancel{(w+2)}}$$

$$= \frac{w-2}{w-3}$$

Method II

1. Multiply the numerator and denominator of the complex fraction by the LCD of all individual fractions within the expression.
2. Apply the distributive property, and simplify the result.
3. Simplify to lowest terms, if necessary.

Example 2

Simplify. $$\frac{1 - \frac{4}{w^2}}{1 - \frac{1}{w} - \frac{6}{w^2}} = \frac{w^2\left(1 - \frac{4}{w^2}\right)}{w^2\left(1 - \frac{1}{w} - \frac{6}{w^2}\right)}$$

$$= \frac{w^2 - 4}{w^2 - w - 6} = \frac{(w-2)\cancel{(w+2)}}{(w-3)\cancel{(w+2)}}$$

$$= \frac{w-2}{w-3}$$

Section 7.6 Rational Equations

Key Concepts

An equation with one or more rational expressions is called a **rational equation**.

Steps to Solve a Rational Equation

1. Factor the denominators of all rational expressions.
2. Identify the LCD of all expressions in the equation.
3. Multiply both sides of the equation by the LCD.
4. Solve the resulting equation.
5. Check each potential solution in the original equation.

Examples

Example 1

Solve. $$\frac{1}{w} - \frac{1}{2w-1} = \frac{-2w}{2w-1}$$

The LCD is $w(2w-1)$.

$$\cancel{w}(2w-1)\frac{1}{\cancel{w}} - w\cancel{(2w-1)}\frac{1}{\cancel{2w-1}} = w\cancel{(2w-1)}\frac{-2w}{\cancel{2w-1}}$$

$$(2w-1)(1) - w(1) = w(-2w)$$

$$2w - 1 - w = -2w^2 \quad \text{Quadratic equation}$$

$$2w^2 + w - 1 = 0$$

$$(2w-1)(w+1) = 0$$

$w \cancel{=} \frac{1}{2}$ Does not check. or $w = -1$ Checks.

Example 2

Solve for I. $$q = \frac{VQ}{I}$$

$$I \cdot q = \frac{VQ}{\cancel{I}} \cdot \cancel{I}$$

$$Iq = VQ$$

$$I = \frac{VQ}{q}$$

Section 7.7 Applications of Rational Equations and Proportions

Key Concepts

Solving Proportions

An equation that equates two rates or ratios is called a **proportion**:

$$\frac{a}{b} = \frac{c}{d} \quad (b \neq 0, d \neq 0)$$

To solve a proportion, multiply both sides of the equation by the LCD.

Examples 2 and 3 give applications of rational equations.

1. Applications involving $d = rt$ (distance = rate · time) and work.

Example 2

Two cars travel from Los Angeles to Las Vegas. One car travels an average of 8 mph faster than the other car. If the faster car travels 189 miles in the same time as the slower car travels 165 miles, what is the average speed of each car?

Let r represent the speed of the slower car.
Let $r + 8$ represent the speed of the faster car.

	Distance	Rate	Time
Slower car	165	r	$\frac{165}{r}$
Faster car	189	$r + 8$	$\frac{189}{r+8}$

$$\frac{165}{r} = \frac{189}{r+8}$$

$$165(r + 8) = 189r$$

$$165r + 1320 = 189r$$

$$1320 = 24r$$

$$55 = r$$

The slower car travels 55 mph, and the faster car travels $55 + 8 = 63$ mph.

Examples

Example 1

A 90-g serving of a particular ice cream contains 10 g of fat. How much fat does 400 g of the same ice cream contain?

$$\frac{10 \text{ g fat}}{90 \text{ g ice cream}} = \frac{x \text{ grams fat}}{400 \text{ g ice cream}}$$

$$\frac{10}{90} = \frac{x}{400}$$

$$3600 \cdot \left(\frac{10}{90}\right) = \left(\frac{x}{400}\right) \cdot 3600$$

$$400 = 9x$$

$$x = \frac{400}{9} \approx 44.4 \text{ g}$$

Example 3

Beth and Cecelia have a house cleaning business. Beth can clean a particular house in 5 hr by herself. Cecelia can clean the same house in 4 hr. How long would it take if they cleaned the house together?

Let x be the number of hours it takes for both Beth and Cecelia to clean the house.

Beth can clean $\frac{1}{5}$ of the house in an hour and $\frac{1}{5}x$ of the house in x hours.

Cecelia can clean $\frac{1}{4}$ of the house in an hour and $\frac{1}{4}x$ of the house in x hours.

$\frac{1}{5}x + \frac{1}{4}x = 1$ Together they clean one whole house.

$$20\left(\frac{1}{5}x + \frac{1}{4}x\right) = (1)20$$

$$4x + 5x = 20$$

$$9x = 20$$

$x = \frac{20}{9}$, or $2\frac{2}{9}$ hr working together.

Section 7.8 Variation (Optional)

Key Concepts

Direct Variation:

y varies directly as x.
y is directly proportional to x. } $y = kx$

Inverse Variation:

y varies inversely as x.
y is inversely proportional to x. } $y = \frac{k}{x}$

Joint Variation:

y varies jointly as w and z.
y is jointly proportional to w and z. } $y = kwz$

Steps to Find a Variation Model:

1. Write a general variation model that relates the variables given in the problem. Let k represent the constant of variation.
2. Solve for k by substituting known values of the variables into the model from step 1.
3. Substitute the value of k into the original variation model from step 1.

Examples

Example 1

t varies directly as the square root of x.

$t = k\sqrt{x}$

Example 2

W is inversely proportional to the cube of x.

$W = \frac{k}{x^3}$

Example 3

y is jointly proportional to x and the square of z.

$y = kxz^2$

Example 4

C varies directly as the square root of d and inversely as t. If $C = 12$ when d is 9 and t is 6, find C if d is 16 and t is 12.

Step 1. $C = \frac{k\sqrt{d}}{t}$

Step 2. $12 = \frac{k\sqrt{9}}{6} \Rightarrow 12 = \frac{k \cdot 3}{6} \Rightarrow k = 24$

Step 3. $C = \frac{24\sqrt{d}}{t} \Rightarrow C = \frac{24\sqrt{16}}{12} \Rightarrow C = 8$

Chapter 7 Review Exercises

Section 7.1

1. For the rational expression $\frac{t-2}{t+9}$

   a. Evaluate the expression (if possible) for $t = 0, 1, 2, -3, -9$

   b. Write the domain of the expression in set-builder notation.

2. For the rational expression $\frac{k+1}{k-5}$

   a. Evaluate the expression for $k = 0, 1, 5, -1, -2$

   b. Write the domain of the expression in set-builder notation.

3. Which of the rational expressions are equal to -1 for all values of x for which the expressions are defined?

   a. $\frac{2-x}{x-2}$

   b. $\frac{x-5}{x+5}$

   c. $\frac{-x-7}{x+7}$

   d. $\frac{x^2-4}{4-x^2}$

For Exercises 4–13, write the domain in set-builder notation. Then simplify the expressions to lowest terms.

4. $\dfrac{x-3}{(2x-5)(x-3)}$

5. $\dfrac{h+7}{(3h+1)(h+7)}$

6. $\dfrac{4a^2+7a-2}{a^2-4}$

7. $\dfrac{2w^2+11w+12}{w^2-16}$

8. $\dfrac{z^2-4z}{8-2z}$

9. $\dfrac{15-3k}{2k^2-10k}$

10. $\dfrac{2b^2+4b-6}{4b+12}$

11. $\dfrac{3m^2-12m-15}{9m+9}$

12. $\dfrac{n+3}{n^2+6n+9}$

13. $\dfrac{p+7}{p^2+14p+49}$

Section 7.2

For Exercises 14–27, multiply or divide as indicated.

14. $\dfrac{3y^3}{3y-6}\cdot\dfrac{y-2}{y}$

15. $\dfrac{2u+10}{u}\cdot\dfrac{u^3}{4u+20}$

16. $\dfrac{11}{v-2}\cdot\dfrac{2v^2-8}{22}$

17. $\dfrac{8}{x^2-25}\cdot\dfrac{3x+15}{16}$

18. $\dfrac{4c^2+4c}{c^2-25}\div\dfrac{8c}{c^2-5c}$

19. $\dfrac{q^2-5q+6}{2q+4}\div\dfrac{2q-6}{q+2}$

20. $\left(\dfrac{-2t}{t+1}\right)(t^2-4t-5)$

21. $(s^2-6s+8)\left(\dfrac{4s}{s-2}\right)$

22. $\dfrac{\dfrac{a^2+5a+1}{7a-7}}{\dfrac{a^2+5a+1}{a-1}}$

23. $\dfrac{\dfrac{n^2+n+1}{n^2-4}}{\dfrac{n^2+n+1}{n+2}}$

24. $\dfrac{5h^2-6h+1}{h^2-1}\div\dfrac{16h^2-9}{4h^2+7h+3}\cdot\dfrac{3-4h}{30h-6}$

25. $\dfrac{3m-3}{6m^2+18m+12}\cdot\dfrac{2m^2-8}{m^2-3m+2}\div\dfrac{m+3}{m+1}$

26. $\dfrac{x-2}{x^2-3x-18}\cdot\dfrac{6-x}{x^2-4}$

27. $\dfrac{4y^2-1}{1+2y}\div\dfrac{y^2-4y-5}{5-y}$

Section 7.3

For Exercises 28–33, convert each expression into an equivalent expression with the indicated denominator.

28. $\dfrac{x+1}{x-2}=\dfrac{\quad}{5x-10}$

29. $\dfrac{y+2}{y-3}=\dfrac{\quad}{2y-6}$

30. $\dfrac{6}{w}=\dfrac{\quad}{w^2-4w}$

31. $\dfrac{2}{r}=\dfrac{\quad}{r^2+3r}$

32. $\dfrac{s-2}{s+4}=\dfrac{\quad}{s^2-16}$

33. $\dfrac{u+1}{u+6}=\dfrac{\quad}{u^2-36}$

For Exercises 34–41, identify the LCD.

34. $\dfrac{2}{a^2bc^2};\dfrac{5}{ab^3}$

35. $\dfrac{6x}{y^2z};\dfrac{3}{xy^2z^4}$

36. $\dfrac{5}{p+2};\dfrac{p}{p-4}$

37. $\dfrac{6}{q};\dfrac{1}{q+8}$

38. $\dfrac{8}{m^2-16};\dfrac{7}{m^2-m-12}$

39. $\dfrac{6}{n^2-9};\dfrac{5}{n^2-n-6}$

40. $\dfrac{4}{2t-5};\dfrac{5}{5-2t}$

41. $\dfrac{-2}{3k-1};\dfrac{6}{1-3k}$

42. State two possible LCDs that could be used to add the fractions.

$$\frac{7}{c-2}+\frac{4}{2-c}$$

43. State two possible LCDs that could be used to subtract the fractions.

$$\frac{10}{3-x}-\frac{5}{x-3}$$

Section 7.4

For Exercises 44–55, add or subtract as indicated.

44. $\dfrac{h+3}{h+1}+\dfrac{h-1}{h+1}$

45. $\dfrac{b-6}{b-2}+\dfrac{b+2}{b-2}$

46. $\dfrac{a^2}{a-5}-\dfrac{25}{a-5}$

47. $\dfrac{x^2}{x+7}-\dfrac{49}{x+7}$

48. $\dfrac{y}{y^2-81}+\dfrac{2}{9-y}$

49. $\dfrac{3}{4-t^2}+\dfrac{t}{2-t}$

50. $\dfrac{4}{3m}-\dfrac{1}{m+2}$

51. $\dfrac{5}{2r+12}-\dfrac{1}{r}$

52. $\dfrac{4p}{p^2+6p+5}-\dfrac{3p}{p^2+5p+4}$

53. $\dfrac{3q}{q^2+7q+10}-\dfrac{2q}{q^2+6q+8}$

54. $\dfrac{1}{h}+\dfrac{h}{2h+4}-\dfrac{2}{h^2+2h}$

55. $\dfrac{x}{3x+9}-\dfrac{3}{x^2+3x}+\dfrac{1}{x}$

Section 7.5

For Exercises 56–63, simplify the complex fractions.

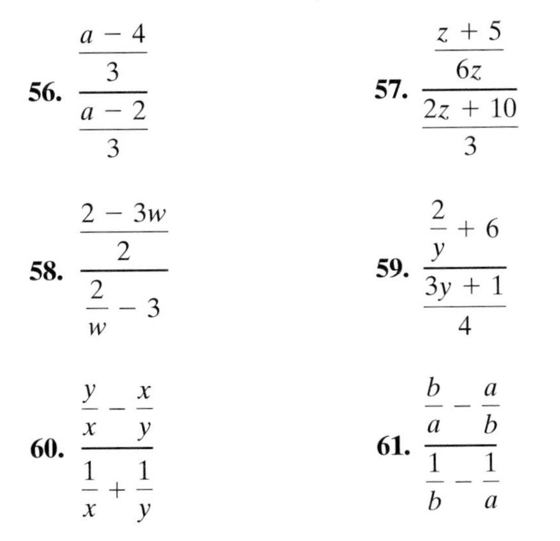

56. $\dfrac{\frac{a-4}{3}}{\frac{a-2}{3}}$

57. $\dfrac{\frac{z+5}{6z}}{\frac{2z+10}{3}}$

58. $\dfrac{\frac{2-3w}{2}}{\frac{2}{w}-3}$

59. $\dfrac{\frac{2}{y}+6}{\frac{3y+1}{4}}$

60. $\dfrac{\frac{y}{x}-\frac{x}{y}}{\frac{1}{x}+\frac{1}{y}}$

61. $\dfrac{\frac{b}{a}-\frac{a}{b}}{\frac{1}{b}-\frac{1}{a}}$

62. $\dfrac{\frac{6}{p+2}+4}{\frac{8}{p+2}-4}$

63. $\dfrac{\frac{25}{k+5}+5}{\frac{5}{k+5}-5}$

Section 7.6

For Exercises 64–71, solve the equations.

64. $\dfrac{2}{x}+\dfrac{1}{2}=\dfrac{1}{4}$

65. $\dfrac{1}{y}+\dfrac{3}{4}=\dfrac{1}{4}$

66. $\dfrac{2}{h-2}+1=\dfrac{h}{h+2}$

67. $\dfrac{w}{w-1}=\dfrac{3}{w+1}+1$

68. $\dfrac{t+1}{3}-\dfrac{t-1}{6}=\dfrac{1}{6}$

69. $\dfrac{4p-4}{p^2+5p-14}+\dfrac{2}{p+7}=\dfrac{1}{p-2}$

70. $\dfrac{y+1}{y+3}=\dfrac{y^2-11y}{y^2+y-6}-\dfrac{y-3}{y-2}$

71. $\dfrac{1}{z+2}=\dfrac{4}{z^2-4}-\dfrac{1}{z-2}$

72. Four times a number is added to 5. The sum is then divided by 6. The result is $\frac{7}{2}$. Find the number.

73. Solve the formula $\dfrac{V}{h}=\dfrac{\pi r^2}{3}$ for h.

74. Solve the formula $\dfrac{A}{b}=\dfrac{h}{2}$ for b.

Section 7.7

For Exercises 75–76, solve the proportions.

75. $\dfrac{m+2}{8}=\dfrac{m}{3}$

76. $\dfrac{12}{a}=\dfrac{5}{8}$

77. A bag of popcorn states that it contains 4 g of fat per serving. If a serving is 2 oz, how many grams of fat are in a 5-oz bag?

78. Bud goes 10 mph faster on his Harley Davidson motorcycle than Ed goes on his Honda motorcycle. If Bud travels 105 miles in the same time that Ed travels 90 miles, what are the rates of the two bikers?

79. There are two pumps set up to fill a small swimming pool. One pump takes 24 min by itself to fill the pool, but the other takes 56 min by itself. How long would it take if both pumps work together?

80. Consider the similar triangles shown here. Find the values of x and y.

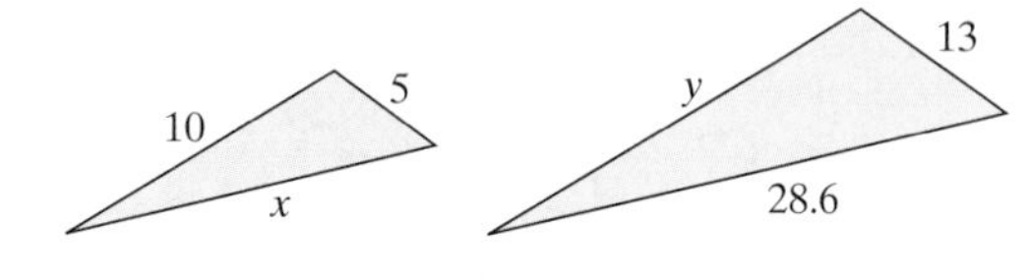

Section 7.8

81. The force applied to a spring varies directly with the distance that the spring is stretched. When 6 lb of force is applied, the spring stretches 2 ft.

a. Write a variation model using k as the constant of variation.

b. Find k.

c. How many pounds of force is required to stretch the spring $1\frac{2}{3}$ ft?

82. Suppose y varies inversely with the cube of x, and $y = 9$ when $x = 2$. Find y when $x = 3$.

83. Suppose y varies jointly with x and the square root of z, and $y = 3$ when $x = 3$ and $z = 4$. Find y when $x = 8$ and $z = 9$.

84. The distance, d, that one can see to the horizon varies directly as the square root of the height above sea level. If a person 25 m above sea level can see 30 km, how far can a person see if she is 64 m above sea level?

Chapter 7 Test

For Exercises 1–2,

a. Write the domain in set-builder notation.

b. Reduce the rational expression to lowest terms.

1. $\dfrac{5(x-2)(x+1)}{30(2-x)}$

2. $\dfrac{7a^2 - 42a}{a^3 - 4a^2 - 12a}$

3. Identify the rational expressions that are equal to -1 for all values of x for which the expression is defined.

a. $\dfrac{x+4}{x-4}$

b. $\dfrac{7-2x}{2x-7}$

c. $\dfrac{9x^2+16}{-9x^2-16}$

d. $-\dfrac{x+5}{x+5}$

For Exercises 4–9, perform the indicated operation.

4. $\dfrac{2}{y^2 + 4y + 3} + \dfrac{1}{3y + 9}$

5. $\dfrac{9 - b^2}{5b + 15} \div \dfrac{b - 3}{b + 3} \div (9b^2 + 28b + 3)$

6. $\dfrac{w^2 - 4w}{w^2 - 8w + 16} \cdot \dfrac{w - 4}{w^2 + w}$

7. $\dfrac{t}{t - 2} - \dfrac{8}{t^2 - 4}$

8. $\dfrac{1}{3x + 4} + \dfrac{2}{3x^2 - 2x - 8} + \dfrac{x}{x - 2}$

9. $\dfrac{1 - \dfrac{4}{m}}{m - \dfrac{16}{m}}$

For Exercises 10–13, solve the equation.

10. $\dfrac{3}{a} + \dfrac{5}{2} = \dfrac{7}{a}$

11. $\dfrac{p}{p - 1} - \dfrac{1}{p} = \dfrac{2p^2 - 1}{p^2 - p}$

12. $\dfrac{3}{c - 2} - \dfrac{1}{c + 1} = \dfrac{7}{c^2 - c - 2}$

13. $\dfrac{4x}{x - 4} = 3 + \dfrac{16}{x - 4}$

14. Solve the formula $\dfrac{C}{2} = \dfrac{A}{r}$ for r.

15. If $\frac{3}{2}$ is added to the reciprocal of a number the result is $\frac{2}{5}$ times the reciprocal of that number. Find the number.

16. Solve the proportion.

$$\frac{y + 7}{-4} = \frac{1}{4}$$

17. A recipe for vegetable soup calls for $\frac{1}{2}$ cup of carrots for six servings. How many cups of carrots are needed to prepare 15 servings?

18. A motorboat can travel 28 miles downstream in the same amount of time as it can travel 18 miles upstream. Find the speed of the current if the boat can travel 23 mph in still water.

19. One printer requires 3 hr to do a job and a second printer requires 6 hr to do the same job. If they worked together, how long would it take to complete the task?

20. Consider the similar triangles shown here. Find the values of a and b.

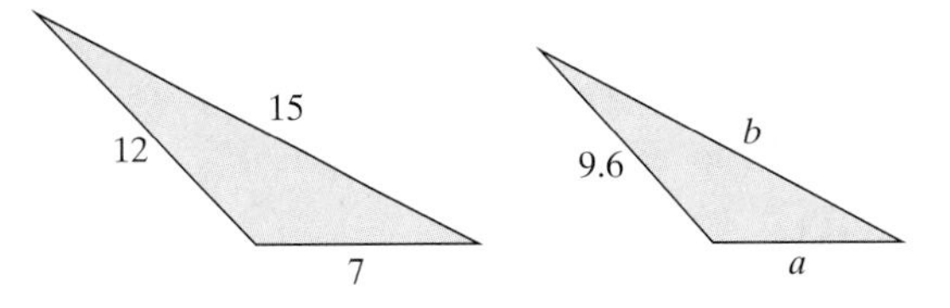

21. Find the LCD of the following pairs of rational expressions.

a. $\dfrac{x}{3(x + 3)}; \dfrac{7}{5(x + 3)}$ **b.** $\dfrac{-2}{3x^2y}; \dfrac{4}{xy^2}$

22. The period of a pendulum, P, varies directly as the square root of the length, L, of the pendulum. The period is 2.2 seconds when the length is 4 feet. Find the period when the length is 9 feet.

Chapters 1–7 Cumulative Review Exercises

For Exercises 1–2, simplify completely.

1. $\left(\dfrac{1}{2}\right)^{-4} + 2^4$

2. $|3 - 5| + |-2 + 7|$

3. Solve for y: $\dfrac{1}{2} - \dfrac{3}{4}(y - 1) = \dfrac{5}{12}$

4. Complete the table.

Set-Builder Notation	Graph	Interval Notation
$\{x \mid x \geq -1\}$	⟶	
	⟶	$(-\infty, 5)$

5. The perimeter of a rectangular swimming pool is 104 m. The length is 1 m more than twice the width. Find the length and width.

6. The height of a triangle is 2 in. less than the base. The area is 40 in.^{2} Find the base and height of the triangle.

7. Simplify. $\left(\dfrac{4x^{-1}y^{-2}}{z^4}\right)^{-2}(2y^{-1}z^3)^3$

8. The length and width of a rectangle are given in terms of x.

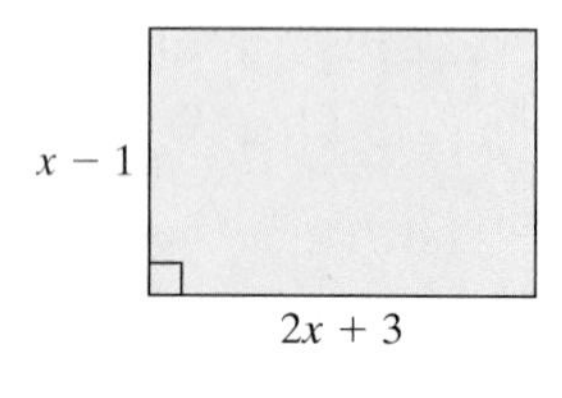

a. Write a polynomial that represents the perimeter of the rectangle.

b. Write a polynomial that represents the area of the rectangle.

9. Factor completely: $25x^2 - 30x + 9$

10. Factor. $10cd + 5d - 6c - 3$

11. Determine the domain of the expression.

$$\frac{x + 3}{(x - 5)(2x + 1)}$$

12. Simplify to lowest terms.

$$\frac{x^2 - 9}{x^2 + 8x + 15}$$

13. Divide. $\dfrac{2x - 6}{x^2 - 16} \div \dfrac{10x^2 - 90}{x^2 - x - 12}$

14. Simplify.

$$\frac{\frac{3}{4} - \frac{1}{x}}{\frac{1}{3x} - \frac{1}{4}}$$

15. Solve.

$$\frac{7}{y^2 - 4} = \frac{3}{y - 2} + \frac{2}{y + 2}$$

16. Solve the proportion.

$$\frac{2b - 5}{6} = \frac{4b}{7}$$

17. Determine the x- and y-intercepts.

a. $-2x + 4y = 8$ **b.** $y = 5x$

18. Determine the slope

a. of the line containing the points $(0, -6)$ and $(-5, 1)$

b. of the line $y = -\frac{2}{3}x - 6$

c. of a line parallel to a line having a slope of 4.

d. of a line perpendicular to a line having a slope of 4.

19. Find an equation of a line passing through the point $(1, 2)$ and having a slope of 5. Write the answer in slope-intercept form.

20. A group of teenagers buys 2 large popcorns and 6 drinks at the movie theater for \$16. A couple buys 1 large popcorn and 2 drinks for \$6.50. Find the price for 1 large popcorn and the price for 1 drink.

8

Radicals

Chapter 8 is devoted to the study of radicals and their applications. The techniques to add, subtract, multiply, and divide radical expressions are presented first. We also solve radical equations and applications involving proportions, rates of change, and the Pythagorean theorem.

The following table represents a magic square. The goal is to fill in numbers so that each row and column adds up to 30. While there are many possible solutions, we offer some clues for one possible solution.

Clues

2. The number, x, where $\sqrt{x} = 2$
3. The cube root of 512
6. The square of box 4
7. Beethoven's famous symphony
8. Valentine's Day
9. The square root and cube root of this number are the same.
10. The length of the hypotenuse for this triangle

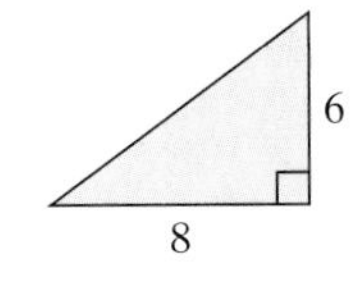

12. One day shy of two weeks
16. A number with only 1 square root

1	2	3	4
5	6	7	8
9	10	11	12
13	14	15	16

Section 8.1 Introduction to Roots and Radicals

Concepts

1. Definition of a Square Root
2. Definition of an *n*th-Root
3. Translations Involving *n*th-Roots
4. Pythagorean Theorem

1. Definition of a Square Root

Recall that to square a number means to multiply the number by itself: $b^2 = b \cdot b$. The reverse operation to squaring a number is to find its square roots. For example, finding a square root of 49 is equivalent to asking: "What number when squared equals 49?"

One obvious answer to this question is 7 because $(7)^2 = 49$. But -7 will also work because $(-7)^2 = 49$.

Definition of a Square Root

b is a **square root** of a if $b^2 = a$.

Example 1 Identifying the Square Roots of a Number

Identify the square roots of

a. 9 **b.** 121 **c.** 0 **d.** -4

Solution:

a. 3 is a square root of 9 because $(3)^2 = 9$.
-3 is a square root of 9 because $(-3)^2 = 9$.

b. 11 is a square root of 121 because $(11)^2 = 121$.
-11 is a square root of 121 because $(-11)^2 = 121$.

c. 0 is a square root of 0 because $(0)^2 = 0$.

d. There are no real numbers that when squared will equal a negative number. Therefore, there are no real-valued square roots of -4.

TIP: All positive real numbers have two real-valued square roots: one positive and one negative. Zero has only one square root, which is 0 itself. Finally, for any negative real number, there are no real-valued square roots.

Skill Practice Identify the square roots of the numbers.

1. 64 **2.** -36 **3.** 36 **4.** $\frac{25}{16}$

Recall from Section 1.2, that the positive square root of a real number can be denoted with a radical sign, $\sqrt{}$.

Notation for Positive and Negative Square Roots

Let a represent a positive real number. Then,

1. $\sqrt{a}$ is the **positive square root** of a. The positive square root is also called the **principal square root**.
2. $-\sqrt{a}$ is the **negative square root** of a.
3. $\sqrt{0} = 0$

Skill Practice Answers

1. 8; -8
2. There are no real-valued square roots.
3. 6; -6
4. $\frac{5}{4}; -\frac{5}{4}$

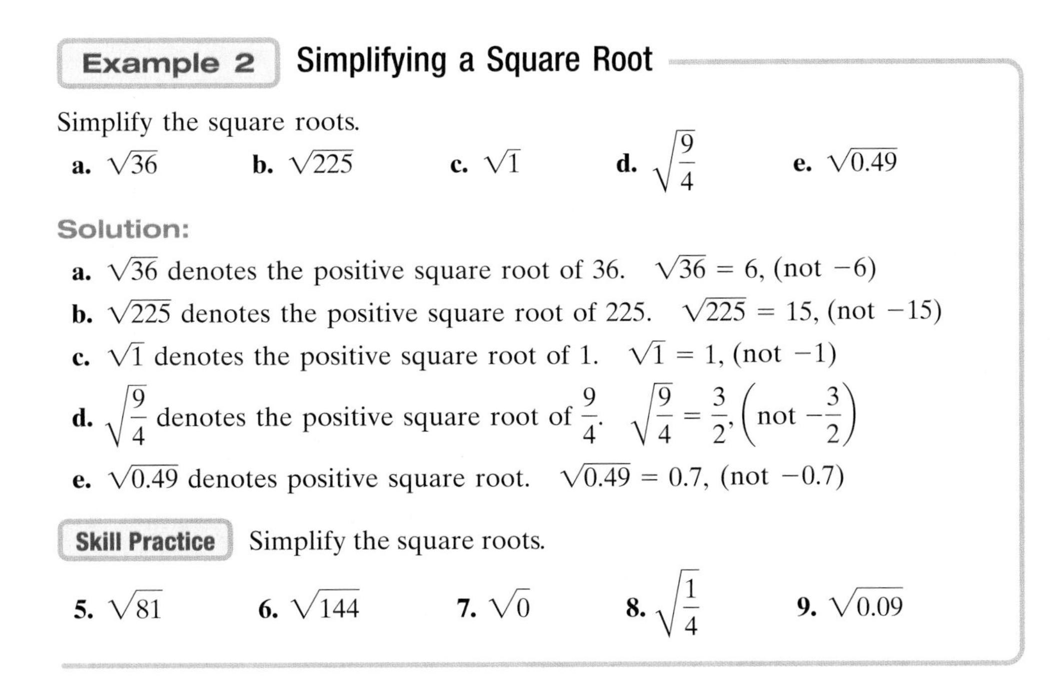

Example 2 Simplifying a Square Root

Simplify the square roots.

a. $\sqrt{36}$ **b.** $\sqrt{225}$ **c.** $\sqrt{1}$ **d.** $\sqrt{\frac{9}{4}}$ **e.** $\sqrt{0.49}$

Solution:

a. $\sqrt{36}$ denotes the positive square root of 36. $\sqrt{36} = 6$, (not -6)

b. $\sqrt{225}$ denotes the positive square root of 225. $\sqrt{225} = 15$, (not -15)

c. $\sqrt{1}$ denotes the positive square root of 1. $\sqrt{1} = 1$, (not -1)

d. $\sqrt{\frac{9}{4}}$ denotes the positive square root of $\frac{9}{4}$. $\sqrt{\frac{9}{4}} = \frac{3}{2}$, $\left(\text{not } -\frac{3}{2}\right)$

e. $\sqrt{0.49}$ denotes positive square root. $\sqrt{0.49} = 0.7$, (not -0.7)

Skill Practice Simplify the square roots.

5. $\sqrt{81}$ **6.** $\sqrt{144}$ **7.** $\sqrt{0}$ **8.** $\sqrt{\frac{1}{4}}$ **9.** $\sqrt{0.09}$

The numbers 36, 225, 1, and $\frac{9}{4}$ are **perfect squares** because their square roots are rational numbers. Radicals that cannot be simplified to rational numbers are irrational numbers. Recall that an irrational number cannot be written as a terminating or repeating decimal. For example, the symbol $\sqrt{13}$ is used to represent the exact value of the square root of 13. The symbol $\sqrt{42}$ is used to represent the exact value of the square root of 42. These values are irrational numbers but can be approximated by rational numbers by using a calculator.

$$\sqrt{13} \approx 3.605551275 \qquad \sqrt{42} \approx 6.480740698$$

Note: The only way to denote the *exact* values of the square root of 13 and the square root of 42 is $\sqrt{13}$ and $\sqrt{42}$.

TIP: Before using a calculator to evaluate a square root, try estimating the value first.

$\sqrt{13}$ must be a number between 3 and 4 because $\sqrt{9} < \sqrt{13} < \sqrt{16}$.

$\sqrt{42}$ must be a number between 6 and 7 because $\sqrt{36} < \sqrt{42} < \sqrt{49}$.

A negative number cannot have a real number as a square root because no real number when squared is negative. For example, $\sqrt{-25}$ is *not a real number* because there is no real number, b, for which $(b)^2 = -25$.

Example 3 Evaluating Square Roots if Possible

Simplify the square roots if possible.

a. $\sqrt{-100}$ **b.** $-\sqrt{100}$ **c.** $\sqrt{-64}$

Solution:

a. $\sqrt{-100}$ Not a real number

b. $-\sqrt{100}$

$-1 \cdot \sqrt{100}$ The expression $-\sqrt{100}$ is equivalent to $-1 \cdot \sqrt{100}$.

$-1 \cdot 10 = -10$

c. $\sqrt{-64}$ Not a real number

Skill Practice Simplify the square roots if possible.

10. $-\sqrt{25}$ **11.** $\sqrt{16}$ **12.** $\sqrt{-4}$

Skill Practice Answers

5. 9 **6.** 12 **7.** 0
8. $\frac{1}{2}$ **9.** 0.3 **10.** -5
11. 4 **12.** Not a real number

2. Definition of an *n*th-Root

Finding a square root of a number is the reverse process of squaring a number. This concept can be extended to finding a third root (called a cube root), a fourth root, and in general, an *n*th-root.

Definition of an *n*th-Root

b is an ***n*th-root** of a if $b^n = a$.

The radical sign, $\sqrt{\ }$, is used to denote the principal square root of a number. The symbol, $\sqrt[n]{\ }$, is used to denote the principal *n*th-root of a number.

In the expression $\sqrt[n]{a}$, n is called the **index** of the radical, and a is called the **radicand**. For a square root, the index is 2, but it is usually not written ($\sqrt[2]{a}$ is denoted simply as $\sqrt{a}$). A radical with an index of three is called a **cube root**, $\sqrt[3]{a}$.

Definition of $\sqrt[n]{a}$

1. If n is a positive *even* integer and $a > 0$, then $\sqrt[n]{a}$ is the principal (positive) *n*th-root of a.
2. If $n > 1$ is a positive *odd* integer, then $\sqrt[n]{a}$ is the *n*th-root of a.
3. If $n > 1$ is a positive integer, then $\sqrt[n]{0} = 0$.

For the purpose of simplifying radicals, it is helpful to know the following patterns:

Perfect cubes	Perfect fourth powers	Perfect fifth powers
$1^3 = 1$	$1^4 = 1$	$1^5 = 1$
$2^3 = 8$	$2^4 = 16$	$2^5 = 32$
$3^3 = 27$	$3^4 = 81$	$3^5 = 243$
$4^3 = 64$	$4^4 = 256$	$4^5 = 1024$
$5^3 = 125$	$5^4 = 625$	$5^5 = 3125$

Example 4 Simplifying *n*th-Roots

Simplify the expressions, if possible.

a. $\sqrt[3]{8}$ **b.** $\sqrt[4]{16}$ **c.** $\sqrt[5]{32}$ **d.** $\sqrt[3]{-64}$

e. $\sqrt[3]{\dfrac{125}{27}}$ **f.** $\sqrt{0.01}$ **g.** $\sqrt[4]{-81}$

Solution:

a. $\sqrt[3]{8} = 2$ Because $(2)^3 = 8$

b. $\sqrt[4]{16} = 2$ Because $(2)^4 = 16$

c. $\sqrt[5]{32} = 2$ Because $(2)^5 = 32$

d. $\sqrt[3]{-64} = -4$ Because $(-4)^3 = -64$

e. $\sqrt[3]{\dfrac{125}{27}} = \dfrac{5}{3}$ Because $\left(\dfrac{5}{3}\right)^3 = \dfrac{125}{27}$

f. $\sqrt{0.01} = 0.1$ Because $(0.1)^2 = 0.01$

Note: $\sqrt{0.01}$ is equivalent to $\sqrt{\frac{1}{100}} = \frac{1}{10}$, or 0.1.

g. $\sqrt[4]{-81}$ is not a real number because no real number raised to the fourth power equals -81.

Skill Practice Simplify the expressions if possible.

13. $\sqrt[3]{27}$ **14.** $\sqrt[4]{1}$ **15.** $\sqrt[3]{216}$ **16.** $\sqrt[5]{-32}$

17. $\sqrt[4]{\frac{16}{625}}$ **18.** $\sqrt{0.25}$ **19.** $\sqrt[4]{-1}$

Avoiding Mistakes:

When evaluating $\sqrt[n]{a}$, where n is *even,* always choose the principal (positive) root. Hence:

$\sqrt[4]{16} = 2$ (not -2)

$\sqrt{0.01} = 0.1$ (not -0.1)

Example 4(g) illustrates that an nth-root of a negative number is not a real number if the index is even because no real number raised to an even power is negative.

Finding an nth-root of a variable expression is similar to finding an nth-root of a numerical expression. However, for roots with an even index, particular care must be taken to obtain a nonnegative solution.

Definition of $\sqrt[n]{a^n}$

1. If n is a positive odd integer, then $\sqrt[n]{a^n} = a$
2. If n is a positive even integer, then $\sqrt[n]{a^n} = |a|$

The absolute value bars are necessary for roots with an even index because the variable, a, may represent a positive quantity or a negative quantity. By using absolute value bars, we ensure that $\sqrt[n]{a^n} = |a|$ is nonnegative and represents the principal nth-root of a.

Example 5 Simplifying Expressions of the Form $\sqrt[n]{a^n}$

Simplify the expressions.

a. $\sqrt{(-3)^2}$ **b.** $\sqrt{x^2}$ **c.** $\sqrt[3]{x^3}$ **d.** $\sqrt[4]{x^4}$ **e.** $\sqrt[5]{x^5}$

Solution:

a. $\sqrt{(-3)^2} = |-3| = 3$ Because the index is *even*, absolute value bars are necessary to make the answer nonnegative.

b. $\sqrt{x^2} = |x|$ Because the index is *even*, absolute value bars are necessary to make the answer nonnegative.

c. $\sqrt[3]{x^3} = x$ Because the index is *odd*, no absolute value bars are necessary.

d. $\sqrt[4]{x^4} = |x|$ Because the index is *even*, absolute value bars are necessary to make the answer nonnegative.

e. $\sqrt[5]{x^5} = x$ Because the index is *odd*, no absolute value bars are necessary.

Skill Practice Simplify.

20. $\sqrt{(-6)^2}$ **21.** $\sqrt[4]{a^4}$ **22.** $\sqrt[3]{w^3}$ **23.** $\sqrt[6]{p^6}$ **24.** $\sqrt[3]{(-2)^3}$

Skill Practice Answers

13. 3 **14.** 1 **15.** 6
16. -2 **17.** $\frac{2}{5}$ **18.** 0.5
19. Not a real number.
20. 6 **21.** $|a|$ **22.** w
23. $|p|$ **24.** -2

If n is an even integer, then $\sqrt[n]{a^n} = |a|$. However, if the variable a is assumed to be nonnegative, then the absolute value bars may be dropped, that is, $\sqrt[n]{a^n} = a$ provided $a \geq 0$. In many examples and exercises, we will make the assumption that the variables within a radical expression are positive real numbers. In such a case, the absolute value bars are not needed to evaluate $\sqrt[n]{a^n}$.

It is helpful to become familiar with the patterns associated with perfect squares and perfect cubes involving variable expressions.

TIP: Any expression raised to an even power (multiple of 2) is a perfect square.

The following powers of x are perfect squares:

Perfect squares

$(x^1)^2 = x^2$
$(x^2)^2 = x^4$
$(x^3)^2 = x^6$
$(x^4)^2 = x^8$
...

TIP: Any expression raised to a power that is a multiple of 3 is a perfect cube.

The following powers of x are perfect cubes:

Perfect cubes

$(x^1)^3 = x^3$
$(x^2)^3 = x^6$
$(x^3)^3 = x^9$
$(x^4)^3 = x^{12}$
...

Example 6 Simplifying *n*th-Roots

Simplify the expressions. Assume that the variables are positive real numbers.

a. $\sqrt{c^6}$ **b.** $\sqrt[3]{d^{15}}$ **c.** $\sqrt[4]{y^8}$ **d.** $\sqrt{a^2b^2}$ **e.** $\sqrt[3]{64z^6}$

Solution:

a. $\sqrt{c^6}$ The expression c^6 is a perfect square.
$\sqrt{c^6} = c^3$ Because $\sqrt{(c^3)^2} = c^3$

b. $\sqrt[3]{d^{15}}$ The expression d^{15} is a perfect cube.
$\sqrt[3]{d^{15}} = d^5$ Because $\sqrt[3]{(d^5)^3} = d^5$

c. $\sqrt[4]{y^8}$ The expression y^8 is a fourth power.
$\sqrt[4]{y^8} = y^2$ Because $\sqrt[4]{(y^2)^4} = y^2$

d. $\sqrt{a^2b^2} = ab$ Because $\sqrt{a^2b^2} = \sqrt{(ab)^2} = ab$

e. $\sqrt[3]{64z^6} = 4z^2$ Because $\sqrt[3]{(4z^2)^3} = 4z^2$

Skill Practice Simplify the expressions. Assume the variables represent positive real numbers.

25. $\sqrt{y^{10}}$ **26.** $\sqrt[3]{x^{12}}$ **27.** $\sqrt[4]{a^{28}}$ **28.** $\sqrt{x^4y^2}$ **29.** $\sqrt{25c^4}$

Skill Practice Answers

25. y^5 **26.** x^4 **27.** a^7
28. x^2y **29.** $5c^2$

3. Translations Involving *n*th-Roots

It is important to understand the vocabulary and language associated with *n*th-roots. For instance, you must be able to distinguish between "squaring a number" and "taking the square *root* of a number." The following example offers practice translating between English form and algebraic form.

Example 7 Translating from English Form to Algebraic Form

Translate each English phrase into an algebraic expression.

a. The difference of the square of x and the principal square root of seven

b. The quotient of one and the cube root of z

Solution:

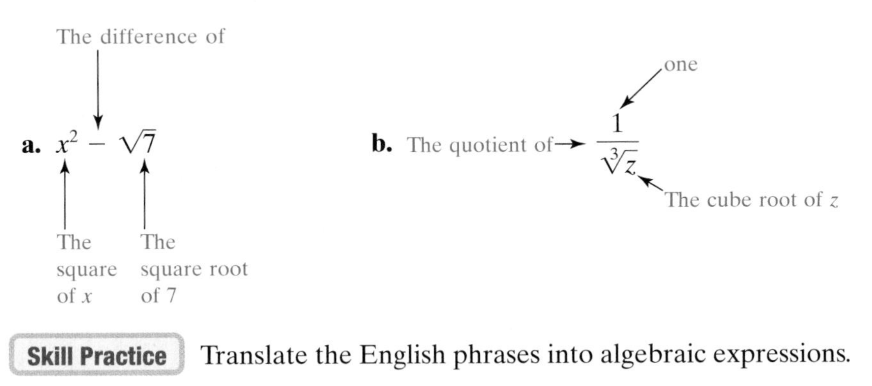

Skill Practice Translate the English phrases into algebraic expressions.

30. The product of the square of y and the principal square root of x

31. The sum of two and the cube root of y.

4. Pythagorean Theorem

Recall that the **Pythagorean theorem** relates the lengths of the three sides of a right triangle (Figure 8-1).

$$a^2 + b^2 = c^2$$

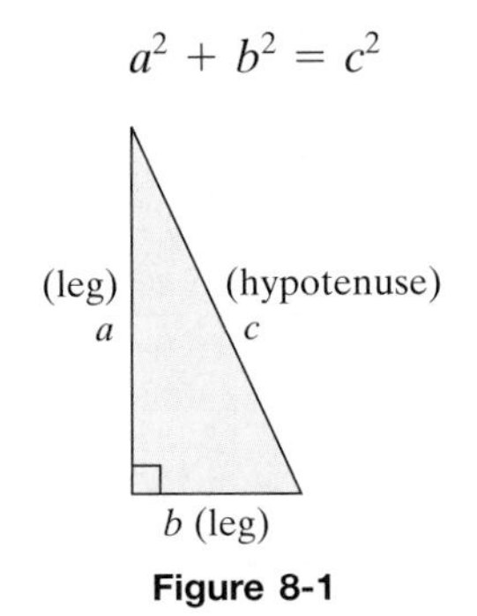

Figure 8-1

The principal square root can be used to solve for an unknown side of a right triangle if the lengths of the other two sides are known.

Skill Practice Answers

30. $y^2\sqrt{x}$ **31.** $2 + \sqrt[3]{y}$

Example 8 Applying the Pythagorean Theorem

Use the Pythagorean theorem and the definition of the principal square root of a number to find the length of the unknown side.

Solution:

Label the sides of the triangle.

$a^2 + b^2 = c^2$

$a^2 + (8)^2 = (10)^2$ Apply the Pythagorean theorem.

$a^2 + 64 = 100$ Simplify.

$a^2 = 100 - 64$

$a^2 = 36$ This equation is quadratic. One method for solving the equation is to set the equation equal to zero, factor, and apply the zero product rule. However, we can also use the definition of a square root to solve for a.

$a = \sqrt{36}$

$a = 6$ By definition, a must be one of the square roots of 36 (either 6 or -6). However, because a represents a distance, choose the *positive* (principal) square root of 36.

The third side is 6 in. long.

Skill Practice

32. Use the Pythagorean theorem to find the length of the unknown side.

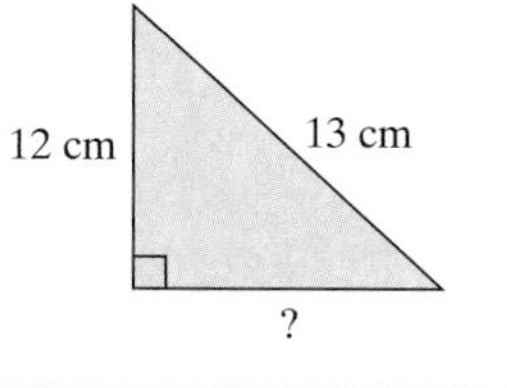

Example 9 Applying the Pythagorean Theorem

A bridge across a river is 600 yd long. A boat ramp is 200 yd due north of point P on the bridge, such that the line segments $\overline{PQ}$ and $\overline{PR}$ form a right angle (Figure 8-2). How far does a kayak travel if it leaves from the boat ramp and paddles to point Q? Use a calculator and round the answer to the nearest yard.

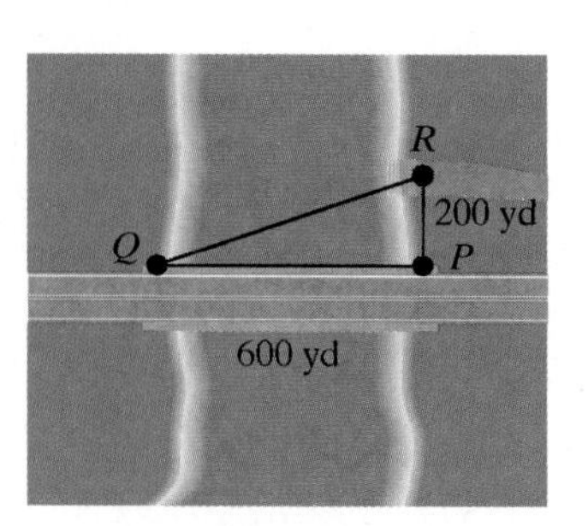

Figure 8-2

Skill Practice Answers

32. 5 cm

Solution:

Label the triangle:

c $a = 200$ yd $b = 600$ yd

$a^2 + b^2 = c^2$	
$(200)^2 + (600)^2 = c^2$	Apply the Pythagorean theorem.
$40{,}000 + 360{,}000 = c^2$	Simplify.
$400{,}000 = c^2$	By definition, c must be one of the square roots of 400,000. Because the value of c is a distance, choose the positive square root of 400,000.
$c = \sqrt{400{,}000}$	
$c \approx 632$	Use a calculator to approximate the positive square root of 400,000.

The kayak must travel approximately 632 yd.

Skill Practice

33. A rope is attached to the top of a 20-ft pole. How long is the rope if it reaches a point on the ground 15 ft from the base of the pole?

? 20 ft 15 ft

A calculator can be used to approximate the value of a radical expression. To evaluate a square root, use the $\sqrt{\ }$ key. For example, evaluate: $\sqrt{25}$, $\sqrt{60}$, $\sqrt{\frac{13}{3}}$

Scientific Calculator

Enter:	2 5 √	**Result:**	5
Enter:	6 0 √	**Result:**	7.745966692
Enter:	1 3 ÷ 3 = √	**Result:**	2.081665999

Graphing Calculator

```
√(25)
                5
√(60)
      7.745966692
√(13/3)
      2.081665999
```

To evaluate cube roots, your calculator may have a $\sqrt[3]{\ }$ key. Otherwise, for cube roots and roots of higher index (fourth roots, fifth roots, and so on), try using the $\sqrt[x]{y}$ key or $\sqrt[x]{\ }$ key. For example, evaluate $\sqrt[3]{64}$, $\sqrt[4]{81}$, and $\sqrt[3]{162}$:

Scientific Calculator

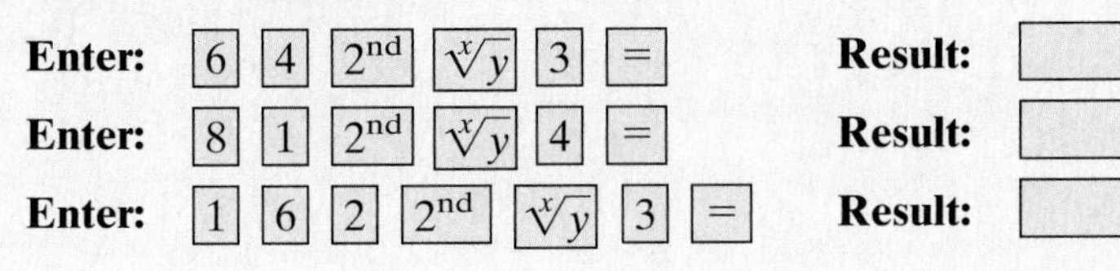

TIP: The values $\sqrt{60}$ and $\sqrt{\frac{13}{3}}$ are approximated on the calculator to 10 digits. However, $\sqrt{60}$ and $\sqrt{\frac{13}{3}}$ are actually irrational numbers. Their decimal forms are nonterminating and nonrepeating. The only way to represent the exact answers is by writing the radical forms, $\sqrt{60}$ and $\sqrt{\frac{13}{2}}$.

Skill Practice Answers

33. The rope is 25 ft long.

Graphing Calculator

On a graphing calculator, the index is usually entered first.

```
3 ˣ√(64)
                4
4 ˣ√(81)
                3
3 ˣ√(162)
      5.451361778
```

Calculator Exercises

Estimate the value of each radical. Then use a calculator to approximate the radical to three decimal places. **(See Tip on page 607.)**

1. $\sqrt{5}$ **2.** $\sqrt{17}$ **3.** $\sqrt{50}$

4. $\sqrt{96}$ **5.** $\sqrt{33}$ **6.** $\sqrt{145}$

7. $\sqrt{80}$ **8.** $\sqrt{170}$ **9.** $\sqrt[3]{7}$

10. $\sqrt[3]{28}$ **11.** $\sqrt[3]{65}$ **12.** $\sqrt[3]{124}$

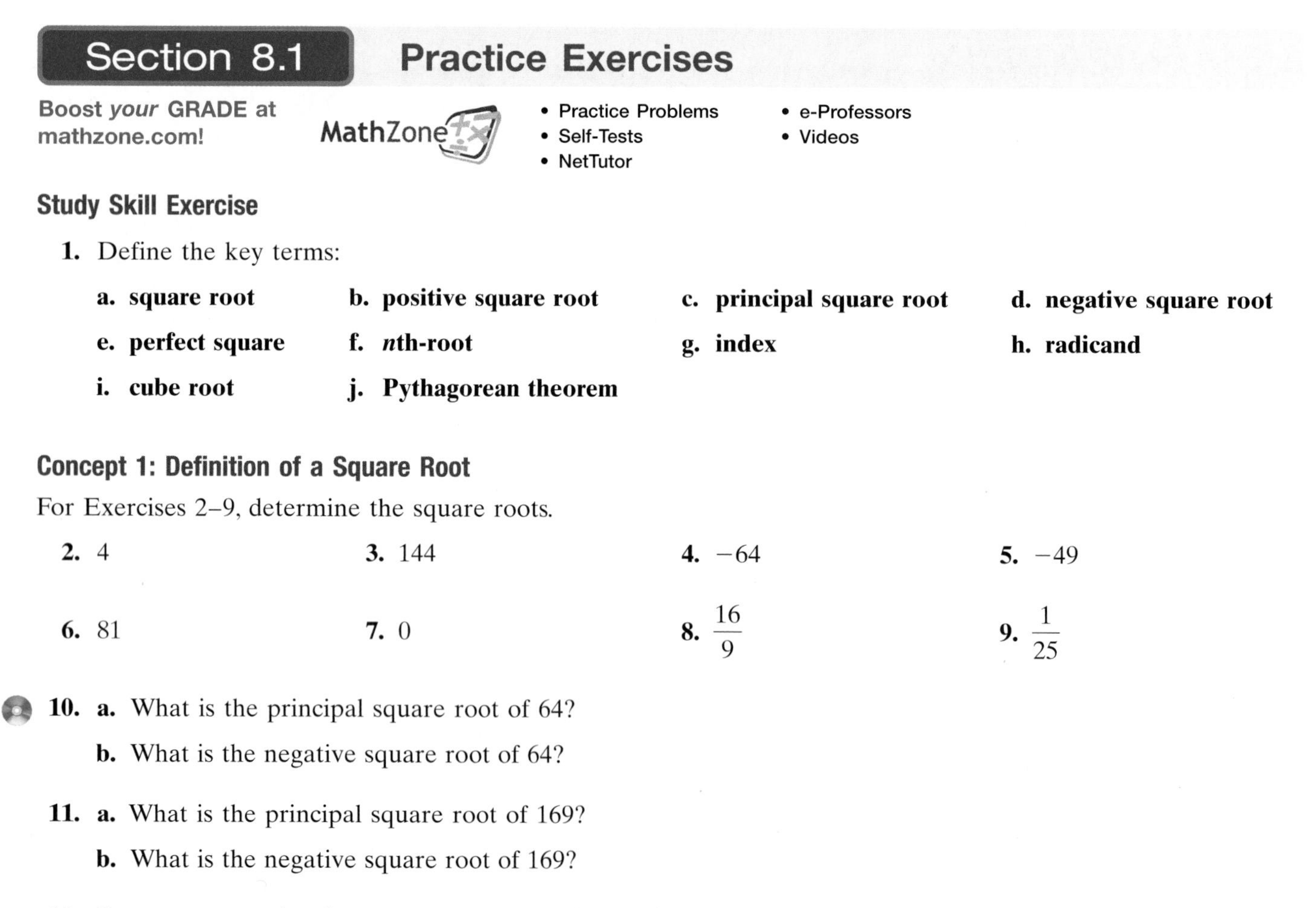

Section 8.1 Practice Exercises

Study Skill Exercise

1. Define the key terms:

a. **square root** **b.** **positive square root** **c.** **principal square root** **d.** **negative square root**

e. **perfect square** **f.** ***n*th-root** **g.** **index** **h.** **radicand**

i. **cube root** **j.** **Pythagorean theorem**

Concept 1: Definition of a Square Root

For Exercises 2–9, determine the square roots.

2. 4 **3.** 144 **4.** -64 **5.** -49

6. 81 **7.** 0 **8.** $\frac{16}{9}$ **9.** $\frac{1}{25}$

10. a. What is the principal square root of 64?

b. What is the negative square root of 64?

11. a. What is the principal square root of 169?

b. What is the negative square root of 169?

12. Does every number have two square roots? Explain.

13. Which number has only one square root?

14. Which of the following are perfect squares?
$\{0, 1, 4, 15, 30, 49, 72, 81, 144, 300, 625, 900\}$

15. Which of the following are perfect squares?
$\{8, 9, 12, 16, 25, 36, 42, 64, 95, 121, 140, 169\}$

For Exercises 16–31, evaluate the square roots.

16. $\sqrt{16}$ **17.** $\sqrt{4}$ **18.** $\sqrt{36}$ **19.** $\sqrt{49}$

20. $\sqrt{169}$ **21.** $\sqrt{81}$ **22.** $\sqrt{225}$ **23.** $\sqrt{625}$

24. $\sqrt{0.25}$ **25.** $\sqrt{0.16}$ **26.** $\sqrt{0.64}$ **27.** $\sqrt{0.09}$

28. $\sqrt{\frac{1}{9}}$ **29.** $\sqrt{\frac{25}{16}}$ **30.** $\sqrt{\frac{49}{121}}$ **31.** $\sqrt{\frac{1}{144}}$

32. Explain the difference between $\sqrt{-16}$ and $-\sqrt{16}$.

33. Using the definition of a square root, explain why $\sqrt{-16}$ does not have a real-valued square root.

34. Evaluate. $-\sqrt{|-25|}$

For Exercises 35–46, evaluate the square roots, if possible.

35. $-\sqrt{4}$ **36.** $-\sqrt{1}$ **37.** $\sqrt{-4}$ **38.** $\sqrt{-1}$

39. $\sqrt{-\frac{4}{49}}$ **40.** $-\sqrt{-\frac{9}{25}}$ **41.** $-\sqrt{-\frac{1}{36}}$ **42.** $-\sqrt{\frac{1}{36}}$

43. $-\sqrt{400}$ **44.** $-\sqrt{121}$ **45.** $\sqrt{-900}$ **46.** $\sqrt{-169}$

Concept 2: Definition of an *n*th-Root

47. Which of the following are perfect cubes?
$\{0, 1, 3, 9, 27, 36, 42, 90, 125\}$

48. Which of the following are perfect cubes?
$\{6, 8, 16, 20, 30, 64, 111, 150, 216\}$

49. Does $\sqrt[3]{-27}$ have a real-valued cube root?

50. Does $\sqrt[3]{-8}$ have a real-valued cube root?

For Exercises 51–66, evaluate the *n*-th roots, if possible.

51. $\sqrt[3]{27}$ **52.** $\sqrt[3]{-27}$ **53.** $\sqrt[3]{64}$ **54.** $\sqrt[3]{-64}$

55. $-\sqrt[4]{16}$ **56.** $-\sqrt[4]{81}$ **57.** $\sqrt[4]{-1}$ **58.** $\sqrt[4]{0}$

59. $\sqrt[4]{-256}$ **60.** $\sqrt[4]{-625}$ **61.** $\sqrt[5]{-32}$ **62.** $-\sqrt[5]{32}$

63. $-\sqrt[6]{1}$ **64.** $\sqrt[6]{64}$ **65.** $\sqrt[6]{0}$ **66.** $\sqrt[6]{-1}$

For Exercises 67–88, simplify the expressions.

67. $\sqrt{(4)^2}$ **68.** $\sqrt{(8)^2}$ **69.** $\sqrt{(-4)^2}$ **70.** $\sqrt{(-8)^2}$

71. $\sqrt[3]{(5)^3}$ **72.** $\sqrt[3]{(7)^3}$ **73.** $\sqrt[3]{(-5)^3}$ **74.** $\sqrt[3]{(-7)^3}$

75. $\sqrt[4]{(2)^4}$ **76.** $\sqrt[4]{(10)^4}$ **77.** $\sqrt[4]{(-2)^4}$ **78.** $\sqrt[4]{(-10)^4}$

79. $\sqrt{a^2}$ **80.** $\sqrt{b^2}$ **81.** $\sqrt[3]{y^3}$ **82.** $\sqrt[3]{z^3}$

83. $\sqrt[4]{w^4}$ **84.** $\sqrt[4]{p^4}$ **85.** $\sqrt[5]{x^5}$ **86.** $\sqrt[5]{y^5}$

87. $\sqrt[6]{m^6}$ **88.** $\sqrt[6]{n^6}$

89. Determine which of the expressions are perfect squares. Then state a rule for determining perfect squares based on the exponent of the expression.

$\{x^2, a^3, y^4, z^5, (ab)^6, (pq)^7, w^8x^8, c^9d^9, m^{10}, n^{11}\}$

90. Determine which of the expressions are perfect cubes. Then state a rule for determining perfect cubes based on the exponent of the expression.

$\{a^2, b^3, c^4, d^5, e^6, (xy)^7, (wz)^8, (pq)^9, t^{10}s^{10}, m^{11}n^{11}, u^{12}v^{12}\}$

91. Determine which of the expressions are perfect fourth powers. Then state a rule for determining perfect fourth powers based on the exponent of the expression.

$\{m^2, n^3, p^4, q^5, r^6, s^7, t^8, u^9, v^{10}, (ab)^{11}, (cd)^{12}\}$

92. Determine which of the expressions are perfect fifth powers. Then state a rule for determining perfect fifth powers based on the exponent of the expression.

$\{a^2, b^3, c^4, d^5, e^6, k^7, w^8, x^9, y^{10}, z^{11}\}$

For Exercises 93–110, simplify the expressions. Assume the variables represent positive real numbers.

93. $\sqrt{y^{12}}$ **94.** $\sqrt{z^{20}}$ **95.** $\sqrt{a^8b^{30}}$ **96.** $\sqrt{t^{50}s^{60}}$

97. $\sqrt[3]{q^{24}}$ **98.** $\sqrt[3]{x^{33}}$ **99.** $\sqrt[5]{c^{45}}$ **100.** $\sqrt[5]{d^{15}}$

101. $\sqrt[3]{8w^6}$ **102.** $\sqrt[3]{-27x^{27}}$ **103.** $\sqrt{(5x)^2}$ **104.** $\sqrt{(6w)^2}$

105. $-\sqrt{25x^2}$ **106.** $-\sqrt{36w^2}$ **107.** $\sqrt[3]{(5p^2)^3}$ **108.** $\sqrt[3]{(2k^4)^3}$

109. $\sqrt[3]{125p^6}$ **110.** $\sqrt[3]{8k^{12}}$

Concept 3: Translations Involving *n*th-Roots

For Exercises 111–114, translate the English phrase to an algebraic expression.

111. The sum of the principal square root of q and the square of p

112. The product of the principal square root of eleven and the cube of x

113. The quotient of six and the principal fourth root of x

114. The difference of the square of y and one

Concept 4: Pythagorean Theorem

For Exercises 115–120, find the length of the third side of each triangle using the Pythagorean theorem. Round the answer to the nearest tenth if necessary.

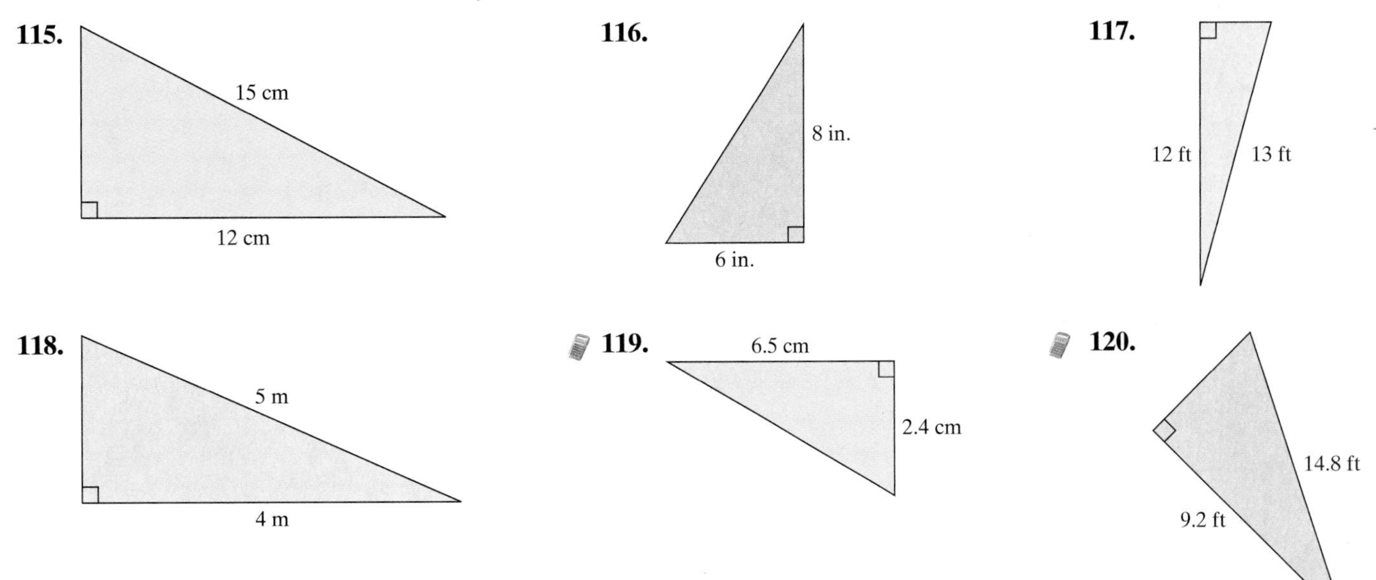

121. Find the length of the diagonal of the square tile shown in the figure. Round the answer to the nearest tenth of an inch.

12 in.

12 in.

122. A baseball diamond is 90 ft on a side. Find the distance between home plate and second base. Round the answer to the nearest tenth of a foot.

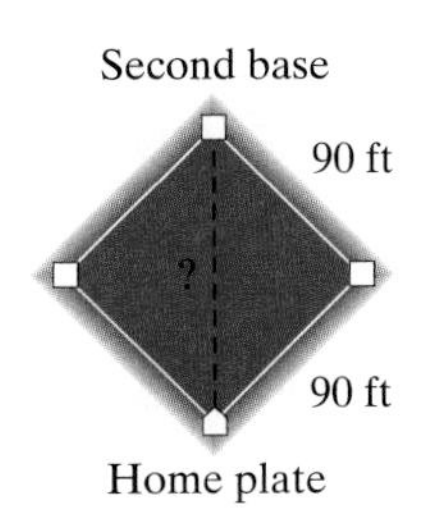

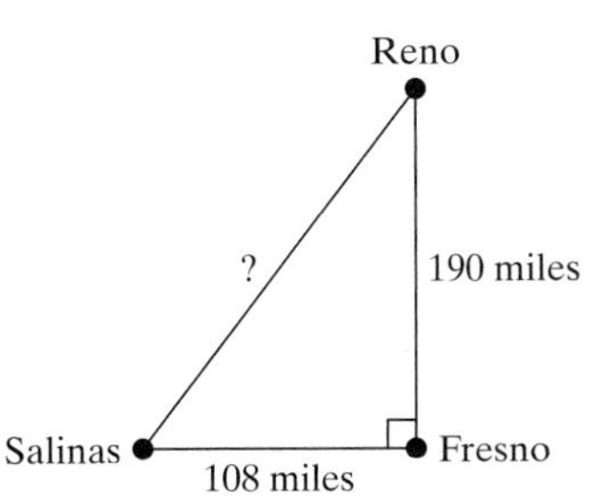

123. On a map, Fresno, California, is 108 miles east of Salinas, California. Reno, Nevada, is 190 miles north of Fresno. Approximate the distance between Reno and Salinas to the nearest mile.

124. Washington, D.C., is north of Richmond, Virginia, a distance of 98 miles. Louisville, Kentucky, is west of Richmond, a distance of 454 miles. How far is it from Washington to Louisville? Round the answer to the nearest mile.

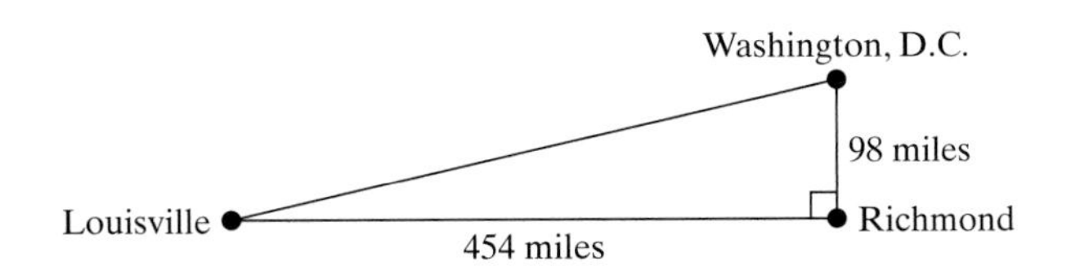

125. On a map, the cities Asheville, North Carolina, Roanoke, Virginia, and Greensboro, North Carolina, form a right triangle (see the figure). The distance between Asheville and Roanoke is 300 km. The distance between Roanoke and Greensboro is 134 km. How far is it from Greensboro to Asheville? Round the answer to the nearest kilometer.

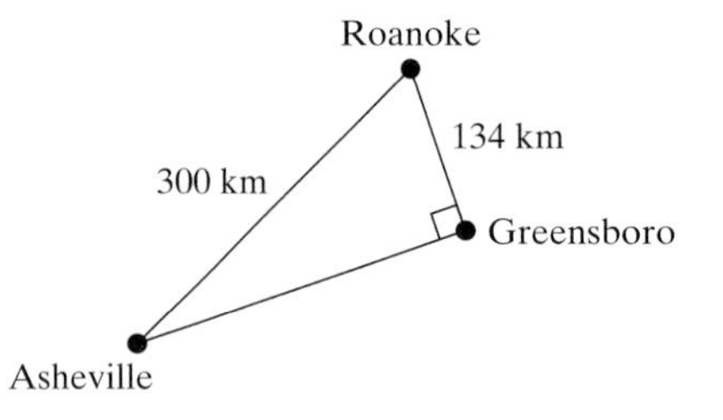

126. Jackson, Mississippi, is west of Meridian, Mississippi, a distance of 141 km. Tupelo, Mississippi, is north of Meridian, a distance of 209 km. How far is it from Jackson to Tupelo? Round the answer to the nearest kilometer.

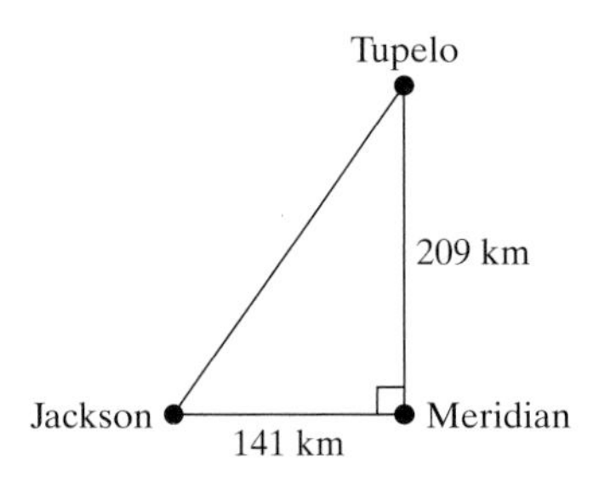

127. Since 1950 there have been 17 hurricanes of category 3 or higher that have hit the Gulf states of Texas, Louisiana, Mississippi, Alabama, and Florida. Before a hurricane, homeowners often apply strips of masking tape over windows to prevent shards of glass from flying into a room.

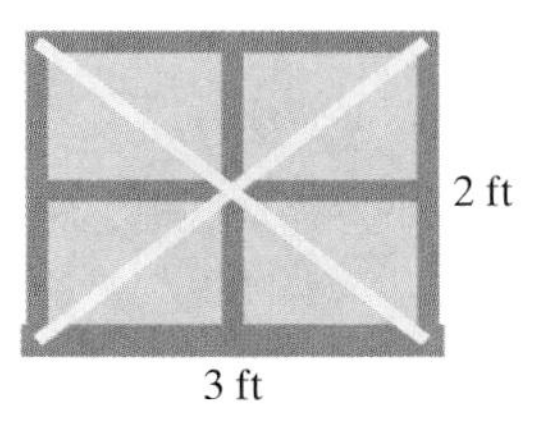

a. How much masking tape is needed to tape one diagonal in the window shown in the figure? (Round to the nearest tenth of a foot.)

b. How much tape is needed to tape both diagonals?

c. How much tape is needed to tape 20 such windows?

Expanding Your Skills

128. For what values of x will $\sqrt{x}$ be a real number?

129. For what values of x will $\sqrt{-x}$ be a real number?

Simplifying Radicals

Section 8.2

Concepts

1. Multiplication and Division Properties of Radicals

You may have already recognized certain properties of radicals involving a product or quotient.

Multiplication and Division Properties of Radicals

Let a and b represent real numbers such that $\sqrt[n]{a}$ and $\sqrt[n]{b}$ are both real. Then,

1. $\sqrt[n]{ab} = \sqrt[n]{a} \cdot \sqrt[n]{b}$ **Multiplication property of radicals**
2. $\sqrt[n]{\frac{a}{b}} = \frac{\sqrt[n]{a}}{\sqrt[n]{b}}$ $b \neq 0$ **Division property of radicals**

The multiplication and division properties of radicals indicate that a product or quotient within a radicand can be written as a product or quotient of radicals provided the roots are real numbers.

$$\sqrt{100} = \sqrt{25} \cdot \sqrt{4}$$

$$\sqrt{\frac{4}{9}} = \frac{\sqrt{4}}{\sqrt{9}}$$

The reverse process is also true. A product or quotient of radicals can be written as a single radical provided the roots are real numbers and they have the same indices.

Same index

$$\sqrt{2} \cdot \sqrt{18} = \sqrt{36}$$

Same index

$$\frac{\sqrt[3]{125}}{\sqrt[3]{8}} = \sqrt[3]{\frac{125}{8}}$$

2. Simplified Form of a Radical

In algebra, it is customary to simplify radical expressions as much as possible.

Simplified Form of a Radical

Consider any radical expression where the radicand is written as a product of prime factors. The expression is in **simplified form** if all of the following conditions are met:

1. The radicand has no factor raised to a power greater than or equal to the index.
2. There are no radicals in the denominator of a fraction.
3. The radicand does not contain a fraction.

3. Simplifying Radicals Using the Multiplication Property of Radicals

The expression $\sqrt{x^2}$ is not simplified because it fails condition 1. Because x^2 is a perfect square, $\sqrt{x^2}$ is easily simplified.

$$\sqrt{x^2} = x \quad (\text{for } x \geq 0)$$

However, how is an expression such as $\sqrt{x^7}$ simplified? This and many other radical expressions are simplified using the multiplication property of radicals. The following examples illustrate how nth powers can be removed from the radicands of nth-roots.

Example 1 Using the Multiplication Property to Simplify a Radical Expression

Use the multiplication property of radicals to simplify the expression $\sqrt{x^7}$. Assume $x \geq 0$.

Solution:

The expression $\sqrt{x^7}$ is equivalent to $\sqrt{x^6 \cdot x}$. By applying the multiplication property of radicals, we have

$\sqrt{x^6 \cdot x} = \sqrt{x^6} \cdot \sqrt{x}$ $\quad$ x^6 is a perfect square because $(x^3)^2 = x^6$

$= x^3 \cdot \sqrt{x}$ $\quad$ Simplify.

$= x^3\sqrt{x}$

In Example 1, the expression x^7 is not a perfect square. Therefore, to simplify $\sqrt{x^7}$, it was necessary to write the expression as the product of the largest perfect square and a remaining, or "leftover," factor: $\sqrt{x^7} = \sqrt{x^6 \cdot x}$.

Example 2 Using the Multiplication Property to Simplify Radicals

Use the multiplication property of radicals to simplify the expressions. Assume the variables represent positive real numbers.

a. $\sqrt{a^{15}}$ $\quad$ **b.** $\sqrt{x^2y^5}$ $\quad$ **c.** $\sqrt{s^9t^{11}}$

Solution:

The goal is to rewrite each radicand as the product of the largest perfect square and a leftover factor.

a. $\sqrt{a^{15}}$

$= \sqrt{a^{14} \cdot a}$ $\quad$ a^{14} is the largest perfect square in the radicand.

$= \sqrt{a^{14}} \cdot \sqrt{a}$ $\quad$ Apply the multiplication property of radicals.

$= a^7\sqrt{a}$ $\quad$ Simplify.

b. $\sqrt{x^2y^5}$

$= \sqrt{x^2y^4 \cdot y}$ $\quad$ x^2y^4 is the largest perfect square in the radicand.

$= \sqrt{x^2y^4} \cdot \sqrt{y}$ $\quad$ Apply the multiplication property of radicals.

$= xy^2\sqrt{y}$ $\quad$ Simplify.

c. $\sqrt{s^9t^{11}}$

$= \sqrt{s^8t^{10} \cdot st}$ — s^8t^{10} is the largest perfect square in the radical.

$= \sqrt{s^8t^{10}} \cdot \sqrt{st}$ — Apply the multiplication property of radicals.

$= s^4t^5\sqrt{st}$ — Simplify.

Skill Practice Use the multiplication property of radicals to simplify the expressions. Assume all the variables are positive.

1. $\sqrt{y^{11}}$ **2.** $\sqrt{x^{20}y^{31}}$ **3.** $\sqrt{u^3w^9}$

Each expression in Example 2 involved a radicand that is a product of variable factors. If a numerical factor is present, sometimes it is necessary to factor the coefficient before simplifying the radical.

Example 3 Using the Multiplication Property to Simplify Radicals

Use the multiplication property of radicals to simplify the expressions. Assume the variables represent positive real numbers.

a. $\sqrt{50}$ **b.** $5\sqrt{24a^6}$ **c.** $\sqrt{81x^4y^3}$

Solution:

The goal is to rewrite each radicand as the product of the largest perfect square and a leftover factor.

a. Write the radicand as a product of prime factors. From the prime factorization, the largest perfect square is easily identified.

$\sqrt{50} = \sqrt{5^2 \cdot 2}$ — Factor the radicand. 5^2 is the largest perfect square.

$$\begin{array}{r} 2\underline{)50} \\ 5\underline{)25} \\ 5 \end{array}$$

$= \sqrt{5^2} \cdot \sqrt{2}$ — Apply the multiplication property of radicals.

$= 5\sqrt{2}$ — Simplify.

b. $5\sqrt{24a^6} = 5\sqrt{2^3 \cdot 3 \cdot a^6}$ — Write the radicand as a product of prime factors: $24 = 2^3 \cdot 3$.

$= 5\sqrt{2^2a^6 \cdot 2 \cdot 3}$ — 2^2a^6 is the largest perfect square in the radicand.

$= 5\sqrt{2^2a^6} \cdot \sqrt{2 \cdot 3}$ — Apply the multiplication property of radicals.

$= 5 \cdot 2a^3\sqrt{6}$ — Simplify the radical.

$= 10a^3\sqrt{6}$ — Simplify the coefficient of the radical.

TIP: The expression $\sqrt{50}$ can also be written as:

$$\sqrt{25 \cdot 2} = \sqrt{25} \cdot \sqrt{2} = 5\sqrt{2}$$

Skill Practice Answers

1. $y^5\sqrt{y}$ **2.** $x^{10}y^{15}\sqrt{y}$
3. $uw^4\sqrt{uw}$

c. $\sqrt{81x^4y^3} = \sqrt{3^4x^4y^3}$ — Write the radical as a product of the prime factors. *Note:* $81 = 3^4$.

$= \sqrt{3^4x^4y^2 \cdot y}$ — $3^4x^4y^2$ is the largest square in the radicand.

$= \sqrt{3^4x^4y^2} \cdot \sqrt{y}$ — Apply the multiplication property of radicals.

$= 3^2x^2y \cdot \sqrt{y}$ — Simplify the radical.

$= 9x^2y\sqrt{y}$ — Simplify the coefficient of the radical.

Skill Practice Use the multiplication property of radicals to simplify the expressions. Assume the variables represent positive real numbers.

4. $\sqrt{12}$ **5.** $\sqrt{60x^2}$ **6.** $7\sqrt{18t^{10}}$

The multiplication property of radicals allows us to simplify a product of factors within a radical. For example,

$$\sqrt{x^2y^2} = \sqrt{x^2} \cdot \sqrt{y^2} = xy \qquad (\text{for } x \geq 0 \text{ and } y \geq 0)$$

However, this rule does not apply to *terms* that are added or subtracted *within* the radical. For example,

$$\sqrt{x^2 + y^2} \quad \text{and} \quad \sqrt{x^2 - y^2}$$

cannot be simplified.

4. Simplifying Radicals Using the Division Property of Radicals

The division property of radicals indicates that a radical of a quotient can be written as the quotient of the radicals and vice versa provided all roots are real numbers.

Example 4 Simplifying Radicals Using the Division Property of Radicals

Simplify the expressions. Assume the variables represent positive real numbers.

a. $\sqrt{\dfrac{a^5}{a^3}}$ **b.** $\dfrac{\sqrt{6}}{\sqrt{96}}$ **c.** $\sqrt{\dfrac{27x^5}{3x}}$

Solution:

a. $\sqrt{\dfrac{a^5}{a^3}}$ — The radical contains a fraction. However, the fraction can be simplified.

$= \sqrt{a^2}$ — Reduce the fraction to lowest terms.

$= a$ — Simplify the radical.

Skill Practice Answers

4. $2\sqrt{3}$ **5.** $2x\sqrt{15}$
6. $21t^5\sqrt{2}$

b. $\dfrac{\sqrt{6}}{\sqrt{96}}$ Notice that the radicands have a common factor.

$= \sqrt{\dfrac{6}{96}}$ Apply the division property of radicals to write a single radical.

$= \sqrt{\dfrac{1}{16}}$ Reduce the fraction to lowest terms.

$= \dfrac{1}{4}$ Simplify.

c. $\sqrt{\dfrac{27x^5}{3x}}$ The fraction within the radicand can be simplified.

$= \sqrt{9x^4}$ Reduce to lowest terms.

$= 3x^2$ Simplify.

Skill Practice Use the division property of radicals to simplify the expressions. Assume the variables represent positive real numbers.

7. $\sqrt{\dfrac{y^{11}}{y^3}}$ **8.** $\dfrac{\sqrt{8}}{\sqrt{50}}$ **9.** $\sqrt{\dfrac{32z^3}{2z}}$

Example 5 **Simplifying Radicals Using the Division Property of Radicals**

Simplify the expression.

$$\frac{5\sqrt{20}}{2}$$

Solution:

$\dfrac{5\sqrt{20}}{2} = \dfrac{5\sqrt{2^2 \cdot 5}}{2}$ 2^2 is the largest perfect square in the radicand.

$= \dfrac{5\sqrt{2^2} \cdot \sqrt{5}}{2}$ Apply the multiplication property of radicals.

$= \dfrac{5 \cdot 2\sqrt{5}}{2}$ Simplify the radical.

$= \dfrac{5 \cdot 2\sqrt{5}}{2}$ Simplify to lowest terms.

$= 5\sqrt{5}$

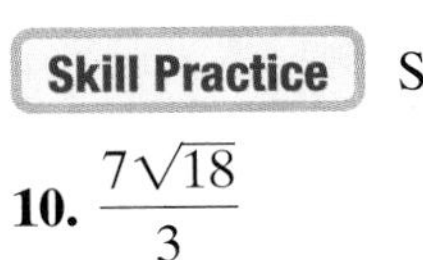

Skill Practice Simplify the expression.

10. $\dfrac{7\sqrt{18}}{3}$

Avoiding Mistakes:

$\dfrac{5\sqrt{20}}{2}$ cannot be simplified as written because 20 is under the radical and 2 is not. To reduce to lowest terms, the radical must be simplified first, $\dfrac{5 \cdot 2\sqrt{5}}{2}$. Then factors outside the radical can be simplified.

Skill Practice Answers

7. y^4 **8.** $\dfrac{2}{5}$

9. $4z$ **10.** $7\sqrt{2}$

5. Simplifying Cube Roots

Example 6 Simplifying Cube Roots

Use the properties of radicals to simplify the expressions.

a. $\sqrt[3]{z^5}$ **b.** $\sqrt[3]{-80}$ **c.** $\sqrt[3]{\dfrac{a^6}{b^3}}$ **d.** $\dfrac{\sqrt[3]{2}}{\sqrt[3]{16}}$

Solution:

a. $\sqrt[3]{z^5}$

$= \sqrt[3]{z^3 \cdot z^2}$ — z^3 is the largest perfect cube in the radicand.

$= \sqrt[3]{z^3} \cdot \sqrt[3]{z^2}$ — Apply the multiplication property of radicals.

$= z\sqrt[3]{z^2}$ — Simplify.

b. $\sqrt[3]{-80}$

$= \sqrt[3]{-1 \cdot 2^4 \cdot 5}$ — Factor the radicand.

$= \sqrt[3]{-1 \cdot 2^3 \cdot 2 \cdot 5}$ — -1 and 2^3 are perfect cubes.

$= \sqrt[3]{-1} \cdot \sqrt[3]{2^3} \cdot \sqrt[3]{2 \cdot 5}$ — Apply the multiplication property of radicals.

$= -1 \cdot 2 \cdot \sqrt[3]{10}$ — Simplify.

$= -2\sqrt[3]{10}$

$$\begin{array}{r} 2\overline{)80} \\ 2\overline{)40} \\ 2\overline{)20} \\ 2\overline{)10} \\ 5 \end{array}$$

c. $\sqrt[3]{\dfrac{a^6}{b^3}}$ — The radical contains an irreducible fraction.

$= \dfrac{\sqrt[3]{a^6}}{\sqrt[3]{b^3}}$ — Apply the division property of radicals.

$= \dfrac{a^2}{b}$ — Simplify.

d. $\dfrac{\sqrt[3]{2}}{\sqrt[3]{16}}$ — Notice that the radicands have a common factor.

$= \sqrt[3]{\dfrac{2}{16}}$ — Apply the division property of radicals to write the radicals as a single radical.

$= \sqrt[3]{\dfrac{1}{8}}$ — Reduce to lowest terms.

$= \dfrac{1}{2}$ — Simplify.

Skill Practice Simplify.

11. $\sqrt[3]{y^4}$ **12.** $\sqrt[3]{24}$ **13.** $\sqrt[3]{\dfrac{x^{12}}{y^6}}$ **14.** $\dfrac{\sqrt[3]{81}}{\sqrt[3]{3}}$

Skill Practice Answers

11. $y\sqrt[3]{y}$ **12.** $2\sqrt[3]{3}$

13. $\dfrac{x^4}{y^2}$ **14.** 3

Calculator Connections

A calculator can support the multiplication property of radicals. For example, use a calculator to evaluate $\sqrt{50}$ and its simplified form $5\sqrt{2}$.

Scientific Calculator

Enter: [5] [0] [√] **Result:** 7.071067812

Enter: [2] [√] [×] [5] [=] **Result:** 7.071067812

Graphing Calculator

```
√(50)
          7.071067812
5*√(2)
          7.071067812
```

A calculator can support the division property of radicals. For example, use a calculator to evaluate $\sqrt{\dfrac{21}{5}}$ and its equivalent form $\dfrac{\sqrt{21}}{\sqrt{5}}$.

Scientific Calculator

Enter: [2] [1] [÷] [5] [=] [√] **Result:** 2.049390153

Enter: [2] [1] [√] [÷] [5] [√] [=] **Result:** 2.049390153

Graphing Calculator

```
√(21/5)
          2.049390153
√(21)/√(5)
          2.049390153
```

Calculator Exercises

Simplify the radical expressions algebraically. Then use a calculator to approximate the original expression and its simplified form.

1. $\sqrt{125}$ **2.** $\sqrt{18}$

3. $\sqrt[3]{54}$ **4.** $\sqrt[3]{108}$

Use a calculator to find a decimal approximation for both the right- and left-hand sides of the expressions.

5. $\sqrt{\dfrac{40}{3}} = \dfrac{\sqrt{40}}{\sqrt{3}}$ **6.** $\sqrt[3]{\dfrac{128}{2}} = \dfrac{\sqrt[3]{128}}{\sqrt[3]{2}}$

TIP: The decimal approximation for $\sqrt{50}$ and $5\sqrt{2}$ agree for the first 10 digits. This in itself does not make $\sqrt{50} = 5\sqrt{2}$. It is the multiplication property of radicals that guarantees that the expressions are equal.

TIP: The calculator approximations agree to the first 10 digits. However, it is the division property of radicals that guarantees that

$$\sqrt{\frac{21}{5}} = \frac{\sqrt{21}}{\sqrt{5}}$$

Section 8.2 Practice Exercises

Boost *your* GRADE at mathzone.com! MathZone

- Practice Problems
- Self-Tests
- NetTutor
- e-Professors
- Videos

Study Skills Exercise

1. Define the key terms:

a. simplified form of a radical **b. multiplication property of radicals**

c. division property of radicals

Review Exercises

2. Which of the following are perfect squares? $\{2, 4, 6, 16, 20, 25, x^2, x^3, x^{15}, x^{20}, x^{25}\}$

3. Which of the following are perfect cubes? $\{3, 6, 8, 9, 12, 27, y^3, y^8, y^9, y^{12}, y^{27}\}$

4. Which of the following are perfect fourth powers? $\{4, 16, 20, 25, 81, w^4, w^{16}, w^{20}, w^{25}, w^{81}\}$

For Exercises 5–12, simplify the expressions, if possible. Assume the variables represent positive real numbers.

5. $-\sqrt{25}$ **6.** $\sqrt{-25}$ **7.** $-\sqrt[3]{27}$ **8.** $\sqrt[3]{-27}$

9. $\sqrt[4]{a^8}$ **10.** $\sqrt[5]{b^{15}}$ **11.** $\sqrt{4x^2y^4}$ **12.** $\sqrt{9p^{10}}$

13. On a map, Seattle, Washington, is 378 km west of Spokane, Washington. Portland, Oregon, is 236 km south of Seattle. Approximate the distance between Portland and Spokane to the nearest kilometer.

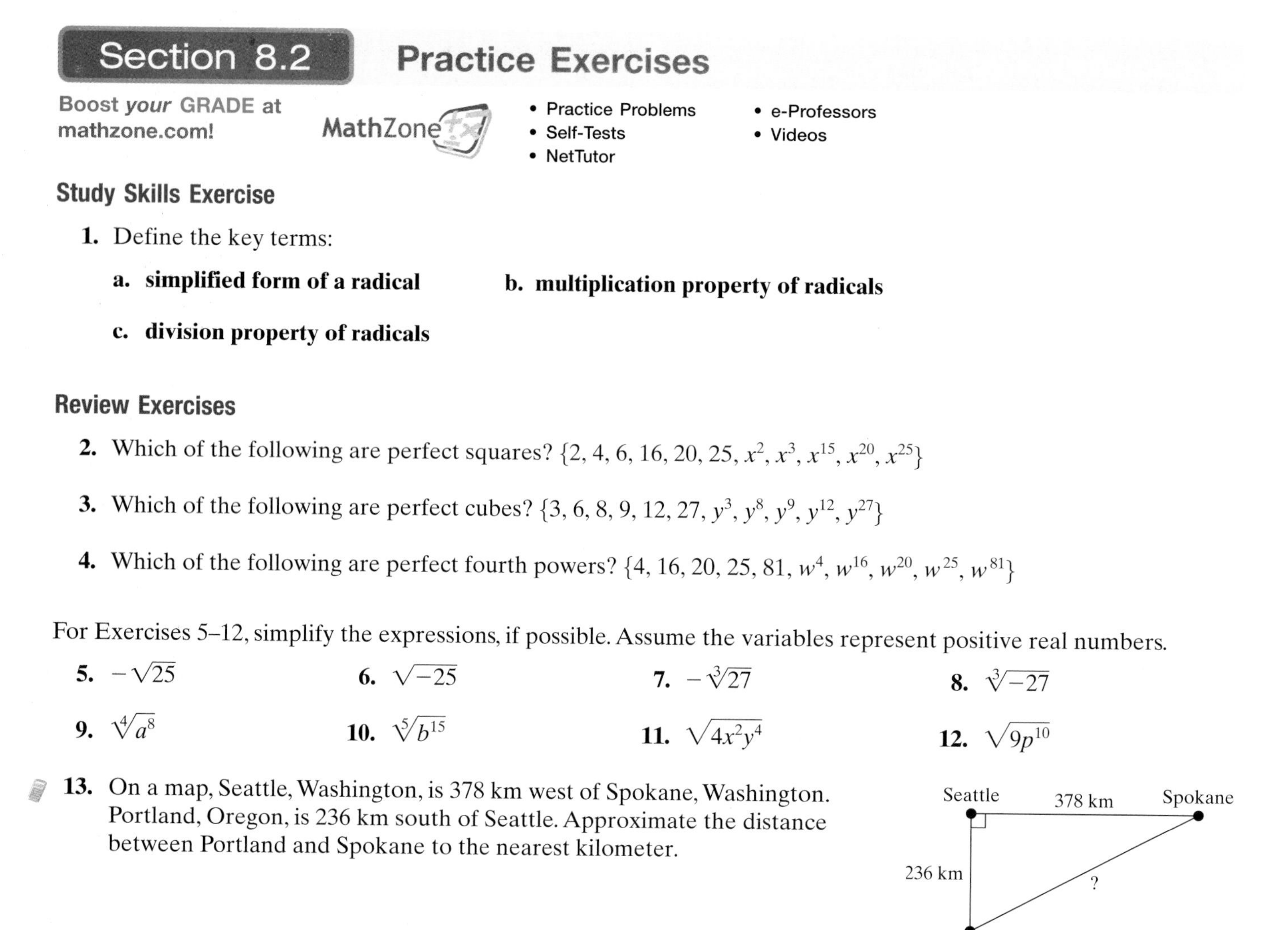

14. A new roof is needed on a shed. How many square feet of tar paper would be needed to cover the top of the roof?

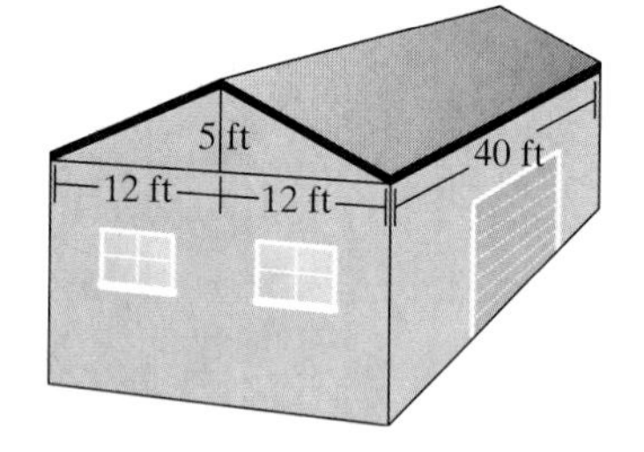

Concept 3: Simplifying Radicals Using the Multiplication Property of Radicals

For Exercises 15–52, use the multiplication property of radicals to simplify the expressions. Assume the variables represent positive real numbers.

15. $\sqrt{18}$ **16.** $\sqrt{75}$ **17.** $\sqrt{28}$ **18.** $\sqrt{40}$

19. $6\sqrt{20}$ **20.** $10\sqrt{27}$ **21.** $-2\sqrt{50}$ **22.** $-11\sqrt{8}$

23. $\sqrt{a^5}$ **24.** $\sqrt{b^9}$ **25.** $\sqrt{w^{22}}$ **26.** $\sqrt{p^{18}}$

27. $\sqrt{m^4n^5}$
28. $\sqrt{c^2d^9}$
29. $x\sqrt{x^{13}y^{10}}$
30. $v\sqrt{u^{10}v^7}$
31. $3\sqrt{t^{11}}$
32. $-4\sqrt{m^9n^4}$
33. $\sqrt{8x^3}$
34. $\sqrt{27y^5}$
35. $\sqrt{16z^3}$
36. $\sqrt{9y^5}$
37. $-\sqrt{45w^6}$
38. $-\sqrt{56v^8}$
39. $\sqrt{z^{25}}$
40. $\sqrt{25p^{49}}$
41. $-\sqrt{15z^{11}}$
42. $-\sqrt{6k^{15}}$
43. $5\sqrt{104a^2b^7}$
44. $3\sqrt{88m^4n^{11}}$
45. $\sqrt{26pq}$
46. $\sqrt{15a}$
47. $m\sqrt{m^{11}n^{16}}$
48. $c^2\sqrt{c^4d^{13}}$
49. $\sqrt{48a^3b^5c^4}$
50. $\sqrt{18xy^4z^3}$
51. $\sqrt{75u^4v^5}$
52. $\sqrt{96p^5q^2}$

Concept 4: Simplifying Radicals Using the Division Property of Radicals

For Exercises 53–74, use the division property of radicals, if necessary, to simplify the expressions. Assume the variables represent positive real numbers.

53. $\sqrt{\dfrac{3}{16}}$
54. $\sqrt{\dfrac{7}{25}}$
55. $\sqrt{\dfrac{a^4}{b^4}}$
56. $\sqrt{\dfrac{y^6}{z^2}}$
57. $\sqrt{\dfrac{a^9}{a}}$ (*Hint:* Simplify the radicand first.)
58. $\sqrt{\dfrac{x^5}{x}}$ (*Hint:* Simplify the radicand first.)
59. $\sqrt{\dfrac{9}{36}}$
60. $\sqrt{\dfrac{4}{64}}$
61. $\sqrt{\dfrac{c^3}{4}}$
62. $\sqrt{\dfrac{d^5}{9}}$
63. $\sqrt{\dfrac{a^9}{b^4}}$
64. $\sqrt{\dfrac{y^3}{z^{10}}}$
65. $\sqrt{\dfrac{200}{81}}$
66. $\sqrt{\dfrac{80}{49}}$
67. $\dfrac{\sqrt{8}}{\sqrt{50}}$ (*Hint:* Write the expression as a single radical and simplify.)
68. $\dfrac{\sqrt{21}}{\sqrt{12}}$ (*Hint:* Write the expression as a single radical and simplify.)
69. $\dfrac{\sqrt{p}}{\sqrt{4p^3}}$
70. $\dfrac{\sqrt{9t}}{\sqrt{t^5}}$
71. $\dfrac{3\sqrt{20}}{2}$
72. $\dfrac{5\sqrt{18}}{3}$
73. $\dfrac{\sqrt{60}}{10}$
74. $\dfrac{\sqrt{40}}{6}$

For Exercises 75–78, find the exact length of the third side of each triangle using the Pythagorean theorem. Simplify the radicals, if possible.

75.
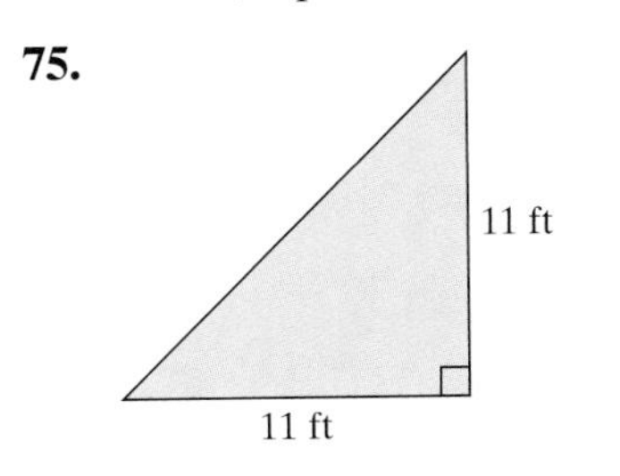

76.
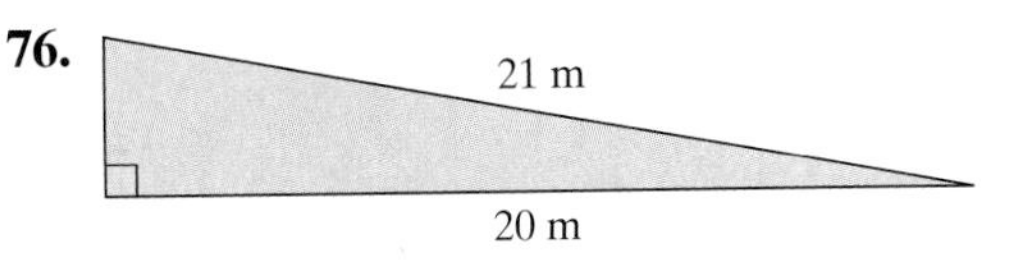

77. **78.**

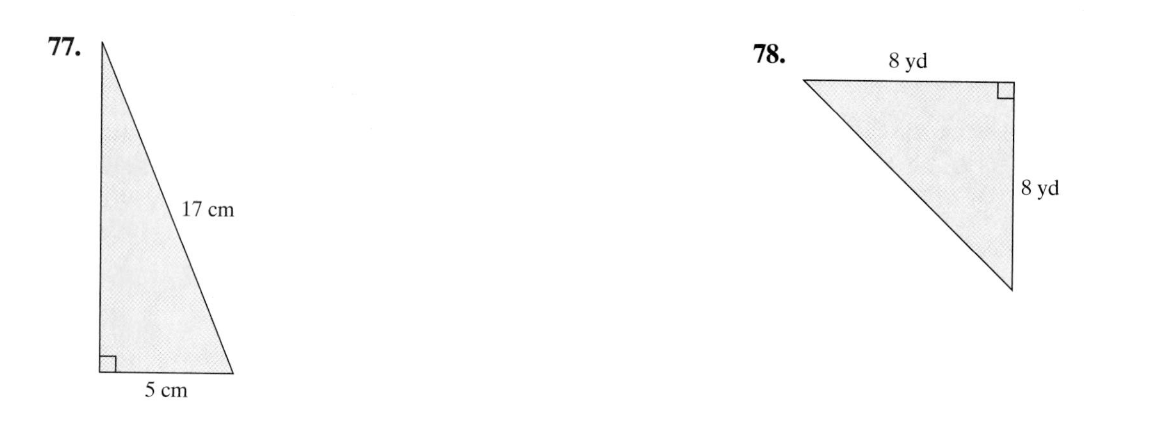

Concept 5: Simplifying Cube Roots

For Exercises 79–90, simplify the cube roots.

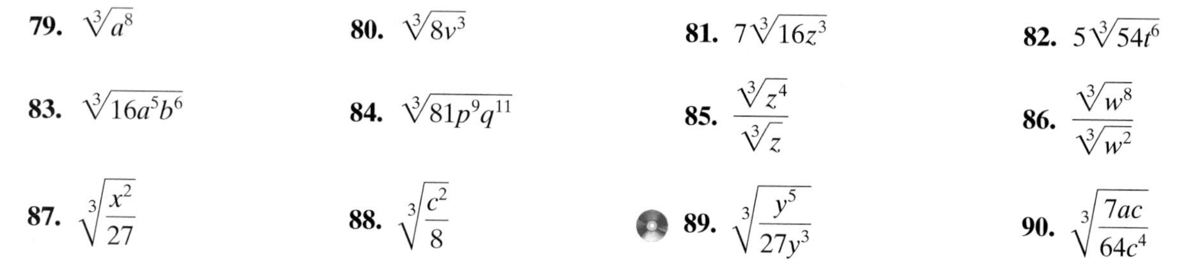

79. $\sqrt[3]{a^8}$ **80.** $\sqrt[3]{8v^3}$ **81.** $7\sqrt[3]{16z^3}$ **82.** $5\sqrt[3]{54t^6}$

83. $\sqrt[3]{16a^5b^6}$ **84.** $\sqrt[3]{81p^9q^{11}}$ **85.** $\dfrac{\sqrt[3]{z^4}}{\sqrt[3]{z}}$ **86.** $\dfrac{\sqrt[3]{w^8}}{\sqrt[3]{w^2}}$

87. $\sqrt[3]{\dfrac{x^2}{27}}$ **88.** $\sqrt[3]{\dfrac{c^2}{8}}$ **89.** $\sqrt[3]{\dfrac{y^5}{27y^3}}$ **90.** $\sqrt[3]{\dfrac{7ac}{64c^4}}$

Mixed Exercises

For Exercises 91–114, simplify the expressions. Assume the variables represent positive real numbers.

91. $\sqrt[3]{16a^3}$ **92.** $\sqrt[3]{125x^6}$ **93.** $\sqrt{16a^3}$ **94.** $\sqrt{125x^6}$

95. $\sqrt{\dfrac{4x^3}{y^2}}$ **96.** $\sqrt{\dfrac{9z^5}{w^2}}$ **97.** $\sqrt[3]{\dfrac{b^4}{27b}}$ **98.** $\sqrt[3]{\dfrac{k^6}{64}}$

99. $\sqrt{32}$ **100.** $\sqrt{64}$ **101.** $\sqrt{52u^4v^7}$ **102.** $\sqrt{44p^8q^{10}}$

103. $\sqrt{216}$ **104.** $\sqrt{250}$ **105.** $\sqrt[3]{216}$ **106.** $\sqrt[3]{250}$

107. $\dfrac{\sqrt{3}}{\sqrt{27}}$ **108.** $\dfrac{\sqrt{5}}{\sqrt{125}}$ **109.** $\sqrt[3]{\dfrac{x^5}{x^2}}$ **110.** $\sqrt[3]{\dfrac{y^{11}}{y^2}}$

111. $\sqrt[3]{15m^4n^{22}}$ **112.** $\sqrt[3]{20s^{15}t^{11}}$ **113.** $\sqrt{8p^2q}$ **114.** $\sqrt{6cd^3}$

Expanding Your Skills

For Exercises 115–118, simplify the expressions. Assume the variables represent positive real numbers.

115. $\sqrt{(-2-5)^2+(-4+3)^2}$ **116.** $\sqrt{(-1-7)^2+[1-(-1)]^2}$

117. $\sqrt{x^2+10x+25}$ **118.** $\sqrt{x^2+6x+9}$

Section 8.3

Concepts

1. Definition of *Like* Radicals
2. Addition and Subtraction of Radicals

Addition and Subtraction of Radicals

1. Definition of *Like* Radicals

Definition of *Like* Radicals

Two radical terms are said to be ***like* radicals** if they have the same index and the same radicand.

The following are pairs of *like* radicals:

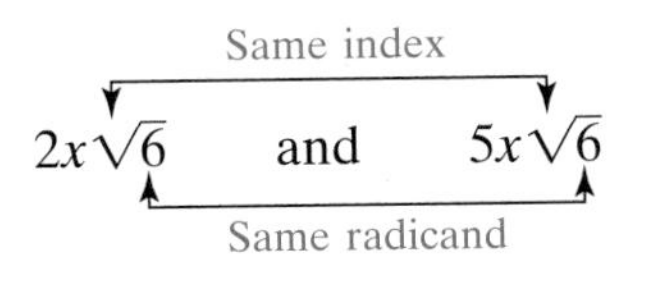

Indices and radicands are the same.
Both indices are 2. Both radicands are 6.

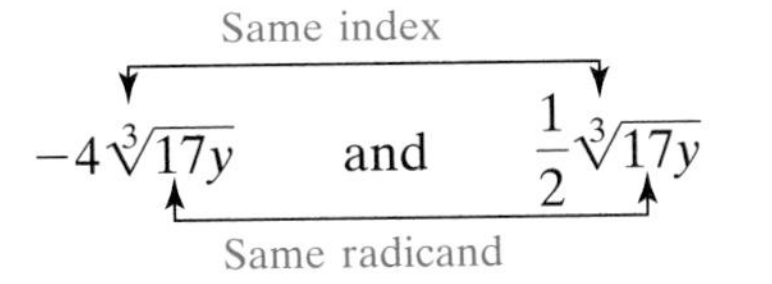

Indices and radicands are the same.
Both indices are 3. Both radicands are $17y$.

These pairs are *not like* radicals:

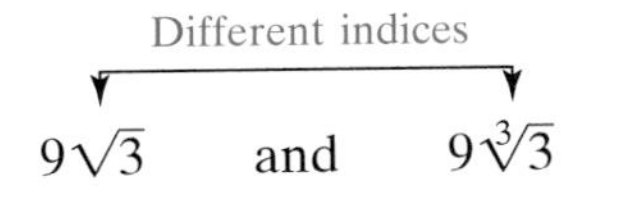

Radicals have different indices.

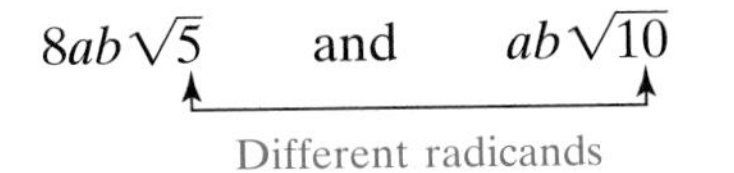

Radicals have different radicands.

2. Addition and Subtraction of Radicals

To add or subtract *like* radicals, use the distributive property. For example,

$$\begin{aligned} 3\sqrt{7} + 5\sqrt{7} &= (3 + 5)\sqrt{7} \\ &= 8\sqrt{7} \end{aligned}$$

$$\begin{aligned} 9\sqrt{2y} - 4\sqrt{2y} &= (9 - 4)\sqrt{2y} \\ &= 5\sqrt{2y} \end{aligned}$$

Example 1 **Adding and Subtracting Radicals**

Add or subtract the radicals as indicated. Assume all variables represent positive real numbers.

a. $\sqrt{5} + \sqrt{5}$ **b.** $6\sqrt{15} + 3\sqrt{15} + \sqrt{15}$

c. $\sqrt{xy} - 6\sqrt{xy} + 4\sqrt{xy}$ **d.** $4\sqrt{2} - 7\sqrt{2}$

Avoiding Mistakes:

The process of adding *like* radicals with the distributive property is similar to adding *like* terms. The end result is that the numerical coefficients are added and the radical factor is unchanged.

$\sqrt{5} + \sqrt{5}$

$= 1\sqrt{5} + 1\sqrt{5}$

$= 2\sqrt{5}$ Correct

Be careful: $\sqrt{5} + \sqrt{5} \neq \sqrt{10}$
In general,

$$\sqrt{x} + \sqrt{y} \neq \sqrt{x + y}$$

Solution:

a. $\sqrt{5} + \sqrt{5}$

$= 1\sqrt{5} + 1\sqrt{5}$ — *Note:* $\sqrt{5} = 1\sqrt{5}$.

$= (1 + 1)\sqrt{5}$ — Apply the distributive property.

$= 2\sqrt{5}$ — Simplify.

b. $6\sqrt{15} + 3\sqrt{15} + \sqrt{15}$ — The radicals have the same radicand and same index.

$= 6\sqrt{15} + 3\sqrt{15} + 1\sqrt{15}$ — *Note:* $\sqrt{15} = 1\sqrt{15}$.

$= (6 + 3 + 1)\sqrt{15}$ — Apply the distributive property.

$= 10\sqrt{15}$

c. $\sqrt{xy} - 6\sqrt{xy} + 4\sqrt{xy}$ — The radicals have the same radicand and same index.

$= 1\sqrt{xy} - 6\sqrt{xy} + 4\sqrt{xy}$ — *Note:* $\sqrt{xy} = 1\sqrt{xy}$.

$= (1 - 6 + 4)\sqrt{xy}$ — Apply the distributive property.

$= -1\sqrt{xy}$ — Simplify.

$= -\sqrt{xy}$

d. $4\sqrt{2} - 7\sqrt{2}$ — The radicals have the same radicand and same index.

$= (4 - 7)\sqrt{2}$ — Apply the distributive property.

$= -3\sqrt{2}$

Skill Practice Add or subtract the radicals as indicated. Assume the variables represent positive real numbers.

1. $3\sqrt{2} + 7\sqrt{2}$
2. $8\sqrt{x} - \sqrt{x}$
3. $4\sqrt{ab} - 2\sqrt{ab} - 9\sqrt{ab}$
4. $6\sqrt{5} - 10\sqrt{5} + \sqrt{5}$

From Example 1, you can see that the process of adding or subtracting *like* radicals is similar to combining *like* terms.

$$x + x \quad \text{similarly,} \quad \sqrt{5} + \sqrt{5}$$
$$= 1x + 1x \qquad\qquad = 1\sqrt{5} + 1\sqrt{5}$$
$$= 2x \qquad\qquad = 2\sqrt{5}$$

The end result is that the numerical coefficients are added or subtracted and the radical factor is unchanged.

Sometimes it is necessary to simplify radicals before adding or subtracting.

Example 2 **Simplifying Radicals before Adding or Subtracting**

Add or subtract the radicals as indicated. Assume the variables represent positive real numbers.

a. $\sqrt{20} + 7\sqrt{5}$
b. $\sqrt{50} - \sqrt{8}$
c. $-4\sqrt{3x^2} - x\sqrt{27} + 5x\sqrt{3}$
d. $a\sqrt{8a^5} + 6\sqrt{2a^7} + \sqrt{9a}$

Skill Practice Answers

1. $10\sqrt{2}$ **2.** $7\sqrt{x}$
3. $-7\sqrt{ab}$ **4.** $-3\sqrt{5}$

Solution:

a. $\sqrt{20} + 7\sqrt{5}$ — Because the radicands are different, try simplifying the radicals first.

$= \sqrt{2^2 \cdot 5} + 7\sqrt{5}$ — Factor the radicand.

$= 2\sqrt{5} + 7\sqrt{5}$ — The terms are *like* radicals.

$= (2 + 7)\sqrt{5}$ — Apply the distributive property.

$= 9\sqrt{5}$ — Simplify.

b. $\sqrt{50} - \sqrt{8}$ — Because the radicands are different, try simplifying the radicals first.

$= \sqrt{5^2 \cdot 2} - \sqrt{2^2 \cdot 2}$ — Factor the radicands.

$= 5\sqrt{2} - 2\sqrt{2}$ — The terms are *like* radicals.

$= (5 - 2)\sqrt{2}$ — Apply the distributive property.

$= 3\sqrt{2}$ — Simplify.

c. $-4\sqrt{3x^2} - x\sqrt{27} + 5x\sqrt{3}$ — Simplify each radical.

$= -4\sqrt{3x^2} - x\sqrt{3^2 \cdot 3} + 5x\sqrt{3}$ — Factor the radicands.

$= -4x\sqrt{3} - 3x\sqrt{3} + 5x\sqrt{3}$ — The terms are *like* radicals.

$= (-4x - 3x + 5x)\sqrt{3}$ — Apply the distributive property.

$= -2x\sqrt{3}$ — Simplify.

d. $a\sqrt{8a^5} + 6\sqrt{2a^7} + \sqrt{9a}$ — Simplify each radical.

$= a\sqrt{2^3a^5} + 6\sqrt{2a^7} + \sqrt{3^2a}$

$= a\sqrt{2^2a^4 \cdot 2a} + 6\sqrt{a^6 \cdot 2a} + \sqrt{3^2 \cdot a}$ — Factor the radicals.

$= a \cdot 2a^2\sqrt{2a} + 6 \cdot a^3\sqrt{2a} + 3\sqrt{a}$ — The terms are *like* radicals.

$= 2a^3\sqrt{2a} + 6a^3\sqrt{2a} + 3\sqrt{a}$ — The first two terms are *like* radicals.

$= (2a^3 + 6a^3)\sqrt{2a} + 3\sqrt{a}$ — Apply the distributive property.

$= 8a^3\sqrt{2a} + 3\sqrt{a}$

Skill Practice Add or subtract the radicals as indicated. Assume the variables represent positive real numbers.

5. $4\sqrt{18} + \sqrt{8}$

6. $\sqrt{50y} - \sqrt{98y}$

7. $4x\sqrt{12} - \sqrt{27x^2}$

8. $\surd\sqrt{28y^3} - y\sqrt{63y} + \sqrt{700}$

It is important to realize that only *like* radicals may be added or subtracted. The next example provides extra practice for recognizing *unlike* radicals.

Skill Practice Answers

5. $14\sqrt{2}$ **6.** $-2\sqrt{2y}$

7. $5x\sqrt{3}$ **8.** $-y\sqrt{7y} + 10\sqrt{7}$

Example 3 **Recognizing *Unlike* Radicals**

The following radicals cannot be simplified further by adding or subtracting. Why?

a. $2\sqrt{7} - 5\sqrt{3}$ **b.** $7 + 4\sqrt{5}$

Solution:

a. $2\sqrt{7} - 5\sqrt{3}$ Radicands are not the same.

b. $7 + 4\sqrt{5}$ One term has a radical, and one does not.

Skill Practice Simplify the expressions as much as possible. Assume the variables represent positive real numbers.

9. $12 - 7\sqrt{5}$ **10.** $2\sqrt{3} - 3\sqrt{2}$

Calculator Connections

A calculator can be used to evaluate a radical expression and its simplified form. For example, use a calculator to evaluate the expression on the left and its simplified form on the right: $2\sqrt{5} + 6\sqrt{5} = 8\sqrt{5}$

Scientific Calculator

Enter: 5 √ × 2 = + 5 √ × 6 = **Result:** 17.88854382

Enter: 5 √ × 8 = **Result:** 17.88854382

Graphing Calculator

```
2√(5)+6√(5)
      17.88854382
8√(5)
      17.88854382
```

A calculator can help you determine when a rule has been applied *incorrectly.* For example, use a calculator to show that $\sqrt{3} + \sqrt{5} \neq \sqrt{8}$

Scientific Calculator

Enter: 3 √ + 5 √ = **Result:** 3.968118785

Enter: 8 √ **Result:** 2.828427125

Values are not equal.

Graphing Calculator

```
√(3)+√(5)
      3.968118785
√(8)
      2.828427125
```

Values are not equal.

Skill Practice Answers

9. Cannot be simplified.
10. Cannot be simplified.

Calculator Exercises

Simplify the radical expression algebraically. Then use a calculator to approximate the original expression and its simplified form.

1. $2\sqrt{3} + 4\sqrt{3}$

2. $-\sqrt{5} - 4\sqrt{5} + 3\sqrt{5}$

3. $\sqrt{20} + \sqrt{5}$

4. $4\sqrt{6} - 7\sqrt{6}$

Section 8.3 Practice Exercises

Study Skills Exercise

1. Define the key term ***like* radicals**.

Review Exercises

For Exercises 2–9, simplify the expressions. Assume the variables represent positive real numbers.

2. $\sqrt{25w^2}$

3. $\sqrt[3]{8y^3}$

4. $\sqrt[3]{4z^4}$

5. $\sqrt{36x^3}$

6. $\sqrt{\dfrac{3a^5}{b^4}}$

7. $\dfrac{\sqrt{5c^6}}{\sqrt{16}}$

8. $\dfrac{\sqrt{2x^3}}{\sqrt{x}}$

9. $\sqrt{-25}$

Concept 1: Definition of *Like* Radicals

10. How do you determine whether two radicals are *like* or *unlike*?

11. Write two radicals that are considered *unlike*.

12. From the three pairs of radicals, identify the pair of *like* radicals:

i. $2\sqrt{x}$ and $8\sqrt[3]{x}$

ii. $\sqrt{5}$ and $-3\sqrt{5}$

iii. $3a\sqrt{3}$ and $3a\sqrt{2}$

13. From the three pairs of radicals, identify the pair of *like* radicals:

i. $13\sqrt{5b}$ and $13b\sqrt{5}$

ii. $\sqrt[4]{x^2y}$ and $\sqrt[3]{x^2y}$

iii. $-2\sqrt[3]{y^2}$ and $6\sqrt[3]{y^2}$

Concept 2: Addition and Subtraction of Radicals

For Exercises 14–27, add or subtract the expressions, if possible. Assume the variables represent positive real numbers.

14. $8\sqrt{6} + 2\sqrt{6}$

15. $3\sqrt{2} + 5\sqrt{2}$

16. $4\sqrt{3} - 2\sqrt{3} + 5\sqrt{3}$

17. $5\sqrt{7} - 3\sqrt{7} + 2\sqrt{7}$

18. $\sqrt{11} + \sqrt{11}$

19. $\sqrt{10} + \sqrt{10}$

20. $12\sqrt{x} - 3\sqrt{x}$

21. $15\sqrt{y} - 4\sqrt{y}$

22. $-3\sqrt{a} + 2\sqrt{a} + \sqrt{a}$

23. $5\sqrt{c} - 6\sqrt{c} + \sqrt{c}$

24. $7x\sqrt{11} - 9x\sqrt{11}$

25. $8y\sqrt{15} - 3y\sqrt{15}$

26. $9\sqrt{2} - 9\sqrt{5}$

27. $x\sqrt{y} - y\sqrt{x}$

For Exercises 28–31, translate the English phrase into an algebraic expression. Then simplify the expression.

28. The sum of three times the cube root of six and eight times the cube root of six.

29. The difference of negative two times the cube root of w and five times the cube root of w.

30. Four times the principal square root of five, minus six times the principal square root of five.

31. Eight times the principal square root of two, plus the principal square root of two.

For Exercises 32–61, simplify. Then add or subtract the expressions, if possible. Assume the variables represent positive real numbers.

32. $2\sqrt{12} + \sqrt{48}$

33. $5\sqrt{32} + 2\sqrt{50}$

34. $4\sqrt{45} - 6\sqrt{20}$

35. $8\sqrt{54} - 4\sqrt{24}$

36. $\frac{1}{2}\sqrt{8} + \frac{1}{3}\sqrt{18}$

37. $\frac{1}{4}\sqrt{32} - \frac{1}{5}\sqrt{50}$

38. $6p\sqrt{20p^2} + p^2\sqrt{80}$

39. $2q\sqrt{48} + \sqrt{27q^2}$

40. $-2\sqrt{2k} + 6\sqrt{8k}$

41. $5\sqrt{27x} - 4\sqrt{12x}$

42. $11\sqrt{a^4b} - a^2\sqrt{b} - 9a\sqrt{a^2b}$

43. $-7\sqrt{x^4y} + 5x^2\sqrt{y} - 6x\sqrt{x^2y}$

44. $4\sqrt{5} - \sqrt{5}$

45. $-3\sqrt{10} - \sqrt{10}$

46. $\frac{5}{6}z\sqrt{6} + \frac{7}{9}z\sqrt{6}$

47. $\frac{3}{4}a\sqrt{b} + \frac{1}{6}a\sqrt{b}$

48. $1.1\sqrt{10} - 5.6\sqrt{10} + 2.8\sqrt{10}$

49. $0.25\sqrt{x} + 1.50\sqrt{x} - 0.75\sqrt{x}$

50. $4\sqrt{x^3} - 2x\sqrt{x}$

51. $8\sqrt{y^9} - 2y^2\sqrt{y^5}$

52. $4\sqrt{7} + \sqrt{63} - 2\sqrt{28}$

53. $8\sqrt{3} - 2\sqrt{27} + \sqrt{75}$

54. $\sqrt{16w} + \sqrt{24w} + \sqrt{40w}$

55. $\sqrt{54y} + \sqrt{81y} - \sqrt{12y}$

56. $\sqrt{x^6y} + 5x^2\sqrt{x^2y}$

57. $7\sqrt{a^5b^2} - a^2\sqrt{ab^2}$

58. $4\sqrt{6} + 2\sqrt{3} - 8\sqrt{6}$

59. $-7\sqrt{y} - \sqrt{z} + 2\sqrt{z}$

60. $x\sqrt{8} - 2\sqrt{18x^2} + \sqrt{2x}$

61. $5\sqrt{p^5} - 2p\sqrt{p} + p\sqrt{16p^3}$

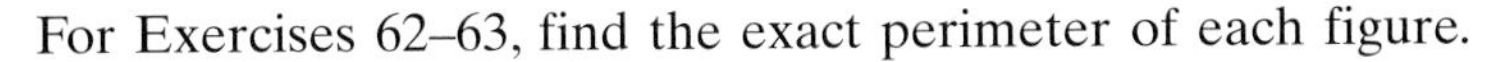

For Exercises 62–63, find the exact perimeter of each figure.

62. **63.**

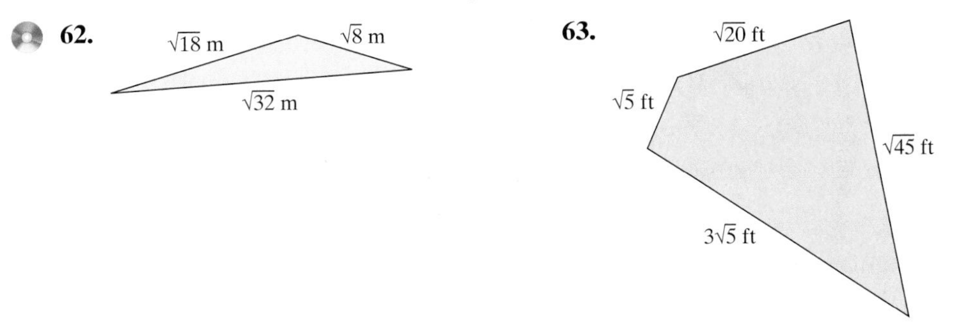

64. Find the exact perimeter of a rectangle whose width is $2\sqrt{3}$ in. and whose length is $3\sqrt{12}$ in.

65. Find the exact perimeter of a square whose side length is $5\sqrt{8}$ cm.

For Exercises 66–71, determine the reason why the following radical expressions cannot be combined by addition or subtraction.

66. $\sqrt{5} + 5\sqrt{2}$ **67.** $3\sqrt{10} + 10\sqrt{3}$ **68.** $3 + 5\sqrt{7}$

69. $-2 + 5\sqrt{11}$ **70.** $-4 + 10\sqrt{6}$ **71.** $8 - 3\sqrt{2}$

Expanding Your Skills

72. Find the slope of the line through the points: $(4, 2\sqrt{3})$ and $(1, \sqrt{3})$.

73. Find the slope of the line through the points: $(7, 4\sqrt{5})$ and $(2, 3\sqrt{5})$.

74. A golfer hits a golf ball at an angle of 30° with an initial velocity of 46.0 meters/second (m/sec). The horizontal position of the ball, x (measured in meters), depends on the number of seconds, t, after the ball is struck according to the equation:

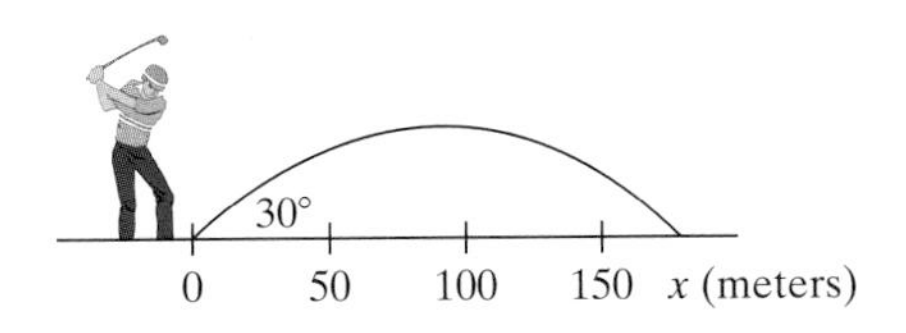

$$x = 23\sqrt{3}t$$

a. What is the horizontal position of the ball after 2 sec? Round the answer to the nearest meter.

b. What is the horizontal position of the ball after 4 sec? Round the answer to the nearest meter.

75. A long-jumper leaves the ground at an angle of 30° at a speed of 9 m/sec. The horizontal position of the long jumper, x (measured in meters), depends on the number of seconds, t, after he leaves the ground according to the equation:

$$x = 4.5\sqrt{3}t$$

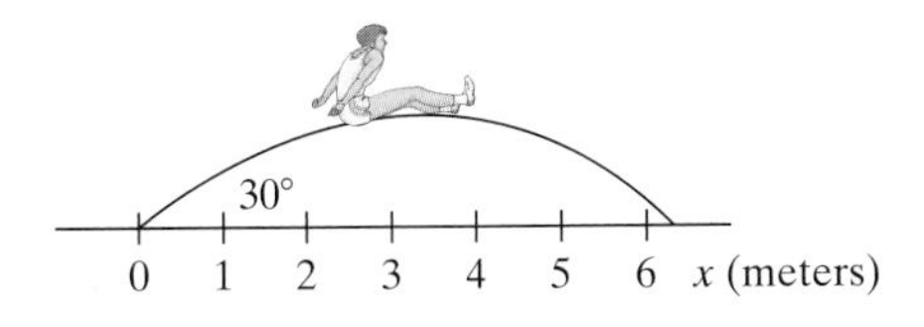

a. What is the horizontal position of the long-jumper after 0.5 sec? Round the answer to the nearest hundredth of a meter.

b. What is the horizontal position of the long-jumper after 0.75 sec? Round the answer to the nearest hundredth of a meter.

Section 8.4 Multiplication of Radicals

Concepts

1. Multiplication Property of Radicals
2. Expressions of the Form $(\sqrt[n]{a})^n$
3. Special Case Products
4. Multiplying Conjugate Radical Expressions

1. Multiplication Property of Radicals

In this section, we will learn how to multiply radicals that have the same index. Recall from Section 8.2 the multiplication property of radicals.

Multiplication Property of Radicals

Let a and b represent real numbers such that $\sqrt[n]{a}$ and $\sqrt[n]{b}$ are both real. Then,

$$\sqrt[n]{a} \cdot \sqrt[n]{b} = \sqrt[n]{ab}$$

To multiply two radical expressions, use the multiplication property of radicals along with the commutative and associative properties of multiplication.

Example 1 Multiplying Radical Expressions

Multiply the expressions and simplify the result. Assume the variables represent positive real numbers.

a. $(5\sqrt{3})(4\sqrt{2})$ **b.** $(2x\sqrt{3})(2\sqrt{15})$ **c.** $(6a\sqrt{ab})\left(\frac{1}{3}a\sqrt{a}\right)$

Solution:

a. $(5\sqrt{3})(4\sqrt{2}) = (5 \cdot 4)(\sqrt{3} \cdot \sqrt{2})$ Commutative and associative properties of multiplication

$= 20\sqrt{6}$ Multiplication property of radicals

b. $(2x\sqrt{3})(2\sqrt{15}) = (2x \cdot 2)(\sqrt{3} \cdot \sqrt{15})$ Commutative and associative properties of multiplication

$= 4x\sqrt{45}$ Multiplication property of radicals

$= 4x\sqrt{3^2 \cdot 5}$ Simplify the radical.

$= 4x \cdot 3\sqrt{5}$

$= 12x\sqrt{5}$

c. $(6a\sqrt{ab})\left(\frac{1}{3}a\sqrt{a}\right) = \left(6a \cdot \frac{1}{3}a\right)(\sqrt{ab} \cdot \sqrt{a})$ Commutative and associative properties of multiplication

$= 2a^2\sqrt{a^2b}$ Multiplication property of radicals

$= 2a^2 \cdot a\sqrt{b}$ Simplify the radical.

$= 2a^3\sqrt{b}$

Skill Practice Multiply the expressions and simplify the result. Assume the variables represent positive real numbers.

1. $(2\sqrt{3})(5\sqrt{2})$ **2.** $(-5z\sqrt{6})(4z\sqrt{2})$ **3.** $(9y\sqrt{x})\left(\frac{1}{3}y\sqrt{xy}\right)$

Skill Practice Answers

1. $10\sqrt{6}$ **2.** $-40z^2\sqrt{3}$
3. $3xy^2\sqrt{y}$

When multiplying radical expressions with more than one term, we use the distributive property.

Example 2 **Multiplying Radical Expressions with Multiple Terms**

Multiply the expressions. Assume the variables represent positive real numbers.

a. $\sqrt{5}(4 + 3\sqrt{5})$ **b.** $(\sqrt{x} - 10)(\sqrt{y} + 4)$ **c.** $(2\sqrt{3} - \sqrt{5})(\sqrt{3} + 6\sqrt{5})$

Solution:

a. $\sqrt{5}(4 + 3\sqrt{5})$

$= \sqrt{5}(4) + \sqrt{5}(3\sqrt{5})$	Apply the distributive property.
$= 4\sqrt{5} + 3\sqrt{5^2}$	Multiplication property of radicals
$= 4\sqrt{5} + 3 \cdot 5$	Simplify the radical.
$= 4\sqrt{5} + 15$	

b. $(\sqrt{x} - 10)(\sqrt{y} + 4)$

$= \sqrt{x}(\sqrt{y}) + \sqrt{x}(4) - 10(\sqrt{y}) - 10(4)$	Apply the distributive property.
$= \sqrt{xy} + 4\sqrt{x} - 10\sqrt{y} - 40$	Simplify.

c. $(2\sqrt{3} - \sqrt{5})(\sqrt{3} + 6\sqrt{5})$

$= 2\sqrt{3}(\sqrt{3}) + 2\sqrt{3}(6\sqrt{5}) - \sqrt{5}(\sqrt{3}) - \sqrt{5}(6\sqrt{5})$	Apply the distributive property.
$= 2\sqrt{3^2} + 12\sqrt{15} - \sqrt{15} - 6\sqrt{5^2}$	Multiplication property of radicals
$= 2 \cdot 3 + 11\sqrt{15} - 6 \cdot 5$	Simplify radicals. Combine *like* radicals.
$= 6 + 11\sqrt{15} - 30$	
$= -24 + 11\sqrt{15}$	Combine *like* terms.

Skill Practice Multiply the expressions and simplify the result. Assume the variables represent positive real numbers.

4. $\sqrt{7}(2\sqrt{7} - 4)$ **5.** $(\sqrt{x} + 2)(\sqrt{x} - 3)$ **6.** $(2\sqrt{a} + 4\sqrt{6})(\sqrt{a} - 3\sqrt{6})$

2. Expressions of the Form $(\sqrt[n]{a})^n$

The multiplication property of radicals can be used to simplify an expression of the form $(\sqrt{a})^2$, where $a \geq 0$.

$$(\sqrt{a})^2 = \sqrt{a} \cdot \sqrt{a} = \sqrt{a^2} = a, \text{ where } a \geq 0$$

This logic can be applied to nth-roots. If $\sqrt[n]{a}$ is a real number, then $(\sqrt[n]{a})^n = a$.

Skill Practice Answers

4. $14 - 4\sqrt{7}$ **5.** $x - \sqrt{x} - 6$
6. $2a - 2\sqrt{6a} - 72$

Example 3 Simplifying Radical Expressions

Simplify the expressions. Assume the variables represent positive real numbers.

a. $(\sqrt{7})^2$ **b.** $(\sqrt[4]{x})^4$ **c.** $(3\sqrt{2})^2$

Solution:

a. $(\sqrt{7})^2 = 7$ **b.** $(\sqrt[4]{x})^4 = x$ **c.** $(3\sqrt{2})^2 = 3^2 \cdot (\sqrt{2})^2 = 9 \cdot 2 = 18$

Skill Practice Simplify the expressions. Assume the variables represent positive real numbers.

7. $(\sqrt{13})^2$ **8.** $(\sqrt[3]{x})^3$ **9.** $(2\sqrt{11})^2$

3. Special Case Products

From Example 2, you may have noticed a similarity between multiplying radical expressions and multiplying polynomials.

Recall from Section 5.6 that the square of a binomial results in a perfect square trinomial.

$$(a + b)^2 = a^2 + 2ab + b^2$$
$$(a - b)^2 = a^2 - 2ab + b^2$$

The same patterns occur when squaring a radical expression with two terms.

Example 4 Squaring a Two-Term Radical Expression

Square the radical expressions as indicated. Assume the variables represent positive real numbers.

a. $(\sqrt{x} + \sqrt{y})^2$ **b.** $(\sqrt{2} - 4\sqrt{3})^2$

Solution:

a. $(\sqrt{x} + \sqrt{y})^2$ This expression is in the form $(a + b)^2$, where $a = \sqrt{x}$ and $b = \sqrt{y}$.

$a^2 + 2ab + b^2$

$= (\sqrt{x})^2 + 2(\sqrt{x})(\sqrt{y}) + (\sqrt{y})^2$ Apply the formula $(a + b)^2 = a^2 + 2ab + b^2$.

$= x + 2\sqrt{xy} + y$ Simplify.

TIP: The product $(\sqrt{x} + \sqrt{y})^2$ can also be found using the distributive property.

$$(\sqrt{x} + \sqrt{y})^2 = (\sqrt{x} + \sqrt{y})(\sqrt{x} + \sqrt{y}) = \sqrt{x} \cdot \sqrt{x} + \sqrt{x} \cdot \sqrt{y} + \sqrt{y} \cdot \sqrt{x} + \sqrt{y} \cdot \sqrt{y}$$
$$= \sqrt{x^2} + \sqrt{xy} + \sqrt{xy} + \sqrt{y^2}$$
$$= x + 2\sqrt{xy} + y$$

Skill Practice Answers

7. 13 **8.** x **9.** 44

b. $(\sqrt{2} - 4\sqrt{3})^2$ This expression is in the form $(a - b)^2$, where $a = \sqrt{2}$ and $b = 4\sqrt{3}$.

$a^2 - 2ab + b^2$

$(\sqrt{2})^2 - 2(\sqrt{2})(4\sqrt{3}) + (4\sqrt{3})^2$ Apply the formula $(a - b)^2 = a^2 - 2ab + b^2$.

$= 2 - 8\sqrt{6} + 16 \cdot 3$ Simplify.

$= 2 - 8\sqrt{6} + 48$

$= 50 - 8\sqrt{6}$

Skill Practice Square the radical expressions. Assume the variables represent positive real numbers.

10. $(\sqrt{p} + 3)^2$ **11.** $(\sqrt{5} - 3\sqrt{2})^2$

4. Multiplying Conjugate Radical Expressions

Recall from Section 5.6 that the product of two conjugate binomials results in a difference of squares.

$$(a + b)(a - b) = a^2 - b^2$$

The same pattern occurs when multiplying two conjugate radical expressions.

Example 5 Multiplying Conjugate Radical Expressions

Multiply the radical expressions. Assume the variables represent positive real numbers.

a. $(\sqrt{5} + 4)(\sqrt{5} - 4)$ **b.** $(2\sqrt{c} - 3\sqrt{d})(2\sqrt{c} + 3\sqrt{d})$

Solution:

a. $(\sqrt{5} + 4)(\sqrt{5} - 4)$ This expression is in the form $(a + b)(a - b)$, where $a = \sqrt{5}$ and $b = 4$.

$a^2 - b^2$

$= (\sqrt{5})^2 - (4)^2$ Apply the formula $(a + b)(a - b) = a^2 - b^2$.

$= 5 - 16$ Simplify.

$= -11$

TIP: The product $(\sqrt{5} + 4)(\sqrt{5} - 4)$ can also be found using the distributive property.

$(\sqrt{5} + 4)(\sqrt{5} - 4) = \sqrt{5} \cdot (\sqrt{5}) + \sqrt{5} \cdot (-4) + 4 \cdot (\sqrt{5}) + 4 \cdot (-4)$

$= 5 - 4\sqrt{5} + 4\sqrt{5} - 16$

$= 5 - 16$

$= -11$

Skill Practice Answers

10. $p + 6\sqrt{p} + 9$
11. $23 - 6\sqrt{10}$

b. $(2\sqrt{c} - 3\sqrt{d})(2\sqrt{c} + 3\sqrt{d})$ This expression is in the form $(a - b)(a + b)$, where $a = 2\sqrt{c}$ and $b = 3\sqrt{d}$.

$$a^2 - b^2$$

$= (2\sqrt{c})^2 - (3\sqrt{d})^2$ Apply the formula $(a + b)(a - b) = a^2 - b^2$.

$= 4c - 9d$

Skill Practice Multiply the radical expressions.

12. $(\sqrt{6} - 3)(\sqrt{6} + 3)$ **13.** $(5\sqrt{a} + \sqrt{b})(5\sqrt{a} - \sqrt{b})$

TIP: The decimal approximations for $(3\sqrt{5})(4\sqrt{2})$ and $12\sqrt{10}$ agree for the first 10 digits. This in itself does not make $(3\sqrt{5})(4\sqrt{2}) = 12\sqrt{10}$. It is the multiplication property of radicals that guarantees that the expressions are equal.

Calculator Connections

A calculator can support the multiplication property of radicals. For example, use a calculator to evaluate $(3\sqrt{5})(4\sqrt{2})$ and its simplified form $12\sqrt{10}$.

Scientific Calculator

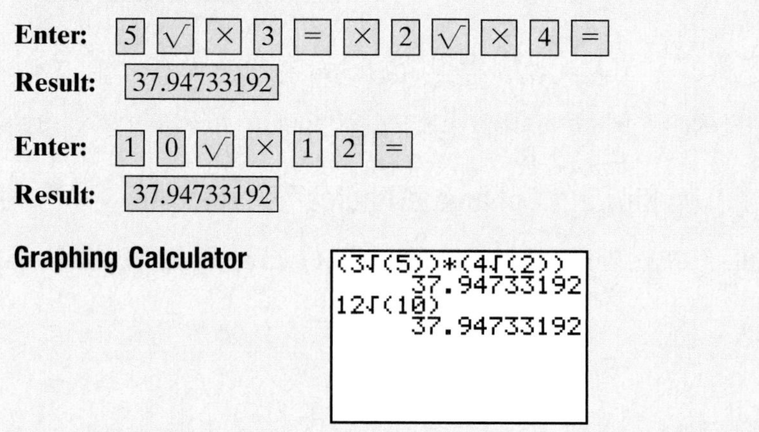

Calculator Exercises

Simplify the radical expressions algebraically. Then use a calculator to approximate the original expression and its simplified form.

1. $(3\sqrt{5})(4\sqrt{10})$

2. $(4\sqrt{6})(7\sqrt{10})$

3. $(\sqrt{2} - \sqrt{3})(\sqrt{2} + \sqrt{3})$

4. $(\sqrt{5} + \sqrt{7})(\sqrt{5} - \sqrt{7})$

5. $(2 + \sqrt{11})^2$

6. $(\sqrt{5} - 4)^2$

Skill Practice Answers

12. -3 **13.** $25a - b$

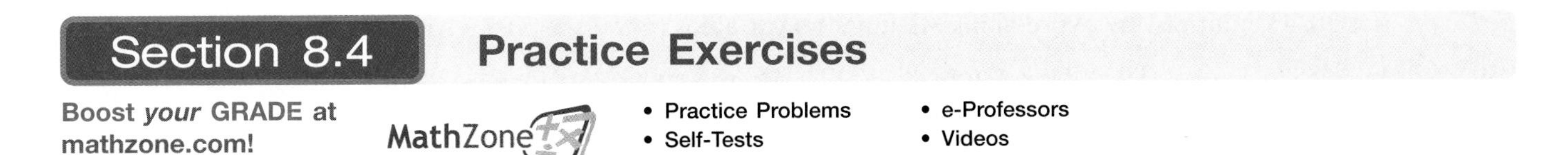

Section 8.4 Practice Exercises

Boost *your* GRADE at mathzone.com! MathZone

- Practice Problems
- Self-Tests
- NetTutor
- e-Professors
- Videos

Study Skills Exercise

1. When writing a radical expression, be sure to note the difference between an exponent on a coefficient and an index to a radical. Write an algebraic expression for each of the following:

x cubed times the square root of y

x times the cube root of y

Review Exercises

For Exercises 2–5, perform the indicated operations and simplify. Assume the variables represent positive real numbers.

2. $\sqrt{25} + \sqrt{16} - \sqrt{36}$ **3.** $\sqrt{100} - \sqrt{4} + \sqrt{9}$ **4.** $6x\sqrt{18} + 2\sqrt{2x^2}$ **5.** $10\sqrt{zw^4} - w^2\sqrt{49z}$

Concept 1: Multiplication Property of Radicals

For Exercises 6–25, multiply the expressions.

6. $\sqrt{5} \cdot \sqrt{3}$ **7.** $\sqrt{7} \cdot \sqrt{6}$ **8.** $\sqrt{47} \cdot \sqrt{47}$ **9.** $\sqrt{59} \cdot \sqrt{59}$

10. $\sqrt{b} \cdot \sqrt{b}$ **11.** $\sqrt{t} \cdot \sqrt{t}$ **12.** $2\sqrt{15} \cdot 3\sqrt{p}$ **13.** $4\sqrt{2} \cdot 5\sqrt{q}$

14. $\sqrt{10} \cdot \sqrt{5}$ **15.** $\sqrt{2} \cdot \sqrt{10}$ **16.** $-\sqrt{7} \cdot (-2\sqrt{14})$ **17.** $-6\sqrt{2} \cdot (-\sqrt{22})$

18. $\sqrt{2} \cdot \sqrt{18}$ **19.** $\sqrt{3} \cdot \sqrt{27}$ **20.** $\frac{1}{3}\sqrt{27} \cdot \sqrt{50}$ **21.** $\sqrt{45} \cdot \frac{1}{6}\sqrt{32}$

22. $6\sqrt{5} \cdot \sqrt{5}$ **23.** $\sqrt{x} \cdot 4\sqrt{x}$ **24.** $-2\sqrt{3} \cdot 4\sqrt{5}$ **25.** $-\sqrt{7} \cdot 2\sqrt{3}$

For Exercises 26–27, find the exact perimeter and exact area of the rectangles.

26. **27.**

For Exercises 28–29, find the exact area of the triangles.

28. **29.**

For Exercises 30–43, multiply the expressions. Assume the variables represent positive real numbers.

30. $\sqrt{3w} \cdot \sqrt{3w}$ **31.** $\sqrt{6p} \cdot \sqrt{6p}$ **32.** $(8\sqrt{5y})(-2\sqrt{2})$ **33.** $(4\sqrt{5x})(7\sqrt{3})$

34. $\sqrt{2}(\sqrt{6} - \sqrt{3})$ **35.** $\sqrt{5}(\sqrt{10} + \sqrt{7})$ **36.** $4\sqrt{x}(\sqrt{x} + 5)$ **37.** $2\sqrt{y}(3 - \sqrt{y})$

38. $(\sqrt{3} + 2\sqrt{10})(4\sqrt{3} - \sqrt{10})$ **39.** $(8\sqrt{7} - \sqrt{5})(\sqrt{7} + 3\sqrt{5})$

40. $(\sqrt{a} - 3b)(9\sqrt{a} - b)$ **41.** $(11\sqrt{m} + 4n)(\sqrt{m} + n)$

42. $(p + 2\sqrt{p})(8p + 3\sqrt{p} - 4)$ **43.** $(5s - \sqrt{s})(s + 5\sqrt{s} + 6)$

Concept 2: Expressions of the Form $(\sqrt[n]{a})^n$

For Exercises 44–51, simplify the expressions. Assume the variables represent positive real numbers.

44. $(\sqrt{10})^2$ **45.** $(\sqrt{23})^2$ **46.** $(\sqrt[3]{4})^3$ **47.** $(\sqrt[3]{29})^3$

48. $(\sqrt[4]{t})^4$ **49.** $(\sqrt[4]{xy})^4$ **50.** $(4\sqrt{c})^2$ **51.** $(10\sqrt{2pq})^2$

Concept 3: Special Case Products

For Exercises 52–61, multiply the radical expressions. Assume the variables represent positive real numbers.

52. $(\sqrt{13} + 4)^2$ **53.** $(6 - \sqrt{11})^2$ **54.** $(\sqrt{a} - 2)^2$ **55.** $(\sqrt{p} + 3)^2$

56. $(2\sqrt{a} - 3)^2$ **57.** $(3\sqrt{w} + 4)^2$ **58.** $(\sqrt{x} - 2\sqrt{y})^2$ **59.** $(5\sqrt{c} + 2\sqrt{d})^2$

60. $(\sqrt{10} - \sqrt{11})^2$ **61.** $(\sqrt{3} - \sqrt{2})^2$

Concept 4: Multiplying Conjugate Radical Expressions

For Exercises 62–73, multiply the radical expressions. Assume the variables represent positive real numbers.

62. $(\sqrt{5} + 2)(\sqrt{5} - 2)$ **63.** $(\sqrt{3} - 4)(\sqrt{3} + 4)$ **64.** $(\sqrt{x} + \sqrt{y})(\sqrt{x} - \sqrt{y})$

65. $(\sqrt{a} + \sqrt{b})(\sqrt{a} - \sqrt{b})$ **66.** $(\sqrt{10} - \sqrt{11})(\sqrt{10} + \sqrt{11})$ **67.** $(\sqrt{3} - \sqrt{2})(\sqrt{3} + \sqrt{2})$

68. $(\sqrt{6} + \sqrt{2})(\sqrt{6} - \sqrt{2})$ **69.** $(\sqrt{15} - \sqrt{5})(\sqrt{15} + \sqrt{5})$ **70.** $(8\sqrt{x} - 2\sqrt{y})(8\sqrt{x} + 2\sqrt{y})$

71. $(4\sqrt{s} + 11\sqrt{t})(4\sqrt{s} - 11\sqrt{t})$ **72.** $(5\sqrt{3} - \sqrt{2})(5\sqrt{3} + \sqrt{2})$ **73.** $(2\sqrt{7} - 4\sqrt{3})(2\sqrt{7} + 4\sqrt{3})$

For Exercises 74–79, multiply the expressions in parts (a) and (b) and compare the process used. Assume the variables represent positive real numbers.

74. **a.** $3(x + 2)$ **b.** $\sqrt{3}(\sqrt{x} + \sqrt{2})$ **75.** **a.** $-5(6 + y)$ **b.** $-\sqrt{5}(\sqrt{6} + \sqrt{y})$

76. **a.** $(2a + 3)^2$ **b.** $(2\sqrt{a} + 3)^2$ **77.** **a.** $(6 - z)^2$ **b.** $(\sqrt{6} - z)^2$

78. **a.** $(b - 5)(b + 5)$ **b.** $(\sqrt{b} - 5)(\sqrt{b} + 5)$ **79.** **a.** $(3w - 1)(3w + 1)$ **b.** $(3\sqrt{w} - 1)(3\sqrt{w} + 1)$

Rationalization

Section 8.5

Concepts

1. Simplified Form of a Radical
2. Rationalizing the Denominator: One Term
3. Rationalizing the Denominator: Two Terms
4. Simplifying a Radical Quotient to Lowest Terms

1. Simplified Form of a Radical

Recall the conditions for a radical to be simplified.

Simplified Form of a Radical

Consider any radical expression where the radicand is written as a product of prime factors. The expression is in simplified form if all of the following conditions are met:

1. The radicand has no factor raised to a power greater than or equal to the index.
2. There are no radicals in the denominator of a fraction.
3. The radicand does not contain a fraction.

The basis of the second and third conditions, which restrict radicals from the denominator of an expression, are largely historical. In some cases, removing a radical from the denominator of a fraction will create an expression that is computationally simpler. For example, we will show that

$$\frac{2}{\sqrt{2}} = \sqrt{2} \qquad \text{and} \qquad \frac{2}{\sqrt{5} + \sqrt{3}} = \sqrt{5} - \sqrt{3}$$

The process to remove a radical from the denominator is called **rationalizing the denominator**. In this section, we will rationalize the denominator in two cases:

1. When the denominator contains a single radical term
2. When the denominator contains two terms involving square roots

2. Rationalizing the Denominator: One Term

To begin the first case, recall that the nth-root of a perfect nth power is easily simplified. For example,

$$\sqrt{x^2} = x \quad x \geq 0$$

Example 1 Rationalizing the Denominator: One Term

Simplify the expressions. Assume the variables represent positive real numbers.

a. $\frac{2}{\sqrt{2}}$ **b.** $\sqrt{\frac{x}{5}}$ **c.** $\frac{14\sqrt{w}}{\sqrt{7}}$ **d.** $\sqrt{\frac{w}{12}}$

Solution:

a. A square root of a perfect square is needed in the denominator to remove the radical.

$\frac{2}{\sqrt{2}} = \frac{2}{\sqrt{2}} \cdot \frac{\sqrt{2}}{\sqrt{2}}$ Multiply the numerator and denominator by $\sqrt{2}$ because $\sqrt{2} \cdot \sqrt{2} = \sqrt{2^2}$.

$= \frac{2\sqrt{2}}{\sqrt{2^2}}$ Multiply the radicals.

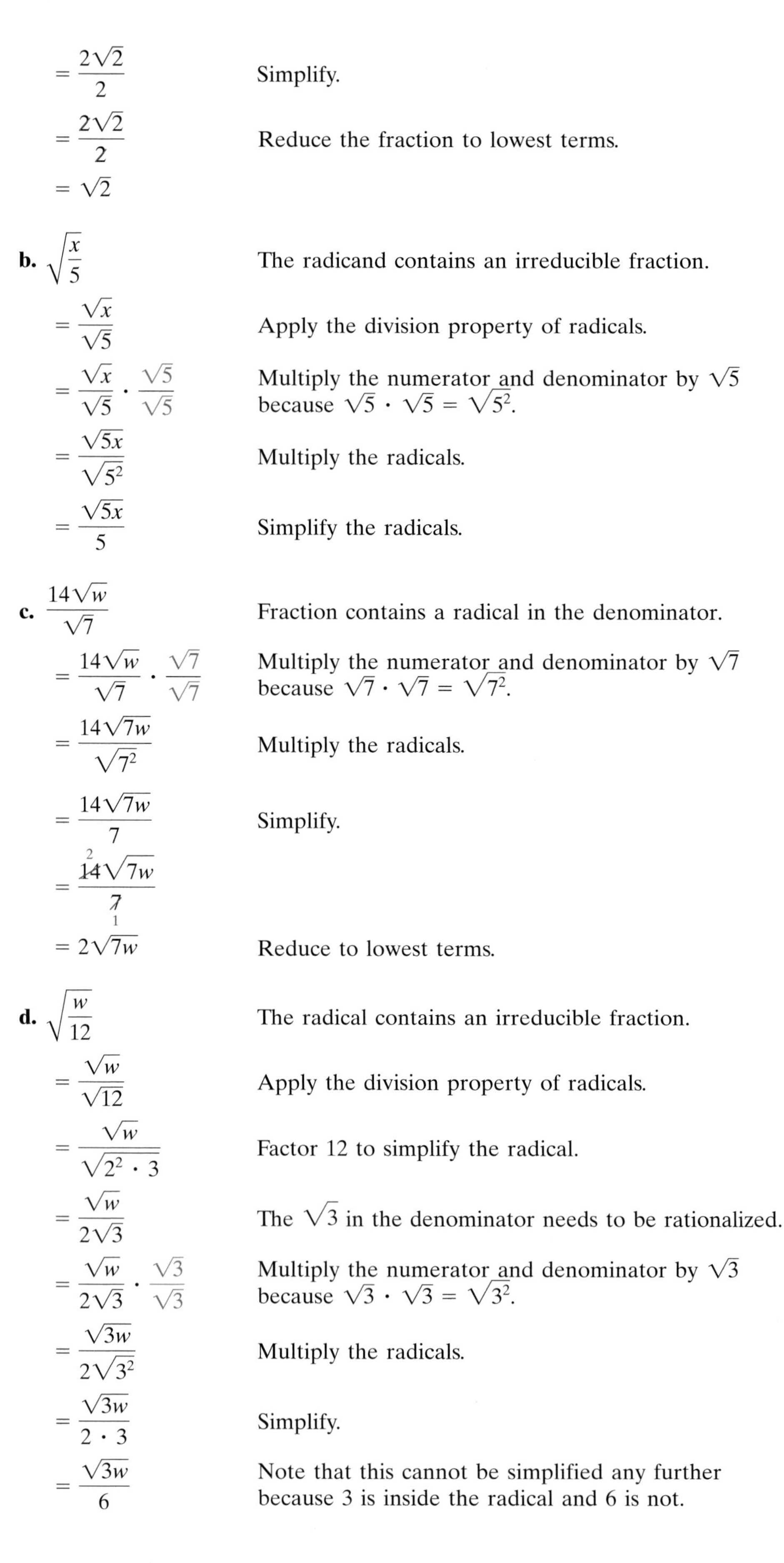

$= \dfrac{2\sqrt{2}}{2}$ Simplify.

$= \dfrac{2\sqrt{2}}{2}$ Reduce the fraction to lowest terms.

$= \sqrt{2}$

b. $\sqrt{\dfrac{x}{5}}$ The radicand contains an irreducible fraction.

$= \dfrac{\sqrt{x}}{\sqrt{5}}$ Apply the division property of radicals.

$= \dfrac{\sqrt{x}}{\sqrt{5}} \cdot \dfrac{\sqrt{5}}{\sqrt{5}}$ Multiply the numerator and denominator by $\sqrt{5}$ because $\sqrt{5} \cdot \sqrt{5} = \sqrt{5^2}$.

$= \dfrac{\sqrt{5x}}{\sqrt{5^2}}$ Multiply the radicals.

$= \dfrac{\sqrt{5x}}{5}$ Simplify the radicals.

Avoiding Mistakes:

In the expression $\dfrac{\sqrt{5x}}{5}$, do not try to "cancel" the factor of $\sqrt{5}$ from the numerator with the factor of 5 in the denominator.

c. $\dfrac{14\sqrt{w}}{\sqrt{7}}$ Fraction contains a radical in the denominator.

$= \dfrac{14\sqrt{w}}{\sqrt{7}} \cdot \dfrac{\sqrt{7}}{\sqrt{7}}$ Multiply the numerator and denominator by $\sqrt{7}$ because $\sqrt{7} \cdot \sqrt{7} = \sqrt{7^2}$.

$= \dfrac{14\sqrt{7w}}{\sqrt{7^2}}$ Multiply the radicals.

$= \dfrac{14\sqrt{7w}}{7}$ Simplify.

$= \dfrac{\overset{2}{\cancel{14}}\sqrt{7w}}{\underset{1}{\cancel{7}}}$

$= 2\sqrt{7w}$ Reduce to lowest terms.

TIP: In the expression

$$\frac{14\sqrt{7w}}{7}$$

the factor of 14 and the factor of 7 may be reduced because both are outside the radical.

$$\frac{14\sqrt{7w}}{7} = \frac{14}{7} \cdot \sqrt{7w}$$
$$= 2\sqrt{7w}$$

d. $\sqrt{\dfrac{w}{12}}$ The radical contains an irreducible fraction.

$= \dfrac{\sqrt{w}}{\sqrt{12}}$ Apply the division property of radicals.

$= \dfrac{\sqrt{w}}{\sqrt{2^2 \cdot 3}}$ Factor 12 to simplify the radical.

$= \dfrac{\sqrt{w}}{2\sqrt{3}}$ The $\sqrt{3}$ in the denominator needs to be rationalized.

$= \dfrac{\sqrt{w}}{2\sqrt{3}} \cdot \dfrac{\sqrt{3}}{\sqrt{3}}$ Multiply the numerator and denominator by $\sqrt{3}$ because $\sqrt{3} \cdot \sqrt{3} = \sqrt{3^2}$.

$= \dfrac{\sqrt{3w}}{2\sqrt{3^2}}$ Multiply the radicals.

$= \dfrac{\sqrt{3w}}{2 \cdot 3}$ Simplify.

$= \dfrac{\sqrt{3w}}{6}$ Note that this cannot be simplified any further because 3 is inside the radical and 6 is not.

Skill Practice Simplify the expression by rationalizing the denominator.

1. $\dfrac{3}{\sqrt{5}}$ **2.** $\sqrt{\dfrac{7}{10}}$ **3.** $\dfrac{6y}{\sqrt{3}}$ **4.** $\sqrt{\dfrac{z}{18}}$

3. Rationalizing the Denominator: Two Terms

Recall from the multiplication of polynomials that the product of two conjugates results in a difference of squares.

$$(a + b)(a - b) = a^2 - b^2$$

If either a or b has a square root factor, the expression will simplify without a radical; that is, the expression is rationalized. For example,

$$\begin{aligned}(\sqrt{5} - \sqrt{3})(\sqrt{5} + \sqrt{3}) &= (\sqrt{5})^2 - (\sqrt{3})^2 \\ &= 5 - 3 \\ &= 2\end{aligned}$$

Multiplying a binomial by its conjugate is the basis for rationalizing a denominator with two terms involving square roots.

Example 2 **Rationalizing the Denominator: Two Terms**

Simplify the expression by rationalizing the denominator.

$$\frac{2}{\sqrt{5} + \sqrt{3}}$$

Solution:

$\dfrac{2}{\sqrt{5} + \sqrt{3}}$ To rationalize a denominator with two terms, multiply the numerator and denominator by the conjugate of the denominator.

$= \dfrac{2}{(\sqrt{5} + \sqrt{3})} \cdot \dfrac{(\sqrt{5} - \sqrt{3})}{(\sqrt{5} - \sqrt{3})}$ (Conjugates) The denominator is in the form $(a + b)(a - b)$, where $a = \sqrt{5}$ and $b = \sqrt{3}$.

$= \dfrac{2(\sqrt{5} - \sqrt{3})}{(\sqrt{5})^2 - (\sqrt{3})^2}$ In the denominator, apply the formula $(a + b)(a - b) = a^2 - b^2$.

$= \dfrac{2(\sqrt{5} - \sqrt{3})}{5 - 3}$ Simplify.

$= \dfrac{2(\sqrt{5} - \sqrt{3})}{2}$

$= \dfrac{2(\sqrt{5} - \sqrt{3})}{2}$ Reduce to lowest terms.

$= \sqrt{5} - \sqrt{3}$

Skill Practice Answers

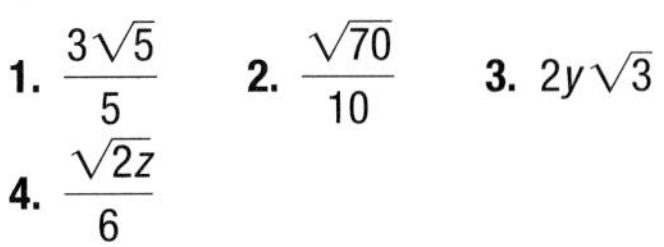

1. $\dfrac{3\sqrt{5}}{5}$ **2.** $\dfrac{\sqrt{70}}{10}$ **3.** $2y\sqrt{3}$ **4.** $\dfrac{\sqrt{2z}}{6}$

Skill Practice Simplify the expression by rationalizing the denominator.

5. $\dfrac{6}{\sqrt{3} - 1}$

Example 3 Rationalizing the Denominator: Two Terms

Simplify the expression by rationalizing the denominator.

$$\frac{\sqrt{x} + \sqrt{2}}{\sqrt{x} - \sqrt{2}}$$

Solution:

$$\frac{\sqrt{x} + \sqrt{2}}{\sqrt{x} - \sqrt{2}} = \frac{(\sqrt{x} + \sqrt{2})}{(\sqrt{x} - \sqrt{2})} \cdot \frac{(\sqrt{x} + \sqrt{2})}{(\sqrt{x} + \sqrt{2})}$$

Conjugates

Multiply the numerator and denominator by the conjugate of the denominator.

$$= \frac{(\sqrt{x} + \sqrt{2})^2}{(\sqrt{x} - \sqrt{2})(\sqrt{x} + \sqrt{2})}$$

$$= \frac{(\sqrt{x})^2 + 2(\sqrt{x})(\sqrt{2}) + (\sqrt{2})^2}{(\sqrt{x})^2 - (\sqrt{2})^2}$$

Simplify using special case products.

$$= \frac{x + 2\sqrt{2x} + 2}{x - 2}$$

Simplify the radicals.

Skill Practice Simplify the expression by rationalizing the denominators.

6. $\dfrac{\sqrt{y} - \sqrt{5}}{\sqrt{y} + \sqrt{5}}$

4. Simplifying a Radical Quotient to Lowest Terms

Sometimes a radical expression within a quotient must be reduced to lowest terms. This is demonstrated in the next example.

Example 4 Simplifying a Radical Quotient to Lowest Terms

Simplify the expression $\dfrac{4 - \sqrt{20}}{10}$.

Solution:

$$\frac{4 - \sqrt{20}}{10}$$

First simplify $\sqrt{20}$ by writing the radicand as a product of prime factors.

$$= \frac{4 - \sqrt{2^2 \cdot 5}}{10}$$

$$= \frac{4 - 2\sqrt{5}}{10}$$

Simplify the radical.

Avoiding Mistakes:

Remember that it is not correct to reduce terms within a rational expression. In the expression

$$\frac{4 - 2\sqrt{5}}{10}$$

do not try to "cancel" the 4 and the 10.

Skill Practice Answers

5. $3\sqrt{3} + 3$

6. $\dfrac{y - 2\sqrt{5y} + 5}{y - 5}$

$$= \frac{2(2 - \sqrt{5})}{2 \cdot 5}$$ Factor out the GCF.

$$= \frac{\cancel{2}(2 - \sqrt{5})}{\cancel{2} \cdot 5}$$ Reduce to lowest terms.

$$= \frac{2 - \sqrt{5}}{5}$$

Skill Practice Simplify the expression.

7. $\dfrac{6 - \sqrt{24}}{12}$

Calculator Connections

After simplifying a radical, a calculator can be used to support your solution. For example, use a calculator to approximate the right- and left-hand sides of each expression.

$$\frac{2}{\sqrt{2}} = \sqrt{2} \quad \text{and} \quad \frac{2}{\sqrt{5} + \sqrt{3}} = \sqrt{5} - \sqrt{3}$$

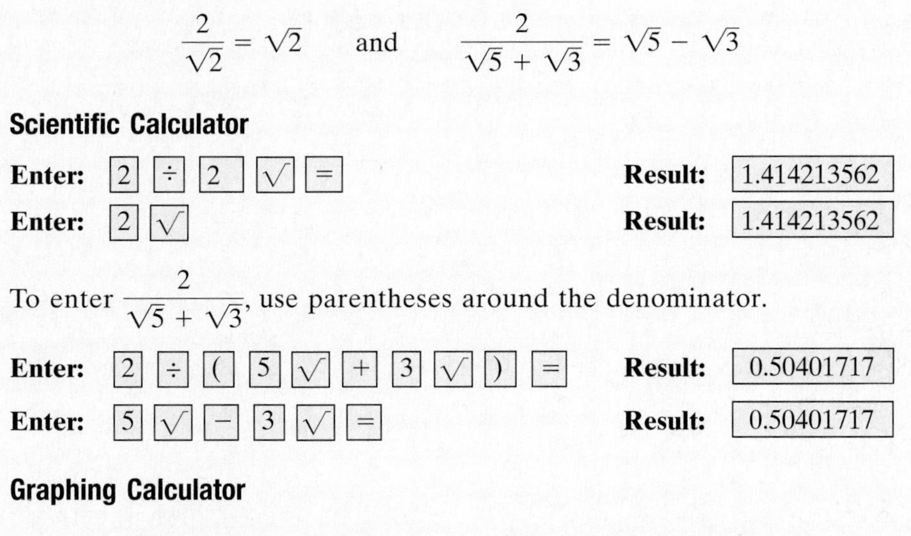

Scientific Calculator

Enter: [2] [÷] [2] [√] [=] **Result:** 1.414213562

Enter: [2] [√] **Result:** 1.414213562

To enter $\dfrac{2}{\sqrt{5} + \sqrt{3}}$, use parentheses around the denominator.

Enter: [2] [÷] [(] [5] [√] [+] [3] [√] [)] [=] **Result:** 0.50401717

Enter: [5] [√] [−] [3] [√] [=] **Result:** 0.50401717

Graphing Calculator

```
2/√(2)
          1.414213562
√(2)
          1.414213562
```

```
2/(√(5)+√(3))
          .5040171699
√(5)-√(3)
          .5040171699
```

The calculator approximation of each expression and its simplified form agree to 10 decimal places.

Calculator Exercises

Simplify each expression. Then use a calculator to approximate the original expression and its simplified form.

1. $\dfrac{5}{\sqrt{11}}$
2. $\dfrac{4}{\sqrt{3}}$
3. $\dfrac{6}{\sqrt{2}}$
4. $\dfrac{10}{\sqrt{3} + \sqrt{2}}$
5. $\dfrac{4}{\sqrt{5} + 1}$
6. $\dfrac{\sqrt{2} + \sqrt{7}}{\sqrt{2} - \sqrt{7}}$

Skill Practice Answers

7. $\dfrac{3 - \sqrt{6}}{6}$

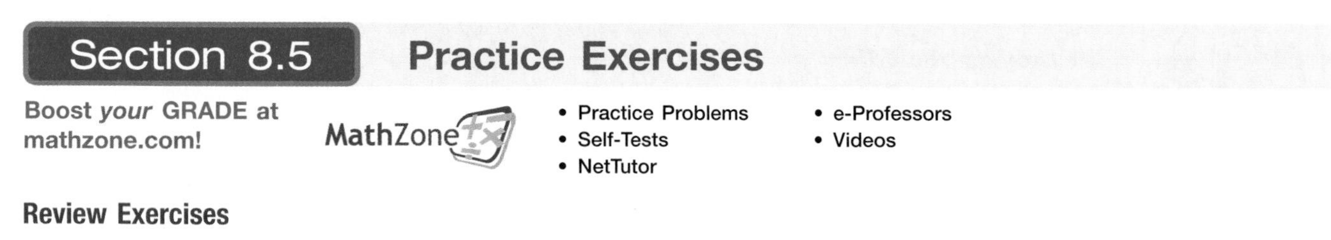

Review Exercises

For Exercises 1–10, perform the indicated operations. Assume the variables represent positive real numbers.

1. $x\sqrt{45} + 4\sqrt{20x^2}$

2. $5b\sqrt{72b} - 3\sqrt{50b^3}$

3. $(2\sqrt{y} + 3)(3\sqrt{y} + 7)$

4. $(4\sqrt{w} - 2)(2\sqrt{w} - 4)$

5. $4\sqrt{3} + \sqrt{5} \cdot \sqrt{15}$

6. $\sqrt{7} \cdot \sqrt{21} + 2\sqrt{27}$

7. $(5 - \sqrt{a})^2$

8. $(\sqrt{z} + 3)^2$

9. $(\sqrt{2} + \sqrt{7})(\sqrt{2} - \sqrt{7})$

10. $(\sqrt{3} + 5)(\sqrt{3} - 5)$

Concept 2: Rationalizing the Denominator: One Term

For Exercises 11–30, rationalize the denominators. Assume the variable expressions represent positive real numbers.

11. $\dfrac{1}{\sqrt{6}}$

12. $\dfrac{5}{\sqrt{2}}$

13. $\dfrac{15}{\sqrt{5}}$

14. $\dfrac{14}{\sqrt{7}}$

15. $\dfrac{6}{\sqrt{x+1}}$

16. $\dfrac{8}{\sqrt{y-3}}$

17. $\sqrt{\dfrac{6}{x}}$

18. $\sqrt{\dfrac{8}{y}}$

19. $\sqrt{\dfrac{3}{7}}$

20. $\sqrt{\dfrac{5}{11}}$

21. $\dfrac{10}{\sqrt{6y}}$

22. $\dfrac{15}{\sqrt{3w}}$

23. $\dfrac{9}{2\sqrt{6}}$

24. $\dfrac{15}{4\sqrt{10}}$

25. $\sqrt{\dfrac{p}{27}}$

26. $\sqrt{\dfrac{x}{32}}$

27. $\dfrac{5}{\sqrt{20}}$

28. $\dfrac{8}{\sqrt{24}}$

29. $\sqrt{\dfrac{x^2}{y^3}}$

30. $\sqrt{\dfrac{a}{b^5}}$

Concept 3: Rationalizing the Denominator: Two Terms

For Exercises 31–32, multiply the conjugates.

31. $(\sqrt{2} + 3)(\sqrt{2} - 3)$

32. $(\sqrt{3} + \sqrt{7})(\sqrt{3} - \sqrt{7})$

33. What is the conjugate of $\sqrt{5} - \sqrt{3}$? Multiply $\sqrt{5} - \sqrt{3}$ by its conjugate.

34. What is the conjugate of $\sqrt{7} + \sqrt{2}$? Multiply $\sqrt{7} + \sqrt{2}$ by its conjugate.

35. What is the conjugate of $\sqrt{x} + 10$? Multiply $\sqrt{x} + 10$ by its conjugate.

36. What is the conjugate of $12 - \sqrt{y}$? Multiply $12 - \sqrt{y}$ by its conjugate.

For Exercises 37–50, rationalize the denominators. Assume the variable expressions represent positive real numbers.

37. $\dfrac{4}{\sqrt{2} + 3}$

38. $\dfrac{6}{4 - \sqrt{3}}$

39. $\dfrac{1}{\sqrt{5} - \sqrt{2}}$

40. $\dfrac{2}{\sqrt{3} + \sqrt{7}}$

41. $\dfrac{\sqrt{8}}{\sqrt{3}+1}$ 42. $\dfrac{\sqrt{18}}{1-\sqrt{2}}$ 43. $\dfrac{1}{\sqrt{x}-\sqrt{3}}$ 44. $\dfrac{1}{\sqrt{y}+\sqrt{5}}$

45. $\dfrac{2+\sqrt{3}}{2-\sqrt{3}}$ 46. $\dfrac{\sqrt{3}-\sqrt{2}}{\sqrt{3}+\sqrt{2}}$ 47. $\dfrac{3}{\sqrt{11}-\sqrt{5}}$ 48. $\dfrac{4}{\sqrt{10}+\sqrt{2}}$

49. $\dfrac{\sqrt{5}+4}{2-\sqrt{5}}$ 50. $\dfrac{3+\sqrt{2}}{\sqrt{2}-5}$

Concept 4: Simplifying a Radical Quotient to Lowest Terms

For Exercises 51–58, simplify the expression.

51. $\dfrac{10-\sqrt{50}}{5}$ 52. $\dfrac{4+\sqrt{12}}{2}$ 53. $\dfrac{21+\sqrt{98}}{14}$ 54. $\dfrac{3-\sqrt{18}}{6}$

55. $\dfrac{2-\sqrt{28}}{2}$ 56. $\dfrac{5+\sqrt{75}}{5}$ 57. $\dfrac{14+\sqrt{72}}{6}$ 58. $\dfrac{15-\sqrt{125}}{10}$

Recall that a radical is simplified if

1. The radicand has no factor raised to a power greater than or equal to the index.
2. There are no radicals in the denominator of a fraction.
3. The radicand does not contain a fraction.

For Exercises 59–62, state which condition(s) fails. Then simplify the radical.

59. a. $\sqrt{8x^9}$ b. $\dfrac{5}{\sqrt{5x}}$ c. $\sqrt{\dfrac{1}{3}}$

60. a. $\sqrt{\dfrac{7}{2}}$ b. $\sqrt{18y^6}$ c. $\dfrac{2}{\sqrt{4x}}$

61. a. $\dfrac{3}{\sqrt{x}+1}$ b. $\sqrt{\dfrac{9w^2}{t}}$ c. $\sqrt{24a^5b^9}$

62. a. $\sqrt{\dfrac{12}{z^3}}$ b. $\dfrac{4}{\sqrt{a}-\sqrt{b}}$ c. $\sqrt[3]{27m^3n^7}$

Mixed Exercises

For Exercises 63–74, simplify the radical expressions, if possible. Assume the variables represent positive real numbers.

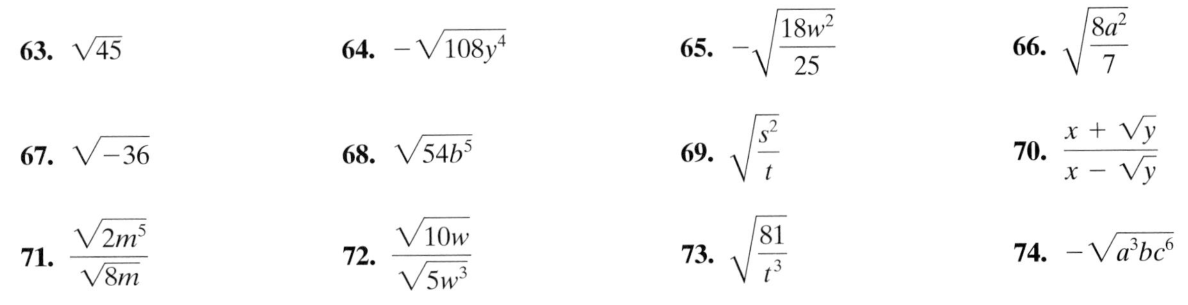

63. $\sqrt{45}$ 64. $-\sqrt{108y^4}$ 65. $-\sqrt{\dfrac{18w^2}{25}}$ 66. $\sqrt{\dfrac{8a^2}{7}}$

67. $\sqrt{-36}$ 68. $\sqrt{54b^5}$ 69. $\sqrt{\dfrac{s^2}{t}}$ 70. $\dfrac{x+\sqrt{y}}{x-\sqrt{y}}$

71. $\dfrac{\sqrt{2m^5}}{\sqrt{8m}}$ 72. $\dfrac{\sqrt{10w}}{\sqrt{5w^3}}$ 73. $\sqrt{\dfrac{81}{t^3}}$ 74. $-\sqrt{a^3bc^6}$

Expanding Your Skills

75. Find the slope of the line through the points: $(5\sqrt{2}, 3)$ and $(\sqrt{2}, 6)$.

76. Find the slope of the line through the points: $(4\sqrt{5}, -1)$ and $(6\sqrt{5}, -5)$.

77. Find the slope of the line through the points: $(\sqrt{3}, -1)$ and $(4\sqrt{3}, 0)$.

78. Find the slope of the line through the points: $(-2\sqrt{7}, -5)$ and $(\sqrt{7}, 2)$.

Chapter 8 Problem Recognition Exercises—Operations on Radicals

Perform the indicated operations and simplify if possible. Assume that all variables represent positive real numbers.

1. $(\sqrt{3})(\sqrt{6})$

2. $(\sqrt{2})(\sqrt{14})$

3. $\sqrt{3} + \sqrt{6}$

4. $\sqrt{2} + \sqrt{14}$

5. $\dfrac{\sqrt{6}}{\sqrt{3}}$

6. $\dfrac{\sqrt{14}}{\sqrt{2}}$

7. $(3 + \sqrt{z})(3 - \sqrt{z})$

8. $(4 - \sqrt{y})(4 + \sqrt{y})$

9. $(2\sqrt{5} + 1)(\sqrt{5} - 2)$

10. $(4\sqrt{3} - 5)(\sqrt{3} + 4)$

11. $2\sqrt{x^2y} - 3x\sqrt{y}$

12. $8\sqrt{a^3b^2} + 3a\sqrt{ab^2}$

13. $-3\sqrt{2}(4\sqrt{2} + 2\sqrt{3} + 1)$

14. $-8\sqrt{5}(2\sqrt{5} - \sqrt{3} - 2)$

15. $\dfrac{2}{\sqrt{x} - 7}$

16. $\dfrac{5}{\sqrt{y} + 4}$

17. $\dfrac{9}{\sqrt{3}}$

18. $\dfrac{15}{\sqrt{5}}$

19. $\sqrt{\dfrac{7}{x}}$

20. $\sqrt{\dfrac{11}{y}}$

21. $\sqrt{y^4z^{11}}$

22. $\sqrt{8q^6}$

23. $\sqrt[3]{27p^8}$

24. $\sqrt[3]{125u^{11}v^{12}}$

25. $\dfrac{\sqrt{10x^3}}{\sqrt{x}}$

26. $\dfrac{\sqrt{15y^3}}{\sqrt{5y}}$

27. $6\sqrt{75} - 5\sqrt{12}$

28. $\sqrt{90} - \sqrt{40}$

29. $\sqrt[4]{\dfrac{1}{81}}$

30. $\sqrt[3]{\dfrac{125}{27}}$

31. $\dfrac{x - 5}{\sqrt{x} + \sqrt{5}}$

32. $\dfrac{y - 7}{\sqrt{y} + \sqrt{7}}$

33. $(4\sqrt{x} + \sqrt{y})(\sqrt{x} - 3\sqrt{y})$

34. $(\sqrt{2} + 7)^2$

35. $(\sqrt{3} + \sqrt{5})^2$

36. $(\sqrt{5} - \sqrt{11})^2$

37. $(\sqrt{x} - 6)^2$

38. $(2\sqrt{3} - 10)(2\sqrt{3} + 10)$

39. $(\sqrt{u} - 3\sqrt{v})(\sqrt{u} + 3\sqrt{v})$

40. $2\sqrt{6} - 5\sqrt{6}$

41. $5\sqrt{a} + 7\sqrt{a} - \sqrt{a}$

42. $x\sqrt{18} + \sqrt{2x^2}$

43. $4\sqrt{75} - 20\sqrt{3}$

44. $\sqrt{5}(\sqrt{5} + \sqrt{7})$

45. $\sqrt{a}(\sqrt{a} + 2)$

46. $(3\sqrt{2} - 4)(5\sqrt{2} + 1)$

Radical Equations

Section 8.6

Concepts

1. Solving Radical Equations
2. Translations Involving Radical Equations
3. Applications of Radical Equations

1. Solving Radical Equations

Definition of a Radical Equation

An equation with one or more radicals containing a variable is called a **radical equation**.

For example, $\sqrt{x} = 5$ is a radical equation. Recall that $(\sqrt[n]{a})^n = a$ provided $\sqrt[n]{a}$ is a real number. The basis to solve a radical equation is to eliminate the radical by raising both sides of the equation to a power equal to the index of the radical.

To solve the equation $\sqrt{x} = 5$, square both sides of the equation.

$$\sqrt{x} = 5$$
$$(\sqrt{x})^2 = (5)^2$$
$$x = 25$$

By raising each side of a radical equation to a power equal to the index of the radical, a new equation is produced. However, it is important to note that the new equation may have **extraneous solutions**; that is, some or all of the solutions to the new equation may *not* be solutions to the original radical equation. For this reason, it is necessary to check *all* potential solutions in the original equation. For example, consider the equation $x = 4$. By squaring both sides we produce a quadratic equation.

Square both sides.

$$x = 4$$
$$(x)^2 = (4)^2$$ Squaring both sides produces a quadratic equation.
$$x^2 = 16$$
$$x^2 - 16 = 0$$
$$(x - 4)(x + 4) = 0$$ Solving this equation, we find two solutions. However, $x = -4$ does not check. The value -4 is an extraneous solution because it is not a solution to the original equation, $x = 4$.
$$x = 4 \quad \text{or} \quad \cancel{x = -4}$$

Steps to Solving a Radical Equation

1. Isolate the radical. If an equation has more than one radical, choose one of the radicals to isolate.
2. Raise each side of the equation to a power equal to the index of the radical.
3. Solve the resulting equation.
4. Check the potential solutions in the original equation.*

*Extraneous solutions can only arise when both sides of the equation are raised to an *even power.* Therefore, an equation with odd-index roots will not have an extraneous solution. However, it is still recommended that you check *all* potential solutions regardless of the type of root.

Example 1 Solving Radical Equations

Solve the equations.

a. $\sqrt{2x+1}+5=8$ **b.** $8+\sqrt{x+2}=7$

Solution:

a. $\sqrt{2x+1}+5=8$

$\sqrt{2x+1}=8-5$ Isolate the radical.

$\sqrt{2x+1}=3$

$(\sqrt{2x+1})^2=(3)^2$ Raise both sides to a power equal to the index of the radical.

$2x+1=9$ Simplify both sides.

$2x=8$ Solve the resulting equation (the equation is linear).

$x=4$

Check: Check $x=4$ as a potential solution.

$\sqrt{2x+1}+5=8$

$\sqrt{2(4)+1}+5\stackrel{?}{=}8$

$\sqrt{8+1}+5\stackrel{?}{=}8$

$\sqrt{9}+5\stackrel{?}{=}8$

$3+5=8$ ✔ The answer checks.

$x=4$ is the solution.

TIP: After isolating the radical in Example 1(b), the equation shows a square root equated to a negative number.

$$\sqrt{x+2}=-1$$

By definition, a principal square root of any real number must be nonnegative. Therefore, there can be no solution to this equation.

b. $8+\sqrt{x+2}=7$

$\sqrt{x+2}=7-8$ Isolate the radical.

$\sqrt{x+2}=-1$

$(\sqrt{x+2})^2=(-1)^2$ Raise both sides to a power equal to the index of the radical.

$x+2=1$ Simplify.

$x=-1$ Solve the resulting equation.

Check: Check $x=-1$ as a potential solution.

$8+\sqrt{x+2}=7$

$8+\sqrt{(-1)+2}\stackrel{?}{=}7$

$8+\sqrt{1}\stackrel{?}{=}7$

$8+1\neq 7$ The value $x=-1$ does not check. It is an extraneous solution.

There is no solution to the equation.

Skill Practice Solve the equations.

1. $\sqrt{p-4}-2=4$ **2.** $\sqrt{2y+5}+7=4$

Skill Practice Answers

1. $p=40$

2. No solution

Example 2 Solving Radical Equations

Solve the equations.

a. $p + 4 = \sqrt{p + 6}$ **b.** $2\sqrt[3]{2x - 3} - \sqrt[3]{x + 6} = 0$

Solution:

a.

$p + 4 = \sqrt{p + 6}$	The radical is already isolated.
$(p + 4)^2 = (\sqrt{p + 6})^2$	Raise both sides to a power equal to the index.
$p^2 + 8p + 16 = p + 6$	
$p^2 + 7p + 10 = 0$	Solve the resulting equation (the equation is quadratic).
$(p + 5)(p + 2) = 0$	Set the equation equal to zero and factor.
$p + 5 = 0$ or $p + 2 = 0$	Set each factor equal to zero.
$p = -5$ or $p = -2$	Solve for p.

Check: $p = -5$

$p + 4 = \sqrt{p + 6}$

$(-5) + 4 \stackrel{?}{=} \sqrt{(-5) + 6}$

$-1 \stackrel{?}{=} \sqrt{1}$

$-1 \neq 1$ Does not check.

Check: $p = -2$

$p + 4 = \sqrt{p + 6}$

$(-2) + 4 \stackrel{?}{=} \sqrt{(-2) + 6}$

$2 \stackrel{?}{=} \sqrt{4}$

$2 = 2$ ✔ The solution checks.

The only solution is $p = -2$ ($p = -5$ does not check).

b. $2\sqrt[3]{2x - 3} - \sqrt[3]{x + 6} = 0$

$2\sqrt[3]{2x - 3} = \sqrt[3]{x + 6}$	Isolate one of the radicals.
$\left(2\sqrt[3]{2x - 3}\right)^3 = \left(\sqrt[3]{x + 6}\right)^3$	Raise both sides to a power equal to the index.
$(2)^3\left(\sqrt[3]{2x - 3}\right)^3 = \left(\sqrt[3]{x + 6}\right)^3$	On the left-hand side, be sure to cube both factors, $(2)^3$ and $\left(\sqrt[3]{2x - 3}\right)^3$.
$8(2x - 3) = x + 6$	Solve the resulting equation.
$16x - 24 = x + 6$	
$15x = 30$	
$x = 2$	

Check:

$2\sqrt[3]{2x - 3} - \sqrt[3]{x + 6} = 0$	Check the potential solution, $x = 2$.
$2\sqrt[3]{2(2) - 3} - \sqrt[3]{2 + 6} \stackrel{?}{=} 0$	
$2\sqrt[3]{4 - 3} - \sqrt[3]{8} \stackrel{?}{=} 0$	
$2\sqrt[3]{1} - 2 \stackrel{?}{=} 0$	
$2 - 2 = 0$ ✔	The solution checks.

TIP: Recall that

$(a + b)^2 = a^2 + 2ab + b^2$.

Hence,

$(p + 4)^2$

$= (p)^2 + 2(p)(4) + (4)^2$

$= p^2 + 8p + 16$

Skill Practice Solve the equations.

3. $\sqrt{x + 34} = x + 4$ **4.** $\sqrt[3]{4p + 1} - \sqrt[3]{p + 16} = 0$

Skill Practice Answers

3. $x = 2$, ($x = -9$ does not check)

4. $p = 5$

2. Translations Involving Radical Equations

Example 3 Translating English Form into Algebraic Form

The principal square root of the sum of a number and three is equal to seven. Find the number.

Solution:

Let x represent the number. Label the variable.

$\sqrt{x+3} = 7$ Translate the verbal model into an algebraic equation.

$(\sqrt{x+3})^2 = (7)^2$ The radical is already isolated. Square both sides.

$x + 3 = 49$ The resulting equation is linear.

$x = 46$ Solve for x.

Check: Check $x = 46$ as a potential solution.

$\sqrt{x+3} = 7$

$\sqrt{46+3} \stackrel{?}{=} 7$

$\sqrt{49} \stackrel{?}{=} 7$

$7 = 7$ ✔ The solution checks.

The number is 46.

Skill Practice

5. The principal square root of the sum of a number and 5 is 2. Find the number.

3. Applications of Radical Equations

Example 4 Using a Radical Equation in an Application

For a small company, the weekly sales, y, of its product are related to the money spent on advertising, x, according to the equation:

$$y = 100\sqrt{x}$$

a. Find the amount in sales if the company spends \$100 on advertising.

b. Find the amount in sales if the company spends \$625 on advertising.

c. Find the amount the company spent on advertising if its sales for 1 week totaled \$2000.

Solution:

a. $y = 100\sqrt{x}$

$= 100\sqrt{100}$ Substitute $x = 100$.

$= 100(10)$

$= 1000$

The amount in sales is \$1000.

Skill Practice Answers

5. The number is -1.

b. $y = 100\sqrt{x}$

$= 100\sqrt{625}$ Substitute $x = 625$.

$= 100(25)$

$= 2500$

The amount in sales is \$2500.

c. $y = 100\sqrt{x}$

$2000 = 100\sqrt{x}$ Substitute $y = 2000$.

$\frac{2000}{100} = \frac{\cancel{100}\sqrt{x}}{\cancel{100}}$ Isolate the radical. Divide both sides by 100.

$20 = \sqrt{x}$ Simplify.

$(20)^2 = (\sqrt{x})^2$ Raise both sides to a power equal to the index.

$400 = x$ Simplify both sides.

Check: Check $x = 400$ as a potential solution.

$y = 100\sqrt{x}$

$2000 \stackrel{?}{=} 100\sqrt{400}$

$2000 \stackrel{?}{=} 100(20)$

$2000 = 2000$ ✔ The solution checks.

The amount spent on advertising was \$400.

Skill Practice

6. If the small company mentioned in Example 4, changes its advertising media, the equation relating money spent on advertising, x, to weekly sales, y, is:

$$y = 100\sqrt{2x}$$

a. Use the given equation to find the amount in sales if the company spends \$200 on advertising.

b. Find the amount spent on advertising if the sales for 1 week totaled \$3000.

Skill Practice Answers

6. a. \$2000 **b.** \$450

Section 8.6 Practice Exercises

Boost *your* GRADE at mathzone.com!

- Practice Problems
- Self-Tests
- NetTutor
- e-Professors
- Videos

Study Skills Exercise

1. Define the key terms:

a. radical equation **b. extraneous solution**

Review Exercises

For Exercises 2–5, rationalize the denominators.

2. $\dfrac{1}{\sqrt{3} - \sqrt{7}}$
3. $\dfrac{1}{\sqrt{2} + \sqrt{10}}$
4. $\dfrac{6}{\sqrt{6}}$
5. $\dfrac{2\sqrt{2}}{\sqrt{3}}$

For Exercises 6–7, simplify the expression.

6. $\dfrac{10 - \sqrt{75}}{5}$
7. $\dfrac{6 + \sqrt{32}}{4}$

For Exercises 8–9, square the binomials.

8. $(x + 4)^2$
9. $(3 - y)^2$

For Exercises 10–13, simplify the expressions. Assume the variable expressions represent positive real numbers.

10. $(\sqrt{2x - 3})^2$
11. $(\sqrt{m + 6})^2$
12. $(\sqrt[3]{t - 9})^3$
13. $(\sqrt[4]{5y - 4})^4$

Concept 1: Solving Radical Equations

For Exercises 14–44, solve the equations. Be sure to check all of the potential answers.

14. $\sqrt{t} = 6$
15. $\sqrt{x + 1} = 4$
16. $\sqrt{x - 3} = 7$
17. $\sqrt{y - 4} = -5$
18. $\sqrt{p + 6} = -1$
19. $\sqrt{5 - t} = 0$
20. $\sqrt{13 + m} = 0$
21. $\sqrt{2n + 10} = 3$
22. $\sqrt{1 - q} = 15$
23. $\sqrt{6w} - 8 = -2$
24. $\sqrt{2z} - 11 = -3$
25. $\sqrt{5a - 4} - 2 = 4$
26. $\sqrt{3b + 4} - 3 = 2$
27. $\sqrt{2x - 3} + 7 = 3$
28. $\sqrt{8y + 1} + 5 = 1$
29. $5\sqrt{c} = \sqrt{10c + 15}$
30. $4\sqrt{x} = \sqrt{10x + 6}$
31. $\sqrt{x^2 - x} = \sqrt{12}$
32. $\sqrt{x^2 + 5x} = \sqrt{150}$
33. $\sqrt{9y^2 - 8y + 1} = 3y + 1$
34. $\sqrt{4x^2 + 2x + 20} = 2x$
35. $\sqrt{x^2 + 3x - 2} = 4$
36. $\sqrt{2k^2 - 3k - 4} = k$
37. $\sqrt{6t + 7} = t + 2$
38. $\sqrt{y + 1} = y + 1$
39. $\sqrt{3p + 3} + 5 = p$
40. $\sqrt{2m + 1} + 7 = m$
41. $\sqrt[3]{3y + 7} = \sqrt[3]{2y - 1}$
42. $\sqrt[3]{p - 5} - \sqrt[3]{2p + 1} = 0$
43. $\sqrt[3]{2x - 8} - \sqrt[3]{-x + 1} = 0$
44. $\sqrt[3]{a - 3} = \sqrt[3]{5a + 1}$

Concept 2: Translations Involving Radical Equations

For Exercises 45–50, translate the English sentence to a radical equation and solve the equation.

45. The principal square root of the sum of a number and 8 equals 12. Find the number.

46. The principal square root of the sum of a number and 10 equals 1. Find the number.

47. The principal square root of a number is 2 less than the number. Find the number.

48. The principal square root of twice a number is 4 less than the number. Find the number.

49. The cube root of the sum of a number and 4 is -5. Find the number.

50. The cube root of the sum of a number and 1 is 2. Find the number.

Concept 3: Applications of Radical Equations

51. Ignoring air resistance, the time, t (in seconds), required for an object to fall x feet is given by the equation:

$$t = \frac{\sqrt{x}}{4}$$

a. Find the time required for an object to fall 64 ft.

b. Find the distance an object will fall in 4 sec.

52. Ignoring air resistance, the velocity, v (in feet per second: ft/sec), of an object in free fall depends on the distance it has fallen, x (in feet), according to the equation:

$$v = 8\sqrt{x}$$

a. Find the velocity of an object that has fallen 100 ft.

b. Find the distance that an object has fallen if its velocity is 136 ft/sec.

53. The speed of a car, s (in miles per hour), before the brakes were applied can be approximated by the length of its skid marks, x (in feet), according to the equation:

$$s = 4\sqrt{x}$$

a. Find the speed of a car before the brakes were applied if its skid marks are 324 ft long.

b. How long would you expect the skid marks to be if the car had been traveling the speed limit of 60 mph?

54. The height of a sunflower plant, y (in inches), can be determined by the time, t (in weeks), after the seed has germinated according to the equation:

$$y = 8\sqrt{t} \quad 0 \leq t \leq 40$$

a. Find the height of the plant after 4 weeks.

b. In how many weeks will the plant be 40 in. tall?

Expanding Your Skills

For Exercises 55–58, solve the equations. First isolate one of the radical terms. Then square both sides. The resulting equation will still have a radical. Repeat the process by isolating the radical and squaring both sides again.

55. $\sqrt{t + 8} = \sqrt{t} + 2$

56. $\sqrt{5x - 9} = \sqrt{5x} - 3$

57. $\sqrt{z + 1} + \sqrt{2z + 3} = 1$

58. $\sqrt{2m + 6} = 1 + \sqrt{7 - 2m}$

Section 8.7

Rational Exponents

Concepts

1. Definition of $a^{1/n}$
2. Definition of $a^{m/n}$
3. Converting between Rational Exponents and Radical Notation
4. Properties of Rational Exponents
5. Applications of Rational Exponents

1. Definition of $a^{1/n}$

In Sections 5.1–5.3, the properties for simplifying expressions with integer exponents were presented. In this section, the properties are expanded to include expressions with rational exponents. We begin by defining expressions of the form $a^{1/n}$.

Definition of $a^{1/n}$

Let a be a real number, and let n be an integer such that $n > 1$. If $\sqrt[n]{a}$ is a real number, then

$$a^{1/n} = \sqrt[n]{a}$$

Note: $(\sqrt{a})^2 = a$ for $a > 0$ and $(a^{1/2})^2 = a^{2/2} = a$, so $\sqrt{a} = a^{1/2}$.

Example 1 Evaluating Expressions of the Form $a^{1/n}$

Convert the expression to radical notation and simplify, if possible.

a. $9^{1/2}$ **b.** $125^{1/3}$ **c.** $16^{1/4}$ **d.** $-25^{1/2}$ **e.** $(-25)^{1/2}$ **f.** $25^{-1/2}$

Solution:

a. $9^{1/2} = \sqrt{9} = 3$

b. $125^{1/3} = \sqrt[3]{125} = 5$

c. $16^{1/4} = \sqrt[4]{16} = 2$

d. $-25^{1/2}$ is equivalent to $-1(25^{1/2})$

$= -1 \cdot \sqrt{25}$

$= -5$

e. $(-25)^{1/2}$ is not a real number because $\sqrt{-25}$ is not a real number.

f. $25^{-1/2} = \dfrac{1}{25^{1/2}} = \dfrac{1}{\sqrt{25}} = \dfrac{1}{5}$

Skill Practice Convert the expression to radical notation and simplify.

1. $36^{1/2}$ **2.** $(-27)^{1/3}$ **3.** $81^{1/4}$ **4.** $(-16)^{1/4}$

5. $(16)^{-1/4}$ **6.** $-16^{1/4}$

2. Definition of $a^{m/n}$

If $\sqrt[n]{a}$ is a real number, then we can define an expression of the form $a^{m/n}$ in such a way that the multiplication property of exponents holds true. For example,

$$16^{3/4} = \begin{cases} (16^{1/4})^3 = (\sqrt[4]{16})^3 = (2)^3 = 8 \\ (16^3)^{1/4} = \sqrt[4]{16^3} = \sqrt[4]{4096} = 8 \end{cases}$$

Skill Practice Answers

1. 6 **2.** -3 **3.** 3
4. Not a real number.
5. $\dfrac{1}{2}$ **6.** -2

Definition of $a^{m/n}$

Let a be a real number, and let m and n be positive integers such that m and n share no common factors and $n > 1$. If $\sqrt[n]{a}$ is a real number, then

$$a^{m/n} = (a^{1/n})^m = (\sqrt[n]{a})^m \quad \text{and} \quad a^{m/n} = (a^m)^{1/n} = \sqrt[n]{a^m}$$

The rational exponent in the expression $a^{m/n}$ is essentially performing two operations. The numerator of the exponent raises the base to the mth-power. The denominator takes the nth-root.

Example 2 Evaluating Expressions of the Form $a^{m/n}$

Convert each expression to radical notation and simplify.

a. $125^{2/3}$ **b.** $100^{-3/2}$ **c.** $(81)^{3/4}$

Solution:

a. $125^{2/3} = (\sqrt[3]{125})^2$ Take the cube root of 125, and square the result.

$= (5)^2$ Simplify.

$= 25$

b. $100^{-3/2} = \dfrac{1}{100^{3/2}}$ Take the reciprocal of the base.

$= \dfrac{1}{(\sqrt{100})^3}$ Take the square root of 100, and cube the result.

$= \dfrac{1}{(10)^3}$ Simplify.

$= \dfrac{1}{1000}$

c. $(81)^{3/4} = (\sqrt[4]{81})^3$ Take the fourth root of 81, and cube the result.

$= (3)^3$ Simplify.

$= 27$

Skill Practice Convert each expression to radical notation and simplify.

7. $16^{3/4}$ **8.** $8^{-2/3}$ **9.** $9^{3/2}$

3. Converting between Rational Exponents and Radical Notation

Example 3 Using Radical Notation and Rational Exponents

Convert the expressions to radical notation. Assume the variables represent positive real numbers.

a. $x^{3/5}$ **b.** $(2a^2)^{1/3}$ **c.** $5y^{1/4}$ **d.** $p^{-1/2}$

Solution:

a. $x^{3/5} = \sqrt[5]{x^3}$ or $(\sqrt[5]{x})^3$

b. $(2a^2)^{1/3} = \sqrt[3]{2a^2}$

Skill Practice Answers

7. $(\sqrt[4]{16})^3$; 8 **8.** $\dfrac{1}{(\sqrt[3]{8})^2}$; $\dfrac{1}{4}$

9. $(\sqrt{9})^3$; 27

c. $5y^{1/4} = 5\sqrt[4]{y}$ The exponent $\frac{1}{4}$ applies only to y.

d. $p^{-1/2} = \dfrac{1}{\sqrt{p}}$

Skill Practice Convert each expression to radical notation. Write the answers with positive exponents only. Assume the variables represent positive real numbers.

10. $y^{4/3}$ **11.** $(5x)^{1/2}$ **12.** $10a^{3/5}$ **13.** $z^{-2/3}$

Example 4 Using Radical Notation and Rational Exponents

Convert each expression to an equivalent expression using rational exponents. Assume that the variables represent positive real numbers.

a. $\sqrt[4]{c^3}$ **b.** $\sqrt{11p}$ **c.** $11\sqrt{p}$

Solution:

a. $\sqrt[4]{c^3} = c^{3/4}$ **b.** $\sqrt{11p} = (11p)^{1/2}$ **c.** $11\sqrt{p} = 11p^{1/2}$

Skill Practice Convert each expression to an equivalent expression using rational exponents.

14. $\sqrt[5]{y^2}$ **15.** $\sqrt{2x}$ **16.** $2\sqrt{x}$

4. Properties of Rational Exponents

In Sections 5.1–5.3, several properties and definitions were introduced to simplify expressions with integer exponents. These properties also apply to rational exponents.

Properties of Exponents and Definitions

Let a and b be real numbers. Let m and n be rational numbers such that a^m, a^n, and b^n are real numbers. Then,

Description	Property	Example
1. Multiplying like bases	$a^m a^n = a^{m+n}$	$x^{1/3} \cdot x^{4/3} = x^{5/3}$
2. Dividing like bases	$\dfrac{a^m}{a^n} = a^{m-n}$	$\dfrac{x^{3/5}}{x^{1/5}} = x^{2/5}$
3. The power rule	$(a^m)^n = a^{mn}$	$(2^{1/3})^{1/2} = 2^{1/6}$
4. Power of a product	$(ab)^m = a^m b^m$	$(xy)^{1/2} = x^{1/2}y^{1/2}$
5. Power of a quotient	$\left(\dfrac{a}{b}\right)^m = \dfrac{a^m}{b^m}$ $(b \neq 0)$	$\left(\dfrac{4}{25}\right)^{1/2} = \dfrac{4^{1/2}}{25^{1/2}} = \dfrac{2}{5}$

Description	Definition	Example
1. Negative exponents	$a^{-m} = \left(\dfrac{1}{a}\right)^m = \dfrac{1}{a^m}$ $(a \neq 0)$	$(8)^{-1/3} = \left(\dfrac{1}{8}\right)^{1/3} = \dfrac{1}{2}$
2. Zero exponent	$a^0 = 1$ $(a \neq 0)$	$5^0 = 1$

Skill Practice Answers

10. $(\sqrt[3]{y})^4$ **11.** $\sqrt{5x}$
12. $10(\sqrt[5]{a})^3$ **13.** $\dfrac{1}{(\sqrt[3]{z})^2}$
14. $y^{2/5}$ **15.** $(2x)^{1/2}$ **16.** $2x^{1/2}$

Example 5 Simplifying Expressions with Rational Exponents

Use the properties of exponents to simplify the expressions. Write the final answers with positive exponents only. Assume the variables represent positive real numbers.

a. $x^{2/3}x^{1/3}$ **b.** $\dfrac{y^{4/5}}{y^{1/10}}$ **c.** $(z^4)^{1/2}$

d. $(s^4t^8)^{1/4}$ **e.** $\left(\dfrac{x^{-2/3}}{y^{-1/2}}\right)^6(x^{-1/5})^{10}$

Solution:

a. $x^{2/3}x^{1/3} = x^{(2/3)+(1/3)}$ Add exponents.

$= x^{3/3}$ Simplify.

$= x$

b. $\dfrac{y^{4/5}}{y^{1/10}} = y^{(4/5)-(1/10)}$ Subtract exponents.

$= y^{(8/10)-(1/10)}$ Get a common denominator.

$= y^{7/10}$ Simplify.

c. $(z^4)^{1/2} = z^{(4)\cdot(1/2)}$ Multiply exponents.

$= z^2$ Simplify.

d. $(s^4t^8)^{1/4} = s^{4/4}t^{8/4}$ Multiply exponents.

$= st^2$

e. $\left(\dfrac{x^{-2/3}}{y^{-1/2}}\right)^6(x^{-1/5})^{10} = \dfrac{x^{(-2/3)(6)}}{y^{(-1/2)(6)}} \cdot x^{(-1/5)(10)}$ Multiply exponents.

$= \left(\dfrac{x^{-4}}{y^{-3}}\right)(x^{-2})$ Simplify exponents.

$= \dfrac{x^{-4}}{y^{-3}} \cdot \dfrac{x^{-2}}{1}$ Write x^{-2} as $\dfrac{x^{-2}}{1}$.

$= \dfrac{x^{-6}}{y^{-3}}$ Add exponents in the numerator.

$= \dfrac{y^3}{x^6}$ Simplify negative exponents.

Skill Practice Use the properties of exponents to simplify the expressions. Write the answers with positive exponents only. Assume the variables represent positive real numbers.

17. $a^{3/4} \cdot a^{5/4}$ **18.** $\dfrac{t^{2/3}}{t^2}$ **19.** $(w^{1/3})^{-12}$

20. $(y^9z^{15})^{1/3}$ **21.** $\left(\dfrac{p^{-1/4}}{m^{1/2}}\right)^8(m^{-1/3})^6$

Skill Practice Answers

17. a^2 **18.** $\dfrac{1}{t^{4/3}}$ **19.** $\dfrac{1}{w^4}$

20. y^3z^5 **21.** $\dfrac{1}{m^6p^2}$

5. Applications of Rational Exponents

Example 6 Using Rational Exponents in an Application

Suppose P dollars in principal is invested in an account that earns interest annually. If after t years the investment grows to A dollars, then the annual rate of return, r, on the investment is given by

$$r = \left(\frac{A}{P}\right)^{1/t} - 1$$

Find the annual rate of return on $8000 that grew to $11,220.41 after 5 years (round to the nearest tenth of a percent).

Solution:

$r = \left(\frac{A}{P}\right)^{1/t} - 1$ where $A = \$11{,}220.41$, $P = \$8000$, and $t = 5$. Hence,

$$\begin{aligned} r &= \left(\frac{11220.41}{8000}\right)^{1/5} - 1 \\ &= (1.40255125)^{1/5} - 1 \\ &\approx 1.070 - 1 \\ &\approx 0.070 \text{ or } 7.0\% \end{aligned}$$

There is a 7.0% annual rate of return.

Skill Practice

22. The formula for finding the radius of a circle given the area is

$$r = \left(\frac{A}{\pi}\right)^{1/2}$$

Find the radius of a circle given that the area is 12.56 in^2. Use 3.14 for π.

Calculator Connections

Rational exponents provide another method to evaluate nth-roots on a calculator. For example, use rational exponents and the y^x key (or ^ key) to evaluate: $125^{2/3}$, $100^{3/2}$, and $81^{3/4}$.

Scientific Calculator

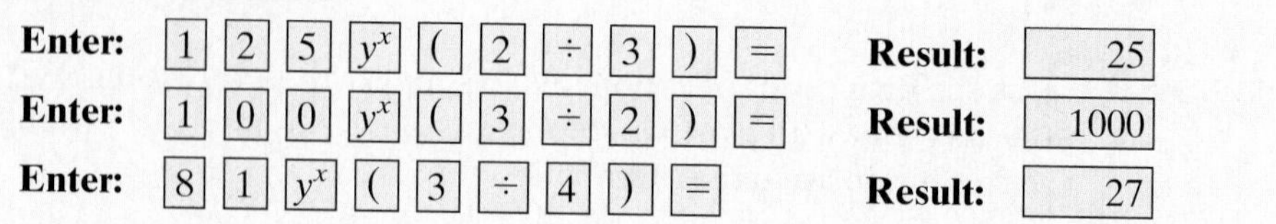

Notice that parentheses are used around the rational exponent to ensure that the base is raised to the entire quotient.

Skill Practice Answers

22. ≈ 2 in.

Graphing Calculator

```
125^(2/3)
                25
100^(3/2)
              1000
81^(3/4)
                27
```

Calculator Exercises

Simplify the expressions algebraically, and then check your answers on a calculator.

1. $169^{1/2}$

2. $729^{1/3}$

3. $1024^{1/5}$

4. $-10{,}000^{1/4}$

5. $8^{-2/3}$

6. $-27^{2/3}$

7. $100{,}000^{4/5}$

8. $16^{-3/4}$

Evaluate the expressions with a calculator.

9. $984^{1/4}$

10. $14.8^{1/3}$

11. $\sqrt[3]{56^2}$

12. $\sqrt[4]{24^3}$

13. $\sqrt[3]{\dfrac{45}{59}}$

14. $\sqrt[3]{\dfrac{3}{104}}$

15. $\sqrt[4]{\left(\dfrac{4}{5}\right)^3}$

16. $\sqrt[3]{\left(\dfrac{5}{6}\right)^2}$

Section 8.7 Practice Exercises

Review Exercises

For the exercises in this set, assume that the variables represent positive real numbers unless otherwise stated.

1. Given $\sqrt[3]{125}$

a. Identify the index. **b.** Identify the radicand.

2. Given $\sqrt{12}$

a. Identify the index. **b.** Identify the radicand.

For Exercises 3–6, simplify the radicals.

3. $(\sqrt[4]{81})^3$

4. $(\sqrt[4]{16})^3$

5. $\sqrt[3]{(a+1)^3}$

6. $\sqrt[5]{(x+y)^5}$

Concept 1: Definition of $a^{1/n}$

For Exercises 7–18, simplify the expression.

7. $81^{1/2}$ **8.** $25^{1/2}$ **9.** $125^{1/3}$ **10.** $8^{1/3}$

11. $81^{1/4}$ **12.** $16^{1/4}$ **13.** $(-8)^{1/3}$ **14.** $(-9)^{1/2}$

15. $-8^{1/3}$ **16.** $-9^{1/2}$ **17.** $36^{-1/2}$ **18.** $16^{-1/2}$

For Exercises 19–30, write the expressions in radical notation.

19. $x^{1/3}$ **20.** $y^{1/4}$ **21.** $(4a)^{1/2}$ **22.** $(36x)^{1/2}$

23. $(yz)^{1/5}$ **24.** $(cd)^{1/4}$ **25.** $(u^2)^{1/3}$ **26.** $(v^3)^{1/4}$

27. $5q^{1/2}$ **28.** $6p^{1/2}$ **29.** $\left(\frac{x}{9}\right)^{1/2}$ **30.** $\left(\frac{y}{8}\right)^{1/3}$

Concept 2: Definition of $a^{m/n}$

31. Explain how to interpret the expression $a^{m/n}$ as a radical.

32. Explain why $(\sqrt[3]{8})^4$ is easier to evaluate than $\sqrt[3]{8^4}$.

For Exercises 33–42, convert the expressions to radical form and simplify.

33. $8^{4/3}$ **34.** $25^{3/2}$ **35.** $16^{3/4}$ **36.** $32^{2/5}$

37. $27^{-2/3}$ **38.** $4^{-5/2}$ **39.** $(-8)^{5/3}$ **40.** $(-27)^{2/3}$

41. $\left(\frac{1}{4}\right)^{-1/2}$ **42.** $\left(\frac{1}{9}\right)^{3/2}$

Concept 3: Converting between Rational Exponents and Radical Notation

For Exercises 43–52, convert each expression to radical notation.

43. $x^{5/3}$ **44.** $a^{2/5}$ **45.** $y^{9/2}$ **46.** $b^{4/9}$

47. $(c^5d)^{1/3}$ **48.** $(a^2b)^{1/8}$ **49.** $(qr)^{-1/5}$ **50.** $(3x)^{-1/4}$

51. $6y^{2/3}$ **52.** $2q^{5/6}$

For Exercises 53–64, write the expressions using rational exponents rather than radical notation.

53. $\sqrt[3]{x}$ **54.** $\sqrt[4]{a}$ **55.** $\sqrt[3]{xy}$

56. $\sqrt[5]{ab}$ **57.** $5\sqrt{x}$ **58.** $7\sqrt[3]{z}$

59. $\sqrt[3]{y^2}$ **60.** $\sqrt[4]{b^3}$ **61.** $\sqrt[4]{m^3n}$

62. $\sqrt[5]{u^3v^4}$ **63.** $4\sqrt[3]{k^3}$ **64.** $6\sqrt[4]{t^4}$

Concept 4: Properties of Rational Exponents

For Exercises 65–84, simplify the expressions using the properties of rational exponents. Write the final answers with positive exponents only.

65. $x^{1/4}x^{3/4}$

66. $2^{2/3}2^{1/3}$

67. $(y^{1/5})^{10}$

68. $(x^{1/2})^8$

69. $6^{-1/5}6^{6/5}$

70. $a^{-1/3}a^{2/3}$

71. $(a^{1/3}a^{1/4})^{12}$

72. $(x^{2/3}x^{1/2})^6$

73. $\dfrac{y^{5/3}}{y^{1/3}}$

74. $\dfrac{z^{5/2}}{z^{1/2}}$

75. $\dfrac{2^{4/3}}{2^{1/3}}$

76. $\dfrac{5^{6/5}}{5^{1/5}}$

77. $(5a^2c^{-1/2}d^{1/2})^2$

78. $(2x^{-1/3}y^2z^{5/3})^3$

79. $\left(\dfrac{x^{-2/3}}{y^{-3/4}}\right)^{12}$

80. $\left(\dfrac{m^{-1/4}}{n^{-1/2}}\right)^{-4}$

81. $\left(\dfrac{16w^{-2}z}{2wz^{-8}}\right)^{1/3}$

82. $\left(\dfrac{50p^{-1}q}{2pq^{-3}}\right)^{1/2}$

83. $(25x^2y^4z^6)^{1/2}$

84. $(8a^6b^3c^9)^{2/3}$

Concept 5: Applications of Rational Exponents

If the area, A, of a square is known, then the length of its sides, s, can be computed by the formula: $s = A^{1/2}$.

85. **a.** Compute the length of the sides of a square having an area of 100 in.^{2}

b. Compute the length of the sides of a square having an area of 72 in.^{2} Round your answer to the nearest 0.01 in.

86. The radius, r, of a sphere of volume, V, is given by

$$r = \left(\frac{3V}{4\pi}\right)^{1/3}$$

Find the radius of a spherical ball having a volume of 55 in.^{3} Round your answer to the nearest 0.01 in.

For Exercises 87–88, use the following information.

If P dollars in principal grows to A dollars after t years with annual interest, then the rate of return is given by

$$r = \left(\frac{A}{P}\right)^{1/t} - 1$$

87. **a.** In one account, \$10,000 grows to \$16,802 after 5 years. Compute the interest rate to the nearest tenth of a percent.

b. In another account \$10,000 grows to \$18,000 after 7 years. Compute the interest rate to the nearest tenth of a percent.

c. Which account produced a higher average yearly return?

88. **a.** In one account, \$5000 grows to \$23,304.79 in 20 years. Compute the interest rate to the nearest whole percent.

b. In another account, \$6000 grows to \$34,460.95 in 30 years. Compute the interest rate to the nearest whole percent.

c. Which account produced a higher average yearly return?

Expanding Your Skills

89. Is $(a + b)^{1/2}$ the same as $a^{1/2} + b^{1/2}$? Why or why not?

For Exercises 90–95, simplify the expressions. Write the final answer with positive exponents only.

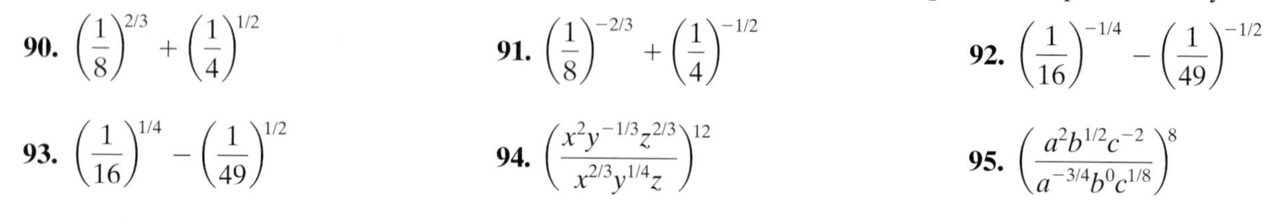

90. $\left(\frac{1}{8}\right)^{2/3} + \left(\frac{1}{4}\right)^{1/2}$

91. $\left(\frac{1}{8}\right)^{-2/3} + \left(\frac{1}{4}\right)^{-1/2}$

92. $\left(\frac{1}{16}\right)^{-1/4} - \left(\frac{1}{49}\right)^{-1/2}$

93. $\left(\frac{1}{16}\right)^{1/4} - \left(\frac{1}{49}\right)^{1/2}$

94. $\left(\frac{x^2y^{-1/3}z^{2/3}}{x^{2/3}y^{1/4}z}\right)^{12}$

95. $\left(\frac{a^2b^{1/2}c^{-2}}{a^{-3/4}b^0c^{1/8}}\right)^8$

Chapter 8 SUMMARY

Section 8.1 Introduction to Roots and Radicals

Key Concepts

b is a **square root** of a if $b^2 = a$.

The expression $\sqrt{a}$ represents the **principal square root** of a.

b is an nth-root of a if $b^n = a$.

1. If n is a positive *even* integer and $a > 0$, then $\sqrt[n]{a}$ is the principal (positive) nth-root of a.
2. If $n > 1$ is a positive *odd* integer, then $\sqrt[n]{a}$ is the nth-root of a.
3. If $n > 1$ is any positive integer, then $\sqrt[n]{0} = 0$.

$\sqrt[n]{a^n} = |a|$ if n is even.

$\sqrt[n]{a^n} = a$ if n is odd.

$\sqrt[n]{a}$ is not a real number if a is *negative* and n is even.

Pythagorean Theorem:

$a^2 + b^2 = c^2$

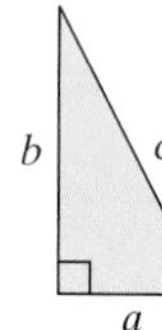

Examples

Example 1

The square roots of 16 are 4 and −4 because $(4)^2 = 16$ and $(-4)^2 = 16$.

$\sqrt{16} = 4$ Because $4^2 = 16$

$\sqrt[4]{16} = 2$ Because $2^4 = 16$

$\sqrt[3]{125} = 5$ Because $5^3 = 125$

Example 2

$\sqrt{y^2} = |y|$ $\sqrt[3]{y^3} = y$ $\sqrt[4]{y^4} = |y|$

Example 3

$\sqrt[4]{-16}$ is not a real number.

Example 4

Find the length of the unknown side.

$a^2 + b^2 = c^2$

$(8)^2 + b^2 = (17)^2$

$64 + b^2 = 289$

$b^2 = 289 - 64$

$b^2 = 225$

$b = \sqrt{225}$

$b = 15$

Because b denotes a length, b must be the positive square root of 225.

b, c = 17 cm, a = 8 cm

The third side is 15 cm.

Section 8.2 Simplifying Radicals

Key Concepts

Multiplication Property of Radicals:

If $\sqrt[n]{a}$ and $\sqrt[n]{b}$ are both real, then

$\sqrt[n]{ab} = \sqrt[n]{a} \cdot \sqrt[n]{b}$

Division Property of Radicals:

$\sqrt[n]{\frac{a}{b}} = \frac{\sqrt[n]{a}}{\sqrt[n]{b}} \quad b \neq 0$

Consider a radical expression whose radicand is written as a product of prime factors. Then the radical is in simplified form if each of the following criteria are met:

1. The radicand has no factor raised to a power greater than or equal to the index.
2. There are no radicals in the denominator of a fraction.
3. The radicand does not contain a fraction.

Examples

Example 1

$\sqrt{3} \cdot \sqrt{5} = \sqrt{3 \cdot 5} = \sqrt{15}$

Example 2

$\sqrt{\frac{x}{9}} = \frac{\sqrt{x}}{\sqrt{9}} = \frac{\sqrt{x}}{3}$

Example 3

$\sqrt[3]{16x^5y^7} = \sqrt[3]{2^4x^5y^7}$

$= \sqrt[3]{2^3x^3y^6 \cdot 2x^2y}$

$= \sqrt[3]{2^3x^3y^6} \cdot \sqrt[3]{2x^2y}$

$= 2xy^2\sqrt[3]{2x^2y}$

Example 4

$\sqrt{\frac{2x^5}{8x}} = \sqrt{\frac{x^4}{4}}$

$= \frac{\sqrt{x^4}}{\sqrt{4}}$

$= \frac{x^2}{2}$

Section 8.3 Addition and Subtraction of Radicals

Key Concepts

Two radical terms are *like* radicals if they have the same index and the same radicand.

Use the distributive property to add or subtract *like* radicals.

Examples

Example 1

Like radicals. $\sqrt[3]{5z}, \quad 6\sqrt[3]{5z}$

Example 2

$3\sqrt{7} - 10\sqrt{7} + \sqrt{7}$

$= (3 - 10 + 1)\sqrt{7}$

$= -6\sqrt{7}$

Section 8.4 Multiplication of Radicals

Key Concepts

Multiplication Property of Radicals:

$\sqrt[n]{a} \cdot \sqrt[n]{b} = \sqrt[n]{ab}$ provided $\sqrt[n]{a}$ and $\sqrt[n]{b}$ are both real.

Special Case Products:

$(a + b)(a - b) = a^2 - b^2$

$(a + b)^2 = a^2 + 2ab + b^2$

$(a - b)^2 = a^2 - 2ab + b^2$

Examples

Example 1

$$(6\sqrt{5})(4\sqrt{3}) = 6 \cdot 4\sqrt{5 \cdot 3} = 24\sqrt{15}$$

Example 2

$$3\sqrt{2}(\sqrt{2} + 5\sqrt{7} - \sqrt{6}) = 3\sqrt{2^2} + 15\sqrt{14} - 3\sqrt{2^2 \cdot 3} = 3 \cdot 2 + 15\sqrt{14} - 3 \cdot 2\sqrt{3} = 6 + 15\sqrt{14} - 6\sqrt{3}$$

Example 3

$$(4\sqrt{x} + \sqrt{2})(4\sqrt{x} - \sqrt{2}) = (4\sqrt{x})^2 - (\sqrt{2})^2 = 16x - 2$$

Example 4

$$(\sqrt{x} - \sqrt{5y})^2 = (\sqrt{x})^2 - 2(\sqrt{x})(\sqrt{5y}) + (\sqrt{5y})^2 = x - 2\sqrt{5xy} + 5y$$

Section 8.5 Rationalization

Key Concepts

Rationalizing the Denominator with One Term:

Multiply the numerator and denominator by an appropriate expression to create an nth-root of an nth-power in the denominator.

Rationalizing a Two-Term Denominator Involving Square Roots:

Multiply the numerator and denominator by the conjugate of the denominator.

Examples

Example 1

$$\frac{10}{\sqrt{5}} = \frac{10}{\sqrt{5}} \cdot \frac{\sqrt{5}}{\sqrt{5}} = \frac{10\sqrt{5}}{\sqrt{5^2}} = \frac{10\sqrt{5}}{5} = 2\sqrt{5}$$

Example 2

$$\frac{\sqrt{2}}{\sqrt{x} - \sqrt{3}} = \frac{\sqrt{2}}{(\sqrt{x} - \sqrt{3})} \cdot \frac{(\sqrt{x} + \sqrt{3})}{(\sqrt{x} + \sqrt{3})} = \frac{\sqrt{2x} + \sqrt{6}}{x - 3}$$

Section 8.6 Radical Equations

Key Concepts

An equation with one or more radicals containing a variable is a **radical equation**.

Steps for Solving a Radical Equation:

1. Isolate the radical. If an equation has more than one radical, choose one of the radicals to isolate.
2. Raise each side of the equation to a power equal to the index of the radical.
3. Solve the resulting equation.
4. Check the potential solutions in the original equation.

Examples

Example 1

Solve. $\sqrt{2x-4}+3=7$

Step 1: $\sqrt{2x-4}=4$

Step 2: $(\sqrt{2x-4})^2=(4)^2$

Step 3: $2x-4=16$

$2x=20$

$x=10$

Step 4:

Check:

$\sqrt{2x-4}+3=7$

$\sqrt{2(10)-4}+3\stackrel{?}{=}7$

$\sqrt{20-4}+3\stackrel{?}{=}7$

$\sqrt{16}+3\stackrel{?}{=}7$

$4+3=7$ ✔ The solution checks.

The solution is $x=10$.

Section 8.7 Rational Exponents

Key Concepts

If $\sqrt[n]{a}$ is a real number, then

$a^{1/n}=\sqrt[n]{a}$

$a^{m/n}=(\sqrt[n]{a})^m=\sqrt[n]{a^m}$

Examples

Example 1

$121^{1/2}=\sqrt{121}=11$

Example 2

$27^{2/3}=(\sqrt[3]{27})^2=(3)^2=9$

Example 3

$8^{-1/3}=\dfrac{1}{8^{1/3}}=\dfrac{1}{\sqrt[3]{8}}=\dfrac{1}{2}$

Chapter 8 Review Exercises

Section 8.1

For Exercises 1–4, state the principal square root and the negative square root.

1. 196 **2.** 1.44

3. 0.64 **4.** 225

5. Explain why $\sqrt{-64}$ is *not* a real number.

6. Explain why $\sqrt[3]{-64}$ *is* a real number.

For Exercises 7–18, simplify the expressions, if possible.

7. $-\sqrt{144}$ **8.** $-\sqrt{25}$ **9.** $\sqrt{-144}$

10. $\sqrt{-25}$ **11.** $\sqrt{y^2}$ **12.** $\sqrt[3]{y^3}$

13. $\sqrt[4]{y^4}$ **14.** $-\sqrt[3]{125}$ **15.** $-\sqrt[4]{625}$

16. $\sqrt[3]{p^{12}}$ **17.** $\sqrt[4]{\dfrac{81}{t^8}}$ **18.** $\sqrt[3]{\dfrac{-27}{w^3}}$

19. The radius, r, of a circle can be found from the area of the circle according to the formula:

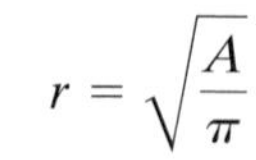

$$r = \sqrt{\frac{A}{\pi}}$$

a. What is the radius of a circular garden whose area is 160 m^2? Round to the nearest tenth of a meter.

b. What is the radius of a circular fountain whose area is 1600 ft^2? Round to the nearest tenth of a foot.

20. Suppose a ball is thrown with an initial velocity of 76 ft/sec at an angle of 30° (see figure). Then the horizontal position of the ball, x (measured in feet), depends on the number of seconds, t, after the ball is thrown according to the equation:

$$x = 38\sqrt{3}t$$

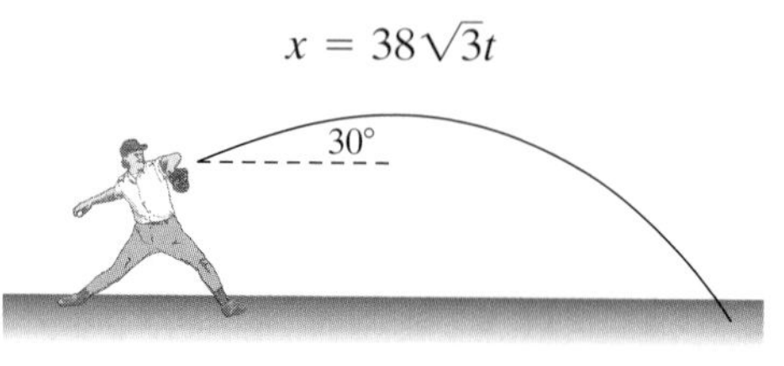

a. What is the horizontal position of the ball after 1 sec? Round your answer to the nearest tenth of a foot.

b. What is the horizontal position of the ball after 2 sec? Round your answer to the nearest tenth of a foot.

For Exercises 21–22, translate the English phrases into algebraic expressions.

21. The square of b plus the principal square root of 5.

22. The difference of the cube root of y and the fourth root of x.

For Exercises 23–24, translate the algebraic expressions into English phrases. (Answers may vary.)

23. $\dfrac{2}{\sqrt{p}}$ **24.** $8\sqrt{q}$

25. A hedge extends 5 ft from the wall of a house. A 13-ft ladder is placed at the edge of the hedge. How far up the house is the tip of the ladder?

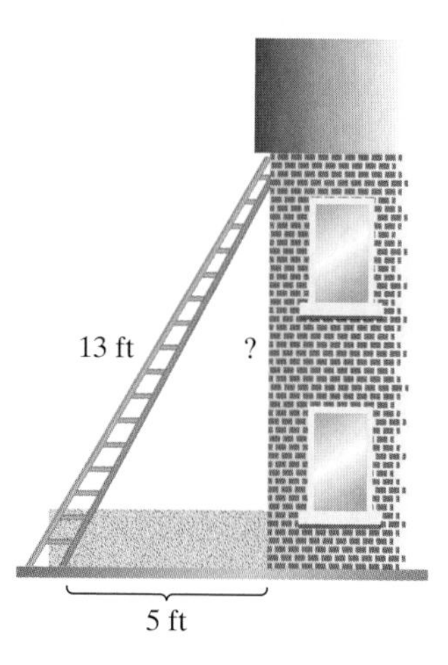

26. Nashville, TN, is north of Birmingham, AL, a distance of 182 miles. Augusta, GA, is east of Birmingham, a distance of 277 miles. How far is it from Augusta to Nashville? Round the answer to the nearest mile.

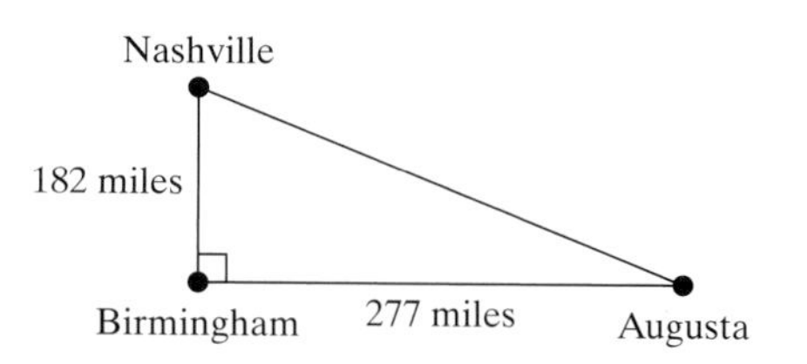

Section 8.2

For Exercises 27–32, use the multiplication property of radicals to simplify. Assume the variables represent positive real numbers.

27. $\sqrt{x^{17}}$
28. $\sqrt[3]{40}$
29. $\sqrt{28}$
30. $5\sqrt{18x^3}$
31. $2\sqrt{27y^{10}}$
32. $\sqrt[3]{27y^{10}}$

For Exercises 33–42, use the division property of radicals to simplify. Assume the variables represent positive real numbers.

33. $\dfrac{\sqrt{50}}{\sqrt{49}}$
34. $\sqrt{\dfrac{a^8}{25}}$
35. $\sqrt{\dfrac{2x^3}{y^6}}$
36. $\dfrac{\sqrt{18x}}{\sqrt{2x^3}}$
37. $\dfrac{\sqrt{200y^5}}{\sqrt{2y^3}}$
38. $\sqrt{\dfrac{w^5}{9z^4}}$
39. $\sqrt[3]{\dfrac{30p^2}{q^6}}$
40. $\dfrac{\sqrt{3n^5}}{\sqrt{48n^7}}$
41. $\dfrac{\sqrt{5t^6}}{\sqrt{121}}$
42. $\dfrac{\sqrt{196}}{\sqrt{36}}$

Section 8.3

For Exercises 43–48, add or subtract as indicated. Assume the variables represent positive real numbers.

43. $8\sqrt{6} - \sqrt{6}$
44. $1.6\sqrt{y} - 1.4\sqrt{y} + 0.6\sqrt{y}$
45. $x\sqrt{20} - 2\sqrt{45x^2}$
46. $y\sqrt{64y} + 3\sqrt{y^3}$
47. $3\sqrt{112} - 4\sqrt{28} + \sqrt{7}$
48. $2\sqrt{50} - 4\sqrt{18} - 6\sqrt{2}$

For Exercises 49–50, translate the English phrases into algebraic expressions and simplify.

49. The sum of the principal square root of the fourth power of x and the square of $5x$.
50. The difference of the principal square root of 128 and the square root of 2.

51. Find the exact perimeter of the triangle.

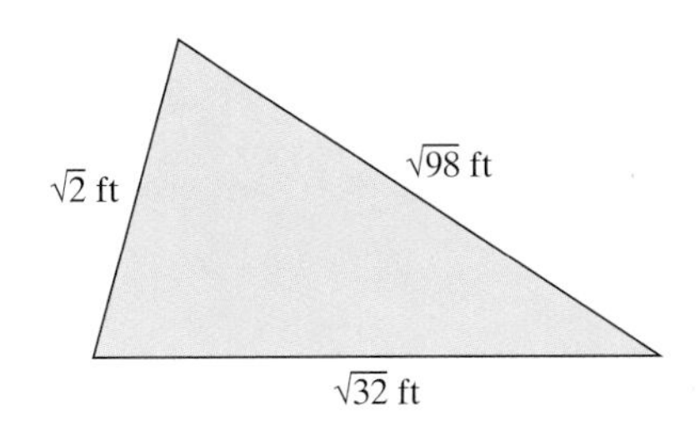

52. Find the exact perimeter of a square whose sides are $3\sqrt{48}$ m.

Section 8.4

For Exercises 53–62, multiply the expressions. Assume the variables represent positive real numbers.

53. $\sqrt{5} \cdot \sqrt{125}$
54. $\sqrt{10p} \cdot \sqrt{6}$
55. $(5\sqrt{6})(7\sqrt{2x})$
56. $(3\sqrt{y})(-2z\sqrt{11y})$
57. $8\sqrt{m}(\sqrt{m} + 3)$
58. $\sqrt{2}(\sqrt{7} + 8)$
59. $(5\sqrt{2} + \sqrt{13})(-\sqrt{2} - 3\sqrt{13})$
60. $(\sqrt{p} + 2\sqrt{q})(4\sqrt{p} - \sqrt{q})$
61. $(8\sqrt{w} - \sqrt{z})(8\sqrt{w} + \sqrt{z})$
62. $(2x - \sqrt{y})^2$
63. Find the exact volume of the box.

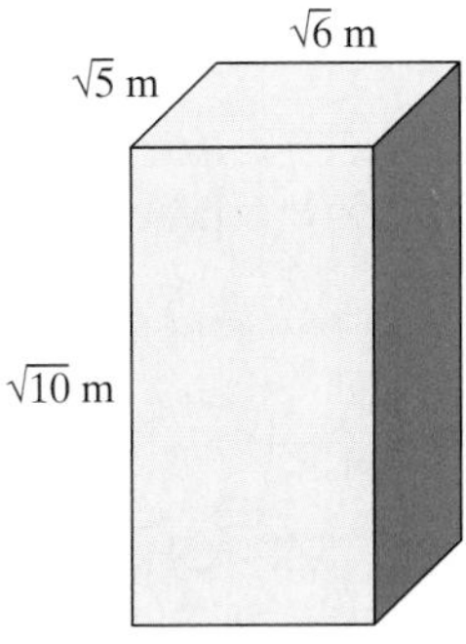

Section 8.5

64. To rationalize the denominator in the expression

$$\frac{6}{\sqrt{a} + 5}$$

which quantity would you multiply by in the numerator and denominator?

a. $\sqrt{a} + 5$
b. $\sqrt{a} - 5$
c. $\sqrt{a}$
d. -5

65. To rationalize the denominator in the expression

$$\frac{w}{\sqrt{w}-4}$$

which quantity would you multiply by in the numerator and denominator?

a. $\sqrt{w}-4$ **b.** $\sqrt{w}+4$

c. $\sqrt{w}$ **d.** 4

For Exercises 66–71, rationalize the denominators. Assume the variables represent positive real numbers.

66. $\dfrac{11}{\sqrt{7}}$ **67.** $\sqrt{\dfrac{18}{y}}$ **68.** $\dfrac{\sqrt{24}}{\sqrt{6x^7}}$

69. $\dfrac{10}{\sqrt{7}-\sqrt{2}}$ **70.** $\dfrac{6}{\sqrt{w}+2}$ **71.** $\dfrac{\sqrt{7}+3}{\sqrt{7}-3}$

72. The velocity of an object, v, (in meters per second: m/sec) depends on the kinetic energy, E (in joules: J), and mass, m (in kilograms: kg), of the object according to the formula:

$$v=\sqrt{\frac{2E}{m}}$$

a. What is the exact velocity of a 3-kg object whose kinetic energy is 100 J?

b. What is the exact velocity of a 5-kg object whose kinetic energy is 162 J?

Section 8.6

For Exercises 73–81, solve the equations. Be sure to check the potential solutions.

73. $\sqrt{p+6}=12$ **74.** $\sqrt{k+1}=-7$

75. $\sqrt{3x-17}-10=0$

76. $\sqrt{14n+10}=4\sqrt{n}$

77. $\sqrt{2z+2}=\sqrt{3z-5}$

78. $\sqrt{5y-5}-\sqrt{4y+1}=0$

79. $\sqrt{2m+5}=m+1$

80. $\sqrt{3n-8}-n+2=0$

81. $\sqrt[3]{2y+13}=-5$

82. The length of the sides of a cube is related to the volume of the cube according to the formula: $x=\sqrt[3]{V}$.

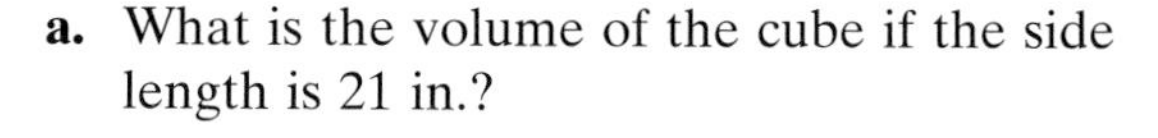

a. What is the volume of the cube if the side length is 21 in.?

b. What is the volume of the cube if the side length is 15 cm?

Section 8.7

For Exercises 83–88, simplify the expressions.

83. $(-27)^{1/3}$ **84.** $121^{1/2}$ **85.** $-16^{1/4}$

86. $(-16)^{1/4}$ **87.** $4^{-3/2}$ **88.** $\left(\dfrac{1}{9}\right)^{-3/2}$

For Exercises 89–92, write the expression in radical notation. Assume the variables represent positive real numbers.

89. $z^{1/5}$ **90.** $q^{2/3}$

91. $(w^3)^{1/4}$ **92.** $\left(\dfrac{b}{121}\right)^{1/2}$

For Exercises 93–96, write the expression using rational exponents rather than radical notation. Assume the variables represent positive real numbers.

93. $\sqrt[5]{a^2}$ **94.** $5\sqrt[3]{m^2}$

95. $\sqrt[5]{a^2b^4}$ **96.** $\sqrt{6}$

For Exercises 97–102, simplify using the properties of rational exponents. Write the answer with positive exponents only. Assume the variables represent positive real numbers.

97. $y^{2/3}y^{4/3}$ **98.** $\dfrac{6^{4/5}}{6^{1/5}}$

99. $a^{1/3}a^{1/2}$ **100.** $(5^{1/2})^{3/2}$

101. $(64a^3b^6)^{1/3}$ **102.** $\left(\dfrac{b^4b^0}{b^{1/4}}\right)^4$

103. The radius, r, of a right circular cylinder can be found if the volume, V, and height, h, are known. The radius is given by

$$r=\left(\frac{V}{\pi h}\right)^{1/2}$$

Find the radius of a right circular cylinder whose volume is 150.8 cm³ and whose height is 12 cm. Round the answer to the nearest tenth of a centimeter.

Chapter 8 Test

1. State the conditions for a radical expression to be in simplified form.

For Exercises 2–7, simplify the radicals, if possible. Assume the variables represent positive real numbers.

2. $\sqrt{242x^2}$ 3. $\sqrt[3]{48y^4}$ 4. $\sqrt{-64}$

5. $\sqrt{\dfrac{5a^6}{81}}$ 6. $\dfrac{9}{\sqrt{6}}$ 7. $\dfrac{2}{\sqrt{5}+6}$

8. Translate the English phrases into algebraic expressions and simplify.

   a. The sum of the principal square root of twenty-five and the cube of five.

   b. The difference of the square of four and the principal square root of 16.

9. A baseball player hits the ball at an angle of 30° with an initial velocity of 112 ft/sec. The horizontal position of the ball, x (measured in feet), depends on the number of seconds, t, after the ball is struck according to the equation

$$x = 56\sqrt{3}t$$

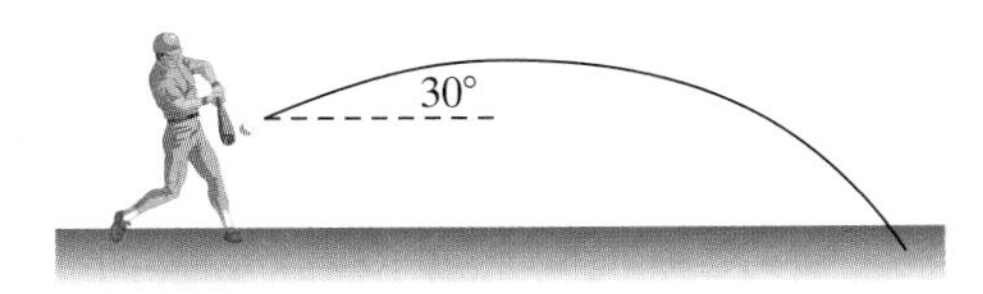

   a. What is the horizontal position of the ball after 1 sec? Round the answer to the nearest foot.

   b. What is the horizontal position of the ball after 3.5 sec? Round the answer to the nearest foot.

For Exercises 10–19, perform the indicated operations. Assume the variables represent positive real numbers.

10. $6\sqrt{z} - 3\sqrt{z} + 5\sqrt{z}$

11. $\sqrt{3}(4\sqrt{2} - 5\sqrt{3})$

12. $\sqrt{50t^2} - t\sqrt{288}$

13. $\sqrt{360} + \sqrt{250} - \sqrt{40}$

14. $(3\sqrt{5} - 1)^2$

15. $(6\sqrt{2} - \sqrt{5})(\sqrt{2} + 4\sqrt{5})$ 16. $\dfrac{\sqrt{2m^3n}}{\sqrt{72m^5}}$

17. $(4 - 3\sqrt{x})(4 + 3\sqrt{x})$ 18. $\sqrt{\dfrac{2}{11}}$

19. $\dfrac{6}{\sqrt{7} - \sqrt{3}}$

20. A triathlon consists of a swim, followed by a bike ride, followed by a run. The swim begins on a beach at point A. The swimmers must swim 50 yd to a buoy at point B, then 200 yd to a buoy at point C, and then return to point A on the beach. How far is the distance from point C to point A? (Round to the nearest yard.)

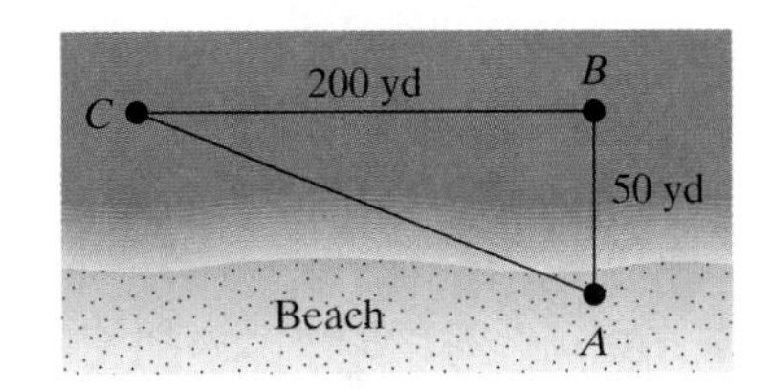

For Exercises 21–23, solve the equations.

21. $\sqrt{2x + 7} + 6 = 2$

22. $\sqrt{1 - 7x} = 1 - x$

23. $\sqrt[3]{x + 6} = \sqrt[3]{2x - 8}$

24. The height, y (in inches), of a tomato plant can be approximated by the time, t (in weeks), after the seed has germinated according to the equation

$$y = 6\sqrt{t}$$

   a. Use the equation to find the height of the plant after 4 weeks.

   b. Use the equation to find the height of the plant after 9 weeks.

c. Use the equation to find the time required for the plant to reach a height of 30 in. Verify your answer from the graph.

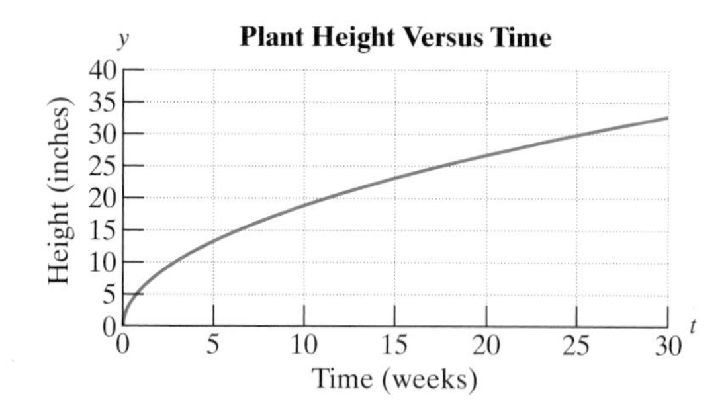

For Exercises 25–26, simplify the expression.

25. $10{,}000^{3/4}$

26. $\left(\frac{1}{8}\right)^{-1/3}$

For Exercises 27–28, write the expressions in radical notation. Assume the variables represent positive real numbers.

27. $x^{3/5}$

28. $5y^{1/2}$

29. Write the expression using rational exponents: $\sqrt[4]{ab^3}$. (Assume $a \geq 0$ and $b \geq 0$.)

For Exercises 30–32, simplify using the properties of rational exponents. Write the final answer with positive exponents only. Assume the variables represent positive real numbers.

30. $p^{1/4} \cdot p^{2/3}$

31. $\dfrac{5^{4/5}}{5^{1/5}}$

32. $(9m^2n^4)^{1/2}$

Chapters 1–8 Cumulative Review Exercises

1. Simplify. $\dfrac{|-3 - 12 \div 6 + 2|}{\sqrt{5^2 - 4^2}}$

2. Solve for y:
$2 - 5(2y + 4) - (-3y - 1) = -(y + 5)$

3. Simplify. Write the final answer with positive exponents only.

$$\left(\frac{1}{3}\right)^0 - \left(\frac{1}{4}\right)^{-2}$$

4. Perform the indicated operations:
$2(x - 3) - (3x + 4)(3x - 4)$

5. Divide:

$$\frac{14x^3y - 7x^2y^2 + 28xy^2}{7x^2y^2}$$

6. Factor completely. $50c^2 + 40c + 8$

7. Solve for x: $10x^2 = x + 2$

8. Perform the indicated operations:

$$\frac{5a^2 + 2ab - 3b^2}{10a + 10b} \div \frac{25a^2 - 9b^2}{50a + 30b}$$

9. Solve for z: $\dfrac{1}{5} + \dfrac{z}{z - 5} = \dfrac{5}{z - 5}$

10. Simplify:

$$\frac{\frac{5}{4} + \frac{2}{x}}{\frac{4}{x} - \frac{4}{x^2}}$$

11. Graph. $3y = 6$

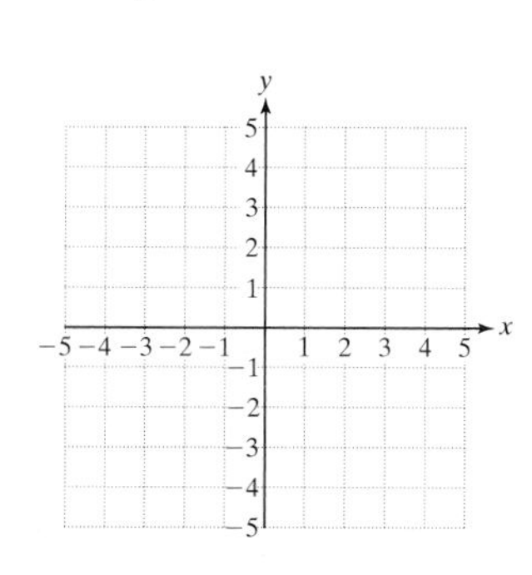

12. The equation $y = 210x + 250$ represents the cost, y (in dollars), of renting office space for x months.

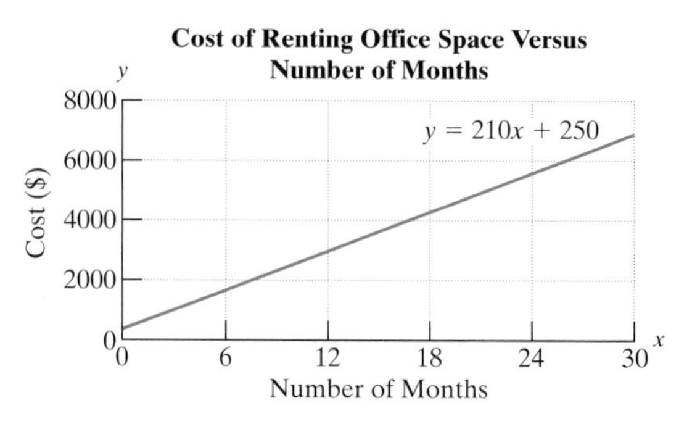

a. Find y when x is 3. Interpret the result in the context of the problem.

b. Find x when y is $2770. Interpret the result in the context of the problem.

c. What is the slope of the line? Interpret the meaning of the slope in the context of the problem.

d. What is the y-intercept? Interpret the meaning of the y-intercept in the context of the problem.

13. Write an equation of the line passing through the points $(2, -1)$ and $(-3, 4)$. Write the answer in slope-intercept form.

14. Solve the system of equations using the addition method. If the system has no solution or infinitely many solutions, so state:

$$3x - 5y = 23$$
$$2x + 4y = -14$$

15. Graph the solution to the inequality: $-2x - y > 3$

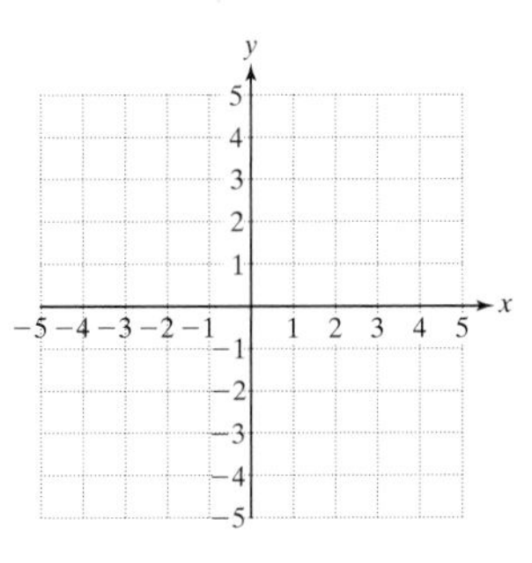

16. How many liters (L) of 20% acid solution must be mixed with a 50% acid solution to obtain 12 L of a 30% acid solution?

17. Simplify. $\sqrt{99}$

18. Perform the indicated operation.

$$5x\sqrt{3} + \sqrt{12x^2}$$

19. Rationalize the denominator. $\dfrac{\sqrt{x}}{\sqrt{x} - \sqrt{y}}$

20. Solve for y. $\sqrt{2y - 1} - 4 = -1$

Functions, Complex Numbers, and Quadratic Equations

9

In this chapter we explore quadratic functions by finding the vertex and intercepts, and by graphing parabolas. We also introduce another set of numbers called the complex numbers.

Many of the key terms found in this chapter can be found in the word search puzzle. Try and find all of the words in the list. They can be written up, down, left, right, or diagonally. Good luck.

Key Terms

complex
conjugates
domain
function
imaginary
linear
parabola
quadratic
range
relation
square root
standard
symmetry
vertex

```
                    M
                  N D X
                Q B L C N
              Z T U I U B D
            X L Z F N I O C H
          E F Y R T E M M Y S Z
        L M Y P H Q A X Q R S V H
      P G D D Q O I R E L A T I O N
    M D S M F U N C T I O N E R S I N
  O P R R T P F E E X S X I Q Q Y L N G
C O V Q E S A Q G F U O E G U W E N G V O
  O M P N Q R N M Y C I T A R D A U Q Z
    N U N S A A D J Q F R M K A C O B
      J Z R B D T A E E E I E I R S
        U U O W B H R B V G W A Q
          G L S S O G D T S D H
            A Y O T D H M J W
              T S F P N C P
                E W C Q O
                  S S N
                    H
```

Section 9.1 Introduction to Functions

Concepts

1. Definition of a Relation
2. Definition of a Function
3. Vertical Line Test
4. Function Notation
5. Domain and Range of a Function
6. Applications of Functions

1. Definition of a Relation

The number of points scored by Shaquille O'Neal during the first six games of a recent basketball season is shown in Table 9-1.

Table 9-1

Game, x	Number of Points, y		Ordered Pair
1	10	⟶	(1, 10)
2	20	⟶	(2, 20)
3	18	⟶	(3, 18)
4	19	⟶	(4, 19)
5	16	⟶	(5, 16)
6	24	⟶	(6, 24)

Each ordered pair from Table 9-1 shows a correspondence, or relationship, between the game number and the number of points scored by Shaquille O'Neal. The set of ordered pairs: $\{(1, 10), (2, 20), (3, 18), (4, 19), (5, 16), (6, 24)\}$ defines a relation between the game number and the number of points scored.

Definition of a Relation in x and y

Any set of ordered pairs, (x, y), is called a **relation** in x and y. Furthermore:

- The set of first components in the ordered pairs is called the **domain** of the relation.
- The set of second components in the ordered pairs is called the **range** of the relation.

Example 1 Finding the Domain and Range of a Relation

Find the domain and range of the relation linking the game number to the number of points scored by O'Neal in the first six games of the season:

$$\{(1, 10), (2, 20), (3, 18), (4, 19), (5, 16), (6, 24)\}$$

Solution:

Domain: $\{1, 2, 3, 4, 5, 6\}$ (Set of first coordinates)

Range: $\{10, 20, 18, 19, 16, 24\}$ (Set of second coordinates)

The domain consists of the game numbers for the first six games of the season. The range represents the corresponding number of points.

Skill Practice

1. Find the domain and range of the relation. $\{(0, 1), (4, 5), (-6, 8), (4, 13), (-8, 8)\}$

Skill Practice Answers

1. Domain: $\{0, 4, -6, -8\}$; Range: $\{1, 5, 8, 13\}$

Example 2 Finding the Domain and Range of a Relation

The three women represented in Figure 9-1 each have children. Molly has one child, Peggy has two children, and Joanne has three children.

a. If the set of mothers is given as the domain and the set of children is the range, write a set of ordered pairs defining the relation given in Figure 9-1.

b. Write the domain and range of the relation.

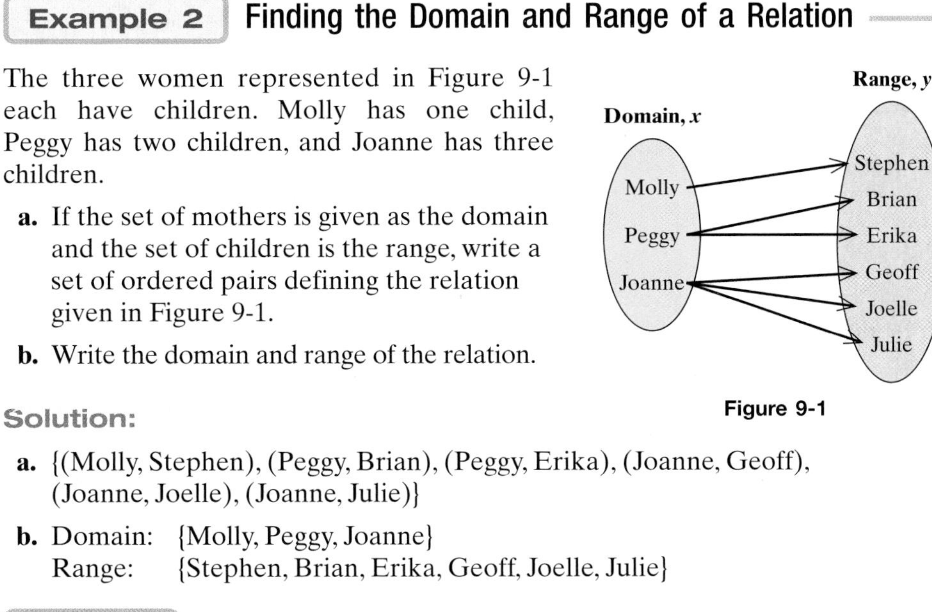

Figure 9-1

Solution:

a. {(Molly, Stephen), (Peggy, Brian), (Peggy, Erika), (Joanne, Geoff), (Joanne, Joelle), (Joanne, Julie)}

b. Domain: {Molly, Peggy, Joanne}
Range: {Stephen, Brian, Erika, Geoff, Joelle, Julie}

Skill Practice Given the relation represented by the figure:

2. Write the relation as a set of ordered pairs.

3. Write the domain and range of the relation.

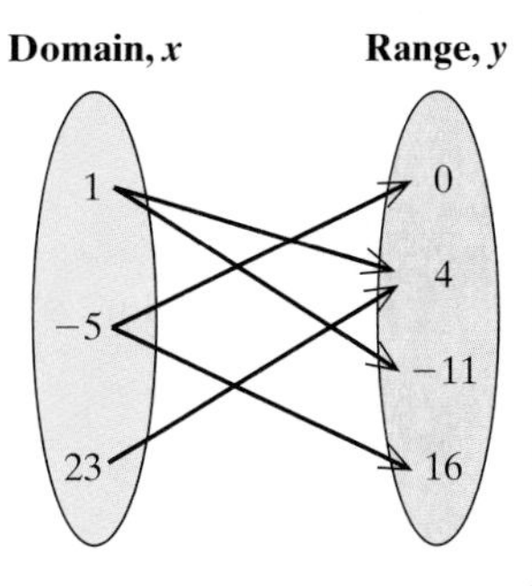

2. Definition of a Function

In mathematics, a special type of relation, called a function, is used extensively.

Definition of a Function

Given a relation in x and y, we say "y is a **function** of x" if for every element x in the domain, there corresponds exactly one element y in the range.

To understand the difference between a relation that is a function and one that is not a function, consider Example 3.

Example 3 Determining Whether a Relation Is a Function

Determine whether the following relations are functions:

a. $\{(2, -3), (4, 1), (3, -1), (2, 4)\}$

b. $\{(-3, 1), (0, 2), (4, -3), (1, 5), (-2, 1)\}$

Skill Practice Answers

2. $\{(1, 4), (1, -11), (-5, 0), (-5, 16), (23, 4)\}$

3. Domain: $\{1, -5, 23\}$; Range: $\{0, 4, -11, 16\}$

Solution:

a. This relation is defined by the set of ordered pairs.

same x-values

$$\{(2, -3), (4, 1), (3, -1), (2, 4)\}$$

different y-values

When $x = 2$, there are two possibilities for y: $y = -3$ and $y = 4$.

This relation is *not* a function because when $x = 2$, there is more than one corresponding element in the range.

b. This relation is defined by the set of ordered pairs: $\{(-3, 1), (0, 2), (4, -3), (1, 5), (-2, 1)\}$. Notice that each value in the domain has only one y-value in the range. Therefore, this relation *is* a function.

When $x = -3$, there is only one possibility for y: $y = 1$.

When $x = 0$, there is only one possibility for y: $y = 2$.

When $x = 4$, there is only one possibility for y: $y = -3$.

When $x = 1$, there is only one possibility for y: $y = 5$.

When $x = -2$, there is only one possibility for y: $y = 1$.

Skill Practice Determine whether the following relations are functions. If the relation is not a function, state why.

4. $\{(0, -7), (4, 9), (-2, -7), \left(\frac{1}{3}, \frac{1}{2}\right), (4, 10)\}$

5. $\{(-8, -3), (4, -3), (-12, 7), (-1, -1)\}$

In Example 2, the relation linking the set of mothers with their respective children is *not* a function. The domain elements, "Peggy" and "Joanne" each have more than one child. Because these x-values in the domain have more than one corresponding y-value in the range, the relation is not a function.

3. Vertical Line Test

A relation that is not a function has at least one domain element, x, paired with more than one range element, y. For example, the ordered pairs (2, 1) and (2, 4) do not make a function. On a graph, these two points are aligned vertically in the xy-plane, and a vertical line drawn through one point also intersects the other point (Figure 9-2). Thus, if a vertical line drawn through a graph of a relation intersects the graph in more than one point, the relation cannot be a function. This idea is stated formally as the **vertical line test**.

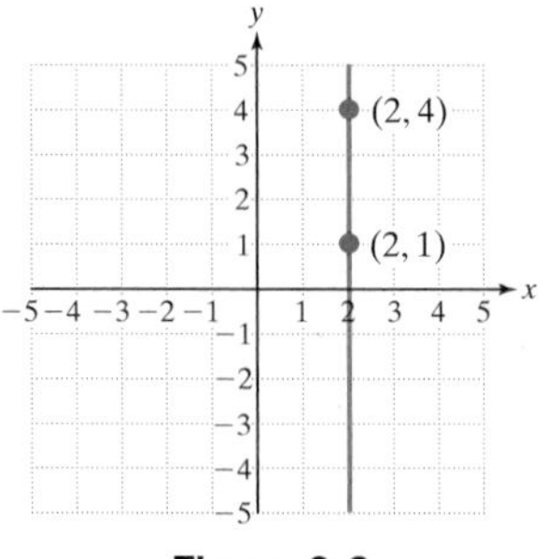

Figure 9-2

Skill Practice Answers

4. Not a function because the domain element, 4, has two different y-values: (4, 9) and (4, 10).

5. Function

Vertical Line Test

Consider a relation defined by a set of points (x, y) on a rectangular coordinate system. Then the graph defines y as a function of x if no vertical line intersects the graph in more than one point.

The vertical line test also implies that if any vertical line drawn through the graph of a relation intersects the relation in more than one point, then the relation does *not* define y as a function of x.

The vertical line test can be demonstrated by graphing the ordered pairs from the relations in Example 3 (Figure 9-3 and Figure 9-4).

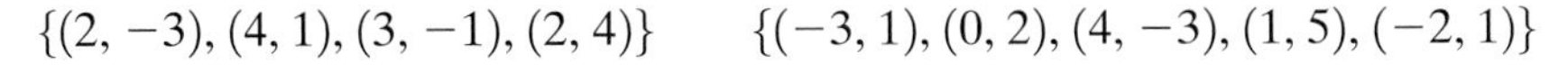

$\{(2, -3), (4, 1), (3, -1), (2, 4)\}$ $\quad\quad$ $\{(-3, 1), (0, 2), (4, -3), (1, 5), (-2, 1)\}$

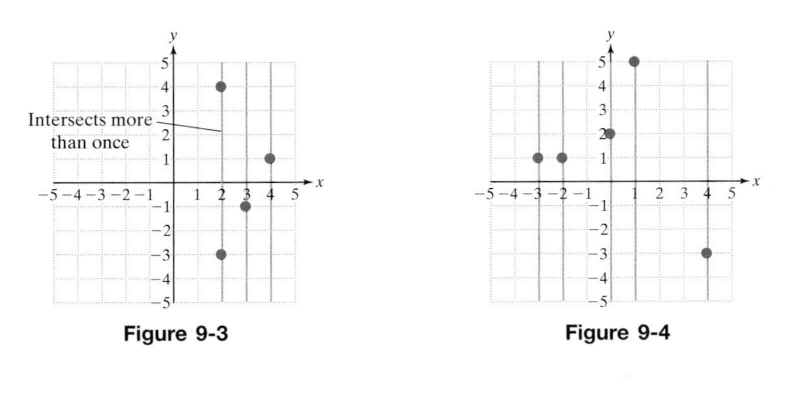

Figure 9-3

Figure 9-4

Not a Function
A vertical line intersects in more than one point.

Function
No vertical line intersects more than once.

The relations in Examples 1, 2, and 3 consist of a finite number of ordered pairs. A relation may, however, consist of an *infinite* number of points defined by an equation or by a graph. For example, the equation $y = x + 1$ defines infinitely many ordered pairs whose y-coordinate is one more than its x-coordinate. These ordered pairs cannot all be listed but can be depicted in a graph.

The vertical line test is especially helpful in determining whether a relation is a function based on its graph.

Example 4 **Using the Vertical Line Test**

Use the vertical line test to determine whether the following relations are functions.

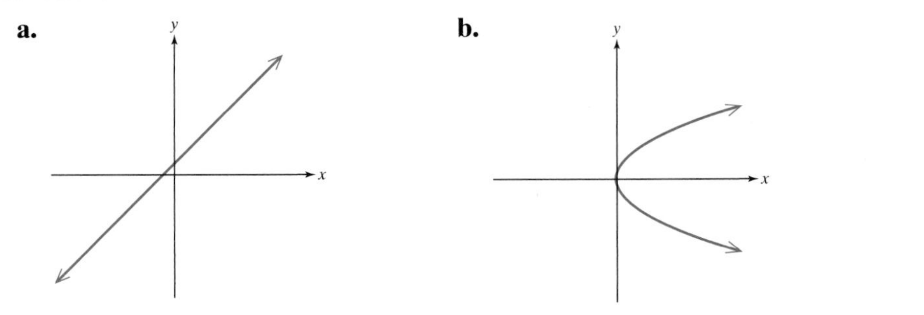

Solution:

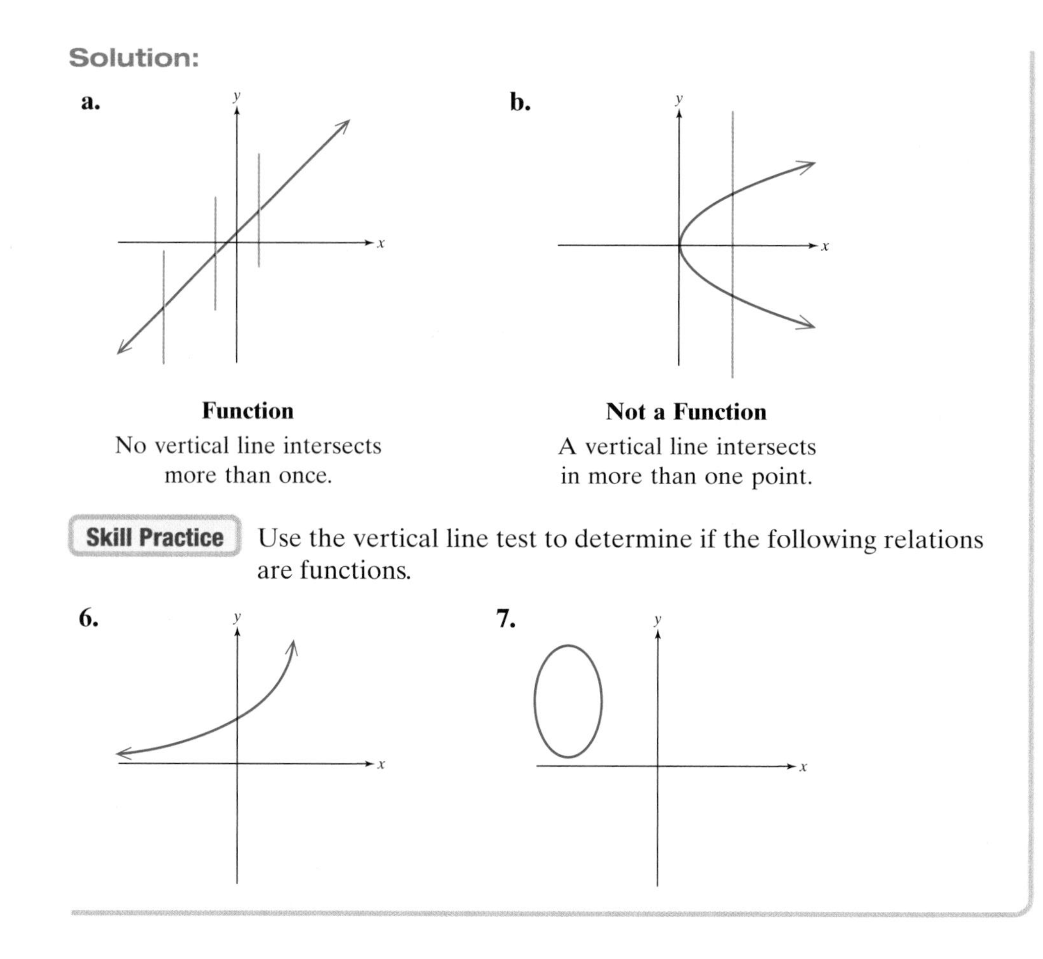

4. Function Notation

Avoiding Mistakes:

The notation $f(x)$ is read as "f of x" and does *not* imply multiplication.

A function is defined as a relation with the added restriction that each value of the domain has only one value in the range. In mathematics, functions are often given by rules or equations to define the relationship between two or more variables. For example, the equation, $y = x + 1$ defines the set of ordered pairs such that the y-value is one more than the x-value.

When a function is defined by an equation, we often use **function notation**. For example, the equation $y = x + 1$ can be written in function notation as

$$f(x) = x + 1$$

where f is the name of the function, x is an input value from the domain of the function, and $f(x)$ is the function value (or y-value) corresponding to x.

The notation $f(x)$ is read as "f of x" or "the value of the function, f, at x."

A function can be evaluated at different values of x by substituting values of x from the domain into the function. For example, for the function defined by $f(x) = x + 1$ we can evaluate f at $x = 3$ by using substitution.

$$f(x) = x + 1$$
$$\downarrow \quad \downarrow$$
$$f(3) = (3) + 1$$
$$f(3) = 4 \qquad \text{This is read as "} f \text{ of 3 equals 4."}$$

Thus, when $x = 3$, the function value is 4. This can also be interpreted as an ordered pair: (3, 4)

The names of functions are often given by either lowercase letters or uppercase letters such as f, g, h, p, k, M, and so on.

Skill Practice Answers

6. Function **7.** Not a function

Example 5 Evaluating a Function

Given the function defined by $h(x) = x^2 - 2$, find the function values.

a. $h(0)$ **b.** $h(1)$ **c.** $h(2)$ **d.** $h(-1)$ **e.** $h(-2)$

Solution:

a. $h(x) = x^2 - 2$

$h(0) = (0)^2 - 2$ Substitute $x = 0$ into the function.

$= 0 - 2$

$= -2$ $h(0) = -2$ means that when $x = 0, y = -2$, yielding the ordered pair $(0, -2)$.

b. $h(x) = x^2 - 2$

$h(1) = (1)^2 - 2$ Substitute $x = 1$ into the function.

$= 1 - 2$

$= -1$ $h(1) = -1$ means that when $x = 1, y = -1$, yielding the ordered pair $(1, -1)$.

c. $h(x) = x^2 - 2$

$h(2) = (2)^2 - 2$ Substitute $x = 2$ into the function.

$= 4 - 2$

$= 2$ $h(2) = 2$ means that when $x = 2, y = 2$, yielding the ordered pair $(2, 2)$.

d. $h(x) = x^2 - 2$

$h(-1) = (-1)^2 - 2$ Substitute $x = -1$ into the function.

$= 1 - 2$

$= -1$ $h(-1) = -1$ means that when $x = -1, y = -1$, yielding the ordered pair $(-1, -1)$.

e. $h(x) = x^2 - 2$

$h(-2) = (-2)^2 - 2$ Substitute $x = -2$ into the function.

$= 4 - 2$

$= 2$ $h(-2) = 2$ means that when $x = -2, y = 2$, yielding the ordered pair $(-2, 2)$.

Figure 9-5

The rule $h(x) = x^2 - 2$ is equivalent to the equation $y = x^2 - 2$. The function values $h(0), h(1), h(2), h(-1)$, and $h(-2)$ are the y-values in the ordered pairs $(0, -2), (1, -1), (2, 2), (-1, -1)$, and $(-2, 2)$, respectively. These points can be used to sketch a graph of the function (Figure 9-5).

Skill Practice Given the function defined by $f(x) = x^2 - 5x$, find the function values.

8. $f(1)$ **9.** $f(0)$ **10.** $f(-3)$ **11.** $f(-1)$ **12.** $f(3)$

Example 6 Evaluating a Function

Given the functions defined by $f(x) = 2x - 3$ and $g(x) = x^3 - 1$, find the function values.

a. $f(1) + g(3)$ **b.** $g(t)$ **c.** $f(a + b)$

Solution:

a. $f(x) = 2x - 3$ $\quad g(x) = x^3 - 1$

$f(1) = 2(1) - 3 \quad g(3) = (3)^3 - 1$

$= 2 - 3 \quad = 27 - 1$

$= -1 \quad = 26$

Therefore,

$f(1) + g(3) = -1 + 26$

$= 25$

b. $g(x) = x^3 - 1$

$g(t) = (t)^3 - 1$ Substitute $x = t$ into the function.

$= t^3 - 1$

c. $f(x) = 2x - 3$

$f(a + b) = 2(a + b) - 3$ Substitute $x = (a + b)$ into the function.

$= 2a + 2b - 3$ Simplify.

Skill Practice Given the functions defined by $g(x) = 4x - 2$ and $h(x) = \frac{4}{x}$, find the function values.

13. $g(5) + h(2)$ **14.** $h(a)$ **15.** $g(p + q)$

5. Domain and Range of a Function

A function is a relation, and it is often necessary to determine its domain and range. Consider a function defined by the equation $y = f(x)$. The domain of f is the set of all x-values that when substituted into the function produce a real number. The range of f is the set of all y-values corresponding to the values of x in the domain.

For the examples encountered in this section of the text, the domain of a function will be all real numbers unless restricted by the following condition:

- The domain must exclude values of the variable that make the denominator zero.

Skill Practice Answers

8. -4 **9.** 0 **10.** 24
11. 6 **12.** -6 **13.** 20
14. $\frac{4}{a}$ **15.** $4p + 4q - 2$

Example 7 Finding the Domain of a Function

Find the domain of the functions defined by the following equations.

a. $f(x) = \dfrac{1}{x - 2}$ **b.** $g(x) = 2x^2 - 6x$

Solution:

a. The function defined by $f(x) = \dfrac{1}{x - 2}$ will be a real number as long as the denominator does not equal zero. The denominator will equal zero only if $x = 2$.

Therefore, the domain of f is $\{x \mid x \text{ is a real number and } x \neq 2\}$.

b. The function defined by $g(x) = 2x^2 - 6x$ does not contain an expression with a variable in the denominator. Therefore, the function has no restrictions on its domain.

The domain of g is the set of all real numbers.

Skill Practice Determine the domain of the functions.

16. $f(x) = 10x^2 + x + 4$ **17.** $g(x) = \dfrac{x + 8}{x - 4}$

Example 8 Finding the Domain and Range of a Function

Find the domain and range of the functions based on the graph of the function. Express the answers in interval notation.

a.

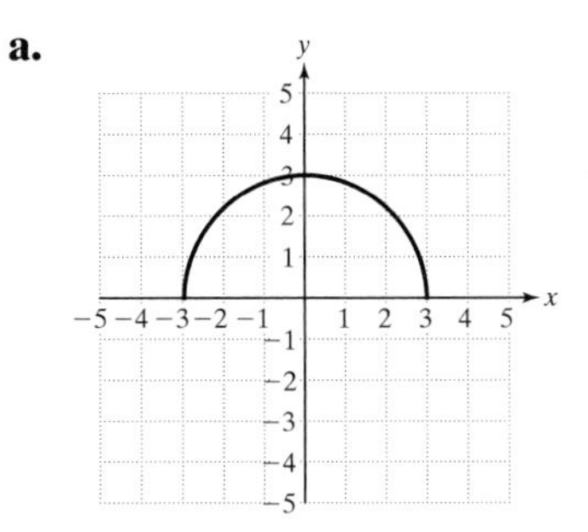

b.

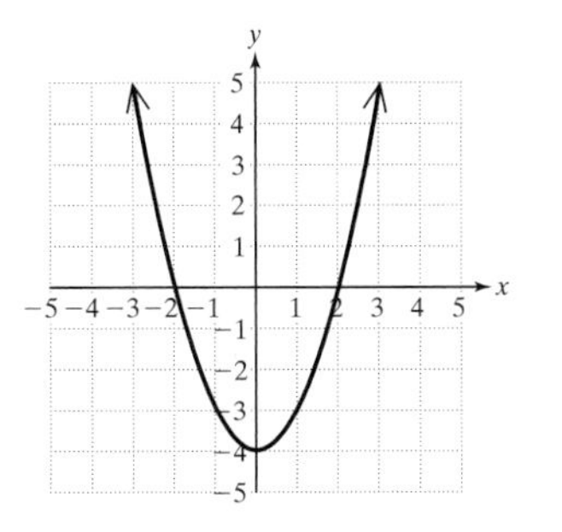

Solution:

a.

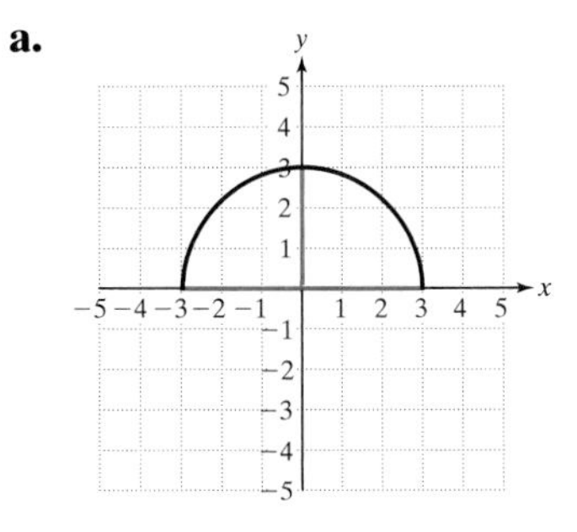

The x-values in the domain are bounded between -3 and 3. (Shown in blue.)

Domain: $[-3, 3]$

The y-values in the range are bounded between 0 and 3. (Shown in red.)

Range: $[0, 3]$

Skill Practice Answers

16. The set of all real numbers

17. $\{x \mid x \text{ is a real number and } x \neq 4\}$

b.

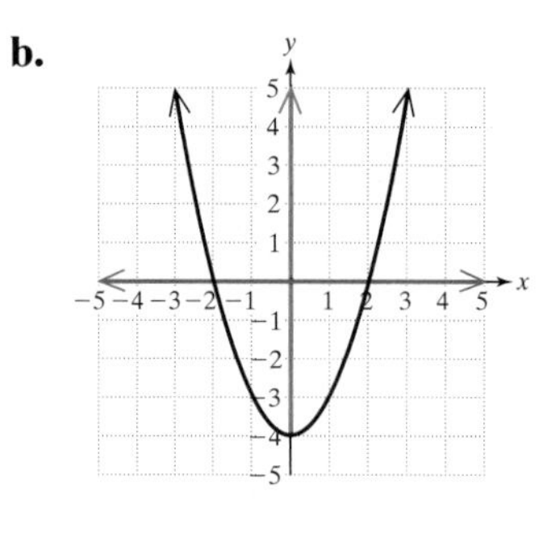

The function extends infinitely far to the left and right. The domain is shown in blue.

Domain: $(-\infty, \infty)$

The y-values extend infinitely far in the positive direction, but are bounded below at $y = -4$. (Shown in red.)

Range: $[-4, \infty)$

Skill Practice Find the domain and range of the functions based on the graph of the function.

18. **19.**

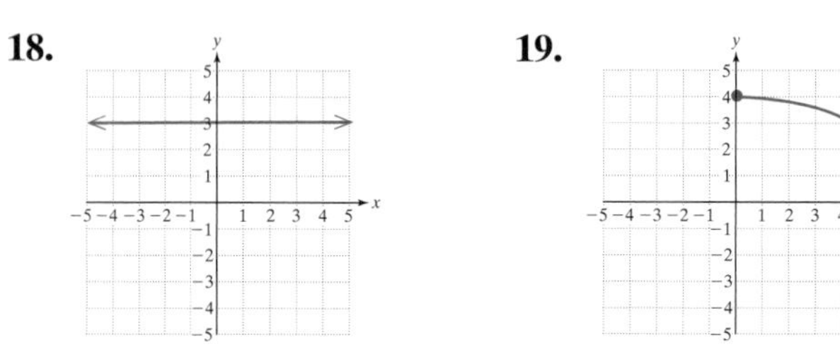

6. Applications of Functions

Example 9 Using a Function in an Application

The score a student receives on an exam is a function of the number of hours the student spends studying. The function defined by

$$P(x) = \frac{100x^2}{40 + x^2} \quad (x \geq 0)$$

indicates that a student will achieve a score of $P\%$ after studying for x hours.

a. Evaluate $P(0)$, $P(10)$, and $P(20)$.

b. Interpret the function values from part (a) in the context of this problem.

Solution:

a. $P(x) = \dfrac{100x^2}{40 + x^2}$

$$P(0) = \frac{100(0)^2}{40 + (0)^2} = \frac{0}{40} = 0$$

$$P(10) = \frac{100(10)^2}{40 + (10)^2} = \frac{10{,}000}{140} = \frac{500}{7} \approx 71.4$$

$$P(20) = \frac{100(20)^2}{40 + (20)^2} = \frac{40{,}000}{440} = \frac{1000}{11} \approx 90.9$$

b. $P(0) = 0$ means that if a student spends 0 hours (hr) studying, the student will score 0% on the exam.

$$P(10) = \frac{500}{7} \approx 71.4$$

means that if a student spends 10 hr studying, the student will score approximately 71.4% on the exam.

Skill Practice Answers

18. Domain: $(-\infty, \infty)$
Range: $\{3\}$

19. Domain: $[0, 5]$
Range: $[2, 4]$

$$P(20) = \frac{1000}{11} \approx 90.9$$

means that if a student spends 20 hr studying, the student will score approximately 90.9% on the exam.

The graph of the function defined by

$$P(x) = \frac{100x^2}{40 + x^2}$$

is shown in Figure 9-6.

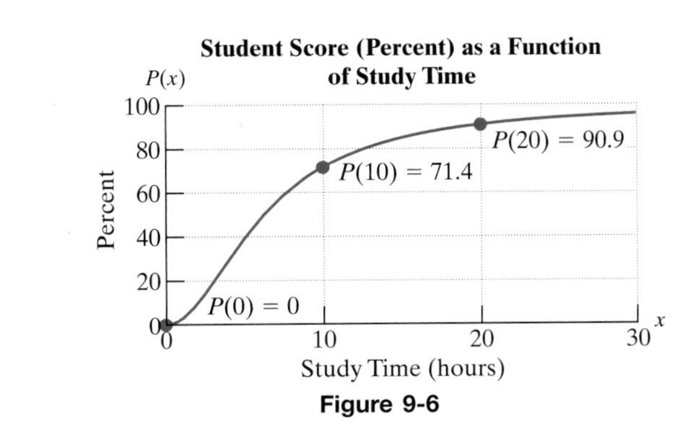

Figure 9-6

Skill Practice The function defined by

$$S(x) = 6x^2 \ (x \geq 0)$$

indicates the surface area of the cube whose side is length x (in inches).

20. Evaluate $S(5)$.

21. Interpret the function value, $S(5)$.

Calculator Connections

A graphing calculator can be used to graph a function. For example,

$$f(x) = \frac{1}{4}x^3 - x^2 - x + 4$$

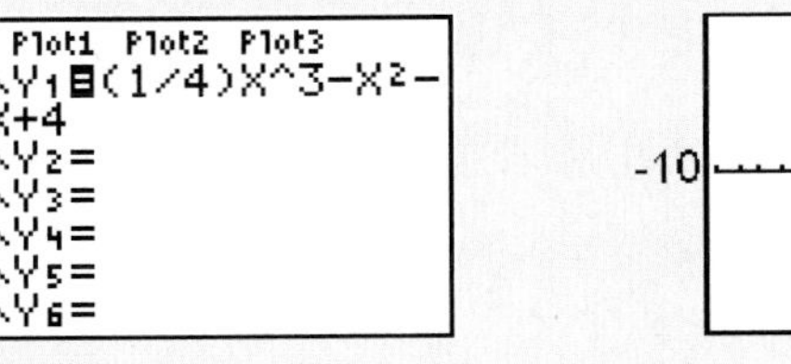

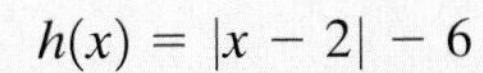

$$h(x) = |x - 2| - 6$$

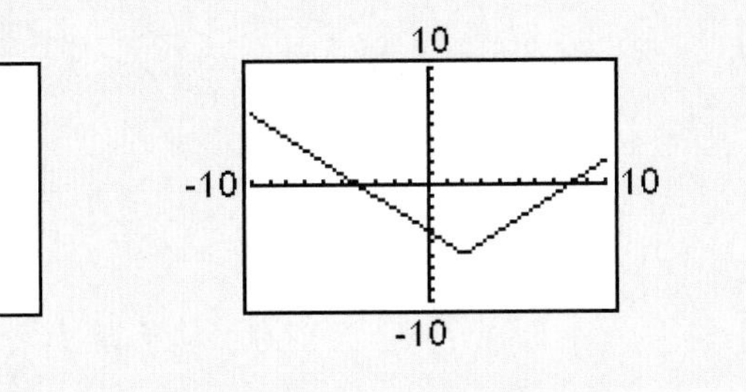

Skill Practice Answers

20. 150

21. For a cube 5 in. on a side, the surface area is 150 in.^{2}

Calculator Exercises

Use a graphing calculator to graph the following functions.

1. $f(x) = x^2 - 5x + 2$

2. $g(x) = 5|x| - x^2$

3. $h(x) = |x + 3| - 4$

4. $y = x^3 - 9x$

5. $y = -4|x| + x^2$

6. $y = -|x| + 8$

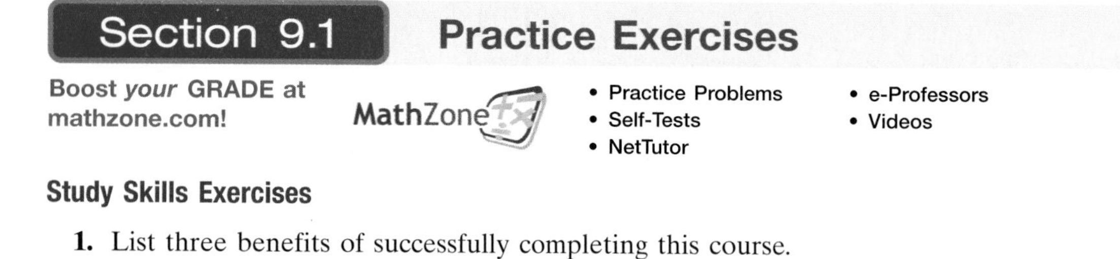

Study Skills Exercises

1. List three benefits of successfully completing this course.

2. Define the key terms:

a. domain **b. function** **c. function notation**

d. range **e. relation** **f. vertical line test**

Concept 1: Definition of a Relation

For Exercises 3–10, determine the domain and range of the relation.

3. $\{(4, 2), (3, 7), (4, 1), (0, 6)\}$

4. $\{(-3, -1), (-2, 6), (1, 3), (1, -2)\}$

5. $\{(\frac{1}{2}, 3), (0, 3), (1, 3)\}$

6. $\{(9, 6), (4, 6), (-\frac{1}{3}, 6)\}$

7.

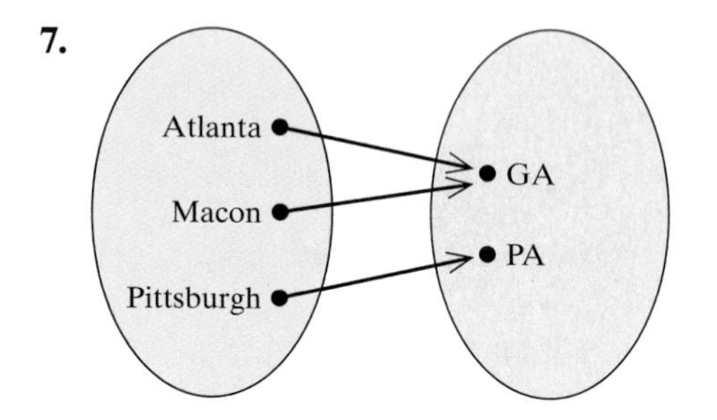

8.

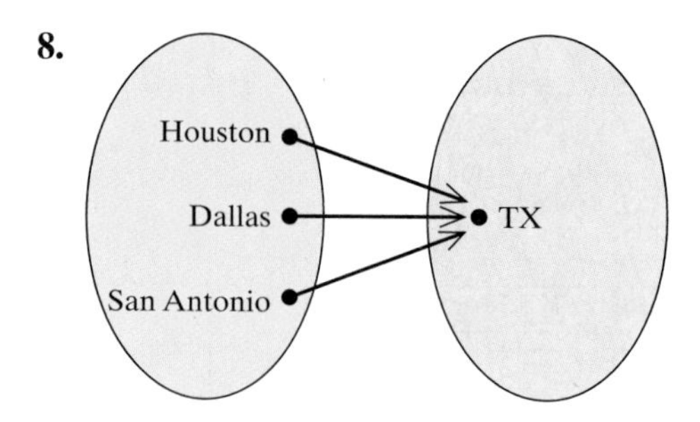

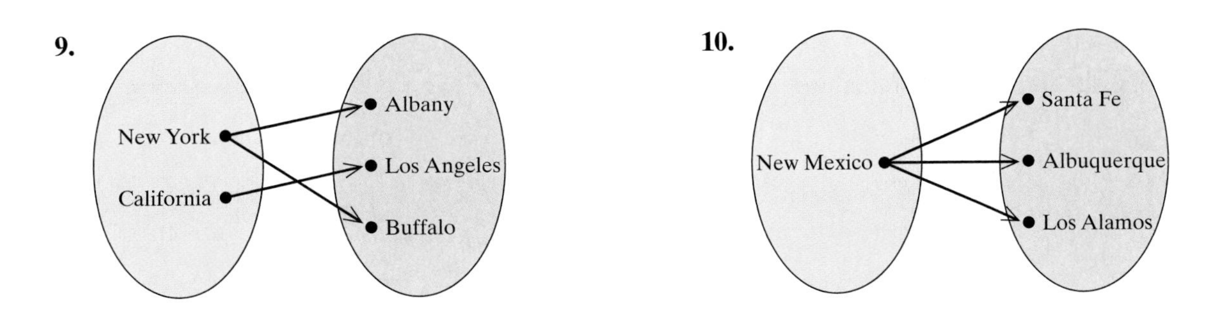

Concept 2: Definition of a Function

11. How can you tell if a set of ordered pairs represents a function?

12. Refer back to Exercises 4, 6, 8, and 10. Identify which relations are functions.

13. Refer back to Exercises 3, 5, 7, and 9. Identify which relations are functions.

Concept 3: Vertical Line Test

14. How can you tell from the graph of a relation if the relation is a function?

For Exercises 15–22, determine if the relation defines y as a function of x.

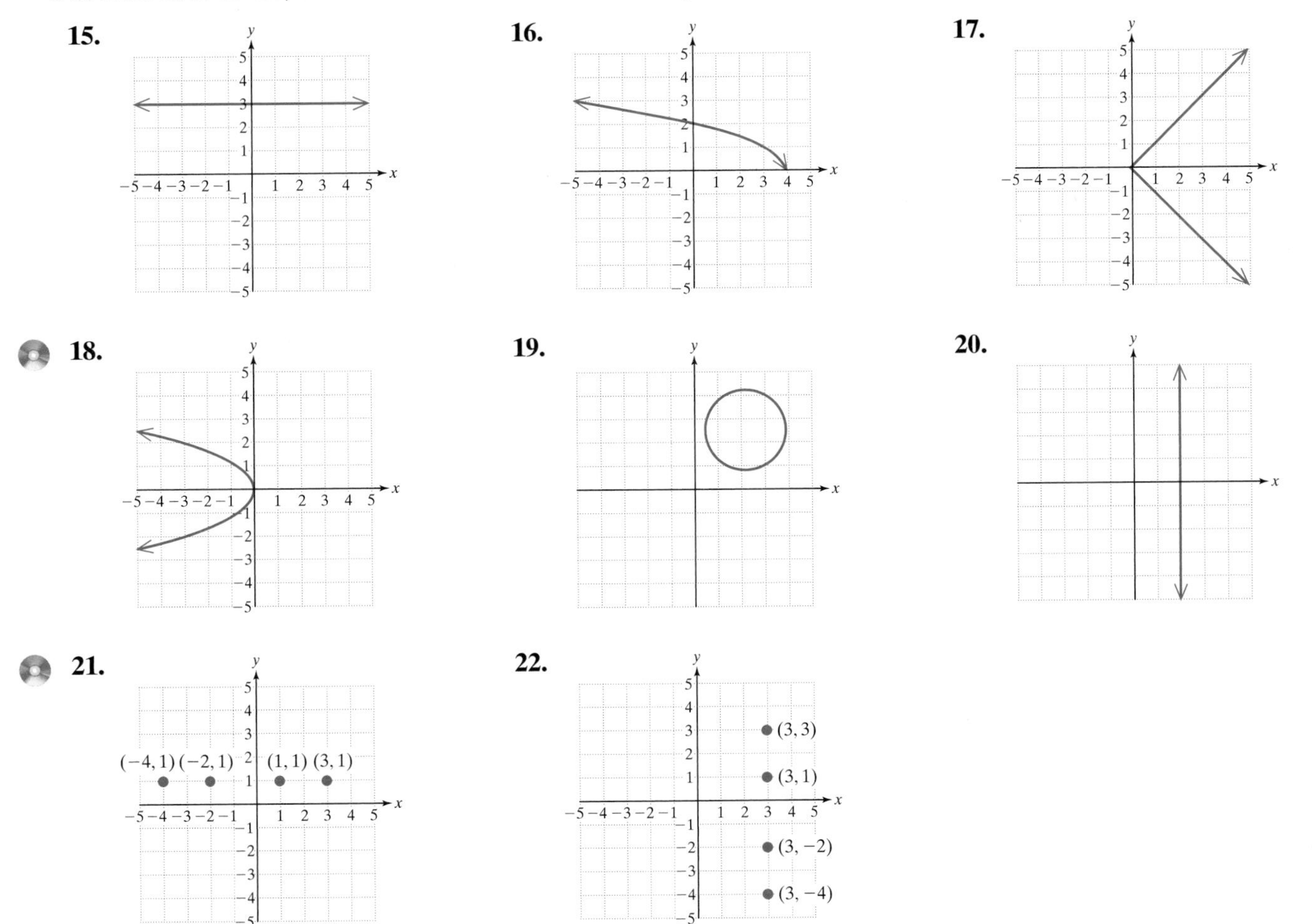

Concept 4: Function Notation

For Exercises 23–26, evaluate the given functions.

23. Let $f(x) = 2x - 5$.
Find: $f(0)$, $f(2)$, $f(-3)$

24. Let $g(x) = x^2 + 1$.
Find: $g(0)$, $g(-1)$, $g(3)$

25. Let $h(x) = \dfrac{1}{x + 4}$.
Find: $h(1)$, $h(0)$, $h(-2)$

26. Let $p(x) = \sqrt{x + 4}$.
Find: $p(0)$, $p(-4)$, $p(5)$

For Exercises 27–40, let the functions f and g be defined by $f(x) = x^2$ and $g(x) = 3x$. Find the function values.

27. $f(2) - g(1)$

28. $f(6) + g(-3)$

29. $g(2) \cdot f(-1)$

30. $f(5) \cdot g(-1)$

31. $\dfrac{g(-2)}{f(-2)}$

32. $\dfrac{f(4)}{g(4)}$

33. $g(5) - f(5)$

34. $f(6) - g(6)$

35. $f(m)$

36. $g(n^2)$

37. $f(a + b)$

38. $f(c + d)$

39. $g(4y)$

40. $g(3t)$

Concept 5: Domain and Range of a Function

For Exercises 41–52, find the domain.

41. $g(x) = 5x - 1$

42. $g(x) = 2x - 2$

43. $L(x) = |x|$

44. $F(x) = x^2 + 2x + 1$

45. $p(x) = \dfrac{1}{x + 6}$

46. $p(x) = \dfrac{1}{x - 3}$

47. $r(x) = \dfrac{x + 2}{x - 4}$

48. $s(x) = \dfrac{x - 6}{x + 2}$

49. $P(x) = \dfrac{x - 10}{x}$

50. $A(x) = \dfrac{x + 11}{x}$

51. $k(x) = \dfrac{4}{9x - 5}$

52. $L(x) = \dfrac{-3}{2x + 5}$

For Exercises 53–56, match the domain and range given with a possible graph. For each graph the relation is shown in black, the domain is shown in red, and the range is shown in blue.

53. Domain: All real numbers
Range: $[1, \infty)$

54. Domain: $[-4, 4]$
Range: $[-2, 2]$

55. Domain: $[-2, \infty)$
Range: The set of all real numbers

56. Domain: $\{x \mid x \text{ is a real number and } x \neq 2\}$
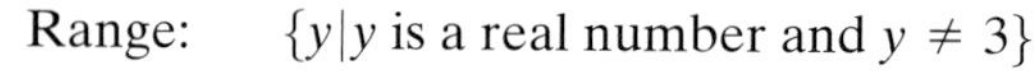
Range: $\{y \mid y \text{ is a real number and } y \neq 3\}$

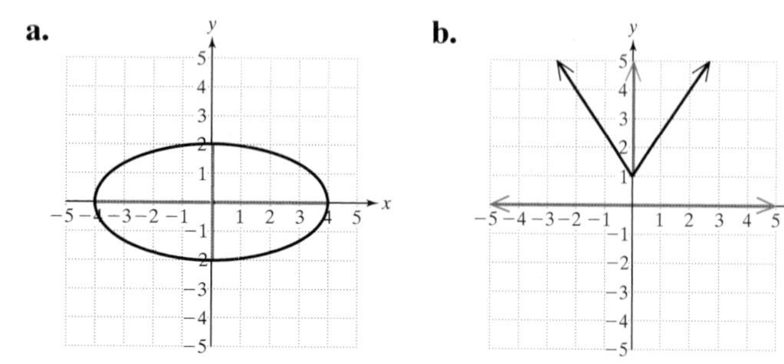

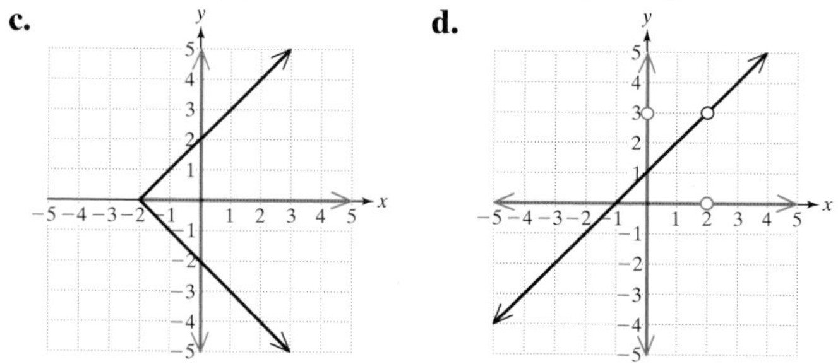

For Exercises 57–60, translate the expressions into English phrases.

57. $f(6) = 2$ **58.** $f(-2) = -14$ **59.** $g\left(\frac{1}{2}\right) = \frac{1}{4}$ **60.** $h(k) = k^2$

61. The function value $f(2) = 7$ corresponds to what ordered pair?

62. The function value $f(-3) = -4$ corresponds to what ordered pair?

63. The function value $f(0) = 8$ corresponds to what ordered pair?

64. The function value $f(4) = 0$ corresponds to what ordered pair?

Concept 6: Applications of Functions

65. Ice cream is \$1.50 for one scoop plus \$0.50 for each topping. The total cost, C (in dollars), can be found using the function defined by

$$C(x) = 1.50 + 0.50x \qquad \text{where } x \text{ is the number of toppings.}$$

a. Find $C(0)$, and interpret the meaning of this function value in terms of cost and the number of toppings.

b. Find $C(1)$, and interpret the meaning in terms of cost and the number of toppings.

c. Find $C(4)$, and interpret the meaning in terms of cost and the number of toppings.

66. Ignoring air resistance, the speed, s (in feet per second: ft/sec), of an object in free fall is a function of the number of seconds, t, after it was dropped:

$$s(t) = 32t$$

a. Find $s(1)$, and interpret the meaning of this function value in terms of speed and time.

b. Find $s(2)$, and interpret the meaning in terms of speed and time.

c. Find $s(10)$, and interpret the meaning in terms of speed and time.

d. A ball dropped from the top of the Sears Tower in Chicago falls for approximately 9.2 sec. How fast was the ball going the instant before it hit the ground?

67. Ignoring air resistance, the speed, s (in meters per second: m/sec), of an object in free fall is a function of the number of seconds, t, after it was dropped:

$$s(t) = 9.8t$$

a. Find $s(0)$, and interpret the meaning of this function value in terms of speed and time.

b. Find $s(4)$, and interpret the meaning in terms of speed and time.

c. Find $s(6)$, and interpret the meaning in terms of speed and time.

d. A penny dropped from the top of the Texas Commerce Tower in Houston falls for approximately 7.9 sec. How fast was it going the instant before it hit the ground?

68. A punter kicks a football straight up with an initial velocity of 64 ft/sec. The height of the ball, h (in feet), is a function of the number of seconds, t, after the ball is kicked:

$$h(t) = -16t^2 + 64t + 3$$

a. Find $h(0)$, and interpret the meaning of the function value in terms of time and height.

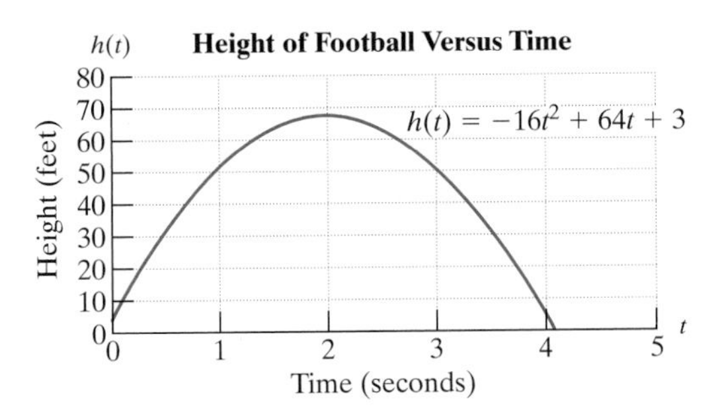

b. Find $h(1)$, and interpret the meaning in terms of time and height.

c. Find $h(2)$, and interpret the meaning in terms of time and height.

d. Find $h(4)$, and interpret the meaning in terms of time and height.

69. A punter kicks a football straight up with an initial velocity of 64 ft/sec. The velocity of the ball, v (in feet per second: ft/sec), is a function of the number of seconds, t, after the ball is kicked:

$$v(t) = -32t + 64$$

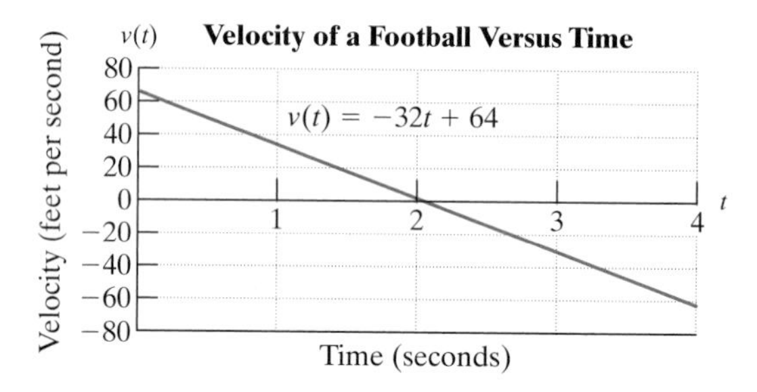

a. Find $v(0)$, and interpret the meaning of the function value in terms of time and velocity.

b. Find $v(1)$, and interpret the meaning in terms of time and velocity.

c. Find $v(2)$, and interpret the meaning in terms of time and velocity.

d. Find $v(4)$, and interpret the meaning in terms of time and velocity.

70. For adults, the maximum recommended heart rate, M (in beats per minute: beats/min), is a function of a person's age, x (in years).

$$M(x) = 220 - x \text{ for } x \geq 18$$

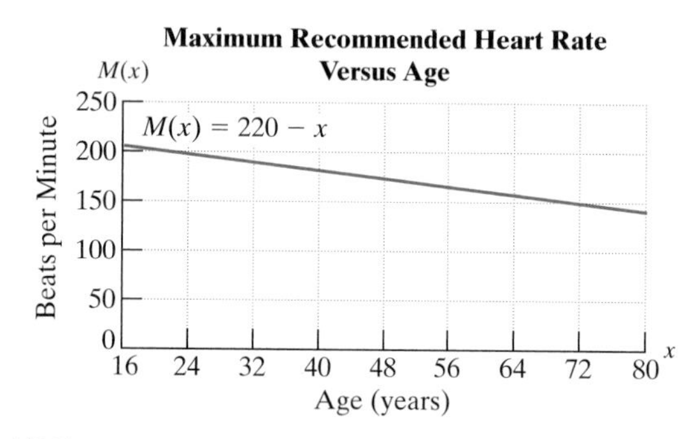

a. Find $M(20)$, and interpret the meaning in terms of maximum recommended heart rate and age.

b. Find $M(30)$, and interpret the meaning in terms of maximum recommended heart rate and age.

c. Find $M(60)$, and interpret the meaning in terms of maximum recommended heart rate and age.

d. Find your own maximum recommended heart rate.

Complex Numbers

Section 9.2

Concepts

1. Definition of i
2. Simplifying Expressions in Terms of i
3. Simplifying Powers of i
4. Definition of a Complex Number
5. Addition, Subtraction, and Multiplication of Complex Numbers
6. Division of Complex Numbers

1. Definition of i

In Section 8.1, we learned that there are no real-valued square roots of a negative number. For example, $\sqrt{-9}$ is not a real number because no real number when squared equals -9. However, the square roots of a negative number are defined over another set of numbers called the *imaginary numbers*. The foundation of the set of imaginary numbers is the definition of the imaginary number, i.

Definition of i

$$i = \sqrt{-1}$$

Note: From the definition of i, it follows that $i^2 = -1$

2. Simplifying Expressions in Terms of i

Using the imaginary number i, we can define the square root of any negative real number.

Definition of $\sqrt{-b}$, $b > 0$

Let b be a real number such that $b > 0$, then $\sqrt{-b} = i\sqrt{b}$

Example 1 Simplifying Expressions in Terms of i

Simplify the expressions in terms of i.

a. $\sqrt{-25}$ **b.** $\sqrt{-81}$ **c.** $\sqrt{-13}$

Solution:

a. $\sqrt{-25} = 5i$

b. $\sqrt{-81} = 9i$

c. $\sqrt{-13} = i\sqrt{13}$

Skill Practice Simplify in terms of i.

1. $\sqrt{-144}$ **2.** $\sqrt{-100}$ **3.** $\sqrt{-7}$

Avoiding Mistakes:

In an expression such as $i\sqrt{13}$ the i is usually written in front of the square root. The expression $\sqrt{13}\,i$ is also correct but may be misinterpreted as $\sqrt{13i}$ (with i incorrectly placed under the square root).

The multiplication and division properties of radicals were presented in Sections 8.2 and 8.4 as follows:

If a and b represent real numbers such that $\sqrt[n]{a}$ and $\sqrt[n]{b}$ are both real, then

$$\sqrt[n]{ab} = \sqrt[n]{a} \cdot \sqrt[n]{b} \quad \text{and} \quad \sqrt[n]{\frac{a}{b}} = \frac{\sqrt[n]{a}}{\sqrt[n]{b}} \quad b \neq 0$$

The conditions that $\sqrt[n]{a}$ and $\sqrt[n]{b}$ must both be real numbers prevent us from applying the multiplication and division properties of radicals for square roots with negative radicands. Therefore, to multiply or divide radicals with negative radicands, write the radicals in terms of the imaginary number i first. This is demonstrated in Example 2.

Skill Practice Answers

1. $12i$ **2.** $10i$ **3.** $i\sqrt{7}$

Example 2 **Simplifying Expressions in Terms of *i***

Simplify the expressions.

a. $\dfrac{\sqrt{-100}}{\sqrt{-25}}$ **b.** $\sqrt{-16} \cdot \sqrt{-4}$

Solution:

a. $\dfrac{\sqrt{-100}}{\sqrt{-25}}$

$= \dfrac{10i}{5i}$ Simplify each radical in terms of i *before* dividing.

$= 2$ Reduce.

b. $\sqrt{-16} \cdot \sqrt{-4}$

$= (4i)(2i)$ Simplify each radical in terms of i *first* before multiplying.

$= 8i^2$

$= 8(-1)$

$= -8$

Avoiding Mistakes:

In Example 2(b), the radical expressions were written in terms of i first before multiplying. If we had mistakenly applied the multiplication property first we would obtain the incorrect answer.

Correct: $\sqrt{-16} \cdot \sqrt{-4}$
$= (4i)(2i) = 8i^2 = -8$

Be careful:
$\sqrt{-16} \cdot \sqrt{-4} \neq \sqrt{64}$

Skill Practice

4. $\dfrac{\sqrt{-36}}{\sqrt{-4}}$ **5.** $\sqrt{-1} \cdot \sqrt{-9}$

3. Simplifying Powers of *i*

From the definition of $i = \sqrt{-1}$, it follows that

$i = i$

$i^2 = -1$

$i^3 = -i$ because $i^3 = i^2 \cdot i = (-1)i = -i$

$i^4 = 1$ because $i^4 = i^2 \cdot i^2 = (-1)(-1) = 1$

$i^5 = i$ because $i^5 = i^4 \cdot i = (1)i = i$

$i^6 = -1$ because $i^6 = i^4 \cdot i^2 = (1)(-1) = -1$

This pattern of values $i, -1, -i, 1, i, -1, -i, 1, \ldots$ continues for all subsequent powers of i. Here is a list of several powers of i.

Powers of *i*

$i^1 = i$	$i^5 = i$	$i^9 = i$
$i^2 = -1$	$i^6 = -1$	$i^{10} = -1$
$i^3 = -i$	$i^7 = -i$	$i^{11} = -i$
$i^4 = 1$	$i^8 = 1$	$i^{12} = 1$

To simplify higher powers of i, we can decompose the expression into multiples of i^4 ($i^4 = 1$) and write the remaining factors as i, i^2, or i^3.

Skill Practice Answers

4. 3 **5.** -3

Example 3 Simplifying Powers of i

Simplify the powers of i.

a. i^9 **b.** i^{14} **c.** i^{103} **d.** i^{28}

Solution:

a.
$$\begin{aligned} i^9 &= (i^8) \cdot i \\ &= (i^4)^2 \cdot (i) \\ &= (1)^2(i) && \text{Recall that } i^4 = 1. \\ &= i && \text{Simplify.} \end{aligned}$$

b.
$$\begin{aligned} i^{14} &= (i^{12}) \cdot i^2 \\ &= (i^4)^3(i^2) \\ &= (1)^3(-1) && i^4 = 1 \text{ and } i^2 = -1. \\ &= -1 && \text{Simplify.} \end{aligned}$$

c.
$$\begin{aligned} i^{103} &= (i^{100}) \cdot i^3 \\ &= (i^4)^{25}(i^3) \\ &= (1)^{25}(-i) && i^4 = 1 \text{ and } i^3 = -i. \\ &= -i && \text{Simplify.} \end{aligned}$$

d.
$$\begin{aligned} i^{28} &= (i^4)^7 \\ &= (1)^7 && i^4 = 1. \\ &= 1 && \text{Simplify.} \end{aligned}$$

Skill Practice Simplify the powers of i.

6. i^{11} **7.** i^{18} **8.** i^{101} **9.** i^{32}

4. Definition of a Complex Number

We have already learned the definitions of the integers, rational numbers, irrational numbers, and real numbers. In this section, we define the complex numbers.

Definition of a Complex Number

A **complex number** is a number of the form $a + bi$, where a and b are real numbers and $i = \sqrt{-1}$.

Notes:

- If $b = 0$, then the complex number, $a + bi$ is a real number.
- If $b \neq 0$, then we say that $a + bi$ is an **imaginary number**.
- The complex number $a + bi$ is said to be written in **standard form**. The quantities a and b are called the **real** and **imaginary parts**, respectively.
- The complex numbers $(a - bi)$ and $(a + bi)$ are called **conjugates**.

Skill Practice Answers

6. $-i$ **7.** -1
8. i **9.** 1

From the definition of a complex number, it follows that all real numbers are complex numbers and all imaginary numbers are complex numbers. Figure 9-7 illustrates the relationship among the sets of numbers we have learned so far.

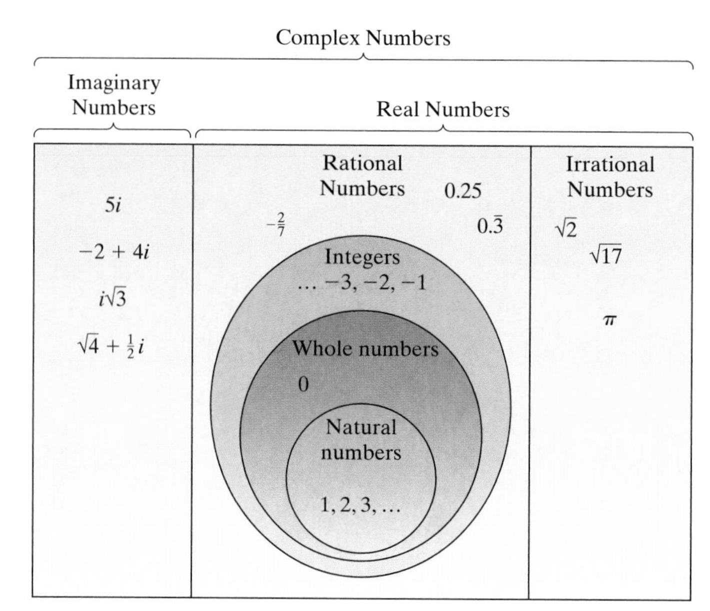

Figure 9-7

Example 4 Identifying the Real and Imaginary Parts of a Complex Number

Identify the real and imaginary parts of the complex numbers.

a. $7 + 4i$ **b.** -6 **c.** $-\dfrac{1}{2}i$

Solution:

a. $7 + 4i$ The real part is 7, and the imaginary part is 4.

b. -6

$= -6 + 0i$ Rewrite -6 in the form $a + bi$.
The real part is -6, and the imaginary part is 0.

c. $-\dfrac{1}{2}i$

$= 0 + -\dfrac{1}{2}i$ Rewrite $-\frac{1}{2}i$ in the form $a + bi$.
The real part is 0, and the imaginary part is $-\frac{1}{2}$.

TIP: Example 4(b) illustrates that a real number is also a complex number.

TIP: Example 4(c) illustrates that an imaginary number is also a complex number.

Skill Practice Identify the real and the imaginary part.

10. $-3 + 2i$ **11.** $6i$ **12.** $-\dfrac{3}{4}$

Skill Practice Answers

10. real part: -3; imaginary part: 2
11. real part: 0; imaginary part: 6
12. real part: $-\dfrac{3}{4}$; imaginary part: 0

5. Addition, Subtraction, and Multiplication of Complex Numbers

The operations for addition, subtraction, and multiplication of real numbers also apply to imaginary numbers. To add or subtract complex numbers, combine the real parts together and combine the imaginary parts together. The commutative, associative, and distributive properties that apply to real numbers also apply to complex numbers.

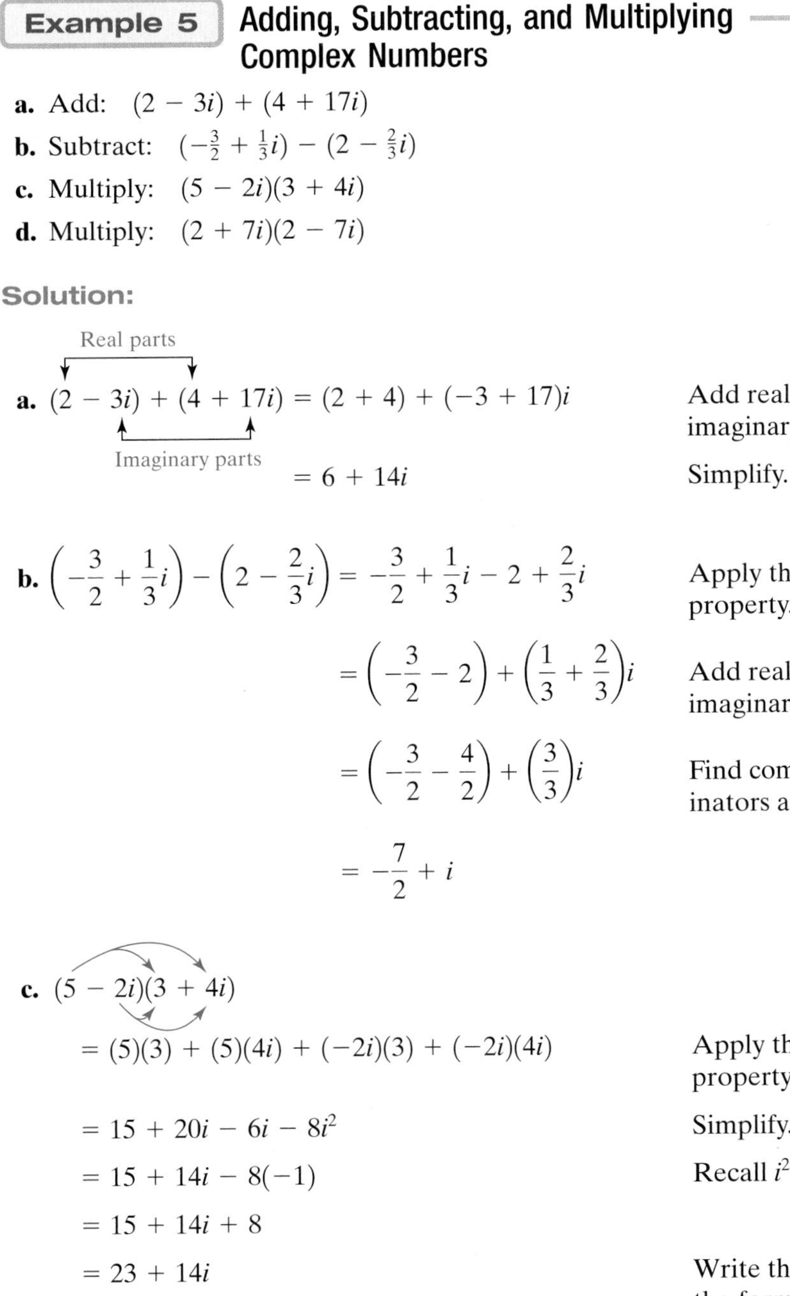

Example 5 **Adding, Subtracting, and Multiplying Complex Numbers**

a. Add: $(2 - 3i) + (4 + 17i)$

b. Subtract: $(-\frac{3}{2} + \frac{1}{3}i) - (2 - \frac{2}{3}i)$

c. Multiply: $(5 - 2i)(3 + 4i)$

d. Multiply: $(2 + 7i)(2 - 7i)$

Solution:

Real parts

a. $(2 - 3i) + (4 + 17i) = (2 + 4) + (-3 + 17)i$ — Add real parts. Add imaginary parts.

Imaginary parts

$= 6 + 14i$ — Simplify.

b. $\left(-\frac{3}{2} + \frac{1}{3}i\right) - \left(2 - \frac{2}{3}i\right) = -\frac{3}{2} + \frac{1}{3}i - 2 + \frac{2}{3}i$ — Apply the distributive property.

$= \left(-\frac{3}{2} - 2\right) + \left(\frac{1}{3} + \frac{2}{3}\right)i$ — Add real parts. Add imaginary parts.

$= \left(-\frac{3}{2} - \frac{4}{2}\right) + \left(\frac{3}{3}\right)i$ — Find common denominators and simplify.

$= -\frac{7}{2} + i$

c. $(5 - 2i)(3 + 4i)$

$= (5)(3) + (5)(4i) + (-2i)(3) + (-2i)(4i)$ — Apply the distributive property.

$= 15 + 20i - 6i - 8i^2$ — Simplify.

$= 15 + 14i - 8(-1)$ — Recall $i^2 = -1$.

$= 15 + 14i + 8$

$= 23 + 14i$ — Write the answer in the form $a + bi$.

d. $(2 + 7i)(2 - 7i)$ The expressions $(2 + 7i)$ and $(2 - 7i)$ are conjugates. The product is a difference of squares.

$(a + b)(a - b) = a^2 - b^2$

$(2 + 7i)(2 - 7i) = (2)^2 - (7i)^2$ Apply the formula, where $a = 2$ and $b = 7i$.

$= 4 - 49i^2$ Simplify.

$= 4 - 49(-1)$ Recall $i^2 = -1$.

$= 4 + 49$

$= 53$

TIP: The complex numbers $2 + 7i$ and $2 - 7i$ can also be multiplied by using the distributive property.

$$
\begin{aligned}
(2 + 7i)(2 - 7i) &= 4 - \cancel{14i} + \cancel{14i} - 49i^2 \\
&= 4 - 49(-1) \\
&= 4 + 49 \\
&= 53
\end{aligned}
$$

Skill Practice

13. Add: $(-3 + 4i) + (-5 - 6i)$

14. Subtract: $\left(\frac{1}{2} - \frac{4}{5}i\right) - \left(\frac{1}{3} + \frac{7}{10}i\right)$

15. Multiply: $(2 - 7i)(3 + 5i)$

16. Multiply: $(5 - i)(5 + i)$

6. Division of Complex Numbers

Example 5(d) illustrates that the product of a complex number and its conjugate produces a real number. Consider the complex numbers $a + bi$ and $a - bi$, where a and b are real numbers. Then,

$$
\begin{aligned}
(a + bi)(a - bi) &= (a)^2 - (bi)^2 \\
&= a^2 - b^2i^2 \\
&= a^2 - b^2(-1) \\
&= a^2 + b^2 \quad \text{(real number)}
\end{aligned}
$$

To divide by a complex number, multiply the numerator and denominator by the conjugate of the denominator. This produces a real number in the denominator so that the resulting expression can be written in the form $a + bi$.

Example 6 Dividing by a Complex Number

Divide the complex numbers. Write the answer in the form $a + bi$.

a. $\dfrac{17}{3 + 5i}$ **b.** $\dfrac{2 + 3i}{4 - 5i}$

Skill Practice Answers

13. $-8 - 2i$ **14.** $\frac{1}{6} - \frac{3}{2}i$

15. $41 - 11i$ **16.** 26

Solution:

a. $\dfrac{17}{3 + 5i}$

$= \dfrac{17}{(3 + 5i)} \cdot \dfrac{(3 - 5i)}{(3 - 5i)}$ Multiply the numerator and denominator by the conjugate of the denominator.

$= \dfrac{17(3 - 5i)}{(3)^2 - (5i)^2}$ In the denominator, apply the formula $(a + b)(a - b) = a^2 - b^2$.

$= \dfrac{17(3 - 5i)}{9 - 25i^2}$ Simplify the denominator.

$= \dfrac{17(3 - 5i)}{9 - 25(-1)}$ Recall $i^2 = -1$.

$= \dfrac{17(3 - 5i)}{9 + 25}$ Simplify.

$= \dfrac{17(3 - 5i)}{34}$

$= \dfrac{\overset{1}{\cancel{17}}(3 - 5i)}{\underset{2}{\cancel{34}}}$ Reduce to lowest terms.

$= \dfrac{3}{2} - \dfrac{5}{2}i$ Write in the form $a + bi$.

b. $\dfrac{2 + 3i}{4 - 5i}$

$\dfrac{(2 + 3i)}{(4 - 5i)} \cdot \dfrac{(4 + 5i)}{(4 + 5i)} = \dfrac{(2)(4) + (2)(5i) + (3i)(4) + (3i)(5i)}{(4)^2 - (5i)^2}$ Multiply the numerator and denominator by the conjugate of the denominator.

$= \dfrac{8 + 10i + 12i + 15i^2}{16 - 25i^2}$ Simplify the numerator and denominator.

$= \dfrac{8 + 22i + 15(-1)}{16 - 25(-1)}$ Recall $i^2 = -1$.

$= \dfrac{8 + 22i - 15}{16 + 25}$

$= \dfrac{-7 + 22i}{41}$ Simplify.

$= -\dfrac{7}{41} + \dfrac{22}{41}i$ Write in the form $a + bi$.

Skill Practice Divide.

17. $\dfrac{4}{3 - 5i}$ **18.** $\dfrac{3 - i}{7 + 5i}$

Skill Practice Answers

17. $\dfrac{6}{17} + \dfrac{10}{17}i$ **18.** $\dfrac{8}{37} - \dfrac{11}{37}i$

Section 9.2 Practice Exercises

Study Skills Exercise

1. Define the key terms.

a. i **b. complex number** **c. imaginary number** **d. standard form**

e. real part **f. imaginary part** **g. conjugates**

Concept 1: Definition of *i*

For Exercises 2–8, simplify the expressions in terms of i.

2. $\sqrt{-49}$ **3.** $\sqrt{-36}$ **4.** $\sqrt{-15}$ **5.** $\sqrt{-21}$

6. $\sqrt{-12}$ **7.** $\sqrt{-48}$ **8.** $\sqrt{-1}$

Concept 2: Simplifying Expressions in Terms of *i*

For Exercises 9–22, perform the indicated operations. Remember to write the radicals in terms of i first.

9. $\sqrt{-100} \cdot \sqrt{-4}$ **10.** $\sqrt{-9} \cdot \sqrt{-25}$ **11.** $\sqrt{-3} \cdot \sqrt{-12}$

12. $\sqrt{-8} \cdot \sqrt{-2}$ **13.** $\dfrac{\sqrt{-81}}{\sqrt{-9}}$ **14.** $\dfrac{\sqrt{-64}}{\sqrt{-16}}$

15. $\dfrac{\sqrt{-50}}{\sqrt{-2}}$ **16.** $\dfrac{\sqrt{-45}}{\sqrt{-5}}$ **17.** $\sqrt{-9} + \sqrt{-121}$

18. $\sqrt{-36} - \sqrt{-49}$ **19.** $\sqrt{-1} - \sqrt{-144} - \sqrt{-169}$ **20.** $\sqrt{-4} + \sqrt{-64} + \sqrt{-81}$

21. $-\sqrt{25} + \sqrt{-25}$ **22.** $\sqrt{-100} - \sqrt{100}$

Concept 3: Simplifying Powers of *i*

For Exercises 23–30, simplify the powers of i.

23. i^6 **24.** i^{17} **25.** i^{12} **26.** i^{27}

27. i^{121} **28.** i^{92} **29.** i^{87} **30.** i^{66}

Concept 4: Definition of a Complex Number

For Exercises 31–36, identify the real part and the imaginary part of the complex number.

31. $-3 - 2i$ **32.** $5 + i$ **33.** 4

34. -6 **35.** $\frac{2}{7}i$ **36.** $0.52i$

Concept 5: Addition, Subtraction, and Multiplication of Complex Numbers

37. Explain how to add or subtract complex numbers.

38. Explain how to multiply complex numbers.

For Exercises 39–68, perform the indicated operations. Write the answers in standard form, $a + bi$.

39. $(2 + 7i) + (-8 + i)$

40. $(6 - i) - (4 + 2i)$

41. $(3 - 4i) + (7 - 6i)$

42. $(-4 - 15i) - (-3 - 17i)$

43. $4i - (9 + i) + 15$

44. $10i - (1 - 5i) - 8$

45. $(5 - 6i) - (9 - 8i) - (3 - i)$

46. $(1 - i) - (5 - 19i) - (24 + 19i)$

47. $(2 - i)(7 - 7i)$

48. $(1 + i)(8 - i)$

49. $(13 - 5i) - (2 + 4i)$

50. $(1 + 8i) + (-6 + 3i)$

51. $(5 + 3i)(3 + 2i)$

52. $(9 + i)(8 + 2i)$

53. $\left(\frac{1}{2} + \frac{1}{5}i\right) - \left(\frac{3}{4} + \frac{2}{5}i\right)$

54. $\left(\frac{5}{6} + \frac{1}{8}i\right) + \left(\frac{1}{3} - \frac{3}{8}i\right)$

55. $8.4i - (3.5 - 9.7i)$

56. $(4.25 - 3.75i) - (10.5 - 18.25i)$

57. $(3 - 2i)(3 + 2i)$

58. $(18 + i)(18 - i)$

59. $(10 - 2i)(10 + 2i)$

60. $(3 - 5i)(3 + 5i)$

61. $\left(\frac{1}{2} - i\right)\left(\frac{1}{2} + i\right)$

62. $\left(\frac{1}{3} - i\right)\left(\frac{1}{3} + i\right)$

63. $(6 - i)^2$

64. $(4 + 3i)^2$

65. $(5 + 2i)^2$

66. $(7 - 6i)^2$

67. $(4 - 7i)^2$

68. $(3 - i)^2$

69. What is the conjugate of $7 - 4i$? Multiply $7 - 4i$ by its conjugate.

70. What is the conjugate of $-3 - i$? Multiply $-3 - i$ by its conjugate.

71. What is the conjugate of $\frac{3}{2} + \frac{2}{5}i$? Multiply $\frac{3}{2} + \frac{2}{5}i$ by its conjugate.

72. What is the conjugate of $-1.3 + 5.7i$? Multiply $-1.3 + 5.7i$ by its conjugate.

73. What is the conjugate of $4i$? Multiply $4i$ by its conjugate.

74. What is the conjugate of $-8i$? Multiply $-8i$ by its conjugate.

Concept 6: Division of Complex Numbers

For Exercises 75–86, divide the complex numbers. Write the answers in standard form, $a + bi$.

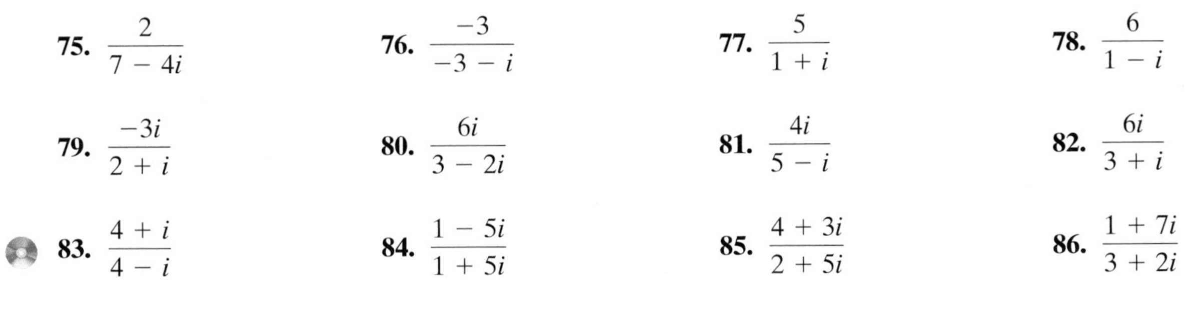

75. $\dfrac{2}{7 - 4i}$

76. $\dfrac{-3}{-3 - i}$

77. $\dfrac{5}{1 + i}$

78. $\dfrac{6}{1 - i}$

79. $\dfrac{-3i}{2 + i}$

80. $\dfrac{6i}{3 - 2i}$

81. $\dfrac{4i}{5 - i}$

82. $\dfrac{6i}{3 + i}$

83. $\dfrac{4 + i}{4 - i}$

84. $\dfrac{1 - 5i}{1 + 5i}$

85. $\dfrac{4 + 3i}{2 + 5i}$

86. $\dfrac{1 + 7i}{3 + 2i}$

Expanding Your Skills

For Exercises 87–99, answer true or false. If an answer is false, explain why.

87. Every complex number is a real number.

88. Every real number is a complex number.

89. Every imaginary number is a complex number.

90. $\sqrt{-64}$ is an imaginary number.

91. $\sqrt[3]{-64}$ is an imaginary number.

92. The product $(2 + 3i)(2 - 3i)$ is a real number.

93. The product $(1 + 4i)(1 - 4i)$ is an imaginary number.

94. The imaginary part of the complex number $2 - 3i$ is 3.

95. The imaginary part of the complex number $4 - 5i$ is -5.

96. i^2 is a real number.

97. i^4 is an imaginary number.

98. i^3 is a real number.

99. i^4 is a real number.

Section 9.3 The Square Root Property and Completing the Square

Concepts

1. Review of the Zero Product Rule
2. Solving Quadratic Equations Using the Square Root Property
3. Completing the Square
4. Solving Quadratic Equations by Completing the Square

1. Review of the Zero Product Rule

In Section 6.7, we learned that an equation that can be written in the form $ax^2 + bx + c = 0$, where $a \neq 0$, is a quadratic equation. One method to solve a quadratic equation is to factor the equation and apply the zero product rule. Recall that the zero product rule states that if $a \cdot b = 0$, then $a = 0$ or $b = 0$. This is reviewed in Example 1.

Example 1 Solving a Quadratic Equation Using the Zero Product Rule

Solve the equations by factoring and applying the zero product rule.

a. $2x(x + 4) = x^2 - 15$ **b.** $x^2 = 25$

Solution:

a. $2x(x + 4) = x^2 - 15$

$2x^2 + 8x = x^2 - 15$ Clear parentheses and combine like terms.

$x^2 + 8x + 15 = 0$ Set the equation equal to zero. The equation is now in the form $ax^2 + bx + c = 0$.

$(x + 5)(x + 3) = 0$ Factor the equation.

$x + 5 = 0$ or $x + 3 = 0$ Set each factor equal to zero.

$x = -5$ or $x = -3$ Solve each equation.

The solutions are $x = -5$ and $x = -3$.

TIP: The solutions to an equation can be checked in the original equation.

Check: $x = -5$

$$2x(x + 4) = x^2 - 15$$
$$2(-5)(-5 + 4) \stackrel{?}{=} (-5)^2 - 15$$
$$-10(-1) \stackrel{?}{=} 25 - 15$$
$$10 = 10 \checkmark$$

Check: $x = -3$

$$2x(x + 4) = x^2 - 15$$
$$2(-3)(-3 + 4) \stackrel{?}{=} (-3)^2 - 15$$
$$-6(1) \stackrel{?}{=} 9 - 15$$
$$-6 = -6 \checkmark$$

b. $x^2 = 25$

$x^2 - 25 = 0$ Set the equation equal to zero.

$(x - 5)(x + 5) = 0$ Factor the equation.

$x - 5 = 0$ or $x + 5 = 0$ Set each factor equal to zero.

$x = 5$ or $x = -5$

The solutions are $x = 5$ and $x = -5$.

Skill Practice Solve the quadratic equations using the zero product rule.

1. $y^2 = 3y + 10$ **2.** $t^2 = 49$

2. Solving Quadratic Equations Using the Square Root Property

In Example 1, the quadratic equations were both factorable. In this section and the next, we learn two techniques to solve *all* quadratic equations, factorable and nonfactorable. The first technique uses the **square root property**.

Square Root Property

For any real number, k, if $x^2 = k$, then $x = \sqrt{k}$ or $x = -\sqrt{k}$.

Note: The solution may also be written as $x = \pm\sqrt{k}$, read as "x equals plus or minus the square root of k."

Example 2 Solving Quadratic Equations Using the Square Root Property

Use the square root property to solve the equations.

a. $x^2 = 25$ **b.** $2x^2 + 18 = 0$ **c.** $(t - 4)^2 = 12$

Solution:

a. $x^2 = 25$ The equation is in the form $x^2 = k$.

$x = \pm\sqrt{25}$ Apply the square root property.

$x = \pm 5$

The solutions are $x = 5$ and $x = -5$.

TIP: The equation $x^2 = 25$ was also solved in Example 1(b) by factoring and using the zero product rule. Notice that we get the same solutions by applying the square root property.

Skill Practice Answers

1. $y = 5, y = -2$
2. $t = 7, t = -7$

Avoiding Mistakes:

A common mistake when applying the square root property is to forget the $\pm$ symbol.

b. $2x^2 + 18 = 0$ Rewrite the equation to fit the form $x^2 = k$.

$2x^2 = -18$

$x^2 = -9$ The equation is now in the form $x^2 = k$.

$x = \pm\sqrt{-9}$ Apply the square root property.

$x = \pm 3i$

The solutions are $x = 3i$ and $x = -3i$.

Check: $x = 3i$

$$2x^2 + 18 = 0$$
$$2(3i)^2 + 18 \stackrel{?}{=} 0$$
$$2(9i^2) + 18 \stackrel{?}{=} 0$$
$$2(-9) + 18 \stackrel{?}{=} 0$$
$$-18 + 18 = 0 \checkmark$$

Check: $x = -3i$

$$2x^2 + 18 = 0$$
$$2(-3i)^2 + 18 \stackrel{?}{=} 0$$
$$2(9i^2) + 18 \stackrel{?}{=} 0$$
$$2(-9) + 18 \stackrel{?}{=} 0$$
$$-18 + 18 = 0 \checkmark$$

c. $(t - 4)^2 = 12$ The equation is in the form $x^2 = k$, where $x = (t - 4)$.

$t - 4 = \pm\sqrt{12}$ Apply the square root property.

$t - 4 = \pm\sqrt{2^2 \cdot 3}$ Simplify the radical.

$t - 4 = \pm 2\sqrt{3}$

$t = 4 \pm 2\sqrt{3}$ Solve for t.

The solutions are $t = 4 + 2\sqrt{3}$ and $t = 4 - 2\sqrt{3}$.

Check: $t = 4 + 2\sqrt{3}$

$$(t - 4)^2 = 12$$
$$(4 + 2\sqrt{3} - 4)^2 \stackrel{?}{=} 12$$
$$(2\sqrt{3})^2 \stackrel{?}{=} 12$$
$$4 \cdot 3 \stackrel{?}{=} 12$$
$$12 = 12 \checkmark$$

Check: $t = 4 - 2\sqrt{3}$

$$(t - 4)^2 = 12$$
$$(4 - 2\sqrt{3} - 4)^2 \stackrel{?}{=} 12$$
$$(-2\sqrt{3})^2 \stackrel{?}{=} 12$$
$$4 \cdot 3 \stackrel{?}{=} 12$$
$$12 = 12 \checkmark$$

Skill Practice Use the square root property to solve the equations.

3. $c^2 = 64$ **4.** $3x^2 + 48 = 0$ **5.** $(p + 3)^2 = 8$

3. Completing the Square

In Example 2(c), we used the square root property to solve an equation in which the square of a binomial was equal to a constant.

$$\underbrace{(t - 4)^2}_{\text{square of a binomial}} = \underset{\text{constant}}{\underset{\uparrow}{12}}$$

Skill Practice Answers

3. $c = \pm 8$
4. $x = \pm 4i$
5. $p = -3 \pm 2\sqrt{2}$

Furthermore, any equation $ax^2 + bx + c = 0$ $(a \neq 0)$ can be rewritten as the square of a binomial equal to a constant by using a process called **completing the square**.

We begin our discussion of completing the square with some vocabulary. For a trinomial $ax^2 + bx + c$ $(a \neq 0)$, the term ax^2 is called the **quadratic term**. The term bx is called the **linear term**, and the term, c, is called the **constant term**.

Next, notice that the square of a binomial is the factored form of a perfect square trinomial.

Perfect Square Trinomial		**Factored Form**
$x^2 + 10x + 25$	$\longrightarrow$	$(x + 5)^2$
$t^2 - 6t + 9$	$\longrightarrow$	$(t - 3)^2$
$p^2 - 14p + 49$	$\longrightarrow$	$(p - 7)^2$

Furthermore, for a perfect square trinomial with a leading coefficient of 1, the constant term is the square of half the coefficient of the linear term. For example:

$x^2 + 10x + 25$ $\qquad$ $t^2 - 6t + 9$ $\qquad$ $p^2 - 14p + 49$

$\left[\frac{1}{2}(10)\right]^2 = [5]^2 = 25$ $\qquad$ $\left[\frac{1}{2}(-6)\right]^2 = [-3]^2 = 9$ $\qquad$ $\left[\frac{1}{2}(-14)\right]^2 = [-7]^2 = 49$

In general, an expression of the form $x^2 + bx$ will result in a perfect square trinomial if the square of half the linear term coefficient, $(\frac{1}{2}b)^2$, is added to the expression.

Example 3 Completing the Square

Complete the square for each expression. Then factor the expression as the square of a binomial.

a. $x^2 + 12x$ $\qquad$ **b.** $x^2 - 22x$ $\qquad$ **c.** $x^2 + 5x$ $\qquad$ **d.** $x^2 - \frac{3}{5}x$

Solution:

The expressions are in the form $x^2 + bx$. Add the square of half the linear term coefficient, $(\frac{1}{2}b)^2$.

a. $x^2 + 12x$

$x^2 + 12x + 36$ $\qquad$ Add $\frac{1}{2}$ of 12, squared. $[\frac{1}{2}(12)]^2 = (6)^2 = 36$

$(x + 6)^2$ $\qquad$ Factored form

b. $x^2 - 22x$

$x^2 - 22x + 121$ $\qquad$ Add $\frac{1}{2}$ of -22, squared. $[\frac{1}{2}(-22)]^2 = (-11)^2 = 121$

$(x - 11)^2$ $\qquad$ Factored form

c. $x^2 + 5x$

$x^2 + 5x + \frac{25}{4}$ $\qquad$ Add $\frac{1}{2}$ of 5, squared. $[\frac{1}{2}(5)]^2 = (\frac{5}{2})^2 = \frac{25}{4}$

$\left(x + \frac{5}{2}\right)^2$ $\qquad$ Factored form

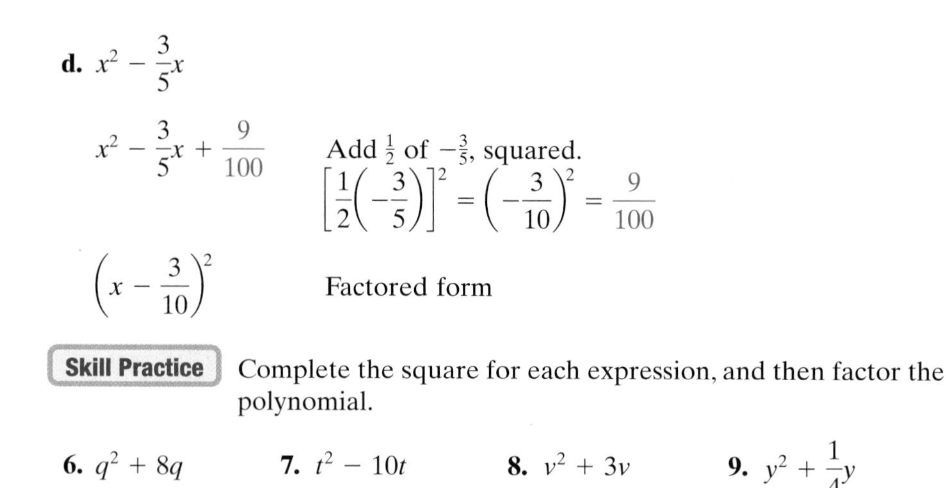

d. $x^2 - \frac{3}{5}x$

$x^2 - \frac{3}{5}x + \frac{9}{100}$ Add $\frac{1}{2}$ of $-\frac{3}{5}$, squared. $\left[\frac{1}{2}\left(-\frac{3}{5}\right)\right]^2 = \left(-\frac{3}{10}\right)^2 = \frac{9}{100}$

$\left(x - \frac{3}{10}\right)^2$ Factored form

Skill Practice Complete the square for each expression, and then factor the polynomial.

6. $q^2 + 8q$ **7.** $t^2 - 10t$ **8.** $v^2 + 3v$ **9.** $y^2 + \frac{1}{4}y$

4. Solving Quadratic Equations by Completing the Square

A quadratic equation can be solved by completing the square and applying the square root property. The following steps outline the procedure.

Solving a Quadratic Equation in the Form $ax^2 + bx + c = 0$ $(a \neq 0)$ by Completing the Square and Applying the Square Root Property

1. Divide both sides by a to make the leading coefficient 1.
2. Isolate the variable terms on one side of the equation.
3. Complete the square (add the square of one-half the linear term coefficient to both sides of the equation. Then factor the resulting perfect square trinomial).
4. Apply the square root property and solve for x.

Example 4 Solving Quadratic Equations by Completing the Square and Applying the Square Root Property

Solve the quadratic equations by completing the square and applying the square root property.

a. $x^2 + 6x - 8 = 0$ **b.** $2x^2 - 16x + 40 = 0$ **c.** $x(2x - 5) = 3$

Solution:

a. $x^2 + 6x - 8 = 0$ The equation is in the form $ax^2 + bx + c = 0$.

Step 1: The leading coefficient is already 1.

$x^2 + 6x = 8$ **Step 2:** Isolate the variable terms on one side.

Skill Practice Answers

6. $q^2 + 8q + 16; (q + 4)^2$
7. $t^2 - 10t + 25; (t - 5)^2$
8. $v^2 + 3v + \frac{9}{4}; \left(v + \frac{3}{2}\right)^2$
9. $y^2 + \frac{1}{4}y + \frac{1}{64}; \left(y + \frac{1}{8}\right)^2$

$x^2 + 6x + 9 = 8 + 9$	**Step 3:** To complete the square, add $[\frac{1}{2}(6)]^2 = (3)^2 = 9$ to both sides.
$(x + 3)^2 = 17$	Factor the perfect square trinomial.
$x + 3 = \pm\sqrt{17}$	**Step 4:** Apply the square root property.
$x = -3 \pm \sqrt{17}$	Solve for x.

The solutions are $x = -3 - \sqrt{17}$ and $x = -3 + \sqrt{17}$.

b. $2x^2 - 16x + 40 = 0$	The equation is in the form $ax^2 + bx + c = 0$.
$\frac{2x^2}{2} - \frac{16x}{2} + \frac{40}{2} = \frac{0}{2}$	**Step 1:** Divide by the leading coefficient.
$x^2 - 8x + 20 = 0$	
$x^2 - 8x = -20$	**Step 2:** Isolate the variable terms on one side.
$x^2 - 8x + 16 = -20 + 16$	**Step 3:** To complete the square, add $[\frac{1}{2}(-8)]^2 = (-4)^2 = 16$ to both sides.
$(x - 4)^2 = -4$	Factor the perfect square trinomial.
$x - 4 = \pm\sqrt{-4}$	**Step 4:** Apply the square root property.
$x - 4 = \pm 2i$	Simplify and solve for x.
$x = 4 \pm 2i$	

The solutions are $x = 4 + 2i$ and $x = 4 - 2i$.

c. $x(2x - 5) = 3$	
$2x^2 - 5x = 3$	Write the equation in the form $ax^2 + bx + c = 0$.
$2x^2 - 5x - 3 = 0$	
$\frac{2x^2}{2} - \frac{5x}{2} - \frac{3}{2} = \frac{0}{2}$	**Step 1:** Divide both sides by the leading coefficient, 2.
$x^2 - \frac{5}{2}x - \frac{3}{2} = 0$	
$x^2 - \frac{5}{2}x = \frac{3}{2}$	**Step 2:** Isolate the variable terms on one side.
$x^2 - \frac{5}{2}x + \frac{25}{16} = \frac{3}{2} + \frac{25}{16}$	**Step 3:** Add $[\frac{1}{2}(-\frac{5}{2})]^2 = (-\frac{5}{4})^2 = \frac{25}{16}$ to both sides.
$\left(x - \frac{5}{4}\right)^2 = \frac{24}{16} + \frac{25}{16}$	Factor the perfect square trinomial. Rewrite the right-hand side with a common denominator.
$\left(x - \frac{5}{4}\right)^2 = \frac{49}{16}$	

TIP: Since the solutions to the equation $x(2x - 5) = 3$ are rational numbers, the equation could have been solved by factoring and using the zero product rule:

$$x(2x - 5) = 3$$
$$2x^2 - 5x - 3 = 0$$
$$(2x + 1)(x - 3) = 0$$
$$2x + 1 = 0 \quad \text{or} \quad x - 3 = 0$$
$$x = -\frac{1}{2} \quad \text{or} \quad x = 3$$

$$x - \frac{5}{4} = \pm\sqrt{\frac{49}{16}}$$ **Step 4:** Apply the square root property.

$$x - \frac{5}{4} = \pm\frac{7}{4}$$ Simplify the radical.

$$x = \frac{5}{4} \pm \frac{7}{4}$$ Solve for x.

The solutions are

$$x = \frac{5}{4} + \frac{7}{4} = \frac{12}{4} = 3 \quad \text{and} \quad x = \frac{5}{4} - \frac{7}{4} = -\frac{2}{4} = -\frac{1}{2}$$

Skill Practice Answers

10. $x = -2 \pm \sqrt{3}$
11. $x = 5 \pm 3i$
12. $x = \frac{2}{3}, x = -2$

Skill Practice Solve the equations by completing the square and then applying the square root property.

10. $x^2 + 4x + 1 = 0$ **11.** $3x^2 - 30x + 102 = 0$ **12.** $x(3x + 4) = 4$

Section 9.3 Practice Exercises

Boost *your* GRADE at mathzone.com! MathZone

- Practice Problems
- Self-Tests
- NetTutor
- e-Professors
- Videos

Study Skills Exercise

1. Define the key terms:

a. square root property **b. completing the square** **c. quadratic term**

d. linear term **e. constant term**

Concept 1: Review of the Zero Product Rule

2. Identify the equations as linear or quadratic.

a. $2x - 5 = 3(x + 2) - 1$ **b.** $2x(x - 5) = 3(x + 2) - 1$

c. $ax^2 + bx + c = 0$ (a, b, and c are real numbers, and $a \neq 0$)

3. Identify the equations as linear or quadratic.

a. $ax + b = 0$ (a and b are real numbers, and $a \neq 0$)

b. $\frac{1}{2}p - \frac{3}{4}p^2 = 0$ **c.** $\frac{1}{2}(p - 3) = 5$

4. Complete the statement of the zero product rule. If $a \cdot b = 0$, then . . .

For Exercises 5–14, solve using the zero product rule.

5. $(3z - 2)(4z + 5) = 0$ **6.** $(t + 5)(2t - 1) = 0$ **7.** $r^2 + 7r + 12 = 0$ **8.** $y^2 - 2y - 35 = 0$

9. $10x^2 = 13x - 4$ **10.** $6p^2 = -13p - 2$ **11.** $2m(m - 1) = 3m - 3$ **12.** $2x^2 + 10x = -7(x + 3)$

13. $x^2 = 4$ **14.** $c^2 = 144$

Concept 2: Solving Quadratic Equations Using the Square Root Property

15. The symbol "±" is read as . . .

For Exercises 16–37, solve the equations using the square root property.

16. $x^2 = 49$ **17.** $x^2 = 16$ **18.** $k^2 - 100 = 0$ **19.** $m^2 - 64 = 0$

20. $p^2 = -24$ **21.** $q^2 = -50$ **22.** $3w^2 + 9 = 0$ **23.** $4v^2 - 5 = 0$

24. $(a - 5)^2 = 16$ **25.** $(b + 3)^2 = 1$ **26.** $(y - 5)^2 = 36$ **27.** $(y + 4)^2 = 4$

28. $(x - 11)^2 = 5$ **29.** $(z - 2)^2 = 7$ **30.** $(a + 1)^2 = -18$ **31.** $(b - 1)^2 = -12$

32. $\left(t - \frac{1}{4}\right)^2 = \frac{7}{16}$ **33.** $\left(t - \frac{1}{3}\right)^2 = \frac{1}{9}$ **34.** $\left(x - \frac{1}{2}\right)^2 + 5 = 20$ **35.** $\left(x + \frac{5}{2}\right)^2 - 3 = 18$

36. $(p - 3)^2 = -16$ **37.** $(t + 4)^2 = -9$

38. Check the solution $x = -3 + \sqrt{5}$ in the equation $(x + 3)^2 = 5$.

39. Check the solution $p = -5 - \sqrt{7}$ in the equation $(p + 5)^2 = 7$.

For Exercises 40–41, answer true or false. If a statement is false, explain why.

40. The only solution to the equation $x^2 = 64$ is $x = 8$.

41. There are two real solutions to every quadratic equation of the form $x^2 = k$, where $k \geq 0$ is a real number.

Concept 3: Completing the Square

For Exercises 42–53, what constant should be added to each expression to make it a perfect square trinomial?

42. $y^2 + 4y$ **43.** $w^2 - 6w$ **44.** $p^2 - 12p$ **45.** $q^2 + 16q$

46. $x^2 - 9x$ **47.** $a^2 - 5a$ **48.** $d^2 + \frac{5}{3}d$ **49.** $t^2 + \frac{1}{4}t$

50. $m^2 - \frac{1}{5}m$ **51.** $n^2 - \frac{5}{7}n$ **52.** $u^2 + u$ **53.** $v^2 - v$

Concept 4: Solving Quadratic Equations by Completing the Square

For Exercises 54–67, solve the equations by completing the square and applying the square root property.

54. $x^2 + 4x = 12$ **55.** $x^2 - 2x = 8$ **56.** $y^2 + 6y = -5$ **57.** $t^2 + 10t = 11$

58. $x^2 = 2x + 1$ **59.** $x^2 = 6x - 2$ **60.** $3x^2 - 6x - 15 = 0$ **61.** $5x^2 + 10x - 30 = 0$

62. $4p^2 + 16p = -4$

63. $2t^2 - 12t = 12$

64. $w^2 + w - 3 = 0$

65. $z^2 - 3z - 5 = 0$

66. $x(x + 2) = 48$

67. $y(y - 4) = 45$

68. Ignoring air resistance, the distance, d (in feet), that an object drops in t seconds is given by the equation

$$d = 16t^2$$

a. Find the distance traveled in 2 sec.

b. Find the time required for the object to fall 200 ft. Round to the nearest tenth of a second.

c. Find the time required for an object to fall from the top of the Empire State Building in New York City if the building is 1250 ft high. Round to the nearest tenth of a second.

69. Ignoring air resistance, the distance, d (in meters), that an object drops in t seconds is given by the equation

$$d = 4.9t^2$$

a. Find the distance traveled in 5 sec.

b. Find the time required for the object to fall 50 m. Round to the nearest tenth of a second.

c. Find the time required for an object to fall from the top of the Canada Trust Tower in Toronto, Canada, if the building is 261 m high. Round to the nearest tenth of a second.

70. An isosceles right triangle has legs of equal length. If the hypotenuse is 10 m long, find the length (in meters) of each leg. Round the answer to the nearest tenth of a meter.

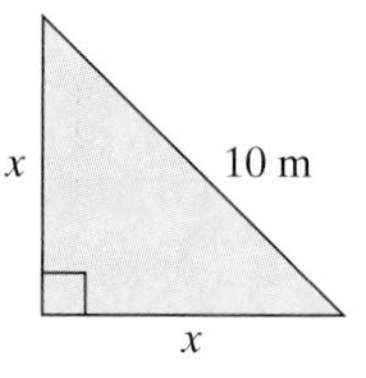

71. The diagonal of a square television screen is 24 in. long. Find the length of the sides to the nearest tenth of an inch.

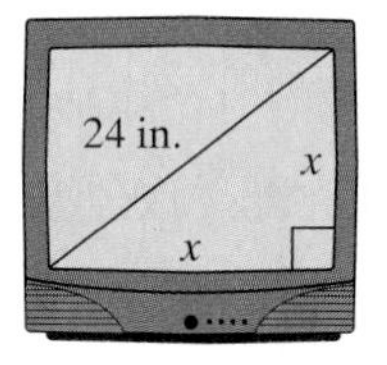

72. The area of a circular wading pool is approximately 200 ft^2. Find the radius to the nearest tenth of a foot.

73. A sprinkler system covers a circular area of approximately 1700 ft^2. Find the radius of the area to the nearest tenth of a foot.

Quadratic Formula

Section 9.4

Concepts

1. Derivation of the Quadratic Formula
2. Solving Quadratic Equations Using the Quadratic Formula
3. Review of the Methods for Solving a Quadratic Equation
4. Applications of Quadratic Equations

1. Derivation of the Quadratic Formula

If we solve a general quadratic equation $ax^2 + bx + c = 0\ (a \neq 0)$ by completing the square and using the square root property, the result is a formula that gives the solutions for x in terms of a, b, and c.

$ax^2 + bx + c = 0$ Begin with a quadratic equation in standard form.

$x^2 + \frac{b}{a}x + \frac{c}{a} = 0$ Divide by the leading coefficient.

$x^2 + \frac{b}{a}x = -\frac{c}{a}$ Isolate the terms containing x.

$x^2 + \frac{b}{a}x + \left(\frac{1}{2} \cdot \frac{b}{a}\right)^2 = \left(\frac{1}{2} \cdot \frac{b}{a}\right)^2 - \frac{c}{a}$ Add the square of $\frac{1}{2}$ the linear term coefficient to both sides of the equation.

$\left(x + \frac{b}{2a}\right)^2 = \frac{b^2}{4a^2} - \frac{c}{a}$ Factor the left side as a perfect square.

$\left(x + \frac{b}{2a}\right)^2 = \frac{b^2 - 4ac}{4a^2}$ Combine fractions on the right side by getting a common denominator.

$x + \frac{b}{2a} = \pm\sqrt{\frac{b^2 - 4ac}{4a^2}}$ Apply the square root property.

$x + \frac{b}{2a} = \frac{\pm\sqrt{b^2 - 4ac}}{2a}$ Simplify the denominator.

$x = -\frac{b}{2a} \pm \frac{\sqrt{b^2 - 4ac}}{2a}$ Subtract $\frac{b}{2a}$ from both sides.

$= \frac{-b \pm \sqrt{b^2 - 4ac}}{2a}$ Combine fractions.

Quadratic Formula

For any quadratic equation of the form $ax^2 + bx + c = 0, (a \neq 0)$ the solutions are

$$x = \frac{-b \pm \sqrt{b^2 - 4ac}}{2a}$$

2. Solving Quadratic Equations Using the Quadratic Formula

Example 1 **Solving Quadratic Equations Using the Quadratic Formula**

Solve the quadratic equation using the quadratic formula: $3x^2 - 7x = -2$

Solution:

$$3x^2 - 7x = -2$$

$$3x^2 - 7x + 2 = 0$$ Write the equation in the form $ax^2 + bx + c = 0$.

$a = 3, b = -7, c = 2$ Identify a, b, and c.

$$x = \frac{-b \pm \sqrt{b^2 - 4ac}}{2a}$$

$$x = \frac{-(-7) \pm \sqrt{(-7)^2 - 4(3)(2)}}{2(3)}$$ Apply the quadratic formula.

$$x = \frac{7 \pm \sqrt{49 - 24}}{6}$$ Simplify.

$$= \frac{7 \pm \sqrt{25}}{6}$$

$$= \frac{7 \pm 5}{6}$$

There are two rational solutions.

$$x = \frac{7 + 5}{6} = \frac{12}{6} = 2 \qquad x = 2$$

$$x = \frac{7 - 5}{6} = \frac{2}{6} = \frac{1}{3} \qquad x = \frac{1}{3}$$

TIP: Because the solutions to the equation $3x^2 - 7x = -2$ are rational numbers, the equation could have been solved by factoring and using the zero product rule.

$$3x^2 - 7x = -2$$

$$3x^2 - 7x + 2 = 0$$

$$(3x - 1)(x - 2) = 0$$

$$3x - 1 = 0 \quad \text{or} \quad x - 2 = 0$$

$$x = \frac{1}{3} \quad \text{or} \quad x = 2$$

Skill Practice Solve by using the quadratic formula.

1. $5x^2 - 9x + 4 = 0$

Skill Practice Answers

1. $x = 1$; $x = \frac{4}{5}$

Example 2 Solving a Quadratic Equation Using the Quadratic Formula

Solve the quadratic equation using the quadratic formula. Then approximate the solutions to three decimal places.

$$2x(x - 1) - 6 = 0$$

Solution:

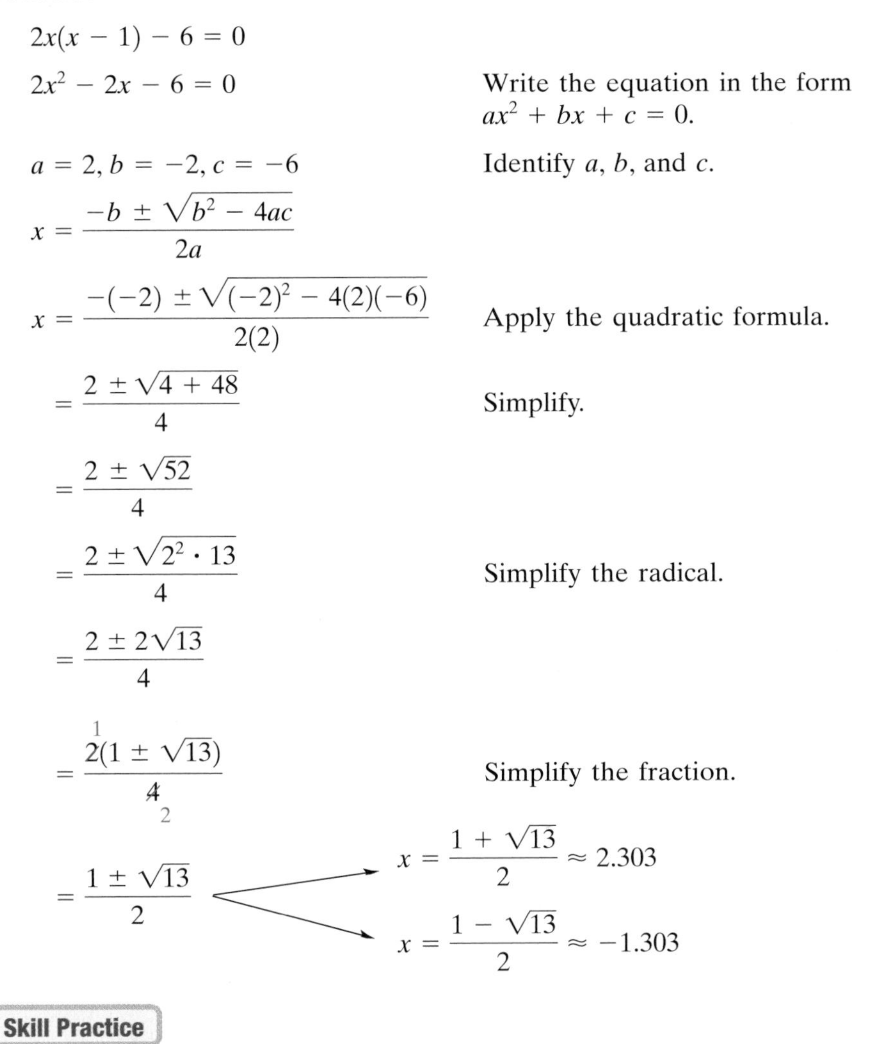

$2x(x - 1) - 6 = 0$

$2x^2 - 2x - 6 = 0$ Write the equation in the form $ax^2 + bx + c = 0$.

$a = 2, b = -2, c = -6$ Identify a, b, and c.

$$x = \frac{-b \pm \sqrt{b^2 - 4ac}}{2a}$$

$$x = \frac{-(-2) \pm \sqrt{(-2)^2 - 4(2)(-6)}}{2(2)}$$ Apply the quadratic formula.

$$= \frac{2 \pm \sqrt{4 + 48}}{4}$$ Simplify.

$$= \frac{2 \pm \sqrt{52}}{4}$$

$$= \frac{2 \pm \sqrt{2^2 \cdot 13}}{4}$$ Simplify the radical.

$$= \frac{2 \pm 2\sqrt{13}}{4}$$

$$= \frac{\overset{1}{\cancel{2}}(1 \pm \sqrt{13})}{\underset{2}{\cancel{4}}}$$ Simplify the fraction.

$$= \frac{1 \pm \sqrt{13}}{2}$$

$$x = \frac{1 + \sqrt{13}}{2} \approx 2.303$$

$$x = \frac{1 - \sqrt{13}}{2} \approx -1.303$$

Skill Practice

2. Solve by using the quadratic formula. Then approximate the solutions to three decimal places.

$$y(y + 2) = 5$$

Example 3 Solving a Quadratic Equation Using the Quadratic Formula

Solve the quadratic equation using the quadratic formula: $\frac{1}{4}w^2 - \frac{1}{2}w + \frac{5}{4} = 0$

Skill Practice Answers

2. $y = -1 \pm \sqrt{6}$; $y \approx 1.449$, $y \approx -3.449$

Solution:

$$\frac{1}{4}w^2 - \frac{1}{2}w + \frac{5}{4} = 0$$ To simplify the equation, multiply both sides by 4.

$$4\left(\frac{1}{4}w^2 - \frac{1}{2}w + \frac{5}{4}\right) = 4(0)$$ Clear fractions.

$$w^2 - 2w + 5 = 0$$ The equation is in the form $ax^2 + bx + c = 0$.

$$a = 1, b = -2, c = 5$$ Identify a, b, and c.

$$x = \frac{-b \pm \sqrt{b^2 - 4ac}}{2a}$$

$$x = \frac{-(-2) \pm \sqrt{(-2)^2 - 4(1)(5)}}{2(1)}$$ Apply the quadratic formula.

$$= \frac{2 \pm \sqrt{4 - 20}}{2}$$ Simplify.

$$= \frac{2 \pm \sqrt{-16}}{2}$$

$$= \frac{2 \pm 4i}{2}$$ The solutions are imaginary numbers.

$$= \frac{2(1 \pm 2i)}{2}$$ Factor the numerator and reduce.

$$= 1 \pm 2i$$

There are two imaginary solutions: $x = 1 + 2i$ and $x = 1 - 2i$.

Skill Practice Solve.

3. $\frac{1}{8}x^2 - \frac{1}{2}x = -\frac{13}{8}$

3. Review of the Methods for Solving a Quadratic Equation

Three methods have been presented for solving quadratic equations.

Methods for Solving a Quadratic Equation

- Factor and use the zero product rule (Section 6.7).
- Use the square root property. Complete the square if necessary (Section 9.3).
- Use the quadratic formula (Section 9.4).

Using the zero product rule only works if you can factor the equation. The square root property and the quadratic formula can be used to solve any quadratic equation. Before solving a quadratic equation, take a minute to analyze it first. Each problem must be evaluated individually before choosing the most efficient method to find its solutions.

Skill Practice Answers

3. $x = 2 \pm 3i$

Example 4 Solving Quadratic Equations Using Any Method

Solve the quadratic equations using any method.

a. $(x + 1)^2 = -4$ **b.** $t^2 - t - 30 = 0$ **c.** $2x^2 + 5x + 1 = 0$

Solution:

a. $(x + 1)^2 = -4$

$x + 1 = \pm\sqrt{-4}$ Apply the square root property.

$x + 1 = \pm 2i$ Simplify the radical.

$x = -1 \pm 2i$ Isolate x.

b. $t^2 - t - 30 = 0$ This equation factors easily.

$(t - 6)(t + 5) = 0$ Factor and apply the zero product rule.

$t = 6$ or $t = -5$

c. $2x^2 + 5x + 1 = 0$ The equation does not factor. Because it is already in the form $ax^2 + bx + c = 0$, use the quadratic formula.

$a = 2, b = 5, c = 1$ Identify a, b, and c.

$$x = \frac{-b \pm \sqrt{b^2 - 4ac}}{2a}$$

$$x = \frac{-(5) \pm \sqrt{(5)^2 - 4(2)(1)}}{2(2)}$$ Apply the quadratic formula.

$$x = \frac{-5 \pm \sqrt{25 - 8}}{4}$$ Simplify.

$$x = \frac{-5 \pm \sqrt{17}}{4}$$

Skill Practice Solve the equations using any method.

4. $(p - 3)^2 + 5 = 7$ **5.** $3x^2 - 7x + 10 = 0$ **6.** $3t^2 - 14t = 5$

4. Applications of Quadratic Equations

Example 5 Solving a Quadratic Equation in an Application

The length of a box is 2 in. longer than the width. The height of the box is 4 in. and the volume of the box is 200 in.^{3} Find the exact dimensions of the box. Then use a calculator to approximate the dimensions to the nearest tenth of an inch.

Skill Practice Answers

4. $p = 3 \pm \sqrt{2}$

5. $x = \dfrac{7 \pm i\sqrt{71}}{6}$

6. $t = -\dfrac{1}{3}, t = 5$

Solution:

Label the box as follows (Figure 9-8):

Width $= x$

Length $= x + 2$

Height $= 4$

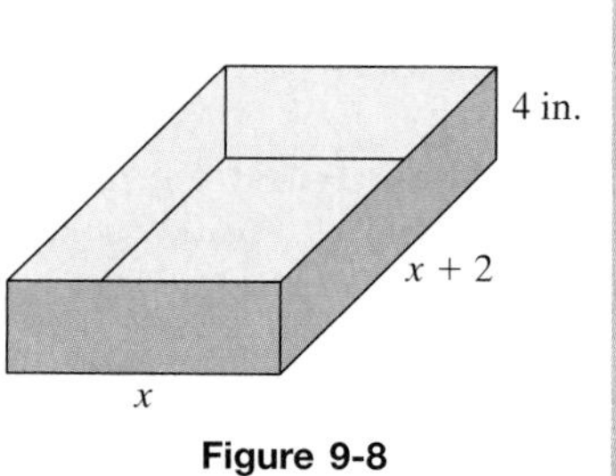

Figure 9-8

The volume of a box is given by the formula $V = lwh$.

$$V = l \cdot w \cdot h$$

$$200 = (x + 2)(x)(4)$$ Substitute $V = 200$, $l = x + 2$, $w = x$, and $h = 4$.

$$200 = (x + 2)4x$$

$$200 = 4x^2 + 8x$$

$$0 = 4x^2 + 8x - 200$$

$$4x^2 + 8x - 200 = 0$$ The equation is in the form $ax^2 + bx + c = 0$.

$$\frac{4x^2}{4} + \frac{8x}{4} - \frac{200}{4} = \frac{0}{4}$$ The coefficients are all divisible by 4. Dividing by 4 will create smaller values of a, b, and c to be used in the quadratic formula.

$$x^2 + 2x - 50 = 0$$ $a = 1, b = 2, c = -50$

$$x = \frac{-2 \pm \sqrt{(2)^2 - 4(1)(-50)}}{2(1)}$$ Apply the quadratic formula.

$$= \frac{-2 \pm \sqrt{4 + 200}}{2}$$ Simplify.

$$= \frac{-2 \pm \sqrt{204}}{2}$$

$$= \frac{-2 \pm 2\sqrt{51}}{2}$$ Simplify the radical. $\sqrt{204} = \sqrt{2^2 \cdot 51} = 2\sqrt{51}$

$$= \frac{\overset{1}{\cancel{2}}(-1 \pm \sqrt{51})}{\underset{1}{\cancel{2}}}$$ Factor and simplify.

$$= -1 \pm \sqrt{51}$$

Because the width of the box must be positive, use $x = -1 + \sqrt{51}$.

The width is $(-1 + \sqrt{51})$ in. ≈ 6.1 in.

The length is $x + 2$: $(-1 + \sqrt{51} + 2)$ in. or $(1 + \sqrt{51})$ in. ≈ 8.1 in.

The height is 4 in.

Avoiding Mistakes:

We do not use the solution $x = -1 - \sqrt{51}$ because it is a negative number; that is,

$$-1 - \sqrt{51} \approx -8.1$$

The width of an object cannot be negative.

Skill Practice

7. The length of a rectangle is 2 in. longer than the width. The area is 10 in.^{2} Find the exact values of the length and width. Then use a calculator to approximate the dimensions to the nearest tenth of an inch.

Skill Practice Answers

7. The width is $(-1 + \sqrt{11})$ in., or approximately 2.3 in. The length is $(1 + \sqrt{11})$ in., or approximately 4.3 in.

Calculator Connections

A calculator can be used to obtain decimal approximations for the real solutions to a quadratic equation. From Example 2, the solutions to the equation $2x(x - 1) - 6 = 0$ are:

$$x = \frac{1 + \sqrt{13}}{2} \quad \text{and} \quad x = \frac{1 - \sqrt{13}}{2}$$

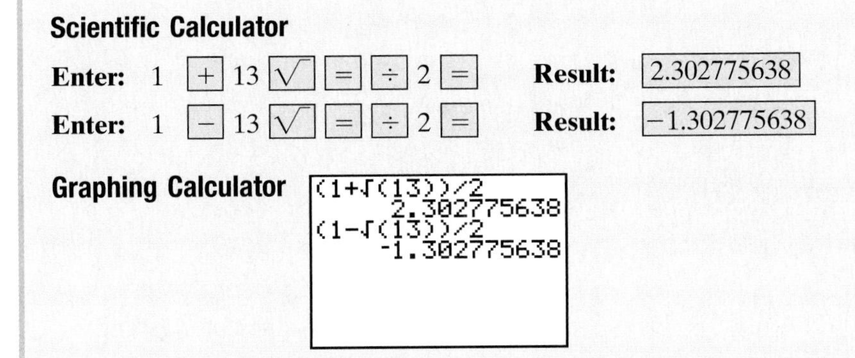

Now consider another example. The solutions to the equation $3x^2 + 6x - 2 = 0$ are:

$$x = \frac{-(6) + \sqrt{(6)^2 - 4(3)(-2)}}{2(3)} \approx 0.2909944487$$

```
(-6+√((6)²-4(3)(
-2)))/(2*3)
        .2909944487   ← Solution
3*Ans²+6*Ans-2
                  0   ← Check
```

and

$$x = \frac{-(6) - \sqrt{(6)^2 - 4(3)(-2)}}{2(3)} \approx -2.290994449$$

```
(-6-√((6)²-4(3)(
-2)))/(2*3)
       -2.290994449   ← Solution
3*Ans²+6*Ans-2
                  0   ← Check
```

TIP: A graphing calculator can be used to apply the quadratic formula directly. The result of the calculation is stored in memory in a variable such as *Ans*. The solution can be checked by substituting the value of *Ans* into the original equation.

Calculator Exercises

Use a calculator to obtain a decimal approximation of each expression.

1. $\frac{-5 + \sqrt{17}}{4}$ and $\frac{-5 - \sqrt{17}}{4}$

2. $\frac{-40 + \sqrt{1920}}{-32}$ and $\frac{-40 - \sqrt{1920}}{-32}$

3. $\frac{-17 - \sqrt{(17)^2 - 4(4)(-3)}}{2(4)}$

4. $\frac{5.2 + \sqrt{(5.2)^2 - 4(2.1)(1.7)}}{2(2.1)}$

Section 9.4 Practice Exercises

Review Exercises

For Exercises 1–4, apply the square root property to solve the equation.

1. $z^2 = 169$ **2.** $p^2 = 1$ **3.** $(x - 4)^2 = 28$ **4.** $(y + 3)^2 = 7$

For Exercises 5–6, solve the equations by completing the square.

5. $x^2 - 5x + 1 = 0$ **6.** $3a^2 - 12a - 12 = 0$

Concept 1: Derivation of the Quadratic Formula

7. State the quadratic formula from memory.

For Exercises 8–13, write the equations in the form $ax^2 + bx + c = 0$. Then identify the values of a, b, and c.

8. $2x^2 - x = 5$ **9.** $5(x^2 + 2) = -3x$

10. $-3x(x - 4) = -2x$ **11.** $x(x - 2) = 3(x + 1)$

12. $x^2 - 9 = 0$ **13.** $x^2 + 25 = 0$

Concept 2: Solving Quadratic Equations Using the Quadratic Formula

For Exercises 14–34, solve the equations using the quadratic formula.

14. $6k^2 - k - 2 = 0$ **15.** $3n^2 + 5n - 2 = 0$ **16.** $5t^2 - t = 3$

17. $2a^2 + 5a = 1$ **18.** $x(x - 2) = 1$ **19.** $2y(y - 3) = -1$

20. $2p^2 = -10p - 11$ **21.** $z^2 = 4z + 1$ **22.** $-4y^2 - y + 1 = 0$

23. $-5z^2 - 3z + 4 = 0$ **24.** $3c^2 = c - 2$ **25.** $5p^2 = 2p - 3$

26. $0.2y^2 = -1.5y - 1$ **27.** $0.2t^2 = t + 0.5$ **28.** $-2.5t(t - 4) = 1.5$

29. $1.6p(p - 2) = 0.8$ **30.** $\frac{2}{3}x^2 + \frac{4}{9}x = \frac{1}{3}$ **31.** $\frac{1}{2}x^2 + \frac{1}{6}x = 1$

32. $(m - 3)(m + 2) = 9$ **33.** $(h - 6)(h - 1) = 12$ **34.** $k^2 + 2k + 5 = 0$

Concept 3: Review of the Methods for Solving a Quadratic Equation

For Exercises 35–52, choose any method to solve the quadratic equations.

35. $16x^2 - 9 = 0$ **36.** $\frac{1}{4}x^2 + 5x + 13 = 0$ **37.** $(x - 5)^2 = -21$

38. $2x^2 + x + 5 = 0$ **39.** $\frac{1}{9}x^2 + \frac{8}{3}x + 11 = 0$ **40.** $7x^2 = 12x$

41. $2x^2 - 6x - 3 = 0$

42. $4(x + 1)^2 = -16$

43. $9x^2 = 11x$

44. $25x^2 - 4 = 0$

45. $(2y - 3)^2 = 5$

46. $(6z + 1)^2 = 7$

47. $0.4x^2 = 0.2x + 1$

48. $0.6x^2 = 0.1x + 0.8$

49. $9z^2 - z = 0$

50. $16p^2 - p = 0$

51. $r^2 - 52 = 0$

52. $y^2 - 32 = 0$

Concept 4: Applications of Quadratic Equations

53. In a triangle, the height is 2 cm more than the base. The area is 72 cm^2. Find the exact values of the base and height of the triangle. Then use a calculator to approximate the base and height to the nearest tenth of a centimeter.

54. In a rectangle, the length is 1 m less than twice the width and the area is 100 m^2. Find the exact dimensions of the rectangle. Then use a calculator to approximate the dimensions to the nearest tenth of a meter.

55. The volume of a rectangular storage area is 240 ft^3. The length is 2 ft more than the width. The height is 6 ft. Find the exact dimensions of the storage area. Then use a calculator to approximate the dimensions to the nearest tenth of a foot.

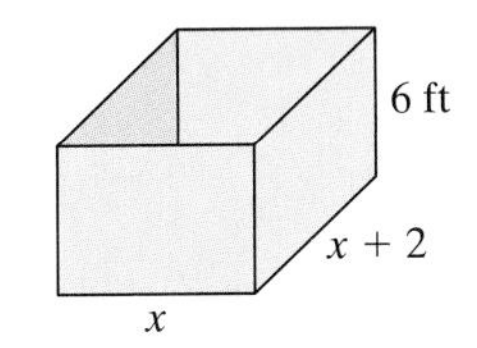

56. In a right triangle, one leg is 2 ft shorter than the other leg. The hypotenuse is 12 ft. Find the exact lengths of the legs of the triangle. Then use a calculator to approximate the legs to the nearest tenth of a foot.

57. In a rectangle, the length is 4 ft longer than the width. The area is 72 ft^2. Find the exact dimensions of the rectangle. Then use a calculator to approximate the dimensions to the nearest tenth of a foot.

58. In a triangle, the base is 4 cm less than twice the height. The area is 60 cm^2. Find the exact values of the base and height of the triangle. Then use a calculator to approximate the base and height to the nearest tenth of a centimeter.

59. In a right triangle, one leg is 3 m longer than the other leg. The hypotenuse is 13 m. Find the exact values of the legs of the triangle. Then use a calculator to approximate the values to the nearest tenth of a meter.

Expanding Your Skills

From Examples 1–3, we see that the solutions to a quadratic equation may be rational numbers, irrational numbers, or imaginary numbers. The *number* and *type* of solution can be determined by noting the value of the square root term in the quadratic formula. The radicand of the square root, $b^2 - 4ac$, is called the discriminant.

Using the Discriminant to Determine the Number and Type of Solutions of a Quadratic Equation

Consider the equation, $ax^2 + bx + c = 0$, where a, b, and c are rational numbers and $a \neq 0$. The expression $b^2 - 4ac$, is called the **discriminant**. Furthermore,

1. If $b^2 - 4ac > 0$ then there will be two real solutions. Moreover,
   a. If $b^2 - 4ac$ is a perfect square, the solutions will be rational numbers.
   b. If $b^2 - 4ac$ is not a perfect square, the solutions will be irrational numbers.

2. If $b^2 - 4ac < 0$ then there will be two imaginary solutions.

3. If $b^2 - 4ac = 0$ then there will be one rational solution.

Using the Discriminant

Use the discriminant to determine the type and number of solutions for each equation.

a. $3x^2 - 4x + 7 = 0$ **b.** $2x^2 + x = 5$

c. $2(x^2 + 10) = 13x$ **d.** $0.4x^2 - 1.2x + 0.9 = 0$

Solution:

For each equation, first write the equation in standard form, $ax^2 + bx + c = 0$. Then determine the value of the discriminant.

Equation	Discriminant	Solution Type and Number
a. $3x^2 - 4x + 7 = 0$	$b^2 - 4ac$	Because $-68 < 0$, there will be two imaginary solutions.
$a = 3, b = -4, c = 7$	$(-4)^2 - 4(3)(7)$	
	$16 - 84$	
	-68	
b. $2x^2 + x = 5$		
$2x^2 + x - 5 = 0$	$b^2 - 4ac$	Because $41 > 0$ but 41 is not a perfect square, there will be two irrational solutions.
$a = 2, b = 1, c = -5$	$(1)^2 - 4(2)(-5)$	
	$1 + 40$	
	41	
c. $2(x^2 + 10) = 13x$		
$2x^2 + 20 = 13x$		
$2x^2 - 13x + 20 = 0$	$b^2 - 4ac$	Because $9 > 0$ and 9 is a perfect square, there will be two rational solutions.
$a = 2, b = -13, c = 20$	$(-13)^2 - 4(2)(20)$	
	$169 - 160$	
	9	
d. $0.4x^2 - 1.2x + 0.9 = 0$		
	$b^2 - 4ac$	Because the discriminant equals 0, there will be only one rational solution.
$a = 0.4, b = -1.2, c = 0.9$	$(-1.2)^2 - 4(0.4)(0.9)$	
	$1.44 - 1.44$	
	0	

For Exercises 60–67, find the discriminant and use the discriminant to determine the number and type of solution to each equation. There may be two rational solutions, two irrational solutions, two imaginary solutions, or one rational solution.

60. $x^2 + 16x + 64 = 0$ **61.** $x^2 - 10x + 25 = 0$ **62.** $6y^2 - y - 2 = 0$ **63.** $3m^2 + 5m - 2 = 0$

64. $2p^2 + 5p = 1$ **65.** $9a^2 + 3a = 1$ **66.** $3b^2 = b - 2$ **67.** $5q^2 = 2q - 3$

Graphing Quadratic Functions

Section 9.5

Concepts

1. Definition of a Quadratic Function
2. Vertex of a Parabola
3. Graphing a Parabola
4. Applications of Quadratic Functions

1. Definition of a Quadratic Function

In Chapter 3, we learned how to graph the solutions to linear equations in two variables. Now suppose we want to graph the *nonlinear* equation, $y = x^2$. To begin, we create a table of points representing several solutions to the equation (Table 9-2). These points form the curve shown in Figure 9-9.

Table 9-2

x	y
−3	9
−2	4
−1	1
0	0
1	1
2	4
3	9

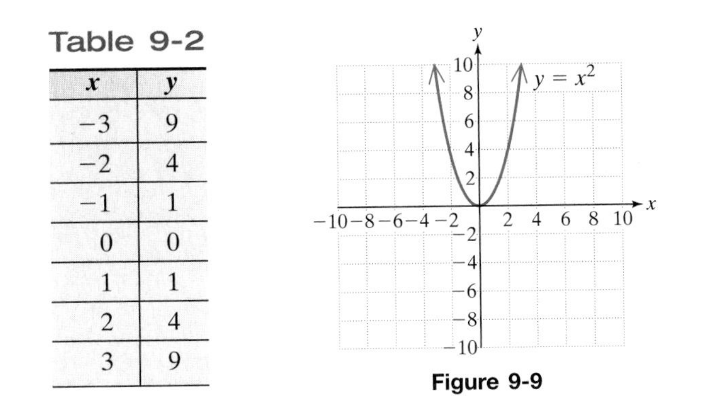

Figure 9-9

The equation $y = x^2$ is a special type of function called a quadratic function, and its graph is in the shape of a **parabola**.

Definition of a Quadratic Function

Let a, b, and c represent real numbers such that $a \neq 0$. Then a function in the form, $y = ax^2 + bx + c$ is called a **quadratic function**.

The graph of a quadratic function is a parabola that opens upward or downward. The leading coefficient, a, determines the direction of the parabola. For the quadratic function defined by $y = ax^2 + bx + c$:

If $a > 0$, the parabola opens *upward*. For example: $y = x^2$.

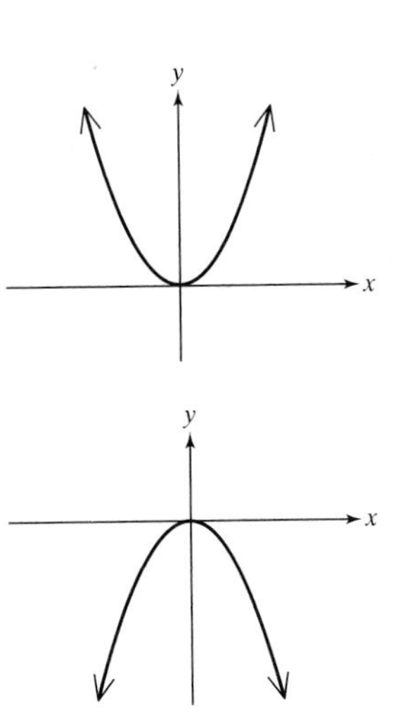

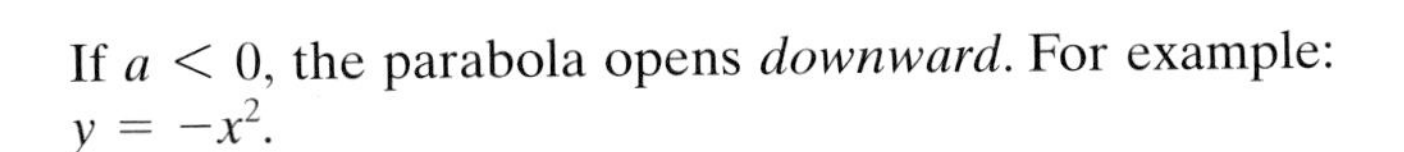

If $a < 0$, the parabola opens *downward*. For example: $y = -x^2$.

If a parabola opens upward, the **vertex** is the lowest point on the graph. If a parabola opens downward, the **vertex** is the highest point on the graph. For a quadratic function, the **axis of symmetry** is the vertical line that passes through the vertex. Notice that the graph of the parabola is its own mirror image to the left and right of the axis of symmetry.

$y = 0.5x^2 + 2x + 3$
$a > 0$
Vertex $(-2, 1)$
Axis of symmetry: $x = -2$

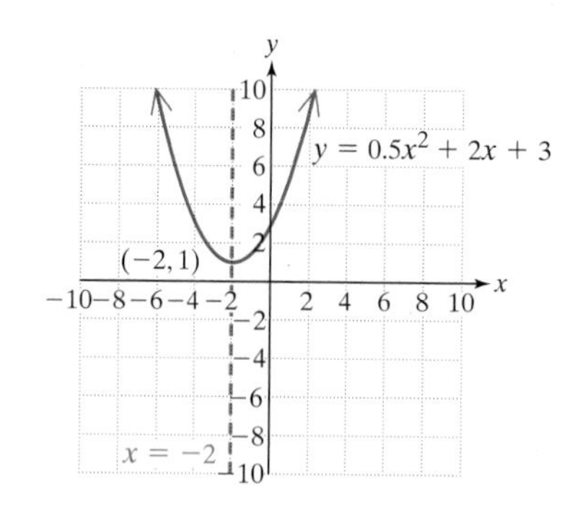

$y = x^2 - 6x + 9$
$a > 0$
Vertex $(3, 0)$
Axis of symmetry: $x = 3$

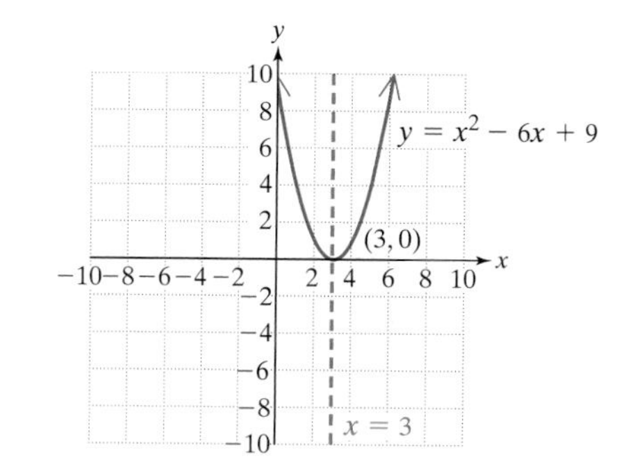

$y = -x^2 + 4x$
$a < 0$
Vertex $(2, 4)$
Axis of symmetry: $x = 2$

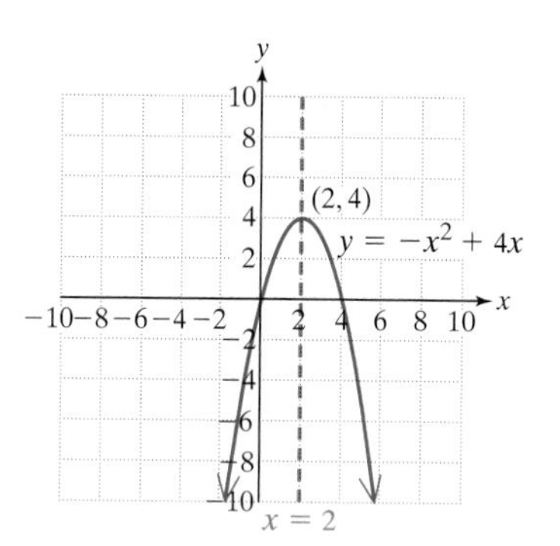

$y = -2x^2 - 4$
$a < 0$
Vertex $(0, -4)$
Axis of symmetry: $x = 0$

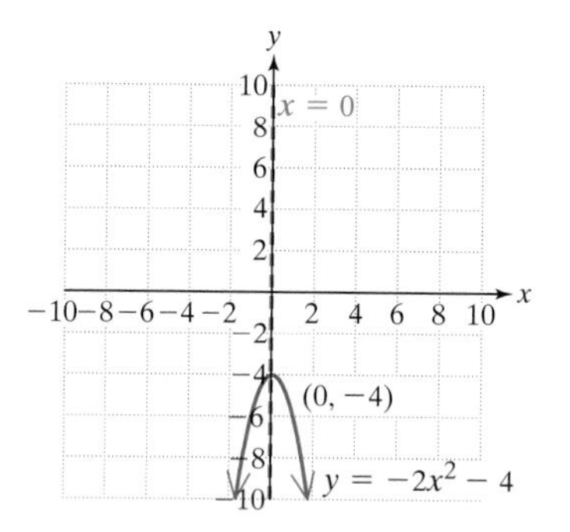

2. Vertex of a Parabola

Quadratic functions arise in many applications of mathematics and applied sciences. For example, an object thrown through the air follows a parabolic path. The mirror inside a reflecting telescope is parabolic in shape. In applications, it is often advantageous to analyze the graph of a parabola. In particular, we want to find the location of the x- and y-intercepts and the vertex.

To find the vertex of a parabola defined by $y = ax^2 + bx + c$ $(a \neq 0)$, we use the following steps:

Vertex of a Parabola

1. The x-coordinate of the vertex of the parabola defined by $y = ax^2 + bx + c$ $(a \neq 0)$ is given by

$$x = \frac{-b}{2a}$$

2. To find the corresponding y-coordinate of the vertex, substitute the value of the x-coordinate found in step 1 and solve for y.

Example 1 Analyzing a Quadratic Function

Given the function defined by $y = -x^2 + 4x - 3$,

a. Determine whether the function opens upward or downward.

b. Find the vertex of the parabola.

c. Find the x-intercept(s).

d. Find the y-intercept.

e. Sketch the parabola.

f. Write the domain of the function in interval notation.

g. Write the range of the function in interval notation.

Solution:

a. The function $y = -x^2 + 4x - 3$ is written in the form $y = ax^2 + bx + c$, where $a = -1$, $b = 4$, and $c = -3$. Because the value of a is negative, the parabola opens downward.

b. The x-coordinate of the vertex is given by $x = \dfrac{-b}{2a}$.

$$x = \frac{-b}{2a} = \frac{-(4)}{2(-1)} \qquad \text{Substitute } b = 4 \text{ and } a = -1.$$

$$= \frac{-4}{-2} \qquad \text{Simplify.}$$

$$= 2$$

The y-coordinate of the vertex is found by substituting $x = 2$ into the equation and solving for y.

$$y = -x^2 + 4x - 3$$

$$= -(2)^2 + 4(2) - 3 \qquad \text{Substitute } x = 2.$$

$$= -4 + 8 - 3$$

$$= 1$$

The vertex is $(2, 1)$. Because the parabola opens downward, the vertex is the maximum point on the graph of the parabola.

c. To find the x-intercept(s), substitute $y = 0$ and solve for x.

$$y = -x^2 + 4x - 3$$

$$0 = -x^2 + 4x - 3 \qquad \text{Substitute } y = 0. \text{ The resulting equation is quadratic.}$$

$$0 = -1(x^2 - 4x + 3) \qquad \text{Factor out } -1.$$

$$0 = -1(x - 3)(x - 1) \qquad \text{Factor the trinomial.}$$

$$x - 3 = 0 \quad \text{or} \quad x - 1 = 0 \qquad \text{Apply the zero product rule.}$$

$$x = 3 \quad \text{or} \quad x = 1$$

The x-intercepts are $(3, 0)$ and $(1, 0)$.

d. To find the y-intercept, substitute $x = 0$ and solve for y.

$$y = -x^2 + 4x - 3$$

$$= -(0)^2 + 4(0) - 3 \qquad \text{Substitute } x = 0.$$

$$= -3$$

The y-intercept is $(0, -3)$.

e. Using the results of parts (a)–(d), we have a parabola that opens downward with vertex $(2, 1)$, x-intercepts at $(3, 0)$ and $(1, 0)$, and y-intercept at $(0, -3)$ (Figure 9-10).

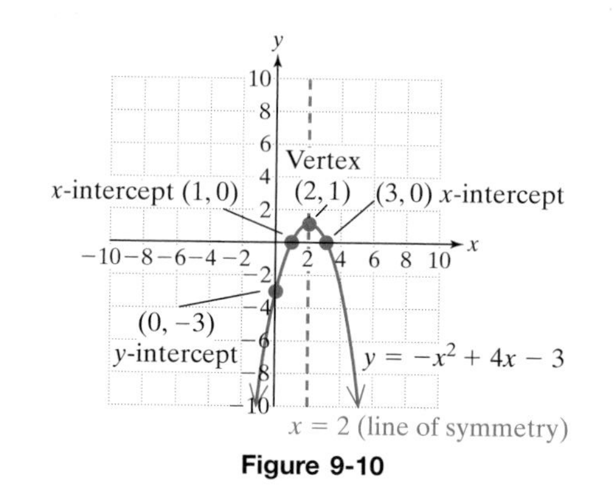

Figure 9-10

f. Because any real number, x, when substituted into the equation $y = -x^2 + 4x - 3$ produces a real number, the domain is all real numbers.

$$\text{Domain: } (-\infty, \infty)$$

g. From the graph, we see that the vertex is the maximum point on the graph. Therefore, the maximum y-value is $y = 1$. The range is restricted to $y \leq 1$.

$$\text{Range: } (-\infty, 1]$$

Skill Practice

1. Given $y = -x^2 - 4x$, perform parts (a)–(g) as in Example 1.

3. Graphing a Parabola

Example 1 illustrates a process to sketch a quadratic function by finding the location of the defining characteristics of the function. These include the vertex and the x- and y-intercepts. Furthermore, notice that the parabola defining the graph of a quadratic function is symmetric with respect to the axis of symmetry.

To analyze the graph of a parabola, we recommend the following guidelines.

Skill Practice Answers

1a. downward
b. $(-2, 4)$
c. $(0, 0)$ and $(-4, 0)$
d. $(0, 0)$
e.

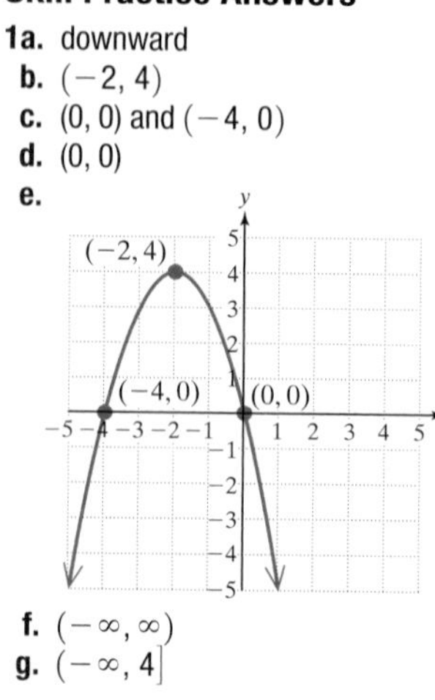

f. $(-\infty, \infty)$
g. $(-\infty, 4]$

Graphing a Parabola

Given a quadratic function defined by $y = ax^2 + bx + c$ $(a \neq 0)$, consider the following guidelines to graph the function.

1. Determine whether the function opens upward or downward.
   - If $a > 0$, the parabola opens upward.
   - If $a < 0$, the parabola opens downward.
2. Find the vertex.
   - The x-coordinate is given by $x = \dfrac{-b}{2a}$
   - To find the y-coordinate, substitute the x-coordinate of the vertex into the equation and solve for y.
3. Find the x-intercept(s) by substituting $y = 0$ and solving the equation for x.
   - *Note:* If the solutions to the equation in step 3 are imaginary numbers, then the function has no x-intercepts.
4. Find the y-intercept by substituting $x = 0$ and solving the equation for y.
5. Plot the vertex and x- and y-intercepts. If necessary, find and plot additional points near the vertex. Then use the symmetry of the parabola to sketch the curve through the points.

Example 2 Graphing a Parabola

Graph the function defined by $y = x^2 - 6x + 9$.

Solution:

1. The function $y = x^2 - 6x + 9$ is written in the form $y = ax^2 + bx + c$, where $a = 1$, $b = -6$, and $c = 9$. Because the value of a is positive, the parabola opens upward.
2. The x-coordinate of the vertex is given by

$$x = \frac{-b}{2a} = \frac{-(-6)}{2(1)} = 3$$

Substituting $x = 3$ into the equation, we have

$$\begin{aligned} y &= (3)^2 - 6(3) + 9 \\ &= 9 - 18 + 9 \\ &= 0 \end{aligned}$$

The vertex is $(3, 0)$.

3. To find the x-intercept(s), substitute $y = 0$ and solve for x.

Substitute $y = 0$

$$y = x^2 - 6x + 9 \longrightarrow 0 = x^2 - 6x + 9$$

$$0 = (x - 3)^2 \qquad \text{Factor.}$$

$$x = 3 \qquad \text{Apply the zero product rule.}$$

The x-intercept is $(3, 0)$.

4. To find the y-intercept, substitute $x = 0$ and solve for y.

Substitute $x = 0$

$$y = x^2 - 6x + 9 \longrightarrow y = (0)^2 - 6(0) + 9$$
$$= 9$$

The y-intercept is $(0, 9)$.

5. Sketch the parabola through the x- and y-intercepts and vertex (Figure 9-11).

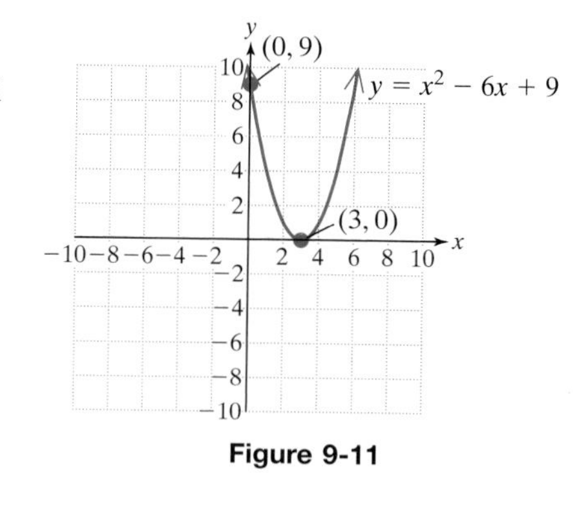

Figure 9-11

TIP: Using the symmetry of the parabola, we know that the points to the right of the vertex must mirror the points to the left of the vertex.

Skill Practice

2. Graph $y = x^2 - 2x + 1$.

Example 3 Graphing a Parabola

Graph the function defined by $y = -x^2 - 4$.

Solution:

1. The function $y = -x^2 - 4$ is written in the form $y = ax^2 + bx + c$, where $a = -1$, $b = 0$, and $c = -4$. Because the value of a is negative, the parabola opens downward.
2. The x-coordinate of the vertex is given by

$$x = \frac{-b}{2a} = \frac{-(0)}{2(-1)} = 0$$

Substituting $x = 0$ into the equation, we have

$$y = -(0)^2 - 4$$
$$= -4$$

The vertex is $(0, -4)$.

3. Because the vertex is below the x-axis and the parabola opens downward, the function cannot have an x-intercept.

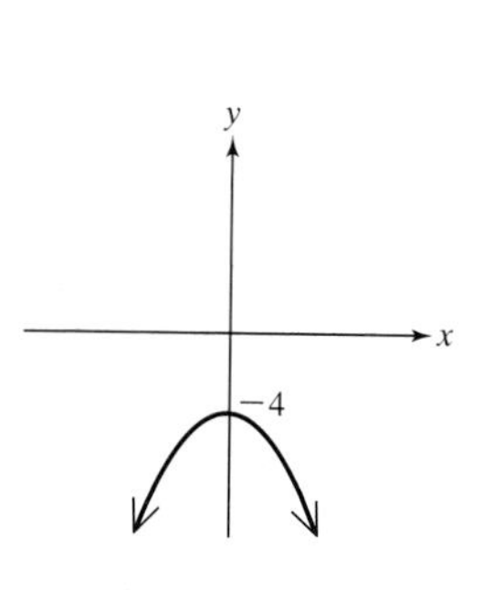

4. The vertex is $(0, -4)$. This is also the y-intercept.

Skill Practice Answers

2.
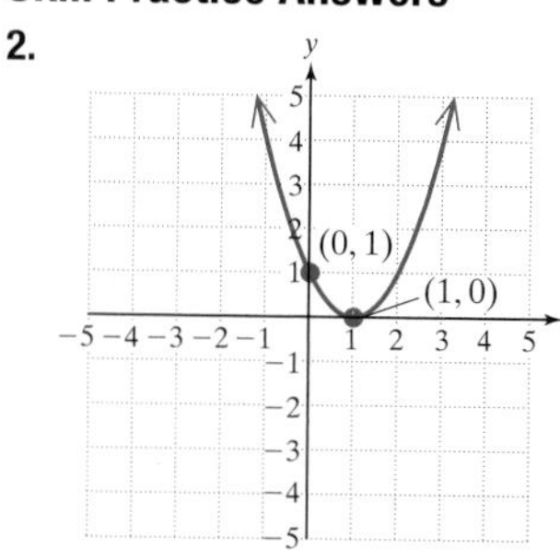

5. Sketch the parabola through the y-intercept and vertex (Figure 9-12).

To verify the proper shape of the graph, find additional points to the right or left of the vertex and use the symmetry of the parabola to sketch the curve.

x	y
1	−5
2	−8
3	−13

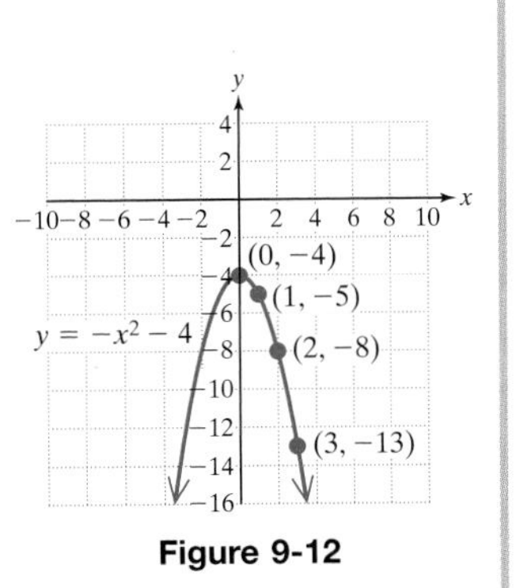

Figure 9-12

TIP: Substituting $y = 0$ into the equation $y = -x^2 - 4$ results in an equation with no real solutions. Therefore, the function $y = -x^2 - 4$ has no x-intercepts.

$$y = -x^2 - 4$$
$$0 = -x^2 - 4$$
$$x^2 = -4$$
$$x = \pm\sqrt{-4}$$
$$x = \pm 2i \quad \text{Not a real number}$$

Skill Practice

3. Graph $y = x^2 - 1$.

4. Applications of Quadratic Functions

Example 4 Using a Quadratic Function in an Application

A golfer hits a ball at an angle of 30°. The height of the ball, y, (in feet) can be represented by

$y = -16x^2 + 60x$ where x is the time in seconds after the ball was hit (Figure 9-13).

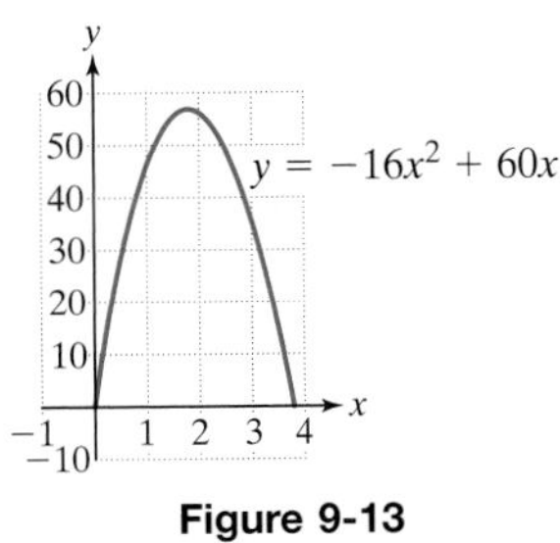

Figure 9-13

Find the maximum height of the ball. In how many seconds will the ball reach its maximum height?

Skill Practice Answers

3. (−1, 0), (1, 0), (0, −1)

Solution:

The function is written in the form $y = ax^2 + bx + c$, where $a = -16$, $b = 60$, and $c = 0$. Because a is negative, the function opens downward. Therefore, the maximum height of the ball occurs at the vertex of the parabola.

The x-coordinate of the vertex is given by

$$x = \frac{-b}{2a} = \frac{-(60)}{2(-16)} = \frac{-60}{-32} = \frac{15}{8}$$

Substituting $x = \frac{15}{8}$ into the equation, we have

$$\begin{aligned} y &= -16\left(\frac{15}{8}\right)^2 + 60\left(\frac{15}{8}\right) \\ &= -16\left(\frac{225}{64}\right) + \frac{900}{8} \\ &= -\frac{225}{4} + \frac{450}{4} \\ &= \frac{225}{4} \end{aligned}$$

The vertex is at the point $\left(\frac{15}{8}, \frac{225}{4}\right)$ or equivalently (1.875, 56.25).

The ball reaches its maximum height of 56.25 ft after 1.875 s.

Skill Practice

4. A basketball player shoots a basketball at an angle of 45°. The height of the ball, y, (in feet) is given by $y = -16t^2 + 40t + 6$. Find the maximum height of the ball and the time required to reach that height.

Calculator Connections

Some graphing calculators have *Minimum* and *Maximum* features that enable the user to approximate the minimum and maximum values of a function. Otherwise, *Zoom* and *Trace* can be used.

For example, the maximum value of the function from Example 4, $y = -16x^2 + 60x$, can be found using the *Maximum* feature.

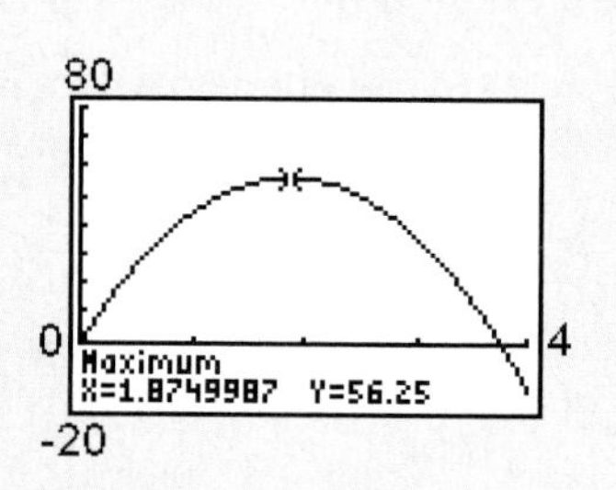

The minimum value of the function from Example 2, $y = x^2 - 6x + 9$, can be found using the *Minimum* feature.

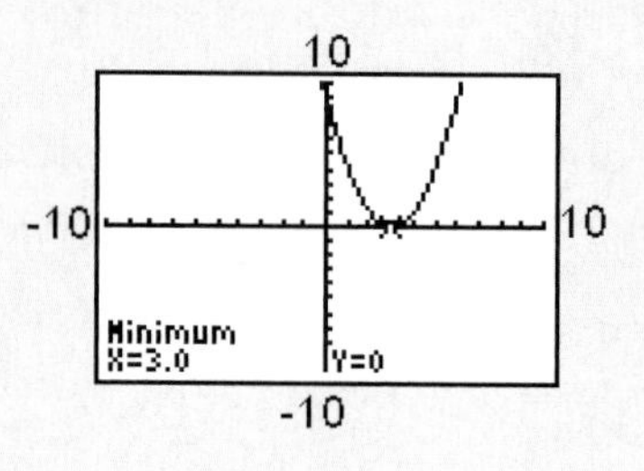

Skill Practice Answers

4. The ball reaches a maximum height of 31 ft in 1.25 sec.

Calculator Exercises

Without using a calculator, find the location of the vertex for each function. Then verify the answer using a graphing calculator.

1. $y = x^2 + 4x + 7$

2. $y = x^2 - 20x + 105$

3. $y = -x^2 - 3x - 4.85$

4. $y = -x^2 + 3.5x - 0.5625$

5. $y = 2x^2 - 10x + \dfrac{25}{2}$

6. $y = 3x^2 + 16x + \dfrac{64}{3}$

7. Use a graphing calculator to graph the functions on the same screen. Describe the relationship between the graph of $y = x^2$ and the graphs in parts (b) and (c).

a. $y = x^2$

b. $y = x^2 + 4$

c. $y = x^2 - 3$

8. Use a graphing calculator to graph the functions on the same screen. Describe the relationship between the graph of $y = x^2$ and the graphs in parts (b) and (c).

a. $y = x^2$

b. $y = (x - 3)^2$

c. $y = (x + 2)^2$

9. Use a graphing calculator to graph the functions on the same screen. Describe the relationship between the graph of $y = x^2$ and the graphs in parts (b) and (c).

a. $y = x^2$

b. $y = 2x^2$

c. $y = \frac{1}{2}x^2$

10. Use a graphing calculator to graph the functions on the same screen. Describe the relationship between the graph of $y = x^2$ and the graphs in parts (b) and (c).

a. $y = x^2$

b. $y = -2x^2$

c. $y = -\frac{1}{2}x^2$

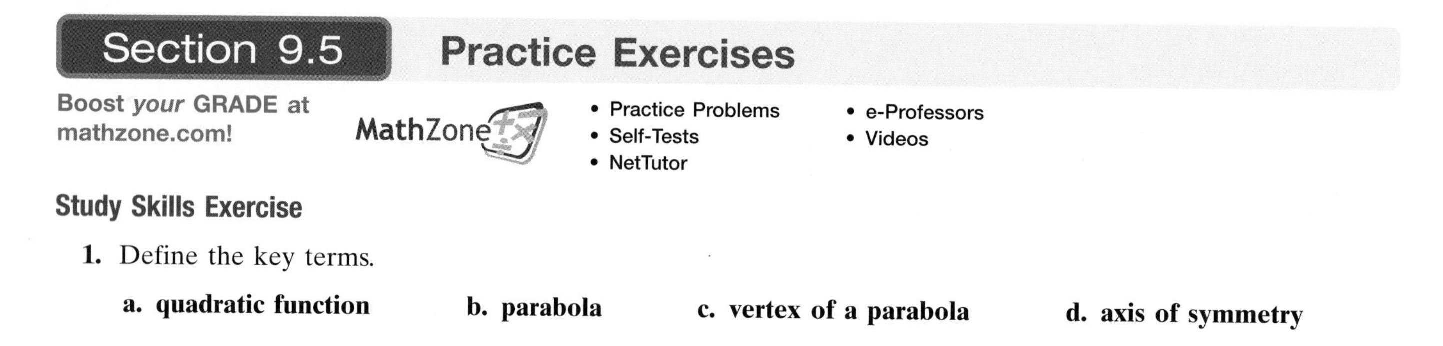

Study Skills Exercise

1. Define the key terms.

a. quadratic function **b. parabola** **c. vertex of a parabola** **d. axis of symmetry**

Review Exercises

For Exercises 2–8, solve the quadratic equations using any one of the following methods: factoring, the square root property, or the quadratic formula.

2. $3(y^2 + 1) = 10y$ **3.** $3 + a(a + 2) = 18$ **4.** $4t^2 - 7 = 0$ **5.** $9p^2 = 5$

6. $(b + 1)^2 = 6$ **7.** $(x - 2)^2 = 8$ **8.** $z^2 + 2z + 7 = 0$

Concept 1: Definition of a Quadratic Function

For Exercises 9–18, identify the equations as linear, quadratic, or neither.

9. $y = -8x + 3$ **10.** $y = 5x - 12$ **11.** $y = 4x^2 - 8x + 22$ **12.** $y = x^2 + 10x - 3$

13. $y = -5x^3 - 8x + 14$ **14.** $y = -3x^4 + 7x - 11$ **15.** $y = 15x$ **16.** $y = -9x$

17. $y = -21x^2$ **18.** $y = 3x^2$

Concept 2: Vertex of a Parabola

19. How do you determine whether the graph of a function $y = ax^2 + bx + c$ $(a \neq 0)$ opens upward or downward?

For Exercises 20–25, identify a and determine if the parabola opens upward or downward.

20. $y = -5x^2 - x + 10$ **21.** $y = -7x^2 + 3x - 1$ **22.** $y = x^2 - 15$ **23.** $y = 2x^2 + 23$

24. $y = -3x^2 + x - 18$ **25.** $y = -10x^2 - 6x - 20$

26. How do you find the vertex of a parabola?

For Exercises 27–34, find the vertex of the parabola.

27. $y = 2x^2 + 4x - 6$ **28.** $y = x^2 - 4x - 4$ **29.** $y = -x^2 + 2x - 5$

30. $y = 2x^2 - 4x - 6$ **31.** $y = x^2 - 2x + 3$ **32.** $y = -x^2 + 4x - 2$

33. $y = x^2 - 4$ **34.** $y = x^2 - 1$

Concept 3: Graphing a Parabola

For Exercises 35–38, find the x- and y-intercepts of the function. Then match each function with a graph.

35. $y = x^2 - 7$

36. $y = x^2 - 9$

37. $y = (x + 3)^2 - 4$

38. $y = (x - 2)^2 - 1$

a.

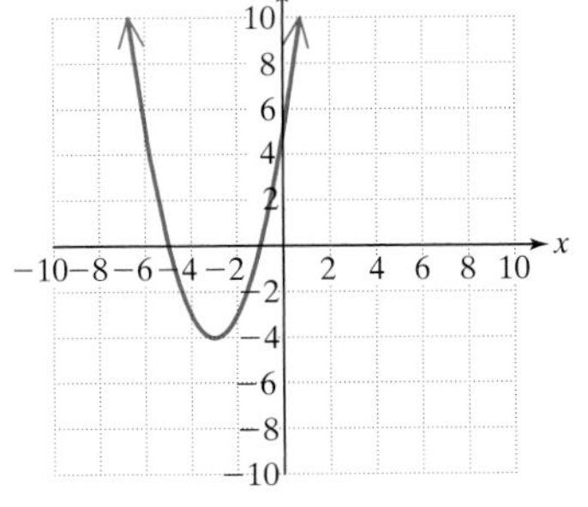

b.

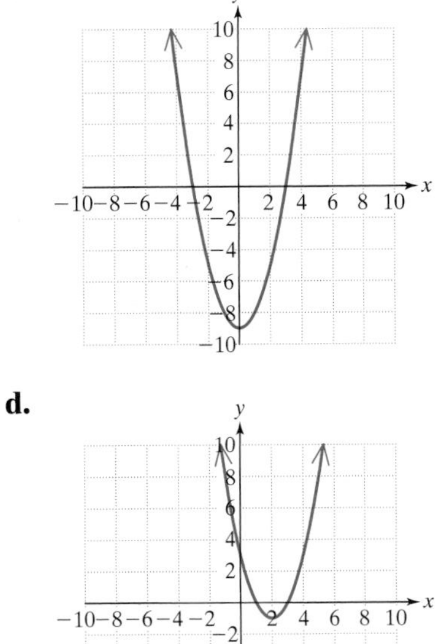

c.

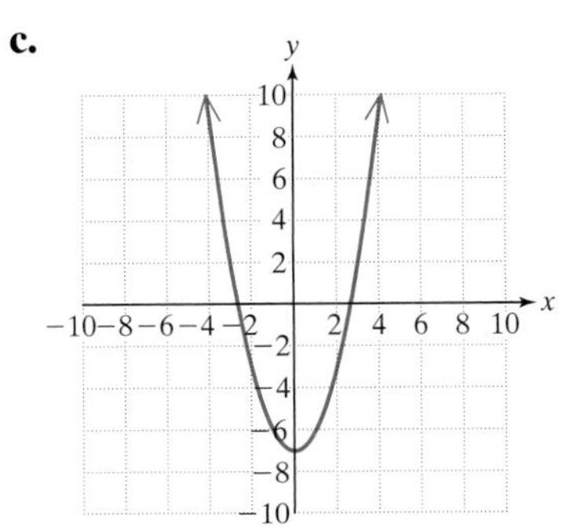

d.

39. What is the y-intercept of the function defined by $y = ax^2 + bx + c$?

40. What is the y-intercept of the function defined by $y = 4x^2 - 3x + 5$?

41. What is the domain of the quadratic function defined by $y = x^2 - x - 12$?

42. What is the domain of the quadratic function defined by $y = x^2 + 10x + 9$?

For Exercises 43–54,

- **a.** Determine whether the graph of the parabola opens upward or downward.
- **b.** Find the vertex.
- **c.** Find the x-intercept(s), if possible.
- **d.** Find the y-intercept.
- **e.** Sketch the function.
- **f.** Identify the domain of the function.
- **g.** Identify the range of the function.

43. $y = x^2 - 9$

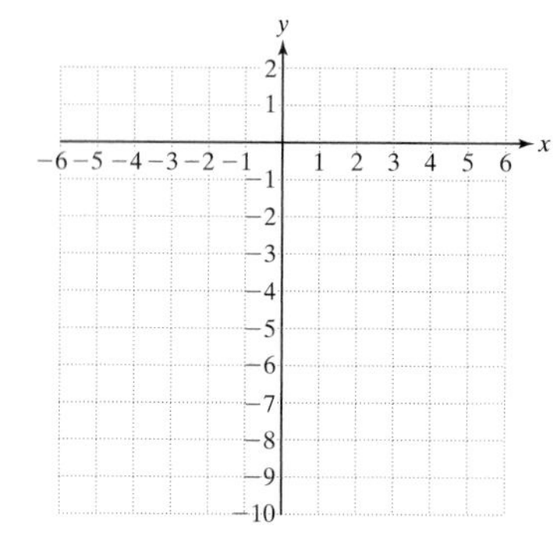

44. $y = x^2 - 4$

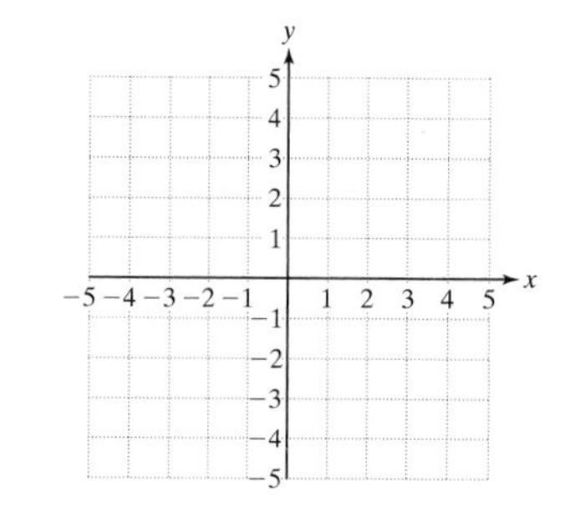

45. $y = x^2 - 2x - 8$

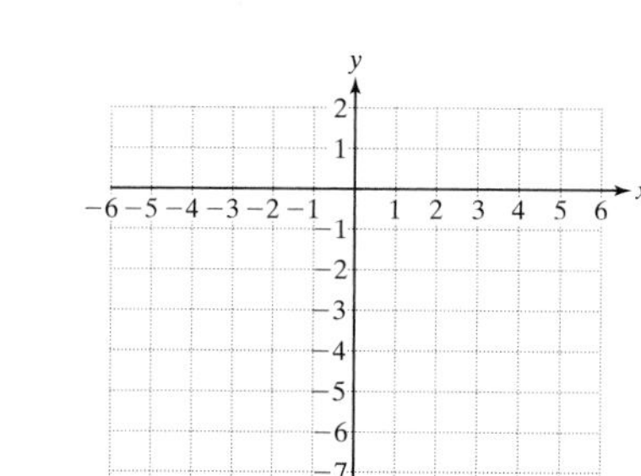

46. $y = x^2 + 2x - 24$

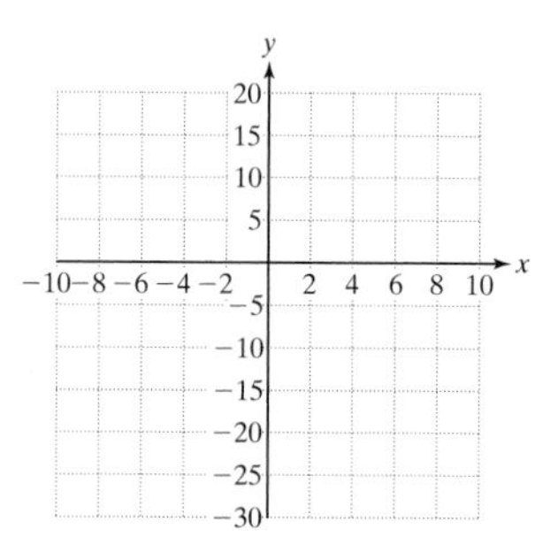

47. $y = -x^2 + 6x - 9$

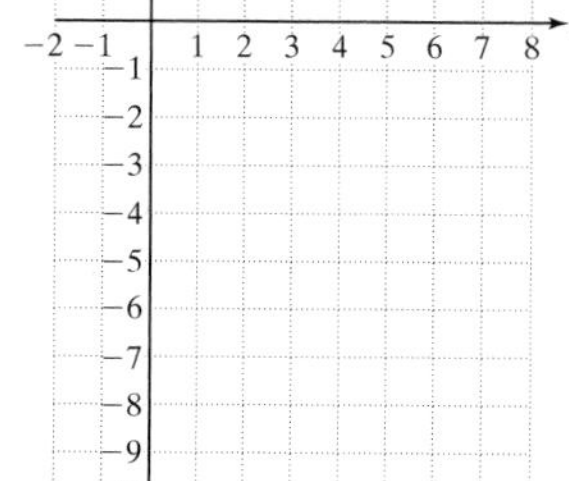

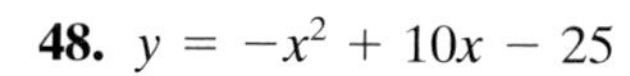

48. $y = -x^2 + 10x - 25$

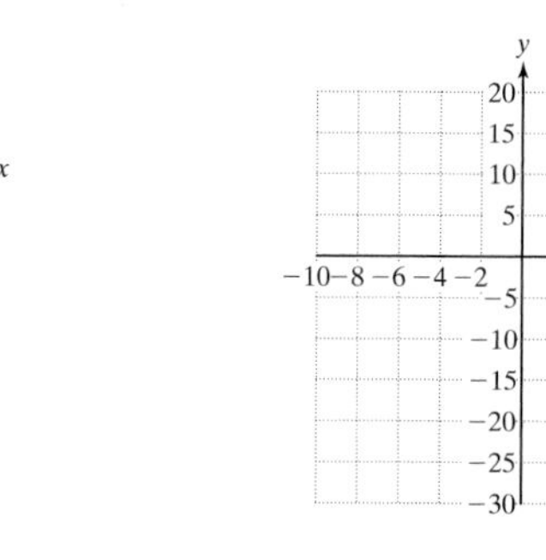

49. $y = -x^2 + 8x - 15$

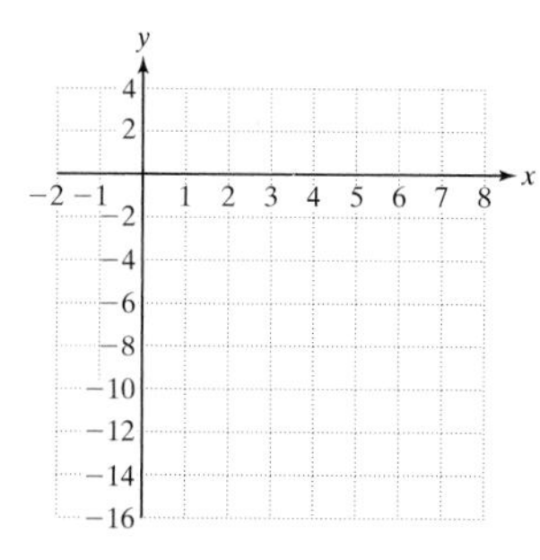

50. $y = -x^2 - 4x + 5$

51. $y = x^2 + 6x + 10$

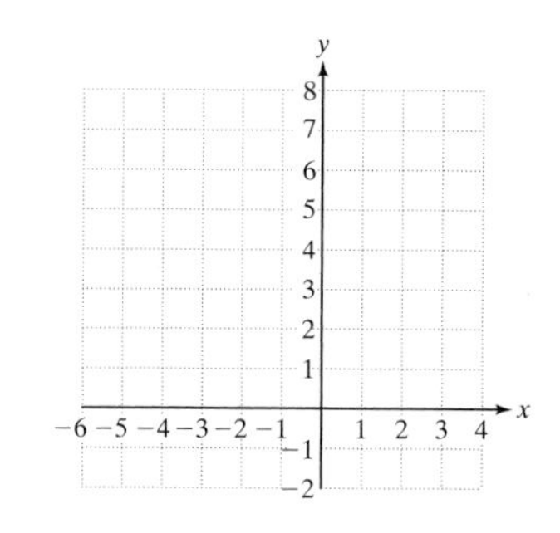

52. $y = x^2 + 4x + 5$

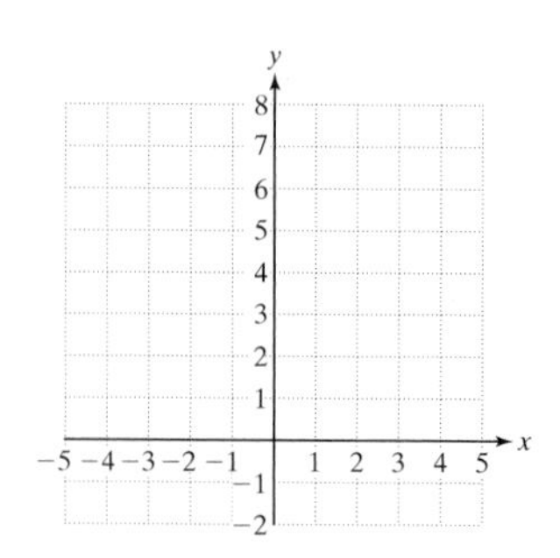

53. $y = -2x^2 - 2$

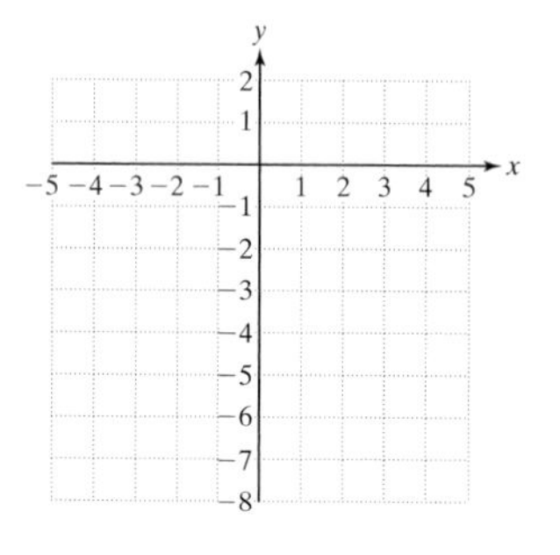

54. $y = -x^2 - 5$

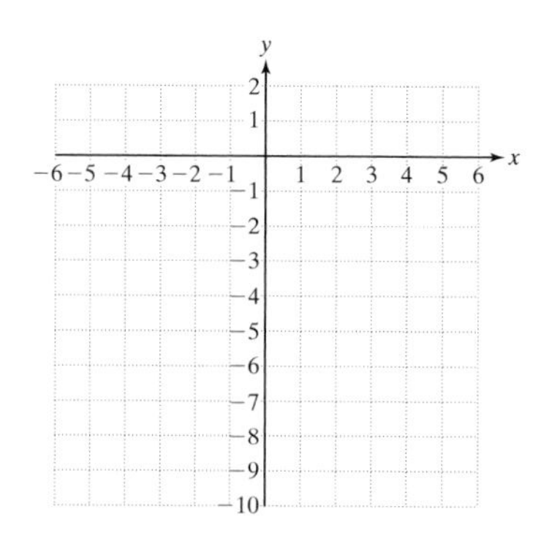

55. True or False: The function $y = -5x^2$ has a maximum value but no minimum value.

56. True or False: The graph of $y = -4x^2 + 9x - 6$ opens upward.

57. True or False: The graph of $y = 1.5x^2 - 6x - 3$ opens downward.

58. True or False: The function $y = 2x^2 - 5x + 4$ has a maximum value but no minimum value.

Concept 4: Applications of Quadratic Functions

59. A concession stand at the Arthur Ashe Tennis Center sells a hamburger/drink combination dinner for $5. The profit, y, (in dollars) can be approximated by

$$y = -0.001x^2 + 3.6x - 400 \quad \text{where } x \text{ is the number of dinners prepared.}$$

a. Find the number of dinners that should be prepared to maximize profit.

b. What is the maximum profit?

60. For a fund-raising activity, a charitable organization produces calendars to sell in the community. The profit, y, (in dollars) can be approximated by

$$y = -\frac{1}{40}x^2 + 10x - 500 \quad \text{where } x \text{ is the number of calendars produced.}$$

a. Find the number of calendars that should be produced to maximize profit.

b. What is the maximum profit?

61. The pressure, x, in an automobile tire can affect its wear. Both over-inflated and under-inflated tires can lead to poor performance and poor mileage. For one particular tire, the number of miles that a tire lasts, y, (in thousands) is given by

$$y = -0.875x^2 + 57.25x - 900 \quad \text{where } x \text{ is the tire pressure in pounds per square inch (psi).}$$

a. Find the tire pressure that will yield the maximum number of miles that a tire will last. Round to the nearest whole unit.

b. Find the maximum number of miles that a tire will last if the proper tire pressure is maintained. Round to the nearest whole unit.

62. A child kicks a ball into the air, and the height of the ball, y, (in feet) can be approximated by

$$y = -16t^2 + 40t + 3 \quad \text{where } t \text{ is the number of seconds after the ball was kicked.}$$

a. Find the maximum height of the ball.

b. How long will it take the ball to reach its maximum height?

Chapter 9 SUMMARY

Section 9.1 Introduction to Functions

Key Concepts

Any set of ordered pairs, (x, y), is called a **relation** in x and y.

The **domain** of a relation is the set of first components in the ordered pairs in the relation. The **range** of a relation is the set of second components in the ordered pairs.

Given a relation in x and y, we say "y is a **function** of x" if for every element x in the domain, there corresponds exactly one element y in the range.

Vertical Line Test for Functions

Consider any relation defined by a set of points (x, y) on a rectangular coordinate system. Then the graph defines y as a function of x if no vertical line intersects the graph in more than one point.

Function Notation

$f(x)$ is the value of the function, f, at x.

The domain of a function defined by $y = f(x)$ is the set of x-values that make the function a real number. In particular,

- Exclude values of x that make the denominator of a fraction zero.

Examples

Example 1

Find the domain and range of the relation.

$\{(0, 0), (1, 1), (2, 4), (3, 9), (-1, 1), (-2, 4), (-3, 9)\}$

Domain: $\{0, 1, 2, 3, -1, -2, -3\}$

Range: $\{0, 1, 4, 9\}$

Example 2

Function: $\{(1, 3), (2, 5), (6, 3)\}$

Nonfunction: $\{(1, 3), (2, 5), (1, -2)\}$

different y-values for the same x-value

Example 3

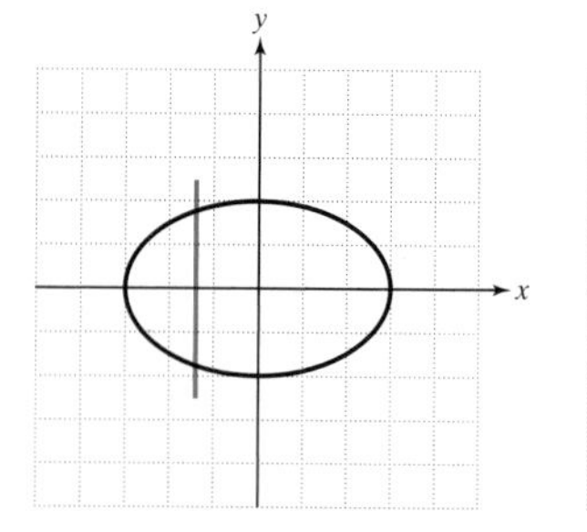

Not a Function
Vertical line intersects more than once.

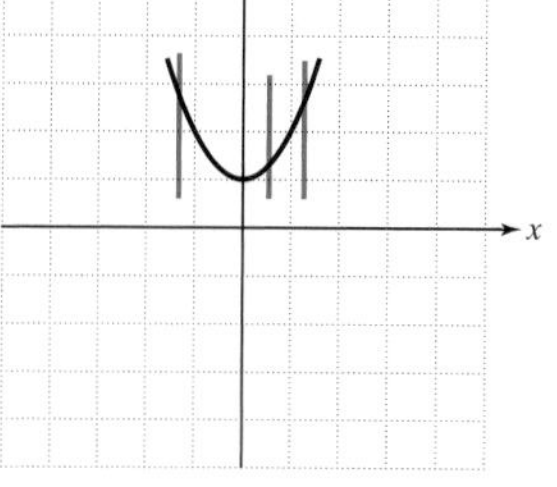

Function
No vertical line intersects more than once.

Example 4

Given $f(x) = -3x^2 + 5x$, find $f(-2)$.

$$\begin{aligned} f(-2) &= -3(-2)^2 + 5(-2) \\ &= -12 - 10 \\ &= -22 \end{aligned}$$

Example 5

Find the domain of the functions.

1. $f(x) = \dfrac{x + 4}{x - 5}$; $\{x \mid x \text{ is a real number and } x \neq 5\}$
2. $g(x) = -x^3 + 5x$; The set of all real numbers

Section 9.2 Complex Numbers

Key Concepts

$i = \sqrt{-1}$ and $i^2 = -1$

For a real number $b > 0$, $\sqrt{-b} = i\sqrt{b}$

A complex number is in the form $a + bi$, where a and b are real numbers. The value a is called the real part, and the value b is called the imaginary part.

To add or subtract complex numbers, combine the real parts and combine the imaginary parts.

Multiply complex numbers by using the distributive property.

Divide complex numbers by multiplying the numerator and denominator by the conjugate of the denominator.

Examples

Example 1

$$\sqrt{-4} \cdot \sqrt{-9} = (2i)(3i) = 6i^2 = -6$$

Example 2

$$(3 - 5i) - (2 + i) + (3 - 2i) = 3 - 5i - 2 - i + 3 - 2i = 4 - 8i$$

Example 3

$$(1 + 6i)(2 + 4i) = 2 + 4i + 12i + 24i^2 = 2 + 16i + 24(-1) = -22 + 16i$$

Example 4

$$\frac{3}{2 - 5i} = \frac{3}{2 - 5i} \cdot \frac{(2 + 5i)}{(2 + 5i)} = \frac{6 + 15i}{4 - 25i^2} = \frac{6 + 15i}{29} \quad \text{or} \quad \frac{6}{29} + \frac{15}{29}i$$

Section 9.3 The Square Root Property and Completing the Square

Key Concepts

Square Root Property

If $x^2 = k$, then $x = \pm\sqrt{k}$.

The square root property can be used to solve a quadratic equation written as a square of a binomial equal to a constant.

Examples

Example 1

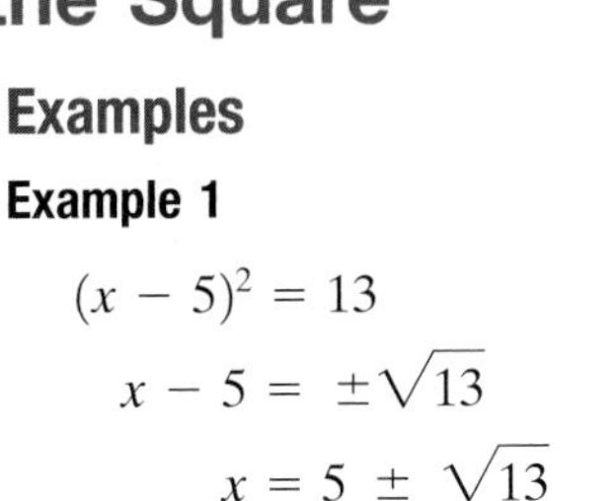

$(x - 5)^2 = 13$

$x - 5 = \pm\sqrt{13}$ Square root property

$x = 5 \pm \sqrt{13}$ Solve for x.

Solving a Quadratic Equation of the Form $ax^2 + bx + c = 0$ ($a \neq 0$) by Completing the Square and Applying the Square Root Property

1. Divide both sides by a to make the leading coefficient 1.
2. Isolate the variable terms on one side of the equation.
3. Complete the square (add the square of $\frac{1}{2}$ the linear term coefficient to both sides of the equation. Then factor the resulting perfect square trinomial).
4. Apply the square root property and solve for x.

Example 2

$2x^2 - 8x - 6 = 0$

Step 1: $\frac{2x^2}{2} - \frac{8x}{2} - \frac{6}{2} = \frac{0}{2}$

$x^2 - 4x - 3 = 0$

Step 2: $x^2 - 4x = 3$

Step 3: $x^2 - 4x + 4 = 3 + 4$ Note that $[\frac{1}{2}(-4)]^2 = (-2)^2 = 4$

$(x - 2)^2 = 7$

Step 4: $x - 2 = \pm\sqrt{7}$

$x = 2 \pm \sqrt{7}$

Section 9.4 Quadratic Formula

Key Concepts

The solutions to a quadratic equation of the form $ax^2 + bx + c = 0$ ($a \neq 0$) is given by the **quadratic formula**

$$x = \frac{-b \pm \sqrt{b^2 - 4ac}}{2a}$$

Three Methods for Solving a Quadratic Equation

1. Factoring and applying the zero product rule
2. Completing the square and applying the square root property
3. Using the quadratic formula

Examples

Example 1

$3x^2 = 2x + 4$

$3x^2 - 2x - 4 = 0 \qquad a = 3, b = -2, c = -4$

$$x = \frac{-(-2) \pm \sqrt{(-2)^2 - 4(3)(-4)}}{2(3)}$$

$$= \frac{2 \pm \sqrt{4 + 48}}{6}$$

$$= \frac{2 \pm \sqrt{52}}{6}$$

$$= \frac{2 \pm 2\sqrt{13}}{6}$$ Simplify the radical.

$$= \frac{2(1 \pm \sqrt{13})}{6}$$ Factor.

$$= \frac{1 \pm \sqrt{13}}{3}$$ Simplify.

Section 9.5 Graphing Quadratic Functions

Key Concepts

Let a, b, and c represent real numbers such that $a \neq 0$. Then a function in the form $y = ax^2 + bx + c$ is called a **quadratic function**.

The graph of a quadratic function is called a **parabola**.

The leading coefficient, a, of a quadratic function, $y = ax^2 + bx + c$, determines if the parabola will open upward or downward. If $a > 0$, then the parabola opens upward. If $a < 0$, then the parabola opens downward.

Finding the Vertex of a Parabola

1. The x-coordinate of the vertex of the parabola defined by $y = ax^2 + bx + c$ $(a \neq 0)$ is given by

$$x = \frac{-b}{2a}$$

2. To find the corresponding y-coordinate of the vertex, substitute the value of the x-coordinate found in step 1 and solve for y.

If a parabola opens upward, the vertex is the lowest point on the graph. If a parabola opens downward, the vertex is the highest point on the graph.

Examples

Example 1

$y = x^2 - 4x - 3$ is a quadratic function. Its graph is in the shape of a parabola.

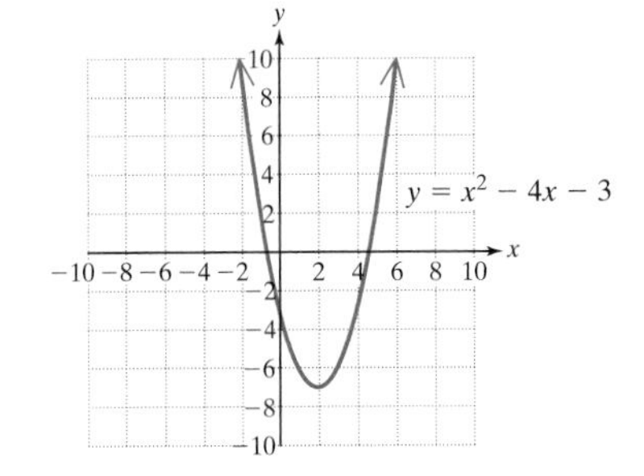

Example 2

Find the vertex of the parabola defined by $y = 3x^2 + 6x - 1$.

$$x = \frac{-b}{2a} = \frac{-6}{2 \cdot 3} = -1$$

$y = 3(-1)^2 + 6(-1) - 1 = -4$

The vertex is the point $(-1, -4)$.

For the equation $y = 3x^2 + 6x - 1$, $a = 3 > 0$. Therefore, the parabola opens upward so the vertex $(-1, -4)$ represents the lowest point of the graph.

Chapter 9 Review Exercises

Section 9.1

For Exercises 1–6, state the domain and range of each relation. Then determine whether the relation is a function.

1. $\{(6, 3), (10, 3), (-1, 3), (0, 3)\}$

2. $\{(2, 0), (2, 1), (2, -5), (2, 2)\}$

3.

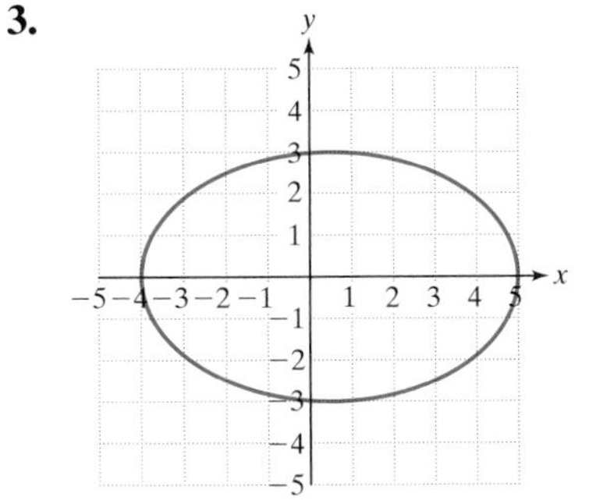

4.

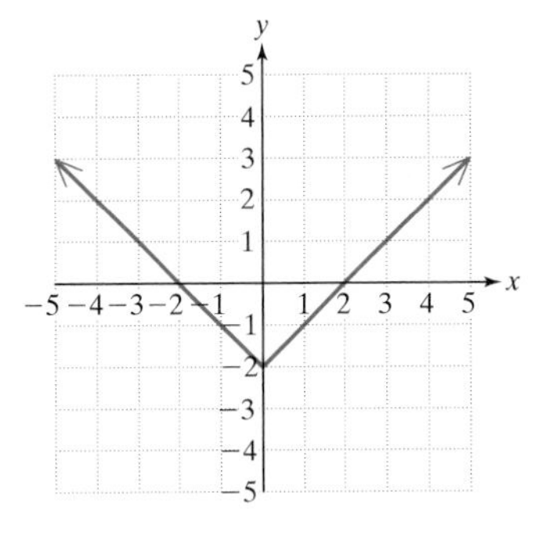

5. $\{(4, 23), (3, -2), (-6, 5), (4, 6)\}$

6. $\{(3, 0), (-4, \frac{1}{2}), (0, 3), (2, -12)\}$

7. Given the function defined by $f(x) = x^3$, find

a. $f(0)$ **b.** $f(2)$ **c.** $f(-3)$

d. $f(-1)$ **e.** $f(4)$ **f.** Write the domain in interval notation.

8. Given the function defined by $g(x) = \frac{x}{5 - x}$, find

a. $g(0)$ **b.** $g(4)$ **c.** $g(-1)$

d. $g(3)$ **e.** $g(-5)$ **f.** Write the domain in interval notation.

For Exercises 9–14, let h and k be defined by $h(x) = x^3 + 2x - 1$ and $k(x) = -4x$. Evaluate the following:

9. $h(-2) + k(8)$ **10.** $h(1) \cdot k(0)$

11. $k(m)$ **12.** $h(m)$

13. $h(2y)$ **14.** $k(-3a)$

15. The landing distance that a certain plane will travel on a runway is determined by the initial landing speed at the instant the plane touches down. The following function relates landing distance, $D(x)$, to initial landing speed, $x \geq 15$:

$$D(x) = \frac{1}{10}x^2 - 3x + 22$$ where D is in feet and x is in feet per second.

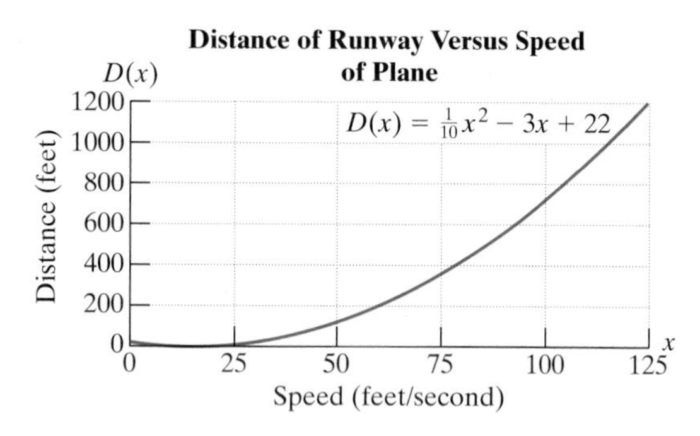

a. Find $D(90)$ and interpret the meaning of the function value in terms of landing speed and length of the runway.

b. Find $D(110)$ and interpret the meaning in terms of landing speed and length of the runway.

For Exercises 16–21, write the domain of each function in set-builder notation.

16. $f(x) = \frac{x + 1}{x - 8}$ **17.** $p(x) = x^4 - 16$

18. $h(x) = \frac{1}{x - 4}$ **19.** $k(x) = \frac{3}{(x - 1)(x + 2)}$

20. $g(x) = x^2 - 3$ **21.** $q(x) = \frac{2}{x + 1}$

Section 9.2

For Exercises 22–27, write the expression in terms of i and simplify.

22. $\sqrt{-36}$ **23.** $\sqrt{-5}$

24. $\sqrt{-32}$ **25.** $\sqrt{-10} \cdot \sqrt{10}$

26. $\sqrt{-25} \cdot \sqrt{-1}$ **27.** $\frac{\sqrt{-64}}{\sqrt{-4}}$

For Exercises 28–31, simplify the powers of i.

28. i^7 **29.** i^2 **30.** i^{20} **31.** i^{41}

For Exercises 32–43, perform the indicated operations. Write the answers in standard form, $a + bi$.

32. $(8 - 3i) - 9$ **33.** $(-2 - 4i) + (6 - 8i)$

34. $7i + (6 - 2i)$ **35.** $-2i \cdot 11i$

36. $-3i(-1 + 9i)$ **37.** $(3 + i)(2 + 4i)$

38. $(4 - 2i)^2$ **39.** $(5 - i)(5 + i)$

40. $\frac{9}{1 - 2i}$ **41.** $\frac{6i}{1 + 2i}$

42. $\frac{3 + 5i}{1 + i}$ **43.** $\frac{2 + 3i}{2 - 3i}$

Section 9.3

For Exercises 44–47, identify the equations as linear or quadratic.

44. $5x - 10 = 3x - 6$ **45.** $(x + 6)^2 = 6$

46. $x(x - 4) = 5x - 2$

47. $3(x + 6) = 18(x - 1)$

For Exercises 48–55, solve the equations using the square root property.

48. $x^2 = 25$ **49.** $x^2 - 19 = 0$

50. $x^2 + 49 = 0$ **51.** $x^2 = -48$

52. $(x + 1)^2 = 14$

53. $(x - 2)^2 = -64$

54. $\left(x - \frac{1}{8}\right)^2 = -\frac{3}{64}$

55. $(2x - 3)^2 = 20$

For Exercises 56–59, find the constant that should be added to each expression to make it a perfect square trinomial.

56. $x^2 + 12x$

57. $x^2 - 18x$

58. $x^2 - \frac{2}{3}x$

59. $x^2 + \frac{1}{7}x$

For Exercises 60–63, solve the quadratic equations by completing the square and applying the square root property.

60. $x^2 + 8x + 3 = 0$

61. $x^2 - 2x - 4 = 0$

62. $2x^2 - 6x + 5 = 0$

63. $3x^2 - 7x - 3 = 0$

64. An isosceles right triangle has legs of equal length. If the hypotenuse is 15 ft long, find the length of each leg. Round the answer to the nearest tenth of a foot.

65. A can in the shape of a right circular cylinder holds approximately 362 cm^3 of liquid. If the height of the can is 12.1 cm, find the radius of the can. Round to the nearest tenth of a centimeter. (*Hint:* The volume of a right circular cylinder is given by: $V = \pi r^2 h$)

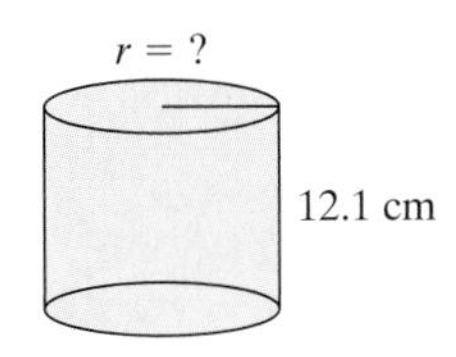

Section 9.4

66. Write the quadratic formula from memory.

For Exercises 67–76, solve the quadratic equations using the quadratic formula.

67. $5x^2 + x - 7 = 0$

68. $x^2 + 4x + 4 = 0$

69. $3x^2 - 2x + 2 = 0$

70. $2x^2 - x - 3 = 0$

71. $\frac{1}{8}x^2 + x = \frac{5}{2}$

72. $\frac{1}{6}x^2 + x + \frac{1}{3} = 0$

73. $0.01x^2 - 0.06x + 0.09 = 0$

74. $1.2x^2 + 6x = 7.2$

75. $(x + 6)(x + 2) = 10$

76. $(x - 1)(x - 7) = -18$

77. One number is two more than another number. Their product is 11.25. Find the numbers.

78. The base of a parallelogram is 1 cm longer than the height, and the area is 24 cm^2. Find the exact values of the base and height of the parallelogram. Then use a calculator to approximate the values to the nearest tenth of a centimeter.

79. An astronaut on the moon tosses a rock upward with an initial velocity of 25 ft/sec. The height of the rock, $h(t)$ (in feet), is determined by the number of seconds, t, after the rock is released according to the equation

$$h(t) = -2.7t^2 + 25t + 5$$

Find the time required for the rock to hit the ground. (*Hint:* At ground level, $h(t) = 0$.) Round to the nearest tenth of a second.

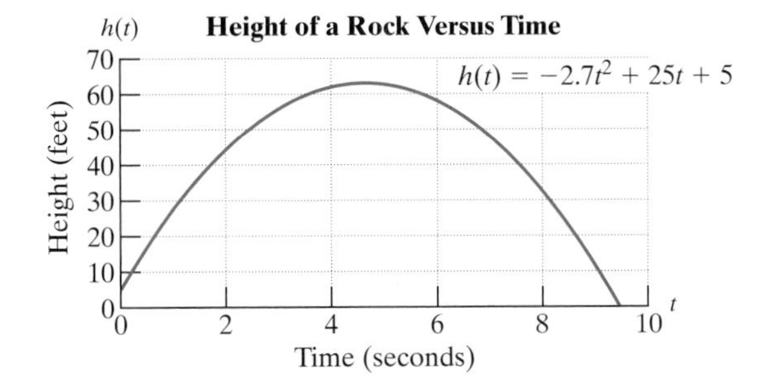

Section 9.5

For Exercises 80–85, find the x- and y-intercepts and match the equation with a graph. Round to the nearest hundredth where necessary.

80. $y = x^2 - 12x + 25$ **81.** $y = x^2 + 6x + 4$

82. $y = -5x^2 + 5$ **83.** $y = -3x^2 + 3$

84. $y = 2x^2 + 8x$ **85.** $y = x^2 - 2x + 4$

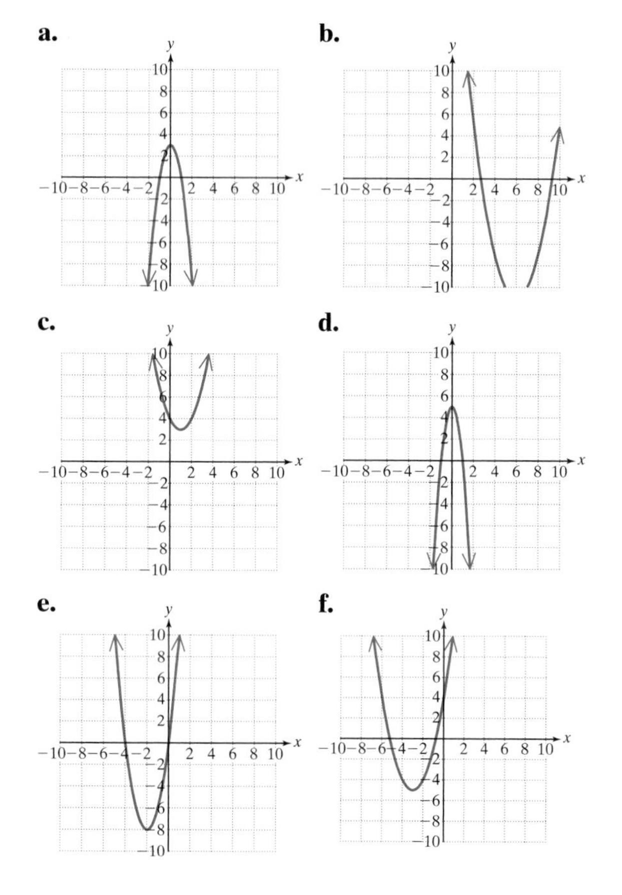

For Exercises 86–89, identify a and determine if the parabola opens upward or downward.

86. $y = x^2 - 3x + 1$ **87.** $y = -x^2 + 8x + 2$

88. $y = -2x^2 + x - 12$ **89.** $y = 5x^2 - 2x - 6$

For Exercises 90–93, find the vertex.

90. $y = 3x^2 + 6x + 4$ **91.** $y = -x^2 + 8x + 3$

92. $y = -2x^2 + 12x - 5$ **93.** $y = 2x^2 + 2x - 1$

For Exercises 94–97,

a. Determine whether the graph of the parabola opens upward or downward.

b. Find the vertex.

c. Find the x-intercept(s), if possible. Round to two decimal places if necessary.

d. Find the y-intercept.

e. Sketch the function.

94. $y = 2x^2 + 4x - 1$

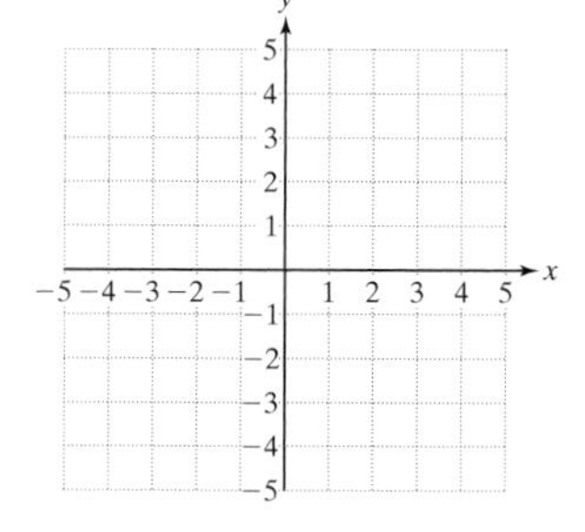

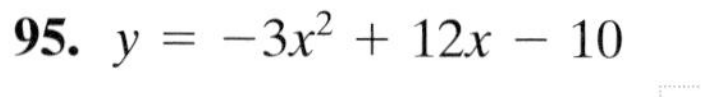

95. $y = -3x^2 + 12x - 10$

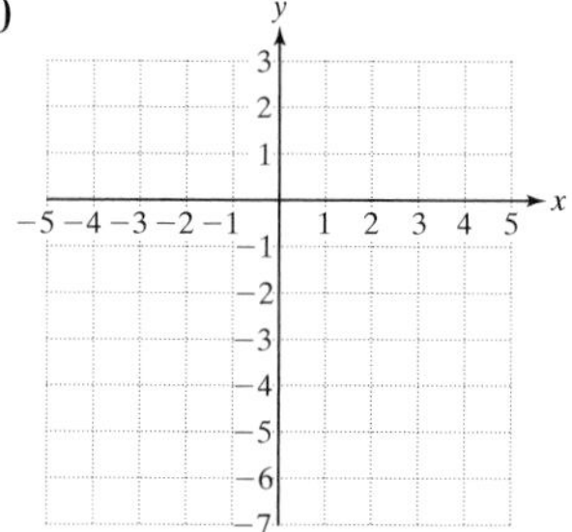

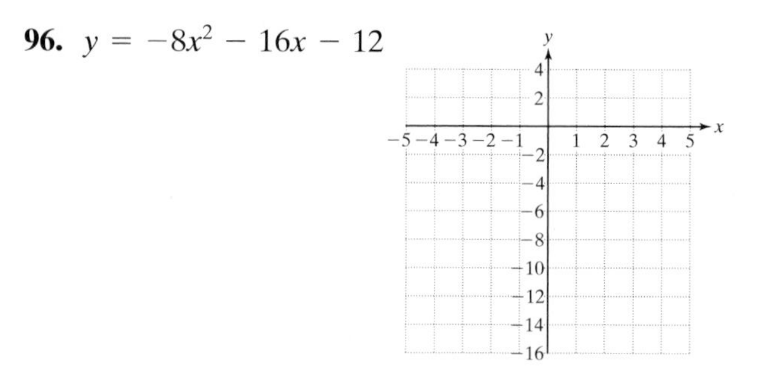

96. $y = -8x^2 - 16x - 12$

97. $y = 3x^2 + 12x + 9$

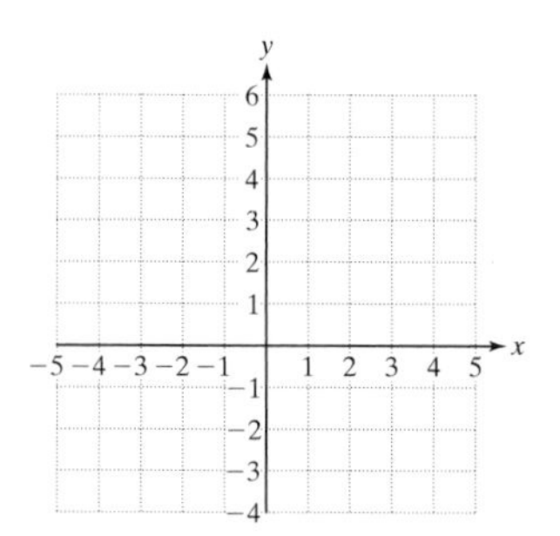

98. An object is launched into the air from ground level with an initial velocity of 256 ft/sec. The height of the object, y (in feet), can be approximated by the function

$y = -16t^2 + 256t$ where t is the number of seconds after launch.

a. Find the maximum height of the object.

b. Find the time required for the object to reach its maximum height.

Chapter 9 Test

1. Write the domain and range for each relation in interval notation. Then determine if the relation is a function.

a.

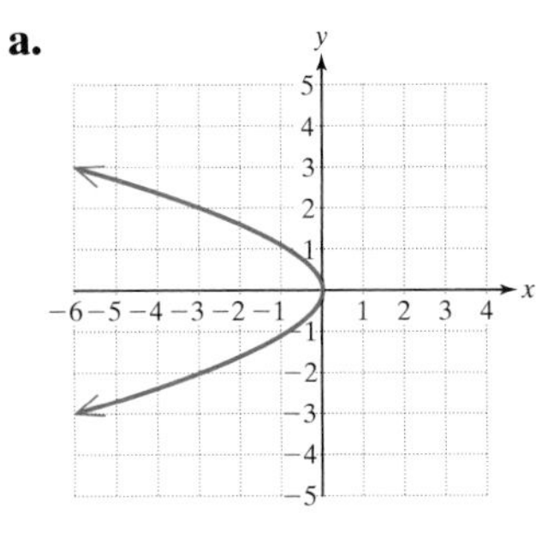

b.

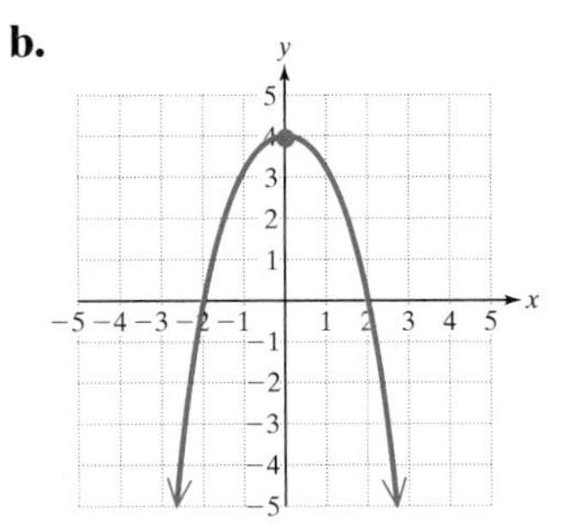

2. For the function defined by $f(x) = \dfrac{1}{x + 2}$

a. Find the function values: $f(0), f(-2), f(6)$.

b. Write the domain of f.

3. The number of diagonals, D, of a polygon is a function of the number of sides, x, of the polygon according to the equation

$$D(x) = \frac{1}{2}x(x - 3)$$

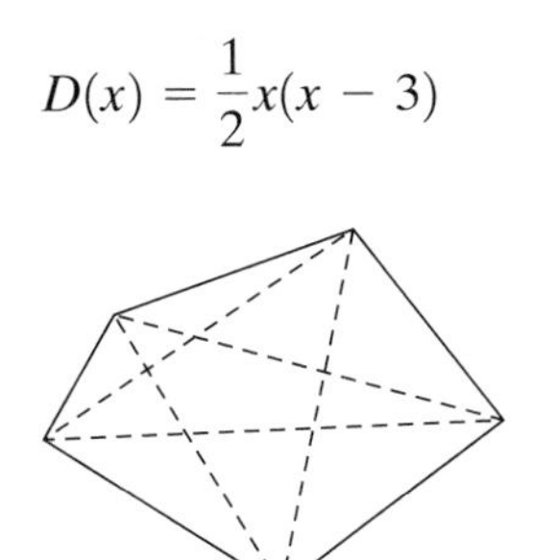

a. Find $D(5)$ and interpret the meaning of the function value. Verify your answer by counting the number of diagonals in the pentagon in the figure.

b. Find $D(10)$ and interpret its meaning.

c. If a polygon has 20 diagonals, how many sides does it have? (*Hint:* Substitute $D(x) = 20$ and solve for x. Try clearing fractions first.)

For Exercises 4–6, simplify the expressions in terms of i.

4. $\sqrt{-100}$ **5.** $\sqrt{-23}$ **6.** $\sqrt{-9} \cdot \sqrt{-49}$

For Exercises 7–8, simplify the powers of i.

7. i^{13} **8.** i^{35}

For Exercises 9–12, perform the indicated operation. Write the answer in standard form, $a + bi$.

9. $(2 - 7i) - (-3 - 4i)$ **10.** $(8 + i)(-2 - 3i)$

11. $(10 - 11i)(10 + 11i)$ **12.** $\dfrac{1}{10 - 11i}$

13. Solve the equation by applying the square root property.

$$(3x + 1)^2 = -14$$

14. Solve the equation by completing the square and applying the square root property.

$$x^2 - 8x - 5 = 0$$

15. Solve the equation by using the quadratic formula.

$$3x^2 - 5x = -1$$

For Exercises 16–22, solve the equations using any method.

16. $5x^2 + x - 2 = 0$ **17.** $(c - 12)^2 = 12$

18. $y^2 + 14y - 1 = 0$ **19.** $3t^2 = 30$

20. $4x(3x + 2) = 15$ **21.** $6p^2 - 11p = 0$

22. $\frac{1}{4}x^2 - \frac{3}{2}x = \frac{11}{4}$

23. The surface area, S, of a sphere is given by the formula $S = 4\pi r^2$, where r is the radius of the sphere. Find the radius of a sphere whose surface area is 201 in.^{2} Round to the nearest tenth of an inch.

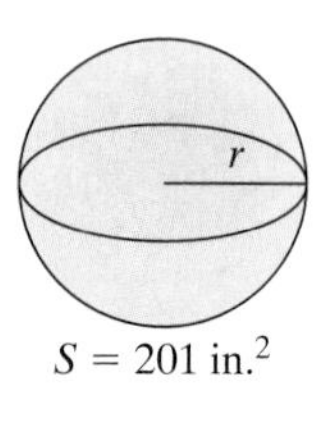

24. The height of a triangle is 2 m longer than twice the base, and the area is 24 m^2. Find the exact values of the base and height. Then use a calculator to approximate the base and height to the nearest tenth of a meter.

25. Explain how to determine if a parabola opens upward or downward.

For Exercises 26–28, find the vertex of the parabola.

26. $y = x^2 - 10x + 25$ **27.** $y = 3x^2 - 6x + 8$

28. $y = -x^2 - 16$

29. Suppose a parabola opens upward and the vertex is located at (24, 3). How many x-intercepts does the parabola have?

30. Given the parabola, $y = x^2 + 6x + 8$

a. Determine whether the parabola opens upward or downward.

b. Find the vertex of the parabola.

c. Find the x-intercepts.

d. Find the y-intercept.

e. Graph the parabola.

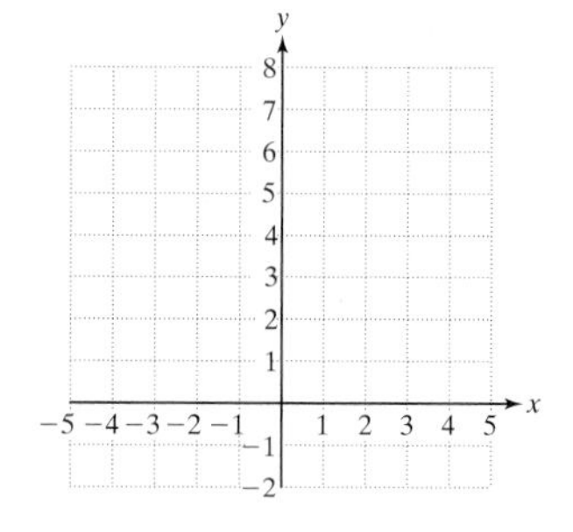

31. Graph the parabola and label the vertex, x-intercepts, and y-intercept.

$$y = -x^2 + 25$$

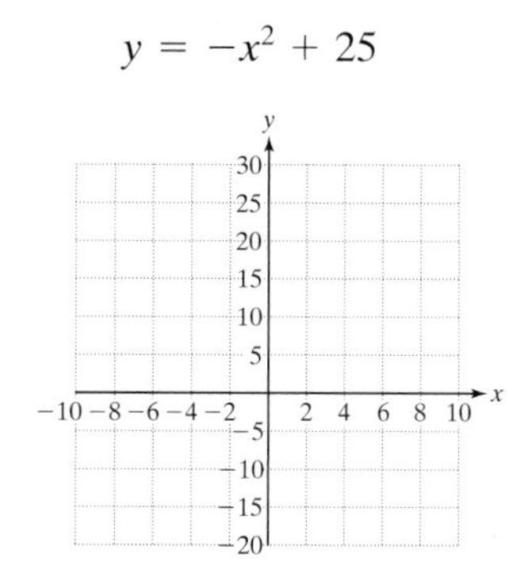

32. The Phelps Arena in Atlanta holds 20,000 seats. If the Atlanta Hawks charge x dollars per ticket for a game, then the total revenue, y (in dollars), can be approximated by

$y = -400x^2 + 20{,}000x$ where x is the price per ticket.

a. Find the ticket price that will produce the maximum revenue.

b. What is the maximum revenue?

Chapters 1–9 Cumulative Review Exercises

1. Solve for x: $3x - 5 = 2(x - 2)$

2. Solve for h: $A = \frac{1}{2}bh$

3. Solve for y: $\frac{1}{2}y - \frac{5}{6} = \frac{1}{4}y + 2$

4. Determine whether $x = 2$ is a solution to the inequality: $-3x + 4 < x + 8$

5. Graph the solution to the inequality: $-3x + 4 < x + 8$. Then write the solution in set-builder notation and in interval notation.

6. The graph depicts the death rate from 60 U.S. cities versus the median education level of the people living in that city. The death rate, y, is measured in number of deaths per 100,000 people. The median education level, x, is a type of "average" and is measured by grade level. (*Source:* U.S. Bureau of the Census)

   The death rate can be predicted from the median education level according to the equation

$$y = -37.6x + 1353 \qquad \text{where } 8 \le x \le 13$$

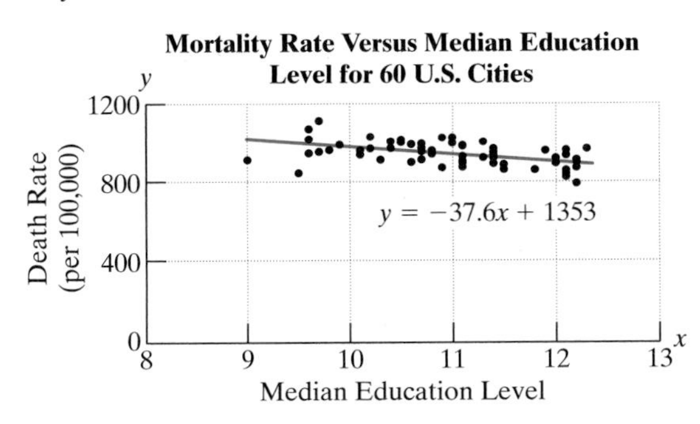

   a. From the graph, does it appear that the death rate increases or decreases as the median education level increases?

   b. What is the slope of the line? Interpret the slope in the context of the death rate and education level.

   c. For a city in the United States with a median education level of 12, what would be the expected death rate?

   d. If the death rate of a certain city is 977 per 100,000 people, what would be the approximate median education level?

7. Simplify completely. Write the final answer with positive exponents only:

$$\left(\frac{2a^2b^{-3}}{c}\right)^{-1} \cdot \left(\frac{4a^{-1}}{b^2}\right)^2$$

8. Approximately 5.2×10^7 disposable diapers are thrown into the trash each day in the United States and Canada. How many diapers are thrown away each year?

9. The distance between Earth and the star, Polaris, is approximately 3.83×10^{15} miles. If 1 light-year is approximately 5.88×10^{12} miles, how many light-years is Polaris from Earth?

10. Perform the indicated operations.
    $(2x - 3)^2 - 4(x - 1)$

11. Divide using long division.
    $(2y^4 - 4y^3 + y - 5) \div (y - 2)$

12. Factor. $2x^2 - 9x - 35$

13. Factor completely. $2xy + 8xa - 3by - 12ab$

14. The base of a triangle is 1 m more than the height. If the area is 36 m^2, find the base and height.

15. Multiply.

$$\frac{x^2 + 10x + 9}{x^2 - 81} \cdot \frac{18 - 2x}{x^2 + 2x + 1}$$

16. Find the domain of the expression

$$\frac{5}{x^2 - 4}?$$

17. Perform the indicated operations.

$$\frac{x^2}{x - 5} - \frac{10x - 25}{x - 5}$$

18. Simplify completely.

$$\frac{\dfrac{1}{x + 1} - \dfrac{1}{x - 1}}{\dfrac{x}{x^2 - 1}}$$

19. Solve for y.

$$1 - \frac{1}{y} = \frac{12}{y^2}$$

20. Solve for b.

$$4b^4 - 2b^3 - 4b^2 + 2b = 0$$

21. Write an equation of the line passing through the point $(-2, 3)$ and having a slope of $\frac{1}{2}$. Write the final answer in slope-intercept form.

For Exercises 22–23,

a. Find the x-intercept (if it exists).

b. Find the y-intercept (if it exists).

c. Find the slope (if it exists).

d. Graph the line.

22. $2x - 4y = 12$

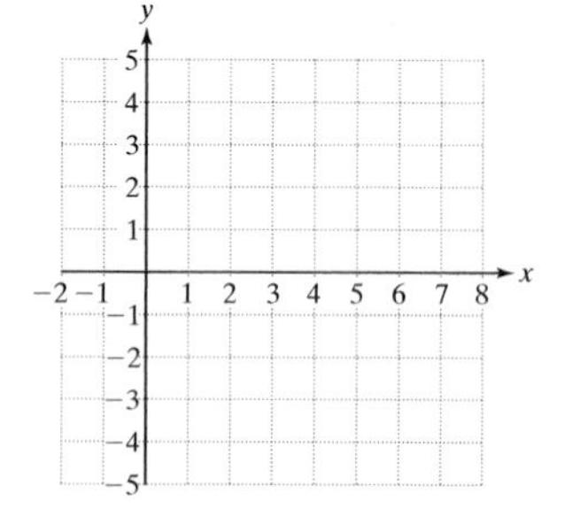

23. $4x + 12 = 0$

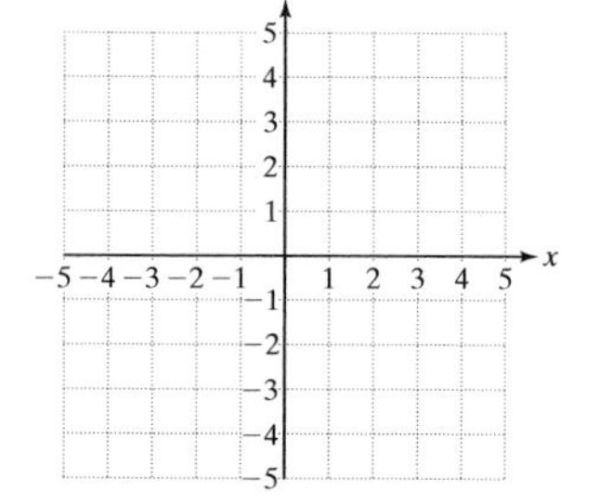

24. Solve the system by using the addition method. If the system has no solution or infinitely many solutions, so state.

$$\frac{1}{2}x - \frac{1}{4}y = \frac{1}{6}$$

$$12x - 6y = 8$$

25. Solve the system by using the substitution method. If the system has no solution or infinitely many solutions, so state.

$$2x - y = 8$$

$$4x - 4y = 3x - 3$$

26. In a right triangle, one acute angle is 2° more than three times the other acute angle. Find the measure of each angle.

27. A bank of 27 coins contains only dimes and quarters. The total value of the coins is \$4.80. Find the number of dimes and the number of quarters.

28. Sketch the inequality. $x - y \leq 4$

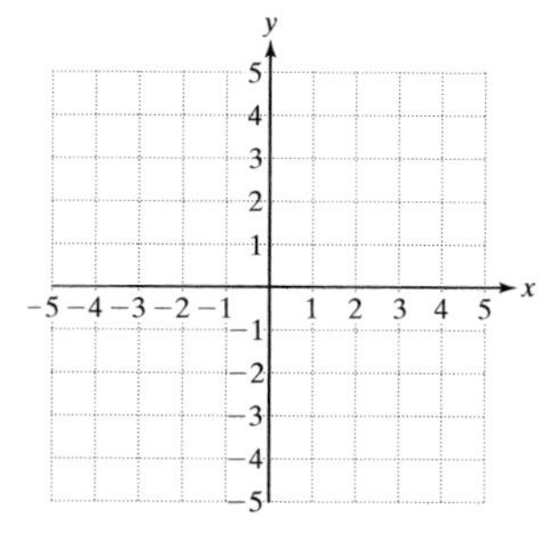

29. Which of the following are irrational numbers? $\{0, -\frac{2}{3}, \pi, \sqrt{7}, 1.2, \sqrt{25}\}$

For Exercises 30–31, simplify the radicals.

30. $\sqrt{\dfrac{1}{7}}$

31. $\dfrac{\sqrt{16x^4}}{\sqrt{2x}}$

32. Perform the indicated operation. $(4\sqrt{3} + \sqrt{x})^2$

33. Add the radicals. $-3\sqrt{2x} + \sqrt{50x}$

34. Rationalize the denominator.

$$\frac{4}{2 - \sqrt{a}}$$

35. Solve for x. $\sqrt{x + 11} = x + 5$

36. Factor completely. $8c^3 - y^3$

37. Which graph defines y as a function of x?

a. **b.**

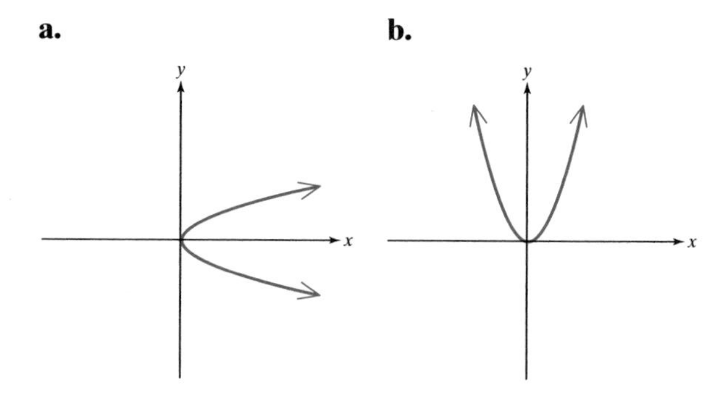

38. Given the functions defined by $f(x) = -\frac{1}{2}x + 4$ and $g(x) = x^2$, find

a. $f(6)$

b. $g(-2)$

c. $f(0) + g(3)$

39. Find the domain and range of the function. $\{(2, 4), (-1, 3), (9, 2), (-6, 8)\}$

40. Find the slope of the line passing through the points $(3, -1)$ and $(-4, -6)$.

41. Find the slope of the line defined by $-4x - 5y = 10$.

42. What value of k would make the expression a perfect square trinomial?

$$x^2 + 10x + k$$

43. Solve the quadratic equation by completing the square and applying the square root property. $2x^2 + 12x + 6 = 0$.

44. Solve the quadratic equation by using the quadratic formula. $2x^2 + 12x + 6 = 0$

45. Graph the parabola defined by the equation. Label the vertex, x-intercepts, and y-intercept.

$$y = x^2 + 4x + 4$$

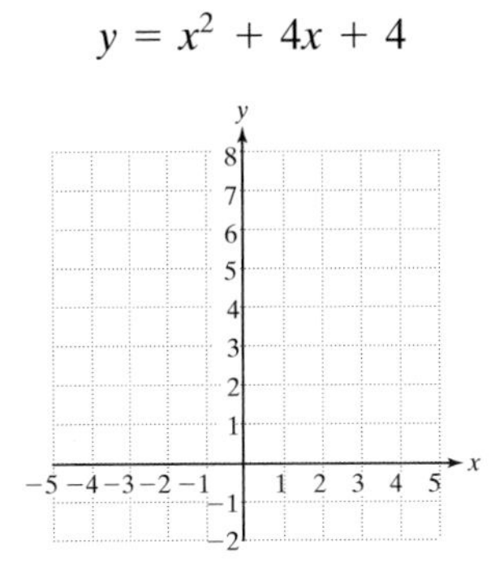

Student Answer Appendix

Chapter R

Chapter Opener Puzzles

p	o	i	n	t	l	e	s	s
5	3	8	1	6	9	2	7	4

Section R.2 Practice Exercises, pp. 18–22

1. Numerator: 7; denominator: 8; proper
3. Numerator: 9; denominator: 5; improper
5. Numerator: 6; denominator: 6; improper
7. Numerator: 12; denominator: 1; improper
9. $\frac{3}{4}$ **11.** $\frac{4}{3}$ **13.** $\frac{1}{6}$ **15.** $\frac{2}{2}$ **17.** $\frac{5}{2}$ or $2\frac{1}{2}$
19. $\frac{6}{2}$ or 3 **21.** The set of whole numbers includes the number 0 and the set of natural numbers does not.
23. For example: $\frac{2}{4}$ **25.** Prime **27.** Composite
29. Composite **31.** Prime **33.** $2 \times 2 \times 3 \times 3$
35. $2 \times 3 \times 7$ **37.** $2 \times 5 \times 11$ **39.** $3 \times 3 \times 3 \times 5$
41. $\frac{1}{5}$ **43.** $\frac{3}{8}$ **45.** $\frac{7}{8}$ **47.** $\frac{3}{4}$ **49.** $\frac{5}{8}$ **51.** $\frac{3}{4}$
53. False: When adding or subtracting fractions, it is necessary to have a common denominator. **55.** $\frac{4}{3}$
57. $\frac{2}{3}$ **59.** $\frac{9}{2}$ **61.** $\frac{3}{5}$ **63.** $\frac{5}{3}$ **65.** $\frac{90}{13}$
67. \$300 **69.** 4 graduated with honors **71.** 8 aprons
73. 8 jars **75.** $\frac{3}{7}$ **77.** $\frac{1}{2}$ **79.** 24 **81.** 40
83. 90 **85.** $\frac{7}{8}$ **87.** $\frac{3}{40}$ **89.** $\frac{3}{26}$ **91.** $\frac{29}{36}$
93. $\frac{7}{10}$ **95.** $\frac{35}{48}$ **97.** $\frac{37}{24}$ or $1\frac{13}{24}$ **99.** $\frac{51}{28}$ or $1\frac{23}{28}$
101. 46 **103.** $\frac{14}{5}$ or $2\frac{4}{5}$ **105.** $\frac{11}{54}$ **107.** $8\frac{19}{24}$ in.
109. $\frac{16}{7}$ or $2\frac{2}{7}$ **111.** $\frac{3}{2}$ or $1\frac{1}{2}$ **113.** $\frac{43}{6}$ or $7\frac{1}{6}$
115. $\frac{11}{7}$ or $1\frac{4}{7}$ **117.** $1\frac{7}{12}$ hr **119.** $2\frac{1}{4}$ lb
121. $\frac{3}{8}$ in. **123.** $4\frac{3}{4}$ in.

Section R.3 Calculator Connections, p. 29

1. $0.\overline{4}$ **2.** $0.\overline{63}$ **3.** $0.1\overline{36}$ **4.** $0.\overline{384615}$

1.–2.
```
4/9
         .4444444444
7/11
         .6363636364
```

3.–4.
```
3/22
         .1363636364
5/13
         .3846153846
```

Section R.3 Practice Exercises, pp. 29–31

1. Tens **3.** Hundreds **5.** Tenths **7.** Hundredths
9. Ones **11.** Ten-thousandths **13.** No, the symbols I, V, X, and so on each represent certain values but the values are not dependent on the position of the symbol within the number. **15.** 0.9 **17.** 0.12 **19.** $1.\overline{7}$
21. $0.\overline{27}$ **23.** $\frac{13}{20}$ **25.** $\frac{273}{1000}$ **27.** $\frac{301}{50}$ or $6\frac{1}{50}$
29. $\frac{12{,}003}{1000}$ or $12\frac{3}{1000}$ **31.** $\frac{8}{9}$ **33.** $\frac{7}{3}$ or $2\frac{1}{3}$
35. $0.8, \frac{4}{5}$ **37.** $0.25, \frac{1}{4}$ **39.** $0.045, \frac{9}{200}$
41. $0.058, \frac{29}{500}$ **43.** $1.40, \frac{7}{5}$
45. Replace the % symbol with $\div$ 100, by $\times \frac{1}{100}$, or by $\times$ 0.01. **47.** 6% **49.** 70% **51.** 480%
53. 930% **55.** 53.6% **57.** 0.2% **59.** 46%
61. 175% **63.** 12.5% **65.** 43.75% **67.** $26.\overline{6}$%
69. $27.\overline{7}$% **71.** 12,096 students **73.** \$3375
75. 7% **77.** \$792 **79.** \$192 **81.** \$67,500
83. 12 questions

Section R.4 Practice Exercises, pp. 41–48

1. b, e, i **3.** 108 cm **5.** 1 ft **7.** $11\frac{1}{2}$ in.
9. 31.4 ft **11.** a, f, g **13.** 40 ft^2 **15.** 37.21 in.^{2}
17. 0.0004 m^2 **19.** 40 mi^2 **21.** 132.665 cm^2
23. 66 in.^{2} **25.** 6 km^2 **27.** 212.00652 cm^3
29. 39 in.^{3} **31.** 113.04 cm^3 **33.** 1695.6 cm^3
35. 3052.08 in.^{3} **37.** 113.04 cm^3 **39.** 32.768 ft^3
41. a. \$0.31/ft^2 **b.** \$129 **43.** Perimeter
45. a. 50.24 in.^{2} **b.** 113.04 in.^{2} **c.** One 12-in. pizza
47. 289.3824 cm^3 **49.** True **51.** True **53.** True
55. True **57.** 45° **59.** Not possible **61.** For example, 100°, 80° **63. a.** $\angle 1$ and $\angle 3$, $\angle 2$ and $\angle 4$
b. $\angle 1$ and $\angle 2$, $\angle 2$ and $\angle 3$, $\angle 3$ and $\angle 4$, $\angle 1$ and $\angle 4$
c. $m(\angle 1) = 100°, m(\angle 2) = 80°, m(\angle 3) = 100°$
65. 57° **67.** 78° **69.** 147° **71.** 58° **73.** 7
75. 1 **77.** 1 **79.** 5
81. $m(\angle a) = 45°, m(\angle b) = 135°, m(\angle c) = 45°, m(\angle d) = 135°, m(\angle e) = 45°, m(\angle f) = 135°, m(\angle g) = 45°$
83. Scalene **85.** Isosceles
87. No, a 90° angle plus an angle greater than 90° would make the sum of the angles greater than 180°.
89. 40° **91.** 37°
93. $m(\angle a) = 80°$, $m(\angle b) = 80°$, $m(\angle c) = 100°$, $m(\angle d) = 100°$, $m(\angle e) = 65°$, $m(\angle f) = 115°$, $m(\angle g) = 115°$, $m(\angle h) = 35°$, $m(\angle i) = 145°$, $m(\angle j) = 145°$
95. $m(\angle a) = 70°$, $m(\angle b) = 65°$, $m(\angle c) = 65°$, $m(\angle d) = 110°$, $m(\angle e) = 70°$, $m(\angle f) = 110°$, $m(\angle g) = 115°$, $m(\angle h) = 115°$, $m(\angle i) = 65°$, $m(\angle j) = 70°$, $m(\angle k) = 65°$
97. 82 ft **99.** 36 in.^{2} **101.** 15.2464 cm^2

Chapter 1

Chapter Opener Puzzle

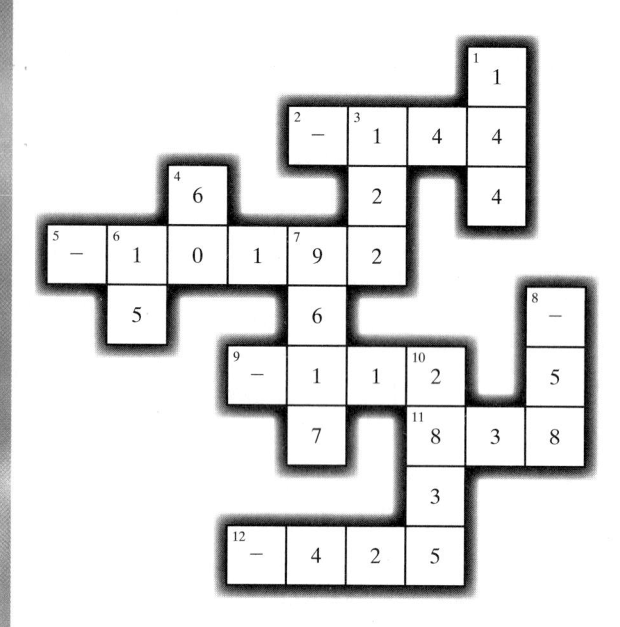

Section 1.1 Practice Exercises, pp. 58–61

3. $-\pi$, $-\frac{5}{2}$, -2, 0, 1, 5.1

−6 −5 −4 −3 −2 −1 0 1 2 3 4 5 6

5. a **7.** b **9.** a **11.** c **13.** a **15.** a **17.** b **19.** c

21. $0.29, 3.8, \frac{1}{9}, \frac{1}{3}, \frac{1}{8}, \frac{1}{5}, -0.125, -3.24, -3, -6, 0.\overline{2}, 0.\overline{6}$

23. For example: $\pi, -\sqrt{2}, \sqrt{3}$

25. For example: $-4, -1, 0$

27. For example: $-\frac{3}{4}, \frac{1}{2}, 0.206$ **29.** $-\frac{3}{2}, -4, 0.\overline{6}, 0, 1$

31. 1 **33.** $-4, 0, 1$

35. a. $>$ **b.** $>$ **c.** $<$ **d.** $>$

37. -18 **39.** 6.1 **41.** $\frac{5}{8}$ **43.** $-\frac{7}{3}$ **45.** 3

47. $-\frac{7}{3}$ **49.** 8 **51.** 2 **53.** 1.5 **55.** -1.5

57. $\frac{3}{2}$ **59.** False, $|n|$ is never negative.

61. True **63.** False **65.** True **67.** False **69.** False **71.** False **73.** True **75.** True **77.** False **79.** True **81.** True **83.** True

85. For all $a < 0$

Section 1.2 Calculator Connections, pp. 67–68

1. 2 **2.** 91 **3.** 84

1.–3.
```
(4+6)/(8-3)
                2
110-5*(2+1)-4
               91
100-2*(5-3)^3
               84
```

4. 12 **5.** 49 **6.** 18 **7.** 4 **8.** 27 **9.** 0.5

4.–6.
```
3+(4-1)²
               12
(12-6+1)²
               49
3*8-√(32+2²)
               18
```

7.–9.
```
√(18-2)
                4
(4*3-3*3)^3
               27
(20-3²)/(26-2²)
               .5
```

Section 1.2 Practice Exercises, pp. 68–71

3. $-4, 5.\overline{6}, 0, 4.02, \frac{7}{9}$ **5.** 9.2 **7.** 34.2

9. 15 **11.** 20 **13.** 11 **15.** 8 **17.** $\frac{2}{9}$

19. 4.8 **21.** 1040 ft **23.** 1000 yd^3 **25.** 10^6

27. $7x^2y^2$ **29.** $3wz^4$ **31.** $\left(\frac{2}{3}t\right)^3$

33. a. y **b.** Yes, 1 **35.** $y \cdot y \cdot y \cdot y$ **37.** $8c \cdot 8c$

39. $x \cdot x \cdot y \cdot y \cdot y$ **41.** $3 \cdot a \cdot a \cdot a \cdot b$ **43.** 64

45. $\frac{1}{32}$ **47.** 0.64 **49.** 169 **51.** 8 **53.** 3

55. 7 **57.** 6 **59.** $\frac{1}{8}$ **61.** $\frac{7}{10}$ **63.** 19

65. 40 **67.** 7 **69.** $\frac{7}{6}$ **71.** $\frac{4}{3}$ **73.** 20

75. 21 **77.** 7 **79.** 17 **81.** 2 **83.** $\frac{7}{2}$

85. $\frac{9}{37}$ **87.** 225 **89.** 44 **91.** $b + 6$

93. $\frac{4}{k}$ or $4 \div k$ **95.** $t - 3$ **97.** $9 - 3p$

99. $2(x - 3)$ **101.** $t - 14$ **103.** 300 **105.** 1

107. 5 **109.** 2 **111.** 32

113. The difference of 18 and x

115. The sum of y and 12

117. One more than the product of 7 and x

119. 6 cubed **121.** The square root of 10

123. 10 squared

125. Multiplication or division are performed in order from left to right. Addition or subtraction are performed in order from left to right.

127. 1 **129.** $\frac{1}{4}$

Section 1.3 Practice Exercises, pp. 77–79

3. $>$ **5.** $>$ **7.** $>$ **9.** -8 **11.** 7

13. 6 **15.** 4 **17.** -7 **19.** 6 **21.** -35

23. -14 **25.** -4 **27.** -8 **29.** 0 **31.** 0

33. -3 **35.** -7 **37.** 0 **39.** -23 **41.** 0

43. -27 **45.** 1.3 **47.** $\frac{3}{16}$ **49.** $-\frac{1}{4}$ **51.** $-\frac{13}{9}$

53. $-\frac{7}{24}$ **55.** -9.2 or $-\frac{46}{5}$ **57.** $-\frac{21}{20}$ or -1.05

59. 0 **61.** $-\frac{21}{22}$ **63.** $\frac{1}{4}$ **65.** -0.0124

67. -6.17 **69.** -5630.15

71. To add two numbers with the same sign, add their absolute values and apply the common sign.

73. 27 **75.** -6 **77.** 15 **79.** 15 **81.** $-3 + 5; 2$

83. $21 + 4; 25$ **85.** $-7 + 24; 17$ **87.** $2(-6 + 10); 8$ **89.** $(4 + (-1)) + (-6); -3$ **91.** $4 + (-9) + 2; -3°F$ **93.** $3 + (-5) + 14$; 12-yd gain **95. a.** $40.02 + (-40.96)$ **b.** Yes **97. a.** $-50{,}000 + (-32{,}000) + (-5000) + 13{,}000 + 26{,}000$ **b.** $-\$48{,}000$

Section 1.4 Calculator Connections, p. 85

1. -13 **2.** -2 **3.** 711 1.–3.

```
-8+(-5)
                -13
4+(-5)+(-1)
                 -2
627-(-84)
                711
```

4. -0.18 **5.** -17.7 **6.** -990 **7.** -17 **8.** 38

4.–6.

```
-0.06-0.12
                -.18
-3.2+(-14.5)
               -17.7
-472+(-518)
                -990
```

7.–8.

```
-12-9+4
                 -17
209-108+(-63)
                  38
```

Section 1.4 Practice Exercises, pp. 85–88

3. x^2 **5.** $-b + 2$ **7.** 9 **9.** 5 **11.** -7 **13.** -9 **15.** 4 **17.** 10 **19.** -3 **21.** 21 **23.** -21 **25.** -4 **27.** 12 **29.** 16 **31.** -16 **33.** -12 **35.** -21 **37.** -29 **39.** 25 **41.** -4 **43.** $-\frac{17}{9}$ **45.** $\frac{23}{15}$ **47.** $\frac{1}{14}$ **49.** $-\frac{17}{24}$ **51.** 9.1 **53.** -7.79 **55.** -34 **57.** -45 **59.** -196.37 **61.** -149.11 **63.** 0.00258 **65.** $18 - (-1); 19$ **67.** $8 - 21; -13$ **69.** $-2 - (-18); 16$ **71.** $-19 - (-31); 12$ **73.** $-3 - 7; -10$ **75.** $1200 - 500 + 800$; \$1500 **77.** 134°F **79.** 20,602 ft **81.** -5 **83.** 8 **85.** 13 **87.** -18 **89.** -18 **91.** $\frac{4}{7}$ **93.** $-\frac{5}{12}$ **95.** 3 **97.** -7 **99.** -9 **101.** -9 **103.** -2 **105.** 9 **107.** 1

Chapter 1 Problem Recognition Exercises—Addition and Subtraction of Signed Numbers, p. 88

1. Add their absolute values and apply a negative sign. **2.** Subtract the smaller absolute value from the larger absolute value. Apply the sign of the number with the larger absolute value. **3.** 41 **4.** 13 **5.** 31 **6.** 46 **7.** -1.3 **8.** -3.6 **9.** -16 **10.** -7 **11.** $-\frac{1}{12}$ **12.** $\frac{7}{24}$ **13.** -36 **14.** -59 **15.** -12 **16.** -50 **17.** $-\frac{19}{6}$ **18.** $-\frac{8}{5}$ **19.** -5 **20.** -32 **21.** 0 **22.** 0 **23.** -7.7 **24.** -10.5 **25.** -114 **26.** -56 **27.** -32 **28.** -46 **29.** -60 **30.** -70 **31.** -30 **32.** -400 **33.** 8 **34.** 2 **35.** $-8 + 20; 12$ **36.** $-11 - (-2); -9$

Section 1.5 Calculator Connections, p. 96

1. -30 **2.** -2 **3.** 625 1.–3.

```
-6(5)
                -30
-5.2/2.6
                 -2
(-5)(-5)(-5)(-5)
                625
```

4. 625 **5.** -625 **6.** -5.76 **7.** 5.76 **8.** -1

4.–6.

```
(-5)^4
                625
-5^4
               -625
-2.4²
              -5.76
```

7.–8.

```
(-2.4)²
               5.76
(-1)(-1)(-1)
                 -1
```

9. 4 **10.** -36 9.–10.

```
-8.4/-2.1
                  4
90/(-5)(2)
                -36
```

Section 1.5 Practice Exercises, pp. 96–99

3. True **5.** True **7.** False **9.** $6 + 6$ **11.** $(-6) + (-6) + (-6) + (-6) + (-6)$ **13.** -12 **15.** -45 **17.** 130 **19.** 128 **21.** 100 **23.** -100 **25.** $-\frac{125}{8}$ **27.** 0.0001 **29.** $(-6)(3) = -18$ **31.** $-4 \cdot 0 = 0$ **33.** No number multiplied by zero equals -4. **35.** -9 **37.** 3 **39.** 7 **41.** $-\frac{1}{11}$ **43.** 48 **45.** -48 **47.** -48 **49.** 48 **51.** 26 **53.** -26 **55.** -26 **57.** 26 **59.** 0 **61.** Undefined **63.** 0 **65.** 0 **67.** $-\frac{3}{2}$ **69.** $\frac{1}{4}$ **71.** -13 **73.** -5383.37 **75.** 0.129 **77.** -8 **79.** 0.3125, or $\frac{5}{16}$ **81.** 49 **83.** -49 **85.** $-\frac{1}{125}$ **87.** 0.000001 **89.** -0.0001 **91.** -16 **93.** 55 **95.** -2.163 **97.** 410 **99.** $-\frac{5}{36}$ **101.** $\frac{6}{23}$ **103.** -60 **105.** 90 **107.** 5 **109.** 3 **111.** $\frac{17}{16}$ or $1\frac{1}{16}$ **113.** $\frac{7}{52}$ **115.** 21 **117.** $\frac{4}{5}$ **119.** 13 **121.** 6 **123.** $\frac{1}{7}$ **125.** Undefined **127.** 6 **129.** -16 **131.** 1 **133.** 42 **135.** -60 **137.** -8 **139.** $\frac{1}{4}$ **141.** Yes, the parentheses indicate that the divisor is the quantity $5x$. **143.** $(-0.4)(-1.258); 0.5032$ **145.** $-\frac{3}{14} \div \frac{1}{7}; -\frac{3}{2}$ **147.** $0.5 + (-2)(0.125); 0.25$ **149.** $-5 - \left(-\frac{5}{6}\right)\frac{3}{8}; -\frac{75}{16}$

151. $-2(6) + 5 = -7$; loss of \$7
153. a. -34 **b.** 5040 **c.** In part (a), we subtract; in part (b), we multiply.

Section 1.6 Practice Exercises, pp. 109–113

3. 8 **5.** -8 **7.** $-\frac{9}{2}$, or -4.5 **9.** 0 **11.** $\frac{7}{8}$
13. $\frac{1}{3}$ **15.** $-\frac{4}{45}$ **17.** $\frac{11}{15}$ **19.** $-8 + 5$
21. $x + 8$ **23.** $4(5)$ **25.** $-12x$
27. $x + (-3)$; $-3 + x$ **29.** $4p + (-9)$; $-9 + 4p$
31. $4(p \cdot p)$; $4p^2$ **33.** $(-5 \cdot 3)x$; $-15x$
35. $\left(\frac{6}{11} \cdot \frac{11}{6}\right)x$; x **37.** $\left(-4 \cdot -\frac{1}{4}\right)t$; t **39.** Reciprocal
41. 0 **43.** $30x + 6$ **45.** $-2a - 16$
47. $15c - 3d$ **49.** $-7y + 14$ **51.** $-\frac{2}{3}x + 4$
53. $\frac{1}{3}m - 1$ **55.** $\frac{3}{2} + 3s$ **57.** $-2p - 10$
59. $3w + 5z$ **61.** $4x + 8y - 4z$ **63.** $6w - x + 3y$
65. $6 + 2x$ **67.** $24z$ **69.** b **71.** i **73.** g
75. d **77.** h **79.**

Term	Coefficient
$2x$	2
$-y$	-1
$18xy$	18
5	5

81.

Term	Coefficient
$-x$	-1
$8y$	8
$-9x^2y$	-9
-3	-3

83. The variable factors are different.
85. The variables are the same and raised to the same power.
87. For example: $5y$, $-2x$, 6 **89.** $2p - 12$
91. $-6y^2 - 7y$ **93.** $7x^2 + x - 4$ **95.** $3t - \frac{7}{5}$
97. $-10.4w + 3.3$ **99.** $-8a - 20$ **101.** $10r$
103. $-6x - 12$ **105.** $-16y - 27$ **107.** 23
109. $-3q + 1$ **111.** $-314p + 107$ **113.** $-7b + 8$
115. $-\frac{9}{5}p + \frac{5}{2}$ **117.** $3k + 11$ **119.** $-y + 25$
121. $-6.12q + 29.72$ **123.** Equivalent
125. Not equivalent. The terms are not *like* terms and cannot be combined.
127. Not equivalent; subtraction is not commutative.
129. Equivalent
131. Not equivalent; coefficients for the variable terms and constants are different.
133. $5\frac{1}{8} + (18\frac{2}{5} + 1\frac{3}{5})$ is easier.

Chapter 1 Review Exercises, pp. 117–119

1. a. 1, 7 **b.** 7, -4, 0, 1 **c.** 7, 0, 1 **d.** 7, $\frac{1}{3}$, -4, 0, $-0.\overline{2}$, 1
e. $-\sqrt{3}$, π **f.** 7, $\frac{1}{3}$, -4, 0, $-\sqrt{3}$, $-0.\overline{2}$, π, 1
2. $\frac{1}{2}$ **3.** 6 **4.** $\sqrt{7}$ **5.** 0 **6.** False **7.** False
8. True **9.** True **10.** True **11.** True **12.** False
13. True **14.** $x \cdot \frac{2}{3}$, or $\frac{2}{3}x$ **15.** $\frac{7}{y}$, or $7 \div y$
16. $2 + 3b$ **17.** $a - 5$ **18.** $5k + 2$ **19.** $13z - 7$
20. $\frac{6}{x} - 18$ **21.** $3y + 12$ **22.** $z - \frac{3}{8}$ **23.** $2p - 5$
24. 216 **25.** 225 **26.** 6 **27.** $\frac{1}{16}$ **28.** $\frac{1}{10}$
29. $\frac{27}{8}$ **30.** 13 **31.** 11 **32.** 7 **33.** 10
34. 2 **35.** 4 **36.** 15 **37.** -17 **38.** $\frac{11}{63}$
39. $-\frac{5}{22}$ **40.** $-\frac{14}{15}$ **41.** $-\frac{27}{10}$ **42.** -2.15
43. -4.28 **44.** 3 **45.** 8 **46.** 7
47. When a and b are both negative or when a and b have different signs and the number with the larger absolute value is negative
48. -1.9°C **49.** -12 **50.** 33 **51.** -1
52. -17 **53.** $-\frac{29}{18}$ **54.** $-\frac{19}{24}$ **55.** -1.2
56. -4.25 **57.** -10.2 **58.** -12.09 **59.** $\frac{10}{3}$
60. $-\frac{17}{20}$ **61.** -1 **62.** If $a < b$
63. $-7 - (-18)$; 11 **64.** $-6 - 41$; -47
65. $7 - 13$; -6 **66.** $[20 - (-7)] - 5$; 22
67. $[6 + (-12)] - 21$; -27 **68.** $125 - (-50)$; 175°F
69. -170 **70.** -91 **71.** -2 **72.** 3 **73.** $-\frac{1}{6}$
74. $-\frac{8}{11}$ **75.** 0 **76.** Undefined **77.** 0 **78.** 0
79. 2.25 **80.** 30 **81.** $-\frac{3}{2}$ **82.** $\frac{1}{4}$ **83.** -30
84. 450 **85.** $\frac{1}{4}$ **86.** $-\frac{1}{7}$ **87.** -2 **88.** $\frac{18}{7}$
89. 17 **90.** 6 **91.** $-\frac{7}{120}$ **92.** 4.4 **93.** -2
94. 26 **95.** -23 **96.** 80 **97.** 70.6 **98.** True
99. False, any nonzero real number raised to an even power is positive.
100. True **101.** True
102. False, the product of two negative numbers is positive.
103. True **104.** True **105.** For example: $2 + 3 = 3 + 2$
106. For example: $(2 + 3) + 4 = 2 + (3 + 4)$
107. For example: $5 + (-5) = 0$
108. For example: $7 + 0 = 7$
109. For example: $5 \cdot 2 = 2 \cdot 5$
110. For example: $(8 \cdot 2)10 = 8(2 \cdot 10)$
111. For example: $3 \cdot \frac{1}{3} = 1$ **112.** For example: $8 \cdot 1 = 8$
113. $5x - 2y = 5x + (-2y)$, then use the commutative property of addition.

114. $3a - 9y = 3a + (-9y)$, then use the commutative property of addition.
115. $3y, 10x, -12, xy$ **116.** $3, 10, -12, 1$
117. a. $8a - b - 10$ **b.** $-7p - 11q + 16$
118. a. $-8z - 18$ **b.** $20w - 40y + 5$
119. $p - 2$ **120.** $-h + 14$ **121.** $-14q - 1$
122. $-5.7b + 2.4$ **123.** $4x + 24$ **124.** $50y + 105$

Chapter 1 Test, p. 119–120

1. Rational, all repeating decimals are rational numbers.
2.

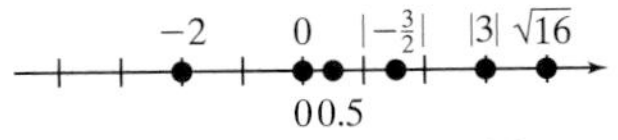

3. a. False **b.** True **c.** True **d.** True
4. a. $(4x)(4x)(4x)$ **b.** $4 \cdot x \cdot x \cdot x$
5. a. Twice the difference of a and b **b.** The difference of twice a and b
6. $\frac{\sqrt{c}}{d^2}$ or $\sqrt{c} \div d^2$ **7.** 6 **8.** −19 **9.** −12
10. 28 **11.** $-\frac{7}{8}$ **12.** 4.66 **13.** −32
14. −12 **15.** Undefined **16.** −28 **17.** 0
18. 96 **19.** $\frac{2}{3}$ **20.** −8 **21.** 9 **22.** $\frac{1}{3}$
23. $-\frac{3}{5}$
24. a. Commutative property of multiplication **b.** Identity property of addition **c.** Associative property of addition **d.** Inverse property of multiplication **e.** Associative property of multiplication
25. $-12x + 2y - 4$ **26.** $-12m - 24p + 21$
27. $40x - 49$ **28.** $-4p - 23$ **29.** $4p - \frac{4}{3}$
30. 5 **31.** 18 **32.** −6 **33.** −32
34. $12 - (-4); 16$ **35.** $6 - 8; -2$

Chapter 2

Chapter Opener Puzzle

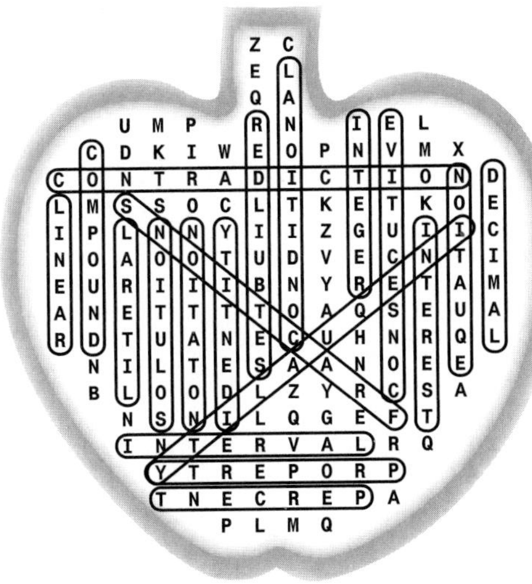

Section 2.1 Practice Exercises, pp. 130–132

3. Expression **5.** Equation
7. Substitute the value into the equation and determine if the right-hand side is equal to the left-hand side.
9. No **11.** Yes **13.** Yes **15.** $x = 12$
17. $w = -8$ **19.** $n = 16$ **21.** $b = 12$ **23.** $b = 0$
25. $y = 7.6$ **27.** $p = -\frac{23}{3}$ or $-7\frac{2}{3}$ **29.** $x = 3$
31. $p = -1.155$ **33.** $k = 4.0629$ **35.** $w = 4$
37. $q = -3$ **39.** $k = 0$ **41.** $z = -7$
43. $h = \frac{3}{7}$ **45.** $b = -24$ **47.** $w = -\frac{4}{3}$
49. $y = -32$ **51.** $q = -5.5$ **53.** $x = 0.014$
55. $31 + x = 13; x = -18$ **57.** $-12 + x = -15; x = -3$
59. $-3x = 24; x = -8$ **61.** $18 = \frac{x}{2}; x = 36$
63. $x - \frac{2}{3} = \frac{1}{3}; x = 1$ **65.** $b = -2$ **67.** $k = 2$
69. $p = 20$ **71.** $y = \frac{57}{4}$ or $14\frac{1}{4}$ **73.** $d = -20$
75. $t = -9$ **77.** $t = 2$ **79.** $s = -\frac{10}{3}$ or $-3\frac{1}{3}$
81. $s = 13$ **83.** $m = -25$ **85.** $p = \frac{11}{4}$ or $2\frac{3}{4}$
87. $w = 2$ **89.** $w = 12.9$ **91.** $m = -1.9834$
93. No **95.** Yes **97.** No **99.** Yes
101. For example: $1 + x = 3$ **103.** For example: $-4t = 40$
105. For example: $6k + 1 = 7$ **107.** $y = 4$
109. $t = -2$

Section 2.2 Practice Exercises, pp. 139–141

3. $-5z + 2$ **5.** $10p - 10$
7. To simplify an expression, clear parentheses and combine *like* terms. To solve an equation, use the addition, subtraction, multiplication, and division properties of equality to isolate the variable.
9. $y = -3$ **11.** $z = -5$ **13.** $z = 2$ **15.** $y = 6$
17. $y = \frac{5}{2}$ or $2\frac{1}{2}$ **19.** $m = 1$ **21.** $x = -\frac{17}{8}$ or $-2\frac{1}{8}$
23. $x = -42$ **25.** $p = -\frac{3}{4}$ **27.** $w = 5$
29. $h = -4$ **31.** $a = -26$ **33.** $r = \frac{50}{3}$ or $16\frac{2}{3}$
35. $z = -8$ **37.** $x = -\frac{7}{3}$ or $-2\frac{1}{3}$ **39.** $y = 0$
41. $p = \frac{9}{2}$ or $4\frac{1}{2}$ **43.** $x = -\frac{1}{3}$ **45.** $y = 12$
47. $s = -6$ **49.** $t = 0$ **51.** $x = -2$
53. $p = \frac{10}{3}$ or $3\frac{1}{3}$ **55.** $w = 6$ **57.** $n = -0.25$
59. Contradiction; no solution
61. Conditional equation; $x = -15$
63. Identity; all real numbers
65. One solution **67.** Infinitely many solutions
69. $t = 10$ **71.** $y = 2$ **73.** $m = 0.6$
75. $n = -\frac{3}{4}$ **77.** No solution **79.** $x = 28$

81. $p = \frac{5}{4}$ **83.** $r = 7$ **85.** $y = -20$
87. All real numbers **89.** $g = \frac{4}{3}$ **91.** $w = 0$
93. $a = -18$ **95.** $a = 4.5$
97. For example: $4x - 3 = 4x + 1$

Section 2.3 Practice Exercises, pp. 147–149

3. $x = -2$ **5.** $y = -5$ **7.** No solution **9.** 18, 36
11. 100; 1000; 10,000 **13.** 30, 60 **15.** $x = 4$
17. $y = -\frac{19}{2}$ **19.** $q = -\frac{15}{4}$ **21.** $w = 8$
23. $m = 3$ **25.** $s = 15$ **27.** No solution
29. All real numbers **31.** $x = 5$ **33.** $w = 2$
35. $x = -15$ **37.** $y = 6$ **39.** $w = 3$
41. All real numbers **43.** $x = 67$ **45.** $p = 90$
47. $y = -2$ **49.** $x = -3.8$ **51.** No solution
53. $x = -0.25$ **55.** $b = 7$ **57.** $y = -8$
59. $a = 20$ **61.** $x = 0$ **63.** $h = 3$
65. $x = \frac{8}{3}$ or $2\frac{2}{3}$ **67.** No solution **69.** $w = \frac{25}{2}$ or $12\frac{1}{2}$
71. All real numbers **73.** $a = -6$ **75.** $h = \frac{1}{3}$
77. $t = -6$ **79.** $x = \frac{23}{12}$ **81.** $k = \frac{1}{10}$
83. The number is $\frac{2}{3}$. **85.** The number is $\frac{13}{24}$.
87. $c = -2$ **89.** $h = 2$

Section 2.4 Practice Exercises, pp. 159–163

3. $x + 16 = -31; x = -47$ **5.** $x - 6 = -3; x = 3$
7. $x - 16 = -1; x = 15$ **9.** The number is −4.
11. The number is −3. **13.** The number is 5.
15. The number is −5. **17.** The number is 9.
19. a. $x + 1, x + 2$ **b.** $x - 1, x - 2$
21. The integers are −34 and −33.
23. The integers are 13 and 15.
25. The page numbers are 470 and 471.
27. The sides are 14, 15, 16, 17, and 18 in.
29. Karen's age is 35, and Clarann's age is 23.
31. There were 165 Republicans and 269 Democrats.
33. The lengths of the pieces are 33 cm and 53 cm.
35. 4.698 million watch *The Dr. Phil Show*.
37. The Congo River is 4370 km long, and the Nile River is 6825 km.
39. The area of Africa is 30,065,000 km^2. The area of Asia is 44,579,000 km^2.
41. She hikes 3 mph to the lake.
43. The plane travels 600 mph in still air.
45. The slower car travels 48 mph and the faster car travels 52 mph.
47. The speeds of the vehicles are 40 mph and 50 mph.
49. The rates of the boats are 20 mph and 40 mph.
51. The number is −80. **53.** 42, 43, and 44
55. The speed of the current is 4 mph.
57. The number is 11.
59. Jennifer Lopez made \$37 million, and U2 made \$69 million.
61. The boats will meet in $\frac{3}{4}$ hr (45 min).
63. The number is 10.
65. The deepest point in the Arctic Ocean is 5122 m.

Section 2.5 Practice Exercises, pp. 168–171

3. 1. Read the problem carefully. 2. Assign labels to unknown quantities. 3. Develop a verbal model. 4. Write a mathematical equation. 5. Solve the equation. 6. Interpret the results, and write the final answer in words.
5. The pyramid at Saqqara is 60 m high, and the Great Pyramid is 137 m high.
7. 65% **9.** 72% **11.** 58.52 **13.** 436.8
15. 840 **17.** 1320 **19.** Approximately 91,000 cases
21. 3% **23. a.** \$89.90 **b.** \$809.10
25. The original price was \$21.95.
27. The markup rate is 88%.
29. Patrick will have to pay \$317.96.
31. The price was \$24.00.
33. The tax rate is 7.5%.
35. The price of a ticket is \$67.
37. The original cost of one CD is \$18.
39. Roxanne will have to pay \$160.
41. Mike borrowed \$3250. **43.** The rate is 8%.
45. Rafael will have \$3300. **47.** Sherica invested \$5500.
49. Dan earned \$26,000.
51. Anna sold \$10,400 worth of appliances.
53. Jessica's rate is 8%.

Section 2.6 Calculator Connections, p. 177

1. 140.056 **2.** 31.831 **3.** −80 **4.** 2

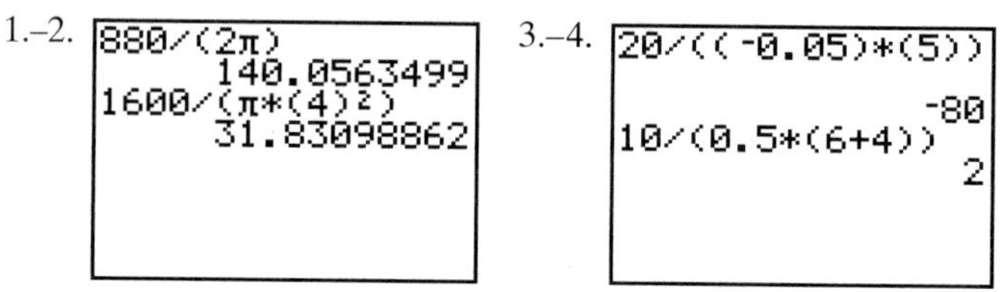

Section 2.6 Practice Exercises, pp. 178–181

3. $y = -5$ **5.** $x = 0$ **7.** $y = -2$
9. $b = P - a - c$ **11.** $d = e - c$
13. $x = y - 35$ **15.** $r = \frac{d}{t}$ **17.** $V_1 = \frac{P_2V_2}{P_1}$
19. $y = -2 - x$ **21.** $x = 6y - 10$
23. $y = \frac{-5x + 10}{2}$ or $y = -\frac{5}{2}x + 5$
25. $x = \frac{y - 13}{3}$ or $x = \frac{1}{3}y - \frac{13}{3}$
27. $y = \frac{-6x + 4}{-3}$ or $y = 2x - \frac{4}{3}$
29. $x = \frac{-by + c}{a}$ or $x = -\frac{b}{a}y + \frac{c}{a}$
31. $L = \frac{P - 2w}{2}$ or $L = \frac{P}{2} - w$

33. $x = \frac{z - 3y}{3}$ or $x = \frac{z}{3} - y$

35. $a = 2Q + b$ **37.** $c = 3A - a - b$ **39.** $m = \frac{Fd^2}{GM}$

41. The length is 12 cm, and the width is 5 cm.
43. The length is 40 yd, and the width is 30 yd.
45. The sides are 4 ft, 5 ft, and 7 ft.
47. The measures of the angles are 20°, 50°, and 110°.
49. The measures of the angles are 38°, 28°, and 114°.
51. $y = 26.5$; the measures of the angles are 28.5° and 61.5°.
53. The angles are 55° and 35°.
55. The angles are 23.5° and 66.5°.
57. The angles are 60° and 120°.
59. $x = 20$; the vertical angles measure 37°.

61. a. $A = lw$ **b.** $w = \frac{A}{l}$ **c.** The width is 29.5 ft.

63. a. $P = 2l + 2w$ **b.** $l = \frac{P - 2w}{2}$ **c.** The length is 103 m.

65. a. $A = \frac{1}{2}bh$ **b.** $h = \frac{2A}{b}$ **c.** The height is 4 km.

67. a. 415.48 m^2 **b.** 10,386.89 m^3

Section 2.7 Practice Exercises, pp. 192–196

3. $x = -3$

Set-Builder Notation	Graph	Interval Notation
5. $\{x \mid x \geq 6\}$	6	$[6, \infty)$
7. $\{x \mid x \leq 2.1\}$	2.1	$(-\infty, 2.1]$
9. $\{x \mid -2 < x \leq 7\}$	−2, 7	$(-2, 7]$
11. $\left\{x \mid x > \frac{3}{4}\right\}$	$\frac{3}{4}$	$\left(\frac{3}{4}, \infty\right)$
13. $\{x \mid -1 < x < 8\}$	−1, 8	$(-1, 8)$
15. $\{x \mid x < -14\}$	−14	$(-\infty, -14)$
17. $\{x \mid x \geq 18\}$	18	$[18, \infty)$
19. $\{x \mid x < -0.6\}$	−0.6	$(-\infty, -0.6)$
21. $\{x \mid -3.5 \leq x < 7.1\}$	−3.5, 7.1	$[-3.5, 7.1)$

23. a. $x = 3$ **b.** $x > 3$ (graph: 3)

25. a. $p = 13$ **b.** $p \leq 13$ (graph: 13)

27. a. $c = -3$ **b.** $c < -3$ (graph: −3)

29. a. $z = -\frac{3}{2}$ **b.** $z \geq -\frac{3}{2}$ (graph: $-\frac{3}{2}$)

31. (graph: −1, 4)

33. (graph: −3, 5) $-3 < x < 5$

35. (graph: 2, 6) $2 \leq x \leq 6$

37. (graph: 1) $(-\infty, 1]$

39. (graph: 10) $(10, \infty)$

41. (graph: 3) $(3, \infty)$

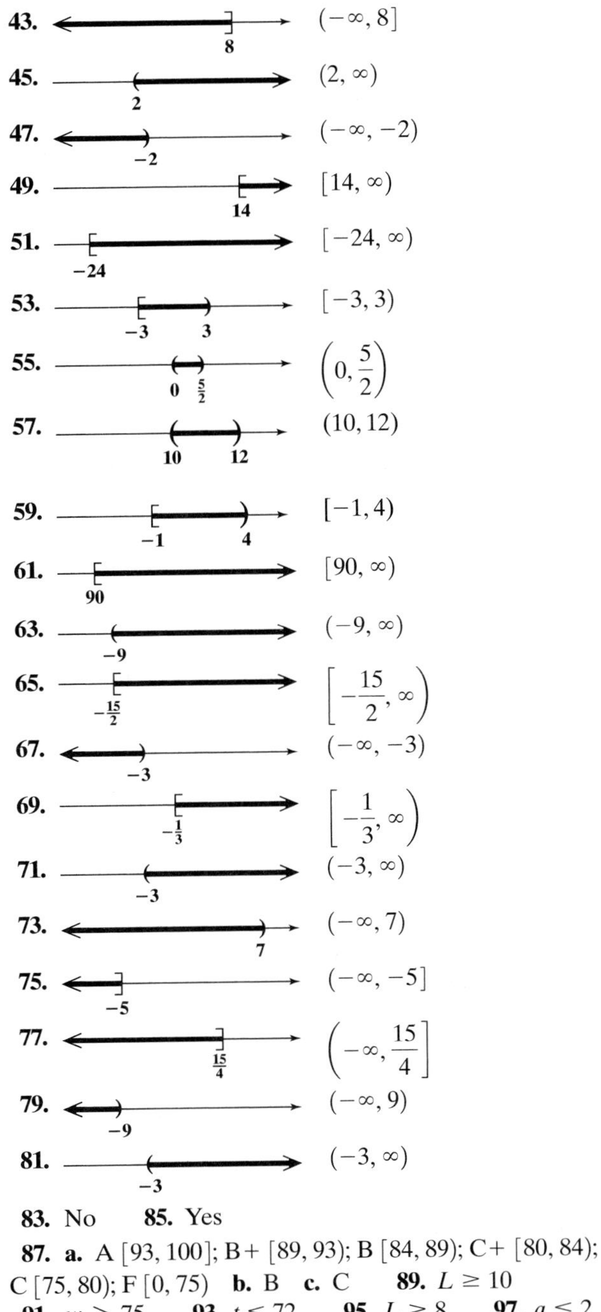

83. No **85.** Yes

87. a. A [93, 100]; B+ [89, 93); B [84, 89); C+ [80, 84); C [75, 80); F [0, 75) **b.** B **c.** C **89.** $L \geq 10$
91. $w > 75$ **93.** $t \leq 72$ **95.** $L \geq 8$ **97.** $a \leq 2$
99. $300 < c < 400$ **101.** More than 10.1 in. of snow is needed. **103. a.** \$384 **b.** It costs \$1112.96 for 148 shirts and \$1104.00 for 150 shirts. 150 shirts cost less than 148 shirts because the discount is greater.
105. a. $14.95 + 0.22x < 18.95 + 0.18x$ **b.** $x < 100$; Company A costs less than Company B if the mileage is less than 100 miles.

107. (graph: −6) $(-\infty, -6)$

109. (graph: −7) $(-\infty, -7)$

111. (graph: $-1.\overline{3}$) $(-\infty, -1.\overline{3}]$

Chapter 2 Review Exercises, pp. 203–206

1. a. Equation **b.** Expression **c.** Equation **d.** Equation
2. A linear equation can be written in the form $ax + b = 0$, $a \neq 0$. **3. a.** Nonlinear **b.** Linear **c.** Nonlinear **d.** Linear **4. a.** No **b.** No **c.** Yes **d.** No

5. $a = -8$ **6.** $z = 15$ **7.** $k = \frac{21}{4}$ **8.** $r = 70$

9. $x = -\frac{21}{5}$ **10.** $t = -60$ **11.** $k = -\frac{10}{7}$

12. $m = 27$ **13.** The number is 60.

14. The number is $\frac{7}{24}$. **15.** The number is -8.

16. The number is -2. **17.** $d = 1$ **18.** $c = -\frac{3}{5}$

19. $c = \frac{9}{4}$ **20.** $w = -6$ **21.** $b = -3$ **22.** $h = 18$

23. $p = \frac{3}{4}$ **24.** $t = \frac{11}{8}$ **25.** $a = 0$ **26.** $c = \frac{1}{8}$

27. $b = \frac{3}{8}$ **28.** $x = \frac{17}{3}$ **29.** A contradiction has no solution and an identity is true for all real numbers.
30. a. Identity **b.** Conditional equation **c.** Contradiction **d.** Identity **e.** Contradiction **31.** $x = 6$ **32.** $y = 22$
33. $z = -3$ **34.** $y = -9$ **35.** $p = -10$

36. $y = -7$ **37.** $t = \frac{5}{3}$ **38.** $w = -\frac{9}{4}$ **39.** $q = 2.5$

40. $z = -4$ **41.** $a = -4.2$ **42.** $t = 2.5$
43. $x = -312$ **44.** $x = 200$ **45.** No solution
46. No solution **47.** All real numbers
48. All real numbers **49.** The number is 30.
50. The number is 11. **51.** The number is -7.
52. The number is -10.
53. The integers are 66, 68, and 70.
54. The integers are 27, 28, and 29.
55. The sides are 25, 26, and 27 in.
56. The sides are 36, 37, 38, 39, and 40 cm.
57. The minimum salary was $30,000 in 1980.
58. Indiana has 6.2 million people and Kentucky has 4.1 million. **59.** The Mooney travels 190 mph, and the Cessna travels 140 mph. **60.** The jogger runs 6 mph and the bicyclist travels 12 mph. **61.** 23.8 **62.** 28.8
63. 12.5% **64.** 95% **65.** 160 **66.** 1750
67. The sale price is $26.39. **68.** The total price is $28.24. **69. a.** $840 **b.** $3840 **70.** $12,000

71. $K = C + 273$ **72.** $C = K - 273$ **73.** $s = \frac{P}{4}$

74. $s = \frac{P}{3}$ **75.** $x = \frac{y - b}{m}$ **76.** $x = \frac{c - a}{b}$

77. $y = \frac{-2 - 2x}{5}$ **78.** $b = \frac{Q - 4a}{4}$ or $b = \frac{Q}{4} - a$

79. a. $h = \frac{3V}{\pi r^2}$ **b.** The height is 5.1 in.

80. The height is 7 m. **81.** The measures of the angles are 22°, 78°, and 80°. **82.** The measures of the angles are 32° and 58°. **83.** The length is 5 ft, and the width is 4 ft.
84. $x = 20$. The angle measure is 65°.
85. The measure of angle y is 53°.
86. a. $(-2, \infty)$

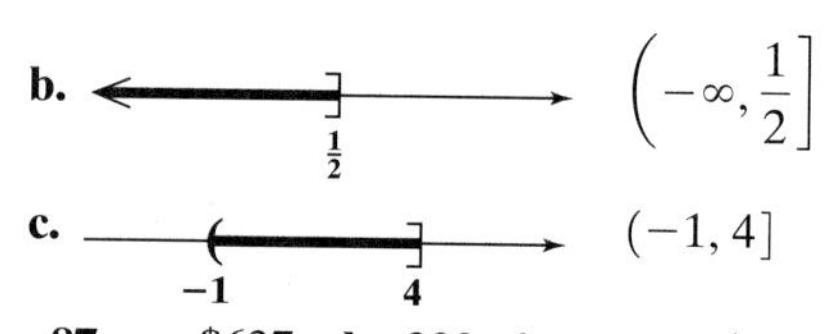

b. $\left(-\infty, \frac{1}{2}\right]$

c. $(-1, 4]$

87. a. $637 **b.** 300 plants cost $1410, and 295 plants cost $1416. It is cheaper to buy 300 plants because there is a greater discount.

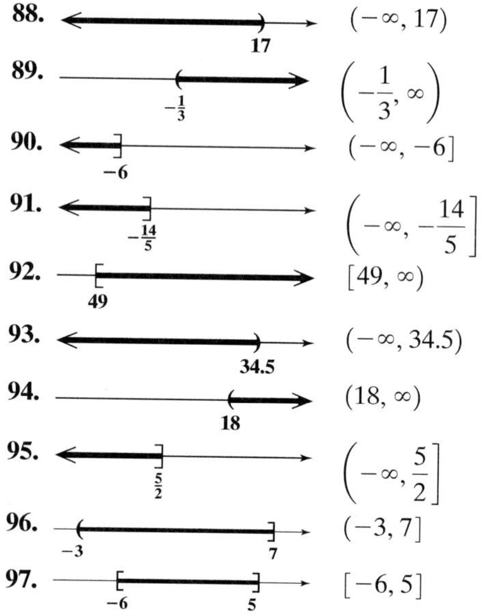

88. $(-\infty, 17)$

89. $\left(-\frac{1}{3}, \infty\right)$

90. $(-\infty, -6]$

91. $\left(-\infty, -\frac{14}{5}\right]$

92. $[49, \infty)$

93. $(-\infty, 34.5)$

94. $(18, \infty)$

95. $\left(-\infty, \frac{5}{2}\right]$

96. $(-3, 7]$

97. $[-6, 5]$

98. More than 2.5 in. is required.
99. a. $1.50x > 33 + 0.4x$ **b.** $x > 30$; a profit is realized if more than 30 hot dogs are sold.

Chapter 2 Test, p. 206–207

1. b, d **2. a.** $5x + 7$ **b.** $x = 9$ **3.** $t = -16$

4. $p = 12$ **5.** $t = -\frac{16}{9}$ **6.** $x = \frac{7}{3}$ **7.** $p = 15$

8. $d = \frac{13}{4}$ **9.** $x = \frac{20}{21}$ **10.** No solution

11. $x = -3$ **12.** $c = -47$ **13.** All real numbers

14. $y = -3x - 4$ **15.** $r = \frac{C}{2\pi}$ **16.** 90

17. The numbers are 18 and 13. **18.** The sides are 61, 62, 63, 64, and 65 in. **19.** Each basketball ticket was $36.32, and each hockey ticket was $40.64.
20. The cost was $82.00. **21.** Clarita originally borrowed $5000. **22.** The field is 110 m long and 75 m wide.
23. $y = 30$; The measures of the angles are 30°, 39°, and 111°.
24. One family travels 55 mph and the other travels 50 mph.
25. The measures of the angles are 32° and 58°.
26. a. $(-\infty, 0)$

b. $[-2, 5)$

27. $(-2, \infty)$

28. $(-\infty, -4]$

29. $[-5, 1]$

30. More than 26.5 in. is required.

Chapters 1–2 Cumulative Review Exercises, p. 207–208

1. $\frac{1}{2}$ **2.** -7 **3.** $-\frac{5}{12}$ **4.** 16 **5.** 4 **6.** $\sqrt{5^2 - 9}$; 4 **7.** $-14 + 12$; -2 **8.** $-7x^2y, 4xy, -6$ **9.** $9x + 13$ **10.** $t = 4$ **11.** $x = -7.2$ **12.** All real numbers **13.** $x = \frac{24}{7}$ **14.** $x = -\frac{4}{7}$ **15.** $w = -80$ **16.** The numbers are 77 and 79. **17.** The cost before tax was \$350.00. **18.** The height is $\frac{41}{6}$ cm or $6\frac{5}{6}$ cm. **19.** $(-2, \infty)$ **20.** $[-1, 9]$

Chapter 3

Chapter Opener Puzzle

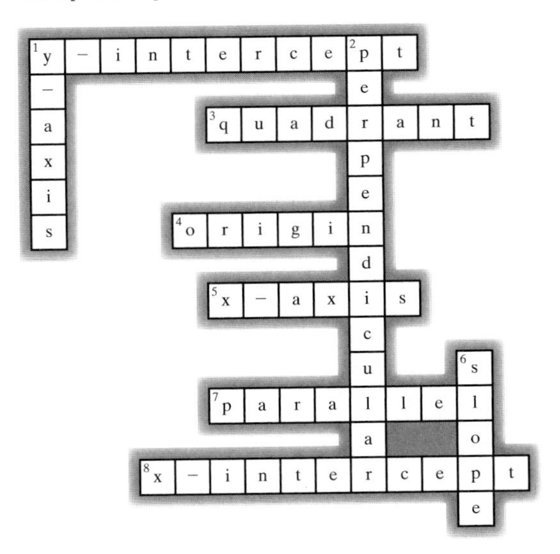

Section 3.1 Practice Exercises, pp. 215–220

3. a. Month 10 **b.** 30 **c.** Between months 3 and 5 and between months 10 and 12 **d.** Months 8 and 9 **e.** Month 3 **f.** 80 **5. a.** On day 1 the price per share was \$89.25. **b.** \$1.75 **c.** $-\$2.75$

7.

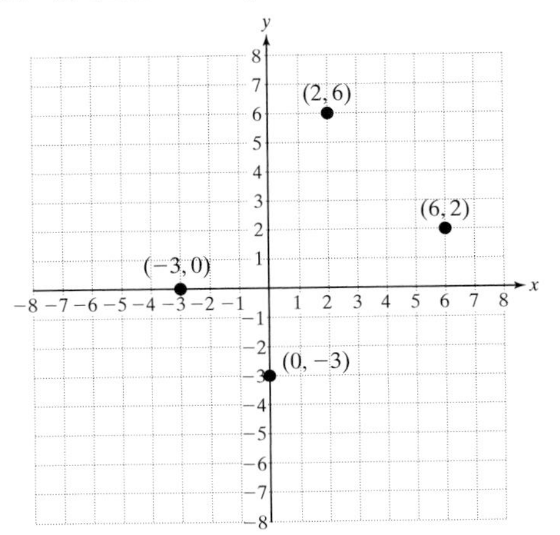

9. **11.**

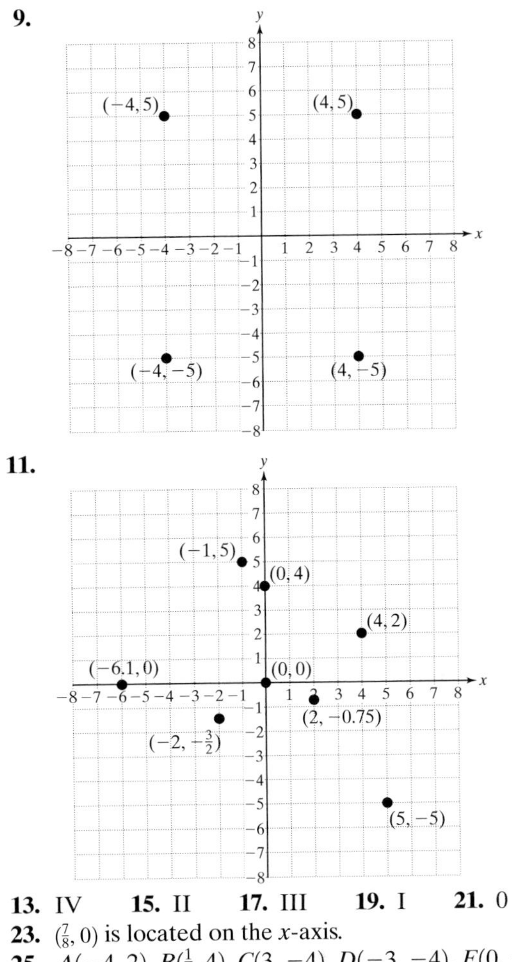

13. IV **15.** II **17.** III **19.** I **21.** 0 **23.** $(\frac{7}{8}, 0)$ is located on the x-axis. **25.** $A(-4, 2)$, $B(\frac{1}{2}, 4)$, $C(3, -4)$, $D(-3, -4)$, $E(0, -3)$, $F(5, 0)$ **27. a.** (250, 225), (175, 193), (315, 330), (220, 209), (450, 570), (400, 480), (190, 185); the ordered pair (250, 225) means that 250 people produce \$225 in popcorn sales.

b.

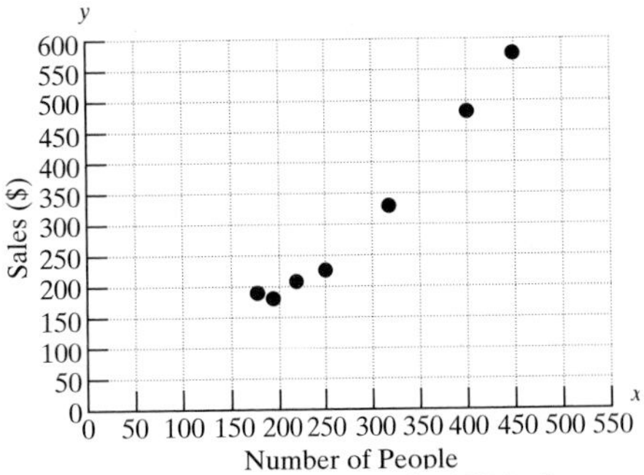

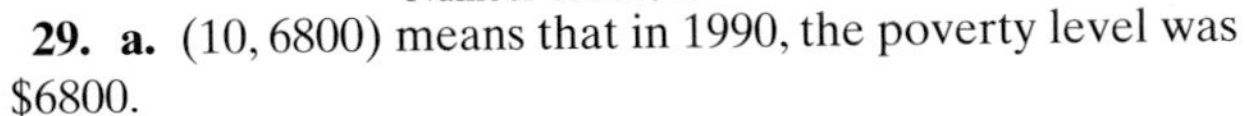

29. a. (10, 6800) means that in 1990, the poverty level was \$6800.

b.

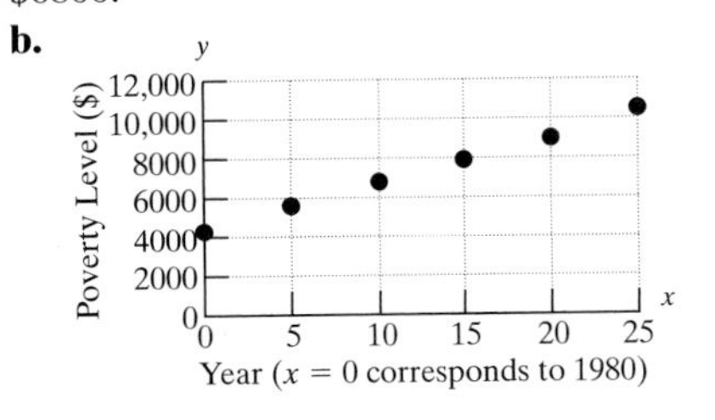

31. a. $(1, -10.2), (2, -9.0), (3, -2.5), (4, 5.7), (5, 13.0), (6, 18.3), (7, 20.9), (8, 19.6), (9, 14.8), (10, 8.7), (11, 2.0), (12, -6.9)$.
b.

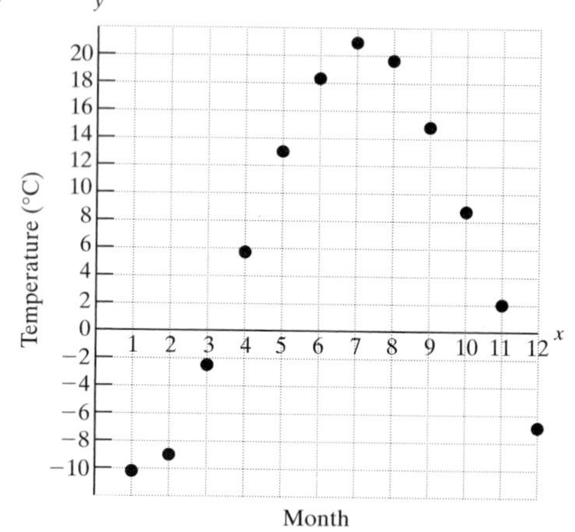

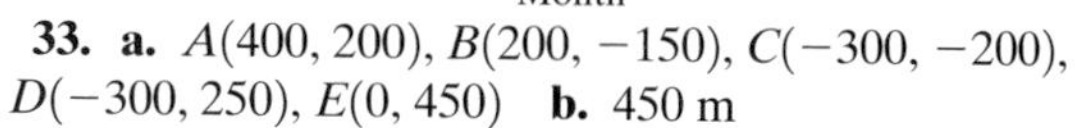

33. a. $A(400, 200)$, $B(200, -150)$, $C(-300, -200)$, $D(-300, 250)$, $E(0, 450)$ **b.** 450 m

Section 3.2 Calculator Connections, pp. 229–230

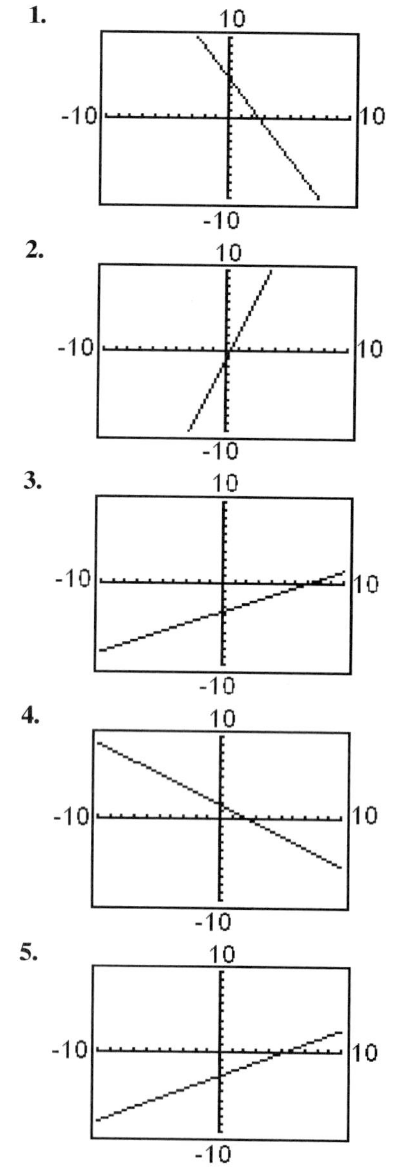

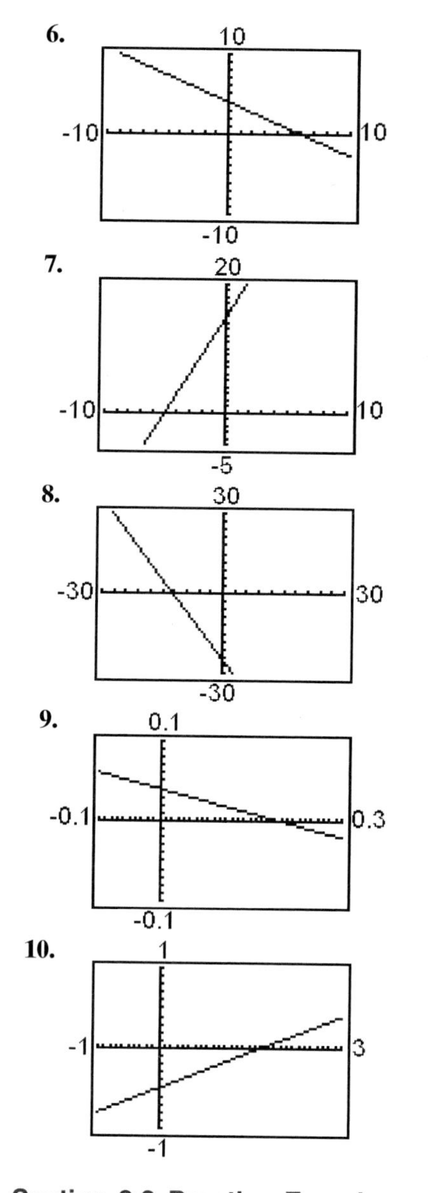

Section 3.2 Practice Exercises, pp. 231–238

3. $(2, 4)$; quadrant I **5.** $(0, -1)$; y-axis **7.** $(3, -4)$; quadrant IV **9.** Yes **11.** Yes **13.** No **15.** No **17.** Yes

19.

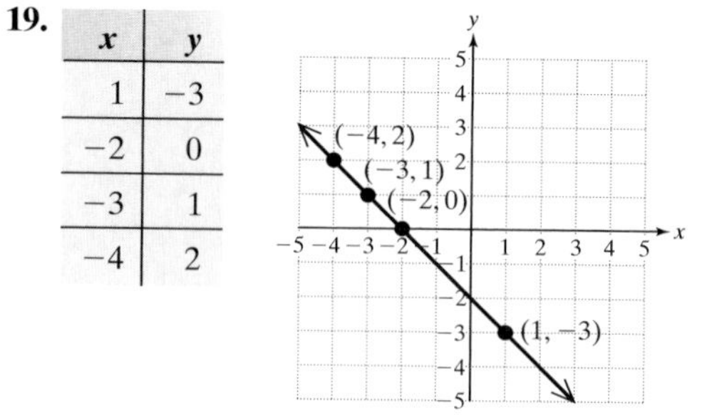

x	y
1	−3
−2	0
−3	1
−4	2

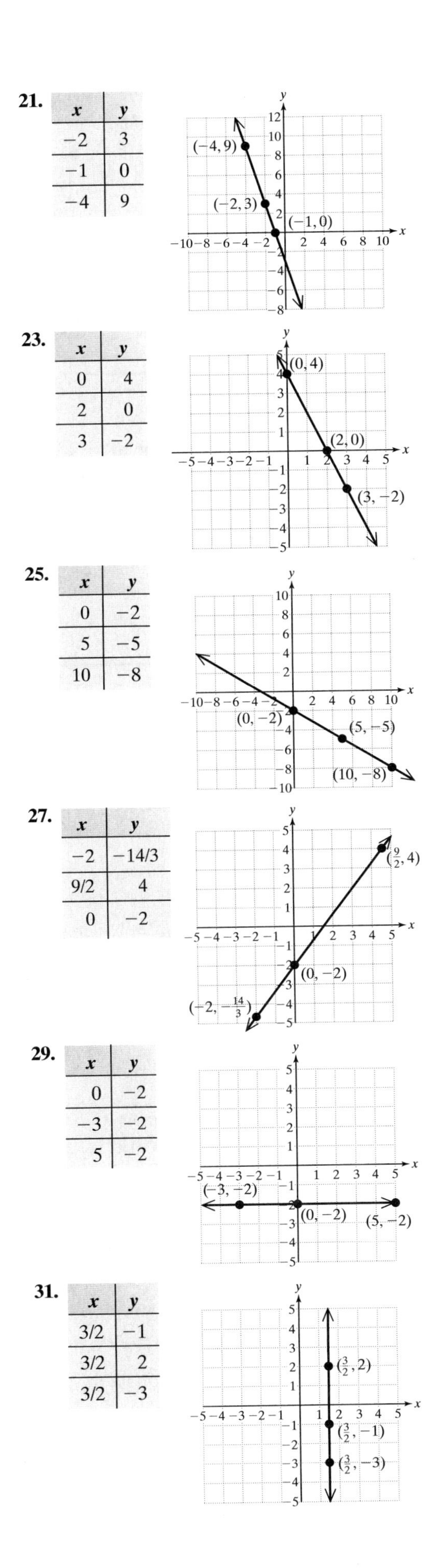

21.

x	y
−2	3
−1	0
−4	9

23.

x	y
0	4
2	0
3	−2

25.

x	y
0	−2
5	−5
10	−8

27.

x	y
−2	−14/3
9/2	4
0	−2

29.

x	y
0	−2
−3	−2
5	−2

31.

x	y
3/2	−1
3/2	2
3/2	−3

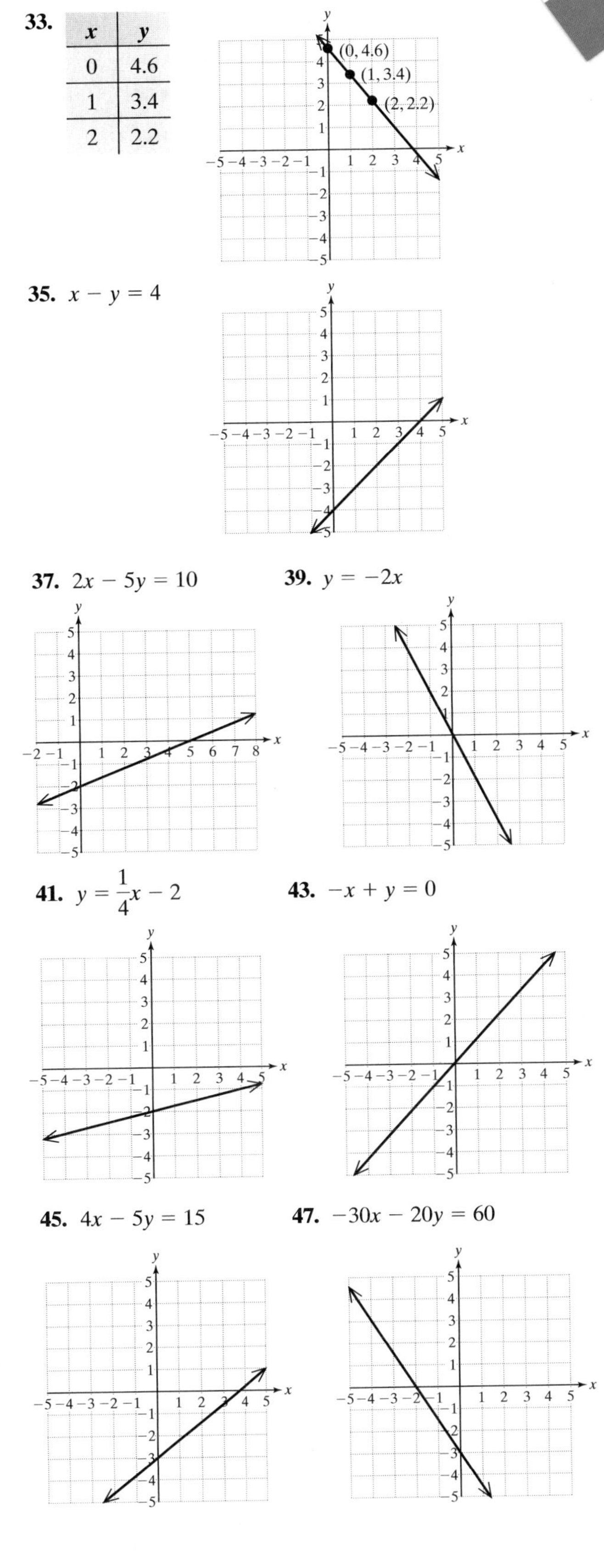

33.

x	y
0	4.6
1	3.4
2	2.2

35. $x - y = 4$

37. $2x - 5y = 10$

39. $y = -2x$

41. $y = \frac{1}{4}x - 2$

43. $-x + y = 0$

45. $4x - 5y = 15$

47. $-30x - 20y = 60$

49. y-axis **51.** x-intercept: $(-1, 0)$; y-intercept: $(0, -3)$
53. x-intercept: $(-4, 0)$; y-intercept: $(0, 1)$

55. x-intercept: $(-9, 0)$; y-intercept: $(0, 3)$
$x - 3y = -9$

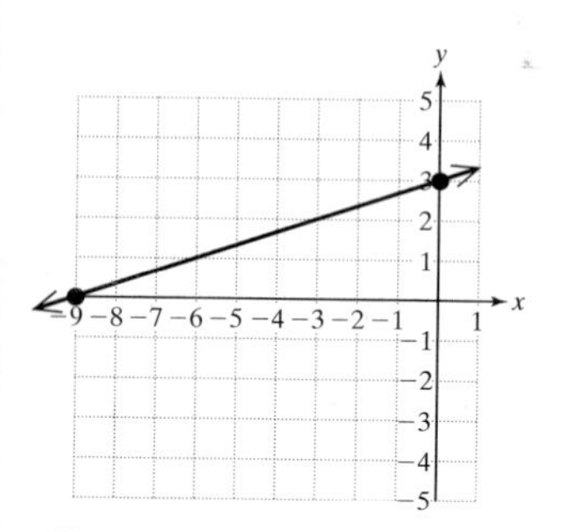

57. x-intercept: $\left(\frac{8}{3}, 0\right)$; y-intercept: $(0, 2)$
$y = -\frac{3}{4}x + 2$

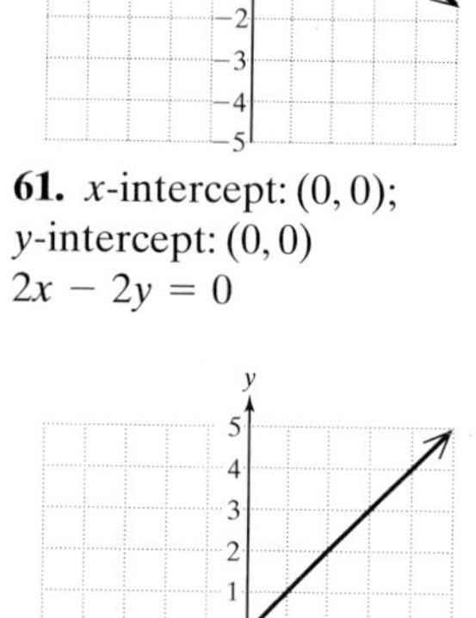

59. x-intercept: $(-4, 0)$; y-intercept: $(0, 8)$
$2x + 8 = y$

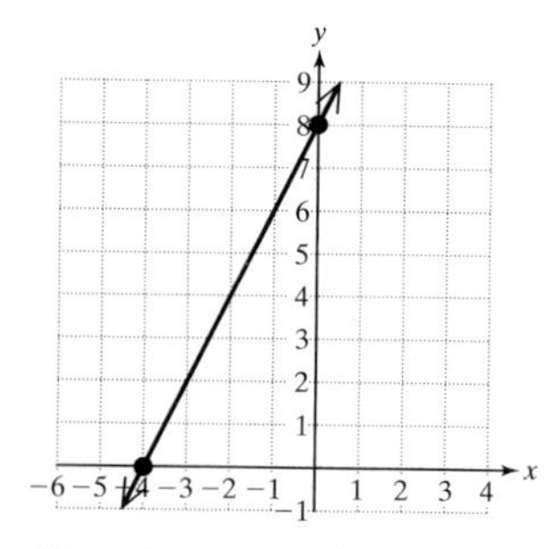

61. x-intercept: $(0, 0)$; y-intercept: $(0, 0)$
$2x - 2y = 0$

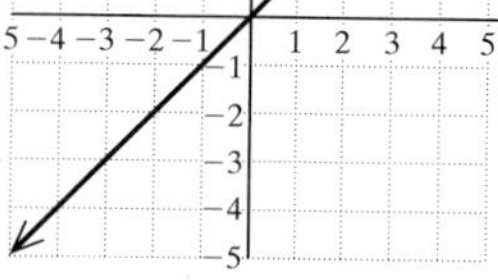

63. x-intercept: $(10, 0)$; y-intercept: $(0, 5)$
$20x = -40y + 200$

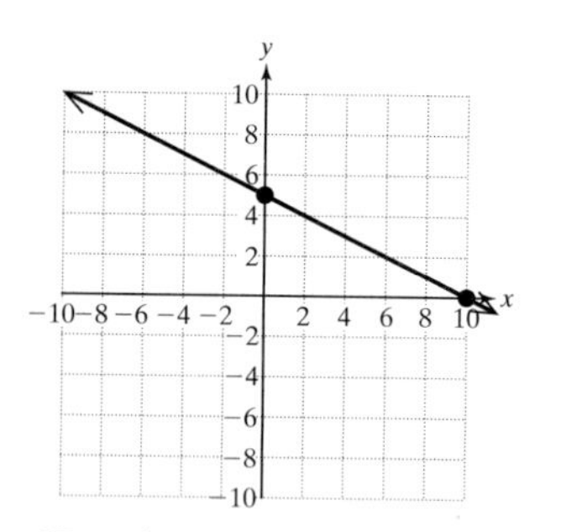

65. x-intercept: $(-2, 0)$; y-intercept: $(0, -1.5)$
$-8.1x - 10.8y = 16.2$

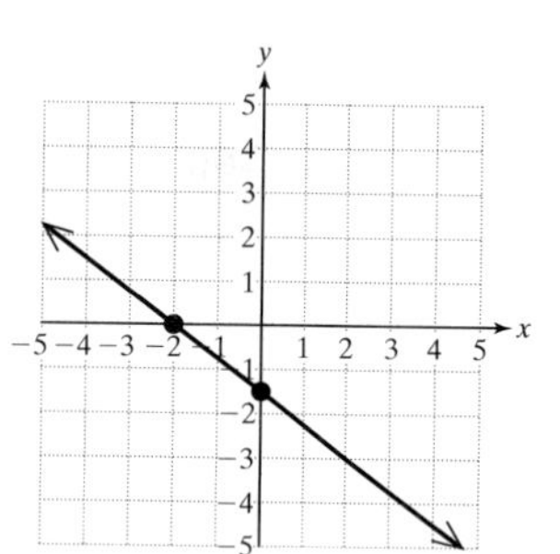

67. x-intercept: $(0, 0)$; y-intercept: $(0, 0)$ $x = -5y$

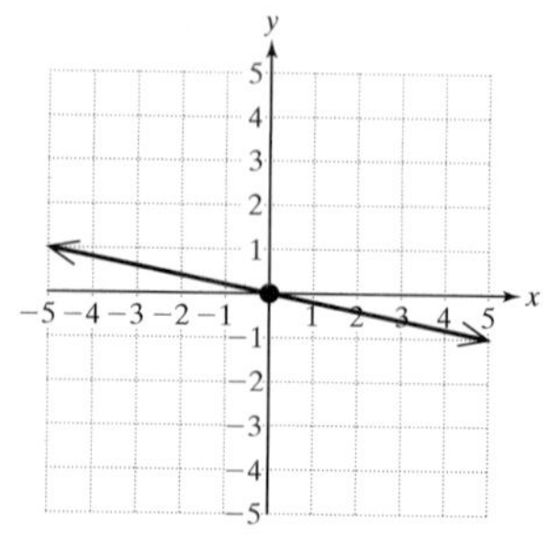

69. b, c, d **71.** False, $x = 3$ is vertical **73.** True

75. a. Vertical
c. x-intercept: $(3, 0)$; no y-intercept

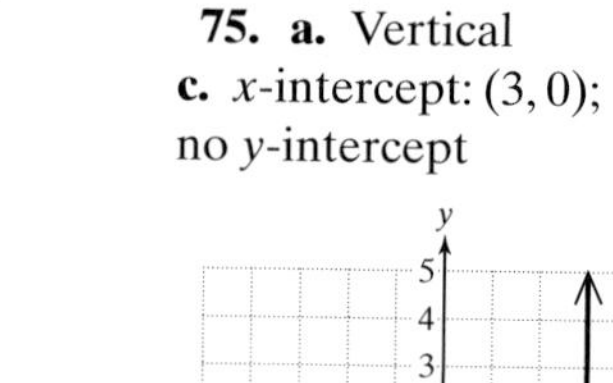

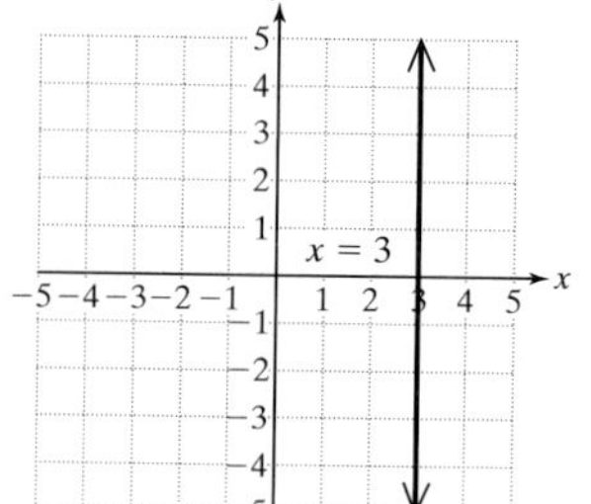

77. a. Horizontal
c. no x-intercept; y-intercept: $(0, -4)$

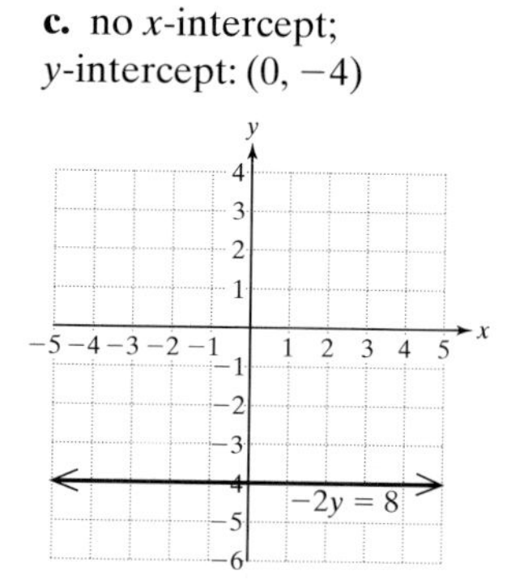

79. a. Vertical
c. x-intercept: $(4, 0)$; no y-intercept

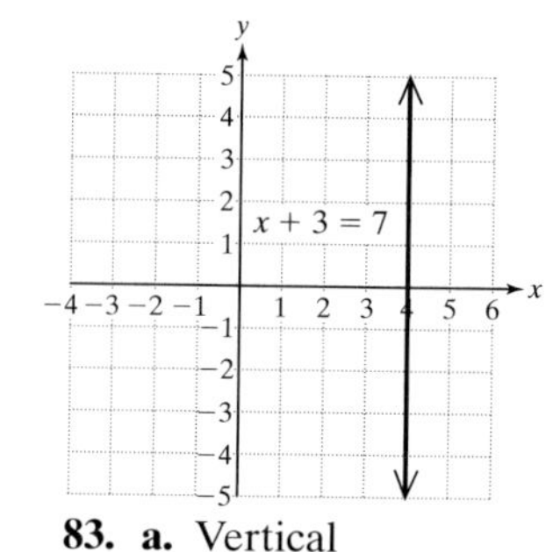

81. a. Horizontal
c. All points on the x-axis are x-intercepts; y-intercept: $(0, 0)$

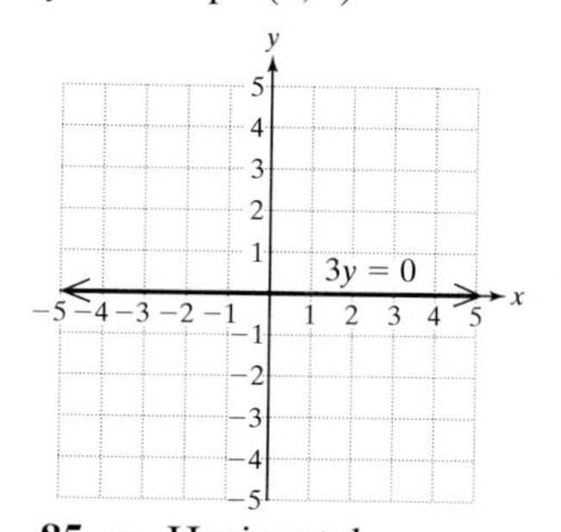

83. a. Vertical
c. x-intercept: $(\frac{3}{2}, 0)$; no y-intercept

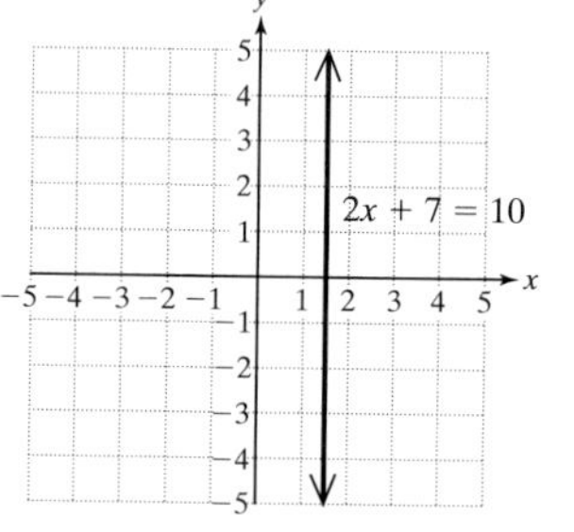

85. a. Horizontal
c. no x-intercept; y-intercept: $(0, \frac{3}{2})$

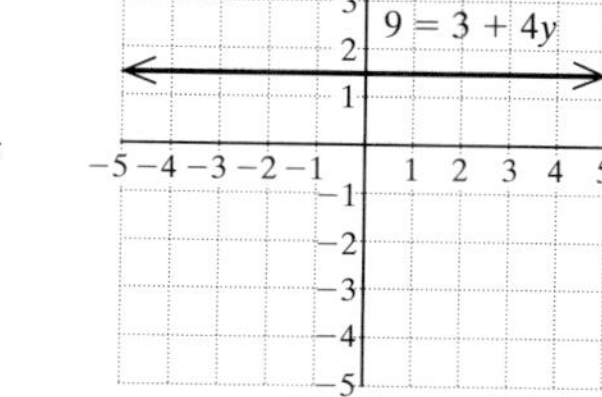

87. a. $y = 17.95$ **b.** $x = 145$
c. $(55, 17.95)$ Collecting 55 lb of cans yields \$17.95. $(145, 80.05)$ Collecting 145 lb of cans yields \$80.05.
d.

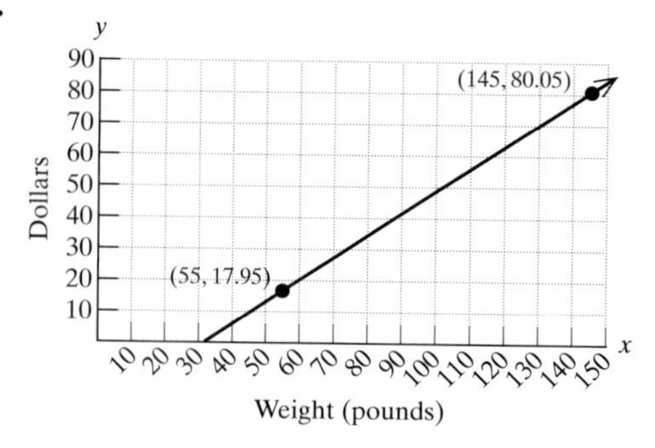

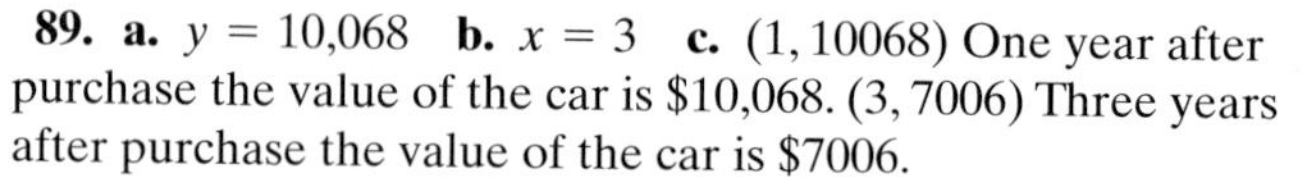

89. a. $y = 10{,}068$ **b.** $x = 3$ **c.** $(1, 10068)$ One year after purchase the value of the car is \$10,068. $(3, 7006)$ Three years after purchase the value of the car is \$7006.

Section 3.3 Practice Exercises, pp. 246–253

3. x-intercept: $(6, 0)$; y-intercept: $(0, -2)$ $x - 3y = 6$

5. x-intercept: $(0, 0)$; y-intercept: $(0, 0)$ $y = \frac{2}{3}x$

7. x-intercept: $(2, 0)$; y-intercept: $(0, 8)$ $4x + y = 8$

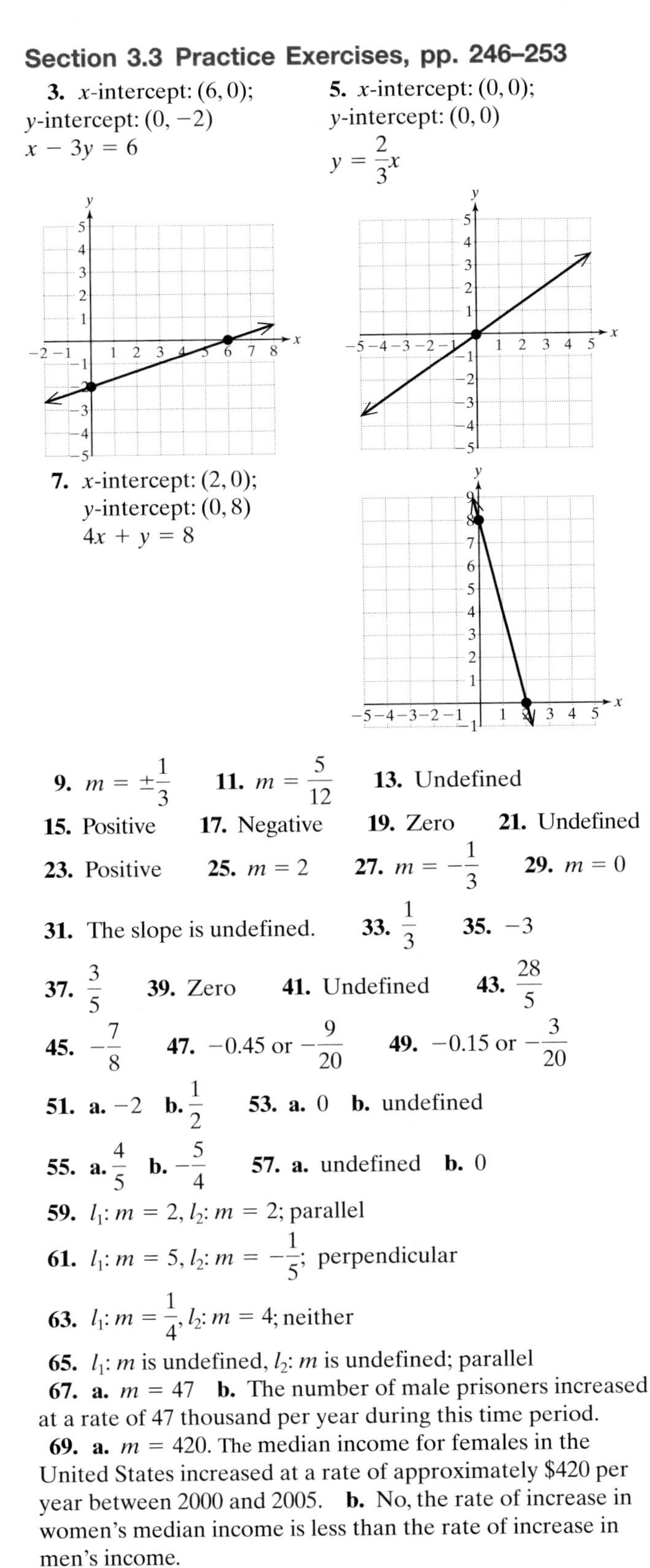

9. $m = \pm\frac{1}{3}$ **11.** $m = \frac{5}{12}$ **13.** Undefined **15.** Positive **17.** Negative **19.** Zero **21.** Undefined **23.** Positive **25.** $m = 2$ **27.** $m = -\frac{1}{3}$ **29.** $m = 0$ **31.** The slope is undefined. **33.** $\frac{1}{3}$ **35.** -3 **37.** $\frac{3}{5}$ **39.** Zero **41.** Undefined **43.** $\frac{28}{5}$ **45.** $-\frac{7}{8}$ **47.** -0.45 or $-\frac{9}{20}$ **49.** -0.15 or $-\frac{3}{20}$ **51. a.** -2 **b.** $\frac{1}{2}$ **53. a.** 0 **b.** undefined **55. a.** $\frac{4}{5}$ **b.** $-\frac{5}{4}$ **57. a.** undefined **b.** 0

59. l_1: $m = 2$, l_2: $m = 2$; parallel

61. l_1: $m = 5$, l_2: $m = -\frac{1}{5}$; perpendicular

63. l_1: $m = \frac{1}{4}$, l_2: $m = 4$; neither

65. l_1: m is undefined, l_2: m is undefined; parallel

67. a. $m = 47$ **b.** The number of male prisoners increased at a rate of 47 thousand per year during this time period.

69. a. $m = 420$. The median income for females in the United States increased at a rate of approximately \$420 per year between 2000 and 2005. **b.** No, the rate of increase in women's median income is less than the rate of increase in men's income.

71. a. 1 mile **b.** 2 miles **c.** 3 miles **d.** $m = 0.2$; The distance between a lightning strike and an observer increases by 0.2 miles for every additional second between seeing lightning and hearing thunder.

73. a. $m = -\frac{2}{5}$ **75.** $m = 1$ **77.** The slope is undefined.

79. For example: $(2, 0)$ and $(-6, -6)$

81. For example: $(0, 1)$ and $(-2, 5)$

83. **85.** **87.** **89.**

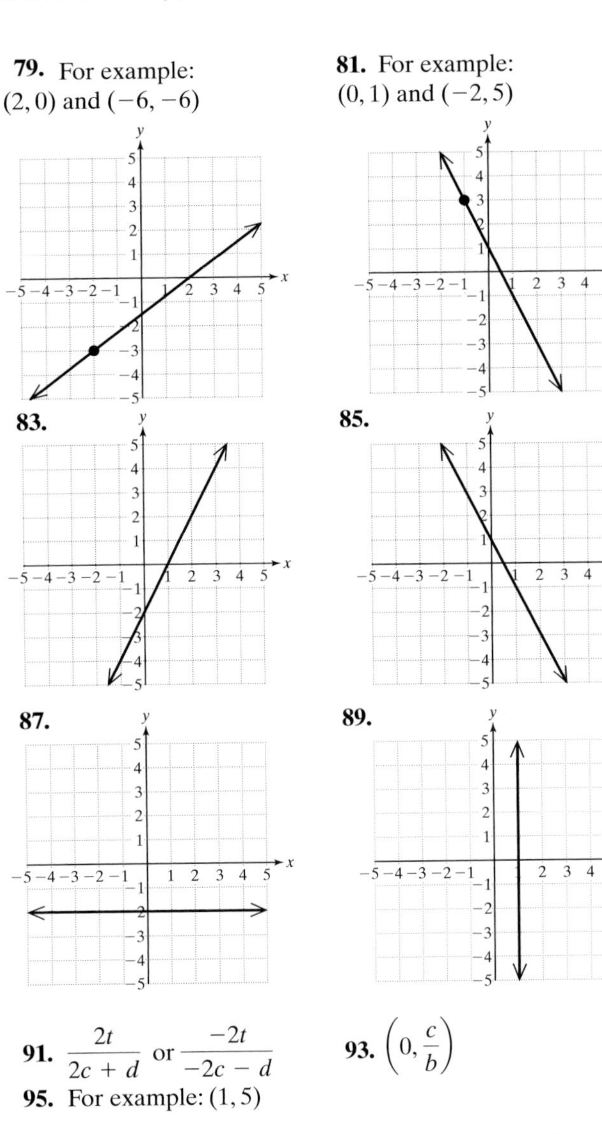

91. $\frac{2t}{2c + d}$ or $\frac{-2t}{-2c - d}$ **93.** $\left(0, \frac{c}{b}\right)$ **95.** For example: $(1, 5)$

Section 3.4 Calculator Connections, p. 258

1. Perpendicular

2. Parallel

3. Neither

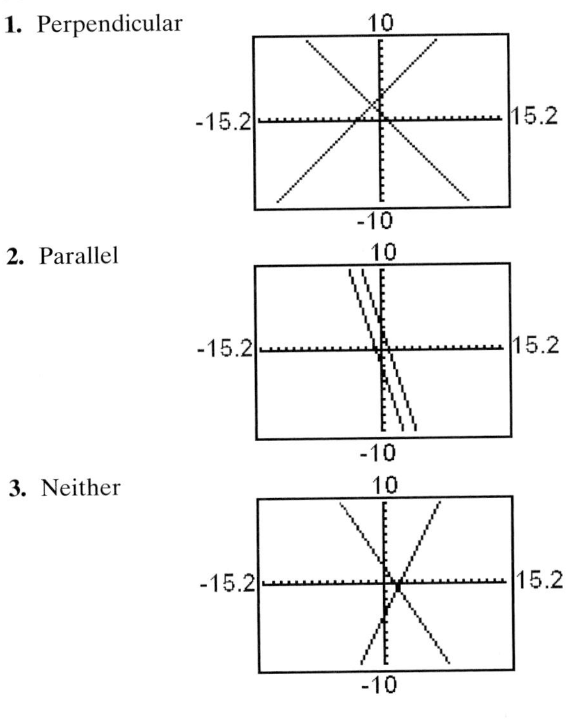

4. The lines may appear parallel; however, they are not parallel because the slopes are different.
5. The lines may appear to coincide on a graph; however, they are not the same line because the y-intercepts are different.

Section 3.4 Practice Exercises, pp. 259–263

3. x-intercept: $(10, 0)$; y-intercept: $(0, -2)$
5. x-intercept: none; y-intercept: $(0, -3)$
7. x-intercept: $(0, 0)$; y-intercept: $(0, 0)$
9. x-intercept: $(4, 0)$; y-intercept: none
11. $m = \frac{2}{3}$; y-intercept: $(0, 5)$
13. $m = -1$; y-intercept: $(0, 6)$
15. $m = -4$; y-intercept: $(0, 0)$
17. $m = 1$; y-intercept: $\left(0, -\frac{5}{3}\right)$
19. $m = -\frac{3}{2}$; y-intercept: $\left(0, \frac{9}{2}\right)$
21. $m = \frac{7}{3}$; y-intercept: $(0, 2)$
23. $m = 1$; y-intercept: $(0, -1)$
25. Undefined slope; no y-intercept
27. $m = 0$; y-intercept: $(0, -8)$
29. $m = \frac{5}{6}$; y-intercept: $(0, 0)$

31. **33.**

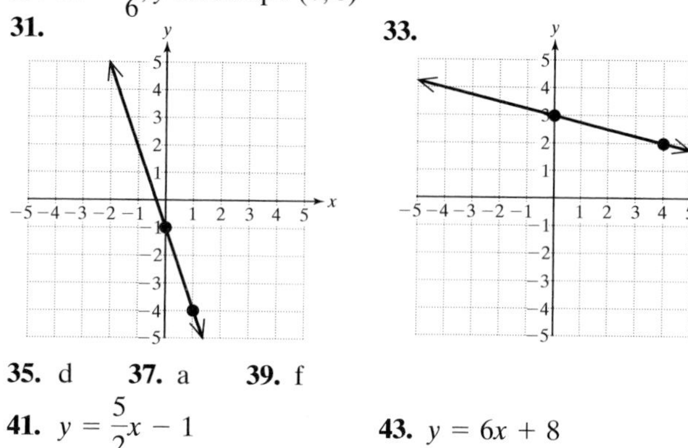

35. d **37.** a **39.** f

41. $y = \frac{5}{2}x - 1$

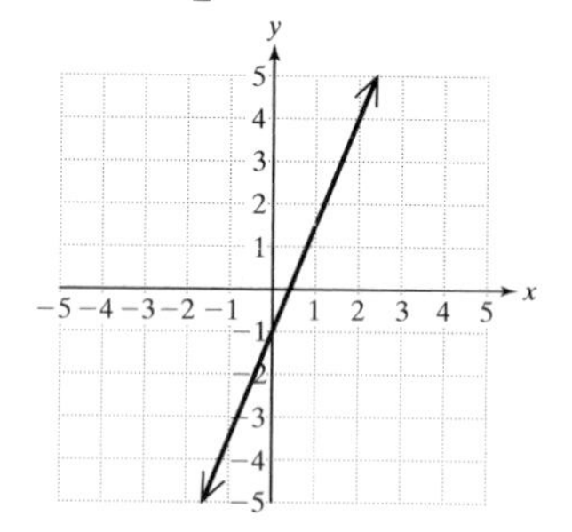

43. $y = 6x + 8$

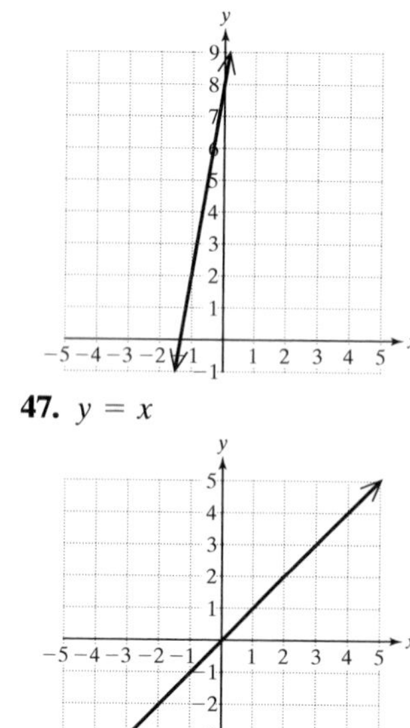

45. $y = 3x + 7$

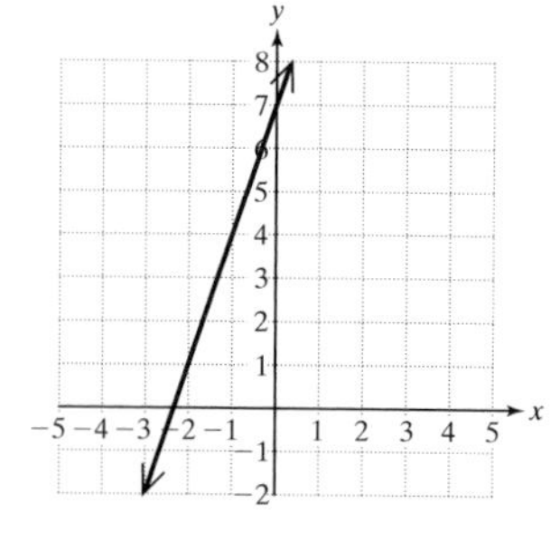

47. $y = x$

49. $y = -\frac{2}{5}x$ **51.** $y = 1$

53. $x = -3$ **55.** $y = 2x - 3$

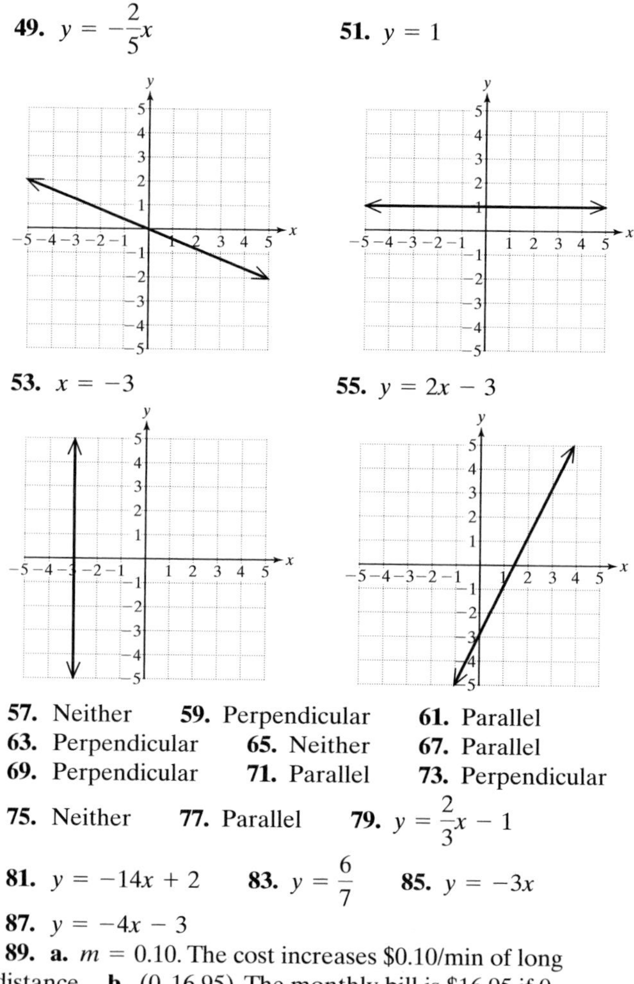

57. Neither **59.** Perpendicular **61.** Parallel
63. Perpendicular **65.** Neither **67.** Parallel
69. Perpendicular **71.** Parallel **73.** Perpendicular
75. Neither **77.** Parallel **79.** $y = \frac{2}{3}x - 1$
81. $y = -14x + 2$ **83.** $y = \frac{6}{7}$ **85.** $y = -3x$
87. $y = -4x - 3$
89. **a.** $m = 0.10$. The cost increases \$0.10/min of long distance. **b.** $(0, 16.95)$. The monthly bill is \$16.95 if 0 minutes of long distance are used. **c.** The cost is \$40.35
91. $m = -\frac{2}{5}$ **93.** $m = \frac{4}{3}$

Section 3.5 Practice Exercises, pp. 269–273

3. $2x - 3y = -3$ **5.** $3 - y = 9$

7. 9 **9.** 0 **11.** $y = 3x + 7$ or $3x - y = -7$
13. $y = -4x - 14$ or $4x + y = -14$
15. $y = -\frac{1}{2}x - \frac{1}{2}$ or $x + 2y = -1$
17. $y = \frac{1}{4}x + 8$ or $x - 4y = -32$
19. $y = 4.5x - 25.6$ or $45x - 10y = 256$
21. $y = -2$

23. $y = -2x + 1$ **25.** $y = \frac{1}{2}x - 1$

27. $y = 2x - 2$ or $2x - y = 2$

29. $y = -\frac{5}{8}x - \frac{19}{8}$ or $5x + 8y = -19$

31. $y = -0.2x - 2.86$ or $20x + 100y = -286$

33. $y = 4x + 13$ or $4x - y = -13$

35. $y = -\frac{3}{2}x + 6$ or $3x + 2y = 12$

37. $y = -2x - 8$ or $2x + y = -8$

39. $y = -\frac{1}{5}x - 6$ or $x + 5y = -30$

41. $y = 3x - 8$ or $3x - y = 8$

43. $y = -\frac{1}{4}x - \frac{3}{2}$ or $x + 4y = -6$

45. iv **47.** vi **49.** iii **51.** $y = 1$ **53.** $x = 2$

55. $x = \frac{5}{2}$ **57.** $y = 2$ **59.** $x = -6$ **61.** $x = -4$

63. a. -0.35 **b.** $y = -0.35x + 33.9$ **c.** In 2000, approximately 21.65% of women smoked.

Section 3.6 Calculator Connections, pp. 278–279

1. 13.3 **2.** -42.3 **3.** 345 **4.** 95

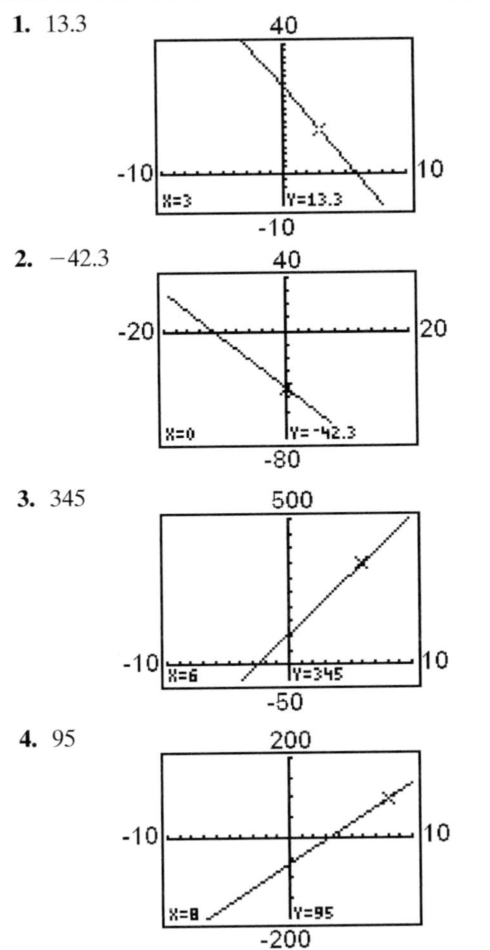

Section 3.6 Practice Exercises, pp. 279–283

3. x-intercept: $(6, 0)$; y-intercept: $(0, 5)$

5. x-intercept: $(-2, 0)$; y-intercept: $(0, -4)$

7. x-intercept: none; y-intercept: $(0, -9)$

9. a. x, year **b.** y, minimum hourly wage **c.** \$3.32 per hour **d.** \$5.82 per hour **e.** The y-intercept is $(0, 0.82)$ and indicates that the minimum hourly wage in the year 1960 ($x = 0$) was approximately \$0.82/hour. **f.** The slope is 0.10 and indicates that minimum hourly wage rises an average of \$0.10 per year.

11. a. 0.75 or equivalently $\frac{3}{4}$. O'Neal would expect to score 0.75 point for each minute played. This is equivalent to scoring 3 points for 4 minutes played. **b.** 0.50 or equivalently $\frac{1}{2}$. Iverson would expect to score 0.5 point for each minute played. This is equivalent to scoring 1 point for 2 minutes played. **c.** O'Neal: 27 points; Iverson: 18 points

13. a. y, temperature **b.** x, latitude **c.** 30.7° **d.** 13.4° **e.** $m = -2.333$. The average temperature in January decreases 2.333° per 1° of latitude. **f.** $(53.2, 0)$. At 53.2° latitude, the average temperature in January is 0°.

15. a. $y = 2.5x + 31$ **b.** 43.5 in.

17. a. $y = 2x + 2$ **b.** $m = 2$. For each additional mile, the time is increased by 2 min. **c.** 38 min

19. a. $y = 0.08x + 18.95$ **b.** \$25.91

21. a. $y = 90x + 105$ **b.** \$1185.00

23. a. $y = 25x + 5000$ **b.** \$8750.00

25. a. $y = 0.5x + 35$ **b.** \$210.00

Chapter 3 Review Exercises, pp. 288–292

1.

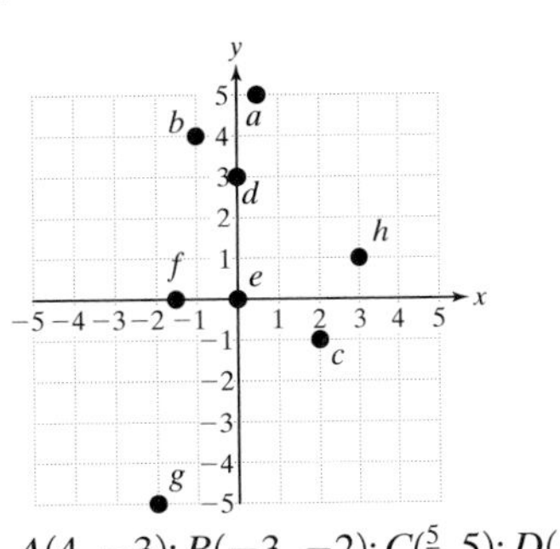

2. $A(4, -3)$; $B(-3, -2)$; $C(\frac{5}{2}, 5)$; $D(-4, 1)$; $E(-\frac{1}{2}, 0)$; $F(0, -5)$

3. III **4.** II **5.** IV **6.** I **7.** IV **8.** III

9. x-axis **10.** y-axis

11. a. On day 1, the price was \$26.25. **b.** Day 2 **c.** \$2.25

12. a. In 2003 (8 years after 1995), there was only 1 space shuttle launch (this was the year that the Columbia and its crew were lost).

b.

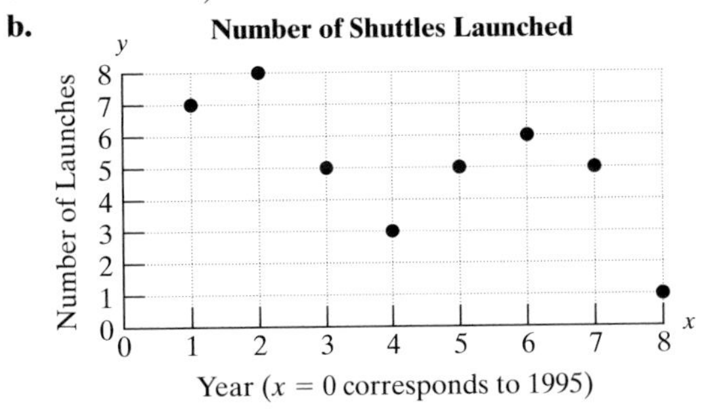

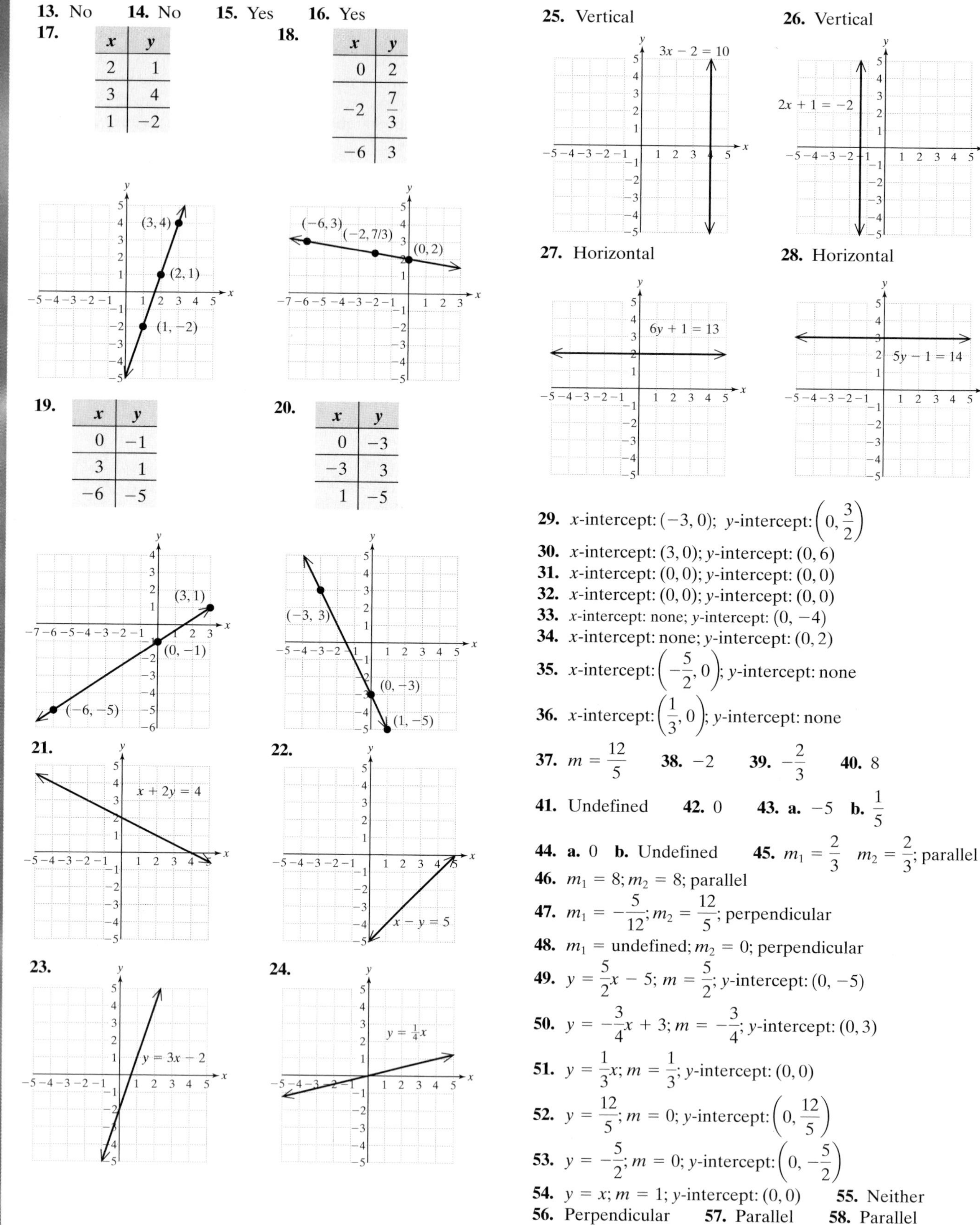

13. No **14.** No **15.** Yes **16.** Yes

17.

x	y
2	1
3	4
1	−2

18.

x	y
0	2
−2	$\frac{7}{3}$
−6	3

19.

x	y
0	−1
3	1
−6	−5

20.

x	y
0	−3
−3	3
1	−5

21.

22.

23.

24.

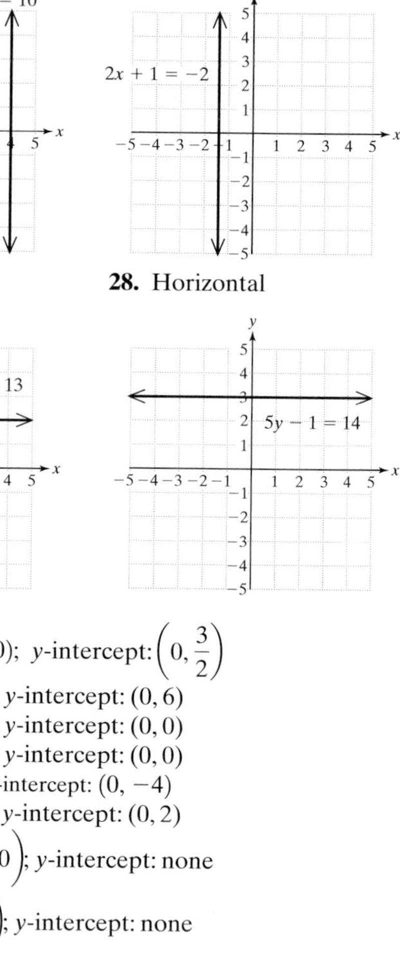

25. Vertical

26. Vertical

27. Horizontal

28. Horizontal

29. x-intercept: $(-3, 0)$; y-intercept: $\left(0, \frac{3}{2}\right)$

30. x-intercept: $(3, 0)$; y-intercept: $(0, 6)$

31. x-intercept: $(0, 0)$; y-intercept: $(0, 0)$

32. x-intercept: $(0, 0)$; y-intercept: $(0, 0)$

33. x-intercept: none; y-intercept: $(0, -4)$

34. x-intercept: none; y-intercept: $(0, 2)$

35. x-intercept: $\left(-\frac{5}{2}, 0\right)$; y-intercept: none

36. x-intercept: $\left(\frac{1}{3}, 0\right)$; y-intercept: none

37. $m = \frac{12}{5}$ **38.** -2 **39.** $-\frac{2}{3}$ **40.** 8

41. Undefined **42.** 0 **43. a.** -5 **b.** $\frac{1}{5}$

44. a. 0 **b.** Undefined **45.** $m_1 = \frac{2}{3}$ $m_2 = \frac{2}{3}$; parallel

46. $m_1 = 8$; $m_2 = 8$; parallel

47. $m_1 = -\frac{5}{12}$; $m_2 = \frac{12}{5}$; perpendicular

48. $m_1 =$ undefined; $m_2 = 0$; perpendicular

49. $y = \frac{5}{2}x - 5$; $m = \frac{5}{2}$; y-intercept: $(0, -5)$

50. $y = -\frac{3}{4}x + 3$; $m = -\frac{3}{4}$; y-intercept: $(0, 3)$

51. $y = \frac{1}{3}x$; $m = \frac{1}{3}$; y-intercept: $(0, 0)$

52. $y = \frac{12}{5}$; $m = 0$; y-intercept: $\left(0, \frac{12}{5}\right)$

53. $y = -\frac{5}{2}$; $m = 0$; y-intercept: $\left(0, -\frac{5}{2}\right)$

54. $y = x$; $m = 1$; y-intercept: $(0, 0)$ **55.** Neither

56. Perpendicular **57.** Parallel **58.** Parallel

59. Perpendicular **60.** $y = -\frac{4}{3}x - 1$ or $4x + 3y = -3$

61. $y = 5x$ or $5x - y = 0$ **62.** For example: $y = 3x + 2$

63. For example: $5x + 2y = -4$ **64.** $m = \frac{y_2 - y_1}{x_2 - x_1}$

65. $y - y_1 = m(x - x_1)$ **66.** For example: $x = 6$

67. For example: $y = -5$

68. $y = -6x + 2$ or $6x + y = 2$

69. $y = \frac{2}{3}x + \frac{5}{3}$ or $2x - 3y = -5$

70. $y = \frac{1}{4}x - 4$ or $x - 4y = 16$ **71.** $y = -5$

72. $y = \frac{6}{5}x + 6$ or $6x - 5y = -30$

73. $y = 4x + 31$ or $4x - y = -31$

74. a. x, age **b.** y, height **c.** 47.8 in. **d.** The slope is 2.4 and indicates that the average height for girls increases at a rate of 2.4 in. per year. **75. a.** $m = 137$ **b.** The number of prescriptions increased by 137 million per year during this time period. **c.** $y = 137x + 2140$ **d.** 3921 million

76. a. $y = 30.6x + 264$ **b.** $814.8 billion

77. a. $y = 20x + 55$ **b.** $235

78. a. $y = 8x + 700$ **b.** $1340

Chapter 3 Test, pp. 292–294

1. a. II **b.** IV **c.** III **2.** 0 **3.** 0

4. a. (5, 46) At age 5 the boy's height was 46 in. (7, 50) (9, 55) (11, 60) **b.**

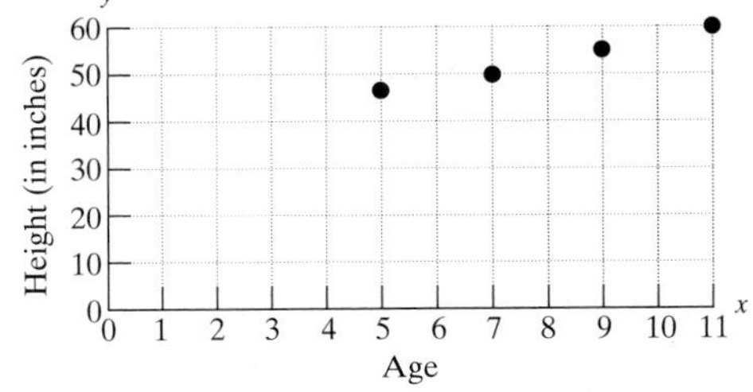

c. 57.5 in. **d.** No, his height will maximize in his teen years.

5. a. No **b.** Yes **c.** Yes **d.** Yes

6.

x	y
0	-2
4	-1
6	$-\frac{1}{2}$

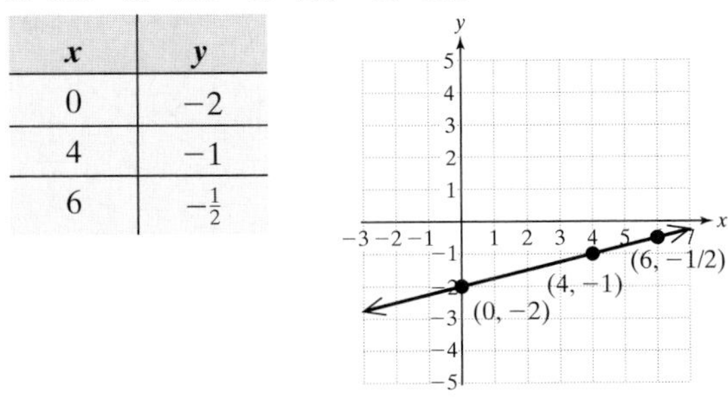

7. a. 202 beats per minute **b.** (20, 200) (30, 190) (40, 180) (50, 170) (60, 160)

8. Horizontal **9.** Vertical

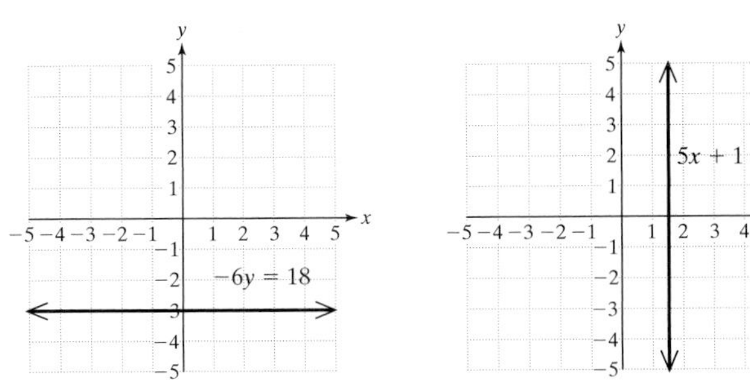

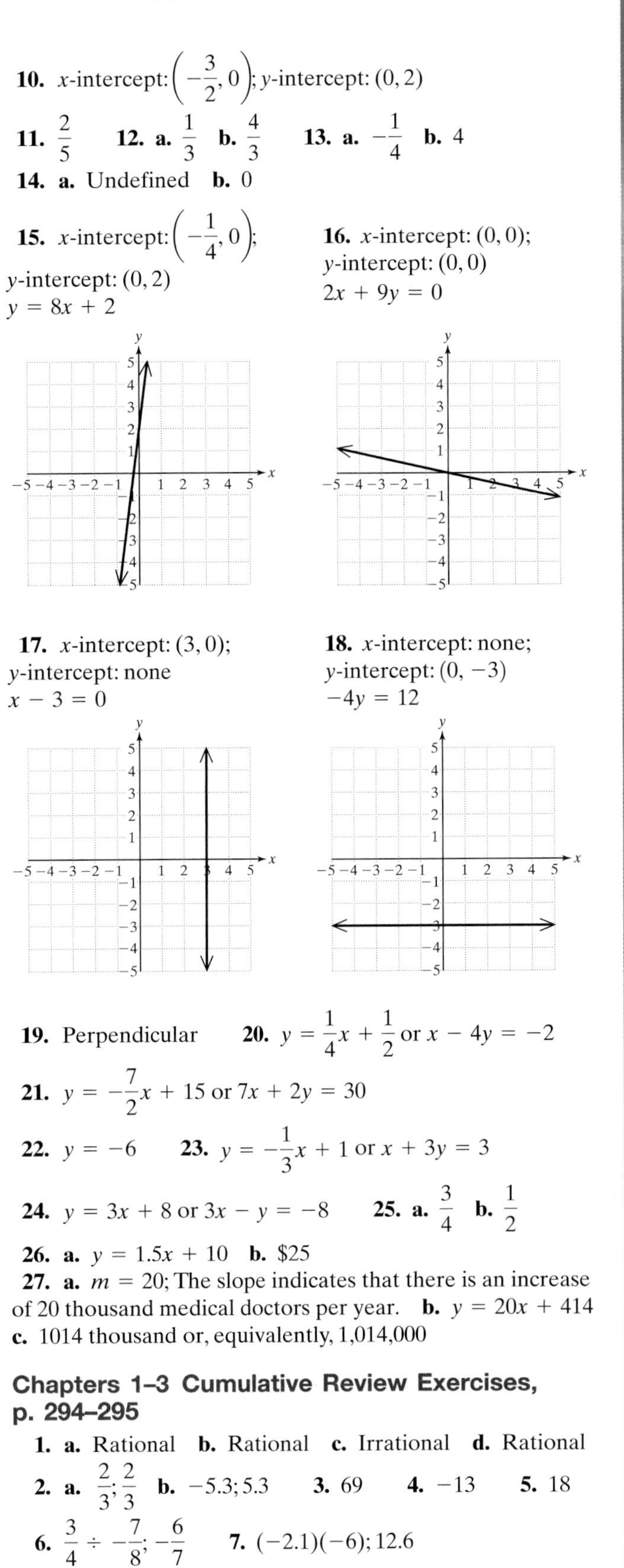

10. x-intercept: $\left(-\frac{3}{2}, 0\right)$; y-intercept: (0, 2)

11. $\frac{2}{5}$ **12. a.** $\frac{1}{3}$ **b.** $\frac{4}{3}$ **13. a.** $-\frac{1}{4}$ **b.** 4

14. a. Undefined **b.** 0

15. x-intercept: $\left(-\frac{1}{4}, 0\right)$; y-intercept: (0, 2) $y = 8x + 2$

16. x-intercept: (0, 0); y-intercept: (0, 0) $2x + 9y = 0$

17. x-intercept: (3, 0); y-intercept: none $x - 3 = 0$

18. x-intercept: none; y-intercept: (0, −3) $-4y = 12$

19. Perpendicular **20.** $y = \frac{1}{4}x + \frac{1}{2}$ or $x - 4y = -2$

21. $y = -\frac{7}{2}x + 15$ or $7x + 2y = 30$

22. $y = -6$ **23.** $y = -\frac{1}{3}x + 1$ or $x + 3y = 3$

24. $y = 3x + 8$ or $3x - y = -8$ **25. a.** $\frac{3}{4}$ **b.** $\frac{1}{2}$

26. a. $y = 1.5x + 10$ **b.** $25

27. a. $m = 20$; The slope indicates that there is an increase of 20 thousand medical doctors per year. **b.** $y = 20x + 414$ **c.** 1014 thousand or, equivalently, 1,014,000

Chapters 1–3 Cumulative Review Exercises, p. 294–295

1. a. Rational **b.** Rational **c.** Irrational **d.** Rational

2. a. $\frac{2}{3}; \frac{2}{3}$ **b.** −5.3; 5.3 **3.** 69 **4.** −13 **5.** 18

6. $\frac{3}{4} \div -\frac{7}{8}; -\frac{6}{7}$ **7.** (−2.1)(−6); 12.6

8. The associative property of addition **9.** $x = 4$

10. $m = 5$ **11.** $y = -\frac{9}{2}$ **12.** $z = -2$ **13.** 9241 mi^2

14. $a = \dfrac{c - b}{3}$ **15.**

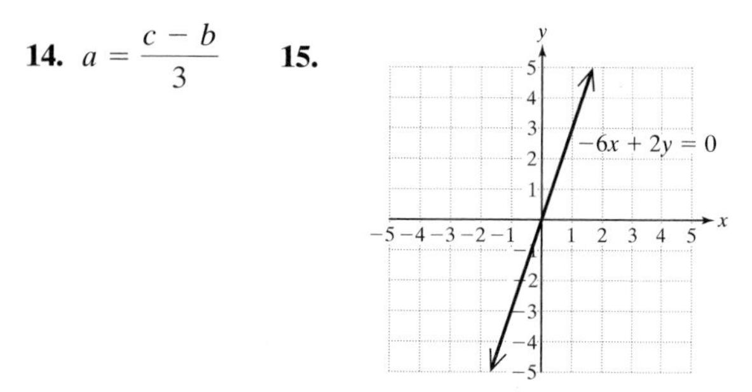

16. x-intercept: $(-2, 0)$; y-intercept: $(0, 1)$

17. $y = -\frac{3}{2}x - 6$; slope: $-\frac{3}{2}$; y-intercept: $(0, -6)$

18. $2x + 3 = 5$ can be written as $x = 1$, which represents a vertical line. A vertical line of the form $x = k$ $(k \neq 0)$ has an x-intercept of $(k, 0)$ and no y-intercept. **19.** $y = -3x + 1$ or $3x + y = 1$ **20.** $y = \frac{2}{3}x + 6$ or $2x - 3y = -18$

Chapter 4

Chapter Opener Puzzle

One drink costs \$2.00 and one small popcorn costs \$3.50.

Section 4.1 Calculator Connections, pp. 303–304

1. $(2, 1)$ **2.** $(6, -1)$ **3.** $(3, 1)$ **4.** $(-2, 0)$

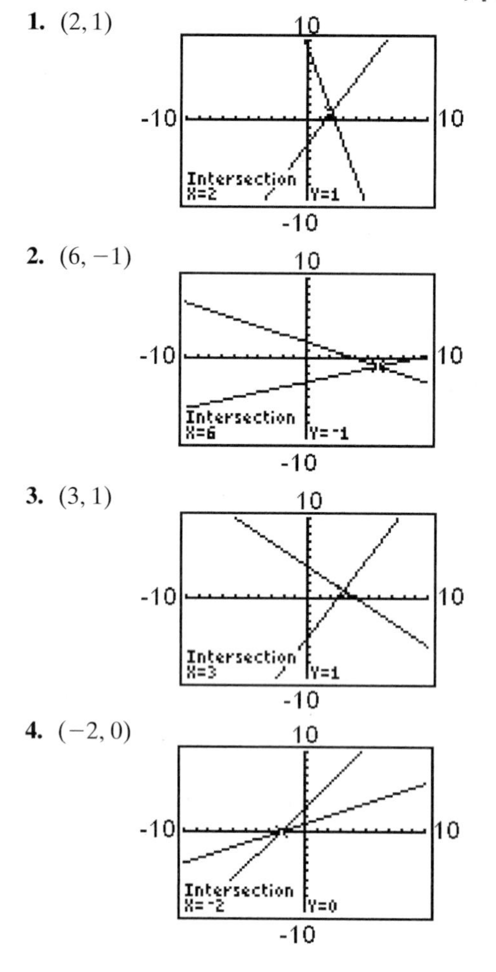

5. No solution, inconsistent system

6. Dependent system

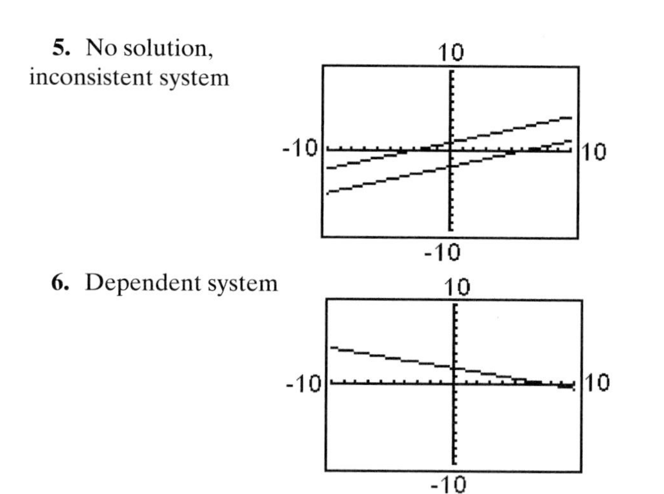

Section 4.1 Practice Exercises, pp. 304–309

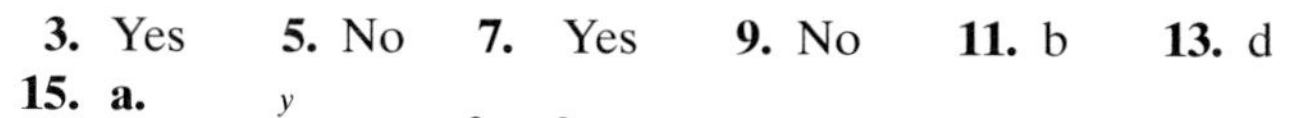
3. Yes **5.** No **7.** Yes **9.** No **11.** b **13.** d

15. a. **b.** **c.**

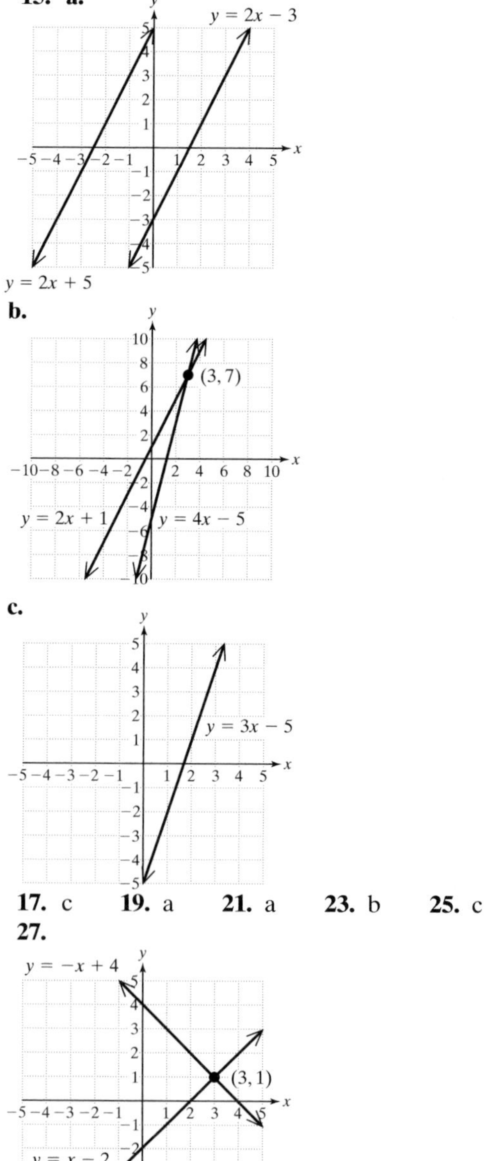

17. c **19.** a **21.** a **23.** b **25.** c

27.

29.

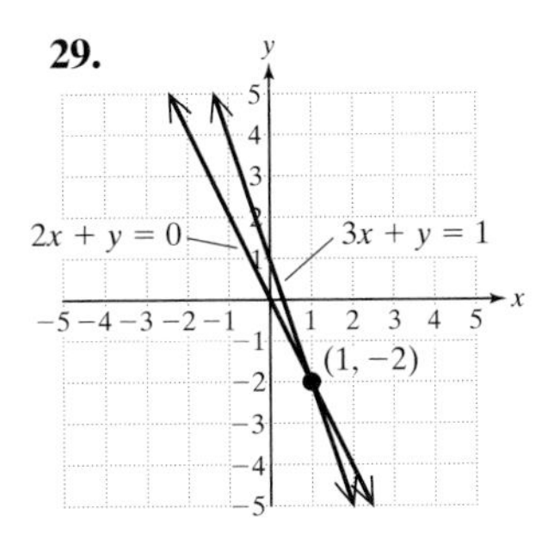

33. No solution; Inconsistent system

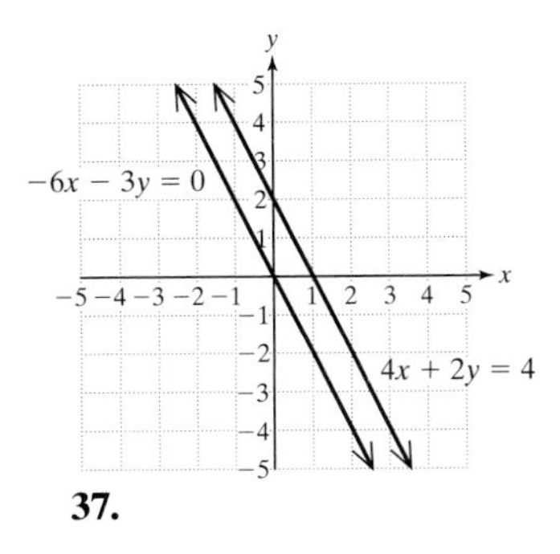

37.

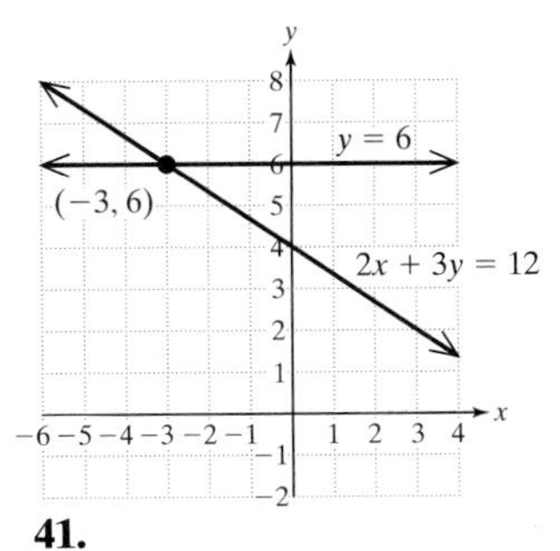

41.

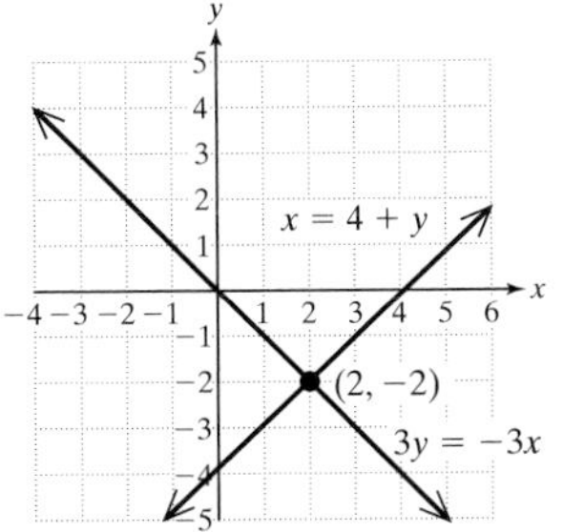

45.

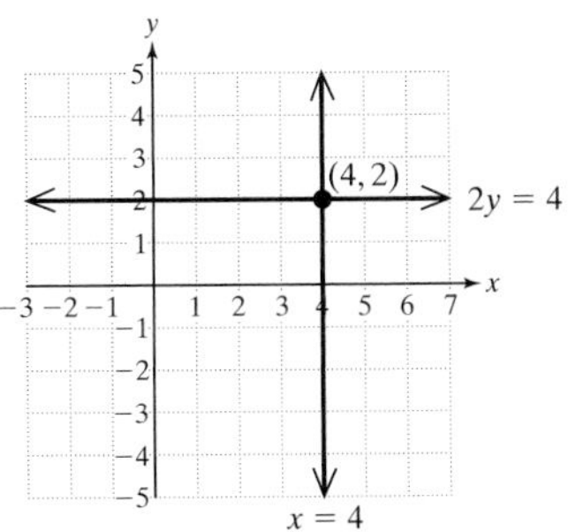

31.

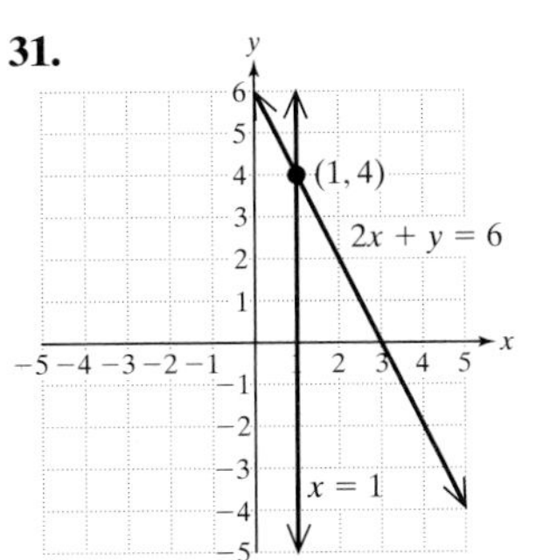

35. Infinitely many solutions $\{(x, y) \mid -2x + y = 3\}$; Dependent system

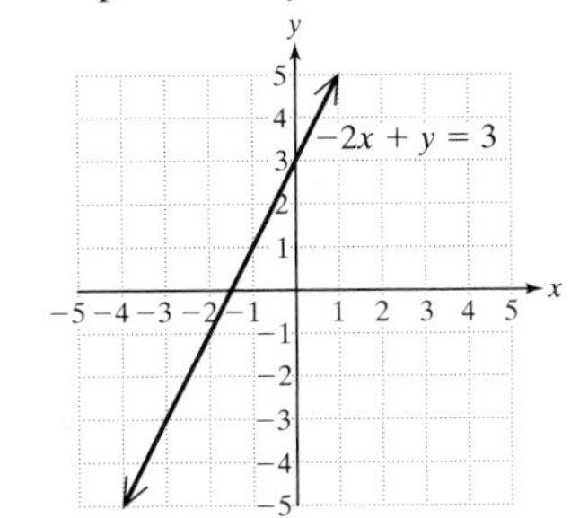

39. Infinitely many solutions $\{(x, y) \mid y = \frac{5}{3}x - 3\}$; Dependent system

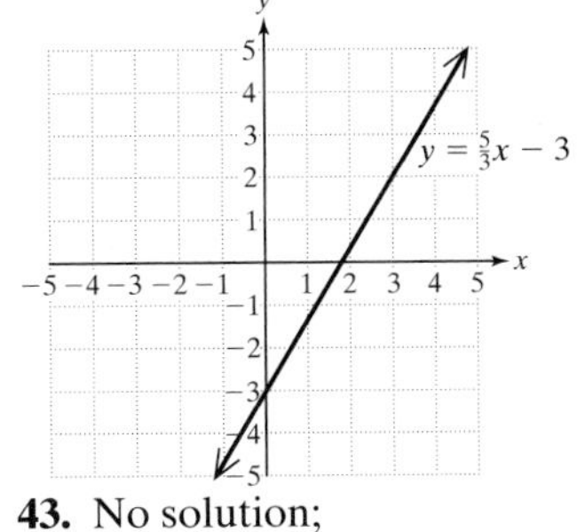

43. No solution; Inconsistent system

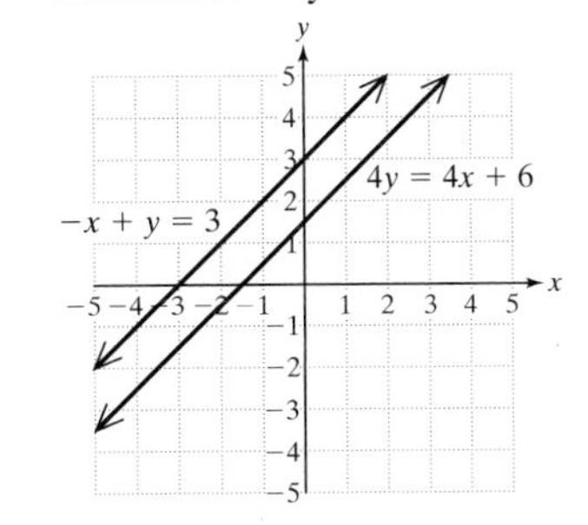

47. No solution; Inconsistent system

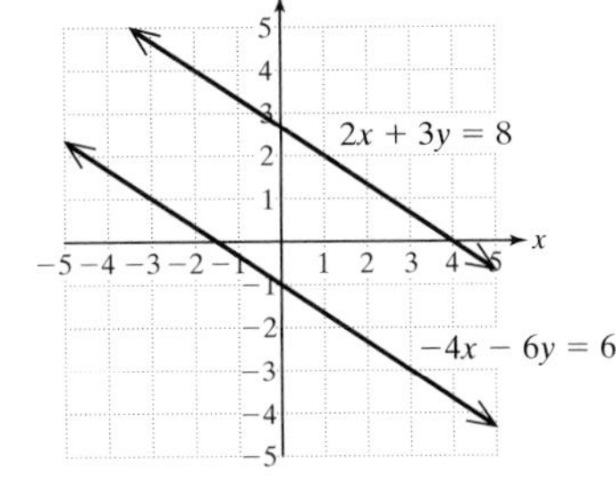

49. **51.** Infinitely many solutions $\{(x, y) \mid y = 0.5x + 2\}$; Dependent system

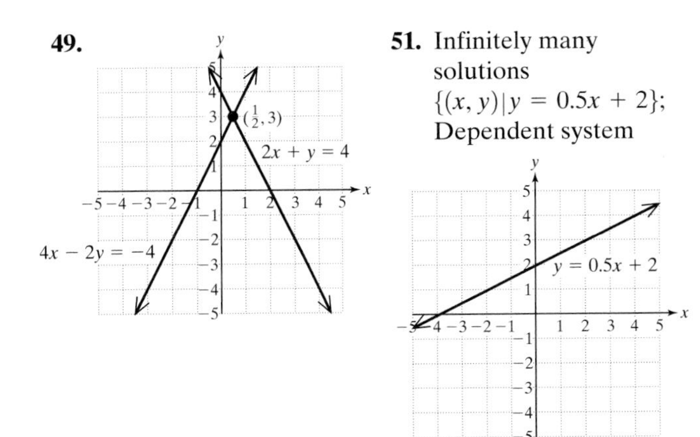

53. 4 lessons will cost \$120 for each instructor. **55.** The point of intersection is below the x-axis and cannot have a positive y-coordinate. **57.** For example: $4x + y = 9$; $-2x - y = -5$ **59.** For example: $2x + 2y = 1$

Section 4.2 Practice Exercises, pp. 317–319

1. $y = 2x - 4$; $y = 2x - 4$; coinciding lines **3.** $y = -\frac{2}{3}x + 2$; $y = x - 5$; intersecting lines **5.** $y = 4x - 4$; $y = 4x - 13$; parallel lines **7.** $(3, -6)$ **9.** $(0, 4)$ **11. a.** y in the second equation is easiest to solve for because its coefficient is 1. **b.** $(1, 5)$ **13.** $(5, 2)$ **15.** $(10, 5)$ **17.** $\left(\frac{1}{2}, 3\right)$ **19.** $(5, 3)$ **21.** $(1, 0)$ **23.** $(1, 4)$ **25.** No solution; Inconsistent system **27.** Infinitely many solutions $\{(x, y) \mid 2x - 6y = -2\}$; Dependent system **29.** $(5, -7)$ **31.** $\left(-5, \frac{3}{2}\right)$ **33.** $(2, -5)$ **35.** $(-4, 6)$ **37.** $(0, 2)$ **39.** Infinitely many solutions $\{(x, y) \mid y = 0.25x + 1\}$; Dependent system **41.** $(1, 1)$ **43.** No solution; Inconsistent system **45.** $(-1, 5)$ **47.** $(-6, -4)$ **49.** The numbers are 48 and 58. **51.** The numbers are 13 and 39. **53.** The angles are 165° and 15°. **55.** The angles are 70° and 20°. **57.** The angles are 42° and 48°. **59.** For example, $(0, 3), (1, 5), (-1, 1)$

Section 4.3 Practice Exercises, pp. 327–330

3. No **5.** Yes **7. a.** True **b.** False, multiply the second equation by 5. **9. a.** x would be easier. **b.** $(0, -3)$ **11.** $(4, -1)$ **13.** $(4, 3)$ **15.** $(2, 3)$ **17.** $(1, -4)$ **19.** $(1, -1)$ **21.** $(-4, -6)$ **23.** There are infinitely many solutions. The lines coincide. **25.** The system will have no solution. The lines are parallel. **27.** No solution; Inconsistent system **29.** Infinitely many solutions; $\{(x, y) \mid 4x - 3y = 6\}$; Dependent system **31.** $(2, -2)$ **33.** $(0, 3)$ **35.** $(5, 2)$ **37.** No solution; Inconsistent system **39.** $(-5, 0)$ **41.** $(2.5, -0.5)$ **43.** $\left(\frac{1}{2}, 0\right)$ **45.** $(-3, 2)$ **47.** $\left(\frac{7}{4}, 3\right)$ **49.** $(0, 1)$ **51.** No solution; Inconsistent system **53.** $(0, -5)$ **55.** $(4, -2)$ **57.** Infinitely many solutions; $\{(a, b) \mid a = 5 + 2b\}$; Dependent system **59.** The numbers are 17 and 19.

61. The numbers are -1 and 3. **63.** $\left(\frac{7}{9}, \frac{5}{9}\right)$

65. $\left(\frac{7}{16}, -\frac{7}{8}\right)$ **67.** $(1, 3)$

69. One line within the system of equations would have to "bend" for the system to have exactly two points of intersection. This is not possible. **71.** $A = -5, B = 2$

Section 4.4 Practice Exercises, pp. 338–341

1. $(-1, 4)$ **3.** $\left(\frac{5}{2}, 1\right)$ **5.** The numbers are 4 and 16.

7. The angles are 80° and 10°.

9. Tapes are \$10.50 each, and CDs are \$15.50 each.

11. Technology stock costs \$16 per share, and the mutual fund costs \$11 per share.

13. Shanelle invested \$3500 in the 10% account and \$6500 in the 7% account.

15. \$9000 is borrowed at 6%, and \$3000 is borrowed at 9%.

17. 15 gal of the 50% mixture should be mixed with 10 gal of the 40% mixture.

19. 12 gal of the 45% disinfectant solution should be mixed with 8 gal of the 30% disinfectant solution.

21. The speed of the boat in still water is 6 mph, and the speed of the current is 2 mph.

23. The speed of the plane in still air is 300 mph, and the wind is 20 mph. **25.** There are 17 dimes and 22 nickels.

27. a. 835 free throws and 1597 field goals **b.** 4029 points **c.** Approximately 50 points per game

29. The speed of the plane in still air is 160 mph, and the wind is 40 mph.

31. 12 lb of candy should be mixed with 8 lb of nuts.

33. \$15,000 is invested in the 5.5% account, and \$45,000 is invested in the 6.5% account.

35. 20 oz of Miracle-Gro should be mixed with 40 oz of Green Light.

37. Dallas scored 30 points, and Buffalo scored 13 points.

39. There were 300 women and 200 men in the survey.

41. There are six 15-sec commercials and sixteen 30-sec commercials.

Section 4.5 Practice Exercises, pp. 346–349

3. **5.**

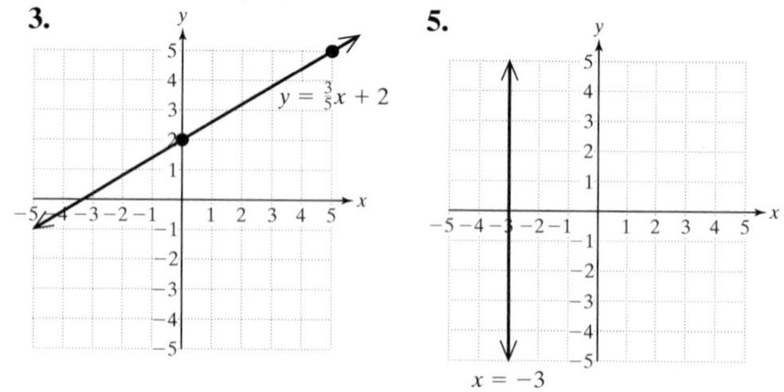

7. When the inequality symbol is $\leq$ or $\geq$

9. All of the points in the shaded region are solutions to the inequality. **11.** a

13. For example:

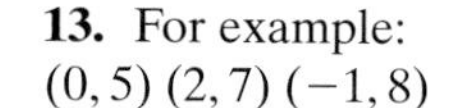

$(0, 5)$ $(2, 7)$ $(-1, 8)$

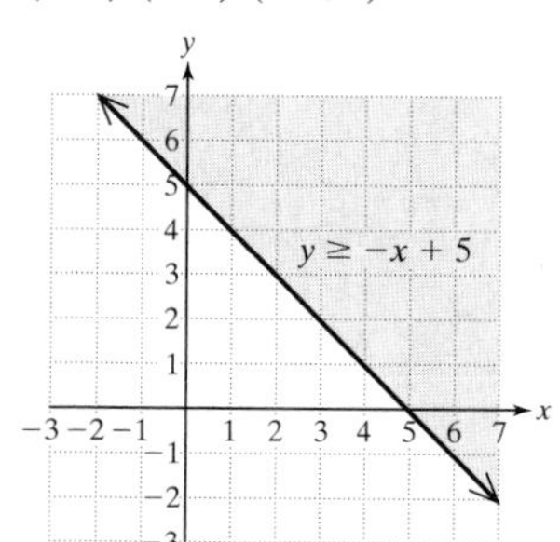

15. For example: $(1, -1)$ $(3, 0)$ $(-2, -9)$

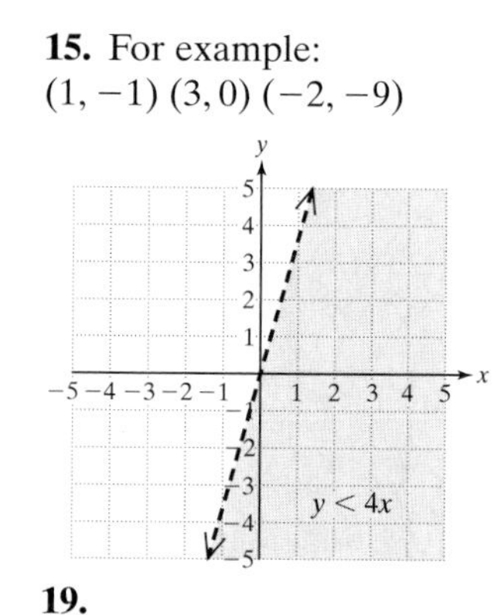

17. For example: $(0, 0)$ $(0, 2)$ $(-1, -3)$

$3x + 7y \leq 14$

19.

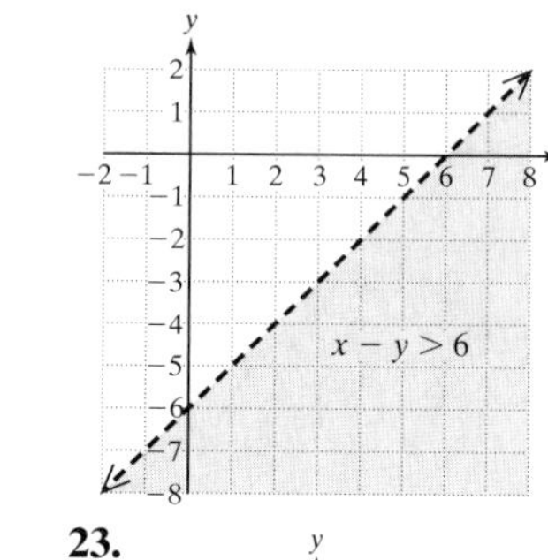

21.

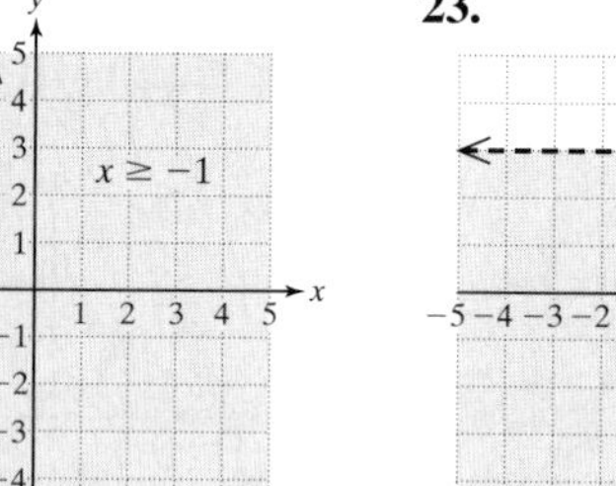

23.

$y < 3$

25.

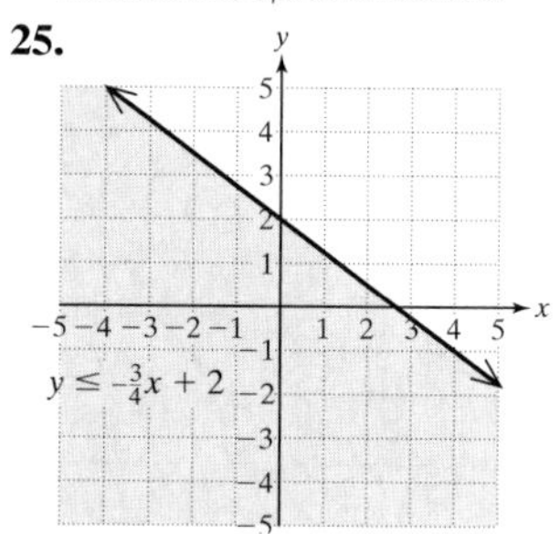

27.

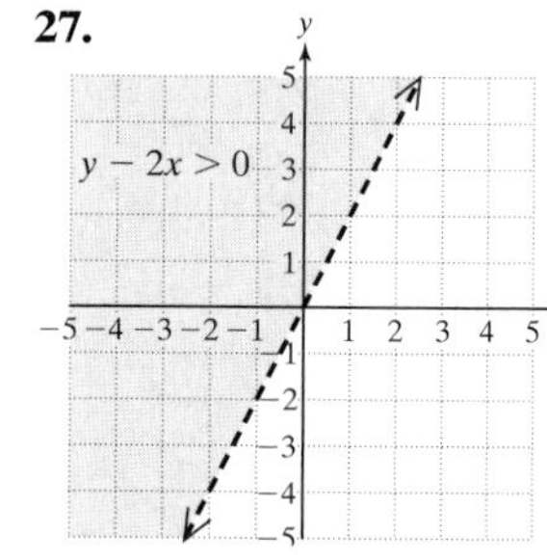

29.

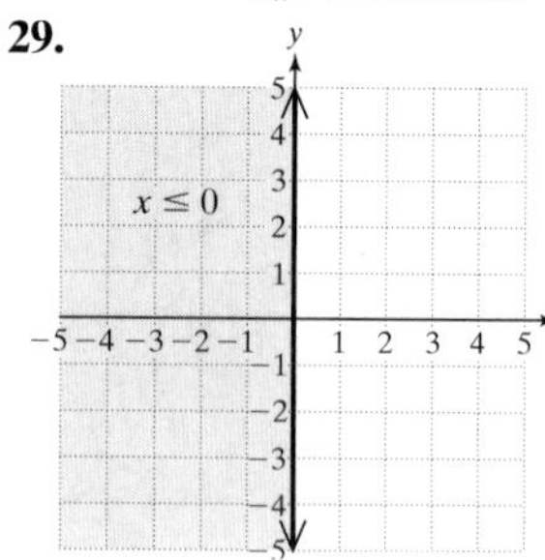

31.

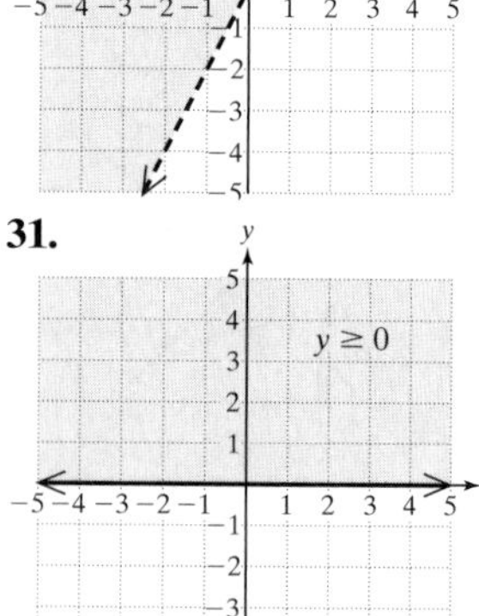

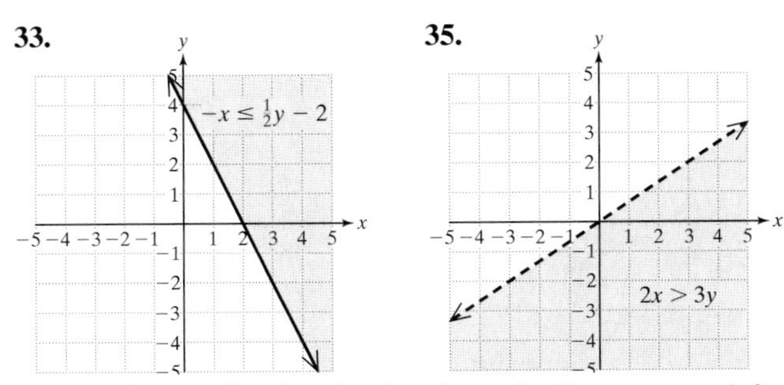

37. a. The set of ordered pairs above the line $x + y = 4$, for example, $(6, 3)(-2, 8)(0, 5)$ **b.** The set of ordered pairs on the line $x + y = 4$, for example, $(0, 4)(4, 0)(2, 2)$
c. The set of ordered pairs below the line $x + y = 4$, for example, $(0, 0)(-2, 1)(3, 0)$

Section 4.6 Practice Exercises, pp. 353–356

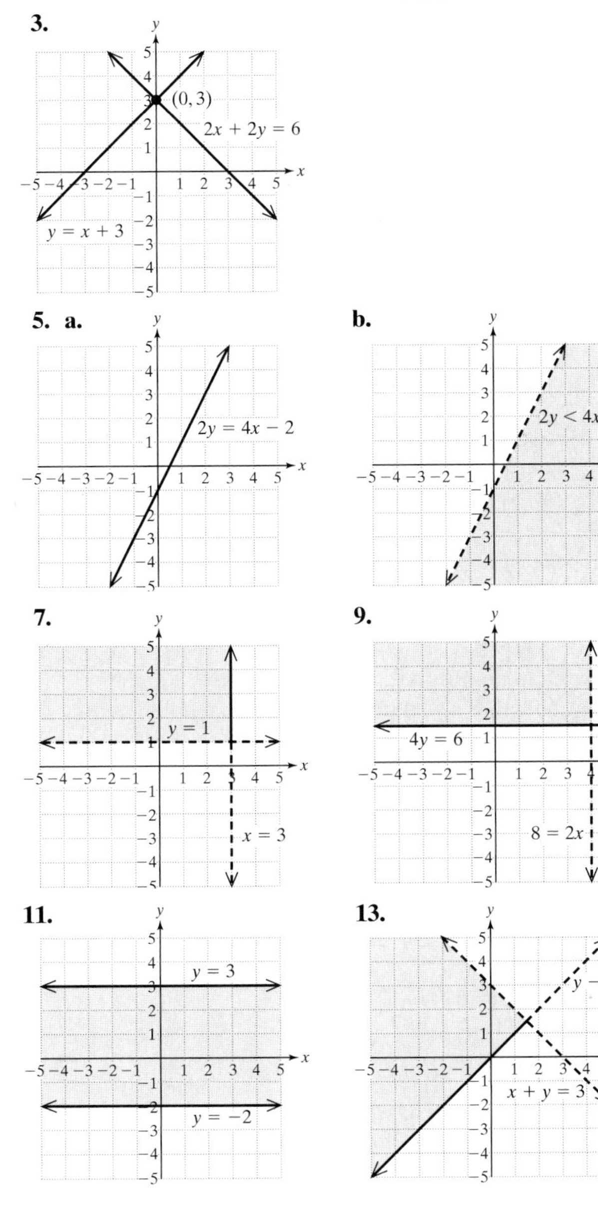

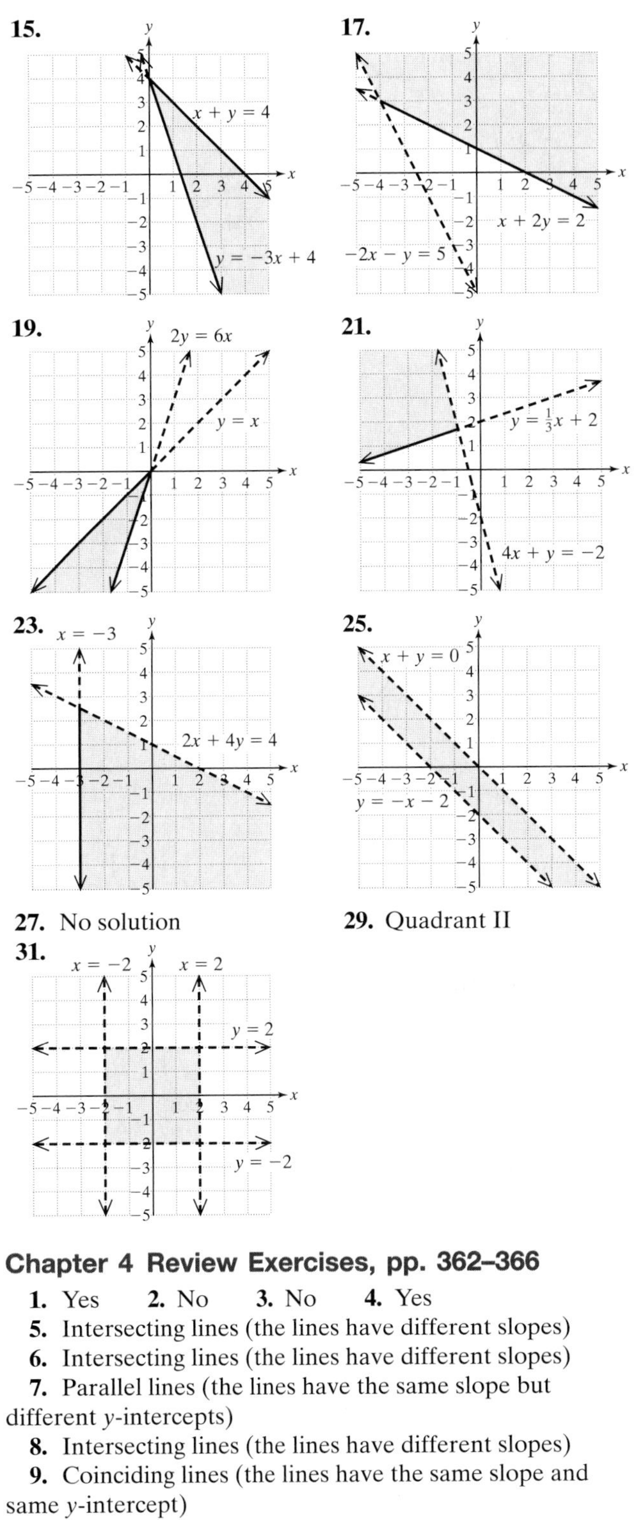

27. No solution **29.** Quadrant II

Chapter 4 Review Exercises, pp. 362–366

1. Yes **2.** No **3.** No **4.** Yes
5. Intersecting lines (the lines have different slopes)
6. Intersecting lines (the lines have different slopes)
7. Parallel lines (the lines have the same slope but different y-intercepts)
8. Intersecting lines (the lines have different slopes)
9. Coinciding lines (the lines have the same slope and same y-intercept)
10. Intersecting lines (the lines have different slopes)

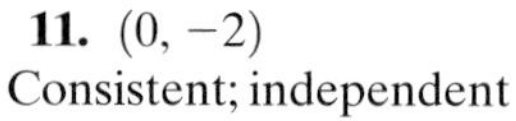

11. (0, −2)
Consistent; independent

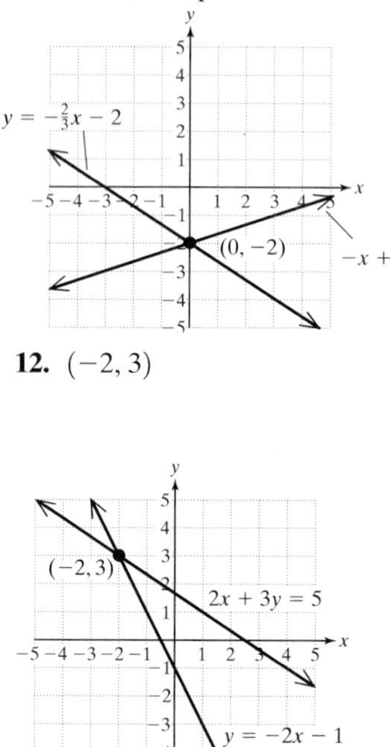

12. (−2, 3)

13. Infinitely many solutions $\{(x, y) \mid 2x + y = 5\}$; Dependent system

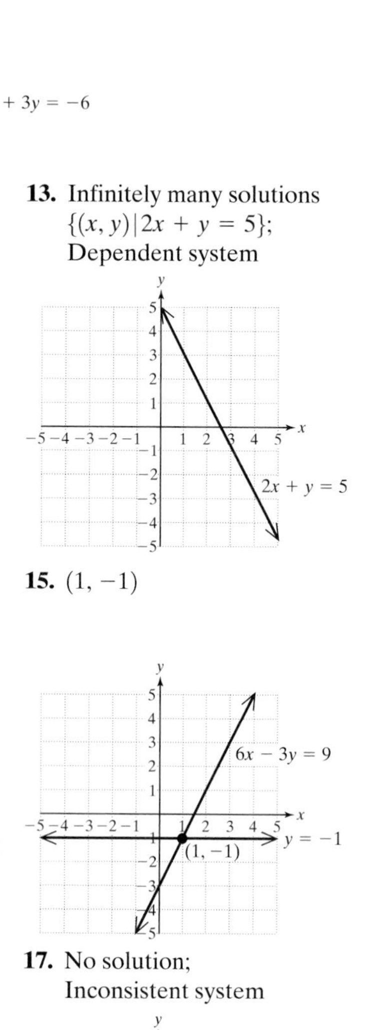

14. Infinitely many solutions $\{(x, y) \mid y = \frac{1}{5}x - 1\}$; Dependent system

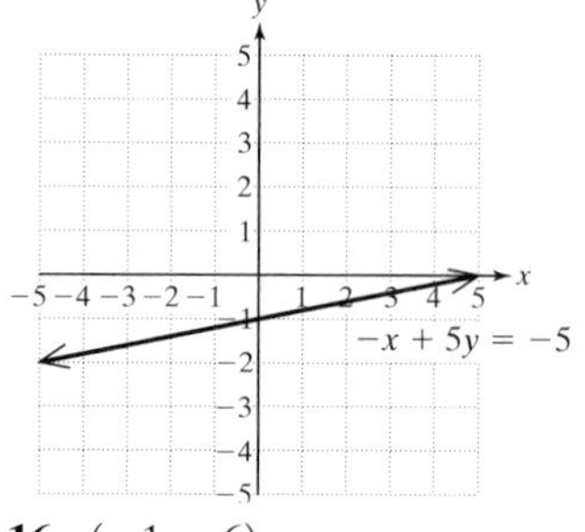

15. (1, −1)

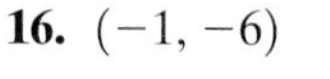

16. (−1, −6)

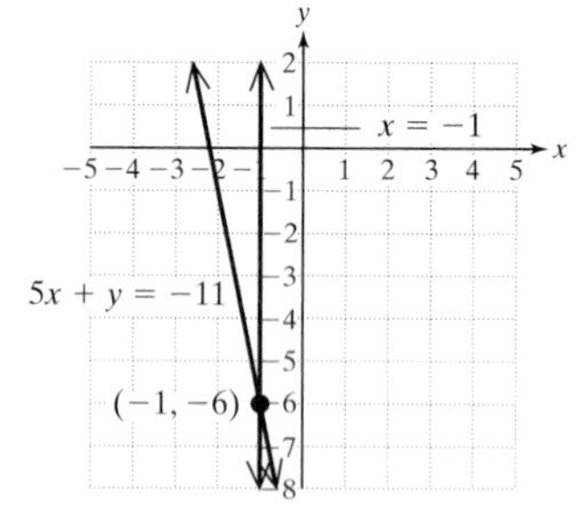

17. No solution; Inconsistent system

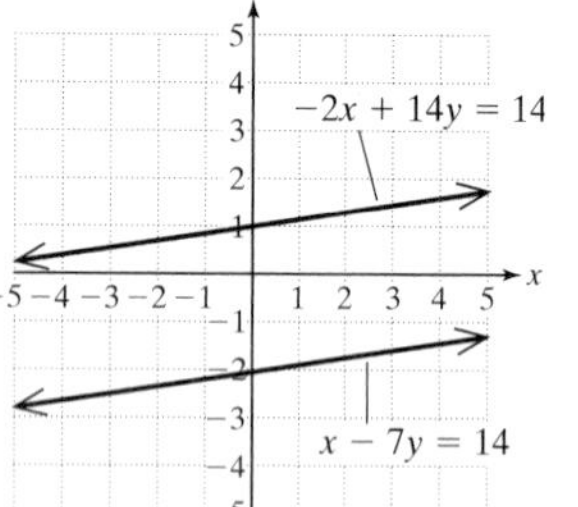

18. No solution; Inconsistent system

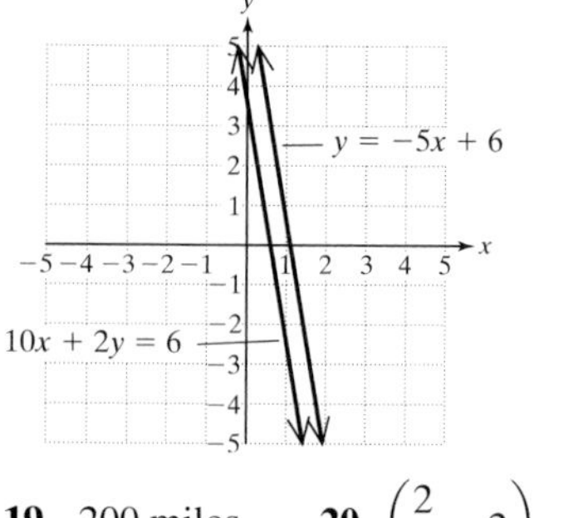

19. 200 miles **20.** $\left(\frac{2}{3}, -2\right)$ **21.** (−4, 1)

22. No solution; Inconsistent system **23.** Infinitely many solutions; $\{(x, y) \mid y = -2x + 2\}$; Dependent system
24. a. x in the first equation is easiest to solve for because its coefficient is 1. **b.** $\left(6, \frac{5}{2}\right)$
25. a. y in the second equation is easiest to solve for because its coefficient is 1. **b.** $\left(\frac{9}{2}, 3\right)$ **26.** (5, −4)
27. (0, 4) **28.** Infinitely many solutions; $\{(x, y) \mid x - 3y = 9\}$; Dependent system
29. No solution; Inconsistent system
30. The numbers are 50 and 8.
31. The angles are 42° and 48°.
32. The angles are $115\frac{1}{3}°$ and $64\frac{2}{3}°$.
33. See page 321. **34. b.** (−3, −2) **35. b.** (2, 2)
36. (2, −1) **37.** (−6, 2) **38.** $\left(-\frac{1}{2}, \frac{1}{3}\right)$ **39.** $\left(\frac{1}{4}, -\frac{2}{5}\right)$
40. Infinitely many solutions; $\{(x, y) \mid -4x - 6y = -2\}$; Dependent system **41.** No solution; Inconsistent system
42. (−4, −2) **43.** (1, 0) **44. b.** (5, −3)
45. b. (−2, −1)
46. There were 8 adult tickets and 52 children's tickets sold.
47. Emillo invested \$2500 in the 5% account and \$17,500 in the 8% account. **48.** 5 liters of the 20% solution should be mixed with 10 L of the 14% solution. **49.** The speed of the boat is 18 mph, and that of the current is 2 mph.
50. Suzanne has seven quarters and three dollar coins.
51. A hot dog costs \$4.50 and a drink costs \$3.50.
52. 3000 women and 2700 men **53.** The score was 72 on the first round and 82 on the second round.

54. For example: $(1, -1)(0, -4)(2, 0)$

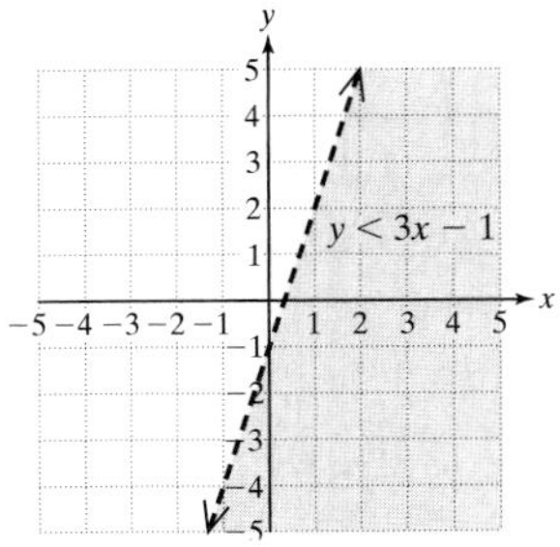

55. For example: $(5, 5)(4, 0)(0, 7)$

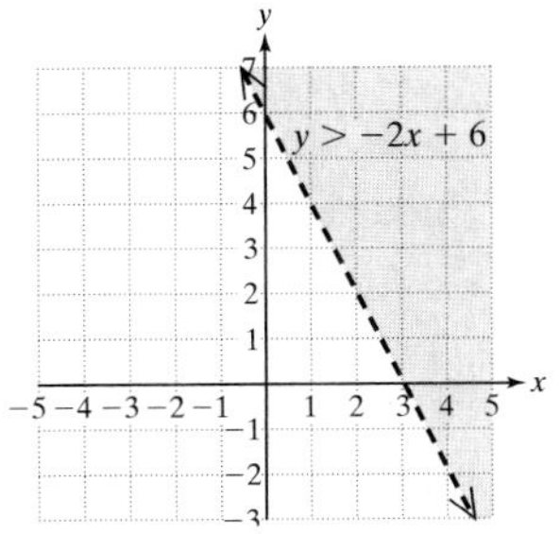

56. For example: $(-4, 0)(-2, -2)(1, -4)$

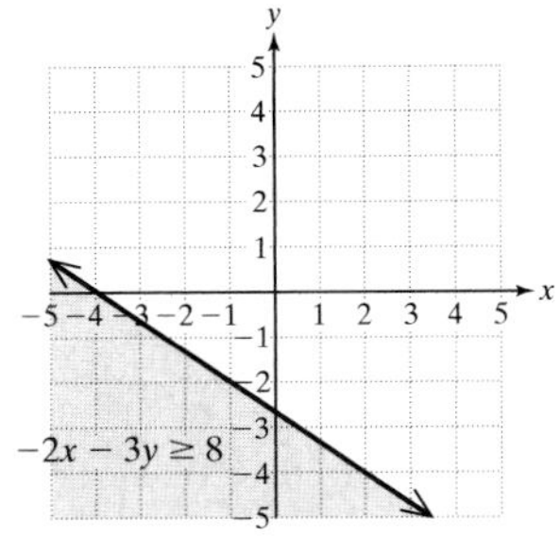

57. For example: $(0, 0)(0, -5)(-1, 1)$

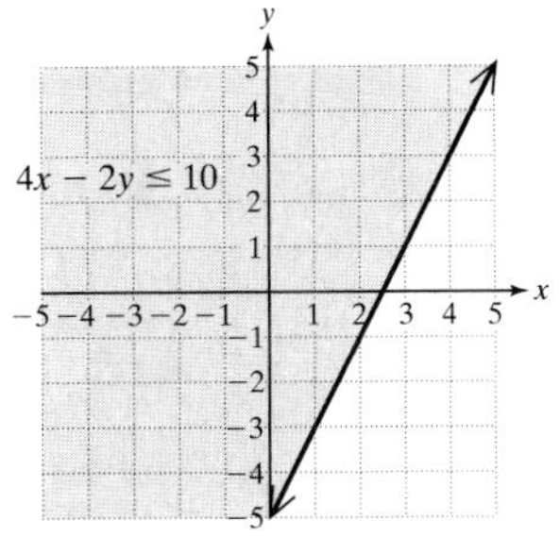

58.

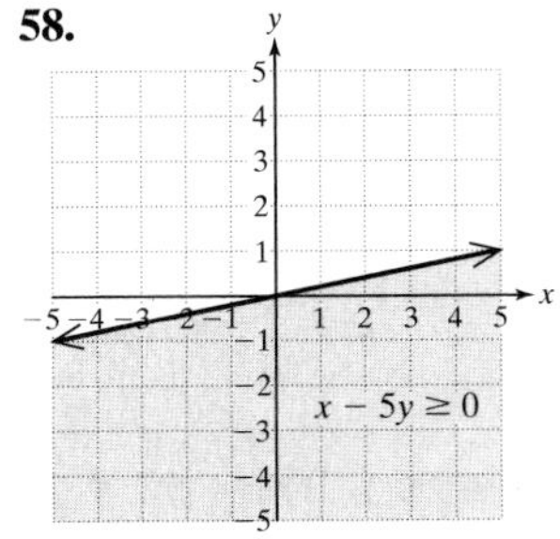

59.

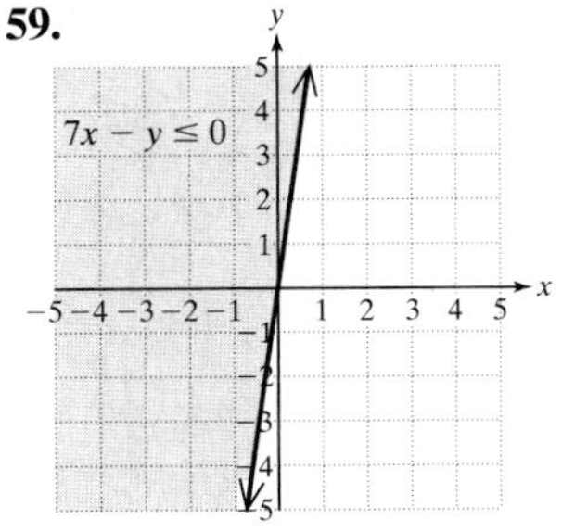

60.

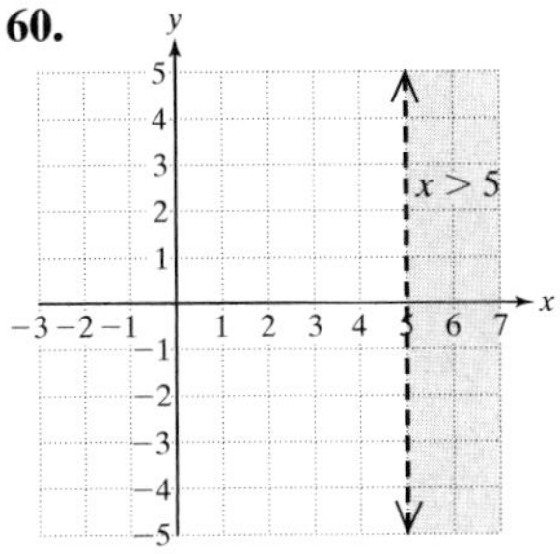

61.

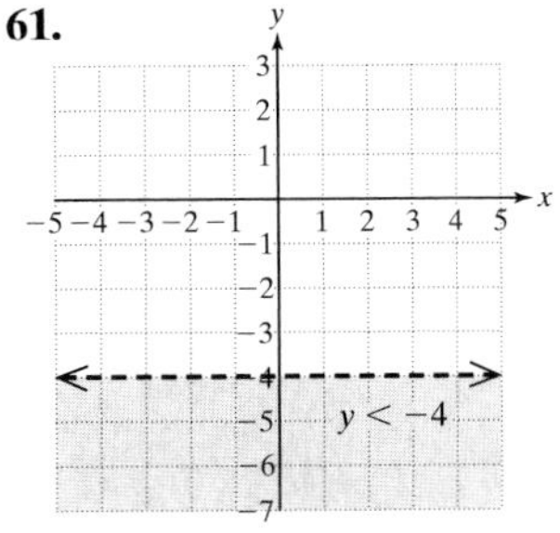

62.

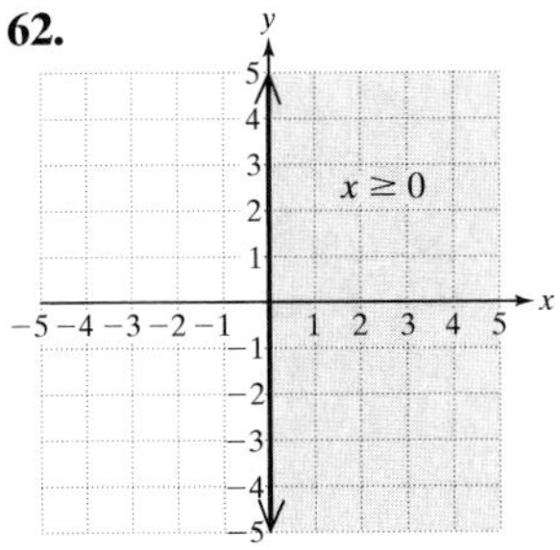

63.

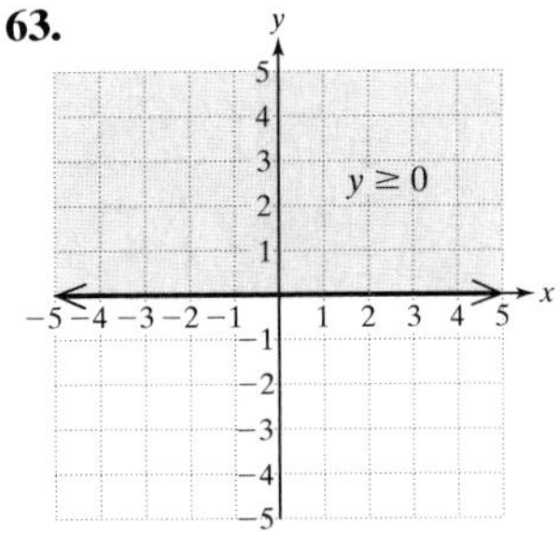

64.

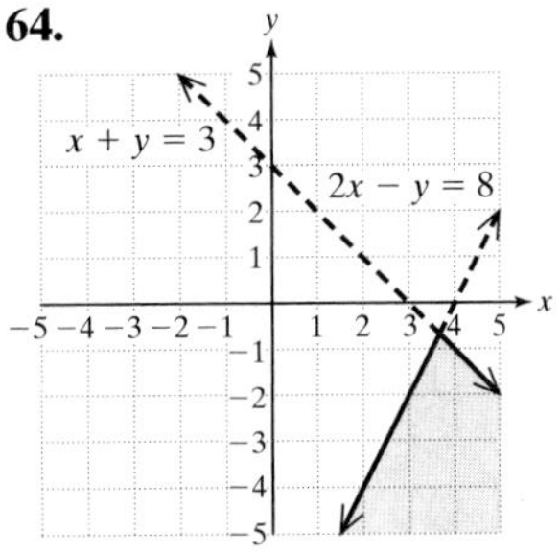

65.

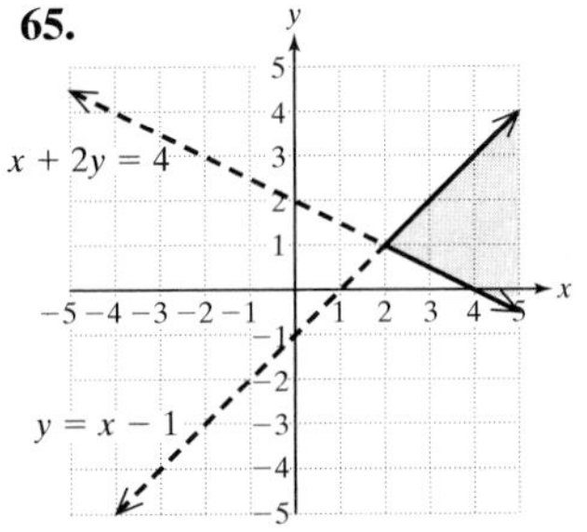

66.

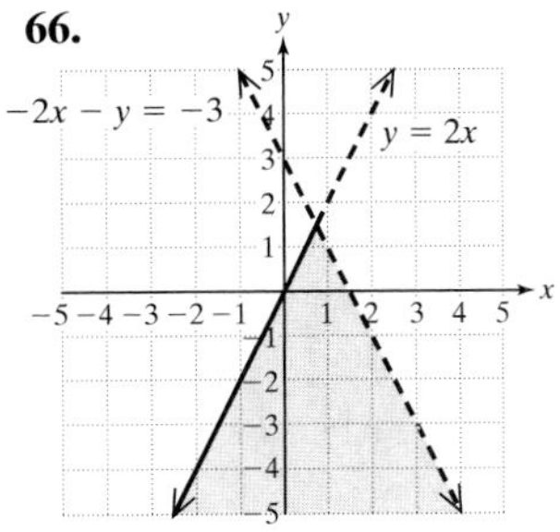

67.

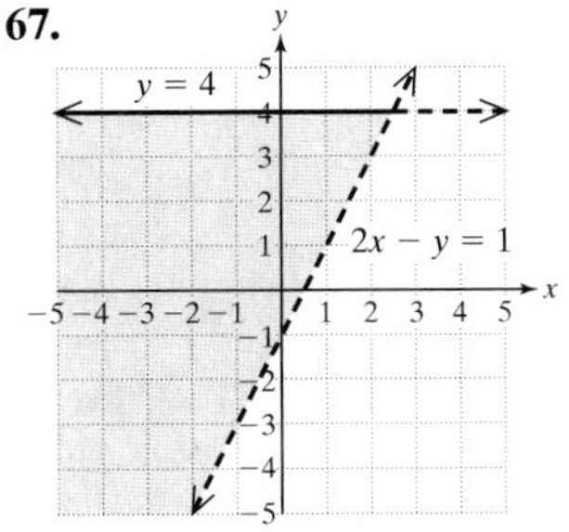

68.

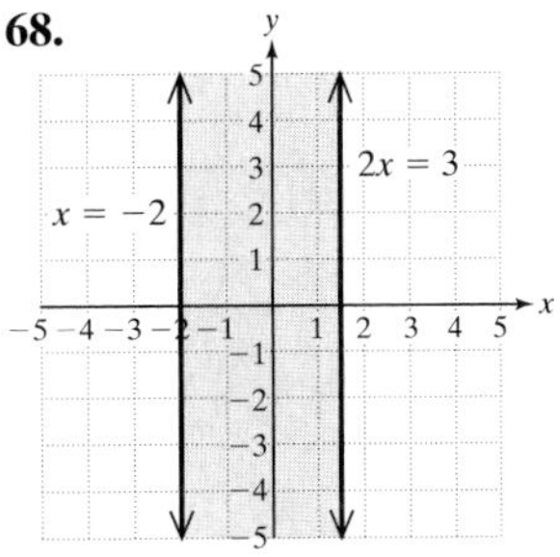

69.

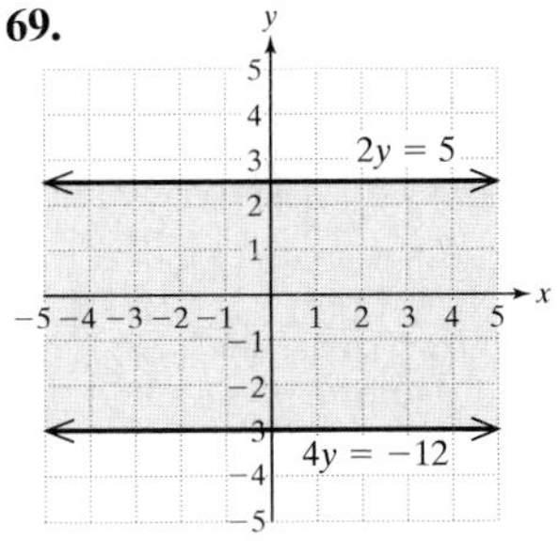

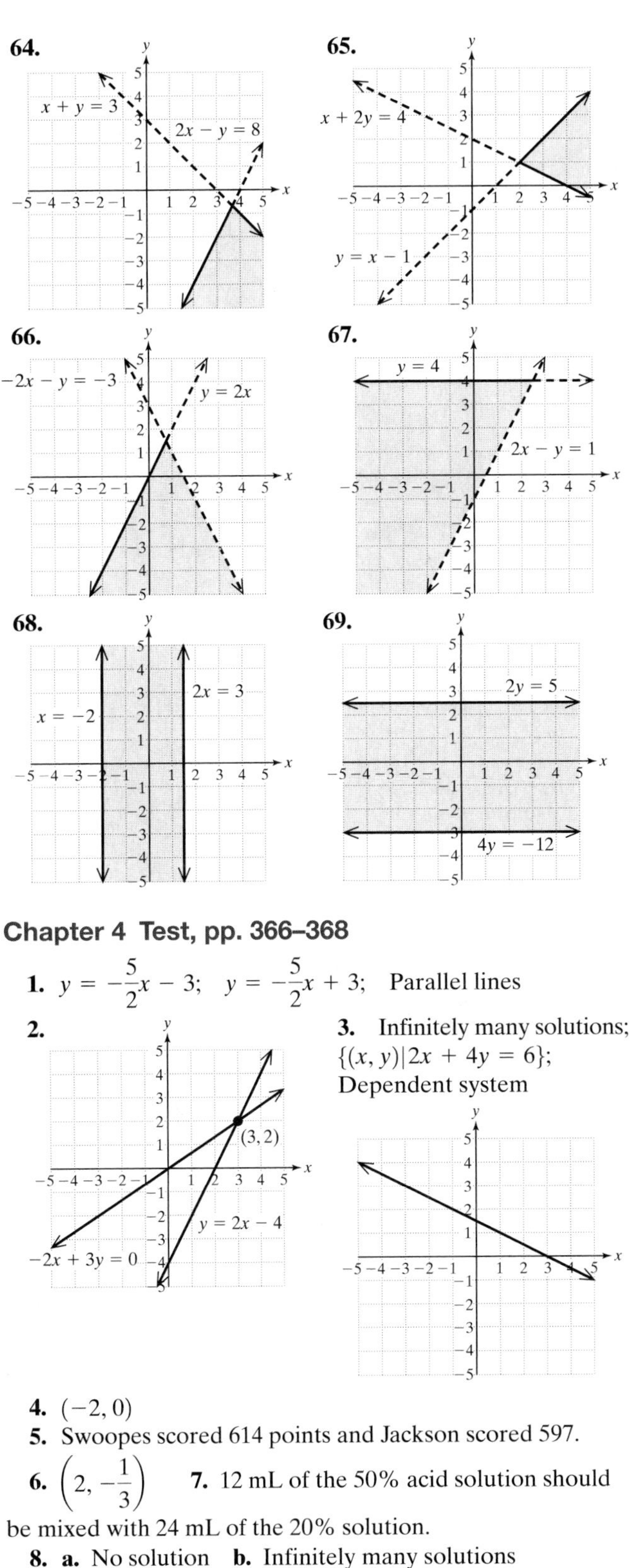

Chapter 4 Test, pp. 366–368

1. $y = -\frac{5}{2}x - 3$; $y = -\frac{5}{2}x + 3$; Parallel lines

2.

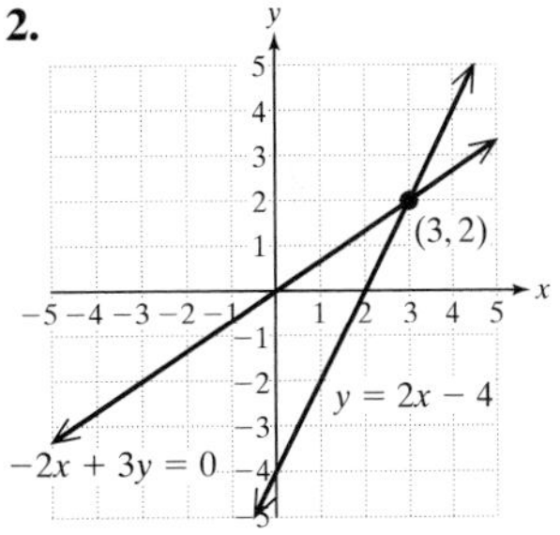

3. Infinitely many solutions; $\{(x, y) | 2x + 4y = 6\}$; Dependent system

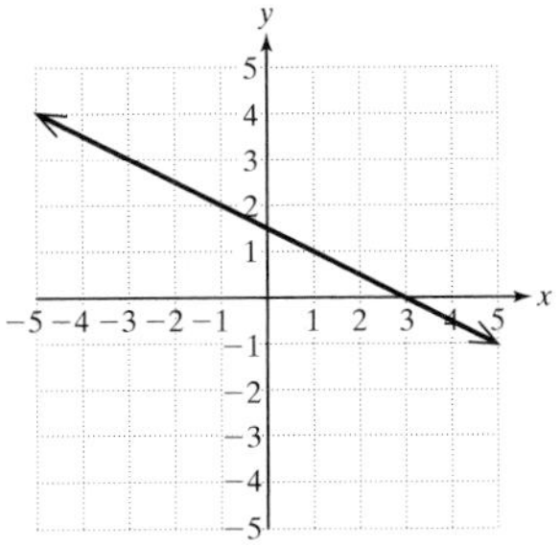

4. $(-2, 0)$
5. Swoopes scored 614 points and Jackson scored 597.
6. $\left(2, -\frac{1}{3}\right)$ **7.** 12 mL of the 50% acid solution should be mixed with 24 mL of the 20% solution.
8. a. No solution **b.** Infinitely many solutions **c.** One solution **9.** $(-5, 4)$ **10.** No solution
11. $(3, -5)$ **12.** $(-1, 2)$ **13.** Infinitely many solutions; $\{(x, y) | 10x + 2y = -8\}$ **14.** $(1, -2)$
15. CDs cost \$8 each and DVDs cost \$11 each.

16. a. \$18 was required. **b.** They used 24 quarters and 12 \$1 bills. **17.** \$1200 was borrowed at 10%, and \$3800 was borrowed at 8%. **18.** He scored 155 receiving touchdowns and 10 touchdowns rushing. **19.** The plane travels 470 mph in still air, and the wind speed is 30 mph. **20.** The cake has 340 calories, and the ice cream has 120 calories.
21. There are 178 men and 62 women on the force.
22. **23.**

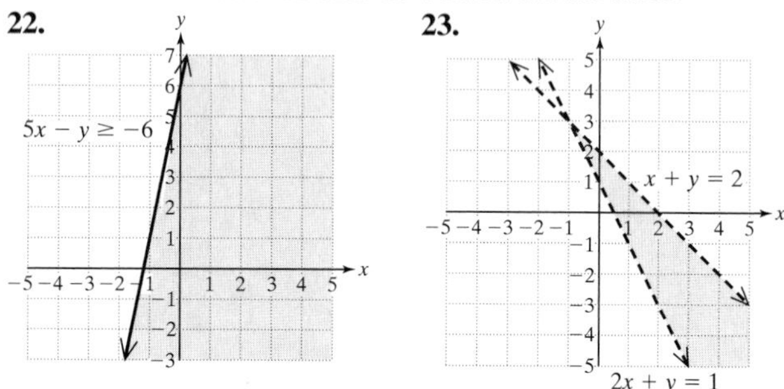

Chapters 1–4 Cumulative Review Exercises, pp. 368–369

1. $\frac{11}{6}$ **2.** $x = -\frac{21}{2}$ **3.** No solution

4. $y = \frac{3}{2}x - 3$ **5.** $\left[\frac{3}{11}, \infty\right)$

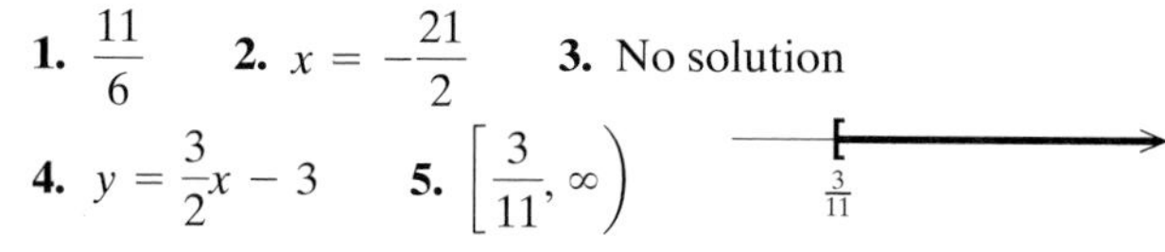

6. The angles are 37°, 33°, and 110°. **7.** The rates of the hikers are 2 mph and 4 mph. **8.** Jesse Ventura received approximately 762,200 votes. **9.** 36% of the goal has been achieved. **10.** The angles are 36.5° and 53.5°.

11. $x = 5z + m$ **12.** $y = \frac{2}{3}x - 2$

13. a. $-\frac{2}{3}$ **b.** $\frac{3}{2}$ **14.** $y = -3x + 3$

15. c. (0, 2) **16.** (0, 2)

17. a. **b.**

c. Part (a) represents the solutions to an equation. Part (b) represents the solutions to a strict inequality.
18. 20 gal of the 15% solution should be mixed with 40 gal of the 60% solution. **19.** x is 27°; y is 63° **20. a.** 1.4
b. Between 1920 and 1990, the winning speed in the Indianapolis 500 increased on average by 1.4 mph per year.

Chapter 5

Chapter Opener Puzzle

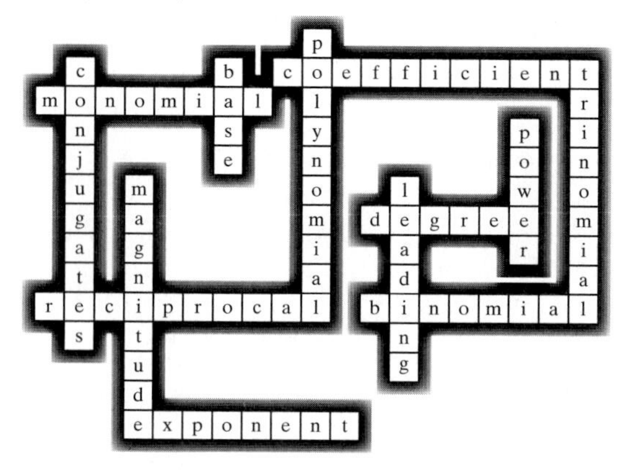

Section 5.1 Calculator Connections, pp. 378

1.–3. **4.–6.**

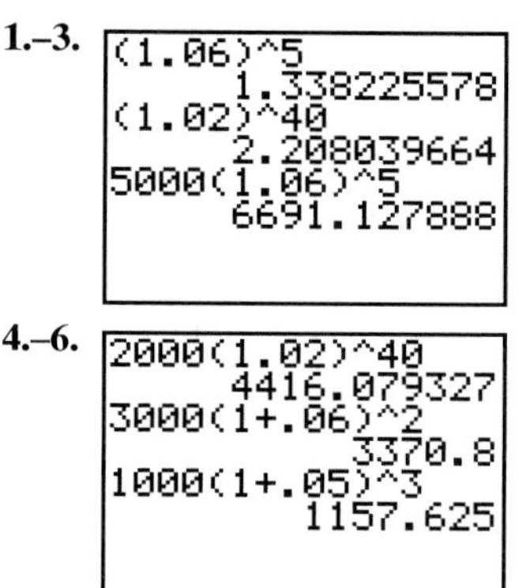

Section 5.1 Practice Exercises, pp. 379–382

3. Base: x; exponent: 4 **5.** Base: 3; exponent: 5
7. Base: -1; exponent: 4 **9.** Base: q; exponent: 1
11. v **13.** y **15.** $(-6b)^2$ **17.** $-6b^2$
19. $(y + 2)^4$ **21.** $\frac{-2}{t^3}$
23. No; $-3^4 = -81$ and $(-3)^4 = 81$
25. Yes; $-5^3 = -125$ and $(-5)^3 = -125$
27. Yes; $\left(\frac{1}{5}\right)^2 = \frac{1}{25}$ and $\frac{1}{5^2} = \frac{1}{25}$
29. Yes; $\left(\frac{7}{10}\right)^3 = \frac{343}{1000}$ and $(0.7)^3 = 0.343$ **31.** 20
33. 1 **35.** 0 **37.** $-\frac{1}{64}$ **39.** $-\frac{9}{25}$ **41.** 0
43. -3 **45.** 72 **47.** 33 **49.** 80 **51.** 400
53. 81 **55.** 81 **57.** 1920 **59.** -80
61. a. $(y \cdot y)(y \cdot y \cdot y \cdot y) = y^6$
b. $(3 \cdot 3)(3 \cdot 3 \cdot 3 \cdot 3) = 3^6$ **63.** w^{11} **65.** p^5
67. 6^{12} **69.** $\left(\frac{1}{x}\right)^3$ **71.** b^{17} **73.** z^{79}
75. a. $\frac{w \cdot w \cdot w \cdot w \cdot w}{w \cdot w} = w^3$ **b.** $\frac{4 \cdot 4 \cdot 4 \cdot 4 \cdot 4}{4 \cdot 4} = 4^3$
77. z **79.** b^{11} **81.** 2^2 **83.** 3^4 **85.** w **87.** n^7

89. 5^{10} **91.** 3^3 **93.** w^5 **95.** $30x^5y^4$ **97.** $42p^7q^{11}$ **99.** $7c^8d^{13}$ **101.** $4h^2k$ **103.** $\frac{w^6z^2}{2}$ **105.** $\frac{15m^4np^3}{4}$ **107.** $50a^6b^2c$ **109.** $-wx^9y^3$ **111.** $2433.31 **113.** $12,155.06 **115.** 1963 ft^2 **117.** 268 in.^{3} **119.** y^{3a} **121.** q **123.** w^{3n+3} **125.** t^3

Section 5.2 Practice Exercises, pp. 385–386

1. 4^9 **3.** a^{20} **5.** d^9 **7.** 7^6 **9.** When multiplying expressions with the same base, add the exponents. When dividing expressions with the same base, subtract the exponents. **11.** 5^{12} **13.** 12^6 **15.** y^{14} **17.** w^{25} **19.** a^{36} **21.** y^{14} **23.** They are both equal to 2^6. **25.** $2^{(2^4)} = 2^{16}$ $(2^2)^4 = 2^8$; the expression $2^{(2^4)}$ is greater than $(2^2)^4$. **27.** $25w^2$ **29.** $s^4r^4t^4$ **31.** $\frac{16}{r^4}$ **33.** $\frac{x^5}{y^5}$ **35.** $81a^4$ **37.** $-27a^3b^3c^3$ **39.** $-\frac{64}{x^3}$ **41.** $\frac{a^2}{b^2}$ **43.** $6^3u^6v^{12}$ or $216u^6v^{12}$ **45.** $5x^8y^4$ **47.** $-h^{28}$ **49.** m^{12} **51.** $\frac{4^5}{r^5s^{20}}$ or $\frac{1024}{r^5s^{20}}$ **53.** $\frac{3^5p^5}{q^{15}}$ or $\frac{243p^5}{q^{15}}$ **55.** y^{14} **57.** x^{31} **59.** $a^{26}b^{18}$ **61.** $16p^8q^{16}$ **63.** $-m^{35}n^{15}$ **65.** $25a^{18}b^6$ **67.** $3x^{12}y^4$ **69.** $\frac{4c^2d^6}{9}$ **71.** $\frac{c^{27}d^{31}}{2}$ **73.** $-\frac{27a^9b^3}{c^6}$ **75.** $16b^{26}$ **77.** x^{2m} **79.** $125a^{6n}$ **81.** $\frac{m^{2b}}{n^{3b}}$ **83.** $\frac{3^na^{3n}}{5^nb^{4n}}$

Section 5.3 Practice Exercises, pp. 393–395

3. c^9 **5.** y **7.** 3^6 or 729 **9.** $7^4w^{28}z^8$ or $2401w^{28}z^8$ **11.** $\frac{25k^6}{h^{14}}$ **13. a.** 1 **b.** 1 **15.** 1 **17.** 1 **19.** -1 **21.** 1 **23.** 1 **25.** 6 **27.** $-p$ **29. a.** $\frac{1}{4^3}$ **b.** $\frac{1}{4^3}$ **31.** $\frac{4}{5}$ **33.** -27 **35.** $\frac{1}{c^5}$ **37.** $\frac{1}{16}$ **39.** $\frac{1}{3z}$ **41.** $\frac{7}{y}$ **43.** $-\frac{1}{64}$ **45.** $\frac{1}{w^2}$ **47.** $-\frac{1}{r^5}$ **49.** b^6 **51.** $\frac{y^5}{y^{-3}} = y^{5-(-3)} = y^8$ **53.** $5b^{-2} = 5 \cdot \frac{1}{b^2} = \frac{5}{b^2}$ **55.** $\frac{1}{s}$ **57.** 1 **59.** b^6 **61.** $\frac{d^{10}}{c^8}$ **63.** $\frac{y^{33}}{6^3x^3}$ or $\frac{y^{33}}{216x^3}$ **65.** $\frac{1}{q^8}$ **67.** u^4 **69.** p^8 **71.** $\frac{1}{s^7}$ **73.** $\frac{1}{9}$ **75.** 1 **77.** $\frac{1}{k^{10}h^2}$ **79.** $\frac{1}{p^{11}}$ **81.** $\frac{1}{25}$ **83.** 1 **85.** $\frac{1}{27u^6}$ **87.** $\frac{-10q^4}{p^6}$ **89.** $\frac{5}{3x^6}$ **91.** $9s^4t^4$ **93.** $\frac{x^{12}}{25}$ **95.** $\frac{m^2}{2n^6}$ **97.** $\frac{3}{16}$ **99.** $\frac{8}{9}$ **101.** $\frac{3}{8}$ **103.** $\frac{5}{4}$ **105.** $\frac{5}{9}$

Section 5.4 Calculator Connections, pp. 400

1.–2.
```
(5.2E6)*(4.6E-3)
               23920
(2.19E-8)*(7.84E
-4)
         1.71696E-11
```

3.–4.
```
(4.76E-5)/(2.38E
9)
               2E-14
(8.5E4)/(4.0E-1)
              212500
```

5.
```
((9.6E7)*(4.0E-3
))/(2.0E-2)
            19200000
```

6.
```
((5.0E-12)*(6.4E
-5))/((1.6E-8)*(
4.0E2))
               5E-11
```

Section 5.4 Practice Exercises, pp. 401–403

3. b^{13} **5.** 10^{13} **7.** $\frac{1}{y^5}$ **9.** $\frac{1}{10^5}$ **11.** w^4 **13.** 10^4 **15.** 100,000 **17.** $\frac{1}{10,000}$ or 0.0001 **19.** 10 **21.** 100 **23.** Move the decimal point between the 2 and 3 and multiply by 10^{-10}; 2.3×10^{-10}. **25.** 5×10^4 **27.** 2.08×10^5 **29.** 6.01×10^6 **31.** 8×10^{-6} **33.** 1.25×10^{-4} **35.** 6.708×10^{-3} **37.** 1.7×10^{-24} g **39.** $\$2.7 \times 10^{10}$ **41.** 6.8×10^7 gal; 1.0×10^2 miles **43.** Move the decimal point nine places to the left; 0.000 000 0031 **45.** 0.00005 **47.** 2800 **49.** 0.000603 **51.** 2,400,000 **53.** 0.019 **55.** 7032 **57.** 0.000 000 000 001 g **59.** 1600 calories and 2800 calories **61.** 5.0×10^4 **63.** 3.6×10^{11} **65.** 2.2×10^4 **67.** 2.25×10^{-13} **69.** 3.2×10^{14} **71.** 2.432×10^{-10} **73.** 3.0×10^{13} **75.** 6.0×10^5 **77.** 1.38×10^1 **79.** 5.0×10^{-14} **81.** 3.75 in. **83.** $\$2.97 \times 10^{10}$ **85. a.** 6.5×10^7 **b.** 2.3725×10^{10} days **c.** 5.694×10^{11} hr **d.** 2.04984×10^{15} sec **87.** Each person would have to pay $25,000. **89.** 1.25×10^{-2} ft or 0.0125 ft

Chapter 5 Problem Recognition Exercises—Properties of Exponents, p. 404

1. t^8 **2.** 2^8 or 256 **3.** y^5 **4.** p^6 **5.** r^4s^8 **6.** $a^3b^9c^6$ **7.** w^6 **8.** $\frac{1}{m^{16}}$ **9.** $\frac{x^4z^3}{y^7}$ **10.** $\frac{a^3c^8}{b^6}$ **11.** 1.25×10^3 **12.** 1.24×10^5 **13.** 8.0×10^8 **14.** 6.0×10^{-9} **15.** p^{15} **16.** p^{15} **17.** $\frac{1}{v^2}$

18. $c^{50}d^{40}$ **19.** 3 **20.** -4 **21.** $\frac{b^9}{2^{15}}$ **22.** $\frac{81}{y^6}$ **23.** $\frac{16y^4}{81x^4}$ **24.** $\frac{25d^6}{36c^2}$ **25.** $3a^7b^5$ **26.** $64x^7y^{11}$ **27.** $\frac{y^4}{x^8}$ **28.** $\frac{1}{a^{10}b^{10}}$ **29.** $\frac{1}{t^2}$ **30.** $\frac{1}{p^7}$ **31.** $\frac{8w^6x^9}{27}$ **32.** $\frac{25b^8}{16c^6}$ **33.** $\frac{q^3s}{r^2t^5}$ **34.** $\frac{m^2p^3q}{n^3}$ **35.** $\frac{1}{y^{13}}$ **36.** w^{10} **37.** $-\frac{1}{8a^{18}b^6}$ **38.** $\frac{4x^{18}}{9y^{10}}$ **39.** $\frac{k^8}{5h^6}$ **40.** $\frac{6n^{10}}{m^{12}}$

Section 5.5 Practice Exercises, pp. 411–415

3. $\frac{15}{x^3}$ **5.** $\frac{2}{t^4}$ **7.** $\frac{1}{3^{12}}$
9. 4.0×10^{-2} is scientific notation in which 10 is raised to the -2 power. 4^{-2} is not scientific notation and 4 is being raised to the -2 power. **11.** $-7x^4 + 7x^2 + 9x + 6$
13. Binomial; 10; 2 **15.** Monomial; 6; 2
17. Binomial; -1; 4 **19.** Trinomial; 12; 4
21. Monomial; 23; 0 **23.** Monomial; -32; 3
25. **a.** 134 ft, 114 ft, and 86 ft **b.** 150 ft
27. **a.** \$740, \$1190, and \$1140 **b.** $-$\$60; when no candles are produced, the business will lose \$60.
29. The exponents on the x-factors are different.
31. $35x^2y$ **33.** $10y$ **35.** $8b^2 - 9$ **37.** $4y^2 + y - 9$
39. $4a - 8c$ **41.** $a - \frac{1}{2}b - 2$ **43.** $\frac{4}{3}z^2 - \frac{5}{3}$
45. $7.9t^3 - 3.4t^2 + 6t - 4.2$ **47.** $-4h + 5$
49. $2m^2 - 3m + 15$ **51.** $-3v^3 - 5v^2 - 10v - 22$
53. $9t^4 + 8t + 39$ **55.** $-8a^3b^2$ **57.** $-53x^3$
59. $-5a - 3$ **61.** $16k + 9$ **63.** $2s + 14$
65. $3t^2 - 4t - 3$ **67.** $-2r - 3s + 3t$
69. $\frac{3}{4}x + \frac{1}{3}y - \frac{3}{10}$ **71.** $-\frac{2}{3}h^2 + \frac{3}{5}h - \frac{5}{2}$
73. $2.4x^4 - 3.1x^2 - 4.4x - 6.7$
75. $4b^3 + 12b^2 - 5b - 12$ **77.** $-3x^3 - 2x^2 + 11x - 31$
79. $4y^3 + 2y^2 + 2$ **81.** $3a^2 - 3a + 5$
83. $9ab^2 - 3ab + 16a^2b$ **85.** $4z^5 + z^4 + 9z^3 - 3z - 2$
87. $2x^4 + 11x^3 - 3x^2 + 8x - 4$ **89.** $-2w^2 - 7w + 18$
91. $-p^2q - 4pq^2 + 3pq$ **93.** 0 **95.** $-5ab + 6ab^2$
97. $11y^2 - 10y - 4$ **99.** For example, $x^3 + 6$
101. For example, $8x^5$ **103.** For example, $-6x^2 + 2x + 5$

Section 5.6 Practice Exercises, pp. 422–425

3. $-2y^2$ **5.** $-8y^4$ **7.** $8uvw^2$ **9.** $7u^2v^2w^4$
11. $3t^3 + 3t$ **13.** $9t^4$ **15.** $-12y$ **17.** $21p$
19. $12a^{14}b^8$ **21.** $-2c^{10}d^{12}$
23. $16p^2q^2 - 24p^2q + 40pq^2$ **25.** $-4k^3 + 52k^2 + 24k$
27. $-45p^3q - 15p^4q^3 + 30pq^2$ **29.** $y^2 - y - 90$
31. $m^2 - 14m + 24$ **33.** $p^2 - p - 2$
35. $w^2 + 11w + 24$ **37.** $p^2 - 14p + 33$
39. $12x^2 + 28x - 5$ **41.** $8a^2 - 22a + 9$
43. $9t^2 - 18t - 7$ **45.** $3x^2 + 28x + 32$
47. $5s^3 + 8s^2 - 7s - 6$ **49.** $27w^3 - 8$
51. $p^4 + 5p^3 - 2p^2 - 21p + 5$ **53.** $9a^2 - 16b^2$
55. $81k^2 - 36$ **57.** $\frac{1}{4} - t^2$ **59.** $u^6 - 25v^2$
61. $4 - 9a^2$ **63.** $\frac{4}{9} - p^2$ **65.** $a^2 + 2ab + b^2$
67. $x^2 - 2xy + y^2$ **69.** $4c^2 + 20c + 25$
71. $9t^4 - 24st^2 + 16s^2$ **73.** $t^2 - 14t + 49$
75. $16q^2 + 24q + 9$
77. **a.** 36 **b.** 20 **c.** $(a + b)^2 \neq a^2 + b^2$ in general.
79. **a.** $9x^2 + 6xy + y^2$ **b.** $9x^2y^2$ **c.** $(a + b)^2$ is the square of a binomial, $a^2 + 2ab + b^2$; $(ab)^2$ is the square of a monomial, a^2b^2. **81.** $36 - y^2$
83. $49q^2 - 42q + 9$ **85.** $3t^3 - 12t$
87. $r^3 - 15r^2 + 63r - 49$ **89.** $81w^2 - 16z^2$
91. $25s^2 - 30st + 9t^2$ **93.** $10a^2 - 13ab + 4b^2$
95. $s^2 + \frac{147}{5}s - 18$ **97.** $4k^3 - 4k^2 - 5k - 25$
99. $u^3 + u^2 - 10u + 8$ **101.** $w^3 + w^2 + 4w + 12$
103. $\frac{4}{25}p^2 - \frac{4}{5}pq + q^2$ **105.** $4v^4 + 48v^3$
107. $4h^2 - 7.29$ **109.** $5k^6 - 19k^3 + 18$
111. $6.25y^2 + 5.5y + 1.21$ **113.** $h^3 + 9h^2 + 27h + 27$
115. $8a^3 - 48a^2 + 96a - 64$
117. $6w^4 + w^3 - 15w^2 - 11w - 5$
119. $30x^3 + 55x^2 - 10x$ **121.** $2y^3 - y^2 - 15y + 18$
123. $x + 6$ **125.** $k = 6$ or -6

Section 5.7 Practice Exercises, pp. 431–434

1. $6z^5 - 10z^4 - 4z^3 - z^2 - 6$ **3.** $10x^2 - 29xy - 3y^2$
5. $11x - 2y$ **7.** $y^2 - \frac{3}{4}y + \frac{1}{2}$ **9.** $a^3 + 27$
11. Use long division when the divisor is a polynomial with two or more terms.
13. **a.** $5t^2 + 6t$ **15.** $3a^2 + 2a - 7$ **17.** $x^2 + 4x - 1$
19. $3p^2 - p$ **21.** $1 + \frac{2}{m}$ **23.** $-2y^2 + y - 3$
25. $x^2 - 6x - \frac{1}{4} + \frac{2}{x}$ **27.** $a - 1 + \frac{b}{a}$
29. $3t - 1 + \frac{3}{2t} - \frac{1}{2t^2} + \frac{2}{t^3}$ **31.** **a.** $z + 2 + \frac{1}{z + 5}$
33. $t + 3$ **35.** $7b + 4$ **37.** $k - 6$
39. $2p^2 + 3p - 4$ **41.** $k - 2 + \frac{-4}{k + 1}$
43. $2x^2 - x + 6 + \frac{2}{2x - 3}$ **45.** $a - 3 + \frac{18}{a + 3}$
47. $4x^2 + 8x + 13$ **49.** $w^2 + 5w - 2 + \frac{1}{w^2 - 3}$
51. $n^2 + n - 6$ **53.** $x - 1 + \frac{-8}{5x^2 + 5x + 1}$
55. Multiply $(x - 2)(x^2 + 4) = x^3 - 2x^2 + 4x - 8$, which does not equal $x^3 - 8$. **57.** Monomial division; $3a^2 + 4a$
59. Long division; $p + 2$
61. Long division; $t^3 - 2t^2 + 5t - 10 + \frac{4}{t + 2}$
63. Long division; $w^2 + 3 + \frac{1}{w^2 - 2}$
65. Long division; $n^2 + 4n + 16$
67. Monomial division; $-3r + 4 - \frac{3}{r^2}$ **69.** $x + 1$
71. $x^3 + x^2 + x + 1$ **73.** $x + 1 + \frac{1}{x - 1}$
75. $x^3 + x^2 + x + 1 + \frac{1}{x - 1}$

Chapter 5 Problem Recognition Exercises—Operations on Polynomials p. 434

1. $2x^3 - 8x^2 + 14x - 12$ **2.** $-3y^4 - 20y^2 - 32$
3. $x^2 - 1$ **4.** $4y^2 + 12$ **5.** $36y^2 - 84y + 49$
6. $9z^2 + 12z + 4$ **7.** $36y^2 - 49$ **8.** $9z^2 - 4$
9. $-x^2 - 3x + 4$ **10.** $5m^2 - 4m + 1$
11. $16x^2 + 8xy + y^2$ **12.** $4a^2 + 4ab + b^2$ **13.** $16x^2y^2$
14. $4a^2b^2$ **15.** $-7m^2 - 16m$
16. $-4n^5 + n^4 + 6n^2 - 7n + 2$
17. $8x^2 + 16x + 34 + \frac{74}{x - 2}$
18. $-4x^2 - 10x - 30 + \frac{-95}{x - 3}$
19. $6x^3 + 5x^2y - 6xy^2 + y^3$ **20.** $6a^3 - a^2b + 5ab^2 + 2b^3$
21. $x^3 + y^6$ **22.** $m^6 + 1$ **23.** $4b$ **24.** $-12z$
25. $a^4 - 4b^2$ **26.** $y^6 - 36z^2$ **27.** $64u^2 + 48uv + 9v^2$
28. $4p^2 - 4pt + t^2$ **29.** $4p + 4 + \frac{-2}{2p - 1}$
30. $2v - 7 + \frac{29}{2v + 3}$ **31.** $4x^2y^2$ **32.** $-9pq$
33. $10a^2 - 57a + 54$ **34.** $28a^2 - 17a - 3$
35. $\frac{9}{49}x^2 - \frac{1}{4}$ **36.** $\frac{4}{25}y^2 - \frac{16}{9}$
37. $-\frac{11}{9}x^3 + \frac{5}{9}x^2 - \frac{1}{2}x - 4$ **38.** $-\frac{13}{10}y^2 - \frac{9}{10}y + \frac{4}{15}$
39. $1.3x^2 - 0.3x - 0.5$ **40.** $5w^3 - 4.1w^2 + 2.8w - 1.2$

Chapter 5 Review Exercises, pp. 438–441

1. Base: 5; exponent: 3 **2.** Base: x; exponent: 4
3. Base: -2; exponent: 0 **4.** Base: y; exponent: 1
5. a. 36 **b.** 36 **c.** -36
6. a. 64 **b.** -64 **c.** -64
7. 5^{13} **8.** a^{11} **9.** x^9 **10.** 6^9 **11.** 10^3
12. y^6 **13.** b^8 **14.** 7^7 **15.** k **16.** 1
17. 2^8 **18.** q^6
19. Exponents are added only when multiplying factors with the same base. In such a case, the base does not change.
20. Exponents are subtracted only when dividing factors with the same base. In such a case, the base does not change.
21. $7146.10 **22.** $22,050 **23.** 7^{12} **24.** c^{12}
25. p^{18} **26.** 9^{28} **27.** $\frac{a^2}{b^2}$ **28.** $\frac{1}{3^4}$ **29.** $\frac{5^2}{c^4d^{10}}$
30. $-\frac{m^{10}}{4^5n^{30}}$ **31.** $2^4a^4b^8$ **32.** $x^{14}y^2$
33. $-\frac{3^3x^9}{5^3y^6z^3}$ **34.** $\frac{r^{15}}{s^{10}t^{30}}$ **35.** a^{11} **36.** 8^2
37. $4h^{14}$ **38.** $2p^{14}q^{13}$ **39.** $\frac{x^6y^2}{4}$ **40.** a^9b^6
41. 1 **42.** 1 **43.** 1 **44.** -1 **45.** 2 **46.** 1
47. $\frac{1}{z^5}$ **48.** $\frac{1}{10^4}$ **49.** $\frac{1}{36a^2}$ **50.** $\frac{6}{a^2}$ **51.** $\frac{17}{16}$
52. $\frac{10}{9}$ **53.** $\frac{1}{t^8}$ **54.** $\frac{1}{r}$ **55.** $\frac{2y^7}{x^6}$ **56.** $\frac{4a^6bc}{5}$
57. $\frac{n^{16}}{16m^8}$ **58.** $\frac{u^{15}}{27v^6}$ **59.** $\frac{k^{21}}{5}$ **60.** $\frac{h^9}{9}$ **61.** $\frac{1}{2}$
62. $\frac{5}{4}$ **63. a.** 9.7×10^7 **b.** 4.2×10^{-3} in.
c. 1.66241×10^8 km² **64. a.** 0.000 000 0001
b. 257,300,000 **c.** $256,000 **65.** 9.43×10^5
66. 1.55×10^{10} **67.** 2.5×10^8 **68.** 1.638×10^3
69. $\approx 9.5367 \times 10^{13}$. This number is too big to fit on most calculator displays.
70. $\approx 1.1529 \times 10^{-12}$. This number is too small to fit on most calculator displays.
71. a. $\approx 5.84 \times 10^8$ miles **b.** $\approx 6.67 \times 10^4$ mph
72. a. $\approx 2.26 \times 10^8$ miles **b.** $\approx 1.07 \times 10^5$ mph
73. a. Trinomial **b.** 4 **c.** 7
74. a. Binomial **b.** 7 **c.** -5
75. $7x - 3$ **76.** $-y^2 - 14y - 2$ **77.** $14a^2 - 2a - 6$
78. $10w^4 + 2w^3 - 7w + 4$ **79.** $\frac{15}{2}x^3 + \frac{1}{4}x^2 + \frac{1}{2}x + 2$
80. $0.01b^5 + b^4 - 0.1b^3 + 0.3b + 0.33$
81. $-2x^2 - 9x - 6$ **82.** $-5x^2 - 9x - 12$
83. For example, $-5x^2 + 2x - 4$
84. For example, $6x^6 + 8$ **85.** $6w + 6$
86. $-75x^6y^4$ **87.** $18a^8b^4$ **88.** $15c^4 - 35c^2 + 25c$
89. $-2x^3 - 10x^2 + 6x$ **90.** $5k^2 + k - 4$
91. $20t^2 + 3t - 2$ **92.** $6q^2 + 47q - 8$
93. $2a^2 + 4a - 30$ **94.** $49a^2 + 7a + \frac{1}{4}$
95. $b^2 - 8b + 16$ **96.** $8p^3 - 27$
97. $-2w^3 - 5w^2 - 5w + 4$ **98.** $b^2 - 16$
99. $\frac{1}{9}r^8 - s^4$ **100.** $49z^4 - 84z^2 + 36$
101. $2h^5 + h^4 - h^3 + h^2 - h + 3$ **102.** $2x^2 + 3x - 20$
103. $4y^2 - 2y$ **104.** $2a^2b - a - 3b$
105. $-3x^2 + 2x - 1$ **106.** $-\frac{z^5w^3}{2} + \frac{3zw}{4} + \frac{1}{z}$
107. $m^2 + \frac{5}{6}m - 1$ **108.** $6n^2 - 2n + 4$
109. $x + 2$ **110.** $2t + 5$ **111.** $p - 3 + \frac{5}{2p + 7}$
112. $a + 6 + \frac{-4}{5a - 3}$ **113.** $b^2 + 5b + 25$
114. $z + 4$ **115.** $y^2 - 4y + 2 + \frac{9y - 4}{y^2 + 3}$
116. $t^2 - 3t + 1 + \frac{-2t - 6}{3t^2 + t + 1}$ **117.** $w^2 + w - 1$

Chapter 5 Test, pp. 441–442

1. $\frac{(3 \cdot 3 \cdot 3 \cdot 3) \cdot (3 \cdot 3 \cdot 3)}{3 \cdot 3 \cdot 3 \cdot 3 \cdot 3 \cdot 3} = 3$ **2.** 9^6 **3.** q^8
4. $27a^6b^3$ **5.** $\frac{16x^4}{y^{12}}$ **6.** 1 **7.** $\frac{1}{c^3}$ **8.** 14
9. $49s^{18}t$ **10.** $\frac{4}{b^{12}}$ **11.** $\frac{16a^{12}}{9b^6}$
12. a. 4.3×10^{10} **b.** 0.000 0056

13. a. $2.4192 \times 10^8\ \text{m}^3$ **b.** $8.83008 \times 10^{10}\ \text{m}^3$
14. $5x^3 - 7x^2 + 4x + 11$ **a.** 3 **b.** 5
15. $24w^2 - 3w - 4$ **16.** $15x^3 - 7x^2 - 2x + 1$
17. $-10x^5 - 2x^4 + 30x^3$ **18.** $8a^2 - 10a + 3$
19. $4y^3 - 25y^2 + 37y - 15$ **20.** $4 - 9b^2$
21. $25z^2 - 60z + 36$ **22.** $100 - 9w^2$
23. $y^3 - 11y^2 + 32y - 12$
24. Perimeter: $12x - 2$; area: $5x^2 - 13x - 6$
25. $-3x^6 + \dfrac{x^4}{4} - 2x$ **26.** $2y - 7$
27. $w^2 - 4w + 5 + \dfrac{-10}{2w + 3}$

Chapters 1–5 Cumulative Review Exercises, pp. 442–443

1. $-\dfrac{35}{2}$ **2.** 4 **3.** $5^2 - \sqrt{4}$; 23 **4.** $x = \dfrac{28}{3}$
5. No solution **6.** Quadrant III **7.** y-axis
8. The measures are 31°, 54°, and 95°.
9.

Time (h) x	Snow Depth (in.) y
0	6
2	9
4	12
6	15
8	18
9	19.5

a. 12 in. **b.** 19.5 in. **c.** 5.5 hr
d.

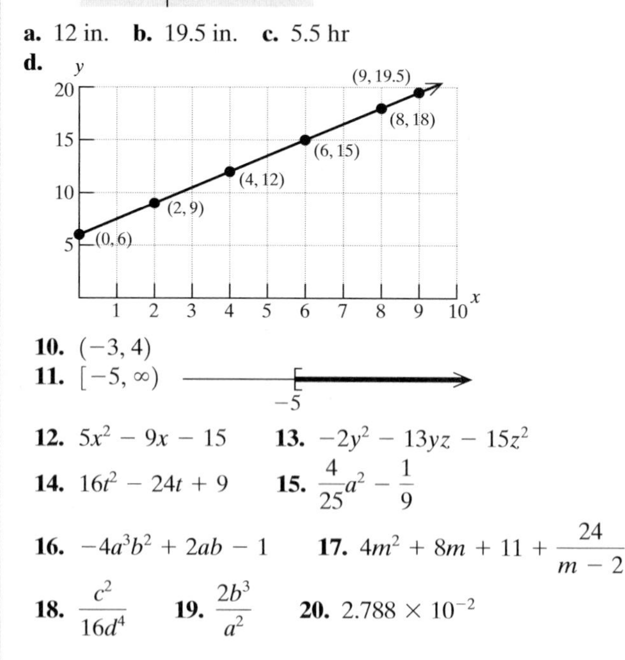

10. $(-3, 4)$
11. $[-5, \infty)$ -5
12. $5x^2 - 9x - 15$ **13.** $-2y^2 - 13yz - 15z^2$
14. $16t^2 - 24t + 9$ **15.** $\dfrac{4}{25}a^2 - \dfrac{1}{9}$
16. $-4a^3b^2 + 2ab - 1$ **17.** $4m^2 + 8m + 11 + \dfrac{24}{m - 2}$
18. $\dfrac{c^2}{16d^4}$ **19.** $\dfrac{2b^3}{a^2}$ **20.** 2.788×10^{-2}

Chapter 6

Chapter Opener Puzzle

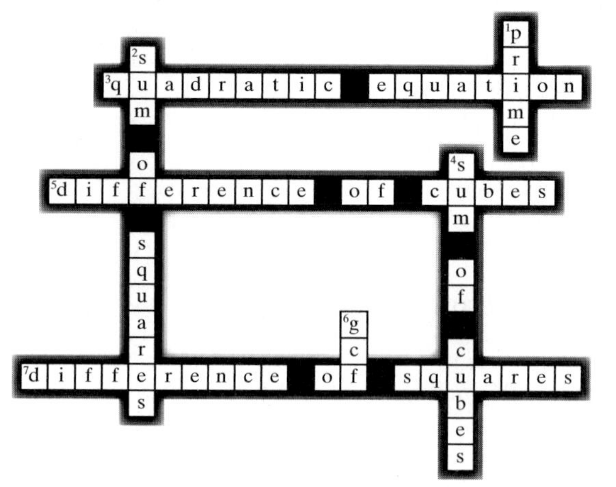

Section 6.1 Practice Exercises, pp. 453–455

3. 7 **5.** 6 **7.** y **9.** 4 **11.** ab **13.** $4w^2z$
15. $2xy^4z^2$ **17.** $(x - y)$ **19. a.** $3x - 6y$
b. $3(x - 2y)$ **21.** $4(p + 3)$ **23.** $5(c^2 - 2c + 3)$
25. $x^3(x^2 + 1)$ **27.** $t(t^3 - 4 + 8t)$ **29.** $2ab(1 + 2a^2)$
31. $19x^2y(2 - y^3)$ **33.** $6xy^5(x^2 - 3y^4z)$
35. The expression is prime because it is not factorable.
37. $7pq^2(6p^2 + 2 - p^3q^2)$ **39.** $t^2(t^3 + 2rt - 3t^2 + 4r^2)$
41. a. $-2x(x^2 + 2x - 4)$ **b.** $2x(-x^2 - 2x + 4)$
43. $-1(8t^2 + 9t + 2)$ **45.** $-15p^2(p + 2)$
47. $-q(q^3 - 2q + 9)$ **49.** $-1(7x + 6y + 2z)$
51. $(a + 6)(13 - 4b)$ **53.** $(w^2 - 2)(8v + 1)$
55. $7x(x + 3)^2$ **57.** $(2a - b)(4a + 3c)$
59. $(q + p)(3 + r)$ **61.** $(2x + 1)(3x + 2)$
63. $(t + 3)(2t - 5)$ **65.** $(3y - 1)(2y - 3)$
67. $(b + 1)(b^3 - 4)$ **69.** $(j^2 + 5)(3k + 1)$
71. $(2x^6 + 1)(7w^6 - 1)$ **73.** $(y + x)(a + b)$
75. $(vw + 1)(w - 3)$ **77.** $5x(x^2 + y^2)(3x + 2y)$
79. $4b(a - b)(x - 1)$ **81.** $6t(t - 3)(s - t^2)$
83. $P = 2(l + w)$ **85.** $S = 2\pi r(r + h)$
87. $\dfrac{1}{7}(x^2 + 3x - 5)$ **89.** $\dfrac{1}{4}(5w^2 + 3w + 9)$
91. For example, $6x^2 + 9x$
93. For example, $16p^4q^2 + 8p^3q - 4p^2q$

Section 6.2 Practice Exercises, pp. 459–461

3. $4xy^5(x^2y^2 - 3x^3 + 2y^3)$ **5.** $(a + 2b)(x - 5)$
7. $(x + 8)(x + 2)$ **9.** $(z - 9)(z - 2)$
11. $(z - 6)(z - 3)$ **13.** $(p + 8)(p - 5)$
15. $(t + 10)(t - 4)$ **17.** Prime
19. $(n + 4)^2$ **21.** a **23.** c
25. They are both correct because multiplication of polynomials is a commutative operation.
27. The expressions are equal and both are correct.
29. Descending order **31.** $(y + 20)(y - 8)$
33. $(t + 8)(t + 9)$ **35.** $(q - 30)(q + 40)$

37. $3(x - 12)(x + 2)$ **39.** $8p(p - 1)(p - 4)$ **41.** $y^2z^2(y - 6)^2$ **43.** $-(x - 4)(x - 6)$ **45.** $-(m + 2)(m - 3)$ **47.** $-2(c + 2)(c + 1)$ **49.** $xy^3(x - 4)(x - 15)$ **51.** $12(p - 7)(p - 1)$ **53.** $-2(m - 10)(m - 1)$ **55.** $(c + 5d)(c + d)$ **57.** $(a - 2b)(a - 7b)$ **59.** Prime **61.** $(q - 7)(q + 9)$ **63.** $(x + 10)^2$ **65.** $(t + 20)(t - 2)$ **67.** The student forgot to factor out the GCF before factoring the trinomial further. The polynomial is not factored completely, because $(2x - 4)$ has a common factor of 2. **69.** $x^2 + 9x - 52$ **71.** $7, 5, -7, -5$ **73.** For example: $c = -16$

Section 6.3 Practice Exercises, pp. 471–473

3. $3ab(7ab + 4b - 5a)$ **5.** $(n - 1)(m - 2)$ **7.** $6(a - 7)(a + 2)$ **9.** a **11.** b **13.** $(2w - 1)(w + 3)$ **15.** $(2a + 3)(a + 2)$ **17.** $(7y - 5)(y + 2)$ **19.** $(3z - 4)(2z + 3)$ **21.** $(2p + 5)(5p - 2)$ **23.** Prime **25.** $(7x - 4)(x - 2)$ **27.** $(8q - 1)(q + 4)$ **29.** $(12y - z)(y - 6z)$ **31.** $3(c - 8)(c - 3)$ **33.** $u^6(3u - 1)(u - 4)$ **35.** $-(x + 2)(x + 5)$ **37.** $-3z^2(3z - 4)(2z + 1)$ **39.** $10(w + 10z)(6w - 5z)$ **41.** $4y^2 - 28y + 49$ **43. a.** $x^2 + 12x + 36$ is a perfect square trinomial. **b.** $x^2 + 13x + 36 = (x + 9)(x + 4)$; $x^2 + 12x + 36 = (x + 6)^2$ **45.** $(y - 4)^2$ **47.** $(6p + 5)^2$ **49.** $(5m - 3n)^2$ **51.** $(w - 2)^2$ **53.** $4(3p - 1)^2$ **55.** $(2y + 9)(2y + 1)$ **57.** $-5(t - 2)(t - 3)$ **59.** $(w - 8)(w + 3)$ **61.** $(4p - 1)(p - 2)$ **63.** $(4r + 3)^2$ **65.** Prime **67.** $(3p - q)(5p + 2q)$ **69.** $(4a - 3b)(a + 2b)$ **71.** $(6x + y)(3x - 2y)$ **73.** Prime **75.** $(8n + 3)(2n + 1)$ **77.** $(6x - 5)(2x - 1)$ **79.** $(y + 4)(y - 10)$ **81.** $(b + 7)(b - 1)$ **83.** $(t + 10)^2$ **85.** $(p - 9q)(p - 4q)$ **87.** Prime **89.** $(x - 18y)(x + y)$ **91.** $(z - 12)(z - 3)$ **93.** $y(3y^2 - y + 12)$ **95.** $2(w + 7)(w + 3)$ **97.** $q^2(p - 3)(p - 11)$ **99. a.** $(x - 15)(x + 2)$ **b.** $(x - 10)(x - 3)$ **101. a.** $(x - 9)(x - 1)$ **b.** $(x + 9)(x + 1)$ **103.** $(y^2 - 3)(y^2 + 7)$ **105.** $(p^2 - 8)(p^2 - 5)$

Section 6.4 Practice Exercises, pp. 480–481

3. $(y + 5)(8 + 9y)$ **5.** $12, 1$ **7.** $-8, -1$ **9.** $5, -4$ **11.** $9, -2$ **13.** $(x + 4)(3x + 1)$ **15.** $(w - 2)(4w - 1)$ **17.** $(m + 3)(2m - 1)$ **19.** $(4k + 3)(2k - 3)$ **21.** $(2k - 5)^2$ **23.** Prime **25.** $(3z - 5)(3z - 2)$ **27.** $2(7y + 4)(y + 3)$ **29.** $(6y - 5z)(2y + 3z)$ **31.** $25y^2 - 70y + 49$ **33. a.** $4x^2 - 20x + 25$ is a perfect square trinomial. **b.** $4x^2 - 25x + 25 = (4x - 5)(x - 5)$; $4x^2 - 20x + 25 = (2x - 5)^2$ **35.** $(t - 8)^2$ **37.** $(7q - 2)^2$ **39.** $(2x + 9)^2$ **41.** $2(4x + 5y)^2$ **43.** $4(c + 1)^2$ **45.** $(3y + 4)(3y + 1)$ **47.** $(5p - 1)(4p - 3)$ **49.** $(3u - 2v)(2u - 5v)$ **51.** $(4a + 5b)(3a - b)$ **53.** $(h + 7k)(3h - 2k)$ **55.** Prime **57.** $(8w + 1)(2w + 1)$ **59.** $(q - 1)^2$ **61.** $-2(a - 1)(a - 9)$ **63.** $(m + 3)(m - 2)$ **65.** $2(10y + 1)(y - 4)$ **67.** $w^3(w - 4)(w - 7)$ **69.** $r(4r - 5)(r + 2)$ **71.** $4q(q - 5)(q + 4)$ **73.** $b^2(a + 10)(a + 3)$ **75.** $-1(m - 2)(m + 17)$ **77.** $-2(h - 9)(h - 5)$ **79.** $(m^2 + 3)(m^2 + 7)$ **81.** No. $(5x - 10)$ contains a common factor of 5.

Section 6.5 Practice Exercises, pp. 489–490

3. $(3x - 1)(2x - 5)$ **5.** $5xy^5(3x - 2y)$ **7.** $(x + b)(a - 6)$ **9.** $(5x + 3)^2$ **11.** $x^2 - 25$ **13.** $4p^2 - 9q^2$ **15.** $(x - 6)(x + 6)$ **17.** $(w - 10)(w + 10)$ **19.** $(2a - 11b)(2a + 11b)$ **21.** $(7m - 4n)(7m + 4n)$ **23.** Prime **25.** $(y + 2z)(y - 2z)$ **27.** $(a - b^2)(a + b^2)$ **29.** $(5pq - 1)(5pq + 1)$ **31.** $(c^3 - 5)(c^3 + 5)$ **33.** $(5 - 4t)(5 + 4t)$ **35.** $x^3, 8, y^6, 27q^3, w^{12}, r^3s^6$ **37.** $(a + b)(a^2 - ab + b^2)$ **39.** $(y - 2)(y^2 + 2y + 4)$ **41.** $(1 - p)(1 + p + p^2)$ **43.** $(w + 4)(w^2 - 4w + 16)$ **45.** $(x - 10)(x^2 + 10x + 100)$ **47.** $(4t + 1)(16t^2 - 4t + 1)$ **49.** $(10a + 3)(100a^2 - 30a + 9)$ **51.** $\left(n - \frac{1}{2}\right)\left(n^2 + \frac{1}{2}n + \frac{1}{4}\right)$ **53.** $(a + b^2)(a^2 - ab^2 + b^4)$ **55.** $(x^3 + 4y)(x^6 - 4x^3y + 16y^2)$ **57.** Prime **59.** $(x^2 - 2)(x^2 + 2)$ **61.** Prime **63.** $(t + 4)(t^2 - 4t + 16)$ **65.** Prime **67.** $4(b + 3)(b^2 - 3b + 9)$ **69.** $5(p - 5)(p + 5)$ **71.** $(\frac{1}{4} - 2h)(\frac{1}{16} + \frac{1}{2}h + 4h^2)$ **73.** $(x - 2)(x + 2)(x^2 + 4)$ **75.** $(q - 2)(q^2 + 2q + 4)(q + 2)(q^2 - 2q + 4)$ **77.** $\left(\frac{2x}{3} - w\right)\left(\frac{2x}{3} + w\right)$ **79.** $(2x + 3)(x - 1)(x + 1)$ **81.** $(2x - y)(2x + y)(4x^2 + y^2)$ **83.** $(3y - 2)(3y + 2)(9y^2 + 4)$ **85.** $(27k + 1)(3k + 1)$ **87.** $(k + 4)(k - 3)(k + 3)$ **89.** $2(t - 5)(t - 1)(t + 1)$ **91.** $y(y - 6)$ **93.** $(2p - 5)(2p + 7)$ **95.** $(-t + 2)(t + 6)$ or $-1(t - 2)(t + 6)$ **97.** $(-2b + 15)(2b + 5)$ or $-1(2b - 15)(2b + 5)$

Section 6.6 Practice Exercises, pp. 493–495

3. Look for a perfect square trinomial: $a^2 + 2ab + b^2$ or $a^2 - 2ab + b^2$.
5. Look for a difference of squares: $a^2 - b^2$, a difference of cubes: $a^3 - b^3$, or a sum of cubes: $a^3 + b^3$.
7. a. Difference of squares **b.** $2(a - 9)(a + 9)$
9. a. None of these **b.** $6w(w - 1)$
11. a. Nonperfect square trinomial **b.** $(3t + 1)(t + 4)$
13. a. Four terms-grouping **b.** $(3c + d)(a - b)$
15. a. Sum of cubes **b.** $(y + 2)(y^2 - 2y + 4)$
17. a. Nonperfect square trinomial **b.** $3(q - 4)(q + 1)$
19. a. None of these **b.** $6a(3a + 2)$
21. a. Difference of squares **b.** $4(t - 5)(t + 5)$
23. a. Nonperfect square trinomial **b.** $10(c^2 + c + 1)$
25. a. Sum of cubes **b.** $(x + 0.1)(x^2 - 0.1x + 0.01)$
27. a. Perfect square trinomial **b.** $(8 + k)^2$
29. a. Four terms-grouping **b.** $(x + 1)(2x - y)$
31. a. Difference of cubes **b.** $(a - c)(a^2 + ac + c^2)$
33. a. Nonperfect square trinomial **b.** Prime
35. a. Perfect square trinomial **b.** $(b + 5)^2$
37. a. Nonperfect square trinomial **b.** $-p(p + 4)(p + 1)$
39. a. Nonperfect square trinomial **b.** $3(2x + 3)(x - 5)$
41. a. None of these **b.** $abc^2(5ac - 7)$
43. a. Nonperfect square trinomial **b.** $(t + 9)(t - 7)$
45. a. Four terms-grouping **b.** $(b + y)(a - b)$
47. a. Nonperfect square trinomial **b.** $(7u - 2v)(2u - v)$
49. a. Nonperfect square trinomial **b.** $2(2q^2 - 4q - 3)$
51. a. Sum of squares **b.** Prime

53. **a.** Nonperfect square trinomial **b.** $(3r + 1)(2r + 3)$
55. **a.** Difference of squares **b.** $(2a - 1)(2a + 1)(4a^2 + 1)$
57. **a.** Perfect square trinomial **b.** $(9u - 5v)^2$
59. **a.** Nonperfect square trinomial **b.** $(x - 6)(x + 1)$
61. **a.** Four terms-grouping **b.** $2(x - 3y)(a + 2b)$
63. **a.** Nonperfect square trinomial **b.** $x^2y(3x + 5)(7x + 2)$
65. **a.** Four terms-grouping **b.** $(4v - 3)(2u + 3)$
67. **a.** Perfect square trinomial **b.** $3(2x - 1)^2$
69. **a.** Nonperfect square trinomial **b.** $n(2n - 1)(3n + 4)$
71. **a.** Difference of squares **b.** $(8 - y)(8 + y)$
73. **a.** Nonperfect square trinomial **b.** Prime
75. $\left(\frac{4}{5}p - \frac{1}{2}q\right)\left(\frac{16}{25}p^2 + \frac{2}{5}pq + \frac{1}{4}q^2\right)$
77. $(a^4 + b^4)(a^8 - a^4b^4 + b^8)$
79. **a.** The quotient is $x^2 + 2x + 4$. **b.** $(x - 2)(x^2 + 2x + 4)$
81. $x^2 + 2x + 4$ **83.** $2x + 1$
85. **a.** $a^2 - b^2$ **b.** $(a - b)(a + b)$ **87.** 4891

Section 6.7 Practice Exercises, pp. 504–508

3. $4(b - 5)(b - 6)$ **5.** $(3x - 2)(x + 4)$
7. $4(x^2 + 4y^2)$ **9.** Neither **11.** Quadratic
13. Linear **15.** $x = -3, x = 1$ **17.** $x = \frac{7}{2}, x = -\frac{7}{2}$
19. $x = -5$ **21.** $x = 0, x = -\frac{1}{3}, x = -1$
23. The equation must have one side equal to zero and the other side factored completely. **25.** $y = 8, y = -1$
27. $w = 8, w = 2$ **29.** $x = -\frac{1}{4}, x = 3$
31. $a = \frac{7}{2}, a = -\frac{7}{2}$ **33.** $t = -5$
35. $n = 0, n = -\frac{1}{3}, n = -1$ **37.** $x = \frac{1}{2}, x = 10, x = -7$
39. $t = 0, t = 6$, or $t = -6$ **41.** $y = -3, y = \frac{3}{2}$
43. $n = \frac{1}{3}, n = -\frac{1}{3}$ **45.** $d = 0, d = -2, d = 4$
47. $h = -13$ **49.** $q = -1, q = 4$ **51.** $x = 0, x = 6$
53. $k = \frac{3}{4}, k = -3$ **55.** $p = 3, p = -2$ **57.** $z = -\frac{4}{7}$
59. $w = 0, w = \frac{2}{3}$ **61.** $d = 0, d = -\frac{1}{4}$
63. $t = -2, t = 4, t = -4$ **65.** $w = -7, w = 5$
67. $k = 4, k = 2$ **69.** The numbers are $-\frac{9}{2}$ and 4.
71. The numbers are 5 and −4.
73. The numbers are 6 and 8, or −8 and −6.
75. The numbers are 0 and 1, or 9 and 10.
77. The painting has length 12 in. and width 10 in.
79. **a.** The picture is 13 in. by 6 in. **b.** 38 in.
81. The base is 10 cm and the height is 25 cm. **83.** 4 sec
85. 0 sec and 4 sec **87.** Given a right triangle with legs a and b and hypotenuse c, then $a^2 + b^2 = c^2$. **89.** $c = 5$ m
91. $b = 12$ yd **93.** The height is 9 km. **95.** 19 yd
97. 10 m

Chapter 6 Review Exercises, pp. 513–515

1. $3a^2b$ **2.** $x + 5$ **3.** $2c(3c - 5)$ **4.** $-2yz$ or $2yz$
5. $2x(3x + x^3 - 4)$ **6.** $11w^2y^3(w - 4y^2)$
7. $t(-t + 5)$ or $-t(t - 5)$ **8.** $u(-6u - 1)$ or $-u(6u + 1)$
9. $(b + 2)(3b - 7)$ **10.** $2(5x + 9)(1 + 4x)$
11. $(w + 2)(7w + b)$ **12.** $(b - 2)(b + y)$
13. $3(4y - 3)(5y - 1)$ **14.** $a(2 - a)(3 - b)$
15. $(x - 3)(x - 7)$ **16.** $(y - 8)(y - 11)$
17. $(z - 12)(z + 6)$ **18.** $(q - 13)(q + 3)$
19. $3w(p + 10)(p + 2)$ **20.** $2m^2(m + 8)(m + 5)$
21. $-(t - 8)(t - 2)$ **22.** $-(w - 4)(w + 5)$
23. $(a + b)(a + 11b)$ **24.** $(c - 6d)(c + 3d)$
25. Different **26.** Both negative **27.** Both positive
28. Different **29.** $(2y + 3)(y - 4)$
30. $(4w + 3)(w - 2)$ **31.** $(2z + 5)(5z + 2)$
32. $(4z - 3)(2z + 3)$ **33.** Prime **34.** Prime
35. $10(w - 9)(w + 3)$ **36.** $3(y - 8)(y + 2)$
37. $(3c - 5d)^2$ **38.** $(x + 6)^2$ **39.** $(v^2 + 1)(v^2 - 3)$
40. $(x^2 + 5)(x^2 + 2)$ **41.** The trinomials in Exercises 37 and 38. **42.** 5, −1 **43.** −3, −5 **44.** $(c - 2)(3c + 1)$
45. $(y + 3)(4y + 1)$ **46.** $(t + 12)(t + 1)$
47. $x(x + 5)(4x - 3)$ **48.** $w(w + 5)(w - 1)$
49. $(p - 3q)(p - 5q)$ **50.** $2(4v + 3)(5v - 1)$
51. $10(4s - 5)(s + 2)$ **52.** $ab(a - 6b)(a - 4b)$
53. $2z^4(z + 7)(z - 3)$ **54.** $(3m - 1)(3m + 2)$
55. $(3p + 2)(2p + 5)$ **56.** $(7x + 10)^2$ **57.** $(3w - z)^2$
58. The trinomials in Exercises 56 and 57.
59. $(a - b)(a + b)$ **60.** Prime
61. $(a + b)(a^2 - ab + b^2)$ **62.** $(a - b)(a^2 + ab + b^2)$
63. $(a - 7)(a + 7)$ **64.** $(d - 8)(d + 8)$
65. $(10 - 9t)(10 + 9t)$ **66.** $(2 - 5k)(2 + 5k)$
67. Prime **68.** Prime **69.** $(4 + a)(16 - 4a + a^2)$
70. $(5 - b)(25 + 5b + b^2)$ **71.** $(p^2 + 2)(p^4 - 2p^2 + 4)$
72. $\left(q^2 - \frac{1}{3}\right)\left(q^4 + \frac{1}{3}q^2 + \frac{1}{9}\right)$ **73.** $6(x - 2)(x^2 + 2x + 4)$
74. $7(y + 1)(y^2 - y + 1)$ **75.** $2(c^2 - 3)(c^2 + 3)$
76. $2(6x - y)(6x + y)$ **77.** $(p + 3)(p - 4)(p + 4)$
78. $(k - 2)(2 - k)(2 + k)$ or $-1(k - 2)^2(2 + k)$
79. $(6y + 1)(y - 2)$ **80.** $3(p - 1)^2$
81. $x(x - 6)(x + 6)$ **82.** $(k - 7)(k - 6)$
83. $(c - 2d)(7a - b)$ **84.** $q(q - 4)(q^2 + 4q + 16)$
85. $4(2h^2 + 5)$ **86.** Prime **87.** $m(m - 8)$
88. $(x + 4)(x + 1)(x - 1)$ **89.** $3st(4s + t)(s - 4t)$
90. $5q(p^2 - 2q)(p^2 + 2q)$ **91.** $3(3a - 1)(2a + 5)$
92. $w^2(w + 8)(w - 7)$ **93.** $n(2 + n)(4 - 2n + n^2)$
94. $14(m - 1)(m^2 + m + 1)$
95. $(x - 3)(2x + 1) = 0$ can be solved directly by the zero product rule because it is a product of factors set equal to zero.
96. $x = \frac{1}{4}, x = -\frac{2}{3}$ **97.** $a = 9, a = \frac{1}{2}$
98. $w = 0, w = -3, w = -\frac{2}{5}$ **99.** $u = 0, u = 7, u = \frac{9}{4}$
100. $k = -\frac{5}{7}, k = 2$ **101.** $h = -\frac{1}{4}, h = 6$
102. $q = 12, q = -12$ **103.** $r = 5, r = -5$
104. $v = 0, v = \frac{1}{5}$ **105.** $x = 4, x = 2$

106. $t = -\frac{5}{6}$ **107.** $s = -\frac{2}{3}$ **108.** $y = \frac{2}{3}, y = 6$

109. $p = \frac{11}{2}, p = -12$ **110.** $y = 0, y = 7, y = 2$

111. $x = 0, x = 2, x = -2$
112. The height is 6 ft, and the base is 13 ft.
113. The ball is at ground level at 0 and 1 sec.
114. The ramp is 13 ft long.
115. The legs are 6 ft and 8 ft; the hypotenuse is 10 ft.
116. The numbers are -8 and 8.
117. The numbers are 29 and 30, or -2 and -1.
118. The height is 4 m, and the base is 9 m.

Chapter 6 Test, p. 515

1. $3x(5x^3 - 1 + 2x^2)$ **2.** $(a - 5)(7 - a)$
3. $(6w - 1)(w - 7)$ **4.** $(13 - p)(13 + p)$
5. $(q - 8)^2$ **6.** $(2 + t)(4 - 2t + t^2)$
7. $(a + 4)(a + 8)$ **8.** $(x + 7)(x - 6)$
9. $(2y - 1)(y - 8)$ **10.** $(2z + 1)(3z + 8)$
11. $(3t - 10)(3t + 10)$ **12.** $(v + 9)(v - 9)$
13. $3(a + 6b)(a + 3b)$ **14.** $(c - 1)(c + 1)(c^2 + 1)$
15. $(y - 7)(x + 3)$ **16.** Prime
17. $-10(u - 2)(u - 1)$ **18.** $3(2t - 5)(2t + 5)$
19. $5(y - 5)^2$ **20.** $7q(3q + 2)$
21. $(2x + 1)(x - 2)(x + 2)$ **22.** $(y - 5)(y^2 + 5y + 25)$
23. $(mn - 9)(mn + 9)$ **24.** $16(a - 2b)(a + 2b)$
25. $(4x - 3y^2)(16x^2 + 12xy^2 + 9y^4)$ **26.** $x = \frac{3}{2}, x = -5$
27. $x = 0, x = 7$ **28.** $x = 8, x = -2$
29. $x = \frac{1}{5}, x = -1$ **30.** $y = 3, y = -3, y = -10$
31. The tennis court is 12 yd by 26 yd. **32.** The two integers are 5 and 7 or -5 and -7. **33.** The base is 12 in., and the height is 7 in. **34.** The shorter leg is 5 ft.

Chapters 1–6 Cumulative Review Exercises, p. 516

1. $\frac{7}{5}$ **2.** $t = -3$ **3.** $y = \frac{3}{2}x - 4$
4. There are 10 quarters, 12 nickels, and 7 dimes.
5. $[-4, \infty)$

6. a. Yes **b.** 1 **c.** $(0, 4)$ **d.** $(-4, 0)$
e.

7. a. Vertical line **b.** Undefined **c.** $(5, 0)$
d. Does not exist **8.** $y = 3x + 14$ **9.** $(5, 2)$
10. $-\frac{7}{2}y^2 - 5y - 14$ **11.** $8p^3 - 22p^2 + 13p + 3$
12. $4w^2 - 28w + 49$ **13.** $r^3 + 5r^2 + 15r + 40 + \frac{121}{r - 3}$
14. c^4 **15.** 1.6×10^3 **16.** $(w - 2)(w + 2)(w^2 + 4)$
17. $(a + 5b)(2x - 3y)$ **18.** $(2x - 5)(2x + 1)$
19. $(y - 3)(y^2 + 3y + 9)$ **20.** $x = 0, x = \frac{1}{2}, x = -5$

Chapter 7

Chapter Opener Puzzle

Column 1	Column 2
$\frac{3}{2} \boxed{+} \frac{1}{2} = 2$	$\frac{3}{10} \boxed{\cdot} \frac{5}{6} \boxed{-} \frac{1}{8} = \frac{1}{8}$
$\frac{3}{4} \boxed{\div} \frac{1}{8} = 6$	$\frac{1}{2} \boxed{+} \frac{1}{2} \boxed{\div} \frac{3}{4} = \frac{7}{6}$
$\frac{5}{7} \boxed{\cdot} \frac{14}{3} = \frac{10}{3}$	
$\frac{2}{3} \boxed{-} \frac{1}{2} = \frac{1}{6}$	

Section 7.1 Practice Exercises, pp. 525–528

3. a. A number $\frac{p}{q}$, where p and q are integers and $q \neq 0$. **b.** An expression $\frac{p}{q}$, where p and q are polynomials and $q \neq 0$. **5.** $-\frac{1}{8}$ **7.** $-\frac{1}{2}$ **9.** 0 **11.** Undefined
13. a. \$196 **b.** \$168 **c.** \$145.60
15. $\{h \mid h \text{ is a real number and } h \neq 4\}$
17. $\left\{y \mid y \text{ is a real number and } y \neq -\frac{7}{3}, y \neq -3\right\}$
19. $\{c \mid c \text{ is a real number and } c \neq 6, c \neq -1\}$ **21.** The set of all real numbers **23.** The set of all real numbers
25. For example: $\frac{1}{x - 5}$
27. For example: $\frac{1}{(x + 1)(x - 4)}$
29. a. Undefined **b.** $\frac{2}{3}$ **31. a.** $\frac{9}{5}$ **b.** $\frac{9}{5}$
33. a. $\{x \mid x \text{ is a real number and } x \neq 1\}$ **b.** 2
35. a. $\{r \mid r \text{ is a real number and } r \neq 2\}$ **b.** $r + 2$
37. a. $\left\{a \mid a \text{ is a real number and } a \neq 0, a \neq \frac{3}{4}\right\}$ **b.** $\frac{2a}{4a - 3}$
39. a. $\left\{b \mid b \text{ is a real number and } b \neq \frac{5}{2}, b \neq -\frac{5}{2}\right\}$
b. $\frac{4}{2b + 5}$
41. a. $\{t \mid t \text{ is a real number and } t \neq -5,\ t \neq 4\}$ **b.** $\frac{t - 2}{t - 4}$
43. $\frac{5}{c^2}$ **45.** $\frac{4a^3}{5}$ **47.** $-\frac{5s^2}{r^3t}$ **49.** 2 **51.** $\frac{c - 1}{c + 2}$

53. $\frac{1}{9(n+2)}$ **55.** $\frac{(p-3)^2}{p+2}$ **57.** $\frac{1}{2c-3}$ **59.** $\frac{3}{x-5}$
61. $\frac{3(p+2)}{q-2}$ **63.** $\frac{5}{z-7}$ **65.** $b+8$
67. Cannot simplify **69.** $\frac{h+3}{h+4}$ **71.** $\frac{x-7}{2x-5}$
73. $\frac{4}{(t-2)(t+2)}$ **75.** $\frac{p-q}{p+q}$ **77.** $\frac{m}{m+1}$
79. $\frac{3}{2(x-y)}$ **81.** $\frac{1}{ax-b}$ **83.** $\frac{2c-d}{5c-2d}$
85. They are opposites. **87.** -1 **89.** -1
91. $-\frac{1}{3}$ **93.** -1 **95.** Cannot simplify
97. $\frac{t-4}{4+t}$ **99.** $-\frac{7+b}{b-3}$ **101.** $y+3$
103. $\frac{x-5}{x^2-5x+25}$

Section 7.2 Practice Exercises, pp. 532–533

1. To multiply fractions, we multiply the numerators and multiply the denominators. To divide the fractions, we would multiply the first fraction by the reciprocal of the second fraction.
3. $\frac{5}{14}$ **5.** 9 **7.** $\frac{7}{5}$ **9.** 6 **11.** $\frac{1}{2s}$ **13.** $\frac{10}{a}$
15. $\frac{3(a+4)}{2}$ **17.** $\frac{p}{6}$ **19.** $-\frac{10}{3}$ **21.** $-\frac{x-y}{x+y}$
23. $\frac{3}{2c^2}$ **25.** $6a^3b$ **27.** $\frac{16}{15}$ **29.** $\frac{16a^2}{15}$
31. $-\frac{b(3+b)}{3(b+1)}$ **33.** $\frac{5(x+y)}{2x-y}$ **35.** $\frac{2t-1}{t}$
37. $\frac{5t+1}{5t-1}$ **39.** 4 **41.** $\frac{4r}{2r+3}$ **43.** $\frac{1}{5}$
45. $\frac{z-4}{z-1}$ **47.** $\frac{p+1}{p-3}$ **49.** $\frac{q-7}{q+4}$ **51.** $\frac{y+6}{y+1}$
53. $-\frac{(p-2)(p+1)}{4(4+p)}$ **55.** $\frac{4h-1}{3h-2}$ **57.** $\frac{2}{y}$
59. $\frac{x+2}{2}$ **61.** $\frac{2m+n}{2}$ **63.** $\frac{1}{3}$ **65.** $\frac{t}{5(t-1)}$
67. r^2-rs+s^2

Section 7.3 Practice Exercises, pp. 539–542

3. $\{x \mid x$ is a real number and $x \neq 1, x \neq -1\}$; $\frac{3}{5(x-1)}$
5. $\frac{a+5}{a+7}$ **7.** $\frac{2}{3y}$ **9.** 36 **11.** 6
13. $15p$ **15.** $12xyz^3$ **17.** $w^2+8w+12$ **19.** -20
21. $5x-10$ **23.** -6 **25.** a, b, c, d **27.** x^5 is the lowest power of x that has x^3, x^5, x^4 as factors.
29. The product of unique factors is $(x+3)(x-2)$.
31. Because $(b-1)$ and $(1-b)$ are opposites; they differ by a factor of -1. **33.** 45 **35.** 16 **37.** 63
39. $9x^2y^3$ **41.** w^2y **43.** $(p+3)(p-1)(p+2)$
45. $9t(t+1)^2$ **47.** $(y-2)(y+2)(y+3)$
49. $3-x$ or $x-3$ **51.** $\frac{6}{5x^2}; \frac{5x}{5x^2}$
53. $\frac{24x}{30x^3}; \frac{5y}{30x^3}$ **55.** $\frac{10}{12a^2b}; \frac{a^3}{12a^2b}$
57. $\frac{6m-6}{(m+4)(m-1)}; \frac{3m+12}{(m+4)(m-1)}$
59. $\frac{6x+18}{(2x-5)(x+3)}; \frac{2x-5}{(2x-5)(x+3)}$
61. $\frac{6w+6}{(w+3)(w-8)(w+1)}; \frac{w^2+3w}{(w+3)(w-8)(w+1)}$
63. $\frac{6p^2+12p}{(p-2)(p+2)^2}; \frac{3p-6}{(p-2)(p+2)^2}$
65. $\frac{1}{a-4}; \frac{-a}{a-4}$ or $\frac{-1}{4-a}; \frac{a}{4-a}$
67. $\frac{8}{2(x-7)}; \frac{-y}{2(x-7)}$ or $\frac{-8}{2(7-x)}; \frac{y}{2(7-x)}$
69. $\frac{1}{a+b}; \frac{-6}{a+b}$ or $\frac{-1}{-a-b}; \frac{6}{-a-b}$
71. $\frac{-9}{24(3y+1)}; \frac{20}{24(3y+1)}$ **73.** $\frac{3z+12}{5z(z+4)}; \frac{5z}{5z(z+4)}$
75. $\frac{z^2+3z}{(z+2)(z+7)(z+3)}; \frac{-3z^2-6z}{(z+2)(z+7)(z+3)}; \frac{5z+35}{(z+2)(z+7)(z+3)}$
77. $\frac{3p+6}{(p-2)(p^2+2p+4)(p+2)}; \frac{p^3+2p^2+4p}{(p-2)(p^2+2p+4)(p+2)}; \frac{5p^3-20p}{(p-2)(p^2+2p+4)(p+2)}$

Section 7.4 Practice Exercises, pp. 548–551

1. $\{x \mid x$ is a real number and $x \neq 6, x \neq -6\}$
3. a. $\frac{1}{6}$, 0, undefined, $-\frac{1}{4}$, undefined
b. $(a-6)(a+2)$; $\{a \mid a$ is a real number and $a \neq 6, a \neq -2\}$
c. $\frac{a-1}{a-6}$ **5.** $\frac{6t-1}{(t+1)(t-6)}$ **7.** $\frac{8}{3}$ **9.** $\frac{2}{3}$ **11.** $\frac{b+9}{b-3}$
13. 6 **15.** $\frac{6p+5}{2p+1}$ **17.** $2w+1$ **19.** $k-3$
21. 5 **23.** $y+7$ **25.** $\frac{1}{x+3}$ **27.** $\frac{1}{x+4}$
29. $\frac{36}{t+3}$ **31.** $\frac{1}{12p}$ **33.** $\frac{10b-7a^2}{6a^2b^2}$ **35.** $\frac{q^2-2p}{p^2q^3}$
37. $\frac{1}{2}$ **39.** $\frac{11}{5(c-4)}$ **41.** $\frac{5h-32}{(h-5)(h+5)}$ **43.** $\frac{5}{24}$
45. $\frac{4}{2w-1}$ or $\frac{-4}{1-2w}$ **47.** $\frac{10x+3y}{(x+y)^2}$
49. $\frac{2r-5}{7}$ or $\frac{-2r+5}{-7}$ **51.** $\frac{4m+1}{m-2}$ or $\frac{-4m-1}{2-m}$
53. $\frac{3(5y-3)}{y(y-1)}$ **55.** $\frac{2y^2+35x}{14xy^2}$ **57.** $\frac{12-5q}{3q(q-2)}$
59. $\frac{3}{z+4}$ **61.** $\frac{2b}{(b+1)(b-1)}$
63. $\frac{-x(x+11)}{(x+1)(x+4)(x-3)}$ **65.** $\frac{8y^2-7y+5}{(3y+1)(2y-3)(y+1)}$
67. $\frac{-1}{3q-2}$ **69.** $\frac{x-y}{x+y}$ **71.** $\frac{-2}{x-y}$ **73.** $\frac{13t+2}{2t^2}$

75. $\frac{1}{n+6}$ **77.** $\frac{12}{p}$ **79.** $n + \left(5 \cdot \frac{1}{n}\right); \frac{n^2+5}{n}$
81. $\frac{1}{m} - \frac{3m}{7}; \frac{7-3m^2}{7m}$ **83.** $\frac{2m^2+1}{(m-1)^2(m^2+m+1)}$
85. $\frac{-17t^2-6t+2}{(2t+1)(4t-1)(t-5)}$
87. $\frac{-5n+17}{2(3n+1)(n-3)}$ or $\frac{5n-17}{2(3n+1)(3-n)}$ **89.** $\frac{2p+3}{3p+4}$
91. $\frac{2}{m}$

Chapter 7, Problem Recognition Exercises—Operations on Rational Expressions p. 552

1. $\frac{-2x+9}{3x+1}$ **2.** $\frac{1}{w-4}$ **3.** $\frac{y-5}{2y-3}$
4. $\frac{7}{(x+3)(2x-1)}$ **5.** $-\frac{1}{x}$ **6.** $\frac{1}{3}$ **7.** $\frac{c+3}{c}$
8. $\frac{x+3}{5}$ **9.** $\frac{a}{12b^4c}$ **10.** $\frac{2a-b}{a-b}$ **11.** $\frac{p-q}{5}$
12. 4 **13.** $\frac{10}{2x+1}$ **14.** $\frac{w+2z}{w+z}$ **15.** $\frac{3}{2x+5}$
16. $\frac{y+7}{x+a}$ **17.** $\frac{1}{2(a+3)}$ **18.** $\frac{2(3y+10)}{(y-6)(y+6)(y+2)}$
19. $(t+8)^2$ **20.** $6b+5$
21. $\{x|x \text{ is a real number and } x \neq -1\}$
22. $\{x|x \text{ is a real number and } x \neq 3, x \neq -3\}$

Section 7.5 Practice Exercises, pp. 558–560

3. $\left\{a \mid a \text{ is a real number and } a \neq \frac{3}{2}, a \neq -5\right\}; \frac{1}{2a-3}$
5. $\frac{3(2k-5)}{5(k-2)}$ **7.** $\frac{1}{3xy(x+y)}$ **9.** $\frac{2b-9a^2}{3a(5-4ab)}$
11. $\frac{4a}{5}$ **13.** $\frac{3}{2}$ **15.** $\frac{27y}{10x^4}$ **17.** $\frac{p^2q^2}{2w^2}$
19. $\frac{10}{23}$ **21.** $1+b$ **23.** $\frac{5(k+5)}{k+1}$ **25.** 6
27. $\frac{3(4+p^2)}{2(p-3)(p+3)}$ **29.** $\frac{2(3+2m^2)}{m(6-5m)}$ **31.** $\frac{3w+5}{4w+7}$
33. $\frac{p+3}{p+2}$ **35.** $\frac{3}{2m}$ **37.** $\frac{10}{\frac{2}{5}-\frac{1}{4}}; \frac{200}{3}$ **39.** $\frac{\frac{3}{5}-\frac{1}{2}}{4}; \frac{1}{40}$

41. a. 51.4 mph average **b.** 51.4 mph **c.** Because the rates going to and leaving from the destination are the same, the average rate is unchanged. The average rate is not affected by the distance traveled.

43. $\frac{-3w+20}{4(w+5)}$ **45.** $-\frac{(y+3)(y-2)}{(y+1)(y-3)}$ **47.** $\frac{5}{3}$

Section 7.6 Practice Exercises, pp. 568–571

3. $\frac{2}{4x-1}$ **5.** $5(h+1)$ **7.** $\frac{(x+4)(x-3)}{x^2}$
9. $b = 3$ **11.** $k = \frac{5}{11}$ **13.** $y = \frac{1}{3}$
15. a. $\{z \mid z \text{ is a real number and } z \neq 0\}$ **b.** $5z$ **c.** $z = 5$
17. a. $\{x \mid x \text{ is a real number and } x \neq 2, x \neq -3\}$
b. $(x-2)(x+3)$ **c.** $x = 7$

19. $x = -\frac{21}{8}$ **21.** $b = 4$ **23.** $x = -\frac{1}{6}$
25. $m = 4, m = -2$ **27.** $b = 8$
29. $t = -4$; ($t = 0$ does not check.)
31. $n = 3$ **33.** No solution; ($q = -5$ does not check.)
35. $w = 5$ **37.** $x = -9; x = -1$
39. $x = -7$; ($x = \frac{3}{2}$ does not check.)
41. No solution; ($x = -3$ does not check.)
43. $y = 2$; ($y = -4$ does not check.) **45.** $a = 60$
47. $w = -3$ **49.** The number is 7.
51. The number is −3. **53.** $a = \frac{FK}{m}$ **55.** $R = \frac{KE}{I}$
57. $r = \frac{E-IR}{I}$ or $r = \frac{E}{I} - R$ **59.** $r = \frac{C}{2\pi}$
61. $w = \frac{V}{lh}$ **63.** $m = \frac{T}{g-f}$ or $m = \frac{-T}{f-g}$
65. $w = \frac{n}{1-Pn}$ or $w = \frac{-n}{Pn-1}$ **67.** $P = \frac{A}{1+rt}$
69. $b = \frac{fa}{a-f}$ or $b = \frac{-fa}{f-a}$

Chapter 7, Problem Recognition Exercises—Comparing Rational Equations and Rational Expressions, p. 571

1. $\frac{y-2}{2y}$ **2.** $x = 6$ **3.** $t = 2$ **4.** $\frac{3a-17}{a-5}$
5. $\frac{4p+27}{18p^2}$ **6.** $\frac{b(b-5)}{(b-1)(b+1)}$ **7.** $h = 5$
8. $\frac{2w+5}{(w+1)^2}$ **9.** $x = 7$ **10.** $m = 5$
11. $\frac{3x+14}{4(x+1)}$ **12.** $t = \frac{11}{3}$

Section 7.7 Practice Exercises, pp. 580–584

3. Expression; $\frac{m^2+m+2}{(m-1)(m+3)}$ **5.** Expression; $\frac{3}{10}$
7. Equation; $p = 2$ **9.** $p = 95$ **11.** $z = 1$
13. $a = \frac{40}{3}$ **15.** $x = 40$ **17.** $y = 3$ **19.** $z = -1$
21. $a = 1$ **23. a.** $V_f = \frac{V_iT_f}{T_i}$ **b.** $T_f = \frac{T_iV_f}{V_i}$
25. Toni can drive 297 mi on 9 gallons of gas.
27. They would produce 1536 lb.
29. The opponent received 4336 votes.
31. This represents 262.5 miles. **33. a.** 4 cm **b.** 5 cm
35. $x = 3.75$ cm; $y = 4.5$ cm
37. The height of the pole is 7 m.
39. The light post is 24 ft high.
41. The speed of the current is 2 mph.
43. The wind speed is 35 mph.
45. Shanelle skis 10 km/hr and Devon skis 15 km/hr.
47. The speeds are 45 mph and 60 mph.
49. Kendra flew 180 mph to Cincinnati and 150 mph back.
51. Sergio rode 12 mph and walked 3 mph.
53. $\frac{1}{2}$ of the room **55.** $5\frac{5}{11}$ ($5.\overline{45}$) minutes
57. $22\frac{2}{9}$ ($22.\overline{2}$) min **59.** $3\frac{1}{3}$ ($3.\overline{3}$) days

61. 48 hr **63.** 16 min
65. There are 40 smokers and 140 nonsmokers.
67. There are 240 men and 200 women.

Section 7.8 Practice Exercises, pp. 590–593

3. $y = 12$ **5.** $a = 6; a = 2$ **7.** $\frac{b+1}{1-b}$
9. a. Decrease **b.** Increase **11.** $P = \frac{k}{r}$
13. $Y = k\sqrt{z}$ **15.** $M = \frac{kp^2}{n^3}$ **17.** $X = kwy^2$
19. $k = \frac{11}{4}$ **21.** $k = 8000$ **23.** $k = 44$
25. $Q = 9$ **27.** $Y = 4$ **29.** $R = 84$ **31.** 64 ft^2
33. 18.75 lumens **35.** 300 W **37.** 336 cycles/sec
39. 150 ft^2 **41.** 2 sec
43. a. Direct **b.** $d = 1.25F$ or $F = 0.8d$

Chapter 7, Review Exercises, pp. 599–602

1. a. $-\frac{2}{9}, -\frac{1}{10}, 0, -\frac{5}{6}$, undefined
b. $\{t \mid t$ is a real number and $t \neq -9\}$
2. a. $-\frac{1}{5}, -\frac{1}{2}$, undefined, $0, \frac{1}{7}$
b. $\{k \mid k$ is a real number and $k \neq 5\}$
3. a, c, d
4. $\{x \mid x$ is a real number and $x \neq \frac{5}{2}, x \neq 3\}$; $\frac{1}{2x-5}$
5. $\{h \mid h$ is a real number and $h \neq -\frac{1}{3}, h \neq -7\}$; $\frac{1}{3h+1}$
6. $\{a \mid a$ is a real number and $a \neq 2, a \neq -2\}$; $\frac{4a-1}{a-2}$
7. $\{w \mid w$ is a real number and $w \neq 4, w \neq -4\}$; $\frac{2w+3}{w-4}$
8. $\{z \mid z$ is a real number and $z \neq 4\}$; $-\frac{z}{2}$
9. $\{k \mid k$ is a real number and $k \neq 0, k \neq 5\}$; $-\frac{3}{2k}$
10. $\{b \mid b$ is a real number and $b \neq -3\}$; $\frac{b-1}{2}$
11. $\{m \mid m$ is a real number and $m \neq -1\}$; $\frac{m-5}{3}$
12. $\{n \mid n$ is a real number and $n \neq -3\}$; $\frac{1}{n+3}$
13. $\{p \mid p$ is a real number and $p \neq -7\}$; $\frac{1}{p+7}$
14. y^2 **15.** $\frac{u^2}{2}$ **16.** $v + 2$ **17.** $\frac{3}{2(x-5)}$
18. $\frac{c(c+1)}{2(c+5)}$ **19.** $\frac{q-2}{4}$ **20.** $-2t(t-5)$
21. $4s(s-4)$ **22.** $\frac{1}{7}$ **23.** $\frac{1}{n-2}$ **24.** $-\frac{1}{6}$
25. $\frac{1}{m+3}$ **26.** $\frac{-1}{(x+3)(x+2)}$ **27.** $-\frac{2y-1}{y+1}$
28. $5x + 5$ **29.** $2y + 4$ **30.** $6w - 24$
31. $2r + 6$ **32.** $s^2 - 6s + 8$ **33.** $u^2 - 5u - 6$
34. $a^2b^3c^2$ **35.** xy^2z^4 **36.** $(p+2)(p-4)$
37. $q(q+8)$ **38.** $(m-4)(m+4)(m+3)$
39. $(n-3)(n+3)(n+2)$ **40.** $2t - 5$ or $5 - 2t$
41. $3k - 1$ or $1 - 3k$ **42.** $c - 2$ or $2 - c$
43. $3 - x$ or $x - 3$ **44.** 2 **45.** 2 **46.** $a + 5$
47. $x - 7$ **48.** $\frac{-y-18}{(y-9)(y+9)}$ or $\frac{y+18}{(9-y)(y+9)}$
49. $\frac{t^2+2t+3}{(2-t)(2+t)}$ **50.** $\frac{m+8}{3m(m+2)}$ **51.** $\frac{3(r-4)}{2r(r+6)}$
52. $\frac{p}{(p+4)(p+5)}$ **53.** $\frac{q}{(q+5)(q+4)}$ **54.** $\frac{1}{2}$
55. $\frac{1}{3}$ **56.** $\frac{a-4}{a-2}$ **57.** $\frac{1}{4z}$ **58.** $\frac{w}{2}$
59. $\frac{8}{y}$ **60.** $y - x$ **61.** $-(b+a)$ **62.** $-\frac{2p+7}{2p}$
63. $-\frac{k+10}{k+4}$ **64.** $x = -8$ **65.** $y = -2$
66. $h = 0$ **67.** $w = 2$ **68.** $t = -2$ **69.** $p = 3$
70. $y = -11, y = 1$
71. No solution; ($z = 2$ does not check.)
72. The number is 4. **73.** $h = \frac{3V}{\pi r^2}$ **74.** $b = \frac{2A}{h}$
75. $m = \frac{6}{5}$ **76.** $a = 19.2$ **77.** 10 g
78. Ed travels at 60 mph, and Bud travels at 70 mph.
79. Together the pumps would fill the pool in 16.8 min.
80. $x = 11; y = 26$
81. a. $F = kd$ **b.** $k = 3$ **c.** $F = 5$ lb **82.** $y = \frac{8}{3}$
83. $y = 12$ **84.** 48 km

Chapter 7, Test, pp. 602–603

1. a. $\{x \mid x$ is a real number and $x \neq 2\}$ **b.** $-\frac{x+1}{6}$
2. a. $\{a \mid a$ is a real number and $a \neq 0, a \neq 6, a \neq -2\}$
b. $\frac{7}{a+2}$
3. b, c, d **4.** $\frac{y+7}{3(y+3)(y+1)}$ **5.** $-\frac{1}{5(9b+1)}$
6. $\frac{1}{w+1}$ **7.** $\frac{t+4}{t+2}$ **8.** $\frac{x(3x+5)}{(3x+4)(x-2)}$ **9.** $\frac{1}{m+4}$
10. $a = \frac{8}{5}$ **11.** $p = -2$; ($p = 1$ does not check.)
12. $c = 1$ **13.** No solution. ($x = 4$ does not check.)
14. $r = \frac{2A}{C}$ **15.** The number is $-\frac{2}{5}$. **16.** $y = -8$
17. $1\frac{1}{4}$ (1.25) cups of carrots
18. The speed of the current is 5 mph. **19.** 2 hr
20. $a = 5.6; b = 12$ **21. a.** $15(x+3)$ **b.** $3x^2y^2$
22. 3.3 sec

Chapters 1–7, Cumulative Review Exercises, pp. 603–604

1. 32 **2.** 7 **3.** $y = \frac{10}{9}$
4.

Set-Builder Notation	Graph	Interval Notation
$\{x \mid x \geq -1\}$	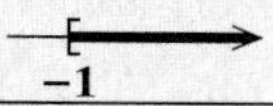 −1	$[-1, \infty)$
$\{x \mid x < 5\}$	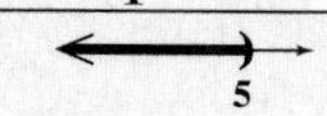 5	$(-\infty, 5)$

5. The width is 17 m and the length is 35 m.
6. The base is 10 in. and the height is 8 in.
7. $\frac{x^2yz^{17}}{2}$ **8. a.** $6x + 4$ **b.** $2x^2 + x - 3$
9. $(5x-3)^2$ **10.** $(2c+1)(5d-3)$
11. $\left\{x \mid x \text{ is a real number and } x \neq 5, x \neq -\frac{1}{2}\right\}$
12. $\frac{x-3}{x+5}$ **13.** $\frac{1}{5(x+4)}$ **14.** -3 **15.** $y = 1$
16. $b = -\frac{7}{2}$

17. a. x-intercept: $(-4, 0)$; y-intercept: $(0, 2)$ **b.** x-intercept: $(0, 0)$; y-intercept: $(0, 0)$

18. a. $m = -\frac{7}{5}$ **b.** $m = -\frac{2}{3}$ **c.** $m = 4$ **d.** $m = -\frac{1}{4}$

19. $y = 5x - 3$

20. One large popcorn costs \$3.50, and one drink costs \$1.50.

Chapter 8

Chapter Opener Puzzle

1: 15	2: 4	3: 8	4: 3
5: 2	6: 9	7: 5	8: 14
9: 1	10: 10	11: 6	12: 13
13: 12	14: 7	15: 11	16: 0

Section 8.1 Calculator Connections, pp. 613–614

1. 2.236 **2.** 4.123 **3.** 7.071 **4.** 9.798
5. 5.745 **6.** 12.042 **7.** 8.944 **8.** 13.038
9. 1.913 **10.** 3.037 **11.** 4.021 **12.** 4.987

Section 8.1 Practice Exercises, pp. 614–618

3. 12; -12
5. There are no real-valued square roots of -49.
7. 0 **9.** $\frac{1}{5}; -\frac{1}{5}$ **11. a.** 13 **b.** -13
13. 0 **15.** 9, 16, 25, 36, 64, 121, 169 **17.** 2 **19.** 7
21. 9 **23.** 25 **25.** 0.4 **27.** 0.3 **29.** $\frac{5}{4}$
31. $\frac{1}{12}$
33. There is no real value of b for which $b^2 = -16$.
35. -2 **37.** Not a real number **39.** Not a real number
41. Not a real number **43.** -20
45. Not a real number **47.** 0, 1, 27, 125 **49.** Yes, -3
51. 3 **53.** 4 **55.** -2 **57.** Not a real number
59. Not a real number **61.** -2 **63.** -1
65. 0 **67.** 4 **69.** 4 **71.** 5 **73.** -5
75. 2 **77.** 2 **79.** $|a|$ **81.** y **83.** $|w|$
85. x **87.** $|m|$
89. $x^2, y^4, (ab)^6, w^8x^8, m^{10}$. The expression is a perfect square if the exponent is even.
91. $p^4, t^8, (cd)^{12}$. The expression is a perfect fourth power if the exponent is a multiple of 4.
93. y^6 **95.** a^4b^{15} **97.** q^8 **99.** c^9 **101.** $2w^2$
103. $5x$ **105.** $-5x$ **107.** $5p^2$ **109.** $5p^2$
111. $\sqrt{q} + p^2$ **113.** $\frac{6}{\sqrt[4]{x}}$ **115.** 9 cm **117.** 5 ft
119. 6.9 cm **121.** 17.0 in. **123.** 219 miles
125. 268 km **127. a.** 3.6 ft **b.** 7.2 ft **c.** 144 ft
129. $x \leq 0$

Section 8.2 Calculator Connections, p. 625

1.
```
√(125)
         11.18033989
5*√(5)
         11.18033989
```
2.
```
√(18)
         4.242640687
3*√(2)
         4.242640687
```
3.
```
³√(54)
          3.77976315
3*³√(2)
          3.77976315
```
4.
```
³√(108)
         4.762203156
3*³√(4)
         4.762203156
```
5.
```
√(40/3)
         3.651483717
√(40)/√(3)
         3.651483717
```
6.
```
³√(128/2)
                   4
³√(128)/³√(2)
                   4
```

Section 8.2 Practice Exercises, pp. 626–628

3. $8, 27, y^3, y^9, y^{12}, y^{27}$ **5.** -5 **7.** -3 **9.** a^2
11. $2xy^2$ **13.** 446 km **15.** $3\sqrt{2}$ **17.** $2\sqrt{7}$
19. $12\sqrt{5}$ **21.** $-10\sqrt{2}$ **23.** $a^2\sqrt{a}$ **25.** w^{11}
27. $m^2n^2\sqrt{n}$ **29.** $x^7y^5\sqrt{x}$ **31.** $3t^5\sqrt{t}$ **33.** $2x\sqrt{2x}$
35. $4z\sqrt{z}$ **37.** $-3w^3\sqrt{5}$ **39.** $z^{12}\sqrt{z}$
41. $-z^5\sqrt{15z}$ **43.** $10ab^3\sqrt{26b}$ **45.** $\sqrt{26pq}$
47. $m^6n^8\sqrt{m}$ **49.** $4ab^2c^2\sqrt{3ab}$ **51.** $5u^2v^2\sqrt{3v}$
53. $\frac{\sqrt{3}}{4}$ **55.** $\frac{a^2}{b^2}$ **57.** a^4 **59.** $\frac{1}{2}$ **61.** $\frac{c\sqrt{c}}{2}$
63. $\frac{a^4\sqrt{a}}{b^2}$ **65.** $\frac{10\sqrt{2}}{9}$ **67.** $\frac{2}{5}$ **69.** $\frac{1}{2p}$ **71.** $3\sqrt{5}$
73. $\frac{\sqrt{15}}{5}$ **75.** $11\sqrt{2}$ ft **77.** $2\sqrt{66}$ cm **79.** $a^2\sqrt[3]{a^2}$
81. $14z\sqrt[3]{2}$ **83.** $2ab^2\sqrt[3]{2a^2}$ **85.** z **87.** $\frac{\sqrt[3]{x^2}}{3}$
89. $\frac{\sqrt[3]{y^2}}{3}$ **91.** $2a\sqrt[3]{2}$ **93.** $4a\sqrt{a}$ **95.** $\frac{2x\sqrt{x}}{y}$
97. $\frac{b}{3}$ **99.** $4\sqrt{2}$ **101.** $2u^2v^3\sqrt{13v}$ **103.** $6\sqrt{6}$
105. 6 **107.** $\frac{1}{3}$ **109.** x **111.** $mn^7\sqrt[3]{15mn}$
113. $2p\sqrt{2q}$ **115.** $5\sqrt{2}$ **117.** $x + 5$

Section 8.3 Calculator Connections, pp. 632–633

1.
```
2√(3)+4√(3)
         10.39230485
6√(3)
         10.39230485
```
2.
```
-√(5)-4√(5)+3√(5
)
        -4.472135955
-2√(5)
        -4.472135955
```

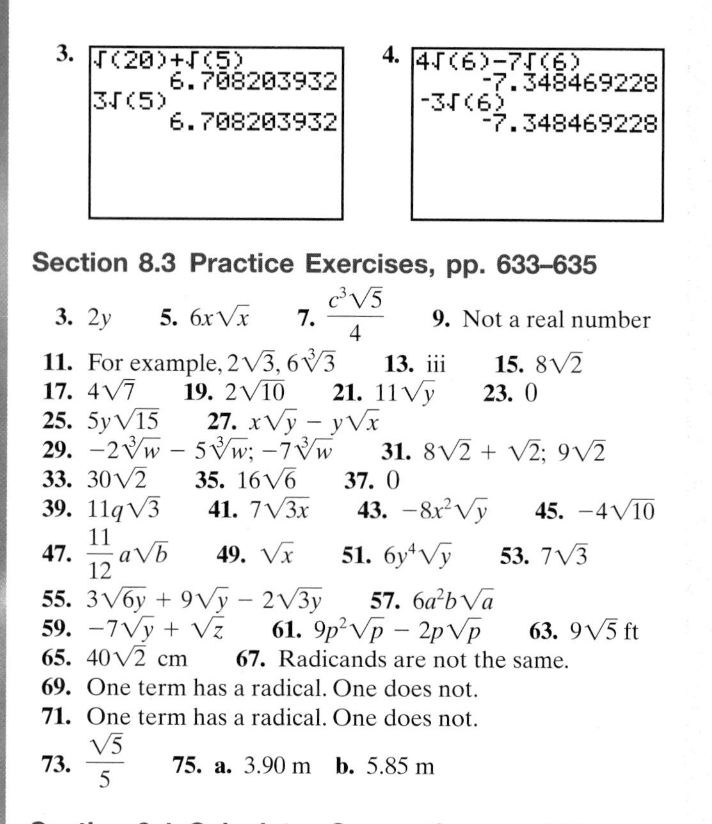

Section 8.3 Practice Exercises, pp. 633–635

3. $2y$ **5.** $6x\sqrt{x}$ **7.** $\dfrac{c^3\sqrt{5}}{4}$ **9.** Not a real number
11. For example, $2\sqrt{3}, 6\sqrt[3]{3}$ **13.** iii **15.** $8\sqrt{2}$
17. $4\sqrt{7}$ **19.** $2\sqrt{10}$ **21.** $11\sqrt{y}$ **23.** 0
25. $5y\sqrt{15}$ **27.** $x\sqrt{y} - y\sqrt{x}$
29. $-2\sqrt[3]{w} - 5\sqrt[3]{w}; -7\sqrt[3]{w}$ **31.** $8\sqrt{2} + \sqrt{2}; 9\sqrt{2}$
33. $30\sqrt{2}$ **35.** $16\sqrt{6}$ **37.** 0
39. $11q\sqrt{3}$ **41.** $7\sqrt{3x}$ **43.** $-8x^2\sqrt{y}$ **45.** $-4\sqrt{10}$
47. $\dfrac{11}{12}a\sqrt{b}$ **49.** $\sqrt{x}$ **51.** $6y^4\sqrt{y}$ **53.** $7\sqrt{3}$
55. $3\sqrt{6y} + 9\sqrt{y} - 2\sqrt{3y}$ **57.** $6a^2b\sqrt{a}$
59. $-7\sqrt{y} + \sqrt{z}$ **61.** $9p^2\sqrt{p} - 2p\sqrt{p}$ **63.** $9\sqrt{5}$ ft
65. $40\sqrt{2}$ cm **67.** Radicands are not the same.
69. One term has a radical. One does not.
71. One term has a radical. One does not.
73. $\dfrac{\sqrt{5}}{5}$ **75. a.** 3.90 m **b.** 5.85 m

Section 8.4 Calculator Connections, p. 640

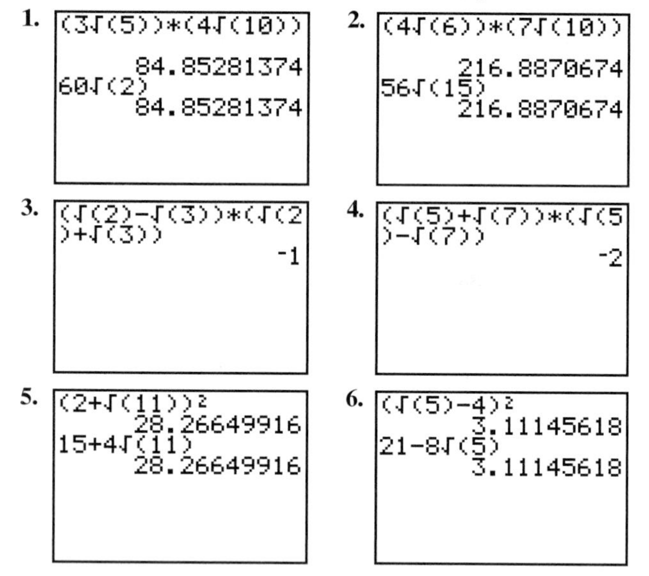

Section 8.4 Practice Exercises, pp. 641–642

3. 11 **5.** $3w^2\sqrt{z}$ **7.** $\sqrt{42}$ **9.** 59 **11.** t
13. $20\sqrt{2q}$ **15.** $2\sqrt{5}$ **17.** $12\sqrt{11}$ **19.** 9
21. $2\sqrt{10}$ **23.** $4x$ **25.** $-2\sqrt{21}$
27. Perimeter: $6\sqrt{2}$ in.; area: 4 in.^{2} **29.** 7 m^2
31. $6p$ **33.** $28\sqrt{15x}$ **35.** $5\sqrt{2} + \sqrt{35}$
37. $6\sqrt{y} - 2y$ **39.** $41 + 23\sqrt{35}$
41. $11m + 15n\sqrt{m} + 4n^2$
43. $5s^2 + 24s\sqrt{s} + 25s - 6\sqrt{s}$ **45.** 23 **47.** 29
49. xy **51.** $200pq$ **53.** $47 - 12\sqrt{11}$
55. $p + 6\sqrt{p} + 9$ **57.** $9w + 24\sqrt{w} + 16$
59. $25c + 20\sqrt{cd} + 4d$ **61.** $5 - 2\sqrt{6}$
63. -13 **65.** $a - b$ **67.** 1 **69.** 10
71. $16s - 121t$ **73.** -20
75. a. $-30 - 5y$ **b.** $-\sqrt{30} - \sqrt{5y}$
77. a. $36 - 12z + z^2$ **b.** $6 - 2z\sqrt{6} + z^2$
79. a. $9w^2 - 1$ **b.** $9w - 1$

Section 8.5 Calculator Connections, p. 647

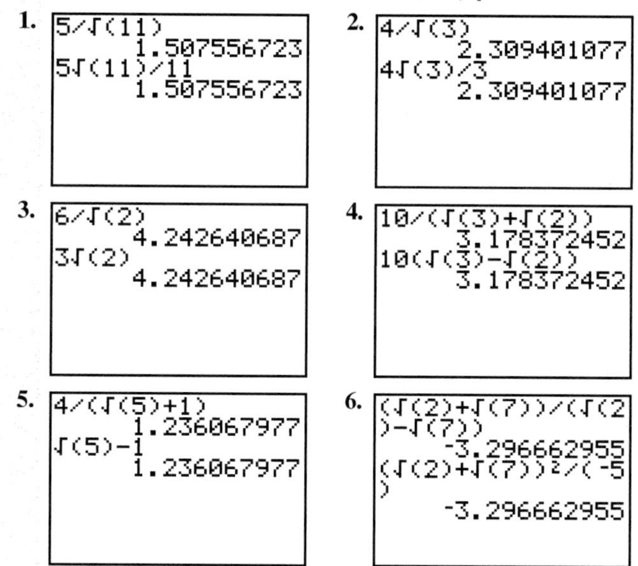

Section 8.5 Practice Exercises, pp. 648–650

1. $11x\sqrt{5}$ **3.** $6y + 23\sqrt{y} + 21$ **5.** $9\sqrt{3}$
7. $25 - 10\sqrt{a} + a$ **9.** -5 **11.** $\dfrac{\sqrt{6}}{6}$
13. $3\sqrt{5}$ **15.** $\dfrac{6\sqrt{x+1}}{x+1}$ **17.** $\dfrac{\sqrt{6x}}{x}$ **19.** $\dfrac{\sqrt{21}}{7}$
21. $\dfrac{5\sqrt{6y}}{3y}$ **23.** $\dfrac{3\sqrt{6}}{4}$ **25.** $\dfrac{\sqrt{3p}}{9}$ **27.** $\dfrac{\sqrt{5}}{2}$
29. $\dfrac{x\sqrt{y}}{y^2}$ **31.** -7 **33.** $\sqrt{5} + \sqrt{3}; 2$
35. $\sqrt{x} - 10; x - 100$ **37.** $\dfrac{4\sqrt{2} - 12}{-7}$ or $\dfrac{12 - 4\sqrt{2}}{7}$
39. $\dfrac{\sqrt{5} + \sqrt{2}}{3}$ **41.** $\sqrt{6} - \sqrt{2}$ **43.** $\dfrac{\sqrt{x} + \sqrt{3}}{x - 3}$
45. $7 + 4\sqrt{3}$ **47.** $\dfrac{\sqrt{11} + \sqrt{5}}{2}$ **49.** $-13 - 6\sqrt{5}$
51. $2 - \sqrt{2}$ **53.** $\dfrac{3 + \sqrt{2}}{2}$ **55.** $1 - \sqrt{7}$
57. $\dfrac{7 + 3\sqrt{2}}{3}$ **59. a.** Condition 1 fails; $2x^4\sqrt{2x}$
b. Condition 2 fails; $\dfrac{\sqrt{5x}}{x}$ **c.** Condition 3 fails; $\dfrac{\sqrt{3}}{3}$
61. a. Condition 2 fails; $\dfrac{3\sqrt{x} - 3}{x - 1}$ **b.** Conditions 1 and 3 fail; $\dfrac{3w\sqrt{t}}{t}$ **c.** Condition 1 fails; $2a^2b^4\sqrt{6ab}$
63. $3\sqrt{5}$ **65.** $-\dfrac{3w\sqrt{2}}{5}$ **67.** Not a real number
69. $\dfrac{s\sqrt{t}}{t}$ **71.** $\dfrac{m^2}{2}$ **73.** $\dfrac{9\sqrt{t}}{t^2}$ **75.** $-\dfrac{3\sqrt{2}}{8}$ **77.** $\dfrac{\sqrt{3}}{9}$

Chapter 8, Problem Recognition Exercises—Operations on Radicals, p. 650

1. $3\sqrt{2}$ **2.** $2\sqrt{7}$ **3.** Cannot be simplified further **4.** Cannot be simplified further **5.** $\sqrt{2}$ **6.** $\sqrt{7}$ **7.** $9 - z$ **8.** $16 - y$ **9.** $8 - 3\sqrt{5}$ **10.** $-8 + 11\sqrt{3}$ **11.** $-x\sqrt{y}$ **12.** $11ab\sqrt{a}$ **13.** $-24 - 6\sqrt{6} - 3\sqrt{2}$ **14.** $-80 + 8\sqrt{15} + 16\sqrt{5}$ **15.** $\frac{2\sqrt{x} + 14}{x - 49}$ **16.** $\frac{5\sqrt{y} - 20}{y - 16}$ **17.** $3\sqrt{3}$ **18.** $3\sqrt{5}$ **19.** $\frac{\sqrt{7x}}{x}$ **20.** $\frac{\sqrt{11y}}{y}$ **21.** $y^2z^5\sqrt{z}$ **22.** $2q^3\sqrt{2}$ **23.** $3p^2\sqrt[3]{p^2}$ **24.** $5u^3v^4\sqrt[3]{u^2}$ **25.** $x\sqrt{10}$ **26.** $y\sqrt{3}$ **27.** $20\sqrt{3}$ **28.** $\sqrt{10}$ **29.** $\frac{1}{3}$ **30.** $\frac{5}{3}$ **31.** $\sqrt{x} - \sqrt{5}$ **32.** $\sqrt{y} - \sqrt{7}$ **33.** $4x - 11\sqrt{xy} - 3y$ **34.** $51 + 14\sqrt{2}$ **35.** $8 + 2\sqrt{15}$ **36.** $16 - 2\sqrt{55}$ **37.** $x - 12\sqrt{x} + 36$ **38.** -88 **39.** $u - 9v$ **40.** $-3\sqrt{6}$ **41.** $11\sqrt{a}$ **42.** $4x\sqrt{2}$ **43.** 0 **44.** $5 + \sqrt{35}$ **45.** $a + 2\sqrt{a}$ **46.** $26 - 17\sqrt{2}$

Section 8.6 Practice Exercises, pp. 655–657

3. $\frac{\sqrt{2} - \sqrt{10}}{-8}$ or $\frac{\sqrt{10} - \sqrt{2}}{8}$ **5.** $\frac{2\sqrt{6}}{3}$ **7.** $\frac{3 + 2\sqrt{2}}{2}$ **9.** $9 - 6y + y^2$ **11.** $m + 6$ **13.** $5y - 4$ **15.** $x = 15$ **17.** No solution **19.** $t = 5$ **21.** $n = -\frac{1}{2}$ **23.** $w = 6$ **25.** $a = 8$ **27.** No solution **29.** $c = 1$ **31.** $x = 4, x = -3$ **33.** $y = 0$ **35.** $x = -6, x = 3$ **37.** $t = 3, t = -1$ **39.** $p = 11$, ($p = 2$ does not check) **41.** $y = -8$ **43.** $x = 3$ **45.** $\sqrt{x + 8} = 12$; 136 **47.** $\sqrt{x} = x - 2$; 4 **49.** $\sqrt[3]{x + 4} = -5$; -129 **51. a.** 2 sec **b.** 256 ft **53. a.** 72 mph **b.** 225 ft **55.** $t = 1$ **57.** $z = -1$, ($z = 3$ does not check)

Section 8.7 Calculator Connections, pp. 662–663

1.–3.

```
169^(1/2)
                13
729^(1/3)
                 9
1024^(1/5)
                 4
```

4.–6.

```
-10000^(1/4)
               -10
8^-(2/3)
               .25
-27^(2/3)
                -9
```

7.–9.

```
100000^(4/5)
             10000
16^-(3/4)
              .125
984^(1/4)
       5.600783363
```

10.–12.

```
14.8^(1/3)
       2.455202052
56^(2/3)
       14.63722284
24^(3/4)
       10.84322404
```

13.–15.

```
(45/59)^(1/3)
       .9136646748
(3/104)^(1/3)
       .3066874269
(4/5)^(3/4)
       .8458970108
```

16.

```
(5/6)^(2/3)
       .8855488077
```

Section 8.7 Practice Exercises, pp. 663–666

1. a. 3 **b.** 125 **3.** 27 **5.** $a + 1$ **7.** 9 **9.** 5 **11.** 3 **13.** -2 **15.** -2 **17.** $\frac{1}{6}$ **19.** $\sqrt[3]{x}$ **21.** $\sqrt{4a}$ or $2\sqrt{a}$ **23.** $\sqrt[3]{yz}$ **25.** $\sqrt[3]{u^2}$ **27.** $5\sqrt{q}$ **29.** $\sqrt{\frac{x}{9}}$ or $\frac{\sqrt{x}}{3}$ **31.** $a^{m/n} = \sqrt[n]{a^m}$ or $(\sqrt[n]{a})^m$, provided the roots exist. **33.** 16 **35.** 8 **37.** $\frac{1}{9}$ **39.** -32 **41.** 2 **43.** $(\sqrt[3]{x})^5$ **45.** $(\sqrt{y})^9$ **47.** $\sqrt[3]{c^5d}$ **49.** $\frac{1}{\sqrt[5]{qr}}$ **51.** $6(\sqrt[4]{y})^2$ **53.** $x^{1/3}$ **55.** $(xy)^{1/3}$ **57.** $5x^{1/2}$ **59.** $y^{2/3}$ **61.** $(m^3n)^{1/4}$ **63.** $4k^{3/3}$ or $4k$ **65.** x **67.** y^2 **69.** 6 **71.** a^7 **73.** $y^{4/3}$ **75.** 2 **77.** $\frac{25a^4d}{c}$ **79.** $\frac{y^9}{x^8}$ **81.** $\frac{2z^3}{w}$ **83.** $5xy^2z^3$ **85. a.** 10 in. **b.** 8.49 in. **87. a.** 10.9% **b.** 8.8% **c.** The account in part (a) **89.** No, for example, $(36 + 64)^{1/2} \neq 36^{1/2} + 64^{1/2}$ **91.** 6 **93.** $\frac{5}{14}$ **95.** $\frac{a^{22}b^4}{c^{17}}$

Chapter 8, Review Exercises, pp. 670–672

1. Principal square root: 14; negative square root: -14 **2.** Principal square root: 1.2; negative square root: -1.2 **3.** Principal square root: 0.8; negative square root: -0.8 **4.** Principal square root: 15; negative square root: -15 **5.** There is no real number b such that $b^2 = -64$. **6.** $\sqrt[3]{-64} = -4$ because $(-4)^3 = -64$. **7.** -12 **8.** -5 **9.** Not a real number **10.** Not a real number **11.** $|y|$ **12.** y **13.** $|y|$ **14.** -5 **15.** -5 **16.** p^4 **17.** $\frac{3}{t^2}$ **18.** $-\frac{3}{w}$ **19. a.** 7.1 m **b.** 22.6 ft **20. a.** 65.8 ft **b.** 131.6 ft **21.** $b^2 + \sqrt{5}$ **22.** $\sqrt[3]{y} - \sqrt[4]{x}$ **23.** The quotient of 2 and the principal square root of p **24.** The product of 8 and the principal square root of q **25.** 12 ft **26.** 331 miles **27.** $x^8\sqrt{x}$ **28.** $2\sqrt[3]{5}$ **29.** $2\sqrt{7}$ **30.** $15x\sqrt{2x}$ **31.** $6y^5\sqrt{3}$ **32.** $3y^3\sqrt{y}$ **33.** $\frac{5\sqrt{2}}{7}$ **34.** $\frac{a^4}{5}$ **35.** $\frac{x\sqrt{2x}}{y^3}$ **36.** $\frac{3}{x}$ **37.** $10y$ **38.** $\frac{w^2\sqrt{w}}{3z^2}$ **39.** $\frac{\sqrt[3]{30p^2}}{q^2}$ **40.** $\frac{1}{4n}$ **41.** $\frac{t^3\sqrt{5}}{11}$ **42.** $\frac{7}{3}$ **43.** $7\sqrt{6}$ **44.** $0.8\sqrt{y}$ **45.** $-4x\sqrt{5}$ **46.** $11y\sqrt{y}$ **47.** $5\sqrt{7}$ **48.** $-8\sqrt{2}$ **49.** $\sqrt{x^4} + (5x)^2$; $26x^2$ **50.** $\sqrt{128} - \sqrt{2}$; $7\sqrt{2}$ **51.** $12\sqrt{2}$ ft **52.** $48\sqrt{3}$ m **53.** 25 **54.** $2\sqrt{15p}$ **55.** $70\sqrt{3x}$ **56.** $-6yz\sqrt{11}$ **57.** $8m + 24\sqrt{m}$ **58.** $\sqrt{14} + 8\sqrt{2}$ **59.** $-49 - 16\sqrt{26}$ **60.** $4p + 7\sqrt{pq} - 2q$ **61.** $64w - z$ **62.** $4x^2 - 4x\sqrt{y} + y$ **63.** $10\sqrt{3}\ \text{m}^3$ **64.** b **65.** b **66.** $\frac{11\sqrt{7}}{7}$ **67.** $\frac{3\sqrt{2y}}{y}$ **68.** $\frac{2\sqrt{x}}{x^4}$ **69.** $2\sqrt{7} + 2\sqrt{2}$ **70.** $\frac{6\sqrt{w} - 12}{w - 4}$ **71.** $-8 - 3\sqrt{7}$ **72. a.** $\frac{10\sqrt{6}}{3}$ m/sec **b.** $\frac{18\sqrt{5}}{5}$ m/sec

73. $p = 138$ **74.** No solution **75.** $x = 39$
76. $n = 5$ **77.** $z = 7$ **78.** $y = 6$
79. $m = 2$, ($m = -2$, does not check)
80. $n = 3, n = 4$ **81.** $y = -69$
82. a. 9261 in.^{3} **b.** 3375 cm^3
83. -3 **84.** 11 **85.** -2 **86.** Not a real number
87. $\frac{1}{8}$ **88.** 27 **89.** $\sqrt[5]{z}$ **90.** $\sqrt[3]{q^2}$ **91.** $\sqrt[4]{w^3}$
92. $\sqrt{\frac{b}{121}} = \frac{\sqrt{b}}{11}$ **93.** $a^{2/5}$ **94.** $5m^{2/3}$
95. $(a^2b^4)^{1/5}$ **96.** $6^{1/2}$ **97.** y^2 **98.** $6^{3/5}$ **99.** $a^{5/6}$
100. $5^{3/4}$ **101.** $4ab^2$ **102.** b^{15} **103.** 2.0 cm

Chapter 8, Test, pp. 673–674

1. 1. The radicand has no factor raised to a power greater than or equal to the index. 2. There are no radicals in the denominator of a fraction. 3. The radicand does not contain a fraction.
2. $11x\sqrt{2}$ **3.** $2y\sqrt[3]{6y}$ **4.** Not a real number
5. $\frac{a^3\sqrt{5}}{9}$ **6.** $\frac{3\sqrt{6}}{2}$ **7.** $\frac{2\sqrt{5} - 12}{-31}$ or $\frac{12 - 2\sqrt{5}}{31}$
8. a. $\sqrt{25} + 5^3$; 130 **b.** $4^2 - \sqrt{16}$; 12
9. a. 97 ft **b.** 339 ft **10.** $8\sqrt{z}$ **11.** $4\sqrt{6} - 15$
12. $-7t\sqrt{2}$ **13.** $9\sqrt{10}$ **14.** $46 - 6\sqrt{5}$
15. $-8 + 23\sqrt{10}$ **16.** $\frac{\sqrt{n}}{6m}$ **17.** $16 - 9x$
18. $\frac{\sqrt{22}}{11}$ **19.** $\frac{3\sqrt{7} + 3\sqrt{3}}{2}$ **20.** 206 yd
21. No solution **22.** $x = 0, x = -5$ **23.** $x = 14$
24. a. 12 in. **b.** 18 in. **c.** 25 weeks
25. 1000 **26.** 2 **27.** $\sqrt[5]{x^3}$ or $(\sqrt[5]{x})^3$ **28.** $5\sqrt{y}$
29. $(ab^3)^{1/4}$ **30.** $p^{11/12}$ **31.** $5^{3/5}$ **32.** $3mn^2$

Chapter 8, Cumulative Review Exercises, pp. 674–675

1. 1 **2.** $y = -2$ **3.** -15 **4.** $-9x^2 + 2x + 10$
5. $\frac{2x}{y} - 1 + \frac{4}{x}$ **6.** $2(5c + 2)^2$ **7.** $x = -\frac{2}{5}, x = \frac{1}{2}$
8. 1 **9.** No solution, ($z = 5$ does not check)
10. $\frac{x(5x + 8)}{16(x - 1)}$
11.

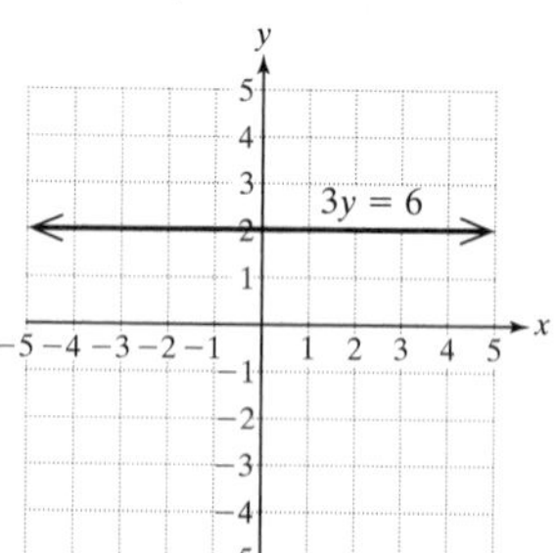

12. a. $y = 880$; the cost of renting the office space for 3 months is \$880. **b.** $x = 12$; the cost of renting office space for 12 months is \$2770. **c.** $m = 210$; the increase in cost is \$210 per month. **d.** (0, 250); the down payment of renting the office space is \$250.
13. $y = -x + 1$ **14.** $(1, -4)$
15.

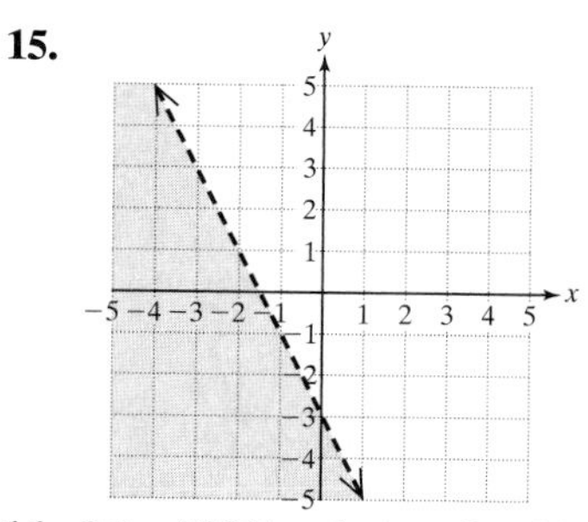

16. 8 L of 20% solution should be mixed with 4 L of 50% solution.
17. $3\sqrt{11}$ **18.** $7x\sqrt{3}$ **19.** $\frac{x + \sqrt{xy}}{x - y}$ **20.** $y = 5$

Chapter 9

Chapter Opener Puzzle

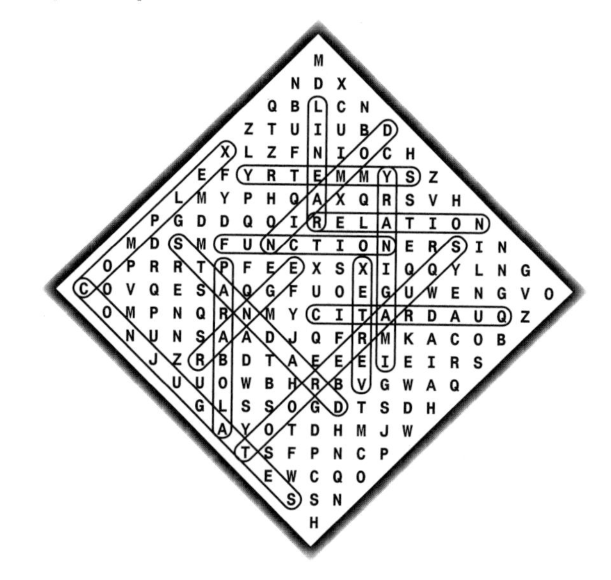

Section 9.1 Calculator Connections, pp. 687–688

1. **2.**

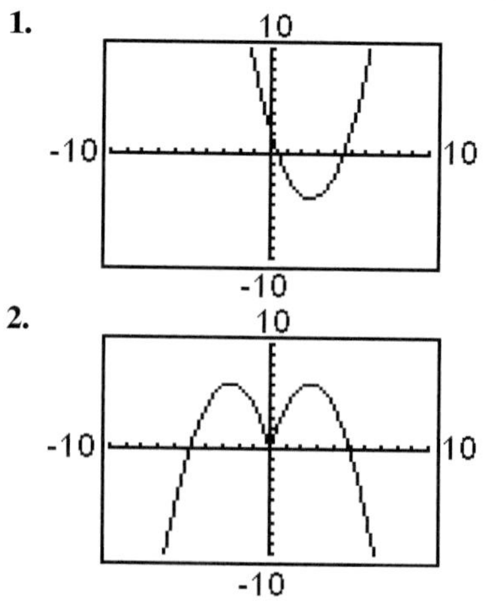

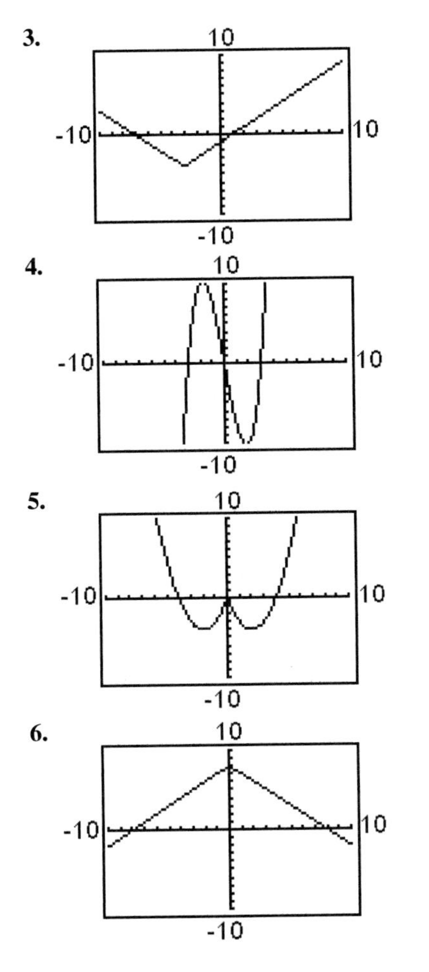

Section 9.1 Practice Exercises, pp. 688–692

3. domain: {4, 3, 0}; range: {2, 7, 1, 6} **5.** domain: {$\frac{1}{2}$, 0, 1}; range: {3} **7.** domain: {Atlanta, Macon, Pittsburgh}; range: {GA, PA} **9.** domain: {New York, California}; range: {Albany, Los Angeles, Buffalo} **11.** The relation is a function if each element in the domain has exactly one corresponding element in the range. **13.** The relations in Exercises 5 and 7 are functions. **15.** Yes **17.** No **19.** No **21.** Yes **23.** –5, –1, –11 **25.** $\frac{1}{5}, \frac{1}{4}, \frac{1}{2}$ **27.** 1 **29.** 6 **31.** $-\frac{3}{2}$ **33.** –10 **35.** m^2 **37.** $(a + b)^2$ or $a^2 + 2ab + b^2$ **39.** $12y$ **41.** The set of all real numbers **43.** The set of all real numbers **45.** $\{x \mid x$ is a real number and $x \neq -6\}$ **47.** $\{x \mid x$ is a real number and $x \neq 4\}$ **49.** $\{x \mid x$ is a real number and $x \neq 0\}$ **51.** $\{x \mid x$ is a real number and $x \neq \frac{5}{9}\}$ **53.** b **55.** c **57.** The function value at $x = 6$ is 2. **59.** The function value at $x = \frac{1}{2}$ is $\frac{1}{4}$. **61.** (2, 7) **63.** (0, 8) **65. a.** $C(0) = 1.50$. The cost of a scoop of ice cream with 0 toppings is \$1.50. **b.** $C(1) = 2.00$. The cost of a scoop of ice cream with 1 topping is \$2.00. **c.** $C(4) = 3.50$. The cost of a scoop of ice cream with four toppings is \$3.50. **67. a.** $s(0) = 0$. The initial speed is 0 m/sec. **b.** $s(4) = 39.2$. The speed of an object 4 sec after being dropped is 39.2 m/sec. **c.** $s(6) = 58.8$. The speed of an object 6 sec after being dropped is 58.8 m/sec. **d.** 77.42 m/sec **69. a.** $v(0) = 64$. The initial velocity is 64 ft/sec. **b.** $v(1) = 32$. The velocity after 1 sec is 32 ft/sec. **c.** $v(2) = 0$. The velocity after 2 sec is 0 ft/sec (the ball is at its peak and momentarily stops). **d.** $v(4) = -64$. The velocity after 4 sec is –64 ft/sec (the football is descending).

Section 9.2 Practice Exercises, pp. 700–702

3. $6i$ **5.** $i\sqrt{21}$ **7.** $4i\sqrt{3}$ **9.** -20 **11.** -6 **13.** 3 **15.** 5 **17.** $14i$ **19.** $-24i$ **21.** $-5 + 5i$ **23.** -1 **25.** 1 **27.** i **29.** $-i$ **31.** Real part: –3; Imaginary part: –2 **33.** Real part: 4; Imaginary part: 0 **35.** Real part: 0; Imaginary part: $\frac{2}{7}$ **37.** Add or subtract the real parts. Add or subtract the imaginary parts. **39.** $-6 + 8i$ **41.** $10 - 10i$ **43.** $6 + 3i$ **45.** $-7 + 3i$ **47.** $7 - 21i$ **49.** $11 - 9i$ **51.** $9 + 19i$ **53.** $-\frac{1}{4} - \frac{1}{5}i$ **55.** $-3.5 + 18.1i$ **57.** 13 **59.** 104 **61.** $\frac{5}{4}$ **63.** $35 - 12i$ **65.** $21 + 20i$ **67.** $-33 - 56i$ **69.** $7 + 4i$; 65 **71.** $\frac{3}{2} - \frac{2}{5}i$; $\frac{241}{100}$ **73.** $-4i$; 16 **75.** $\frac{14}{65} + \frac{8}{65}i$ **77.** $\frac{5}{2} - \frac{5}{2}i$ **79.** $-\frac{3}{5} - \frac{6}{5}i$ **81.** $-\frac{2}{13} + \frac{10}{13}i$ **83.** $\frac{15}{17} + \frac{8}{17}i$ **85.** $\frac{23}{29} - \frac{14}{29}i$ **87.** False. For example: $2 + 3i$ is not a real number. **89.** True **91.** False. $\sqrt[3]{-64} = -4$. **93.** False. $(1 + 4i)(1 - 4i) = 17$. **95.** True **97.** False. $i^4 = 1$. **99.** True

Section 9.3 Practice Exercises, pp. 708–710

3 a. Linear **b.** Quadratic **c.** Linear **5.** $z = \frac{2}{3}, z = -\frac{5}{4}$ **7.** $r = -4, r = -3$ **9.** $x = \frac{1}{2}, x = \frac{4}{5}$ **11.** $m = \frac{3}{2}, m = 1$ **13.** $x = 2, x = -2$ **15.** Plus or minus **17.** $x = \pm 4$ **19.** $m = \pm 8$ **21.** $q = \pm 5i\sqrt{2}$ **23.** $v = \frac{\pm\sqrt{5}}{2}$ **25.** $b = -2, b = -4$ **27.** $y = -2, y = -6$ **29.** $z = 2 \pm \sqrt{7}$ **31.** $b = 1 \pm 2i\sqrt{3}$ **33.** $t = \frac{2}{3}, t = 0$ **35.** $x = -\frac{5}{2} \pm \sqrt{21}$ **37.** $t = -4 \pm 3i$ **39.** The solution checks. **41.** False. If $k = 0$, there is only one solution. **43.** 9 **45.** 64 **47.** $\frac{25}{4}$ **49.** $\frac{1}{64}$ **51.** $\frac{25}{196}$ **53.** $\frac{1}{4}$ **55.** $x = 4, x = -2$ **57.** $t = 1, t = -11$ **59.** $x = 3 \pm \sqrt{7}$ **61.** $x = -1 \pm \sqrt{7}$ **63.** $t = 3 \pm \sqrt{15}$ **65.** $z = \frac{3}{2} \pm \frac{\sqrt{29}}{2}$

67. $y = -5, y = 9$
69. **a.** 122.5 m **b.** 3.2 sec **c.** 7.3 sec
71. 17.0 in. **73.** 23.3 ft

Section 9.4 Calculator Connections, p. 717

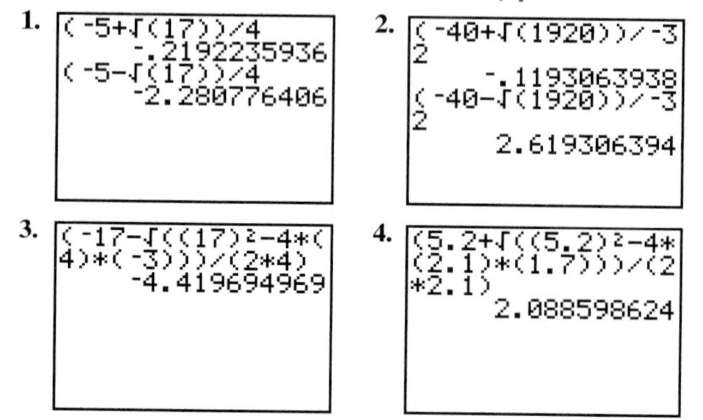

Section 9.4 Practice Exercises, pp. 718–720

1. $z = \pm 13$ **3.** $x = 4 \pm 2\sqrt{7}$ **5.** $x = \dfrac{5 \pm \sqrt{21}}{2}$

7. For $ax^2 + bx + c = 0$, $x = \dfrac{-b \pm \sqrt{b^2 - 4ac}}{2a}$

9. $5x^2 + 3x + 10 = 0; a = 5, b = 3, c = 10$

11. $x^2 - 5x - 3 = 0; a = 1, b = -5, c = -3$

13. $x^2 + 0x + 25 = 0; a = 1, b = 0, c = 25$

15. $n = -2, n = \dfrac{1}{3}$ **17.** $a = \dfrac{-5 \pm \sqrt{33}}{4}$

19. $y = \dfrac{3 \pm \sqrt{7}}{2}$ **21.** $z = 2 \pm \sqrt{5}$

23. $z = \dfrac{3 \pm \sqrt{89}}{-10}$ or $\dfrac{-3 \pm \sqrt{89}}{10}$ **25.** $p = \dfrac{1 \pm i\sqrt{14}}{5}$

27. $t = \dfrac{5 \pm \sqrt{35}}{2}$ **29.** $p = \dfrac{2 \pm \sqrt{6}}{2}$

31. $x = \dfrac{-1 \pm \sqrt{73}}{6}$ **33.** $h = \dfrac{7 \pm \sqrt{73}}{2}$

35. $x = \dfrac{3}{4}, x = -\dfrac{3}{4}$ **37.** $x = 5 \pm i\sqrt{21}$

39. $x = -12 \pm 3\sqrt{5}$ **41.** $x = \dfrac{3 \pm \sqrt{15}}{2}$

43. $x = 0, x = \dfrac{11}{9}$ **45.** $y = \dfrac{3 \pm \sqrt{5}}{2}$

47. $x = \dfrac{1 \pm \sqrt{41}}{4}$ **49.** $z = 0, z = \dfrac{1}{9}$

51. $r = \pm 2\sqrt{13}$

53. The base is $-1 + \sqrt{145} \approx 11.0$ cm. The height is $1 + \sqrt{145} \approx 13.0$ cm

55. The length is $1 + \sqrt{41} \approx 7.4$ ft. The width is $-1 + \sqrt{41} \approx 5.4$ ft. The height is 6 ft.

57. The width is $-2 + 2\sqrt{19} \approx 6.7$ ft. The length is $2 + 2\sqrt{19} \approx 10.7$ ft.

59. The legs are $\dfrac{3 + \sqrt{329}}{2} \approx 10.6$ m and $\dfrac{-3 + \sqrt{329}}{2} \approx 7.6$ m.

61. Discriminant is 0; one rational solution
63. Discriminant is 49; two rational solutions
65. Discriminant is 45; two irrational solutions
67. Discriminant is -56; two imaginary solutions

Section 9.5 Calculator Connections, pp. 728–729

1. $(-2, 3)$
2. $(10, 5)$
3. $(-1.5, -2.6)$
4. $(1.75, 2.5)$
5. $\left(\dfrac{5}{2}, 0\right)$
6. $\left(-\dfrac{8}{3}, 0\right)$

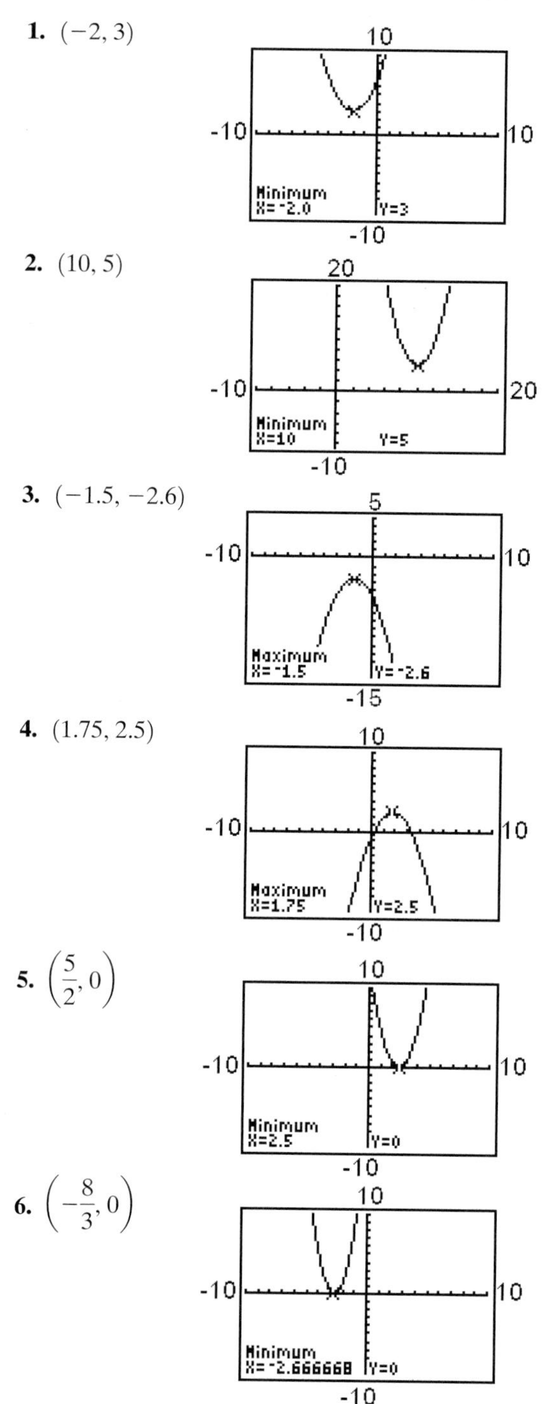

7. The graph in part (b) is shifted up 4 units. The graph in part (c) is shifted down 3 units.

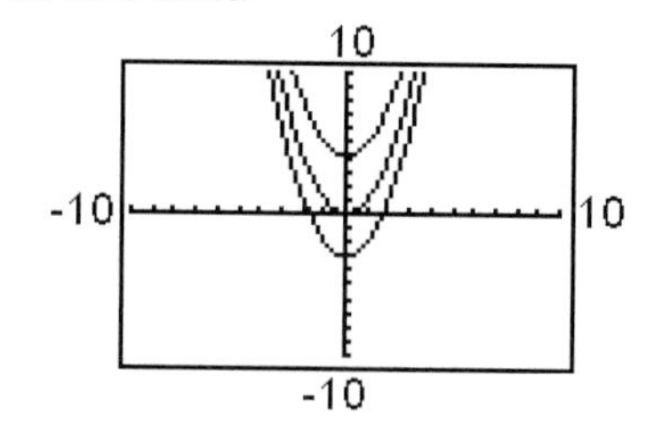

8. In part (b), the graph is shifted to the right 3 units. In part (c), the graph is shifted to the left 2 units.

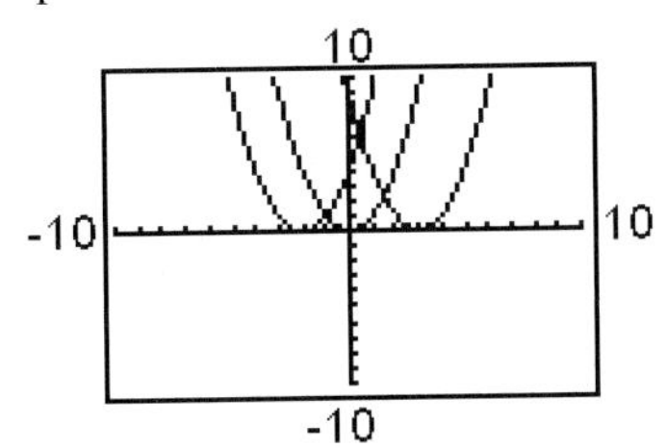

9. In part (b), the graph is stretched vertically by a factor of 2. In part (c), the graph is shrunk vertically by a factor of $\frac{1}{2}$.

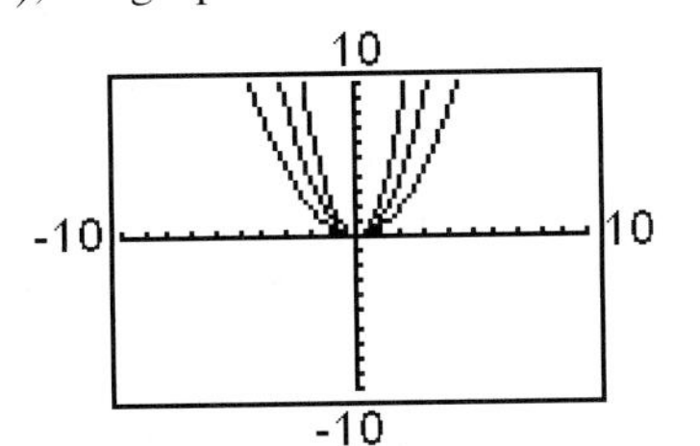

10. In part (b), the graph has been stretched vertically and reflected across the x-axis. In part (c), the graph has been shrunk vertically and reflected across the x-axis.

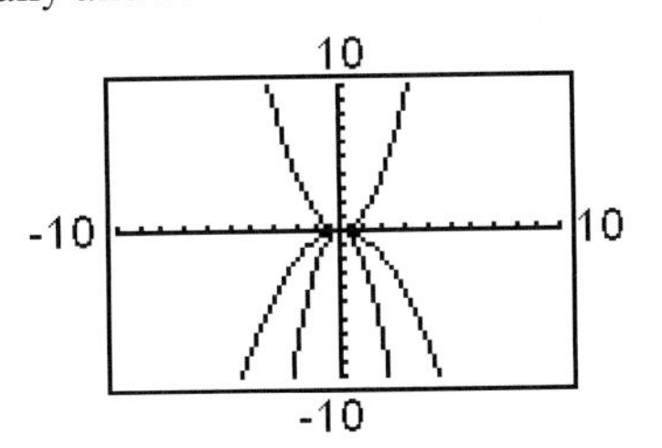

Section 9.5 Practice Exercises, pp. 730–733

3. $a = -5, a = 3$ **5.** $p = \dfrac{\pm\sqrt{5}}{3}$
7. $x = 2 \pm 2\sqrt{2}$ **9.** Linear **11.** Quadratic
13. Neither **15.** Linear **17.** Quadratic
19. If $a > 0$ the graph opens upward; if $a < 0$ the graph opens downward. **21.** $a = -7$; downward
23. $a = 2$; upward **25.** $a = -10$; downward
27. $(-1, -8)$ **29.** $(1, -4)$ **31.** $(1, 2)$ **33.** $(0, -4)$
35. c; x-intercepts: $(\sqrt{7}, 0)(-\sqrt{7}, 0)$; y-intercept: $(0, -7)$
37. a; x-intercepts: $(-1, 0)(-5, 0)$; y-intercept: $(0, 5)$
39. $(0, c)$ **41.** $(-\infty, \infty)$
43. a. Upward **b.** $(0, -9)$ **c.** $(3, 0)(-3, 0)$ **d.** $(0, -9)$
e.

$y = x^2 - 9$

f. $(-\infty, \infty)$ **g.** $[-9, \infty)$

45. a. Upward **b.** $(1, -9)$ **c.** $(4, 0)(-2, 0)$ **d.** $(0, -8)$
e.

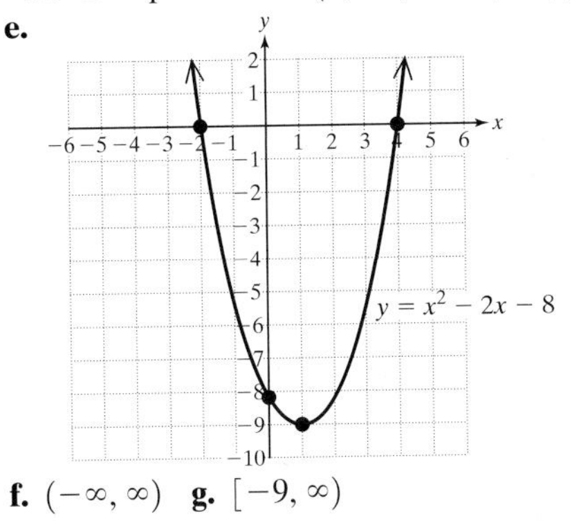

f. $(-\infty, \infty)$ **g.** $[-9, \infty)$

47. a. Downward **b.** $(3, 0)$ **c.** $(3, 0)$ **d.** $(0, -9)$
e.

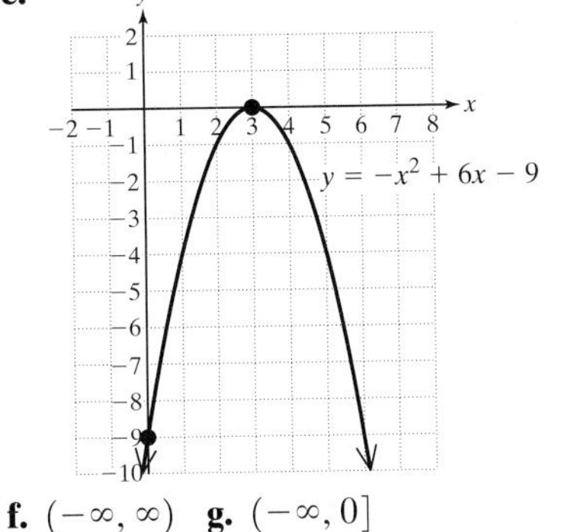

f. $(-\infty, \infty)$ **g.** $(-\infty, 0]$

49. a. Downward **b.** $(4, 1)$ **c.** $(3, 0)(5, 0)$ **d.** $(0, -15)$
e.

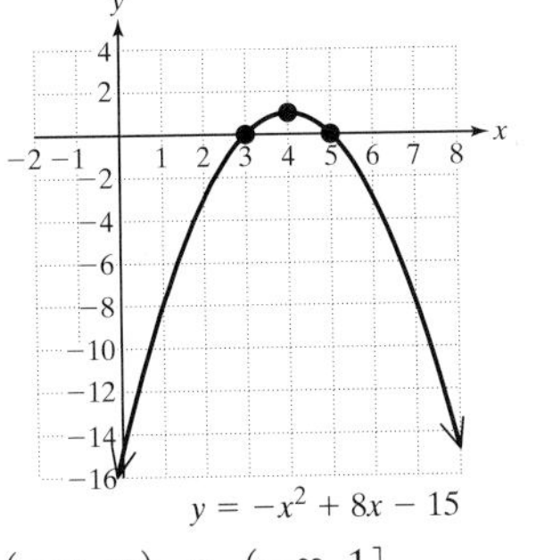

f. $(-\infty, \infty)$ **g.** $(-\infty, 1]$

51. a. Upward **b.** $(-3, 1)$ **c.** none **d.** $(0, 10)$
e.

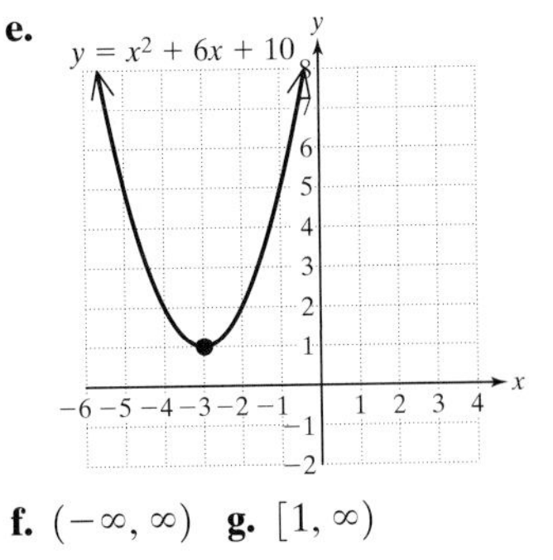

f. $(-\infty, \infty)$ **g.** $[1, \infty)$

53. a. Downward **b.** (0, −2) **c.** none **d.** (0, −2)
e.

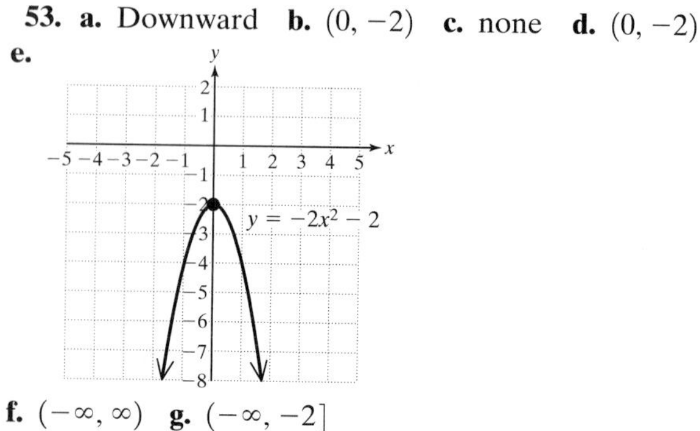

f. $(-\infty, \infty)$ **g.** $(-\infty, -2]$
55. True **57.** False **59. a.** 1800 dinners **b.** $2840
61. a. 33 psi **b.** 36,000 miles

Chapter 9 Review Exercises, pp. 737–741

1. Domain: {6, 10, −1, 0}; range: {3}; function
2. Domain: {2}; range: {0, 1, −5, 2}; not a function
3. Domain: [−4, 5]; Range: [−3, 3]; not a function
4. Domain: $(-\infty, \infty)$; Range: $[-2, \infty)$; function
5. Domain: {4, 3, −6}; range: {23, −2, 5, 6}; not a function
6. Domain: {3, −4, 0, 2}; range: $\{0, \frac{1}{2}, 3, -12\}$; function
7. a. 0 **b.** 8 **c.** −27 **d.** −1 **e.** 64 **f.** $(-\infty, \infty)$
8. a. 0 **b.** 4 **c.** $-\frac{1}{6}$ **d.** $\frac{3}{2}$ **e.** $-\frac{1}{2}$ **f.** $(-\infty, 5) \cup (5, \infty)$
9. −45 **10.** 0 **11.** $-4m$ **12.** $m^3 + 2m - 1$
13. $8y^3 + 4y - 1$ **14.** $12a$
15. a. $D(90) = 562$. A plane traveling 90 ft/s when it touches down will require 562 ft of runway.
b. $D(110) = 902$. A plane traveling 110 ft/s when it touches down will require 902 ft of runway.
16. $\{x \mid x \text{ is a real number and } x \neq 8\}$
17. $\{x \mid x \text{ is a real number}\}$
18. $\{x \mid x \text{ is a real number and } x \neq 4\}$
19. $\{x \mid x \text{ is a real number and } x \neq 1, x \neq -2\}$
20. $\{x \mid x \text{ is a real number}\}$
21. $\{x \mid x \text{ is a real number and } x \neq -1\}$
22. $6i$ **23.** $i\sqrt{5}$ **24.** $4i\sqrt{2}$ **25.** $10i$ **26.** −5
27. 4 **28.** $-i$ **29.** −1 **30.** 1 **31.** i
32. $-1 - 3i$ **33.** $4 - 12i$ **34.** $6 + 5i$
35. 22 **36.** $27 + 3i$ **37.** $2 + 14i$ **38.** $12 - 16i$
39. 26 **40.** $\frac{9}{5} + \frac{18}{5}i$ **41.** $\frac{12}{5} + \frac{6}{5}i$
42. $4 + i$ **43.** $-\frac{5}{13} + \frac{12}{13}i$ **44.** Linear
45. Quadratic **46.** Quadratic **47.** Linear
48. $x = \pm 5$ **49.** $x = \pm\sqrt{19}$ **50.** $x = \pm 7i$
51. $x = \pm 4i\sqrt{3}$ **52.** $x = -1 \pm \sqrt{14}$
53. $x = 2 \pm 8i$ **54.** $x = \frac{1}{8} \pm \frac{i\sqrt{3}}{8}$
55. $x = \frac{3 \pm 2\sqrt{5}}{2}$ **56.** 36 **57.** 81 **58.** $\frac{1}{9}$
59. $\frac{1}{196}$ **60.** $x = -4 \pm \sqrt{13}$ **61.** $x = 1 \pm \sqrt{5}$
62. $\frac{3}{2} \pm \frac{1}{2}i$ **63.** $x = \frac{7 \pm \sqrt{85}}{6}$ **64.** 10.6 ft
65. 3.1 cm
66. For $ax^2 + bx + c = 0$, $x = \frac{-b \pm \sqrt{b^2 - 4ac}}{2a}$
67. $x = \frac{-1 \pm \sqrt{141}}{10}$ **68.** $x = -2$
69. $x = \frac{1 \pm i\sqrt{5}}{3}$ **70.** $x = \frac{3}{2}, x = -1$
71. $x = -10, x = 2$ **72.** $x = -3 \pm \sqrt{7}$
73. $x = 3$ **74.** $x = 1, x = -6$
75. $x = -4 \pm \sqrt{14}$ **76.** $x = 4 \pm 3i$
77. The numbers are −2.5 and −4.5, or 2.5 and 4.5.
78. The height is $\frac{-1 + \sqrt{97}}{2} \approx 4.4$ cm. The base is $\frac{1 + \sqrt{97}}{2} \approx 5.4$ cm.
79. 9.5 sec
80. b; x-intercepts: (9.32, 0)(2.68, 0); y-intercept: (0, 25)
81. f; x-intercepts: (−0.76, 0)(−5.24, 0); y-intercept: (0, 4)
82. d; x-intercepts: (1, 0)(−1, 0); y-intercept: (0, 5)
83. a; x-intercepts: (1, 0)(−1, 0); y-intercept: (0, 3)
84. e; x-intercepts: (0, 0)(−4, 0); y-intercept: (0, 0)
85. c; x-intercepts: none; y-intercept: (0, 4)
86. $a = 1$; upward **87.** $a = -1$; downward
88. $a = -2$; downward **89.** $a = 5$; upward
90. (−1, 1) **91.** (4, 19) **92.** (3, 13)
93. $\left(-\frac{1}{2}, -\frac{3}{2}\right)$
94. a. Upward **b.** (−1, −3) **c.** Approximately (0.22, 0)(−2.22, 0) **d.** (0, −1)
e.

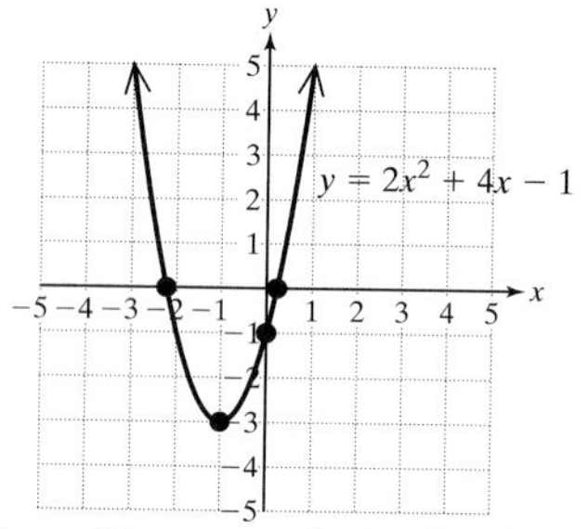

95. a. Downward **b.** (2, 2) **c.** Approximately (2.82, 0) (1.18, 0) **d.** (0, −10)
e.

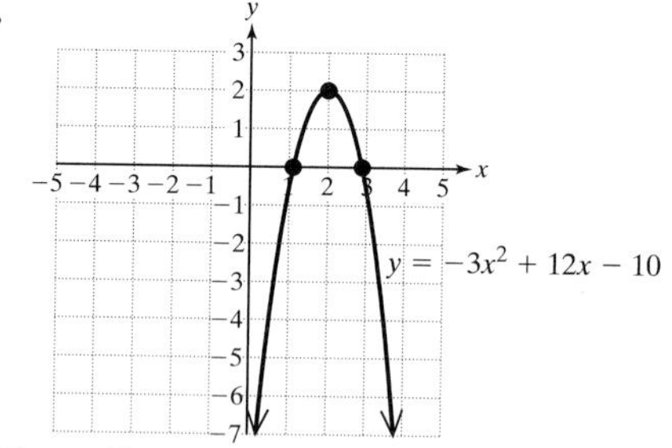

96. a. Downward **b.** (−1, −4) **c.** No x-intercepts
d. (0, −12)
e.

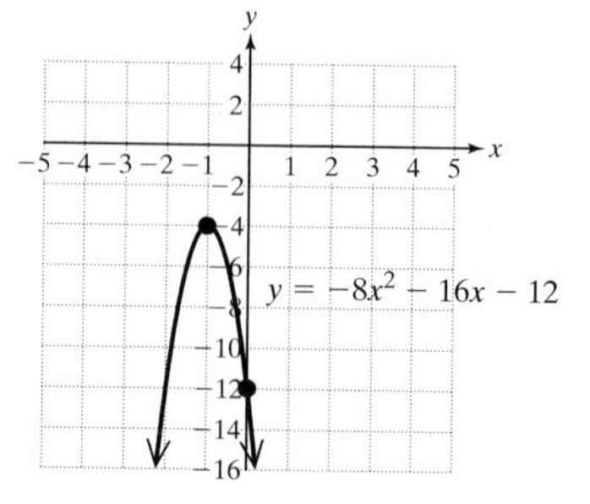

97. a. Upward **b.** $(-2, -3)$ **c.** $(-3, 0)(-1, 0)$
d. $(0, 9)$ **e.**

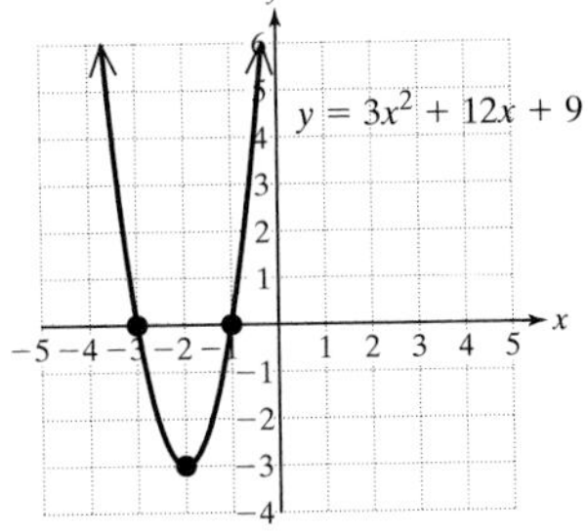

98. a. 1024 ft **b.** 8 sec

Chapter 9 Test, pp. 741–742

1. a. Domain: $(-\infty, 0]$; Range: $(-\infty, \infty)$; not a function
b. Domain: $(-\infty, \infty)$; Range: $(-\infty, 4]$; function

2. a. $f(0) = \frac{1}{2}$, $f(-2)$ is undefined, $f(6) = \frac{1}{8}$
b. $\{x \mid x \text{ is a real number and } x \neq -2\}$

3. a. $D(5) = 5$; a five-sided polygon has five diagonals.
b. $D(10) = 35$; a 10-sided polygon has 35 diagonals.
c. 8 sides

4. $10i$ **5.** $i\sqrt{23}$ **6.** -21 **7.** i **8.** $-i$
9. $5 - 3i$ **10.** $-13 - 26i$ **11.** 221
12. $\frac{10}{221} + \frac{11}{221}i$ **13.** $x = \frac{-1 \pm i\sqrt{14}}{3}$
14. $x = 4 \pm \sqrt{21}$ **15.** $x = \frac{5 \pm \sqrt{13}}{6}$
16. $x = \frac{-1 \pm \sqrt{41}}{10}$ **17.** $c = 12 \pm 2\sqrt{3}$
18. $y = -7 \pm 5\sqrt{2}$ **19.** $t = \pm\sqrt{10}$
20. $x = \frac{5}{6}, x = -\frac{3}{2}$ **21.** $p = 0, p = \frac{11}{6}$
22. $x = 3 \pm 2\sqrt{5}$ **23.** 4.0 in.
24. The base is $\frac{-1 + \sqrt{97}}{2} \approx 4.4$ m. The height is $1 + \sqrt{97} \approx 10.8$ m.
25. For $y = ax^2 + bx + c$, if $a > 0$ the parabola opens upward; if $a < 0$ the parabola opens downward.
26. $(5, 0)$ **27.** $(1, 5)$ **28.** $(0, -16)$
29. The parabola has no x-intercepts.
30. a. Opens upward **b.** Vertex: $(-3, -1)$
c. x-intercepts: $(-2, 0)$ and $(-4, 0)$ **d.** y-intercept: $(0, 8)$
e.

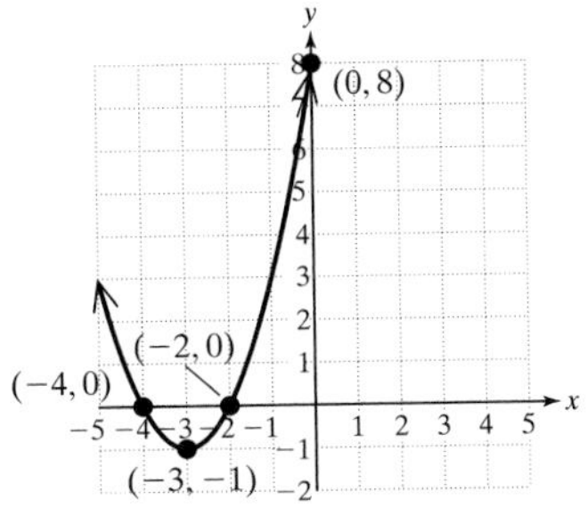

31. Vertex: $(0, 25)$; x-intercepts: $(-5, 0)(5, 0)$; y-intercept: $(0, 25)$

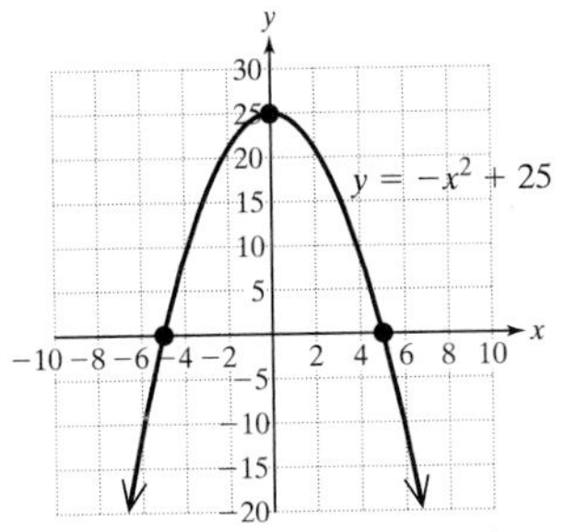

32. a. \$25 per ticket **b.** \$250,000

Chapters 1–9 Cumulative Review Exercises, pp. 742–745

1. $x = 1$ **2.** $h = \frac{2A}{b}$ **3.** $y = \frac{34}{3}$
4. Yes, $x = 2$ is a solution.
5. $\{x \mid x > -1\}$; $(-1, \infty)$

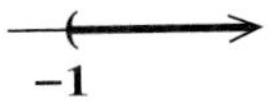

6. a. Decreases **b.** $m = -37.6$. For each additional increase in education level, the death rate decreases by approximately 38 deaths per 100,000 people. **c.** 901.8 per 100,000 **d.** 10th grade
7. $\frac{8c}{a^4 b}$ **8.** 1.898×10^{10} diapers
9. Approximately 651 light-years **10.** $4x^2 - 16x + 13$
11. $2y^3 + 1 - \frac{3}{y - 2}$ **12.** $(2x + 5)(x - 7)$
13. $(y + 4a)(2x - 3b)$ **14.** The base is 9 m, and the height is 8 m.
15. $-\frac{2}{x + 1}$
16. $\{x \mid x \text{ is a real number and } x \neq 2, x \neq -2\}$
17. $x - 5$ **18.** $-\frac{2}{x}$ **19.** $y = 4, y = -3$
20. $b = 0, b = \frac{1}{2}, b = 1, b = -1$ **21.** $y = \frac{1}{2}x + 4$
22. a. $(6, 0)$ **b.** $(0, -3)$ **c.** $\frac{1}{2}$
d.

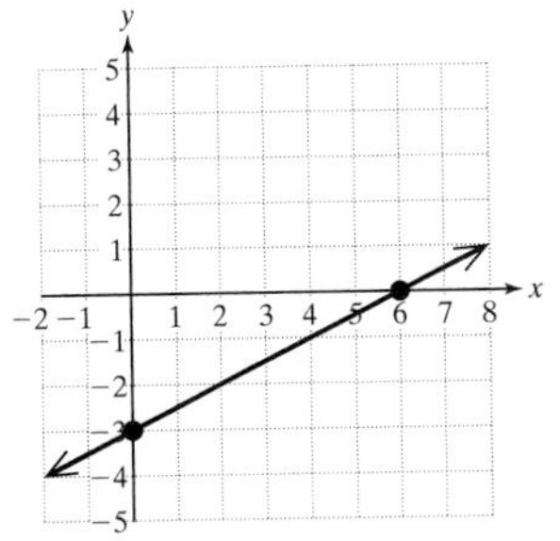

23. a. $(-3, 0)$ **b.** No y-intercept **c.** Slope is undefined.
d.

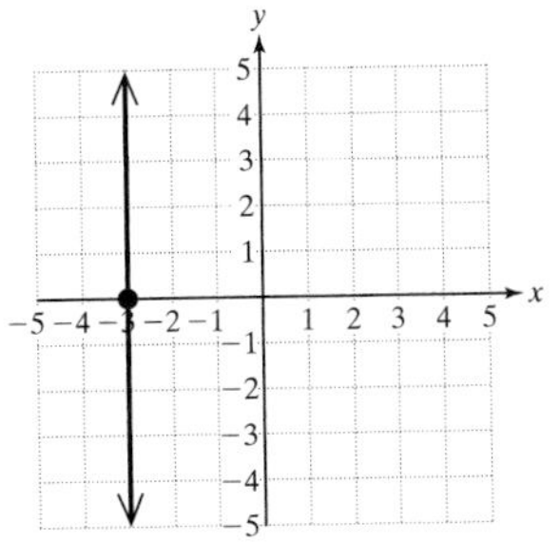

24. No solution **25.** (5, 2)
26. The angles are 22° and 68°.
27. There are 13 dimes and 14 quarters.
28.

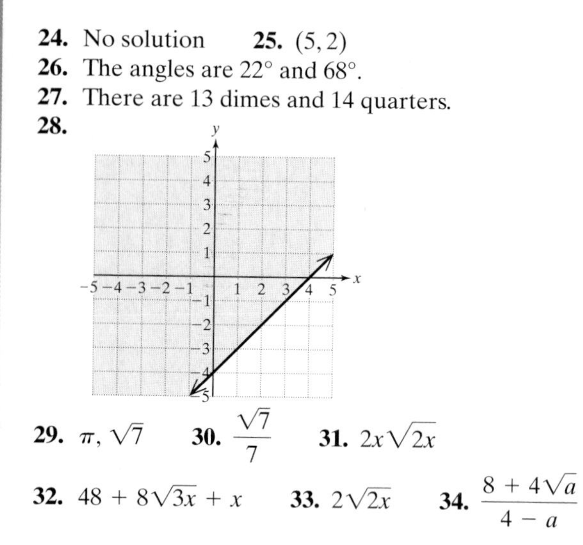

29. $\pi, \sqrt{7}$ **30.** $\frac{\sqrt{7}}{7}$ **31.** $2x\sqrt{2x}$

32. $48 + 8\sqrt{3x} + x$ **33.** $2\sqrt{2x}$ **34.** $\frac{8 + 4\sqrt{a}}{4 - a}$

35. $x = -2$, ($x = -7$ does not check)
36. $(2c - y)(4c^2 + 2cy + y^2)$ **37.** b
38. **a.** 1 **b.** 4 **c.** 13
39. Domain: {2, −1, 9, −6}; range: {4, 3, 2, 8}
40. $m = \frac{5}{7}$ **41.** $m = -\frac{4}{5}$ **42.** 25
43. $x = -3 \pm \sqrt{6}$ **44.** $x = -3 \pm \sqrt{6}$
45.

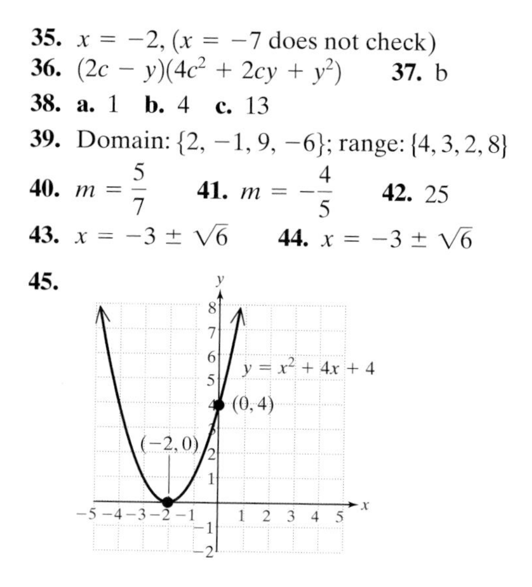

Application Index

BIOLOGY/HEALTH/ LIFE SCIENCES

BUSINESS

CHEMISTRY

CONSTRUCTION

CONSUMER APPLICATIONS

DISTANCE/SPEED/TIME

ENVIRONMENT

INVESTMENT

POLITICS

SCHOOL

SCIENCE

SPORTS

STATISTICS/DEMOGRAPHICS

Subject Index

A

B

C

D

E

F

S